STUDENT SOLUTIONS MANUAL

Laurel Technical Services

EXPERIENCING ALGEBRA

JoAnne Thomasson

Bob Pesut

PRENTICE HALL, Upper Saddle River, NJ 07458

Senior Editor: Karin Wagner
Supplement Editor: Kate Marks
Special Projects Manager: Barbara A. Murray
Production Editor: Mark Marsden
Supplement Cover Manager: Paul Gourhan
Supplement Cover Designer: Liz Nemeth
Manufacturing Buyer: Alan Fischer

Printed in the United States of America

10 9 8 7 6 5 4 3 2 1

ISBN 0-13-799982-8

Prentice-Hall International (UK) Limited, *London*
Prentice-Hall of Australia Pty. Limited, *Sydney*
Prentice-Hall Canada, Inc., *London*
Prentice-Hall Hispanoamericana, S.A., *Mexico*
Prentice-Hall of India Private Limited, *New Delhi*
Prentice-Hall of Japan, Inc., *Tokyo*
Simon & Schuster Asia Pte. Ltd., *Singapore*
Editora Prentice-Hall do Brazil, Ltda., *Rio de Janeiro*

Table of Contents

Discovery Answers

Chapter 1

Discovery 1

1. a. 8

   b. 8

2. a. 12

   b. 12

3. a. 3.7

   b. 3.7

4. a. $\frac{3}{4}$

   b. $\frac{3}{4}$

The sum of two positive rational numbers is a positive number that is the sum of the absolute values of the addends.

Discovery 2

1. a. −8

   b. 8

2. a. −12

   b. 12

3. a. −3.7

   b. 3.7

4. a. $-\frac{3}{4}$

   b. $\frac{3}{4}$

The sum of two negative rational numbers is a negative number that is the opposite of the sum of the absolute values of the addends.

Discovery 3

1. a. 4

   b. 4

2. a. −6

   b. 6

3. a. −1.3

   b. 1.3

4. a. $\frac{1}{4}$

   b. $\frac{1}{4}$

The sum of a positive and a negative rational number has the same sign as the addend with the larger absolute value and is the difference of the absolute values of the addends.

Discovery 4

1. a. −4

   b. 4

2. a. 6

   b. 6

3. a. 1.3

   b. 1.3

4. a. $-\frac{1}{4}$

b. $\frac{1}{4}$

The sum of a negative and a positive rational number has the same sign as the addend with the larger absolute value and is the difference of the absolute values of the addends.

Discovery 5

1. a. 4

b. 4

2. a. –6

b. –6

3. a. –1.3

b. –1.3

4. a. $\frac{1}{4}$

b. $\frac{1}{4}$

The difference of two positive rational numbers is the sum of the minuend and the opposite of the subtrahend.

Discovery 6

1. a. –4

b. –4

2. a. 6

b. 6

3. a. 1.3

b. 1.3

4. a. $-\frac{1}{4}$

b. $-\frac{1}{4}$

The difference of two negative rational numbers is the sum of the minuend and the opposite of the subtrahend.

Discovery 7

1. a. 8

b. 8

2. a. –12

b. –12

3. a. 3.7

b. 3.7

4. a. $-\frac{3}{4}$

b. $-\frac{3}{4}$

The difference of a positive and negative rational number is the sum of the minuend and the opposite of the subtrahend.

Discovery 8

1. a. 12

b. 12

2. a. 12

b. 12

3. a. 3

b. 3

4. a. $\frac{1}{8}$

b. $\frac{1}{8}$

The product of two rational numbers with like signs is a positive number that is the product of the absolute values of the factors.

Discovery 9

1. a. -12

   b. 12

2. a. -12

   b. 12

3. a. -3

   b. 3

4. a. $-\frac{1}{8}$

   b. $\frac{1}{8}$

The product of two rational numbers with unlike signs is a negative number that is the opposite of the product of the absolute values of the factors.

Discovery 10

1. a. -72

   b. 72

2. a. 72

   b. 72

3. a. -72

   b. 72

4. a. 72

   b. 72

The product of three or more rational numbers is a positive number when the number of negative factors is even. The product of three or more rational numbers is a negative number when the number of negative factors is odd. The absolute value of the product is the product of the absolute values of the factors.

Discovery 11

1. 0

2. 0

3. 0

The product of zero and any rational number (or numbers) is zero.

Discovery 12

1. a. 4

   b. 4

2. a. 4

   b. 4

3. a. 0.48

   b. 0.48

4. a. 2

   b. 2

The quotient of two rational numbers with like signs is a positive number that is the quotient of the absolute values of the dividend and divisor.

Discovery 13

1. a. -4

   b. 4

2. a. -4

   b. 4

3. a. -0.48

   b. 0.48

4. a. -2

   b. 2

The quotient of two rational numbers with unlike signs is a negative number that is the opposite of the quotient of the absolute values of the dividend and divisor.

Discovery 14

1. This results in a calculator error.
2. This results in a calculator error.
3. This results in a calculator error.
4. This results in a calculator error.

The quotient of a rational number divided by zero is undefined.

Discovery 15

1. 0
2. 0
3. 0
4. 0
5. This results in a calculator error.

The quotient of zero divided by any rational number other than zero is zero. The quotient of zero divided by zero is indeterminate.

Chapter 2

Discovery 1

1. 4
2. –8
3. 16
4. –32

The sign of an exponential expression with a negative base is negative if the exponent is an odd number (such as 3 and 5) and is positive if the exponent is an even number (such as 2 and 4).

Discovery 2

1. 3
2. –3
3. 2.4
4. $\frac{1}{2}$
5. 0

An exponential expression with an exponent of one results in the base number.

Discovery 3

1. 1
2. 1
3. 1
4. 1
5. indeterminate

An exponential expression with an exponent of zero and a nonzero base will be 1 and with a zero base will be indeterminate.

Discovery 4

1. **a.** 0.5
   b. 0.5
2. **a.** 0.25
   b. 0.25
3. **a.** 0.125
   b. 0.125
4. **a.** –0.5
   b. –0.5
5. **a.** –0.25

b. –0.25

6. a. –0.125

b. –0.125

An exponential expressions with a nonzero base and a negative integer exponent is equal to an exponential expression having a base that is the reciprocal of the original base and an exponent that is the opposite of the original exponent.

Discovery 5

1. a. 2

b. 2

2. a. 4

b. 4

3. a. 8

b. 8

4. a. –2

b. –2

5. a. 4

b. 4

6. a. –8

b. –8

An exponential expression with a fraction base and a negative integer exponent is equal to an exponential expression having a base that is the reciprocal of the original base and an exponent that is the opposite of the original exponent.

Discovery 6

1. $(1.1)^2 = 1.21$
$(1.2)^2 = 1.44$
$(1.3)^2 = 1.69$
$(1.4)^2 = 1.96$
$(1.5)^2 = 2.25$

Therefore, $\sqrt{2}$ must be between 1.4 and 1.5 .

2. $(1.41)^2 = 1.9881$
$(1.42)^2 = 2.0164$

Therefore, $\sqrt{2}$ must be between 1.41 and 1.42 .

Discovery 7

1. a. 1.732050808

b. 1.732050808

2. a. 1.44224957

b. 1.44224957

3. a. 1.316074013

b. 1.316074013

4. a. error

b. error

5. a. –1.44224957

b. –1.44224957

6. a. error

b. error

A radical expression is equal to an exponential expression having a base that is the radicand of the radical expression and an exponent that is the reciprocal of the index of the radical expression.

Discovery 8

1. a. –8

b. –8

2. a. –4

b. –4

3. **a.** –4

   b. 4

4. **a.** –8

   b. 8

5. **a.** 12

   b. 12

6. **a.** –12

   b. –12

7. **a.** 3

   b. $\frac{1}{3}$

8. **a.** –3

   b. $-\frac{1}{3}$

Changing the order of real numbers when addition and multiplication operations are performed results in equivalent values. However, changing the order of real numbers when subtraction and division operations are performed results in different values.

Discovery 9

1. **a.** –5

   b. –5

2. **a.** –1

   b. –1

3. **a.** –7

   b. –1

4. **a.** –5

   b. –11

5. **a.** 36

   b. 36

6. **a.** –36

   b. –36

7. **a.** 1

   b. 9

8. **a.** –9

   b. –1

Changing the grouping of real numbers when addition and multiplication operations are performed results in equivalent values. However, changing the grouping of real numbers when subtraction and division operations are performed results in different values.

Discovery 10

1. **a.** 20

   b. 20

2. **a.** 4

   b. 4

The product of a real number factor and a sum of real number addends is the same as the sum of the products of the factor and each addend.

Discovery 11

1. 7
2. –4
3. 1
4. 1

Perform all multiplication and division in order from left to right before performing addition and subtraction in order left to right.

Chapter 4

Discovery 1

Ordered pairs may vary.
In quadrant I, both coordinates are positive; in quadrant II, the x-coordinate is negative and the y-coordinate is positive; in quadrant III, both coordinates are negative; and in quadrant IV, the x-coordinate is positive and the y-coordinate is negative.

Discovery 2

Ordered pairs may vary.
On the x-axis, the y-coordinate is always 0, and on the y-axis, the x-coordinate is always 0.

Discovery 3

1.
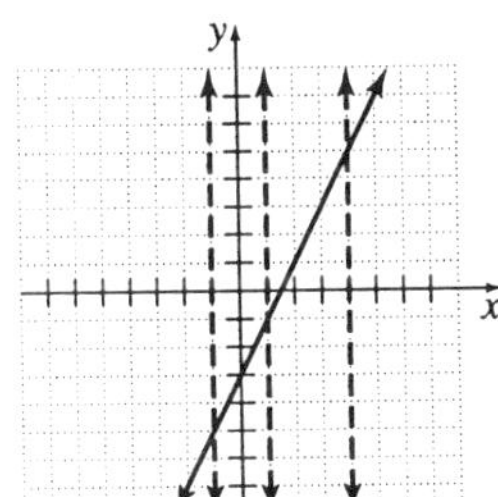

2.
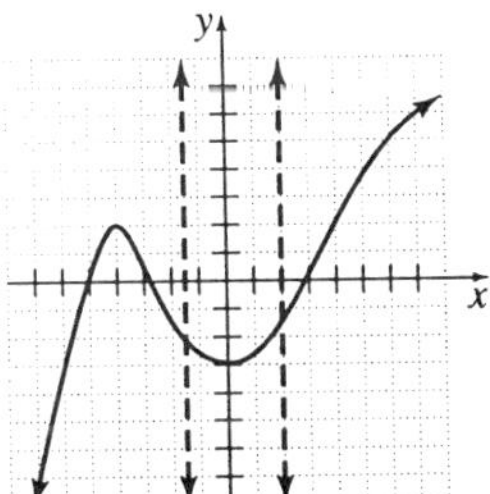

3.
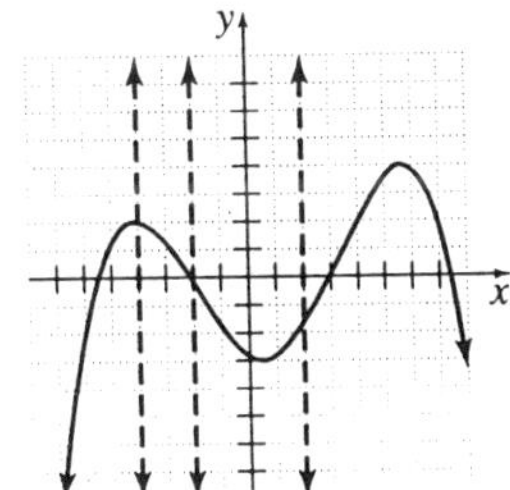

A vertical line does not cross the graph of a function more than once.

4.
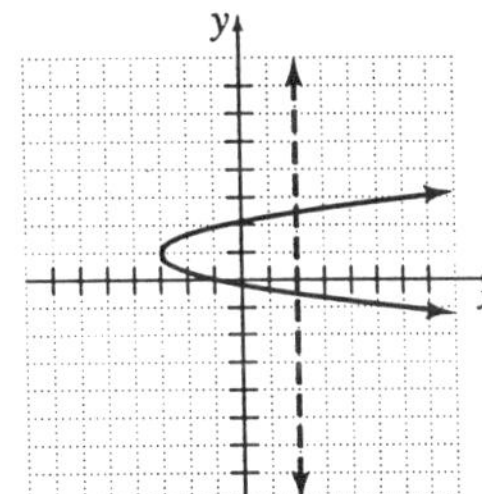

5.
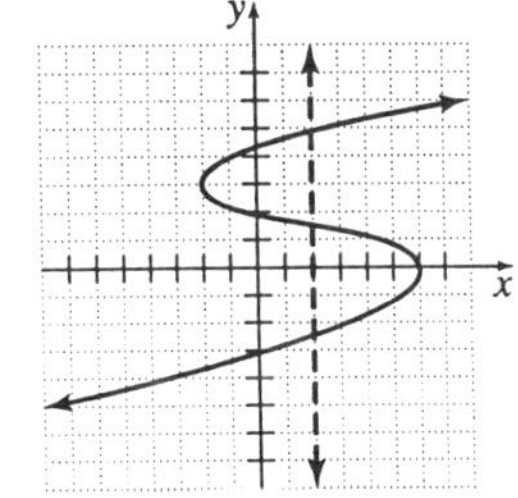

Discovery 4

1.
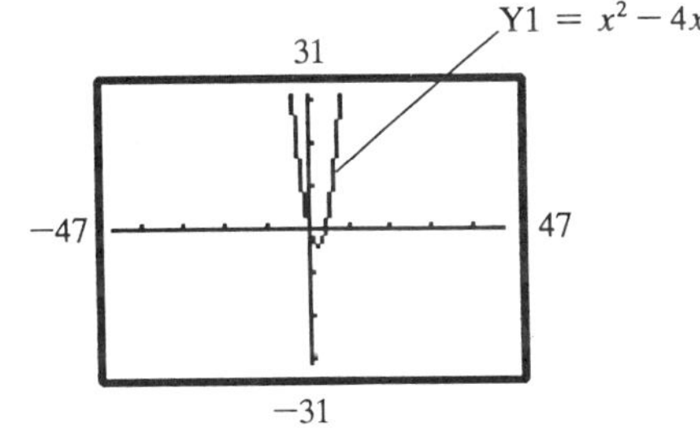

2. falls; rises

3. 12; 5; 0; –3; –4; –3; 0; 5; 12

4. decrease; increase

A function is increasing if the graph is rising. A function is decreasing if the graph is falling.

Chapter 5

Discovery 1

1.

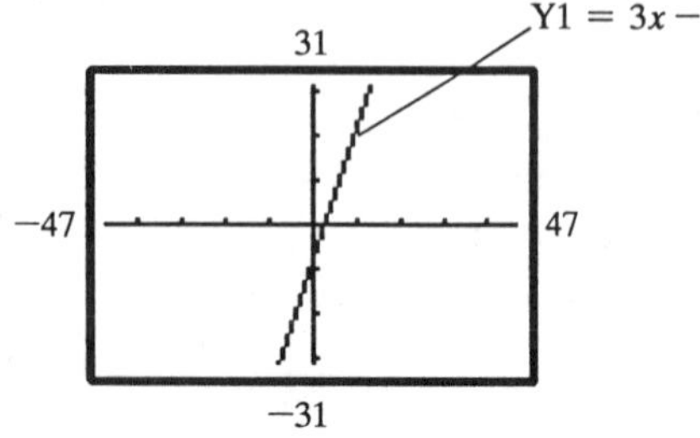

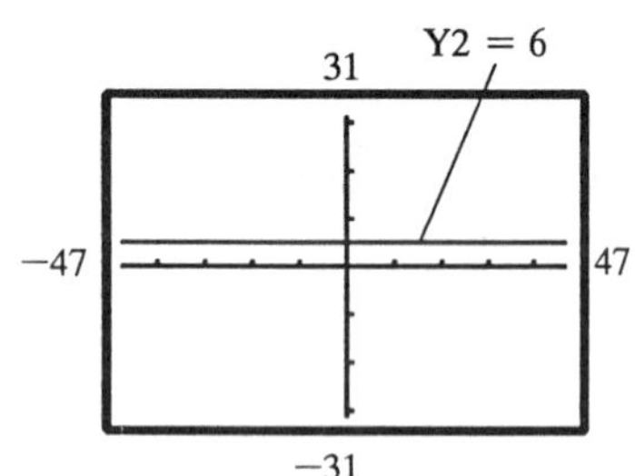

2.

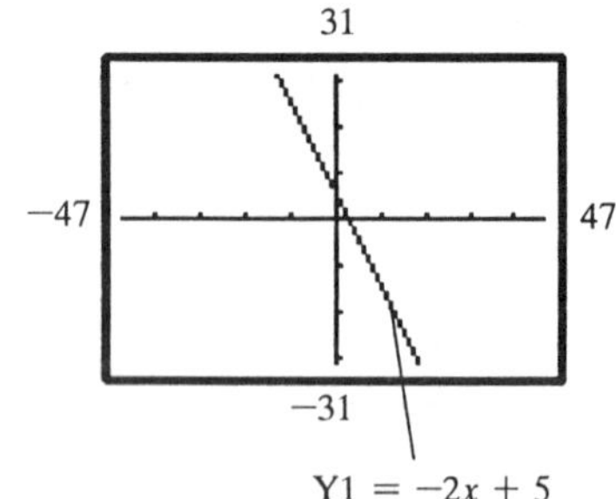

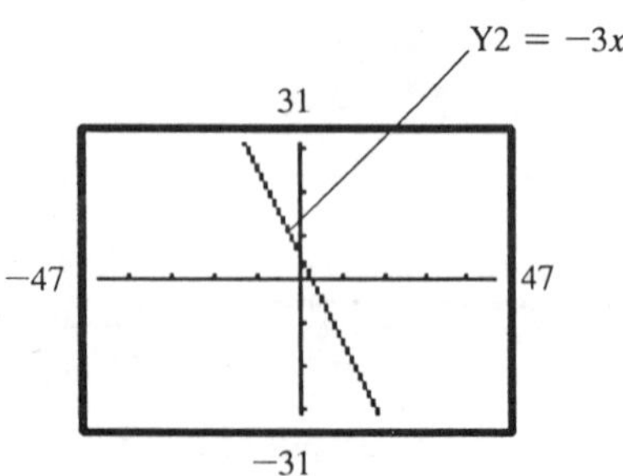

The graphs of the functions Y1 and Y2 are straight lines.

3.

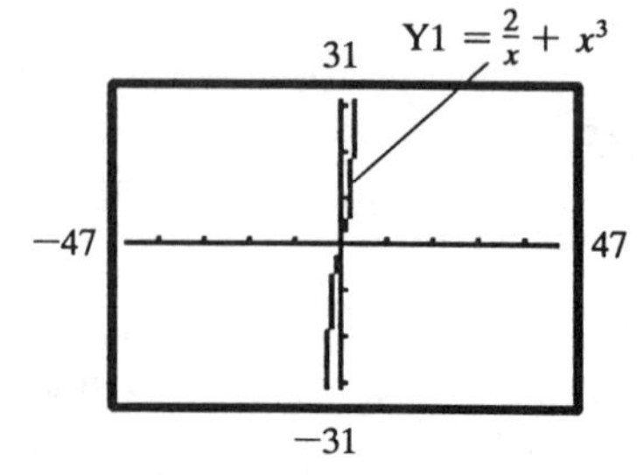

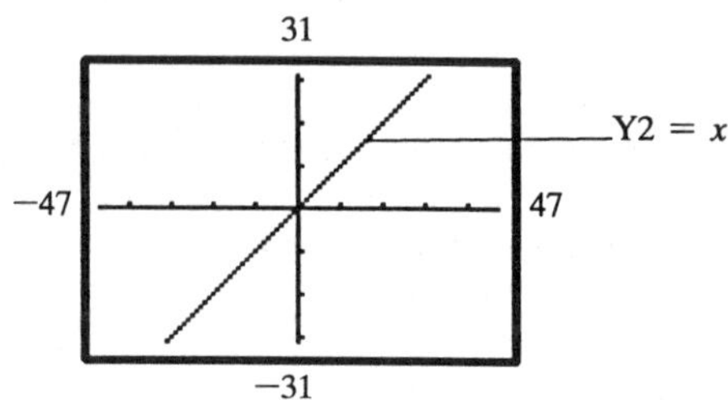

The graph of the function Y1 is not a straight line but a curve.
The graph of the function Y2 is a straight line.

Discovery 2

x	$2x+3$	$= x+5$	
0	3	5	$3<5$
1	5	6	$5<6$
2	7	7	$7=7$
3	9	8	$9>8$
4	11	9	$11>9$

The solution of an equation is the value for the variable that determines equal values for the expression on the left and the expression on the right of the equation.

Discovery 3

x	$(5x+4)-2(3x+1)$	$=2(x-7)$	
3	−1	−8	$-1>-8$
4	−2	−6	$-2>-6$
5	−3	−4	$-3>-4$
6	−4	−2	$-4<-2$
7	−5	0	$-5<0$

The solution of an equation is between two integer values when the order of the values for the expression on the left and the expression on the right changes from less than to greater than or greater than to less than.

Discovery 4

x	$2x+5$	$=2x+10$		
0	5	10	$5\neq 10$	$10-5=5$
1	7	12	$7\neq 12$	$12-7=5$
2	9	14	$9\neq 14$	$14-9=5$
3	11	16	$11\neq 16$	$16-11=5$

If the values of the expression on the left and the expression on the right are never equal and their differences remain the same, then the equation has no solution.

Discovery 5

x	$2x+5$	$=(x+3)+(x+2)$	
0	5	5	$5=5$
1	7	7	$7=7$
2	9	9	$9=9$
3	11	11	$11=11$

If the values of the expression on the left and the expression on the right are always equal, then the equation has many solutions.

Discovery 6

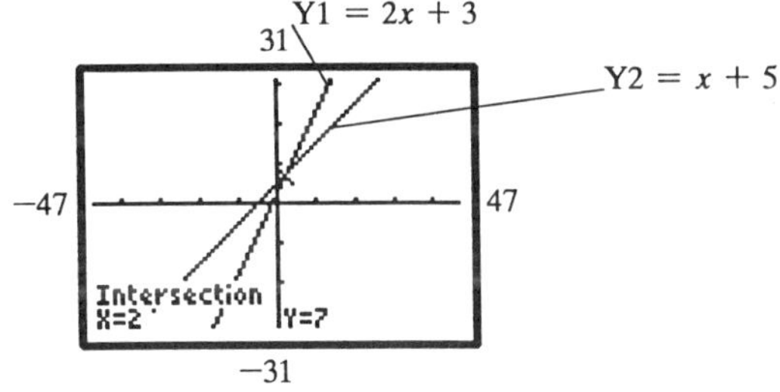

The value of the x-coordinate of the point of intersection of the two graphs is the solution of the equation.
The value of the y-coordinate of the point of intersection of the two graphs is the value obtained when both expressions are evaluated with the solution.

Discovery 7

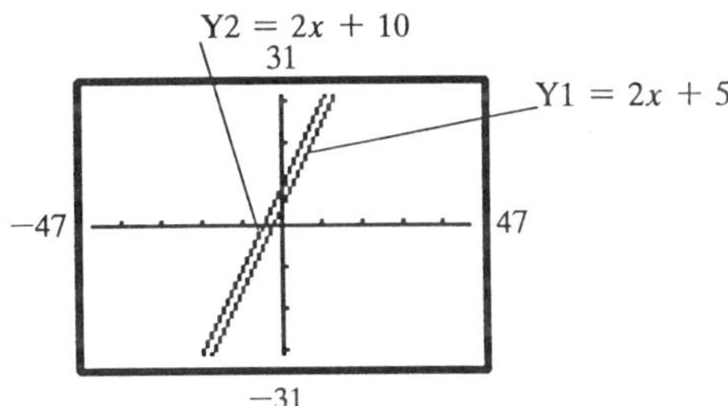

If the graphs of the functions defined by the expression on the left and the expression on the right do not intersect (are parallel), then the equation has no solution.

Discovery 8

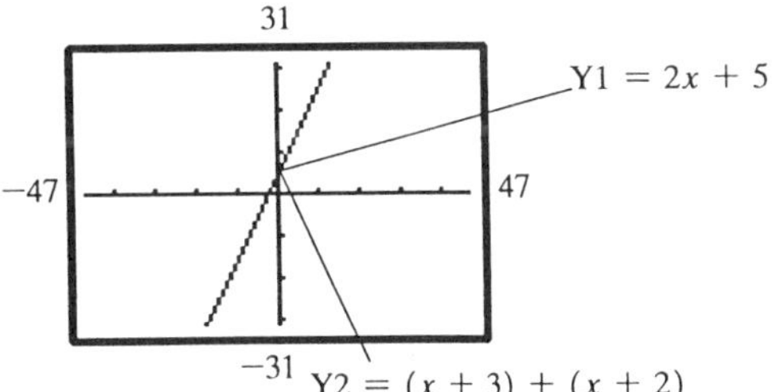

If the graphs of the functions defined by the expression on the left and the expression on the right intersect at all times (are coinciding), then the equation has many solutions.

Discovery 9

1. $7 = 7$

$7 + (-2)$	$7 + (-2)$
5	5

2. $6 + 1 = 4 + 3$

$6 + 1 + 2$	$4 + 3 + 2$
9	9

3. $6 + 1 = 4 + 3$

$6 + 1 + (-2)$	$4 + 3 + (-2)$
5	5

If the same number is added to both expressions in an equation, the result is an equivalent equation.

Discovery 10

1. $7 = 7$

$7(-2)$	$7(-2)$
-14	-14

2. $6 + 1 = 4 + 3$

$(6 + 1)2$	$(4 + 3)2$
14	14

3. $6 + 1 = 4 + 3$

$(6 + 1)(-2)$	$(4 + 3)(-2)$
-14	-14

If both expressions in an equation are multiplied by the same number, the result is an equivalent equation.

Discovery 11

$2x + 5 = 2x + 10$

$2x + 5 - 2x = 2x + 10 - 2x$ Subtract $2x$ from both sides.

$5 = 10$ Simplify.

When isolating the variable results in an equation without a variable and the equation is false (a contradiction), then the equation has no solution.

Discovery 12

$2x + 5 = (x + 3) + (x + 2)$

$2x + 5 = 2x + 5$ Simplify the right side.

$2x + 5 - 2x = 2x + 5 - 2x$ Subtract $2x$ from both sides.

$5 = 5$ Simplify.

When isolating the variable results in an equation without a variable and the equation is true (an identity), then the equation has many solutions.

Discovery 13

1.

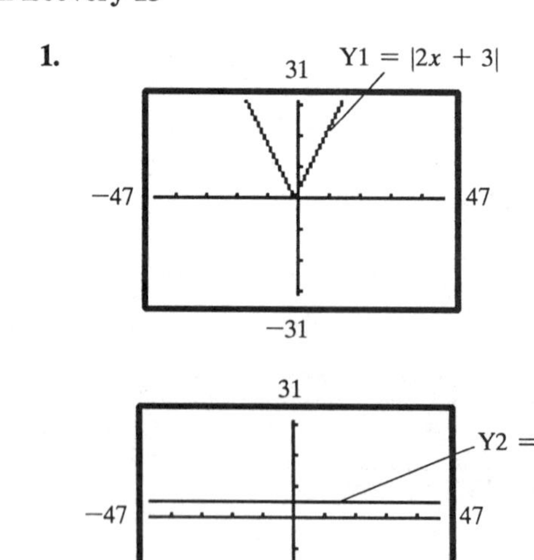

2.

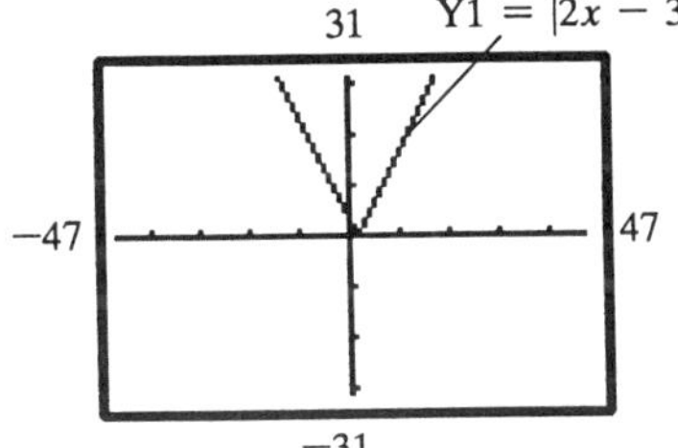

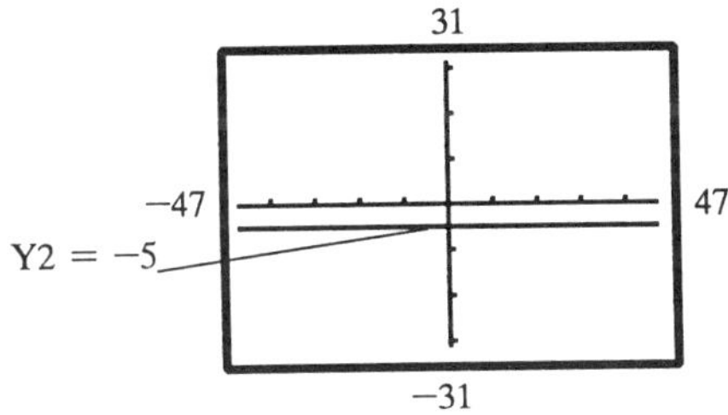

The graph of the function defined by the absolute value expression is a V-shaped graph made up of two lines and the graph of the constant function is a line.

3.

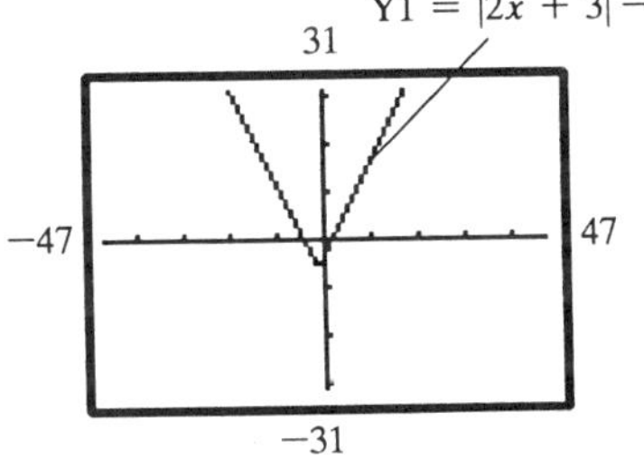

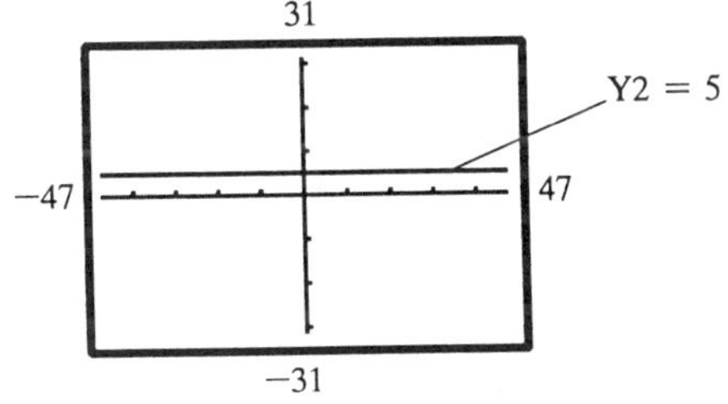

4.

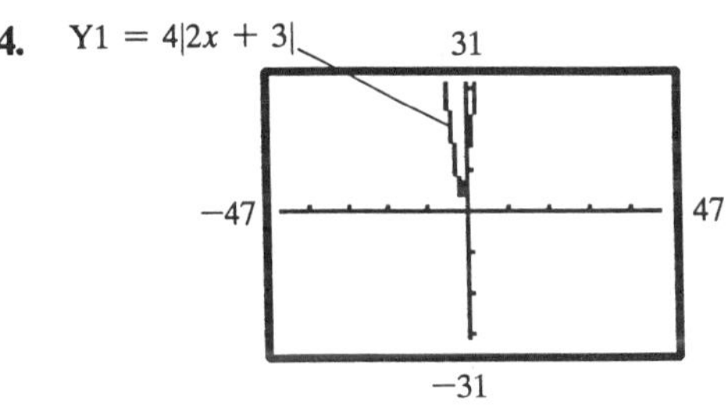

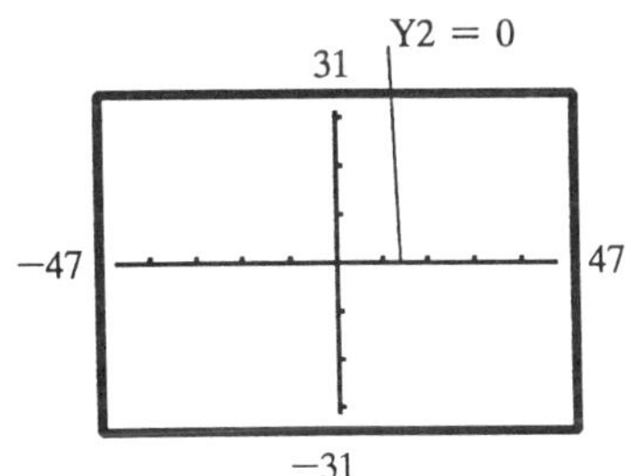

The graph of the function defined by the expression containing the absolute value expression is a V-shaped graph made up of lines, and the graph of the constant function is a line.

Discovery 14

1.

x	$\lvert x\rvert$	3
–5	5	3
–4	4	3
–3	3	3
–2	2	3
–1	1	3
0	0	3
1	1	3
2	2	3
3	3	3
4	4	3
5	5	3

solution: –3 and 3

2.

x	$\lvert x\rvert$	–3
–5	5	–3
–4	4	–3
–3	3	–3
–2	2	–3
–1	1	–3
0	0	–3
1	1	–3
2	2	–3
3	3	–3
4	4	–3
5	5	–3

solution: no solution

3.

x	$\lvert x\rvert$	0
–5	5	0
–4	4	0
–3	3	0
–2	2	0
–1	1	0
0	0	0
1	1	0
2	2	0
3	3	0
4	4	0
5	5	0

solution: 0

The number of solutions that can be found for a linear absolute value equation is determined by the number the absolute value expression is equated to. That is, if the number is positive, two solutions can be found. If the number is negative, no solution can be found. If the number is zero, then one solution can be found.

Discovery 15

1.

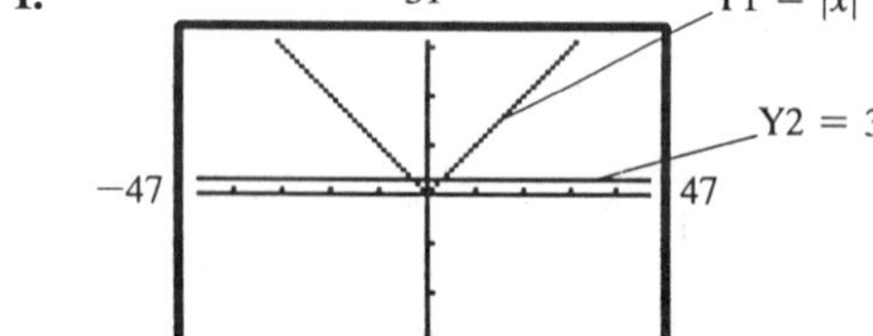

solution: –3 and 3

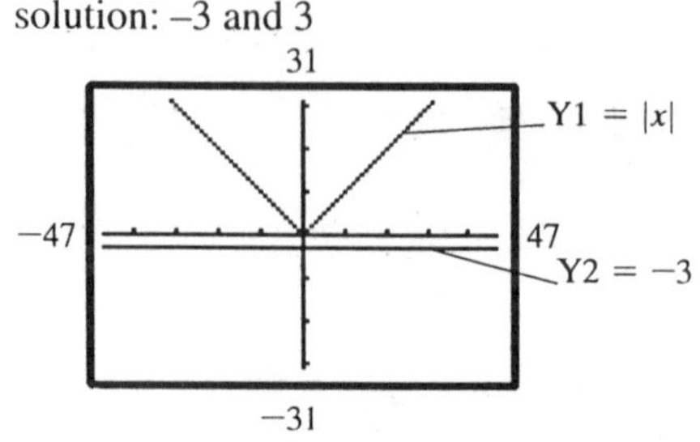

solution: no solution

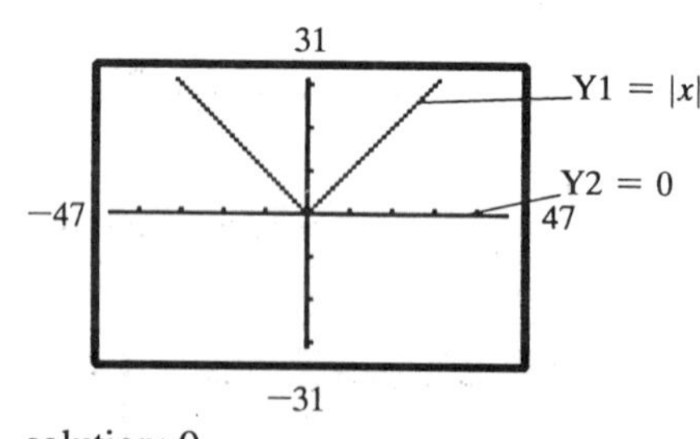

solution: 0

The number of solutions that can be found for a linear absolute value equation is determined by the number the absolute value expression is equated to. That is, if the number is positive, two solutions can be found. If the number is negative, no solution can be found. If the number is zero, then one solution can be found.

Chapter 6

Discovery 1

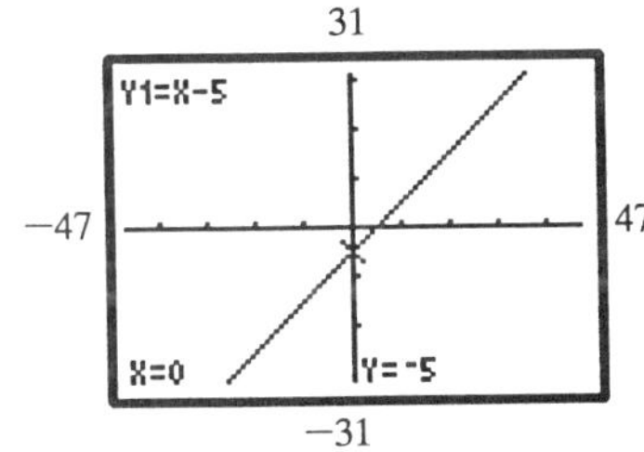

2.

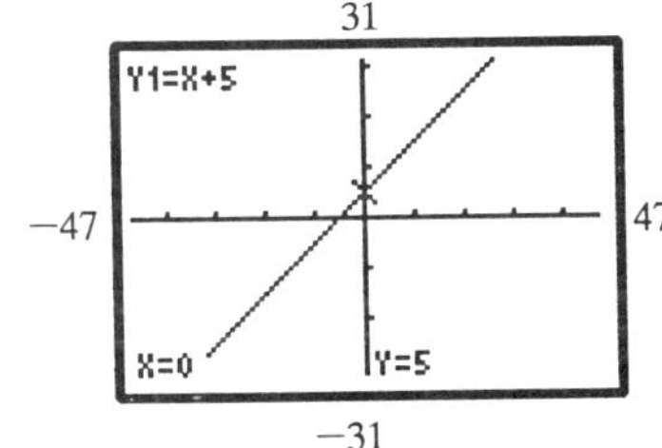

3.

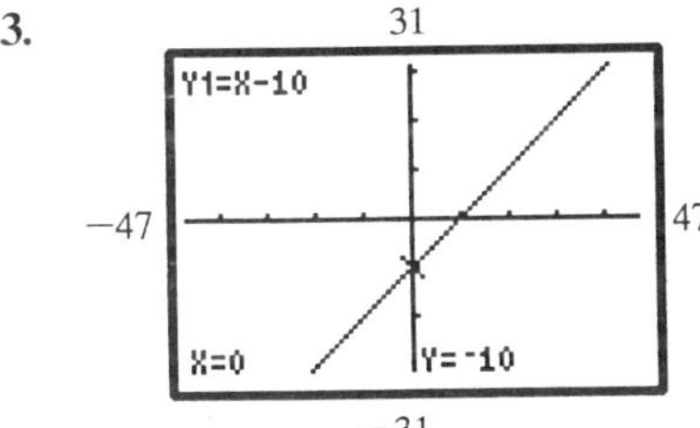

4.

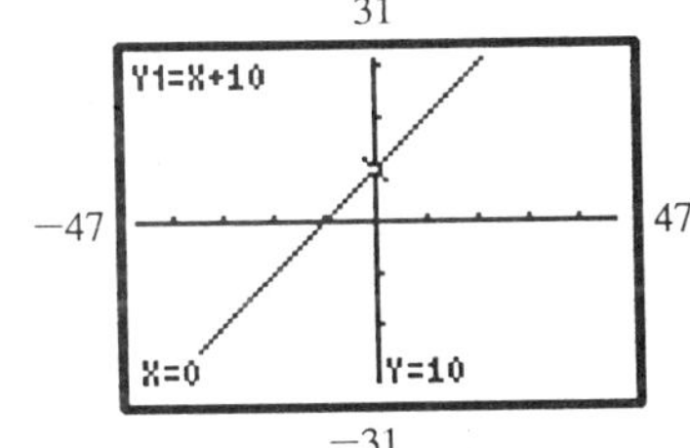

Given an equation solved for y, the y-coordinate of the y-intercept is the constant term of the expression in the equation.

Discovery 2

1.

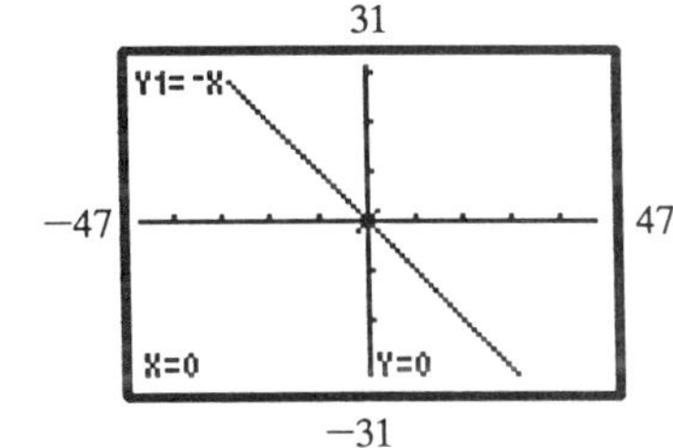

2.

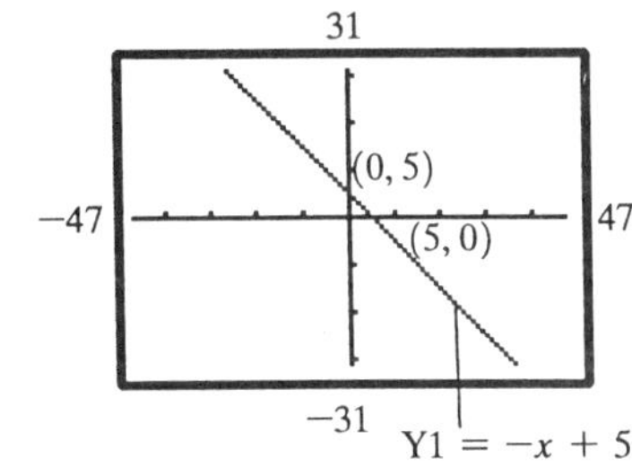

3.

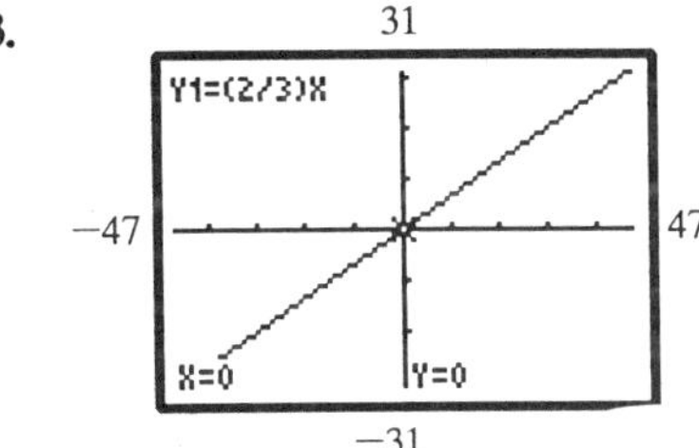

4.

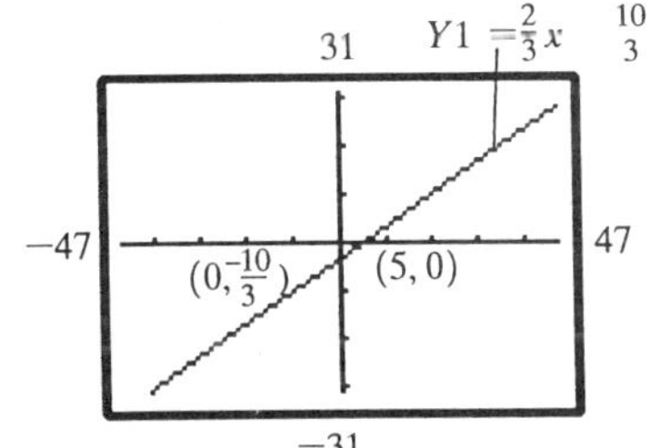

The graph of a linear equation in two variables written in standard form has one point that is both the x-intercept and y-intercept if the constant term is zero. The intercept is the origin, (0, 0).

Discovery 3

1. $\frac{1}{2}$

2. 2

3. $-\frac{1}{3}$

4. –3

5. 0

6. undefined

7. positive; rise

8. negative; fall

9. zero; horizontal

10. undefined; vertical

11. more

Discovery 4

1.

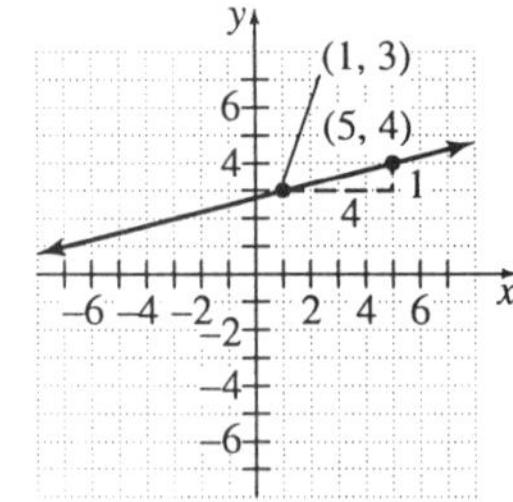

2. 1

3. 4

4. 1

5. 4

6. $\frac{1}{4}$

The slope of the graph is the ratio of the difference of the ordered pairs' y-coordinates to the difference in the ordered pairs' x-coordinates.

Discovery 5

Possible integer coordinate points are labeled.

1. a.

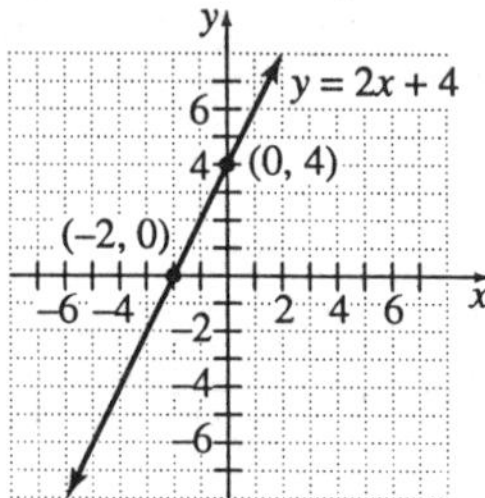

b.

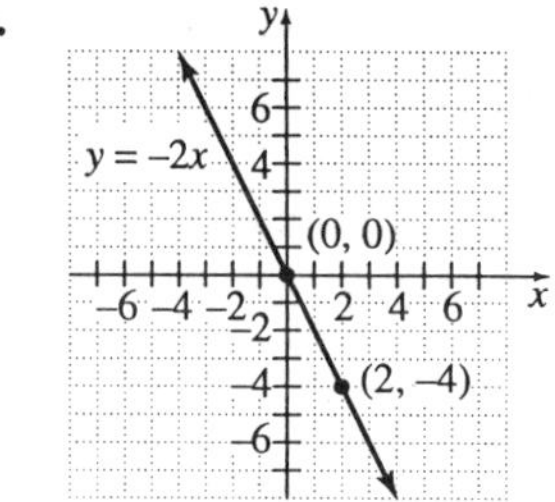

c.

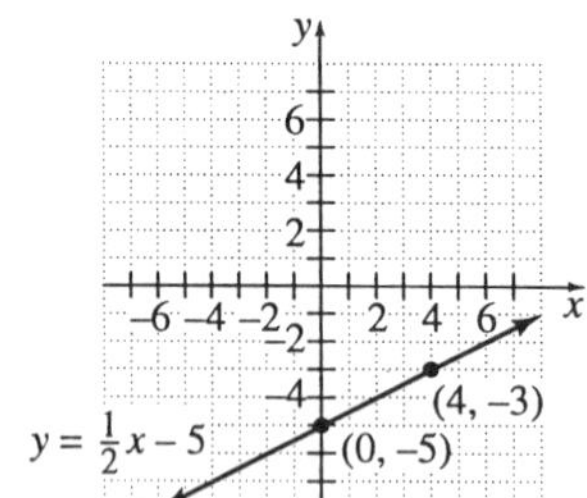

2. a. 2

b. –2

c. $\frac{1}{2}$

3 a. 2

b. –2

c. $\frac{1}{2}$

Given a linear equation solved for y, the slope of its graph is the coefficient of the x-term.

Discovery 6

1. a.

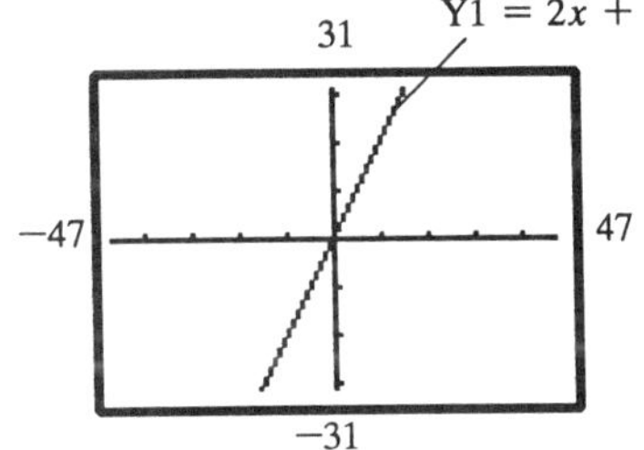

b.

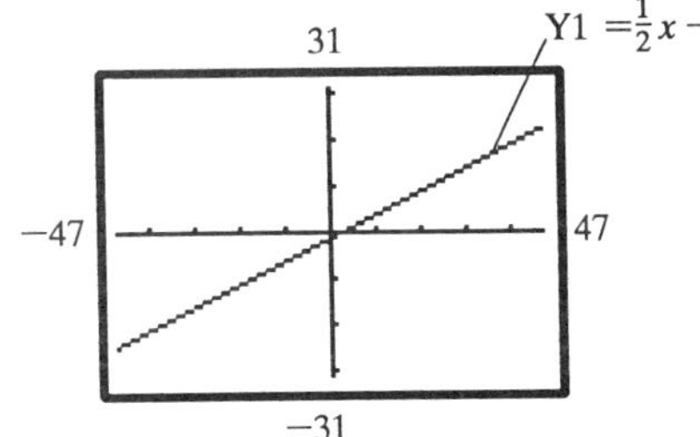

c.

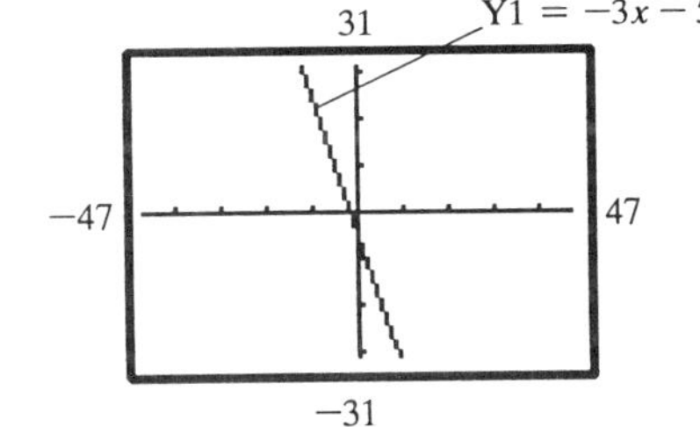

2. a. $m = 2$ $b = 1$ $m = 2$ $b = 1$

b. $m = \frac{1}{2}$ $b = -\frac{3}{2}$ $m = \frac{1}{2}$ $b = -\frac{3}{2}$

c. $m = -3$ $b = -5$ $m = -3$ $b = -5$

3. coinciding

4. equal

5. equal

The graphs of two linear equations in two variables are coinciding if their corresponding equations written in slope-intercept form have equal slopes (m) and equal y-intercepts (b).

Discovery 7

1. a.

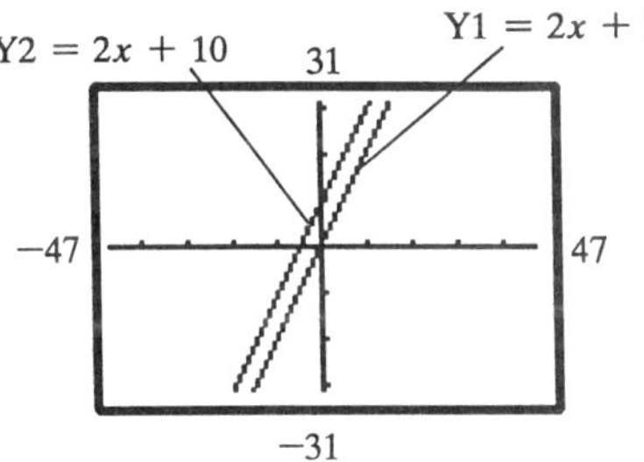

b.

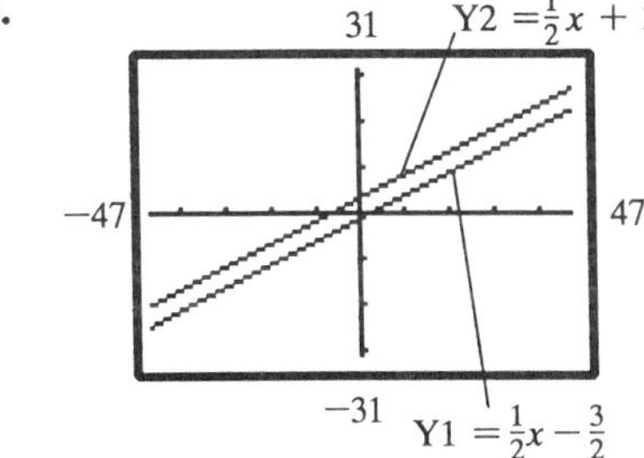

c.

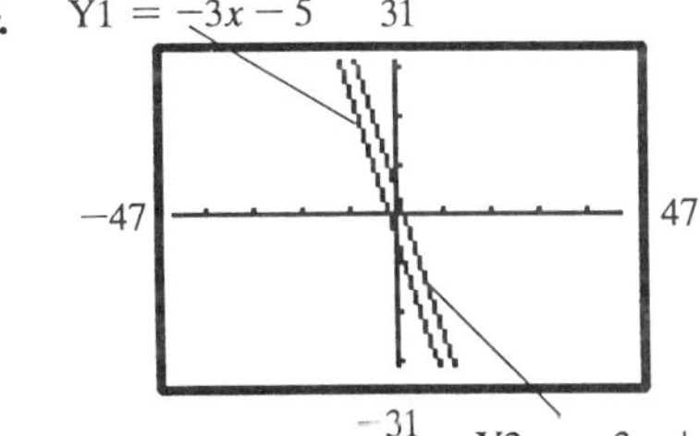

2 a. $m = 2$ $b = 1$ $m = 2$ $b = 10$

b. $m = \frac{1}{2}$ $b = -\frac{3}{2}$ $m = \frac{1}{2}$ $b = 3$

c. $m = -3$ $b = -5$ $m = -3$ $b = 5$

3. parallel

4. equal

5. not equal

The graphs of two linear equations in two variables are parallel if their corresponding equations written in slope-intercept form have equal slopes (m) and non-equal y-intercepts (b).

Discovery 8

1. a.

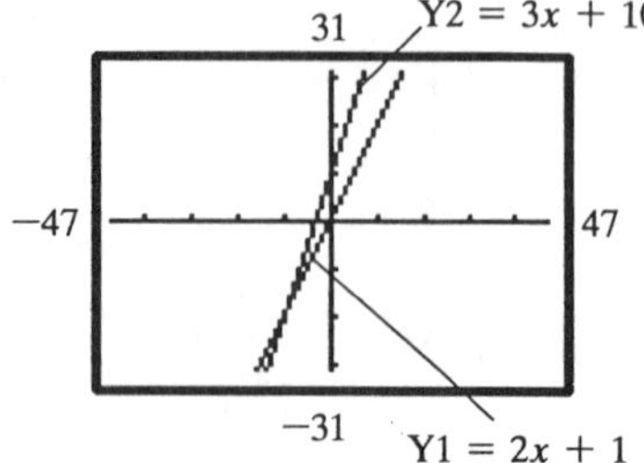

b.

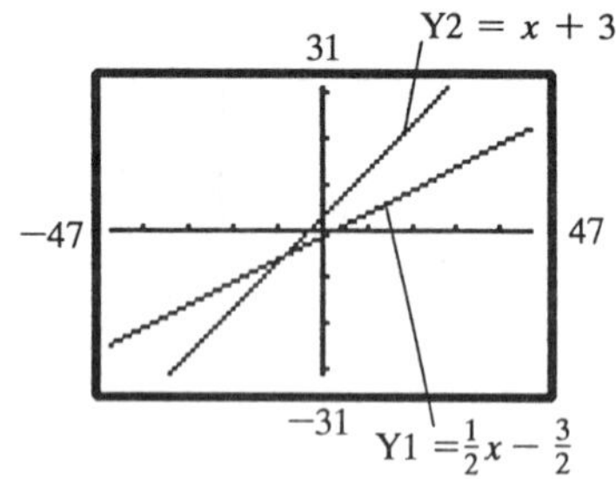

c.

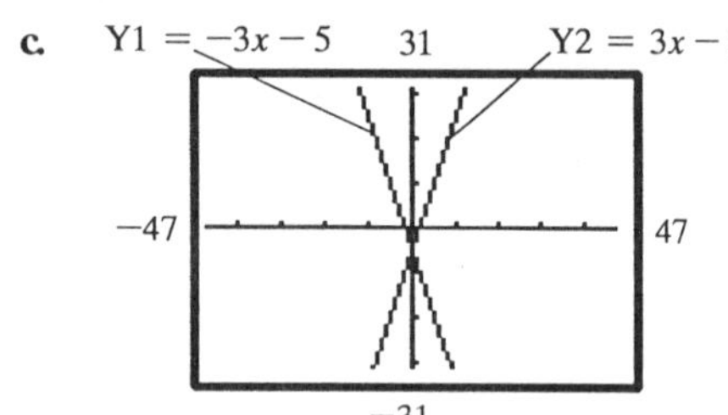

2. a. $m=2$ $b=1$ $m=3$ $b=10$

   b. $m=\frac{1}{2}$ $b=-\frac{3}{2}$ $m=1$ $b=3$

   c. $m=-3$ $b=-5$ $m=3$ $b=-5$

3. intersecting

4. not equal

The graphs of two linear equations in two variables are intersecting if their corresponding equations written in slope-intercept form have non-equal slopes (m).

Discovery 9

1. a.

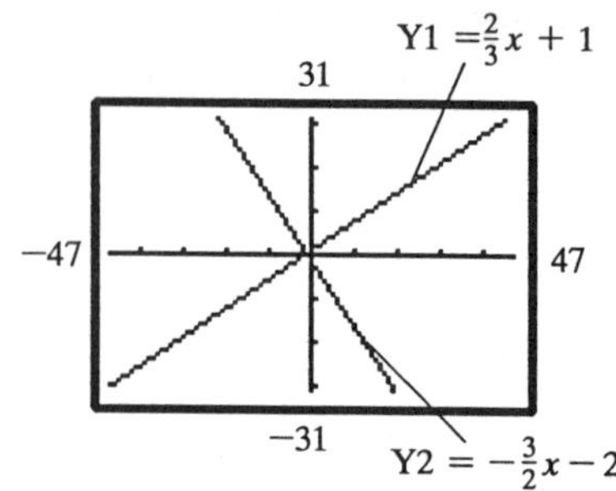

b.

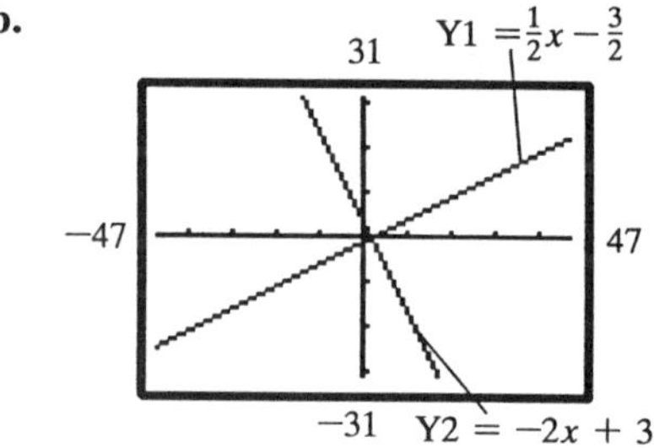

c.

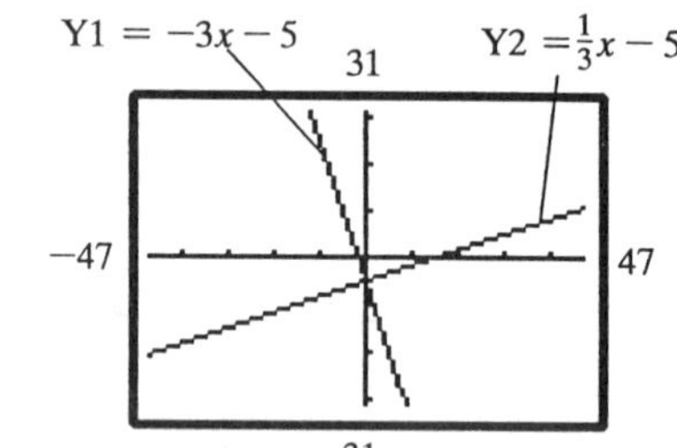

2. a. $m=\frac{2}{3}$ $b=1$ $m=-\frac{3}{2}$ $b=$ -2

   b. $m=\frac{1}{2}$ $b=-\frac{3}{2}$ $m=-2$ $b=3$

   c. $m=-3$ $b=-5$ $m=\frac{1}{3}$ $b=-5$

3. intersecting and perpendicular

4. not equal

5. opposite

The graphs of two linear equations in two variables are intersecting and perpendicular if their corresponding equations written in slope-intercept form have non-equal slopes (m) which are opposite reciprocals.

Chapter 7

Discovery 1

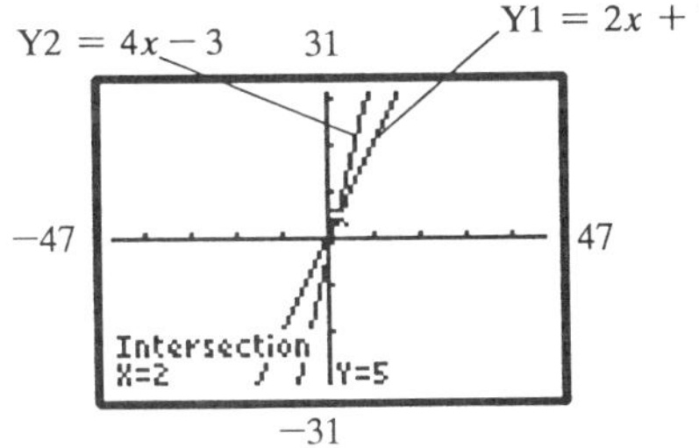

1. (2, 5)

2. (2, 5)

If the graphs of the two equations of a system intersect, then the coordinates of the point of intersection is the ordered pair solution of the system.

Discovery 2

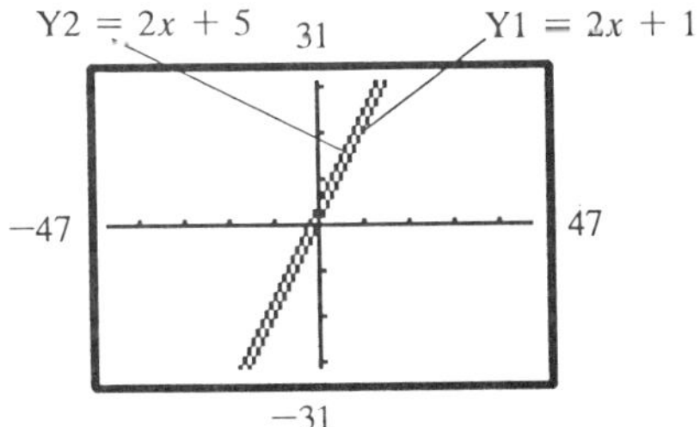

If the graphs of the equations of a system do not intersect, then the system of linear equations is inconsistent or has no solution.

Discovery 3

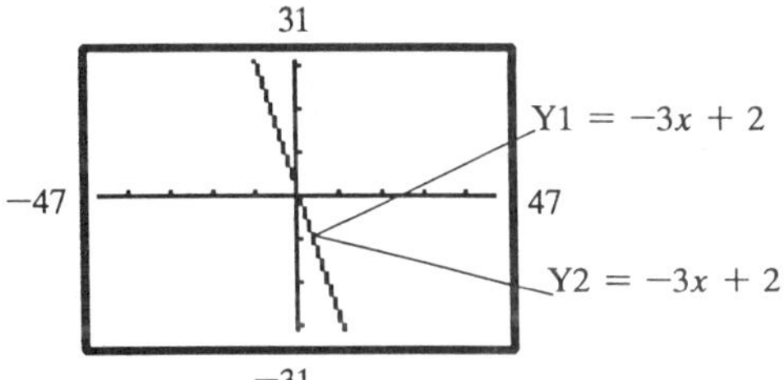

If the graphs of the equations of a system are coinciding, then the system of linear equations consists of dependent equations or has an infinite number of solutions.

Discovery 4

Solve the first equation for y.
$y = 2x + 1$
Substitute $(2x + 1)$ for y in the second equation and solve for x.
$-4x + 2(2x + 1) = 3$
$-4x + 4x + 2 = 3$
$2 = 3$ (contradiction)

If the substitution method yields a contradiction, then the system of linear equations is inconsistent or has no solution.

Discovery 5

Substitute $(-3x + 2)$ for y in the first equation.
$-3x - (-3x + 2) = -2$
$-3x + 3x - 2 = -2$
$-2 = -2$ (identity)
If the substitution method yields an identity, then the system of linear equations has dependent equations or an infinite number of solutions.

Discovery 6

Multiply both sides of the first equation by −2.
$-2(-2x + y) = -2(1)$
The new system is:
$4x - 2y = -2$
$-4x + 2y = 3$
Add corresponding members.
$0 = 1$ (contradiction)

If the elimination method yields a contradiction, then the system of linear equations is inconsistent or has no solution.

Discovery 7

Write both equations in standard form.

$-3x - y = -2$

$3x + y = 2$

Add corresponding members.

$0 = 0$ (identity)

If the elimination method yields an identity, then the system of linear equations has dependent equations or an infinite number of solutions.

Chapter 8

Discovery 1

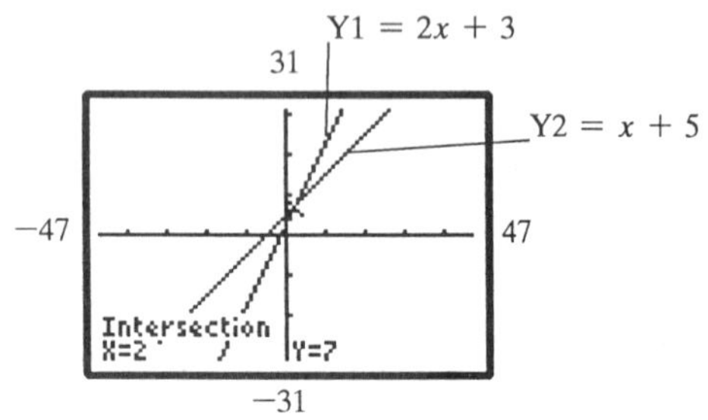

1. (2, 7)
2. 2
3. below; left; less than
4. $x \leq 2$

Discovery 2

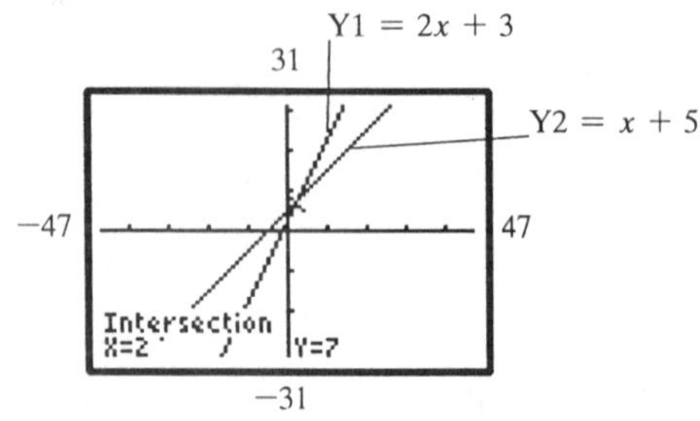

1. (2, 7)
2. 2
3. above; right; greater than
4. $x \geq 2$

Discovery 3

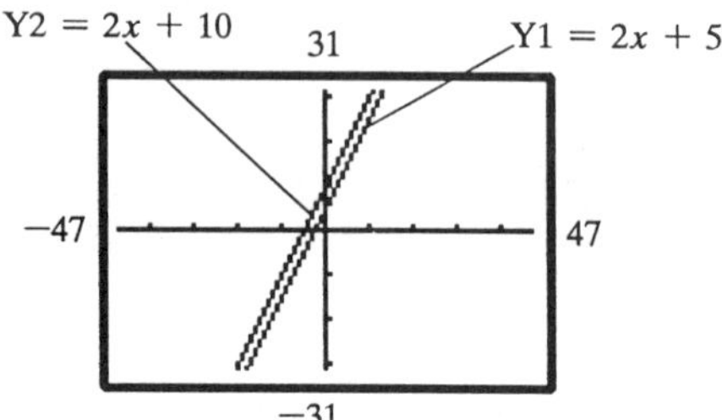

A "less than" inequality Y1 < Y2 has no solution if the graphs of the two functions defined by Y1 and Y2 coincide or if the graphs do not intersect and the graph of Y1 lies above the graph of Y2.

A "greater than" inequality Y1 > Y2 has no solution if the graphs of the two functions defined by Y1 and Y2 coincide or if the graphs do not intersect and the graph of Y1 lies below the graph of Y2.

A "less than or equal to" inequality Y1 ≤ Y2 has no solution if the graphs of the two functions defined by Y1 and Y2 do not intersect and the graph of Y1 lies above the graph of Y2.

A "greater than or equal to" inequality Y1 ≥ Y2 has no solution if the graphs of the two functions defined by Y1 and Y2 do not intersect and the graph of Y1 lies below the graph of Y2.

Discovery 4

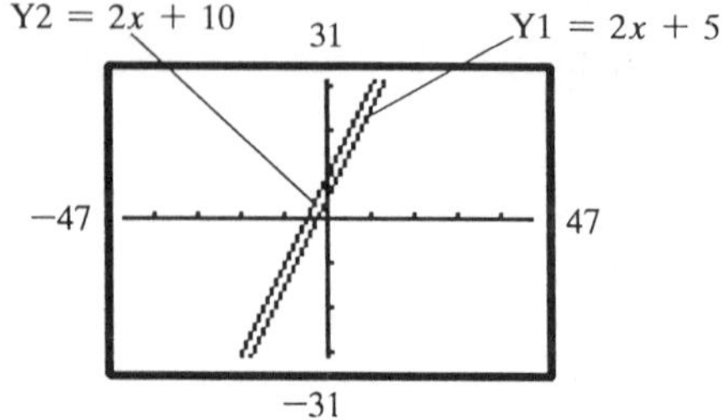

A "less than" inequality Y1 < Y2 has a solution set of all real numbers if the graphs of the two functions defined by Y1 and Y2 do not intersect and the graph of Y1 lies below the graph of Y2.

A "greater than" inequality Y1 > Y2 has a solution set of all real numbers if the graphs of the two functions defined by Y1 and Y2 do not intersect and the graph of Y1 lies above the graph of Y2.

Also, the inequality Y1 ≤ Y2 or Y1 ≥ Y2 has a solution set of all real numbers if the graphs of the two functions defined by Y1 and Y2 are the same.

Discovery 5

1. $10 < 12$

$10 + 2$	$12 + 2$
12	14

2. $10 < 12$

$10 + (-2)$	$12 + (-2)$
8	10

3. $10 < 12$

$10 - 2$	$12 - 2$
8	10

4. $10 < 12$

$10 - (-2)$	$12 - (-2)$
12	14

If a number is added to (or subtracted from) both expressions in a true inequality the resulting inequality is also true.

Discovery 6

1. $10 < 12$

$10 \cdot 2$	$12 \cdot 2$
20	24

2. $10 < 12$

$10 \cdot (-2)$	$12 \cdot (-2)$
−20	−24

3. $10 < 12$

$10 \div 2$	$12 \div 2$
5	6

4. $10 < 12$

$10 \div (-2)$	$12 \div (-2)$
−5	−6

If the expressions in a true inequality are both multiplied (or divided) by a positive number the resulting inequality is also true.
If the expressions in a true inequality are both multiplied (or divided) by a negative number the resulting inequality is not true.

Discovery 7

$2x + 5 > 2x + 10$
$2x + 5 - 2x > 2x + 10 - 2x$ Subtract $2x$ from both sides.
$5 > 10$ Simplify.

A linear inequality has no solution if in the process of solving the variable term is deleted and the result is a false inequality.

Discovery 8

$2x + 5 < 2x + 10$
$2x + 5 - 2x < 2x + 10 - 2x$ Subtract $2x$ from both sides.
$5 < 10$ Simplify.

A linear inequality has a solution set of all real numbers if in the process of solving the variable term is deleted and the result is a true inequality.

Discovery 9

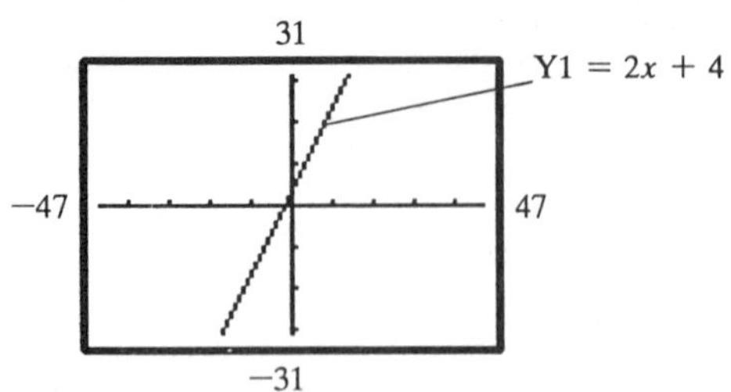

1. a. An infinite number of ordered pairs can be found.

   b. yes

2. a. An infinite number of ordered pairs can be found.

   b. no

3. a. An infinite number of ordered pairs can be found.

   b. yes

The solution set of a "less than or equal to" inequality that is solved for y may be illustrated by graphing the line for the related equation and shading all points below the line.

Discovery 10

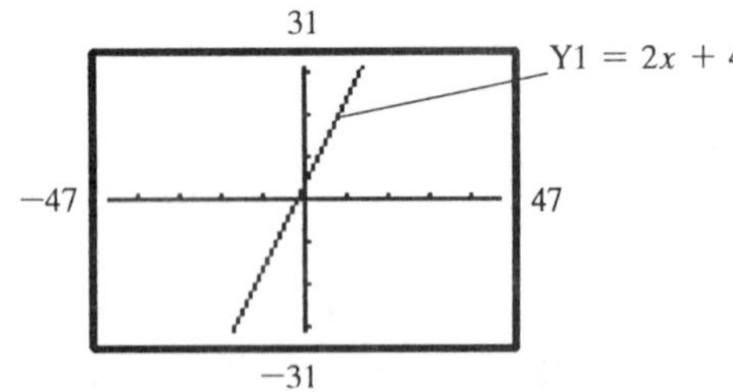

1. a. An infinite number of ordered pairs can be found.

   b. no

2. a. An infinite number of ordered pairs can be found.

   b. yes

3. a. An infinite number of ordered pairs can be found.

   b. no

The solution set of a "greater than" inequality that is solved for y may be illustrated by graphing the line for the related equation with a dashed line and shading all points above the line.

Discovery 11

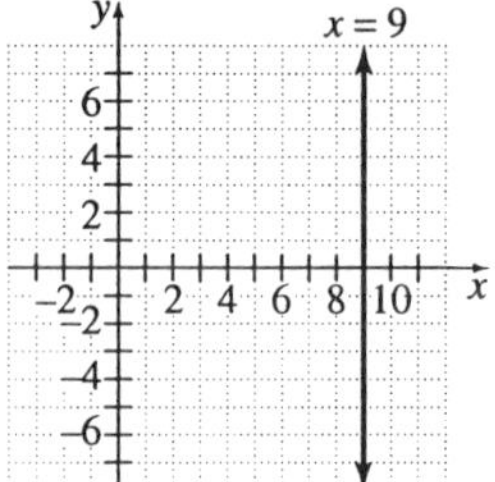

1. a. An infinite number of ordered pairs can be found.

   b. yes

   c. no

   d. yes

   e. no

2. a. An infinite number of ordered pairs can be found.

   b. yes

   c. yes

   d. no

   e. no

3. a. An infinite number of ordered pairs can be found.

   b. no

   c. no

d. yes

e. yes

The solution set of a linear inequality containing no y-term that is solved for x may be illustrated by graphing the line for the related equation. If the inequality symbol is $<$, then the line is dashed and all points to the left of the line are shaded. If the inequality symbol is $>$, then the line is dashed and all points to the right of the line are shaded. If the inequality symbol is $\leq$, then the line is solid and all points to the left of the line is shaded. If the inequality symbol is $\geq$, then the line is solid and all points to the right of the line is shaded.

Chapter 9

Discovery 1

1.

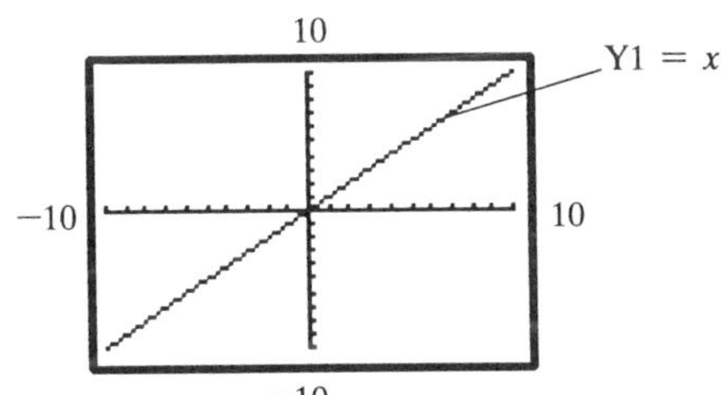

2.

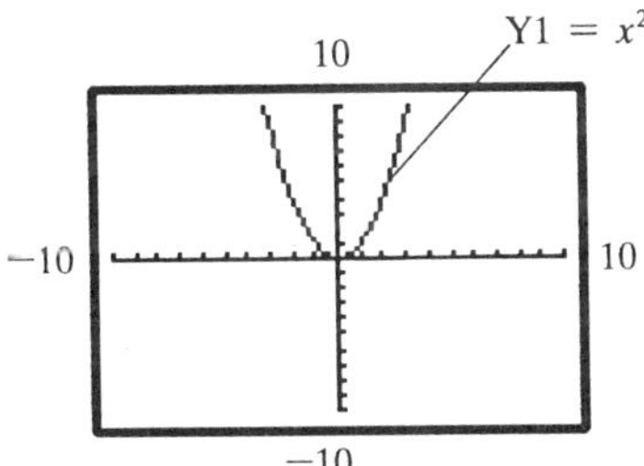

3.

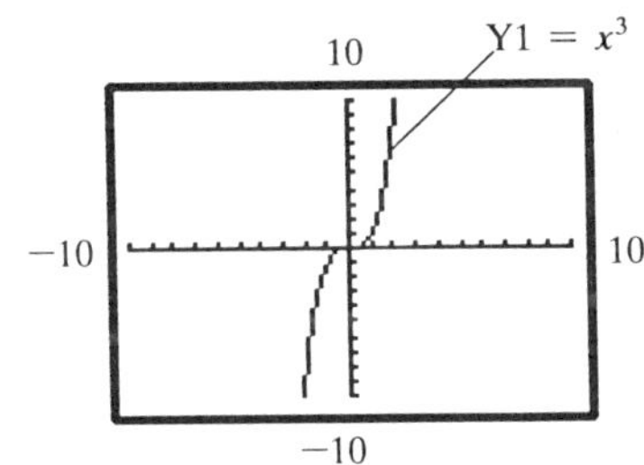

4.

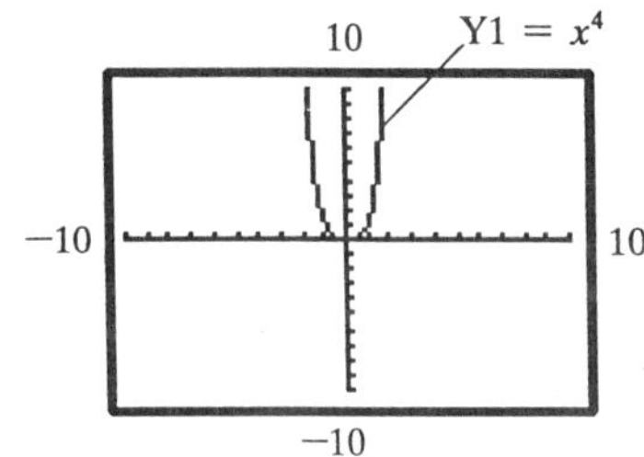

Each polynomial relation passes the vertical line test.

Discovery 2

1.–6.

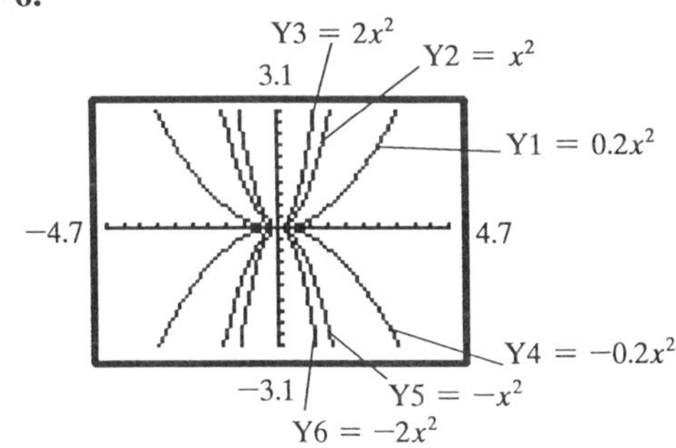

7. positive; upward

8. negative; downward

9. narrower

10. wider

Discovery 3

1.

x	y	
−3	9	
−2	4	
−1	1	
0	0	← vertex
1	1	
2	4	
3	9	

2.

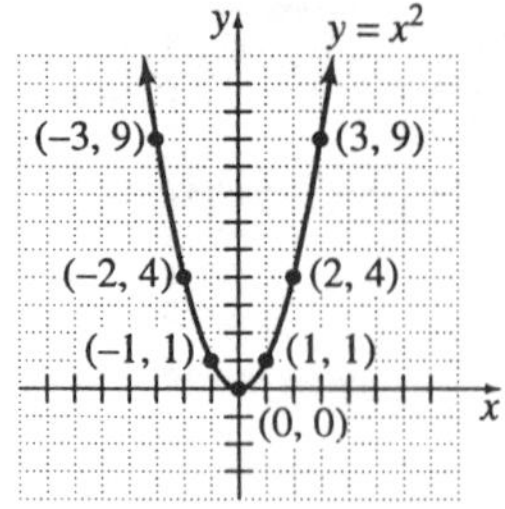

3. 1

4. 4

5. 9

Discovery 4

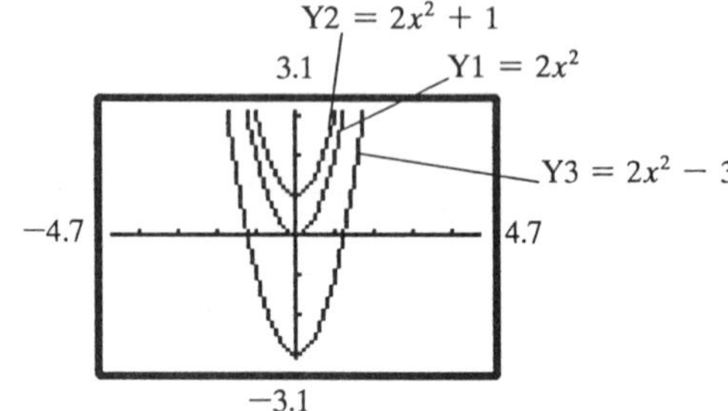

The *y*-coordinate of the *y*-intercept is the constant term.

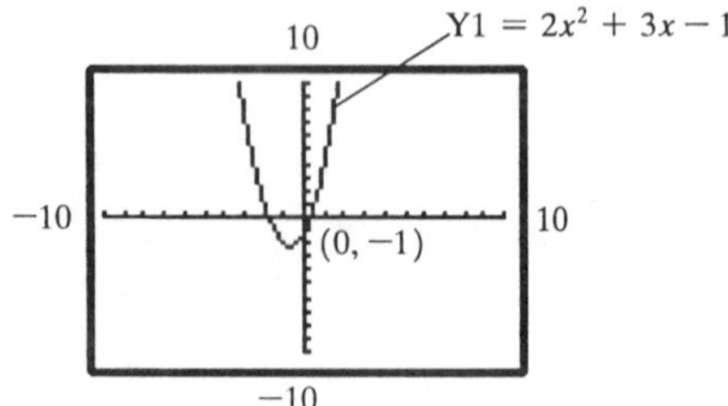

Chapter 10

Discovery 1

1. a. 1024

b. 1024

2. a. 64

b. 64

3. a. $\frac{27}{64}$

b. $\frac{27}{64}$

To multiply exponential expressions with like bases, add the exponents.

4. x^7

Discovery 2

1. a. 64

b. 64

2. a. 4

b. 4

3. a. $\frac{9}{16}$

b. $\frac{9}{16}$

4. a. indeterminate

b. indeterminate

To divide exponential expressions with like bases, subtract the exponent in the denominator from the exponent in the numerator.

5. x^4

Discovery 3

1. a. 4096

b. 4096

2. a. 256

b. 256

3. a. $\frac{729}{4096}$

b. $\frac{729}{4096}$

To raise an exponential expression to a power, multiply the exponents.

4. x^{12}

Discovery 4

1. a. 512

b. 512

2. a. 36

b. 36

3. a. $\frac{27}{1000}$

b. $\frac{27}{1000}$

Raising a product to a power is equivalent to the product of the factors raised to a power.

4. x^4y^4

Discovery 5

1. $x^2 - 25$

2. $9x^2 - 1$

3. $x^2 - y^2$

The product of the sum and difference of the same two terms is the difference of the square of the first term and the square of the second term.

Discovery 6

1. $x^2 + 10x + 25$

2. $9x^2 - 6x + 1$

3. $x^2 + 2xy + y^2$

4. $x^2 - 2xy + y^2$

The square of a binomial that is a sum of two terms is the square of the first term, plus two times the product of the first and last terms, plus the square of the last term.

The square of a binomial that is a difference of two terms is the square of the first term, minus two times the product of the first and last terms, plus the square of the last term.

Discovery 7

1. $x \cdot x \cdot x$; $x \cdot x \cdot y$; $x \cdot x \cdot x \cdot x$

2. x

3. 2

4. x^2

The GCF for a set of monomials with variable factors is the common variable with the smallest exponent common in the set of factors.

Discovery 8

1. a. $x^2 + 5x + 6$

b. 6; 5

2. a. $x^2 - 5x + 6$

b. 6; –5

The first term in each binomial factor is x, or the square root of the quadratic term, x^2. The second term in each of the binomial factors must be the factors of the constant term, c. The sum of the factors is the coefficient b.

Discovery 9

1. $(x + 3)(x + 2)$; positive
2. $(x - 3)(x - 2)$; negative
3. $(x + 3)(x - 2)$; positive
4. $(x - 3)(x + 2)$; negative

The signs of the factors are determined in the following way:

If c is positive, then the factors both have the same sign as b.
If c is negative, then the factors have different signs (the factor with the larger absolute value has the same sign as b).

Discovery 10

1. $x^3 + 125$
2. $x^3 - 125$

The first product is a sum of two cubes. The second product is a difference of two cubes. In order to factor these polynomials, we turn these results around and write a binomial and trinomial factor.

Chapter 11

Discovery 1

x	$4x^2 - x - 3$	$3x$	
–3	36	–9	$36 > -9$
–2	15	–6	$15 > -6$
–1	2	–3	$2 > -3$
0	–3	0	$-3 < 0$
1	0	3	$0 < 3$
2	11	6	$11 > 6$
3	30	9	$30 > 9$

The expression on the left is greater than the expression on the right for x-values of –3, –2, and –1. The expression on the left is less than the expression on the right for x-values of 0 and 1. Since each function is defined for all real numbers between –1 and 0, the expression on the left is equal to the expression on the right at some x-value between –1 and 0.

Also, the expression on the left is less than the expression on the right for x-values of 0 and 1 and the expression on the left is greater than the expression on the right for x-values of 2 and 3. Since each function is defined for all real numbers between 1 and 2, the expression on the left is equal to the expression on the right at some x-value between 1 and 2.

Discovery 2

x	$x^2 + x + 10$	$x^2 + x - 10$		
0	10	–10	$10 \neq -10$	$10 - (-10) = 20$
1	12	–8	$12 \neq -8$	$12 - (-8) = 20$
2	16	–4	$16 \neq -4$	$16 - (-4) = 20$
3	22	2	$22 \neq 2$	$22 - 2 = 20$

The expression on the right is always 20 less than the expression on the left. Therefore, it appears that the two expressions are never equal. The equation does not have a solution.

Discovery 3

x	$2x^3+4x^2+5$	$(x^3+3x^2+4)+(x^3+x^2+1)$	
0	5	5	$5=5$
1	11	11	$11=11$
2	37	37	$37=37$
3	95	95	$95=95$

The two expressions are equal for all values of the independent variable in the table. If we could view this table with all real-number values for x, the expressions would always be equal. Therefore, the solution of the equation is the set of all real numbers (the permissible replacements for the independent variable).

Discovery 4

x	x^2+x+1	$4x^2+4x+4$
0	1	4
1	3	12
2	7	28
3	13	52

The expression on the left is always less than the expression on the right. Therefore, the two expressions are never equal. It appears the equation does not have a solution.

Discovery 5

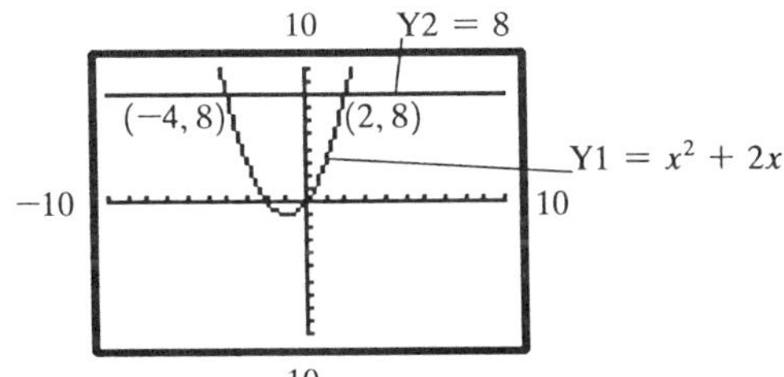

The values of the x-coordinates of the points of intersection of the two graphs are the solutions of the equation.

The values of the y-coordinates of the points of intersection of the two graphs are the values obtained when both expressions are evaluated with the solution.

Discovery 6

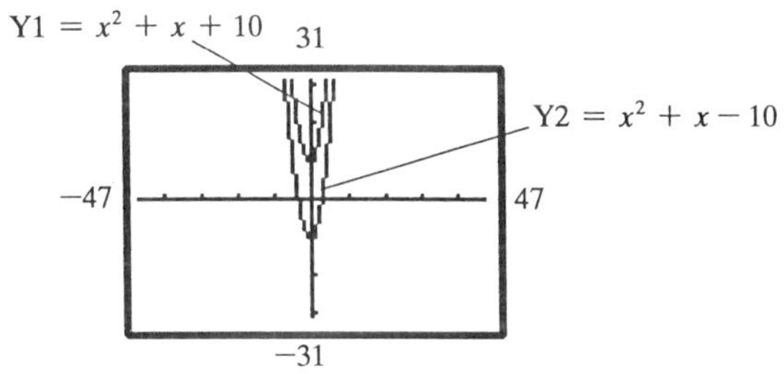

The two graphs do not appear to intersect. Therefore, there is no ordered pair common to both functions.

Discovery 7

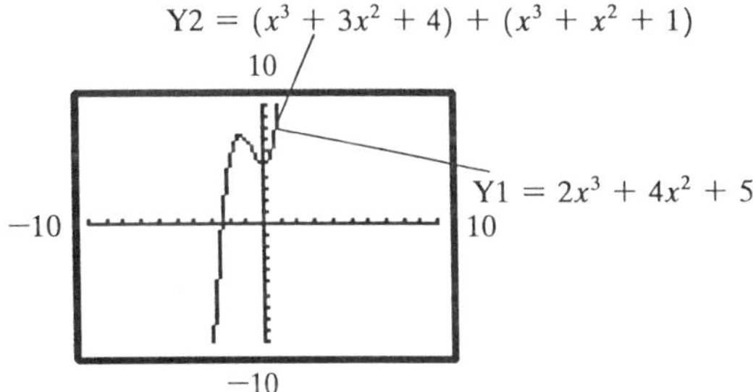

There is only one graph on the screen. (Actually, both graphs are the same.) Therefore, all ordered pairs on the graph are common to both functions, or all x-coordinates in the domain of the functions are solutions of the equation.

Discovery 8

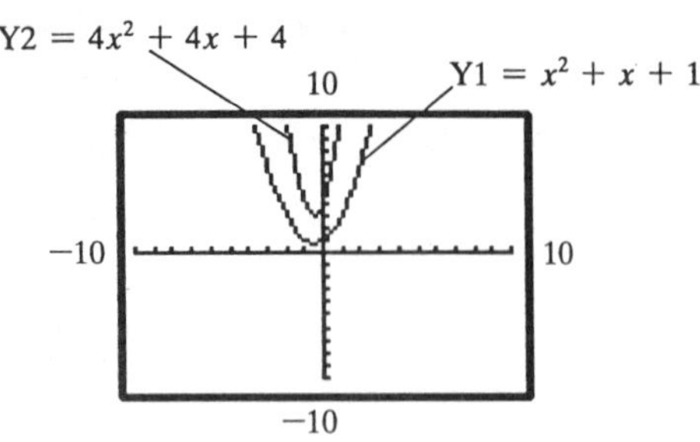

The two graphs do not appear to intersect. Therefore, there is no ordered pair common to both graphed functions. We need more information to determine the solution.

Discovery 9

1. a. 4.583

   b. 4.583

2. a. 2.45

   b. 2.45

To multiply square roots, we multiply the radicands and then take the square root of the product.

Discovery 10

1. a. 1.732

   b. 1.732

2. a. 1.414

   b. 1.414

To divide square roots, we divide the radicands and then take the square root of the quotient.

Discovery 11

1. 49

2. 41

3. 0

4. −7

The radicand, $b^2 - 4ac$, determines the characteristics of the quadratic equation solutions. If the radicand is a perfect square ($\neq 0$), there will be two rational-number solutions. If the radicand is 0, there will be one rational-number solution. If the radicand is a positive number that is not a perfect square, then there will be two irrational-number solutions. If the radicand is a negative number, there will be no real-number solutions.

Discovery 12

1.

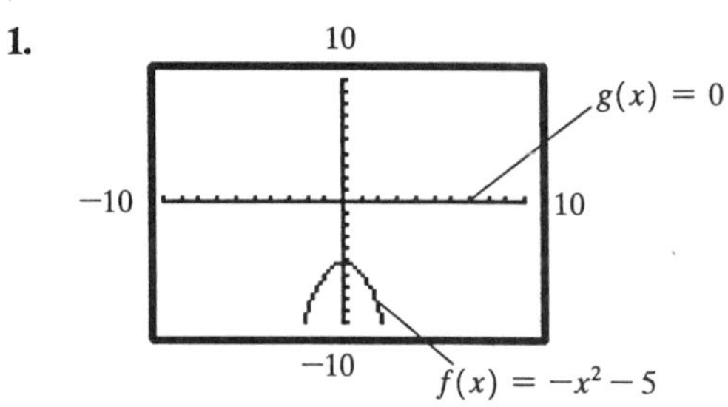

   a. no solution

   b. no solution

   c. all real numbers

   d. all real numbers

2.

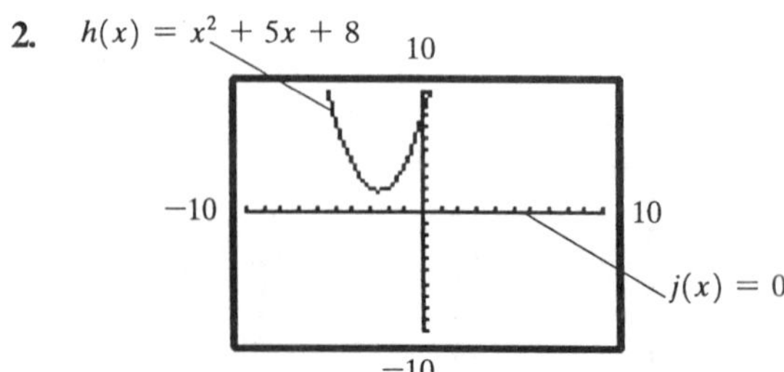

   a. all real numbers

   b. all real numbers

   c. no solution

   d. no solution

3.

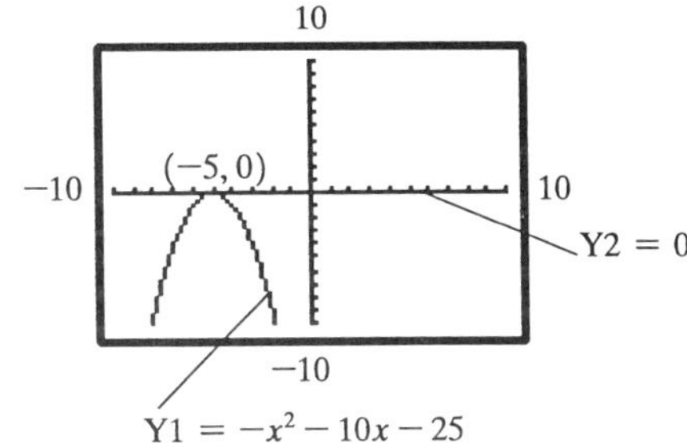

a. no solution

b. $x = -5$

c. $x \neq -5$

d. all real numbers

4.

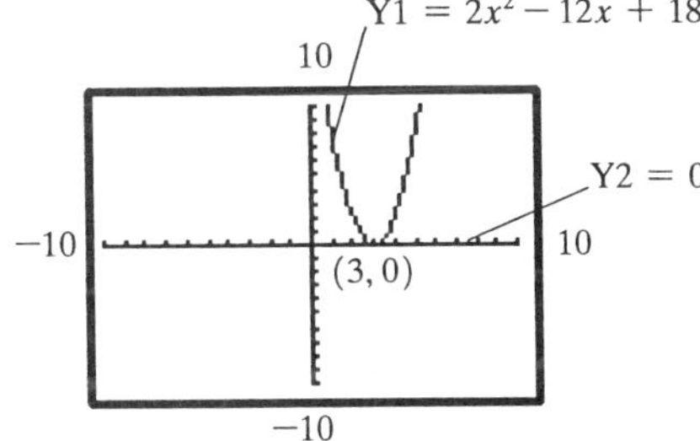

a. $x \neq 3$

b. all real numbers

c. no solution

d. $x = 3$

Chapter 12

Discovery 1

1. a. $f(x) = \dfrac{6}{x}$

x	$f(x)$
–3	–2
–2	–3
–1	–6
0	undefined
1	6
2	3
3	2

b. $g(x) = \dfrac{x+2}{x-1}$

x	$g(x)$
–3	$\frac{1}{4}$
–2	0
–1	$-\frac{1}{2}$
0	–2
1	undefined
2	4
3	$2\frac{1}{2}$

c. $h(x)=\dfrac{x^2-x-2}{x-2}$

x	$h(x)$
−3	−2
−2	−1
−1	0
0	1
1	2
2	undefined
3	4

2 a. $x=0$

b. $x=1$

c. $x=2$

3. There is a "hole" in the graph.

The restrictions on the domain of a rational function are the numbers that result in undefined values for the rational expression.

The restricted values (x-values) of the function do not have corresponding function values.

To determine the restricted values algebraically, set the expression in the denominator of the rational expression equal to zero and solve for x.

Discovery 2

1. In viewing the table, the original expression is undefined when $x = 0$ and the simplified expression is defined for all integers.

2. The original expression has a restriction of $x = 0$. The simplified expression has no restrictions.

3. When simplifying the original expression, we divided out an x in the numerator and denominator. Therefore, a factor of x in the original expression is deleted from the simplified expression.

Discovery 3

x	$\frac{x+4}{x-1}$	$\frac{x-4}{x-1}$
−3	−0.25	1.75
−2	−0.6667	2
−1	−1.5	2.5
0	−4	4
1	undefined	undefined
2	6	−2
3	3.5	−0.5

It appears that the expression on the left, Y1, is never equal to the expression on the right, Y2. The x-values less than the restricted value of 1 correspond to Y1 < Y2. The x-values greater than the restricted value of 1 correspond to Y1 > Y2. Since the restricted value is between this change, an alternative method is needed to confirm that there is no solution of the equation.

Discovery 4

x	$\frac{3}{x}$	$\frac{6}{2x}$
−3	−1	−1
−2	−1.5	−1.5
−1	−3	−3
0	undefined	undefined
1	3	3
2	1.5	1.5
3	1	1

The two expressions are equal for every value of the independent variable in the table. The domain for the independent variable is all real numbers not equal to 0. Therefore, it appears that the solution of the equation is the set of all real numbers not equal to 0.

Discovery 5

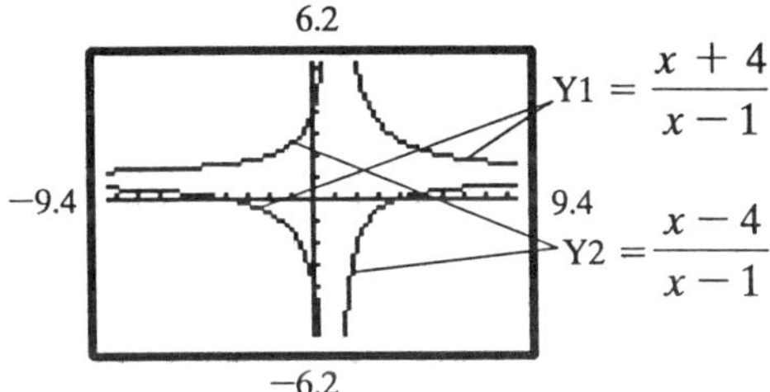

The two functions do not appear to intersect. There is no ordered pair common to both functions; therefore, there is no solution of the equation.

Discovery 6

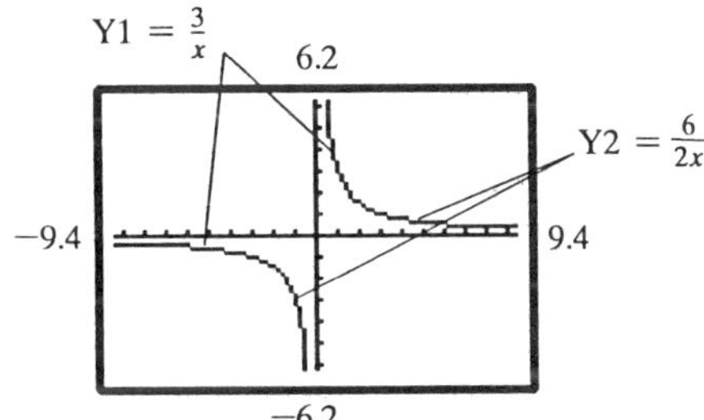

It appears that there is only one graph on the screen. Actually, both graphs are the same. Therefore, all ordered pairs on the graph are common to both functions, or all x-coordinates are solutions of the equation. The solution is the set of all real numbers not equal to 0 (the domain of the graphed functions).

Chapter 13

Discovery 1

1. a. 8

b. 8

c. 8

2. a. 16

b. 16

c. 16

3. a. 16

b. 16

c. 16

4. a. 16

b. 16

c. 16

The base of the expression having a rational exponent is the radicand of the radical expression. The denominator of the rational exponent is the index of the radical expression. The numerator of the rational exponent is the power the radical expression is raised to, or the power the radicand is raised to.

Discovery 2

1. a. $f(x) = \sqrt{x}$

x	$f(x)$
–3	no real number
–2	no real number
–1	no real number
0	0
1	1

b. $g(x) = \sqrt[4]{x+1}$

x	$g(x)$
–3	no real number
–2	no real number
–1	0
0	1
1	1.1892

c. $h(x)=\sqrt[3]{x}$

x	$h(x)$
−3	−1.442
−2	−1.26
−1	−1
0	0
1	1

d. $j(x)=\sqrt[5]{x+1}$

x	$j(x)$
−3	−1.149
−2	−1
−1	0
0	1
1	1.1487

2 **a.** $x=-3,\ x=-2,\ x=-1$

b. $x=-3,\ x=-2$

c. none

d. none

3. The even roots have restricted values.

4. There appears to be a "hole" in the graph at the restricted values.

In viewing a table of values, the numbers restricted from the domain of a radical function are the numbers that result in nonreal numbers when evaluating the radical expression.

When viewing the function's graph, the restricted values of the function do not have corresponding function values.

Since these restricted values occur when the function is evaluated for a nonreal value, and this is when the radicand of an even root is negative, or less than zero, we can determine the restricted values of even roots algebraically by setting the radicand less than zero and solving for the independent variable.

Discovery 3

1. a. −2.884

b. −2.884

2. a. 2.632

b. 2.632

To multiply radicals with like indices, we multiply the radicands and then take the root of the product.

Discovery 4

Divide the exponent of the variable in the radicand by the index of the radical. The result is a mixed numeral. The simplification is the product of the variable raised to a power of the whole number part of the mixed number and the root of the variable raised to the power of the numerator of the fractional part of the mixed number.

Discovery 5

1. a. 1.587

b. 1.587

2. a. 1.682

b. 1.682

To divide radicals with like indices, we divide the radicands and then take the root of the quotient.

Discovery 6

1. a. 2.828

b. 2.828

2. a. 4.243

b. 4.243

3. a. 1.414

b. 1.414

4. a. 2.52

b. 2.52

5. a. 3.78

b. 3.78

6. a. 1.26

b. 1.26

To add (or subtract) like radicals, we add (or subtract) the coefficients of each like radical.

Discovery 7

1. -1

2. $4 - x$

The factors contained square roots but the product did not contain a square root.

Discovery 8

x	$\sqrt{x+2}-5$	$\sqrt{x+2}+8$
−3	no real number	no real number
−2	−5	8
−1	−4	9
0	−3.586	9.4142
1	−3.268	9.7321
2	−3	10
3	−2.764	10.236

It appears that the expression on the left is never equal to the expression on the right. In fact, the expression on the left is always 13 less than the expression on the right. Therefore, the equation does not appear to have a solution.

Discovery 9

x	$\sqrt{x}$	$\sqrt[6]{x^3}$
0	0	0
1	1	1
2	1.4142	1.4142
3	1.7321	1.7321

The two expressions are equal for every value of the independent variable in the table. The domain for the independent variable is all real numbers greater than or equal to 0. Therefore, it appears that the solution of the equation is the set of all real numbers greater than or equal to 0.

Discovery 10

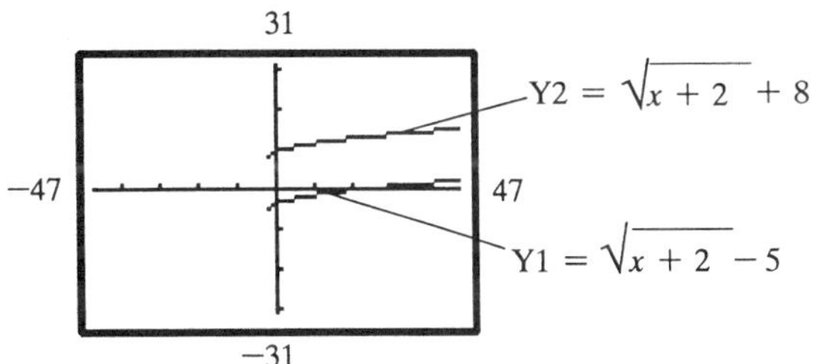

The two functions do not appear to intersect. Therefore, there is no ordered pair common to both functions. There appears to be no solution to the equation.

Discovery 11

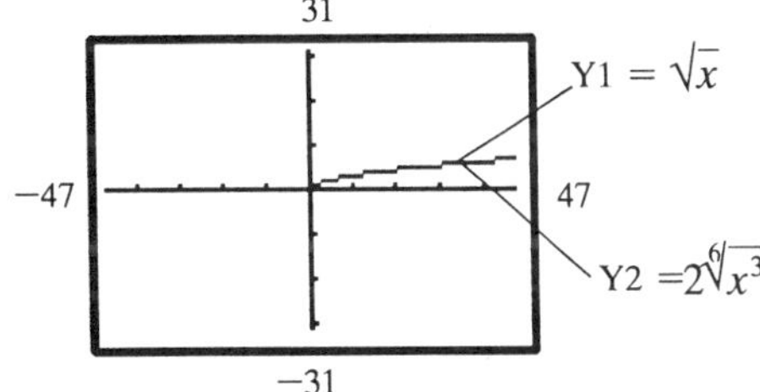

It appears that there is only one graph on the screen. Actually, both graphs are the same. You can verify this by tracing and using the arrow keys to move from one graph to the another. Therefore, all ordered pairs on the graph are in common to both functions, or all x-coordinates are solutions of the equation. The solution set is all real numbers greater than or equal to 0 (the domain of the graphed functions).

Discovery 12

1. true
2. true
3. true
4. true
5. true
6. true
7. true
8. true
9. true

If we begin with an equation that is true and we raise both expressions contained in the equation to the same power, the result is a true equation.

Chapter 14

Discovery 1

1. **a.** $-i$

   b. 1

   c. i

   d. -1

2. $-i$

3. 1

Divide the power of i by 4, if the remainder is 1 the value of the expression is i; if the remainder is 2 the value of the expression is -1; if the remainder is 3 the value of the expression is $-i$; if the remainder is 0 the value of the expression is 1.

Discovery 2

1. 13
2. 17

The product of $(a+bi)(a-bi)$ is the sum of the squares of a and b.

Chapter 15

Discovery 1

1. **a.**

Domain	Range
0	2
1	5
2	8
3	11

   b.

Domain	Range
0	2
1	3
−1	3
−2	6

2. **a.** $\{(2,0), (5,1), (8,2), (11,3)\}$

   b. $\{(2,0), (3,1), (3,-1), (6,-2)\}$

3. **a.**

Domain	Range
2	0
5	1
8	2
11	3

   b.

Domain	Range
2	0
3	1
3	−1
6	−2

4. **a.** yes

   b. no

If each element in the range of a function is paired with only one element in the domain of the function, the inverse of the function is a function.

Discovery 2

A sample horizontal line is drawn for each figure.

1.

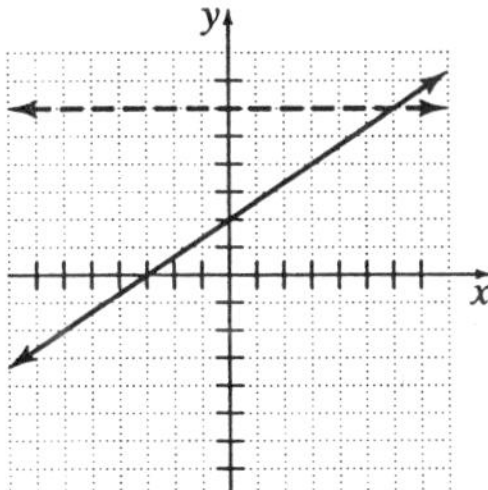

2.

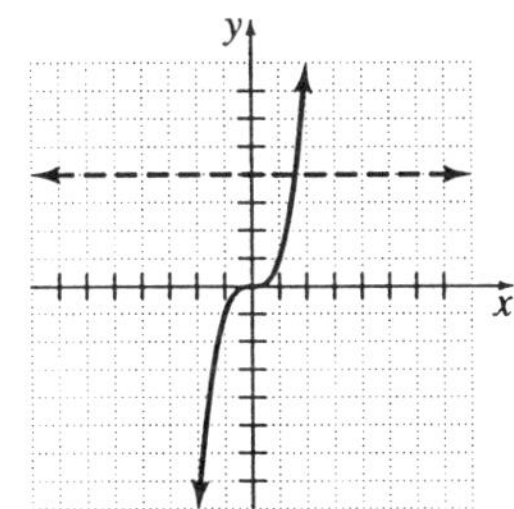

3.

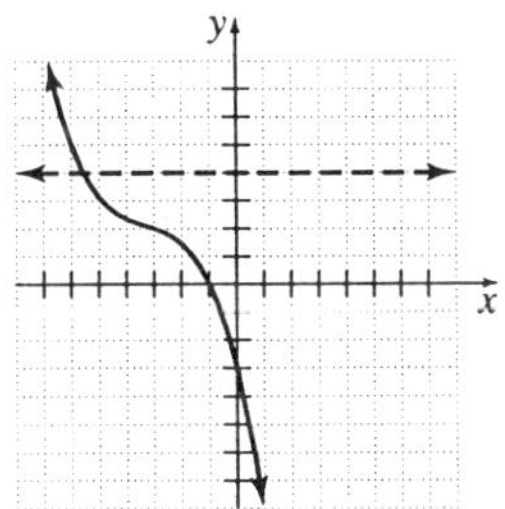

4.

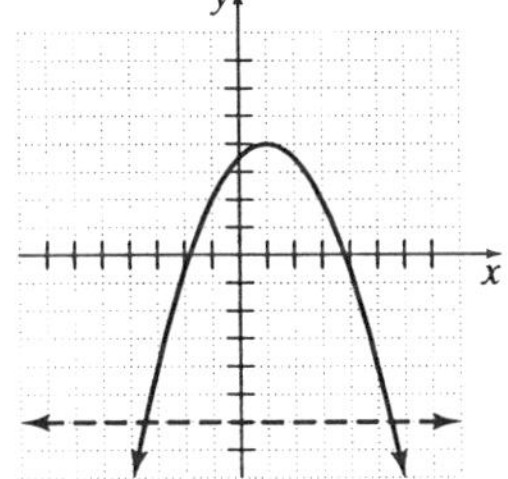

Discovery 3

1. a.

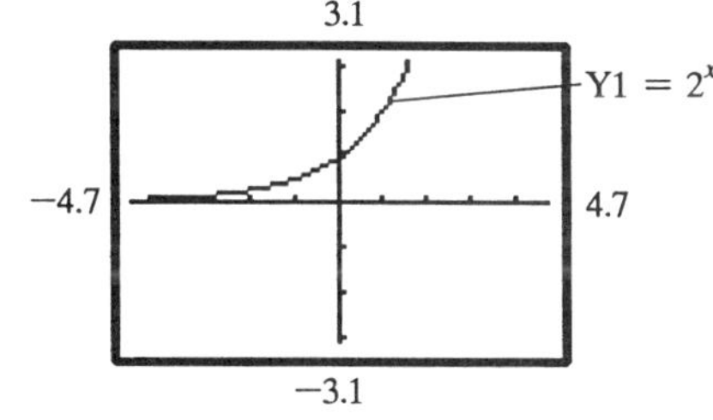

b.

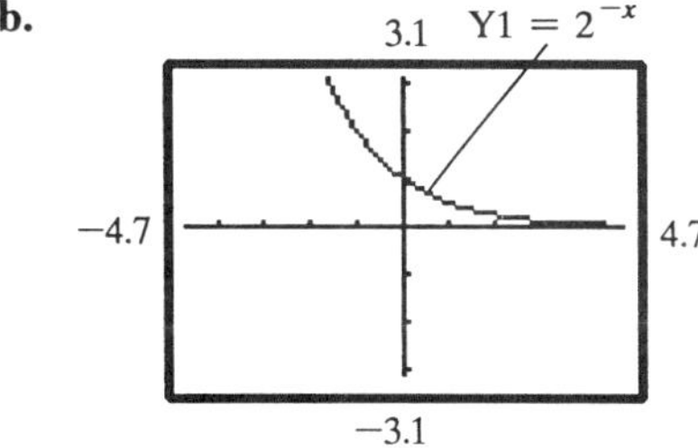

c.

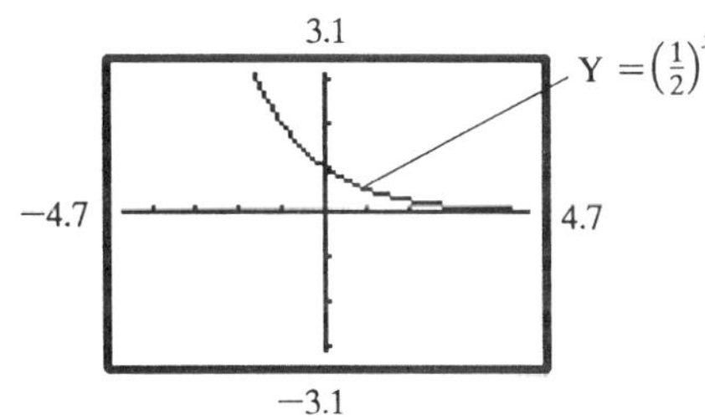

d.

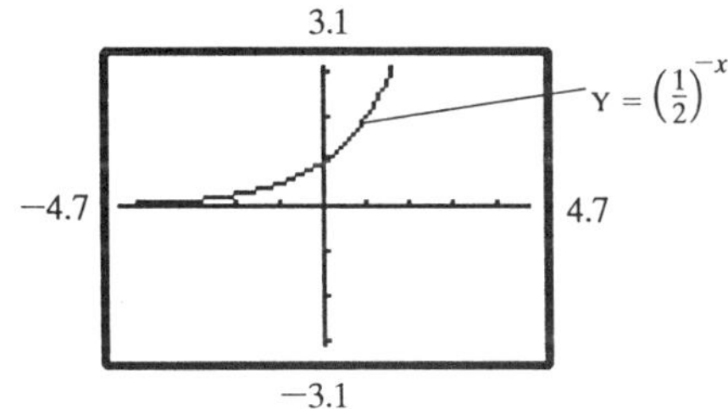

2. a and d; b and c

An exponential function $f(x) = a^{-x}$, where $a > 1$, is equivalent to the function $f(x) = \left(\frac{1}{a}\right)^x$.

Discovery 4

1.

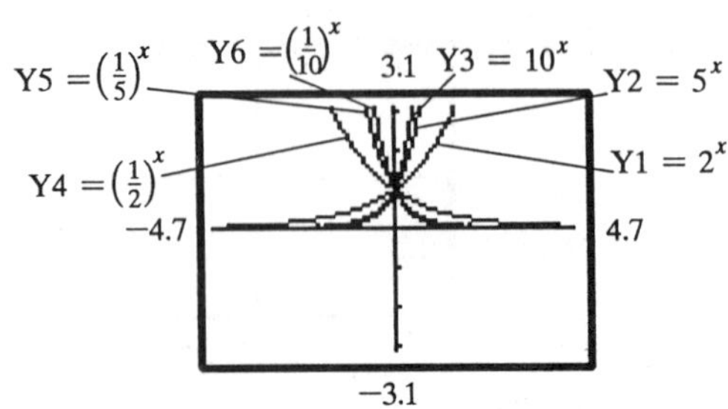

2. all real numbers
3. all real numbers greater than 0
4. yes
5. (0,1)
6. There is no x-intercept.
7. increasing; steeper
8. decreasing; steeper

Discovery 5

1.

x	y
1	2
10	2.5937
100	2.7048
1000	2.7169
10,000	2.7181
100,000	2.7183
1,000,000	2.7183

2. 2.7183

Discovery 6

1.

x	$y=2^x$	$y=x^2$
−5	0.03125	25
−4	0.0625	16
−3	0.125	9
−2	0.25	4
−1	0.5	1
0	1	0
1	2	1
2	4	4
3	8	9
4	16	16
5	32	25
6	64	36
7	128	49
8	256	64
9	512	81
10	1024	100
11	2048	121
12	4096	144
13	8192	169
14	16,384	196
15	32,768	225

2. $y=2^x$

Discovery 7

1. a. 0

   b. 0

2. a. 1

b. 1

3. a. 2

b. 2

4. a. undefined

b. undefined

5. a. undefined

b. undefined

6. 0

7. 1

8. 2

9. undefined

10. undefined

Discovery 8

1. a. 1.0792

b. 1.0792

c. 2.3026

d. 2.3026

2. a. –0.1249

b. –0.1249

c. –0.9163

d. –0.9163

3. a. 1.9085

b. 1.9085

c. 3.4657

d. 3.4657

The logarithm of a product is the sum of the logarithms of the factors.

The logarithm of a quotient is the difference of the logarithms of the numerator and the denominator.

The logarithm of a base raised to a power is the product of the power and the logarithm of the base.

Discovery 9

1. a. true

b. true

c. true

Raise both sides to the same base of a true equation to obtain a true equation.

2. a. true

b. true

c. true

Take the logarithm with the same base of both sides of a true equation to obtain a true equation.

Chapter 1

1.1 Experiencing Algebra the Exercise Way

1. **a.** 15: integer, rational

   b. 29: natural, whole, integer, rational

   c. 1 million: natural, whole, integer, rational

   d. $\frac{3}{7}$: rational

3. **a.** 0: whole, integer, rational

   b. 12.75: rational

   c. $2\frac{4}{9}$: rational

   d. –8.35: rational

5. wins \$3 = 3
   wins \$1.75 = 1.75
   loses \$2.00 = –2
   loses \$1.25 = –1.25

7. in the red \$345.67 = –345.67

9. correct = 5
   incorrect = –5
   unanswered = –2

11. $\frac{17}{3} = 5\frac{2}{3}$
    Graph between 5 and 6.

13. $-3.5 \quad -2 \quad -1\frac{1}{4} \quad -\frac{4}{5} \quad \frac{1}{2} \quad 1.5 \quad \frac{9}{4} \quad 4$
    –4 –3 –2 –1 0 1 2 3 4

15. $-9 < -5$

17. $0 > -6$

19. $5 > 3$

21. $-\frac{1}{2} < -\frac{2}{5}$

23. $\frac{3}{7} < \frac{7}{10}$

25. $2\frac{3}{5} > -2\frac{3}{5}$

27. $1\frac{4}{5} = 1.8$

29. $-3.7 > -5.8$

31. $1.7 < 3.2$

33. T

35. F

37. F

39. F

41. F

43. F

45. F

47. T

49. F

51. $0.295 < 0.7$

53. $-1 > -5$

55. $0.4 = \frac{2}{5}$

57. $|15.34| = 15.34$

59. $|-15.34| = 15.34$

61. $\left|-\left(-3\frac{1}{3}\right)\right| = \left|3\frac{1}{3}\right| = 3\frac{1}{3}$

63. $\left|-\left(-\left(-3\frac{1}{3}\right)\right)\right| = \left|-3\frac{1}{3}\right| = 3\frac{1}{3}$

65. $-|23| = -23$

67. $-|-23| = -|23| = -23$

69. $-(15) = -15$

71. $-(-25) = 25$

73. $-\left(-\left(-\frac{3}{2}\right)\right) = -\left(\frac{3}{2}\right) = -\frac{3}{2}$

75. loss of $712 million = –712

1.1 Experiencing Algebra the Calculator Way

1. F

2. F

3. F

```
-193987< -194879
               0
123/579>296/978
               0
15+11/19<15+9/16
               0
```

4. F

5. F

```
119/250>0.476
               0
-149/275< -356/56
7
               0
```

6. F

7. F

```
-13.0987>-13.087
9
               0
-129/500=-0.2585
               0
```

8. T

9. T

10. F

```
12.368=1546/125
               1
abs( -1.5)=1.5
               1
abs( -1.5)=-1.5
               0
```

11. F

12. T

```
abs( -3/2)=-1.5
               0
abs( -3/2)=1.5
               1
```

13. F

14. T

```
abs( -4)=-4
               0
-abs(4)=-4
               1
```

1.2 Experiencing Algebra the Exercise Way

1. $-7 + 9 = 2$

3. $0 + (-13) = -13$

5. $-19 + (-12) = -31$

7. $15 + (-15) = 0$

9. $52 + (-13) = 39$

11. $32 + (-579) = -547$

13. $2.7 + 3.96 = 6.66$

15. $1.2 + (-2.5) = -1.3$

17. $-2.73 + 4.1 = 1.37$

19. $-1.1 + (-2.27) = -3.37$

21. $-1.25 + 0 = -1.25$

23. $1.23 + (-1.23) = 0$

25. $-\frac{3}{5}+\left(-\frac{1}{2}\right)$
$=-\frac{6}{10}+\left(-\frac{5}{10}\right)$
$=-\frac{11}{10}$

27. $-\frac{7}{9}+\frac{1}{6}$
$=-\frac{14}{18}+\frac{3}{18}$
$=-\frac{11}{18}$

29. $\frac{2}{3}+\left(-\frac{2}{9}\right)$
$=\frac{6}{9}+\left(-\frac{2}{9}\right)$
$=\frac{4}{9}$

31. $\frac{2}{3}+0=\frac{2}{3}$

33. $\frac{1}{3}+\left(-\frac{1}{3}\right)=0$

35. $-2\frac{3}{4}+3\frac{2}{3}$
$=-2\frac{9}{12}+3\frac{8}{12}$
$=-\frac{33}{12}+\frac{44}{12}$
$=\frac{11}{12}$

37. $1+\left(-4\frac{3}{4}\right)=-3\frac{3}{4}$

39. $-13 + (-25) + (-26)$
$= -38 + (-26)$
$= -64$

41. $17 + (-23) + (-16) + 13$
$= -6 + (-16) + 13$
$= -22 + 13$
$= -9$

43. $1 + (-2.3) + (-5.71)$
$= -1.3 + (-5.71)$
$= -7.01$

45. $-\frac{1}{2}+\frac{1}{3}+\left(-\frac{1}{4}\right)$
$=-\frac{6}{12}+\frac{4}{12}+\left(-\frac{3}{12}\right)$
$=-\frac{2}{12}+\left(-\frac{3}{12}\right)$
$=-\frac{5}{12}$

47. $1124 + (-923) + 2305 + (-1156) + (-109)$
$= 201 + 2305 + (-1156) + (-109)$
$= 2506 + (-1156) + (-109)$
$= 1350 + (-109)$
$= 1241$

49. $1500 + 150 + 150 + 150 + (-75) + (-500) + (-200) + 12$
$= 1650 + 150 + 150 + (-75) + (-500) + (-200) + 12$
$= 1800 + 150 + (-75) + (-500) + (-200) + 12$
$= 1950 + (-75) + (-500) + (-200) + 12$
$= 1875 + (-500) + (-200) + 12$
$= 1375 + (-200) + 12$
$= 1175 + 12$
$= 1187$
The net balance of Lindsay's account is \$1187.

51. $-148{,}561 + 5{,}253{,}628 = 5{,}105{,}067$
The census for 1990 of Cook County, Illinois was 5,105,067.

53. $8+\left(-4\frac{3}{4}\right)+22\frac{1}{2}$
$=8+\left(-4\frac{3}{4}\right)+22\frac{2}{4}$
$=\frac{32}{4}+\left(-\frac{19}{4}\right)+\frac{90}{4}$
$=\frac{13}{4}+\frac{90}{4}$
$=\frac{103}{4}$
$=25\frac{3}{4}$
Heath had a net gain of $25\frac{3}{4}$ yards.

55. $-255 + 375 + (-575) + 1525$
$= 120 + (-575) + 1525$
$= -455 + 1525$
$= 1070$
Karin is \$1070 above her quota, because 1070 is a positive number.

57. $-2.5 + 1.25 + (-1.8) + (-2.5)$
$= -1.25 + (-1.8) + (-2.5)$
$= -3.05 + (-2.5)$
$= -5.55$
Beth lost 5.5 pounds in four weeks.

59. $378 + 322 + 218$
$= 700 + 218$
$= 918$
Liu drove 918 miles

61. $25 + 8 + (-12) + 17$
$= 33 + (-12) + 17$
$= 21 + 17$
$= 38$
Julia had 38 chips at the end of the third hand.

63. $\frac{1}{2}+\frac{1}{4}+1+1+\frac{1}{2}$
$=\frac{2}{4}+\frac{1}{4}+\frac{4}{4}+\frac{4}{4}+\frac{2}{4}$
$=\frac{2+1+4+4+2}{4}$
$=\frac{13}{4}$
$=3\frac{1}{4}$
The recipe requires $3\frac{1}{4}$ cups of dry ingredients.

1.2 Experiencing Algebra the Calculator Way

A.

1. $\frac{55}{7}=7\frac{6}{7}$

```
55/7
          7.857142857
Ans-7
          .8571428571
Ans▸Frac
                  6/7
```

2. $-\frac{295}{113}=-2\frac{69}{113}$

```
-295/113
         -2.610619469
Ans+2
          -.610619469
Ans▸Frac
              -69/113
```

3. $\frac{1227}{487}=2\frac{253}{487}$

```
1227/487
          2.519507187
Ans-2
          .5195071869
Ans▸Frac
              253/487
```

4. $-\frac{108}{19}=-5\frac{13}{19}$

```
-108/19
          -5.684210526
Ans+5
          -.6842105263
Ans▸Frac
                -13/19
```

6. $15\frac{17}{19}+\left(-21\frac{13}{17}\right)=-5\frac{281}{323}$

```
(15+17/19)-(21+1
3/17)
            -5.86996904
Ans+5
           -.8699690402
Ans▸Frac
               -281/323
```

B.

1. $-\frac{171}{92}+\left(-\frac{170}{33}\right)=-7\frac{31}{3036}$

```
-(171/92)+(-170/
33)
           -7.010210804
Ans+7
           -.0102108037
Ans▸Frac
              -31/3036
```

2. $\frac{171}{92}+\left(-\frac{170}{33}\right)=-3\frac{889}{3036}$

```
(171/92)+(-170/3
3)
           -3.292819499
Ans+3
           -.2928194993
Ans▸Frac
             -889/3036
```

3. $-\frac{171}{92}+\left(\frac{170}{33}\right)=3\frac{889}{3036}$

```
-(171/92)+(170/3
3)
            3.292819499
Ans-3
            .2928194993
Ans▸Frac
              889/3036
```

4. $\frac{171}{92}+\left(\frac{170}{33}\right)=7\frac{31}{3036}$

```
(171/92)+(170/33
)
            7.010210804
Ans-7
            .0102108037
Ans▸Frac
               31/3036
```

5. $15\frac{17}{19}+21\frac{13}{17}=37\frac{213}{323}$

```
(15+17/19)+(21+1
3/17)
            37.65944272
Ans-37
            .6594427245
Ans▸Frac
               213/323
```

7. $-15\frac{17}{19}+21\frac{13}{17}=5\frac{281}{323}$

```
-(15+17/19)+(21+
13/17)
             5.86996904
Ans-5
            .8699690402
Ans▸Frac
               281/323
```

8. $-15\frac{17}{19}+\left(-21\frac{13}{17}\right)=-37\frac{213}{323}$

```
-(15+17/19)-(21+
13/17)
           -37.65944272
Ans+37
           -.6594427245
Ans▸Frac
              -213/323
```

9. $437.925 + (-108.0065) = 329.9185$

10. $-437.925 + (-108.0065) = -545.9315$

```
437.925+-108.006
5
               329.9185
-437.925+-108.00
65
              -545.9315
```

11. $-437.925 + 108.0065 = -329.9185$

12. $437.925 + 108.0065 = 545.9315$

```
-437.925+108.006
5
              -329.9185
437.925+108.0065
               545.9315
```

13. 893,475 + 1,093,007 = 1,986,482

14. –893,475 + 1,093,007= 199,532

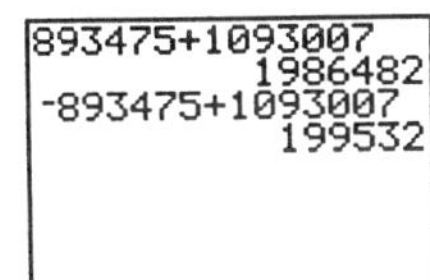

15. 893,475 + (–1,093,007) = –199,532

16. –893,475 + (–1,093,007) = –1,986,482

```
893475+-1093007
        -199532
-893475+-1093007
       -1986482
```

1.3 Experiencing Algebra the Exercise Way

1. –7 –9 = –7 + (–9) = –16

3. 0 – (–13) = 0 + 13 = 13

5. –19 – (–12) = –19 + 12 = –7

7. 15 – (–15) = 15 + 15 = 30

9. 52 – (–13) = 52 + 13 = 65

11. 32 – 579 = 32 + (–579) = –547

13. 1.2 – (–2.5) = 1.2 + 2.5 = 3.7

15. –2.73 – 4.1 = –2.73 + (–4.1) = –6.83

17. –1.1 – (–2.27) = –1.1 + 2.27 = 1.17

19. –1.25 – 0 = –1.25

21. 1.23 – (–1.23) = 1.23 + 1.23 = 2.46

23. 2.7 – 3.96 = 2.7 + (–3.96) = –1.26

25. $-\frac{3}{5}-\left(-\frac{1}{2}\right)=-\frac{3}{5}+\frac{1}{2}=-\frac{6}{10}+\frac{5}{10}=-\frac{1}{10}$

27. $-\frac{7}{9}-\frac{1}{6}$
$=-\frac{7}{9}+\left(-\frac{1}{6}\right)$
$=-\frac{14}{18}+\left(-\frac{3}{18}\right)$
$=-\frac{17}{18}$

29. $\frac{2}{3}-\left(-\frac{7}{9}\right)=\frac{2}{3}+\frac{7}{9}=\frac{6}{9}+\frac{7}{9}=\frac{13}{9}$

31. $\frac{2}{3}-0=\frac{2}{3}$

33. $\frac{1}{3}-\left(-\frac{1}{3}\right)=\frac{1}{3}+\frac{1}{3}=\frac{2}{3}$

35. $\frac{3}{7}-3=\frac{3}{7}+(-3)=\frac{3}{7}+\left(-\frac{21}{7}\right)=-\frac{18}{7}$

37. $1\frac{5}{6}-\frac{1}{3}=1\frac{5}{6}+\left(-\frac{1}{3}\right)=\frac{11}{6}+\left(-\frac{2}{6}\right)=\frac{9}{6}=\frac{3}{2}$

39. $-5-\left(-1\frac{4}{5}\right)=-5+1\frac{4}{5}=-\frac{25}{5}+\frac{9}{5}=-\frac{16}{5}$

41. $3\frac{2}{3}-5=3\frac{2}{3}+(-5)=\frac{11}{3}+\left(-\frac{15}{3}\right)=-\frac{4}{3}$

43. –13 – (–25) – (–26)
= –13 + 25 + 26
= 12 + 26
= 38

45. 17 – (–23) – (–16) – 13
= 17 + 23 + 16 + (–13)
= 40 + 16 + (–13)
= 56 + (–13)
= 43

47. 1.2 – (–2.31) – (–5.7)
= 1.2 + 2.31 + 5.7
= 3.51 + 5.7
= 9.21

49. $5-(-7)-(-2.3)$
$=5+7+2.3$
$=12+2.3$
$=14.3$

51. $-\frac{1}{2}-\frac{1}{3}-\left(-\frac{1}{4}\right)$
$=-\frac{1}{2}+\left(-\frac{1}{3}\right)+\frac{1}{4}$
$=-\frac{6}{12}+\left(-\frac{4}{12}\right)+\frac{3}{12}$
$=-\frac{10}{12}+\frac{3}{12}$
$=-\frac{7}{12}$

53. $1124-(-924)-2305-(-1156)-(-109)$
$=1124+924+(-2305)+1156+109$
$=2048+(-2305)+1156+109$
$=-257+1156+109$
$=899+109$
$=1008$

55. $23+56-34+(-12)-68-(-31)$
$=23+56+(-34)+(-12)+(-68)+31$
$=79+(-34)+(-12)+(-68)+31$
$=45+(-12)+(-68)+31$
$=33+(-68)+31$
$=-35+31$
$=-4$

57. $1\frac{1}{5}-2\frac{3}{10}+\frac{4}{5}-\left(-\frac{7}{10}\right)+\left(-\frac{3}{5}\right)$
$=1\frac{1}{5}+\left(-2\frac{3}{10}\right)+\frac{4}{5}+\frac{7}{10}+\left(-\frac{3}{5}\right)$
$=1\frac{2}{10}+\left(-2\frac{3}{10}\right)+\frac{8}{10}+\frac{7}{10}+\left(-\frac{6}{10}\right)$
$=\frac{12}{10}+\left(-\frac{23}{10}\right)+\frac{8}{10}+\frac{7}{10}+\left(-\frac{6}{10}\right)$
$=-\frac{11}{10}+\frac{8}{10}+\frac{7}{10}+\left(-\frac{6}{10}\right)$
$=-\frac{3}{10}+\frac{7}{10}+\left(-\frac{6}{10}\right)$
$=\frac{4}{10}+\left(-\frac{6}{10}\right)$
$=-\frac{2}{10}$
$=-\frac{1}{5}$

59. $3.75-1.2+(-1.09)-(-0.76)+13.13$
$=3.75+(-1.2)+(-1.09)+0.76+13.13$
$=2.55+(-1.09)+0.76+13.13$
$=1.46+0.76+13.13$
$=2.22+13.13$
$=15.35$

61. $98-(-90)=98+90=188$
The range is 188°F.

63. $130-(-180)=130+180=310$
The change of the mean surface temperature is 310°C.

65. $20{,}320-(-282)=20{,}320+282=20{,}602$
The difference is 20,602 feet.

67. $420.35+185.00+75.00+(-50.00)+(-120.00)+(-12.55)+(-110.76)+(-5.50)$
$=680.35+(-298.81)$
$=381.54$
Rolanda's current balance is \$381.54.

69. $20+\left(-2\frac{1}{2}\right)+\left(-1\frac{2}{3}\right)+(-1)+\left(-2\frac{1}{4}\right)$
$=\frac{240}{12}+\left(-2\frac{6}{12}\right)+\left(-1\frac{8}{12}\right)+\left(-\frac{12}{12}\right)+\left(-2\frac{3}{12}\right)$
$=\frac{240}{12}+\left(-\frac{30}{12}\right)+\left(-\frac{20}{12}\right)+\left(-\frac{12}{12}\right)+\left(-\frac{27}{12}\right)$
$=\frac{210}{12}+\left(-\frac{20}{12}\right)+\left(-\frac{12}{12}\right)+\left(-\frac{27}{12}\right)$
$=\frac{190}{12}+\left(-\frac{12}{12}\right)+\left(-\frac{27}{12}\right)$
$=\frac{178}{12}+\left(-\frac{27}{12}\right)$
$=\frac{151}{12}$
$=12\frac{7}{12}$
Betty has $12\frac{7}{12}$ cups of flour left.

71. 25 + (–5.75) + 15 + (–12.50) + (–4.50) + 2.35
= 19.25 + 15 + (–12.50) + (–4.50) + 2.35
= 34.25 + (–12.50) + (–4.50) + 2.35
= 21.75 + (–4.50) + 2.35
= 17.25 + 2.35
= 19.60
There is \$19.60 left in the petty cash fund.

73. 19.95 + 19.95 + 19.95 + (–25) + (–59.27) + (–19.95)
= 59.85 + (–104.22)
= –44.37
Rosie lost \$44.37 that morning.

1.3 Experiencing Algebra the Calculator Way

1. $-\frac{171}{92}-\left(-\frac{170}{33}\right)=3\frac{889}{3036}$

```
-171/92--170/33
      3.292819499
Ans-3
     .2928194993
Ans▸Frac
        889/3036
```

2. $\frac{171}{92}-\left(-\frac{170}{33}\right)=7\frac{31}{3036}$

```
171/92--170/33
      7.010210804
Ans-7
     .0102108037
Ans▸Frac
         31/3036
```

3. $-\dfrac{171}{92}-\left(\dfrac{170}{33}\right)=-7\dfrac{31}{3036}$

```
-171/92-170/33
     -7.010210804
Ans+7
     -.0102108037
Ans▸Frac
        -31/3036
```

4. $\dfrac{171}{92}-\dfrac{170}{33}=-3\dfrac{889}{3036}$

```
171/92-170/33
     -3.292819499
Ans+3
     -.2928194993
Ans▸Frac
       -889/3036
```

5. $15\dfrac{17}{19}-21\dfrac{13}{17}=-5\dfrac{281}{323}$

```
(15+17/19)-(21+1
3/17)
      -5.86996904
Ans+5
     -.8699690402
Ans▸Frac
         -281/323
```

6. $15\dfrac{17}{19}-\left(-21\dfrac{13}{17}\right)=37\dfrac{213}{323}$

```
(15+17/19)-(-21-
13/17)
      37.65944272
Ans-37
      .6594427245
Ans▸Frac
          213/323
```

7. $-15\dfrac{17}{19}-21\dfrac{13}{17}=-37\dfrac{213}{323}$

```
-(15+17/19)-(21+
13/17)
     -37.65944272
Ans+37
     -.6594427245
Ans▸Frac
         -213/323
```

8. $-15\dfrac{17}{19}-\left(-21\dfrac{13}{17}\right)=5\dfrac{281}{323}$

```
-(15+17/19)-(-21
-13/17)
       5.86996904
Ans-5
      .8699690402
Ans▸Frac
          281/323
```

9. 437.925 – (–108.0065) = 545.9315

10. –437.925 – (–108.0065) = –329.9185

```
437.925--108.006
5
        545.9315
-437.925--108.00
65
       -329.9185
```

11. –437.925 – 108.0065 = –545.9315

12. 437.925 – 108.0065 = 329.9185

```
-437.925-108.006
5
       -545.9315
437.925-108.0065
        329.9185
```

13. 893,475 – 1,093,007 = –199,532

14. –893,475 – 1,093,007 = –1,986,482

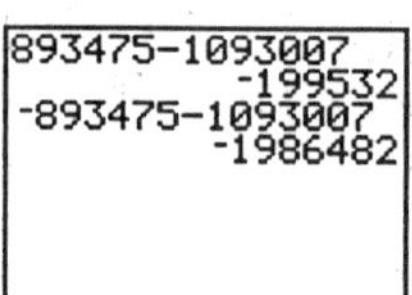

```
893475-1093007
          -199532
-893475-1093007
         -1986482
```

15. 893,475 – (–1,093,007) = 1,986,482

16. –893,475 – (–1,093,007) = 199,532

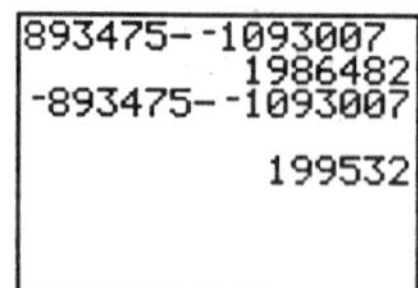

```
893475--1093007
          1986482
-893475--1093007
           199532
```

1.4 Experience Algebra the Exercise Way

1. $45\cdot(-3)=-135$

3. $-32\cdot(-4)=128$

5. $-25\cdot 5=-125$

7. $51\cdot 3=153$

9. $0\cdot(-15)=0$

11. $(0.88)(-1.1)=-0.968$

13. $(-1.7)(-0.2)=0.34$

15. $(24.3)(0.3)=7.29$

17. $(-0.25)(50) = -12.5$

19. $\left(\frac{2}{5}\right)\left(\frac{25}{48}\right) = \frac{50}{240} = \frac{5}{24}$

21. $\left(1\frac{2}{3}\right)\left(-\frac{3}{4}\right) = \left(\frac{5}{3}\right)\left(-\frac{3}{4}\right) = -\frac{15}{12} = -\frac{5}{4}$

23. $\left(-\frac{1}{3}\right)\left(-\frac{3}{7}\right) = \frac{3}{21} = \frac{1}{7}$

25. $\left(-\frac{4}{7}\right)\left(\frac{3}{16}\right) = -\frac{12}{112} = -\frac{3}{28}$

27. $(0)\left(-3\frac{5}{19}\right) = 0$

29. $(-1.11)(0) = 0$

31. four negative factors: positive product

33. five negative factors: negative product

35. two negative factors: positive product

37. three negative factors: negative product

39. $(-2)(-3)(-4)(-10)(20) = 4800$

41. $\left(-\frac{1}{5}\right)\left(-\frac{2}{3}\right)\left(-\frac{4}{5}\right)\left(\frac{1}{2}\right) = -\frac{8}{150} = -\frac{4}{75}$

43. $(5.2)(-0.1)(-2.2) = 1.144$

45. $\left(\frac{1}{3}\right)(-2)(4.2)(5) = -\frac{42}{3} = -14$

47. $(5)(-6.8)\left(\frac{1}{2}\right)(0) = 0$

49. $(14)(0)(-35)(0)(-312) = 0$

51. $12 \cdot 225 = 2700$
Steve pays $2700 in rent each year.

53. 22% of 645
$(0.22)(-645) = -141.9$
Sara has $141.90 deducted every 2 weeks.
$(6)(-141.9) = -851.4$
This amounts to $851.40 every 12 weeks.

55. $(4)\left(-4\frac{1}{2}\right) = -18$
Heath lost 18 yards on these four sacks.

57. $(4.5)(4)(7) = 126$
The track athlete will run 126 miles in four weeks.

59. $(20)(25) = 500$
There are 500 soldirers in each group.

61. $(10)(3)(4) = 120$
There are 120 people riding in the cars.

63. $(0.05)(40)\left(1\frac{1}{2}\right)(3) = 9$
Sammy earns $9 for his work.

65. $(4)(3)(20) = 240$
Nellie will be able to store 240 CD's.

67. $(50)\left(8\frac{1}{2}\right)(4) = 1700$
George drove approximately 1700 miles.

69. $(24)(0.59)(14) = 198.24$
The total retail value for the vegetables is $198.24.

71. $(100)\left(2\frac{3}{4}\right)(12) = 3300$
Drucilla will pay her parents $3300.

73. $(8)(6)(4)(24) = 4608$
4608 ounces will be needed to fill the order.

1.4 Experiencing Algebra the Calculator Way

1. $\left(-\frac{171}{92}\right)\left(-\frac{170}{33}\right) = 9\frac{291}{506}$

```
(-171/92)*(-170/
33)
          9.575098814
Ans-9
          .5750988142
Ans▸Frac
              291/506
```

2. $\left(\frac{171}{92}\right)\left(-\frac{170}{33}\right)=-9\frac{291}{506}$

```
(171/92)*(-170/3
3)
      -9.575098814
Ans+9
      -.5750988142
Ans▸Frac
          -291/506
```

3. $\left(-\frac{171}{92}\right)\left(\frac{170}{33}\right)=-9\frac{291}{506}$

```
(-171/92)*(170/3
3)
      -9.575098814
Ans+9
      -.5750988142
Ans▸Frac
          -291/506
```

4. $\left(\frac{171}{92}\right)\left(\frac{170}{33}\right)=9\frac{291}{506}$

```
(171/92)*(170/33
)
       9.575098814
Ans-9
       .5750988142
Ans▸Frac
           291/506
```

5. $\left(15\frac{17}{19}\right)\left(21\frac{13}{17}\right)=345\frac{305}{323}$

```
(15+17/19)*(21+1
3/17)
       345.9442724
Ans-345
       .9442724458
Ans▸Frac
           305/323
```

6. $\left(15\frac{17}{19}\right)\left(-21\frac{13}{17}\right)=-345\frac{305}{323}$

```
(15+17/19)*(-21-
13/17)
      -345.9442724
Ans+345
      -.9442724458
Ans▸Frac
          -305/323
```

7. $\left(-15\frac{17}{19}\right)\left(21\frac{13}{17}\right)=-345\frac{305}{323}$

```
(-15-17/19)*(21+
13/17)
      -345.9442724
Ans+345
      -.9442724458
Ans▸Frac
          -305/323
```

8. $\left(-15\frac{17}{19}\right)\left(-21\frac{13}{17}\right)=345\frac{305}{323}$

```
(-15-17/19)*(-21
-13/17)
       345.9442724
Ans-345
       .9442724458
Ans▸Frac
           305/323
```

9. (437.925)(–108.0065) = –47,298.74651

10. (–437.925)(–108.0065) = 47,298.74651

```
437.925*-108.006
5
      -47298.74651
-437.925*-108.00
65
       47298.74651
```

11. (–437.925)(108.0065) = –47,298.74651

12. (437.925)(108.0065) = 47,298.74651

```
-437.925*108.006
5
      -47298.74651
437.925*108.0065
       47298.74651
```

13. (893)(1,093) = 976,049

14. (–893)(1,093) = –976,049

```
893*1093
            976049
-893*1093
           -976049
```

15. (893)(–1,093) = –976,049

16. (–893)(–1,093) = 976,049

```
893*-1093
           -976049
-893*-1093
            976049
```

1.5 Experiencing Algebra the Exercise Way

1. 45 ÷ (–3) = –15

3. –32 ÷ (–4) = 8

5. $51 \div 3 = 17$

7. $0 \div (-15) = 0$

9. $18 \div (-18) = -1$

11. $26 \div (-0.13) = -200$

13. $-1.7 \div (-0.2) = 8.5$

15. $24.3 \div 0.3 = 81$

17. $-2.7 \div (-2.7) = 1$

19. $\frac{-25}{5} = -5$

21. $\frac{-16}{0} = \text{undefined}$

23. $\frac{0.88}{-1.1} = -0.8$

25. $\frac{-5.7}{19} = -0.3$

27. $\frac{2}{5} \div \frac{25}{48} = \frac{2}{5} \cdot \frac{48}{25} = \frac{96}{125}$

29. $\left(1\frac{2}{3}\right) \div \left(-\frac{3}{4}\right) = \left(\frac{5}{3}\right)\left(-\frac{4}{3}\right) = -\frac{20}{9}$

31. $\frac{6}{7} \div \left(-\frac{18}{19}\right) = \frac{6}{7} \cdot \left(-\frac{19}{18}\right) = -\frac{114}{126} = -\frac{19}{21}$

33. $-\frac{1}{3} \div \left(-\frac{3}{7}\right) = -\frac{1}{3} \cdot \left(-\frac{7}{3}\right) = \frac{7}{9}$

35. $\left(-1\frac{1}{4}\right) \div \left(-\frac{4}{5}\right) = \left(-\frac{5}{4}\right) \cdot \left(-\frac{5}{4}\right) = \frac{25}{16}$

37. $-\frac{4}{7} \div \frac{3}{16} = -\frac{4}{7} \cdot \frac{16}{3} = -\frac{64}{21}$

39. $0 \div \left(-3\frac{5}{19}\right) = 0$

41. $\frac{3}{5} \div 0 = \text{undefined}$

43. $-\frac{2}{3} \div \left(-\frac{2}{3}\right) = -\frac{2}{3} \cdot \left(-\frac{3}{2}\right) = \frac{6}{6} = 1$

45. $\frac{8}{9} \div 4 = \frac{8}{9} \cdot \frac{1}{4} = \frac{8}{36} = \frac{2}{9}$

47. $14 \div \left(-\frac{1}{3}\right) = 14 \cdot \left(-\frac{3}{1}\right) = -42$

49. $\frac{0}{1.3} = 0$

51.
$$\begin{aligned}
&(-15)(4) \div (-3)(12) \div 3(-10) \\
&= -60 \div (-3)(12) \div 3(-10) \\
&= 20(12) \div 3(-10) \\
&= 240 \div 3(-10) \\
&= 80(-10) \\
&= -800
\end{aligned}$$

53.
$$\begin{aligned}
&(-3.3)(2.7) \div (-11)(0.6) \\
&= -8.91 \div (-11)(0.6) \\
&= 0.81(0.6) \\
&= 0.486
\end{aligned}$$

55.
$$\begin{aligned}
&\left(\frac{2}{3}\right)\left(-\frac{5}{8}\right) \div \left(-\frac{5}{16}\right) \\
&= -\frac{10}{24} \div \left(-\frac{5}{16}\right) \\
&= -\frac{10}{24} \cdot \left(-\frac{16}{5}\right) \\
&= \frac{160}{120} \\
&= \frac{4}{3}
\end{aligned}$$

57. $\frac{20 + \left(-3\frac{1}{2}\right) + 8\frac{3}{4} + 12}{4} = \frac{37\frac{1}{4}}{4} = 9\frac{5}{16}$

Heath averaged $9\frac{5}{16}$ yards per play.

59. $\frac{3.5+2.0+0.5+(-1.5)+(-2.5)+3.5+4.0+(-4.0)+3.5+2.5+3.5}{11}$
$=\frac{15}{11}$
$=1.\overline{36}$
Professor Chip's average rating was approximately 1.36 points.

61. $364 \div 26 = 14$
There will be 14 soldiers in each row.

63. $479 \div 60 = 7.98\overline{3}$
It will take Michaela approximately 8 hours to drive to New Orleans.

65. $3200 \div 12 = 266.\overline{6}$
The distributor will have 266 full packs.

67. $659 \div 19.4 \approx 33.97$
Al will use approximately 34 gtallons of gas on his trip.

69. $47{,}500 \div 40 = 1187.5$
1187 bottles can be filled.

71. $16{,}000{,}000 \text{ seconds} \times \frac{1 \text{ minute}}{60 \text{ seconds}} \times \frac{1 \text{ hour}}{60 \text{ minutes}} \times \frac{1 \text{ day}}{24 \text{ hours}}$
≈ 185.2 days
It will take a little over 185 days.

73. $850 \div 35 \approx 24.3$
It will take Anita 25 weeks to pay her mother.

75. $335 \div (3 \times 20) = 335 \div 60 \approx 5.6$
Billie will need 6 CD cabinets to store her collection.

77. $1{,}035{,}900{,}000 \div 4 = 258{,}975{,}000$
The estimated expenditure for a quarter was $258,975,000.

1.5 Experiencing Algebra the Calculator Way

1. $-\frac{171}{92} \div \left(-\frac{17}{3}\right) = \frac{513}{1564}$

2. $\frac{171}{92} \div \left(-\frac{17}{3}\right) = -\frac{513}{1564}$

```
(-171/92)/(-17/3
)▸Frac
           513/1564
(171/92)/(-17/3)
▸Frac
          -513/1564
```

3. $-\frac{171}{92} \div \frac{17}{3} = -\frac{513}{1564}$

4. $\frac{171}{92} \div \frac{17}{3} = \frac{513}{1564}$

```
(-171/92)/(17/3)
▸Frac
          -513/1564
(171/92)/(17/3)▸
Frac
           513/1564
```

5. $\left(15\frac{17}{19}\right)\div\left(21\frac{13}{17}\right)=\frac{2567}{3515}$

6. $\left(15\frac{17}{19}\right)\div\left(-21\frac{13}{17}\right)=-\frac{2567}{3515}$

```
(15+17/19)/(21+1
3/17)▸Frac
       2567/3515
(15+17/19)/-(21+
13/17)▸Frac
      -2567/3515
```

7. $\left(-15\frac{17}{19}\right)\div\left(21\frac{13}{17}\right)=-\frac{2567}{3515}$

8. $\left(-15\frac{17}{19}\right)\div\left(-21\frac{13}{17}\right)=\frac{2567}{3515}$

```
-(15+17/19)/(21+
13/17)▸Frac
      -2567/3515
-(15+17/19)/-(21
+13/17)▸Frac
       2567/3515
```

9. $\frac{437.925}{-108.0065}\approx -4.055$

10. $\frac{-437.925}{-108.0065}\approx 4.055$

```
437.925/-108.006
5
     -4.054617083
-437.925/-108.00
65
      4.054617083
```

11. $\frac{-437.925}{108.0065}\approx -4.055$

12. $\frac{437.925}{108.0065}\approx 4.055$

```
-437.925/108.006
5
     -4.054617083
437.925/108.0065
      4.054617083
```

13. $893 \div 1093 = 0.817$

14. $-893 \div 1093 \approx -0.817$

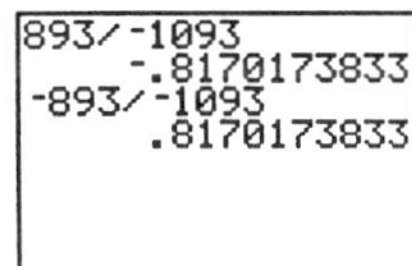

15. $893 \div (-1093) \approx -0.817$

16. $-893 \div (-1093) \approx 0.817$

```
893/-1093
    -.8170173833
-893/-1093
     .8170173833
```

Chapter 1 Review

Reflections

1.–7. Answers will vary.

Exercises

1. –15: integer, rational
1000: natural, whole, integer, rational
0: whole, integer, rational
–2000: integer, rational
13: natural, whole, integer, rational
$-\frac{15}{17}$: rational
12.97: rational
$3\frac{5}{8}$: rational
$\frac{12}{4}=3$: natural, whole, integer, rational

2. 3 above par = 3
even par = 0
2 below par= –2
1 above par = 1

3. Number line with points: -3, $-2\frac{1}{2}$, -1.1, 0, $\frac{3}{4}$, $1\frac{1}{2}$, 2.5, 4

Axis: –4 –3 –2 –1 0 1 2 3 4

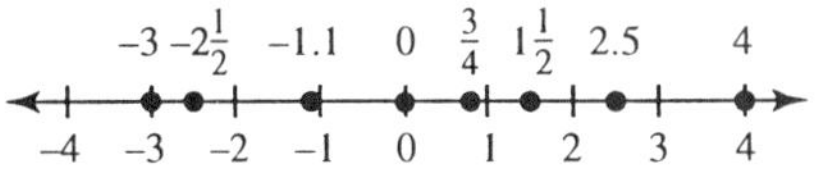

4. $-7 > -9$

5. $-0.54 > -0.65$

6. $\frac{3}{11} < \frac{1}{3}$

7. $-329 < -115$

8. $\frac{13}{45} < \frac{4}{9}$

9. $\frac{27}{75} = 0.36$

10. $2035 > 491$

11. $12.304 < 12.344$

12. $-28 < 13$

13. $-\frac{23}{79} < \frac{13}{47}$

14. $0.056 > -0.197$

15. $331 > -331$

16. $\frac{23}{17} > -\frac{17}{23}$

17. $-0.34 < 0.17$

18. $-1.34 < -1.04$

19. $-\frac{3}{4} < -\frac{5}{8}$

20. $-4\frac{1}{3} > -5\frac{1}{2}$
True

21. $\frac{13}{23} < \frac{51}{91}$
False

22. $-408 > -513$
True

23. $-23 > -49$
$-49 < -23$

24. $\frac{252}{529} < \frac{13}{17}$
$\frac{13}{17} > \frac{252}{529}$

25. $-34.58 = -\frac{1729}{50}$
$-\frac{1729}{50} = -34.58$

26. $-\left|-\frac{17}{33}\right| = -\frac{17}{33}$

27. $|-(-67)| = |67| = 67$

28. $-|-16.85| = -16.85$

29. $-\left(-\left(-\left(-5\frac{1}{2}\right)\right)\right)$
$= -\left(-\left(5\frac{1}{2}\right)\right)$
$= 5\frac{1}{2}$

30. $-(-32.698) = 32.698$

31. $-(-(-(257))) = -257$

32. $-27 + 13 = -14$

33. $-16 + (-22) = -38$

34. $35 + (-19) = 16$

35. $-1123 + (-3406) = -4529$

36. $82.56 + (-43.7) = 38.86$

37. $-5\frac{2}{9} + 4\frac{5}{9} = -\frac{47}{9} + \frac{41}{9} = -\frac{6}{9} = -\frac{2}{3}$

38. $18 + \left(-1\frac{1}{3}\right) = \frac{54}{3} + \left(-\frac{4}{3}\right) = \frac{50}{3}$

39. $\frac{102}{43}+\left(-\frac{29}{75}\right)$
$=\frac{7650}{3225}+\left(-\frac{1247}{3225}\right)$
$=\frac{6403}{3225}$
$=1\frac{3178}{3225}$

40. $235{,}407+(-571{,}004)=-335{,}597$

41. $12.097+1.92=14.017$

42. $\frac{32}{77}+\frac{50}{99}=\frac{288}{693}+\frac{350}{693}=\frac{638}{693}=\frac{58}{63}$

43. $-3.57+(-41.098)-44.668$

44. $23{,}596+143{,}271=166{,}867$

45. $-49.071+105.399=56.328$

46. $-\frac{103}{345}+\left(-\frac{21}{115}\right)=-\frac{103}{345}+\left(-\frac{63}{345}\right)$
$=-\frac{166}{345}$

47. $0+(-123)=-123$

48. $53+(-97)+33+50+(-83)+(-101)$
$=-44+33+50+(-83)+(-101)$
$=-11+50+(-83)+(-101)$
$=39+(-83)+(-101)$
$=-44+(-101)$
$=-145$

49. $\frac{1}{8}+\frac{5}{6}+\left(-\frac{2}{3}\right)+\left(-\frac{7}{12}\right)+\left(-\frac{5}{2}\right)$
$=\frac{3}{24}+\frac{20}{24}+\left(-\frac{16}{24}\right)+\left(-\frac{14}{24}\right)+\left(-\frac{60}{24}\right)$
$=\frac{23}{24}+\left(-\frac{16}{24}\right)+\left(-\frac{14}{24}\right)+\left(-\frac{60}{24}\right)$
$=\frac{7}{24}+\left(-\frac{14}{24}\right)+\left(-\frac{60}{24}\right)$
$=-\frac{7}{24}+\left(-\frac{60}{24}\right)$
$=-\frac{67}{24}$

50. $650+(-250)+1200+(-700)=900$
Willy is \$900 above quota at the end of four weeks.

51. $65+(-100)=-35$
The resultant will be pointing 35 km west.

52. $-35-(-61)=-35+61=26$

53. $-\frac{3}{8}-\frac{4}{7}=-\frac{3}{8}+\left(-\frac{4}{7}\right)=-\frac{21}{56}+\left(-\frac{32}{56}\right)=-\frac{53}{56}$

54. $15.6 - 18 = 15.6 + (-18) = -2.4$

55. $0 - 3.97 = 0 + (-3.97) = -3.97$

56. $-2.3 - (-2.3) = -2.3 + 2.3 = 0$

57. $3\frac{5}{9} - 5\frac{2}{9} = \frac{32}{9} + \left(-\frac{47}{9}\right) = -\frac{15}{9} = -\frac{5}{3}$

58. $\frac{23}{112} - \frac{19}{24} = \frac{69}{336} + \left(-\frac{266}{336}\right) = -\frac{197}{336}$

59. $4{,}079{,}321 - 7{,}056{,}902$
$= 4{,}079{,}321 + (-7{,}056{,}902)$
$= -2{,}977{,}581$

60. $-473 - 2091 = -473 + (-2091) = -2564$

61. $-\frac{34}{75} - \left(-\frac{203}{275}\right) = -\frac{374}{825} + \frac{609}{825} = \frac{235}{825} = \frac{47}{165}$

62. $101.02 - (-23.9) = 101.02 + 23.9 = 124.92$

63. $553 - (-392) = 553 + 392 = 945$

64. $2\frac{26}{27} - \left(-5\frac{1}{27}\right) = \frac{80}{27} + \frac{136}{27} = \frac{216}{27} = 8$

65. $1\frac{61}{125} - 1.488 = 1.488 - 1.488 = 0$

66. $3.5 - 4.7 - (-8.2) - (-10.1) - 4.9 - (-16.7)$
$= 3.5 + (-4.7) + 8.2 + 10.1 + (-4.9) + 16.7$
$= 28.9$

67. $1 - 2 - (-3) - (-4) - 5 - 6 - (-7) - 8 - (-9)$
$= 1 + (-2) + 3 + 4 + (-5) + (-6) + 7 + (-8) + 9$
$= 3$

68. $\frac{3}{2} - \frac{1}{4} - \left(-\frac{8}{3}\right) - \frac{5}{6} - \left(-\frac{7}{2}\right) - 3$
$= \frac{18}{12} + \left(-\frac{3}{12}\right) + \frac{32}{12} + \left(-\frac{10}{12}\right) + \frac{42}{12} + \left(-\frac{36}{12}\right)$
$= \frac{43}{12}$
$= 3\frac{7}{12}$

69. $31 - 16 + (-23) - 45 - (-37) + 52 + (-83)$
$= 31 + (-16) + (-23) + (-45) + 37 + 52 + (-83)$
$= -47$

70. $3.7 - 6.83 + 5.5 + (-9.02) - 0.8 - (-15.2)$
$= 3.7 + (-6.83) + 5.5 + (-9.02) + (-0.8) + 15.2$
$= 7.75$

71. $\frac{4}{5} - \frac{7}{3} + \frac{8}{9} - \left(-\frac{4}{15}\right) + \left(-\frac{2}{3}\right) + 3$
$= \frac{36}{45} + \left(-\frac{105}{45}\right) + \frac{40}{45} + \frac{12}{45} + \left(-\frac{30}{45}\right) + \frac{135}{45}$
$= \frac{88}{45}$
$= 1\frac{43}{45}$

72. $735.66 + (-276.12) + (-187.05) + (-68.57) + 75.00 + 185.00 + 50 + (-4.65) + (-12.00)$
$= 1045.66 + (-548.39)$
$= 497.27$
Cleta has \$497.29 left in her account at the end of the month.

73. $14{,}494 - (-282) = 14{,}494 + 282 = 14{,}776$
The difference in elevation between these two points is 14,776 feet.

74. $28\frac{1}{4} - 22\frac{1}{2}$
$= \frac{113}{4} + \left(-\frac{45}{2}\right)$
$= \frac{113}{4} + \left(-\frac{90}{4}\right)$
$= \frac{23}{4}$
John's gain per share was $\$5\frac{3}{4}$.

75. $(-13)(-6) = 78$

76. $(21)(-5) = -105$

77. $\left(-7\frac{2}{5}\right)\left(-\frac{5}{7}\right) = \left(-\frac{37}{5}\right)\left(-\frac{5}{7}\right) = \frac{185}{35} = \frac{37}{7}$

78. $\left(-3\frac{1}{5}\right)\left(\frac{5}{16}\right) = \left(-\frac{16}{5}\right)\left(\frac{5}{16}\right) = -\frac{80}{80} = -1$

79. $(23.05)(-0.04) = -0.922$

80. $0 \cdot (-11) = 0$

81. $(-1) \cdot 25 = -25$

82. $(-34)(20) = -680$

83. $(-2357)(1037) = -2{,}444{,}209$

84. $\left(\frac{23}{171}\right)\left(\frac{33}{230}\right) = \frac{759}{39{,}330} = \frac{11}{570}$

85. $(-2.04)(-4.12) = 8.4048$

86. $2905 \cdot 1197 = 3{,}477{,}285$

87. $(13.902)(4.387) \approx 60.988$

88. $(-763)(-924) = 705{,}012$

89. $-\frac{13}{29}\cdot\frac{58}{169}=-\frac{754}{4901}=-\frac{2}{13}$

90. $(-2.101)(3.337)\approx -7.011$

91. $765\cdot(-835)=-638{,}775$

92. $(43.996)(0)=0$

93. $(-32)(20)(-1)(5)(-2)(10)=-64{,}000$

94. $(-1.1)(0.2)(-4)(-10)(-0.8)=7.04$

95. $\left(-\frac{1}{3}\right)\left(\frac{6}{7}\right)\left(-\frac{14}{15}\right)\left(-\frac{25}{8}\right)=-\frac{2100}{2520}=-\frac{5}{6}$

96. $(-54)(21)(0)(32)(0)(-25)=0$

97. $(14)(36)(16)=8064$
8064 ounces of cleanser will be required to process the order.

98. $(22.5)(45)=1012.5$
The bill for labor will be $1012.50.

99. $220\div(-4)=-55$

100. $(-78)\div(-3)=26$

101. $-26{,}320\div(-112)=235$

102. $\frac{5}{9}\div\left(-\frac{2}{3}\right)=\frac{5}{9}\cdot\left(-\frac{3}{2}\right)=-\frac{15}{18}=-\frac{5}{6}$

103. $-10.557\div 2.3=-4.59$

104. $0\div 25=0$

105. $-2\div 0=$ undefined

106. $(-13.7)\div(-1)=13.7$

107. $749.82645\div(-35.65)=-21.033$

108. $-\frac{23}{356}\div\frac{46}{89}=-\frac{23}{356}\cdot\frac{89}{46}=-\frac{2047}{16{,}376}=-\frac{1}{8}$

109. $-1{,}363{,}443\div 2539=-537$

110. $\frac{143{,}883}{657}=219$

111. $-\frac{87}{121}\div\left(-\frac{58}{99}\right)=-\frac{87}{121}\cdot\left(-\frac{99}{58}\right)$
$=\frac{8613}{7018}=1\frac{5}{22}$

112. $\frac{65}{323}\div\frac{91}{247}=\frac{65}{323}\cdot\frac{247}{91}=\frac{16{,}055}{29{,}393}=\frac{65}{119}$

113. $7.32864\div 1.056=6.94$

114. $90{,}258\div(-307)=-294$

115. $(-25)\div(-5)(12)(-3)\div(-9)(-5)$
$=5(12)(-3)\div(-9)(-5)$
$=60(-3)\div(-9)(-5)$
$=-180\div(-9)(-5)$
$=20(-5)$
$=-100$

116. $(4.2)(-3.2)\div(1.6)(0.2)\div(-0.4)(-2.2)$
$=-13.44\div(1.6)(0.2)\div(-0.4)(-2.2)$
$=-8.4(0.2)\div(-0.4)(-2.2)$
$=-1.68\div(-0.4)(-2.2)$
$=4.2(-2.2)$
$=-9.24$

117. $\left(\frac{5}{12}\right)\left(-\frac{6}{25}\right)\div\left(-\frac{3}{10}\right)\left(\frac{15}{17}\right)\div\left(-\frac{5}{13}\right)$
$=\left(\frac{5}{12}\right)\left(-\frac{6}{25}\right)\left(-\frac{10}{3}\right)\left(\frac{15}{17}\right)\left(-\frac{13}{5}\right)$
$=-\frac{58{,}500}{76{,}500}$
$=-\frac{13}{17}$

118. $\frac{10+(-5)+8+1+(-12)}{5}=\frac{2}{5}=0.4$
David's score is 0.4 points above the class average.

119. $\dfrac{625+690+620+590+630+660}{6}$
$=\dfrac{3815}{6}\approx 635.83$
A fair monthly rent would be \$635.83.

120. $345 \div 18.7 \approx 18.45$
Clarence's average mileage is approximately 18.45 mpg.

Chapter 1 Mixed Reveiw

1. $-\left|\dfrac{9}{25}\right| = -\dfrac{9}{25}$

2. $|-(-102)| = |102| = 102$

3. $-|-(-3.7)| = -|3.7| = -3.7$

4. $-\left(-\left(-\left(4\dfrac{6}{7}\right)\right)\right) = -4\dfrac{6}{7}$

5. $-(0.085) = -0.085$

6. $-(-(-(-7))) = -(-7) = 7$

7. $-549 + (-908) = -1457$

8. $3.07 + (-2.9) = 0.17$

9. $-1\dfrac{4}{13} + 2\dfrac{1}{2} = -1\dfrac{8}{26} + 2\dfrac{13}{26}$
$= -\dfrac{34}{26} + \dfrac{65}{26} = \dfrac{31}{26}$
$= 1\dfrac{5}{26}$

10. $-\dfrac{1}{8} + \left(-1\dfrac{1}{4}\right)$
$= -\dfrac{1}{8} + \left(-1\dfrac{2}{8}\right)$
$= -\dfrac{1}{8} + \left(-\dfrac{10}{8}\right)$
$= -\dfrac{11}{8} = -1\dfrac{3}{8}$

11. $1260 + 1111 = 2371$

12. $-67{,}853 + 80{,}000 = 12{,}147$

13. $0.005 + 0.05 = 0.055$

14. $\dfrac{17}{65} + \dfrac{8}{91} = \dfrac{119}{455} + \dfrac{40}{455} = \dfrac{159}{455}$

15. $-9.678 + (-9.678) = -19.356$

16. $-34.07 - (-31.97) = -34.07 + 31.97 = -2.1$

17. $-576 - (-394) = -576 + 394 = -182$

18. $-\dfrac{4}{13} - \dfrac{7}{39} = -\dfrac{12}{39} + \left(-\dfrac{7}{39}\right) = -\dfrac{19}{39}$

19. $0.52 - 3 = 0.52 + (-3) = -2.48$

20. $0 - \dfrac{3}{11} = 0 + \left(-\dfrac{3}{11}\right) = -\dfrac{3}{11}$

21. $-4.07 - (-5.1) = -4.07 + 5.1 = 1.03$

22. $\dfrac{13}{25} - \dfrac{2}{5} = \dfrac{13}{25} + \left(-\dfrac{2}{5}\right) = \dfrac{13}{25} + \left(-\dfrac{10}{25}\right) = \dfrac{3}{25}$

23. $\dfrac{3}{17} - \left(-\dfrac{5}{34}\right) = \dfrac{3}{17} + \dfrac{5}{34} = \dfrac{6}{34} + \dfrac{5}{34} = \dfrac{11}{34}$

24. $1{,}000{,}000 - 1{,}000 = 1{,}000{,}000 + (-1{,}000)$
$= 999{,}000$

25. $-21 - 453 = -21 + (-453) = -474$

26. $-\dfrac{4}{9} - \left(-\dfrac{47}{81}\right) = -\dfrac{4}{9} + \dfrac{47}{81} = -\dfrac{36}{81} + \dfrac{47}{81} = \dfrac{11}{81}$

27. $15.07 - (-0.907) = 15.07 + 0.907 = 15.977$

28. $4500 - (-45) = 4500 + 45 = 4545$

29. $\left(-\dfrac{12}{55}\right)\left(-\dfrac{5}{6}\right) = \dfrac{60}{330} = \dfrac{2}{11}$

30. $\left(-4\dfrac{3}{7}\right)\left(\dfrac{14}{31}\right) = \left(-\dfrac{31}{7}\right)\left(\dfrac{14}{31}\right) = -\dfrac{434}{217} = -2$

31. $(0.09)(-0.06) = -0.0054$

32. $0 \cdot (-8.05) = 0$

33. $(-45) \cdot 65 = -2925$

34. $(-1)(-88) = 88$

35. $(-767)(-11) = 8437$

36. $\left(\frac{17}{72}\right)\left(\frac{9}{34}\right) = \frac{153}{2448} = \frac{1}{16}$

37. $(-31.6)(-1.001) = 31.6316$

38. $4000 \cdot 1200 = 4{,}800{,}000$

39. $(121.3)(12.1) = 1467.73$

40. $(-111)(0) = 0$

41. $\frac{-1175}{-5} = 235$

42. $\frac{4}{13} \div \left(-\frac{12}{13}\right) = \frac{4}{13} \cdot \left(-\frac{13}{12}\right) = -\frac{52}{156} = -\frac{1}{3}$

43. $-18.87 \div 5.1 = -3.7$

44. $0 \div 0 =$ indeterminate

45. $-21 \div 0 =$ undefined

46. $(56.8) \div (-1) = -56.8$

47. $-9.02 \div (-1.1) = 8.2$

48. $-\frac{8}{55} \div \left(-\frac{4}{17}\right) = -\frac{8}{55} \cdot \left(-\frac{17}{4}\right) = \frac{136}{220} = \frac{34}{55}$

49. $\frac{-16{,}281}{243} = -67$

50. $5.5 \div 2.2 = 2.5$

51. $\frac{184{,}008}{902} = 204$

52. $\frac{2}{5} \div \frac{56}{75} = \frac{2}{5} \cdot \frac{75}{56} = \frac{150}{280} = \frac{15}{28}$

53. $-\frac{1}{5} - \left(-\frac{2}{9}\right) = -\frac{1}{5} + \frac{2}{9} = -\frac{9}{45} + \frac{10}{45} = \frac{1}{45}$

54. $75 > 59$

55. $3.54 < 3.65$

56. $\frac{31}{51} > \frac{10}{17}$

57. $-142 < -105$

58. $-\frac{21}{59} > -\frac{29}{59}$

59. $\frac{3}{8} = 0.375$

60. $-35 < 71$

61. $-9.05 > -9.07$

62. $-3.7 < 0.53$

63. $-\frac{6}{7} < \frac{3}{7}$

64. $0 > -197$

65. $31 > 0$

66. $2.56 + (-3.78) + 0.5 + 9.882 + (-1.05) + (-0.009)$
$= 8.103$

67. $\frac{5}{7}+\frac{3}{5}+\left(-\frac{1}{2}\right)+\left(-\frac{23}{35}\right)+\left(-\frac{7}{10}\right)$
$=\frac{50}{70}+\frac{42}{70}+\left(-\frac{35}{70}\right)+\left(-\frac{46}{70}\right)+\left(-\frac{49}{70}\right)$
$=-\frac{38}{70}$
$=-\frac{19}{35}$

68. $17.6 - 0.9 - (-12.7) - (-28.4) - 25.8 - (-21.4)$
$= 17.6 + (-0.9) + 12.7 + 28.4 + (-25.8) + 21.4$
$= 53.4$

69. $-45 - 88 - (-74) - (-90) - 65 - 71 - (-27) - 53 - (-48)$
$= -45 + (-88) + 74 + 90 + (-65) + (-71) + 27 + (-53) + 48$
$= -83$

70. $\frac{5}{6}-\frac{11}{30}-\left(-\frac{3}{5}\right)-\frac{14}{15}-\left(-\frac{9}{2}\right)-1$
$=\frac{25}{30}+\left(-\frac{11}{30}\right)+\frac{18}{30}+\left(-\frac{28}{30}\right)+\frac{135}{30}+\left(-\frac{30}{30}\right)$
$=\frac{109}{30}=3\frac{19}{30}$

71. $57 - 61 + (-32) - 54 - (-73) + 25 + (-38)$
$= 57 + (-61) + (-32) + (-54) + 73 + 25 + (-38)$
$= -30$

72. $7.3 - 38.6 + 5.5 + (-2.09) - 8 - (-2.51)$
$= 7.3 + (-38.6) + 5.5 + (-2.09) + (-8) + 2.51$
$= -33.38$

73. $\frac{3}{7}-\frac{17}{21}+\frac{11}{14}-\left(-\frac{5}{6}\right)+\left(-\frac{35}{42}\right)+7$
$=\frac{18}{42}+\left(-\frac{34}{42}\right)+\frac{33}{42}+\frac{35}{42}+\left(-\frac{35}{42}\right)+\frac{294}{42}$
$=\frac{311}{42}$
$=7\frac{17}{42}$

74. $(-55)(12)(-2)(9)(-3)(1) = -35{,}640$

75. $(-3.2)(0.6)(-5)(-100)(-0.1) = 96$

76. $\left(-\frac{14}{22}\right)\left(\frac{8}{21}\right)\left(-\frac{11}{4}\right)\left(-\frac{3}{5}\right)=-\frac{3696}{9240}=-\frac{2}{5}$

77. $(-11.2)(3.1)(0)(-9.4)(-1)(-7.5) = 0$

78. $(-15) \div (-3)(22)(-4) \div (-8)(-7)$
$= 5(22)(-4) \div (-8)(-7)$
$= 110(-4) \div (-8)(-7)$
$= -440 \div (-8)(-7)$
$= 55(-7)$
$= -385$

79. $(13.1)(-4.2) \div (2.62)(0.5) \div (-0.7)(-1.1)$
$= -55.02 \div (2.62)(0.5) \div (-0.7)(-1.1)$
$= -21(0.5) \div (-0.7)(-1.1)$
$= -10.5 \div (-0.7)(-1.1)$
$= 15(-1.1)$
$= -16.5$

80. $\left(\frac{9}{14}\right)\left(-\frac{7}{18}\right)\div\left(-\frac{3}{4}\right)\left(\frac{6}{7}\right)\div\left(-\frac{1}{21}\right)$
$=\left(\frac{9}{14}\right)\left(-\frac{7}{18}\right)\left(-\frac{4}{3}\right)\left(\frac{6}{7}\right)\left(-\frac{21}{1}\right)$
$=-\frac{31,752}{5292}$
$=-6$

81. $-3\frac{2}{3}$ -2.7 $-\frac{2}{3}$ 0 $2\frac{1}{2}$ 3.6

-4 -3 -2 -1 0 1 2 3 4

82. $76 > 75$
$75 < 76$

83. $-\frac{2}{5} < \frac{1}{7}$
$\frac{1}{7} > -\frac{2}{5}$

84. $0.52 = \frac{13}{25}$
$\frac{13}{25} = 0.52$

85. a. $275,000 - 183,000$
$= 275,000 + (-183,000)$
$= 92,000$
The business made a profit of \$92,000.

b. $695,000 - 710,000$
$= 695,000 + (-710,000)$
$= -15,000$
The business lost \$15,000.

86. $\frac{5\frac{3}{8}+\left(-1\frac{5}{8}\right)+\left(-3\frac{3}{4}\right)+0}{4}$
$=\frac{\frac{43}{8}+\left(-\frac{13}{8}\right)+\left(-\frac{30}{8}\right)+0}{4}$
$=\frac{0}{4}$
$=0$
The average weekly change is 0 for the four weeks.

87. $763 \div \left(6\frac{1}{2}\right)$
$= 763 \div \left(\frac{13}{2}\right)$
$= 763 \cdot \left(\frac{2}{13}\right)$
$= \frac{1526}{13} \approx 117.4$
Approximately 117 credits can run per minute.

88. $10 \times \frac{3}{4} = \frac{30}{4} = 7\frac{1}{2}$

It takes $7\frac{1}{2}$ gallons of water to fill ten 10-gallon hats.

89. $1012 - 430 = 1012 + (-430) = 582$
Captain Picard had 582 more people under his command.

90. $24 \times 28 \times 8 = 5376$
5376 footballs will be needed.

Chapter 1 Test

1. natural numbers: 15

2. whole numbers: 0, 15

3. integers: 0, 15, –17

4. rational numbers:
$0, -\frac{1}{3}, 15, -4\frac{1}{5}, 0.546, -17$

5. $\frac{4}{9} < \frac{5}{6}$

6. $-18 > -23$

7. $\frac{17}{25} = 0.68$

8. $\frac{5}{16} > \frac{1}{4}$
$\frac{1}{4} < \frac{5}{16}$

9. $-3.2 < -2.3$
$-2.3 > -3.2$

10. $\frac{3}{50} = 0.06$
$0.06 = \frac{3}{50}$

11. $|-13.37| = 13.37$

12. $-(-20) = 20$

13. $-|-20| = -20$

14. $59 + (-95) = -36$

15. $-\frac{17}{95} + \frac{4}{19} = -\frac{17}{95} + \frac{20}{95} = \frac{3}{95}$

16. $4.378 - 7.98 = 4.378 + (-7.98) = -3.602$

17. $\left(-\frac{3}{4}\right)\left(-\frac{8}{15}\right) = \frac{24}{60} = \frac{2}{5}$

18. $-6985 - (-2576) = -6985 + 2576 = -4409$

19. $45.78 \cdot (-1) = -45.78$

20. $(-37{,}562)(456) = -17{,}128{,}272$

21. $-413.9 + (-597.65) = -1011.55$

22. $-819 \div (-9) = 91$

23. $-\frac{44}{57} \div \frac{11}{19} = -\frac{44}{57} \cdot \frac{19}{11} = -\frac{836}{627} = -\frac{4}{3}$

24. $15.9 \div 0 =$ undefined

25. $0 \div (-53) = 0$

26. $2\frac{3}{7} \div 1\frac{3}{14} = \frac{17}{7} \div \frac{17}{14} = \frac{17}{7} \cdot \frac{14}{17} = \frac{238}{119} = 2$

27. $-12.05 - 2.4 = -12.05 + (-2.4) = -14.45$

28. $(-23)(-4)(0)(-17)(0)(-45) = 0$

29. $\left(-\frac{5}{6}\right)\left(-\frac{3}{7}\right)\left(\frac{1}{2}\right)\left(-\frac{2}{15}\right) = -\frac{30}{1260} = -\frac{1}{42}$

30. $4.3 + (-0.1) - (-2) + (-1.1)$
$= 4.3 + (-0.1) + 2 + (-1.1)$
$= 5.1$

31. $3\frac{2}{3} \div 4\frac{1}{9} \cdot \left(-7\frac{4}{11}\right)$
$= \frac{11}{3} \div \frac{37}{9} \cdot \left(-\frac{81}{11}\right)$
$= \frac{11}{3} \cdot \frac{9}{37} \cdot \left(-\frac{81}{11}\right)$
$= -\frac{8019}{1221}$
$= -6\frac{21}{37}$

32. $(-42)(22) \div 77(-4) \div (-16)$
$= -924 \div 77(-4) \div (-16)$
$= -12(-4) \div (-16)$
$= 48 \div (-16)$
$= -3$

33.

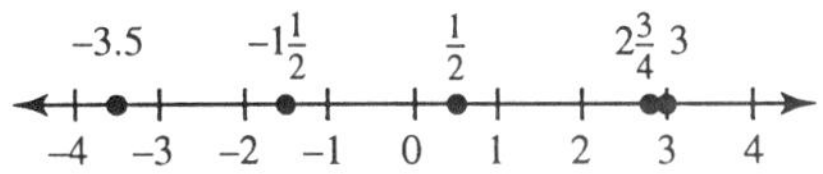

34. $500 + (-123.75) + (-56.80) + (-95.87) + 250$
$= 473.58$
Sally's current account balance is $473.58.

35. 24% of 675
$0.24 \times 675 = 162$
Regina will have $162 per week deducted.

36. If the signs are the same, the product is positive. If the signs are different, the product is negative.

37. If the sum of the negative signs is even, the product is positive. If the sum of the negative signs is odd, the product is negative.

Chapter 2

2.1 Experiencing Algebra the Exercise Way

1. $15 \cdot 15 \cdot 15 \cdot 15 \cdot 15 \cdot 15 = 15^6$

3. $\left(-\frac{1}{5}\right)\left(-\frac{1}{5}\right)\left(-\frac{1}{5}\right)\left(-\frac{1}{5}\right)=\left(-\frac{1}{5}\right)^4$

5. $(3.7)(3.7)(3.7) = (3.7)^3$

7. $0 \cdot 0 \cdot 0 \cdot 0 \cdot 0 \cdot 0 \cdot 0 = 0^7$

9. $(-55)^8$, positive

11. -55^8, negative

13. $(-55)^5$, negative

15. -55^5, negative

17. $3^4 = 3 \cdot 3 \cdot 3 \cdot 3 = 81$

19. $(-3)^4 = (-3)(-3)(-3)(-3) = 81$

21. $-3^4 = -(3 \cdot 3 \cdot 3 \cdot 3) = -81$

23. $-(-3)^4 = -[(-3)(-3)(-3)(-3)] = -81$

25. $(-4)^3 = (-4)(-4)(-4) = -64$

27. $-(-4)^3 = -[(-4)(-4)(-4)] = -(-64) = 64$

29. $(-2.5)^2 = (-2.5)(-2.5) = 6.25$

31. $-(0.5)^3 = -[(0.5)(0.5)(0.5)] = -0.125$

33. $-\left(-\frac{3}{7}\right)^2 = -\left[\left(-\frac{3}{7}\right)\left(-\frac{3}{7}\right)\right] = -\frac{9}{49}$

35. $\left(1\frac{1}{3}\right)^3 = \left(\frac{4}{3}\right)^3 = \left(\frac{4}{3}\right)\left(\frac{4}{3}\right)\left(\frac{4}{3}\right) = \frac{64}{27}$

37. $0^8 = 0$

39. $1^{32} = 1$

41. $(-1)^{29} = -1$

43. $(-1)^{122} = 1$

45. $(-79)^4 = 38{,}950{,}081$

47. $(-1.08)^5 \approx -1.469$

49. $\left(-\frac{2}{3}\right)^6 = \frac{64}{729}$

51. $\left(\frac{5}{6}\right)^5 = \frac{3125}{7776}$

53. $(-37)^5 = -69{,}343{,}957$

55. $(2.24)^6 \approx 126.325$

57. $-\left(3\frac{2}{11}\right)^4 = -\left(\frac{35}{11}\right)^4 \approx -102.495$

59. $-\left(7\frac{21}{29}\right)^7 = -\left(\frac{224}{29}\right)^7 \approx -1{,}640{,}401.445$

61. $1256^1 = 1256$

63. $-(13.06)^1 = -13.06$

65. $1256^0 = 1$

67. $-4721^0 = -1$

69. $7^{-2} = \left(\frac{1}{7}\right)^2 = \frac{1}{49}$

71. $\left(\frac{2}{3}\right)^{-4} = \left(\frac{3}{2}\right)^4 = \frac{81}{16}$

73. $(-0.2)^{-2}$
$=\left(-\frac{2}{10}\right)^{-2}$
$=\left(-\frac{10}{2}\right)^{2}$
$=(-5)^2$
$=25$

75. $-11^{-2}=-\left(\frac{1}{11}\right)^2=-\frac{1}{121}$

77. $(-12)^{-1}=\left(-\frac{1}{12}\right)^1=-\frac{1}{12}$

79. $-(-97)^{-1}=-\left(-\frac{1}{97}\right)^1=\frac{1}{97}$

81. $-\left(-\frac{5}{6}\right)^{-4}=-\left(-\frac{6}{5}\right)^4=-\frac{1296}{625}$

83. $-(2.06)^{-4}=-\left(\frac{1}{2.06}\right)^4\approx 0.056$

85. $-\left(2\frac{2}{9}\right)^{-2}=-\left(\frac{20}{9}\right)^{-2}=-\left(\frac{9}{20}\right)^2=-\frac{81}{400}$

87. $\left(-\frac{2}{7}\right)^{-3}=\left(-\frac{7}{2}\right)^3=-\frac{343}{8}$

89. $\left(1\frac{1}{5}\right)^{-3}=\left(\frac{6}{5}\right)^{-3}=\left(\frac{5}{6}\right)^3=\frac{125}{216}$

91. $-\left(-5\frac{4}{11}\right)^{-3}$
$=-\left(-\frac{59}{11}\right)^{-3}$
$=-\left(-\frac{11}{59}\right)^{3}$
≈ -0.006

93. $(3.5)^3=42.875$
The volume of the pit is 42.875 ft^3.

95. $(6.5)^2=42.25$
Aladdin's magic carpet will cover 42.25 ft^2.

97. $(755)^2=570,025$
The Great Pyramid covers 570,025 ft^2 of ground.

2.1 Experiencing Algebra the Calculator Way

1. $17^6=24,137,569$

```
17^6
            24137569
```

2. $-17^6=-24,137,569$

```
-17^6
           -24137569
```

3. $(-17)^6=24,137,569$

```
(-17)^6
            24137569
```

4. $17^5=1,419,857$

```
17^5
             1419857
```

5. $-17^5=-1,419,857$

```
-17^5
            -1419857
```

6. $(-17)^5 = -1,419,857$

```
(-17)^5
        -1419857
```

7. $-(-24)^6 = -191,102,976$

```
-(-24)^6
       -191102976
```

8. $-\left(\frac{2}{9}\right)^4 = -\frac{16}{6561}$

```
-(2/9)^4▸Frac
         -16/6561
```

9. $\left(-\frac{2}{9}\right)^4 = \frac{16}{6561}$

```
(-2/9)^4▸Frac
          16/6561
```

10. $(12.89)^4 \approx 27,606.520$

```
(12.89)^4
      27606.52033
```

11. $-(12.89)^4 \approx -27,606.520$

```
-(12.89)^4
     -27606.52033
```

12. $(-0.56)^4 \approx 0.098$

```
(-0.56)^4
        .09834496
```

13. $-(-0.56)^4 \approx -0.098$

```
-(-0.56)^4
       -.09834496
```

14. $-\left(3\frac{5}{6}\right)^4 = -\left(\frac{23}{6}\right)^4 = -\frac{279,841}{1296}$

```
-(23/6)^4▸Frac
     -279841/1296
```

15. $\left(-4\frac{1}{4}\right)^4 = \left(-\frac{17}{4}\right)^4 = \frac{83,521}{256}$

```
(-17/4)^4▸Frac
        83521/256
```

16. $-\left(-4\frac{1}{4}\right)^4 = -\left(-\frac{17}{4}\right)^4 = -\frac{83,521}{256}$

```
-(-17/4)^4▸Frac
       -83521/256
```

17. $\left(-\frac{4}{7}\right)^7 \approx -0.020$

```
(-4/7)^7
    -.0198945289
```

18. $-\left(-\frac{4}{7}\right)^7 \approx 0.020$

```
-(-4/7)^7
          .0198945289
```

19. $(-4.076)^7 \approx -18{,}691.288$

```
(-4.076)^7
          -18691.28792
```

20. $-(-4.076)^7 \approx 18{,}691.288$

```
-(-4.076)^7
          18691.28792
```

21. $\left(2\frac{2}{7}\right)^5 = \left(\frac{16}{7}\right)^5 \approx 62.389$

```
(16/7)^5
          62.38924258
```

22. $-\left(-3\frac{2}{3}\right)^7 = -\left(-\frac{11}{3}\right)^7 = \frac{19{,}487{,}171}{2187}$

```
-(-11/3)^7▸Frac
       19487171/2187
```

23. $\left(-7\frac{19}{21}\right)^3 = \left(-\frac{166}{21}\right)^3 = -\frac{4{,}574{,}296}{9261}$

```
(-166/21)^3▸Frac
       -4574296/9261
```

24. $\left(-17\frac{21}{37}\right)^6 = \left(-\frac{650}{37}\right)^6 \approx 29{,}394{,}751.66$

```
(-650/37)^6
          29394751.66
```

2.2 Experiencing Algebra the Exercise Way

1. $23{,}450{,}000{,}000 = 2.345 \times 10^{10}$

3. $-203{,}415{,}000{,}000{,}000 = -2.03415 \times 10^{14}$

5. $0.0000000176 = 1.76 \times 10^{-8}$

7. $-0.000006591 = -6.591 \times 10^{-6}$

9. $3.6943 = 3.6943 \times 10^0$

11. $-4.7502 = -4.7502 \times 10^0$

13. 6.3 E17
$= 6.3 \times 10^{17}$
$= 630{,}000{,}000{,}000{,}000{,}000$

15. -7.1103 E5 $= -7.1103 \times 10^5 = -711{,}030$

17. -3.7 E-5 $= -3.7 \times 10^{-5} = -0.000037$

19. 1.966 E-2 $= 1.966 \times 10^{-2} = 0.01966$

21. 4.356 E0 $= 4.356 \times 10^0 = 4.356$

23. -9.95 E0 $= -9.95 \times 10^0 = -9.95$

25. $2.7 \times 10^7 = 27{,}000{,}000$

27. $-4.005 \times 10^6 = -4{,}005{,}000$

29. $4.056 \times 10^{-7} = 0.0000004056$

31. $-3.0303 \times 10^{-4} = -0.00030303$

33. $1.26 \times 10^0 = 1.26$

35. $-4.5\times10^0=-4.5$

37. $\$1.024769\times10^{12}=\$1,024,769,000,000$

39. $1.024769\times10^{12}-2.71638\times10^{11}$
$=7.53131\times10^{11}$
The GNP of the U.K. is $\$7.53131\times10^{11}$ or \$753,131,000,000 more than the GNP of India.

41. $14,438,000=1.4438\times10^7$ crimes
$255,458,000=2.55458\times10^8$ population
$\frac{1.4438\times10^7}{2.55458\times10^8}\approx0.056518$
5.6518×10^{-2} ratio
In 1992, there were about 0.056518 crimes per person in the U.S.

43. $1,878,285$
$=1.878285\times10^6$ Native Americans
$248,700,000=2.487\times10^8$ total population
$\frac{1,878,285}{248,700,000}=0.00755$
0.755% of the population were Native Americans.

45. $258,245,000\times365$
$=9.4259425\times10^{10}$
9.4259425×10^{10} or 94,259,425,000 cups of fruit were consumed in a year.

47. $2.65\times10^9=2,650,000,000$
2,650,000,000 Christmas cards are expected to be sold in 1996.

49. $6,000,000\times2000=12,000,000,000$
$=1.2\times10^{10}$
Each pyramid at Giza would weigh over 1.2×10^{10} pounds.

51. $3,700\times5280=1.9536\times10^7$
The Great Wall of China is 1.9536×10^7 feet long.

2.2 Experiencing Algebra the Calculator Way

1. a. 6.022 E23

```
6.022E23
```

b. 602,600,000,000,000,000,000,000

c. scientific notation; takes up less room

d. 1.2044 E26

```
1.2044E26
```

e. 1.2046×10^{26}

2. a. 5.642151×10^9

b. 5.642151 E9

```
5.642151E9
```

c.

```
1.6926453*10^10
```

1.6926453×10^{10} people will live in the world.

d. 16,926,453,000

2.3 Experiencing Algebra the Exercise Way

1. $\sqrt{36}=6$ because $6^2=36$.

3. $\sqrt{256}=16$ because $16^2=256$.

5. $-\sqrt{25}=-5$ because $5^2=25$.

7. $\sqrt{0.64} = 0.8$ because $0.8^2 = 0.64$.

9. $\sqrt{\dfrac{49}{81}} = \dfrac{7}{9}$ because $7^2 = 49$ and $9^2 = 81$.

11. $-\sqrt{\dfrac{16}{9}} = -\dfrac{4}{3}$ because $4^2 = 16$ and $3^2 = 9$.

13. $\sqrt{1} = 1$ because $1^2 = 1$.

15. $-\sqrt{0} = 0$ because $0^2 = 0$.

17. $\sqrt{-16}$ is not a real number.

19. $\sqrt{10}$ lies between $\sqrt{9} = 3$ and $\sqrt{16} = 4$ or approximately 3.162.

21. $-\sqrt{3}$ lies between
$-\sqrt{4} = -2$ and $-\sqrt{1} = -1$ or approximately -1.732.

23. $-\sqrt{1200} \approx -34.641$

25. $\sqrt{10.5} \approx 3.240$

27. $-\sqrt{\dfrac{1}{10}} \approx -0.316$

29. $\sqrt{5\dfrac{8}{13}} \approx 2.370$

31. $-\sqrt{2.5} \approx -1.581$

33. $\sqrt[3]{64} = 4$

35. $\sqrt[3]{1728} = 12$

37. $\sqrt[3]{1234} \approx 10.726$

39. $\sqrt[3]{-125} = -5$

41. $-\sqrt[4]{1296} = -6$

43. $\sqrt[6]{17} \approx 1.604$

45. $\sqrt[4]{-1296}$ is not a real number

47. $\sqrt[5]{-12.7} \approx -1.662$

49. $\sqrt[4]{11.3} \approx 1.833$

51. $\sqrt[3]{\dfrac{1}{8}} = \dfrac{1}{2}$

53. $\sqrt[5]{-\dfrac{32}{3125}} = -\dfrac{2}{5}$

55. $-\sqrt[3]{2\dfrac{5}{7}} \approx -1.395$

57. $\sqrt[5]{-7\dfrac{19}{32}} = \sqrt[5]{-\dfrac{243}{32}} = -\dfrac{3}{2} = -1.5$

59.

$-\sqrt{\frac{8}{3}}$ $\sqrt{\frac{9}{16}}$
$-\sqrt{45}$ $-\sqrt{16}$ $\sqrt[3]{8}$ $\sqrt{25}$ $\sqrt{75}$
-10 -8 -6 -4 -2 0 2 4 6 8 10

61. $A = s^2$
$182.25 = s^2$
$s = \sqrt{182.25} = 13.5$
Each side is 13.5 feet.

63. $l = 8$ ft
$w = 5$ ft
$\text{diagonal} = \sqrt{8^2 + 5^2}$
$= \sqrt{64 + 25}$
$= \sqrt{89}$
≈ 9.434
Jennie will need approximately 9.434 feet of logs.

65. $\text{Current} = \dfrac{120}{400\sqrt{3}} \approx 0.173$
The current will be approximately 0.173 ampere.

67. Speed $= 2\sqrt{5 \cdot 50}$
$= 2\sqrt{250}$
$= 2(15.8113883)$
≈ 31.623
The speed was approximately 31.6 mph.

69. Speed $= (3.25)\sqrt{300}$
$= (3.25)(17.32050808)$
≈ 56.29
The speed was approximately 56 mph.

71. flow rate $= (2.97)(1.5^2)\sqrt{40}$
$\approx (2.97)(2.25)(6.32455532)$
≈ 42.264
The flow rate is approximately 42.264 gallons per minute.

2.3 Experiencing Algebra the Calculator Way

1. $\sqrt{351,649} = 593$

```
√(351649)
                593
```

2. $\sqrt[3]{-13,312,053} = -237$

```
3√(-13312053)
               -237
```

3. $\sqrt{\dfrac{1681}{2601}} = \dfrac{41}{51}$

```
√(1681/2601)▸Fra
c
              41/51
```

4. $\sqrt{695.798884} = 26.378$

```
√(695.798884)
             26.378
```

5. $\sqrt[3]{-\dfrac{512}{729}} = -\dfrac{8}{9}$

```
3√(-512/729)▸Fra
c
               -8/9
```

6. $\sqrt[5]{1,160,290,625} = 65$

```
5*√1160290625
                 65
```

7. $\sqrt{-24,025}$ is not a real number.

8. $-\sqrt{24,025} = -155$

```
-√(24025)
               -155
```

9. $\sqrt[5]{14539.33568} = 6.8$

```
5*√14539.33568
                6.8
```

10. $\sqrt{\pi} \approx 1.772$

```
√(π)
        1.772453851
```

11. $\sqrt{1999} \approx 44.710$

```
√(1999)
           44.71017781
```

12. $\sqrt{\dfrac{2}{3}} \approx 0.816$

```
√(2/3)
           .8164965809
```

2.4 Experiencing Algebra the Exercise Way

1. $\sqrt[3]{27} = 27^{\frac{1}{3}} = 3$

3. $\sqrt[3]{-27} = (-27)^{\frac{1}{3}} = -3$

5. $-\sqrt[3]{27} = -(27)^{\frac{1}{3}} = -3$

7. $-\sqrt[3]{-27} = -(-27)^{\frac{1}{3}} = 3$

9. $-\sqrt[4]{256} = -(256)^{\frac{1}{4}} = -4$

11. $\sqrt[4]{-256} = (-256)^{\frac{1}{4}}$ is not a real number

13. $\sqrt[5]{250} = (250)^{\frac{1}{5}} \approx 3.017$

15. $\sqrt[3]{-39.304} = (-39.304)^{\frac{1}{3}} = -3.4$

17. $\sqrt[4]{-5.6} = (-5.6)^{\frac{1}{4}}$ is not a real number

19. $\sqrt[5]{75.85} = 75.85^{\frac{1}{5}} \approx 2.377$

21. $\sqrt[3]{\dfrac{64}{343}} = \left(\dfrac{64}{343}\right)^{\frac{1}{3}} = \dfrac{4}{7}$

23. $\sqrt[4]{\dfrac{12}{13}} = \left(\dfrac{12}{13}\right)^{\frac{1}{4}} \approx 0.980$

25. $\sqrt[5]{\dfrac{1}{243}} = \left(\dfrac{1}{243}\right)^{\frac{1}{5}} = \dfrac{1}{3}$

27. $\sqrt{289} = (289)^{\frac{1}{2}} = 17$

29. $\sqrt[5]{-1024} = (-1024)^{\frac{1}{5}} = -4$

31. $\sqrt[4]{187.4161} = (187.4161)^{\frac{1}{4}} = 3.7$

33. $\sqrt[4]{\dfrac{16}{2401}} = \left(\dfrac{16}{2401}\right)^{\frac{1}{4}} = \dfrac{2}{7}$

35. $\sqrt[3]{6859} = (6859)^{\frac{1}{3}} = 19$

37. $-\sqrt[4]{4096} = -(4096)^{\frac{1}{4}} = -8$

39. $-\sqrt[4]{4.85} = -(4.85)^{\frac{1}{4}} \approx -1.484$

41. $\sqrt{\dfrac{841}{1225}} = \left(\dfrac{841}{1225}\right)^{\frac{1}{2}} = \dfrac{29}{35}$

43. $\sqrt[4]{614,656} = (614,656)^{\frac{1}{4}} = 28$

45. $\sqrt[3]{-5832} = (-5832)^{\frac{1}{3}} = -18$

47. $\sqrt[5]{-5.55} = (-5.55)^{\frac{1}{5}} \approx -1.409$

49. $\sqrt[5]{-2\dfrac{1}{7}} = \left(-2\dfrac{1}{7}\right)^{\frac{1}{5}} \approx -1.165$

51. $\sqrt[5]{3125} = (3125)^{\frac{1}{5}} = 5$

53. $-\sqrt{841} = -(841)^{\frac{1}{2}} = -29$

55. $\sqrt[3]{-17.576} = (-17.576)^{\frac{1}{3}} = -2.6$

57. $-\sqrt[4]{3\dfrac{3}{5}} = -\left(3\dfrac{3}{5}\right)^{\frac{1}{4}} \approx -1.377$

59. side $= \sqrt[3]{125} = (125)^{\frac{1}{3}} = 5$
Each side is 5 yards.

61. side $= \sqrt[3]{3} = 3^{\frac{1}{3}} \approx 1.442$
Each side is approximately 1.442 feet.

63. side $= \sqrt{12} = (12)^{\frac{1}{2}} \approx 3.464$
Each side is approximately 3.464 feet.

65. side $= \sqrt{377{,}000} = (377{,}000)^{\frac{1}{2}} \approx 614.003$
Each side is approximately 614.003 feet.

67. radius $= \sqrt{\frac{19{,}400}{\pi}} = \left(\frac{19{,}400}{\pi}\right)^{\frac{1}{2}} \approx 78.583$
The radius is approximately 78.583 feet.

2.4 Experiencing Algebra the Calculator Way

1. $\sqrt[5]{57{,}392} = (57{,}392)^{\frac{1}{5}} \approx 8.949$

```
(57392)^(1/5)
          8.94891271
```

2. $-\sqrt[4]{37{,}652} = -(37{,}652)^{\frac{1}{4}} \approx -13.930$

```
-(37652)^(1/4)
         -13.92986837
```

3. $\sqrt[3]{\pi} = (\pi)^{\frac{1}{3}} \approx 1.465$

```
π^(1/3)
          1.464591888
```

4. $\sqrt{13\frac{15}{19}} = \left(13\frac{15}{19}\right)^{\frac{1}{2}} \approx 3.713$

```
(13+15/19)^(1/2)
          3.713418059
```

5. $\sqrt[7]{-23\frac{15}{17}} = \left(-23\frac{15}{17}\right)^{\frac{1}{7}} \approx -1.574$

```
(-23+15/17)^(1/7
)
         -1.556343874
```

6. $\sqrt[3]{\frac{2197}{3375}} = \left(\frac{2197}{3375}\right)^{1/3} = \frac{13}{15}$

```
(2197/3375)^(1/3
)▸Frac
               13/15
```

7. $\sqrt[4]{299{,}821.9536} = (299{,}821.9536)^{\frac{1}{4}} = 23.4$

```
(299821.9536)^(1
/4)
                23.4
```

8. $\sqrt{0.1} = (0.1)^{\frac{1}{2}} \approx 0.316$

```
(0.1)^(1/2)
          .316227766
```

9. $\sqrt[5]{\frac{1}{5}} = \left(\frac{1}{5}\right)^{\frac{1}{5}} \approx 0.725$

```
(1/5)^(1/5)
         .7247796637
```

2.5 Experiencing Algebra the Exercise Way

1. b

3. c

5. f

7. b

9. a

11. e

13. d

15. g

17. h

19. j

21. i

23.

Number	12	–3	–6.2	3.5	$\frac{5}{9}$	$-\frac{3}{5}$
Opposite	–12	3	6.2	–3.5	$-\frac{5}{9}$	$\frac{3}{5}$
Reciprocal	$\frac{1}{12}$	$-\frac{1}{3}$	$-\frac{1}{6.2}$	$\frac{1}{3.5}$	$\frac{9}{5}$	$-\frac{5}{3}$

25. $-4.33 + 1.9 = 1.9 + (-4.33)$

27. $58(-65) = -65(58)$

29. $5.6[(-3.7)(1.1)] = [5.6(-3.7)](1.1)$

31. $\left(\frac{5}{13}+\frac{8}{13}\right)+\frac{1}{13}=\frac{5}{13}+\left(\frac{8}{13}+\frac{1}{13}\right)$

33. $\left(\frac{3}{8}\right)\left(\frac{5}{7}-\frac{1}{9}\right)=\left(\frac{3}{8}\right)\left(\frac{5}{7}\right)-\left(\frac{3}{8}\right)\left(\frac{1}{9}\right)$

35. $2.7(-1.5 + 3.2) = 2.7(-1.5) + 2.7(3.2)$

37. $\frac{217-175}{7}=\frac{217}{7}-\frac{175}{7}$

39. $-(15 + 19.3) = -15 + (-19.3) = -15 - 19.3$

41. $-\left(-\frac{6}{7}-\frac{5}{9}\right)=-\left(-\frac{6}{7}\right)+\left[-\left(-\frac{5}{9}\right)\right]=\frac{6}{7}+\frac{5}{9}$

43. $-\left(1\frac{1}{7}-2\frac{1}{5}\right)$
$=-1\frac{1}{7}+\left[-\left(-2\frac{1}{5}\right)\right]$
$=-1\frac{1}{7}+2\frac{1}{5}$

45. $-(19.37 + 15.043)$
$= -19.37 + (-15.043)$
$= -19.37 - 15.043$

47. $15(17) + 15(21) = 15(17 + 21)$

49. $-3(14) - 3(21) = -3(14 + 21)$

51. $-6(12) - 6(-13) = -6(12 - 13)$

53. $-3(15) + (-3)(17) = -3(15 + 17)$

55. $-(6^2 - 12) + 5(-3 - 8)$
$= -(36 - 12) + 5(-3 - 8)$
$= -(24) + 5(-11)$
$= -24 + (-55)$
$= -79$

57. $[4(7 - 5) + 2] - 9$
$= [4(2) + 2] - 9$
$= (8 + 2) - 9$
$= 10 - 9$
$= 1$

59. $-\sqrt{36+64}$
$= -\sqrt{100}$
$= -10$

61. $\frac{18-4^2+7}{2+1}$
$=\frac{18-16+7}{2+1}$
$=\frac{2+7}{3}$
$=\frac{9}{3}$
$= 3$

63. $2.7 + 5.6 - 16 \div 4 - 3 \cdot 2$
$= 2.7 + 5.6 - 4 - 3 \cdot 2$
$= 2.7 + 5.6 - 4 - 6$
$= 8.3 - 4 - 6$
$= 4.3 - 6$
$= -1.7$

65. $3(-4) - (5)(6) + 4$
$= -12 - 30 + 4$
$= -42 + 4$
$= -38$

67. $(4.3)3 - 5(1.6) + 42.9 \div 3$
$= 12.9 - 5(1.6) + 42.9 \div 3$
$= 12.9 - 8 + 42.9 \div 3$
$= 12.9 - 8 + 14.3$
$= 4.9 + 14.3$
$= 19.2$

69. $-5(39 - 4^2) - 2(23 - 11)$
$= -5(39 - 16) - 2(23 - 11)$
$= -5(23) - 2(12)$
$= -115 - 24$
$= -139$

71. $4(15 - 8) + 31(14 - 11)$
$= 4(7) + 31(3)$
$= 28 + 93$
$= 121$

73. $-|5.2 - 31.3 + 3.95| = -|-22.15| = -22.15$

75. $6^2 + 12 \div (-2) - 12 \cdot (-4)$
$= 36 + 12 \div (-2) - 12 \cdot (-4)$
$= 36 + (-6) - (-48)$
$= 30 + 48$
$= 78$

77. $\left(\frac{2}{3}\right)^2 \div \left(\frac{1}{3}+\frac{1}{2}\right)\cdot\left(\frac{8}{9}\right)$
$=\frac{4}{9}\div\left(\frac{1}{3}+\frac{1}{2}\right)\cdot\left(\frac{8}{9}\right)$
$=\frac{4}{9}\div\left(\frac{5}{6}\right)\cdot\left(\frac{8}{9}\right)$
$=\frac{4}{9}\cdot\frac{6}{5}\cdot\frac{8}{9}$
$=\frac{64}{135}$

79. $\left(\frac{1}{5}\right)\cdot\left(\frac{15}{22}\right)\div\left(\frac{1}{11}+\frac{1}{33}\right)-\left(\frac{1}{3}\right)^2$
$=\left(\frac{1}{5}\right)\cdot\left(\frac{15}{22}\right)\div\left(\frac{1}{11}+\frac{1}{33}\right)-\left(\frac{1}{9}\right)$
$=\frac{3}{22}\div\left(\frac{4}{33}\right)-\left(\frac{1}{9}\right)$
$=\frac{3}{22}\cdot\left(\frac{33}{4}\right)-\frac{1}{9}$
$=\frac{9}{8}-\frac{1}{9}$
$=\frac{73}{72}$

81. $100-(24+7^2-5)\cdot 3+102\div 2$
$=100-(24+49-5)\cdot 3+102\div 2$
$=100-(68)\cdot 3+102\div 2$
$=100-204+51$
$=-53$

83. $2\{3[5-2(3+4)]+9\div 3\}-9(-8)$
$=2\{3[5-2(7)]+3\}+72$
$=2\{3[5-14]+3\}+72$
$=2\{3[-9]+3\}+72$
$=2\{-27+3\}+72$
$=2\{-24\}+72$
$=-48+72$
$=24$

85. $\frac{29+3-2^3}{2^2+2^3}$
$=\frac{29+3-8}{4+8}$
$=\frac{24}{12}$
$=2$

87. $15-\sqrt{2^2+3\cdot 7}$
$=15-\sqrt{4+21}$
$=15-\sqrt{25}$
$=15-5$
$=10$

89. $\frac{4+3\cdot 9-5^2-2\cdot 3}{6^2-5}$
$=\frac{4+3\cdot 9-25-2\cdot 3}{36-5}$
$=\frac{4+27-25-6}{36-5}$
$=\frac{0}{31}$
$=0$

91. $\frac{8^2+3\cdot 12}{5-4\cdot 6+2\cdot 3^2+1^2}$
$=\frac{64+3\cdot 12}{5-4\cdot 6+2\cdot 9+1}$
$=\frac{64+36}{5-24+18+1}$
$=\frac{100}{0}$
= undefined

93. $\frac{87+80+92+76+100}{5}$
$=\frac{435}{5}$
$=87$
Holly's average grade on the tests was 87.

95. $\frac{675+375+545+390}{4}$
$=\frac{1985}{4}$
= \$496.25
Estrelita's average weekly sales were \$496.25.

97. $\frac{168-136}{12}=\frac{32}{12}=2\frac{2}{3}$ lb

Marilyn lost on average $2\frac{2}{3}$ lb per week.

99. "Can't Buy Me Love" sold 7 million records
" Do They Know It's Christmas" sold
7 million records
"We Are the World" sold 7 million records

$$x = 2(7) - \frac{1}{2}$$
$$= 13.5$$

"I Will Always Love You" sold 13.5 million records

$$x = 4(7) + 2$$
$$= 30$$

"White Christmas" sold 30 million records
Total = 7 + 7 + 7 + 13.5 + 30 = 64.5 million
Total records sold was 64.5 million.

2.5 Experiencing Algebra the Calculator Way

1.
```
1085+57(273(480-
575)-(-1233))-(-
857+654)+56²
          -1403590
```

2.
```
15/17-(-19/23+7/
2)-5/3(8/15(2/3-
7/9)-14/17)+(7/3
)²
       5.124198794
```

3.
```
(11.98644/(-2.36
))²(5.76-3.1³)+(
15.75)^-2
      -619.9054362
```

4.
```
√((2.35)³-(11/19
)²)
       3.555656753
```

5.
```
(57²-31(15+17)-4
7(-53))/(14(31)-
29²+15³)▸Frac
          1187/742
```

6.
```
57^2-43(-59)
              5786
36^2-54(2^2)(6)
                 0
```

$\frac{5786}{0}$ is undefined.

7.
```
15/32√(14(56)³-2
(56-7(2)³)
               735
```

8.
```
(14+2(-3+5))/(5+
2²)+3
                 5
```

9.
```
5*125-6*115
               -65
2*7^2-(5*18+2^3)
                 0
```

$\frac{-65}{0}$ is undefined.

10.
```
(7*35-5*7²)/(425
-4³)
                 0
```

Chapter 2 Review

Reflections

1.–12. Answers will vary.

Exercises

1. $12 \cdot 12 \cdot 12 \cdot 12 \cdot 12 \cdot 12 \cdot 12 \cdot 12 \cdot 12 \cdot 12 = 12^{10}$

2. $\left(\frac{5}{11}\right)\left(\frac{5}{11}\right)\left(\frac{5}{11}\right)\left(\frac{5}{11}\right)\left(\frac{5}{11}\right) = \left(\frac{5}{11}\right)^5$

3. **a.** positive

b. negative

c. negative

d. negative

e. negative

f. positive

4. $2^8 = 2\cdot2\cdot2\cdot2\cdot2\cdot2\cdot2\cdot2 = 256$

5. $(1.2)^2 = (1.2)(1.2) = 1.44$

6. $\left(1\frac{1}{3}\right)^3 = \left(1\frac{1}{3}\right)\left(1\frac{1}{3}\right)\left(1\frac{1}{3}\right) = 2\frac{10}{27}$

7. $0^{10} = 0$

8. $1^{15} = 1$

9. $(-3)^5 = (-3)(-3)(-3)(-3)(-3) = -243$

10. $(-3)^4 = (-3)(-3)(-3)(-3) = 81$

11. $(-0.2)^3 = (-0.2)(-0.2)(-0.2) = -0.008$

12. $\left(-1\frac{1}{3}\right)^2 = \left(-1\frac{1}{3}\right)\left(-1\frac{1}{3}\right) = \frac{16}{9}$

13. $(-1)^{18} = 1$

14. $(-1)^{21} = -1$

15. $(-33)^5 = -39,135,393$

16. $\left(-2\frac{1}{3}\right)^5 = \left(-\frac{7}{3}\right)^5 = -\frac{16,807}{243}$

17. $(0.47)^6 \approx 0.011$

18. $(-56)^4 = 9,834,496$

19. $\left(-\frac{3}{4}\right)^6 = \frac{729}{4096}$

20. $-\left(-2\frac{1}{3}\right)^4 = -\left(-\frac{7}{3}\right)^4 = -\frac{2401}{81}$

21. $15^0 = 1$

22. $(-15)^0 = 1$

23. $-15^0 = -1$

24. $15^1 = 15$

25. $-15^1 = -15$

26. $(-15)^1 = -15$

27. $1^0 = 1$

28. $0^0 =$ indeterminate

29. $1^1 = 1$

30. $0^1 = 0$

31. $(229,384)^0 = 1$

32. $(3.079)^1 = 3.079$

33. $\left(\frac{13}{23}\right)^0 = 1$

34. $(229,384)^1 = 229,384$

35. $(-12)^{-2} = \left(-\frac{1}{12}\right)^2 = \frac{1}{144}$

36. $-12^{-2} = -\left(\frac{1}{12}\right)^2 = -\frac{1}{144}$

37. $56,439^{-3} = \left(\frac{1}{56,439}\right)^3 \approx 5.562\times10^{-15}$

38. $(-9)^{-3}=\left(-\frac{1}{9}\right)^3=-\frac{1}{729}$

39. $-9^{-3}=-\left(\frac{1}{9}\right)^3=-\frac{1}{729}$

40. $\left(-2\frac{3}{8}\right)^{-1}=\left(-\frac{19}{8}\right)^{-1}=\left(-\frac{8}{19}\right)^1=-\frac{8}{19}$

41. $\left(1\frac{3}{4}\right)^{-2}=\left(\frac{7}{4}\right)^{-2}=\left(\frac{4}{7}\right)^2=\frac{16}{49}$

42. $1^{-9}=\left(\frac{1}{1}\right)^9=1$

43. $(-1)^{-9}=\left(-\frac{1}{1}\right)^9=-1$

44. $(1.8)^{-3}=\left(\frac{1}{1.8}\right)^3\approx 0.171$

45. $\left(\frac{7}{8}\right)^{-3}=\left(\frac{8}{7}\right)^3=\frac{512}{343}$

46. $A=s^2$
$A=(7.5)^2=56.25$
Yes, since 56.25 ft^2 is greater than 60 ft^2, there will be enough plants to fill the garden.

47. $A=s^3$
$A=(2.5)^3=15.625$
15.625 ft^3 of mulch will fill the bin.

48. $0.000000189=1.89\times10^{-7}$

49. $189{,}000{,}000{,}000{,}000{,}000=1.89\times10^{17}$

50. $-0.00056=-5.6\times10^{-4}$

51. $-27{,}085{,}000{,}000=-2.7085\times10^{10}$

52. $5.89\text{ E}11=5.89\times10^{11}=589{,}000{,}000{,}000$

53. $4.02\text{ E}-7=4.02\times10^{-7}=0.000000402$

54. $-1.3\text{ E}7=-1.3\times10^7=-13{,}000{,}000$

55. $-7.093\text{ E}-5$
$=-7.093\times10^{-5}=-0.00007093$

56. $5{,}904{,}822{,}000{,}000=\$5.904822\times10^{12}$

57. $2.37\times10^8=\$237{,}000{,}000$
$5.904822\times10^{12}-2.37\times10^8$
$=\$5.904585\times10^{12}$
The U.S. GNP exceeds the Solomon Islands GNP by $\$5.904585\times10^{12}$.

58. $8\times10^{-6}=0.000008$ m

59. $40{,}000{,}000=4.0\times10^7$
Michael Jackson's album sold over 4.0×10^7 copies.

60. $\sqrt{81}=9$

61. $\sqrt{0.64}=0.8$

62. $\sqrt{\frac{9}{25}}=\frac{3}{5}$

63. $-\sqrt{49}=-7$

64. $\sqrt{-16}$ is not a real number

65. $\sqrt{1.2769}=1.13$

66. $-\sqrt{470.89}=-21.7$

67. $-\sqrt{\frac{576}{1369}}=-\frac{24}{37}$

68. $\sqrt{2\frac{6}{11}}\approx 1.595$

69. $\sqrt{15}\approx 3.873$

70. $\sqrt{-1.2}$ is not a real number

71. $\sqrt[4]{81}=3$

72. $\sqrt[3]{0.125}=0.5$

73. $\sqrt[4]{\frac{16}{81}}=\frac{2}{3}$

74. $\sqrt[3]{-27{,}000}=-30$

75. $\sqrt[5]{-1}=-1$

76. $-\sqrt[4]{13.0321}=-1.9$

77. $-\sqrt[5]{10}\approx-1.585$

78.

79. $A=s^2$
$729=s^2$
$s=\sqrt{729}=27$
The diamond is 27 meters on a side.

80. $V=s^3$
$91\frac{1}{8}=s^3$
$s=\sqrt[3]{91\frac{1}{8}}=4.5$
The box is 4.5 in. on a side.

81. $\sqrt[4]{16}=16^{\frac{1}{4}}=2$

82. $\sqrt[4]{-16}=(-16)^{\frac{1}{4}}$ is not a real number

83. $-\sqrt{9}=-(9)^{\frac{1}{2}}=-3$

84. $\sqrt[5]{-3\frac{53}{1024}}=\left(-3\frac{53}{1024}\right)^{\frac{1}{5}}=-\frac{3125}{1024}=-\frac{5}{4}$

85. $\sqrt[7]{-123}=(-123)^{\frac{1}{7}}\approx-1.989$

86. $\sqrt[5]{20}=20^{\frac{1}{5}}\approx1.821$

87. diagonal $=\sqrt{(21.5)^2+(21.5)^2}$
≈ 30.406
The diagonal is approximately 30.406 inches long.

88. c

89. k

90. f

91. a

92. j

93. b

94. d

95. h

96. e

97. g

98. i

99.

Number	8	–12	$\frac{3}{17}$	$-\frac{5}{3}$
Opposite	–8	12	$-\frac{3}{17}$	$\frac{5}{3}$
Reciprocal	$\frac{1}{8}$	$-\frac{1}{12}$	$\frac{17}{3}$	$-\frac{3}{5}$

100. $-33.05 + 12.4 = 12.4 + (-33.05)$

101. $\left(1\frac{9}{10}\right)\left(2\frac{4}{19}\right) = \left(2\frac{4}{19}\right)\left(1\frac{9}{10}\right)$

102. $132 + (-207 + 391) = [132 + (-207)] + 391$

103. $\left(\frac{3}{7}\cdot\frac{14}{15}\right)\cdot\frac{5}{9} = \frac{3}{7}\cdot\left(\frac{14}{15}\cdot\frac{5}{9}\right)$

104. $-2.6(-1.9 + 3.2) = -2.6(-1.9) - 2.6(3.2)$

105. $\frac{5}{6}\left(-\frac{3}{5} - \frac{4}{15}\right) = \frac{5}{6}\left(-\frac{3}{5}\right) + \frac{5}{6}\left(-\frac{4}{15}\right)$

106. $\frac{1687 - 1372}{7} = \frac{1687}{7} - \frac{1372}{7}$

107. $-(2.7 + 3.09) = -2.7 + (-3.09) = -2.7 - 3.09$

108. $-(-296 - 132) = -(-296) - (-132) = 296 + 132$

109. $-[14.2 - (-10.9)] = -(14.2 + 10.9) = -14.2 - 10.9$

110. $-[32 + (-51)] = -32 - (-51) = -32 + 51$

111. $21(18) + 21(47) = 21(18 + 47)$

112. $5(19) - 5(17) = 5(19 - 17)$

113. $17 + 31 - 20 \div 2 - 15 \cdot (-3)$
$= 17 + 31 - 10 - (-45)$
$= 83$

114. $-2(-4.8) - 5(1.7) + 9.2$
$= 9.6 - 8.5 + 9.2$
$= 10.3$

115. $41(-73+65)-(52-46)$
$=41(-8)-(6)$
$=-328-6$
$=-334$

116. $-(27-4^2)-16(-5-3)$
$=-(27-16)-16(-8)$
$=-11+128$
$=117$

117. $\dfrac{191+104-11^2}{5^2+3^3-46}$
$=\dfrac{191+104-121}{25+27-46}$
$=\dfrac{174}{6}$
$=29$

118. $[15+21(14-18)]-7^2$
$=[15+21(-4)]-49$
$=(15-84)-49$
$=-69-49$
$=-118$

119. $22-\sqrt{9^2-45}$
$=22-\sqrt{81-45}$
$=22-\sqrt{36}$
$=22-6$
$=16$

120. $\dfrac{8\cdot 9-2\cdot 6^2}{125-7^2}$
$=\dfrac{8\cdot 9-2\cdot 36}{125-49}$
$=\dfrac{72-72}{125-49}$
$=\dfrac{0}{76}$
$=0$

121. $\dfrac{78-3\cdot 17}{5^2-4\cdot 6-1^3}$
$=\dfrac{78-51}{25-24-1}$
$=\dfrac{27}{0}$
= undefined

122. Area $=12(14)-6^2$
$=168-36$
$=132$
132 ft^2 will not be covered by the rug.

123. $2(7)+2(5)+2\sqrt{7^2+5^2}$
$=14+10+2\sqrt{49+25}$
$=14+10+2\sqrt{74}$
$=41.205$
Mary will need approximately 41.205 feet of lace.

124. $\dfrac{77+(77+4)+68+(68-6)+82}{5}=\dfrac{370}{5}$
$=74$
The daily average temperature was 74°F.

125. $\dfrac{227-185}{20}=\dfrac{42}{20}=2.1$
Richard lost 2.1 pounds per week on average.

Chapter 2 Mixed Review

1. $(-9)(-9)(-9)(-9)(-9)(-9)(-9)$
$=(-9)^7$
$=-4{,}782{,}969$

2. $(-5.5)(-5.5)(-5.5)(-5.5)$
$=(-5.5)^4$
$=915.0625$

3. $0.00000475=4.75\times 10^{-6}=4.75$ E–6

4. $355{,}400{,}000{,}000{,}000{,}000$
$=3.554\times 10^{17}$
$=3.554$ E17

5. $-0.000092 = -9.2 \times 10^{-5} = -9.2$ E–5

6. $-100,505,000$
$= -1.00505 \times 10^{8}$
$= -1.00505$ E8

7. 1.12 E9 $= 1.12 \times 10^{9} = 1,120,000,000$

8. 7.94 E–12
$= 7.94 \times 10^{-12} = 0.00000000000794$

9. -6.876 E4 $= -6.876 \times 10^{4} = -68,760$

10. -2.35 E–7 $= -2.35 \times 10^{-7} = -0.000000235$

11. $\sqrt[3]{2197} = 2197^{\frac{1}{3}} = 13$

12. $\sqrt[5]{32} = 32^{\frac{1}{5}} = 2$

13. $\sqrt[4]{-81} = (-81)^{\frac{1}{4}}$ is not a real number

14. $\sqrt[5]{-125} = (-125)^{\frac{1}{5}} \approx -2.627$

15. $\sqrt[3]{-\frac{64}{125}} = \left(-\frac{64}{125}\right)^{\frac{1}{3}} = -\frac{4}{5}$

16. $-\sqrt{121} = -(121)^{\frac{1}{2}} = -11$

17. $\sqrt[4]{10} = 10^{\frac{1}{4}} \approx 1.778$

18. $-\sqrt[4]{4521.2176} = -(4521.2176)^{\frac{1}{4}} = -8.2$

19. $42(15) - 42(23) = 42(15 - 23)$

20. $-31(12) - 31(42) = -31(12 + 42)$

21. $25 + 63 - 48 \div 3 - 22 \cdot (-8)$
$= 25 + 63 - 16 - (-176)$
$= 248$

22. $-3.4(-1.1) - 7(2.6) + 14.3$
$= 3.74 - 18.2 + 14.3$
$= -14.46 + 14.3$
$= -0.16$

23. $2 - \{4[3 + 2(7 - 5)] - 18 \div 3\}$
$= 2 - \{4[3 + 2(2)] - 6\}$
$= 2 - \{4[3 + 4] - 6\}$
$= 2 - \{4[7] - 6\}$
$= 2 - \{28 - 6\}$
$= 2 - \{22\}$
$= 2 - 22$
$= -20$

24. $-(34 - 7^{2}) - 21(-8 - 4)$
$= -(34 - 49) - 21(-12)$
$= -(-15) + 252$
$= 267$

25. $\frac{412 + 204 - 4^{2}}{5^{2} + 35}$
$= \frac{412 + 204 - 16}{25 + 35}$
$= \frac{600}{60}$
$= 10$

26. $[651 + 18(21 - 19)] - 5^{2}$
$= [651 + 18(2)] - 25$
$= (651 + 36) - 25$
$= 687 - 25$
$= 662$

27. $96 - \sqrt{11^{2} - 21}$
$= 96 - \sqrt{121 - 21}$
$= 96 - \sqrt{100}$
$= 96 - 10$
$= 86$

28. $\frac{10 \cdot 25 - 2 \cdot 5^{3}}{275 - 3^{5}}$
$= \frac{10 \cdot 25 - 2 \cdot 125}{275 - 243}$
$= \frac{250 - 250}{32}$
$= \frac{0}{32}$
$= 0$

29. $\dfrac{35-6\cdot 22}{6^3-4\cdot 50-4^2}$
$=\dfrac{35-132}{216-200-16}$
$=-\dfrac{97}{0}$
= undefined

30. $13^2=169$

31. $(3.4)^7\approx 5252.335$

32. $\left(2\frac{1}{5}\right)^2=\left(\frac{11}{5}\right)^2=\frac{121}{25}$

33. $0^{35}=0$

34. $1^{25}=1$

35. $(-5)^3=-125$

36. $(-5)^4=625$

37. $(-3.34)^5\approx -415.654$

38. $\left(-1\frac{2}{5}\right)^3=\left(-\frac{7}{5}\right)^3=-\frac{343}{125}$

39. $(-1)^{24}=1$

40. $(-1)^{35}=-1$

41. $3^4=81$

42. $-\left(\frac{1}{3}\right)^5=-\frac{1}{243}$

43. $(-2.1)^4=19.4481$

44. $22^0=1$

45. $(-22)^0=1$

46. $-22^0=-1$

47. $22^1=22$

48. $-22^1=-22$

49. $(-22)^1=-22$

50. $-1^0=-1$

51. 0^0 = indeterminate

52. $-1^1=-1$

53. $0^1=0$

54. $(2,333,145)^0=1$

55. $(65.02)^1=65.02$

56. $(-0.84)^0=1$

57. $(2,333,145)^1=2,333,145$

58. $\left(\frac{79}{95}\right)^0=1$

59. $\left(-\frac{11}{65}\right)^1=-\frac{11}{65}$

60. $(-0.0004)^0=1$

61. $20^{-2}=\left(\frac{1}{20}\right)^2=\frac{1}{400}$

62. $(-20)^{-2}=\left(-\frac{1}{20}\right)^2=\frac{1}{400}$

63. $-20^{-2}=-\left(\frac{1}{20}\right)^2=-\frac{1}{400}$

64. $6^{-3}=\left(\frac{1}{6}\right)^3=\frac{1}{216}$

65. $(-6)^{-3} = \left(-\frac{1}{6}\right)^3 = -\frac{1}{216}$

66. $-6^{-3} = -\left(\frac{1}{6}\right)^3 = -\frac{1}{216}$

67. $\left(1\frac{3}{5}\right)^{-2} = \left(\frac{8}{5}\right)^{-2} = \left(\frac{5}{8}\right)^2 = \frac{25}{64}$

68. $1^{-7} = \left(\frac{1}{1}\right)^7 = 1$

69. $(-1)^{-7} = \left(-\frac{1}{1}\right)^7 = -1$

70. $\sqrt{2500} = 50$

71. $\sqrt{-400}$ is not a real number

72. $-\sqrt{400} = -20$

73. $\sqrt{4.5369} = 2.13$

74. $\sqrt{\frac{16}{289}} = \frac{4}{17}$

75. $-\sqrt{\frac{441}{1444}} = -\frac{21}{38}$

76. $\sqrt{5\frac{4}{9}} = \sqrt{\frac{49}{9}} = \frac{7}{3} = 2\frac{1}{3}$

77. $\sqrt{1.44} = 1.2$

78. $-\sqrt{\frac{5}{400}} \approx -0.112$

79. $-\sqrt{2335.6} \approx -47.282$

80. $\sqrt[3]{-64} = -4$

81. $\sqrt[4]{\frac{81}{16}} = \frac{3}{2}$

82. $-\sqrt[3]{274.625} = -6.5$

83. $\sqrt[4]{3\frac{13}{81}} = \sqrt[4]{\frac{256}{81}} = \frac{4}{3} = 1\frac{1}{3}$

84. $-\sqrt[5]{25} \approx -1.904$

85. $\sqrt[3]{\frac{1}{3}} \approx 0.693$

86. $\sqrt[4]{2.56} \approx 1.265$

87. $\sqrt[3]{-2\frac{5}{7}} \approx -1.395$

88. $\sqrt[5]{\frac{243}{1024}} = \frac{3}{4}$

89. $-\sqrt[4]{6.5536} = -1.6$

90. $-\sqrt[4]{2\frac{113}{256}} = -1.25$

91.

$\sqrt{\frac{9}{25}}$

$-\sqrt{64}$ -7 $-\sqrt{7}$ $\sqrt[3]{0}$ $\sqrt{7}$ $\sqrt{25}$ $\sqrt{81}$

-10 -8 -6 -4 -2 0 2 4 6 8 10

92.

Number	24	–15	$\frac{4}{9}$	$-\frac{5}{8}$
Opposite	–24	15	$-\frac{4}{9}$	$\frac{5}{8}$
Reciprocal	$\frac{1}{24}$	$-\frac{1}{15}$	$\frac{9}{4}$	$-\frac{8}{5}$

93. $A = s^2$
$A = (14)^2 = 196$
The porch's area is 196 ft^2.

94. $V = s^3$
$V = (6)^3 = 216$
The volume of the fruit cellar is 216 ft^3.

95. $1{,}176{,}946{,}000 = 1.176946 \times 10^9$

96. $9.87 \times 10^8 = 987{,}000{,}000$

97. $A = s^2$
$300 = s^2$
$s = \sqrt{300} = (300)^{\frac{1}{2}} \approx 17.321$
The dimensions of the tarpaulin are approximately 17.321 feet on each side.

98. $V = s^3$
$343 = s^3$
$s = \sqrt[3]{343} = (343)^{\frac{1}{3}} = 7$
The dimensions of the tree house are 7 feet on each side.

99. $\dfrac{35 + (35 + 8) + 52 + (52 - 16)}{4}$
$= \dfrac{166}{4}$
$= 41.5$
Katie's average cost per frame was $41.50.

100. $\dfrac{32{,}800 - 22{,}675}{6}$
$= \dfrac{10{,}125}{6}$
$= 1687.5$
Cecilia's salary increased on average $1687.50 yearly.

Chapter 2 Test

1. $5{,}239{,}000{,}000{,}000{,}000 = 5.239 \times 10^{15}$

2. $-0.00000203 = -2.03 \times 10^{-6}$

3. $5.07 \text{ E}6 = 5.07 \times 10^6 = 5{,}070{,}000$

4. $-1.04 \text{ E}{-8}$
$= -1.04 \times 10^{-8} = -0.0000000104$

5.

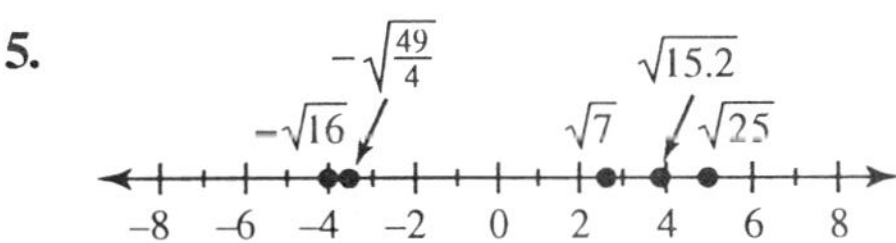

6. $-\sqrt[3]{117{,}649} = -(117{,}649)^{\frac{1}{3}} = -49$

7. $\sqrt[6]{\dfrac{4046}{729}} = \left(\dfrac{4096}{729}\right)^{\frac{1}{6}} = \dfrac{4}{3}$

8. b

9. c

10. a

11. $(1.5)^2 = 2.25$

12. $3^4 = 81$

13. $\left(\frac{4}{3}\right)^4 = \frac{256}{81}$

14. $1^9 = 1$

15. $0^{12} = 0$

16. $(-4)^3 = -64$

17. $(-4.008)^1 = -4.008$

18. $\left(\frac{55}{81}\right)^0 = 1$

19. $1^1 = 1$

20. $0^0 =$ indeterminate

21. $\left(-\frac{3}{8}\right)^{-2} = \left(-\frac{8}{3}\right)^2 = \frac{64}{9}$

22. $(-3)^0 = 1$

23. $-\sqrt{196} = -14$

24. $\sqrt{\frac{36}{121}} = \frac{6}{11}$

25. $-\sqrt{3.6} \approx -1.897$

26. $\sqrt[6]{\frac{262,144}{729}} = \frac{8}{3}$

27. $\sqrt[4]{7\frac{58}{81}} = \sqrt[4]{\frac{625}{81}} = \frac{5}{3} = 1\frac{2}{3}$

28. $-[51.3-(-20.9)]$
$= -(51.3 + 20.9)$
$= -72.2$

29. $\frac{-609+928}{29}$
$= \frac{319}{29}$
$= 11$

30. $\sqrt[3]{17^2 - 6\cdot 12 - 1} + 126 \div 9$
$= \sqrt[3]{289 - 6\cdot 12 - 1} + 126 \div 9$
$= \sqrt[3]{289 - 72 - 1} + 126 \div 9$
$= \sqrt[3]{216} + 14$
$= 6 + 14$
$= 20$

31. $\frac{2(5^2+3^2)-8^2-2^2}{3.65}$
$= \frac{2(25+9)-64-4}{3.65}$
$= \frac{2(34)-64-4}{3.65}$
$= \frac{68-64-4}{3.65}$
$= \frac{0}{3.65}$
$= 0$

32. $(9.5)^2 - 68 = 22.25$
The garden will have 22.25 ft^2 in excess.

33. $2.49\times 10^{11} - 2.49\times 10^{10} = 2.241\times 10^{11}$
The GNP of Switzerland exceeds that of Romania by $\$2.241\times 10^{11}$ or $\$224,100,000,000$.

34. Take the positive power of the reciprocal.

35. Raise the radical expression to the power that is the reciprocal of the index.

Chapter 3

3.1 Experiencing Algebra the Exercise Way

For exercises 1–19 let each x, y, z = some number.

1. $25 - x$

3. $\frac{3}{4}x$

5. $x \div 15$

7. $2.5x - 19.59 \div x$

9. $12x - 25$

11. $x + y + 2xy$

13. $(x + y + z) \div 3$

15. $2x - 5$

17. $\frac{1}{3}x + 6$

19. $2x \div (x + 5)$

21. Let a = number of adult tickets
    c = number of children tickets
    Total money = $6a + 2c$

23. Let
    x = number of hours worked at \$9.50 per hr
    y = number of hours worked at \$12.25 per hr
    Total money = $9.5x + 12.25y$

For exercises 25–40 answers will vary. One possibility is given.

25. $x + 27$
    The sum of a number and 27

27. $\frac{35}{a}$
    The quotient of 35 divided by some number

29. $\frac{3}{5} \cdot z$
    Three-fifths of a number

31. $t - 12.50$
    12.50 subtracted from some number

33. $14 - 5z$
    The difference of 14 and the product of a number times 5

35. $3.14d$
    The product of 3.14 and some number

37. $\frac{1}{2}(L + H)$
    One-half of the sum of two different numbers

39. $(a + b + c) \div 3$
    The sum of three different grades divided by 3

41. $3x + 5$ for $x = -5$
    $3x + 5 = 3(-5) + 5 = -15 + 5 = -10$

43. $3x + 5$ for $x = \frac{2}{3}$
    $3x + 5 = 3\left(\frac{2}{3}\right) + 5 = 2 + 5 = 7$

45. $3x + 5$ for $x = -2.7$
    $3x + 5 = 3(-2.7) + 5 = -8.1 + 5 = -3.1$

47. $18 - 3z$ for $z = -12.07$
    $18 - 3z = 18 - 3(-12.07)$
    $= 18 + 36.21$
    $= 54.21$

49. $18 - 3z$ for $z = 23$
    $18 - 3z = 18 - 3(23) = 18 - 69 = -51$

51. $18-3z$ for $z=-\frac{5}{6}$

$18-3z=18-3\left(-\frac{5}{6}\right)$
$=18+\frac{5}{2}$
$=\frac{36}{2}+\frac{5}{2}$
$=\frac{41}{2}$

53. $\frac{1}{2}bh$ for $b=56, h=14$

$\frac{1}{2}bh=\frac{1}{2}(56)(14)=392$

55. $\frac{1}{2}bh$ for $b=\frac{8}{5}, h=\frac{16}{3}$

$\frac{1}{2}bh=\frac{1}{2}\left(\frac{8}{5}\right)\left(\frac{16}{3}\right)=\frac{64}{15}$

57. $\frac{1}{2}bh$ for $b=6.8, h=4.2$

$\frac{1}{2}bh=\frac{1}{2}(6.8)(4.2)=14.28$

59. x^2+2x+9 for $x=-7$

$x^2+2x+9=(-7)^2+2(-7)+9$
$=49+(-14)+9$
$=44$

61. x^2+2x+9 for $x=2.5$

$x^2+2x+9=(2.5)^2+2(2.5)+9$
$=6.25+5+9=20.25$

63. $\sqrt{4x^2-20x+25}$ for $x=3$

$\sqrt{4x^2-20x+25}=\sqrt{4(3)^2-20(3)+25}$
$=\sqrt{36-60+25}$
$=\sqrt{1}$
$=1$

65. $\sqrt{4x^2-20x+25}$ for $x=-4$

$\sqrt{4x^2-20x+25}=\sqrt{4(-4)^2-20(-4)+25}$
$=\sqrt{64+80+25}$
$=\sqrt{169}$
$=13$

67. $|4.5-3.1b|$ for $b=2$

$|4.5-3.1b|=|4.5-3.1(2)|$
$=|4.5-6.2|$
$=|-1.7|$
$=1.7$

69. $|4.5-3.1b|$ for $b=-2$

$|4.5-3.1b|=|4.5-3.1(-2)|$
$=|4.5+6.2|$
$=|10.7|$
$=10.7$

71. $\frac{x^2+2x+1}{x+1}$ for $x=-3$

$\frac{x^2+2x+1}{x+1}=\frac{(-3)^2+2(-3)+1}{-3+1}$
$=\frac{9-6+1}{-2}$
$=\frac{4}{-2}$
$=-2$

73. $\frac{x^2+2x+1}{x+1}$ for $x=3$

$\frac{x^2+2x+1}{x+1}=\frac{3^2+2(3)+1}{3+1}$
$=\frac{9+6+1}{4}$
$=\frac{16}{4}$
$=4$

75. Let x = number of adult tickets
y = number of children's tickets
Total revenue $= 4x + 0.75y$
$= 4(150) + 0.75(200)$
$= 600 + 150$
$= 750$
The revenue from the tickets was \$750.

77. Let d = diameter
Circumference $= \pi d = \pi(12) \approx 37.699$
The circumference of the pool is approximately 37.7 feet.

3.1 Experiencing Algebra the Calculator Way

1. a.

```
120→A:150→B:(A+1
5)/(B-10)
        .9642857143
```

b.

```
275→A:13→B:(A+15
)/(B-10)
        96.66666667
```

c.

```
16→A:375→B:(A+15
)/(B-10)
        .0849315068
```

d.

```
7/8→R:πR²
        2.405281875
```

e.

```
3+5/9→R:πR²
        39.71593676
```

f.

```
100→R:πR²
        31415.92654
```

2. a.

```
19→R:πR²
        1134.114948
```

b.

```
23.76→R:πR²
        1773.547177
```

c.

```
.05→R:πR²
        .0078539816
```

3. a.

```
120→A:160→B:√(A²
+B²)
                200
```

b.

```
3→A:4→B:√(A²+B²)
                  5
```

c.

```
2.4→A:3.2→B:√(A²
+B²)
                  4
```

d.
```
1→A:1→B:√(A²+B²)
      1.414213562
```

e.
```
3/5→A:4/5→B:√(A²
+B²)
                1
```

f.
```
√(5)→A:√(11)→B:√
(A²+B²)
                4
```

4. a.
```
7→X:(4X²-28X+49)
/(2X-7)
                7
```

b.
```
-2→X:(4X²-28X+49
)/(2X-7)
              -11
```

c.
```
-9/2→X:(4X²-28X+
49)/(2X-7)
              -16
```

d.
```
7.5→X:(4X²-28X+4
9)/(2X-7)
                8
```

e.
```
0→X:(4X²-28X+49)
/(2X-7)
               -7
```

f.
```
7/2→X:(4X²-28X+4
9)/(2X-7)
```

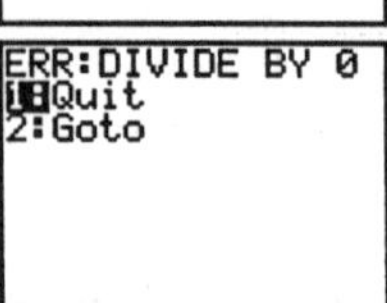

Indeterminate

5. a.
```
10→X:abs((X-250)
/(2X+100))
                2
```

b.
```
100→X:abs((X-250
)/(2X+100)
               .5
```

c.
```
1000→X:abs((X-25
0)/(2X+100))
      .3571428571
```

d.
```
10000→X:abs((X-2
50)/(2X+100))
      .4850746269
```

e.
```
-50→X:abs((X-250
)/(2X+100))
```

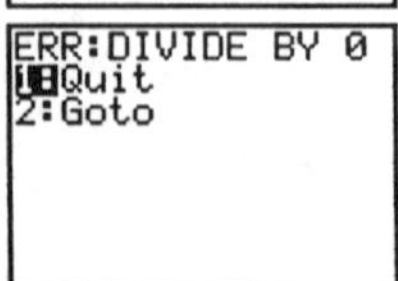

Undefined

f.

```
250→X:abs((X-250
)/(2X+100))
               0
```

3.2 Experiencing Algebra the Exercise Way

1. $12x - 11$

a. number of terms = 2

b. coefficient of each term: 12, −11

c. like terms: none

3. $-15y + 12z - y + 9$

a. number of terms = 4

b. coefficient of each term: −15, 12, −1, 9

c. like terms: $-15y$ and $-y$

5. $2x^2 - 6x + x + 12$

a. number of terms = 4

b. coefficient of each term: 2, −6, 1, 12

c. like terms: $-6x$ and x

7. $3.4a - 11.2b - 0.3a$

a. number of terms = 3

b. coefficient of each term: 3.4, −11.2, −0.3

c. like terms: $3.4a$ and $-0.3a$

9. $2m + 3(n - 5) + 6(n - 5)$

a. number of terms = 3

b. coefficient of each term: 2, 3, 6

c. like terms: $3(n - 5)$ and $6(n - 5)$

11. $x^2 + 3xy - y^2 + 7$

a. number of terms = 4

b. coefficient of each term: 1, 3, −1, 7

c. like terms: none

13. $27x + 44x = 71x$

15. $17a + a = 18a$

17. $5x + 9 - 13x + 17 - 12 + 9x$
$= 5x - 13x + 9x + 9 + 17 - 12$
$= 1x + 14$
$= x + 14$

19. $2x^3 + 7x^2 - 2x + 8 - x^3 - 7x^2 + 3x - 2$
$= 2x^3 - x^3 + 7x^2 - 7x^2 - 2x + 3x + 8 - 2$
$= 1x^3 + 0x^2 + 1x + 6$
$= x^3 + x + 6$

21. $5 - xy + 2yz + 5xz - 17 + xy - 12yz$
$= 5 - 17 - xy + xy + 2yz - 12yz + 5xz$
$= -12 + 0xy - 10yz + 5xz$
$= -12 - 10yz + 5xz$

23. $3.05a + 6.29b - 1.18a + 0.49b$
$= 3.05a - 1.18a + 6.29b + 0.49b$
$= 1.87a + 6.78b$

25. $\frac{1}{6}x + \frac{2}{9} - \frac{2}{3}x + \frac{5}{18} = \frac{1}{6}x - \frac{2}{3}x + \frac{2}{9} + \frac{5}{18}$
$= \frac{1}{6}x - \frac{4}{6}x + \frac{4}{18} + \frac{5}{18}$
$= -\frac{3}{6}x + \frac{9}{18}$
$= -\frac{1}{2}x + \frac{1}{2}$

27. $6x^3+3x^2y-5xy^2+3y^3-5x^2y+xy^2+x^3+6y^3$
$=6x^3+x^3+3x^2y-5x^2y-5xy^2+xy^2+3y^3+6y^3$
$=7x^3-2x^2y-4xy^2+9y^3$

29. $\frac{5}{7}x+\frac{1}{6}y+\frac{5}{6}x-\left(-\frac{2}{7}y\right)=\frac{5}{7}x+\frac{5}{6}x+\frac{1}{6}y+\frac{2}{7}y$
$=\frac{30}{42}x+\frac{35}{42}x+\frac{7}{42}y+\frac{12}{42}y$
$=\frac{65}{42}x+\frac{19}{42}y$

31. $9.35a^2-4.31b^2+2.35ab-1.61+4.39a^2-5.77b^2+10.06$
$=9.35a^2+4.39a^2-4.31b^2-5.77b^2+2.35ab-1.61+10.06$
$=13.74a^2-10.08b^2+2.35ab+8.45$

33. $35a^3-41b^3+5ab-11+4a^2-7b^2+16$
$=35a^3-41b^3+5ab+4a^2-7b^2-11+16$
$=35a^3+41b^3+5ab+4a^2-7b^2+5$

35. $15a+14b-12+3a+9-8b-17a-6b$
$=15a+3a-17a+14b-8b-6b-12+9$
$=1a+0b-3$
$=a-3$

37. $8w+(7w-5)=(8w+7w)-5$
$=15w-5$

39. $5.3x+1.4+(3.4-1.7x)$
$=5.3x+1.4+3.4-1.7x$
$=5.3x-1.7x+1.4+3.4$
$=3.6x+4.8$

41. $(45x-112)+(21x+33)$
$=45x-112+21x+33$
$=45x+21x-112+33$
$=66x-79$

43. $-(29d-7c)=-29d+7c$

45. $4.9z+1.8-(2.6z+0.5)$
$=4.9z+1.8-2.6z-0.5$
$=4.9z-2.6z+1.8-0.5$
$=2.3z+1.3$

47. $(235-12y)-(307+31y)$
$=235-12y-307-31y$
$=235-307-12y-31y$
$=-72-43y$

49. $(x+y+4z)-(2x-5y+z)$
$=x+y+4z-2x+5y-z$
$=x-2x+y+5y+4z-z$
$=-x+6y+3z$

51. $25z-(12z+7)=25z-12z-7=13z-7$

53. $-(1.8y-3.5z)+(4.1y-2.7z)$
$=-1.8y+3.5z+4.1y-2.7z$
$=-1.8y+4.1y+3.5z-2.7z$
$=2.3y+0.8z$

55. $(a+b)-(a+b)+(a+b)-(a+b)$
$= a+b-a-b+a+b-a-b$
$= a-a+a-a+b-b+b-b$
$= 0a+0b$
$= 0$

57. $(-x+y)+(x-y) = -x+y+x-y$
$= -x+x+y-y$
$= 0x+0y$
$= 0$

59. $(a+5b)-(-a+5b) = a+5b+a-5b$
$= a+a+5b-5b$
$= 2a+0b$
$= 2a$

61. $2(250+400)-(65+3) = 2(650)-68$
$= 1300-68$
$= 1232$
Willy needs 1232 feet of fencing.

63. $6(7)^2 = 6(49) = 294$
Rusty needs 294 square inches of balsa wood.

65. Let x = number of square yards of carpet
Profit $= 20 + 1.55x - 0.65x - 6.5$
$= 13.5 + 0.9x$
$= 13.5 + 0.9(250)$
$= 238.5$
Carl will realize $238.50 profit on the job.

67. Let x = number of ounces
Profit $= 5 + 1.5x - 2.25$
$= 2.75 + 1.5x$
$= 2.75 + 1.5(4)$
$= 8.75$
The net profit is $8.75 for a 4-ounce letter.

69. Let x = number of discs
Cost $= 9.99x + 0.05(9.99x)$
$= 9.99x + 0.4995x$
$= 10.4895x$
Change $= 50 - 10.4895x$
Cost for 3 discs $= 10.4895(3) = 31.4685$
Change $= 50 - 31.47 = 18.53$
Laurie will pay $31.47 for the 3 CD's and get $18.53 in change.

71. Let x = number of adults
y = number of children
$ made $= 15x + 7y - 6x - 2.5y$
$= 9x + 4.5y$
$= 9(75) + 4.5(40)$
$= 675 + 180$
$= 855$
$855 was made on the dinner

3.2 Experiencing Algebra the Calculator Way

1. $12.078x + 2.093 - 17.42x - 13.9035$
$= 12.078x - 17.42x + 2.093 - 13.9035$
$= -5.342x - 11.8105$

```
12.078-17.42
              -5.342
2.093-13.9035
            -11.8105
```

2. $(2579x - 4302) - (1087x - 306)$
$= 2579x - 1087x - 4302 + 306$
$= 1492x - 3996$

```
2579-1087
                1492
-4302+306
               -3996
```

3. $\frac{10}{13}x - \frac{5}{52}y - \frac{17}{20}x - \frac{7}{13}y$
$= \frac{10}{13}x - \frac{17}{20}x - \frac{5}{52}y - \frac{7}{13}y$
$= -\frac{21}{260}x - \frac{33}{52}y$

```
10/13-17/20►Frac
             -21/260
-5/52-7/13►Frac
              -33/52
```

4. $\left(2\frac{11}{25}\right)x + 5\frac{17}{30} - \left(3\frac{13}{15}\right)x - 3\frac{23}{75}$
$= \left(2\frac{11}{25}\right)x - \left(3\frac{13}{15}\right)x + 5\frac{17}{30} - 3\frac{23}{75}$
$= -1\frac{32}{75}x + 2\frac{13}{50}$

```
2+11/25-3-13/15►
Frac
             -107/75
5+17/30-3-23/75►
Frac
              113/50
```

5. $(1.0009x + 0.0004) - (0.0909x - 1.0031)$
$= 1.0009x - 0.0909x + 0.0004 + 1.0031$
$= 0.91x + 1.0035$

```
1.0009-.0909
                .91
.0004+1.0031
             1.0035
```

6. $-(935.3376x + 701.315) - (83.027x - 581.9534)$
$= -935.3376x - 701.315 - 83.027x + 581.9534$
$= -935.3376x - 83.027x - 701.315 + 581.9534$
$= -1018.3646x - 119.3616$

```
-935.3376-83.027
          -1018.3646
-701.315+581.953
4
           -119.3616
```

7. $3.995x + 12.083 - 2.995x - 9.083$
$= 3.995x - 2.995x + 12.083 - 9.083$
$= x + 3$

```
3.995-2.995
                   1
12.083-9.083
                   3
```

3.3 Experiencing Algebra the Exercise Way

1. $31(2xy) = (31 \cdot 2)xy = 62xy$

3. $(-120a)(8b) = (-120 \cdot 8)ab = -960ab$

5. $-55(-4mn) = [-55 \cdot (-4)]mn = 220mn$

7. $-4.3(-0.7st) = [-4.3 \cdot (-0.7)]st = 3.01st$

9. $-\frac{16}{27}\left(\frac{15}{24}cd\right) = \left(-\frac{16}{27} \cdot \frac{15}{24}\right)cd$
$= -\frac{240}{648}cd$
$= -\frac{10}{27}cd$

11. $12(4a + 7) = 12(4a) + 12(7)$
$= (12 \cdot 4)a + 12(7)$
$= 48a + 84$

13. $-15(2x + 3) = -15(2x) - 15(3)$
$= (-15 \cdot 2)x - 15(3)$
$= -30x - 45$

15. $-4(-5z - 14) = -4(-5z) - 4(-14)$
$= 20z + 56$

17. $2.2(3.5x - 7.3) = 2.2(3.5x) + 2.2(-7.3)$
$= 7.7x - 16.06$

19. $-8(-4.2z - 5) = -8(-4.2z) - 8(-5)$
$= 33.6z + 40$

21. $\frac{36}{49}\left(-\frac{7}{6}m + \frac{49}{72}\right) = \frac{36}{49}\left(-\frac{7}{6}m\right) + \frac{36}{49}\left(\frac{49}{72}\right)$
$= -\frac{6}{7}m + \frac{1}{2}$

23. $-4\left(-\frac{5}{12}b + \frac{3}{16}\right) = -4\left(-\frac{5}{12}b\right) - 4\left(\frac{3}{16}\right)$
$= \frac{5}{3}b - \frac{3}{4}$

25. $-\frac{2}{3}\left(-\frac{5}{8}d - \frac{9}{16}\right) = -\frac{2}{3}\left(-\frac{5}{8}d\right) - \frac{2}{3}\left(-\frac{9}{16}\right)$
$= \frac{5}{12}d + \frac{3}{8}$

27. $3x(5x+4y)=3x(5x)+3x(4y)$
$=(3\cdot 5)x\cdot x+(3\cdot 4)xy$
$=15x^2+12xy$

29. $-12d(-3c-7d)=-12d(-3c)-12d(-7d)$
$=36cd+84d^2$

31. $-0.9m(-2.2m+3.6n)$
$=-0.9m(-2.2m)-0.9m(3.6n)$
$=1.98m^2-3.24mn$

33. $-\frac{3}{5}a\left(\frac{15}{17}a+b\right)=-\frac{3}{5}a\left(\frac{15}{17}a\right)-\frac{3}{5}a(b)$
$=-\frac{9}{17}a^2-\frac{3}{5}ab$

35. $\frac{3}{8}m(m-4n)=\frac{3}{8}m(m)+\frac{3}{8}m(-4n)$
$=\frac{3}{8}m^2-\frac{3}{2}mn$

37. $\frac{108p}{9}=108p\cdot\frac{1}{9}$
$=\left(108\cdot\frac{1}{9}\right)\cdot p$
$=12p$

39. $\frac{-138x}{46}=\frac{-138}{46}\cdot x=-3x$

41. $\frac{-58z}{-29}=\frac{-58}{-29}\cdot z=2z$

43. $\frac{5.4c}{-2.7}=\frac{5.4}{-2.7}\cdot c=-2c$

45. $\frac{-\frac{1}{2}x}{2}=-\frac{1}{2}x\div 2$
$=-\frac{1}{2}x\cdot\frac{1}{2}$
$=\left(-\frac{1}{2}\cdot\frac{1}{2}\right)x$
$=-\frac{1}{4}x$

47. $\frac{-\frac{2}{3}a}{-4}=-\frac{2}{3}a\div(-4)$
$=-\frac{2}{3}a\cdot\left(-\frac{1}{4}\right)$
$=-\frac{2}{3}\left(-\frac{1}{4}\right)a$
$=\frac{1}{6}a$

49. $\frac{120xy}{24y}=\frac{120}{24}\cdot x\cdot\frac{y}{y}=5\cdot x\cdot 1=5x$

51. $\frac{-1.9rs}{9.5r}=\frac{-1.9}{9.5}\cdot\frac{r}{r}\cdot s=-0.2\cdot 1\cdot s=-0.2s$

53. $\frac{12.321pqr}{-1.11pr}=\frac{12.321}{-1.11}\cdot\frac{p}{p}\cdot q\cdot\frac{r}{r}$
$=-11.1\cdot 1\cdot q\cdot 1$
$=-11.1q$

55. $\frac{-\frac{5}{9}mn}{5n}=-\frac{5}{9}\cdot\frac{1}{5}\cdot m\cdot\frac{n}{n}=-\frac{1}{9}\cdot m\cdot 1=-\frac{1}{9}m$

57. $\frac{36x+60}{12}=\frac{36x}{12}+\frac{60}{12}=3x+5$

59. $\frac{20.4b-3.4c}{-6.8}=\frac{20.4b}{-6.8}-\frac{3.4c}{-6.8}$
$=-3b+\frac{1}{2}c$

61. $\frac{96a+24b-120c}{8}=\frac{96a}{8}+\frac{24b}{8}-\frac{120c}{8}$
$=12a+3b-15c$

63. $11(-3a+2b-4c)-8(5a-7b+2c)$
$=-33a+22b-44c-40a+56b-16c$
$=-73a+78b-60c$

65. $-4.6(2x-5y)+9.9(5x-3y)$
$=-9.2x+23y+49.5x-29.7y$
$=40.3x-6.7y$

67. $\frac{3}{8}\left(-\frac{4}{9}p-\frac{2}{9}q\right)+\frac{2}{3}\left(\frac{7}{8}p-\frac{6}{7}q\right)$
$=-\frac{1}{6}p-\frac{1}{12}q+\frac{7}{12}p-\frac{4}{7}q$
$=-\frac{2}{12}p+\frac{7}{12}p-\frac{7}{84}q-\frac{48}{84}q$
$=\frac{5}{12}p-\frac{55}{84}q$

69. $[15-2(3x+6y-10)+4x]+[6x+2(8y-12)]$
$=(15-6x-12y+20+4x)+(6x+16y-24)$
$=35-2x-12y+6x+6y-24$
$=11+4x+4y$

71. $2[-5a+3(2b-4c)+15]-[7(2a+6b-c)+12]$
$=2(-5a+6b-12c+15)-(14a+42b-7c+12)$
$=-10a+12b-24c+30-14a-42b+7c-12$
$=-24a-30b-17c+18$

73. $6\{2[x+2(3y-4z)]-[x-y+3(y+2z)]\}$
$=6[2(x+6y-8z)-(x-y+3y+6z)]$
$=6(2x+12y-16z-x+y-3y-6z)$
$=6(x+10y-22z)$
$=6x+60y-132z$

75. $\frac{8(5a+7c)-6(2a+4c)}{4}=\frac{40a+56c-12a-24c}{4}$
$=\frac{40a}{4}+\frac{56c}{4}-\frac{12a}{4}-\frac{24c}{4}$
$=10a+14c-3a-6c$
$=7a+8c$

77. $\frac{2.6m+3(1.2m-2.6n)-(4.8m+7.4n)-1.6n}{3m+2(-2m+1)+m}$
$=\frac{2.6m+3.6m-7.8n-4.8m-7.4n-1.6n}{3m-4m+2+m}$
$=\frac{1.4m-16.8n}{2}$
$=\frac{1.4m}{2}-\frac{16.8n}{2}$
$=0.7m-8.4n$

79. Let w = width
$2w$ = length
w = height
Surface area $= 2[w(2w)] + 2[2w(w)] + 2[w \cdot w]$
$= 2(2w^2) + 2(2w^2) + 2w^2$
$= 4w^2 + 4w^2 + 2w^2$
$= 10w^2$
$= 10(2)^2$
$= 10(4)$
$= 40$
The surface area is 40 square feet.

81. Let x = number of days
y = miles driven
Expense $= 35x + 0.2y + 2(120x) + 2(50)$
$= 275x + 0.2y + 100$
$= 275(4) + 0.2(625) + 100$
$= 1325$
\$1325 is the total trip expense

83. Let x = number of days
y = number of miles
Expense $= 30x + 0.22y + 450 + 85(2x) + 2(50)$
$= 200x + 0.22y + 550$
$= 200(2) + 0.22(375) + 550$
$= 1032.5$
\$1032.50 is the total expense

85. Let x = number of hours
Cost $= 2(38x) + 2(22) + 3(16x)$
$= 124x + 44$
$= 124(7) + 44$
$= 912$
The total cost is \$912.

3.3 Experiencing Algebra the Calculator Way

1. $-679(138x - 349y) + 903(287x + 423y)$
$= -679(138x) - 679(-349y) + 903(287x) + 903(423y)$
$= 165{,}459x + 618{,}940y$

2. $\dfrac{1173.04a + 2147.893b}{23.65} = \dfrac{1173.04a}{23.65} + \dfrac{2147.893b}{23.65}$
$= 49.6a + 90.82b$

3. $27.12[39.675(x-41.56)-21.876(y-22.7)]+109.35[30.4(x+33.9)+43.76(y-52.7)]$
$=27.12(39.675x-1648.893-21.876y+496.5852)+109.35(30.4x+1030.56+43.76y-2306.152)$
$=27.12(39.675x-21.876y-1152.3078)+109.35(30.4x+43.76y-1275.592)$
$=1075.986x-593.27712y-31{,}250.58754+3324.24x+4785.156y-139{,}485.9852$
$=4400.226x+4191.87888y-170{,}736.5727$

4. $\left(3\frac{5}{12}\right)\left[\left(4\frac{1}{2}\right)x-\left(3\frac{3}{5}\right)y\right]$
$=\frac{41}{12}\left(\frac{9}{2}\right)x-\frac{41}{12}\left(\frac{18}{5}\right)y$
$=\frac{123}{8}x-\frac{123}{10}y$

5. $\frac{12.0832abc}{2.36b}=5.12ac$

6. $\frac{-2102.17yz}{41.3yz}=-50.9$

7. $\frac{-34.155xyz}{-2.07xz}=16.5y$

8. $\frac{\frac{5}{12}y}{-\frac{15}{16}}=\frac{5}{12}y\cdot\left(-\frac{16}{15}\right)=-\frac{4}{9}y$

9. $\frac{-\frac{2}{5}ab}{\frac{3}{5}ab}=-\frac{2ab}{5}\cdot\left(\frac{5}{3ab}\right)=-\frac{2}{3}$

10. $\frac{\frac{3}{5}xy}{-\frac{2}{5}x}=\frac{3xy}{5}\cdot\left(-\frac{5}{2x}\right)=-\frac{3}{2}y$

11. $\frac{-5bc}{-\frac{5}{7}b}=-5bc\cdot\left(-\frac{7}{5b}\right)=7c$

12. $\frac{-\frac{5}{21}x+\frac{15}{49}y}{-\frac{5}{7}}=\left(-\frac{5x}{21}\right)\left(-\frac{7}{5}\right)+\left(\frac{15y}{49}\right)\left(-\frac{7}{5}\right)$
$=\frac{1}{3}x-\frac{3}{7}y$

3.4 Experiencing Algebra the Exercise Way

1. equation; equality

3. expression; no equality

5. equation; equality

7. equation; equality

9. expression; no equality

11. $7+15=2\cdot 11$

22	22

Since 22 = 22, the equation is true.

13. $4\cdot 7=5+24$

28	29

Since 28 ≠ 29, the equation is false.

15. $\frac{1}{7}+\frac{2}{3}=\frac{6}{7}-\frac{1}{3}$

$\frac{3}{21}+\frac{14}{21}$	$\frac{18}{21}-\frac{7}{21}$
$\frac{17}{21}$	$\frac{11}{21}$

Since $\frac{17}{21}\neq\frac{11}{21}$, the equation is false.

17. $\frac{1}{2}+\frac{1}{3}=\frac{2}{3}+\frac{1}{6}$

$\frac{3}{6}+\frac{2}{6}$	$\frac{4}{6}+\frac{1}{6}$
$\frac{5}{6}$	$\frac{5}{6}$

Since $\frac{5}{6}=\frac{5}{6}$, the equation is true.

19. $3.7-4.8=1.7-0.6$

$$\begin{array}{r|l} -1.1 & 1.1 \end{array}$$

Since $-1.1 \neq 1.1$, the equation is false.

21. $4.9+1.2=8.4-2.2$

$$\begin{array}{r|l} 6.1 & 6.2 \end{array}$$

Since $6.1 \neq 6.2$, the equation is false.

23. $2(16-5)=3(19-12)$

$$\begin{array}{r|l} 2(11) & 3(7) \\ 22 & 21 \end{array}$$

Since $22 \neq 21$, the equation is false.

25. $\frac{2}{3}+\frac{3}{2}=2$

$$\begin{array}{r|l} \frac{4}{6}+\frac{9}{6} & 2 \\ \frac{13}{6} & \end{array}$$

Since $\frac{13}{6} \neq 2$, the equation is false.

27. $2x+4=10$

$$\begin{array}{r|l} 2(3)+4 & 10 \\ 6+4 & \\ 10 & \end{array}$$

Since $10 = 10$, 3 is a solution.

29. $5y-7=9$

$$\begin{array}{r|l} 5(3)-7 & 9 \\ 15-7 & \\ 8 & \end{array}$$

Since $8 \neq 9$, 3 is not a solution.

31. $6a+5=3a+17$

$$\begin{array}{r|l} 6(3)+5 & 3(3)+17 \\ 18+5 & 9+17 \\ 23 & 26 \end{array}$$

Since $23 \neq 26$, 3 is not a solution.

33. $9z-23=6z-29$

$$\begin{array}{r|l} 9(-2)-23 & 6(-2)-29 \\ -18-23 & -12-29 \\ -41 & -41 \end{array}$$

Since $-41=-41$, -2 is a solution.

35. $3(x-5)+9=4(6x-5)-7$

$$\begin{array}{r|l} 3(1-5)+9 & 4(6\cdot 1-5)-7 \\ 3(-4)+9 & 4(1)-7 \\ -12+9 & 4-7 \\ -3 & -3 \end{array}$$

Since $-3=-3$, 1 is a solution.

37. $2[3(x-4)-6]+x=3x$

$$\begin{array}{r|l} 2[3(8-4)-6]+8 & 3(8) \\ 2(3\cdot 4-6)+8 & 24 \\ 2(12-6)+8 & \\ 2(6)+8 & \\ 12+8 & \\ 20 & \end{array}$$

Since $20 \neq 24$, 8 is not a solution.

39. $x^2+5=33-3x$

$$\begin{array}{r|l} 4^2+5 & 33-3(4) \\ 16+5 & 33-12 \\ 21 & 21 \end{array}$$

Since $21 = 21$, 4 is a solution.

41. $7+6(5)=3(7)+4^2$

43. Let x = a number
$x+6=15$

45. Let x = a number
$2x=12$

47. Let x = a number
$x^2-21=100$

49. Let x = a number
$2(x+5^2)=x+100$

51. Let x = a number
$2(x+2)=4+2x$

53. Let x = a number
$17+\frac{x}{2}=4+3x$

55. Let p = perimeter
d = diameter
r = radius
$p=d+\pi r$

57. $I=A-P$

3.4 Experiencing Algebra the Calculator Way

1.

```
4-6*3+28/7
                -10
16/4+2(-16+9)
                -10
```

True

2.

```
14/2+6*3-2²
                 21
5²-2²
                 21
```

True

3.

```
4(8+7*3)-100
                 16
5+3(2*6-20)
                -19
```

False

4.

```
-7(-5+4)+20/5
                 11
-2(3²+1)+19
                 -1
```

False

5.

```
13(1/5+3/7)▸Frac
             286/35
11(3/5+1/7)▸Frac
             286/35
```

True

6.

```
1/2(2/3+4/7)-3/7
▸Frac
               4/21
1-17/21▸Frac
               4/21
```

True

7.

```
(3/4)/(1/2)+2(3/
8)-(1+1/4)▸Frac
                  1
1/2▸Frac
                1/2
```

False

8.

```
1/3-(5/7)/(5/3)+
6(1/7)-(2+1/3)▸F
rac
              -11/7
-(1+4/7)▸Frac
              -11/7
```

False

9.
```
1.2-3(4.7)+6.5/5
             -11.6
-3(3.2)+2(4.3-2.
4-2.9)
             -11.6
```

True

10.
```
-1.2+4.7-6.9/2.3
                .5
3(5.6)-4(4.2-0.5
)
                 2
```

False

11.
```
5.7/3+7(-1.2)
              -6.5
-2(3.3
              -6.6
```

False

12.
```
2(2.7)-5.8/2
               2.5
7.8/3-0.1
               2.5
```

True

13.
```
269.1/7.5
             35.88
13.8(2.6)
             35.88
```

True

14.
```
-6-1/3→Z:5Z+13=2
Z-6
                 1
```

Yes

15.
```
-2/7→X:15X-2=X-7
                 0
```

No

16.
```
27.5→X:15X-13=6(
2X+7)+X
                 1
```

Yes

17.
```
-.5→Z:3Z+20=5-3(
Z-4)
                 1
```

Yes

3.5 Experiencing Algebra the Exercise Way

1. $A = \frac{1}{2}bh = \frac{1}{2}(39)(24)$
   $A = 468 \text{ in}^2$
   $P = a + b + c = 39 + 40 + 25 = 104 \text{ in.}$

3. $A = LW = 6(4) = 24 \text{ cm}^2$
   $P = 2L + 2W = 2(6) + 2(4) = 12 + 8 = 20 \text{ cm}$

5. $A = s^2 = \left(2\frac{1}{2}\right)^2 = \left(\frac{5}{2}\right)^2 = \frac{25}{4} \text{ ft}^2$
   $P = 4s = 4\left(2\frac{1}{2}\right) = 4\left(\frac{5}{2}\right) = 10 \text{ ft}$

7. $A = bh = 68(45) = 3060 \text{ m}^2$
   $P = 2a + 2b$
   $= 2(53) + 2(68)$
   $= 106 + 136$
   $= 242 \text{ m}$

9. $A = \frac{1}{2}h(b + B)$
   $= \frac{1}{2}(8)(2 + 23)$
   $= 4(25)$
   $= 100 \text{ ft}^2$
   $P = a + b + c + B = 10 + 2 + 17 + 23 = 52 \text{ ft}$

11. $A = \pi r^2 = (\pi)\left(5\frac{1}{4}\right)^2 \approx 86.6 \text{ in}^2$

$C = 2\pi r = 2(\pi)\left(5\frac{1}{4}\right) \approx 33.0 \text{ in.}$

13. $V = LWH = 5(3)(1) = 15 \text{ ft}^3$
$S = 2LW + 2WH + 2LH$
$= 2(5)(3) + 2(3)(1) + 2(5)(1)$
$= 30 + 6 + 10$
$= 46 \text{ ft}^2$

15. $V = s^3 = (7.5)^3 = 421.875 \text{ in}^3$
$S = 6s^2 = 6(7.5)^2 = 337.5 \text{ in}^2$

17. $V = \pi r^2 h = (\pi)\left(\frac{3}{2}\right)^2 (5) \approx 35.3 \text{ in}^3$

$S = 2\pi r^2 + 2\pi rh$

$= 2(\pi)\left(\frac{3}{2}\right)^2 + 2(\pi)\left(\frac{3}{2}\right)(5)$

$\approx 61.3 \text{ in}^2$

19. $V = \frac{4}{3}\pi r^3 = \frac{4}{3}(\pi)\left(\frac{20}{2}\right)^3 \approx 4189 \text{ cm}^3$

$S = 4\pi r^2 = 4(\pi)(10)^2 \approx 1257 \text{ cm}^2$

21. $V = \frac{1}{3}\pi r^2 h$

$= \frac{1}{3}(\pi)(0.25)^2(0.75)$

$\approx 0.049 \text{ ft}^3$

$S = \pi r\sqrt{r^2 + h^2} + \pi r^2$

$= \pi(0.25)\sqrt{(0.25)^2 + (0.75)^2} + \pi(0.25)^2$

$\approx 0.817 \text{ ft}^2$

23. $V = \frac{1}{3}Bh = \frac{1}{3}(8)^2(15) = 320 \text{ cm}^3$

25. $90° - 65° = 25°$

27. $180° - 65° = 115°$

29. $180° - (33° + 68°) = 79°$

31. $c = \sqrt{a^2 + b^2} = \sqrt{(40)^2 + (42)^2} = 58 \text{ in.}$

33. $c = \sqrt{a^2 + b^2} = \sqrt{(6)^2 + \left(2\frac{1}{2}\right)^2} = 6\frac{1}{2} \text{ cm}$

35. $c = \sqrt{a^2 + b^2}$
$= \sqrt{(4.81)^2 + (6)^2}$
$= 7.69 \text{ mm}$

37. $A = \frac{1}{2}h(b + B)$

$= \frac{1}{2}(14)(12 + 15)$
$= 7(27)$
$= 189 \text{ ft}^2$

John will need to cover 189 ft^2.

39. $A = s^2 = (52)^2 = 2704 \text{ in}^2$
$P = 4s = 4(52) = 208 \text{ in.}$

The area of coverage is 2704 in^2. Gretchen will need 208 in. of fringe material.

41. $A = LW = 12(8.5) = 102 \text{ ft}^2$
$P = 2L + 2W = 2(12) + 2(8.5) = 41 \text{ ft}$

Linda had 102 ft^2 of sod removed. She needed 41 ft of border.

43. $A = \pi r^2 = \pi(5)^2 \approx 78.5 \text{ ft}^2$
$C = 2\pi r = 2\pi(5) \approx 31.4 \text{ ft}$

The square footage of the area is 78.5 ft^2. The border will be 31.4 ft.

45. $A = bh = 220(250) = 55{,}000 \text{ ft}^2$

The lot is 55,000 ft^2.

47. $V = LWH = 4(2)(2) = 16 \text{ ft}^3$
$S = 2LW + 2WH + 2LH$
$= 2(4)(2) + 2(2)(2) + 2(4)(2)$
$= 16 + 8 + 16$
$= 40 \text{ ft}^2$
$40 \div 20 = 2$

The box will hold 16 ft^3 of toys. Jim painted 40 ft^2 of surface area, using 2 pints of paint.

49. $V = \pi r^2 h = \pi(1)^2(4) \approx 12.6 \text{ ft}^3$
$S = 2\pi r^2 + 2\pi rh$
$= 2\pi(1)^2 + 2\pi(1)(4)$
$\approx 31.4 \text{ ft}^2$
The barrel contains 12.6 ft^3 of space.
31.4 ft^2 of surface area will need to be painted.

51. $V = \frac{4}{3}\pi r^3 = \frac{4}{3}\pi(10)^3 \approx 4188.8 \text{ in}^3$
$S = 4\pi r^2 = 4\pi(10)^2 \approx 1256.6 \text{ in}^2$
The tank will hold 4188.8 in^3 of gas. The surface area is 1256.6 in^2.

53. $V = s^3 = (18)^3 = 5832 \text{ in}^3$
$S = 6s^2 = 6(18)^2 = 1944 \text{ in}^2$
The case contains 5832 in^3, with a surface area of 1944 in^2.

55. $V = \frac{1}{3}Bh = \frac{1}{3}(3.5)^2(7) = 28\frac{7}{12} \text{ in}^3$
The volume of the paperweight is $28\frac{7}{12} \text{ in}^3$.

57. $180° - (30° + 65°) = 85°$
The third angle is 85°.

59. $90° - 50° = 40°$
The pitch of the roof is 40°.

61. $180° - 45° = 135°$
The other angle is 135°.

63. $c = \sqrt{a^2 + b^2} = \sqrt{6^2 + 8^2} = 10 \text{ ft}$
The diagonal must measure 10 ft.

3.5 Experiencing Algebra the Calculator Way

Students should develop other programs for various formulas.

3.6 Experiencing Algebra the Exercise Way

1. $I = Prt = 2500(0.065)(1) \approx 162.50$
JoAnne paid $162.50 in interest.

3. $A = P(1+r)^t$
$= 2500(1+0.006)^{12}$
≈ 2686.06
$I = A - P = 2686.06 - 2500 \approx 186.06$
Simple interest is better.

5. $A = Pe^{nr} = 2500(e)^{1(0.065)} \approx 2667.90$
$I = A - P = 2667.90 - 2500 = 167.90$
The option in exercise 1 is the best choice.

7. $I = Prt = 500(0.07)\left(\frac{1}{12}\right) = 2.91\overline{6}$
$2.92 of interest is earned.

9. $I = Prt = 1200(0.08)\left(\frac{1}{2}\right) = 48$
$48 in interest is paid.

11. $A = P(1+r)^t = 2000(1+0.09)^2 = 2376.2$
$I = A - P = 2376.2 - 2000 = 376.2$
Steve earned $376.20 in interest.

13. $A = P(1+r)^t$
$= 4000(1+0.07)^5$
≈ 5610.2069
$I = A - P$
$= 5610.2069 - 4000$
$= 1610.2069$
Chauncie had $1610.21 in interest.

15. $A = P(1+r)^t$
$= 3000(1+0.04)^3$
$= 3374.592$
$I = 3374.592 - 3000 = 374.592$
The interest earned will be $374.59.

17. $A = Pe^{nr}$
$= 1800e^{2(0.05)}$
≈ 1989.307653
$I = A - P \approx 1989.307653 - 1800$
≈ 189.3076525
The interest is $189.31.

19. $A = Pe^{nr} = 3000e^{.06(4.5)} \approx 3929.893352$
$I = A - P \approx 3929.893352 - 3000$
≈ 929.8933522
The interest is \$929.89.

21. $F = \frac{9}{5}C + 32 = \frac{9}{5}(25) + 32 = 77$
The temperature is 77°F.

23. $C = \frac{5}{9}(F - 32) = \frac{5}{9}(92 - 32) = 33\frac{1}{3}$
The temperature is $33\frac{1}{3}$°C.

25. $F = \frac{9}{5}C + 32 = \frac{9}{5}(100) + 32 = 212$
The temperature is 212°F.

27. $F = \frac{9}{5}C + 32 = \frac{9}{5}(95) + 32 = 203$
The temperature is 203°F.

29. $C = \frac{5}{9}(F - 32)$
$= \frac{5}{9}(-37.97 - 32)$
$= -38.87\overline{2}$
The temperature is –38.87°C.

31. $C = \frac{5}{9}(F - 32) = \frac{5}{9}(84 - 32) = 28.\overline{8}$
The temperature is 28.9°C.

33. $d = rt = 55\left(9\frac{1}{2}\right) = 522.5$
The distance covered was 522.5 miles.

35. $d = rt = 74.602(3) = 223.806$
He drove 223.806 miles.

37. $s = -16t^2 + v_0t + s_0$
$= -16(3)^2 + 0(3) + 200$
$= 56$
The pie is 56 feet high.

39. $s = -16t^2 + v_0t + s_0$
$= -16(3)^2 + 25(3) + 200$
$= 131$
The baseball is 131 feet high.

41. $T = 2\pi\sqrt{\frac{L}{32}} = 2\pi\sqrt{\frac{3}{32}} \approx 1.924$
It will take 1.924 seconds.

43. $T = 2\pi\sqrt{\frac{L}{32}} = 2\pi\sqrt{\frac{2.25}{32}} \approx 1.666$
The period is 1.666 seconds.

45. $T = 2\pi\sqrt{\frac{L}{32}} = 2\pi\sqrt{\frac{0.75}{32}} \approx 0.962$
The period is 0.962 second.

47. $T = 2\pi\sqrt{\frac{L}{32}} = 2\pi\sqrt{\frac{2.5}{32}} \approx 1.756$
The leg takes 1.756 seconds.
$r = \frac{d}{t} = \frac{4}{1.756} = 2.28$
Eydie's speed is 2.28 feet per second.

3.6 Experiencing Algebra the Calculator Way

$A = Pe^{nr}$

$A = 80,000e^{n(0.085)}$

Number of Years (n)	Amount earned (A)	Interest = A – 80,000
5	$122,367	$42,367
10	$187,172	$107,172
15	$286,296	$206,296
20	$437,916	$357,916
25	$669,832	$589,832
30	$1,024,568	$944,568

Chapter 3 Review

Reflections

Answers will vary.

Exercises

1. Let x = a number
   $55x + 4$

2. Let x = a number
   $55(x + 4)$

3. Let x = a number
   $\frac{3}{4}(x+35)$

4. Let x = a number
   $\frac{3}{4}x+35$

5. Let x = a number
   $2x - 20$

6. Let x = a number
   $20 - 2x$

7. $2500 + 275n$

8. $\frac{650}{h}$

9. $200 + 5.5k$

For problems 10–18, answers will vary. One possibility is given.

10. $5 + x$
    The sum of 5 and a number

11. $8 - 6n$
    The difference of 8 and 6 times some number

12. $9x \div 6$
    The product of 9 and a number, divided by 6

13. $\frac{2}{3}(x-75)$
    Two-thirds of the difference of a number and 75

14. $x^2 + y$
    The square of one number added to another number

15. xyz
    The product of three different numbers

16. $5x - 25 = 5(8) - 25 = 40 - 25 = 15$

17. $5x - 25 = 5(-4.6) - 25 = -23 - 25 = -48$

18. $\sqrt{6x+10} = \sqrt{6(9)+10}$
$= \sqrt{54+10}$
$= \sqrt{64}$
$= 8$

19. $\sqrt{6x+10} = \sqrt{6(-1)+10}$
$= \sqrt{-6+10}$
$= \sqrt{4}$
$= 2$

20. $\frac{12y-84}{6y+36} = \frac{12(9)-84}{6(9)+36} = \frac{24}{90} = \frac{4}{15}$

21. $\frac{12y-84}{6y+36} = \frac{12(-6)-84}{6(-6)+36} = \frac{-156}{0} =$ undefined

22. $|2x^2 - 45x - 75| = |2(0.1)^2 - 45(0.1) - 75|$
$= |0.02 - 4.5 - 75|$
$= |-79.48|$
$= 79.48$

23. $|2x^2 - 45x - 75| = |2(10)^2 - 45(10) - 75|$
$= |200 - 450 - 75|$
$= |-325|$
$= 325$

24. $x^2 = (-15)^2 = 225$

25. $-x^2 = -(-15)^2 = -225$

26. $(-x)^2 = [-(-15)]^2 = (15)^2 = 225$

27. Let m = number of months
$1200 + 75m = 1200 + 75(12) = 2100$
Sandi will have \$2100 after 12 months.
Yes, she will have enough money.

	Algebraic Expression	# of terms	terms
28.	$3x^2 - 16x + 35$	3	$3x^2$, $-16x$, 35
29.	$5x + 12 - 3y - 16$	4	$5x$, 12, $-3y$, -16
30.	$2(x+1) - 4(y+2) - 4$	3	$2(x+1)$, $-4(y+2)$, -4
31.	$a^2 + 2ab + b^2$	3	a^2, $2ab$, b^2

	Algebraic Expression	Coefficients of terms	like terms
32.	$3x - 2y + 4x + 9y$	3, −2, 4, 9	$3x$ and $4x$, $-2y$ and $9y$
33.	$2a^2 - a + 3a^2 - 5a^3$	2, −1, 3, −5	$2a^2$ and $3a^2$
34.	$2.4x + 5.1 + 6.2x$	2.4, 5.1, 6.2	$2.4x$ and $6.2x$
35.	$4(a+b) - 2(a+b) + 2a$	4, −2, 2	$4(a+b)$ and $-2(a+b)$

36. $6c - 17c = (6 - 17)c = -11c$

37. $2.4z + 1.7z - 3.9z = (2.4 + 1.7 - 3.9)z = 0.2z$

38. $17x + 51 + 26x - 86 - 19x - 7$
$= 17x + 26x - 19x + 51 - 86 - 7$
$= 24x - 42$

39. $\frac{3}{4}x + \frac{5}{8}y - \frac{2}{3}x + \frac{1}{4}y = \frac{3}{4}x - \frac{2}{3}x + \frac{5}{8}y + \frac{1}{4}y$
$= \frac{9}{12}x - \frac{8}{12}x + \frac{5}{8}y + \frac{2}{8}y$
$= \frac{1}{12}x + \frac{7}{8}y$

40. $15x^2 - 14xy + 12y^2 - 23 + 42xy - 7y^2 + 21 - 6x^2$
$= 15x^2 - 6x^2 - 14xy + 42xy + 12y^2 - 7y^2 - 23 + 21$
$= 9x^2 + 28xy + 5y^2 - 2$

41. $-(19p - 21) = -19p + 21$

42. $4a + 6 - (a - 1) = 4a + 6 - a + 1$
$= 4a - a + 6 + 1$
$= 3a + 7$

43. $(3a + 4b) - (-2a - 6b) + (a - b) - (-a + b)$
$= 3a + 4b + 2a + 6b + a - b + a - b$
$= 3a + 2a + a + a + 4b + 6b - b - b$
$= 7a + 8b$

44. Let x = number of cans of peanuts
y = number of cans of bridge mix
$2.75(x + y) - (1.25x + 1.5y)$
$= 2.75x + 2.75y - 1.25x - 1.5y$
$= 2.75x - 1.25x + 2.75y - 1.5y$
$= 1.5x + 1.25y$
$= 1.5(220) + 1.25(480)$
$= 330 + 600 = 930$
The net profit is \$930.

45. Let x = number of pins
$20 - [5x + 0.06(5x)] = 20 - (5x + 0.3x)$
$= 20 - 5.3x$
$= 20 - 5.3(3)$
$= 20 - 15.9$
$= 4.1$
Katie received \$4.10 in change.

46. Let x = number of frames
$-3.25x + 12x - 52.65 = 8.75x - 52.65$
$= 8.75(32) - 52.65$
$= 280 - 52.65$
$= 227.35$
Margaret's net profit is \$227.35.

47. $14(5ab) = (14 \cdot 5)ab = 70ab$

48. $-2(-4.1m) = [-2 \cdot (-4.1)]m = 8.2m$

49. $-2a(3.8a - 4.7b) = -2a(3.8a) - 2a(-4.7b)$
$= -7.6a + 9.4ab$

50. $-\frac{5}{12}\left(\frac{6}{7}x - \frac{4}{15}y\right) = -\frac{5}{12}\left(\frac{6}{7}x\right) - \frac{5}{12}\left(-\frac{4}{15}y\right)$
$= -\frac{5}{14}x + \frac{1}{9}y$

51. $2x(3x - 17y) = 2x(3x) + 2x(-17y)$
$= 6x^2 - 34xy$

52. $\frac{-123m}{-3} = -123m \div (-3)$
$= -123m \cdot \left(-\frac{1}{3}\right)$
$= -123\left(-\frac{1}{3}\right)m$
$= 41m$

53. $\dfrac{\frac{15}{16}z}{-5} = \dfrac{15}{16}z \div (-5)$
$= \dfrac{15}{16}z \cdot \left(-\dfrac{1}{5}\right)$
$= \dfrac{15}{16}\left(-\dfrac{1}{5}\right) \cdot z$
$= -\dfrac{3}{16}z$

54. $\dfrac{36.6ab}{2} = \dfrac{36.6}{2} \cdot ab = 18.3ab$

55. $\dfrac{-308.88ab}{4b} = \dfrac{-308.88}{4} \cdot a \cdot \dfrac{b}{b}$
$= -77.22 \cdot a \cdot 1$
$= -77.22a$

56. $\dfrac{27a - 36b + 18c}{9} = \dfrac{27a}{9} - \dfrac{36b}{9} + \dfrac{18c}{9}$
$= 3a - 4b + 2c$

57. $12(7x + 9y) + 15(3x - 7y)$
$= 84x + 108y + 45x - 105y$
$= 129x + 3y$

58. $3[-2(x + 3y) - 5] - [3(2x + y) + 16]$
$= 3(-2x - 6y - 5) - (6x + 3y + 16)$
$= -6x - 18y - 15 - 6x - 3y - 16$
$= -12x - 21y - 31$

59. $\dfrac{25(2a - 6b) + 5(4b - 3c) + 75}{4a - 2(2a - 3) - 1}$
$= \dfrac{50a - 150b + 20b - 15c + 75}{4a - 4a + 6 - 1}$
$= \dfrac{50a - 130b - 15c + 75}{5}$
$= \dfrac{50a}{5} - \dfrac{130b}{5} - \dfrac{15c}{5} + \dfrac{75}{5}$
$= 10a - 26b - 3c + 15$

60. Tom: x pushups
Charles: $2x - 5$ pushups
Jim: $x + 7$ pushups
All three: $x + 2x - 5 + x + 7 = 4x + 2$
Total $= 4(45) + 2 = 182$
182 is the pushups total.

61. Let x = number of days
y = number of miles
Total cost $= 45x + 0.2y + 2(125x) + 2(20x)$
$= 45x + 0.2y + 250x + 40x$
$= 335x + 0.2y$
$= 335(4) + 0.2(345)$
$= 1409$
The total cost is \$1409.

62. expression; no equality

63. equation; equality

64. equation; equality

65. expression; no equality

66. $17 - 3 \cdot 4 = 20 \cdot 2 - 6^2$

$17 - 12$	$40 - 36$
5	4

Since $5 \neq 4$, the equation is false.

67. $2(3.5) - 3(0.7) = 2.5 + 2(1.2)$

$7 - 2.1$	$2.5 + 2.4$
4.9	4.9

Since $4.9 = 4.9$, the equation is true.

68. $3 + 5(9 - 8) + 6 \cdot 5 = 4 + 6(12 - 8) + 45 \div 9 + 5$

$3 + 5(1) + 30$	$4 + 6(4) + 5 + 5$
38	38

Since $38 = 38$, the equation is true.

69. $\frac{2}{3}+5\left(\frac{1}{6}-\frac{2}{9}\right)=\frac{4}{9}-3\left(\frac{2}{3}-\frac{17}{18}\right)$

$$\begin{array}{r|l}
\frac{2}{3}+5\left(\frac{3}{18}-\frac{4}{18}\right) & \frac{4}{9}-3\left(\frac{12}{18}-\frac{17}{18}\right) \\
\frac{2}{3}+5\left(-\frac{1}{18}\right) & \frac{4}{9}-3\left(-\frac{5}{18}\right) \\
\frac{12}{18}-\frac{5}{18} & \frac{8}{18}+\frac{15}{18} \\
\frac{7}{18} & \frac{23}{18}
\end{array}$$

Since $\frac{7}{18}\neq\frac{23}{18}$, the equation is false.

70. $15y-35=12$

$$\begin{array}{r|l}
15(3)-35 & 12 \\
45-35 & \\
10 &
\end{array}$$

Since $10\neq 12$, 3 is not a solution.

71. $8a+2(3a-7)=11a-32$

$$\begin{array}{r|l}
8(-6)+23(-6)-7] & 11(-6)-32 \\
-48+2(-18-7) & -66-32 \\
-48+2(-25) & -98 \\
-98 &
\end{array}$$

Since $-98=-98$, -6 is a solution.

72. $x^3-25x=2x^2-32-35$

$$\begin{array}{r|l}
5^3-25(5) & 2(5)^2-3(5)-35 \\
125=125 & 50-15-35 \\
0 & 0
\end{array}$$

Since $0=0$, 5 is a solution.

73. $2(x-3.4)=x-5$

$$\begin{array}{r|l}
2(1.8-3.4) & 1.8-5 \\
2(-1.6) & -3.2 \\
-3.2 &
\end{array}$$

Since $-3.2=-3.2$, 1.8 is a solution.

74. $x-\frac{2}{3}=-\frac{1}{9}$

$$\begin{array}{r|l}
\frac{4}{9}-\frac{2}{3} & -\frac{1}{9} \\
\frac{4}{9}-\frac{6}{9} & \\
-\frac{2}{9} &
\end{array}$$

Since $-\frac{2}{9}\neq-\frac{1}{9}$, $\frac{4}{9}$ is not a solution.

75. $3+2(27-15)=3^3$

76. Let x = a number

$5+4x=65+\frac{x}{4}$

77. $A=\frac{1}{2}bh=\frac{1}{2}(26)(16)=208\text{ m}^2$

$P=a+b+c=22+26+20=68$ m

78. $A=LW=44(20)=880\text{ in}^2$

$P=2L+2W=2(44)+2(20)=88+40$

$=128$ in.

79. $A=s^2=(15)^2=225\text{ cm}^2$

$P=4s=4(15)=60$ cm

80. $A=bh=10.0(6.5)=65\text{ m}^2$

$P=2a+2b=2(7.0)+2(10.0)=24$ m

81. $A = \frac{1}{2}h(b+B)$
$= \frac{1}{2}(70)(110+160)$
$= 9450 \text{ yd}^2$
$P = a + b + c + B$
$= 75 + 110 + 80 + 160$
$= 425 \text{ yd}$

82. $A = \pi r^2 = \pi(12.2)^2 \approx 467.59 \text{ ft}^2$
$C = 2\pi r = 2\pi(12.2) \approx 76.65 \text{ ft}$

83. $V = LWH = 35(9)(21) = 6615 \text{ in}^3$
$S = 2LW + 2WH + 2LH$
$= 2(35)(9) + 2(9)(21) + 2(35)(21)$
$= 630 + 378 + 1470$
$= 2478 \text{ in}^2$

84. $V = s^3 = (14.6)^3 = 3112.136 \text{ mm}^3$
$S = 6s^2 = 6(14.6)^2 = 1278.96 \text{ mm}^2$

85. $V = \pi r^2 h = \pi(18)^2(54) \approx 54{,}965.3 \text{ in}^3$
$S = 2\pi r^2 + 2\pi rh$
$= 2\pi(18)^2 + 2\pi(18)(54)$
$\approx 8143.01 \text{ in}^2$

86. $V = \frac{4}{3}\pi r^3 = \frac{4}{3}\pi(32.6)^3 \approx 145{,}124.7 \text{ cm}^3$
$S = 4\pi r^2 = 4\pi(32.6)^2 \approx 13{,}355.04 \text{ cm}^2$

87. $V = \frac{1}{3}\pi r^2 h = \frac{1}{3}\pi(4)^2(7) \approx 117.3 \text{ cm}^3$
$S = \pi r\sqrt{r^2 + h^2} + \pi r^2$
$= \pi(4)\sqrt{4^2 + 7^2} + \pi(4)^2$
$\approx 151.6 \text{ cm}^2$

88. $V = \frac{1}{3}Bh = \frac{1}{3}(5)^2(8) = 66\frac{2}{3} \text{ in}^3$

89. $180° - 58° = 122°$
The other angle is 122°.

90. $90° - 58° = 32°$
The other angle is 32°.

91. $180° - (67° + 88°) = 25°$
The third angle is 25°.

92. $c = \sqrt{a^2 + b^2} = \sqrt{(20)^2 + (21)^2} = 29 \text{ in.}$
The hypotenuse is 29 inches.

93. $c = \sqrt{a^2 + b^2}$
$= \sqrt{(5.39)^2 + (29.4)^2}$
$= 29.89 \text{ cm}$
The hypotenuse is 29.89 cm.

94. $A = LW = 85(60) = 5100 \text{ ft}^2$
$P = 2L + 2W = 2(85) + 2(60) = 290$
fence $= 290 - 3 - 35 = 252$ ft
Dan should order 5100 ft^2 of sod and 252 feet of fencing.

95. $A = \pi r^2 - \pi\left(\frac{2}{2}\right)^2 = \pi(8)^2 - \pi(1)^2$
$\approx 197.9 \text{ ft}^2$
The garden will have about 197.9 ft^2.

96. Area of 2 10" pizzas $= 2(\pi r^2)$
$= 2(\pi)(5)^2$
$= 50\pi \text{ in}^2$
Area of one 14" pizza $= \pi r^2$
$= \pi(7)^2$
$= 49\pi \text{ in}^2$
The 2 10-inch pizzas are the better deal because 50π is greater than 49π.

97. $V = LWH = 8(5)(2) = 80 \text{ ft}^3$
The truck will hold 80 ft^3.

98. $S = 4\pi r^2 = 4\pi\left(\frac{6}{2}\right)^2 \approx 113.1 \text{ ft}^2$
The surface area is approximately 113.1 ft^2.

99. $90° - 40° = 50°$
The other angle is 50°.

100. $c = \sqrt{a^2 + b^2} = \sqrt{(18)^2 + (13.5)^2} = 22.5 \text{ ft}$
The diagonal must be 22.5 feet.

101. $I = Prt = 850(0.125)(1) = 106.25$
$106.25 interest
$A = P + I = 850 + 106.25 = \$956.25$
The total amount of the loan is $956.25.

102. $A = P(1+r)^t$
$= 15,000(1+0.085)^6$
$\approx 24,472.013$
$I = A - P = 24,472.013 - 15,000 = 9472.013$
The interest is $9472.01.

103. $A = Pe^{nr} = 10,000e^{12(0.055)} \approx 19,347,923$
$I = A - P = 19,347.923 - 10,000 = 9347.923$
The amount of interest is $9347.92.

104. $F = \frac{9}{5}C + 32 = \frac{9}{5}(50) + 32 = 122$
The temperature is 122°F.

105. $C = \frac{5}{9}(F-32) = \frac{5}{9}(80-32) = 26\frac{2}{3}$
The temperature is $26\frac{2}{3}$ °C.

106. $d = rt = 62\left(5\frac{3}{4}\right) = 356.5$
LuAnn traveled 356.5 miles.

107. $s = -16t^2 + v_0t + s_0$
$= -16(3)^2 + 0(3) + 180$
$= 36$
The wedding bouquet is 36 feet above the ground.

108. $s = -16t^2 + v_0t + s_0$
$= -16(2)^2 + 96(2) + 0$
$= 128$
The height is 128 feet after 2 seconds.
$-16(3)^2 + 96(3) + 0 = 144$
The height is 144 feet after 3 seconds.
$-16(4)^2 + 96(4) + 0 = 128$
The height is 128 feet after 4 seconds.
$-16(5)^2 + 96(5) + 0 = 80$
The height is 80 feet after 5 seconds.

109. $T = 2\pi\sqrt{\frac{L}{32}} = 2\pi\sqrt{\frac{6}{32}} \approx 2.72$
The period is 2.72 seconds.

110. $I = Prt = 1000(0.08)(2) = 160$
$160 simple interest
$A = P(1+r)^t = 1000(1+0.07)^2 = 1144.9$
$I = 1144.9 - 1000 = 144.9$
$144.90 compounded interest
The first option is the better choice.

Chapter 3 Mixed Review

	Algebraic Expression	# of terms	terms
1.	$12 + y - z + 23$	4	$12x, y, -z, 23$
2.	$3(a-2) + 5(b-4) + 75$	3	$3(a-2), 5(b-4), 75$
3.	$12 - 7x + 14x - 18 + x$	5	$12, -7x, 14x, -18, x$

	Algebraic Expression	Coefficients of terms	like terms
4.	$x-2y-5+4x-y+6$	1, –2, –5, 4, –1, 6	x and $4x$, $-2y$ and $-y$, –5 and 6
5.	$b^2+2b-3b^2+6b+b^3$	1, 2, –3, 6, 1	b^2 and $-3b^2$, $2b$ and $6b$

6. $3x = 3(-18) = -54$

7. $-3x = -3(-18) = 54$

8. $-(-3x) = -[-3(-18)] = -54$

9. $-(-(-3x)) = -(-54) = 54$

10. $x^2 = (-18)^2 = 324$

11. $-x^2 = -(-18)^2 = -324$

12. $(-x)^2 = [-(-18)]^2 = 324$

13. $-(-x)^2 = -324$

14. $\sqrt{12y+20} = \sqrt{12(8)+20}$
$= \sqrt{116}$
≈ 10.77033

15. $\sqrt{12y+20} = \sqrt{12\left(-\frac{1}{3}\right)+20}$
$= \sqrt{-4+20}$
$= \sqrt{16}$
$= 4$

16. $\sqrt{12y+20} = \sqrt{12(-5)+20} = \sqrt{-40}$
Not a real number

17. $\frac{7x+84}{2x-3} = \frac{7(-12)+84}{2(-12)-3} = \frac{0}{-27} = 0$

18. $\frac{7x+84}{2x-3} = \frac{7(3)+84}{2(3)-3} = \frac{105}{3} = 35$

19. $\frac{7x+84}{2x-3} = \frac{7(1.5)+84}{2(1.5)-3} = \frac{94.5}{0}$
Undefined

20. $3 + 5 \cdot 12 = 7(3 + 6)$

$3 + 60$	$7(9)$
63	63

Since 63 = 63, the equation is true.

21. $38 - 2 \cdot 11 = (1+3)^2$

$38 - 22$	4^2
16	16

Since 16 = 16, the equation is true.

22. $5.9 + 3.6(8.7 - 3.1) + 7(6.3) = 100 - 3(8.52) - 4.3$

$5.9 + 3.6(5.6) + 44.1$	$100 - 25.56 - 4.3$
70.16	70.14

Since $70.16 \neq 70.14$, the equation is false.

23. $\dfrac{25 + 5(14 - 2\cdot 9) - 33(85 - 5\cdot 7)}{100 - 5\cdot 13} = -5\cdot 9 - 2(50 - 7^2)$

$\dfrac{25 + 5(-4) - 33(50)}{100 - 65}$	$-45 - 2(1)$
$\dfrac{-1645}{35}$	$-45 - 2$
-47	-47

Since $-47 = -47$, the equation is true.

24. $3x - 7 = x + 1$

$3(4) - 7$	$4 + 1$
$12 - 7$	5
5	

Since $5 = 5$, 4 is a solution.

25. $5x + 17 = 10x + 5$

$5(3) + 17$	$10(3) + 5$
$15 + 17$	$30 + 5$
32	35

Since $32 \neq 35$, 3 is not a solution.

26. $3x + 17 = 2(x - 5)$

$3(-27) + 17$	$2(-27 - 5)$
$-81 + 17$	$2(-32)$
-64	-64

Since $-64 = -64$, -27 is a solution.

27. $2.1x - 1.9 = 0.6x - 4.6$

$2.1(-1.8) - 1.9$	$0.6(-1.8) - 4.6$
$-3.78 - 1.9$	$-1.08 - 4.6$
-5.68	-5.68

Since $-5.68 = -5.68$, -1.8 is a solution.

28. $\dfrac{3}{4}\left(x - \dfrac{8}{9}\right) = \dfrac{1}{2}\left(x + \dfrac{2}{3}\right)$

$\dfrac{3}{4}\left(5\dfrac{1}{3} - \dfrac{8}{9}\right)$	$\dfrac{1}{2}\left(5\dfrac{1}{3} + \dfrac{2}{3}\right)$
$\dfrac{3}{4}\left(4\dfrac{4}{9}\right)$	$\dfrac{1}{2}(6)$
$3\dfrac{1}{3}$	3

Since $3\dfrac{1}{3} \neq 3$, $5\dfrac{1}{3}$ is not a solution.

29. $3x^2 - 6x - 10 = x^2 + x - 5$

$3(-5)^2 - 6(-5) - 10$	$(-5)^2 - 5 - 5$
$75 + 30 - 10$	$25 - 5 - 5$
95	15

Since $95 \neq 15$, -5 is not a solution.

30. $12h + 9h - 4h = (12 + 9 - 4)h = 17h$

31. $6m + 22 - m - 12 + 3m$
$= 6m - m + 3m + 22 - 12$
$= 8m + 10$

32. $3x-35+4y-5x-6y+7x+27+17y+22x$
$=3x-5x+7x+22x-35+27+4y-6y+17y$
$=27x-8+15y$

33. $3x^4+5x-7x^2+12x^4-17x-34x+x^3-1$
$=3x^4+12x^4+x^3-7x^2+5x-17x-34x-1$
$=15x^4+x^3-7x^2-46x-1$

34. $(6.2a+5.3b)+(4.7a-1.9b)$
$=6.2a+5.3b+4.7a-1.9b$
$=6.2a+4.7a+5.3b-1.9b$
$=10.9a+3.4b$

35. $-(27y-15)=-27y+15$

36. $5g+8-(g+4)=5g+8-g-4$
$=5g-g+8-4$
$=4g+4$

37. $(-2x+4y-7z)-(-x+6y+8z)$
$=-2x+4y-7z+x-6y-8z$
$=-2x+x+4y-6y-7z-8z$
$=-x-2y-15z$

38. $-14x(3y)=(-14)(3)xy=-42xy$

39. $\frac{15}{22}\left(\frac{11}{25}a+\frac{33}{50}b\right)=\frac{165}{550}a+\frac{495}{1100}b$
$=\frac{3}{10}a+\frac{9}{20}b$

40. $\frac{-115.388z}{-4.55}=25.36z$

41. $\frac{104x-156y+221z}{13}=\frac{104x}{13}-\frac{156y}{13}+\frac{221z}{13}$
$=8x-12y+17z$

42. $\frac{\frac{21}{25}uv}{-3u}=\frac{21uv}{25}\cdot\left(-\frac{1}{3u}\right)=-\frac{21uv}{75u}=-\frac{7}{25}v$

43. $\frac{-18x+24y-36z}{-6}=\frac{-18x}{-6}+\frac{24y}{-6}-\frac{36z}{-6}$
$=3x-4y+6z$

44. $4(3.9x-11.1y)+7(2.9x-0.7y)$
$=4(3.9x)+4(-11.1y)+7(2.9x)+7(-0.7y)$
$=15.6x-44.4y+20.3x-4.9y$
$=35.9x-49.3y$

45. $12[-3(2a-5b)+9]+8[-9(a+13)-6(b-12)]$
$=12(-6a+15b+9)+8(-9a-117-6b+72)$
$=-72a+180b+108-72a-936-48b+576$
$=-144a+132b-252$

46. $\frac{14.4(2x+5)-21.6(5x-2)+7.2(x-1)}{3x-4(x+8)+x+34.4}$
$=\frac{28.8x+72-108x+43.2+7.2x-7.2}{3x-4x-32+x+34.4}$
$=\frac{-72x+108}{2.4}$
$=-30x+45$

47. Let x = a number
$2(x+50)-\frac{1}{2}x$

48. Let x = a number
$11+\frac{1}{4}x$

49. Let x = a number
$x^2 + 2x = x + 306$

50. Let x = number of hours
$225 + 45x = 225 + 45(120) = 5625$
Lakeetha will make \$5625 in earnings.

51. Let n = number of weeks
$500 + 145n - 15n = 500 + 130n$
$= 500 + 130(15) = 2450$
Carmen has \$2450 deposited after 15 weeks.
$2450 - 625 = 1825$
She has a \$1825 balance after the withdrawal.

52. a. Beatrice: x appliances
Marie: $2x - 5$ appliances
Ann: $\frac{1}{2}x$ appliances
Magdalene: x appliances

b. Beatrice: $100 + 25x$ dollars
Marie: $100 + 25(2x - 5) = 100 + 50x - 125 = 50x - 25$ dollars
Ann: $100 + 25\left(\frac{1}{2}x\right) = 100 + \frac{25}{2}x$ dollars
Magdalene: $100 + 25x$ dollars

c. $100 + 25x + 100 + 25(2x - 5) + 100 + 25\left(\frac{1}{2}x\right) + 100 + 25x$

d. $100 + 25x + 50x - 25 + 100 + 12.5x + 100 + 25x = 275 + 112.5x$

e. $275 + 112.5(20) = 2525$
\$2525 total money is earned.

53. $A = \frac{1}{2}h(b + B)$
$= \frac{1}{2}(100)(120 + 200)$
$= 50(320)$
$= 16{,}000$
Fuad must buy 16,000 ft^2 of sod.

54. $V = \pi r^2 h = \pi(10)^2(4.5) \approx 1413.7$
Chum's pool will hold 1413.7 ft^3 of water.

55. $C = \frac{5}{9}(F - 32) = \frac{5}{9}(96 - 32) = 35\frac{5}{9}$
The temperature is $35\frac{5}{9}$°C.

56. $A = P(1 + r)^t$
$= 18{,}500(1 + 0.055)^{10}$
$\approx 31{,}600.67$
You will have \$31,600.67 over 10 years.

57. $d = rt = 13(1.25) = 16.25$
Randy bicycled 16.25 miles.

58. $c = \sqrt{a^2 + b^2} = \sqrt{(95)^2 + (168)^2} = 193$
The hypotenuse is 193 mm.

59. $V = \pi r^2 h = \pi(1.625)^2(1.5) \approx 12.44$
The can volume is 12.44 in^3.
$S = 2\pi r^2 + 2\pi rh$
$= 2\pi(1.625)^2 + 2\pi(1.625)(1.5)$
≈ 31.91
The can's surface area is about 31.91 in^2.

60. $s = -16t^2 + v_0 t + s_0$
$= -16(1)^2 + 80(1) + 6$
$= 70$ feet
The ball is 70 ft after 1 second.
$s = -16(2)^2 + 80(2) + 6 = 102$ feet
The ball is 102 ft after 2 seconds.
$s = -16(3)^2 + 80(3) + 6 = 102$ feet
The ball is 102 ft after 3 seconds.
$s = -16(4)^2 + 80(4) + 6 = 70$ feet
The ball is 70 ft after 4 seconds.
$s = -16(5)^2 + 80(5) + 6 = 6$ feet
The ball is 6 ft after 5 seconds.

61. Answer will vary.
Possible answer: The sum of $\frac{2}{3}$ times a number and 25

62. Answer will vary.
Possible answer:
Seven times the difference of a number and 45, added to 15

63. equation; equality

64. expression: no equality

65. expression: no equality

66. equation: equality

67. $90° - 85° = 5°$
The complementary angle is 5°.

68. $180° - 43° = 137°$
The supplementary angle is 137°.

69. $180° - (31° + 58°) = 91°$
The third angle is 91°.

Chapter 3 Test

1. Let x = a number
$\frac{x}{12}$

2. Let x = a number
$\frac{12}{x}$

3. $x + 0.08x = 1.08x = 1.08(15) = 16.2$
\$16.20

4. Answer will vary.
Possible answer: b subtracted from the square of a

5. $\sqrt{4x^2 - 20x + 25} = \sqrt{4(2)^2 - 20(2) + 25}$
$= \sqrt{1}$
$= 1$

6. $\frac{-18x + 54}{x - 8} = \frac{-18(-1) + 54}{-1 - 8}$
$= \frac{18 + 54}{-9}$
$= \frac{72}{-9}$
$= -8$

7. $x^2 = (-6)^2 = 36$

8. $-x^2 = -(-6)^2 = -36$

9. $(-x)^2 = [-(-6)]^2 = 6^2 = 36$

10. 8 terms

11. $y^3, -5y^2, 15y, 7y^2, 4y, 6y^3$

12. $-3, -12$

13. $1, -5, 15, -3, 7, -12, 4, 6$

14. y^3 and $6y^3$, $-5y^2$ and $7y^2$, $15y$ and $4y$, -3 and -12

15. $\frac{2}{3}x+\frac{5}{6}y-\frac{8}{9}+\frac{1}{6}x+\frac{7}{9}y+\frac{1}{3}$
$=\frac{2}{3}x+\frac{1}{6}x+\frac{5}{6}y+\frac{7}{9}y-\frac{8}{9}+\frac{1}{3}$
$=\frac{4}{6}x+\frac{1}{6}x+\frac{15}{18}y+\frac{14}{18}y-\frac{8}{9}+\frac{3}{9}$
$=\frac{5}{6}x+\frac{29}{18}y-\frac{5}{9}$

16. $-(5p+2q)-(-9p+q)+(p+q)-(-p-q)$
$=-5p-2q+9p-q+p+q+p+q$
$=-5p+9p+p+p-2q-q+q+q$
$=6p-q$

17. $\frac{25x-45}{-5}=\frac{25x}{-5}-\frac{45}{-5}=-5x+9$

18. $2a(3a+6b-8c)$
$=2a(3a)+2a(6b)+2a(-8c)$
$=6a^2+12ab-16ac$

19. $-7x-4=6x+9$

$-7(-2)-4$	$6(-2)+9$
$14-4$	$-12+9$
10	-3

Since $10 \neq -3$, -2 is not a solution.

20. $8x^2+40x+45=2x^2-2x-27$

$8(-4)^2+40(-4)+45$	$2(-4)^2-2(-4)-27$
$8(16)-160+45$	$2(16)+8-27$
$128-160+45$	$32+8-27$
13	13

Since $13 = 13$, -4 is a solution.

21. $V=LWH=4(2)(1.5)=12\text{ ft}^3$
The tool box's volume is 12 ft^3.
$S=2LW+2WH+2LH$
$=2(4)(2)+2(2)(1.5)+2(4)(1.5)$
$=34\text{ ft}^2$
The outside surface of the tool box is 34 ft^2.

22. $A=Pe^{nr}=2000e^{40(0.055)}\approx 18{,}050.027$
Tracy's account will have \$18,050.03 in 40 years.

23. $180°-78°=102°$
The supplementary angle is 102°.

24. $c=\sqrt{a^2+b^2}=\sqrt{(20)^2+(48)^2}=52$
The hypotenuse is 52 inches.

25. $F=\frac{9}{5}C+32=\frac{9}{5}(25)+32=77$
The temperature is 77°F.

26. Answers will vary.
Possible answer: The area is the space contained within the boundaries, whereas the perimeter is the distance around the boundaries.

Chapter 4

4.1 Experiencing Algebra the Exercise Way

1.

x	$y=5x+4$	y
–2	$y=5(-2)+4$ $y=-6$	–6
–1	$y=5(-1)+4$ $y=-1$	–1
0	$y=5(0)+4$ $y=4$	4
1	$y=5(1)+4$ $y=9$	9
2	$y=5(2)+4$ $y=14$	14
3	$y=5(3)+4$ $y=19$	19

3.

x	$y=\frac{3}{5}x-2$	y
–15	$y=\frac{3}{5}(-15)-2$ $y=-11$	–11
–10	$y=\frac{3}{5}(-10)-2$ $y=-8$	–8
–5	$y=\frac{3}{5}(-5)-2$ $y=-5$	–5
0	$y=\frac{3}{5}(0)-2$ $y=-2$	–2
5	$y=\frac{3}{5}(5)-2$ $y=1$	1
10	$y=\frac{3}{5}(10)-2$ $y=4$	4
15	$y=\frac{3}{5}(15)-2$ $y=7$	7

5.

x	$y = 2.3x + 1.6$	y
−2	$y = 2.3(-2) + 1.6$ $y = -3$	−3
−1	$y = 2.3(-1) + 1.6$ $y = -0.7$	−0.7
0	$y = 2.3(0) + 1.6$ $y = 1.6$	1.6
1	$y = 2.3(1) + 1.6$ $y = 3.9$	3.9
2	$y = 2.3(2) + 1.6$ $y = 6.2$	6.2

7.

x	$y = \frac{1}{3}(x+7)$	y
−1	$y = \frac{1}{3}(-1+7)$ $y = 2$	2
−4	$y = \frac{1}{3}(-4+7)$ $y = 1$	1
−7	$y = \frac{1}{3}(-7+7)$ $y = 0$	0
−10	$y = \frac{1}{3}(-10+7)$ $y = -1$	−1
−13	$y = \frac{1}{3}(-13+7)$ $y = -2$	−2

For exercises 9–15, answers will vary. One possibility is given.

9.

x	$y = 6x - 8$	y
−2	$y = 6(-2) - 8$ $y = -20$	−20
−1	$y = 6(-1) - 8$ $y = -14$	−14
0	$y = 6(0) - 8$ $y = -8$	−8
1	$y = 6(1) - 8$ $y = -2$	−2
2	$y = 6(2) - 8$ $y = 4$	4

11.

x	$y = \frac{2}{7}x - 2$	y
−14	$y = \frac{2}{7}(-14) - 2$ $y = -6$	−6
−7	$y = \frac{2}{7}(-7) - 2$ $y = -4$	−4
0	$y = \frac{2}{7}(0) - 2$ $y = -2$	−2
7	$y = \frac{2}{7}(7) - 2$ $y = 0$	0
14	$y = \frac{2}{7}(14) - 2$ $y = 2$	2

13.

x	$y=-4.6x+2.1$	y
–2	$y=-4.6(-2)+2.1$ $y=11.3$	11.3
–1	$y=-4.6(-1)+2.1$ $y=6.7$	6.7
0	$y=-4.6(0)+2.1$ $y=2.1$	2.1
1	$y=-4.6(1)+2.1$ $y=-2.5$	–2.5
2	$y=-4.6(2)+2.1$ $y=-7.1$	–7.1

15.

x	$y=\frac{1}{4}(3x-2)$	y
–6	$y=\frac{1}{4}[3(-6)-2]$ $y=-5$	–5
–2	$y=\frac{1}{4}[3(-2)-2]$ $y=-2$	–2
0	$y=\frac{1}{4}(3\cdot 0-2)$ $y=-\frac{1}{2}$	$-\frac{1}{2}$
2	$y=\frac{1}{4}(3\cdot 2-2)$ $y=1$	1
6	$y=\frac{1}{4}(3\cdot 6-2)$ $y=4$	4

17.

x	$y=12x-13$	y
–2	$y=12(-2)-13$ $y=-37$	–37
0	$y=12(0)-13$ $y=-13$	–13
2	$y=12(2)-13$ $y=11$	11

19.

y	$z=\frac{1}{3}y+5$	z
–3	$z=\frac{1}{3}(-3)+5$ $z=4$	4
0	$z=\frac{1}{3}(0)+5$ $z=5$	5
3	$z=\frac{1}{3}(3)+5$ $z=6$	6

21.

b	$a=14.2b+5.7$	a
–1	$a=14.2(-1)+5.7$ $a=-8.5$	–8.5
0	$a=14.2(0)+5.7$ $a=5.7$	5.7
1	$a=14.2(1)+5.7$ $a=19.9$	19.9

23.

x	$y=2x^2+3x+1$	y
–3	$y=2(-3)^2+3(-3)+1$ $y=10$	10
–2	$y=2(-2)^2+3(-2)+1$ $y=3$	3
–1	$y=2(-1)^2+3(-1)+1$ $y=0$	0
0	$y=2(0)^2+3(0)+1$ $y=1$	1
1	$y=2(1)^2+3(1)+1$ $y=6$	6
2	$y=2(2)^2+3(2)+1$ $y=15$	15
3	$y=2(3)^2+3(3)+1$ $y=28$	28

25.

x	$y=(2x-3)(3x+4)$	y
–1	$y=[2(-1)-3][3(-1)+4]$ $y=-5$	–5
0	$y=(2\cdot 0-3)(3\cdot 0+4)$ $y=-12$	–12
1	$y=(2\cdot 1-3)(3\cdot 1+4)$ $y=-7$	–7

27.

x	$y=\dfrac{3x+7}{x-1}$	y
–1	$y=\dfrac{3(-1)+7}{-1-1}$ $y=-2$	–2
1	$y=\dfrac{3(1)+7}{1-1}$ $y=\dfrac{10}{0}$	undefined

29.

H	$V=4(2)H$ $V=8H$	V
1	$V=8(1)$ $V=8$	8
3	$V=8(3)$ $V=24$	24
5	$V=8(5)$ $V=40$	40
7	$V=8(7)$ $V=56$	56
9	$V=8(9)$ $V=72$	72

31.

t	$I = 5000(0.045)t$ $I = 225t$	I
1	$I = 225(1)$ $I = 225$	225
2	$I = 225(2)$ $I = 450$	450
3	$I = 225(3)$ $I = 675$	675
4	$I = 225(4)$ $I = 900$	900
5	$I = 225(5)$ $I = 1125$	1125
6	$I = 225(6)$ $I = 1350$	1350
7	$I = 225(7)$ $I = 1575$	1575
8	$I = 225(8)$ $I = 1800$	1800
9	$I = 225(9)$ $I = 2025$	2025
10	$I = 225(10)$ $I = 2250$	2250
11	$I = 225(11)$ $I = 2475$	2475
12	$I = 225(12)$ $I = 2700$	2700

33.

R	$I = \frac{9}{R}$	I
1	$I = \frac{9}{1}$ $I = 9$	9
2	$I = \frac{9}{2}$ $I = 4.5$	4.5
3	$I = \frac{9}{3}$ $I = 3$	3
4	$I = \frac{9}{4}$ $I = 2.25$	2.25
5	$I = \frac{9}{5}$ $I = 1.8$	1.8
6	$I = \frac{9}{6}$ $I = 1.5$	1.5
7	$I = \frac{9}{7}$ $I \approx 1.2857$	1.2857
8	$I = \frac{9}{8}$ $I = 1.125$	1.125
9	$I = \frac{9}{9}$ $I = 1$	1

35. $y = 1.2x + 4$
$y = 1.2(-2) + 4$
$y = 1.6$
$y = 1.2(-1) + 4$
$y = 2.8$
$y = 1.2(0) + 4$
$y = 4$
$y = 1.2(1) + 4$
$y = 5.2$
$y = 1.2(2) + 4$
$y = 6.4$
(–2, 1.6), (–1, 2.8), (0, 4), (1, 5.2), (2, 6.4)

37. $q = 1 - p$
$q = 1 - \frac{1}{6}$
$q = \frac{5}{6}$
$q = 1 - \frac{1}{5}$
$q = \frac{4}{5}$
$q = 1 - \frac{1}{4}$
$q = \frac{3}{4}$
$q = 1 - \frac{1}{3}$
$q = \frac{2}{3}$
$q = 1 - \frac{1}{2}$
$q = \frac{1}{2}$
$\left(\frac{1}{6}, \frac{5}{6}\right), \left(\frac{1}{5}, \frac{4}{5}\right), \left(\frac{1}{4}, \frac{3}{4}\right), \left(\frac{1}{3}, \frac{2}{3}\right), \left(\frac{1}{2}, \frac{1}{2}\right)$

39. $P = 4s$
$P = 4(2)$
$P = 8$
$P = 4(4)$
$P = 16$
$P = 4(6)$
$P = 24$
$P = 4(8)$
$P = 32$
$P = 4(10)$
$P = 40$
(2, 8), (4, 16), (6, 24), (8, 32), (10, 40)

41. $R = \{(3, 15.8), (5, 17.8), (7, 19.8), (9, 21.8)\}$
domain {3, 5, 7, 9}
range {15.8, 17.8, 19.8, 21.8}

43. $S = \{(4, -3), (4, -1), (4, 1), (4, 3), \ldots\}$
domain {4}
range {–3, –1, 1, 3, ...}

45. $T = \{\ldots, (2, -2), (2, -1), (2, 0), (2, 1), (2, 2), \ldots\}$
domain {2}
range {..., –2, –1, 0, 1, 2, ...}

47. $y = 4x - 5$ for $x = \{2, 4, 6\}$
$y = 4(2) - 5$
$y = 3$
$y = 4(4) - 5$
$y = 11$
$y = 4(6) - 5$
$y = 19$
domain {2, 4, 6}
range {3, 11, 19}

49. $y = 6 - x$
domain all real numbers
range all real numbers

51. $y = x^2 + 1$
The x-value can be any real number, whereas x^2 will always be zero or positive.
domain all real numbers
range all real numbers ≥ 1

53. $y = \sqrt{x-2}$

$x - 2$ must be positive or zero, so $x \geq 2$, which makes y zero or a positive number.
domain all real numbers ≥ 2
range all real numbers ≥ 0

55. $y = \dfrac{6}{x-2}$

$x - 2$ must not equal zero, so $x \neq 2$, which means $y \neq 0$.
domain all real numbers $x \neq 2$
range all real numbers $y \neq 0$

57. $V = s^3$

s must be positive, which makes V positive.
domain all real numbers $s > 0$
range all real numbers $V > 0$

59. $s = -16t^2 + v_0 t + s_0$

$s = -16t^2 + 0t + 50$

$s = -16t^2 + 50$

His position will be s feet high t seconds after the jump.

61. $d = rt$

$d = 65t$

t	$d = 65t$	d
1	$d = 65(1)$ $d = 65$	65
2	$d = 65(2)$ $d = 130$	130
3	$d = 65(3)$ $d = 195$	195
4	$d = 65(4)$ $d = 260$	260

(1, 65), (2, 130), (3, 195), (4, 260)
The distance traveled (in miles) after 1, 2, 3, and 4 hours is 65, 130, 195, and 260 respectively.

4.1 Experiencing Algebra the Calculator Way

1. $y = 2.75x - 15.8$

$Y1 = 2.75x - 15.8$

X	Y1	
-250	-703.3	
-200	-565.8	
-150	-428.3	
-100	-290.8	
-50	-153.3	
0	-15.8	
50	121.7	

Y1=2.75X-15.8

X	Y1	
100	259.2	
150	396.7	
200	534.2	
250	671.7	

Y1=2.75X-15.8

2. $y = 3x^3 + 5x^2 + 7x + 9$

$Y1 = 3x^3 + 5x^2 + 7x + 9$

X	Y1	
-5	-276	
-3.873	-117.4	
-3.142	-56.66	
-3	-48	
-2	-9	
-1.442	.30467	
0	9	

Y1=3X^3+5X²+7X+9

X	Y1	
2	67	
3.1416	173.36	
3.1623	176	

Y1=3X^3+5X²+7X+9

3. $y = (2x+1)^4$

$Y1 = (2x+1)^4$

X	Y1	
25	6.77E6	
30	1.38E7	
35	2.54E7	
40	4.3E7	
45	6.86E7	
50	1.04E8	

Y1=(2X+1)^4

4. $y = (8)2\pi x + 2\pi x^2$

$Y1 = (8)2\pi x + 2\pi x^2$

X	Y1	
2	125.66	
2.5	164.93	
3	207.35	
3.5	252.9	
4	301.59	
4.5	353.43	
5	408.41	

Y1=8*2πX+2πX²

5. $y = 12,000(1+0.075)^x$

$y = 12,000(1.075)^x$

$\text{Y1} = 12000(1.075)^x$

X	Y1
1	12900
2	13868
3	14908
4	16026
5	17228
6	18520
7	19909

Y1=12000(1.075)...

X	Y1
8	21402
9	23007
10	24732
11	26587
12	28581

Y1=12000(1.075)...

6. $y = 12,000e^{(0.075x)}$

$\text{Y1} = 12000e^{(0.075x)}$

X	Y1
1	12935
2	13942
3	15028
4	16198
5	17460
6	18820
7	20286

Y1=12000e^(0.07...

X	Y1
8	21865
9	23568
10	25404
11	27383
12	29515

Y1=12000e^(0.07...

4.2 Experiencing Algebra the Exercise Way

1. $A(-7,-5)$, $B(-5,-7)$

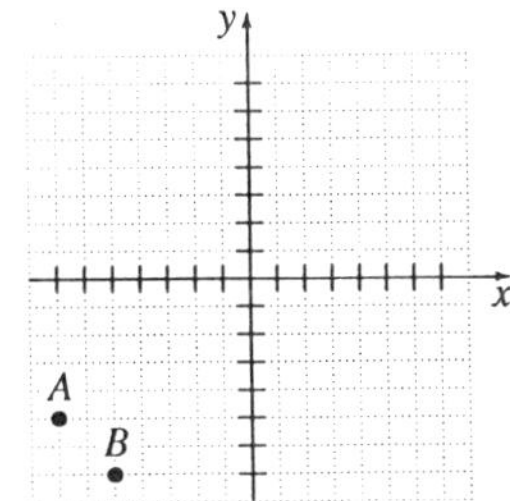

3. $E(4, 9)$, $F(9, 4)$

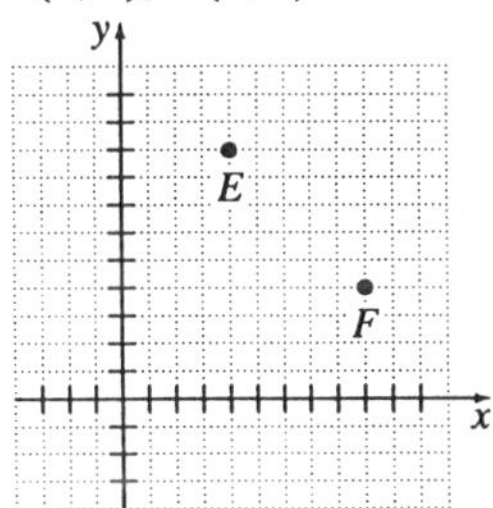

5. $I(-5, 5)$, $J(5, -5)$

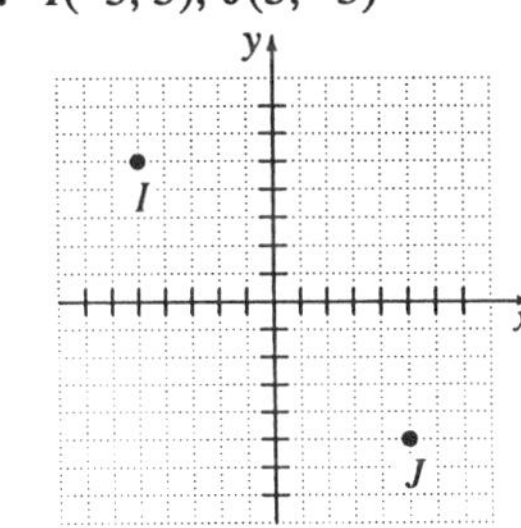

7. $M(2, -1)$, $N(-1, 2)$

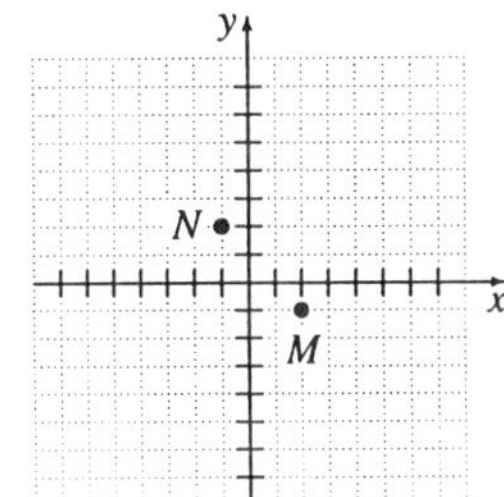

9. $Q(3, -1)$, $R(-3, 1)$

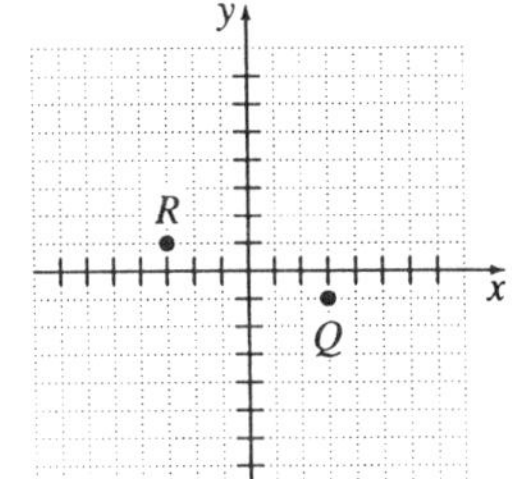

11. $U(-8, -2)$, $V(8, 2)$

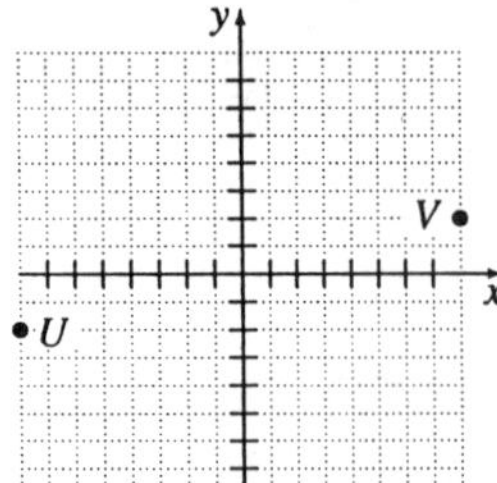

13. $A(1.2, 2.4)$

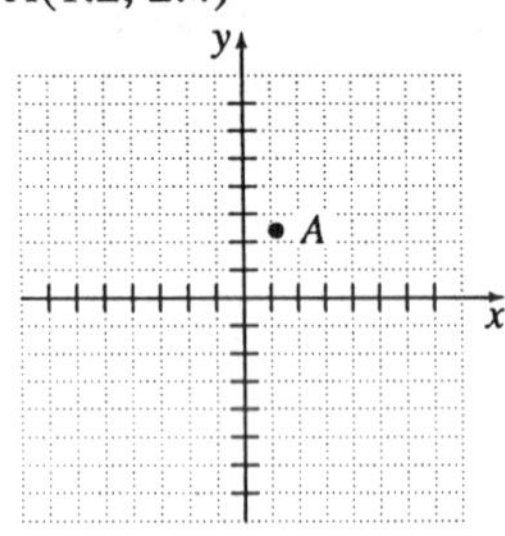

15. $C(-4.5, -2.6)$

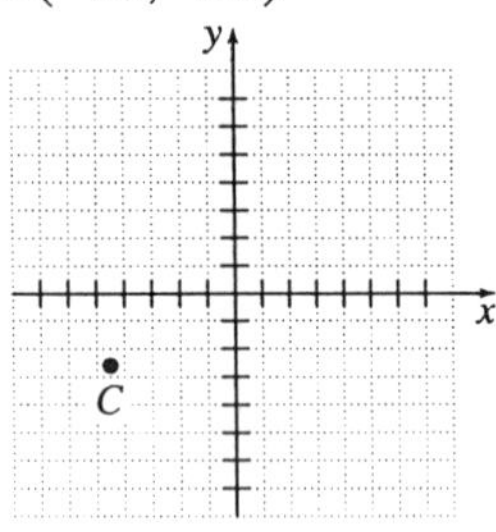

17. $E(-2.4, 2.1)$

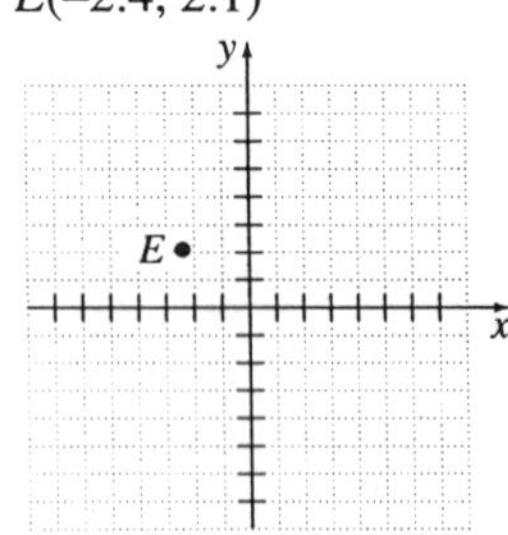

19. $G(1.8, -2.7)$

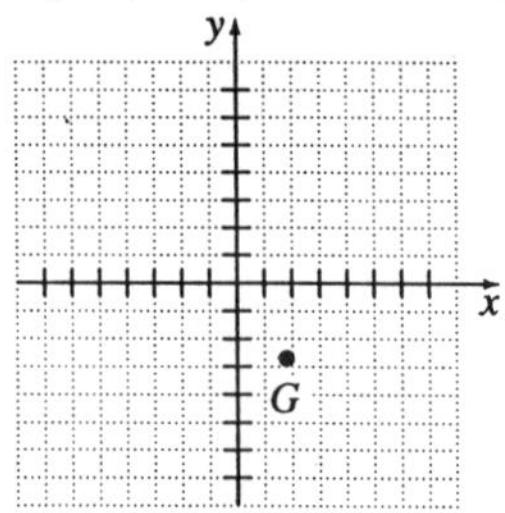

21. $(-21, 35)$
Coordinates are $(-, +)$
quadrant II

23. $(4, 96)$
Coordinates are $(+, +)$
quadrant I

25. $(-3, -19)$
Coordinates are $(-, -)$
quadrant III

27. $(0, -31)$
x-coordinate is 0
y-axis

29. $(90, -100)$
Coordinates are $(+, -)$
quadrant IV

31. $(24, 0)$
y-coordinate is 0
x-axis

33. $(-19, 0)$
y-coordinate is zero
x-axis

35. $(0.05, 1.003)$
Coordinates are $(+, +)$
quadrant I

37. $A(8, 2)$, $B(-9, 7)$, $C(0, 0)$, $D(-2, -3)$, $E(0, 4)$, $F(5, -6)$, $G(-5, 0)$

39. $(0, 3.7)$
x-coordinate is 0
y-axis

41. $\left(\frac{13}{27}, -\frac{11}{19}\right)$
Coordinates are (+, –)
quadrant IV

43. $\left(-\frac{53}{100}, -\frac{39}{100}\right)$
Coordinates are (–, –)
quadrant III

45. $\left(0, \frac{28}{51}\right)$
x-coordinate is zero
y-axis

47. $A = \{(-3, -3), (-2, -2), (-1, -1), (0, 0), (1, 1), (2, 2), (3, 3)\}$

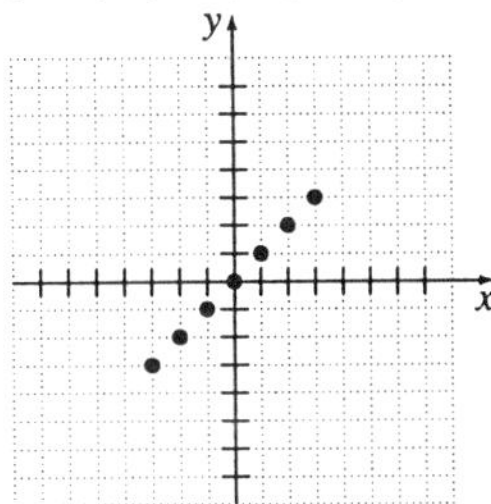

49. $C = \{(-4, 4), (-2, 2), (0, 0), (2, -2), (4, -4)\}$

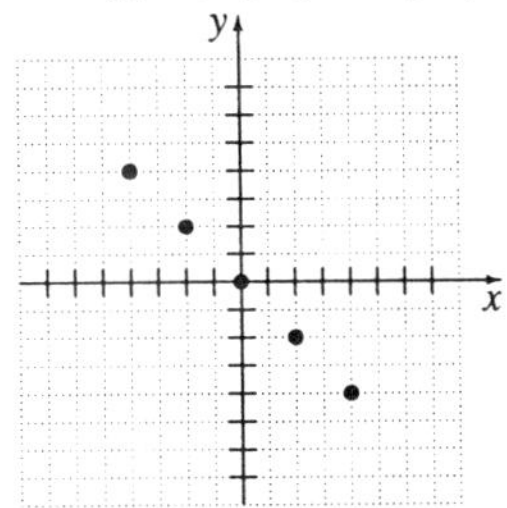

51. $E = \{(5, -3), (5, -1), (5, 1), (5, 3)\}$

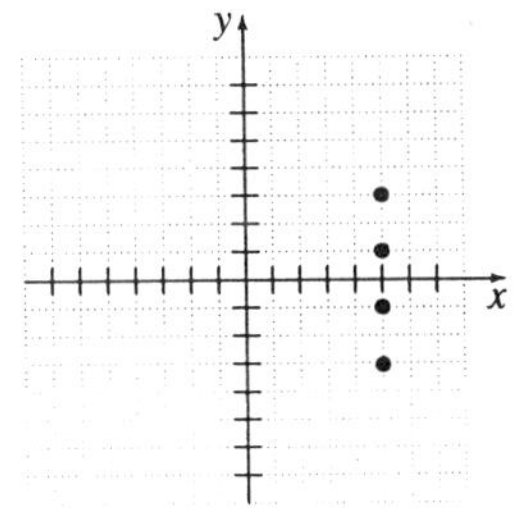

53. $G = \{\ldots, (-2, -1), (-1, 0), (0, 1), (1, 2), (2, 3), \ldots\}$

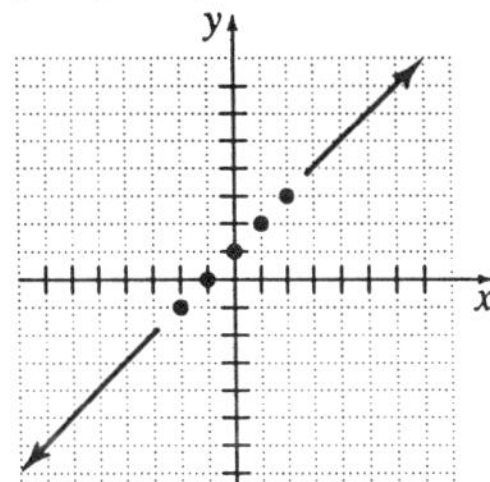

55. $I = \{\ldots, (-2, -6), (-1, -5), (0, -4), (1, -3), (2, -2), (3, -1), \ldots\}$

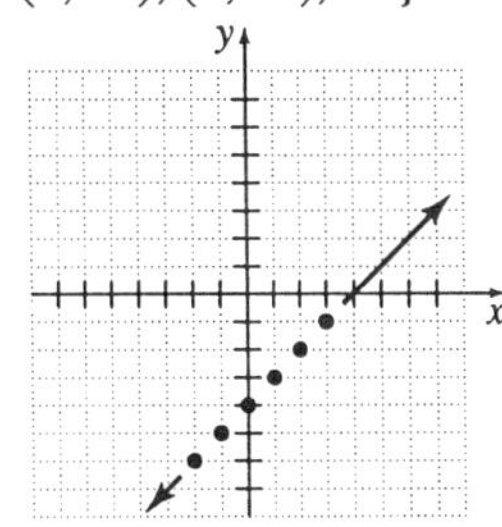

57. $y = 12x - 15$
Possible points:

x	y
$-\frac{1}{2}$	–21
0	–15
$\frac{1}{2}$	–9

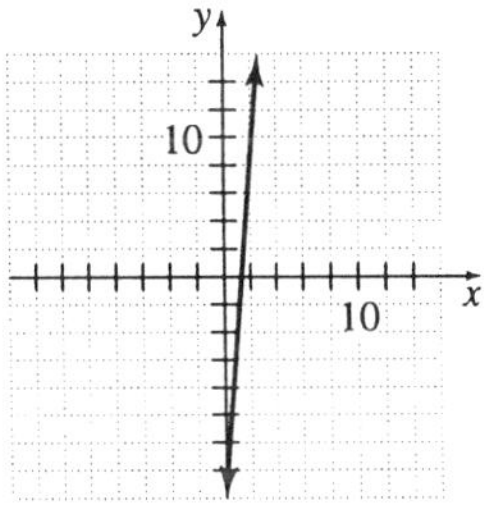

59. $y = -10x + 9$
Possible points:

x	y
–1	19
0	9
1	–1

61. $y = \frac{1}{2}x + 3$
Possible points:

x	y
–2	2
0	3
2	4

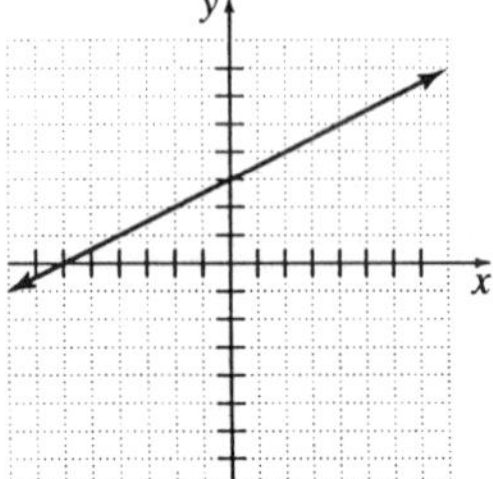

63. $y = -\frac{2}{3}x - 2$
Possible points:

x	y
–3	0
0	–2
3	–4

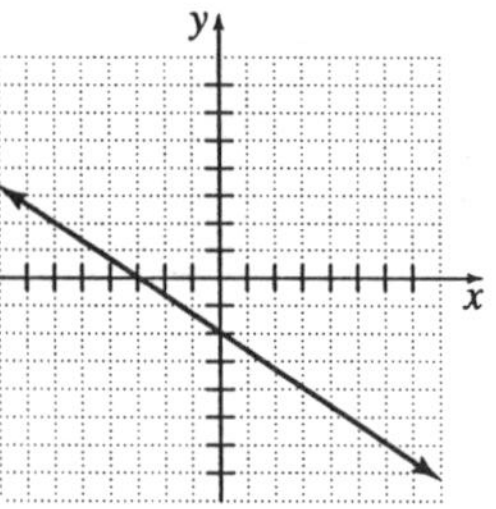

65. $y = 2x^2 - 5$
Possible points:

x	y
–2	3
0	–5
2	3

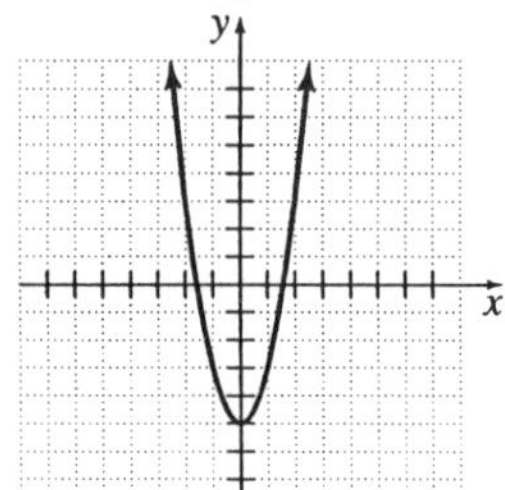

67. $y = |2x|$
Possible points:

x	y
–2	4
0	0
2	4

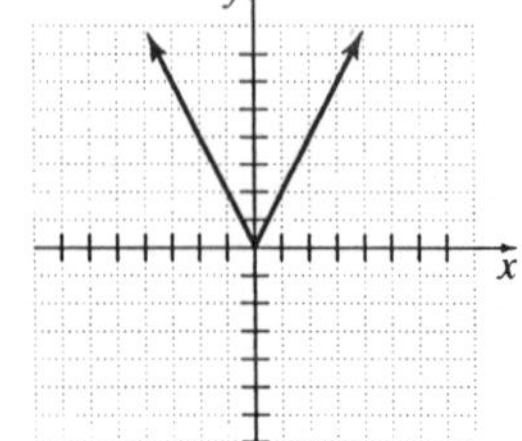

69. domain all real numbers
range all real numbers ≤ -1

71. domain all real numbers
range all real numbers ≥ 3

73. domain 1960 through 1992
range approximately 45% to 62%
The domain represents the time period between the years 1960 and 1992 during which the data was collected.
The range represents the percent of the adult population who identify themselves as Democrats for each year.

75. domain all real numbers $0 \leq x \leq 2.5$
range all real numbers $0 \leq y \leq 100$
The domain represents the time that has passed from 0 seconds to 2.5 seconds. The range represents the height from 100 feet to 0 feet.

4.2 Experiencing Algebra the Calculator Way

Part A

1. Integer setting
2. Decimal setting
3. Standard setting

Part B

1. $Y1 = 0.6x - 1.2$

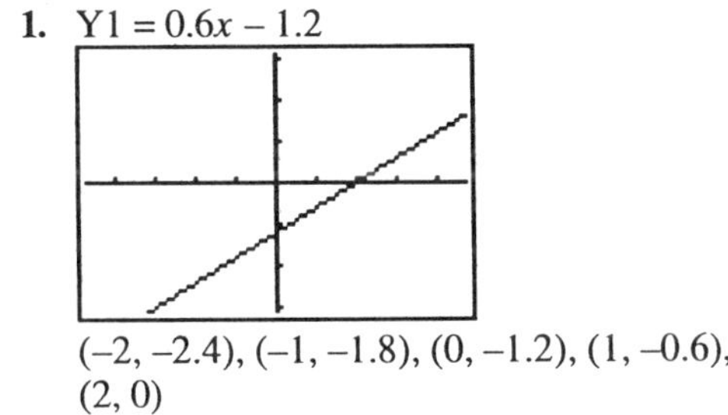

(–2, –2.4), (–1, –1.8), (0, –1.2), (1, –0.6), (2, 0)

2. $Y1 = -0.5x + 2.2$

(–2, 3.2), (–1, 2.7), (0, 2.2), (1, 1.7), (2, 1.2)

3. $Y1 = |x| - 2$

(–2, 0), (–1, –1), (0, –2), (1, –1), (2, 0)

4. $Y1 = |x - 2|$

(–2, 4), (–1, 3), (0, 2), (1, 1), (2, 0)

5. $Y1 = x - 2$

(–2, –4), (–1, –3), (0, –2), (1, –1), (2, 0)

6. Exercise 3:
[MATH] [▶] [1] [X,T,θ,n] [)] [–] [2]
Exercise 4:
[MATH] [▶] [1] [X,T,θ,n] [–] [2] [)]
Exercise 5:
[X,T,θ,n] [–] [2]

4.3 Experiencing Algebra the Exercise Way

1. *A* is not a function because two elements in the domain, –2 and 2, each correspond to two elements in the range, 1 and 3, and –3 and –1, respectively.

3. *C* is a function because every element in the domain corresponds to only one element in the range.

5. *E* is not a function because an element in the domain, 6, corresponds to 4 elements in the range, –1, –3, –5, and –7.

7. *G* is a function because every element in the domain corresponds to only one element in the range.

9. $y = 22x - 11$ is a function. Each element in the domain corresponds to a single element in the range.

11. $y = 13x^2 + 4$ is a function. Each element in the domain corresponds to a single element in the range.

13. $y^2 = 16x + 25$ is not a function. Each element in the domain, except –1.5625, corresponds to two elements in the range.

15. $y = 25x + 125$ is a function. Each element in the domain corresponds to a single element in the range.

17. $y = 6x^3 + 4x^2 + 2x$ is a function. Each element in the domain corresponds to a single element in the range.

19. $y^2 = 25 - x^2$ is not a function. Each element in the domain, except –5 and 5, corresponds to two elements in the range.

21. $y = |-3x - 18|$ is a function. Each element in the domain corresponds to a single element in the range.

23. $y = \sqrt{2x^2 + 4x + 1}$ is a function. Each element in the domain corresponds to a single element in the range.

25. This graph represents a function. All possible vertical lines cross the graph a maximum of one time.

27. This graph does not represent a function. Several vertical lines can be drawn to pass through more than one point.

29. This graph represents a function. All possible vertical lines cross the graph a maximum of one time.

31.
$$f(x) = 20x + 12$$
$$f(5) = 20(5) + 12$$
$$f(5) = 100 + 12$$
$$f(5) = 112$$

33.
$$f(x) = 20x + 12$$
$$f(-7) = 20(-7) + 12$$
$$f(-7) = -140 + 12$$
$$f(-7) = -128$$

35.
$$f(x) = 20x + 12$$
$$f(2.4) = 20(2.4) + 12$$
$$f(2.4) = 48 + 12$$
$$f(2.4) = 60$$

37.
$$f(x) = 20x + 12$$
$$f\left(-\frac{1}{4}\right) = 20\left(-\frac{1}{4}\right) + 12$$
$$f\left(-\frac{1}{4}\right) = -5 + 12$$
$$f\left(-\frac{1}{4}\right) = 7$$

39.
$$f(x) = 20x + 12$$
$$f(a) = 20a + 12$$

41.
$$f(x) = 20x + 12$$
$$f(a + 2) = 20(a + 2) + 12$$
$$f(a + 2) = 20a + 40 + 12$$
$$f(a + 2) = 20a + 52$$

43.
$$f(x) = 20x + 12$$
$$f(a - 4) = 20(a - 4) + 12$$
$$f(a - 4) = 20a - 80 + 12$$
$$f(a - 4) = 20a - 68$$

45.
$$f(x) = 20x + 12$$
$$f(a + b) = 20(a + b) + 12$$
$$f(a + b) = 20a + 20b + 12$$

47. $h(x) = 2x^2 - 4x + 5$
$h(7) = 2(7)^2 - 4(7) + 5$
$h(7) = 98 - 28 + 5$
$h(7) = 75$

49. $h(x) = 2x^2 - 4x + 5$
$h(-4) = 2(-4)^2 - 4(-4) + 5$
$h(-4) = 32 + 16 + 5$
$h(-4) = 53$

51. $h(x) = 2x^2 - 4x + 5$
$h(-1.1) = 2(-1.1)^2 - 4(-1.1) + 5$
$h(-1.1) = 2.42 + 4.4 + 5$
$h(-1.1) = 11.82$

53. $h(x) = 2x^2 - 4x + 5$
$h\left(-\frac{2}{5}\right) = 2\left(-\frac{2}{5}\right)^2 - 4\left(-\frac{2}{5}\right) + 5$
$h\left(-\frac{2}{5}\right) = \frac{8}{25} + \frac{40}{25} + \frac{125}{25}$
$h\left(-\frac{2}{5}\right) = \frac{173}{25}$

55. $h(x) = 2x^2 - 4x + 5$
$h\left(2\frac{3}{5}\right) = 2\left(2\frac{3}{5}\right)^2 - 4\left(2\frac{3}{5}\right) + 5$
$h\left(2\frac{3}{5}\right) = 2\left(\frac{13}{5}\right)^2 - 4\left(\frac{13}{5}\right) + 5$
$h\left(2\frac{3}{5}\right) = \frac{338}{25} - \frac{260}{25} + \frac{5}{25}$
$h\left(2\frac{3}{5}\right) = \frac{83}{25}$

57. $h(x) = 2x^2 - 4x + 5$
$h(b) = 2b^2 - 4b + 5$

59. $g(x) = |-3x + 9|$
$g(5) = |-3(5) + 9|$
$g(5) = |-15 + 9|$
$g(5) = |-6|$
$g(5) = 6$

61. $g(x) = |-3x + 9|$
$g(-5) = |-3(-5) + 9|$
$g(-5) = |15 + 9|$
$g(-5) = |24|$
$g(-5) = 24$

63. $g(x) = |-3x + 9|$
$g(4.5) = |-3(4.5) + 9|$
$g(4.5) = |-13.5 + 9|$
$g(4.5) = |-4.5|$
$g(4.5) = 4.5$

65. $g(x) = |-3x + 9|$
$g(-4.5) = |-3(-4.5) + 9|$
$g(-4.5) = |13.5 + 9|$
$g(-4.5) = |22.5|$
$g(-4.5) = 22.5$

67. $g(x) = |-3x + 9|$
$g\left(\frac{2}{3}\right) = \left|-3\left(\frac{2}{3}\right) + 9\right|$
$g\left(\frac{2}{3}\right) = |-2 + 9|$
$g\left(\frac{2}{3}\right) = |7|$
$g\left(\frac{2}{3}\right) = 7$

69. $g(x) = |-3x + 9|$
$g\left(-4\frac{2}{3}\right) = \left|-3\left(-4\frac{2}{3}\right) + 9\right|$
$g\left(-4\frac{2}{3}\right) = \left|-3\left(-\frac{14}{3}\right) + 9\right|$
$g\left(-4\frac{2}{3}\right) = |14 + 9|$
$g\left(-4\frac{2}{3}\right) = |23|$
$g\left(-4\frac{2}{3}\right) = 23$

71. $F(x)=\sqrt{x+15}$
$F(85)=\sqrt{85+15}$
$F(85)=\sqrt{100}$
$F(85)=10$

73. $F(x)=\sqrt{x+15}$
$F(-6)=\sqrt{-6+15}$
$F(-6)=\sqrt{9}$
$F(-6)=3$

75. $F(x)=\sqrt{x+15}$
$F(-25)=\sqrt{-25+15}$
$F(-25)=\sqrt{-10}$, which is not a real number.

77. $F(x)=\sqrt{x+15}$
$F(5.25)=\sqrt{5.25+15}$
$F(5.25)=\sqrt{20.25}$
$F(5.25)=4.5$

79. $F(x)=\sqrt{x+15}$
$F\left(-2\frac{3}{4}\right)=\sqrt{-2\frac{3}{4}+15}$
$F\left(-2\frac{3}{4}\right)=\sqrt{-\frac{11}{4}+\frac{60}{4}}$
$F\left(-2\frac{3}{4}\right)=\sqrt{\frac{49}{4}}$
$F\left(-2\frac{3}{4}\right)=\frac{7}{2}$

81. Let x = the number of televisions.
$f(x) = 1500 + 35x$
$f(400) = 1500 + 35(400)$
$f(400) = 1500 + 14{,}000$
$f(400) = 15{,}500$
The production run costs $15,500.

83. Let x = the number of players.
$f(x) = 125(x) - 470$
$f(400) = 125(400) - 470$
$f(400) = 50{,}000 - 470$
$f(400) = 49{,}530$
The net revenue is $49,530.

85. Let x = the number of seats sold.
The cost of each seat = $\$175 - 3x$.
$f(x) = 175x - 3x^2$
$f(22) = 175(22) - 3(22)^2$
$f(22) = 3850 - 1452$
$f(22) = 2398$
If 22 customers make reservations for the tour, the company will make $2398.

87. Let x = the number of CD's, that is, 5 or more.
$f(x) = 25 + 4(x - 5)$
$f(x) = 25 + 4x - 20$
$f(x) = 5 + 4x$
$f(12) = 5 + 4(12)$
$f(12) = 5 + 48$
$f(12) = 53$
The cost of 12 used CD's will be $53.

89. Let x = the number of days.
$f(x) = 39 + 25x$
$f(3) = 39 + 25(3)$
$f(3) = 39 + 75$
$f(3) = 114$
The charge for renting a truck for 3 days will be $114.

91. Let x = the number of half-hour increments.
$f(x) = 2.5 + x$
$f(7) = 2.5 + 7$
$f(7) = 9.5$
The charge is $9.50.

4.3 Experiencing Algebra the Calculator Way

1.–6.

X	Y1
65	278916
⁻83	⁻5.6E5
3.1416	45.017
1.4142	7.2426
5634	1.8E11
⁻3.142	⁻23.28

Y1=X^3+X²+X+1

7.–12.

X	Y1
⁻8	4
0	4
⁻4	0
.8	4.8
⁻6.3	2.3
.75	4.75

Y1=√(X²+8X+16)

13.–18.

X	Y_1
10	1
-5	-.5
20	.33333
5.5	10
5	ERROR
.2	-1.042

$Y_1 = 5/(X-5)$

4.4 Experiencing Algebra the Exercise Way

1. **a.** (–2, 0), (6, 0)

b. (0, 3)

c. $x < 2$

d. $x > 2$

e. 4

f. none

3. **a.** (–5, 0), (–1, 0), (4, 0)

b. (0, –1)

c. $x < -3$, $x > 2$

d. $-3 < x < 2$

e. 2

f. –3

5. $y = 3x - 6$

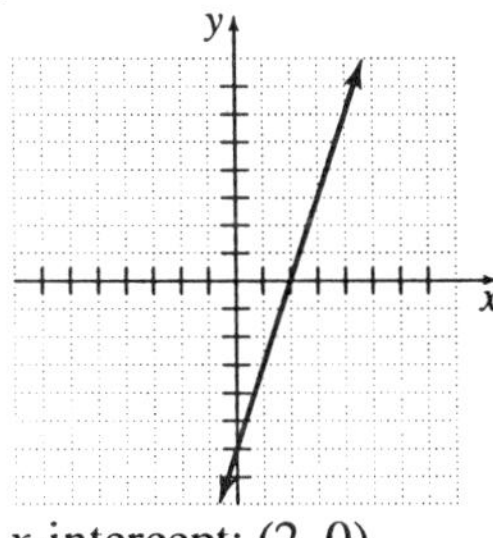

x-intercept: (2, 0)
y-intercept: (0, –6)

7. $y = \frac{1}{2}x + 1$

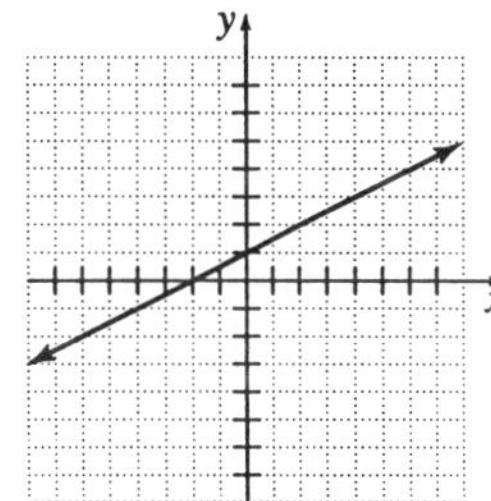

x-intercept: (–2, 0)
y-intercept: (0, 1)

9. $y = 1.2x - 6$

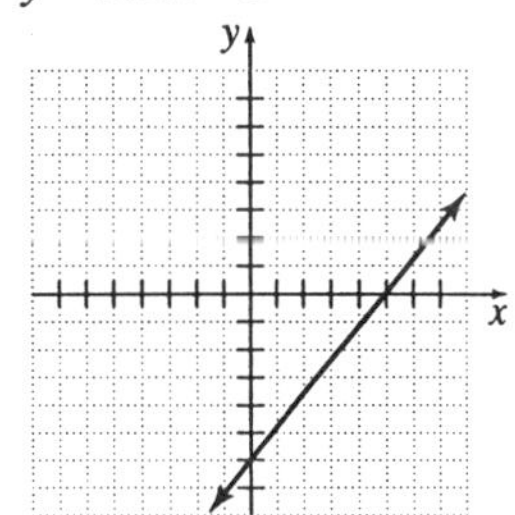

x-intercept: (5, 0)
y-intercept: (0, –6)

11. $f(x) = -12x + 24$

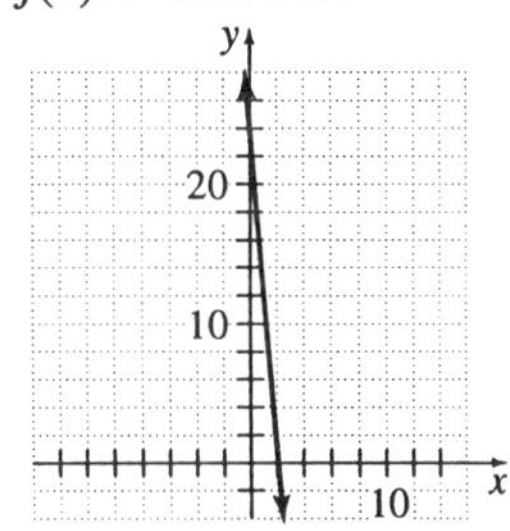

x-intercept: (2, 0)
y-intercept: (0, 24)

13. $f(x) = 9x + 15$

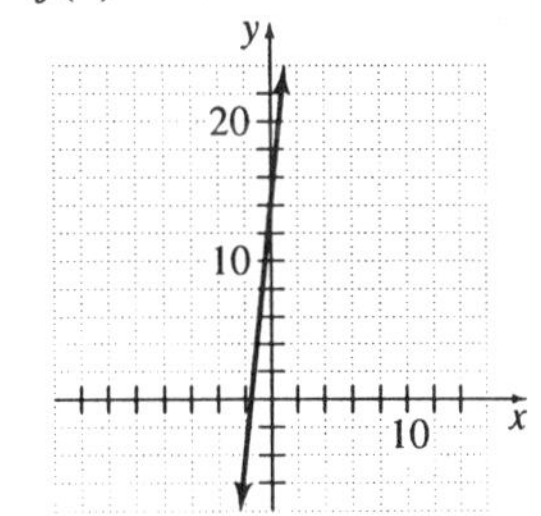

x-intercept: $\left(-\frac{5}{3}, 0\right)$
y-intercept: (0, 15)

15. $y = x^2 - 9$

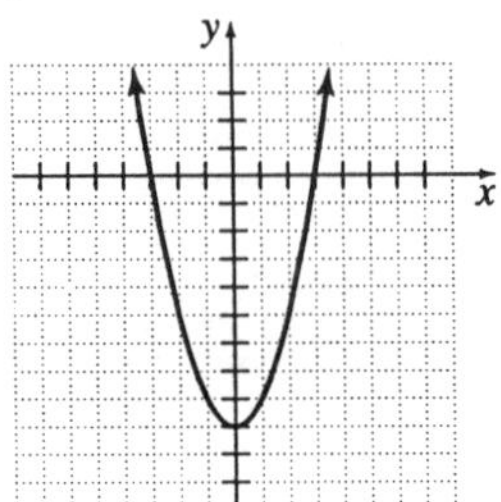

x-intercepts: (3, 0), (–3, 0)
y-intercept: (0, –9)

17. $y = x^2 + 6x + 9$

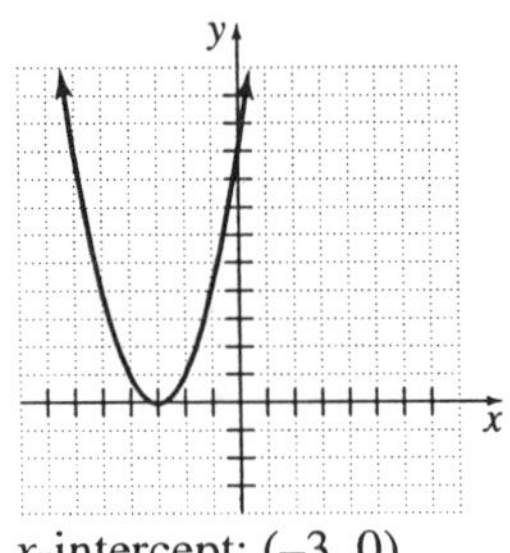

x-intercept: (–3, 0)
y-intercept: (0, 9)

19. $y = 4x^2 + 4x + 1$

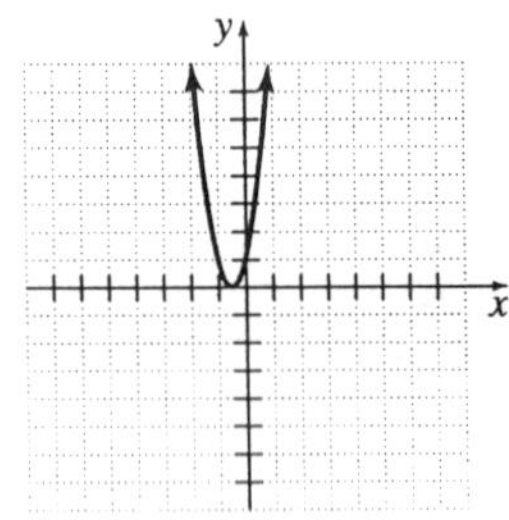

x-intercept: $\left(-\frac{1}{2}, 0\right)$
y-intercept: (0, 1)

21. $g(x) = x^2 + 10x - 3$

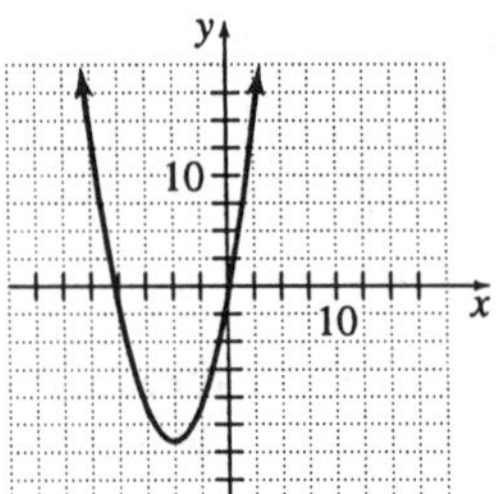

x-intercepts: (–10.29, 0), (0.29, 0)
y-intercept: (0, –3)

23. $H(x) = x^2 - 5x - 24$

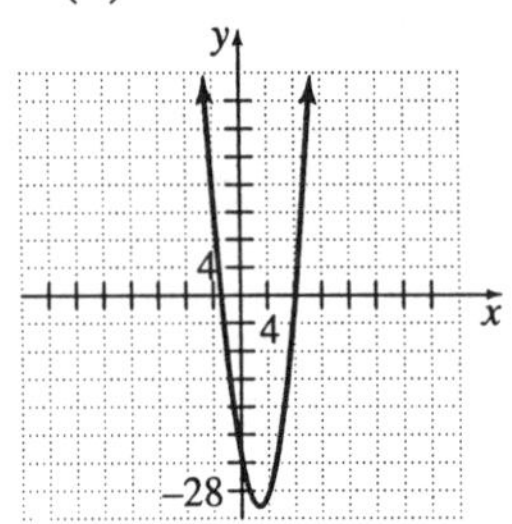

x-intercepts: (–3, 0), (8, 0)
y-intercept: (0, –24)

25. $y = x^3 + x^2 - 2x$

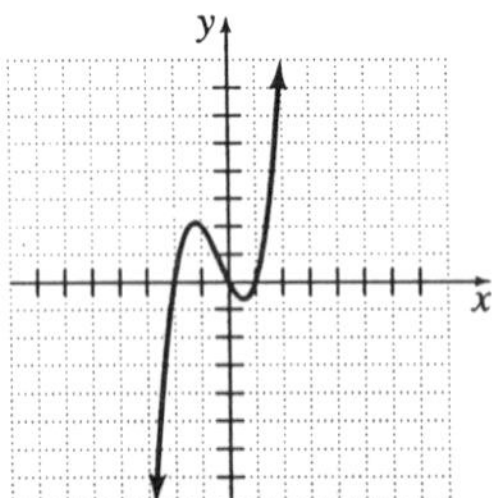

x-intercepts: (–2, 0), (0, 0), (1, 0)
y-intercept: (0, 0)

27. $f(x) = x^3 + 2x^2 - x - 2$

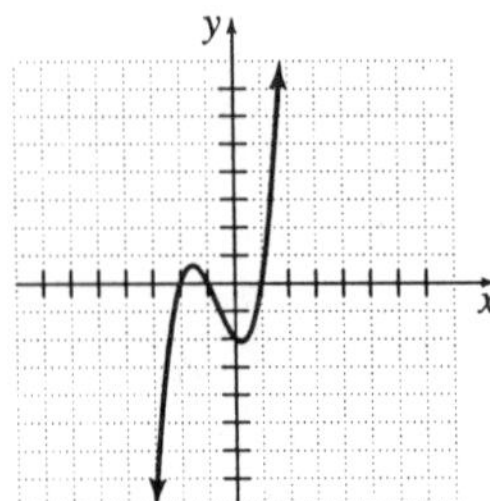

x-intercepts: (–2, 0), (–1, 0), (1, 0)
y-intercept: (0, –2)

29. $h(x) = |x| - 6$

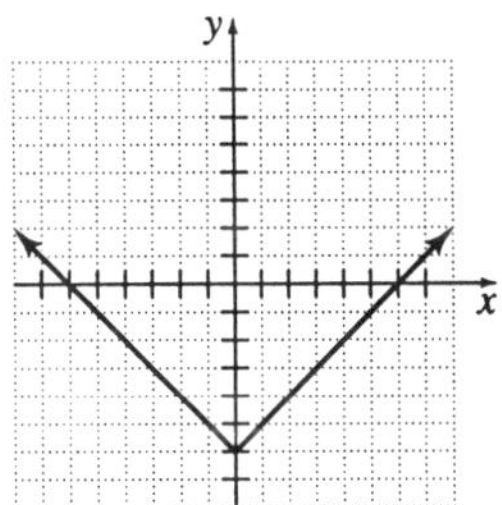

x-intercepts: (–6, 0), (6, 0)
y-intercept: (0, –6)

31. $y = |2x - 3| - 1$

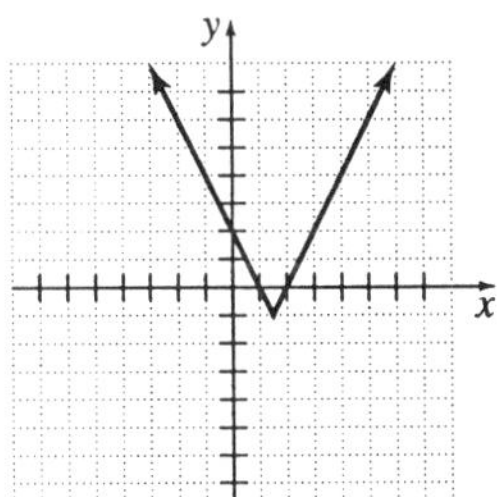

x-intercepts: (1, 0), (2, 0)
y-intercept: (0, 2)

33. $y = |x^2 - 2| - 1$

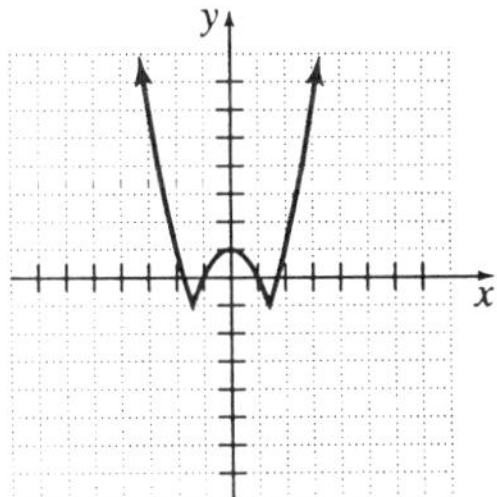

x-intercepts: (–1.73, 0), (–1, 0), (1, 0), (1.73, 0)
y-intercept: (0, 1)

35. $y = 2x + 8$

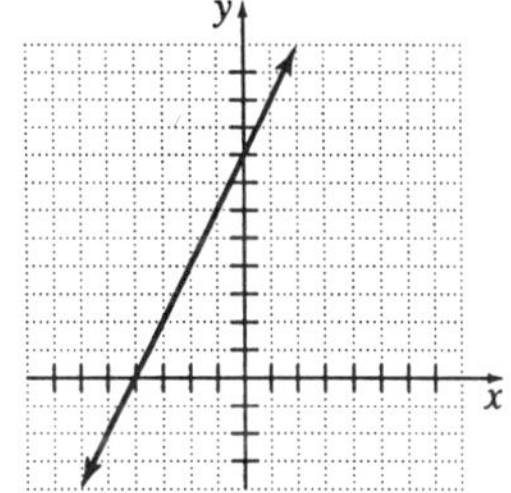

increasing for all x-values

37. $f(x) = 3 - 2x$

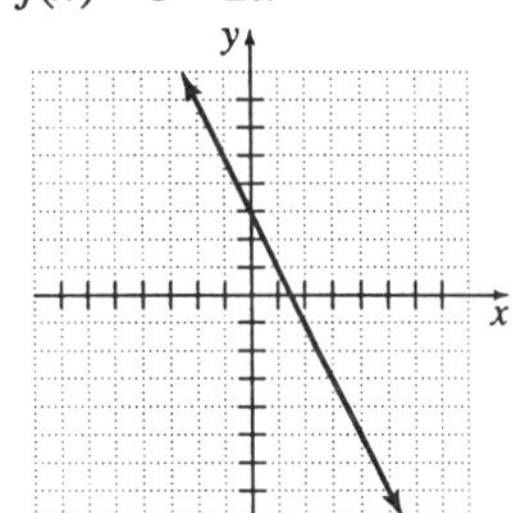

decreasing for all x-values

39. $y = 1 - x^2$

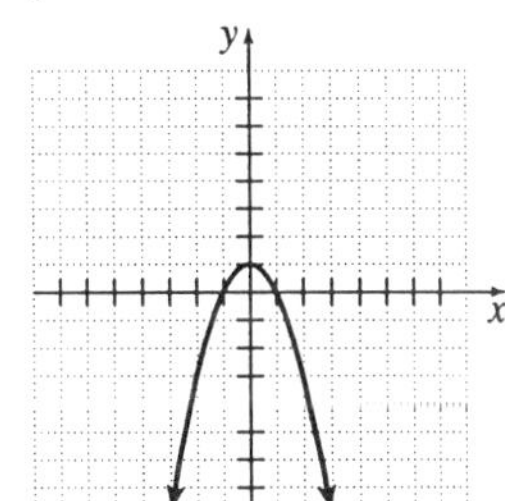

increasing for $x < 0$
decreasing for $x > 0$
relative maximum is 1

41. $g(x) = x^2 + 4x + 3$

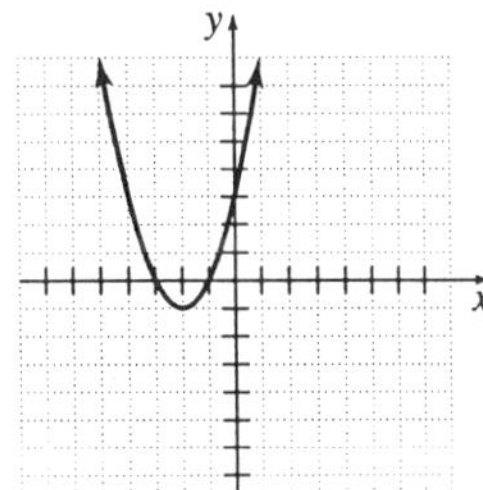

increasing for $x > -2$
decreasing for $x < -2$
relative minimum is –1

43. $y = |x+3|$

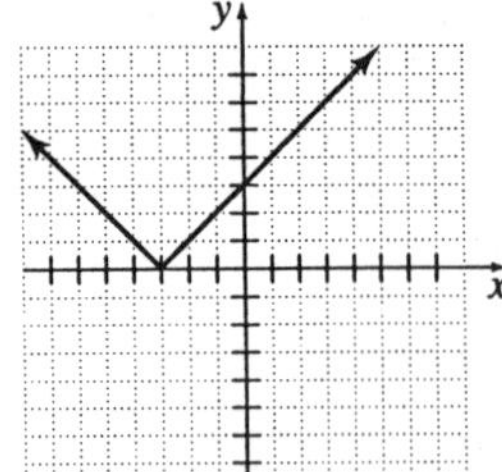

increasing for $x > -3$
decreasing for $x < -3$
relative minimum is 0

45. $f(x) = x^3 + 2x^2 - x - 2$

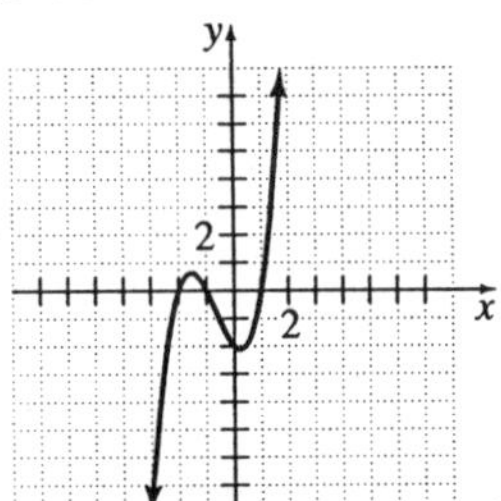

increasing for $x < -1.55$ and $x > 0.22$
decreasing for $-1.55 < x < 0.22$
relative maximum ≈ 0.63
relative minimum ≈ -2.11

47. $y = 3x - 5$ and $y = -2x + 15$

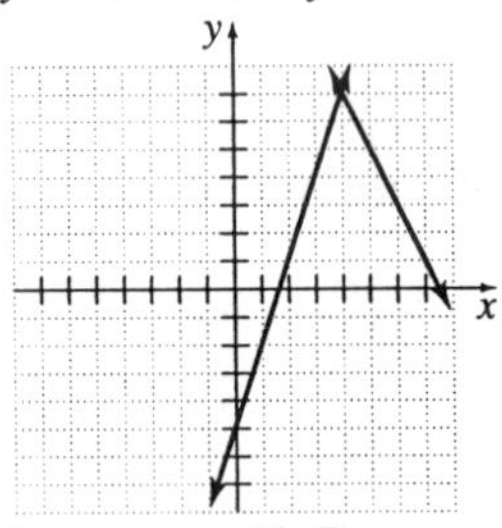

intersect at (4, 7)

49. $f(x) = 2x + 7$ and $g(x) = -x + 1$

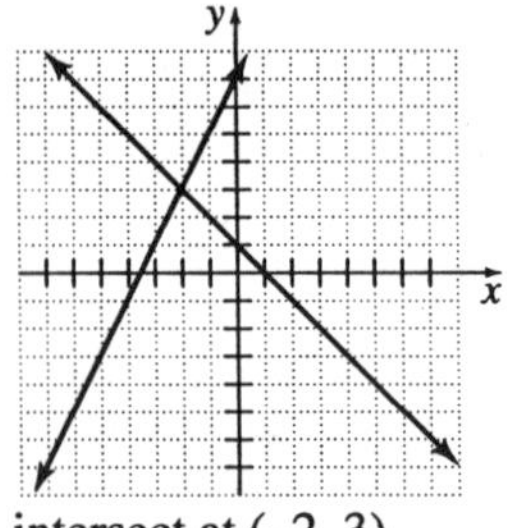

intersect at (–2, 3)

51. $y = -5x + 2$ and $y = 3x + 8$

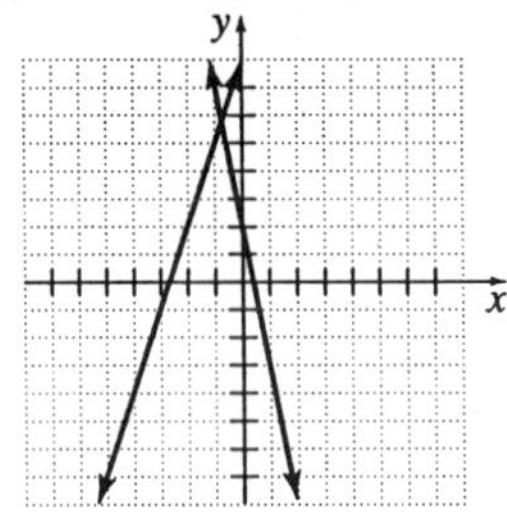

intersect at (–0.75, 5.75)

53. $r(x) = 5x - 7$ and $c(x) = 12$

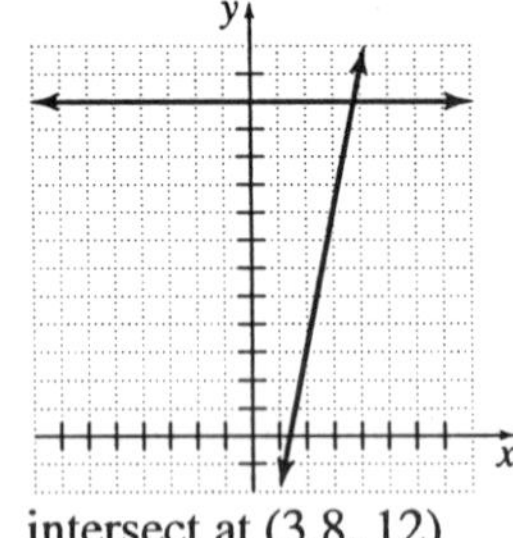

intersect at (3.8, 12)

55. $y = 3$ and $y = -x^2 + 4$

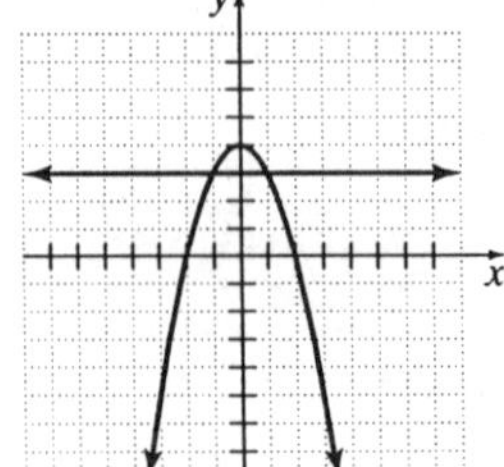

intersect at (–1, 3), (1, 3)

57. $f(x) = 2x^2 - 4x + 5$ and $g(x) = 4x - 1$

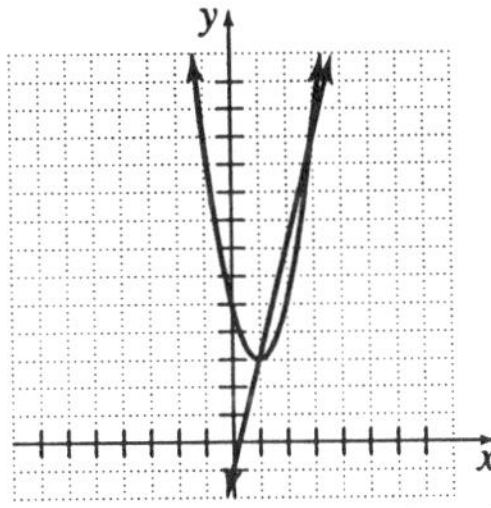

intersect at (1, 3), (3, 11)

59. $y = \frac{1}{4}x^2 - 2$ and $y = \frac{1}{2}x$

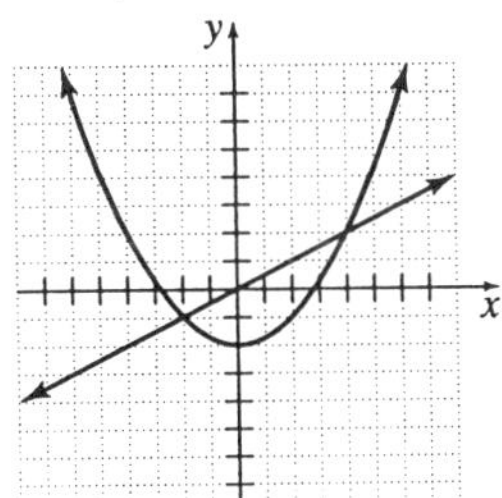

intersect at (–2, –1), (4, 2)

61. $y = |x| - 5$ and $y = 2$

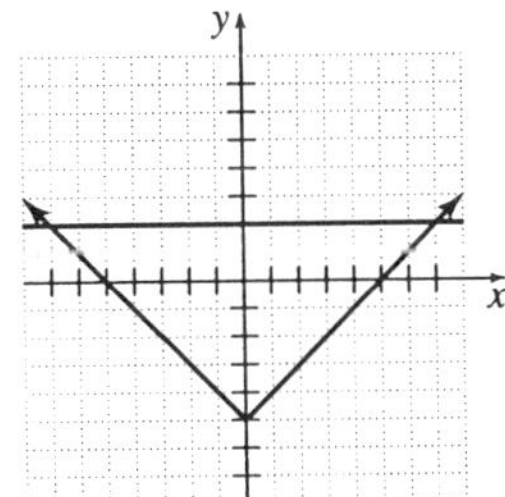

intersect at (–7, 2), (7, 2)

63. Let x = the number of crafts.
$Y1 = 5x - 50$

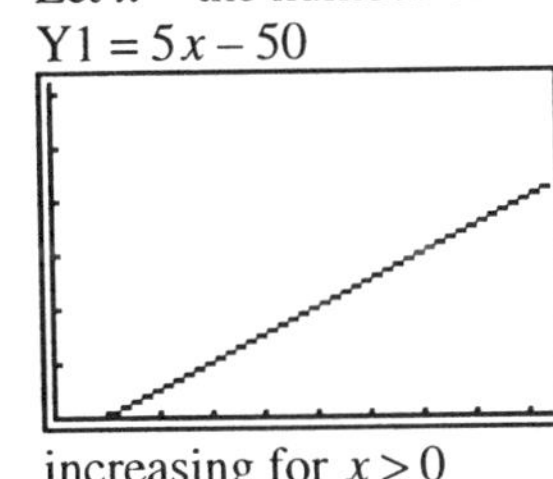

increasing for $x > 0$

65. Let x = the number of pieces of equipment.
$Y1 = 400x - 10x^2$ for $0 < x \le 25$

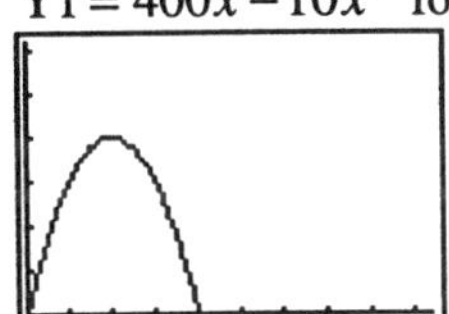

increasing for $0 < x < 20$
dereasing for $20 < x < 25$
relative maximum at $x = 20$

67. a. Let x = number of containers made in one run.
$f(x) = 4x + 50$

b. $g(x) = 10x$

c. 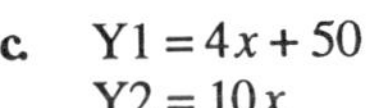
$Y1 = 4x + 50$
$Y2 = 10x$

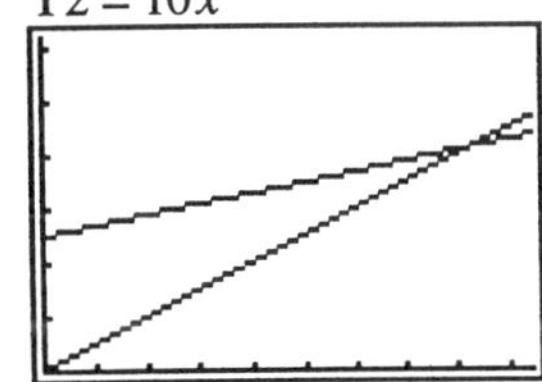

intersection at $(8.\overline{3}, 83.\overline{3})$

d. At $x = 8.3$ containers, the cost for producing and the cost for selling is the same at $83.33. This means that 9 containers must be sold to cover the costs of production.

69. a. $f(x) = 200 + 50x$
$g(x) = 75x$

b. $Y1 = 200 + 50x$
$Y2 = 75x$

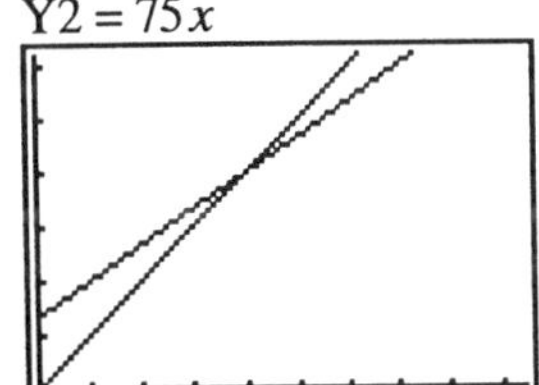

intersect at (8, 600)

c. At 8 credit hours, her pay will be the same, at $600.

4.4 Experiencing Algebra the Calculator Way

For exercises 1–3, graphs will vary.

1. $g(x)=\frac{1}{3}x+3$ and $f(x)=\frac{1}{4}x^2-4$ intersect at $(-4.\overline{6}, 1.\overline{4})$ and (6, 5)

2. $y=|x|-6$ and $y=-|x|+4$ intersect at (–5, –1) and (5, –1)

3. $y=x^2-18$ and $y=-x^2+54$ intersect at (–6, 18) and (6, 18)

Chapter 4 Review

Reflections

Answers will vary.

Exercises

1.

a	$b=-2a+7$	b
–3	$b=-2(-3)+7$ $b=13$	13
–2	$b=-2(-2)+7$ $b=11$	11
–1	$b=-2(-1)+7$ $b=9$	9
0	$b=-2(0)+7$ $b=7$	7
1	$b=-2(1)+7$ $b=5$	5
2	$b=-2(2)+7$ $b=3$	3
3	$b=-2(3)+7$ $b=1$	1

2.

x	$y=\frac{2}{3}x+4$	y
9	$y=\frac{2}{3}(9)+4$ $y=10$	10
6	$y=\frac{2}{3}(6)+4$ $y=8$	8
3	$y=\frac{2}{3}(3)+4$ $y=6$	6
0	$y=\frac{2}{3}(0)+4$ $y=4$	4
–3	$y=\frac{2}{3}(-3)+4$ $y=2$	2
–6	$y=\frac{2}{3}(-6)+4$ $y=0$	0
–9	$y=\frac{2}{3}(-9)+4$ $y=-2$	–2

3.

x	$y = 0.4x - 1.2$	y
–3	$y = 0.4(-3) - 1.2$ $y = -2.4$	–2.4
–2	$y = 0.4(-2) - 1.2$ $y = -2$	–2
–1	$y = 0.4(-1) - 1.2$ $y = -1.6$	–1.6
0	$y = 0.4(0) - 1.2$ $y = -1.2$	–1.2
1	$y = 0.4(1) - 1.2$ $y = -0.8$	–0.8
2	$y = 0.4(2) - 1.2$ $y = -0.4$	–0.4
3	$y = 0.4(3) - 1.2$ $y = 0$	0

4.

b	$a = 3b^2 - 2b + 5$	a
–2	$a = 3(-2)^2 - 2(-2) + 5$ $a = 21$	21
–1	$a = 3(-1)^2 - 2(-1) + 5$ $a = 10$	10
1	$a = 3(1)^2 - 2(1) + 5$ $a = 6$	6
2	$a = 3(2)^2 - 2(2) + 5$ $a = 13$	13

5.

x	$y=5x^3-3x^2+2x-22$	y
–18	$y=5(-18)^3-3(-18)^2+2(-18)-22$ $y=-30,190$	–30,190
–7	$y=5(-7)^3-3(-7)^2+2(-7)-22$ $y=-1898$	–1898
0	$y=5(0)^3-3(0)^2+2(0)-22$ $y=-22$	–22
6	$y=5(6)^3-3(6)^2+2(6)-22$ $y=962$	962
21	$y=5(21)^3-3(21)^2+2(21)-22$ $y=45,002$	45,002
22.5	$y=5(22.5)^3-3(22.5)^2+2(22.5)-22$ $y=55,457.375$	55,457.375

6.

x	$y=\lvert x^2-6x+5\rvert$	y
–2	$y=\lvert(-2)^2-6(-2)+5\rvert$ $y=21$	21
–1	$y=\lvert(-1)^2-6(-1)+5\rvert$ $y=12$	12
0	$y=\lvert 0^2-6\cdot 0+5\rvert$ $y=5$	5
1	$y=\lvert 1^2-6\cdot 1+5\rvert$ $y=0$	0
2	$y=\lvert 2^2-6\cdot 2+5\rvert$ $y=3$	3
3	$y=\lvert 3^2-6\cdot 3+5\rvert$ $y=4$	4

7.

x	$y = 3.6x^2 + 1.5x - 14.2$	y
−2.7	$y = 3.6(-2.7)^2 + 1.5(-2.7) - 14.2$ $y = 7.994$	7.994
−1.9	$y = 3.6(-1.9)^2 + 1.5(-1.9) - 14.2$ $y = -4.054$	−4.054
−0.6	$y = 3.6(-0.6)^2 + 1.5(-0.6) - 14.2$ $y = -13.804$	−13.804
0	$y = 3.6(0)^2 + 1.5(0) - 14.2$ $y = -14.2$	−14.2
0.8	$y = 3.6(0.8)^2 + 1.5(0.8) - 14.2$ $y = -10.696$	−10.696
1.5	$y = 3.6(1.5)^2 + 1.5(1.5) - 14.2$ $y = -3.85$	−3.85
2.4	$y = 3.6(2.4)^2 + 1.5(2.4) - 14.2$ $y = 10.136$	10.136

8. Answers will vary. Possible answer:

x	$y = 12x - 21$	y
−1	$y = 12(-1) - 21$ $y = -33$	−33
0	$y = 12(0) - 21$ $y = -21$	−21
1	$y = 12(1) - 21$ $y = -9$	−9
2	$y = 12(2) - 21$ $y = 3$	3
3	$y = 12(3) - 21$ $y = 15$	15

9. Answers will vary. Possible answer:

x	$y = -1.6x + 4.5$	y
−2	$y = -1.6(-2) + 4.5$ $y = 7.7$	7.7
−1	$y = -1.6(-1) + 4.5$ $y = 6.1$	6.1
0	$y = -1.6(0) + 4.5$ $y = 4.5$	4.5
1	$y = -1.6(1) + 4.5$ $y = 2.9$	2.9
2	$y = -1.6(2) + 4.5$ $y = 1.3$	1.3

10. Answers will vary. Possible answer:

x	$y=\frac{3}{2}x-6$	y
–4	$y=\frac{3}{2}(-4)-6$ $y=-12$	–12
–2	$y=\frac{3}{2}(-2)-6$ $y=-9$	–9
0	$y=\frac{3}{2}(0)-6$ $y=-6$	–6
2	$y=\frac{3}{2}(2)-6$ $y=-3$	–3
4	$y=\frac{3}{2}(4)-6$ $y=0$	0

11. Answers will vary. Possible answer:

x	$y=\lvert 2x-9\rvert$	y
–4	$y=\lvert 2(-4)-9\rvert$ $y=17$	17
–2	$y=\lvert 2(-2)-9\rvert$ $y=13$	13
0	$y=\lvert 2\cdot 0-9\rvert$ $y=9$	9
2	$y=\lvert 2\cdot 2-9\rvert$ $y=5$	5
4	$y=\lvert 2\cdot 4-9\rvert$ $y=1$	1

12.

x	$y=(5x+2)(x-4)$	y
–3	$y=[5(-3)+2](-3-4)$ $y=91$	91
–1	$y=[5(-1)+2](-1-4)$ $y=15$	15
1	$y=(5\cdot 1+2)(1-4)$ $y=-21$	–21
3	$y=(5\cdot 3+2)(3-4)$ $y=-17$	–17

13.

x	$y=\frac{3}{5}x+8$	y
–15	$y=\frac{3}{5}(-15)+8$ $y=-1$	–1
–10	$y=\frac{3}{5}(-10)+8$ $y=2$	2
–5	$y=\frac{3}{5}(-5)+8$ $y=5$	5
0	$y=\frac{3}{5}(0)+8$ $y=8$	8
5	$y=\frac{3}{5}(5)+8$ $y=11$	11
10	$y=\frac{3}{5}(10)+8$ $y=14$	14
15	$y=\frac{3}{5}(15)+8$ $y=17$	17

14.

x	$y = 17.1x - 12.9$	y
–3	$y = 17.1(-3) - 12.9$ $y = -64.2$	–64.2
–2	$y = 17.1(-2) - 12.9$ $y = -47.1$	–47.1
–1	$y = 17.1(-1) - 12.9$ $y = -30$	–30
0	$y = 17.1(0) - 12.9$ $y = -12.9$	–12.9
1	$y = 17.1(1) - 12.9$ $y = 4.2$	4.2
2	$y = 17.1(2) - 12.9$ $y = 21.3$	21.3
3	$y = 17.1(3) - 12.9$ $y = 38.4$	38.4

15.

x	$y = 3x^2 - 5x + 17$	y
–6	$y = 3(-6)^2 - 5(-6) + 17$ $y = 155$	155
–2	$y = 3(-2)^2 - 5(-2) + 17$ $y = 39$	39
0	$y = 3(0)^2 - 5(0) + 17$ $y = 17$	17
3	$y = 3(3)^2 - 5(3) + 17$ $y = 29$	29
8	$y = 3(8)^2 - 5(8) + 17$ $y = 169$	169
11	$y = 3(11)^2 - 5(11) + 17$ $y = 325$	325

16.

r	$A = \pi r^2$	A
4	$A = \pi(4)^2$ $A \approx 50.265$	50.265
6	$A = \pi(6)^2$ $A \approx 113.097$	113.097
8	$A = \pi(8)^2$ $A \approx 201.062$	201.062
10	$A = \pi(10)^2$ $A \approx 314.159$	314.159

17.

r	$V = \pi(6)^2 h$ $V = 36\pi h$	V
3	$V = 36\pi(3)$ $V \approx 339.3$	339.3
6	$V = 36\pi(6)$ $V \approx 678.6$	678.6
9	$V = 36\pi(9)$ $V \approx 1017.9$	1017.9
12	$V = 36\pi(12)$ $V \approx 1357.2$	1357.2

18.

a	$b=90-a$	b
10	$b=90-10$ $b=80$	80
20	$b=90-20$ $b=70$	70
30	$b=90-30$ $b=60$	60
40	$b=90-40$ $b=50$	50
45	$b=90-45$ $b=45$	45

19.

t	$A=2000(1.06)^t$	A
2	$A=2000(1.06)^2$ $A=2247.2$	2247.2
3	$A=2000(1.06)^3$ $A\approx 2382.03$	2382.03
4	$A=2000(1.06)^4$ $A\approx 2524.95$	2524.95

20.

F	$C=\frac{5}{9}(F-32)$	C
−23	$C=\frac{5}{9}(-23-32)$ $C\approx -30.6$	−30.6
−14	$C=\frac{5}{9}(-14-32)$ $C\approx -25.6$	−25.6
0	$C=\frac{5}{9}(0-32)$ $C\approx -17.8$	−17.8
41	$C=\frac{5}{9}(41-32)$ $C=5$	5
50	$C=\frac{5}{9}(50-32)$ $C=10$	10
59	$C=\frac{5}{9}(59-32)$ $C=15$	15
100	$C=\frac{5}{9}(100-32)$ $C\approx 37.8$	37.8

21.

t	$d = 65t$	d
2	$d = 65(2)$ $d = 130$	130
3	$d = 65(3)$ $d = 195$	195
4	$d = 65(4)$ $d = 260$	260
5	$d = 65(5)$ $d = 325$	325
6	$d = 65(6)$ $d = 390$	390
7	$d = 65(7)$ $d = 455$	455

22.

x	$y = 7 - 3x$	y
–10	$y = 7 - 3(-10)$ $y = 37$	37
–5	$y = 7 - 3(-5)$ $y = 22$	22
0	$y = 7 - 3(0)$ $y = 7$	7
5	$y = 7 - 3(5)$ $y = -8$	–8
10	$y = 7 - 3(10)$ $y = -23$	–23

ordered pairs: (–10, 37), (–5, 22), (0, 7), (5, –8), (10, –23)

23.

x	$y = \sqrt{x+8}$	y
–8	$y = \sqrt{-8+8}$ $y = 0$	0
–7	$y = \sqrt{-7+8}$ $y = 1$	1
–4	$y = \sqrt{-4+8}$ $y = 2$	2
1	$y = \sqrt{1+8}$ $y = 3$	3
8	$y = \sqrt{8+8}$ $y = 4$	4

ordered pairs: (–8, 0), (–7, 1), (–4, 2), (1, 3), (8, 4)

24.

c	$d = \frac{2}{3}c - 1$	d
–6	$d = \frac{2}{3}(-6) - 1$ $d = -5$	–5
–3	$d = \frac{2}{3}(-3) - 1$ $d = -3$	–3
0	$d = \frac{2}{3}(0) - 1$ $d = -1$	–1
3	$d = \frac{2}{3}(3) - 1$ $d = 1$	1
6	$d = \frac{2}{3}(6) - 1$ $d = 3$	3

ordered pairs: (–6, –5), (–3, –3), (0, –1), (3, 1), (6, 3)

25.

d	$r=\frac{d}{2}$	r
2	$r=\frac{2}{2}$ $r=1$	1
4	$r=\frac{4}{2}$ $r=2$	2
6	$r=\frac{6}{2}$ $r=3$	3
8	$r=\frac{8}{2}$ $r=4$	4
10	$r=\frac{10}{2}$ $r=5$	5

ordered pairs: (2, 1), (4, 2), (6, 3), (8, 4), (10, 5)

26. domain {1, 3, 5, 7, 9}
range {2, 6, 10, 14, 18}

27. domain {..., –6, –4, –2, 0, 2, 4, 6, ...}
range {..., 6, 4, 2, 0, –2, –4, –6, ...}

28. $y=4(x+5)-1$
$y=4(-5+5)-1$
$y=0-1=-1$
$y=4(-4+5)-1$
$y=4-1=3$
$y=4(-3+5)-1$
$y=8-1=7$
$y=4(-2+5)-1$
$y=12-1=11$
$y=4(-1+5)-1$
$y=16-1=15$
domain {–5, –4, –3, –2, –1}
range {–1, 3, 7, 11, 15}

29. domain all real numbers
range all real numbers ≥ 2.5

30. $5+x\geq 0$
$x\geq -5$
domain all real numbers ≥ -5
range all real numbers ≥ 0

31. $1-x\neq 0$
$x\neq 1$
domain all real numbers $x\neq 1$
range all real numbers $y\neq 0$

32. $y=5+4x$
domain all real numbers
range all real numbers

33. $\frac{L}{32}\geq 0$
$L\geq 0$
domain all real numbers ≥ 0
range all real numbers ≥ 0

34. $A(3, 2)$

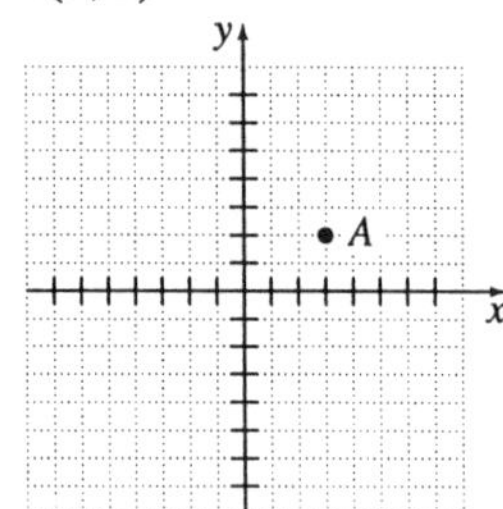

35. $B(4, -3)$

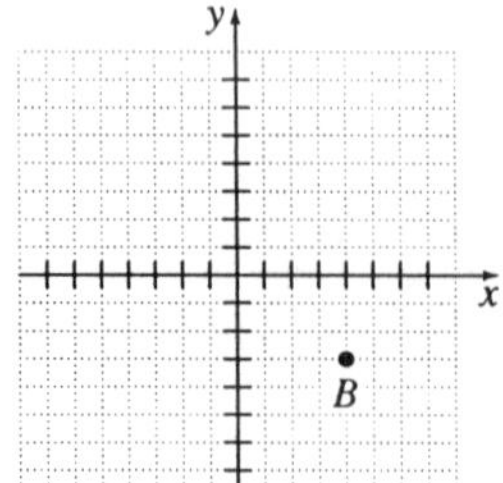

36. $C(-3, 2)$

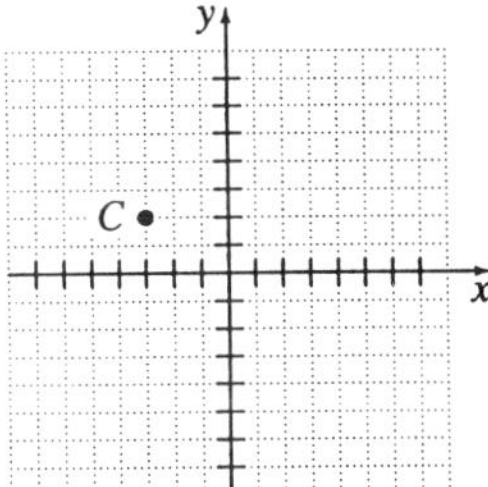

37. $D(-4, -3)$

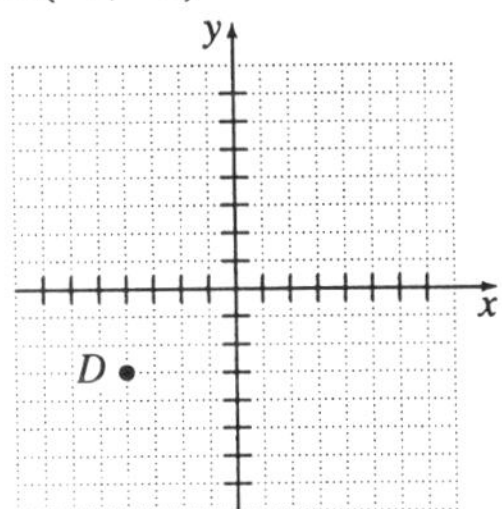

38. $E(0, 5)$

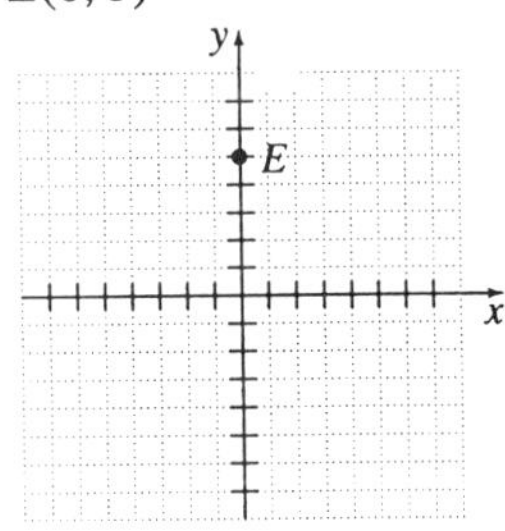

39. $F(-5, 0)$

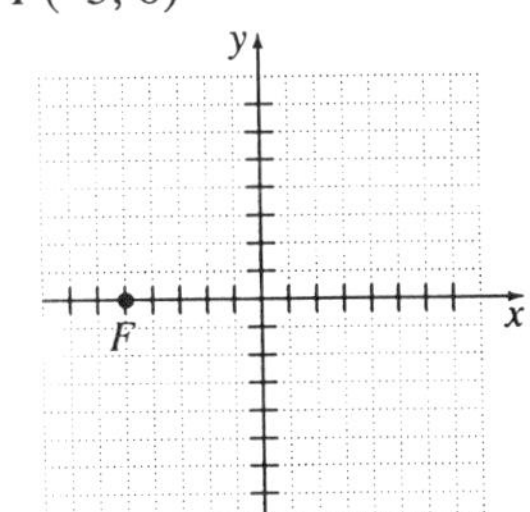

40. $A(5, 3)$

41. $B(-2, -5)$

42. $C(2, -2)$

43. $D(-3, 5)$

44. $E(5, 0)$

45. $F(0, -4)$

46. $G(0, 0)$

47. quadrant I

48. quadrant III

49. quadrant IV

50. quadrant II

51. x-axis

52. y-axis

53. origin

54. $S = \{(0, -4), (1, -3), (2, -2), (3, -1), (4, 0), (5, 1)\}$

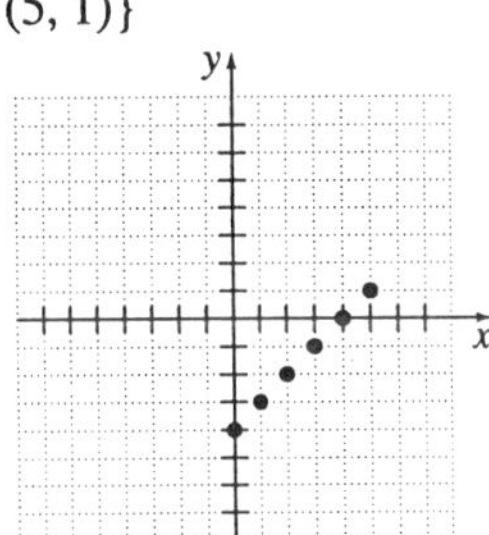

55.

x	$y = 2x - 3$	y
-1	$y = 2(-1) - 3$ $y = -5$	-5
0	$y = 2(0) - 3$ $y = -3$	-3
1	$y = 2(1) - 3$ $y = -1$	-1
2	$y = 2(2) - 3$ $y = 1$	1
3	$y = 2(3) - 3$ $y = 3$	3
4	$y = 2(4) - 3$ $y = 5$	5

(–1, –5), (0, –3), (1, –1), (2, 1), (3, 3), (4, 5)

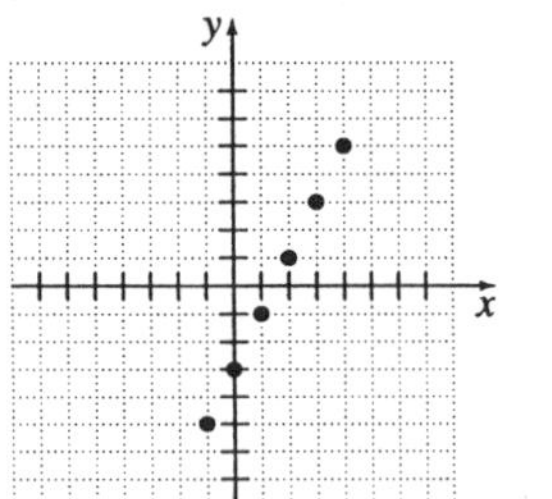

56. $y = 3 - 2x$

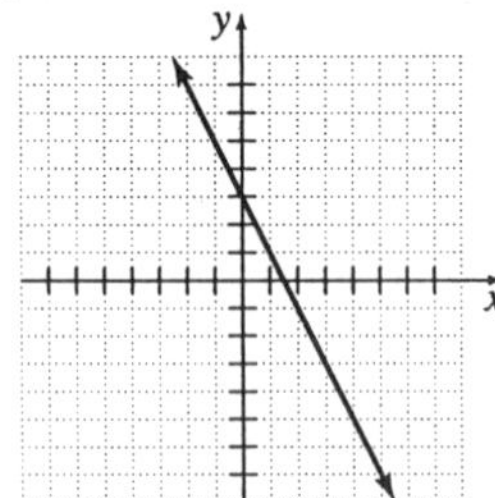

57. $T = \{(-5, 3), (-3, 3), (-1, 3), (1, 3), (3, 3), (5, 3)\}$

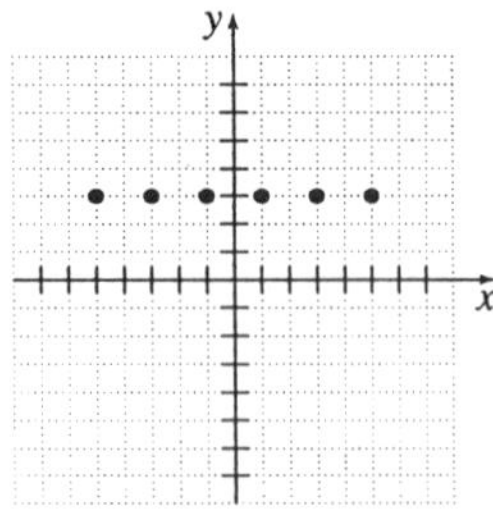

58. domain all real numbers
range all real numbers ≤ 2

59. domain all real numbers
range $\{-3\}$

60. $y = 6x - 5$
domain all real numbers
range all real numbers

61. $y = \sqrt{8 - 4x}$
$8 - 4x \geq 0$
$4x \leq 8$
$x \leq 2$
domain all real numbers ≤ 2
range all real numbers ≥ 0

62. $y = x^2 + 6x + 9$

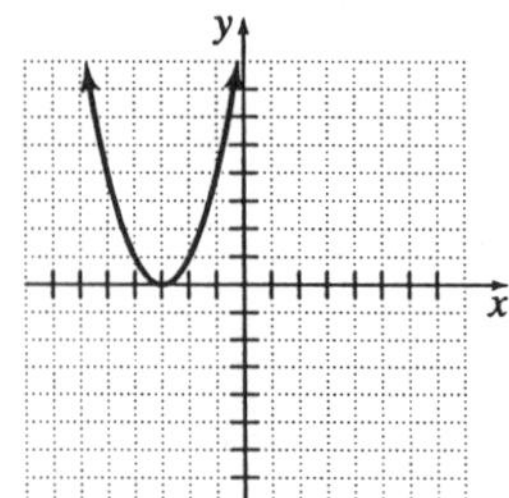

domain all real numbers
range all real numbers ≥ 0

63. $y = -x^2$

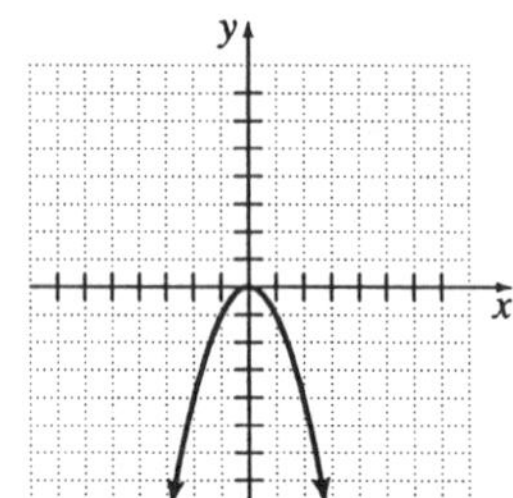

domain all real numbers
range all real numbers ≤ 0

64. domain years 1970 to 1993
range all real numbers $65 \leq y \leq 85$
For a year (independent variable) the energy consumption (dependent variable) is graphed

65. Domain integers 1 to 7
Range {40, 50, 55, 60}
The domain represents the number of children from one family enrolled in the child care center.
The range represents the charge in dollars for a week's care for the children from a family.

66. domain all real numbers > 0
range all real numbers > 0
The length of a side (independent variable) and its corresponding area (dependent variable) is graphed.

67. *S* is not a function. Two elements in the domain, –2 and –1, each correspond to 2 elements in the range, 3 and 11, and 5 and 9, respectively.

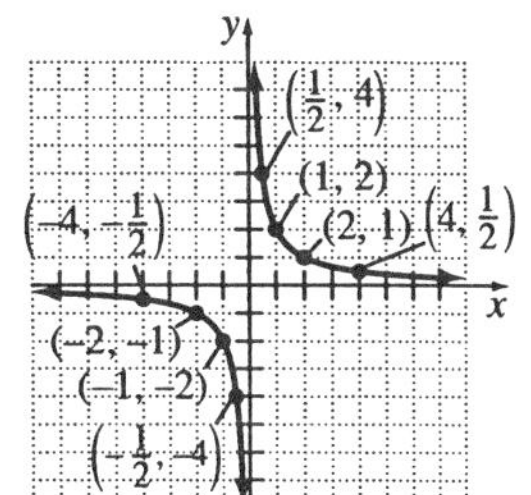

68. T is a function. Every element in the domain corresponds to only one element in the range.

69. $y = x^2 - 10$ is a function. Every element in the domain corresponds to only one element in the range.

70. $y^2 = x - 10$ is not a function. There are two elements in the range for each element in the domain (except $x = 10$).

71. $y = 4$ for all x is a function. Each element in the domain corresponds to only one element in the range.

72. $x = 2$ for all y is not a function. Each element in the domain corresponds to the same element in the domain.

73. No; does not pass the vertical line test.

74. Yes; does pass the vertical line test.

75. $f(x) = -4x + 13$
$f(13) = -4(13) + 13$
$f(13) = -39$

76. $f(x) = -4x + 13$
$f(-21) = -4(-21) + 13$
$f(-21) = 97$

77. $f(x) = -4x + 13$
$f(2.5) = -4(2.5) + 13$
$f(2.5) = 3$

78. $f(x) = -4x + 13$
$f(-3.7) = -4(-3.7) + 13$
$f(-3.7) = 27.8$

79. $f(x) = -4x + 13$
$f(a + 3) = -4(a + 3) + 13$
$f(a + 3) = -4a - 12 + 13$
$f(a + 3) = -4a + 1$

80. $f(x) = -4x + 13$
$f(-b) = -4(-b) + 13$
$f(-b) = 4b + 13$

81. $g(x) = 5x^2 + x - 4$
$g(3) = 5(3)^2 + 3 - 4$
$g(3) = 45 + 3 - 4$
$g(3) = 44$

82. $g(x) = 5x^2 + x - 4$
$g(-2) = 5(-2)^2 + (-2) - 4$
$g(-2) = 20 - 2 - 4$
$g(-2) = 14$

83. $g(x) = 5x^2 + x - 4$
$g(0.5) = 5(0.5)^2 + (0.5) - 4$
$g(0.5) = 1.25 + 0.5 - 4$
$g(0.5) = -2.25$

84. $g(x) = 5x^2 + x - 4$
$g(a) = 5a^2 + a - 4$

85. $g(x) = 5x^2 + x - 4$
$g(-a) = 5(-a)^2 + (-a) - 4$
$g(-a) = 5a^2 - a - 4$

86. $g(x) = 5x^2 + x - 4$
$$g\left(-\frac{1}{4}\right) = 5\left(-\frac{1}{4}\right)^2 + \left(-\frac{1}{4}\right) - 4$$
$$g\left(-\frac{1}{4}\right) = \frac{5}{16} - \frac{1}{4} - 4$$
$$g\left(-\frac{1}{4}\right) = \frac{5}{16} - \frac{4}{16} - \frac{64}{16}$$
$$g\left(-\frac{1}{4}\right) = -\frac{63}{16}$$

87. $S(x) = \sqrt{2x + 3}$
$S(3) = \sqrt{2 \cdot 3 + 3}$
$S(3) = \sqrt{9}$
$S(3) = 3$

88. $S(x) = \sqrt{2x + 3}$
$S(11) = \sqrt{2 \cdot 11 + 3}$
$S(11) = \sqrt{25}$
$S(11) = 5$

89. $S(x) = \sqrt{2x+3}$
$S(59) = \sqrt{2 \cdot 59 + 3}$
$S(59) = \sqrt{121}$
$S(59) = 11$

90. Let x = the number of widgets.
$f(x) = 4500 + 17x$
$f(1200) = 4500 + 17(1200)$
$f(1200) = 4500 + 20{,}400$
$f(1200) = 24{,}900$
The cost of producing 1200 widgets is \$24,900.

91. Let x = the number of days.
$f(x) = 50 + 25x$
$f(4) = 50 + 25(4)$
$f(4) = 50 + 100$
$f(4) = 150$
The cost of a 4-day rental is \$150.

92. Let x = the number of employees.
$f(x) = 1500 + 125x$
$f(20) = 1500 + 125(20)$
$f(20) = 1500 + 2500$
$f(20) = 4000$
The charge for a training session for 20 employees is \$4,000.

93. Let x = the number of faces.
$f(x) = 1.5x - 15$
$f(135) = 1.5(135) - 15$
$f(135) = 202.5 - 15$
$f(135) = 187.5$
He will make a profit of \$187.50.

94. **a.** $(-5, 0), (1, 0), (6, 0)$

b. $(0, 1)$

c. $x < -3, x > 3$

d. $-3 < x < 3$

e. 3

f. -3

95. $y = 3x + 9$

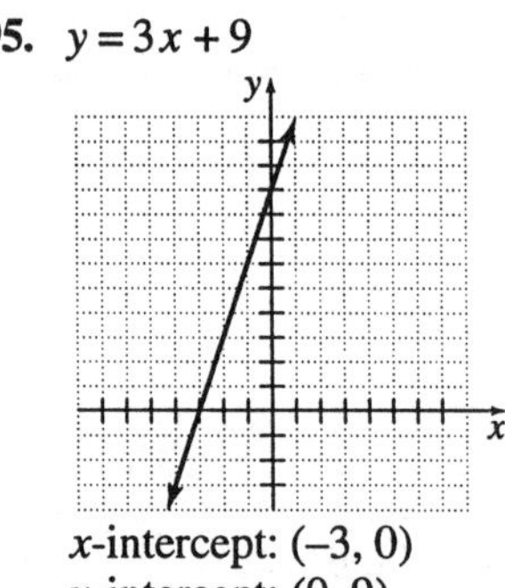

x-intercept: $(-3, 0)$
y-intercept: $(0, 9)$

96. $y = 6 - x$

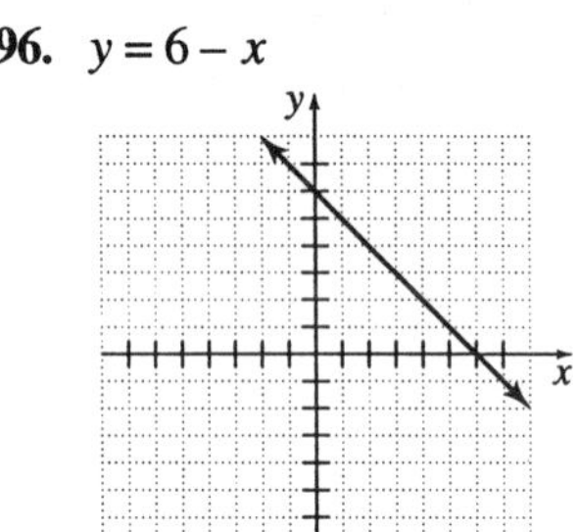

x-intercept: $(6, 0)$
y-intercept: $(0, 6)$

97. $y = \frac{3}{4}x - 9$

x-intercept: $(12, 0)$
y-intercept: $(0, -9)$

98. $y = x^2 - 0.36$

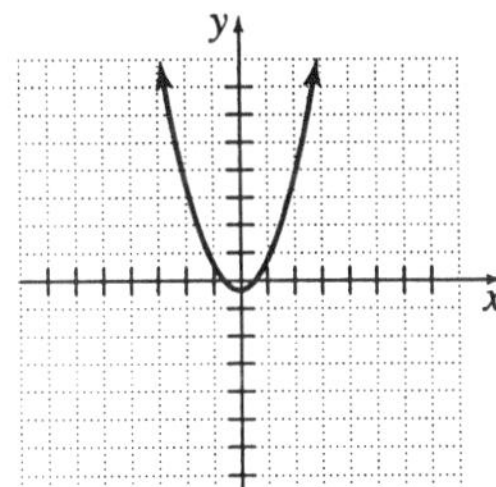

x-intercepts: (–0.6, 0), (0.6, 0)
y-intercept: (0, –0.36)

99. $y = |x| - 4$

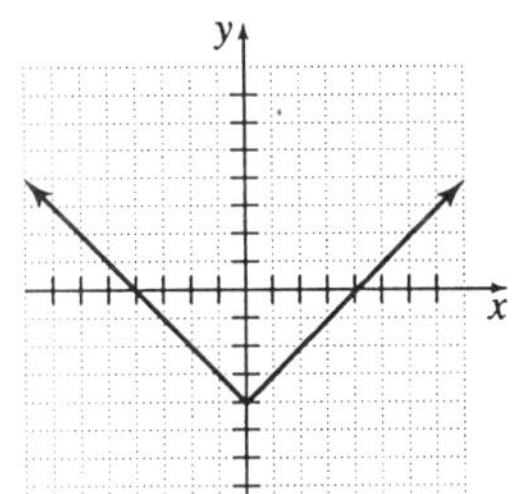

x-intercepts: (–4, 0), (4, 0)
y-intercept: (0, –4)

100. $y = 4x - 3$

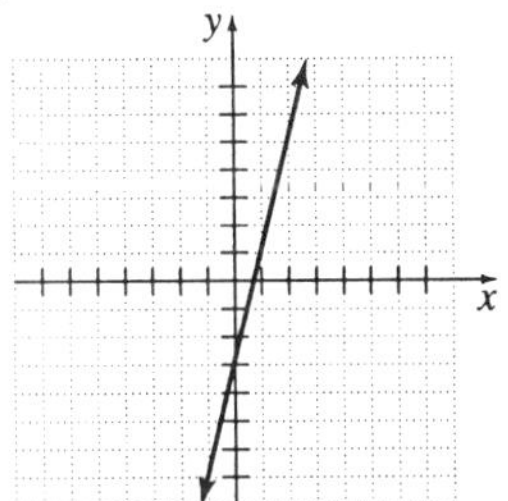

increasing for all values of x

101. $h(x) = 6 - 2x$

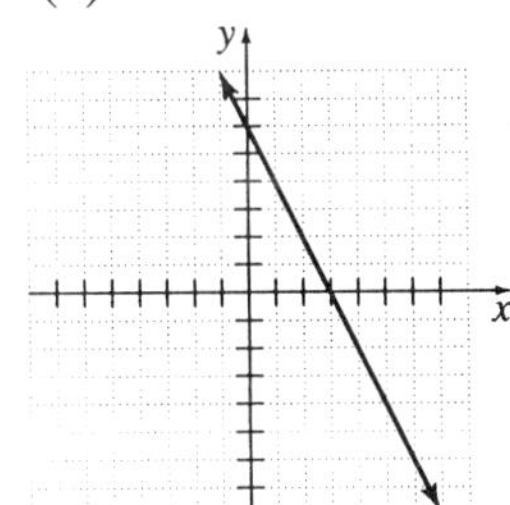

decreasing for all values of x

102. $y = 3 - x^2$

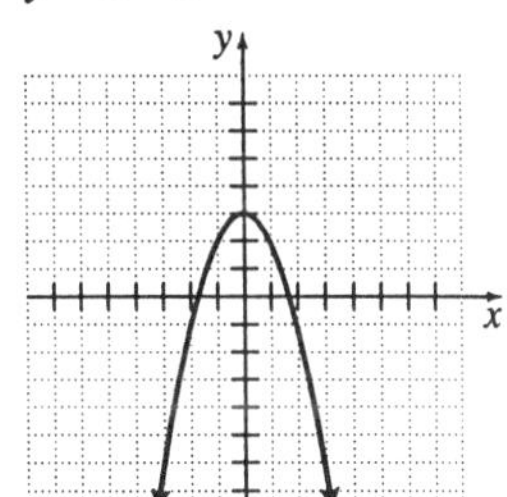

increasing for $x < 0$
decreasing for $x > 0$
relative maximum is 3

103. $y = |x| + 2$

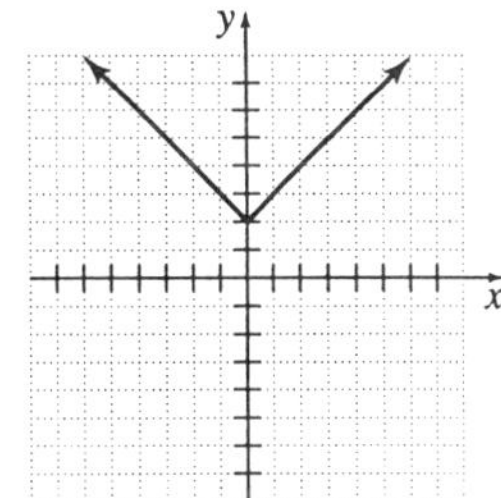

increasing for $x > 0$
decreasing for $x < 0$
relative minimum is 2

104. $y = \left|x^2 - 1\right|$

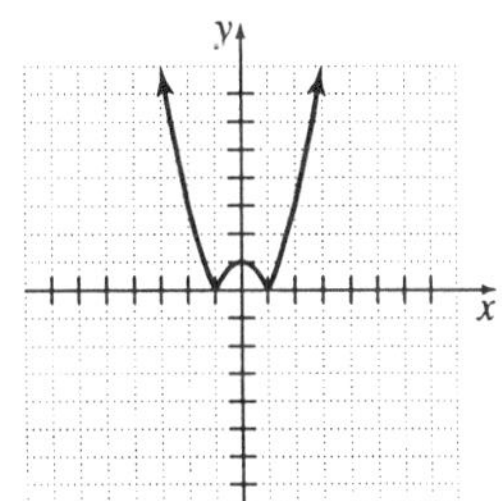

Increasing for $-1 < x < 0$, $x > 1$
decreasing for $x < -1$, $0 < x < 1$
relative maximum is 1
relative minima are 0, 0

105. $y = 2x - 2$ and $y = -\frac{1}{3}x + 5$

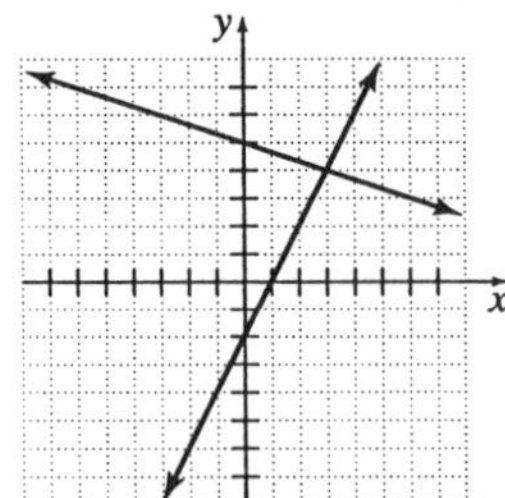

intersect at (3, 4)

106. $y = x^2 - 6$ and $y = x$

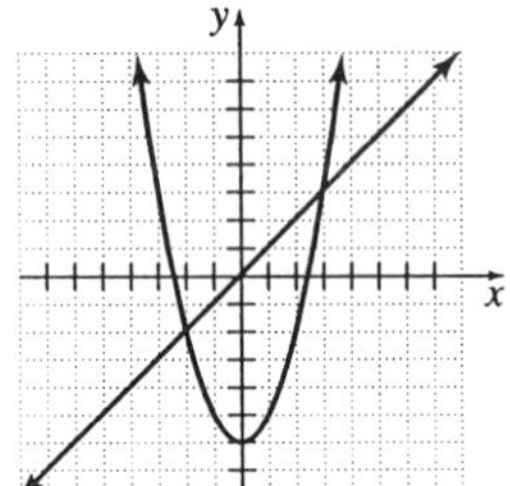

intersect at (–2, –2), (3, 3)

107. $f(x) = |x + 5|$ and $g(x) = 2$

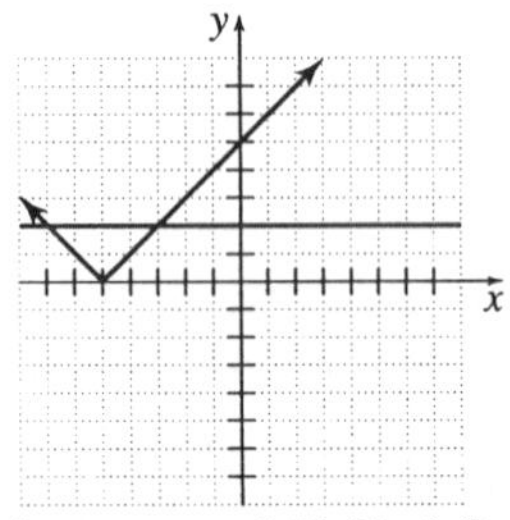

intersect at (–7, 2), (–3, 2)

108. $f(x) = 10.45x$

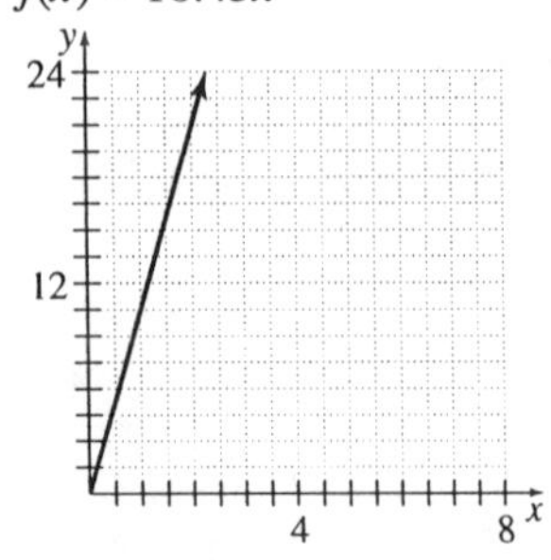

increasing for $x > 0$

109. $f(x) = 216 - 4.5x$

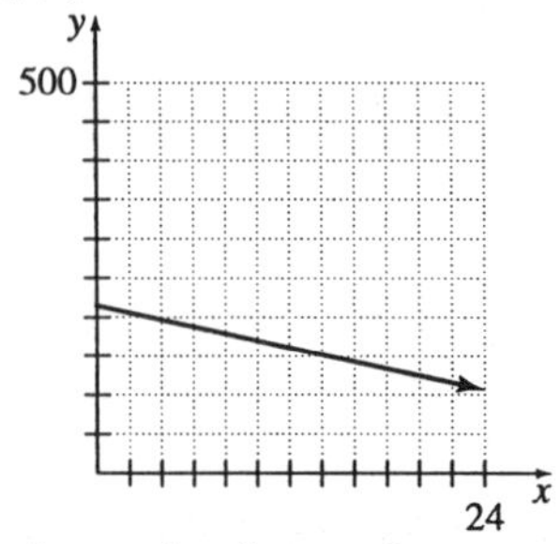

decreasing for $x > 0$

110. $Y1 = 325 + 325x - 15x^2$

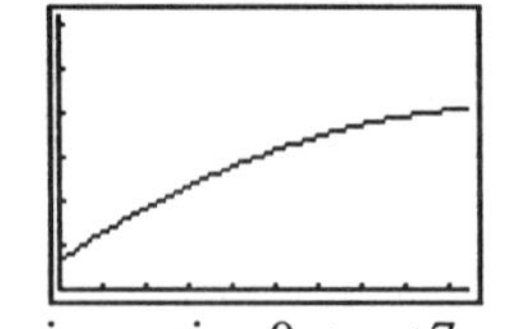

increasing $0 < x < 7$

111. a. Let x = credit hours.
$f(x) = 400 + 65x$
$g(x) = 100x$

b. $Y1 = 400 + 65x$
$Y2 = 100x$

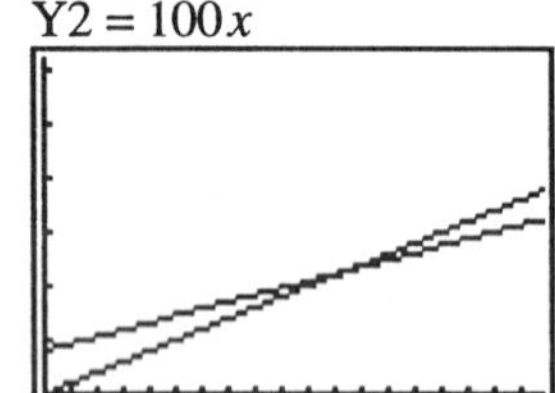

intersect at approximately (11.4, 1143)

c. At about 11.4 credit hours the stipend is the same, at about $1143, for both options.

112. a. Let x = the number of items.
$f(x) = 500 + 12x$
$g(x) = 25x$

b. $Y1 = 500 + 12x$
$Y2 = 25x$

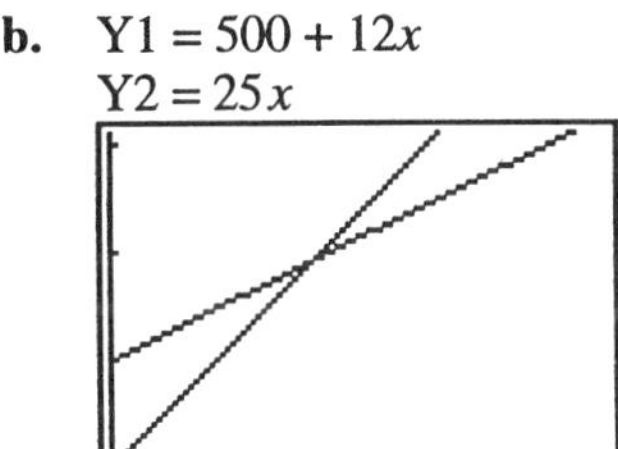

intersect at approximately(38.5, 961.5)

c. At about 38.5 items, the cost to produce and the revenue are equal at about \$961.50 each. This means 39 items must be sold to break even.

Chapter 4 Mixed Review

1. $f(x) = -x + 9$
$f(9) = -9 + 9$
$f(9) = 0$

2. $f(x) = -x + 9$
$f(-9) = -(-9) + 9$
$f(-9) = 9 + 9$
$f(-9) = 18$

3. $f(x) = -x + 9$
$f(1.8) = -1.8 + 9$
$f(1.8) = 7.2$

4. $f(x) = -x + 9$
$f(-2.7) = -(-2.7) + 9$
$f(-2.7) = 2.7 + 9$
$f(-2.7) = 11.7$

5. $f(x) = -x + 9$
$f(-b) = -(-b) + 9$
$f(-b) = b + 9$

6. $f(x) = -x + 9$
$f(h + 1) = -(h + 1) + 9$
$f(h + 1) = -h - 1 + 9$
$f(h + 1) = -h + 8$

7. $g(x) = x^2 - 3x - 4$
$g(4) = 4^2 - 3 \cdot 4 - 4$
$g(4) = 16 - 12 - 4$
$g(4) = 0$

8. $g(x) = x^2 - 3x - 4$
$g(-1) = (-1)^2 - 3(-1) - 4$
$g(-1) = 1 + 3 - 4$
$g(-1) = 0$

9. $g(x) = x^2 - 3x - 4$
$g(1.5) = (1.5)^2 - 3(1.5) - 4$
$g(1.5) = 2.25 - 4.5 - 4$
$g(1.5) = -6.25$

10. $g(x) = x^2 - 3x - 4$
$g(v) = v^2 - 3v - 4$

11. $g(x) = x^2 - 3x - 4$
$g(-v) = (-v)^2 - 3(-v) - 4$
$g(-v) = v^2 + 3v - 4$

12. $g(x) = x^2 - 3x - 4$
$$g\left(-\frac{2}{3}\right) = \left(-\frac{2}{3}\right)^2 - 3\left(-\frac{2}{3}\right) - 4$$
$$g\left(-\frac{2}{3}\right) = \frac{4}{9} + \frac{6}{3} - 4$$
$$g\left(-\frac{2}{3}\right) = \frac{4}{9} + \frac{18}{9} - \frac{36}{9}$$
$$g\left(-\frac{2}{3}\right) = -\frac{14}{9}$$

13. $S(x) = \sqrt{6x - 8}$
$S(4) = \sqrt{6 \cdot 4 - 8}$
$S(4) = \sqrt{16}$
$S(4) = 4$

14. $S(x) = \sqrt{6x - 8}$
$S(12) = \sqrt{6(12) - 8}$
$S(12) = \sqrt{64}$
$S(12) = 8$

15. $S(x) = \sqrt{6x - 8}$
$S(44) = \sqrt{6(44) - 8}$
$S(44) = \sqrt{256}$
$S(44) = 16$

16. Yes; each element in the domain corresponds to a single element in the range.

17. No; an element in the domain, 2, corresponds to 5 elements in the range, –3, –2, –1, 0, and 1.

18. $y = x^2 + 5$

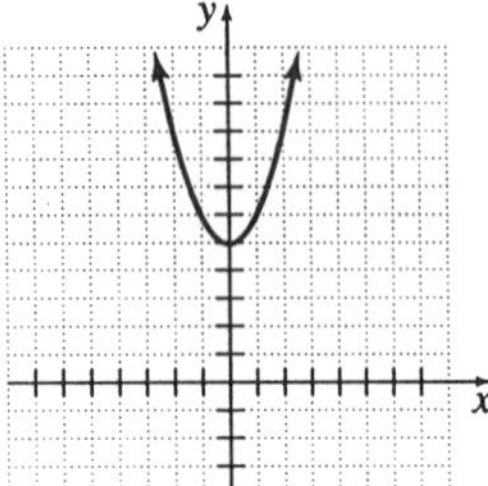

Yes; the graph passes the vertical line test.

19. $y^2 = x + 5$

No; for each element in the domain, except –5, there are two corresponding elements in the range.

20. $y = -4$ for all x.

Yes; each element in the domain corresponds to only one element in the range.

21. $y = 5x - 10$

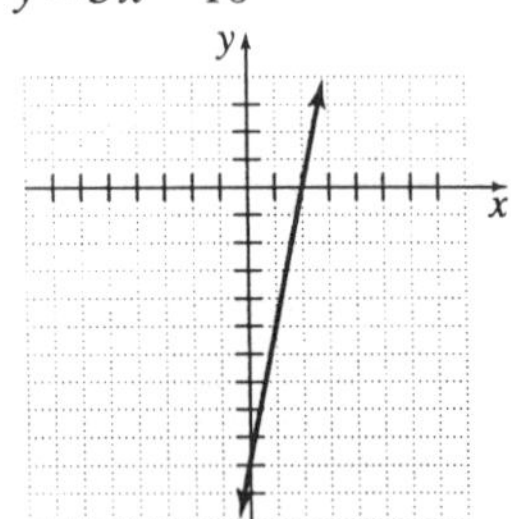

x-intercept at (2, 0)
y-intercept at (0, –10)
increasing for all x-values

22. $y = 8 - 2x$

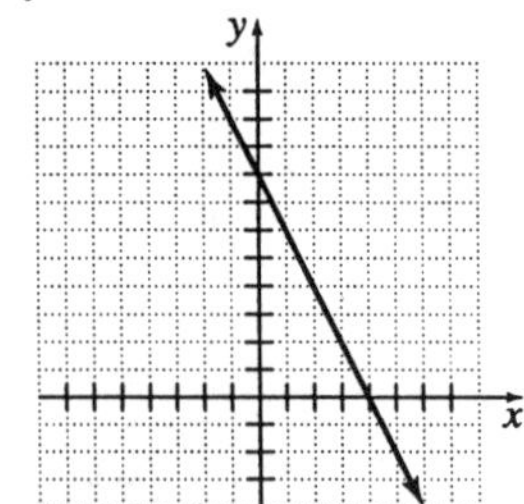

x-intercept at (0, 8)
y-intercept at (4, 0)
decreasing for all x-values

23. $y = 4.8x - 1.2$

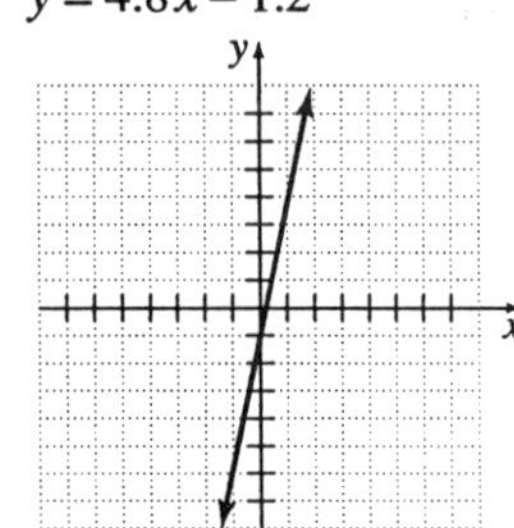

x-intercept: (0.25, 0)
y-intercept: (0, –1.2)
increasing for all x-values

24. $y = \frac{2}{5}x + 4$

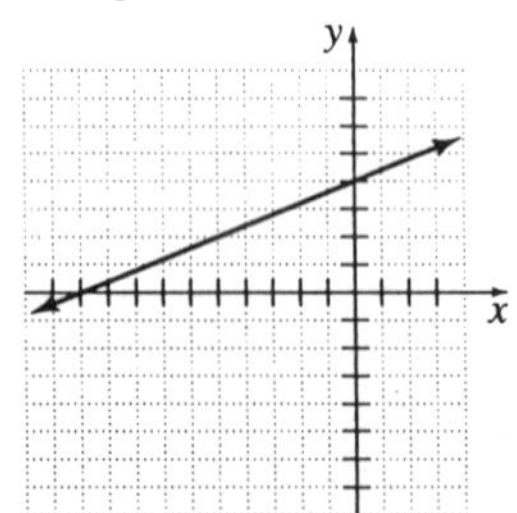

x-intercept at (–10, 0)
y-intercept at (0, 4)
increasing for all x-values

25. $y = x^2 - 1.21$

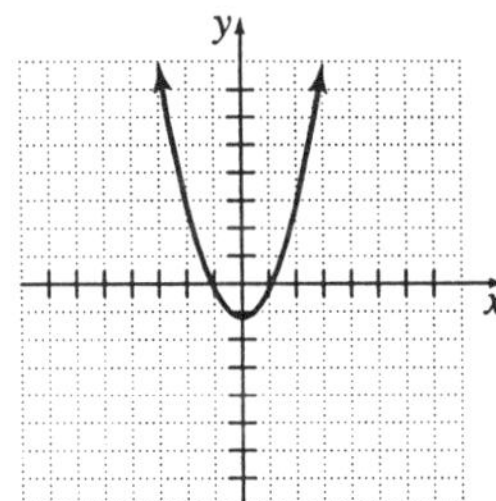

x-intercepts at (–1.1, 0) and (1.1, 0)
y-intercept at (0, –1.21)
increasing for $x > 0$
decreasing for $x < 0$
relative minimum is –1.21

26. $y = 2 - |x|$

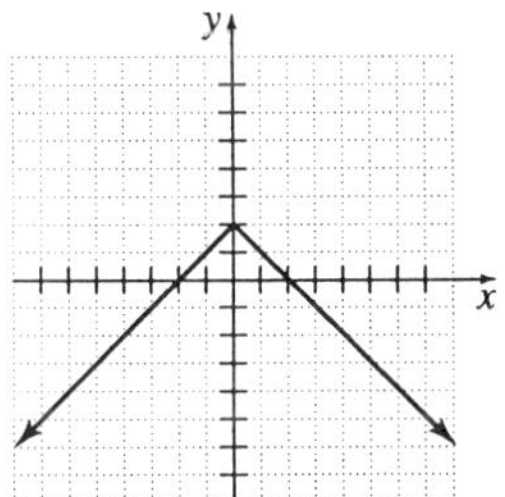

x-intercepts at (–2, 0), (2, 0)
y-intercept at (0, 2)
increasing for $x < 0$
decreasing for $x > 0$
relative maximum is 2

27. $y = 2x + 2$ and $y = -2x - 10$

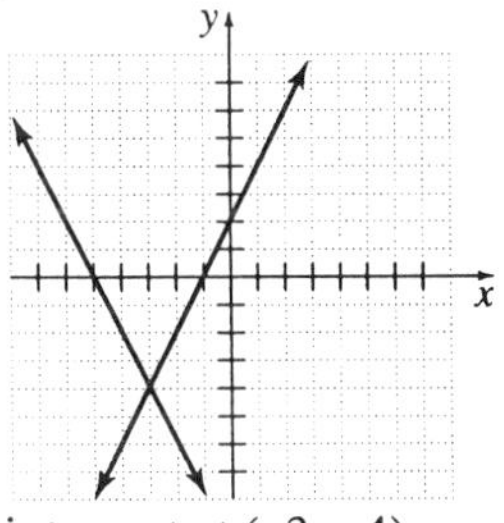

intersect at (–3, –4)

28. $y = x^2$ and $y = 3x$

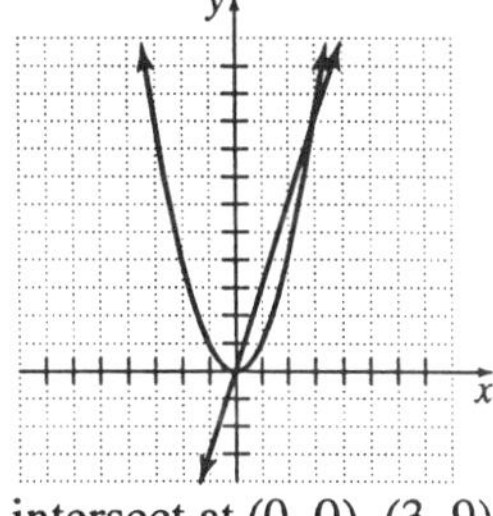

intersect at (0, 0), (3, 9)

29. $f(x) = |2x|$ and $g(x) = x + 3$

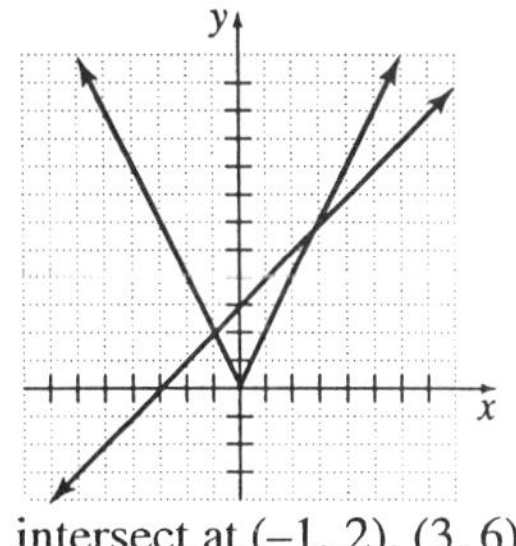

intersect at (–1, 2), (3, 6)

30. domain {2, 4, 6, 8, 10}
range {1, 2, 3, 4, 5}

31. domain {..., –6, –4, –2, 0, 2, 4, 6, ...}
range {3}

32. domain {–5, –4, –3, –2, –1}
rangc {25, 16, 9, 4, 1}

33. domain all real numbers
range all real numbers ≥ -1.5

34. $x - 8 \geq 0$
$x \geq 8$
domain all real numbers ≥ 8
range all real numbers ≥ 0

35. $x^2 \neq 0$
$x \neq 0$
domain all real numbers $\neq 0$
range all real numbers > 0

36. domain all real numbers
range all real numbers

37. $6-2x\geq 0$
$6\geq 2x$
$x\leq 3$
domain all real numbers ≤ 3
range all real numbers ≥ 0

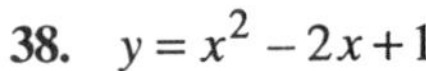
38. $y=x^2-2x+1$

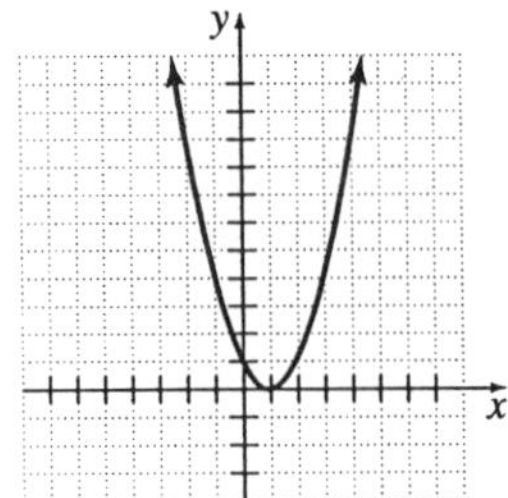

domain all real numbers
range all real numbers ≥ 0

39. $y=-x^2+3$

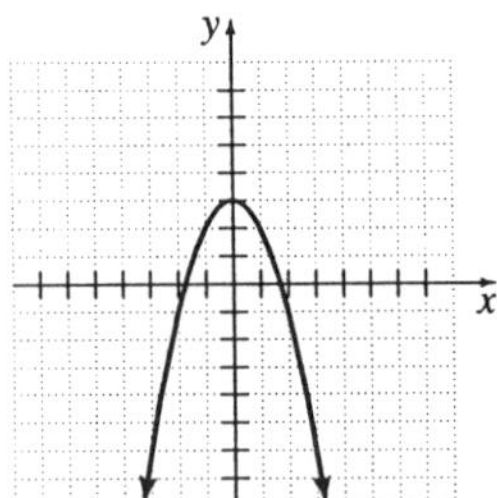

domain all real numbers
range all real numbers ≤ 3

40.

x	$y=12-8x$	y
–6	$y=12-8(-6)$ $y=60$	60
–3	$y=12-8(-3)$ $y=36$	36
0	$y=12-8(0)$ $y=12$	12
3	$y=12-8(3)$ $y=-12$	–12
6	$y=12-8(6)$ $y=-36$	–36

ordered pairs: (–6, 60), (–3, 36), (0, 12), (3, –12), (6, –36)

41.

x	$y=\sqrt{10-3x}$	y
3	$y=\sqrt{10-3\cdot 3}$ $y=1$	1
2	$y=\sqrt{10-3\cdot 2}$ $y=2$	2
1	$y=\sqrt{10-3\cdot 1}$ $y=\sqrt{7}$	$\sqrt{7}$
0	$y=\sqrt{10-3\cdot 0}$ $y=\sqrt{10}$	$\sqrt{10}$
–1	$y=\sqrt{10-3(-1)}$ $y=\sqrt{13}$	$\sqrt{13}$
–2	$y=\sqrt{10-3(-2)}$ $y=4$	4

ordered pairs: (3, 1), (2, 2), $\left(1, \sqrt{7}\right)$, $\left(0, \sqrt{10}\right)$, $\left(-1, \sqrt{13}\right)$, (–2, 4)

42.

s	$t=\frac{4}{7}s+5$	t
–7	$t=\frac{4}{7}(-7)+5$ $t=1$	1
0	$t=\frac{4}{7}(0)+5$ $t=5$	5
7	$t=\frac{4}{7}(7)+5$ $t=9$	9
14	$t=\frac{4}{7}(14)+5$ $t=13$	13
21	$t=\frac{4}{7}(21)+5$ $t=17$	17

ordered pairs: (–7, 1), (0, 5), (7, 9), (14, 13), (21, 17)

43.

x	$y=(3x-5)(2x+1)$	y
–2	$y=[3(-2)-5][2(-2)+1]$ $y=33$	33
0	$y=(3\cdot 0-5)(2\cdot 0+1)$ $y=-5$	–5
2	$y=(3\cdot 2-5)(2\cdot 2+1)$ $y=5$	5

44.

x	$y=\frac{3}{4}x-5$	y
–8	$y=\frac{3}{4}(-8)-5$ $y=-11$	–11
–4	$y=\frac{3}{4}(-4)-5$ $y=-8$	–8
0	$y=\frac{3}{4}\cdot 0-5$ $y=-5$	–5
4	$y=\frac{3}{4}\cdot 4-5$ $y=-2$	–2
8	$y=\frac{3}{4}\cdot 8-5$ $y=1$	1

45.

x	$y=15.8-4.7x$	y
–2	$y=15.8-4.7(-2)$ $y=25.2$	25.2
–1	$y=15.8-4.7(-1)$ $y=20.5$	20.5
0	$y=15.8-4.7(0)$ $y=15.8$	15.8
1	$y=15.8-4.7(1)$ $y=11.1$	11.1
2	$y=15.8-4.7(2)$ $y=6.4$	6.4

46.

x	$y=4x^2-17x-15$	y
-2	$y=4(-2)^2-17(-2)-15$ $y=35$	35
$-\frac{3}{4}$	$y=4\left(-\frac{3}{4}\right)^2-17\left(-\frac{3}{4}\right)-15$ $y=0$	0
0	$y=4\cdot 0^2-17\cdot 0-15$ $y=-15$	-15
$\frac{3}{4}$	$y=4\left(\frac{3}{4}\right)^2-17\left(\frac{3}{4}\right)-15$ $y=-25.5$	-25.5
5	$y=4(5)^2-17(5)-15$ $y=0$	0

47.

x	$y=x^3+x^2+x+1$	y
-15	$y=(-15)^3+(-15)^2+(-15)+1$ $y=-3164$	-316[illegible]
-5	$y=(-5)^3+(-5)^2+(-5)+1$ $y=-104$	-104
0	$y=0^3+0^2+0+1$ $y=1$	1
5	$y=5^3+5^2+5+1$ $y=156$	156
15	$y=(15)^3+(15)^2+15+1$ $y=3616$	3616
25	$y=(25)^3+(25)^2+25+1$ $y=16,276$	16,276

48.

x	$y=\lvert 1-2x-3x^2\rvert$	y
-6	$y=\lvert 1-2(-6)-3(-6)^2\rvert$ $y=95$	95
-3	$y=\lvert 1-2(-3)-3(-3)^2\rvert$ $y=20$	20
0	$y=\lvert 1-2(0)-3(0)^2\rvert$ $y=1$	1
3	$y=\lvert 1-2\cdot 3-3(3)^2\rvert$ $y=32$	32
6	$y=\lvert 1-2(6)-3(6)^2\rvert$ $y=119$	119
9	$y=\lvert 1-2(9)-3(9)^2\rvert$ $y=260$	260

49.

x	$y = 4.6x^2 + 2.8x + 10.4$	y
–3.7	$y = 4.6(-3.7)^2 + 2.8(-3.7) + 10.4$ $y = 63.014$	63.014
–2.2	$y = 4.6(-2.2)^2 + 2.8(-2.2) + 10.4$ $y = 26.504$	26.504
–0.7	$y = 4.6(-0.7)^2 + 2.8(-0.7) + 10.4$ $y = 10.694$	10.694
0	$y = 4.6(0)^2 + 2.8(0) + 10.4$ $y = 10.4$	10.4
0.8	$y = 4.6(0.8)^2 + 2.8(0.8) + 10.4$ $y = 15.584$	15.584
2.3	$y = 4.6(2.3)^2 + 2.8(2.3) + 10.4$ $y = 41.174$	41.174
3.8	$y = 4.6(3.8)^2 + 2.8(3.8) + 10.4$ $y = 87.464$	87.464

50. Answers will vary. Possible answer:

x	$y = 17 - 5x$	y
–1	$y = 17 - 5(-1)$ $y = 22$	22
0	$y = 17 - 5(0)$ $y = 17$	17
1	$y = 17 - 5(1)$ $y = 12$	12

51. Answers will vary. Possible answer:

x	$y = 4.5x - 1.6$	y
–1	$y = 4.5(-1) - 1.6$ $y = -6.1$	–6.1
0	$y = 4.5(0) - 1.6$ $y = -1.6$	–1.6
1	$y = 4.5(1) - 1.6$ $y = 2.9$	2.9

52. Answers will vary. Possible answer:

x	$y=\frac{1}{4}x+3$	y
-4	$y=\frac{1}{4}(-4)+3$ $y=2$	2
0	$y=\frac{1}{4}(0)+3$ $y=3$	3
4	$y=\frac{1}{4}(4)+3$ $y=4$	4

53. Answers will vary. Possible answer:

x	$y=\lvert 3x-10\rvert$	y
-3	$y=\lvert 3(-3)-10\rvert$ $y=19$	19
0	$y=\lvert 3(0)-10\rvert$ $y=10$	10
3	$y=\lvert 3(3)-10\rvert$ $y=1$	1

54.

r	$C=2\pi r$	C
$\frac{1}{4}$	$C=2\pi\left(\frac{1}{4}\right)$ $C\approx 1.57$	1.57
$\frac{1}{2}$	$C=2\pi\left(\frac{1}{2}\right)$ $C\approx 3.14$	3.14
1	$C=2\pi(1)$ $C\approx 6.28$	6.28
$\frac{3}{2}$	$C=2\pi\left(\frac{3}{2}\right)$ $C\approx 9.42$	9.42
2	$C=2\pi(2)$ $C\approx 12.57$	12.57

ordered pairs: $\left(\frac{1}{4}, 1.5708\right)$, $\left(\frac{1}{2}, 3.1416\right)$, $(1, 6.2832)$, $\left(\frac{3}{2}, 9.4248\right)$, $(2, 12.566)$

55.

s	$A=s^2$	A
3	$A=3^2$ $A=9$	9
5	$A=5^2$ $A=25$	25

56.

H	$V=2H$	V
1	$V=2(1)$ $V=2$	2
1.25	$V=2(1.25)$ $V=2.5$	2.5
1.5	$V=2(1.5)$ $V=3$	3
1.75	$V=2(1.75)$ $V=3.5$	3.5
2	$V=2(2)$ $V=4$	4

57.

a	$b=180-a$	b
30	$b=180-30$ $b=150$	150
60	$b=180-60$ $b=120$	120
90	$b=180-90$ $b=90$	90
120	$b=180-120$ $b=60$	60
150	$b=180-150$ $b=30$	30

58.

t	$I=2000(0.06)t$ $I=120t$	I
2	$I=120(2)$ $I=240$	240
3	$I=120(3)$ $I=360$	360
4	$I=120(4)$ $I=480$	480

59.

C	$F=\frac{9}{5}C+32$	F
−10	$F=\frac{9}{5}(-10)+32$ $F=14$	14
−5	$F=\frac{9}{5}(-5)+32$ $F=23$	23
0	$F=\frac{9}{5}(0)+32$ $F=32$	32
5	$F=\frac{9}{5}(5)+32$ $F=41$	41
10	$F=\frac{9}{5}(10)+32$ $F=50$	50
15	$F=\frac{9}{5}(15)+32$ $F=59$	59
20	$F=\frac{9}{5}(20)+32$ $F=68$	68
25	$F=\frac{9}{5}(25)+32$ $F=77$	77

60. Let x = the number of items.
$f(x) = 2500 + 12x$
$f(1650) = 2500 + 12(1650)$
$f(1650) = 2500 + 19{,}800$
$f(1650) = 22{,}300$
The production run will cost \$22,300.

61. Let x = the number of hours of rental.
$f(x) = 15 + 2x$
$f(10) = 15 + 2(10)$
$f(10) = 15 + 20$
$f(10) = 35$
It will cost \$35 to rent the grinder for 10 hours.

62. Let x = the number of employees.
$f(x) = 275 + 9.5x$
$f(135) = 275 + 9.5(135)$
$f(135) = 275 + 1282.5$
$f(135) = 1557.5$
A luncheon for 135 employees will cost \$1557.50.

63. Let x = the number of admissions.
$f(x) = -185 + 4x$
$f(310) = -185 + 4(310)$
$f(310) = -185 + 1240$
$f(310) = 1055$
The net profit for the game will be \$1055.

64. $f(x) = 250 - 3.5x$
$Y1 = 250 - 3.5x$

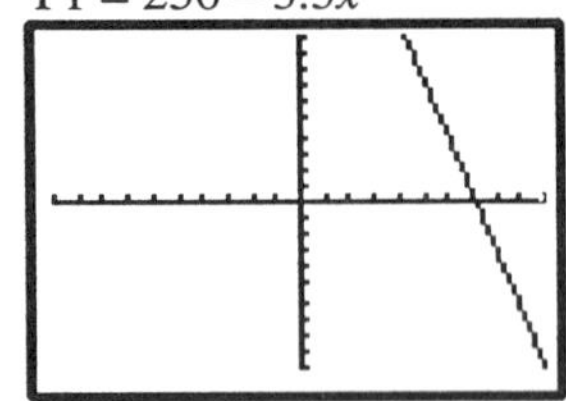

decreasing for $0 < x < 71.43$

65. $f(x) = 1000 + 50x$
$Y1 = 1000 + 50x$

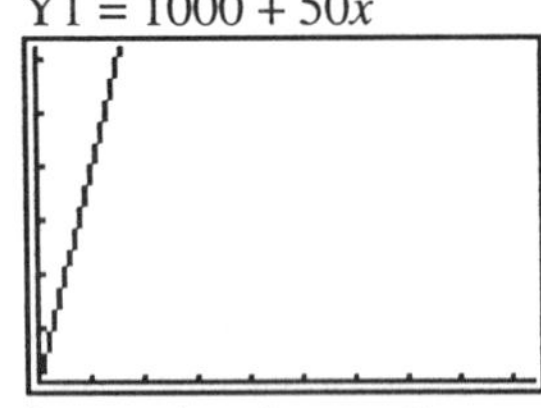

increasing for $x > 0$

66. $A = LW$
$f(x) = x(100 - x)$
$Y1 = 100x - x^2$

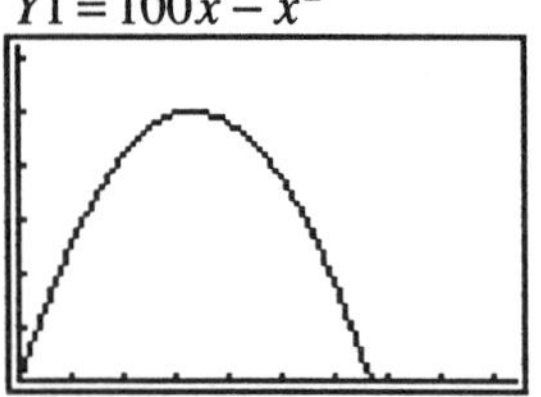

increasing for $0 < x < 50$
decreasing for $50 < x < 100$

67. **a.** $f(x) = 25{,}000 + 5000x$
$g(x) = 6000x$

b. $Y1 = 25{,}000 + 5000x$
$Y2 = 6000x$

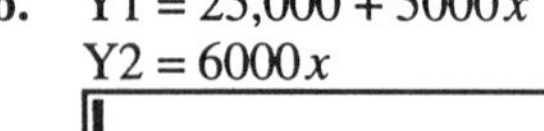

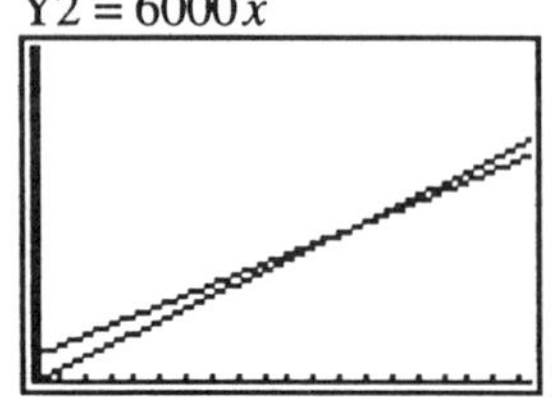

intersect at (25, 150,000)

c. At 25 years, the money received is the same at \$150,000.

68. **a.** $f(x) = 22x + 600$
$g(x) = 75x$

b. $Y1 = 22x + 600$
$Y2 = 75x$

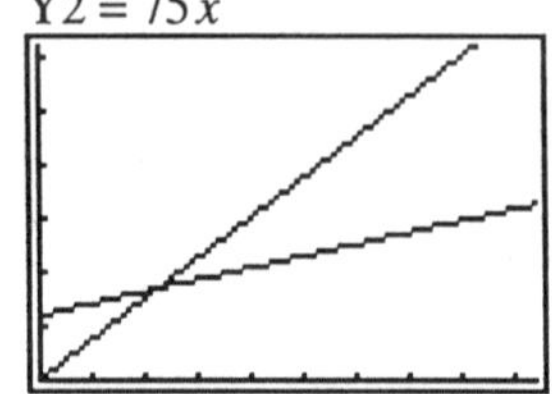

intersect at approximately (11.32, 849.06)

c. At about 11.32 appliances the total acquisition cost and total revenue are equal at about \$849.06. This means 12 appliances must be sold to break even.

Chapter 4 Test

1.

x	$y=2x^2+17x-9$	y
–9	$y=2(-9)^2+17(-9)-9$ $y=0$	0
0	$y=2(0)^2+17(0)-9$ $y=-9$	–9
3	$y=2(3)^2+17(3)-9$ $y=60$	60

2. $y=|2x|$

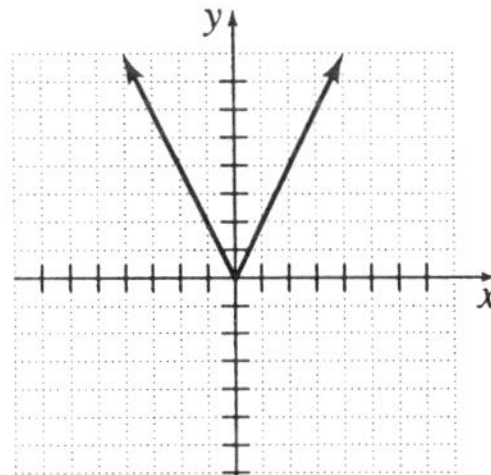

3. domain all real numbers

4. range all real numbers ≥ 0

5. Yes; all possible vertical lines cross the graph a maximum of one time.

6. increasing for $x>0$

7. decreasing for $x<0$

8. none

9. relative minimum is 0

10. x-intercept at (0, 0)

11. y-intercept at (0, 0)

12. $A(1, 2)$, $B(-2, -4)$, $C(-5, 3)$, $D(2, -5)$, $E(0, -2)$

13. quadrant IV

14. $f(x)=\frac{1}{2}x+6$
$f(4)=\frac{1}{2}(4)+6$
$f(4)=2+6$
$f(4)=8$

15. $f(x)=\frac{1}{2}x+6$
$f(-6)=\frac{1}{2}(-6)+6$
$f(-6)=-3+6$
$f(-6)=3$

16. $f(x)=\frac{1}{2}x+6$
$f(2a-2)=\frac{1}{2}(2a-2)+6$
$f(2a-2)=a-1+6$
$f(2a-2)=a+5$

17. $f(x)=\frac{1}{2}x+6$
$f(-b)=\frac{1}{2}(-b)+6$
$f(-b)=-\frac{1}{2}b+6$

18. $f(x)=450+21.5x$

19. $f(250)=450+21.5(250)$
$f(250)=450+5375$
$f(250)=5825$
The cost is $5825.

20. $y=3x-10$ and $y=-x-2$

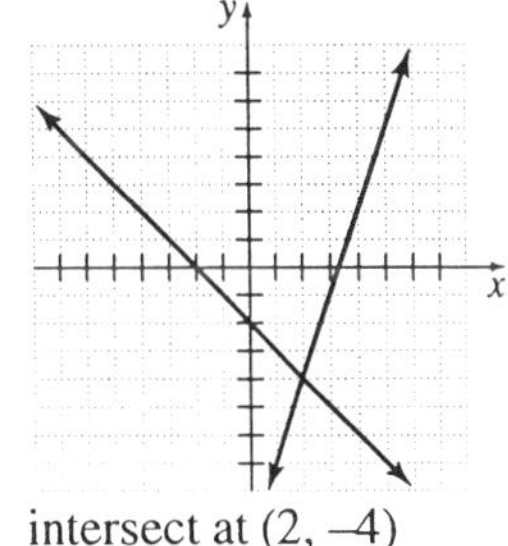

intersect at (2, –4)

21. Answers will vary. Possible answer: Each point in the plane corresponds to an ordered pair. The first number denotes the distance and direction along the x-axis. The second number denotes the distance and direction along the y-axis.

Chapters 1–4 Cumulative Review

1. 12 is the only whole number.

2. 0 and 12 are integers.

3. $-\frac{2}{3}$, 0, 12, $1\frac{4}{5}$, and –0.33 are rational numbers.

4. $\sqrt{7}$ is an irrational number.

5. $\frac{3}{8} > \frac{1}{3}$

6. $\frac{2}{3} > 0.66$

7. $-2.8 < -1.6$

8.

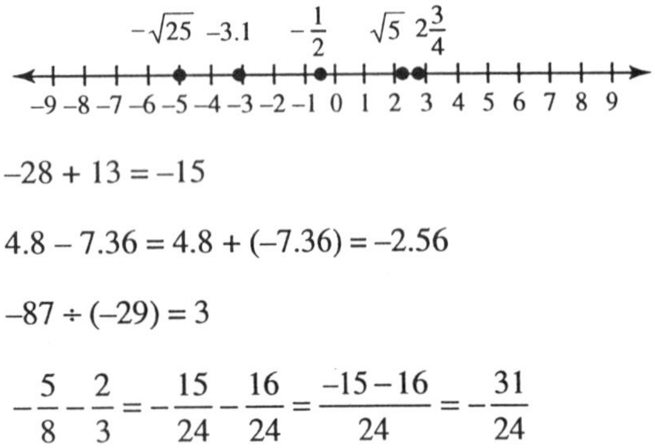

9. $-28 + 13 = -15$

10. $4.8 - 7.36 = 4.8 + (-7.36) = -2.56$

11. $-87 \div (-29) = 3$

12. $-\frac{5}{8} - \frac{2}{3} = -\frac{15}{24} - \frac{16}{24} = \frac{-15-16}{24} = -\frac{31}{24}$

13. $-2\frac{3}{4} \div 1\frac{3}{7} = -\frac{11}{4} \div \frac{10}{7} = -\frac{11}{4} \cdot \frac{7}{10}$
$= \frac{-11 \cdot 7}{4 \cdot 10} = -\frac{77}{40} = -1\frac{37}{40}$

14. $\left(-\frac{2}{3}\right)\left(\frac{3}{8}\right)\left(-\frac{7}{16}\right)\left(\frac{9}{10}\right) = \frac{(-2)(3)(-7)(9)}{(3)(8)(16)(10)}$
$= \frac{(2)(7)(9)}{(8)(16)(10)} = \frac{(7)(9)}{(4)(16)(10)} = \frac{63}{640}$

15. $(12.96)(-4.8) = -62.208$

16. $(14)(0)(5)(-6) = 0$

17. $(-12)(16) \div 4(-2)$
$= (-192) \div 4(-2)$
$= (-48)(-2)$
$= 96$

18. $14 + (-7) + 22 - 16 - (-18)$
$= 7 + 22 - 16 - (-18)$
$= 29 - 16 - (-18)$
$= 13 - (-18)$
$= 13 + 18 = 31$

19. $-[3.8 - (-2.4)] = -[3.8 + 2.4] = -[6.2] = -6.2$

20. $\frac{2(3^2 + 7) - 2^5}{3 \cdot 18} = \frac{2(9+7) - 32}{54}$
$= \frac{2(16) - 32}{54} = \frac{32 - 32}{54} = \frac{0}{54} = 0$

21. $-|12 - 20| = -|-8| = -(8) = -8$

22. $\sqrt{\frac{16}{25}} = \frac{\sqrt{16}}{\sqrt{25}} = \frac{4}{5}$

23. $-\sqrt{1.2} = -\sqrt{4(0.3)} = -2\sqrt{0.3} \approx -1.095$

24. $\sqrt{-16}$ is not a real number.

25. $\sqrt[3]{1\frac{13}{81}} = \sqrt[3]{\frac{94}{81}} = \sqrt[3]{\frac{94}{27 \cdot 3}} = \frac{1}{3}\sqrt[3]{\frac{94}{3}}$
$= \frac{1}{3}\sqrt[3]{\frac{94 \cdot 3^2}{3 \cdot 3^2}} = \frac{1}{3}\left(\frac{1}{3}\right)\sqrt[3]{94 \cdot 9}$
$= \frac{\sqrt[3]{846}}{9} \approx 1.051$

26. $(14)^0 = 1$

27. $1^{12} = 1$

28. 0^0 = indeterminate

29. $-8^4 = -4096$

30. $(-8)^4 = 4096$

31. $\left(\frac{3}{4}\right)^{-2} = \left(\frac{4}{3}\right)^2 = \frac{4^2}{3^2} = \frac{16}{9}$

32. $0.00000305 = 3.05 \times 10^{-6}$

33. $-4,235,600 = -4.2356 \times 10^6$

34. 3.56 E–2 = 0.0356

35. 6.78 E8 = 678,000,000

36. $\sqrt[4]{3125} = 3125^{1/4} \approx 7.4767$

37. $\sqrt{-3^2 + 5(3) - 2} = \sqrt{-9 + 15 - 2} = \sqrt{8}$
$= \sqrt{4(2)} = 2\sqrt{2}$

38. **a.** There are six terms in the expression.

b. The variable terms are a^3, $-2a^2$, a, $-2a^3$, and $7a$.

c. The only constant term is –5.

39. $-(3y + 2z) + (4y - 2z) - (-3y - 5z)$
$= (-3y - 2z) + (4y - 2z) + (3y + 5z)$
$= -3y + 4y + 3y - 2z - 2z + 5z$
$= 4y + z$

40. $\frac{3}{4}x + \frac{5}{8}y - \frac{1}{16} - \frac{3}{4}x + \frac{1}{8}y - \frac{5}{6}$
$= \left(\frac{3}{4} - \frac{3}{4}\right)x + \left(\frac{5}{8} + \frac{1}{8}\right)y - \frac{1}{16} - \frac{5}{6}$
$= 0x + \frac{6}{8}y - \frac{3}{48} - \frac{40}{48}$
$= \frac{3}{4}y - \frac{43}{48}$

41. $-(-2)^2 + 3(-2) + 8 \stackrel{?}{=} -3(-2)$
$-4 + (-6) + 8 \stackrel{?}{=} 6$
$-10 + 8 \stackrel{?}{=} 6$
$-2 \neq 6$
$x = -2$ is not a solution of the equation.

42.

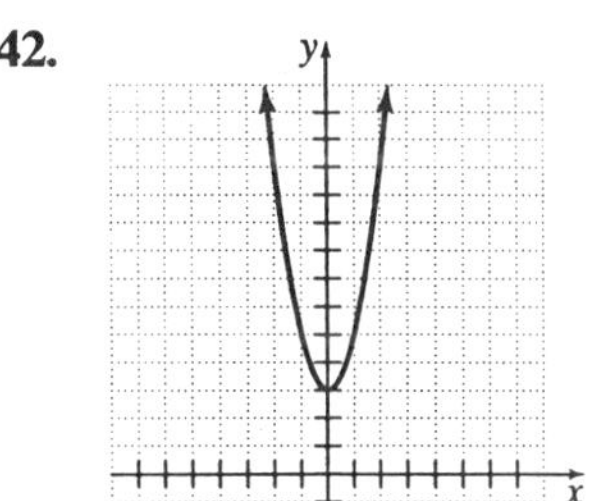

43. The domain of the function is all real numbers. The range of the function is all $y \geq 3$.

44. Since no vertical line will cross the graph more than once, the relation is a function.

45. The relation is increasing for all $x > 0$.

46. The relative minimum value of the relation is $y = 3$, which occurs when $x = 0$.

47. $f(x) = \frac{1}{3}x - 5$

a. $f(9) = \frac{1}{3}(9) - 5$
$f(9) = 3 - 5$
$f(9) = -2$

b. $f(a+h) = \frac{1}{3}(a+h) - 5$
$f(a+h) = \frac{1}{3}a + \frac{1}{3}h - 5$

48. Volume is the product of length, width, and height.
(3.5)(2.25)(1.75) = 13.78125
The volume is 13.78125 ft^3.

49. $A = P(1+r)^t$
$A = 500(1+0.055)^4$
$A = 500(1.055)^4$
$A \approx 619.41$
The interest is the amount by which the compound amount exceed the original balance.
$619.41 - 500 = 119.41$
Kelsie's interest is $119.41.

50. The cost for one production run is the sum of the setup cost and the cost per ornament.
$c(x) = 35 + 2.80x$
$c(150) = 35 + 2.80(150)$
$c(150) = 35 + 420$
$c(150) = 455$
The cost of producing 150 ornaments in one production run is $455.

Chapter 5

5.1 Experiencing Algebra the Exercise Way

1. $6x - 55 = x + 72$ is linear because it is in the form $ax + b = ax + b$, where $a \neq 0$ in at least one of the expressions and the coefficient of a is not the same in each expression.

3. $4x^2 + 5 = 2x - 6$ is nonlinear because x has an exponent of 2.

5. $\frac{7}{9}z - \frac{2}{3} = 0$ is linear because it is in the form $ax + b = 0$.

7. $\sqrt[3]{4x+16} = 27$ is nonlinear because the radical expression has a variable in the radicand.

9. $3(2x - 5) = x + 3(x - 9)$ is linear because it simplifies to $6x - 15 = 4x - 27$.

11. A sample table is shown below.

x	$2x - 7$	$35 - x$	
13	19	22	$19 < 22$
14	21	21	$21 = 21$
15	23	20	$23 > 20$

The solution is 14.

13. A sample table is shown below.

x	$3(2x + 11)$	$3(5 + x)$	
–5	3	0	$3 > 0$
–6	–3	–3	$-3 = -3$
–7	–9	–6	$-9 < -6$

The solution is –6.

15. A sample table is shown below.

a	$6.8a + 4.3$	$2.6a + 33.7$	
6	45.1	49.3	$45.1 < 49.3$
7	51.9	51.9	$51.9 = 51.9$
8	58.7	54.5	$58.7 > 54.5$

The solution is 7.

17. A sample table is shown below.

x	$7(x + 10) + 15$	$6(x + 15) + (x - 5)$	
0	85	85	$85 = 85$
1	92	92	$92 = 92$
2	99	99	$99 = 99$

The expressions are always equal. The equation is an identity. The solution is all real numbers.

19. A sample table is shown below.

a	$(a-4)-(a+4)$	$(a+3)-(a-2)$	
1	–8	5	$-8 \neq 5$
2	–8	5	$-8 \neq 5$
3	–8	5	$-8 \neq 5$

The expressions always equal the same value. The equation is a contradiction. There is no solution.

21. A sample table is shown below.

z	$3.5(z-1)$	$7(0.5z+0.6)+2$	
0	–3.5	6.2	$6.2-(-3.5)=9.7$
1	0	9.7	$9.7-0=9.7$
2	3.5	13.2	$13.2-3.5=9.7$

The values of the left and right expressions are never equal and their differences remain the same. The equation is a contradiction. There is no solution.

23. A sample table is shown below.

x	$\frac{4}{5}(x-1)$	$6\left(\frac{1}{15}x-\frac{1}{10}\right)$	
0	$-\frac{4}{5}$	$-\frac{3}{5}$	$-\frac{4}{5}<-\frac{3}{5}$
1	0	$-\frac{1}{5}$	$0>-\frac{1}{5}$

The solution is a noninteger between 0 and 1.

25. $Y1 = x+6$
$Y2 = 9+2x$

31

–47 (–3, 3) 47

–31

The intersection is (–3, 3).
The solution is –3.

27. $Y1 = (x+4)+(x+2)$
$Y2 = (x-1)+(x-3)$

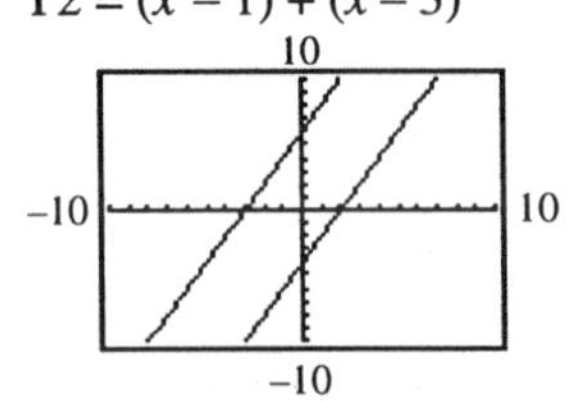

The lines are parallel.
There is no solution.

29. $Y1 = 2(x + 3$
$Y2 = 3(x - 1) - (x - 9)$

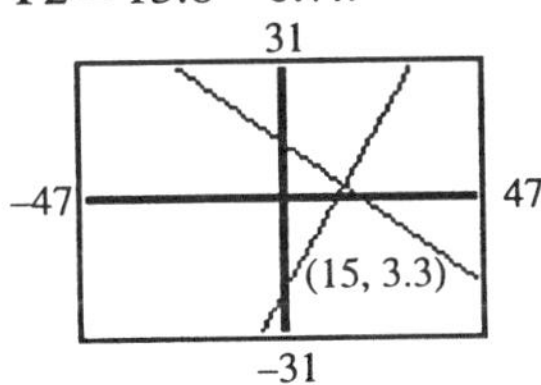

The lines are the same.
The solution is all real numbers.

31. $Y1 = 1.7x - 22.2$
$Y2 = 13.8 - 0.7x$

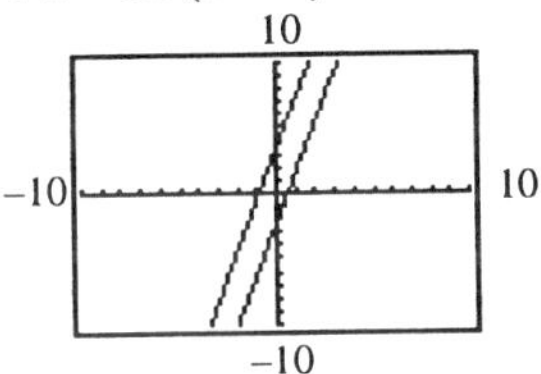

The intersection is (15, 3.3).
The solution is 15.

33. $Y1 = 2.2(x - 1) + 1.7x$
$Y2 = 3.5(x + 1) + 0.4x$

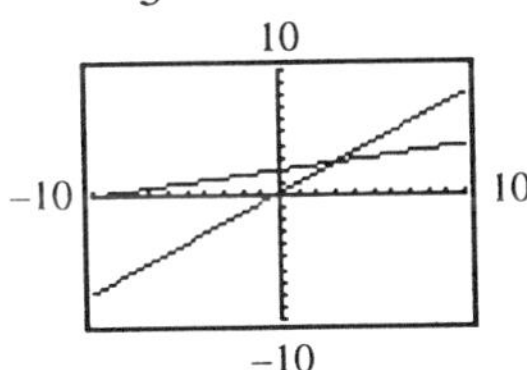

The lines are parallel.
There is no solution.

35. $Y1 = \frac{4}{5}x + \frac{1}{5}$

$Y2 = \frac{1}{5}x + 2$

10
−10
10
−10

The intersection is (3, 2.6).
The solution is 3.

37. $Y1 = \frac{2}{3}(x + 1) - \frac{1}{3}$

$Y2 = \frac{1}{3}(x + 1) + \frac{1}{3}x$

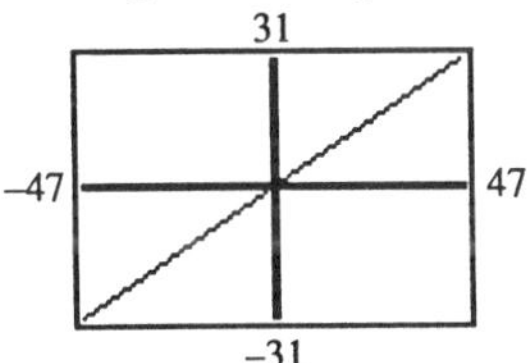

The lines are the same.
The solution is all real numbers.

39. Let x = number of miles
$49.95 = 29.95 + 0.25x$

x	49.95	$29.95 + 0.25x$	
79	49.95	49.7	$49.95 > 49.7$
80	49.95	49.95	$49.95 = 49.95$
81	49.95	50.2	$49.95 < 50.2$

The number of miles is 80.

41. Let x = number of pairs of shoes
$280 + 8x = 22x$
$Y1 = 280 + 8x$
$Y2 = 22x$

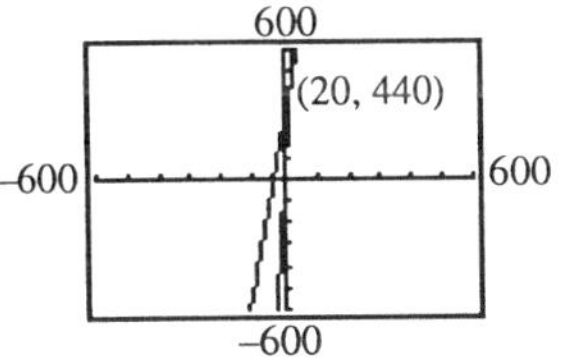

The intersection is (20, 440).
The factory should produce 20 pairs of shoes.

43. Let x = day 5 expenses

$$\frac{28+19+22+27+x}{5}=25$$

x	25	$\frac{96+x}{5}$	
28	25	24.8	25 > 24.8
29	25	25	25 = 25
30	25	25.2	25 < 25.2

She can spend $29.

45. Let x = width,
then $x + 3$ = length
$2x + 2(x + 3) = 26$
$Y1 = 2x + 2(x + 3)$
$Y2 = 26$

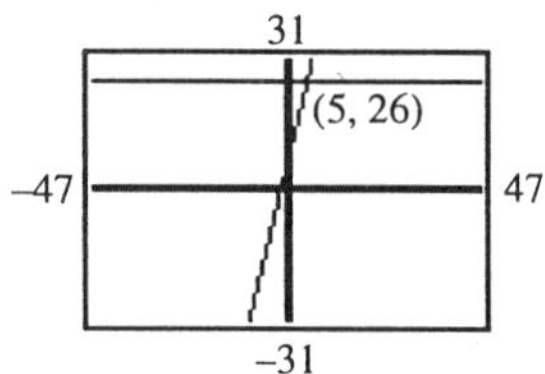

The intersection is (5, 26) so $x = 5$, $x + 3 = 8$.
The dimensions are 5 ft by 8 ft.

47. Let x = number of rolls
$25 + 9x + 5x = 25 + 14x$

x	$25+9x+5x$	$25+14x$	
1	39	39	39 = 39
2	53	53	53 = 53
3	67	67	67 = 67

The values are always equal.
Any number of rolls will have the same charge for both.

5.1 Experiencing Algebra the Calculator Way

A.

1. Graphs will vary.
The intersection is (22, 64).
The solution is 22.

2. Graphs will vary.
The intersection is (–12, –54.6).
The solution is –12.

3. Graphs will vary.
The intersection is (250, 2200).
The solution is 250.

4. Graphs will vary.
The intersection is (–500, –5200).
The solution is –500

B. Students should experiment with some exercises.

5.2 Experiencing Algebra the Exercise Way

1. $x + 33 = 51$
$x + 33 - 33 = 51 - 33$
$x = 18$
The solution is 18.

3. $75 = a - 41$
$75 + 41 = a - 41 + 41$
$116 = a$
The solution is 116.

5. $-4.91 = y + 3.07$
$-4.91 - 3.07 = y + 3.07 - 3.07$
$-7.98 = y$
The solution is –7.98.

7. $a - \frac{13}{18} = \frac{5}{6}$

$a - \frac{13}{18} + \frac{13}{18} = \frac{15}{18} + \frac{13}{18}$

$a = \frac{28}{18} = \frac{14}{9}$

The solution is $\frac{14}{9}$.

9. $27 + (x - 13) = 11$
$27 + x - 13 = 11$
$14 + x = 11$
$14 + x - 14 = 11 - 14$
$x = -3$
The solution is –3.

11. $(13.9 + x) + 0.88 = -2.07$
$13.9 + x + 0.88 = -2.07$
$x + 14.78 = -2.07$
$x + 14.78 - 14.78 = -2.07 - 14.78$
$x = -16.85$
The solution is -16.85.

13. $\left(x - \frac{3}{10}\right) - \frac{2}{5} = -3\frac{1}{2}$
$x - \frac{3}{10} - \frac{4}{10} = -\frac{7}{2}$
$x - \frac{7}{10} = -\frac{35}{10}$
$x - \frac{7}{10} + \frac{7}{10} = -\frac{35}{10} + \frac{7}{10}$
$x = -\frac{28}{10} = -\frac{14}{5}$
The solution is $-\frac{14}{5}$.

15. $(5x - 2) - (4x + 7) = 27$
$5x - 2 - 4x - 7 = 27$
$x - 9 = 27$
$x - 9 + 9 = 27 + 9$
$x = 36$
The solution is 36.

17. $\left(\frac{1}{3}x + \frac{1}{8}\right) + \left(\frac{3}{4} + \frac{2}{3}x\right) = -\frac{3}{16}$
$\frac{1}{3}x + \frac{1}{8} + \frac{6}{8} + \frac{2}{3}x = -\frac{3}{16}$
$\frac{3}{3}x + \frac{7}{8} = -\frac{3}{16}$
$x + \frac{14}{16} - \frac{14}{16} = -\frac{3}{16} - \frac{14}{16}$
$x = -\frac{17}{16}$
The solution is $-\frac{17}{16}$.

19. $(3x + 76) - (2x - 45) = 31$
$3x + 76 - 2x + 45 = 31$
$x + 121 = 31$
$x + 121 - 121 = 31 - 121$
$x = -90$
The solution is -90.

21. $-324 = -4y$
$\frac{-324}{-4} = \frac{-4y}{-4}$
$81 = y$
The solution is 81.

23. $-5.1x = 0.102$
$\frac{-5.1x}{-5.1} = \frac{0.102}{-5.1}$
$x = -0.02$
The solution is -0.02.

25. $-3\frac{1}{3}x = -1\frac{1}{3}$
$-\frac{10}{3}x = -\frac{4}{3}$
$-\frac{3}{10}\left(-\frac{10}{3}x\right) = -\frac{3}{10}\left(-\frac{4}{3}\right)$
$x = \frac{2}{5}$
The solution is $\frac{2}{5}$.

27. $\frac{x}{4} = 1.22$
$4\left(\frac{x}{4}\right) = 4(1.22)$
$x = 4.88$
The solution is 4.88.

29. $-x = 57$
$-1(-x) = -1(57)$
$x = -57$
The solution is -57.

31. $57 = 2x + 17x$
$57 = 19x$
$\frac{57}{19} = \frac{19x}{19}$
$3 = x$
The solution is 3.

33. $18.22x - 12.9x = -12.76$
$5.32x = -12.76$
$\frac{5.32x}{5.32} = \frac{-12.76}{5.32}$
$x \approx -2.398$
The solution is about –2.398.

35. $\frac{5}{14} = \frac{9}{14}a + \frac{3}{7}a$
$\frac{5}{14} = \frac{9}{14}a + \frac{6}{14}a$
$\frac{5}{14} = \frac{15}{14}a$
$\frac{14}{15}\left(\frac{5}{14}\right) = \frac{14}{15}\left(\frac{15}{14}a\right)$
$\frac{1}{3} = a$
The solution is $\frac{1}{3}$.

37. $2(3x + 6) + 3(x - 4) = 126$
$6x + 12 + 3x - 12 = 126$
$9x = 126$
$\frac{9x}{9} = \frac{126}{9}$
$x = 14$
The solution is 14.

39. $2.2(x + 3.7) + 7.4(x - 1.1) = 60.48$
$2.2x + 8.14 + 7.4x - 8.14 = 60.48$
$9.6x = 60.48$
$\frac{9.6x}{9.6} = \frac{60.48}{9.6}$
$x = 6.3$
The solution is 6.3.

41. $3\left(\frac{1}{2}x - \frac{3}{4}\right) - 18\left(x - \frac{1}{8}\right) = 0$
$\frac{3}{2}x - \frac{9}{4} - 18x + \frac{18}{8} = 0$
$\frac{3}{2}x - \frac{36}{2}x - \frac{9}{4} + \frac{9}{4} = 0$
$-\frac{33}{2}x = 0$
$-\frac{2}{33}\left(-\frac{33}{2}x\right) = -\frac{2}{33}(0)$
$x = 0$
The solution is 0.

43. Let x = number of remaining servings
$4 + x = 10$
$4 + x - 4 = 10 - 4$
$x = 6$
6 servings remain in the box.

45. Let x = gross pay
$x - 567.32 = 1784.26$
$x - 567.32 + 567.32 = 1784.26 + 567.32$
$x = 2351.58$
His gross pay was $2351.58.

47. Let x = amount to borrow
$3\frac{3}{4} + x = 5\frac{1}{2}$
$\frac{15}{4} + x = \frac{11}{2}$
$\frac{15}{4} + x - \frac{15}{4} = \frac{22}{4} - \frac{15}{4}$
$x = \frac{7}{4} = 1\frac{3}{4}$
She should borrow $1\frac{3}{4}$ cups.

49. Let x = sales tax
$49.95 + x = 54.32$
$49.95 + x - 49.95 = 54.32 - 49.95$
$x = 4.37$
There was $4.37 sales tax.

51. Let x = number of feet of additional wallpaper
$35 + x = 2(18 + 22)$
$35 + x = 2(40)$
$35 + x = 80$
$35 + x - 35 = 80 - 35$
$x = 45$
She must buy 45 feet.
$2(20) = 40$ feet
$45 - 40 = 5$
No, she will not have enough.

53. Let x = total number of paid admissions
$0.55x = 264$
$\frac{0.55x}{0.55} = \frac{264}{0.55}$
$x = 480$
There were 480 paid admissions.

55. Let x = number of packets they must sell
$2.5x = 1450$
$\frac{2.5x}{2.5} = \frac{1450}{2.5}$
$x = 580$
They must sell 580 packets.

57. Let x = amount of estate
$\frac{3}{5}x = 45,240$
$\frac{5}{3}\left(\frac{3}{5}x\right) = \frac{5}{3}(45,240)$
$x = 75,400$
The estate was worth \$75,400.

59. Let h = height
$20h = 350$
$\frac{20h}{20} = \frac{350}{20}$
$h = 17.5$
The height is 17.5 feet.

61. Let h = height
$\text{Volume} = \pi r^2 h$
$300 = \pi(4)^2 h$
$300 = 16\pi h$
$\frac{300}{16\pi} = \frac{16\pi h}{16\pi}$
$5.97 \approx h$
The height must be about 6 feet.

63. Let P = amount of principal
$864 = P(0.045)3$
$864 = 0.135P$
$\frac{864}{0.135} = \frac{0.135P}{0.135}$
$6400 = P$
You must place \$6400 into savings.

65. Let x = amount of quarterly profits
$\frac{x}{5} = 12,730$
$5\left(\frac{x}{5}\right) = 5(12,730)$
$x = 63,650$
The quarterly profits were \$63,650.

67. Let x = amount of sales
$0.03x = 40.5$
$\frac{0.03x}{0.03} = \frac{40.5}{0.03}$
$x = 1350$
Her sales were \$1350.

5.2 Experiencing Algebra the Calculator Way

1. Let x = linear feet of fencing
$x + 2(5) = 2\pi(75)$
$x + 10 = 150\pi$
$x + 10 - 10 = 150\pi - 10$
$x \approx 461.2$
It will require about 461.2 feet.

2. Let A = total amount paid
$A = 8000(1 + 0.04)^5$
$A \approx 9733.22$
Let I = interest
$I = A - 8000$
$I = 1733.22$
She will pay \$1733.22.

3. Let d = difference
$d + 25.8 = 25.8\pi$
$d + 25.8 - 25.8 = 25.8\pi - 25.8$
$d \approx 55.3$
The difference is about 55.3 inches.

4. Let P = amount of principal
$7325 = Pe^{3(0.045)}$
$\frac{7325}{e^{0.135}} = \frac{Pe^{0.135}}{e^{0.135}}$
$6400 \approx P$
About \$6400 was invested.

5. Let x = amount of increase per share

$250x = 531\frac{1}{4}$

$\frac{250x}{250} = \frac{531.25}{250}$

$x = 2.125 = 2\frac{1}{8}$

The increase was $2\frac{1}{8}$ per share.

6. Let x = number of pieces

$5\frac{3}{8}x = 34\frac{1}{2}$

$\frac{43}{8}x = \frac{69}{2}$

$\frac{8}{43}\left(\frac{43}{8}x\right) = \frac{8}{43}\left(\frac{69}{2}\right)$

$x = \frac{276}{43} \approx 6.4$

You can cut 6 pieces.

7. Let x = gallons of fuel

$18.5x = 220$

$\frac{18.5x}{18.5} = \frac{220}{18.5}$

$x \approx 11.89$

The car will need about 12 gallons.

8. Let x = number of square feet of wall

$6.5x = 800$

$\frac{6.5x}{6.5} = \frac{800}{6.5}$

$x \approx 123.1$

About 123 square feet of wall can be constructed.

9. Let x = daily recommended amount of fat

$0.04x = 2.5$

$\frac{0.04x}{0.04} = \frac{2.5}{0.04}$

$x = 62.5$

The recommended amount of fat is 62.5 grams.

10. Let x = daily recommended amount of fat

$0.05x = 3$

$\frac{0.05x}{0.05} = \frac{3}{0.05}$

$x = 60$

The recommended amount is 60 grams. This does not agree with problem 9. The difference may be round-off error.

11. Let r = speed

$250 = r(15)$

$\frac{250}{15} = \frac{15r}{15}$

$16.7 \approx r$

The speed was about 16.7 feet per second.

5.3 Experiencing Algebra the Exercise Way

1. $4x + 8 = 0$

$4x + 8 - 8 = 0 - 8$

$4x = -8$

$\frac{4x}{4} = \frac{-8}{4}$

$x = -2$

The solution is –2.

3. $-3x + 7 = 7$

$-3x + 7 - 7 = 7 - 7$

$-3x = 0$

$\frac{-3x}{-3} = \frac{0}{-3}$

$x = 0$

The solution is 0.

5. $15.17 = 5.9x - 4.3$

$15.17 + 4.3 = 5.9x - 4.3 + 4.3$

$19.47 = 5.9x$

$\frac{19.47}{5.9} = \frac{5.9x}{5.9}$

$3.3 = x$

The solution is 3.3.

7. $6.1 = -0.55a + 6.1$

$6.1 - 6.1 = -0.55a + 6.1 - 6.1$

$0 = -0.55a$

$\frac{0}{-0.55} = \frac{-0.55a}{-0.55}$

$0 = a$

The solution is 0.

9. $-9.2x - 4.3 = -70.6$
$-9.2x - 4.3 + 4.3 = -70.6 + 4.3$
$-9.2x = -66.3$
$\frac{-9.2x}{-9.2} = \frac{-66.3}{-9.2}$
$x \approx 7.21$
The solution is about 7.21.

11. $-\frac{5}{9}b + \frac{11}{12} = \frac{23}{36}$
$36\left(-\frac{5}{9}b\right) + 36\left(\frac{11}{12}\right) = 36\left(\frac{23}{36}\right)$
$-20b + 33 = 23$
$-20b + 33 - 33 = 23 - 33$
$-20b = -10$
$\frac{-20b}{-20} = \frac{-10}{-20}$
$b = \frac{1}{2}$
The solution is $\frac{1}{2}$.

13. $-2\frac{2}{3}z - 3\frac{1}{2} = -8\frac{5}{6}$
$-\frac{8}{3}z - \frac{7}{2} = -\frac{53}{6}$
$6\left(-\frac{8}{3}z\right) - 6\left(\frac{7}{2}\right) = 6\left(-\frac{53}{6}\right)$
$-16z - 21 = -53$
$-16z - 21 + 21 = -53 + 21$
$-16z = -32$
$\frac{-16z}{-16} = \frac{-32}{-16}$
$z = 2$
The solution is 2.

15. $5x + 6 = x + 126$
$5x + 6 - 6 = x + 126 - 6$
$5x = x + 120$
$5x - x = x + 120 - x$
$4x = 120$
$\frac{4x}{4} = \frac{120}{4}$
$x = 30$
The solution is 30.

17. $27x - 49 = -12x - 10$
$27x - 49 + 49 = -12x - 10 + 49$
$27x = -12x + 39$
$27x + 12x = -12x + 39 + 12x$
$39x = 39$
$\frac{39x}{39} = \frac{39}{39}$
$x = 1$
The solution is 1.

19. $156z - 210 = 47z + 662$
$156z - 210 + 210 = 47z + 662 + 210$
$156z = 47z + 872$
$156z - 47z = 47z + 872 - 47z$
$109z = 872$
$\frac{109z}{109} = \frac{872}{109}$
$z = 8$
The solution is 8.

21. $4x - (3x + 5) = x - 5$
$4x - 3x - 5 = x - 5$
$x - 5 = x - 5$
$x - 5 - x = x - 5 - x$
$-5 = -5$
This is a true equation for all real numbers.

23. $6x - (x + 1) = 5x + 7$
$6x - x - 1 = 5x + 7$
$5x - 1 = 5x + 7$
$5x - 1 - 5x = 5x + 7 - 5x$
$-1 = 7$
This is a false equation.
There is no solution.

25. $5(0.3x + 8.7) = 1.5x + 43.5$
$1.5x + 43.5 = 1.5x + 43.5$
$1.5x + 43.5 - 1.5x = 1.5x + 43.5 - 1.5x$
$43.5 = 43.5$
This is a true equation for all real numbers.

27. $5.5x = 1.2x + 3.3(x - 2)$
$5.5x = 1.2x + 3.3x - 6.6$
$5.5x = 4.5x - 6.6$
$5.5x - 4.5x = 4.5x - 6.6 - 4.5x$
$x = -6.6$
The solution is –6.6.

29. $\frac{3}{4}x+6=\frac{1}{2}x+\frac{1}{4}x$
$\frac{3}{4}x+6=\frac{2}{4}x+\frac{1}{4}x$
$\frac{3}{4}x+6=\frac{3}{4}x$
$\frac{3}{4}x+6-\frac{3}{4}x=\frac{3}{4}x-\frac{3}{4}x$
$6=0$
This is a false equation.
There is no solution.

31. $3x-\frac{1}{4}=\left(x+\frac{1}{2}\right)+\left(2x+\frac{1}{3}\right)$
$3x-\frac{1}{4}=x+\frac{3}{6}+2x+\frac{2}{6}$
$3x-\frac{1}{4}=3x+\frac{5}{6}$
$3x-\frac{1}{4}-3x=3x+\frac{5}{6}-3x$
$-\frac{1}{4}=\frac{5}{6}$
This is a false equation.
There is no solution.

33. $11x-12=7(3x-6)-2(x+9)$
$11x-12=21x-42-2x-18$
$11x-12=19x-60$
$11x-12+12=19x-60+12$
$11x=19x-48$
$11x-19x=19x-48-19x$
$-8x=-48$
$\frac{-8x}{-8}=\frac{-48}{-8}$
$x=6$
The solution is 6.

35. $7x-5(3x+9)=-2(4x+35)$
$7x-15x-45=-8x-70$
$-8x-45=-8x-70$
$-8x-45+8x=-8x-70+8x$
$-45=-70$
This is a false equation.
There is no solution.

37. $\frac{1}{4}x+\frac{5}{9}=\frac{5}{6}$
$36\left(\frac{1}{4}x\right)+36\left(\frac{5}{9}\right)=36\left(\frac{5}{6}\right)$
$9x+20=30$
$9x+20-20=30-20$
$9x=10$
$\frac{9x}{9}=\frac{10}{9}$
$x=\frac{10}{9}$
The solution is $\frac{10}{9}$.

39. $3x+\frac{1}{4}=2x+\frac{7}{36}$
$36(3x)+36\left(\frac{1}{4}\right)=36(2x)+36\left(\frac{7}{36}\right)$
$108x+9=72x+7$
$108x+9-9=72x+7-9$
$108x=72x-2$
$108x-72x=72x-2-72x$
$36x=-2$
$\frac{36x}{36}=\frac{-2}{36}$
$x=-\frac{1}{18}$
The solution is $-\frac{1}{18}$.

41. $\frac{2}{5}b-12=\frac{2}{3}b+20$
$15\left(\frac{2}{5}b\right)-15(12)=15\left(\frac{2}{3}b\right)+15(20)$
$6b-180=10b+300$
$6b-180+180=10b+300+180$
$6b=10b+480$
$6b-10b=10b+480-10b$
$-4b=480$
$\frac{-4b}{-4}=\frac{480}{-4}$
$b=-120$
The solution is -120.

43. $\frac{3}{4}\left(x+\frac{4}{5}\right)=-\frac{7}{8}x-\frac{2}{5}$

$\frac{3}{4}x+\frac{3}{5}=-\frac{7}{8}x-\frac{2}{5}$

$40\left(\frac{3}{4}x\right)+40\left(\frac{3}{5}\right)=40\left(-\frac{7}{8}x\right)-40\left(\frac{2}{5}\right)$

$30x+24=-35x-16$

$30x+24-24=-35x-16-24$

$30x=-35x-40$

$30x+35x=-35x-40+35x$

$65x=-40$

$\frac{65x}{65}=\frac{-40}{65}$

$x=-\frac{8}{13}$

The solution is $-\frac{8}{13}$.

45. $0.05x+10.5=0.15x-0.25$

$100(0.05x)+100(10.5)$
$=100(0.15x)-100(0.25)$

$5x+1050=15x-25$

$5x+1050-1050=15x-25-1050$

$5x=15x-1075$

$5x-15x=15x-1075-15x$

$-10x=-1075$

$\frac{-10x}{-10}=\frac{-1075}{-10}$

$x=\frac{1075}{10}=107.5$

The solution is 107.5.

47. $21.1x+0.46=10.9x+0.46$

$100(21.1x)+100(0.46)$
$=100(10.9x)+100(0.46)$

$2110x+46=1090x+46$

$2110x+46-46=1090x+46-46$

$2110x=1090x$

$2110x-1090x=1090x-1090x$

$1020x=0$

$\frac{1020x}{1020}=\frac{0}{1020}$

$x=0$

The solution is 0.

49. $15.2y-175.43=-2.4y-176.31$

$100(15.2y)-100(175.43)$
$=100(-2.4y)-100(176.31)$

$1520y-17{,}543=-240y-17{,}631$

$1520y-17{,}543+17{,}543$
$=-240y-17{,}631+17{,}543$

$1520y=-240y-88$

$1520y+240y=-240y-88+240y$

$1760y=-88$

$\frac{1760y}{1760}=\frac{-88}{1760}$

$y=-0.05$

The solution is –0.05.

51. Let x = number of miles

$49.95+0.12x=200$

$49.95+0.12x-49.95=200-49.95$

$0.12x=150.05$

$\frac{0.12x}{0.12}=\frac{150.05}{0.12}$

$x\approx 1250.4$

You could drive about 1250 miles.

53. Let x = monthly payment

$24x=2252-200$

$24x=2052$

$\frac{24x}{24}=\frac{2052}{24}$

$x=85.5$

The monthly payments would be \$85.50.

55. Let x = number of liters of 30% solution

$4(0.10)+x(0.30)=0.25(4+x)$

$0.4+0.3x=1+0.25x$

$100(0.4)+100(0.3x)=100(1)+100(0.25)$

$40+30x=100+25x$

$40+30x-40=100+25x-40$

$30x=60+25x$

$30x-25x=60+25x-25x$

$5x=60$

$\frac{5x}{5}=\frac{60}{5}$

$x=12$

There were 12 liters of 30% solution.

57. Let x = brother's earnings
$25+2x=730.10$
$25+2x-25=730.1-25$
$2x=705.1$
$\frac{2x}{2}=\frac{705.1}{2}$
$x=352.55$
Her brother's average weekly earnings are $352.55.

59. Let x = amount of sales
$150+0.10x-0.02x=200+0.06x+0.02x$
$150+0.8x=200+0.08x$
$150+0.08x-0.08x=200+0.08x-0.08x$
$150=200$
This is a false equation. There is no solution. There is no value of sales for which the two are equal.

61. Let x = total sales
$300+0.04x=700+0.04(x-10{,}000)$
$300+0.04x=700+0.04x-400$
$300+0.04x=300+0.04x$
$300+0.04x-0.04x=300+0.04x-0.04x$
$300=300$
This is a true equation. The solution is all real numbers. The plans are the same for any amount of total sales.

5.3 Experiencing Algebra the Calculator Way

1. Let x = total sales
$0.0375x+25=49.48$
$0.0375x+25-25=49.48-25$
$0.0375x=24.48$
$\frac{0.0375x}{0.0375}=\frac{24.48}{0.0375}$
$x=652.8$
The sales were $652.80.

2. Let x = number of 120-grain tablets
$200+120x=620$
$200+120x-200=620-200$
$120x=420$
$\frac{120x}{120}=\frac{420}{120}$
$x=3.5$
She should administer 3.5 tablets.

3. Let x = number of sheets of wood
$2+\frac{3}{4}x=4\frac{1}{4}$
$2+\frac{3}{4}x-2=4\frac{1}{4}-2$
$\frac{3}{4}x=2\frac{1}{4}$
$\frac{3}{4}x=\frac{9}{4}$
$\frac{4}{3}\left(\frac{3}{4}x\right)=\frac{4}{3}\left(\frac{9}{4}\right)$
$x=3$
He needs 3 sheets.

5.4 Experiencing Algebra the Exercise Way

1. $P=4s$, for s
$\frac{P}{4}=\frac{4s}{4}$
$\frac{P}{4}=s$ or $s=\frac{P}{4}$

3. $C=\pi d$, for d
$\frac{C}{\pi}=\frac{\pi d}{\pi}$
$\frac{C}{\pi}=d$ or $d=\frac{C}{\pi}$

5. $V=LWH$, for L
$\frac{V}{WH}=\frac{LWH}{WH}$
$\frac{V}{WH}=L$ or $L=\frac{V}{WH}$

7. $S=2LW+2LH+2WH$, for L
$S-2WH=2LW+2LH+2WH-2WH$
$S-2WH=2LW+2LH$
$S-2WH=L(2W+2H)$
$\frac{S-2WH}{2W+2H}=\frac{L(2W+2H)}{2W+2H}$
$\frac{S-2WH}{2W+2H}=L$ or $L=\frac{S-2WH}{2W+2H}$

9. $V = \pi r^2 h$, for h

$$\frac{V}{\pi r^2} = \frac{\pi r^2 h}{\pi r^2}$$
$$\frac{V}{\pi r^2} = h \text{ or } h = \frac{V}{\pi r^2}$$

11. $F = \frac{9}{5}C + 32$, for C

$$F - 32 = \frac{9}{5}C + 32 - 32$$
$$F - 32 = \frac{9}{5}C$$
$$\frac{5}{9}(F - 32) = \frac{5}{9}\left(\frac{9}{5}C\right)$$
$$\frac{5}{9}(F - 32) = C \text{ or } C = \frac{5}{9}(F - 32)$$

13. $I = Prt$, for P

$$\frac{I}{rt} = \frac{Prt}{rt}$$
$$\frac{I}{rt} = P \text{ or } P = \frac{I}{rt}$$

15. $A = P(1+i)^t$, for P

$$\frac{A}{(1+i)^t} = \frac{P(1+i)^t}{(1+i)^t}$$
$$\frac{A}{(1+i)^t} = P \text{ or } P = \frac{A}{(1+i)^t}$$

17. $v = gt$, for g

$$\frac{v}{t} = \frac{gt}{t}$$
$$\frac{v}{t} = g \text{ or } g = \frac{v}{t}$$

19. $I = \frac{V}{R}$, for R

$$RI = R\left(\frac{V}{R}\right)$$
$$RI = V$$
$$\frac{RI}{I} = \frac{V}{I}$$
$$R = \frac{V}{I}$$

21. $z = \frac{x - m}{s}$, for m

$$zs = s\left(\frac{x - m}{s}\right)$$
$$zs = x - m$$
$$zs - x = x - m - x$$
$$zs - x = -m$$
$$\frac{zs - x}{-1} = \frac{-m}{-1}$$
$$-zs + x = m \text{ or } m = x - zs$$

23. $4x + 3y = 0$

$$4x + 3y - 4x = 0 - 4x$$
$$3y = -4x$$
$$\frac{3y}{3} = \frac{-4x}{3}$$
$$y = -\frac{4}{3}x$$

25. $-5x + 10y = 0$

$$-5x + 10y + 5x = 0 + 5x$$
$$10y = 5x$$
$$\frac{10y}{10} = \frac{5x}{10}$$
$$y = \frac{1}{2}x$$

27. $-x - y = 0$

$$-x - y + x = 0 + x$$
$$-y = x$$
$$\frac{-y}{-1} = \frac{x}{-1}$$
$$y = -x$$

29. $5x + 4y = 20$

$$5x + 4y - 5x = 20 - 5x$$
$$4y = 20 - 5x$$
$$\frac{4y}{4} = \frac{20 - 5x}{4}$$
$$y = \frac{20}{4} - \frac{5}{4}x$$
$$y = 5 - \frac{5}{4}x$$
$$y = -\frac{5}{4}x + 5$$

31. $-x-y=7$
$-x-y+x=7+x$
$-y=7+x$
$\frac{-y}{-1}=\frac{7+x}{-1}$
$y=-7-x$
$y=-x-7$

33. $7x-14y=-28$
$7x-14y-7x=-28-7x$
$-14y=-28-7x$
$\frac{-14y}{-14}=\frac{-28-7x}{-14}$
$y=\frac{-28}{-14}-\frac{7x}{-14}$
$y=2+\frac{1}{2}x$
$y=\frac{1}{2}x+2$

35. $-x+y=-1$
$-x+y+x=-1+x$
$y=-1+x$
$y=x-1$

37. $y-5=4(x-6)$
$y-5=4x-24$
$y-5+5=4x-24+5$
$y=4x-19$

39. $y+6=-2(x-7)$
$y+6=-2x+14$
$y+6-6=-2x+14-6$
$y=-2x+8$

41. $y+2=-1(x+4)$
$y+2=-x-4$
$y+2-2=-x-4-2$
$y=-x-6$

43. $y-4=\frac{2}{3}(x+9)$
$y-4=\frac{2}{3}x+6$
$y-4+4=\frac{2}{3}x+6+4$
$y=\frac{2}{3}x+10$

45. $y+\frac{5}{9}=-\frac{2}{3}\left(x-\frac{1}{3}\right)$
$y+\frac{5}{9}=-\frac{2}{3}x+\frac{2}{9}$
$y+\frac{5}{9}-\frac{5}{9}=-\frac{2}{3}x+\frac{2}{9}-\frac{5}{9}$
$y=-\frac{2}{3}x-\frac{3}{9}$
$y=-\frac{2}{3}x-\frac{1}{3}$

47. $y=5x+15$
$y-15=5x+15-15$
$y-15=5x$
$\frac{y-15}{5}=\frac{5x}{5}$
$\frac{y}{5}-\frac{15}{5}=x$
$\frac{y}{5}-3=x$ or $x=\frac{1}{5}y-3$

49. $y=-2x+8$
$y-8=-2x+8-8$
$y-8=-2x$
$\frac{y-8}{-2}=\frac{-2x}{-2}$
$\frac{y}{-2}-\frac{8}{-2}=x$
$-\frac{1}{2}y+4=x$ or $x=-\frac{1}{2}y+4$

51. $y=-\frac{6}{7}x+8$
$y-8=-\frac{6}{7}x+8-8$
$y-8=-\frac{6}{7}x$
$-\frac{7}{6}(y-8)=-\frac{7}{6}\left(-\frac{6}{7}x\right)$
$-\frac{7}{6}y+\frac{28}{3}=x$ or $x=-\frac{7}{6}y+\frac{28}{3}$

53. $y = \frac{5}{9}x - \frac{2}{3}$
$y + \frac{2}{3} = \frac{5}{9}x - \frac{2}{3} + \frac{2}{3}$
$y + \frac{2}{3} = \frac{5}{9}x$
$\frac{9}{5}\left(y + \frac{2}{3}\right) = \frac{9}{5}\left(\frac{5}{9}x\right)$
$\frac{9}{5}y + \frac{6}{5} = x$ or $x = \frac{9}{5}y + \frac{6}{5}$

55. $y = mx + b$
$y - b = mx + b - b$
$y - b = mx$
$\frac{y - b}{m} = \frac{mx}{m}$
$\frac{y}{m} - \frac{b}{m} = x$
$\frac{1}{m}y - \frac{b}{m} = x$ or $x = \frac{1}{m}y - \frac{b}{m}$

57. $P = 200 + 85m$
$P - 200 = 200 + 85m - 200$
$P - 200 = 85m$
$\frac{P - 200}{85} = \frac{85m}{85}$
$\frac{1}{85}P - \frac{40}{17} = m$ or $m = \frac{1}{85}P - \frac{40}{17}$
$m = \frac{1}{85}(2240) - \frac{40}{17}$
$m = 24$
It will take 24 months to pay off $2240.
$m = \frac{1}{85}(1200) - \frac{40}{17}$
$m \approx 11.76$
It will take 12 months to pay off $1200.

59. $T = 22c + 12.50$
$T - 12.50 = 22c + 12.50 - 12.50$
$T - 12.50 = 22c$
$\frac{T - 12.50}{22} = \frac{22c}{22}$
$\frac{1}{22}T - \frac{12.5}{22} = c$ or $c = \frac{1}{22}T - \frac{25}{44}$
$c = \frac{1}{22}(35) - \frac{25}{44}$
$c \approx 1.02$
She can spend about $1.02 on each student for a $35 party.
$c = \frac{1}{22}(50) - \frac{25}{44}$
$c \approx 1.70$
She can spend about $1.70 on each student for a $50 party.

61. $B = 75 + 3T$
$B - 75 = 75 + 3T - 75$
$B - 75 = 3T$
$\frac{B - 75}{3} = \frac{3T}{3}$
$\frac{1}{3}B - 25 = T$ or $T = \frac{1}{3}B - 25$
$T = \frac{1}{3}(725) - 25$
$T \approx 216.67$
Ted's weekly earnings are about $216.67 when his boss averages $725 per week.
$T = \frac{1}{3}(1275) - 25$
$T = 400$
Ted's weekly earnings are $400 when his boss averages $1275 per week.

63. $C = 75 + 65d$
$C - 75 = 75 + 65d - 75$
$C - 75 = 65d$
$\frac{C - 75}{65} = \frac{65d}{65}$
$\frac{1}{65}C - \frac{15}{13} = d$ or $d = \frac{1}{65}C - \frac{15}{13}$
$d = \frac{1}{65}(250) - \frac{15}{13}$
$d \approx 2.69$
The equipment could be rented for 2 days for under $250.
$d = \frac{1}{65}(400) - \frac{15}{13}$
$d = 5$
The equipment could be rented for 5 days for $400.

65. $V = 3(5)h$
$V = 15h$
$\frac{V}{15} = \frac{15h}{15}$
$\frac{V}{15} = h$ or $h = \frac{V}{15}$
$h = \frac{60}{15}$
$h = 4$
The height should be 4 feet for a volume of 60 cubic feet.
$h = \frac{100}{15}$
$h = 6\frac{2}{3}$
The height should be $6\frac{2}{3}$ feet for a volume of 100 cubic feet.

5.4 Experiencing Algebra the Calculator Way

$A = Pe^{rt}$
$\frac{A}{e^{rt}} = \frac{Pe^{rt}}{e^{rt}}$
$\frac{A}{e^{rt}} = P$ or $P = \frac{A}{e^{rt}}$
$P = \frac{10,000}{e^{5(0.045)}}$
$P \approx 7985.16$
The amount to invest is about $7985.16 if you want $10,000 at the end of 5 years at 4.5% interest.

A	t	r	P
$10,000	5	4.5%	$7985.16
$10,000	7	4.5%	$7297.89
$10,000	10	4.5%	$6376.28
$10,000	12	4.5%	$5827.48
$25,000	5	7%	$17,617.20
$25,000	7	7%	$15,315.66
$25,000	10	7%	$12,414.63
$25,000	12	7%	$10,792.76

5.5 Experiencing Algebra the Exercise Way

Method of solution may vary.

1. Let x = number of grains in first dosage
$x + 2$ = number of grains in second dosage
$x + 4$ = number of grains in third dosage
$x + x + 2 + x + 4 = 24$
$Y1 = x + x + 2 + x + 4$
$Y2 = 24$

X	Y1	Y2
1	9	24
2	12	24
3	15	24
4	18	24
5	21	24
6	24	24
7	27	24

X=7

The solution is 6.
$x = 6$
$x + 2 = 8$
$x + 4 = 10$
The doses are 6 grains, 8 grains, and 10 grains.

3. Let x = number of prizes in first stage
$x + 2$ = number of prizes in second stage
$x + 4$ = number of prizes in third stage
$x + 6$ = number of prizes in fourth stage
$x + x + 2 + x + 4 + x + 6 = 24$
$Y1 = x + x + 2 + x + 4 + x + 6$
$Y2 = 24$

X	Y1	Y2
1	16	24
2	20	24
3	24	24
4	28	24
5	32	24
6	36	24
7	40	24

X=1

$x = 3$
$x + 2 = 5$
$x + 4 = 7$
$x + 6 = 9$
The number awarded at each stage is 3 prizes, 5 prizes, 7 prizes, and 9 prizes.

5. Let x = lowest grade
$x + 1$ = second lowest grade
$\vdots$
$x + 7$ = highest grade
$x + x + 1 + x + 2 + x + 3 + x + 4 + x + 5 + x + 6 + x + 7 = 676$
$8x + 28 = 676$
$8x + 28 - 28 = 676 - 28$
$8x = 648$

$\frac{8x}{8}=\frac{648}{8}$
$x=81$
$x+7=88$
The lowest grade was 81 points and the highest grade was 88 points.

7. Let x = first angle measurement
$x+10$ = second angle measurement
$x+20$ = third angle measurement
$x+x+10+x+20=180$
$3x+30=180$
$3x+30-30=180-30$
$3x=150$
$\frac{3x}{3}=\frac{150}{3}$
$x=50$
$x+10=60$
$x+20=70$
The angles measure 50°, 60°, and 70°.

9. Let x = measure of each side
$x+x+x=29\frac{1}{4}$
$Y1=x+x+x$
$Y2=29\frac{1}{4}$

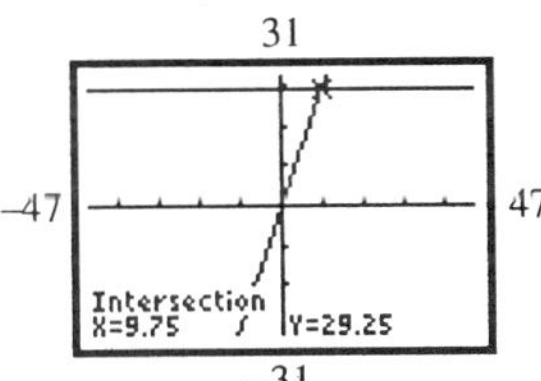

The intersection is (9.75, 29.25).

The solution is 9.75 or $9\frac{3}{4}$. Each side measures $9\frac{3}{4}$ inches.

11. Let x = length of the first side
x = length of the second side
$\frac{2}{3}x$ = length of the third side
$x+x+\frac{2}{3}x=16$
$Y1=\frac{8}{3}x$
$Y2=16$

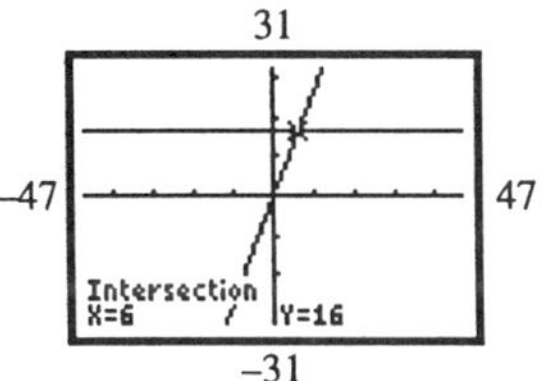

$x=6$
$\frac{2}{3}x=4$
The sides measure 6 feet, 6 feet, and 4 feet.

13. Let x = width
$x+30$ = length
$2x+2(x+30)=400$
$2x+2x+60=400$
$4x+60=400$
$4x+60-60=400-60$
$4x=340$
$\frac{4x}{4}=\frac{340}{4}$
$x=85$
$x+30=115$
The dimensions are 85 yards by 115 yards.

15. Let x = length
$0.55x$ = width
$2x+2(0.55x)=294.5$
$2x+1.1x=294.5$
$3.1x=294.5$
$\frac{3.1x}{3.1}=\frac{294.5}{3.1}$
$x=95$
$0.55x=52.25$
The dimensions are 95 cm by 52.25 cm.

17. Let x = width
$5x$ = length
$2x+2(5x)=96$
$2x+10x=96$
$12x=96$
$\frac{12x}{12}=\frac{96}{12}$
$x=8$
$5x=40$
The dimensions should be 8 feet by 40 feet.
Area = 8(40) = 320
It will cover 320 square feet of yard.

19. Let x = width
$5x$ = length
$2x + 5x = 96$
$7x = 96$
$\frac{7x}{7} = \frac{96}{7}$
$x \approx 13.7$
$5x \approx 68.5$
The dimensions are about 13.7 feet by 68.6 feet.
Area $\approx (13.7)(68.6) \approx 938 \text{ ft}^2$
The area of this run is larger than the run in exercise 17.

21. Let $3x$ = first angle measurement
x = second angle measurement
$x - 5$ = third angle measurement
$3x + x + x - 5 = 180$
$5x - 5 = 180$
$5x - 5 + 5 = 180 + 5$
$5x = 185$
$\frac{5x}{5} = \frac{185}{5}$
$x = 37$
$3x = 111$
$x - 5 = 32$
The angles measure 37°, 111°, 32°.

23. $I = PRT$
$I = P(0.125)(1)$
$I = 0.125P$
$P + 0.125P = 4500$
$1.125P = 4500$
$\frac{1.125P}{1.125} = \frac{4500}{1.125}$
$P = 4000$
$0.125P = 500$
The amount borrowed was $4000.
The interest was $500.

25. $I = PRT$
$I = P(0.09)(1)$
$I = 0.09P$
$P + 0.09P = 5000$
$1.09P = 5000$
$\frac{1.09P}{1.09} = \frac{5000}{1.09}$
$P = 4587.156$
You must invest about $4587.16.

27. $I = PRT$
$I = P(0.10)(1)$
$I = 0.1P$
$P + 0.1P = 500{,}000$
$1.1P = 500{,}000$
$\frac{1.1P}{1.1} = \frac{500{,}000}{1.1}$
$P \approx 454{,}545$
He should invest about $454,545.

29. Let x = amount invested at 8%
$15{,}000 - x$ = amount invested at 6.5%
$0.08x + 0.065(15{,}000 - x) = 1117.50$
$0.08x + 975 - 0.065x = 1117.5$
$975 + 0.015x = 1117.5$
$975 + 0.015x - 975 = 1117.5 - 975$
$0.015x = 142.5$
$\frac{0.015x}{0.015} = \frac{142.5}{0.015}$
$x = 9500$
$15{,}000 - x = 5500$
$9500 was invested at 8%.
$5500 was invested at 6.5%.

31. $A = P + 0.07P$
$A = 1.07P$
$\frac{A}{1.07} = \frac{1.07P}{1.07}$
$\frac{A}{1.07} = P$ or $P = \frac{A}{1.07}$
$P = \frac{1350}{1.07} \approx 1261.68$
About $1261.68 should be invested to have $1350.
$P = \frac{2500}{1.07} \approx 2336.45$
About $2336.45 should be invested to have $2500.

33. Let x = original price
$x - 0.20x = 68$
$0.8x = 68$
$\frac{0.8x}{0.8} = \frac{68}{0.8}$
$x = 85$
The original price was $85.

35. Let x = original cost
$x+0.6x=19.95$
$1.6x=19.95$
$\frac{1.6x}{1.6}=\frac{19.95}{1.6}$
$x\approx 12.47$
The original cost was about $12.47.

37. Let x = markup percentage
$x=\frac{17.10-9.50}{9.50}$
$x=\frac{7.6}{9.5}$
$x=0.8$ or 80%
The markup percentage was 80%.

39. Let x = regular price
$x-0.25x=195$
$0.75x=195$
$\frac{0.75x}{0.75}=\frac{195}{0.75}$
$x=260$
The regular price is $260.

41. Let x = SRP
$x-0.35x=12{,}500$
$0.65x=12{,}500$
$\frac{0.65x}{0.65}=\frac{12{,}500}{0.65}$
$x\approx 19{,}230.77$
The SRP should be about $19,230.77.

43. $y=x-0.1x$
$y=0.9x$
$\frac{y}{0.9}=\frac{0.9x}{0.9}$
$\frac{y}{0.9}=x$ or $x=\frac{y}{0.9}$
$x=\frac{53.96}{0.9}\approx 59.96$
The original price is about $59.96 if the sale price is $53.96.
$x=\frac{98.95}{0.9}\approx 109.94$
The original price is about $109.94 if the sale price is $98.95.

45. Let x = hourly wage before increase
$x+0.022x=13.75$
$1.022x=13.75$
$\frac{1.022x}{1.022}=\frac{13.75}{1.022}$
$x\approx 13.45$
His hourly wage before the increase was about $13.45.

47. Let x = amount of bill before gratuity
$x+0.15x=143.24$
$1.15x=143.24$
$\frac{1.15x}{1.15}=\frac{143.24}{1.15}$
$x\approx 124.56$
The bill before the gratuity was added was about $124.56.

5.5 Experiencing Algebra the Calculator Way

1.

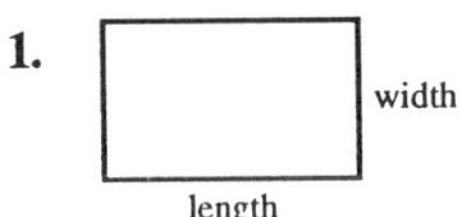

2. Let x = length in inches

3.

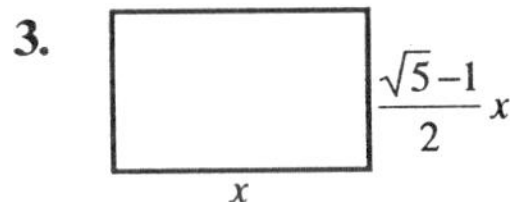

4. $P=2L+2W$
$323.61=2x+2Gx$

5. $323.61=x(2+2G)$
$\frac{323.61}{2+2G}=\frac{x(2+2G)}{2+2G}$
$\frac{323.61}{2+2G}=x$

```
(√(5)-1)/2→G:323
.61/(2+2G)
            100.0009895
```

$x\approx 100$
The length is about 100 inches.

6.
```
(√(5)-1)/2→G:323
.61/(2+2G)
         100.0009895
G*Ans
         61.80401045
```

$Gx \approx 61.804$
The width is about 61.804 inches.

7. $P = 2L + 2W$

323.61	$2L + 2W$
323.61	$2(100) + 2(61.804)$
323.61	323.608

The perimeter is 323.61 inches (after rounding).

5.6 Experiencing Algebra the Exercise Way

1. $|x+6| = 3$
There are two solutions.

x	$\lvert x+6\rvert$	3	
–9	3	3	$3 = 3$
–6	0	3	$0 \neq 3$
–3	3	3	$3 = 3$

The solutions are –9 and –3.

3. $|-x| = 13$
There are two solutions.

x	$\lvert -x\rvert$	13	
–13	13	13	$13 = 13$
0	0	13	$0 \neq 13$
13	13	13	$13 = 13$

The solutions are –13 and 13.

5. $|-x+3| = 2$
There are two solutions.

x	$\lvert -x+3\rvert$	2	
1	2	2	$2 = 2$
3	0	2	$0 \neq 2$
5	2	2	$2 = 2$

The solutions are 1 and 5.

7. $|2x+6| = 4$
There are two solutions.

x	$\lvert 2x+6\rvert$	4	
–5	4	4	$4 = 4$
–3	0	4	$0 \neq 4$
–1	4	4	$4 = 4$

The solutions are –5 and –1.

9. $|-4x-12| = 0$
There is one solution.

x	$\lvert -4x-12\rvert$	0	
–3	0	0	$0 = 0$
0	12	0	$12 \neq 0$

The solution is –3.

11. $|x| = 9$
$Y1 = |x|$
$Y2 = 9$

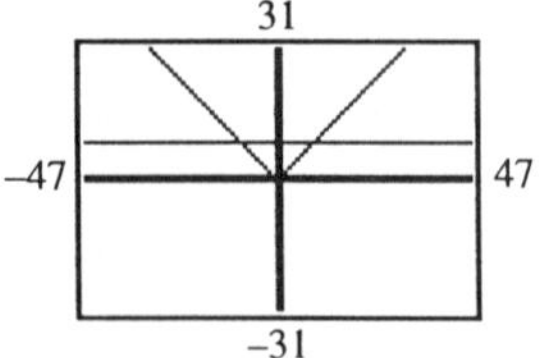

The graphs intersect at (–9, 9) and (9, 9)
The solutions are –9 and 9.

13. $|x+1|=3$
$Y1=|x+1|$
$Y2=3$

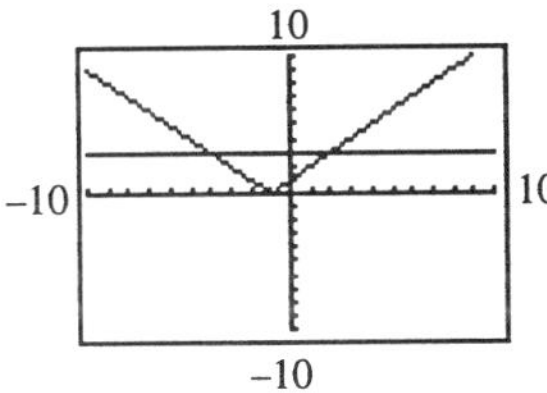

The graphs intersect at (–4, 3) and (2, 3). The solutions are –4 and 2.

15. $|6x+6|=12$
$Y1=|6x+6|$
$Y2=12$

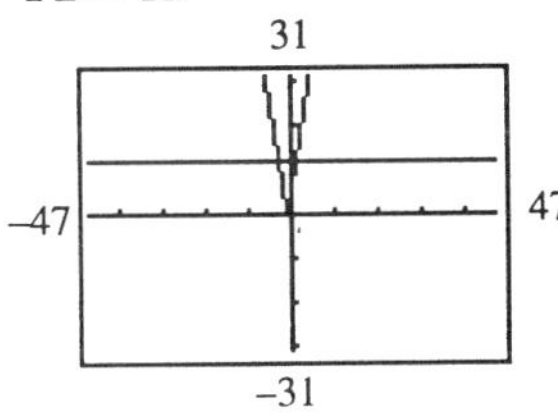

The graphs intersect at (–3, 12) and (1, 12) The solutions are –3 and 1.

17. $|-x-1|=1$
$Y1=|-x-1|$
$Y2=1$

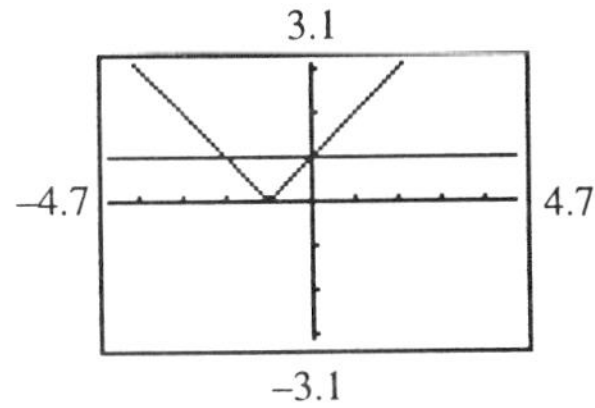

The graphs intersect at (–2, 1) and (0, 1). The solutions are –2 and 0.

19. $|x+21|=26$
$Y1=|x+21|$
$Y2=26$

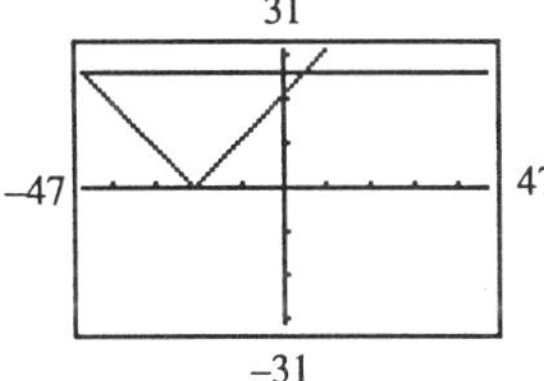

The graphs intersect at (–47, 26) and (5, 26). The solutions are –47 and 5.

21. $|-x-19|=22$
$Y1=|-x-19|$
$Y2=22$

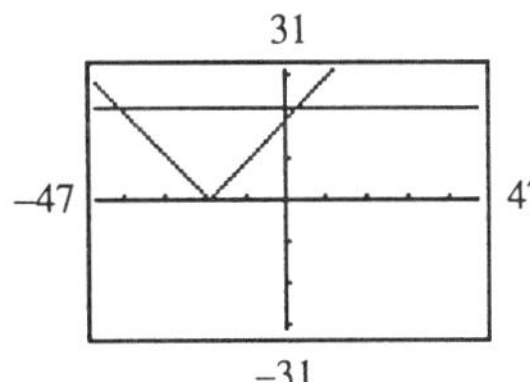

The graphs intersect at (–41, 22) and (3, 22). The solutions are –41 and 3.

23. $|x-2.17|=6.09$
$Y1=|x-2.17|$
$Y2=6.09$

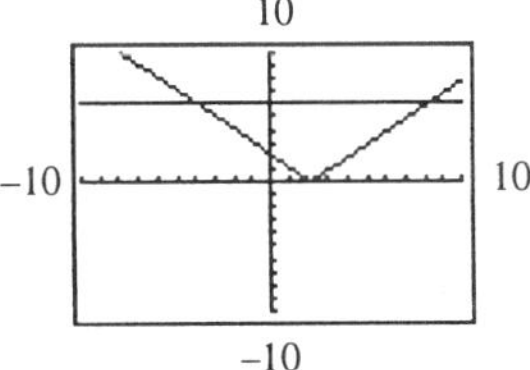

The graphs intersect at (–3.92, 6.09) and (8.26, 6.09). The solutions are –3.92 and 8.26.

25. $\left|x+\frac{2}{5}\right|=\frac{11}{15}$
$Y1=\left|x+\frac{2}{5}\right|$
$Y2=\frac{11}{15}$

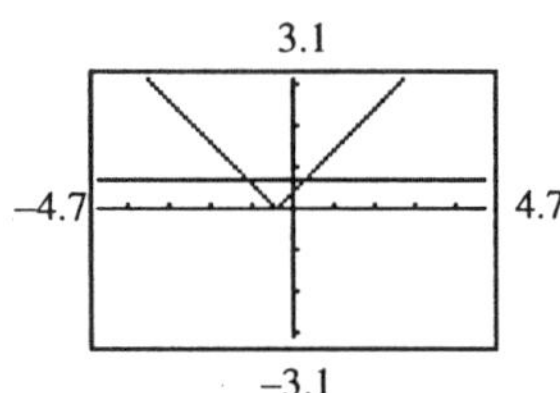

The graphs intersect at (–1.133, 7.333) and (0.333, 7.333). The solutions are –1.133 and 0.333.

27. $2|x-15|+25=15$
$Y1=2|x-15|+25$
$Y2=15$

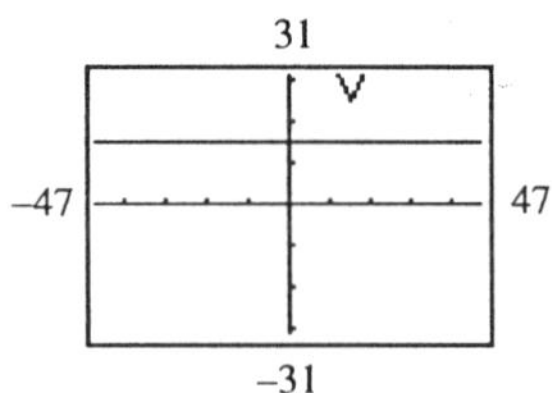

The graphs do not intersect.
There is no solution.

29. $7|x-32|+21=21$
$Y1=7|x-32|+21$
$Y2=21$

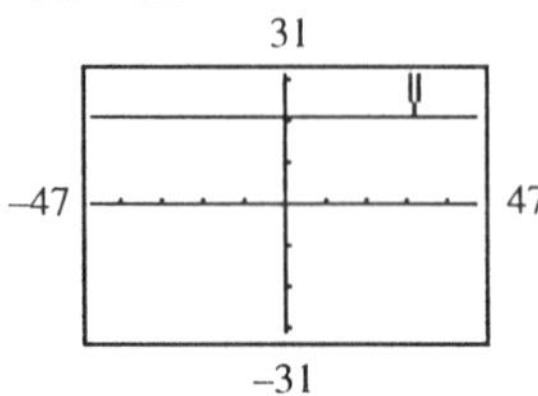

The graphs intersect at (32, 21).
The solution is 32.

31. $3|x+5|-7=4$
$Y1=3|x+5|-7$
$Y2=4$

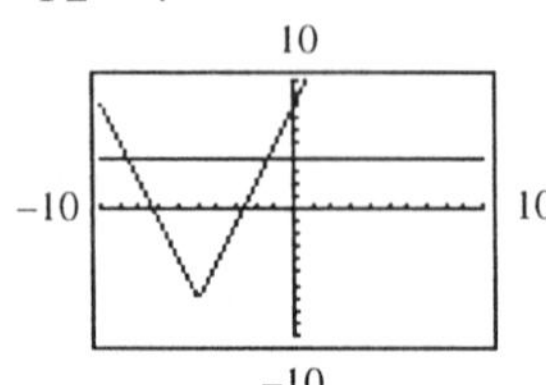

The graphs intersect at (–8.667, 4) and (–1.333, 4). The solutions are –8.667 and –1.333.

33. $|x|=138$
$x=-138$ or $x=138$
The solutions are –138 and 138.

35. $|a|=-512$
There is no solution because the absolute value expression equals a negative number.

37. $|b|=41.67$
$b=-41.67$ or $b=41.67$
The solutions are –41.67 and 41.67.

39. $|c|=14\frac{5}{9}$
$c=-14\frac{5}{9}$ or $c=14\frac{5}{9}$
The solutions are $-14\frac{5}{9}$ and $14\frac{5}{9}$.

41. $|x+578|=286$
$x+578=-286$
$x+578-578=-286-578$
$x=-864$
or
$x+578=286$
$x+578-578=286-578$
$x=-292$
The solutions are –864 and –292.

43. $|x-721|=1942$
$x-721=-1942$
$x-721+721=-1942+721$
$x=-1221$
or
$x-721=1942$
$x-721+721=1942+721$
$x=2663$
The solutions are –1221 and 2663.

45. $5|x-29|-46=74$
$5|x-29|-46+46=74+46$
$5|x-29|=120$
$\frac{5|x-29|}{5}=\frac{120}{5}$
$|x-29|=24$
$x-29=-24$
$x-29+29=-24+29$
$x=5$

or
$x - 29 = 24$
$x - 29 + 29 = 24 + 29$
$x = 53$
The solutions are 5 and 53.

47. $2|x+41|+96 = 48$
$2|x+41|+96-96 = 48-96$
$2|x+41| = -48$
$\frac{2|x+41|}{2} = \frac{-48}{2}$
$|x+41| = -24$
There is no solution because the absolute value expression equals a negative number.

49. $\left|\frac{x-14}{7}\right| = 8$
$\frac{x-14}{7} = -8$
$7\left(\frac{x-14}{7}\right) = 7(-8)$
$x - 14 = -56$
$x - 14 + 14 = -56 + 14$
$x = -42$
or
$\frac{x-14}{7} = 8$
$7\left(\frac{x-14}{7}\right) = 7(8)$
$x - 14 = 56$
$x - 14 + 14 = 56 + 14$
$x = 70$
The solutions are −42 and 70.

51. $|-3x-6|-18 = -6$
$|-3x-6|-18+18 = -6+18$
$|-3x-6| = 12$
$-3x - 6 = -12$ or $-3x - 6 = 12$
$-3x - 6 + 6 = -12 + 6$　　$-3x - 6 + 6 = 12 + 6$
$-3x = -6$　　$-3x = 18$
$\frac{-3x}{-3} = \frac{-6}{-3}$　　$\frac{-3x}{-3} = \frac{18}{-3}$
$x = 2$　　$x = -6$
The solutions are −6 and 2.

53. $-4|x+12| = -16$
$\frac{-4|x+12|}{-4} = \frac{-16}{-4}$
$|x+12| = 4$
$x + 12 = -4$
$x + 12 - 12 = -4 - 12$
$x = -16$
or
$x + 12 = 4$
$x + 12 - 12 = 4 - 12$
$x = -8$
The solutions are −16 and −8.

55. Let x = maximum depth of Lake Superior
$|x-923| = 410$
$x - 923 = -410$
$x - 923 + 923 = -410 + 923$
$x = 513$
or
$x - 923 = 410$
$x - 923 + 923 = 410 + 923$
$x = 1333$
The maximum depth of Lake Superior is either 513 feet or 1333 feet. The maximum depth is 1333 feet because 1333 is greater than 923.

57. Let x = male height in feet.
5 ft 9 in. = 5.75 ft
6 in. = 0.5 ft
$|x-5.75| = |\pm 0.5|$
$|x-5.75| = 0.5$
$x - 5.75 = -0.5$
$x - 5.75 + 5.75 = -0.5 + 5.75$
$x = 5.25$
$x = 5'3''$
or
$x - 5.75 = 0.5$
$x - 5.75 + 5.75 = 0.5 + 5.75$
$x = 6.25$
$x = 6'3''$
The minimum and maximum heights the clothes will fit are 5 feet 3 inches and 6 feet 3 inches.

59. Let x = percentage
$|x-42|=|\pm 3|$
$|x-42|=3$
$x-42=-3$
$x-42+42=-3+42$
$x=39$
or
$x-42=3$
$x-42+42=3+42$
$x=45$
The minimum and maximum percentages are 39% and 45%.

61. Let x = weight in pounds
$|x-132|=|\pm 2|$
$|x-132|=2$
$x-132=-2$
$x-132+132=-2+132$
$x=130$
or
$x-132=2$
$x-132+132=2+132$
$x=134$
The range is from 130 pounds to 134 pounds.

63. Let x = length in inches
$\left|x-5\frac{1}{2}\right|=\left|\pm\frac{1}{4}\right|$
$\left|x-5\frac{1}{2}\right|=\frac{1}{4}$
$x-5\frac{1}{2}=-\frac{1}{4}$
$x-5\frac{1}{2}+5\frac{1}{2}=-\frac{1}{4}+5\frac{1}{2}$
$x=5\frac{1}{4}$
or
$x-5\frac{1}{2}=\frac{1}{4}$
$x-5\frac{1}{2}+5\frac{1}{2}=\frac{1}{4}+5\frac{1}{2}$
$x=5\frac{3}{4}$
The lengths range from $5\frac{1}{4}$ inches to $5\frac{3}{4}$ inches.

65. Let x = score
$|x-71|=|\pm 22|$
$|x-71|=22$
$x-71=-22$
$x-71+71=-22+71$
$x=49$
or
$x-71=22$
$x-71+71=22+71$
$x=93$
The possible values for the extreme score are 49 and 93.

67. Let x = age
$|x-47|=|\pm 12|$
$|x-47|=12$
$x-47=-12$
$x-47+47=-12+47$
$x=35$
or
$x-47=12$
$x-47+47=12+47$
$x=59$
The pairs of ages are 35 and 47 or 47 and 59.

5.6 Experiencing Algebra the Calculator Way

1. $|x+2|=|x-3|$
$Y1=|x+2|$
$Y2=|x-3|$

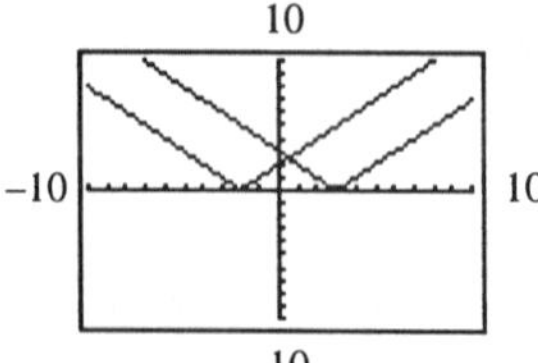

The graphs intersect at (0.5, 2.5).
The solution is 0.5.

2. $3|x+2|-5=3|x+2|+5$
$Y1=3|x+2|-5$
$Y2=3|x+2|+5$

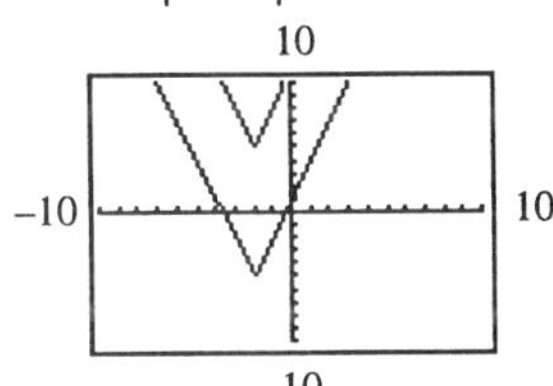

The graphs do not intersect.
There is no solution.

3. $|x+6|-9=-|x+3|+8$
$Y1=|x+6|-9$
$Y2=-|x+3|+8$

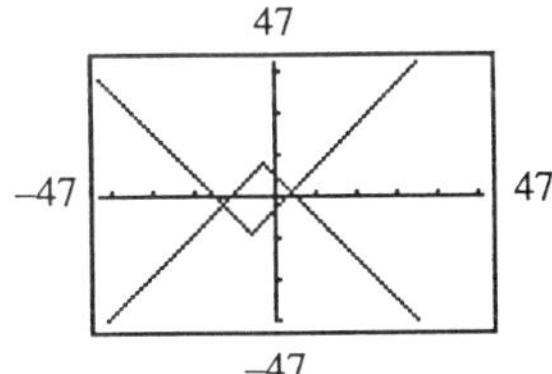

The graphs intersect at (4, 1) and (−13, −2).
The solutions are 4 and −13.

4. $|x+2|=-3x+8$
$Y1=|x+2|$
$Y2=-3x+8$

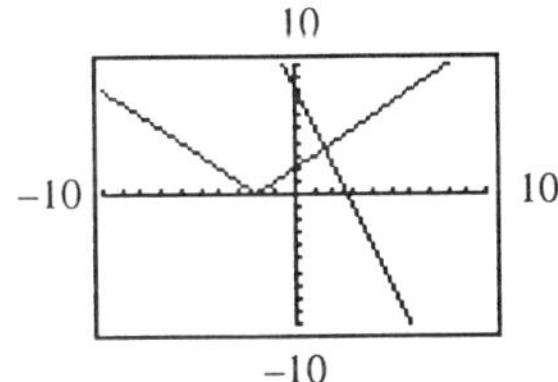

The graphs intersect at (1.5, 3.5).
The solution is 1.5.

5. $|x+3|+|x-3|=15$
$Y1=|x+3|+|x-3|$
$Y2=15$

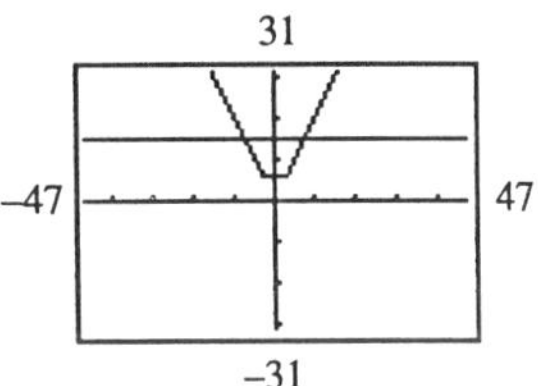

The graphs intersect at (−7.5, 15) and (7.5, 15). The solutions are −7.5 and 7.5.

Chapter 5 Review

Reflections

1.–8. Answers will vary.

Exercises

1. $4x^2-2x+1=0$ is nonlinear because x has an exponent of 2.

2. $5x+3=x-4$ is linear because it is in the form $ax+b=ax+b$.

3. $5.7x-8.2(x+46)=0$ is linear because it simplifies to $-2.5x-37.72=0$ which is in the form $ax+b=c$.

4. $\sqrt{x}+2=3x-6$ is nonlinear because there is a radical expression with a variable in its radicand.

5. $\frac{1}{8}x+\frac{3}{4}=\frac{11}{16}$ is linear because it is in the form $ax+b=c$.

6. $\frac{3}{x}+5=12x$ is nonlinear because there is a variable in the denominator of a term.

7.

x	$4x+7$	$2x-5$	
–7	–21	–19	$-21<-19$
–6	–17	–17	$-17=-17$
–5	–13	–15	$-13>-15$

The solution is –6.

8.

x	$2.4x-9.6$	4.8	
5	2.4	4.8	$2.4<4.8$
6	4.8	4.8	$4.8=4.8$
7	7.2	4.8	$7.2>4.8$

The solution is 6.

9.

x	$\frac{3}{5}x-\frac{7}{10}$	$\frac{1}{5}x+\frac{1}{2}$	
2	0.5	0.9	$0.5<0.9$
3	1.1	1.1	$1.1=1.1$
4	1.7	1.3	$1.7>1.3$

The solution is 3.

10.

x	$14x+12$	$11(x-5)+60$	
–3	–30	–28	$-30<-28$
–2	–16	–17	$-16>-17$

The solution is a noninteger between –3 and –2.

11.

x	$3(x-2)-1$	$4(x-1)-(x+3)$	
0	–7	–7	$-7=-7$
1	–4	–4	$-4=-4$
2	–1	–1	$-1=-1$

The values of the expressions are always equal. The equation is an identity. The solution is all real numbers.

12.

x	$2(2x+1)+x$	$3(2x-1)-(x+1)$	
0	2	–4	$2-(-4)=6$
1	7	1	$7-1=6$
2	12	6	$12-6=6$

The difference in the values of the expressions is a constant, 6. There is no solution.

13. $2x-2=-x+4$
$Y1=2x-2$
$Y2=-x+4$

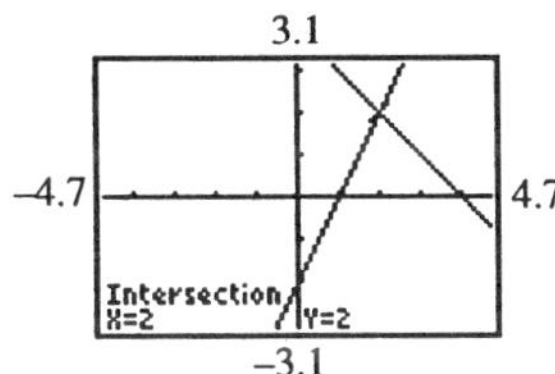

The graphs intersect at (2, 2).
The solution is 2.

14. $\frac{1}{2}x-2=-\frac{1}{3}x-\frac{11}{3}$

$Y1=\frac{1}{2}x-2$

$Y2=-\frac{1}{3}x-\frac{11}{3}$

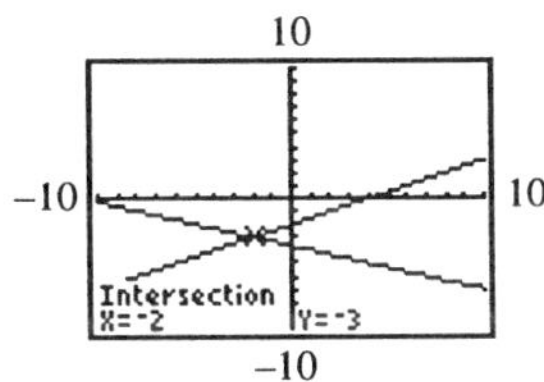

The graphs intersect at (–2, –3).
The solution is –2.

15. $(x+3)+(x+1)=3(x+1)-(x-1)$
$Y1=(x+3)+(x+1)$
$Y2=3(x+1)-(x-1)$

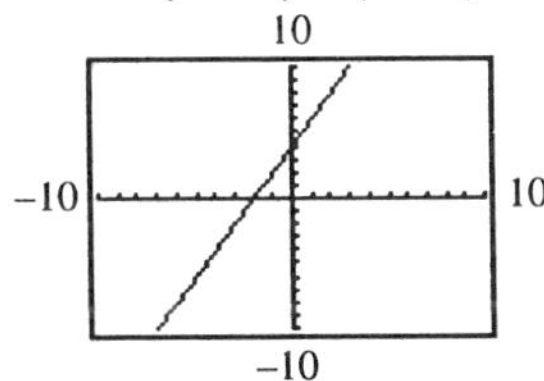

The graphs are the same.
The solution is all real numbers.

16. $(x+6)-3(x+1)=(2x+5)-2(2x+3)$
$Y1=(x+6)-3(x+1)$
$Y2=(2x+5)-2(2x+3)$

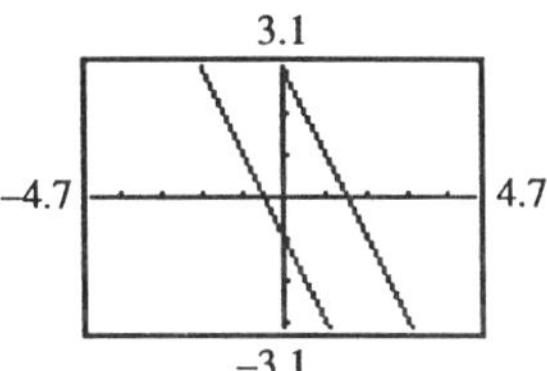

The graphs are parallel. There is no solution.

17. $1.2x+0.72=-2.1x+8.64$
$Y1=1.2x+0.72$
$Y2=-2.1x+8.64$

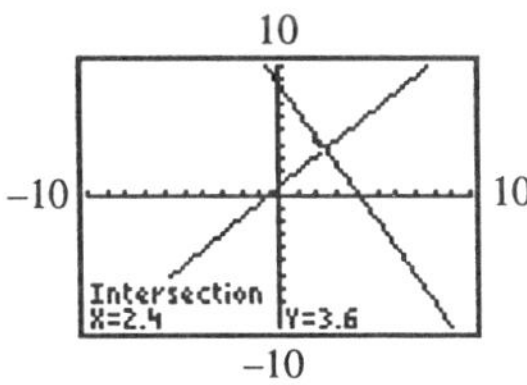

The graphs intersect at (2.4, 3.6). The solution is 2.4.

18. Let x = the number of hours worked
$90=25+6.5x$
$Y1=90$
$Y2=25+6.5x$

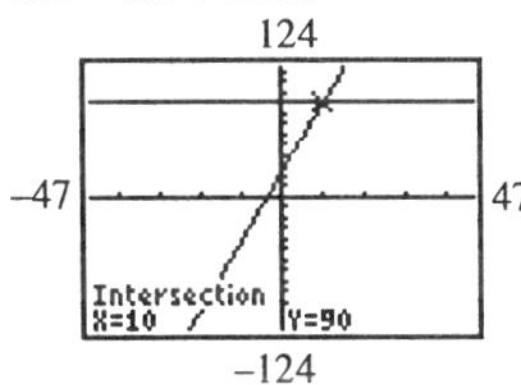

The graphs intersect at (10, 90).
The solution is 10.
The two offers are equivalent at 10 hours.

19. Let x = the amount of the fourth week donation

$$\frac{2200+1750+1885+x}{4}=2000$$

$$Y1=\frac{2200+1750+1885+x}{4}$$

$Y2=2000$

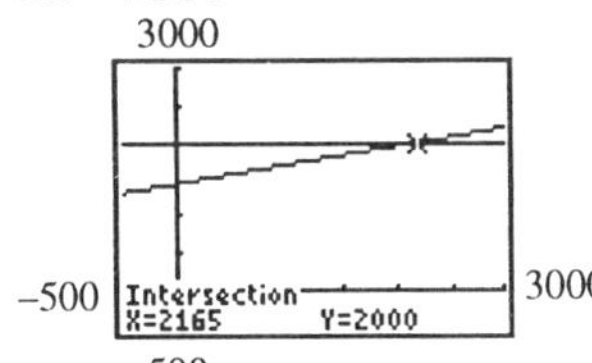

The graphs intersect at (2165, 2000). The solution is 2165. The fourth week donation should be $2165.

20. Let x = length of first side
x = length of second side
$3 + x$ = length of third side
$x + x + 3 + x = 26.25$
$Y1 = x + x + 3 + x$
$Y2 = 26.25$

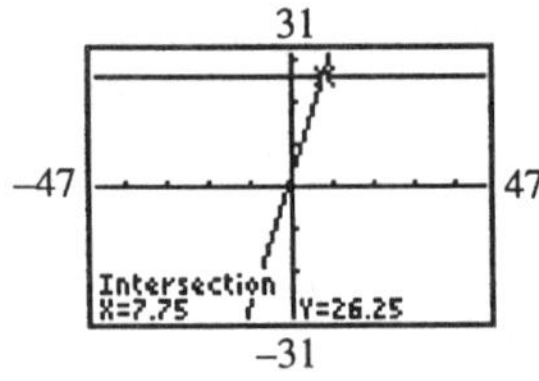

The graphs intersect at (7.75, 26.25). The solution is 7.75, so $3 + x = 10.75$. The sides measure 7.75 feet, 7.75 feet, and 10.75 feet.

21. $41 + x = 67$
$41 + x - 41 = 67 - 41$
$x = 26$
The solution is 26.

22. $y - \frac{7}{13} = \frac{11}{39}$
$y - \frac{7}{13} + \frac{7}{13} = \frac{11}{39} + \frac{7}{13}$
$y = \frac{11}{39} + \frac{21}{39}$
$y = \frac{32}{39}$
The solution is $\frac{32}{39}$.

23. $5 - (2 - x) = 1$
$5 - 2 + x = 1$
$3 + x = 1$
$3 + x - 3 = 1 - 3$
$x = -2$
The solution is –2.

24. $0.59(z - 1) + 0.41(z + 2) = 3.163$
$0.59z - 0.59 + 0.41z + 0.82 = 3.163$
$1z + 0.23 = 3.163$
$z + 0.23 - 0.23 = 3.163 - 0.23$
$z = 2.933$
The solution is 2.933.

25. $-4x = 272$
$\frac{-4x}{-4} = \frac{272}{-4}$
$x = -68$
The solution is –68.

26. $45.86z = -1765.61$
$\frac{45.86z}{45.86} = \frac{-1765.61}{45.86}$
$z = -38.5$
The solution is –38.5.

27. $\frac{a}{7} = 15$
$7\left(\frac{a}{7}\right) = 7(15)$
$a = 105$
The solution is 105.

28. $\frac{4}{5}x = \frac{64}{125}$
$\frac{5}{4}\left(\frac{4}{5}x\right) = \frac{5}{4}\left(\frac{64}{125}\right)$
$x = \frac{16}{25}$
The solution is $\frac{16}{25}$.

29. $15x - 16x = -12$
$-x = -12$
$\frac{-x}{-1} = \frac{-12}{-1}$
$x = 12$
The solution is 12.

30. $-y = 2.98$
$\frac{-y}{-1} = \frac{2.98}{-1}$
$y = -2.98$
The solution is –2.98.

31. $4(2x-6)+6(x+4)=49$
$8x-24+6x+24=49$
$14x=49$
$\frac{14x}{14}=\frac{49}{14}$
$x=3.5$
The solution is 3.5.

32. Let x = the number of passes given
$189+x=247$
$189+x-189=247-189$
$x=58$
There were 58 passes given.

33. Let x = total graduates
$0.55x=154$
$\frac{0.55x}{0.55}=\frac{154}{0.55}$
$x=280$
There were 280 graduates.

34. Let x = the number of books they must sell
$15x=80{,}000$
$\frac{15x}{15}=\frac{80{,}000}{15}$
$x=5333\frac{1}{3}$
They must sell 5334 books.

35. Let x = total proceeds
$\frac{3}{7}x=18{,}270$
$\frac{7}{3}\left(\frac{3}{7}x\right)=\frac{7}{3}(18{,}270)$
$x=42{,}630$
The total proceeds were $42,630.

36. $3x+7=4x+21$
$3x+7-7=4x+21-7$
$3x=4x+14$
$3x-4x=4x+14-4x$
$-x=14$
$\frac{-x}{-1}=\frac{14}{-1}$
$x=-14$
The solution is –14.

37. $14-2x=5x$
$14-2x+2x=5x+2x$
$14=7x$
$\frac{14}{7}=\frac{7x}{7}$
$2=x$
The solution is 2.

38. $8.7x+4.33=-2.4x-33.41$
$8.7x+4.33-4.33=-2.4x-33.41-4.33$
$8.7x=-2.4x-37.74$
$8.7x+2.4x=-2.4x-37.74+2.4x$
$11.1x=-37.74$
$\frac{11.1x}{11.1}=\frac{-37.74}{11.1}$
$x=-3.4$
The solution is –3.4.

39. $6.8z-9.52=0$
$6.8z-9.52+9.52=0+9.52$
$6.8z=9.52$
$\frac{6.8z}{6.8}=\frac{9.52}{6.8}$
$z=1.4$
The solution is 1.4.

40. $\frac{3}{8}x+\frac{1}{2}=\frac{3}{4}$
$8\left(\frac{3}{8}x\right)+8\left(\frac{1}{2}\right)=8\left(\frac{3}{4}\right)$
$3x+4=6$
$3x+4-4=6-4$
$3x=2$
$\frac{3x}{3}=\frac{2}{3}$
$x=\frac{2}{3}$
The solution is $\frac{2}{3}$.

41. $\frac{5}{7}a+\frac{11}{14}=\frac{2}{7}a$
$14\left(\frac{5}{7}a\right)+14\left(\frac{11}{14}\right)=14\left(\frac{2}{7}a\right)$
$10a+11=4a$
$10a-4a+11=4a-4a$
$6a+11=0$
$6a+11-11=0-11$

$6a = -11$
$\frac{6a}{6} = \frac{-11}{6}$
$a = -\frac{11}{6}$
The solution is $-\frac{11}{6}$.

42. $2(x+5)-(x+6)=2(x+2)-x$
$2x+10-x-6=2x+4-x$
$x+4=x+4$
$x+4-x=x+4-x$
$4=4$
This is a true equation.
The solution is all real numbers.

43. $3(x-4)+2(x+1)=5x+10$
$3x-12+2x+2=5x+10$
$5x-10=5x+10$
$5x-10-5x=5x+10-5x$
$-10=10$
This is a false equation. There is no solution.

44. Let x = the number of miles.
$49.95+0.22x=250$
$49.95+0.22x-49.95=250-49.95$
$0.22x=200.05$
$\frac{0.22x}{0.22}=\frac{200.05}{0.22}$
$x \approx 909.3$
You can drive 909 miles.

45. Let x = annual depreciation.
$7x+25{,}000=150{,}000$
$7x+25{,}000-25{,}000=150{,}000-25{,}000$
$7x=125{,}000$
$\frac{7x}{7}=\frac{125{,}000}{7}$
$x \approx 17{,}857.14$
The annual depreciation is about $17,857.

46. Let x = number of hours.
$175+35x=100+40x$
$175+35x-175=100+40x-175$
$35x=-75+40x$
$35x-40x=-75+40x-40x$
$-5x=-75$
$\frac{-5x}{-5}=\frac{-75}{-5}$
$x=15$
The offers are the same at 15 hours.

47. $A=\frac{1}{2}h(b+B)$ for h
$2A=2\left[\frac{1}{2}h(b+B)\right]$
$2A=h(b+B)$
$\frac{2A}{b+B}=\frac{h(b+B)}{b+B}$
$\frac{2A}{b+B}=h$ or $h=\frac{2A}{b+B}$

48. $S=2LW+2WH+2LH$ for W
$S-2LH=2LW+2WH+2LH-2LH$
$S-2LH=2LW+2WH$
$S-2LH=W(2L+2H)$
$\frac{S-2LH}{2L+2H}=\frac{W(2L+2H)}{2L+2H}$
$\frac{S-2LH}{2L+2H}=W$ or $W=\frac{S-2LH}{2L+2H}$

49. $6x+7y=42$ for x
$6x+7y-7y=42-7y$
$6x=42-7y$
$\frac{6x}{6}=\frac{42-7y}{6}$
$x=\frac{42}{6}-\frac{7y}{6}$
$x=7-\frac{7}{6}y$ or $x=-\frac{7}{6}y+7$

50. $\frac{3}{4}x-\frac{5}{8}y=\frac{11}{12}$ for y
$24\left(\frac{3}{4}x\right)-24\left(\frac{5}{8}y\right)=24\left(\frac{11}{12}\right)$
$18x-15y=22$
$18x-15y-18x=22-18x$
$-15y=22-18x$
$\frac{-15y}{-15}=\frac{22-18x}{-15}$
$y=\frac{22}{-15}-\frac{18x}{-15}$
$y=-\frac{22}{15}+\frac{6}{5}x$ or $y=\frac{6}{5}x-\frac{22}{15}$

51. $A = 6000 + 8000n$
$A - 6000 = 6000 + 8000n - 6000$
$\frac{A-6000}{8000} = \frac{8000n}{8000}$
$\frac{A-6000}{8000} = n$ or $n = \frac{A-6000}{8000}$
$n = \frac{78,000-6000}{8000}$
$n = 9$
It will last 9 years if the amount is \$78,000.
$n = \frac{126,000-6000}{8000}$
$n = 15$
It will last 15 years if the amount is \$126,000.

52. Let x = length of first piece
$x + 2$ = length of second piece
$x + 4$ = length of third piece
$x + x + 2 + x + 4 = 3(12)$
$3x + 6 = 36$
$3x + 6 - 6 = 36 - 6$
$3x = 30$
$\frac{3x}{3} = \frac{30}{3}$
$x = 10$
$x + 2 = 12$
$x + 4 = 14$
The lengths should be 10 in., 12 in., and 14 in.

53. Let x = length of first piece
$x + 1$ = length of second piece
$x + 2$ = length of third piece
$x + x + 1 + x + 2 = 3(12)$
$3x + 3 = 36$
$3x + 3 - 3 = 36 - 3$
$3x = 33$
$\frac{3x}{3} = \frac{33}{3}$
$x = 11$
$x + 1 = 12$
$x + 2 = 13$
The lengths should be 11 in., 12 in., and 13 in.

54. Let x = width
$3x$ = length
$2x + 2(3x) - 4 - 6 = 230$
$8x - 10 = 230$
$8x - 10 + 10 = 230 + 10$
$8x = 240$
$\frac{8x}{8} = \frac{240}{8}$
$x = 30$
$3x = 90$
The dimensions are 30 ft by 90 ft.

55. $I = PRT$
$2250 = 5000(0.075)T$
$2250 = 375T$
$\frac{2250}{375} = \frac{375T}{375}$
$6 = T$
It should be invested for 6 years.

56. $I = PRT$
$2250 = P(0.075)(3)$
$2250 = 0.225P$
$\frac{2250}{0.225} = \frac{0.225P}{0.225}$
$10,000 = P$
You should invest \$10,000.

57. Let x = the amount that the retail price should be
$x - 0.2x = 210$
$x(1 - 0.2) = 210$
$0.8x = 210$
$\frac{0.8x}{0.8} = \frac{210}{0.8}$
$x = 262.5$
No, the store was not being honest. The suit's retail price should have been \$262.50.

58. Let x = the artist's price
$x + 0.3x = 32.50$
$x(1 + 0.3) = 32.50$
$1.3x = 32.5$
$\frac{1.3x}{1.3} = \frac{32.5}{1.3}$
$x = 25$
The artist was paid \$25.

59. $|x| = 23$
$x = 23$ or $x = -23$
The solutions are 23 and –23.

60. $|x| = -12$
There is no solution because the absolute value expression equals a negative number.

61. $|c| = 0$
$c = 0$
The solution is 0.

62. $|a-7| = 0$
$a - 7 = 0$
$a - 7 + 7 = 0 + 7$
$a = 7$
The solution is 7.

63. $|-b+12| = 4$
$-b + 12 = 4$
$-b + 12 - 12 = 4 - 12$
$-b = -8$
$\frac{-b}{-1} = \frac{-8}{-1}$
$b = 8$
or
$-b + 12 = -4$
$-b + 12 - 12 = -4 - 12$
$-b = -16$
$\frac{-b}{-1} = \frac{-16}{-1}$
$b = 16$
The solutions are 8 and 16.

64. $|c-2| = -4$
There is no solution because the absolute value expression equals a negative number.

65. $|2x-7| = 8$
$2x - 7 = -8$ or $2x - 7 = 8$
$2x - 7 + 7 = -8 + 7$ $\quad$ $2x - 7 + 7 = 8 + 7$
$2x = -1$ $\quad$ $2x = 15$
$\frac{2x}{2} = \frac{-1}{2}$ $\quad$ $\frac{2x}{2} = \frac{15}{2}$
$x = -\frac{1}{2}$ $\quad$ $x = \frac{15}{2}$
The solutions are $-\frac{1}{2}$ and $\frac{15}{2}$.

66. $2|x-7| - 4 = 8$
$2|x-7| - 4 + 4 = 8 + 4$
$2|x-7| = 12$
$\frac{2|x-7|}{2} = \frac{12}{2}$
$|x-7| = 6$
$x - 7 = -6$ or $x - 7 = 6$
$x - 7 + 7 = -6 + 7$ $\quad$ $x - 7 + 7 = 6 + 7$
$x = 1$ $\quad$ $x = 13$
The solutions are 1 and 13.

67. $5|2x-7| + 10 = 8$
$5|2x-7| + 10 - 10 = 8 - 10$
$5|2x-7| = -2$
$\frac{5|2x-7|}{5} = \frac{-2}{5}$
$|2x-7| = -\frac{2}{5}$
There is no solution because the absolute value expression equals a negative number.

68. $2\left|\frac{x-1}{2}\right| - 5 = -2$
$2\left|\frac{x-1}{2}\right| - 5 + 5 = -2 + 5$
$2\left|\frac{x-1}{2}\right| = 3$
$\frac{2\left|\frac{x-1}{2}\right|}{2} = \frac{3}{2}$
$\left|\frac{x-1}{2}\right| = \frac{3}{2}$
$\frac{x-1}{2} = -\frac{3}{2}$ or $\frac{x-1}{2} = \frac{3}{2}$
$2\left(\frac{x-1}{2}\right) = 2\left(-\frac{3}{2}\right)$ $\quad$ $2\left(\frac{x-1}{2}\right) = 2\left(\frac{3}{2}\right)$
$x - 1 = -3$ $\quad$ $x - 1 = 3$
$x - 1 + 1 = -3 + 1$ $\quad$ $x - 1 + 1 = 3 + 1$
$x = -2$ $\quad$ $x = 4$
The solutions are –2 and 4.

69. Let x = the limit on the part
$|x - 62.79| = |\pm 0.04|$
$|x - 62.79| = 0.04$
$x - 62.79 = -0.04$
$x - 62.79 + 62.79 = -0.04 + 62.79$
$x = 62.75$
or
$x - 62.79 = 0.04$
$x - 62.79 + 62.79 = 0.04 + 62.79$
$x = 62.83$
The permissible limits on the part are 62.75 mm and 62.83 mm.

70. Let x = the limit on the percentage of voters
$|x - 49| = |\pm 4|$
$|x - 49| = 4$
$x - 49 = -4$ or $x - 49 = 4$
$x - 49 + 49 = -4 + 49$ $x - 49 + 49 = 4 + 49$
$x = 45$ $x = 53$
The limits on the percentage of voters are 45% and 53%.

Chapter 5 Mixed Review

1. $3(x + 2) - 2(x - 1) = (2x + 5) - (x - 3)$
$Y1 = 3(x + 2) - 2(x - 1)$
$Y2 = (2x + 5) - (x - 3)$

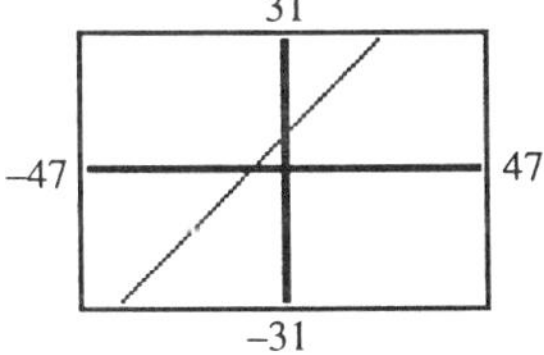

The graphs are the same.
The solution is all real numbers.

2. $(2x + 1) - (3x - 7) = (x + 5) - 2(x - 2)$
$Y1 = (2x + 1) - (3x - 7)$
$Y2 = (x + 5) - 2(x - 2)$

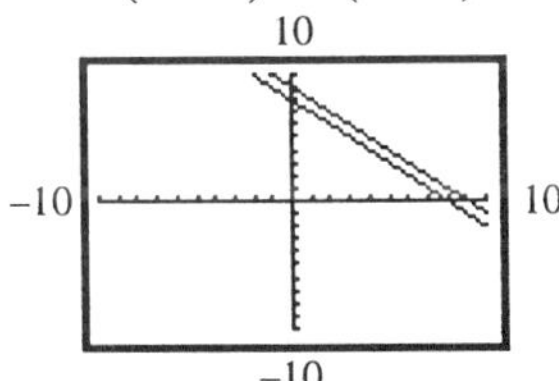

The graphs are parallel.
There is no solution.

3. $x + 1 = 2x + 5$
$Y1 = x + 1$
$Y2 = 2x + 5$

31
−47
(−4, −3)
47
−31

The graphs intersect at (−4, −3).
The solution is −4.

4. $\frac{1}{3}x + 3 = 6 - \frac{2}{3}x$
$Y1 = \frac{1}{3}x + 3$
$Y2 = 6 - \frac{2}{3}x$

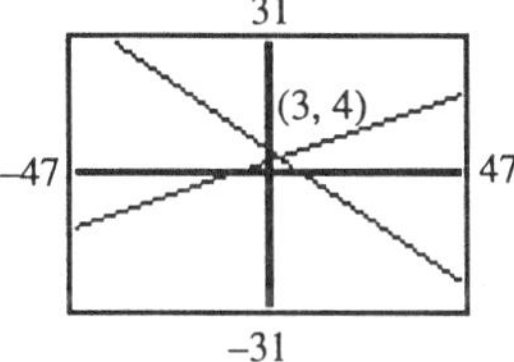

The graphs intersect at (3, 4).
The solution is 3.

5. $1.2x - 6.12 = -2.2x + 4.42$
$Y1 = 1.2x - 6.12$
$Y2 = -2.2x + 4.42$

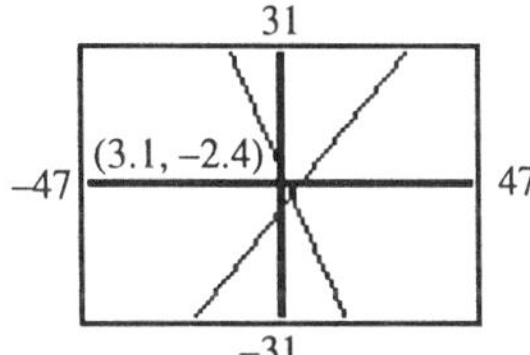

The graphs intersect at (3.1, −2.4).
The solution is 3.1.

6.

x	$4x-5$	$7-2x$	
1	–1	5	$-1<5$
2	3	3	$3=3$
3	7	1	$7>1$

The solution is 2.

7.

x	$14(x+6)-17$	$12(x+5)$	
–4	11	12	$11<12$
–3	25	24	$25>24$

The solution is a noninteger between –4 and –3.

8.

x	$1.5x+5.5$	$-2.4x-6.2$	
–4	–0.5	3.4	$-0.5<3.4$
–3	1	1	$1=1$
–2	2.5	–1.4	$2.5>-1.4$

The solution is –3.

9.

x	$2(2x-3)+3(x+1)$	$6(x-1)+(x+3)$	
0	–3	–3	$-3=-3$
1	4	4	$4=4$
2	11	11	$11=11$

The values of the expressions are always equal. The solution is all real numbers.

10.

x	$\frac{1}{3}x+\frac{14}{3}$	$\frac{5}{2}-\frac{3}{4}x$	
–3	3.67	4.75	$3.67<4.75$
–2	4	4	$4=4$
–1	4.33	3.25	$4.33>3.25$

The solution is –2.

11.

x	$4(x-2)-(x-1)$	$3x+1$	
0	–7	1	$-7-1=-8$
1	–4	4	$-4-4=-8$
2	–1	7	$-1-7=-8$

The difference in the values of the expressions is a constant, –8.
There is no solution.

12. $5-\frac{3}{7}x=\frac{2}{3}x+2$ is linear because it is in the form $b+ax=ax+b$.

13. $2x-4=1+\frac{9}{x}$ is nonlinear because a variable appears in the denominator of a term.

14. $-2x^3+3x=x-5$ is nonlinear because the variable x has an exponent of 3.

15. $15=2(x-7)-(x-2)$ is linear because it simplifies to $15=x-12$.

16. $3.9(x-1.2)-6.7x=0$ is linear because it simplifies to $-2.8x-4.68=0$.

17. $\sqrt[3]{x+1}-5=4x$ is nonlinear because there is a radical expression with a variable in its radicand.

18. $\frac{z}{29}=-12$

$29\left(\frac{z}{29}\right)=29(-12)$

$z=-348$

The solution is –348.

19. $\frac{5}{22}=\frac{25}{33}y$

$\frac{33}{25}\left(\frac{5}{22}\right)=\frac{33}{25}\left(\frac{25}{33}y\right)$

$\frac{3}{10}=y$

The solution is $\frac{3}{10}$.

20. $59a=1888$

$\frac{59a}{59}=\frac{1888}{59}$

$a=32$

The solution is 32.

21. $-174.243=2.41x$

$\frac{-174.243}{2.41}=\frac{2.41x}{2.41}$

$-72.3=x$

The solution is –72.3.

22. $2.3x-3.3x=14$

$-x=14$

$\frac{-x}{-1}=\frac{14}{-1}$

$x=-14$

The solution is –14.

23. $-b=14.59$

$\frac{-b}{-1}=\frac{14.59}{-1}$

$b=-14.59$

The solution is –14.59.

24. $3(x-8)+8(x+3)=77$

$3x-24+8x+24=77$

$11x=77$

$\frac{11x}{11}=\frac{77}{11}$

$x=7$

The solution is 7.

25. $z+193=-251$

$z+193-193=-251-193$

$z=-444$

The solution is –444.

26. $\frac{13}{17}+a=\frac{5}{51}$

$\frac{13}{17}+a-\frac{13}{17}=\frac{5}{51}-\frac{13}{17}$

$a=\frac{5}{51}-\frac{39}{51}$

$a=-\frac{34}{51}=-\frac{2}{3}$

The solution is $-\frac{2}{3}$.

27. $0.92(x-2)+0.08(x+1)=5.73$

$0.92x-1.84+0.08x+0.08=5.73$

$x-1.76=5.73$

$x-1.76+1.76=5.73+1.76$

$x=7.49$

The solution is 7.49.

28. $6.2x + 5.67 = 4.9x + 16.98$
$6.2x + 5.67 - 5.67 = 4.9x + 16.98 - 5.67$
$6.2x = 4.9x + 11.31$
$6.2x - 4.9x = 4.9x + 11.31 - 4.9x$
$1.3x = 11.31$
$\frac{1.3x}{1.3} = \frac{11.31}{1.3}$
$x = 8.7$
The solution is 8.7.

29. $24.96 - 3.9a = 0$
$24.96 - 3.9a + 3.9a = 0 + 3.9a$
$24.96 = 3.9a$
$\frac{24.96}{3.9} = \frac{3.9a}{3.9}$
$6.4 = a$
The solution is 6.4.

30. $\frac{2}{3}x - \frac{3}{4} = -\frac{5}{6}$
$12\left(\frac{2}{3}x\right) - 12\left(\frac{3}{4}\right) = 12\left(-\frac{5}{6}\right)$
$8x - 9 = -10$
$8x - 9 + 9 = -10 + 9$
$8x = -1$
$\frac{8x}{8} = \frac{-1}{8}$
$x = -\frac{1}{8}$
The solution is $-\frac{1}{8}$.

31. $\frac{4}{9}y + \frac{11}{18} = \frac{5}{6}y$
$18\left(\frac{4}{9}y\right) + 18\left(\frac{11}{18}\right) = 18\left(\frac{5}{6}y\right)$
$8y + 11 = 15y$
$8y + 11 - 8y = 15y - 8y$
$11 = 7y$
$\frac{11}{7} = \frac{7y}{7}$
$\frac{11}{7} = y$
The solution is $\frac{11}{7}$.

32. $7x - 4 = 3x + 20$
$7x - 4 + 4 = 3x + 20 + 4$
$7x = 3x + 24$
$7x - 3x = 3x + 24 - 3x$
$4x = 24$
$\frac{4x}{4} = \frac{24}{4}$
$x = 6$
The solution is 6.

33. $7x = 15 - 3x$
$7x + 3x = 15 - 3x + 3x$
$10x = 15$
$\frac{10x}{10} = \frac{15}{10}$
$x = 1.5$
The solution is 1.5.

34. $2(x + 2) + (x + 1) = 5(x + 1) - 2x$
$2x + 4 + x + 1 = 5x + 5 - 2x$
$3x + 5 = 3x + 5$
$3x + 5 - 3x = 3x + 5 - 3x$
$5 = 5$
This is a true equation. The solution is all real numbers.

35. $4(2x - 1) = 3(2x + 1) + 2(x + 1)$
$8x - 4 = 6x + 3 + 2x + 2$
$8x - 4 = 8x + 5$
$8x - 4 - 8x = 8x + 5 - 8x$
$-4 = 5$
This is a false equation. There is no solution.

36. $|2x + 6| = 0$
$2x + 6 = 0$
$2x + 6 - 6 = 0 - 6$
$2x = -6$
$\frac{2x}{2} = \frac{-6}{2}$
$x = -3$
The solution is –3.

37. $|4-z|=4$

$4-z=-4$ or $4-z=4$

$4-z-4=-4-4$ $4-z-4=4-4$

$-z=-8$ $-z=0$

$\frac{-z}{-1}=\frac{-8}{-1}$ $\frac{-z}{-1}=\frac{0}{-1}$

$z=8$ $z=0$

The solutions are 0 and 8.

38. $|2x-1|=-4$

There is no solution because the absolute value expression equals a negative number.

39. $|5x+2|=12$

$5x+2=-12$ or $5x+2=12$

$5x+2-2=-12-2$ $5x+2-2=12-2$

$5x=-14$ $5x=10$

$\frac{5x}{5}=\frac{-14}{5}$ $\frac{5x}{5}=\frac{10}{5}$

$x=-\frac{14}{5}$ $x=2$

The solutions are $-\frac{14}{5}$ and 2.

40. $5|x-7|=15$

$\frac{5|x-7|}{5}=\frac{15}{5}$

$|x-7|=3$

$x-7=-3$ or $x-7=3$

$x-7+7=-3+7$ $x-7+7=3+7$

$x=4$ $x=10$

The solutions are 4 and 10.

41. $3\left|\frac{x+1}{4}\right|-9=-6$

$3\left|\frac{x+1}{4}\right|-9+9=-6+9$

$3\left|\frac{x+1}{4}\right|=3$

$\frac{3\left|\frac{x+1}{4}\right|}{3}=\frac{3}{3}$

$\left|\frac{x+1}{4}\right|=1$

$\frac{x+1}{4}=-1$ or $\frac{x+1}{4}=1$

$4\left(\frac{x+1}{4}\right)=4(-1)$ $4\left(\frac{x+1}{4}\right)=4(1)$

$x+1=-4$ $x+1=4$

$x+1-1=-4-1$ $x+1-1=4-1$

$x=-5$ $x=3$

The solutions are -5 and 3.

42. $|x|=31$

$x=-31$ or $x=31$

The solutions are -31 and 31.

43. $|x|=-29$

There is no solution because the absolute value expression equals a negative number.

44. $|-a|=0$

$-a=0$

$\frac{-a}{-1}=\frac{0}{-1}$

$a=0$

The solution is 0.

45. $1.2x-2.4y=0.36$ for x

$1.2x-2.4y+2.4y=0.36+2.4y$

$1.2x=0.36+2.4y$

$\frac{1.2x}{1.2}=\frac{0.36+2.4y}{1.2}$

$x=\frac{0.36}{1.2}+\frac{2.4y}{1.2}$

$x=0.3+2y$ or $x=2y+0.3$

46. $\frac{1}{9}x+\frac{2}{3}y=\frac{1}{6}$ for y

$18\left(\frac{1}{9}x\right)+18\left(\frac{2}{3}y\right)=18\left(\frac{1}{6}\right)$

$2x+12y=3$

$2x+12y-2x=3-2x$

$12y=3-2x$

$\frac{12y}{12}=\frac{3-2x}{12}$

$y=\frac{3}{12}-\frac{2}{12}x$

$y=\frac{1}{4}-\frac{1}{6}x$ or $y=-\frac{1}{6}x+\frac{1}{4}$

47. $S=2\pi r^2+2\pi rh$ for h

$S-2\pi r^2=2\pi r^2+2\pi rh-2\pi r^2$

$S-2\pi r^2=2\pi rh$

$\frac{S-2\pi r^2}{2\pi r}=\frac{2\pi rh}{2\pi r}$

$\frac{S-2\pi r^2}{2\pi r}=h$ or $h=\frac{S-2\pi r^2}{2\pi r}$

48. $I=Prt$ for P

$\frac{I}{rt}=\frac{Prt}{rt}$

$\frac{I}{rt}=P$ or $P=\frac{I}{rt}$

49. Let x = height of bank

$|x-853|=|\pm170|$

$|x-853|=170$

$x-853=-170$

$x-853+853=-170+853$

$x=683$

or

$x-853=170$

$x-853+853=170+853$

$x=1023$

The two possible heights are 683 feet and 1023 feet. The bank is 1023 feet because 1023 is greater than 853.

50. Let x = limits for C grade

$|x-72|=|\pm5|$

$|x-72|=5$

$x-72=-5$ or $x-72=5$

$x-72+72=-5+72$ $\quad$ $x-72+72=5+72$

$x=67$ $\quad$ $x=77$

The limits are 67 and 77 points.

51. Let x = th65.58

$65.58+5.41=70.99$

The subtotal was about $65.58.

The total bill was about $70.99.

52. Let x = the total amount he expects to receive

$\frac{2}{5}x=5650$

$\frac{5}{2}\left(\frac{2}{5}x\right)=\frac{5}{2}(5650)$

$x=14,125$

He expects to receive $14,125.

53. Let x = the number of hours

$13.25x=16,562.50$

$\frac{13.25x}{13.25}=\frac{16,562.5}{13.25}$

$x=1250$

She worked 1250 hours.

54. Let x = the number of houses

$15+0.25x=45$

$15+0.25x-15=45-15$

$0.25x=30$

$\frac{0.25x}{0.25}=\frac{30}{0.25}$

$x=120$

An employee needs 120 houses.

55. Let x = the annual depreciation

$20x+15,000=250,000$

$20x+15,000-15,000=250,000-15,000$

$20x=235,000$

$\frac{20x}{20}=\frac{235,000}{20}$

$x=11,750$

The annual depreciation is $11,750.

56. Let x = additional money
$x + 985 = 1399$
$x + 985 - 985 = 1399 - 985$
$x = 414$
She needs an additional \$414.

57. Let x = length of first piece
$x + 1$ = length of second piece
$x + 2$ = length of third piece
$x + 3$ = length of fourth piece
$x + x + 1 + x + 2 + x + 3 = 26$
$4x + 6 = 26$
$4x + 6 - 6 = 26 - 6$
$4x = 20$
$\frac{4x}{4} = \frac{20}{4}$
$x = 5$
$x + 1 = 6$
$x + 2 = 7$
$x + 3 = 8$
The pieces should be 5 in., 6 in., 7 in., and 8 in.

58. Let x = length of first piece
$x + 2$ = length of second piece
$x + 4$ = length of third piece
$x + 6$ = length of fourth piece
$x + x + 2 + x + 4 + x + 6 = 26 - 2$
$4x + 12 = 24$
$4x + 12 - 12 = 24 - 12$
$4x = 12$
$\frac{4x}{4} = \frac{12}{4}$
$x = 3$
$x + 2 = 5$
$x + 4 = 7$
$x + 6 = 9$
The pieces should be 3 in., 5 in., 7 in., and 9 in.

59. Let x = price before markdown.
$x - 0.15x = 259.95$
$x(1 - 0.15) = 259.95$
$0.85x = 259.95$
$\frac{0.85x}{0.85} = \frac{259.95}{0.85}$
$x \approx 305.82$
The original price was about \$305.82.

60. $I = PRT$
$1560 = 8000(R)(3)$
$1560 = 24{,}000R$
$\frac{1560}{24{,}000} = \frac{24{,}000R}{24{,}000}$
$0.065 = R$
$R = 6.5\%$
The simple interest rate should be 6.5%.

61. $I = PRT$
$562.50 = P(0.045)(1)$
$562.5 = 0.045P$
$\frac{562.5}{0.045} = \frac{0.045P}{0.045}$
$12{,}500 = P$
You should invest \$12,500.

62. Let x = width
$1.5x$ = length
$2x + 2(1.5x) = 84$
$2x + 3x = 84$
$5x = 84$
$\frac{5x}{5} = \frac{84}{5}$
$x = 16.8$
$1.5x = 25.2$
The dimensions are 16.8 feet by 25.2 feet.

63. Let x = the price before tax
$x + 0.0875x = 325.16$
$1.0875x = 325.16$
$\frac{1.0875x}{1.0875} = \frac{325.16}{1.0875}$
$x \approx 299.00$
$299(0.0875) \approx 26.16$
The price before tax was about \$299.
Sales tax was about \$26.16.

64. $A = 40 + 22.5h$
$A - 40 = 40 + 22.5h - 40$
$A - 40 = 22.5h$
$\frac{A - 40}{22.5} = \frac{22.5h}{22.5}$
$\frac{A - 40}{22.5} = h$ or $h = \frac{A - 40}{22.5}$
$h = \frac{208.75 - 40}{22.5}$
$h = 7.5$
A job that cost \$208.75 lasts 7.5 hours.

$h = \frac{85-40}{22.5}$
$h = 2$
A job that cost \$85 lasts 2 hours.

65. Let x = the measure of the angle made with width
$x + 35 + 90 = 180$
$x + 125 = 180$
$x + 125 - 125 = 180 - 125$
$x = 55$
The angle made with the width of the rug is 55°.

Chapter 5 Practice Test

1. $3.14x + 9.07 = 5.72x$ is linear because it is in the form $ax + b = ax$.

2. $5x = 12 + \frac{19}{x}$ is nonlinear because a variable appears in the denominator of a term.

3. $4x + 21 = 5x^2$ is nonlinear because the variable x has an exponent of 2.

4. $4(x - 6) = 3(5 - x) + 12$ is linear because it simplifies to $4x - 24 = 27 - 3x$.

5. $2(2x - 5) - 2(2 - x) = 6(x + 1) + 1$
$4x - 10 - 4 + 2x = 6x + 6 + 1$
$6x - 14 = 6x + 7$
$6x - 14 - 6x = 6x + 7 - 6x$
$-14 = 7$
This is a false equation. There is no solution.

6. $2(x + 5) = -3(x + 1) - 2$
$2x + 10 = -3x - 3 - 2$
$2x + 10 = -3x - 5$
$2x + 10 - 10 = -3x - 5 - 10$
$2x = -3x - 15$
$2x + 3x = -3x - 15 + 3x$
$5x = -15$
$\frac{5x}{5} = \frac{-15}{5}$
$x = -3$
The solution is –3.

7. $1.41(x + 5.08) + 1.17x + 0.00102 = -3.46x - 5.39334$
$1.41x + 7.1628 + 1.17x + 0.00102 = -3.46x - 5.39334$
$2.58x + 7.16382 = -3.46x - 5.39334$
$2.58x + 7.16382 - 7.16382 = -3.46x - 5.39334 - 7.16382$
$2.58x = -3.46x - 12.55716$
$2.58x + 3.46x = -3.46x - 12.55716 + 3.46x$
$6.04x = -12.55716$
$\frac{6.04x}{6.04} = \frac{-12.55716}{6.04}$
$x = -2.079$
The solution is –2.079.

8. $(x + 1) - 4(x - 1) = 3(2 - x) - 1$
$x + 1 - 4x + 4 = 6 - 3x - 1$
$-3x + 5 = 5 - 3x$
$-3x + 5 + 3x = 5 - 3x + 3x$
$5 = 5$
This is a true equation. The solution is all real numbers.

9. $\frac{4}{5}x + \frac{31}{10} = -\frac{4}{3}x + \frac{41}{6}$
$30\left(\frac{4}{5}x\right) + 30\left(\frac{31}{10}\right) = 30\left(-\frac{4}{3}x\right) + 30\left(\frac{41}{6}\right)$
$24x + 93 = -40x + 205$
$24x + 93 - 93 = -40x + 205 - 93$
$24x = -40x + 112$
$24x + 40x = -40x + 112 + 40x$
$64x = 112$
$\frac{64x}{64} = \frac{112}{64}$
$x = 1\frac{3}{4}$
The solution is $1\frac{3}{4}$.

10. $|x + 3| = 7$
$x + 3 = -7$ or $x + 3 = 7$
$x + 3 - 3 = -7 - 3$ $\quad$ $x + 3 - 3 = 7 - 3$
$x = -10$ $\quad$ $x = 4$
The solutions are –10 and 4.

11. $3|2x-9|+14=5$
$3|2x-9|+14-14=5-14$
$3|2x-9|=-9$
$\frac{3|2x-9|}{3}=\frac{-9}{3}$
$|2x-9|=-3$
There is no solution because the absolute value expression equals a negative number.

12. Let x = length of first piece
$x+1$ = length of second piece
$x+2$ = length of third piece
$x+x+1+x+2=45$
$3x+3=45$
$3x+3-3=45-3$
$3x=42$
$\frac{3x}{3}=\frac{42}{3}$
$x=14$
$x+1=15$
$x+2=16$
The pieces should be cut into sections measuring 14 in., 15in., and 16 in.

13. Let x = his monthly payment
$12x+850=2470$
$12x+850-850=2470-850$
$12x=1620$
$\frac{12x}{12}=\frac{1620}{12}$
$x=135$
His monthly payments will be $135.

14. Let x = the original price
$x-0.25x=179.95$
$x(1-0.25)=179.95$
$0.75x=179.95$
$\frac{0.75x}{0.75}=\frac{179.95}{0.75}$
$x\approx 239.93$
The price before it went on sale was about $239.93.

15. $P=2L+2W$ for W
$P-2L=2L+2W-2L$
$P-2L=2W$
$\frac{P-2L}{2}=\frac{2W}{2}$
$\frac{P-2L}{2}=W$ or $W=\frac{P-2L}{2}$
$W=\frac{44.8-2(14.8)}{2}$
$W=\frac{15.2}{2}$
$W=7.6$
The width is 7.6 inches.

16. Let x = first angle measurement
$x+10$ = second angle measurement
$x+x+10$ = third angle measurement
$x+x+10+x+x+10=180$
$4x+20=180$
$4x+20-20=180-20$
$4x=160$
$\frac{4x}{4}=\frac{160}{4}$
$x=40$
$x+10=50$
$x+x+10=90$
The measures of the angles are 40°, 50°, and 90°.

17. Let x = true percent.
$|x-52|=|\pm 3|$
$|x-52|=3$
$x-52=-3$ or $x-52=3$
$x-52+52=-3+52$ $x-52+52=3+52$
$x=49$ $x=55$
The limits on the true percent are 49% and 55%.

18. Let x = number of liters of 60% apple juice.
$0.60x+0.20(500)=0.50(x+500)$
$0.60x+100=0.50x+250$
$0.60x+100-100=0.50x+250-100$
$0.60x=0.50x+150$
$0.60x-0.50x=0.50x+150-0.50x$
$0.10x=150$
$\frac{0.10x}{0.10}=\frac{150}{0.10}$
$x=1500$
1500 liters of 60% apple juice must be added.

19. Let x = amount of sales.
$1500 = 1200 + 0.4x$
$1500 - 1200 = 1200 + 0.4x - 1200$
$300 = 0.4x$
$\frac{300}{0.4} = \frac{0.4x}{0.4}$
$7500 = x$
The two payment plans will be equal for $7500 of sales.

20. Let x = the number of hours to do the job.
$12x + 50 = 15x$
$12x + 50 - 12x = 15x - 12x$
$50 = 3x$
$\frac{50}{3} = \frac{3x}{3}$
$x = \frac{50}{3} = 16\frac{2}{3}$
The two plans will cost the same if the job lasts $16\frac{2}{3}$ hours.

21. Answers will vary.

Chapter 6

6.1 Experiencing Algebra the Exercise Way

1. $5x + 7y = 35$ is linear and it is written in standard form, $5x + 7y = 35$.
$a = 5,\ b = 7,\ c = 35$

3. $-4\sqrt{x} + y = 8$ is nonlinear because the variable x is in the radicand.

5. $6x^2 + 2y = 12$ is nonlinear because the variable x is squared.

7. $y = \frac{5}{6}x - 2$ is linear.
$y - \frac{5}{6}x = \frac{5}{6}x - 2 - \frac{5}{6}x$
$y - \frac{5}{6}x = -2$
$-\frac{5}{6}x + y = -2$ is standard form.
$a = -\frac{5}{6},\ b = 1,\ c = -2$

9. $x - 2y + 1 = x - 4y + 9$ is linear.
$x - 2y + 1 - 1 = x - 4y + 9 - 1$
$x - 2y = x - 4y + 8$
$x - 2y - x = x - 4y + 8 - x$
$-2y = -4y + 8$
$-2y + 4y = -4y + 8 + 4y$
$2y = 8$
$\frac{2y}{2} = \frac{8}{2}$
$y = 4$ is standard form.
$a = 0,\ b = 1,\ c = 4$

11. $2x - 5 = 0$ is linear.
$2x - 5 + 5 = 0 + 5$
$2x = 5$ is standard form.
$a = 2,\ b = 0,\ c = 5$

13. $4.2x + 0.3y = 1.4$ is linear and it is in standard form, $4.2x + 0.3y = 1.4$.
$a = 4.2,\ b = 0.3,\ c = 1.4$

15.
$$\begin{array}{r|l} \multicolumn{2}{c}{2x - 3y = 11} \\ \hline 2(5) - 3(7) & 11 \\ 10 - 21 & \\ -11 & \end{array}$$
Since $-11 \neq 11$, (5, 7) is not a solution.

17.
$$\begin{array}{r|l} \multicolumn{2}{c}{-2x + y = 2} \\ \hline -2\left(\frac{1}{2}\right) + 3 & 2 \\ -1 + 3 & \\ 2 & \end{array}$$
Since $2 = 2$, $\left(\frac{1}{2}, 3\right)$ is a solution.

19.
$$\begin{array}{r|l} \multicolumn{2}{c}{m(x) = 2.4x + 3.8} \\ \hline 11 & 2.4(3) + 3.8 \\ & 7.2 + 3.8 \\ & 11 \end{array}$$
Since $11 = 11$, (3, 11) is a solution.

21.
$$\begin{array}{r|l} \multicolumn{2}{c}{y = -6.4x + 10.4} \\ \hline -4 & -6.4(4) + 10.4 \\ & -25.6 + 10.4 \\ & -15.2 \end{array}$$
Since $-4 \neq -15.2$, (4, −4) is not a solution.

23.
$$\begin{array}{r|l} \multicolumn{2}{c}{y = 5} \\ \hline 5 & 5 \end{array}$$
Since $5 = 5$, (11, 5) is a solution.

25.
$$\begin{array}{r|l} \multicolumn{2}{c}{-4x = 12} \\ \hline -4(-3) & 12 \\ 12 & \end{array}$$
Since $12 = 12$, (−3, 9) is a solution.

27.
$$\begin{array}{r|l} \multicolumn{2}{c}{y = -2} \\ \hline -5 & -2 \end{array}$$
Since $-5 \neq -2$, (4, −5) is not a solution.

29. $\dfrac{x = 15}{3 \mid 15}$

Since $3 \neq 15$, (3, 15) is not a solution.

31.

$x - 4y = 7$	
$18.6 - 4(-2.9)$	7
$18.6 + 11.6$	
30.2	

Since $30.2 \neq 7$, (18.6, –2.9) is not a solution.

33. Answers will vary. Possible answer:

$x - 6y = 12$
$x - 6y - x = 12 - x$
$-6y = 12 - x$
$\dfrac{-6y}{-6} = \dfrac{12 - x}{-6}$
$\dfrac{-6y}{-6} = \dfrac{12 - x}{-6}$
$y = \dfrac{12}{-6} - \dfrac{x}{-6}$
$y = \dfrac{1}{6}x - 2$

x	$y = \frac{1}{6}x - 2$	y
–6	$y = \frac{1}{6}(-6) - 2 = -3$	–3
0	$y = \frac{1}{6}(0) - 2 = -2$	–2
6	$y = \frac{1}{6}(6) - 2 = -1$	–1

(–6, –3), (0, –2), and (6, –1) are three possible solutions.

35. Answers will vary. Possible answer:

x	$y = -\frac{5}{8}x + 1$	y
–8	$y = -\frac{5}{8}(-8) + 1 = 6$	6
0	$y = -\frac{5}{8}(0) + 1 = 1$	1
8	$y = -\frac{5}{8}(8) + 1 = -4$	–4

(–8, 6), (0, 1), and (8, –4) are three possible solutions.

37. Answers will vary. Possible answer:

x	$p(x) = \frac{4x+1}{3}$	$p(x)$
–1	$p(x) = \frac{4(-1)+1}{3} = -1$	–1
2	$p(x) = \frac{4(2)+1}{3} = 3$	3
5	$p(x) = \frac{4(5)+1}{3} = 7$	7

(–1, –1), (2, 3), and (5, 7) are three possible solutions.

39. Answers will vary. Possible answer:

x	$y = 8$	y
0	$y = 8$	8
1	$y = 8$	8
2	$y = 8$	8

(0, 8), (1, 8), and (2, 8) are three possible solutions.

41. Answers will vary. Possible answer:

x	$y = 2.6x - 4.2$	y
–1	$y = 2.6(-1) - 4.2 = -6.8$	–6.8
0	$y = 2.6(0) - 4.2 = -4.2$	–4.2
1	$y = 2.6(1) - 4.2 = -1.6$	–1.6

(–1, –6.8), (0, –4.2), and (1, –1.6) are three possible solutions.

43. Answers will vary. Possible answer:

$2x + y = 3$
$2x + y - 2x = 3 - 2x$
$y = 3 - 2x$
$y = -2x + 3$

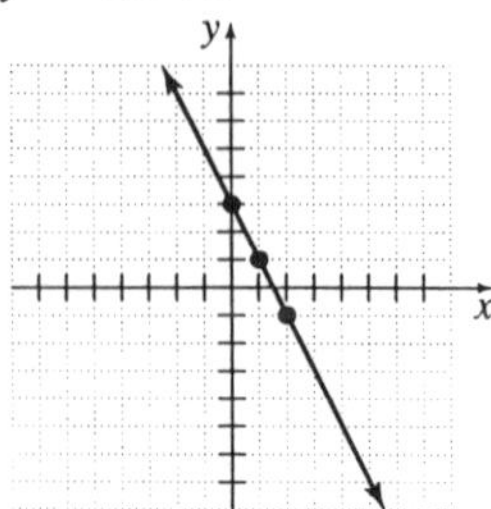

(0, 3), (1, 1), and (2, –1) are three possible solutions.

45. $5x - 3y = 6$
$5x - 3y - 5x = 6 - 5x$
$-3y = 6 - 5x$
$\frac{-3y}{-3} = \frac{6 - 5x}{-3}$
$y = -2 + \frac{5}{3}x$
$y = \frac{5}{3}x - 2$

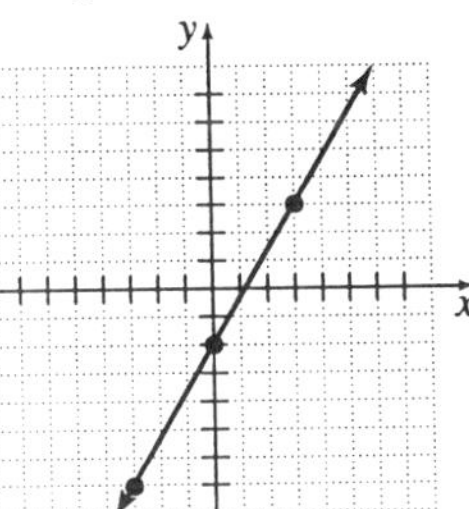

(–3, –7), (0, –2), and (3, 3) are three possible solutions.

47. Answers will vary. Possible answer:
$r(x) = -\frac{3}{4}x + 4$
$y = -\frac{3}{4}x + 4$

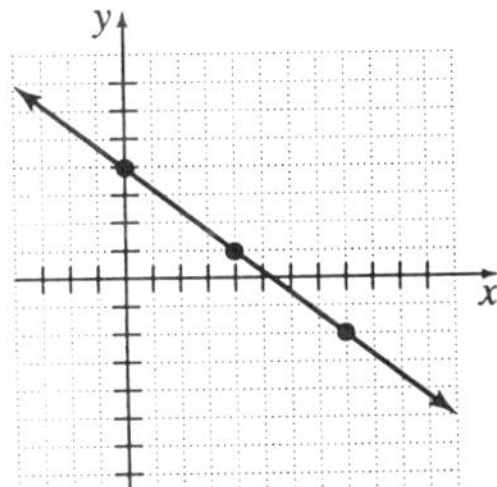

(0, 4), (4, 1), and (8, –2) are three possible solutions.

49. Answers will vary. Possible answer:
$y = 2.8x - 1.6$

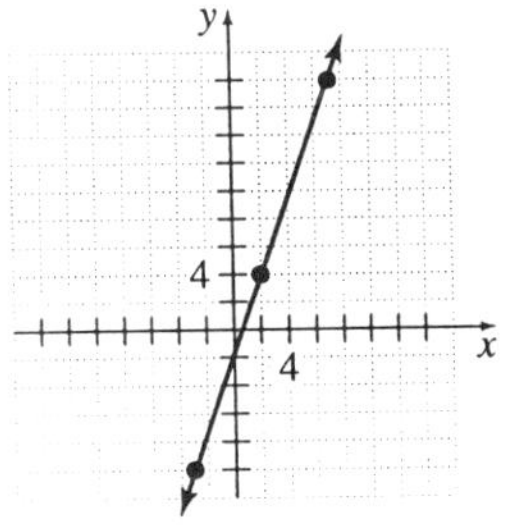

(–3, –10), (2, 4), and (7, 18) are three possible solutions.

51. Answers will vary. Possible answer:
$y = \frac{3x - 5}{2}$
$y = \frac{3}{2}x - \frac{5}{2}$

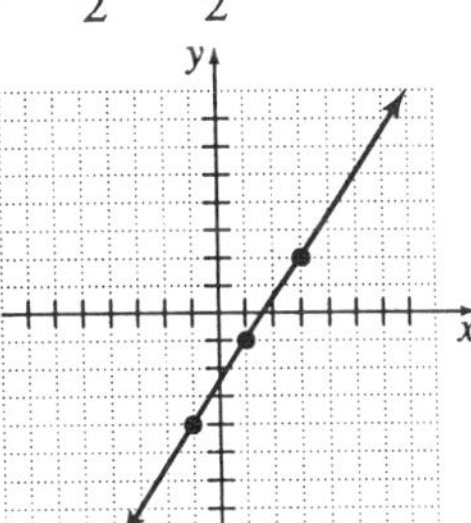

(–1, –4), (1, –1), and (3, 2) are three possible solutions.

53. Answers will vary. Possible answer:
$3x + y - 4 = x + 2y - 3$
$3x - y - 4 - 3x = x + 2y - 3 - 3x$
$y - 4 = -2x + 2y - 3$
$y - 4 + 4 = -2x + 2y - 3 + 4$
$y = -2x + 2y + 1$
$y - 2y = -2x + 2y + 1 - 2y$
$-y = -2x + 1$
$\frac{-y}{-1} = \frac{-2x + 1}{-1}$
$y = 2x - 1$

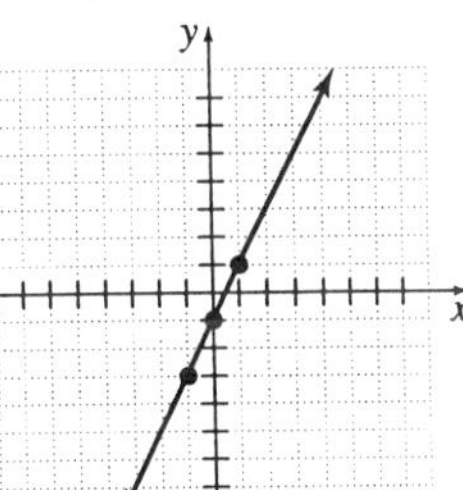

(–1, –3), (0, –1), and (1, 1) are three possible solutions.

55. Answers will vary. Possible answer:
$5y = -20$
$\frac{5y}{5} = \frac{-20}{5}$
$y = -4$

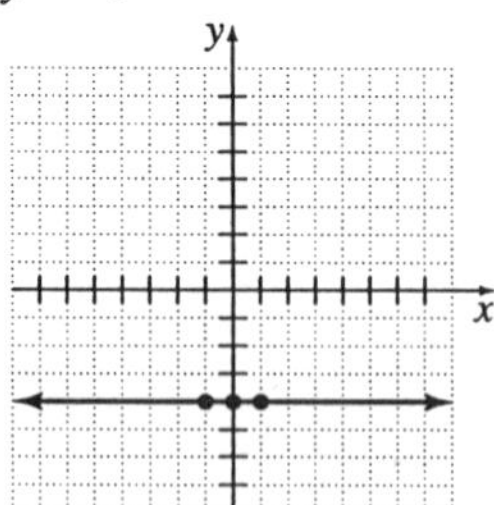

(−1, −4), (0, −4) and (1, −4) are three possible solutions.

57. **a.** (1, 0), (2, 5), (4, 15)

b.

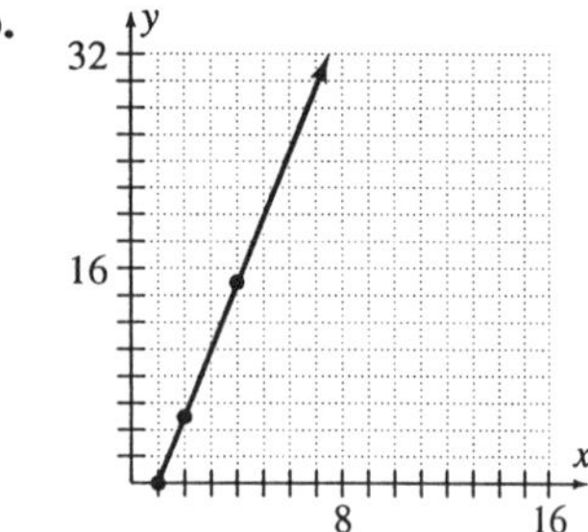

c. At $x = 5$, $y = 20$
A crew of 5 people would pack 20 boxes per minute.

d. Answers will vary. Possible answer: There is probably a maximum number of boxes that can be packed by a large number of people maybe given the size of the packing plant.

59. **a.** $p(x) = 8x + 25$
$p(7) = 8(7) + 25$
$p(7) = 81$
She would earn $81.

b. $p(5) = 8(5) + 25$
$p(5) = 65$
No, she should be paid $65.

c. $p(0) = 8(0) + 25$
$p(0) = 25$
She would be paid $25.

d. Answers will vary. Possible answer:
Domain: 0 to 50
Range: 25 to 425

61. **a.** (2, 6), (3.5, 10.5), (10.5, 31.5)

b.

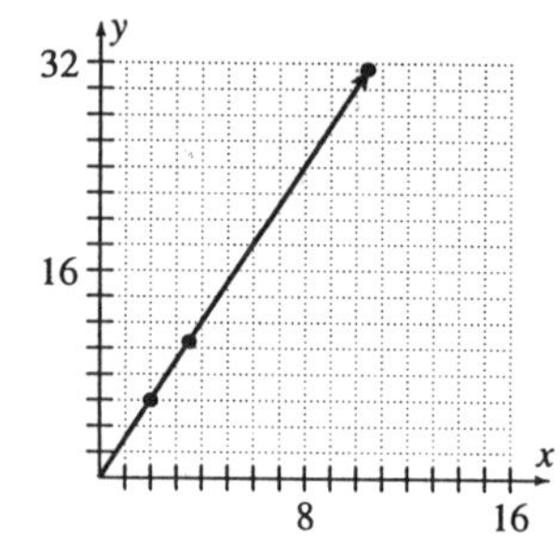

c. At $x = 4$, $y = 12$
The border would be 12 inches.

d. Yes, all equilateral triangles have three sides of equal measure.

63. **a.** $D(x) = 3.75x + 250$
$D(20) = 3.75(20) + 250$
$D(20) = 325$
The cost would be $325.

b. $D(30) = 3.75(30) + 250$
$D(30) = 362.5$
The cost would be $362.50.

c. $D(25) = 3.75(25) + 250$
$D(25) = 343.75$
Yes, $343.75 is less than $350.

d.

x	$D(x) = 3.75x + 250$	$D(x)$
0	$D(0) = 3.75(0) + 250 = 250$	250
5	$D(5) = 3.75(5) + 250 = 268.75$	268.75
10	$D(10) = 3.75(10) + 250 = 287.5$	287.50
15	$D(15) = 3.75(15) + 250 = 306.25$	306.25
20	$D(20) = 3.75(20) + 250 = 325$	325
25	$D(25) = 3.75(25) + 250 = 343.75$	343.75
30	$D(30) = 3.75(30) + 250 = 362.5$	362.50
35	$D(35) = 3.75(35) + 250 = 381.25$	381.25
40	$D(40) = 3.75(40) + 250 = 400$	400
45	$D(45) = 3.75(45) + 250 = 418.75$	418.75
50	$D(50) = 3.75(50) + 250 = 437.5$	437.50

65. Answers will vary. Possible answer:
(5, 225), (10, 350), (15, 475)
The cost of producing 5 items, 10 items, and 15 items is $225, $350, and $475, respectively.

6.1 Experiencing Algebra the Calculator Way

1. $y = 6.871x - 35.711249$

3.287	$6.871(4.719) - 35.711249$
	-3.287

Since $3.287 \neq -3.287$, (4.719, 3.287) is not a solution.

2. $1.05x + 1.45y = -23.75$

$1.05(25) + 1.45(50)$	-23.75
98.75	

Since $98.75 \neq -23.75$, (25, 50) is not a solution.

3. $132x - 297y = 1526$

$132\left(\frac{17}{33}\right) - 297\left(-4\frac{10}{11}\right)$	1526
1526	

Since $1526 = 1526$, $\left(\frac{17}{33}, -4\frac{10}{11}\right)$ is a solution.

4. $y = 3.721x - 1.857$

x	y
1.5	3.7245
1.6	4.0966
1.7	4.4687
1.8	4.8408
1.9	5.2129
2.0	5.585

5. $y=\dfrac{13}{28}x-5$

x	y
0	–5
7	–1.75
14	1.5
21	4.75
28	8
35	11.25

6. $y=\sqrt{3}x-\sqrt{2}$

$\text{Y1}=\sqrt{3}x-\sqrt{2}$

31

–47 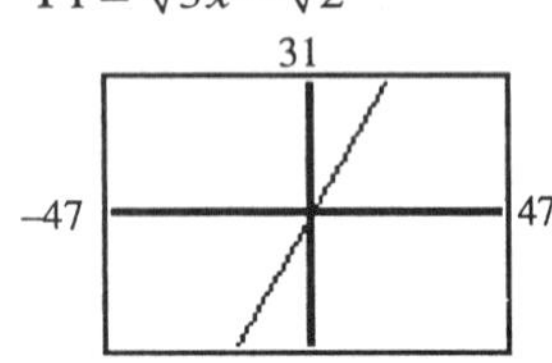 47

–31

(0, –1.41), (2, 2.05), (4, 5.51), (6, 8.98), (8, 12.44)

7. $y=\pi r^2 x$

$y=\pi(0.25)^2 x$

$y=0.0625\pi x$

$\text{Y1}=0.0625\pi x$

31

–47 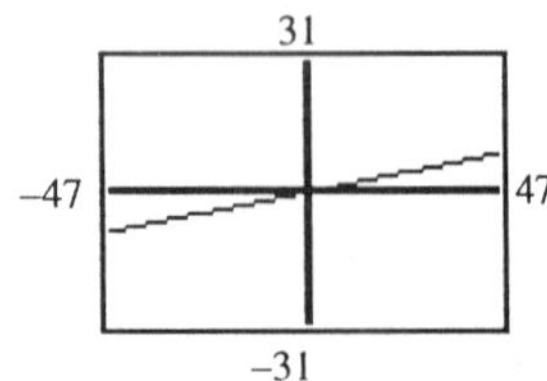 47

–31

(1, 0.20), (2, 0.39), (3, 0.59), (4, 0.79), (5, 0.98)

8. $\text{Y1}=9.64x$

X	Y1
5	48.2
10	96.4
15	144.6
20	192.8
25	241
30	289.2
35	337.4

X=35

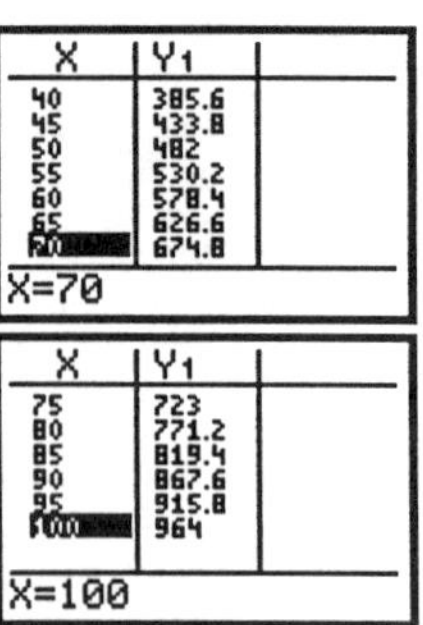

X	Y1
40	385.6
45	433.8
50	482
55	530.2
60	578.4
65	626.6
70	674.8

X=70

X	Y1
75	723
80	771.2
85	819.4
90	867.6
95	915.8
100	964

X=100

6.2 Experiencing Algebra the Exercise Way

1. x-intercept (–2, 0)
y-intercept (0, 4)

3. x-intercept (4, 0)
y-intercept (0, 2)

5. x-intercept (0, 0)
y-intercept (0, 0)

7. x-intercept (3, 0)
y-intercept none

9. x-intercept none
y-intercept (0, 1)

11. $\text{Y1}=5x-24$

31

–47 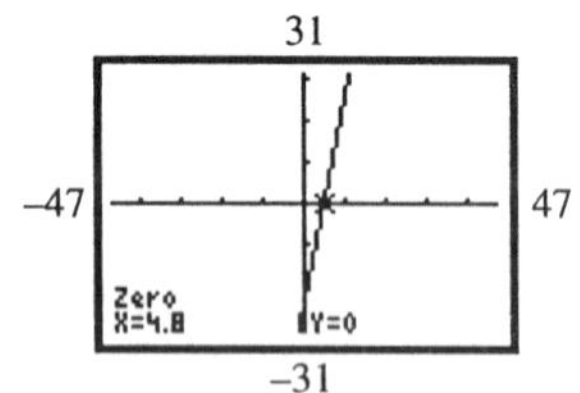 47

–31

x-intercept (4.8, 0)
y-intercept (0, –24)

13. $\text{Y1}=3x+18$

31

–47 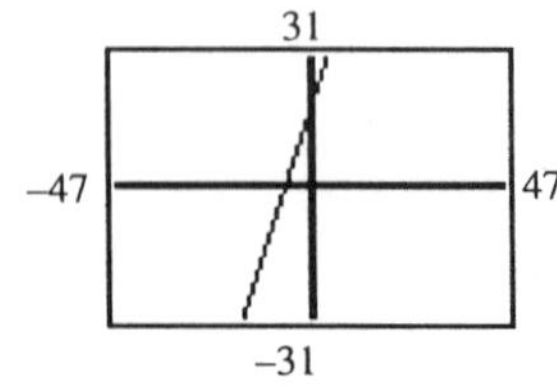 47

–31

x-intercept (–6, 0)
y-intercept (0, 18)

15. $5x+4y=30$
$5x+4y-5x=30-5x$
$4y=30-5x$
$\frac{4y}{4}=\frac{30-5x}{4}$
$y=\frac{30}{4}-\frac{5}{4}x$
$y=-\frac{5}{4}x+\frac{15}{2}$
$Y1=-\frac{5}{4}x+\frac{15}{2}$

31
−47 47
−31

x-intercept (6, 0)
y-intercept (0, 7.5)

17. $5x-7y=28$
$5x-7y-5x=28-5x$
$-7y=28-5x$
$\frac{-7y}{-7}=\frac{28-5x}{-7}$
$y=-\frac{28}{7}+\frac{5}{7}x$
$y=\frac{5}{7}x-4$
$Y1=\frac{5}{7}x-4$

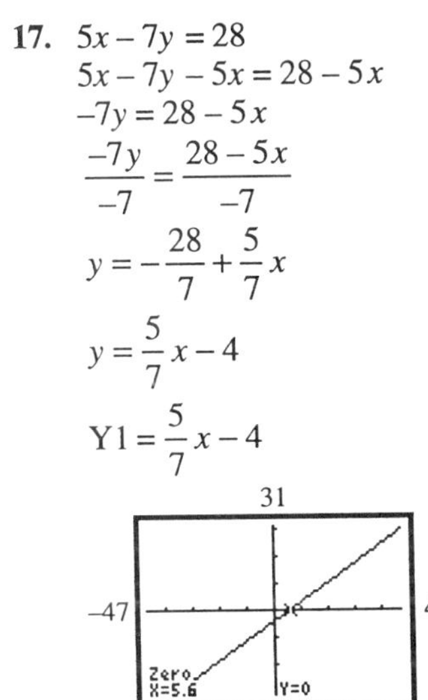

x-intercept (5.6, 0)
y-intercept (0, −4)

19. $-2x-3y=10$
$-2x-3y+2x=10+2x$
$-3y=10+2x$
$\frac{-3y}{-3}=\frac{10+2x}{-3}$
$y=-\frac{10}{3}-\frac{2}{3}x$
$y=-\frac{2}{3}x-\frac{10}{3}$
$Y1=-\frac{2}{3}x-\frac{10}{3}$

31
−47 47
−31

x-intercept (−5, 0)
y-intercept $(0, -3.\overline{3})$

21. $3x+5y=12$
To find the x-intercept, substitute 0 for y.
$3x+5(0)=12$
$3x=12$
$\frac{3x}{3}=\frac{12}{3}$
$x=4$
The x-intercept is (4, 0).
To find the y-intercept, substitute 0 for x.
$3(0)+5y=12$
$5y=12$
$\frac{5y}{5}=\frac{12}{5}$
$y=\frac{12}{5}$
The y-intercept is $\left(0, \frac{12}{5}\right)$.

23. $4x-7y=14$
Substitute 0 for y.
$4x-7(0)=14$
$4x=14$
$\frac{4x}{4}=\frac{14}{4}$
$x=\frac{7}{2}$

The x-intercept is $\left(\frac{7}{2}, 0\right)$.
Substitute 0 for x.
$4(0) - 7y = 14$
$-7y = 14$
$\frac{-7y}{-7} = \frac{14}{-7}$
$y = -2$
The y-intercept is $(0, -2)$.

25. $-2x - 9y = 27$
Substitute 0 for y.
$-2x - 9(0) = 27$
$-2x = 27$
$\frac{-2x}{-2} = \frac{27}{-2}$
$x = -\frac{27}{2}$
The x-intercept is $\left(-\frac{27}{2}, 0\right)$.
Substitute 0 for x.
$-2(0) - 9y = 27$
$-9y = 27$
$\frac{-9y}{-9} = \frac{27}{-9}$
$y = -3$
The y-intercept is $(0, -3)$.

27. $6x + 9y - 36 = 0$
Substitute 0 for y.
$6x + 9(0) - 36 = 0$
$6x - 36 = 0$
$6x - 36 + 36 = 0 + 36$
$6x = 36$
$\frac{6x}{6} = \frac{36}{6}$
$x = 6$
The x-intercept is $(6, 0)$.
Substitute 0 for x.
$6(0) + 9y - 36 = 0$
$9y - 36 = 0$
$9y - 36 + 36 = 0 + 36$
$9y = 36$
$\frac{9y}{9} = \frac{36}{9}$
$y = 4$
The y-intercept is $(0, 4)$.

29. $3x + 7y = 0$
Substitute 0 for y.
$3x + 7(0) = 0$
$3x = 0$
$\frac{3x}{3} = \frac{0}{3}$
$x = 0$
The x-intercept is $(0, 0)$.
Substitute 0 for x.
$3(0) + 7y = 0$
$7y = 0$
$\frac{7y}{7} = \frac{0}{7}$
$y = 0$
The y-intercept is $(0, 0)$.

31. $6x - 8 = 2x + 32$
$6x - 8 - 2x = 2x + 32 - 2x$
$4x - 8 = 32$
$4x - 8 + 8 = 32 + 8$
$4x = 40$
$\frac{4x}{4} = \frac{40}{4}$
$x = 10$
The x-intercept is $(10, 0)$.
There is no y-intercept.

33. $y = 3y - 22$
$y - 3y = 3y - 22 - 3y$
$-2y = -22$
$\frac{-2y}{-2} = \frac{-22}{-2}$
$y = 11$
The y-intercept is $(0, 11)$.
There is no x-intercept.

35. $12x - y = 24$
$12x - y - 12x = 24 - 12x$
$-y = 24 - 12x$
$\frac{-y}{-1} = \frac{24 - 12x}{-1}$
$y = -24 + 12x$
$y = 12x - 24$
The y-coordinate is the constant -24.
The x-coordinate is 0. The y-intercept is $(0, -24)$.

37. $y = 5(x - 3)$
$y = 5x - 15$
The y-intercept is $(0, -15)$.

39. $5x - 15y = 0$
$5x - 15y - 5x = 0 - 5x$
$-15y = -5x$
$\frac{-15y}{-15} = \frac{-5x}{-15}$
$y = \frac{1}{3}x$
The constant is 0.
The y-intercept is $(0, 0)$.

41. $3y = 12y + 18$
$3y - 12y = 12y + 18 - 12y$
$-9y = 18$
$\frac{-9y}{-9} = \frac{18}{-9}$
$y = -2$
The y-intercept is $(0, -2)$.

43. $y = 0$
The y-intercept is $(0, 0)$.

45. $-17.6x + 2.2y = 19.8$
$-17.6x + 2.2y + 17.6x = 19.8 + 17.6x$
$2.2y = 19.8 + 17.6x$
$\frac{2.2y}{2.2} = \frac{19.8 + 17.6x}{2.2}$
$y = \frac{19.8}{2.2} + \frac{17.6x}{2.2}$
$y = 8x + 9$
The y-intercept is $(0, 9)$.

47. $x = 12y$
$\frac{x}{12} = \frac{12y}{12}$
$\frac{x}{12} = y$
$y = \frac{1}{12}x$
The constant is 0.
The y-intercept is $(0, 0)$.

49. $3x + 5y = 30$
Substitute 0 for y.
$3x + 5(0) = 30$
$3x = 30$
$\frac{3x}{3} = \frac{30}{3}$
$x = 10$
The x-intercept is $(10, 0)$.
Substitute 0 for x.
$3(0) + 5y = 30$
$5y = 30$
$\frac{5y}{5} = \frac{30}{5}$
$y = 6$
The y-intercept is $(0, 6)$.

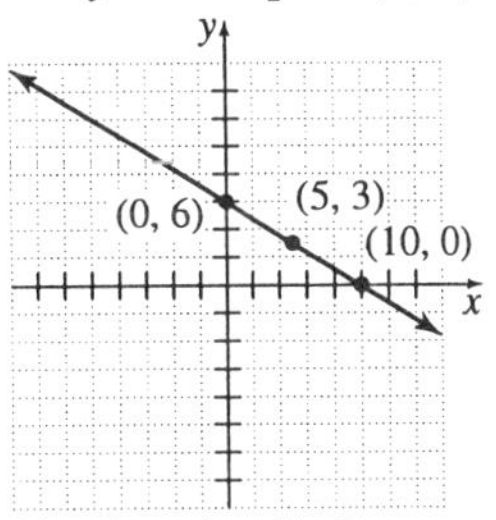

Check point $(5, 3)$.

$3x + 5y = 30$	
$3(5) + 5(3)$	30
$15 + 15$	
30	

Since $30 = 30$, $(5, 3)$ is a solution.

51. $4x - 3y = 24$
Substitute 0 for y.
$4x - 3(0) = 24$
$4x = 24$
$\frac{4x}{4} = \frac{24}{4}$
$x = 6$
The x-intercept is $(6, 0)$.
Substitute 0 for x.
$4(0) - 3y = 24$
$-3y = 24$
$\frac{-3y}{-3} = \frac{24}{-3}$
$y = -8$
The y-intercept is $(0, -8)$.

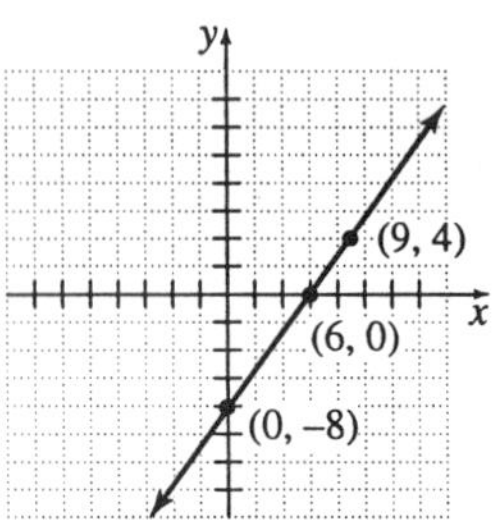

Check point (9, 4).

$4x-3y=24$	
$4(9)-3(4)$	24
$36-12$	
24	

Since 24 = 24, (9, 4) is a solution.

53. $x-y=9$
Substitute 0 for y.
$x-0=9$
$x=9$
The x-intercept is (9, 0).
Substitute 0 for x.
$0-y=9$
$\frac{-y}{-1}=\frac{9}{-1}$
$y=-9$
The y-intercept is (0, –9).

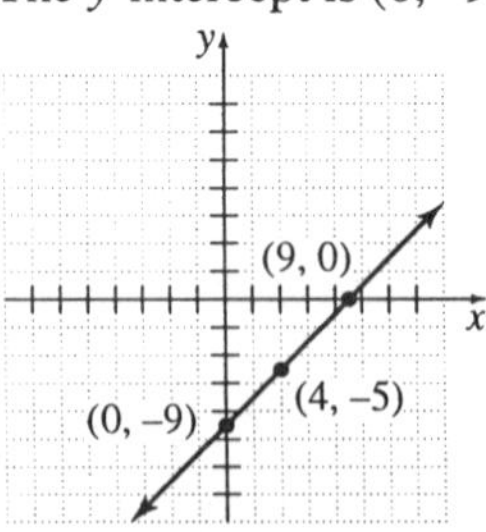

Check point (4, –5).

$x-y=9$	
$4-(-5)$	9
$4+5$	
9	

Since 9 = 9, (4, –5) is a solution.

55. $-x-y=9$
Substitute 0 for y.
$-x-0=9$
$\frac{-x}{-1}=\frac{9}{-1}$
$x=-9$
The x-intercept is (–9, 0).
Substitute 0 for x.
$0-y=9$
$\frac{-y}{-1}=\frac{9}{-1}$
$y=-9$
The y-intercept is (0, –9).

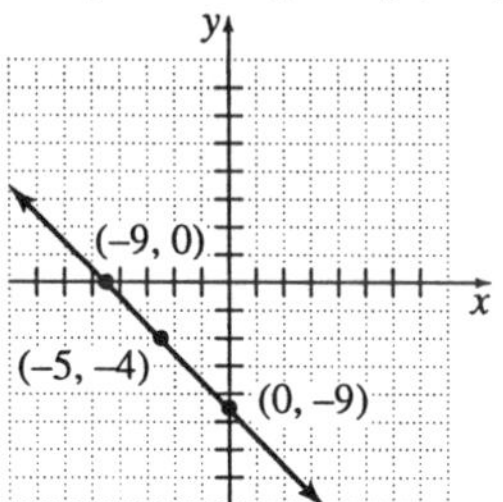

Check point (–5, –4).

$-x-y=9$	
$-(-5)-(-4)$	9
$5+4$	
9	

Since 9 = 9, (–5, –4) is a solution.

57. $3x-7y=-14$
Substitute 0 for y.
$3x-7(0)=-14$
$3x=-14$
$\frac{3x}{3}=\frac{-14}{3}$
$x=-\frac{14}{3}$
The x-intercept is $\left(-\frac{14}{3}, 0\right)$.
Substitute 0 for x.
$3(0)-7y=-14$
$-7y=-14$
$\frac{-7y}{-7}=\frac{-14}{-7}$
$y=2$
The y-intercept is (0, 2).

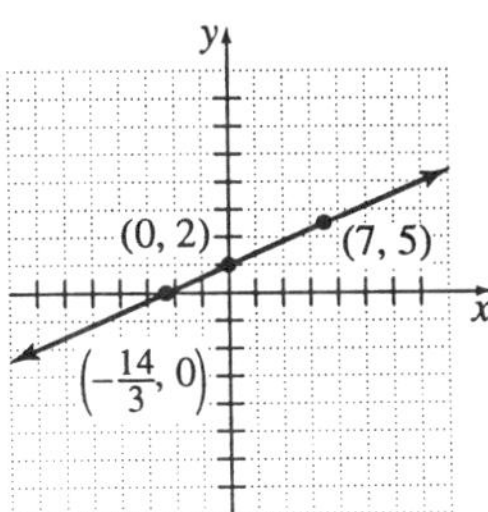

Check point (7, 5).

$3x - 7y$	$= -14$
$3(7) - 7(5)$	-14
$21 - 35$	
-14	

Since $-14 = -14$, (7, 5) is a solution.

59. $y = 6.1x - 23.18$
Substitute 0 for y.
$0 = 6.1x - 23.18$
$0 + 23.18 = 6.1x - 23.18 + 23.18$
$23.18 = 6.1x$
$\frac{23.18}{6.1} = \frac{6.1x}{6.1}$
$3.8 = x$
The x-intercept is (3.8, 0).
The y-intercept is (0, –23.18).

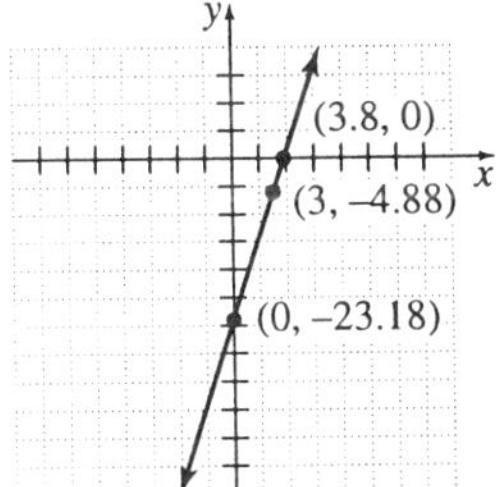

Check point (3, –4.88).

y	$= 6.1x - 23.18$
-4.88	$6.1(3) - 23.18$
	$18.3 - 23.18$
	-4.88

Since $-4.88 = -4.88$, (3, –4.88) is a solution.

61. $y = \frac{5}{3}x + 10$
Substitute 0 for y.
$0 = \frac{5}{3}x + 10$
$0 - 10 = \frac{5}{3}x + 10 - 10$
$-10 = \frac{5}{3}x$
$\frac{3}{5}(-10) = \frac{3}{5}\left(\frac{5}{3}x\right)$
$-6 = x$
The x-intercept is (–6, 0).
The y-intercept is (0, 10).

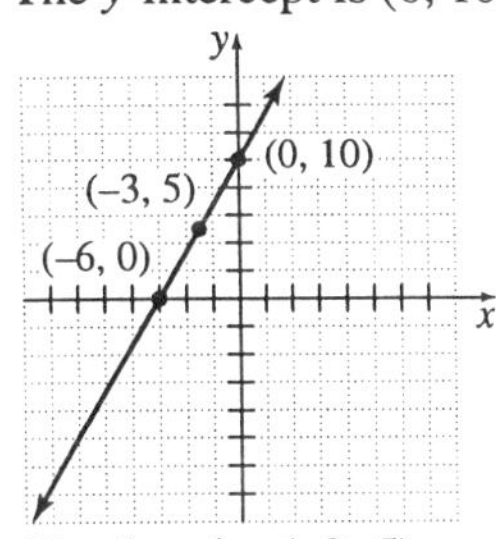

Check point (–3, 5).

y	$= \frac{5}{3}x + 10$
5	$\frac{5}{3}(-3) + 10$
	$-5 + 10$
	5

Since $5 = 5$, (–3, 5) is a solution.

63. $y = -40x + 200$
Substitute 0 for y.
$0 = -40x + 200$
$40x + 0 = -40x + 200 + 40x$
$40x = 200$
$\frac{40x}{40} = \frac{200}{40}$
$x = 5$
Substitute 0 for x.
$y = -40(0) + 200$
$y = 200$
The x-intercept is (5, 0) which represents after 5 hours, there will be 0 miles left. The y-intercept is (0, 200) which represents after 0 hours, there are 200 miles to travel.

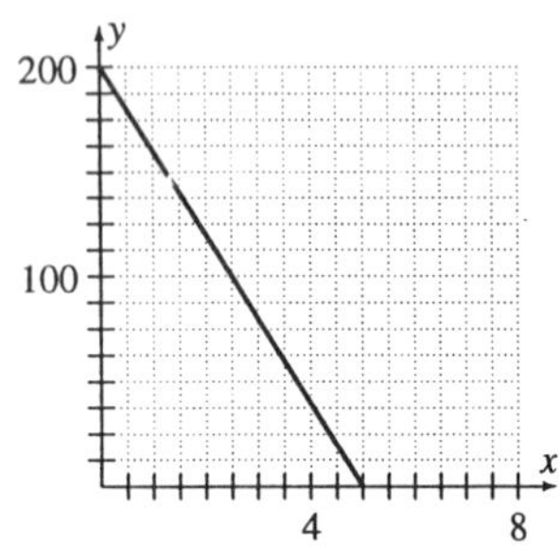

65. $y = -4x + 150$
Substitute 0 for y.
$0 = -4x + 150$
$4x + 0 = -4x + 150 + 4x$
$4x = 150$
$\frac{4x}{4} = \frac{150}{4}$
$x = 37.5$
Substitute 0 for x.
$y = -4(0) + 150$
$y = 150$
The x-intercept is (37.5, 0) which represents after 37.5 minutes, the tank is empty. The y-intercept is (0, 150) which represents after 0 minutes, the tank is full.

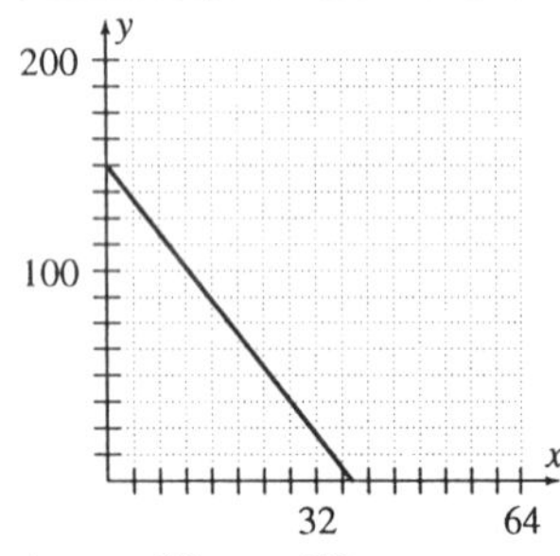

At $x = 20$, $y = 70$.
After 20 minutes there are 70 gallons remaining.

67. The x-intercept is (0, 0).
The y-intercept is (0, 0).
The cost of producing 0 units is $0.
At $x = 50$, $y = 250$.
The cost of producing 50 units is $250.
At $x = 75$, $y = 375$.
The cost of producing 75 units is $375.

6.2 Experiencing Algebra the Calculator Way

1. $\text{Y1} = \frac{45}{88}x - 15$

31

–47 47

Zero
X=29.333333 Y=0

–31

x-intercept $(29.\overline{3}, 0)$
y-intercept (0, –15)

2. $\text{Y1} = -2.56x + 33.71$

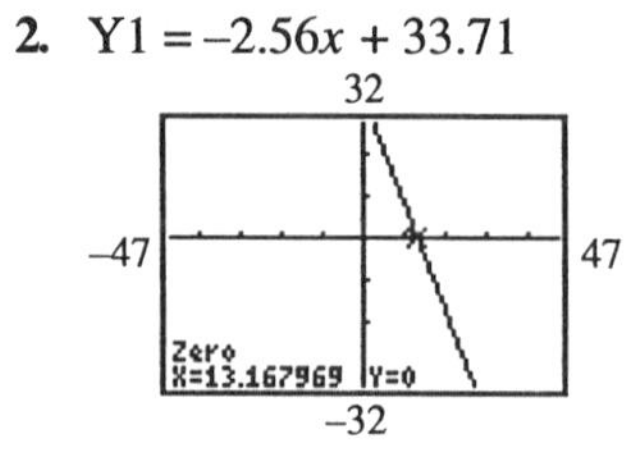

x-intercept (13.168, 0)
y-intercept (0, 33.71)

3. $F = \frac{9}{5}C + 32$

$\text{Y1} = \frac{9}{5}x + 32$

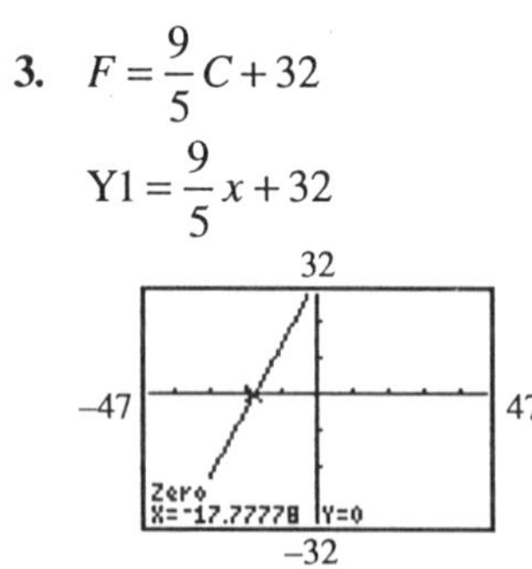

x-intercept $(-17.\overline{7}, 0)$
y-intercept (0, 32)

4. $C = \frac{5}{9}(F - 32)$

$Y1 = \frac{5}{9}(x - 32)$

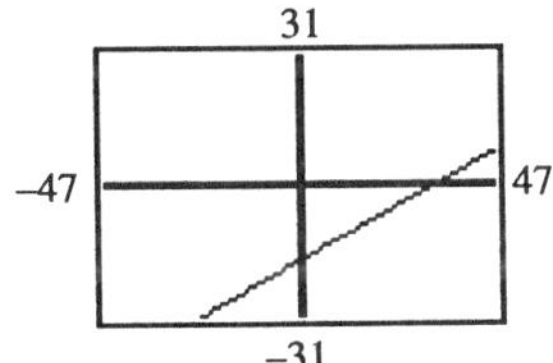

x-intercept (32, 0)
y-intercept $(0, -17.\overline{7})$

5. At 0°C, the Fahrenheit temperature is 32°F.
At 0°F, the Celsius temperature is $-17.\overline{7}$°C.

6.3 Experiencing Algebra the Exercise Way

1.

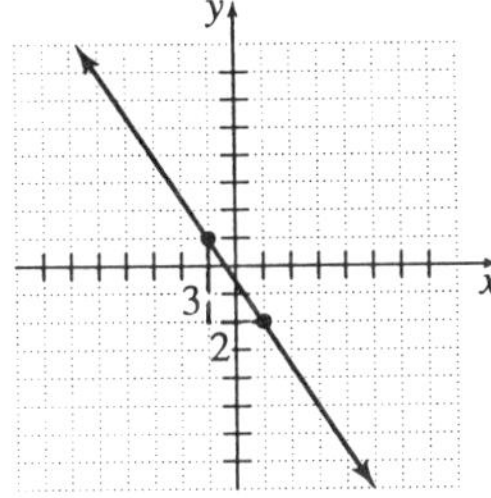

The slope is negative.

The slope is $-\frac{3}{2}$.

3. The slope is zero.

5.

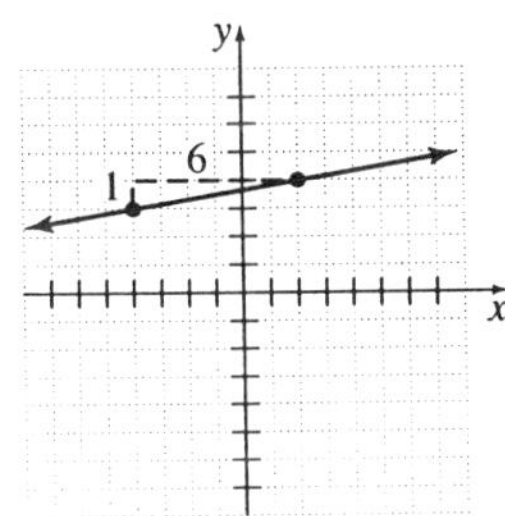

The slope is positive.

The slope is $\frac{1}{6}$.

7. The slope is undefined.

9. (–7, –2) and (5, 6)

$m = \frac{y_2 - y_1}{x_2 - x_1}$

$m = \frac{6-(-2)}{5-(-7)}$

$m = \frac{8}{12}$

$m = \frac{2}{3}$

The slope is $\frac{2}{3}$.

11. (–12, –9) and (4, –9)

$m = \frac{y_2 - y_1}{x_2 - x_1}$

$m = \frac{-9-(-9)}{4-(-12)}$

$m = \frac{0}{16}$

$m = 0$

The slope is 0.

13. (0, 3) and (0, 8)

$m = \frac{y_2 - y_1}{x_2 - x_1}$

$m = \frac{8-3}{0-0}$

$m = \frac{5}{0}$

The slope is undefined.

15. (6, –4) and (7, –6)

$m = \frac{y_2 - y_1}{x_2 - x_1}$

$m = \frac{-6-(-4)}{7-6}$

$m = \frac{-2}{1}$

$m = -2$

The slope is –2.

17. (0, 4) and (5, 0)
$m=\frac{y_2-y_1}{x_2-x_1}$
$m=\frac{0-4}{5-0}$
$m=-\frac{4}{5}$
The slope is $-\frac{4}{5}$.

19. (11.5, –9.2) and (6.9, 18.4)
$m=\frac{y_2-y_1}{x_2-x_1}$
$m=\frac{18.4-(-9.2)}{6.9-11.5}$
$m=\frac{27.6}{-4.6}$
$m=-6$
The slope is –6.

21. $\left(\frac{1}{2},\frac{3}{4}\right)$ and $\left(-\frac{1}{2},-\frac{5}{6}\right)$
$m=\frac{y_2-y_1}{x_2-x_1}$
$m=\frac{-\frac{5}{6}-\frac{3}{4}}{-\frac{1}{2}-\frac{1}{2}}$
$m=\frac{-\frac{19}{12}}{-1}$
$m=\frac{19}{12}$
The slope is $\frac{19}{12}$.

23. $y=21x+15$
$m=21,\ b=15$
The slope is 21 and the y-intercept is (0, 15).

25. $y=\frac{11}{15}x-\frac{21}{25}$
$m=\frac{11}{15},\ b=-\frac{21}{25}$
The slope is $\frac{11}{15}$ and the y-intercept is $\left(0,-\frac{21}{25}\right)$.

27. $4=5.95x-2.01$
$m=5.95,\ b=-2.01$
The slope is 5.95 and the y-intercept is (0, –2.01).

29. $y=85{,}600-1255x$
$y=-1255x+85{,}600$
$m=-1255,\ b=85{,}600$
The slope is –1255 and the y-intercept is (0, 85,600).

31. $16x-4y=64$
$16x-4y-16x=64-16x$
$-4y=64-16x$
$\frac{-4y}{-4}=\frac{64-16x}{-4}$
$y=\frac{64}{-4}-\frac{16x}{-4}$
$y=-16+4x$
$y=4x-16$
$m=4,\ b=-16$
The slope is 4 and the y-intercept is (0, –16).

33. $7y+18=2(y+6)-4$
$7y+18=2y+12-4$
$7y+18=2y+8$
$7y+18-2y=2y+8-2y$
$5y+18=8$
$5y+18-18=8-18$
$5y=-10$
$\frac{5y}{5}=\frac{-10}{5}$
$y=-2$
or $y=0x-2$
$m=0,\ b=-2$
The slope is 0 and the y-intercept is (0, –2).

35. $7.83x - 2.61y = 10.44$
$7.83x - 2.61y - 7.83x = 10.44 - 7.83x$
$-2.61y = 10.44 - 7.83x$
$\frac{-2.61y}{-2.61} = \frac{10.44 - 7.83x}{-2.61}$
$y = \frac{10.44}{-2.61} - \frac{7.83x}{-2.61}$
$y = -4 + 3x$
$y = 3x - 4$
$m = 3,\ b = -4$
The slope is 3 and the y-intercept is $(0, -4)$.

37. $\frac{3}{2}x - \frac{3}{5}y = \frac{21}{10}$
$\frac{3}{2}x - \frac{3}{5}y - \frac{3}{2}x = \frac{21}{10} - \frac{3}{2}x$
$-\frac{3}{5}y = \frac{21}{10} - \frac{3}{2}x$
$-\frac{5}{3}\left(-\frac{3}{5}y\right) = -\frac{5}{3}\left(\frac{21}{10} - \frac{3}{2}x\right)$
$y = -\frac{7}{2} + \frac{5}{2}x$
$y = \frac{5}{2}x - \frac{7}{2}$
$m = \frac{5}{2},\ b = -\frac{7}{2}$
The slope is $\frac{5}{2}$ and the y-intercept is $\left(0, -\frac{7}{2}\right)$.

39. $y = -4\frac{7}{8}$
or $y = 0x - 4\frac{7}{8}$
$m = 0,\ b = -4\frac{7}{8}$
The slope is 0 and the y-intercept is $\left(0, -4\frac{7}{8}\right)$.

41. (8, 3) and $m = \frac{4}{7}$

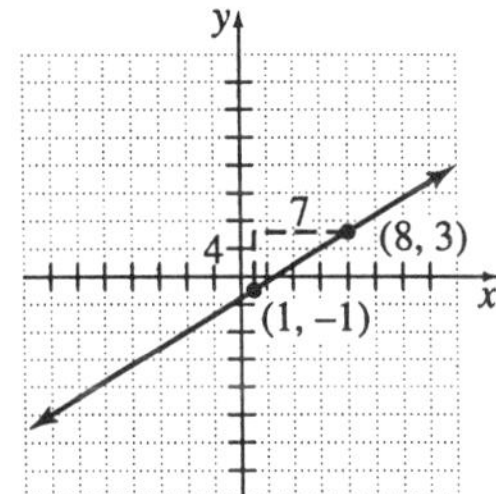

43. (−10, 4) and $m = -\frac{5}{9}$

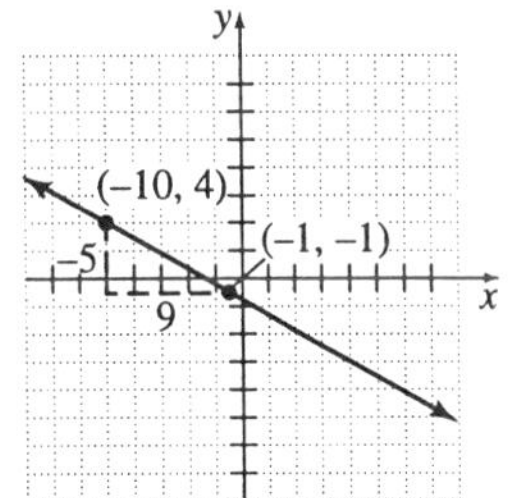

45. (5, 7) and $m = 4$

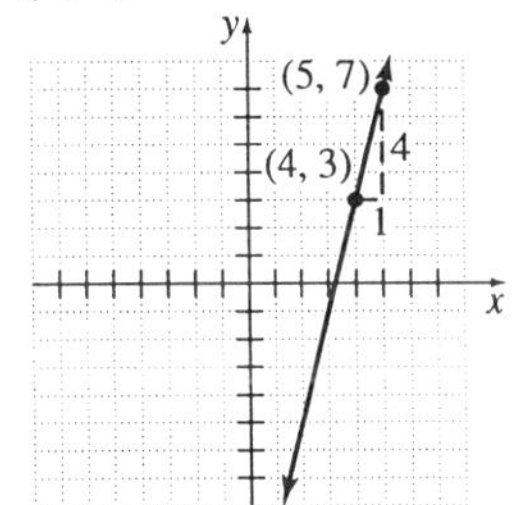

47. (0, 9) and $m = 0$

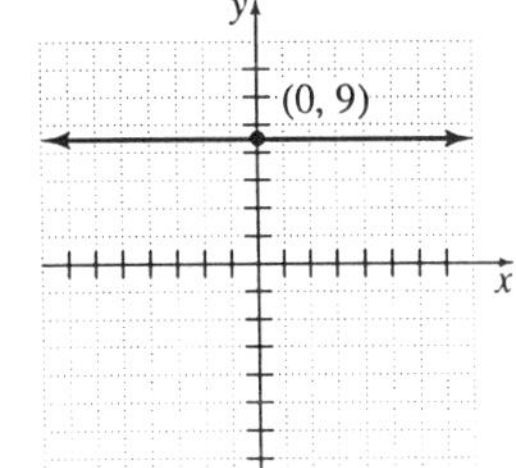

49. (9, 0) and m = undefined

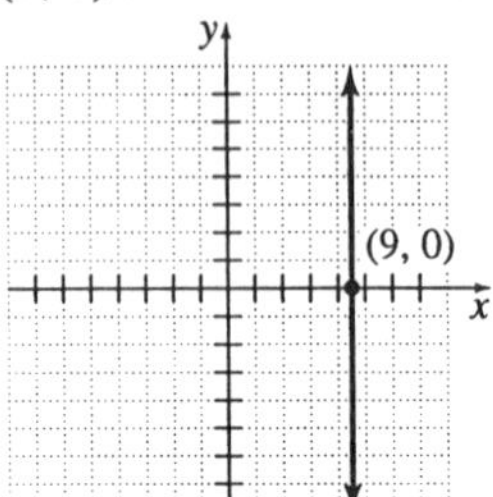

51. $y=\frac{5}{3}x-4$

$m=\frac{5}{3},\ b=-4$

The slope is $\frac{5}{3}$ and the y-intercept is (0, –4).

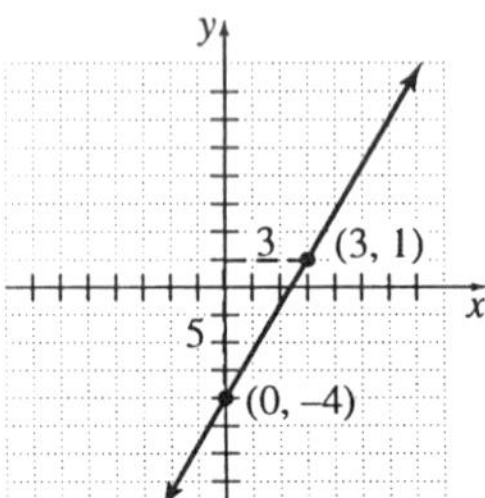

53. $16x-8y=40$

$16x-8y-16x=40-16x$

$-8y=40-16x$

$\frac{-8y}{-8}=\frac{40-16x}{-8}$

$y=\frac{40}{-8}-\frac{16x}{-8}$

$y=-5+2x$

$y=2x-5$

$m=2,\ b=-5$

The slope is 2 and the y-intercept is (0, –5).

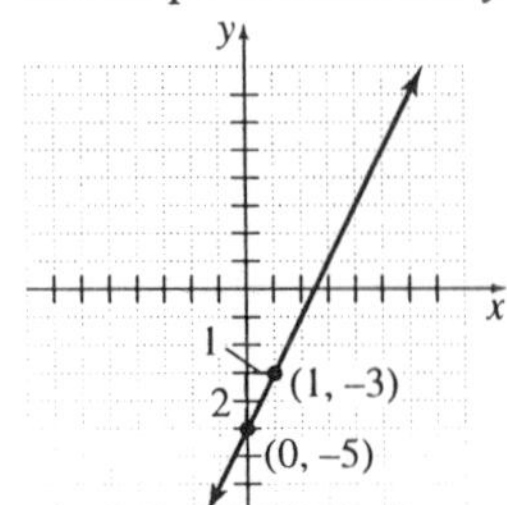

55. $7x+2y=-16$

$7x+2y-7x=-16-7x$

$2y=-16-7x$

$\frac{2y}{2}=\frac{-16-7x}{2}$

$y=\frac{-16}{2}-\frac{7x}{2}$

$y=-\frac{7}{2}x-8$

$m=-\frac{7}{2},\ b=-8$

The slope is $-\frac{7}{2}$ and the y-intercept is (0, –8).

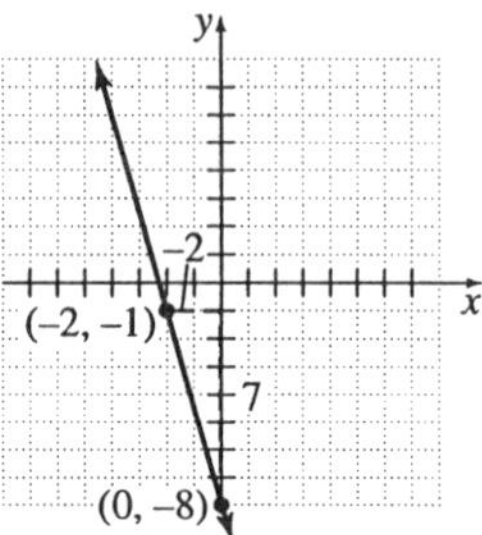

57. $14y+21=6y+5$

$14y+21-6y=6y+5-6y$

$8y+21=5$

$8y+21-21=5-21$

$8y=-16$

$\frac{8y}{8}=\frac{-16}{8}$

$y=-2$

$m=0,\ b=-2$

The slope is 0 and the y-intercept is (0, –2).

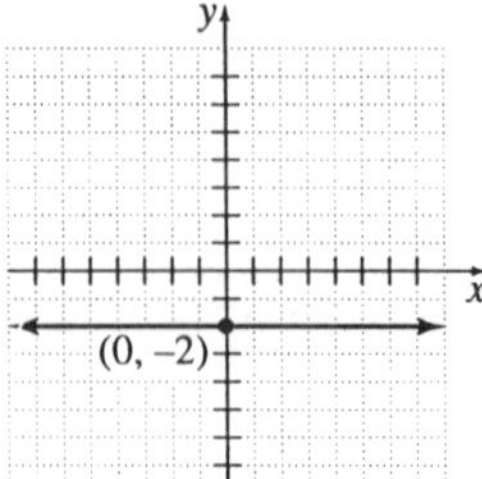

59. $5y = 150x + 350$

$\frac{5y}{5} = \frac{150x + 350}{5}$

$y = 30x + 70$

$m = 30,\ b = 70$

The slope is 30 and the y-intercept is (0, 70).

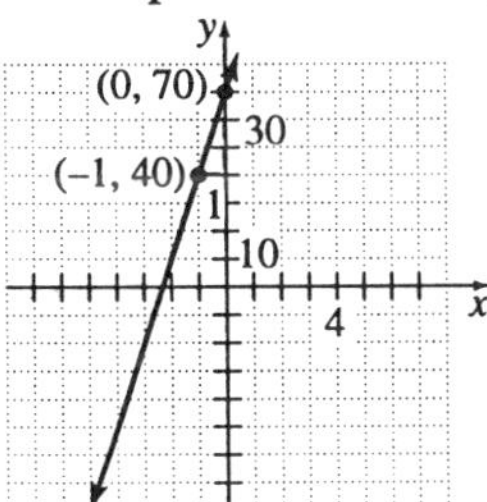

61. $f(x) = 0.3x - 1.2$

$m = 0.3,\ b = -1.2$

The slope is 0.3 or $\frac{3}{10}$ and the y-intercept is (0, –1.2).

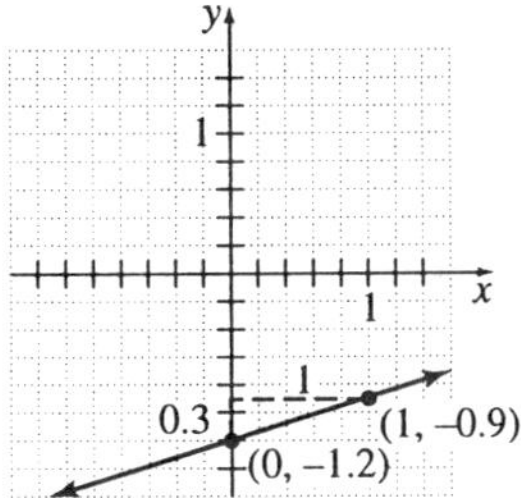

63.

56 ft

160 ft

$\text{grade} = \frac{56}{160} \times 100\%$

$= 35\%$

The grade of the advertised terrain is 35%.

65.

3.3 in.

12 in.

$\text{pitch} = \frac{3.3}{12} \times 100\% = 27.5\%$

The pitch of the roof is 27.5%.

67. $R = \frac{D}{T}$

$R = \frac{10.5}{1.5}$

$R = 7$

Their average speed was 7 miles per hour.

$y = 7x$

$m = 7,\ b = 0$

The slope is 7 and the y-intercept is (0, 0).

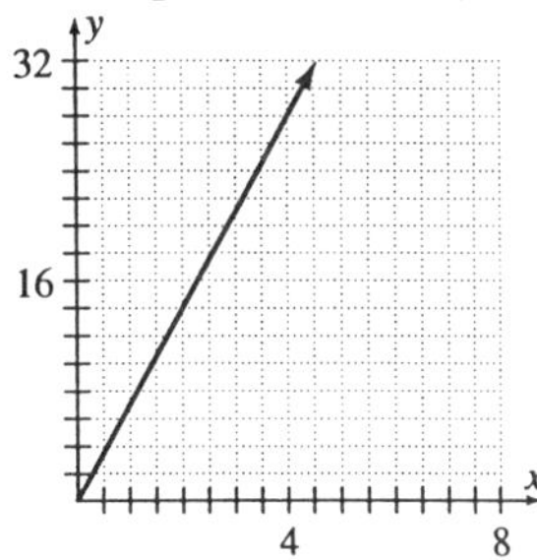

At $x = 2.5$, $y = 17.5$

They would travel 17.5 miles.

69. **a.** $\frac{1045 - 1624}{10 - 1} = \frac{-579}{9}$

b. $y = -\frac{579}{9}x + 1624$

c. $Y1 = -\frac{579}{9}x + 1624$

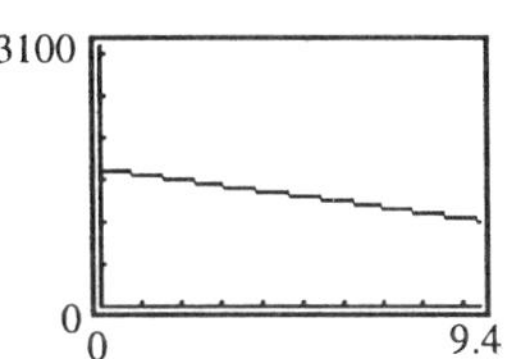

d. $Y1 = -64\frac{1}{3} + 1624$

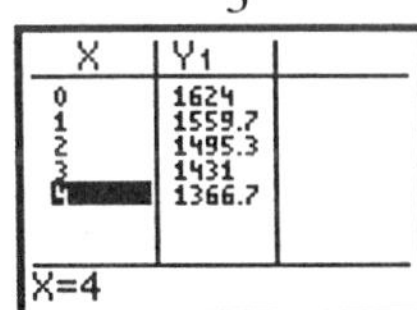

X	Y1
0	1624
1	1559.7
2	1495.3
3	1431
4	1366.7

X=4

X	Y1
5	1302.3
6	1238
7	1173.7
8	1109.3
9	1045

X=9

71. a. 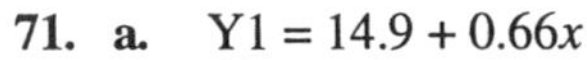
Y1 = 14.9 + 0.66x

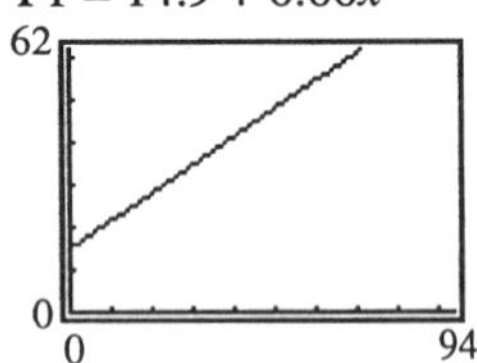

b. $y = 14.9 + 0.66(15)$
$y = 24.8$
The actual value is (15, 25) which is very close.

c.

x	actual y	predicted y	difference
27	30	32.72	–2.72
22	26	29.42	–3.42
15	25	24.8	0.2
35	42	38	4
30	38	34.7	3.3
52	40	49.22	–9.22
35	32	38	–6
55	54	51.2	2.8
40	50	41.3	8.7
40	43	41.3	1.7

73. $D(p) = 80 - \frac{4}{5}p$

a. $y = -\frac{4}{5}x + 80$

$m = -\frac{4}{5}$, $b = 80$

The slope is $-\frac{4}{5}$ and the y-intercept is (0, 80).

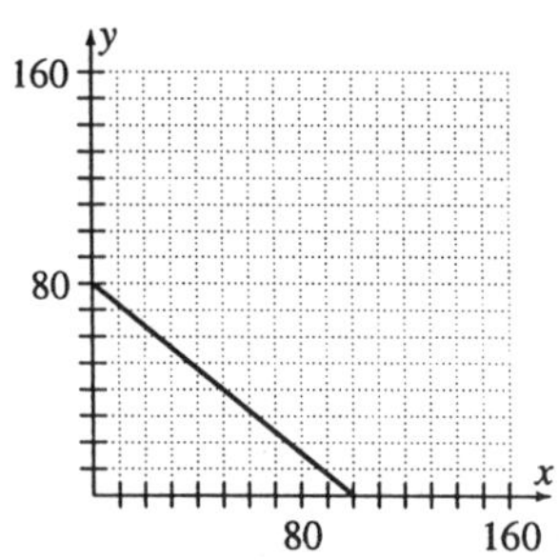

b. At $x = 10$, $y = 72$
$x = 20$, $y = 64$
$x = 40$, $y = 48$
$x = 64$, $y = 28.8$

c. Demand decreases as price increases.

d. At a price of $100 or more, demand is 0.

6.3 Experiencing Algebra the Calculator Way

1. a. $y = 4x$
$m = 4$
$b = 0$

b. $y = 2x$
$m = 2$
$b = 0$

c. $y = \left(\frac{1}{2}\right)x$
$m = \frac{1}{2}$
$b = 0$
$\text{Y1} = 4x$
$\text{Y2} = 2x$
$\text{Y3} = \frac{1}{2}x$

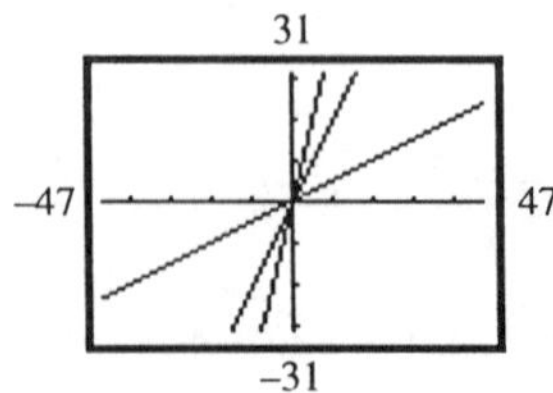

The y-intercept, (0, 0), is the same.
The slopes are different, but all are positive.
The graphs rise at different rates, but all cross at (0, 0).

2. a. $y = -4x$
$m = -4$
$b = 0$

b. $y = -2x$
$m = -2$
$b = 0$

c. $y = \left(-\frac{1}{2}\right)x$
$m = -\frac{1}{2}$
$b = 0$
$\text{Y1} = -4x$
$\text{Y2} = -2x$
$\text{Y3} = -\frac{1}{2}x$

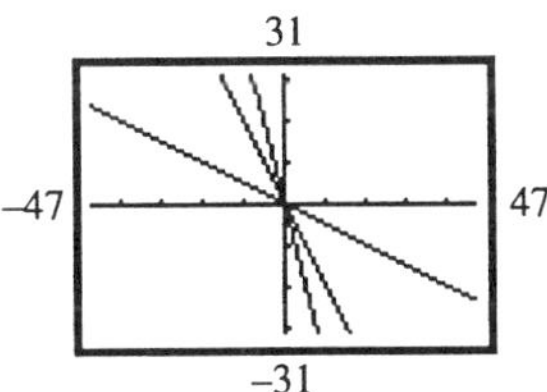

The y-intercept is the same at (0, 0). The slopes are different, but all are negative.
The graphs decline at different rates but all cross at (0, 0).

3. a. $y = 4x + 9$
$m = 4$
$b = 9$

b. $y = 4x$
$m = 4$
$b = 0$

c. $y = 4x - 9$
$m = 4$
$b = -9$
$\text{Y1} = 4x + 9$
$\text{Y2} = 4x$
$\text{Y3} = 4x - 9$

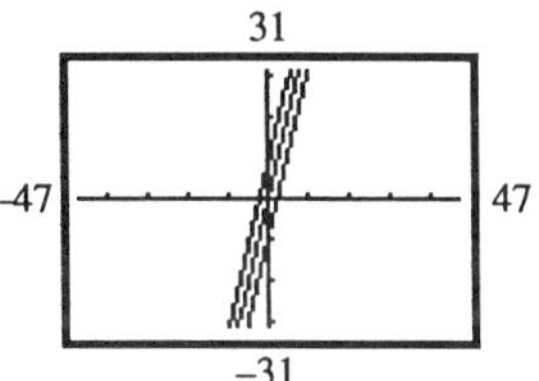

The slopes are the same. The y-intercepts are different. They are all parallel, but cross the y-axis at different points.

4. a. $y = -4x + 9$
$m = -4$
$b = 9$

b. $y = -4x$
$m = -4$
$b = 0$

c. $y = -4x - 9$
$m = -4$
$b = -9$
$\text{Y1} = -4x + 9$
$\text{Y2} = -4x$
$\text{Y3} = -4x - 9$

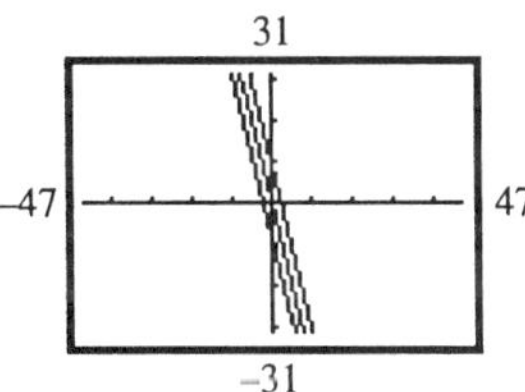

The slopes are the same. The y-intercepts are different. They are parallel, but cross the y-axis at different points.

6.4 Experiencing Algebra the Exercise Way

1. $3x - 2y = 5(y + 7)$
$3x - 2y = 5y + 35$
$3x - 2y - 3x = 5y + 35 - 3x$
$-2y = 5y + 35 - 3x$
$-2y - 5y = 5y + 35 - 3x - 5y$
$-7y = -3x + 35$
$\frac{-7y}{-7} = \frac{-3x + 35}{-7}$
$y = \frac{3}{7}x - 5 \quad m = \frac{3}{7},\ b = -5$

$7x = 3(1 - y)$
$7x = 3 - 3y$
$7x - 3 = 3 - 3y - 3$
$7x - 3 = -3y$
$\frac{7x-3}{-3} = \frac{-3y}{-3}$
$-\frac{7}{3}x + 1 = y$
$y = -\frac{7}{3}x + 1 \quad m = -\frac{7}{3},\ b = 1$
The lines are intersecting and perpendicular because the slopes (m) are opposite reciprocals of each other.

3. $x = 4(y - 3)$
$x = 4y - 12$
$x + 12 = 4y - 12 + 12$
$x + 12 = 4y$
$\frac{x+12}{4} = \frac{4y}{4}$
$y = \frac{1}{4}x + 3 \quad m = \frac{1}{4},\ b = 3$
$x = 4(y + 5)$
$x = 4y + 20$
$x - 20 = 4y + 20 - 20$
$x - 20 = 4y$
$\frac{x-20}{4} = \frac{4y}{4}$
$y = \frac{1}{4}x - 5 \quad m = \frac{1}{4},\ b = -5$
The lines are parallel because the slopes (m) are equal and the y-intercepts (b) are not equal.

5. $4x - y = 6$
$y = 4x - 6 \quad m = 4,\ b = -6$
$2x - y + 3 = 0$
$y = 2x + 3 \quad m = 2,\ b = 3$
The lines are only intersecting because the slopes (m) are not equal and their product is not -1.

7. $x = 2(y - 7)$
$x = 2y - 14$
$2y = x + 14$
$y = \frac{1}{2}x + 7 \quad m = \frac{1}{2},\ b = 7$
$y = \frac{1}{2}x + 7 \quad m = \frac{1}{2},\ b = 7$
The lines are coinciding because the slopes (m) are equal and the y-intercepts (b) are equal.

9. $5x + y = -6$
$y = -5x - 6 \quad m = -5,\ b = -6$
$3x + y = 0$
$y = -3x \quad m = -3,\ b = 0$
The lines are only intersecting because the slopes (m) are not equal and their product is not -1.

11. $4x + y = 8$
$y = -4x + 8 \quad m = -4,\ b = 8$
$4x + y + 2 = 0$
$y = -4x - 2 \quad m = -4,\ b = -2$
The lines are parallel because the slopes (m) are equal and the y-intercepts are not equal.

13. $y - 5 = 0$
$y = 5$
$2x + 6 = x + 9$
$x = 5$
The lines are intersecting and perpendicular because one is vertical and the other is horizontal.

15. $2y - 3 = 13$
$2y = 16$
$y = 8 \quad m = 0, b = 8$
$y + 1 = 4$
$y = 3 \quad m = 0, b = 3$
The lines are parallel because both lines are horizontal with different y-intercepts.

17. $x + 3 = 0$
$x = -3$
$x - 5 = 0$
$x = 5$
The lines are parallel because they are both vertical with different constants.

19. $2x-9=0$
$x=\frac{9}{2}$
$x-4=5-x$
$2x=9$
$x=\frac{9}{2}$
The lines are coinciding because both are vertical with the same constant.

21. $3(y-3)=1$
$3y-9=1$
$3y=10$
$y=\frac{10}{3} \quad m=0,\ b=\frac{10}{3}$

$5y=10+2y$
$3y=10$
$y=\frac{10}{3} \quad m=0,\ b=\frac{10}{3}$
The lines are coinciding because they are both horizontal lines with the same constant.

23. $x=2 \quad m=\text{undefined}$
$y=2x-1 \quad m=2,\ b=-1$
The lines are only intersecting because one is a vertical line and the other is not a vertical or a horizontal line.

25. $y-3=0$
$y=3 \quad m=0, b=3$

$2x+3y=0$
$3y=-2x$
$y=-\frac{2}{3}x \quad m=-\frac{2}{3},\ b=0$
The lines are only intersecting because the slopes are not equal and their product is not -1.

27. **a.** $y=0.15x+2.5+0.1x+1$
$y=0.25x+3.5$

b. $y=0.25x$

c. The lines are parallel. Since they never intersect, there is no break-even point.

d. $y=\frac{1}{3}x$

e. $0.25x+3.5=\frac{1}{3}x$
$\frac{1}{4}x+3.5=\frac{1}{3}x$
$3.5=\frac{1}{12}x$
$x=42$
At 42 candy bars, Brook will break even.

f. total cost $y=0.15x+1$
total revenue $y=0.25x$
$0.15x+1=0.25x$
$0.15x+1-0.15x=0.25x-0.15x$
$1=0.1x$
$x=10$
At 10 candy bars, Brook will break even.

g. $0.15x+1=\frac{1}{3}x$
$\frac{9}{60}x+1=\frac{20}{60}x$
$\frac{9}{60}x+1-\frac{9}{60}x=\frac{20}{60}x-\frac{9}{60}x$
$1=\frac{11}{60}x$
$x\approx 5.45$
At about 5.45 candy bars, Brook will break even. At 6 candy bars, Brook will start making a profit.

h. Answers will vary.

29. $y_1=35+0.25x \quad m=0.25, b=35$
$y_2=60 \quad m=0, b=60$
Their graphs will intersect because the slopes are not equal.

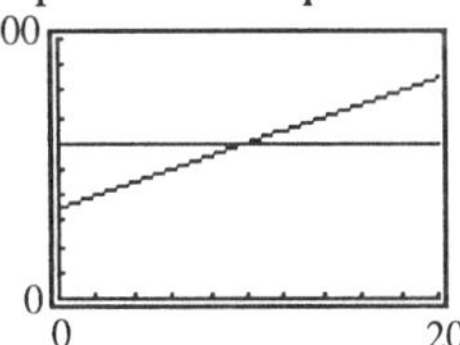

The intersection is (100, 60). At 100 miles, the prices are equal at $60 per day.

31. $y_1 = 50 + 2x \quad m = 2, b = 50$
$y_2 = 75 \quad m = 0, b = 75$
$x = 8 \quad m = \text{undefined}, b = \text{none}$
Yes, y_1 intersects with $x = 8$ and y_2 intersects with $x = 8$ because each has a different slope. $x = 8$ is perpendicular to y_2 because one is vertical and the other is horizontal. The intersection of y_1 and $x = 8$ is (8, 66). At 8 years, she will receive $66. The intersection of y_2 and $x = 8$ is (8, 75). At 8 years, she will receive $75. The intersection of y_1 and y_2 is (12.5, 75). At $12\frac{1}{2}$ years she will receive the same either way, $75.

33. $y_1 = 10x \quad m = 10, b = 0$
$y_2 = 15(x - 4) \quad m = 15, b = 60$
$y_3 = 150 \quad m = 0, b = 150$
y_1 and y_2 intersect at (12, 120). At 12 seconds, Speedie will catch Archie at a distance of 120 feet or before the end zone 150 feet away.

6.4 Experiencing Algebra the Calculator Way

1. Xmin = –10
Xmax = 10
Xscl = 1
Ymin = –10
Ymax = 10
Yscl = 1

2. Xmin = –47
Xmax = 47
Xscl = 10
Ymin = –31
Ymax = 31
Yscl = 10

3. Xmin = 0
Xmax = 94
Xscl = 10
Ymin = 0
Ymax = 62
Yscl = 10

4. Xmin = –470
Xmax = 470
Xscl = 100
Ymin = –310
Ymax = 310
Yscl = 100

6.5 Experiencing Algebra the Exercise Way

1. $m = \frac{3}{2}, \; b = -1$
$y = mx + b$
$y = \frac{3}{2}x - 1$

3. The graph is a vertical line with $c = -5$.
$x = -5$

5. $m = -3, \; b = 0$
$y = mx + b$
$y = -3x + 0$
$y = -3x$

7. The graph is a horizontal line with $b = -\frac{3}{2}$.
$y = -\frac{3}{2}$

9. $m = -\frac{2}{5}, \; b = 4$
$y = mx + b$
$y = -\frac{2}{5}x + 4$

11. $m = \frac{5}{9}, \; b = 0$
$y = mx + b$
$y = \frac{5}{9}x + 0$
$y = \frac{5}{9}x$

13. $m = 4, \; b = -\frac{3}{4}$
$y = mx + b$
$y = 4x - \frac{3}{4}$

15. $m = -4.1,\ b = 0.5$
$y = mx + b$
$y = -4.1x + 0.5$

17. $m = 0,\ b = -33$
$y = mx + b$
$y = 0x - 33$
$y = -33$

19. $m = \frac{2}{3},\ (3, -3)$
$y - (-3) = \frac{2}{3}(x - 3)$
$y + 3 = \frac{2}{3}x - 2$
$y = \frac{2}{3}x - 5$

21. $m = -3,\ (0, 4)$
$y - 4 = -3(x - 0)$
$y - 4 = -3x$
$y = -3x + 4$

23. $m = -1.7,\ (3, -1.5)$
$y - (-1.5) = -1.7(x - 3)$
$y + 1.5 = -1.7x + 5.1$
$y = -1.7x + 3.6$

25. $(-1, 1)$ and $(1, -2)$
$m = \frac{y_2 - y_1}{x_2 - x_1}$
$m = \frac{-2 - 1}{1 - (-1)}$
$m = \frac{-3}{2}$
$y - y_1 = m(x - x_1)$
$y - 1 = -\frac{3}{2}[x - (-1)]$
$y - 1 = -\frac{3}{2}(x + 1)$
$y - 1 = -\frac{3}{2}x - \frac{3}{2}$
$y - 1 + 1 = -\frac{3}{2}x - \frac{3}{2} + \frac{2}{2}$
$y = -\frac{3}{2}x - \frac{1}{2}$

27. $(-1, -2)$ and $(-1, 5)$
$m = \frac{y_2 - y_1}{x_2 - x_1}$
$m = \frac{5 - (-2)}{-1 - (-1)}$
$m = \frac{7}{0}$
m is undefined.
The line is a vertical line.
The constant term is -1.
$x = -1$

29. $(-1, 1)$ and $(-2, -1)$
$m = \frac{y_2 - y_1}{x_2 - x_1}$
$m = \frac{-1 - 1}{-2 - (-1)}$
$m = \frac{-2}{-1}$
$m = 2$
$y - y_1 = m(x - x_1)$
$y - 1 = 2[x - (-1)]$
$y - 1 = 2(x + 1)$
$y - 1 + 1 = 2x + 2 + 1$
$y = 2x + 3$

31. $(-2, 2)$ and $(4, 2)$
$m = \frac{y_2 - y_1}{x_2 - x_1}$
$m = \frac{2 - 2}{4 - (-2)}$
$m = 0$
$y - y_1 = m(x - x_1)$
$y - 2 = 0[x - (-2)]$
$y - 2 + 2 = 0 + 2$
$y = 2$

33. $\left(4\frac{1}{2}, 5\frac{1}{4}\right)$ and (1, 4)

$$m = \frac{y_2 - y_1}{x_2 - x_1}$$
$$m = \frac{4 - 5\frac{1}{4}}{1 - 4\frac{1}{2}}$$
$$m = \frac{-1\frac{1}{4}}{-3\frac{1}{2}}$$
$$m = -\frac{5}{4} \cdot \left(-\frac{2}{7}\right)$$
$$m = \frac{5}{14}$$
$$y - y_2 = m(x - x_2)$$
$$y - 4 = \frac{5}{14}(x - 1)$$
$$y - 4 + 4 = \frac{5}{14}x - \frac{5}{14} + \frac{56}{14}$$
$$y = \frac{5}{14}x + \frac{51}{14}$$

35. $\left(-1\frac{1}{3}, 2\right)$ and (0, 0)

$$m = \frac{y_2 - y_1}{x_2 - x_1}$$
$$m = \frac{0 - 2}{0 - \left(-1\frac{1}{3}\right)}$$
$$m = \frac{-2}{\frac{4}{3}}$$
$$m = -2 \cdot \frac{3}{4}$$
$$m = -\frac{3}{2}$$
$$y - y_2 = m(x - x_2)$$
$$y - 0 = -\frac{3}{2}(x - 0)$$
$$y = -\frac{3}{2}x$$

37. (0.5, 0) and (–0.8, 4.2)

$$m = \frac{y_2 - y_1}{x_2 - x_1}$$
$$m = \frac{4.2 - 0}{-0.8 - 0.5}$$
$$m = \frac{4.2}{-1.3}$$
$$m = -\frac{42}{13}$$
$$y - y_1 = m(x - x_1)$$
$$y - 0 = -\frac{42}{13}(x - 0.5)$$
$$y = -\frac{42}{13}x + \frac{21}{13}$$

39. (2.4, 2.8) and (–2.6, –2.2)

$$m = \frac{y_2 - y_1}{x_2 - x_1}$$
$$m = \frac{-2.2 - 2.8}{-2.6 - 2.4}$$
$$m = \frac{5}{5}$$
$$m = 1$$
$$y - y_1 = m(x - x_1)$$
$$y - 2.8 = 1(x - 2.4)$$
$$y - 2.8 = x - 2.4$$
$$y - 2.8 + 2.8 = x - 2.4 + 2.8$$
$$y = x + 0.4$$

41. $3x - 8y = 32$

$$3x - 8y - 3x = 32 - 3x$$
$$-8x = -3x + 32$$
$$\frac{-8y}{-8} = \frac{-3x + 32}{-8}$$
$$y = \frac{3}{8}x - 4$$
$$m = \frac{3}{8}$$
$$y - y_1 = m(x - x_1)$$
$$y - 7 = \frac{3}{8}(x - 8)$$
$$y - 7 = \frac{3}{8}x - 3$$
$$y - 7 + 7 = \frac{3}{8}x - 3 + 7$$
$$y = \frac{3}{8}x + 4$$

43. $y = 3x - 10$
$m = 3$
A parallel line has the same slope.
$y - y_1 = m(x - x_1)$
$y - \left(-\frac{5}{6}\right) = 3\left(x - \frac{4}{9}\right)$
$y + \frac{5}{6} = 3x - \frac{4}{3}$
$y + \frac{5}{6} - \frac{5}{6} = 3x - \frac{8}{6} - \frac{5}{6}$
$y = 3x - \frac{13}{6}$

45. $y = 3x + 12$
$m = 3$
A perpendicular line has an opposite reciprocal slope. This slope is $-\frac{1}{3}$.
$y - y_1 = m(x - x_1)$
$y - 5.8 = -\frac{1}{3}(x - 3.6)$
$y - 5.8 = -\frac{1}{3}x + 1.2$
$y - 5.8 + 5.8 = -\frac{1}{3}x + 1.2 + 5.8$
$y = -\frac{1}{3}x + 7$

47. $y - 5x - 1$
$m = 5$
A perpendicular line has an opposite reciprocal slope. This slope is $-\frac{1}{5}$.
$y - y_1 = m(x - x_1)$
$y - (-30) = -\frac{1}{5}(x - 15)$
$y + 30 = -\frac{1}{5}x + 3$
$y + 30 - 30 = -\frac{1}{5}x + 3 - 30$
$y = -\frac{1}{5}x - 27$

49. a. $m = \frac{y_2 - y_1}{x_2 - x_1}$
$m = \frac{25.5 - 42.4}{25 - 0}$
$m = \frac{-16.9}{25}$
$m = -0.676$

b. $y - y_1 = m(x - x_1)$
$y - 42.4 = -0.676(x - 0)$
$y - 42.4 = -0.676x$
$y - 42.4 + 42.4 = -0.676x + 42.4$
$y = -0.676x + 42.4$

62
0 94
0

c. In 1988, $x = 23$
At $x = 23$, $y = 26.852$
The percent is predicted to be 26.852%.
It is fairly close.

d. In 2005, $x = 40$
At $x = 40$, $y = 15.36$
The percent is predicted to be 15.36%.

e. At $x = 62$, $y \approx 0$
$1965 + 62 = 2027$
In year 2027, the predicted percentage becomes close to zero.
Answers will vary.

51. $m = \frac{y_2 - y_1}{x_2 - x_1}$
$m = \frac{263 - 273}{-10 - 0}$
$m = \frac{-10}{-10}$
$m = 1$
$b = 273$
$y = mx + b$
$y = 1x + 273$
$y = x + 273$
$y = 100 + 273$
$y = 373$
A Kelvin temperature of 373 corresponds to 100°C.

53. $m = \dfrac{y_2 - y_1}{x_2 - x_1}$

$m = \dfrac{16.96 - 15.92}{1.6 - 1.2}$

$m = \dfrac{1.04}{0.4}$

$m = 2.6$

$y - y_1 = m(x - x_1)$

$y - 15.92 = 2.6(x - 1.2)$

$y - 15.92 = 2.6x - 3.12$

$y - 15.92 + 15.92 = 2.6x - 3.12 + 15.92$

$y = 2.6x + 12.8$

$y = 2.6(1.2) + 12.8 = 15.92$

$y = 2.6(1.6) + 12.8 = 16.96$

$y = 2.6(2) + 12.8 = 18.00$

$y = 2.6(2.4) + 12.8 = 19.04$

$y = 2.6(2.8) + 12.8 = 20.08$

55. (2500, 40) and (3500, 35)

$m = \dfrac{y_2 - y_1}{x_2 - x_1}$

$m = \dfrac{35 - 40}{3500 - 2500}$

$m = \dfrac{-5}{1000}$

$m = -0.005$

$y - y_1 = m(x - x_1)$

$y - 40 = -0.005(x - 2500)$

$y - 40 = -0.005x + 12.5$

$y - 40 + 40 = -0.005x + 12.5 + 40$

$y = -0.005x + 52.5$

57. (6, 5000) and (15, 14,000)

$m = \dfrac{14{,}000 - 5000}{15 - 6}$

$m = \dfrac{9000}{9}$

$m = 1000$

$y - y_1 = m(x - x_1)$

$y - 5000 = 1000(x - 6)$

$y - 5000 = 1000x - 6000$

$y - 5000 + 5000 = 1000x - 6000 + 5000$

$y = 1000x - 1000$

$y = 1000(10) - 1000$

$y = 9000$

The prediction is 9000 sales.

59. $m = 120,\ b = 5470$

$y = mx + b$

$y = 120x + 5470$

$x = 1996 - 1990$

$x = 6$

$y = 120(6) + 5470$

$y = 6190$

The predicted enrollment in 1996 is 6190 students.

Yes, this is a good estimate.

$x = 2000 - 1990$

$x = 10$

$y = 120(10) + 5470$

$y = 6670$

The predicted enrollment in 2000 would be 6670 students.

6.5 Experiencing Algebra the Calculator Way

1. (12, 925) and (72, 4225)

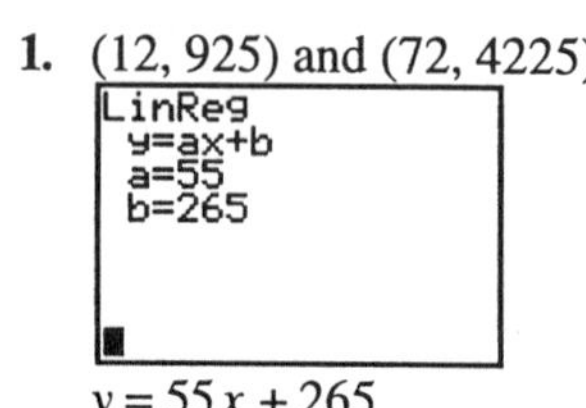

$y = 55x + 265$

2. (16, –351) and (40, 417)

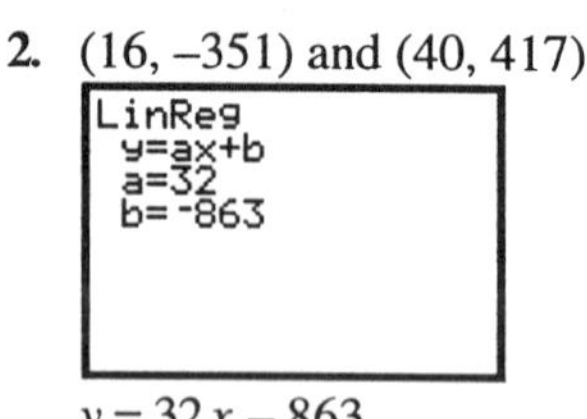

$y = 32x - 863$

3. (0, 6.4) and (36.4, 103.59)

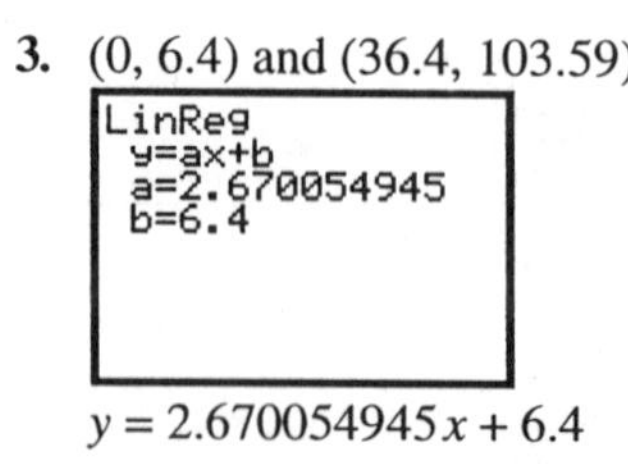

$y = 2.670054945x + 6.4$

4. $\left(1, \frac{17}{4}\right)$ and $\left(3, \frac{57}{4}\right)$

```
LinReg
 y=ax+b
 a=5
 b=-.75
```

$y = 5x - 0.75$

5. (–5, 21) and (–9, 7)

```
LinReg
 y=ax+b
 a=3.5
 b=38.5
```

$y = 3.5x + 38.5$

6. (16, –2) and (7, 4)

```
LinReg
 y=ax+b
 a=-.6666666667
 b=8.666666667
```

$y = -0.\overline{6}x + 8.\overline{6}$

Chapter 6 Review

Reflections

1.–9. Answers will vary.

Exercises

1. $y = 0.6x + 2.3$ is linear.
$y - 2.3 - y = 0.6x + 2.3 - 2.3 - y$
$-2.3 = 0.6x - y$
$0.6x - y = -2.3$ is in standard form.

2. $y = 4x^2 - 2$ is non-linear because the x variable is squared.

3. $3x - 6y + 12 = 0$ is linear.
$3x - 6y + 12 - 12 = 0 - 12$
$3x - 6y = -12$ is in standard form.

4. $5x - 3y + 7 = x - y + 9$ is linear
$5x - 3y + 7 - 7 = x - y + 9 - 7$
$5x - 3y = x - y + 2$
$5x - 3y - x + y = x - y + 2 - x + y$
$4x - 2y = 2$ is in standard form.

5. $3x^2 + y = 1$ is non-linear because the x variable is squared.

6. $5y - 12 = 7 - y$ is linear
$5y - 12 + 12 = 7 - y + 12$
$5y = 19 - y$
$5y + y = 19 - y + y$
$6y = 19$ is in standard form.

7.

$f(x) = 10x - 42$	
-2	$10(4) - 42$
	$40 - 42$
	-2

Since $-2 = -2$, $(4, -2)$ is a solution.

8.

$3x - 2y = 6$	
$3(1) - 2(-1)$	6
$3 + 2$	
5	

Since $5 \neq 6$, $(1, -1)$ is not a solution.

9.

$\frac{3}{5}x - \frac{8}{9}y = 1$	
$\frac{3}{5}(15) - \frac{8}{9}(18)$	1
$9 - 16$	
-7	

Since $-7 \neq 1$, $(15, 18)$ is not a solution.

10.

$45x - 14y = 31$	
$45\left(\frac{5}{12}\right) - 14\left(-\frac{7}{8}\right)$	31
$\frac{225}{12} + \frac{98}{8}$	
31	

Since $31 = 31$, $\left(\frac{5}{12}, -\frac{7}{8}\right)$ is a solution.

11. Answers will vary. Possible answer:
$12x + 6y = 48$
$6y = -12x + 48$
$y = -2x + 8$

x	$y = -2x + 8$	y
–1	$y = -2(-1) + 8 = 10$	10
0	$y = -2(0) + 8 = 8$	8
1	$y = -2(1) + 8 = 6$	6

(–1, 10), (0, 8), and (1, 6) are three possible solutions.

12. Answers will vary. Possible answer:

x	$y = \frac{8}{13}x - 7$	y
–13	$y = \frac{8}{13}(-13) - 7 = -15$	–15
0	$y = \frac{8}{13}(0) - 7 = -7$	–7
13	$y = \frac{8}{13}(13) - 7 = 1$	1

(–13, –15), (0, –7), and (13, 1) are three possible solutions.

13. Answers will vary. Possible answer:

x	$y = -9$	y
0	$y = -9$	–9
1	$y = -9$	–9
2	$y = -9$	–9

(0, –9), (1, –9), and (2, –9) are three possible solutions.

14. Answers will vary. Possible answer:
$9x - y = 12$
$9x - y + y = 12 + y$
$9x = 12 + y$
$9x - 12 = 12 + y - 12$
$9x - 12 = y$
$y = 9x - 12$

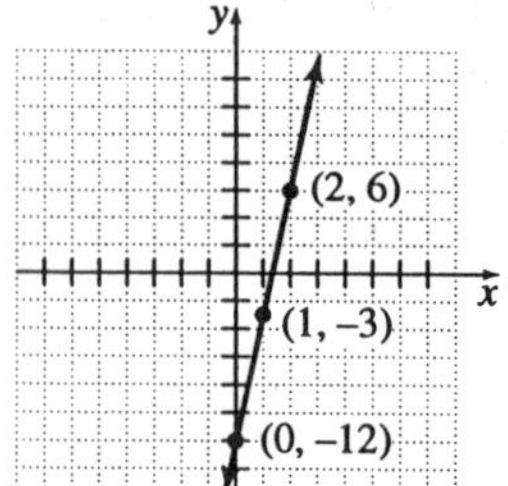

(0, –12), (1, –3), and (2, 6) are three possible solutions.

15. Answers will vary. Possible answer:
$y = -\dfrac{4}{3}x + 2$

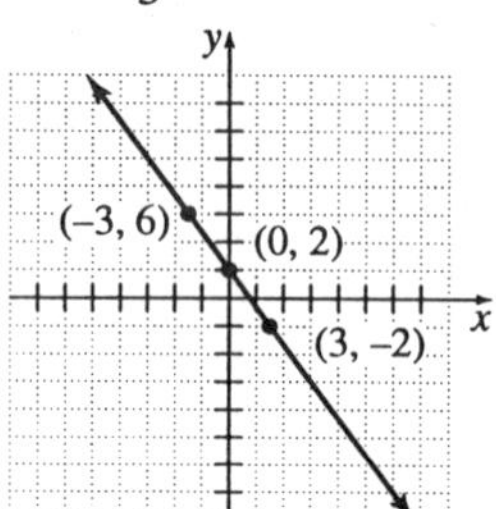

(–3, 6), (0, 2), and (3, –2) are three possible solutions.

16. Answers will vary. Possible answer:
$5y - 2 = y - 10$
$5y - 2 - y = y - 10 - y$
$4y - 2 = -10$
$4y - 2 + 2 = -10 + 2$
$4y = -8$
$\dfrac{4y}{4} = \dfrac{-8}{4}$
$y = -2$

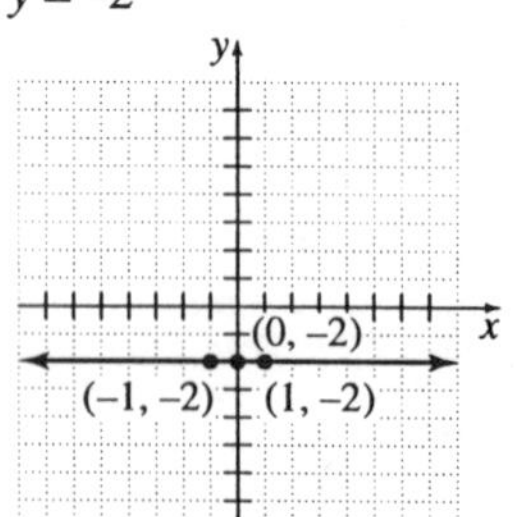

(–1, –2), (0, –2), and (1, –2) are three possible solutions.

17. Answers will vary. Possible answer:
$2x + 16 = x + 18$
$2x + 16 - x = x + 18 - x$
$x + 16 = 18$
$x + 16 - 16 = 18 - 16$
$x = 2$

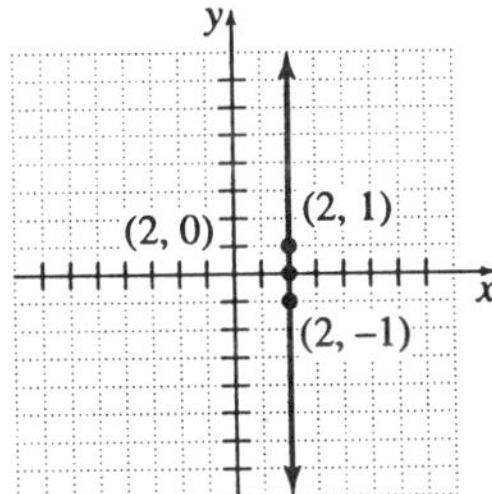

(2, –1), (2, 0), and (2, 1) are three possible solutions.

18. a. (10, 85) and (20, 160)

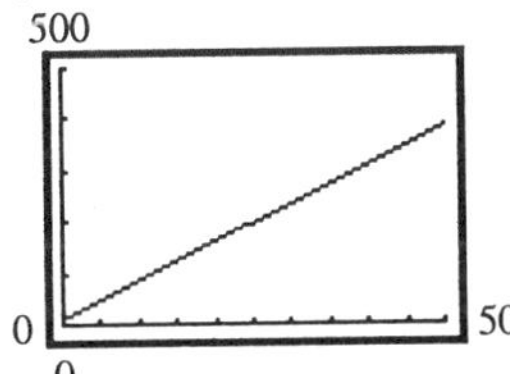

b. At $x = 30$, $y = 235$
She will receive $235 for a job that is 30 pages long.

19. a. $y = 20 + 9x$
$y = 20 + 9(10)$
$y = 110$
He would earn $110.

b. $y = 20 + 9(15)$
$y = 155$
No, he should be paid $155.

c. $y = 20 + 9(0)$
$y = 20$
He would be paid $20.

20. $8x + 12y = -24$
Substitute 0 for y.
$8x + 12(0) = -24$
$8x = -24$
$\frac{8x}{8} = \frac{-24}{8}$
$x = -3$
The x-intercept is (–3, 0).
Substitute 0 for x.
$8(0) + 12y = -24$
$12y = -24$
$\frac{12y}{12} = \frac{-24}{12}$
$y = -2$
The y-intercept is (0, –2).

21. $y = \frac{2x + 20}{5}$
Substitute 0 for y.
$0 = \frac{2x + 20}{5}$
$5(0) = 5\left(\frac{2x + 20}{5}\right)$
$0 = 2x + 20$
$0 - 20 = 2x + 20 - 20$
$-20 = 2x$
$\frac{-20}{2} = \frac{2x}{2}$
$-10 = x$
The x-intercept is (–10, 0).
Substitute 0 for x.
$y = \frac{2(0) + 20}{5}$
$y = \frac{20}{5}$
$y = 4$
The y-intercept is (0, 4).

22. $9x - 3y = -12$
$9x - 3y - 9x = -12 - 9x$
$-3y = -12 - 9x$
$\frac{-3y}{-3} = \frac{-12 - 9x}{-3}$
$y = 4 + 3x$
$y = 3x + 4$
The y-intercept is (0, 4).

23. $6x + 2y = x$
$6x + 2y - 6x = x - 6x$
$2y = -5x$
$\frac{2y}{2} = \frac{-5x}{2}$
$y = -\frac{5}{2}x$
The y-intercept is (0, 0).

24. $7x + 11y = 77$
Substitute 0 for y.
$7x + 11(0) = 77$
$7x = 77$
$\frac{7x}{7} = \frac{77}{7}$
$x = 11$
The x-intercept is (11, 0).
Substitute 0 for x.
$7(0) + 11y = 77$
$11y = 77$
$\frac{11y}{11} = \frac{77}{11}$
$y = 7$
The y-intercept is (0, 7).

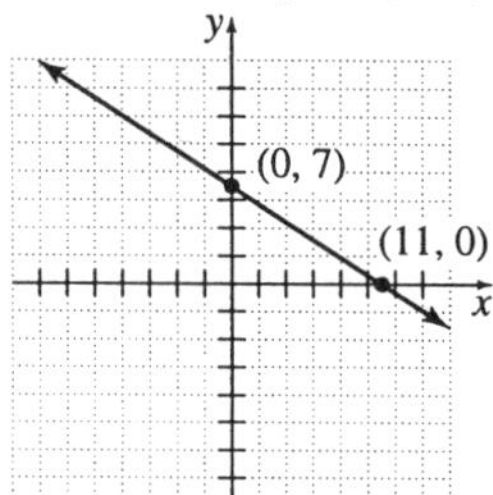

25. $-x - y = 2$
Substitute 0 for y.
$-x - 0 = 2$
$\frac{-x}{-1} = \frac{2}{-1}$
$x = -2$
The x-intercept is (–2, 0).
Substitute 0 for x.
$-0 - y = 2$
$\frac{-y}{-1} = \frac{2}{-1}$
$y = -2$
The y-intercept is (0, –2).

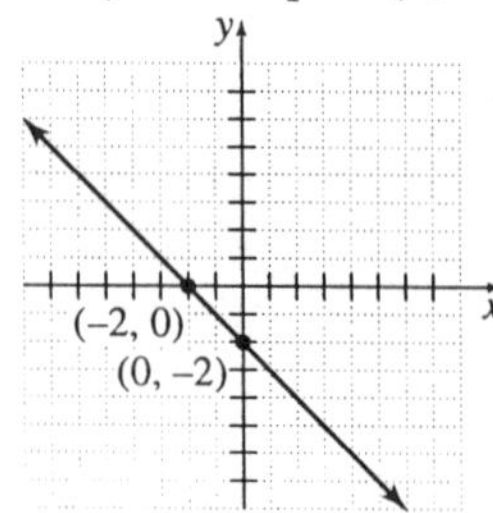

26. $7.4x + 14.8y = 29.6$
Substitute 0 for y.
$7.4x + 14.8(0) = 29.6$
$7.4x = 29.6$
$\frac{7.4x}{7.4} = \frac{29.6}{7.4}$
$x = 4$
The x-intercept is (4, 0).
Substitute 0 for x.
$7.4(0) + 14.8y = 29.6$
$\frac{14.8y}{14.8} = \frac{29.6}{14.8}$
$y = 2$
The x-intercept is (0, 2).

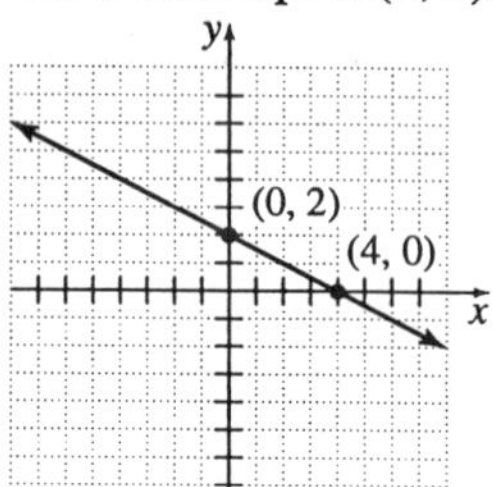

27. a. $y = 5200 - 100x$

b. Substitute 0 for y.
$0 = 5200 - 100x$
$\frac{-5200}{-100} = \frac{-100x}{-100}$
$52 = x$
The x-intercept is (52, 0). At 52 weeks the balance is \$0.
Substitute 0 for x.
$y = 5200 - 100(0)$
$y = 5200$
The y-intercept is (0, 5200). At 0 weeks, the balance is \$5200.

c.

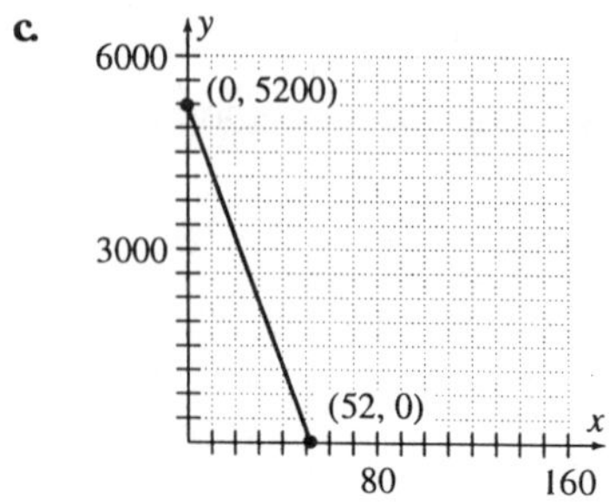

d. At $x = 32$, $y = 2000$
There is $2000 in the account.

28. a. $y = 50 - 0.5x$

b. Substitute 0 for y.
$0 = 50 - 0.5x$
$\frac{-50}{-0.5} = \frac{-0.5x}{-0.5}$
$100 = x$
The x-intercept is (100, 0). After 100 plays, Bryon has $0 left.
Substitute 0 for x.
$y = 50 - 0.5(0)$
$y = 50$
The y-intercept is (0, 50). After 0 plays, Bryon has $50.

c.

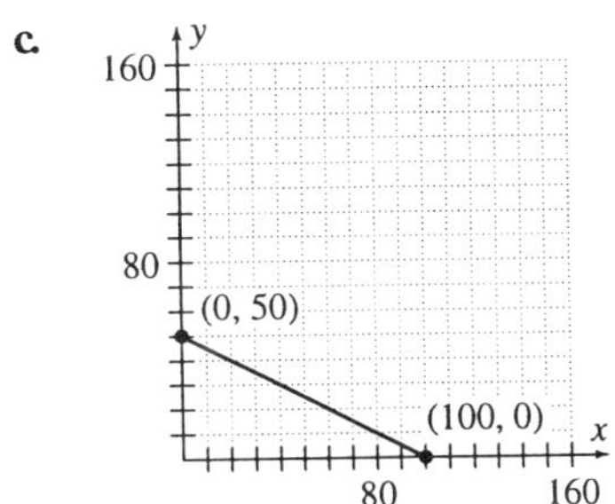

d. At $x = 60$, $y = 20$
He has $20 left.

29. The slope is 0 because the line is horizontal.

30. The slope is $\frac{9}{2}$.

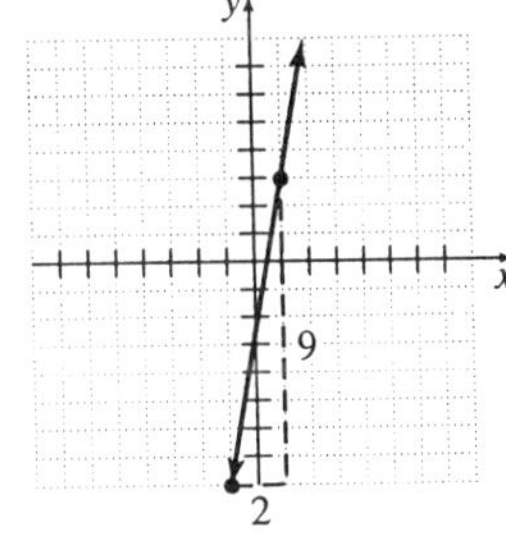

31. The slope is $-\frac{8}{5}$.

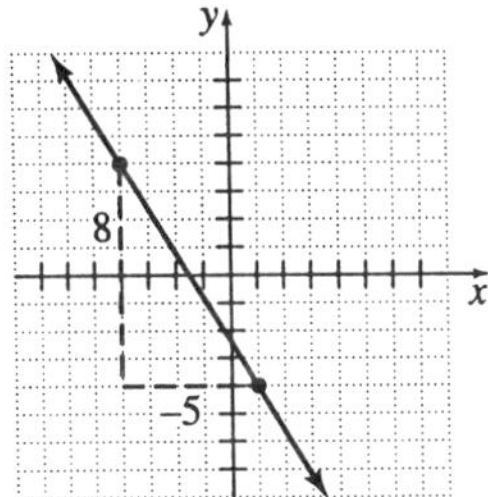

32. The slope is undefined because the line is vertical.

33. (–5, –3) and (1, 0)
$m = \frac{y_2 - y_1}{x_2 - x_1}$
$m = \frac{0 - (-3)}{1 - (-5)}$
$m = \frac{3}{6}$
$m = \frac{1}{2}$
The slope is $\frac{1}{2}$.

34. (–7, 7) and (–4, 2)
$m = \frac{y_2 - y_1}{x_2 - x_1}$
$m = \frac{2 - 7}{-4 - (-7)}$
$m = \frac{-5}{3}$
The slope is $-\frac{5}{3}$.

35. (4, –3) and (10, –3)
$m = \frac{y_2 - y_1}{x_2 - x_1}$
$m = \frac{-3 - (-3)}{10 - 4}$
$m = \frac{0}{6}$
The slope is 0.

36. (4, –3) and (4, 3)

$m = \frac{y_2 - y_1}{x_2 - x_1}$

$m = \frac{3-(-3)}{4-4}$

$m = \frac{6}{0}$

The slope is undefined.

37. $y = 23x - 51$

$m = 23,\ b = -51$

The slope is 23 and the y-intercept is (0, –51).

38. $y = -3.05x + 2.97$

$m = -3.05,\ b = 2.97$

The slope is –3.05 and the y-intercept is (0, 2.97).

39. $6x + 5y = 12$

$6x + 5y - 6x = 12 - 6x$

$5y = -6x + 12$

$\frac{5y}{5} = \frac{-6x+12}{5}$

$y = -\frac{6}{5}x + \frac{12}{5}$

$m = -\frac{6}{5},\ b = \frac{12}{5}$

The slope is $-\frac{6}{5}$ and the y-intercept is $\left(0, \frac{12}{5}\right)$.

40. $4(y-2) = 3(2y-1) + 4$

$4y - 8 = 6y - 3 + 4$

$4y - 8 = 6y + 1$

$4y - 8 - 6y = 6y + 1 - 6y$

$-2y - 8 = 1$

$-2y - 8 + 8 = 1 + 8$

$-2y = 9$

$\frac{-2y}{-2} = \frac{9}{-2}$

$y = -\frac{9}{2}$

$m = 0,\ b = -\frac{9}{2}$

The slope is 0 and the y-intercept is $\left(0, -\frac{9}{2}\right)$.

41. (–2, 3) and $m = \frac{5}{9}$

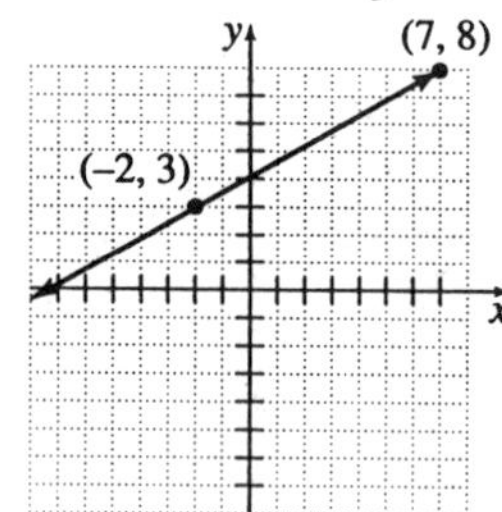

42. (3, –2) and $m = 3$

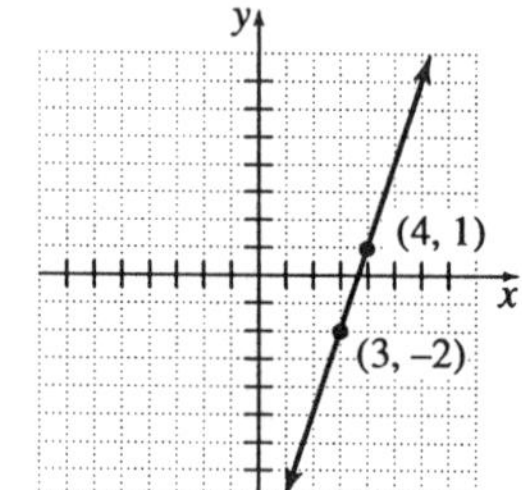

43. (–2, –2) and $m = 0$

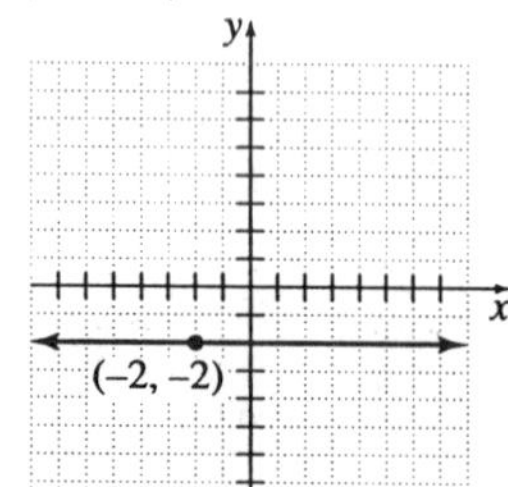

44. (–2, –2) and m = undefined

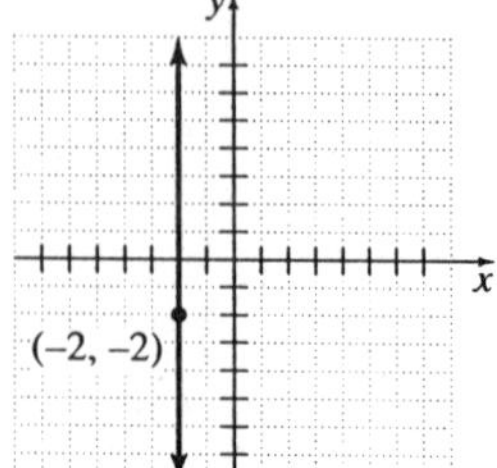

45. $y = -\frac{5}{3}x + 2$

$m = -\frac{5}{3},\ b = 2$

The slope is $-\frac{5}{3}$ and the y-intercept is (0, 2).

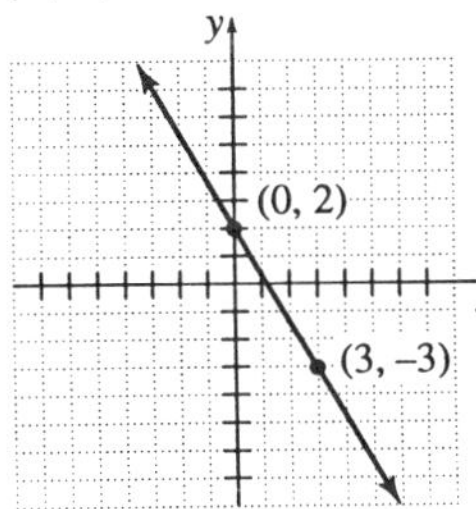

46. $3x + 2y = 12$
$3x + 2y - 3x = 12 - 3x$
$2y = -3x + 12$
$\frac{2y}{2} = \frac{-3x + 12}{2}$
$y = -\frac{3}{2}x + 6$

The slope is $-\frac{3}{2}$ and the y-intercept is (0, 6).

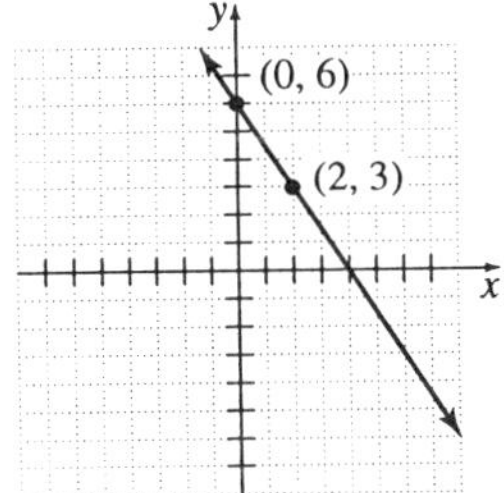

47. $m(x) = \frac{5}{8}x$

The slope is $\frac{5}{8}$ and the y-intercept is (0, 0).

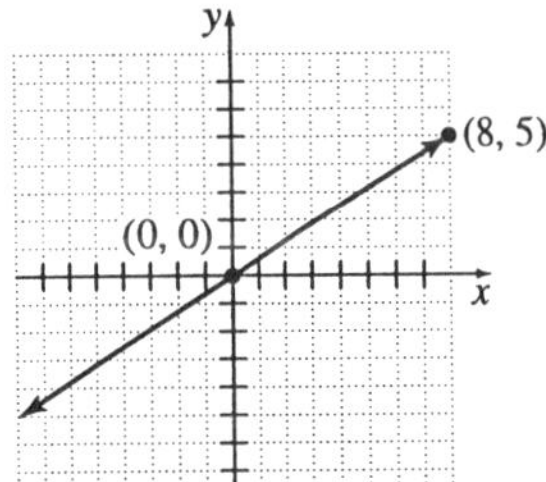

48. $y = -4$
The slope is 0 and the y-intercept is (0, –4).

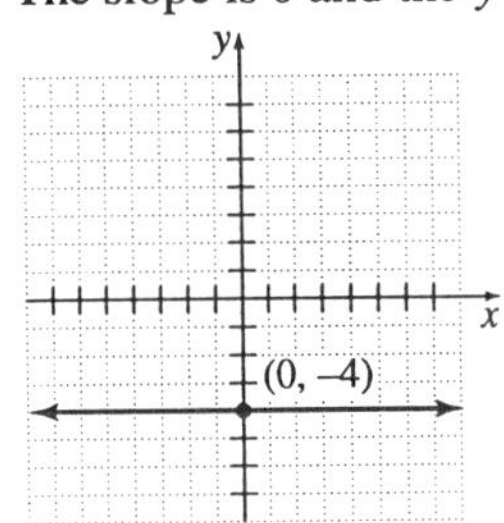

49. a. (0, 62) (4, 100)

$m = \frac{y_2 - y_1}{x_2 - x_1}$

$m = \frac{100 - 62}{4 - 0} = \frac{38}{4} = 9.5$

$m = 9.5$

The slope is 9.5.

b. $y - y_1 = m(x - x_1)$
$y - 100 = 9.5(x - 4)$
$y - 100 = 9.5x - 38$
$y - 100 + 100 = 9.5x - 38 + 100$
$y = 9.5x + 62$

c.

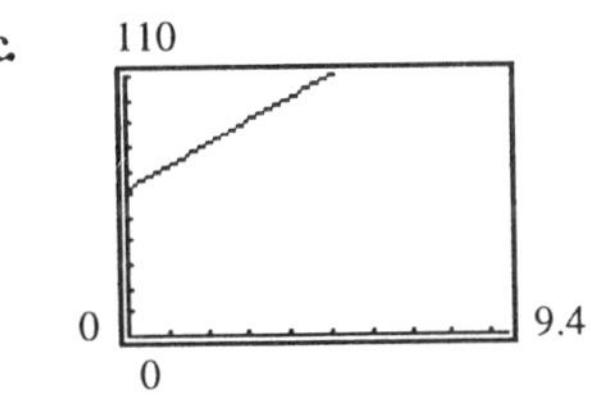

At $x = 3,\ y = 90.5$
The prediction is not very close to 98.
At $x = 2,\ y = 81$
The prediction is fairly close to 80.

d. At $x = 1.5, y = 76.25$
The predicted grade would be about 76 points.

50. $y = 2x + 6 \quad m = 2, b = 6$
$3y - x = 15$
$3y = x + 15$
$y = \frac{1}{3}x + 5 \quad m = \frac{1}{3}, b = 5$
The lines are only intersecting because their slopes are not equal and their product is not –1.

51. $2y - 2x = y + 3x + 2$
$y = 5x + 2 \quad m = 5, b = 2$
$5x - y = -2$
$y = 5x + 2 \quad m = 5, b = 2$
The lines are coinciding because they have the same slopes and the same y-intercepts.

52. $4x - 20 = 0$
$4x = 20$
$x = 5$
$2x + 3 = x + 4$
$x = 1$
The lines are parallel because they are both vertical with different constants.

53. $2x + y = 4$
$y = -2x + 4 \quad m = -2, b = 4$
$y = -2(x + 2)$
$y = -2x - 4 \quad m = -2, b = -4$
The lines are parallel because they have the same slopes and different y-intercepts.

54. $3(y + 2) = 2(y + 4)$
$3y + 6 = 2y + 8$
$y = 2$
$2(x - 2) = 0$
$2x - 4 = 0$
$2x = 4$
$x = 2$
The lines are intersecting and perpendicular because one is a vertical line and the other is a horizontal line.

55. $5y - 4(y + 3) = -10$
$5y - 4y - 12 = -10$
$y - 12 = -10$
$y = 2 \quad m = 0, b = 2$
$y - 1 = -2(x - 1)$
$y - 1 = -2x + 2$
$y = -2x + 3 \quad m = -2, b = 3$
The lines are only intersecting because the slopes are not equal and their product is not –1.

56. $2x - 3y = -9$
$-3y = -2x - 9$
$y = \frac{2}{3}x + 3 \quad m = \frac{2}{3}, b = 3$

$3x + 2y = 6$
$2y = -3x + 6$
$y = -\frac{3}{2}x + 3 \quad m = -\frac{3}{2}, b = 3$
The lines are intersecting and perpendicular because the product of the slopes equals –1.

57. $y = x + 7 \quad m = 1, b = 7$
$y = -x + 7 \quad m = -1, b = 7$
The lines are intersecting and perpendicular because the product of the slopes is –1.

58. **a.** $y = 25 + 35 + 30x + 25 + 5x$
$y = 35x + 85$

b. $y = 35x$

c.

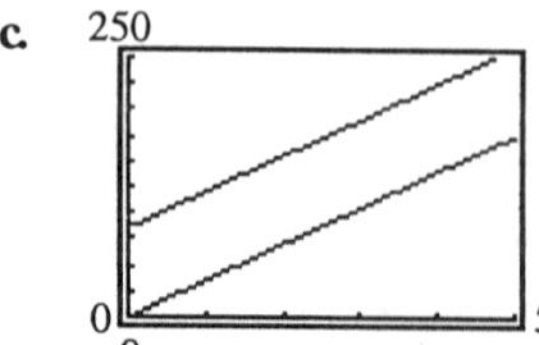

There is no break-even point.

d. $y = 60x$

e.

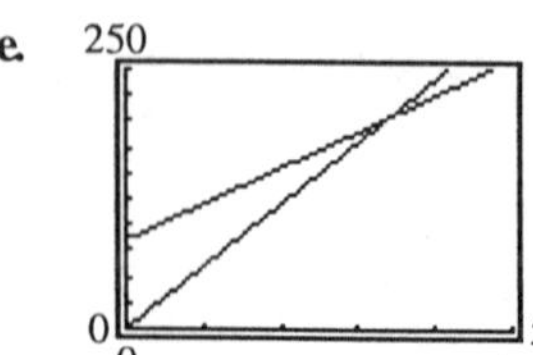

The break-even point is at (3.4, 204). He will start making a profit when he sells the fourth calculator.

f. J.R. should sell the calculators for $60 each.

59. $m=-\frac{1}{4},\ b=1$

$y=-\frac{1}{4}x+1$

60. $m=-2,\ b=3$
$y=-2x+3$

61. $m=\frac{3}{5},\ b=-2$

$y=\frac{3}{5}x-2$

62. $m=0,\ b=-3.5$
$y=0x-3.5$
$y=-3.5$

63. Vertical line, $c=2.6$
$x=2.6$

64. $m=-5,\ (2,-3)$
$y+3=-5(x-2)$
$y+3=-5x+10$
$y=-5x+7$

65. (9, 5) and (–2, 2)

$m=\frac{y_2-y_1}{x_2-x_1}$

$m=\frac{2-5}{-2-9}$

$m=\frac{-3}{-11}$

$m=\frac{3}{11}$

$y-y_1=m(x-x_1)$

$y-5=\frac{3}{11}(x-9)$

$y-5=\frac{3}{11}x-\frac{27}{11}$

$y=\frac{3}{11}x+\frac{28}{11}$

66. $y-y_1=m(x-x_1)$

$y-6=\frac{1}{3}(x-4)$

$y-6=\frac{1}{3}x-\frac{4}{3}$

$y=\frac{1}{3}x+\frac{14}{3}$

67. $y=4x+5$
$m=4$
Parallel lines have equal slopes.
$y-y_1=m(x-x_1)$
$y-1=4(x-1)$
$y-1=4x-4$
$y=4x-3$

68. $y=2x-1$
$m=2$
Perpendicular lines have opposite reciprocal slopes. This line has slope $-\frac{1}{2}$.

$y-y_1=m(x-x_1)$

$y-4=-\frac{1}{2}(x-2)$

$y-4=-\frac{1}{2}x+1$

$y=-\frac{1}{2}x+5$

69. a. (2, 15) and (8, 30)

$m=\frac{y_2-y_1}{x_2-x_1}$

$m=\frac{30-15}{8-2}$

$m=\frac{15}{6}$

$m=\frac{5}{2}$

$y-y_1=m(x-x_1)$

$y-15=\frac{5}{2}(x-2)$

$y-15=\frac{5}{2}x-5$

$y=\frac{5}{2}x+10$

b. $y = \frac{5}{2}(5) + 10$
$y = 22.5$
The prediction is 22.5 thousands of records sold. This is not very close to the actual sales.

Chapter 6 Mixed Review

1. Answers will vary. Possible answer:
$3(y - 5) = y + 7$
$3y - 15 = y + 7$
$3y - 15 - y + 15 = y + 7 - y + 15$
$2y = 22$
$\frac{2y}{2} = \frac{22}{2}$
$y = 11$

x	$y = 11$	y
–1	$y = 11$	11
0	$y = 11$	11
1	$y = 11$	11

(–1, 11), (0, 11), (1, 11) are three possible solutions.

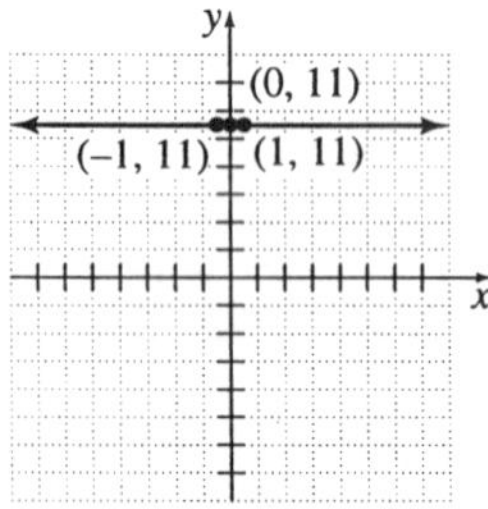

2. Answers will vary. Possible answer.
$14x + 7y = 7$
$14x + 7y - 14x = 7 - 14x$
$7y = -14x + 7$
$\frac{7y}{7} = \frac{-14x + 7}{7}$
$y = -2x + 1$

x	$y = -2x + 1$	y
–2	$y = -2(-2) + 1 = 5$	5
0	$y = -2(0) + 1 = 1$	1
2	$y = -2(2) + 1 = -3$	–3

(–2, 5), (0, 1), and (2, –3) are three possible solutions.

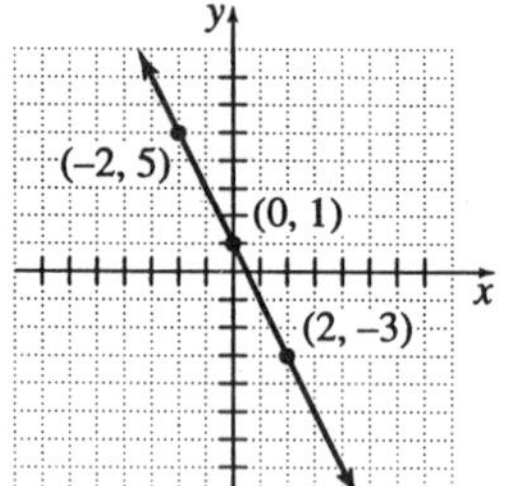

3. Answers will vary. Possible answer:

x	$y = -\frac{7}{5}x - 10$	y
–5	$y = -\frac{7}{5}(-5) - 10 = -3$	–3
0	$y = -\frac{7}{5}(0) - 10 = -10$	–10
5	$y = -\frac{7}{5}(5) - 10 = -17$	–17

(–5, –3), (0, –10), and (5, –17) are three possible solutions.

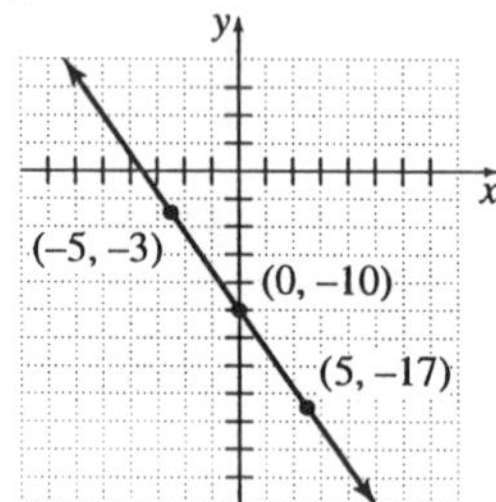

4. (2, 6) and $m = 0$

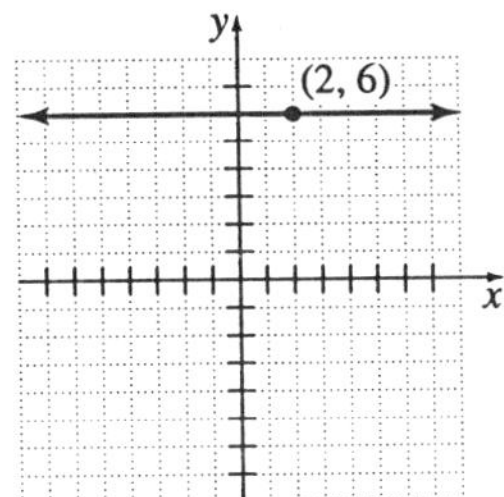

5. (2, 6) and $m =$ undefined

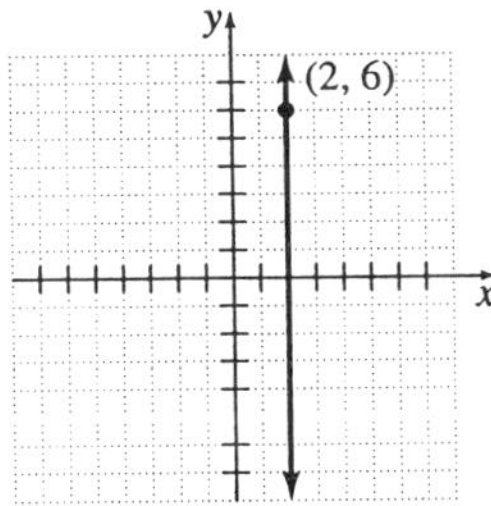

6. (1, –4) and $m = -2$

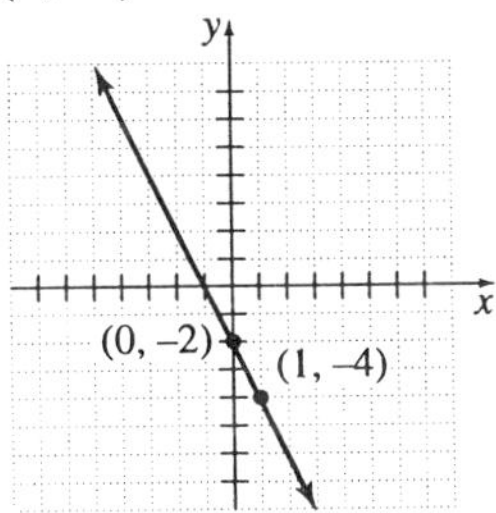

7. (0, 3) and $m = \frac{2}{7}$

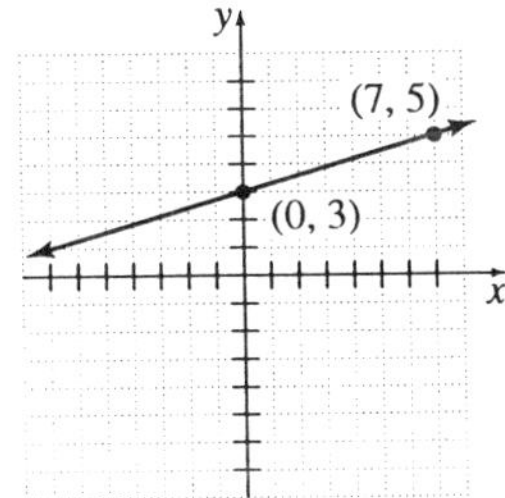

8. $y = -\frac{3}{8}x$

$m = -\frac{3}{8},\ b = 0$

The slope is $-\frac{3}{8}$ and the y-intercept is (0, 0).

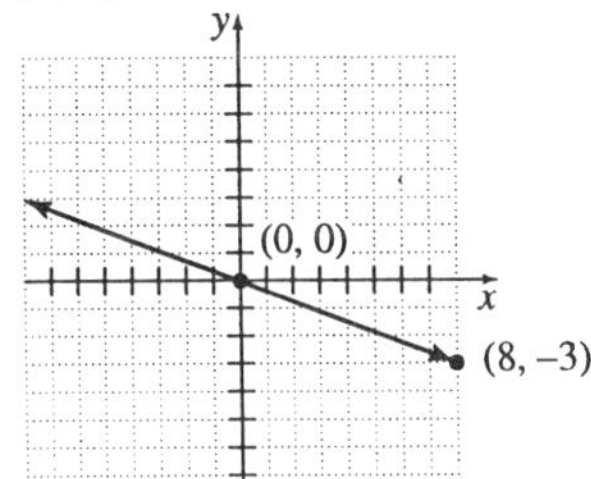

9. $y = 7$

$m = 0,\ b = 7$

The slope is 0 and the y-intercept is (0, 7).

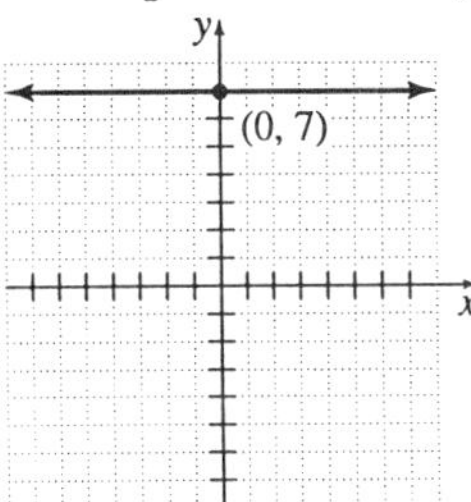

10. $5x - 4y = 12$

$5x - 4y - 5x = 12 - 5x$

$-4y = -5x + 12$

$\frac{-4y}{-4} = \frac{-5x + 12}{-4}$

$y = \frac{5}{4}x - 3$

$m = \frac{5}{4},\ b = -3$

The slope is $\frac{5}{4}$ and the y-intercept is (0, –3).

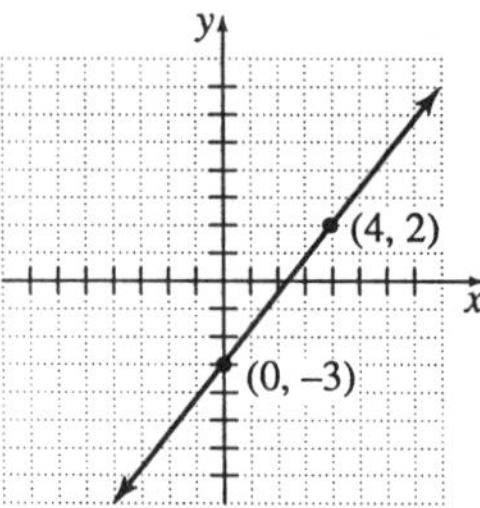

11. $15.5x + 21.7y = 108.5$
Substitute 0 for y.
$15.5x + 21.7(0) = 108.5$
$\frac{15.5x}{15.5} = \frac{108.5}{15.5}$
$x = 7$
The x-intercept is (7, 0).
Substitute 0 for x.
$15.5(0) + 21.7y = 108.5$
$\frac{21.7y}{21.7} = \frac{108.5}{21.7}$
$y = 5$
The y-intercept is (0, 5).

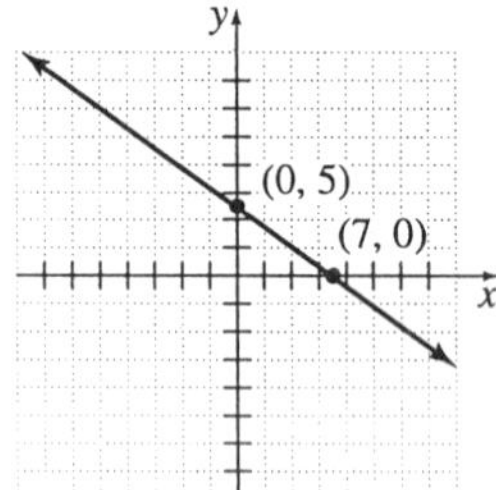

12. $12x - 24y = -48$
Substitute 0 for y.
$12x - 24(0) = -48$
$\frac{12x}{12} = \frac{-48}{12}$
$x = -4$
The x-intercept is (–4, 0).
Substitute 0 for x.
$12(0) - 24y = -48$
$\frac{-24y}{-24} = \frac{-48}{-24}$
$y = 2$
The y-intercept is (0, 2).

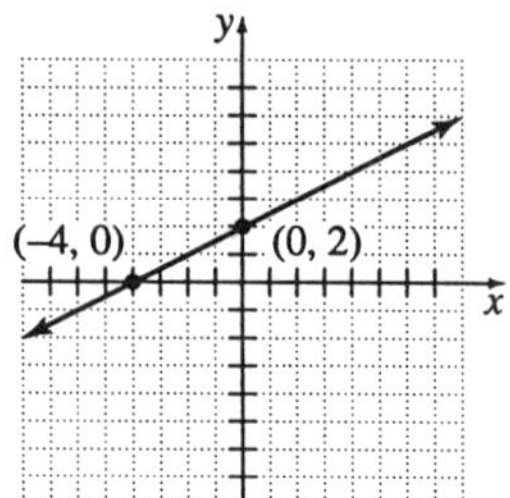

13. $-x + y = -6$
Substitute 0 for y.
$-x + 0 = -6$
$\frac{-x}{-1} = \frac{-6}{-1}$
$x = 6$
The x-intercept is (6, 0).
Substitute 0 for x.
$-0 + y = -6$
$y = -6$
The y-intercept is (0, –6).

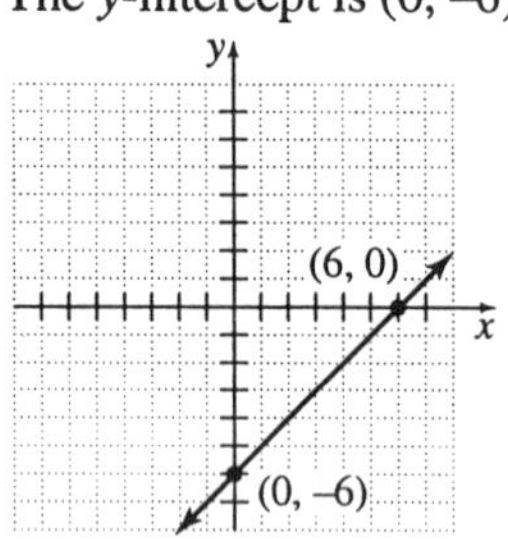

14. $2x - 3 = 1$
$2x = 4$
$x = 2$ m is undefined

$4x - 3y = 6$
$3y = 4x - 6$
$y = \frac{4}{3}x - 2$ $m = \frac{4}{3}$, $b = -2$
The lines are only intersecting because one is vertical and the other has a slope of $\frac{4}{3}$.

15. $5x - 4y = 8$
$4y = 5x - 8$
$y = \frac{5}{4}x - 2$ $m = \frac{5}{4}$, $b = -2$

$4x - 5y = -15$
$5y = 4x + 15$
$y = \frac{4}{5}x + 3 \quad m = \frac{4}{5},\ b = 3$
The lines are only intersecting because they have different slopes and their product is not -1.

16. $y = 5(x - 1)$
$y = 5x - 5 \quad m = 5,\ b = -5$

$2(x - 2) = 7x - y + 3$
$2x - 2 = 7x - y + 3$
$y = 5x + 5 \quad m = 5,\ b = 5$
The lines are parallel because they have the same slopes, but different y-intercepts.

17. $5(x + 1) = 15$
$5x + 5 = 15$
$5x = 10$
$x = 2$

$3y + 1 = 10$
$3y = 9$
$y = 3$
The lines are intersecting and perpendicular because one is vertical and the other is horizontal.

18. $y = -2x + 3 \quad m = -2,\ b = 3$
$2(x + y) = y + 3$
$2x + 2y = y + 3$
$y = -2x + 3 \quad m = -2,\ b = 3$
The lines are coinciding because their slopes are the same and their y-intercepts are the same.

19. $y = x + 3 \quad m = 1,\ b = 3$
$y = -x - 4 \quad m = -1,\ b = -4$
The lines are intersecting and perpendicular because the product of their slopes is -1.

20. $5y = 20$
$y = 4 \quad m = 0, b = 4$
$2(y + 3) = -2$
$2y + 6 = -2$
$2y = -8$
$y = -4 \quad m = 0, b = -4$
The lines are parallel because they are both horizontal with different y-intercepts.

21. Answers will vary. Possible answer:

x	$y = 8$	y
-1	$y = 8$	8
0	$y = 8$	8
1	$y = 8$	8

$(-1, 8)$, $(0, 8)$, and $(1, 8)$ are three possible solutions.

22. Answers will vary. Possible answer:
$8x - 9y = -72$
$8x - 9y - 8x = -72 - 8x$
$-9y = -8x - 72$
$\frac{-9y}{-9} = \frac{-8x - 72}{-9}$
$y = \frac{8}{9}x + 8$

x	$y = \frac{8}{9}x + 8$	y
-9	$y = \frac{8}{9}(-9) + 8 = 0$	0
0	$y = \frac{8}{9}(0) + 8 = 8$	8
9	$y = \frac{8}{9}(9) + 8 = 16$	16

$(-9, 0)$, $(0, 8)$, and $(9, 16)$ are three possible solutions.

23. Answers will vary. Possible answer:

x	$t(x) = \frac{9}{11}x - 8$	$t(x)$
-11	$t(x) = \frac{9}{11}(-11) - 8 = -17$	-17
0	$t(x) = \frac{9}{11}(0) - 8 = -8$	-8
11	$t(x) = \frac{9}{11}(11) - 8 = 1$	1

$(-11, -17)$, $(0, -8)$, and $(11, 1)$ are three possible solutions.

24. $y = x$ is linear.
$y - y = x - y$
$0 = x - y$
$x - y = 0$ is in standard form.

25. $y = 2x^2 - 8$ is non-linear because the variable x is squared.

26. $y = 1.3x - 0.5$ is linear.
$y - y = 1.3x - 0.5 - y$
$0 = 1.3x - 0.5 - y$
$0.5 + 0 = 1.3x - 0.5 - y + 0.5$
$0.5 = 1.3x - y$
$1.3x - y = 0.5$ is in standard form.

27. $y = 15x$ is linear.
$y - y = 15x - y$
$0 = 15x - y$
$15x - y = 0$ is in standard form.

28. $8y - 5x = 21$ is linear.
$-5x + 8y = 21$
$-1(-5x) - 1(8y) = -1(21)$
$5x - 8y = -21$ is in standard form.

29. $2x + 3y - 1 = y^2 + x$ is non-linear because one y-variable is squared.

30.

$\frac{2}{3}x + \frac{3}{4}y = 2$	
$\frac{2}{3}(12) + \frac{3}{4}(-8)$	2
$8 + (-6)$	
2	

Since $2 = 2$, $(12, -8)$ is a solution.

31.

$22x - 5y = 17$	
$22\left(\frac{7}{12}\right) - 5\left(-\frac{5}{6}\right)$	17
$\frac{154}{12} + \frac{25}{6}$	
17	

Since $17 = 17$, $\left(\frac{7}{12}, -\frac{5}{6}\right)$ is a solution.

32.

$u(x) = -8x + 2$	
-14	$-8(2) + 2$
	$-16 + 2$
	-14

Since $-14 = -14$, $(2, -14)$ is a solution.

33.

$7x + 8y = 8$	
$7(0) + 8(-3)$	8
-24	

Since $-24 \neq 8$, $(0, -3)$ is not a solution.

34.

$x = -8$	
-8	-8

Since $-8 = -8$, $(-8, -2)$ is a solution.

35.

$v(x) = 12$	
-2	12

Since $-2 \neq 12$, $(12, -2)$ is not a solution.

36. $x + 4(x - 2) = 2$
$x + 4x - 8 = 2$
$5x - 8 = 2$
$5x - 8 + 8 = 2 + 8$
$5x = 10$
$\frac{5x}{5} = \frac{10}{5}$
$x = 2$
The slope is undefined.
There is no y-intercept.

37. $x - 3(y + 2) = 4(x + 1) - y$
$x - 3y - 6 = 4x + 4 - y$
$x - 3y - 6 - x + 6 = 4x + 4 - y - x + 6$
$-3y = 3x + 10 - y$
$-3y + y = 3x + 10 - y + y$
$-2y = 3x + 10$
$\frac{-2y}{-2} = \frac{3x + 10}{-2}$
$y = -\frac{3}{2}x - 5$
$m = -\frac{3}{2}$, $b = -5$
The slope is $-\frac{3}{2}$ and the y-intercept is $(0, -5)$.

38. $12x - 4y = 8$
$12x - 4y - 12x = 8 - 12x$
$-4y = -12x + 8$
$\frac{-4y}{-4} = \frac{-12x + 8}{-4}$
$y = 3x - 2$
$m = 3$, $b = -2$
The slope is 3 and the y-intercept is $(0, -2)$.

39. $2y-6=4(y-2)+2$
$2y-6=4y-8+2$
$2y-6=4y-6$
$2y-6+6=4y-6+6$
$2y=4y$
$2y-4y=4y-4y$
$-2y=0$
$\frac{-2y}{-2}=\frac{0}{-2}$
$y=0$
$m=0,\ b=0$
The slope is 0 and the y-intercept is (0, 0).

40. $y=13x-15$
$m=13,\ b=-15$
The slope is 13 and the y-intercept is (0, –15).

41. $y=-5.03x+7.92$
$m=-5.03,\ b=7.92$
The slope is –5.03 and the y-intercept is (0, 7.92).

42. (5, 5) and (8, 5)
$m=\frac{y_2-y_1}{x_2-x_1}$
$m=\frac{5-5}{8-5}$
$m=\frac{0}{3}$
The slope is 0.

43. (–3, –3) and (–3, 3)
$m=\frac{y_2-y_1}{x_2-x_1}$
$m=\frac{3-(-3)}{-3-(-3)}$
$m=\frac{6}{0}$
The slope is undefined.

44. (–4, 4) and (–1, –5)
$m=\frac{y_2-y_1}{x_2-x_1}$
$m=\frac{-5-4}{-1-(-4)}$
$m=\frac{-9}{3}$
$m=-3$
The slope is –3.

45. $\left(-\frac{2}{5},\frac{4}{7}\right)$ and $\left(-\frac{4}{7},\frac{1}{5}\right)$
$m=\frac{y_2-y_1}{x_2-x_1}$
$m=\frac{\frac{1}{5}-\frac{4}{7}}{-\frac{4}{7}-\left(-\frac{2}{5}\right)}$
$m=\frac{\frac{7-20}{35}}{\frac{-20+14}{35}}$
$m=\frac{-13}{35}\cdot\frac{35}{-6}$
$m=\frac{-13}{-6}$
$m=\frac{13}{6}$
The slope is $\frac{13}{6}$.

46. $m=\frac{y_2-y_1}{x_2-x_1}$
$m=\frac{2-(-2)}{5-4}$
$m=\frac{4}{1}$
$m=4$
$y-y_1=m(x-x_1)$
$y-(-2)=4(x-4)$
$y+2=4x-16$
$y=4x-18$

47. $y - y_1 = m(x - x_1)$
$y - (-1) = -\frac{1}{4}[x - (-1)]$
$y + 1 = -\frac{1}{4}x - \frac{1}{4}$
$y = -\frac{1}{4}x - \frac{5}{4}$

48. $m = -\frac{2}{3}$, $b = 5$
$y = -\frac{2}{3}x + 5$

49. A vertical line has a constant of 4.1.
$x = 4.1$

50. $y = -3x + 6$
$m = -3$
Parallel lines have equal slopes.
$y - y_1 = m(x - x_1)$
$y - (-2) = -3[x - (-2)]$
$y + 2 = -3x - 6$
$y = -3x - 8$

51. $m = 0$, $b = 8$
$y = 0x + 8$
$y = 8$

52. $m = 4$, $(1.2, 8)$
$y - 8 = 4(x - 1.2)$
$y - 8 = 4x - 4.8$
$y = 4x + 4.8$

53. $m = 3$, $b = -2$
$y = 3x - 2$

54. $y = -\frac{1}{2}x - 1$
$m = -\frac{1}{2}$
Perpendicular lines have opposite reciprocal slopes. The slope is 2.
$m = 2$, $b = 2$

55. **a.** $m = \frac{y_2 - y_1}{x_2 - x_1}$
$m = \frac{99 - 57}{94 - 50}$
$m = \frac{42}{44}$
$m = \frac{21}{22}$
$y - y_1 = m(x - x_1)$
$y - 57 = \frac{21}{22}(x - 50)$
$y - 57 = \frac{21}{22}x - \frac{525}{11}$
$y - 57 + 57 = \frac{21}{22}x - \frac{525}{11} + \frac{627}{11}$
$y = \frac{21}{22}x + \frac{102}{11}$

b. $y = \frac{21}{22}(80) + \frac{102}{11}$
$y \approx 85.6$
The predicted score of about 86 is not very close to the actual score of 78.

56. **a.** (7, 24) and (9, 28)

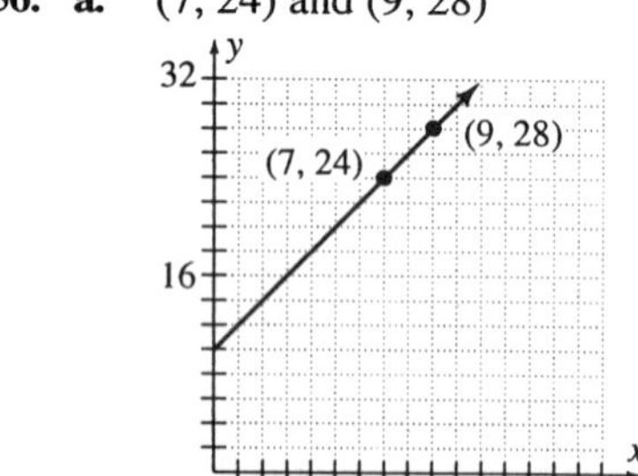

b. See the graph in part a.
At $x = 8$, $y = 26$
Frank would receive $26 for 8 innings.
At $x = 10$, $y = 30$
Frank would receive $30 for 10 innings.

57. **a.** $y = 180 + 10x - 15x$
$y = -5x + 180$

b. Substitute 0 for y.
$0 = -5x + 180$
$5x = 180$
$\frac{5x}{5} = \frac{180}{5}$
$x = 36$
The x-intercept is (36, 0). After 36 hours, the tank is empty.
Substitute 0 for x.
$y = -5(0) + 180$
$y = 180$
The y-intercept is (0, 180). After 0 hours, the tank has 180 gallons.

c.

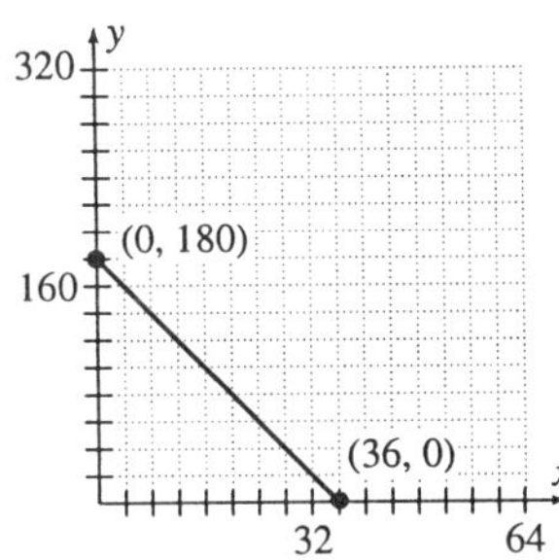

d. At $x = 16$, $y = 100$
After 16 hours, 100 gallons remain in the tank.

58. a. $y = 25x + 15$

b. $y = 25(2) + 15$
$y = 65$
The trainer would earn \$65.

c. $y = 25\left(1\frac{1}{2}\right) + 15$
$y = 52\frac{1}{2}$
He would earn \$52.50.

59. a. $c(x) = 75 + 150 + 0.75x$
$c(x) = 0.75x + 225$

b. $r(x) = 2x$

c.

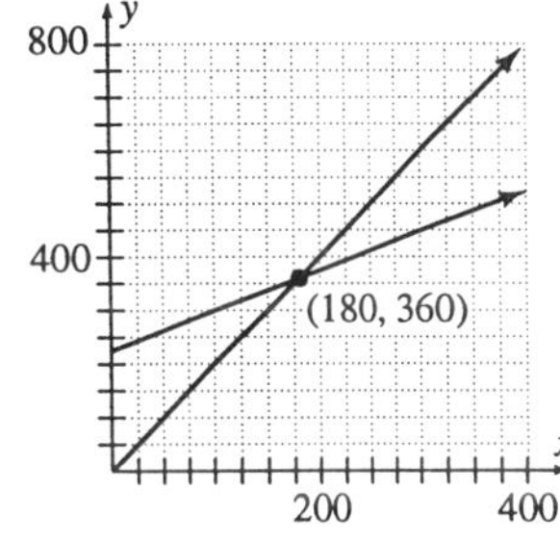

The break-even point is (180, 360). Revenue equals cost at 180 bows.

d. No, 150 is less than 180.
Yes, 200 is greater than 180.
She must sell at least 180 bows.

60. grade $= \frac{60 \text{ feet}}{125 \text{ feet}} = \frac{12}{25} = 48\%$

Chapter 6 Test

1.

$7x = 8(y - 3)$	
7(8)	8(10 − 3)
56	8(7)
	56

Since 56 = 56, (8, 10) is a solution.

2.

$x - 8y = 16$	
−8 − 8(−2)	16
−8 + 16	
8	

Since $8 \neq 16$, (−8, −2) is not a solution.

3. $5x - 7y = 2(x + 1) - 3$ is linear because it simplifies to $3x - 7y = -1$.

4. $y = \frac{11}{12}x - 5$ is linear because it is in slope-intercept form $y = ax + b$.

5. $y^2 - 16 = 0$ is nonlinear because the variable y is squared.

6. $7x + 2y = 3x^2 + 3$ is nonlinear because one of the x-variables is squared.

7. Answers will vary. Possible answer:
$2x - 8y = 0$
$2x - 8y - 2x = 0 - 2x$
$-8y = -2x$
$\frac{-8y}{-8} = \frac{-2x}{-8}$
$y = \frac{1}{4}x$

x	$y = \frac{1}{4}x$	y
−4	$y = \frac{1}{4}(-4) = -1$	−1
0	$y = \frac{1}{4}(0) = 0$	0
4	$y = \frac{1}{4}(4) = 1$	1

(−4, −1), (0, 0), and (4, 1) are three possible solutions.

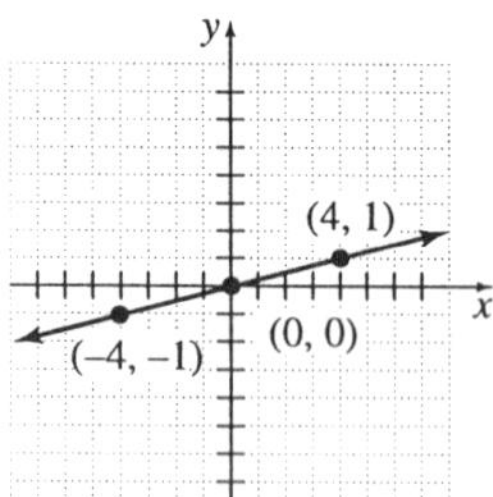

8. $12x + 15y = 60$
Substitute 0 for y.
$12x + 15(0) = 60$
$\frac{12x}{12} = \frac{60}{12}$
$x = 5$
The x-intercept is (5, 0).
Substitute 0 for x.
$12(0) + 15y = 60$
$\frac{15y}{15} = \frac{60}{15}$
$y = 4$
The y-intercept is (0, 4).

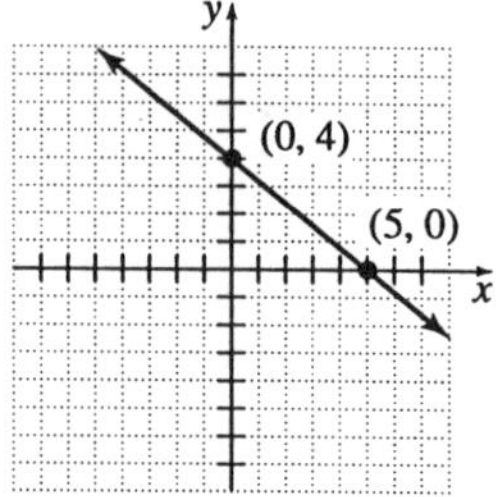

9. $g(x) = 3x - 5$
$m = 3,\ b = -5$
The slope is 3 and the y-intercept is (0, −5).

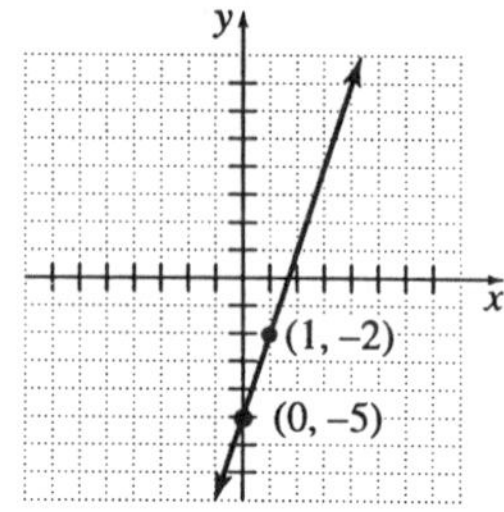

10. (0, −1) and (−1, −1)
$m = \frac{y_2 - y_1}{x_2 - x_1}$
$m = \frac{-1-(-1)}{-1-0}$
$m = \frac{0}{-1}$
The slope is 0.

11. (4.8, 0.6) and (−1.5, 2.4)
$m = \frac{y_2 - y_1}{x_2 - x_1}$
$m = \frac{2.4 - 0.6}{-1.5 - 4.8}$
$m = \frac{1.8}{-6.3}$
$m = -\frac{18}{63}$
$m = -\frac{2}{7}$
The slope is $-\frac{2}{7}$.

12. (–3, –4) and (1, 0)

$m = \frac{y_2 - y_1}{x_2 - x_1}$

$m = \frac{0-(-4)}{1-(-3)}$

$m = \frac{4}{4}$

$m = 1$

The slope is 1.

13. (–2, 2) and (–2, 8)

$m = \frac{y_2 - x_1}{x_2 - x_1}$

$m = \frac{8-2}{-2-(-2)}$

$m = \frac{6}{0}$

The slope is undefined.

14. $y = -8x + 1 \quad m = -8,\ b = 1$

$y = x - 8 \quad m = 1,\ b = -8$

The lines are only intersecting because the slopes are not equal and their product is not equal to –1.

15. $9(y-2) + 2x = 7x$

$9y - 18 + 2x = 7x$

$9y - 18 = 5x$

$9y = 5x + 18$

$y = \frac{5}{9}x + 2 \quad m = \frac{5}{9},\ b = 2$

$4x = 9(y+4) - x$

$4x = 9y + 36 - x$

$9y = 5x - 36$

$y = \frac{5}{9}x - 4 \quad m = \frac{5}{9},\ b = -4$

The lines are parallel because their slopes are equal and their *y*-intercepts are not equal.

16. $3x - y = 2$

$y = 3x - 2 \quad m = 3,\ b = -2$

$y - x = 2(x-1)$

$y - x = 2x - 2$

$y = 3x - 2 \quad m = 3,\ b = -2$

The lines are coinciding because their slopes are the same and their *y*-intercepts are the same.

17. $5y - 2 = 3y + 2$

$2y = 4$

$y = 2$

$3x = 2(x+2)$

$3x = 2x + 4$

$x = 4$

The lines are intersecting and perpendicular because one is horizontal and the other is vertical.

18. $4y = 3x + 12$

$\frac{4y}{4} = \frac{3x+12}{4}$

$y = \frac{3}{4}x + 3$

$y = \frac{3}{4},\ b = 3$

The slope is $\frac{3}{4}$ and the *y*-intercept is (0, 3).

19. $7x + 7y = 21$

$7x + 7y - 7x = 21 - 7x$

$7y = -7x + 21$

$\frac{7y}{7} = \frac{-7x+21}{7}$

$y = -x + 3$

$m = -1,\ b = 3$

The slope is –1 and the *y*-intercept is (0, 3).

20. $y = \frac{3}{5}x$

$m = \frac{3}{5}$

Perpendicular lines have slopes which are opposite reciprocals.

The slope is $-\frac{5}{3}$.

$y - y_1 = m(x - x_1)$

$y - 1 = -\frac{5}{3}(x-2)$

$y - 1 = -\frac{5}{3}x + \frac{10}{3}$

$y = -\frac{5}{3}x + \frac{13}{3}$

21. $m = 9,\ b = 7$

$y = 9x + 7$

22. $y - 2 = 6[x - (-1)]$
$y - 2 = 6(x + 1)$
$y - 2 = 6x + 6$
$y = 6x + 8$

23. (2, 3) (–4, 1)
$m = \dfrac{1-3}{-4-2} = \dfrac{-2}{-6} = \dfrac{1}{3}$
$y - 3 = \dfrac{1}{3}(x - 2)$
$y - 3 = \dfrac{1}{3}x - \dfrac{2}{3}$
$y = \dfrac{1}{3}x + \dfrac{7}{3}$

24. $y = 1450 - 150x$

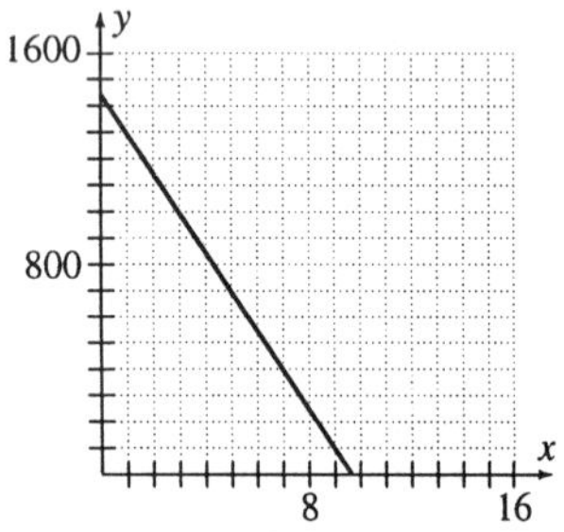

At $x = 4$, $y = 850$
After 4 months, his remaining loan is $850.

25. Answers will vary.

Chapter 7

7.1 Experiencing Algebra the Exercise Way

1.

$y = x + 1$	
4	$3 + 1$
	4

$2y = x + 5$	
2(4)	$3 + 5$
8	8

Both equations are true. Therefore, (3, 4) is a solution of the system.

3. $p(x) = \frac{1}{4}x + 1$

$\frac{6}{5}$	$\frac{1}{4}\left(\frac{4}{5}\right) + 1$
	$\frac{1}{5} + \frac{5}{5}$
	$\frac{6}{5}$

$q(x) = \frac{1}{2}x + \frac{4}{5}$

$\frac{6}{5}$	$\frac{1}{2}\left(\frac{4}{5}\right) + \frac{4}{5}$
	$\frac{2}{5} + \frac{4}{5}$
	$\frac{6}{5}$

Both equations are true. Therefore, $\left(\frac{4}{5}, \frac{6}{5}\right)$ is a solution of the system.

5.

$3y = x + 6$	
3(2.1)	$0.3 + 6$
6.3	6.3

$10y = 25$	
10(2.1)	25
21	

The second equation is false. Therefore (0.3, 2.1) is not a solution.

7.

$7y + 2 = 7$	
$7\left(\frac{5}{7}\right) + 2$	7
$5 + 2$	
7	

$5y + 3 = 0$	
$5\left(\frac{5}{7}\right) + 3$	0
$\frac{25}{7} + \frac{21}{7}$	
$\frac{46}{7}$	

The second equation is false. Therefore $\left(\frac{2}{3}, \frac{5}{7}\right)$ is not a solution.

9.

$2y = -x$	
2(–2)	–5
–4	

$x = 5$	
5	5

The first equation is false. Therefore, (5, –2) is not a solution.

11. $2x + 3y = -6$
$x - 4y = 8$

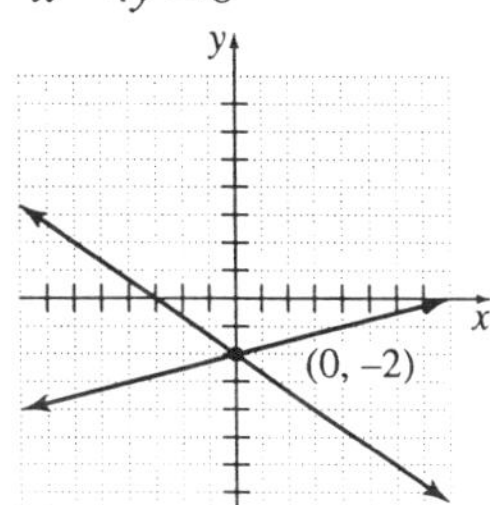

The solution is (0, –2).

13. $a + b = 4$
$a - 2b = 7$

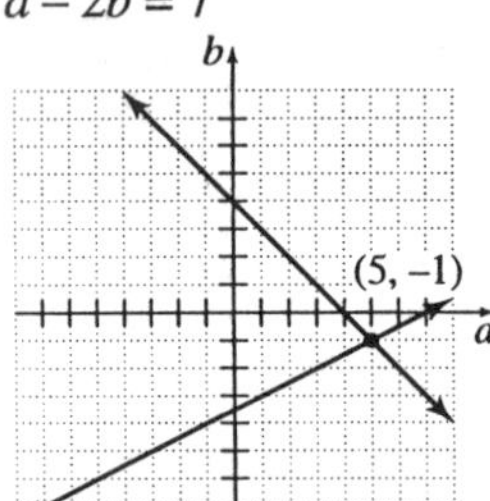

The solution is (5, –1).

15. $x + 2y = 6$
$x + 2y = 2$

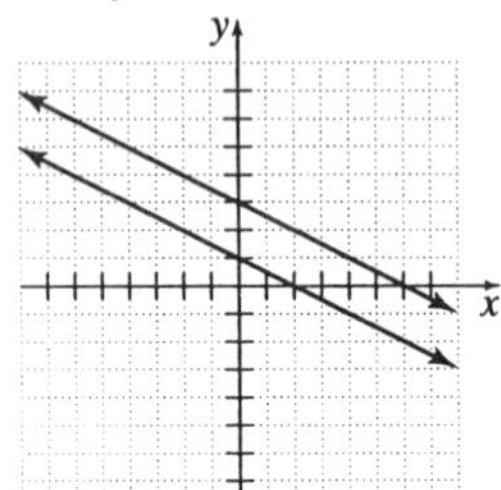

The lines are parallel. There is no solution.

17. $x - y = -1$
$3x - 3y = -3$

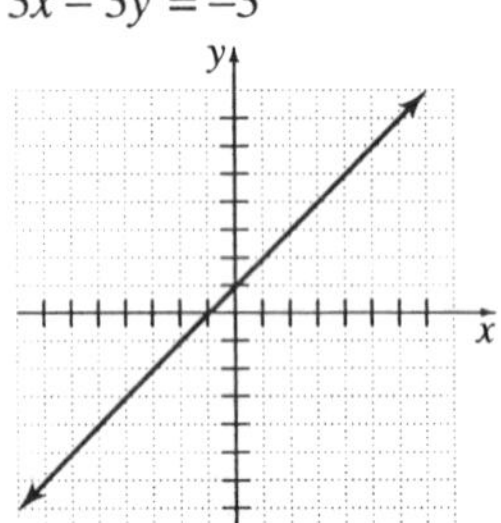

The graphs are coinciding. There is an infinite number of solutions. The solutions are all ordered pairs (x, y) that satisfy $x - y = -1$.

19. $y = \frac{1}{2}x + 3$

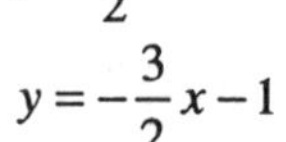

$y = -\frac{3}{2}x - 1$

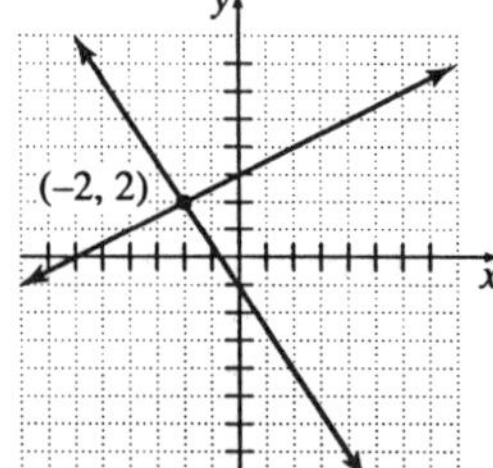

The solution is (–2, 2).

21. $f(x) = 3x - 1$
$g(x) = -2x + 4$

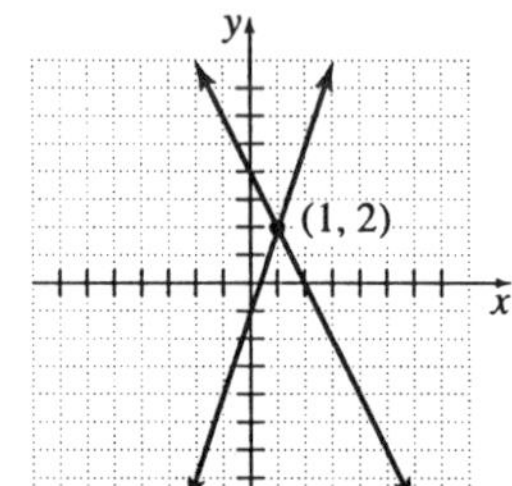

The solution is (1, 2).

23. $F(x) = 2x + 2$
$G(x) = 2x - 3$

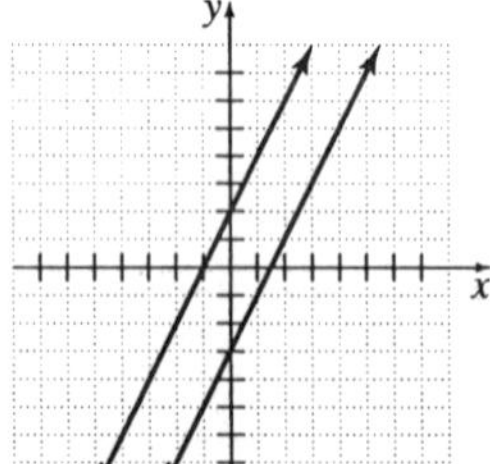

The graphs are parallel. There is no solution.

25. $y = \frac{1}{2}x + 2$
$6y = 3x + 12$

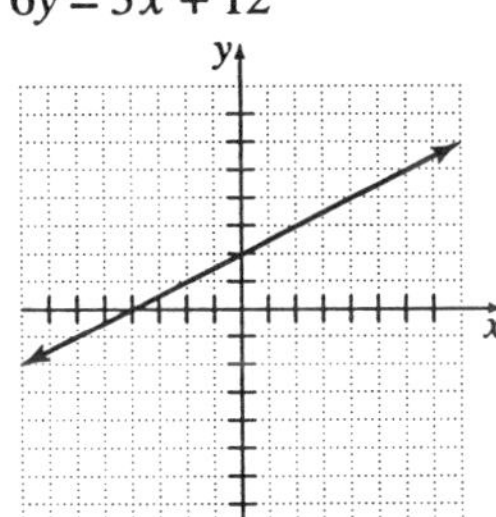

The graphs are coinciding. The solutions are all ordered pairs (x, y) that satisfy
$y = \frac{1}{2}x + 2.$

27. $3x + 2y = 12$
$y - 2 = 1$

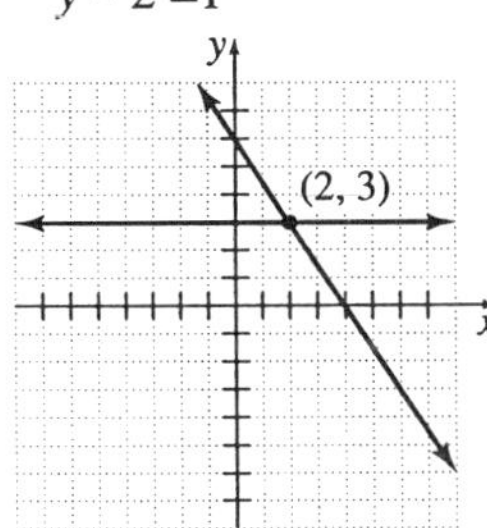

The solution is (2, 3).

29. $3y - 2 = 2y + 2$
$x - 1 = 0$

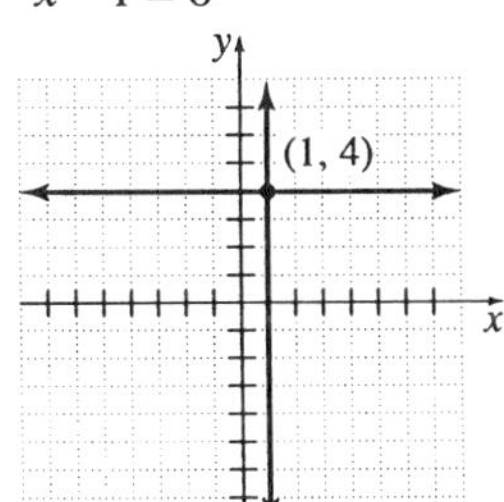

The solution is (1, 4).

31. $y - 1 = 0$
$y + 3 = 0$

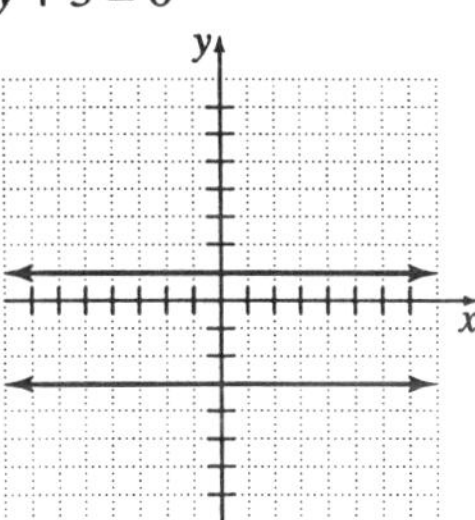

The graphs are parallel. There is no solution.

33. $y = 3x - 5$
$x = 4 - 2y$

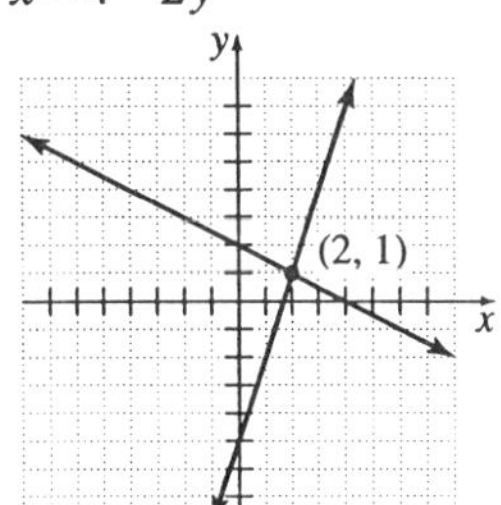

The solution is (2, 1).

35. $y = x - 7$
$x = y + 7$

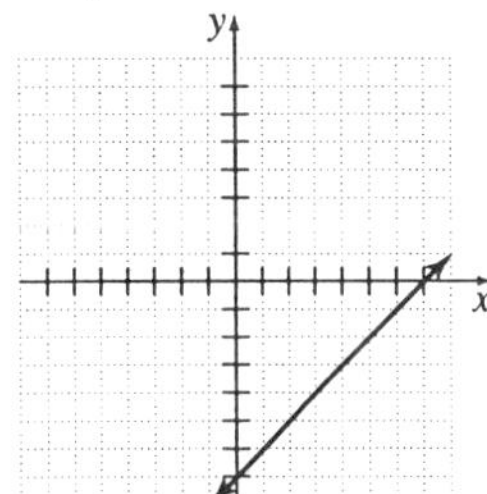

The graphs are coinciding. The solutions are all ordered pairs (x, y) that satisfy $y = x - 7$.

37. $a(x) = 1 + 4x$
$b(x) = 4x - 2$

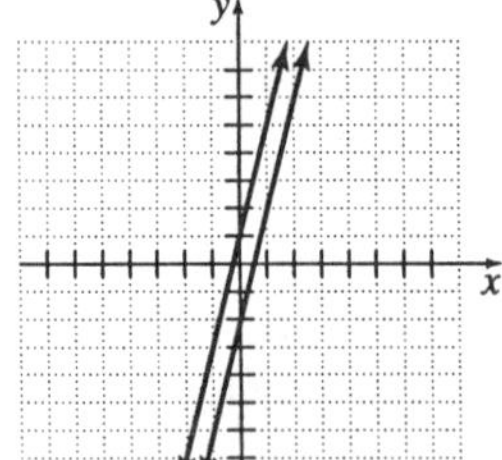

The graphs are parallel. There is no solution.

39. a. Turtle Rental: $y = 39.95 + 0.25(x - 150)$
Snail Rental: $y = 79.95 + 0.15(x - 200)$

b.

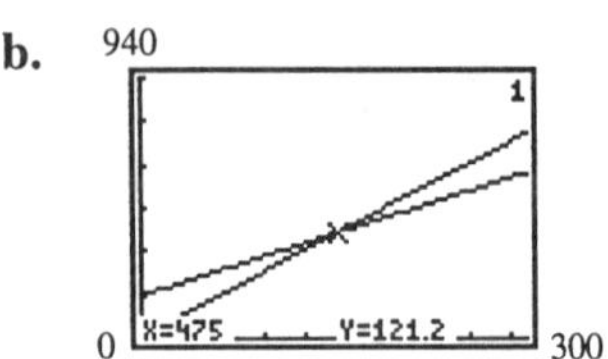

The solution is (475, 121.2).

c. If Ahmet drives less than 475 miles, Turtle Rental is a better deal. If Ahmet drives more than 475 miles, Snail Rental is a better deal.

d. Turtle Rental

e. Snail Rental

41. a. Bulk-rate: $y = 75 + 0.226x$
First-class: $y = 0.32x$

b.

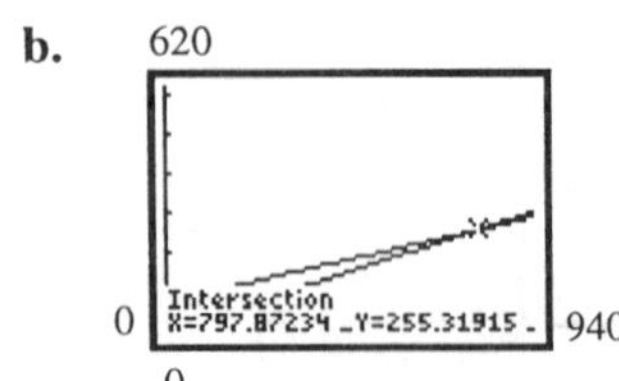

The solution is approximately (798, 255).

c. When there are more than 798 pieces of mail, it is more cost effective to use bulk-rate. The graph representing bulk-rate is lower than that representing first-class for values of x greater than 798.

d. When there are fewer than 798 pieces of mail, it is more cost effective to use first-class. The graph representing first-class is lower than that representing bulk-rate for values of x less than 798.

43. Let x = car payment
y = rent payment
$x + y = 500$
$y = 3x$

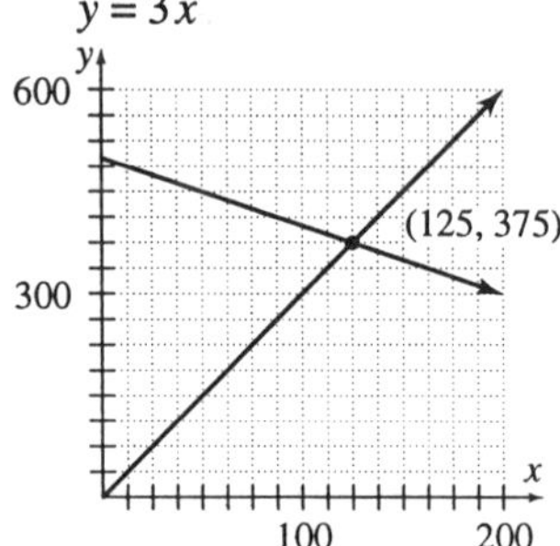

The solution is (125, 375). The car payment is \$125 and the rent payment is \$375.

45. Let x = number of hours at \$7.00 per hour
y = number of hours at \$8.20 per hour
$x + y = 40$
$7x + 8.2y = 298$

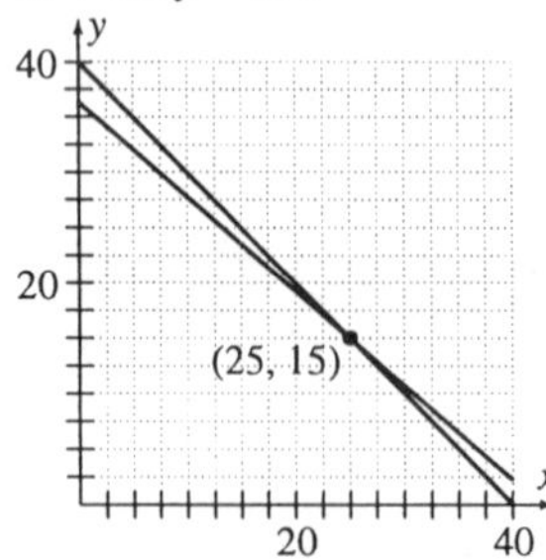

The solution is (25, 15). She should work 15 hours at the job that pays \$8.20 per hour and 25 hours at the job paying \$7 per hour.

47. Let W = width
L = length
$L = 2W - 25$
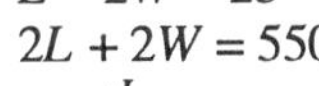
$2L + 2W = 550$

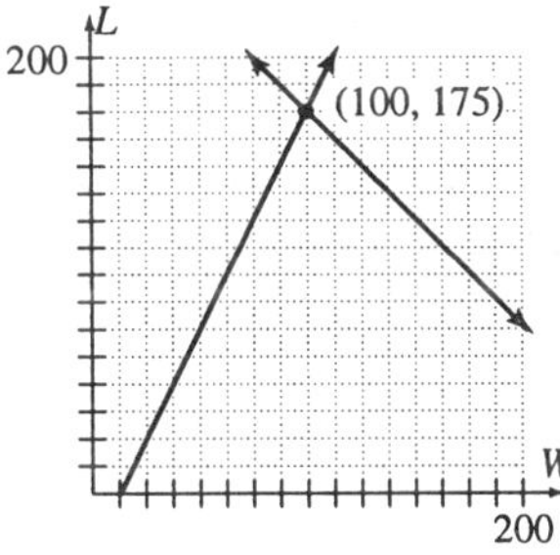

The solution is (100, 175). The width is 100 feet and the length is 175 feet.

49. Let x = number of Shania's records
y = number of Reba's records
$y = 2x + 1$
$x + y = 22$

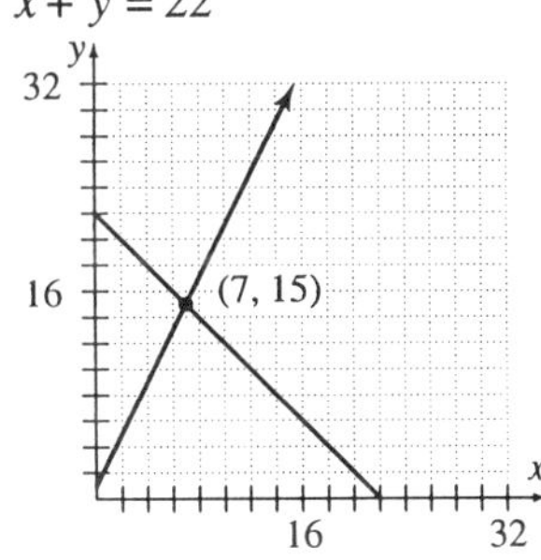

The solution is (7, 15). Shania had 7 records and Reba had 15 records.

7.1 Experiencing Algebra the Calculator Way

1. $y = 2.4x + 1.2$
$y = -1.3x - 2.7$

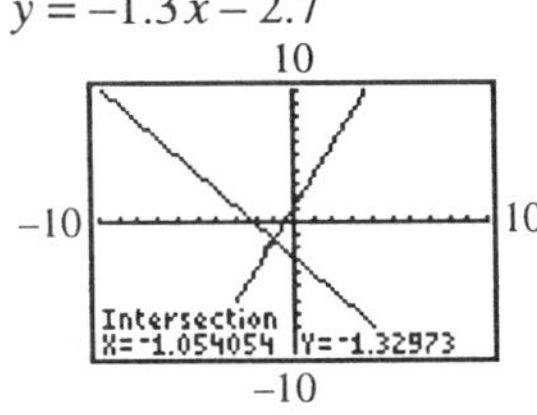

The solution is approximately (−1.05, −1.33).

2. $y = 0.4x - 8.3$
$y = 2.2x + 2.4$

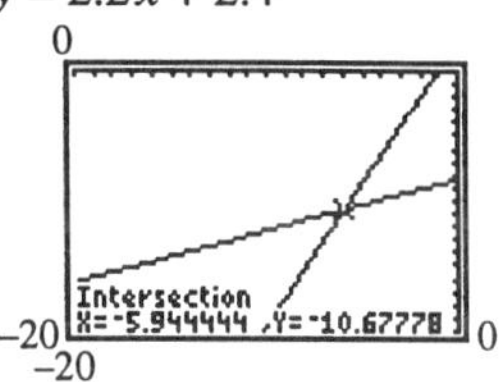

The solution is approximately (−5.94, −10.68).

3. $h(x) = 13.7x$
$k(x) = -115.3$

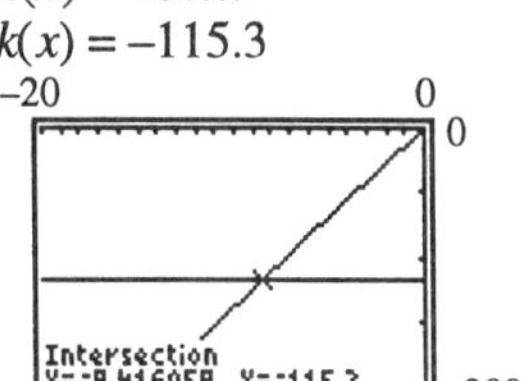

The solution is approximately (−8.42, −115.3).

4. $y = \dfrac{36 - 8x}{12}$
$y = \dfrac{18x - 45}{15}$

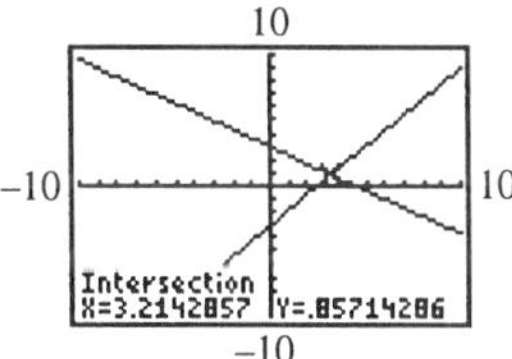

The solution is approximately (3.21, 0.86).

5. $y = \dfrac{15x - 12}{3}$
$y = 5x - 20$

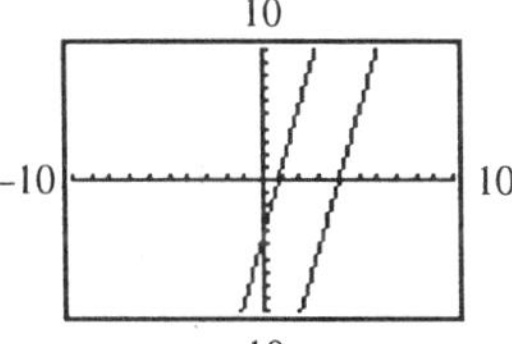

The graphs are parallel. There is no solution.

6. 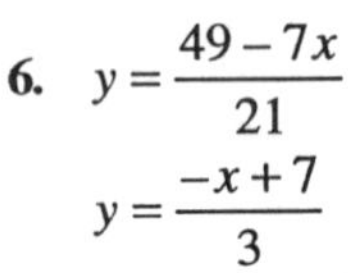

$y = \dfrac{49 - 7x}{21}$

$y = \dfrac{-x + 7}{3}$

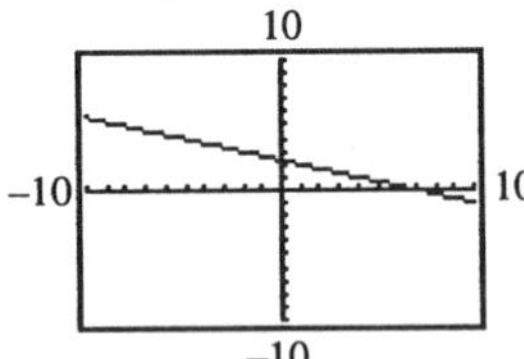

The graphs are coinciding. The solution is all ordered pairs (x, y) that satisfy $y = \dfrac{-x+7}{3}$.

7. $F(x) = 45x + 705$
$G(x) = -35x + 155$

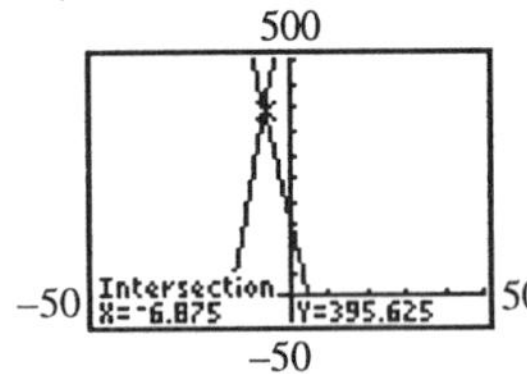

The solution is approximately (−6.88, 395.63).

8. $y = 35x + 25$
$y = 575$

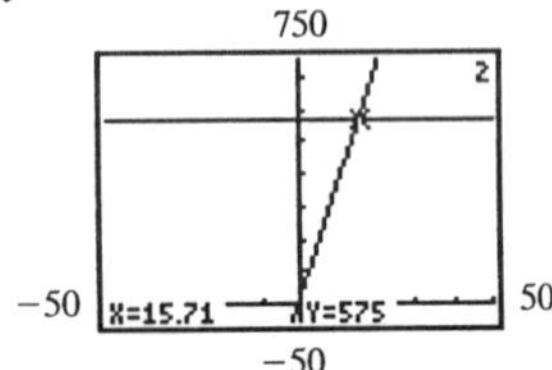

The solution is approximately (15.71, 575).

9. $y = 100x + 6000$
$y = 300x$

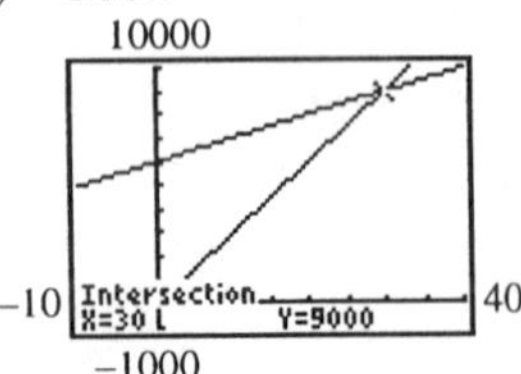

The solution is (30, 9000).

7.2 Experiencing Algebra the Exercise Way

1. $x - 2y = 26$
$5x + 10y = -10$
Solve the first equation for x: $x = 2y + 26$.
Substitute $2y + 26$ for x in the second equation.
$5(2y + 26) + 10y = -10$
$10y + 130 + 10y = -10$
$20y + 130 = -10$
$20y + 130 - 130 = -10 - 130$
$20y = -140$
$\dfrac{20y}{20} = \dfrac{-140}{20}$
$y = -7$
Substitute −7 for y in the first equation.
$x - 2(-7) = 26$
$x + 14 = 26$
$x + 14 - 14 = 26 - 14$
$x = 12$
The solution is (12, −7).

3. $y = 3x - 15$
$x - 5 = 0$
Solve the second equation for x: $x = 5$.
Substitute 5 for x in the first equation.
$y = 3(5) - 15$
$y = 15 - 15$
$y = 0$
The solution is (5, 0).

5. $x - 5y = 20$
$2y + 3 = -7$
Solve the second equation for y: $y = -5$.
Substitute −5 for y in the first equation.
$x - 5(-5) = 20$
$x + 25 = 20$
$x + 25 - 25 = 20 - 25$
$x = -5$
The solution is (−5, −5).

7. $3y = x + 6$
$x - 3y = 9$
Solve the second equation for x: $x = 3y + 9$.
Substitute $3y + 9$ for x in the first equation.
$3y = 3y + 9 + 6$
$3y = 3y + 15$
$3y - 3y = 3y + 15 - 3y$
$0 = 15$
This is a contradiction. There is no solution.

9. $y = x - 3$
$2x - y = 19$
Substitute $x - 3$ for y in the second equation.
$2x - (x - 3) = 19$
$2x - x + 3 = 19$
$x + 3 = 19$
$x + 3 - 3 = 19 - 3$
$x = 16$
Substitute 16 for x in the first equation.
$y = 16 - 3$
$y = 13$
The solution is (16, 13).

11. $y = x + 1$
$5x - 10y = 3$
Substitute $x + 1$ for y in the second equation.
$5x - 10(x + 1) = 3$
$5x - 10x - 10 = 3$
$-5x - 10 = 3$
$-5x - 10 + 10 = 3 + 10$
$-5x = 13$
$\frac{-5x}{-5} = \frac{13}{-5}$
$x = -\frac{13}{5}$
Substitute $-\frac{13}{5}$ for x in the first equation.
$y = -\frac{13}{5} + 1$
$y = -\frac{13}{5} + \frac{5}{5}$
$y = -\frac{8}{5}$
The solution is $\left(-\frac{13}{5}, -\frac{8}{5}\right)$.

13. $4y = x + 4$
$5y = 2x + 11$
Solve the first equation for x: $x = 4y - 4$.
Substitute $4y - 4$ for x in the second equation.
$5y = 2(4y - 4) + 11$
$5y = 8y - 8 + 11$
$5y = 8y + 3$
$5y - 8y = 8y + 3 - 8y$
$-3y = 3$
$\frac{-3y}{-3} = \frac{3}{-3}$
$y = -1$
Substitute -1 for y in the first equation.
$4(-1) = x + 4$
$-4 = x + 4$
$-4 - 4 = x + 4 - 4$
$-8 = x$
The solution is $(-8, -1)$.

15. $3y = x + 6$
$10y = 25$
Solve the second equation for y: $y = \frac{5}{2}$.
Substitute $\frac{5}{2}$ for y in the first equation.
$3\left(\frac{5}{2}\right) = x + 6$
$\frac{15}{2} = x + 6$
$\frac{15}{2} - 6 = x + 6 - 6$
$\frac{15}{2} - \frac{12}{2} = x$
$\frac{3}{2} = x$
The solution is $\left(\frac{3}{2}, \frac{5}{2}\right)$.

17. $y = -3x + 2$
$2y + 2x = 2 + y - x$
Substitute $-3x + 2$ for y in the second equation.
$2(-3x + 2) + 2x = 2 + (-3x + 2) - x$
$-6x + 4 + 2x = 2 - 3x + 2 - x$
$-4x + 4 = 4 - 4x$
$-4x + 4 + 4x = 4 - 4x + 4x$
$4 = 4$
This is an identity. The solutions are all ordered pairs (x, y) that satisfy $y = -3x + 2$.

19. $2x + y = -3$
$y = -0.5x + 3$
Substitute $-0.5x + 3$ for y in the first equation.
$2x + (-0.5x + 3) = -3$
$1.5x + 3 = -3$
$1.5x + 3 - 3 = -3 - 3$
$1.5x = -6$
$\frac{1.5x}{1.5} = \frac{-6}{1.5}$

$x = -4$
Substitute –4 for x in the second equation.
$y = -0.5(-4) + 3$
$y = 5$
The solution is (–4, 5).

21. $5x - 3y = -13$
$4x + y = 27$
Solve the second equation for y: $y = 27 - 4x$.
Substitute $27 - 4x$ for y in the first equation.
$5x - 3(27 - 4x) = -13$
$5x - 81 + 12x = -13$
$17x - 81 = -13$
$17x - 81 + 81 = -13 + 81$
$17x = 68$
$\frac{17x}{17} = \frac{68}{17}$
$x = 4$
Substitute 4 for x in the second equation.
$4(4) + y = 27$
$16 + y - 16 = 27 - 16$
$y = 11$
The solution is (4, 11).

23. $x + 2y = -28$
$3x + y = -9$
Solve the second equation for y: $y = -3x - 9$.
Substitute $-3x - 9$ for y in the first equation.
$x + 2(-3x - 9) = -28$
$x - 6x - 18 = -28$
$-5x - 18 = -28$
$-5x - 18 + 18 = -28 + 18$
$-5x = -10$
$\frac{-5x}{-5} = \frac{-10}{-5}$
$x = 2$
Substitute 2 for x in the second equation:
$3(2) + y = -9$
$6 + y - 6 = -9 - 6$
$y = -15$
The solution is (2, –15).

25. $x - y = -1$
$3x - 3y = -3$
Solve the first equation for x: $x = y - 1$.
Substitute $y - 1$ for x in the second equation.
$3(y - 1) - 3y = -3$
$3y - 3 - 3y = -3$
$-3 = -3$
This is an identity. The solutions are all ordered pairs (x, y) that satisfy $x - y = -1$.

27. $y = 3x - 1$
$2x + y = 4$
Substitute $3x - 1$ for y in the second equation.
$2x + 3x - 1 = 4$
$5x - 1 = 4$
$5x - 1 + 1 = 4 + 1$
$5x = 5$
$\frac{5x}{5} = \frac{5}{5}$
$x = 1$
Substitute 1 for x in the first equation.
$y = 3(1) - 1$
$y = 2$
The solution is (1, 2).

29. $y = \frac{1}{2}x + 3$
$3x + 2y = -2$
Substitute $\frac{1}{2}x + 3$ for y in the second equation.
$3x + 2\left(\frac{1}{2}x + 3\right) = -2$
$3x + x + 6 = -2$
$4x + 6 - 6 = -2 - 6$
$4x = -8$
$\frac{4x}{4} = \frac{-8}{4}$
$x = -2$
Substitute –2 for x in the first equation.
$y = \frac{1}{2}(-2) + 3$
$y = 2$
The solution is (–2, 2).

31. $y = \frac{1}{2}x + 2$
$6y = 3x + 12$
Substitute $\frac{1}{2}x + 2$ for y in the second equation.

$6\left(\frac{1}{2}x+2\right)=3x+12$
$3x+12=3x+12$
$3x+12-3x=3x+12-3x$
$12=12$
This is an identity. The solutions are all ordered pairs (x, y) that satisfy $y=\frac{1}{2}x+2.$

33. $y=2x+2$
$2x-y=3$
Substitute $2x+2$ for y in the second equation.
$2x-(2x+2)=3$
$2x-2x-2=3$
$-2=3$
This is a contradiction. There is no solution.

35. $3x+2y=12$
$y-2=1$
Solve the second equation for y: $y=3$.
Substitute 3 for y in the first equation.
$3x+2(3)=12$
$3x+6-6=12-6$
$3x=6$
$\frac{3x}{3}=\frac{6}{3}$
$x=2$
The solution is (2, 3).

37. $3x-y=5$
$x+2y=4$
Solve the second equation for x: $x=4-2y$.
Substitute $4-2y$ for x in the first equation.
$3(4-2y)-y=5$
$12-6y-y=5$
$12-7y=5$
$12-7y-12=5-12$
$-7y=-7$
$\frac{-7y}{-7}=\frac{-7}{-7}$
$y=1$
Substitute 1 for y in the second equation.
$x+2(1)=4$
$x+2-2=4-2$
$x=2$
The solution is (2, 1).

39. $y=x-7$
$x=y+7$
Substitute $x-7$ for y in the second equation.
$x=x-7+7$
$x=x$
This is an identity. The solutions are all ordered pairs (x, y) that satisfy $y=x-7$.

41. $y=1+4x$
$4x-y=2$
Substitute $1+4x$ for y in the second equation.
$4x-(1+4x)=2$
$4x-1-4x=2$
$-1=2$
This is a contradiction. There is no solution.

43. $5x+7y=35$
$2x-5y=53$
Solve the first equation for x,
$x=7-\frac{7}{5}y.$
Substitute $7-\frac{7}{5}y$ for x in the second equation.
$2\left(7-\frac{7}{5}y\right)-5y=53$
$14-\frac{14}{5}y-5y=53$
$14-\frac{14}{5}y-\frac{25}{5}y=53$
$14-\frac{39}{5}y=53$
$14-\frac{39}{5}y-14=53-14$
$-\frac{39}{5}y=39$
$\left(-\frac{5}{39}\right)\left(-\frac{39}{5}y\right)=-\frac{5}{39}(39)$
$y=-5$

Substitute –5 for y in the second equation.
$2x-5(-5)=53$
$2x+25=53$
$2x+25-25=53-25$
$2x=28$
$\frac{2x}{2}=\frac{28}{2}$
$x=14$
The solution is (14, –5).

45. Let W = width
L = length
$2W+2L=146$
$L=2W-11$
Substitute $2W-11$ for L in the first equation.
$2W+2(2W-11)=146$
$2W+4W-22=146$
$6W-22+22=146+22$
$6W=168$
$\frac{6W}{6}=\frac{168}{6}$
$W=28$
$L=2(28)-11=45$
The width is 28 feet and the length is 45 feet.

47. Let x = measure of small angle
y = measure of large angle
$x+y=90$
$y=4x-10$
Substitute $4x-10$ for y in the first equation.
$x+4x-10=90$
$5x-10+10=90+10$
$5x=100$
$\frac{5x}{5}=\frac{100}{5}$
$x=20$
$y=4(20)-10=70$
The angles measure 20° and 70°.

49. Let x = measure of the two equal angles
y = measure of third angle
$x+x+y=180$
$y=x+x+20$
Substitute $x+x+20$ for y in the first equation.
$x+x+x+x+20=180$
$4x+20-20=180-20$
$4x=160$
$\frac{4x}{4}=\frac{160}{4}$
$x=40$
$y=40+40+20=100$
The angles measure 40°, 40°, and 100°.

51. Let r = radius of small circle
R = radius of large circle
$R=2r+5$
$2\pi R=283$
Substitute $2r+5$ for R in the second equation.
$2\pi(2r+5)=283$
$4\pi r+10\pi=283$
$4\pi r+10\pi-10\pi=283-10\pi$
$4\pi r=283-10\pi$
$\frac{4\pi r}{4\pi}=\frac{283-10\pi}{4\pi}$
$r=\frac{283-10\pi}{4\pi}$
$r\approx 20$
$R\approx 2(20)+5=45$
The radius of each measures 20 inches and 45 inches.

53. Let x = amount borrowed at 7%
y = amount borrowed at 6%
$x+y=100{,}000$
$0.07x+0.06y=6450$
Solve the first equation for x.
$x=100{,}000-y$
Substitute $100{,}000-y$ for x in the second equation.
$0.07(100{,}000-y)+0.06y=6450$
$7000-0.07y+0.06y=6450$
$7000-0.01y-7000=6450-7000$
$-0.01y=-550$
$\frac{-0.01y}{-0.01}=\frac{-550}{-0.01}$
$y=55{,}000$
Substitute 55,000 for y in the second equation.
$x+55{,}000=100{,}000$
$x+55{,}000-55{,}000=100{,}000-55{,}000$
$x=45{,}000$
She borrowed \$45,000 at 7% simple interest and \$55,000 at 6% simple interest.

55. Let x = number of acres of soybeans
y = number of acres of corn
$x+y=15$
$y=25-x$
Substitute $25-x$ for y in the first equation.
$x+25-x=15$
$25=15$
This is a contradiction. There is no solution.

57. Let x = amount from individuals
y = amount from other sources
$x+y=1.4155\times10^{12}$
$y=x+1.687\times10^{11}$
Substitute $x+1.687\times10^{11}$ for y in the first equation.
$x+x+1.687\times10^{11}=1.4155\times10^{12}$
$2x=1.2468\times10^{12}$
$x=6.234\times10^{11}$
$y=6.234\times10^{11}+1.687\times10^{11}$
$y=7.921\times10^{11}$
The amounts are as follows:
$\$6.234\times10^{11}$ from individual income taxes and $\$7.921\times10^{11}$ from other sources.

7.2 Experiencing Algebra the Calculator Way

1. $15.8x+y=2655.10$
$18.4x+73.2y=19361.22$
Solve the first equation for y.
$y=2655.1-15.8x$
Substitute $2655.1-15.8x$ for y in the second equation.

$18.4x+73.2(2655.1-15.8x)=19361.22$
$18.4x+194353.32-1156.56x=19361.22$
$-1138.16x=-174992.1$
$x=153.75$
$y=2655.1-15.8(153.75)=225.85$
The solution is (153.75, 225.85).

2. $x+\frac{9}{17}y=\frac{137}{170}$
$\frac{3}{8}x+\frac{7}{16}y=\frac{2303}{5280}$
Solve the first equation for x.
$x=\frac{137}{170}-\frac{9}{17}y$
Substitute $\frac{137}{170}-\frac{9}{17}y$ in the second equation for x.
$\frac{3}{8}\left(\frac{137}{170}-\frac{9}{17}y\right)+\frac{7}{16}y=\frac{2303}{5280}$
$\frac{411}{1360}-\frac{27}{136}y+\frac{7}{16}y=\frac{2303}{5280}$
$\left(\frac{7}{16}-\frac{27}{136}\right)y=\frac{2303}{5280}-\frac{411}{1360}$
$y=\frac{37}{66}$
$x=\frac{137}{170}-\frac{9}{17}\left(\frac{37}{66}\right)=\frac{28}{55}$
The solution is $\left(\frac{28}{55},\frac{37}{66}\right)$.

3.
```
.0983→X:-1.0768→
Y:4.055X-8.752Y=
9.8277601
                0
```

The first equation is not true. Therefore, (0.0983, –1.0768) is not a solution.

4.
```
-2897.65→A:1725.
85→B:.408A+.526B
=-274.4441
               1
.005A-.802B=-139
8.720
               0
```

The second equation is not true. Therefore, (–2897.65, 1725.85) is not a solution.

5.
```
(13/35)→M:(-17/5
5)→N:(7/39)M+(5/
34)N=7/330
               1
(5/42)M-(11/85)N
=169/7350
               0
```

The second equation is not true. Therefore $\left(\frac{13}{35},-\frac{17}{55}\right)$ is not a solution.

6. 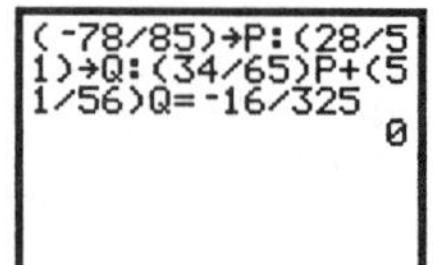

The first equation is not true. Therefore $\left(-\frac{78}{85}, \frac{28}{51}\right)$ is not a solution.

7.3 Experiencing Algebra the Exercise Way

1. $2x - y = -6$
$\underline{5x + y = -8}$
$7x = -14$

$$\frac{7x}{7} = \frac{-14}{7}$$
$x = -2$

Substitute –2 for x in the second equation.
$5(-2) + y = -8$
$-10 + y + 10 = -8 + 10$
$y = 2$
The solution is (–2, 2).

3. $x + 7y = 19$
$2y = x - 1$
Write the second equation in standard from and add to the first equation.
$x + 7y = 19$
$\underline{-x + 2y = -1}$
$9y = 18$
$$\frac{9y}{9} = \frac{18}{9}$$
$y = 2$
Substitute 2 for y in the first equation.
$x + 7(2) = 19$
$x + 14 - 14 = 19 - 14$
$x = 5$
The solution is (5, 2).

5. $5x + y = -24$
$3x - 2y = 9$
Multiply the first equation by 2 and add to the second equation.
$10x + 2y = -48$
$\underline{3x - 2y = 9}$
$13x = -39$
$$\frac{13x}{13} = -\frac{39}{13}$$
$x = -3$

Substitute –3 for x in the first equation.
$5(-3) + y = -24$
$-15 + y + 15 = -24 + 15$
$y = -9$
The solution is (–3, –9).

7. $x + 3y = 2$
$x + 5y = -2$
Multiply the second equation by –1 and add to the first equation.
$x + 3y = 2$
$\underline{-x - 5y = 2}$
$-2y = 4$
$$\frac{-2y}{-2} = \frac{4}{-2}$$
$y = -2$
Substitute –2 for y in the first equation.
$x + 3(-2) = 2$
$x - 6 + 6 = 2 + 6$
$x = 8$
The solution is (8, –2).

9. $2x + 7y = 29$
$4x + 3y = 25$
Multiply the first equation by –2 and add to the second equation.
$-4x - 14y = -58$
$\underline{4x + 3y = 25}$
$-11y = -33$
$$\frac{-11y}{-11} = \frac{-33}{-11}$$
$y = 3$
Substitute 3 for y in the first equation.
$2x + 7(3) = 29$
$2x + 21 - 21 = 29 - 21$
$2x = 8$
$$\frac{2x}{2} = \frac{8}{2}$$
$x = 4$
The solution is (4, 3).

11. $3x - 5y = 66$
$4x + 3y = 1$
Multiply the first equation by 3 and the second equation by 5. Add the new equations.

$$\begin{aligned} 9x - 15y &= 198 \\ 20x + 15y &= 5 \\ \hline 29x &= 203 \end{aligned}$$

$$\frac{29x}{29} = \frac{203}{29}$$

$$x = 7$$

Substitute 7 for x in the first equation.

$3(7) - 5y = 66$

$21 - 5y - 21 = 66 - 21$

$-5y = 45$

$\frac{-5y}{-5} = \frac{45}{-5}$

$y = -9$

The solution is (7, –9).

13. $2x + 9y = 102$

$5x - 11y = -147$

Multiply the first equation by –5 and the second equation by 2. Add the new equations.

$$\begin{aligned} -10x - 45y &= -510 \\ 10x - 22y &= -294 \\ \hline -67y &= -804 \end{aligned}$$

$$\frac{-67y}{-67} = \frac{-804}{-67}$$

$$y = 12$$

Substitute 12 for y in the first equation.

$2x + 9(12) = 102$

$2x + 108 - 108 = 102 - 108$

$2x = -6$

$\frac{2x}{2} = \frac{-6}{2}$

$x = -3$

The solution is (–3, 12).

15. $40x = 23 + 10y$

$50x + 10y = 94$

Write the first equation in standard form and add to the second equation.

$$\begin{aligned} 40x - 10y &= 23 \\ 50x + 10y &= 94 \\ \hline 90x \quad &= 117 \end{aligned}$$

$$\frac{90x}{90} = \frac{117}{90}$$

$$x = \frac{13}{10}$$

Substitute $\frac{13}{10}$ for x in the second equation.

$50\left(\frac{13}{10}\right) + 10y = 94$

$65 + 10y - 65 = 94 - 65$

$10y = 29$

$\frac{10y}{10} = \frac{29}{10}$

$y = \frac{29}{10}$

The solution is $\left(\frac{13}{10}, \frac{29}{10}\right)$.

17. $5x + 40y = 77$

$5x + 15y = 17$

Multiply the second equation by –1 and add to the first equation.

$$\begin{aligned} 5x + 40y &= 77 \\ -5x - 15y &= -17 \\ \hline 25y &= 60 \end{aligned}$$

$$\frac{25y}{25} = \frac{60}{25}$$

$$y = \frac{12}{5}$$

Substitute $\frac{12}{5}$ for y in the first equation.

$5x + 40\left(\frac{12}{5}\right) = 77$

$5x + 96 - 96 = 77 - 96$

$5x = -19$

$\frac{5x}{5} = -\frac{19}{5}$

$x = -\frac{19}{5}$

The solution is $\left(-\frac{19}{5}, \frac{12}{5}\right)$.

19.
$$\begin{aligned} 10x - 4y &= 28 \\ 5x + 4y &= 35 \\ \hline 15x \quad &= 63 \end{aligned}$$

$$\frac{15x}{15} = \frac{63}{15}$$

$$x = \frac{21}{5}$$

Substitute $\frac{21}{5}$ for x in the first equation.

$10\left(\frac{21}{5}\right) - 4y = 28$
$42 - 4y - 42 = 28 - 42$
$-4y = -14$
$\frac{-4y}{-4} = \frac{-14}{-4}$
$y = \frac{7}{2}$

The solution is $\left(\frac{21}{5}, \frac{7}{2}\right)$.

21. $2x + 8y = 29$
$13y = 3x + 39$
Multiply the first equation by 3 and the second equation by 2. Write the second equation in standard form and add to the first equation.

$$\begin{aligned} 6x + 24y &= 87 \\ -6x + 26y &= 78 \\ \hline 50y &= 165 \end{aligned}$$

$\frac{50y}{50} = \frac{165}{50}$
$y = \frac{33}{10}$

Substitute $\frac{33}{10}$ for y in the first equation.

$2x + 8\left(\frac{33}{10}\right) = 29$
$2x + \frac{132}{5} - \frac{132}{5} = 29 - \frac{132}{5}$
$2x = \frac{145}{5} - \frac{132}{5}$
$2x = \frac{13}{5}$
$\frac{1}{2}(2x) = \frac{1}{2}\left(\frac{13}{5}\right)$
$x = \frac{13}{10}$

The solution is $\left(\frac{13}{10}, \frac{33}{10}\right)$.

23. $\frac{1}{4}x + \frac{1}{3}y = \frac{5}{12}$
$\frac{1}{4}x = \frac{1}{3}y - \frac{1}{12}$
Multiply each equation by 12. Write the second equation in standard form and add to the first equation.

$$\begin{aligned} 3x + 4y &= 5 \\ 3x - 4y &= -1 \\ \hline 6x \quad &= 4 \end{aligned}$$

$\frac{6x}{6} = \frac{4}{6}$
$x = \frac{2}{3}$

Substitute $\frac{2}{3}$ for x in the first equation.

$\frac{1}{4}\left(\frac{2}{3}\right) + \frac{1}{3}y = \frac{5}{12}$
$\frac{1}{6} + \frac{1}{3}y - \frac{1}{6} = \frac{5}{12} - \frac{1}{6}$
$\frac{1}{3}y = \frac{1}{4}$
$3\left(\frac{1}{3}y\right) = 3\left(\frac{1}{4}\right)$
$y = \frac{3}{4}$

The solution is $\left(\frac{2}{3}, \frac{3}{4}\right)$.

25. $\frac{1}{3}x + \frac{1}{2}y = -\frac{1}{4}$
$\frac{1}{6}x - \frac{5}{6}y = \frac{11}{16}$
Multiply the first equation by –24 and the second equation by 48. Add the new equations.

$$\begin{aligned} -8x - 12y &= 6 \\ 8x - 40y &= 33 \\ \hline -52y &= 39 \end{aligned}$$

$\frac{-52y}{-52} = \frac{39}{-52}$
$y = -\frac{3}{4}$

Substitute $-\frac{3}{4}$ for y in the first equation.

$\frac{1}{3}x+\frac{1}{2}\left(-\frac{3}{4}\right)=-\frac{1}{4}$
$\frac{1}{3}x-\frac{3}{8}=-\frac{1}{4}$
$\frac{1}{3}x-\frac{3}{8}+\frac{3}{8}=-\frac{2}{8}+\frac{3}{8}$
$\frac{1}{3}x=\frac{1}{8}$
$3\left(\frac{1}{3}x\right)=3\left(\frac{1}{8}\right)$
$x=\frac{3}{8}$
The solution is $\left(\frac{3}{8}, -\frac{3}{4}\right)$.

27. $y=\frac{3}{2}x-9$
$\frac{1}{6}x-\frac{1}{9}y=1$
Multiply the first equation by 2 and write in standard form. Multiply the second equation by 18. Add the new equations.
$$\begin{array}{r} -3x+2y=-18 \\ 3x-2y=\ \ 18 \\ \hline 0=0 \end{array}$$
This is an identity. The solutions are all ordered pairs (x, y) that satisfy $y=\frac{3}{2}x-9$.

29. $\frac{4}{5}x-\frac{3}{5}y=-1$
$\frac{3}{2}x-\frac{9}{8}y=1$
Multiply the first equation by –15 and the second equation by 8. Add the new equations.
$$\begin{array}{r} -12x+9y=15 \\ 12x-9y=\ \ 8 \\ \hline 0=23 \end{array}$$
This is a contradiction. There is no solution.

31. $x-20y=70$
$3x+10y=70$
Multiply the second equation by 2 and add to the first equation.
$$\begin{array}{r} x-20y=\ \ 70 \\ 6x+20y=140 \\ \hline 7x\qquad=210 \end{array}$$
$\frac{7x}{7}=\frac{210}{7}$
$x=30$
Substitute 30 for x in the first equation.
$30-20y=70$
$30-20y-30=70-30$
$\frac{-20y}{-20}=\frac{40}{-20}$
$y=-2$
The solution is (30, –2).

33. $3x+y=40$
$x+y=20$
Multiply the second equation by –1 and add to the first equation.
$$\begin{array}{r} 3x+y=\ \ 40 \\ -x-y=-20 \\ \hline 2x\qquad=\ \ 20 \end{array}$$
$\frac{2x}{2}=\frac{20}{2}$
$x=10$
Substitute 10 for x in the second equation.
$10+y=20$
$10+y-10=20-10$
$y=10$
The solution is (10, 10).

35. $x-y=300$
$2x-y=-100$
Multiply the second equation by –1 and add to the first equation.
$$\begin{array}{r} x-y=300 \\ -2x+y=100 \\ \hline -x\qquad=400 \end{array}$$
$\frac{-x}{-1}=\frac{400}{-1}$
$x=-400$
Substitute –400 for x in the first equation.
$-400-y=300$
$-400-y+400=300+400$
$-y=700$
$\frac{-y}{-1}=\frac{700}{-1}$
$y=-700$
The solution is (–400, –700).

37. $10x - 10y = 22$
$y = 2x - 11$
Multiply the second equation by 5 and write in standard form. Add the result to the first equation.

$$\begin{array}{r} 10x - 10y = 22 \\ -10x + 5y = -55 \\ \hline -5y = -33 \end{array}$$

$$\frac{-5y}{-5} = \frac{-33}{-5}$$

$$y = \frac{33}{5}$$

Substitute $\frac{33}{5}$ for y in the first equation.

$$10x - 10\left(\frac{33}{5}\right) = 22$$
$$10x - 66 = 22$$
$$10x - 66 + 66 = 22 + 66$$
$$10x = 88$$
$$\frac{10x}{10} = \frac{88}{10}$$
$$x = \frac{44}{5}$$

The solution is $\left(\frac{44}{5}, \frac{33}{5}\right)$.

39. $3.2x + 4.2y = 368$
$4.4x - 2.1y = 128$
Multiply the second equation by 2 and add to the first equation.

$$\begin{array}{r} 3.2x + 4.2y = 368 \\ 8.8x - 4.2y = 256 \\ \hline 12x = 624 \end{array}$$

$$\frac{12x}{12} = \frac{624}{12}$$
$$x = 52$$

Substitute 52 for x in the first equation.
$$3.2(52) + 4.2y = 368$$
$$166.4 + 4.2y - 166.4 = 368 - 166.4$$
$$4.2y = 201.6$$
$$\frac{4.2y}{4.2} = \frac{201.6}{4.2}$$
$$y = 48$$

The solution is (52, 48).

41. $0.05x + 0.10y = 0.75$
$x + y = 11$
Multiply the second equation by –0.05 and add to the first equation.

$$\begin{array}{r} 0.05x + 0.10y = 0.75 \\ -0.05x - 0.05y = -0.55 \\ \hline 0.05y = 0.2 \end{array}$$

$$\frac{0.05y}{0.05} = \frac{0.2}{0.05}$$
$$y = 4$$

Substitute 4 for y in the second equation.
$$x + 4 = 11$$
$$x + 4 - 4 = 11 - 4$$
$$x = 7$$

The solution is (7, 4).

43. $0.3x + 0.45y = 43.5$
$x + y = 110$
Multiply the second equation by –0.3 and add to the first equation.

$$\begin{array}{r} 0.3x + 0.45y = 43.5 \\ -0.3x - 0.3y = -33 \\ \hline 0.15y = 10.5 \end{array}$$

$$\frac{0.15y}{0.15} = \frac{10.5}{0.15}$$
$$y = 70$$

Substitute 70 for y in the second equation.
$$x + 70 = 110$$
$$x + 70 - 70 = 110 - 70$$
$$x = 40$$

The solution is (40, 70).

45. $12.50x + 6.50y = 1780$
$x + y = 200$
Multiply the second equation by –6.5 and add to the first equation.

$$\begin{array}{r} 12.5x + 6.5y = 1780 \\ -6.5x - 6.5y = -1300 \\ \hline 6x = 480 \end{array}$$

$$\frac{6x}{6} = \frac{480}{6}$$
$$x = 80$$

Substitute 80 for x in the second equation.
$$80 + y = 200$$
$$80 + y - 80 = 200 - 80$$
$$y = 120$$

The solution is (80, 120).

47. Let x = number of students
y = number of others
$x + y = 683$
$1.5x + 5y = 2645$
Multiply the first equation by –5 and add to the second equation.

$$\begin{array}{r} -5x - 5y = -3415 \\ 1.5x + 5y = 2645 \\ \hline -3.5x = -770 \end{array}$$

$$\frac{-3.5x}{-3.5} = \frac{-770}{-3.5}$$
$$x = 220$$

There were 220 students who attended the game.

49. Let x = number of cookbooks
y = number of calendars
$x - y = 50$
$8.5x + 5y = 3462.5$
Multiply the first equation by 5 and add to the second equation.

$$\begin{array}{r} 5x - 5y = 250 \\ 8.5x + 5y = 3462.5 \\ \hline 13.5x = 3712.5 \end{array}$$

$$\frac{13.5x}{13.5} = \frac{3712.5}{13.5}$$
$$x = 275$$

Substitute 275 for x in the first equation.
$275 - y = 50$
$275 - y - 275 = 50 - 275$
$-y = -225$
$\frac{-y}{-1} = \frac{-225}{-1}$
$y = 225$
They sold 275 cookbooks and 225 calendars.

51. Let x = amount invested at 5%
y = amount invested at 7.25%
$x + y = 16{,}500$
$0.05x + 0.0725y = 1000$
Multiply the first equation by –0.05 and add to the second equation.

$$\begin{array}{r} -0.05x - 0.05y = -825 \\ 0.05x + 0.0725y = 1000 \\ \hline 0.0225y = 175 \end{array}$$

$$\frac{0.0225y}{0.0225} = \frac{175}{0.0225}$$
$$y \approx 7777.78$$

Substitute 7777.78 for y in the first equation.
$x + 7777.78 = 16{,}500$
$x + 7777.78 - 7777.78 = 16{,}500 - 7777.78$
$x = 8722.22$
She should invest \$8722.22 at 5% and \$7777.78 at 7.25%.

53. Let x = number of adults
y = number of children
$x + y = 385$
$2x + 2y = 770$
Multiply the first equation by –2 and add to the second equation.

$$\begin{array}{r} -2x - 2y = -770 \\ 2x + 2y = 770 \\ \hline 0 = 0 \end{array}$$

This is an identity. The solution is any number of adults, x, and any number of children, y, where $x + y = 385$.

55. Let x = width of smaller field
y = length of smaller field
$2x + 2y = 620$
$2x + 2(y + 90) = 800$
Multiply the first equation by –1 and add to the second equation in standard form.

$$\begin{array}{r} -2x - 2y = -620 \\ 2x + 2y = 620 \\ \hline 0 = 0 \end{array}$$

This is an identity. The solutions are any number of yards, x, and any number of yards, y, where $x + y = 310$.

57. Let x = measure of each equal angle
y = measure of third angle
$x + x + y = 180$
$y = 90 - 2x$
Multiply the second equation by –1 and write in standard form. Add the result to the first equation in standard form.

$$\begin{array}{r} 2x + y = 180 \\ -2x - y = -90 \\ \hline 0 = 90 \end{array}$$

This is a contradiction. There is no solution.

59. Let x = distance from Earth to sun
y = distance from Mars to sun
$y = x + 4.864 \times 10^7$
$\frac{x+y}{2} = 1.1728 \times 10^8$
Multiply the second equation by 2 and add to the first equation written in standard form.
$$\begin{array}{r} -x+y = 4.864\times 10^7 \\ x+y = 2.3456\times 10^8 \\ \hline 2y = 2.832\times 10^8 \end{array}$$
$\frac{2y}{2} = \frac{2.832\times 10^8}{2}$
$y = 1.416 \times 10^8$
Substitute 1.416×10^8 for y in the first equation.
$1.416\times 10^8 = x + 4.864\times 10^7$
$1.416\times 10^8 - 4.864\times 10^7$
$= x + 4.864\times 10^7 - 4.864\times 10^7$
$9.296\times 10^7 = x$
The distances from the sun are 9.296×10^7 miles for Earth and 1.416×10^8 miles for Mars.

7.3 Experiencing Algebra the Calculator Way

1.
$$\begin{array}{rcl} 2578x - 4823y &=& 299,758 \\ 7395x + 4823y &=& 3,210,738 \\ \hline 9973x &=& 3,510,496 \end{array}$$
$\frac{9973x}{9973} = \frac{3,510,496}{9973}$
$x = 352$
Substitute 352 for x in the first equation.
$2578(352) - 4823y = 299,758$
$907,456 - 4823y = 299,758$
$-4823y = -607,698$
$y = 126$
The solution is (352, 126).

2. $487x + 182y = 51,567$
$976x + 364y = 103,280$
Multiply the first equation by –2 and add to the second equation.
$$\begin{array}{rcl} -974x - 364y &=& -103,134 \\ 976x + 364y &=& 103,280 \\ \hline 2x &=& 146 \end{array}$$
$x = 73$
Substitute 73 for x in the first equation.
$487(73) + 182y = 51,567$
$35,551 + 182y = 51,567$
$182y = 16,016$
$y = 88$
The solution is (73, 88).

3. $4.376x - 2.659y = 4.479378$
$-2.188x - 5.033y = 19.787286$
Multiply the second equation by 2 and add to the first equation.
$$\begin{array}{rcl} 4.376x - 2.659y &=& 4.479378 \\ -4.376x - 10.066y &=& 39.574572 \\ \hline -12.725y &=& 44.05395 \end{array}$$
$y = -3.462$
Substitute –3.462 for y in the first equation.
$4.376x - 2.659(-3.462) = 4.479378$
$4.376x + 9.205458 = 4.479378$
$4.376x = -4.72608$
$x = -1.08$
The solution is (–1.08, –3.462).

4. $5882x + 0.473y = 2764.366$
$2750x - y = -4508$
Multiply the second equation by 0.473 and add to the first equation.
$$\begin{array}{rcl} 5882x + 0.473y &=& 2764.366 \\ 1300.75x - 0.473y &=& -2132.284 \\ \hline 7182.75x &=& 632.082 \end{array}$$
$x = 0.088$
Substitute 0.088 for x in the second equation.
$2750(0.088) - y = -4508$
$242 - y = -4508$
$-y = -4750$
$y = 4750$
The solution is (0.088, 4750).

5. $\frac{27}{85}x + y = -\frac{37}{95}$
$x + \frac{57}{82}y = \frac{7}{72}$
Multiply the second equation by $-\frac{27}{85}$ and add to the first equation.

$$\begin{aligned} \frac{27}{85}x + \phantom{\frac{1539}{6970}}y &= -\frac{37}{95} \\ -\frac{27}{85}x - \frac{1539}{6970}y &= -\frac{189}{6120} \\ \hline \frac{5431}{6970}y &= -\frac{37}{95} - \frac{189}{6120} \\ y &= -\frac{41}{76} \end{aligned}$$

Substitute $-\frac{41}{76}$ for y in the second equation.

$$x + \frac{57}{82}\left(-\frac{41}{76}\right) = \frac{7}{72}$$
$$x = \frac{17}{36}$$

The solution is $\left(\frac{17}{36}, -\frac{41}{76}\right)$.

6. $\frac{5}{17}x - \frac{4}{19}y = \frac{421}{1615}$
$\frac{8}{19}x + \frac{11}{17}y = -\frac{2}{323}$
Multiply the first equation by $\frac{11}{17}$ and add to the second equation multiplied by $\frac{4}{19}$.

$$\begin{aligned} \frac{55}{289}x - \frac{44}{323}y &= \frac{4631}{27,455} \\ \frac{32}{361}x + \frac{44}{323}y &= -\frac{8}{6137} \\ \hline \frac{29,103}{104,329}x &= \frac{87,309}{521,645} \\ x &= \frac{3}{5} \end{aligned}$$

Substitute $\frac{3}{5}$ for x in the second equation.

$$\frac{8}{19}\left(\frac{3}{5}\right) + \frac{11}{17}y = -\frac{2}{323}$$
$$y = -\frac{2}{5}$$

The solution is $\left(\frac{3}{5}, -\frac{2}{5}\right)$.

7.4 Experiencing Algebra the Exercise Way

1. Let x = number of pounds of peanuts
y = number of pounds of candy
$x + y = 30$
$0.9x + 1.5y = 1.25(30)$
Solve the first equation for x: $x = 30 - y$.
Substitute $30 - y$ for x in the second equation.
$0.9(30 - y) + 1.5y = 1.25(30)$
$27 - 0.9y + 1.5y = 37.5$
$27 + 0.6y - 27 = 37.5 - 27$
$0.6y = 10.5$
$\frac{0.6y}{0.6} = \frac{10.5}{0.6}$
$y = 17.5$
$x = 30 - y$
$x = 30 - 17.5$
$x = 12.5$
She should mix 12.5 pounds of peanuts with 17.5 pounds of candy.

3. Let x = number of \$5 bills
y = number of \$10 bills
$x + y = 65$
$5x + 10y = 365$
Multiply the first equation by –5 and add to the second equation.

$$\begin{aligned} -5x - 5y &= -325 \\ 5x + 10y &= 365 \\ \hline 5y &= 40 \end{aligned}$$
$$\frac{5y}{5} = \frac{40}{5}$$
$$y = 8$$

Substitute 8 for y in the first equation.
$x + 8 = 65$
$x + 8 - 8 = 65 - 8$
$x = 57$
There are 57 \$5 bills and 8 \$10 bills.

5. Let x = number of gallons of concentrate
y = number of gallons of water
$x + y = 130$
$0.65x + 0y = 0.35(130)$
Solve the second equation for x: $x = 70$.
Substitute 70 for x in the first equation.
$70 + y = 130$
$70 + y - 70 = 130 - 70$
$y = 60$
The mixture should have 70 gallons of concentrate and 60 gallons of water.

7. Let x = speed of plane with no wind
y = speed of wind
$450 = 3(x + y)$
$450 = 5(x - y)$
Multiply the first equation by 5 and the second equation by 3. Add the new equations.

$$\begin{array}{l} 2250 = 15x + 15y \\ \underline{1350 = 15x - 15y} \\ 3600 = 30x \end{array}$$

$$\frac{3600}{30} = \frac{30x}{30}$$

$120 = x$

Substitute 120 for x in the first equation.
$450 = 3(120 + y)$
$450 = 360 + 3y$
$450 - 360 = 360 + 3y - 360$
$90 = 3y$

$$\frac{90}{3} = \frac{3y}{3}$$

$30 = y$

The speed of the plane with no wind is 120 mph and the speed of the wind is 30 mph.

9. Let x = number of children
y = number of adults
$4.5x + 7.5y = 3937.5$
$x + y = 625$
Multiply the second equation by –4.5 and add to the first equation.

$$\begin{array}{rl} 4.5x + 7.5y = & 3937.5 \\ \underline{-4.5x - 4.5y =} & \underline{-2812.5} \\ 3y = & 1125 \end{array}$$

$$\frac{3y}{3} = \frac{1125}{3}$$

$y = 375$

Substitute 375 for y in the second equation.
$x + 375 = 625$
$x + 375 - 375 = 625 - 375$
$x = 250$
There were 250 children and 375 adults.

11. Let x = number of pints of 70% solution
y = number of pints of 40% solution
$x + y = 5$
$0.7x + 0.4y = 0.5(5)$
Multiply the first equation by –0.4 and add to the second equation.

$$\begin{array}{l} -0.4x - 0.4y = -2 \\ \underline{0.7x + 0.4y = 2.5} \\ 0.3x \qquad = 0.5 \end{array}$$

$$\frac{0.3x}{0.3} = \frac{0.5}{0.3}$$

$$x = 1\frac{2}{3}$$

Substitute $1\frac{2}{3}$ for x in the first equation.

$$1\frac{2}{3} + y = 5$$

$$1\frac{2}{3} + y - 1\frac{2}{3} = 5 - 1\frac{2}{3}$$

$$y = 3\frac{1}{3}$$

He should mix $1\frac{2}{3}$ pints of 70% weed killer and $3\frac{1}{3}$ pints of 40% weed killer.

13. Let x = number of pounds of French vanilla
y = number of pounds of hazelnut
$9.5x + 7y = 8.5(20)$
$x + y = 20$
Multiply the second equation by –7 and add to the first equation.

$$\begin{array}{rl} 9.5x + 7y = & 170 \\ \underline{-7x - 7y =} & \underline{-140} \\ 2.5x \qquad = & 30 \end{array}$$

$$\frac{2.5x}{2.5} = \frac{30}{2.5}$$

$x = 12$

Substitute 12 for x in the second equation.
$12 + y = 20$
$12 + y - 12 = 20 - 12$
$y = 8$
He should use 12 pounds of French vanilla coffee and 8 pounds of hazelnut coffee.

15. Let x = amount at 8.5%
y = amount at 7%
$x + y = 10{,}000$
$0.085x + 0.07y = 752.5$
Multiply the first equation by –0.07 and add to the second equation.

$$\begin{array}{r} -0.07x - 0.07y = -700 \\ 0.085x + 0.07y = 752.5 \\ \hline 0.015x \quad\quad = 52.5 \end{array}$$

$$\frac{0.015x}{0.015} = \frac{52.5}{0.015}$$

$$x = 3500$$

Substitute 3500 for x in the first equation.

$3500 + y = 10{,}000$

$3500 + y - 3500 = 10{,}000 - 3500$

$y = 6500$

She invested \$3500 at 8.5% and \$6500 at 7%.

17. Let x = time spent walking

y = time spent riding

$x = 2y$

$11 = 3x + 60y$

Substitute $2y$ for x in the second equation.

$11 = 3(2y) + 60y$

$11 = 66y$

$\frac{11}{66} = \frac{66y}{66}$

$\frac{1}{6} = y$

$x = 2\left(\frac{1}{6}\right) = \frac{1}{3}$

He walked for $\frac{1}{3}$ hour and rode for $\frac{1}{6}$ hour.

19. Let x = number of azaleas

y = number of rhododendrons

$x + y = 30$

$5x + 12y = 250$

Multiply the first equation by –5 and add to the second equation.

$$\begin{array}{r} -5x - 5y = -150 \\ 5x + 12y = 250 \\ \hline 7y = 100 \end{array}$$

$$\frac{7y}{7} = \frac{100}{7}$$

$$y \approx 14$$

Substitute 14 for y in the first equation.

$x + 14 = 30$

$x + 14 - 14 = 30 - 14$

$x = 16$

She can buy 16 azaleas and 14 rhododendrons.

21. Let x = wage as waitress

y = wage as cook

$15x + 20y = 320$

$18x + 24y = 384$

Multiply the first equation by 18 and the second equation by –15. Add the new equations.

$$\begin{array}{r} 270x + 360y = 5760 \\ -270x - 360y = -5760 \\ \hline 0 = 0 \end{array}$$

This is an identity. There are an infinite number of solutions. The solutions are any wage, x, and any wage, y, where $15x + 20y = 320$.

7.4 Experiencing Algebra the Calculator Way

1. a. $R(x) = 25x$

b. $C(x) = 10x + 1500$

c.

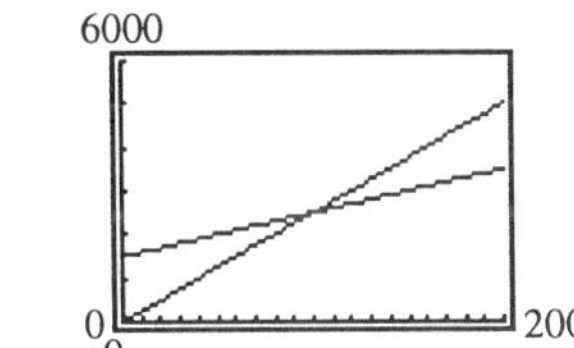

d. The graphs intersect at (125, 5625). The break-even point is at producing and selling 125 lamps.

e.

$R(x) = 25x$	
2500	25(100)
	2500

$C(x) = 10x + 1500$	
2500	10(100) + 1500
	1000 + 1500
	2500

The solution checks.

2. a. $R(x) = 45x$

b. $C(x) = 25x + 2500$

c.

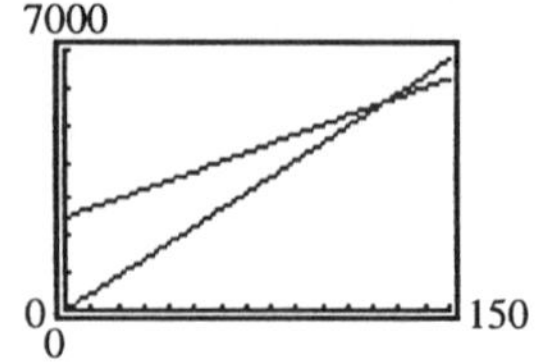

d. The graphs intersect at (125, 5625). The break-even point is at producing and selling 125 tables.

e.

$R(x) = 45x$	
5625	45(125)
	5625

$C(x) = 25x + 2500$	
5625	25(125) + 2500
	3125 + 2500
	5625

The solution checks.

3. $R(x) = 32.49x$
$C(x) = 15.89x + 1256$

5000

Intersection X=75.662651 Y=2458.2795

0 150
0

The graphs intersect at approximately (75.7, 2458.3). The break-even point is at producing and selling 76 purses.

Chapter 7 Review

Reflections

1.–8. Answers may vary.

Exercises

1.

$3x + 2y = -2$	
3(2) + 2(–4)	–2
6 – 8	
–2	

$4x - 3y = 20$	
4(2) – 3(–4)	20
8 + 12	
20	

Both equations are true. Therefore, (2, –4) is a solution.

2.

$2x + y = 6$	
$2\left(\frac{17}{7}\right) + \frac{8}{7}$	6
$\frac{34}{7} + \frac{8}{7}$	
$\frac{42}{7}$	
6	

$-x + 3y = 1$	
$-\frac{17}{7} + 3\left(\frac{8}{7}\right)$	1
$-\frac{17}{7} + \frac{24}{7}$	
$\frac{7}{7}$	
1	

Both equations are true . Therefore, $\left(\frac{17}{7}, \frac{8}{7}\right)$ is a solution.

3.

$4x - 5y = 3$	
4(0.25) – 5(–0.45)	3
1+ 2.25	
3.25	

$8x + 5y = 0$	
8(0.25) + 5(–0.45)	0
2 – 2.25	
–0.25	

Both equations are false. Therefore, (0.25, –0.45) is not a solution.

4.

$x + 3y = 13$	
7 + 3(–2)	13
7 – 6	
1	

$x - y = 5$	
$7 - (-2)$	5
9	

Both equations are false. Therefore (7, –2) is not a solution.

5.

$3x + 6y = -2$	
$3\left(\frac{2}{3}\right) + 6\left(\frac{2}{3}\right)$	-2
$2 + 4$	
6	

$6x - 3y = 6$	
$6\left(\frac{2}{3}\right) - 3\left(\frac{2}{3}\right)$	6
$4 - 2$	
2	

Both equations are false. Therefore, $\left(\frac{2}{3}, \frac{2}{3}\right)$ is not a solution.

6.

$6x + 5y = -3$	
$6(1.5) + 5(-2.4)$	-3
$9 - 12$	
-3	

$2x - 10y = 27$	
$2(1.5) - 10(-2.4)$	27
$3 + 24$	
27	

Both equations are true. Therefore, (1.5, –2.4) is a solution.

7.

$2(x - 3) = 2$	
$2(4 - 3)$	2
$2(1)$	
2	

$3(y + 1) = -3$	
$3(2 + 1)$	-3
$3(3)$	
9	

The second equation is false. Therefore, (4, 2) is not a solution.

8.

$x + 5 = 2$	
$-3 + 5$	2
2	

$2y - 3 = 7$	
$2(5) - 3$	7
$10 - 3$	
7	

Both equations are true. Therefore, (–3, 5) is a solution.

9. $2x + y = 17$
$y = 3x - 18$

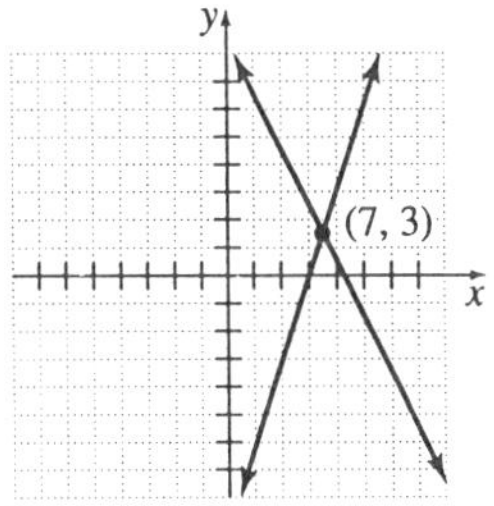

(7, 3)

The solution is (7, 3).

10. $3(x + 2) + 1 = -5$
$2x - y = -10$

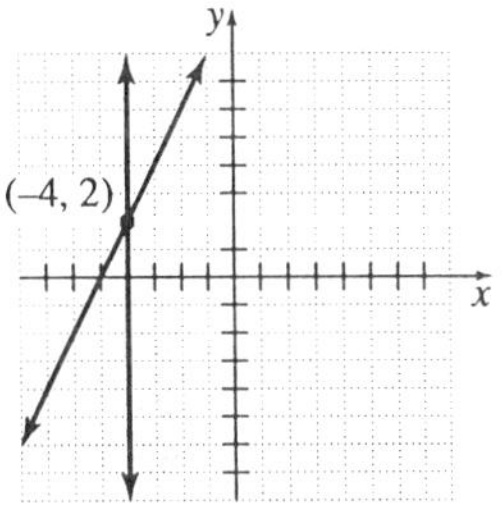

(–4, 2)

The solution is (–4, 2).

11. $x + 6 = 4$
$2(y + 2) = y + 3$

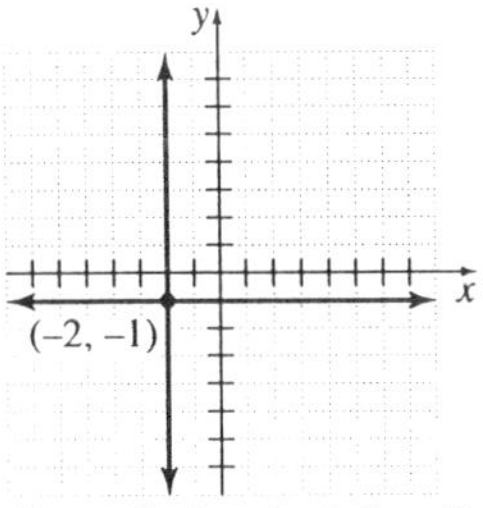

(–2, –1)

The solution is (–2, –1).

12. $x - 2y = 11$
$2x + 3y = -6$

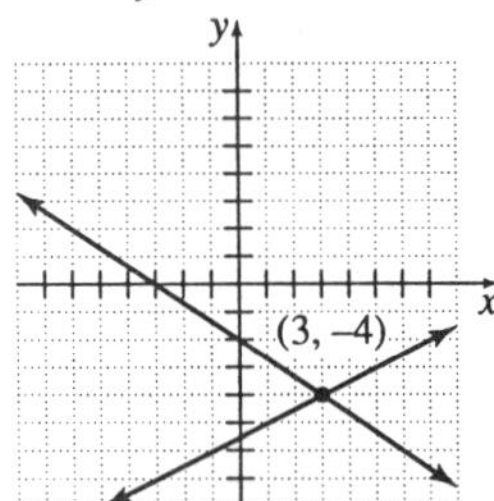

The solution is (3, –4).

13. $y = 2(x + 2)$
$2x - y = 5$

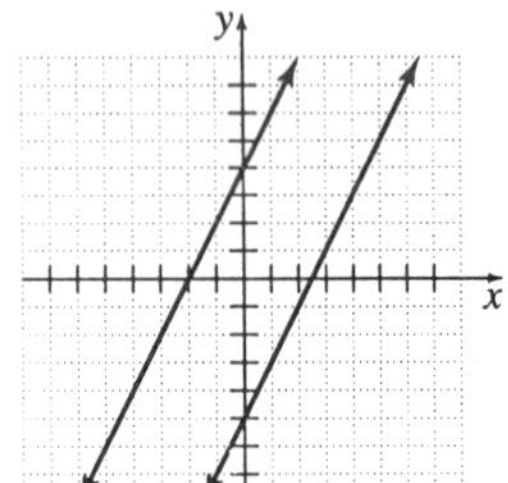

The graphs are parallel. There is no solution.

14. $y = \frac{3}{2}x - 6$
$3x - 2y = 12$

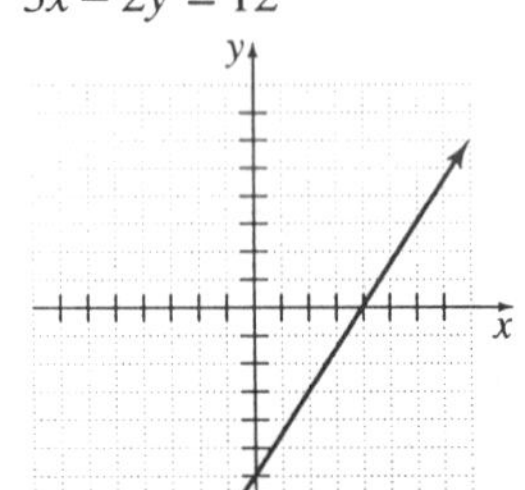

The graphs coincide. The solutions are all ordered pairs (x, y) that satisfy $y = \frac{3}{2}x - 6$.

15. $y = 3x - 6$
$y = -3$

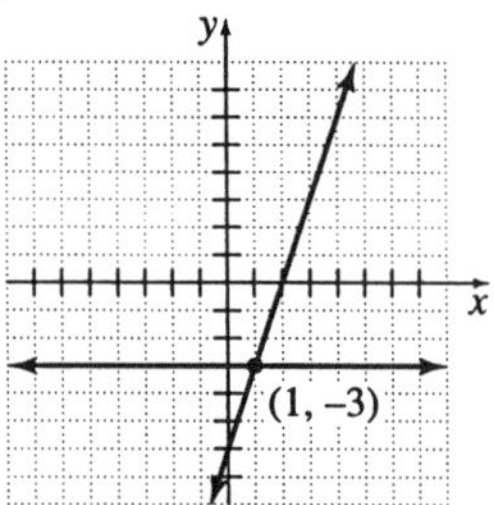

The solution is (1, –3).

16. $2y - 3 = 1$
$5(y - 4) = 10$

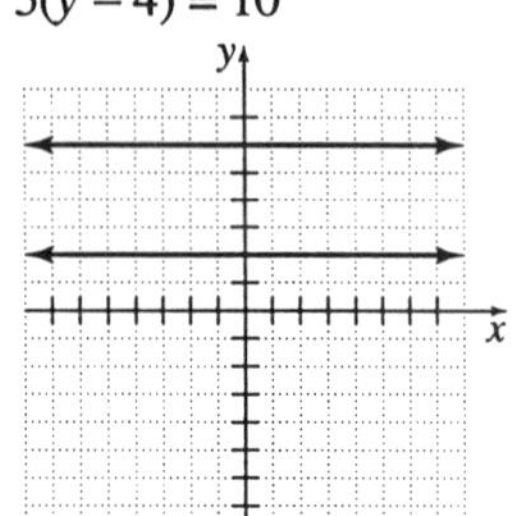

The graphs are parallel. There is no solution.

17. $2x - 11 = 3$
$14 - 2x = x - 7$

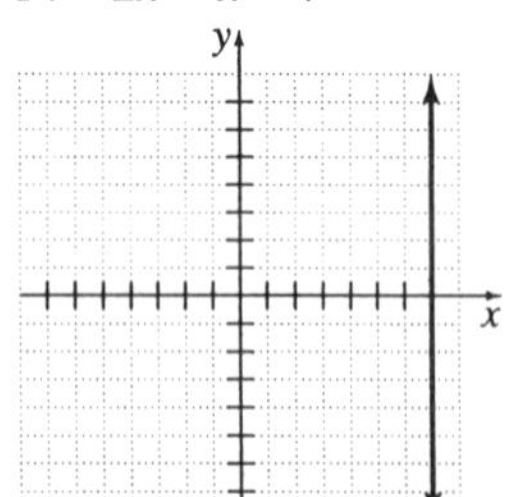

The graphs coincide. The solutions are all ordered pairs (x, y) that satisfy $2x - 11 = 3$ or $x = 7$.

18. $y - 1 = 4$
$3y - 2 = 13$

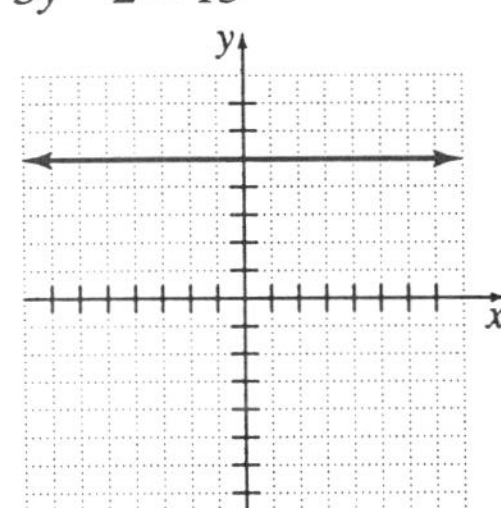

The graphs coincide. The solutions are all ordered pairs (x, y) that satisfy $y - 1 = 4$ or $y = 5$.

19. $2x + 3y = 6$
$x + y = 1$

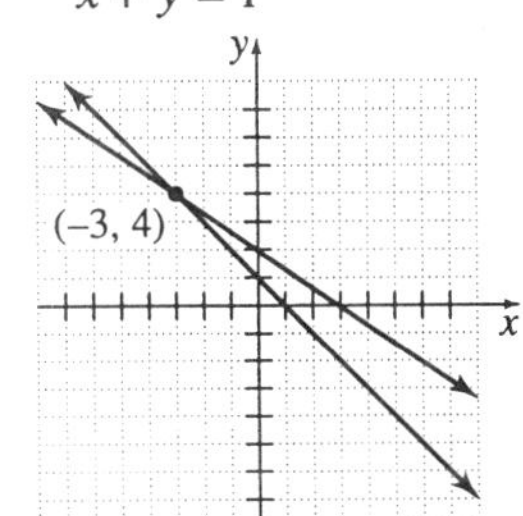

The solution is (–3, 4).

20. $f(x) = 2x - 15$
$g(x) = -3x + 10$

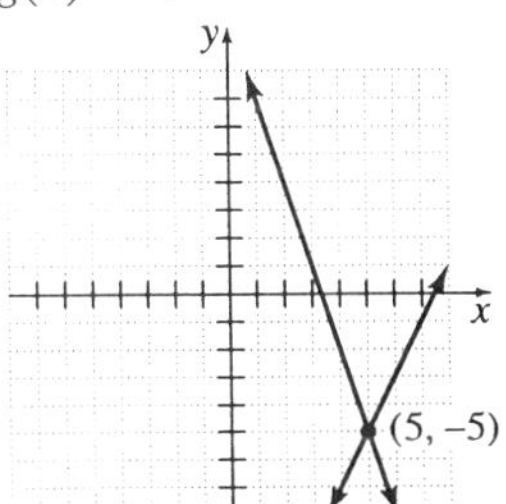

The solution is (5, –5).

21. $y = 2x + 1$
$x = 2y + 7$

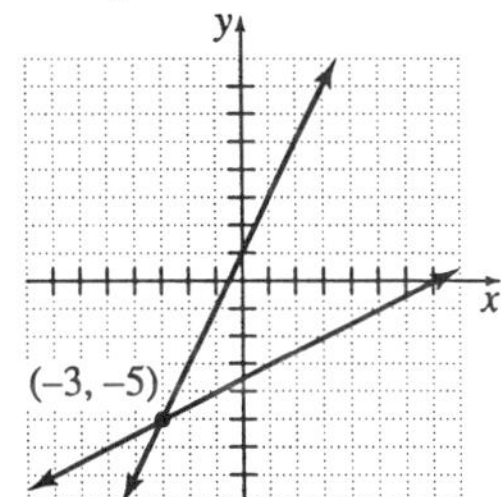

The solution is (–3, –5).

22. $2y + 6 = 3y + 5$
$2(x - 3) + 1 = 5$

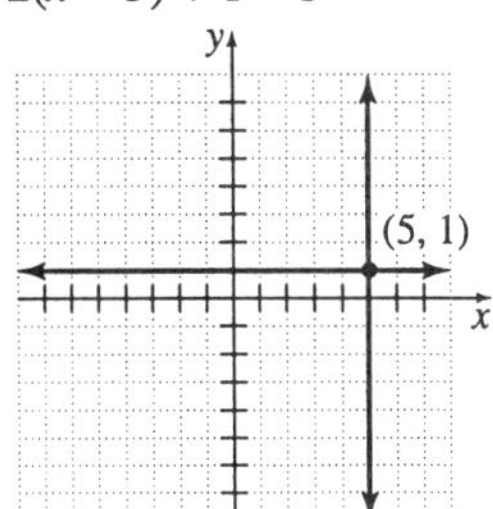

The solution is (5, 1).

23. Let x = number of Democrats
y = number of Republicans
$x + y = 500 - 120$
$y = x + 40$

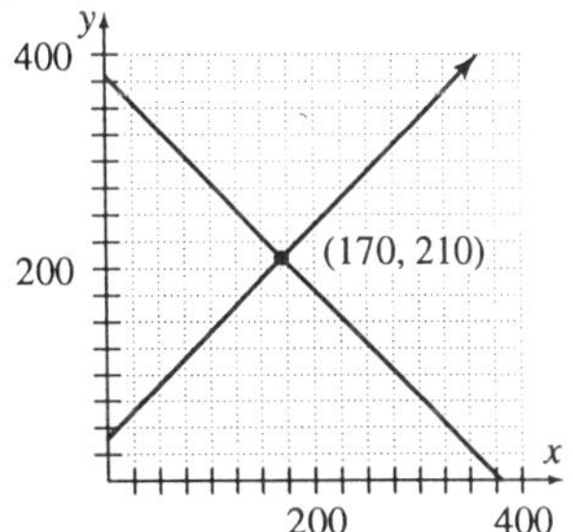

The solution is (170, 210).
There were 170 Democrats.
$170 - 120 = 50$
There were 50 more Democrats than Independents.

24. Let W = width of first towel
L = length of first towel
$2W + 2L = 116.6$
$2(W + 4.7) + 2(L - 2.9) = 120.2$

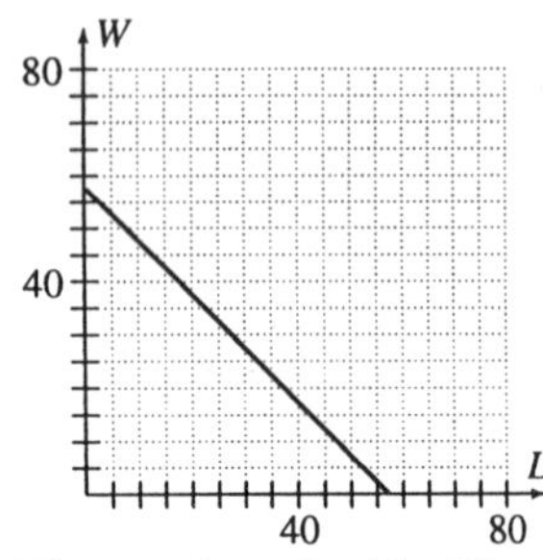

The graphs coincide. The solution is all ordered pairs (W, L) that satisfy $2W + 2L = 116.6$.

25. $8y = 5$
$4x + 8y = 2$

Solve the first equation for y: $y = \frac{5}{8}$.

Substitute $\frac{5}{8}$ for y in the second equation.

$4x + 8\left(\frac{5}{8}\right) = 2$
$4x + 5 = 2$
$4x + 5 - 5 = 2 - 5$
$4x = -3$
$\frac{4x}{4} = \frac{-3}{4}$
$x = -\frac{3}{4}$

The solution is $\left(-\frac{3}{4}, \frac{5}{8}\right)$.

26. $10x - 5y = -14$
$5x - 1 = 0$

Solve the second equation for x: $x = \frac{1}{5}$.

Substitute $\frac{1}{5}$ for x in the first equation.

$10\left(\frac{1}{5}\right) - 5y = -14$
$2 - 5y = -14$
$2 - 5y - 2 = -14 - 2$
$-5y = -16$
$\frac{-5y}{-5} = \frac{-16}{-5}$
$y = \frac{16}{5}$

The solution is $\left(\frac{1}{5}, \frac{16}{5}\right)$.

27. $4x - y = 5$
$y = 4x + 3$
Substitute $4x + 3$ for y in the first equation.
$4x - (4x + 3) = 5$
$4x - 4x - 3 = 5$
$-3 = 5$
This is a contradiction. There is no solution.

28. $x - 2y = 71$
$3x - 7y = 275$
Solve the first equation for x: $x = 2y + 71$.
Substitute $2y + 71$ for x in the second equation.
$3(2y + 71) - 7y = 275$
$6y + 213 - 7y = 275$
$213 - y = 275$
$213 - y - 213 = 275 - 213$
$-y = 62$
$\frac{-y}{-1} = \frac{62}{-1}$
$y = -62$
Substitute -62 for y in the first equation.
$x - 2(-62) = 71$
$x + 124 = 71$
$x + 124 - 124 = 71 - 124$
$x = -53$
The solution is $(-53, -62)$.

29. $2x + 3y = -69$
$2x - 4y = 218$
Solve the second equation for x.
$x = 2y + 109$
Substitute $2y + 109$ for x in the first equation.

$2(2y + 109) + 3y = -69$
$4y + 218 + 3y = -69$
$7y + 218 - 218 = -69 - 218$
$7y = -287$
$\frac{7y}{7} = \frac{-287}{7}$
$y = -41$
Substitute –41 for y in the second equation.
$2x - 4(-41) = 218$
$2x + 164 = 218$
$2x + 164 - 164 = 218 - 164$
$2x = 54$
$\frac{2x}{2} = \frac{54}{2}$
$x = 27$
The solution is (27, –41).

30. $3(x + 2) - y = -1$
$y = 3x + 7$
Substitute $3x + 7$ for y in the first equation.
$3(x + 2) - (3x + 7) = -1$
$3x + 6 - 3x - 7 = -1$
$-1 = -1$
This is an identity. The solutions are all ordered pairs (x, y) that satisfy $y = 3x + 7$.

31. $x + 8y = 5$
$12x + y = 10$
Solve the first equation for x: $x = 5 - 8y$
Substitute $5 - 8y$ for x in the second equation.
$12(5 - 8y) + y = 10$
$60 - 96y + y = 10$
$60 - 95y = 10$
$60 - 95y - 60 = 10 - 60$
$-95y = -50$
$\frac{-95y}{-95} = \frac{-50}{-95}$
$y = \frac{10}{19}$
Substitute $\frac{10}{19}$ for y in the first equation.
$x + 8\left(\frac{10}{19}\right) = 5$
$x + \frac{80}{19} - \frac{80}{19} = 5 - \frac{80}{19}$
$x = \frac{95}{19} - \frac{80}{19}$
$x = \frac{15}{19}$
The solution is $\left(\frac{15}{19}, \frac{10}{19}\right)$.

32. $5x - 3y = 406$
$2x - y = 327$
Solve the second equation for y.
$y = 2x - 327$
Substitute $2x - 327$ for y in the first equation.
$5x - 3(2x - 327) = 406$
$5x - 6x + 981 = 406$
$-x + 981 - 981 = 406 - 981$
$-x = -575$
$\frac{-x}{-1} = \frac{-575}{-1}$
$x = 575$
Substitute 575 for x in the second equation.
$2(575) - y = 327$
$1150 - y - 1150 = 327 - 1150$
$-y = -823$
$\frac{-y}{-1} = \frac{-823}{-1}$
$y = 823$
The solution is (575, 823).

33. $y = \frac{1}{3}x - 4$
$x - 5y = 0$
Substitute $\frac{1}{3}x - 4$ for y in the second equation.

$x - 5\left(\frac{1}{3}x - 4\right) = 0$
$x - \frac{5}{3}x + 20 = 0$
$\frac{3}{3}x - \frac{5}{3}x + 20 - 20 = 0 - 20$
$-\frac{2}{3}x = -20$
$-\frac{3}{2}\left(-\frac{2}{3}x\right) = -\frac{3}{2}(-20)$
$x = 30$
Substitute 30 for x in the first equation.
$y = \frac{1}{3}(30) - 4$
$y = 10 - 4$
$y = 6$
The solution is (30, 6).

34. $y = \frac{1}{2}x + 7$
$y = -\frac{3}{5}x - 4$
Substitute $\frac{1}{2}x + 7$ for y in the second equation.
$\frac{1}{2}x + 7 = -\frac{3}{5}x - 4$
$\frac{1}{2}x + 7 - 7 = -\frac{3}{5}x - 4 - 7$
$\frac{1}{2}x = -\frac{3}{5}x - 11$
$\frac{1}{2}x + \frac{3}{5}x = -\frac{3}{5}x - 11 + \frac{3}{5}x$
$\frac{5}{10}x + \frac{6}{10}x = -11$
$\frac{11}{10}x = -11$
$\frac{10}{11}\left(\frac{11}{10}x\right) = \frac{10}{11}(-11)$
$x = -10$
Substitute −10 for x in the first equation.
$y = \frac{1}{2}(-10) + 7$
$y = -5 + 7$
$y = 2$
The solution is (−10, 2).

35. $3x - y = 10$
$2x - 6y = 26$
Solve the first equation for y: $y = 3x - 10$
Substitute $3x - 10$ for y in the second equation.
$2x - 6(3x - 10) = 26$
$2x - 18x + 60 = 26$
$-16x + 60 - 60 = 26 - 60$
$-16x = -34$
$\frac{-16x}{-16} = \frac{-34}{-16}$
$x = \frac{17}{8}$
Substitute $\frac{17}{8}$ for x in the first equation.
$3\left(\frac{17}{8}\right) - y = 10$
$\frac{51}{8} - y - \frac{51}{8} = 10 - \frac{51}{8}$
$-y = \frac{29}{8}$
$-1(-y) = -1\left(\frac{29}{8}\right)$
$y = -\frac{29}{8}$
The solution is $\left(\frac{17}{8}, -\frac{29}{8}\right)$.

36. $x = 160y$
$3x - 440y = 30$
Substitute $160y$ for x in the second equation.
$3(160y) - 440y = 30$
$480y - 440y = 30$
$40y = 30$
$\frac{40y}{40} = \frac{30}{40}$
$y = \frac{3}{4}$
Substitute $\frac{3}{4}$ for y in the first equation.
$x = 160\left(\frac{3}{4}\right)$
$x = 120$
The solution is $\left(120, \frac{3}{4}\right)$.

37. Let x = measure of smaller angle
y = measure of larger angle
$x + y = 90$
$y = 2x + 12$
Substitute $2x + 12$ for y in the first equation.
$x + 2x + 12 = 90$
$3x + 12 - 12 = 90 - 12$
$3x = 78$
$\frac{3x}{3} = \frac{78}{3}$
$x = 26$
Substitute 26 for x in the second equation.
$y = 2(26) + 12$
$y = 52 + 12$
$y = 64$
$64 - 26 = 38$
The difference in the angles is 38°.

38. Let W = width
L = length
$L = 3W - 2.5$
$2W + 2L = 31.8$
Substitute $3W - 2.5$ for L in the second equation.
$2W + 2(3W - 2.5) = 31.8$
$2W + 6W - 5 = 31.8$
$8W - 5 + 5 = 31.8 + 5$
$8W = 36.8$
$\frac{8W}{8} = \frac{36.8}{8}$
$W = 4.6$
Substitute 4.6 for W in the first equation.
$L = 3(4.6) - 2.5$
$L = 13.8 - 2.5$
$L = 11.3$
Area = (11.3)(4.6) = 51.98
The area is 51.98 cm^2.

39. Let x = amount at 4.5%
y = amount at 6%
$x + y = 10{,}000{,}000$
$0.045x + 0.06y = 487{,}500$
Solve the first equation for y.
$y = 10{,}000{,}000 - x$.
Substitute $10{,}000{,}000 - x$ in the second equation.
$0.045x + 0.06(10{,}000{,}000 - x) = 487{,}500$
$0.045x + 600{,}000 - 0.06x = 487{,}500$
$-0.015x + 600{,}000 - 600{,}000$
$= 487{,}500 - 600{,}000$
$-0.015x = -112{,}500$
$\frac{-0.015x}{-0.015} = \frac{-112{,}500}{-0.015}$
$x = 7{,}500{,}000$
Substitute 7,500,000 for x in the first equation.
$7{,}500{,}000 + y = 10{,}000{,}000$
$7{,}500{,}000 + y - 7{,}500{,}000$
$= 10{,}000{,}000 - 7{,}500{,}000$
$y = 2{,}500{,}000$
She invested \$7,500,000 at 4.5% simple interest and \$2,500,000 at 6% simple interest.

40.
$$\begin{array}{r} 5x + 3y = -10 \\ \underline{5x - 3y = 80} \\ 10x \quad = 70 \end{array}$$
$\frac{10x}{10} = \frac{70}{10}$
$x = 7$
Substitute 7 for x in the first equation.
$5(7) + 3y = -10$
$35 + 3y - 35 = -10 - 35$
$3y = -45$
$\frac{3y}{3} = \frac{-45}{3}$
$y = -15$
The solution is (7, −15).

41. $x = 18 - 2y$
$3x + 2y = 30$
Multiply the first equation by −1 and write in standard form. Add the result to the second equation.
$$\begin{array}{r} -x - 2y = -18 \\ \underline{3x + 2y = 30} \\ 2x \quad = 12 \end{array}$$
$\frac{2x}{2} = \frac{12}{2}$
$x = 6$
Substitute 6 for x in the second equation.

$3(6) + 2y = 30$
$18 + 2y - 18 = 30 - 18$
$2y = 12$
$\frac{2y}{2} = \frac{12}{2}$
$y = 6$
The solution is (6, 6).

42. $\frac{1}{2}x + \frac{1}{3}y = 3$
$\frac{1}{4}x - \frac{2}{5}y = -7$
Multiply the first equation by 6 and the second equation by 5. Add the new equations.

$$\begin{array}{l} 3x + 2y = 18 \\ \underline{\frac{5}{4}x - 2y = -35} \\ \frac{17}{4}x \quad\quad = -17 \end{array}$$

$\frac{4}{17}\left(\frac{17}{4}x\right) = \frac{4}{17}(-17)$
$x = -4$
Substitute –4 for x in the second equation.
$\frac{1}{4}(-4) - \frac{2}{5}y = -7$
$-1 - \frac{2}{5}y + 1 = -7 + 1$
$-\frac{2}{5}y = -6$
$-\frac{5}{2}\left(-\frac{2}{5}y\right) = -\frac{5}{2}(-6)$
$y = 15$
The solution is (–4, 15).

43. $5x + 10y = -55$
$2x - 3y = 6$
Multiply the first equation by 3 and the second equation by 10. Add the new equations.

$$\begin{array}{l} 15x + 30y = -165 \\ \underline{20x - 30y = 60} \\ 35x \quad\quad = -105 \end{array}$$

$\frac{35x}{35} = \frac{-105}{35}$
$x = -3$
Substitute –3 for x in the first equation.
$5(-3) + 10y = -55$
$-15 + 10y = -55$
$-15 + 10y + 15 = -55 + 15$
$10y = -40$
$\frac{10y}{10} = \frac{-40}{10}$
$y = -4$
The solution is (–3, –4).

44. $2x = 3y + 1$
$15y = 10x + 5$
Multiply the first equation by 5 and add both equations written in standard form.

$$\begin{array}{r} 10x - 15y = 5 \\ \underline{-10x + 15y = 5} \\ 0 = 10 \end{array}$$

This is a contradiction. There is no solution.

45. $5x + 7y = 21$
$3x - 2y = 13$
Multiply the first equation by 2 and the second equation by 7. Add the new equations.

$$\begin{array}{l} 10x + 14y = 42 \\ \underline{21x - 14y = 91} \\ 31x \quad\quad = 133 \end{array}$$

$\frac{31x}{31} = \frac{133}{31}$
$x = \frac{133}{31}$
Substitute $\frac{133}{31}$ for x in the first equation.
$5\left(\frac{133}{31}\right) + 7y = 21$
$\frac{665}{31} + 7y - \frac{665}{31} = 21 - \frac{665}{31}$
$7y = \frac{651}{31} - \frac{665}{31}$
$7y = -\frac{14}{31}$
$\frac{1}{7}(7y) = \frac{1}{7}\left(-\frac{14}{31}\right)$
$y = -\frac{2}{31}$
The solution is $\left(\frac{133}{31}, -\frac{2}{31}\right)$.

46. $3x + y = 20$

$y = \frac{2}{3}x + \frac{16}{3}$

Multiply the second equation by –3 and write in standard form. Add the result to the first equation multiplied by 3.

$$\begin{array}{r} 9x + 3y = 60 \\ 2x - 3y = -16 \\ \hline 11x \quad\;\; = 44 \end{array}$$

$\frac{11x}{11} = \frac{44}{11}$

$x = 4$

Substitute 4 for x in the first equation.

$3(4) + y = 20$

$12 + y - 12 = 20 - 12$

$y = 8$

The solution is (4, 8).

47. $7(y - 1) = 5x$

$y = \frac{5}{7}x + 1$

Multiply the second equation by 7. Add both equations written in standard form.

$$\begin{array}{r} -5x + 7y = 7 \\ 5x - 7y = -7 \\ \hline 0 = 0 \end{array}$$

This is an identity. The solutions are all ordered pairs (x, y) that satisfy $y = \frac{5}{7}x + 1$.

48. $3x - 3y = 4$

$9x + 9y = -2$

Multiply the first equation by 3 and add to the second equation.

$$\begin{array}{r} 9x - 9y = 12 \\ 9x + 9y = -2 \\ \hline 18x \quad\;\; = 10 \end{array}$$

$\frac{18x}{18} = \frac{10}{18}$

$x = \frac{5}{9}$

Substitute $\frac{5}{9}$ for x in the second equation.

$9\left(\frac{5}{9}\right) + 9y = -2$

$5 + 9y = -2$

$5 + 9y - 5 = -2 - 5$

$9y = -7$

$\frac{9y}{9} = \frac{-7}{9}$

$y = -\frac{7}{9}$

The solution is $\left(\frac{5}{9}, -\frac{7}{9}\right)$.

49. $3x - y = 0$

$2x + 2y = 7$

Multiply the first equation by 2 and add to the second equation.

$$\begin{array}{r} 6x - 2y = 0 \\ 2x + 2y = 7 \\ \hline 8x \quad\;\; = 7 \end{array}$$

$\frac{8x}{8} = \frac{7}{8}$

$x = \frac{7}{8}$

Substitute $\frac{7}{8}$ for x in the first equation.

$3\left(\frac{7}{8}\right) - y = 0$

$\frac{21}{8} - y - \frac{21}{8} = 0 - \frac{21}{8}$

$-y = -\frac{21}{8}$

$-1(-y) = -1\left(-\frac{21}{8}\right)$

$y = \frac{21}{8}$

The solution is $\left(\frac{7}{8}, \frac{21}{8}\right)$.

50. $0.25x + 0.3y = 4$

$0.5x - 0.2y = 4$

Multiply the second equation by –0.5 and add to the first equation.

$$\begin{array}{r} 0.25x + 0.3y = 4 \\ -0.25x + 0.1y = -2 \\ \hline 0.4y = 2 \end{array}$$
$$\frac{0.4y}{0.4} = \frac{2}{0.4}$$
$$y = 5$$
Substitute 5 for y in the first equation.
$0.25x + 0.3(5) = 4$
$0.25x + 1.5 - 1.5 = 4 - 1.5$
$0.25x = 2.5$
$\frac{0.25x}{0.25} = \frac{2.5}{0.25}$
$x = 10$
The solution is (10, 5).

51. $0.2x + 0.1y = 5$
$0.02x - 0.01y = 13.5$
Multiply the first equation by 0.1 and add to the second equation.
$$\begin{array}{r} 0.02x + 0.01y = 0.5 \\ 0.02x - 0.01y = 13.5 \\ \hline 0.04x \qquad = 14 \end{array}$$
$$\frac{0.04x}{0.04} = \frac{14}{0.04}$$
$$x = 350$$
Substitute 350 for x in the first equation.
$0.2(350) + 0.1y = 5$
$70 + 0.1y - 70 = 5 - 70$
$0.1y = -65$
$\frac{0.1y}{0.1} = \frac{-65}{0.1}$
$y = -650$
The solution is (350, –650).

52. Let x = number of pounds of 25% copper
y = number of pounds of 30% copper
$x + y = 100$
$0.25x + 0.3y = 0.27(100)$
Multiply the first equation by –0.3 and add to the second equation.
$$\begin{array}{r} -0.3x - 0.3y = -30 \\ 0.25x + 0.3y = 27 \\ \hline -0.05x \qquad = -3 \end{array}$$
$$\frac{-0.05x}{-0.05} = \frac{-3}{-0.05}$$
$$x = 60$$
Substitute 60 for x in the first equation.
$60 + y = 100$
$60 + y - 60 = 100 - 60$
$y = 40$

The amounts are as follows:
60 pounds of 25% copper and 40 pounds of 30% copper.

53. Let x = number of small offices
y = number of large offices
$x + y = 40$
$250x + 400y = 12{,}700$
Multiply the first equation by –250 and add to the second equation.
$$\begin{array}{r} -250x - 250y = -10{,}000 \\ 250x + 400y = 12{,}700 \\ \hline 150y = 2700 \end{array}$$
$$\frac{150y}{150} = \frac{2700}{150}$$
$$y = 18$$
Substitute 18 for y in the first equation.
$x + 18 = 40$
$x + 18 - 18 = 40 - 18$
$x = 22$
She will have more of the small offices. She will have 22 small offices and 18 large offices.

54. Let x = mass of Mars
y = mass of Earth
$y = x + 5.328 \times 10^{24}$
$\frac{x + y}{2} = 3.306 \times 10^{24}$
Write each equation in standard form and add them.
$$\begin{array}{r} -x + y = 5.328 \times 10^{24} \\ x + y = 6.612 \times 10^{24} \\ \hline 2y = 1.194 \times 10^{25} \end{array}$$
$$\frac{2y}{2} = \frac{1.194 \times 10^{25}}{2}$$
$$y = 5.97 \times 10^{24}$$
Substitute 5.97×10^{24} for y in the first equation.
$5.97 \times 10^{24} = x + 5.328 \times 10^{24}$
$6.42 \times 10^{23} = x$
The mass of Mars is 6.42×10^{23} kg and the mass of Earth is 5.97×10^{24} kg.

55. Let x = number from A
y = number from B
$25x + 35y = 5500$
$x + y = 200$
Multiply the second equation by –25 and add to the first equation.

$$\begin{array}{r} 25x + 35y = 5500 \\ -25x - 25y = -5000 \\ \hline 10y = 500 \end{array}$$

$$\frac{10y}{10} = \frac{500}{10}$$

$$y = 50$$

Substitute 50 for y in the second equation.
$x + 50 = 200$
$x + 50 - 50 = 200 - 50$
$x = 150$
The plant should order 150 components from Supplier A and 50 components from Supplier B.

56. Let x = number of pounds of 10% nitrogen
y = number of pounds of 5% nitrogen
$x + y = 150$
$0.1x + 0.05y = 0.08(150)$
Multiply the first equation by –0.05 and add to the second equation.

$$\begin{array}{r} -0.05x - 0.05y = -7.5 \\ 0.1x + 0.05y = 12 \\ \hline 0.05x \qquad = 4.5 \end{array}$$

$$\frac{0.05x}{0.05} = \frac{4.5}{0.05}$$

$$x = 90$$

Substitute 90 for x in the first equation.
$90 + y = 150$
$90 + y - 90 = 150 - 90$
$y = 60$
He should use 90 pounds of 10% nitrogen with 60 pounds of 5% nitrogen.

57. Let x = number of pounds of broccoli
y = number of pounds of cauliflower
$0.49x + 0.99y = 0.69(200)$
$x + y = 200$
Multiply the second equation by –0.49 and add to the first equation.

$$\begin{array}{r} 0.49x + 0.99y = 138 \\ -0.49x - 0.49y = -98 \\ \hline 0.5y = 40 \end{array}$$

$$\frac{0.5y}{0.5} = \frac{40}{0.5}$$

$$y = 80$$

Substitute 80 for y in the second equation.
$x + 80 = 200$
$x + 80 - 80 = 200 - 80$
$x = 120$
They should mix 120 pounds of broccoli with 80 pounds of cauliflower.

58. Let x = number of hours of interstate driving
y = number of hours of highway driving
$x + y = 10$
$65x + 45y = 600$
Multiply the first equation by –45 and add to the first equation.

$$\begin{array}{r} -45x - 45y = -450 \\ 65x + 45y = 600 \\ \hline 20x \qquad = 150 \end{array}$$

$$\frac{20x}{20} = \frac{150}{20}$$

$$x = 7.5$$

$7.5(65) = 487.5$
He drove 487.5 miles on the interstate.

59. Let x = speed of current
y = distance of race
$y = 217.26(x + 1.603)$
$y = 217.50(x + 1.598)$
Substitute $217.26(x + 1.603)$ for y in the second equation.
$217.26(x + 1.603) = 217.50(x + 1.598)$
$217.26x + 348.268 = 217.50x + 347.565$
$217.26x - 217.5x = 347.565 - 348.268$
$-0.24x = -0.703$

$$\frac{-0.24x}{-0.24} = \frac{-0.703}{-0.24}$$

$x \approx 2.929$
Substitute 2.929 for x in the first equation.
$y = 217.26(2.929 + 1.603)$
$y \approx 984.62$
The speed of the current was 2.929 mps and the distance of the race was 984.62 meters.

60. $C(x) = 400 + 5x$
$R(x) = 8.5x$

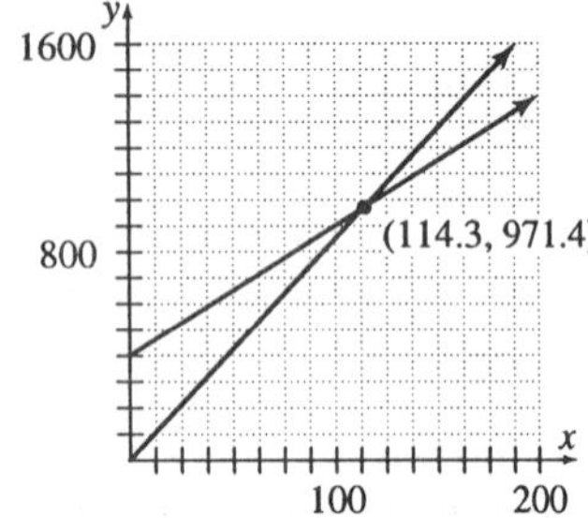

The intersection is approximately at (114.3, 971.4). The break-even point is at about 114 items.
When the company produces and sells less than 114 items, cost exceeds revenue. When the company produces and sells more than 115 items, revenue exceeds cost.

61. Let x = miles driven.
$A(x) = 150 + 0.25x$
$B(x) = 175 + 0.2x$

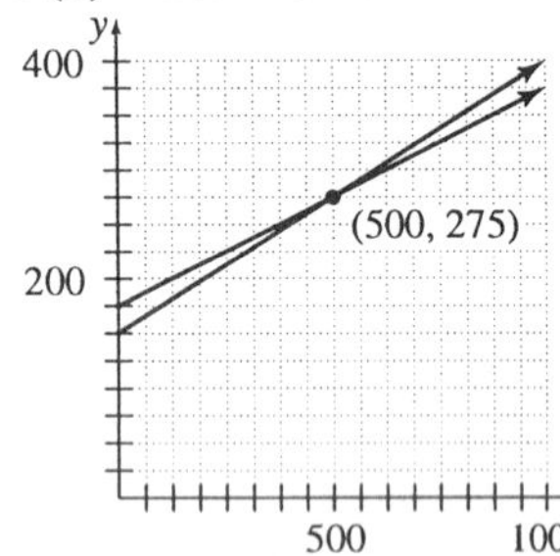

The graphs intersect at (500, 275).
He must drive more than 500 miles.

Chapter 7 Mixed Review

1. $y = \frac{4}{7}x + 2$
$x - 3y = 4$
Multiply the first equation by 7 and the second equation by 4. Add the new equations written in standard form.
$$\begin{array}{r} -4x + 7y = 14 \\ 4x - 12y = 16 \\ \hline -5y = 30 \\ \frac{-5y}{-5} = \frac{30}{-5} \\ y = -6 \end{array}$$
Substitute –6 for y in the second equation.
$x - 3(-6) = 4$
$x + 18 = 4$
$x + 18 - 18 = 4 - 18$
$x = -14$
The solution is (–14, –6).

2. $y = \frac{4}{5}x - 6$
$5y + 2 = 4(x - 7)$
Multiply the first equation by –5. Add the two equations written in standard form.
$$\begin{array}{r} 4x - 5y = 30 \\ -4x + 5y = -30 \\ \hline 0 = 0 \end{array}$$
This is an identity. The solutions are all ordered pairs (x, y) that satisfy $y = \frac{4}{5}x - 6$.

3. $1.25x + 3.5y = -25$
$4.5x + 2.8y = 8$
Multiply the first equation by 2.8 and the second equation by –3.5. Add the new equations.
$$\begin{array}{r} 3.5x + 9.8y = -70 \\ -15.75x - 9.8y = -28 \\ \hline -12.25x = -98 \\ \frac{-12.25x}{-12.25} = \frac{-98}{-12.25} \\ x = 8 \end{array}$$
Substitute 8 for x in the second equation.
$4.5(8) + 2.8y = 8$
$36 + 2.8y - 36 = 8 - 36$
$2.8y = -28$
$\frac{2.8y}{2.8} = \frac{-28}{2.8}$
$y = -10$
The solution is (8, –10).

4.
$$\begin{array}{r} 0.55x - 0.68y = 48 \\ -0.51x + 0.68y = 0 \\ \hline 0.04x = 48 \\ \frac{0.04x}{0.04} = \frac{48}{0.04} \\ x = 1200 \end{array}$$
Substitute 1200 for x in the first equation.

$0.55(1200) - 0.68y = 48$
$660 - 0.68y - 660 = 48 - 660$
$-0.68y = -612$
$\frac{-0.68y}{-0.68} = \frac{-612}{-0.68}$
$y = 900$
The solution is (1200, 900).

5. $8x + 2y = -6$
$6x - 2y = -78$
$14x = -84$
$\frac{14x}{14} = \frac{-84}{14}$
$x = -6$
Substitute –6 for x in the first equation.
$8(-6) + 2y = -6$
$-48 + 2y + 48 = -6 + 48$
$2y = 42$
$\frac{2y}{2} = \frac{42}{2}$
$y = 21$
The solution is (–6, 21).

6. $2x + 8y = -12$
$y = 2x + 3$
Write the second equation in standard form and add to the first equation.
$2x + 8y = -12$
$-2x + y = 3$
$9y = -9$
$\frac{9y}{9} = \frac{-9}{9}$
$y = -1$
Substitute –1 for y in the first equation.
$2x + 8(-1) = -12$
$2x - 8 + 8 = -12 + 8$
$2x = -4$
$\frac{2x}{2} = \frac{-4}{2}$
$x = -2$
The solution is (–2, –1).

7. $3x + y = 4$
$12x + 4y = 9$
Multiply the first equation by –4 and add to the second equation.
$-12x - 4y = -16$
$12x + 4y = 9$
$0 = -7$
This is a contradiction. There is no solution.

8. $3x - 5y = 11$
$4x + 2y = 13$
Multiply the first equation by 2 and the second equation by 5. Add the new equations.
$6x - 10y = 22$
$20x + 10y = 65$
$26x = 87$
$\frac{26x}{26} = \frac{87}{26}$
$x = \frac{87}{26}$
Substitute $\frac{87}{26}$ for x in the second equation.
$4\left(\frac{87}{26}\right) + 2y = 13$
$\frac{174}{13} + 2y - \frac{174}{13} = 13 - \frac{174}{13}$
$2y = \frac{169}{13} - \frac{174}{13}$
$2y = -\frac{5}{13}$
$\frac{1}{2}(2y) = \frac{1}{2}\left(-\frac{5}{13}\right)$
$y = -\frac{5}{26}$
The solution is $\left(\frac{87}{26}, -\frac{5}{26}\right)$.

9. $\frac{1}{3}x - \frac{2}{5}y = 10$
$\frac{1}{2}x + \frac{4}{5}y = -13$
Multiply the first equation by –15 and the second equation by 10. Add the new equations.

$$\begin{array}{r} -5x+6y=-150 \\ 5x+8y=-130 \\ \hline 14y=-280 \end{array}$$

$$\frac{14y}{14}=-\frac{280}{14}$$

$$y=-20$$

Substitute –20 for y in the second equation.

$$\frac{1}{2}x+\frac{4}{5}(-20)=-13$$

$$\frac{1}{2}x-16+16=-13+16$$

$$\frac{1}{2}x=3$$

$$2\left(\frac{1}{2}x\right)=2(3)$$

$$x=6$$

The solution is (6, –20).

10. $4x+8y=68$

$5x-3y=-6$

Multiply the first equation by 3 and the second equation by 8. Add the new equations.

$$\begin{array}{r} 12x+24y=204 \\ 40x-24y=-48 \\ \hline 52x=156 \end{array}$$

$$\frac{52x}{52}=\frac{156}{52}$$

$$x=3$$

Substitute 3 for x in the first equation.

$$4(3)+8y=68$$

$$12+8y-12=68-12$$

$$8y=56$$

$$\frac{8y}{8}=\frac{56}{8}$$

$$y=7$$

The solution is (3, 7).

11. $2x+y=0$

$x+4y=3$

Multiply the second equation by –2 and add to the first equation.

$$\begin{array}{r} 2x+y=0 \\ -2x-8y=-6 \\ \hline -7y=-6 \end{array}$$

$$\frac{-7y}{-7}=\frac{-6}{-7}$$

$$y=\frac{6}{7}$$

Substitute $\frac{6}{7}$ for y in the second equation.

$$x+4\left(\frac{6}{7}\right)=3$$

$$x+\frac{24}{7}-\frac{24}{7}=3-\frac{24}{7}$$

$$x=\frac{21}{7}-\frac{24}{7}$$

$$x=-\frac{3}{7}$$

The solution is $\left(-\frac{3}{7}, \frac{6}{7}\right)$.

12. $5x+3y=9$

$x-y=6$

Multiply the second equation by 3 and add to the first equation.

$$\begin{array}{r} 5x+3y=9 \\ 3x-3y=18 \\ \hline 8x=27 \end{array}$$

$$\frac{8x}{8}=\frac{27}{8}$$

$$x=\frac{27}{8}$$

Substitute $\frac{27}{8}$ for x in the second equation.

$$\frac{27}{8}-y=6$$

$$\frac{27}{8}-y-\frac{27}{8}=6-\frac{27}{8}$$

$$-y=\frac{48}{8}-\frac{27}{8}$$

$$-1(-y)=-1\left(\frac{21}{8}\right)$$

$$y=-\frac{21}{8}$$

The solution is $\left(\frac{27}{8}, -\frac{21}{8}\right)$.

13. $2y + 2 = y$
$y = -x + 1$

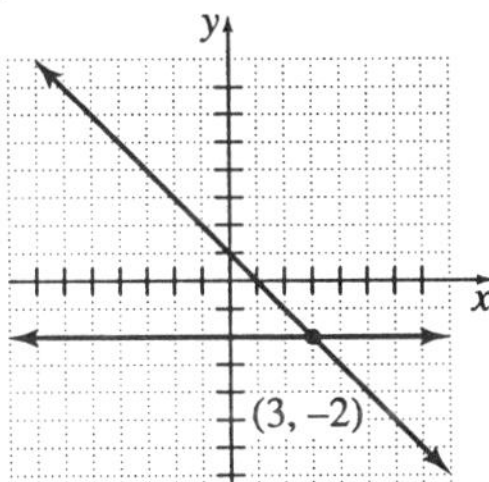

The solution is (3, –2).

14. $2x + 3 = x$
$2x = 5$

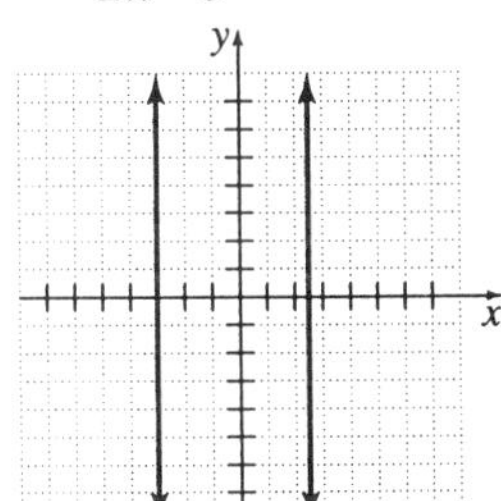

The graphs are parallel. There is no solution.

15. $y = 2x + 10$
$3x + y = -10$

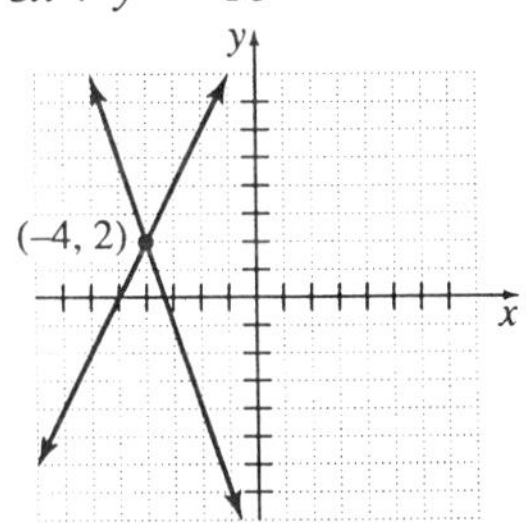

The solution is (–4, 2).

16. $2(x - 1) - 4 = 0$
$2x + y = 3$

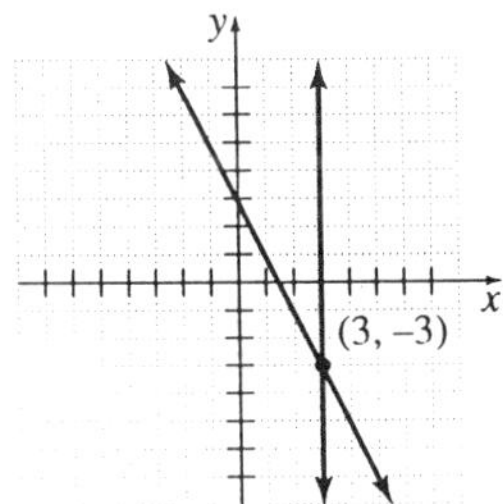

The solution is (3, –3).

17. $x + y = 9 - x$
$2x + y = 5$

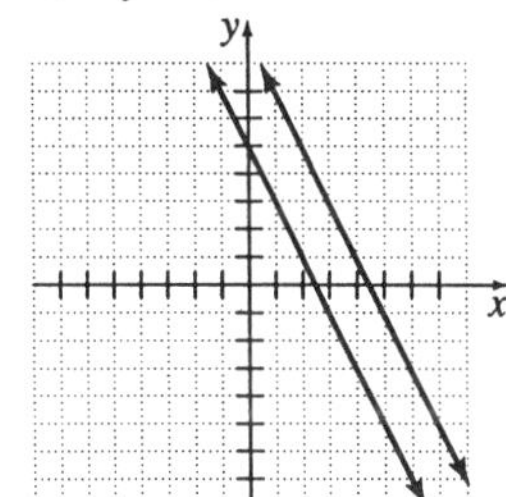

The graphs are parallel. There is no solution.

18. $y = -4x - 3$
$4x + 2y = y - 3$

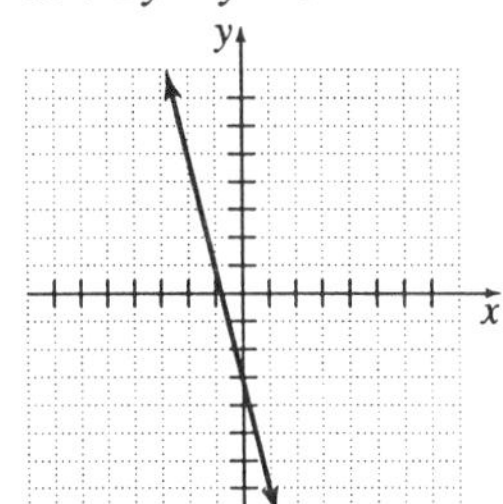

The graphs are coinciding. The solutions are all ordered pairs (x, y) satisfying $y = -4x - 3$.

19. $3x + 4 = 2x$
$x + 7 = 3$

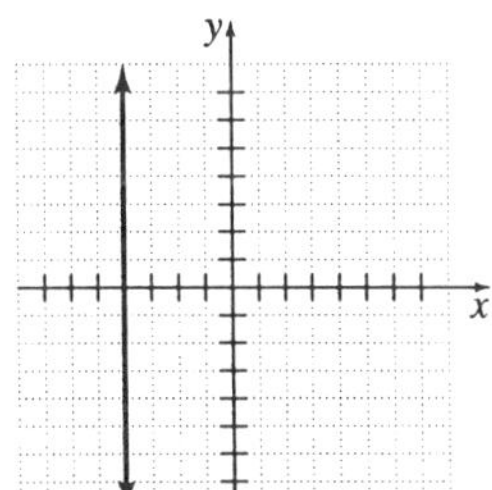

The graphs are coinciding. The solutions are all ordered pairs (x, y) that satisfy $x + 7 = 3$ or $x = -4$.

20. $3y+7=2y+5$
$y+9=7$

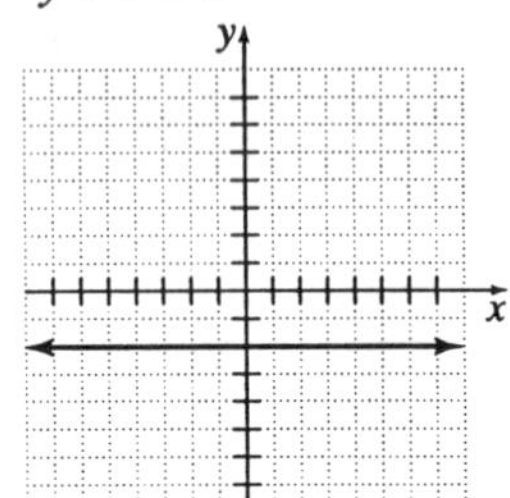

The graphs are coinciding. The solutions are all ordered pairs (x, y) that satisfy $y+9=7$ or $y=-2$.

21. $3(x-3)-1=8$
$3y-10=15-2y$

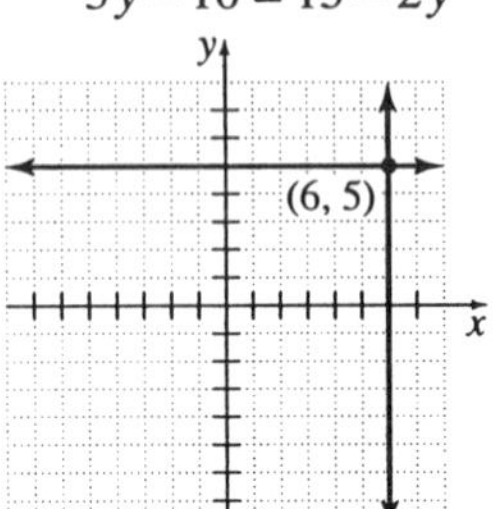

The solution is (6, 5).

22. $y=3x+7$
$x-y=-1$

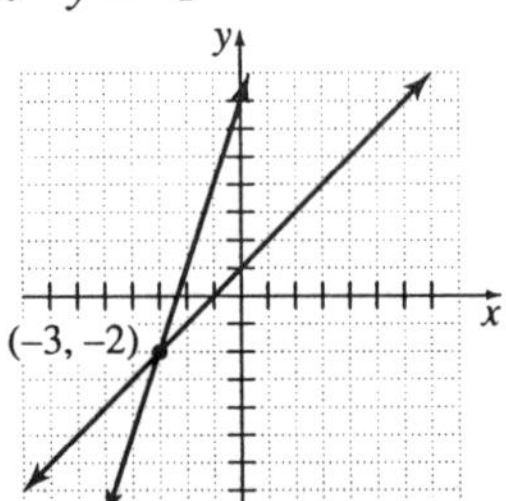

The solution is (–3, –2).

23. $x-5y=15$
$x+5y=-5$

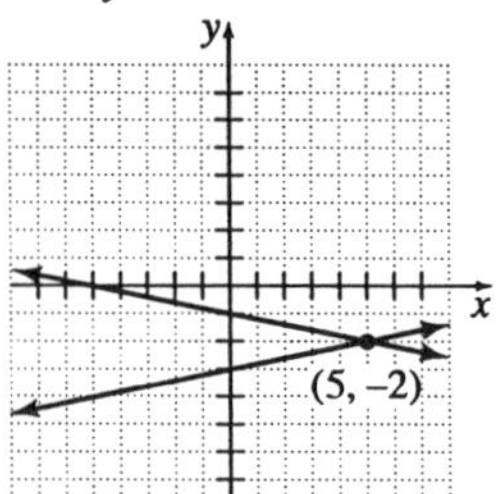

The solution is (5, –2).

24. $f(x)=-4x-6$
$g(x)=3x+8$

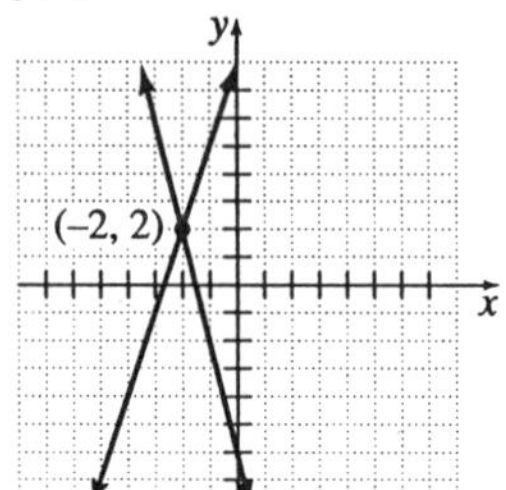

The solution is (–2, 2).

25. $y=x$
$x=2-y$

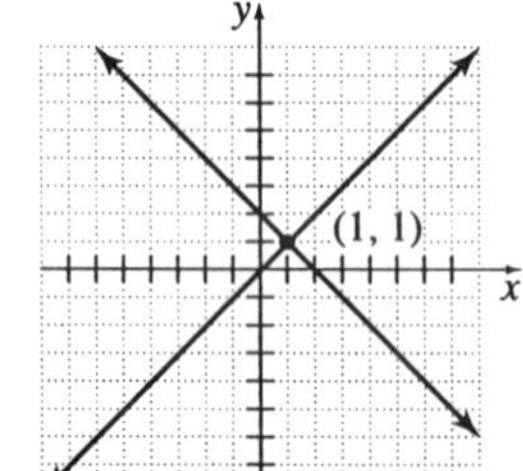

The solution is (1, 1).

26. $x+9=3$
$3y=2y-1$

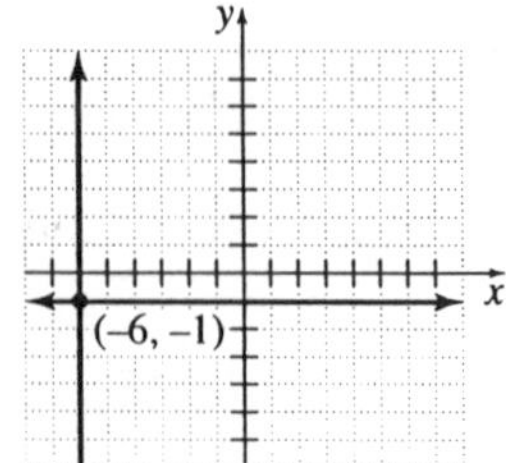

The solution is (–6, –1).

27. $y=\frac{5}{11}x-43$
$x+2y=-2$
Substitute $\frac{5}{11}x-43$ for y in the second equation.
$x+2\left(\frac{5}{11}x-43\right)=-2$
$x+\frac{10}{11}x-86=-2$
$\frac{21}{11}x-86+86=-2+86$
$\frac{21}{11}x=84$
$\frac{11}{21}\left(\frac{21}{11}x\right)=\frac{11}{21}(84)$
$x=44$
Substitute 44 for x in the first equation.
$y=\frac{5}{11}(44)-43$
$y=-23$
The solution is (44, –23).

28. $y=\frac{3}{4}x+14$
$y=-\frac{2}{3}x-3$
Substitute $\frac{3}{4}x+14$ for y in the second equation.
$\frac{3}{4}x+14=-\frac{2}{3}x-3$
$\frac{3}{4}x+\frac{2}{3}x=-3-14$
$\frac{17}{12}x=-17$
$\frac{12}{17}\left(\frac{17}{12}x\right)=\frac{12}{17}(-17)$
$x=-12$
Substitute –12 for x in the first equation.
$y=\frac{3}{4}(-12)+14$
$y=5$
The solution is (–12, 5).

29. $2x+3y=53$
$2x-4y=-248$
Solve the second equation for x, $x=2y-124$. Substitute $2y-124$ for x in the first equation.
$2(2y-124)+3y=53$
$7y-248=53$
$7y-248+248=53+248$
$7y=301$
$\frac{7y}{7}=\frac{301}{7}$
$y=43$
Substitute 43 for y in the first equation.
$2x+3(43)=53$
$2x+129-129=53-129$
$2x=-76$
$\frac{2x}{2}=-\frac{76}{2}$
$x=-38$
The solution is (–38, 43).

30. $y=4x-7$
$2y+12=4x+y+5$
Substitute $4x-7$ for y in the second equation.
$2(4x-7)+12=4x+4x-7+5$
$8x-14+12=8x-2$
$8x-2-8x=8x-2-8x$
$-2=-2$
This is an identity. The solutions are all ordered pairs (x, y) that satisfy $y=4x-7$.

31. $y=-5x$
$x-y=1$
Substitute $-5x$ for y in the second equation.
$x-(-5x)=1$
$6x=1$
$\frac{6x}{6}=\frac{1}{6}$
$x=\frac{1}{6}$
Substitute $\frac{1}{6}$ for x in the first equation.
$y=-5\left(\frac{1}{6}\right)$
$y=-\frac{5}{6}$
The solution is $\left(\frac{1}{6}, -\frac{5}{6}\right)$.

32. $5x - 10y = 21$
$5x = -3$

Solve the second equation for x, $x = -\frac{3}{5}$.

Substitute $-\frac{3}{5}$ for x in the first equation.

$5\left(-\frac{3}{5}\right) - 10y = 21$
$-3 - 10y + 3 = 21 + 3$
$-10y = 24$
$\frac{-10y}{-10} = \frac{24}{-10}$
$y = -\frac{12}{5}$

The solution is $\left(-\frac{3}{5}, -\frac{12}{5}\right)$.

33. $x + 7y = 3$
$4x + y = 9$
Solve the first equation for x, $x = 3 - 7y$.
Substitute $3 - 7y$ for x in the second equation.
$4(3 - 7y) + y = 9$
$12 - 28y + y = 9$
$12 - 27y - 12 = 9 - 12$
$-27y = -3$
$\frac{-27y}{-27} = \frac{-3}{-27}$
$y = \frac{1}{9}$

Substitute $\frac{1}{9}$ for y in the first equation.

$x + 7\left(\frac{1}{9}\right) = 3$
$x + \frac{7}{9} - \frac{7}{9} = 3 - \frac{7}{9}$
$x = \frac{20}{9}$

The solution is $\left(\frac{20}{9}, \frac{1}{9}\right)$.

34. $5x + 6y = 669$
$x - 2y = 1705$
Solve the second equation for x, $x = 2y + 1705$. Substitute $2y + 1705$ for x in the first equation.

$5(2y + 1705) + 6y = 669$
$10y + 8525 + 6y = 669$
$16y + 8525 - 8525 = 669 - 8525$
$16y = -7856$
$\frac{16y}{16} = \frac{-7856}{16}$
$y = -491$
Substitute -491 for y in the second equation.
$x - 2(-491) = 1705$
$x + 982 - 982 = 1705 - 982$
$x = 723$
The solution is $(723, -491)$.

35. $x - y = 5$
$6x + 2y = 9$
Solve the first equation for x, $x = y + 5$.
Substitute $y + 5$ for x in the second equation.
$6(y + 5) + 2y = 9$
$6y + 30 + 2y = 9$
$8y + 30 - 30 = 9 - 30$
$8y = -21$
$\frac{8y}{8} = -\frac{21}{8}$
$y = -\frac{21}{8}$

Substitute $-\frac{21}{8}$ for y in the first equation.

$x - \left(-\frac{21}{8}\right) = 5$
$x + \frac{21}{8} - \frac{21}{8} = \frac{40}{8} - \frac{21}{8}$
$x = \frac{19}{8}$

The solution is $\left(\frac{19}{8}, -\frac{21}{8}\right)$.

36. $x = 324y$
$3x - 810y = 90$
Substitute $324y$ for x in the second equation.
$3(324y) - 810y = 90$
$972y - 810y = 90$
$162y = 90$
$\frac{162y}{162} = \frac{90}{162}$
$y = \frac{5}{9}$

Substitute $\frac{5}{9}$ for y in the first equation.
$x = 324\left(\frac{5}{9}\right)$
$x = 180$
The solution is $\left(180, \frac{5}{9}\right)$.

37. $y = -2x + 7$
$2x + y = 0$
Substitute $-2x + 7$ for y in the second equation.
$2x + (-2x + 7) = 0$
$7 = 0$
This is a contradiction. There is no solution.

38. $2x - y = 60$
$5x - 3y = 60$
Solve the first equation for y, $y = 2x - 60$.
Substitute $2x - 60$ for y in the second equation.
$5x - 3(2x - 60) = 60$
$5x - 6x + 180 = 60$
$-x + 180 - 180 = 60 - 180$
$-x = -120$
$\frac{-x}{-1} = \frac{-120}{-1}$
$x = 120$
Substitute 120 for x in the first equation.
$2(120) - y = 60$
$240 - y - 240 = 60 - 240$
$-y = -180$
$\frac{-y}{-1} = \frac{-180}{-1}$
$y = 180$
The solution is (120, 180).

39.

$4x + 2y = 3$	
$4\left(\frac{5}{6}\right) + 2\left(-\frac{1}{6}\right)$	3
$\frac{10}{3} - \frac{1}{3}$	
3	

$x - y = -1$	
$\frac{5}{6} - \left(-\frac{1}{6}\right)$	-1
$\frac{5}{6} + \frac{1}{6}$	
1	

The second equation is false. Therefore, $\left(\frac{5}{6}, -\frac{1}{6}\right)$ is not a solution.

40.

$3x + y = -1$	
$3(-1.6) + 3.8$	-1
$-4.8 + 3.8$	
-1	

$5x + 5y = 11$	
$5(-1.6) + 5(3.8)$	11
$-8 + 19$	
11	

Both equations are true. Therefore, (–1.6, 3.8) is a solution.

41.

$5x + 7y = -6$	
$5(-4) + 7(2)$	-6
$-20 + 14$	
-6	

$2x - 7y = -22$	
$2(-4) - 7(2)$	-22
$-8 - 14$	
-22	

Both equations are true. Therefore, (–4, 2) is a solution.

42.

$x + y = 1$	
$\frac{13}{9} + \left(-\frac{4}{9}\right)$	1
$\frac{9}{9}$	
1	

$8x - y = 12$	
$8\left(\frac{13}{9}\right) - \left(-\frac{4}{9}\right)$	12
$\frac{104}{9} + \frac{4}{9}$	
$\frac{108}{9}$	
12	

Both equations are true. Therefore, $\left(\frac{13}{9}, -\frac{4}{9}\right)$ is a solution.

43.

$2x - 1 = 5$	
$2(3) - 1$	5
$6 - 1$	
5	

$3y + 5 = 2$	
$3(-2) + 5$	2
$-6 + 5$	
-1	

The second equation is false. Therefore, (3, –2) is not a solution.

44.

$x + 10 = 5$	
$-5 + 10$	5
5	

$3y + 11 = 2$	
$3(-3) + 11$	2
$-9 + 11$	
2	

Both equations are true. Therefore, (–5, –3) is a solution.

45.

$x + 2y = 2$	
$-0.44 + 2(1.22)$	2
$-0.44 + 2.44$	
2	

$2x + 4y = 5$	
$2(-0.44) + 4(1.22)$	5
$-0.88 + 4.88$	
4	

The second equation is false. Therefore, (–0.44, 1.22) is not a solution.

46.

$2x + 5y = -19$	
$2(-2) + 5(-3)$	-19
$-4 - 15$	
-19	

$3x - 7y = 10$	
$3(-2) - 7(-3)$	10
$-6 + 21$	
15	

The second equation is false. Therefore, (–2, –3) is not a solution.

47. Let x = number of pounds of 15%
y = number of pounds of 35%
$x + y = 200$
$0.15x + 0.35y = 0.27(200)$
Multiply the first equation by –0.15 and add to the second equation.
$$\begin{aligned} -0.15x - 0.15y &= -30 \\ 0.15x + 0.35y &= 54 \\ \hline 0.2y &= 24 \\ \frac{0.2y}{0.2} &= \frac{24}{0.2} \\ y &= 120 \end{aligned}$$
Substitute 120 for y in the first equation.
$x + 120 = 200$
$x + 120 - 120 = 200 - 120$
$x = 80$
The amounts should be as follows:
80 pounds of 15% brass
120 pounds of 35% brass
$120 - 80 = 40$
40 pounds more of the 35% alloy will be used.

48. Let x = amount for 1-bedroom
y = amount for 2-bedroom
$8x + 12y = 7300$
$y = x + 150$
Substitute $x + 150$ for y in the first equation.
$8x + 12(x + 150) = 7300$
$8x + 12x + 1800 = 7300$
$20x + 1800 - 1800 = 7300 - 1800$
$20x = 5500$
$\frac{20x}{20} = \frac{5500}{20}$
$x = 275$
Substitute 275 for x in the second equation.
$y = 275 + 150 = 425$
$8x = 8(275) = 2200$
$12y = 12(425) = 5100$
She will receive \$2200 for all of the 1-bedroom apartments and \$5100 for all of the 2-bedroom apartments.

49. Let x = measure of smaller angle
y = measure of larger angle
$x + y = 180$
$y = x + 10$
Substitute $x + 10$ for y in the first equation.
$x + x + 10 = 180$
$2x + 10 - 10 = 180 - 10$
$2x = 170$
$\frac{2x}{2} = \frac{170}{2}$
$x - 85$
The smaller angle measures 85°.

50. Let x = side length of smaller square
y = side length of larger square
$y = x + 5$
$4x + 4y = 100$
Substitute $x + 5$ for y in the second equation.
$4x + 4(x + 5) = 100$
$4x + 4x + 20 = 100$
$8x + 20 - 20 = 100 - 20$
$8x = 80$
$\frac{8x}{8} = \frac{80}{8}$
$x = 10$
$y = 10 + 5 = 15$
$(10)^2 = 100$
$(15)^2 = 225$
The areas of the squares will be 100 in^2 and 225 in^2.

51. Let x = number of women surveyed
y = number of men surveyed
$x + y = 700$
$0.7x + 0.4y = 400$
Multiply the first equation by –0.4 and add to the second equation.
$-0.4x - 0.4y = -280$
$0.7x + 0.4y = 400$
$0.3x = 120$
$\frac{0.3x}{0.3} = \frac{120}{0.3}$
$x = 400$
Substitute 400 for x in the first equation.
$400 + y = 700$
$400 + y - 400 = 700 - 400$
$y = 300$
$400 - 300 = 100$
There were 100 more women surveyed.

52. Let x = measure of smaller angle
y = measure of larger angle
$x + y = 180$
$y = 3x + 12$
Substitute $3x + 12$ for y in the first equation.
$x + 3x + 12 = 180$
$4x + 12 - 12 = 180 - 12$
$4x = 168$
$\frac{4x}{4} = \frac{168}{4}$
$x = 42$
Substitute 42 for x in second equation.
$y = 3(42) + 12$
$y = 138$
The angles measure 42° and 138°.

53. Let x = number of cc of 15%
y = number of cc of 25%
$x + y = 30$
$0.15x + 0.25y = 0.21(30)$
Multiply the first equation by –0.15 and add to the second equation.
$-0.15x - 0.15y = -4.5$
$0.15x + 0.25y = 6.3$
$0.1y = 1.8$
$\frac{0.1y}{0.1} = \frac{1.8}{0.1}$
$y = 18$
Substitute 18 for y in the first equation.

$x + 18 = 30$
$x + 18 - 18 = 30 - 18$
$x = 12$
She should use 12 cc of 15% and 18 cc of 25%.

54. Let x = walking rate
y = jogging rate
$3.5 = \frac{1}{2}x + \frac{1}{3}y$
$4 = \frac{2}{3}x + \frac{1}{3}y$
Multiply the first equation by –6 and the second equation by 6. Add the new equations.
$$\begin{aligned} -21 &= -3x - 2y \\ 24 &= 4x + 2y \\ \hline 3 &= x \end{aligned}$$
Substitute 3 for x in the second equation.
$4 = \frac{2}{3}(3) + \frac{1}{3}y$
$4 - 2 = 2 + \frac{1}{3}y - 2$
$2 = \frac{1}{3}y$
$3(2) = 3\left(\frac{1}{3}y\right)$
$6 = y$
He walked at 3 mph and jogged at 6 mph.

55. Let x = number of pounds of gourmet coffee
y = number of pounds of chocolate
$x + y = 50$
$8.5x + 12.5y = 9.5(50)$
Multiply the first equation by –8.5 and add to the second equation.
$$\begin{aligned} -8.5x - 8.5y &= -425 \\ 8.5x + 12.5y &= 475 \\ \hline 4y &= 50 \end{aligned}$$
$\frac{4y}{4} = \frac{50}{4}$
$y = 12.5$
Substitute 12.5 for y in the first equation.
$x + 12.5 = 50$
$x + 12.5 - 12.5 = 50 - 12.5$
$x = 37.5$
The shop should mix 37.5 pounds of gourmet coffee with 12.5 pounds of dutch chocolate.

56. Let x = number of hours on line
y = number of hours as clerk
$10.75x + 6.5y = 181$
$x + y = 20$
Multiply the second equation by –6.5 and add to the first equation.
$$\begin{aligned} 10.75x + 6.5y &= 181 \\ -6.5x - 6.5y &= -130 \\ \hline 4.25x &= 51 \end{aligned}$$
$\frac{4.25x}{4.25} = \frac{51}{4.25}$
$x = 12$
Substitute 12 for x in the second equation.
$12 + y = 20$
$12 + y - 12 = 20 - 12$
$y = 8$
She should work 12 hours at $10.75 per hour and 8 hours at $6.50 per hour.

57. Let x = number of hours of shorter trip
y = distance
$y = x(400 + 40)$
$y = (x + 1)(400 - 40)$
Substitute $x(400 + 40)$ for y in the second equation.
$x(400 + 40) = (x + 1)(400 - 40)$
$440x = 360x + 360$
$440x - 360x = 360x + 360 - 360x$
$80x = 360$
$\frac{80x}{80} = \frac{360}{80}$
$x = 4.5$
Substitute 4.5 for x in the first equation.
$y = 4.5(400 + 40)$
$y = 1980$
The distance traveled was 1980 miles.

58. Let x = number of miles driven
$U(x) = 39.95 + 0.15x$
$B(x) = 19.95 + 0.22x$

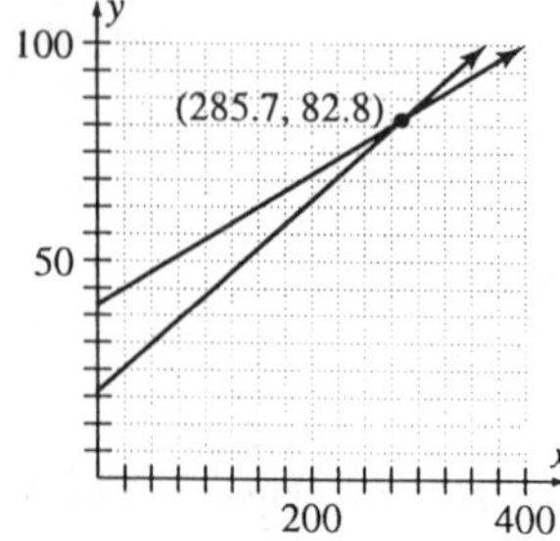

The graphs intersect at approximately (285.7, 82.8). You must drive at least 286 miles in order for U Rent It to be less costly.

59. Let x = number of items
$C(x) = 75 + 2.5x$
$R(x) = 6.5x$

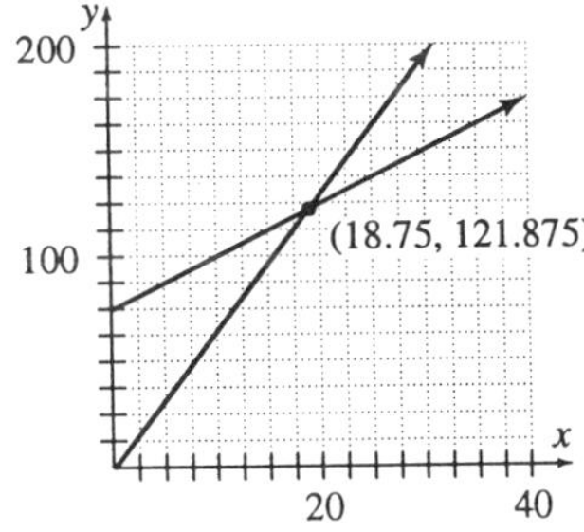

The graphs intersect at (18.75, 121.875). The shopkeeper must produce and sell at least 19 items in order not to lose money.

Chapter 7 Test

1.

	$y = -3.5x + 15.5$
–64.3	$-3.5(22.8) + 15.5$
	$-79.8 + 15.5$
	-64.3

$15x + 10y$	$= -301$
$15(22.8) + 10(-64.3)$	301
$342 - 643$	
-301	

Both equations are true.
Therefore, (22.8, –64.3) is a solution.

2. $2y - 8 = 0$
$y = 4$

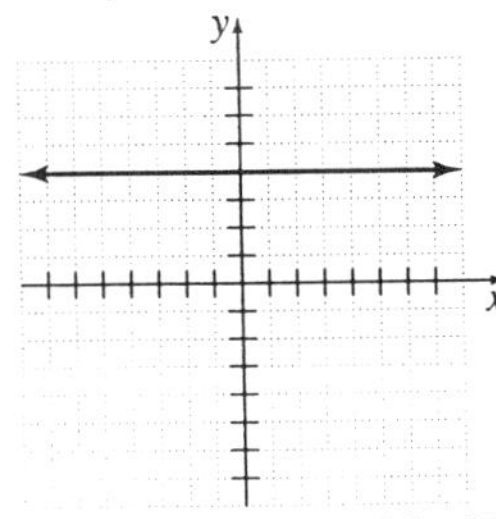

The graphs coincide. The solutions are all ordered pairs (x, y) that satisfy $y = 4$.

3. $y = \frac{1}{2}x + 3$
$y = \frac{1}{2}x - 5$

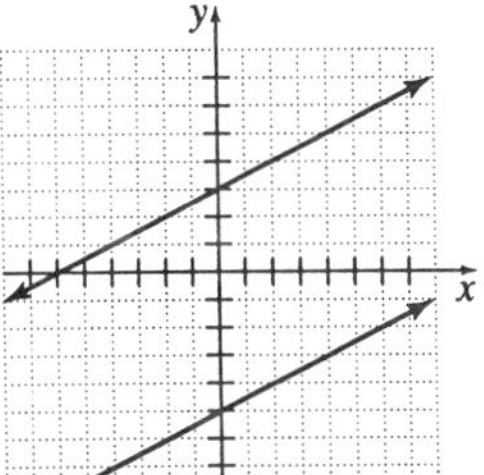

The graphs are parallel. There is no solution.

4. $y = 5x - 9$
$4x + 8y = 16$

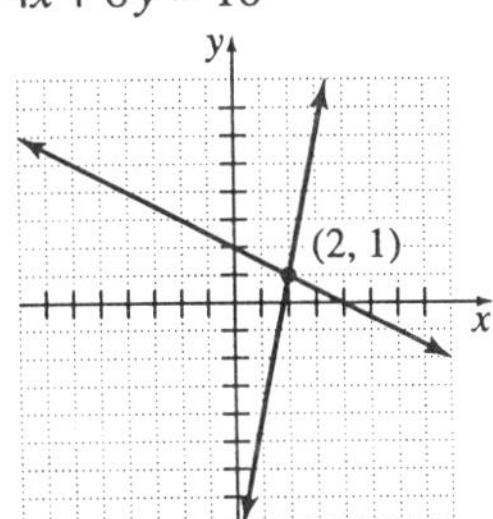

The solution is (2, 1).

5. $x + 2y = 6$
$5x - 11y = -54$
Solve the first equation for x, $x = 6 - 2y$.
Substitute $6 - 2y$ for x in the second equation.
$5(6 - 2y) - 11y = -54$
$30 - 10y - 11y = -54$
$30 - 21y = -54$
$30 - 21y - 30 = -54 - 30$
$-21y = -84$
$\frac{-21y}{-21} = \frac{-84}{-21}$
$y = 4$
Substitute 4 for y in the first equation.
$x + 2(4) = 6$
$x + 8 - 8 = 6 - 8$
$x = -2$
The solution is (–2, 4).

6. $3x+6y=12$
$x+2y=-5$
Solve the second equation for x, $x=-5-2y$.
Substitute $-5-2y$ for x in the first equation.
$3(-5-2y)+6y=12$
$-15-6y+6y=12$
$-15=12$
This is a contradiction. There is no solution.

7. $x+6y=-2$
$x-2y=1$
Solve the second equation for x, $x=2y+1$.
Substitute $2y+1$ for x in the first equation.
$2y+1+6y=-2$
$8y+1-1=-2-1$
$8y=-3$
$\frac{8y}{8}=\frac{-3}{8}$
$y=-\frac{3}{8}$
Substitute $-\frac{3}{8}$ for y in the second equation.
$x-2\left(-\frac{3}{8}\right)=1$
$x+\frac{3}{4}-\frac{3}{4}=1-\frac{3}{4}$
$x=\frac{1}{4}$
The solution is $\left(\frac{1}{4}, -\frac{3}{8}\right)$.

8. $3x+7y=11$
$6x+14y=2$
Multiply the first equation by -2 and add to the second equation.
$-6x-14y=-22$
$6x+14y=2$
$0=-20$
This is a contradiction. There is no solution.

9. $2x+9y=16$
$5y=8-x$
Multiply the second equation by -2 and write in standard form. Add to the first equation.
$2x+9y=16$
$-2x-10y=-16$
$-y=0$
$\frac{-y}{-1}=\frac{0}{-1}$
$y=0$
Substitute 0 for y in the second equation.
$5(0)=8-x$
$0=8-x$
$0-8=8-x-8$
$-8=-x$
$\frac{-8}{-1}=\frac{-x}{-1}$
$8=x$
The solution is (8, 0).

10. $5x+3y=11$
$2x+5y=7$
Multiply the first equation by -2 and the second equation by 5. Add the new equations.
$-10x-6y=-22$
$10x+25y=35$
$19y=13$
$\frac{19y}{19}=\frac{13}{19}$
$y=\frac{13}{19}$
Substitute $\frac{13}{19}$ for y in the first equation.
$5x+3\left(\frac{13}{19}\right)=11$
$5x+\frac{39}{19}-\frac{39}{19}=11-\frac{39}{19}$
$5x=\frac{170}{19}$
$\frac{1}{5}(5x)=\frac{1}{5}\left(\frac{170}{19}\right)$
$x=\frac{34}{19}$
The solution is $\left(\frac{34}{19}, \frac{13}{19}\right)$.

11. $C(x) = 5500 + 65x$
$R(x) = 125x$

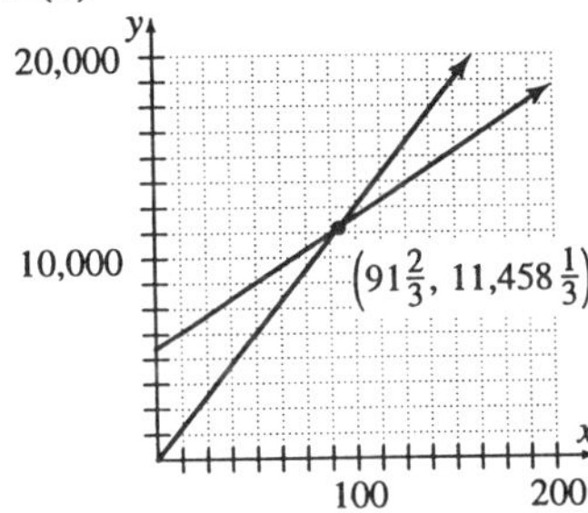

The graphs intersect at $\left(91\frac{2}{3}, 11,458\frac{1}{3}\right)$.

The break-even point is at $91\frac{2}{3}$ items.

After producing and selling 92 items, the factory's revenue will be greater than its cost.

12. Let x = Kenny's rate
y = distance to Dolly
$y = 3x$
$y = 2(x+4)$
Substitute $3x$ for y in the second equation.
$3x = 2(x+4)$
$3x = 2x + 8$
$3x - 2x = 2x + 8 - 2x$
$x = 8$
$y = 3(8) = 24$
Kenny's rate was 8 mph and the distance to Dolly was 24 miles.

13. Let x = number of nanoseconds to perform one addition
y = number of nanoseconds to perform one multiplication
$6x + 8y = 58$
$10x + 5y = 55$
Divide the first equation by 2 and the second equation by 5. Then multiply the new second equation by –4 and add it to the new first equation.
$$\begin{array}{r} 3x + 4y = 29 \\ -8x - 4y = -44 \\ \hline -5x = -15 \end{array}$$
$$\frac{-5x}{-5} = \frac{-15}{-5}$$
$x = 3$

Substitute 3 for x in the first equation.
$6(3) + 8y = 58$
$18 + 8y = 58$
$18 + 8y - 18 = 58 - 18$
$8y = 40$
$$\frac{8y}{8} = \frac{40}{8}$$
$y = 5$
The computer takes 3 nanoseconds per addition and 5 nanoseconds per multiplication.

14. Let x = number of pounds of raisins
y = number of pounds of peanuts
$x + y = 30$
$1.25x + 2y = 1.5(30)$
Multiply the first equation by –1.25 and add to the second equation.
$$\begin{array}{r} -1.25x - 1.25y = -37.5 \\ 1.25x + 2y = 45 \\ \hline 0.75y = 7.5 \end{array}$$
$$\frac{0.75y}{0.75} = \frac{7.5}{0.75}$$
$y = 10$
Substitute 10 for y in the first equation.
$x + 10 = 30$
$x + 10 - 10 = 30 - 10$
$x = 20$
He should mix 20 pounds of raisins with 10 pounds of peanuts.

15. Let x = measure of small angle
y = measure of large angle
$x + y = 90$
$y = 2x$
Substitute $2x$ for y in the first equation.
$x + 2x = 90$
$3x = 90$
$$\frac{3x}{3} = \frac{90}{3}$$
$x = 30$
$y = 2(30) = 60$
The angles measure 30° and 60°.

16. Let x = number of cc of 50% solution
y = number of cc of 10% solution
$x + y = 100$
$0.5x + 0.1y = 0.3(100)$
Multiply the first equation by –0.1 and add to the second equation.

$$\begin{array}{r} -0.1x - 0.1y = -10 \\ 0.5x + 0.1y = 30 \\ \hline 0.4x = 20 \end{array}$$

$$\frac{0.4x}{0.4} = \frac{20}{0.4}$$

$$x = 50$$

Substitute 50 for x in the first equation.
$50 + y = 100$
$50 + y - 50 = 100 - 50$
$y = 50$
She should mix 50 cc of 50% solution and 50 cc of 10% solution.

17. One solution exists.
No solution exists.
An infinite number of solutions exist.
Answers will vary.

Chapter 8

8.1 Experiencing Algebra the Exercise Way

1. $3(x + 1) > -(5x - 4)$ is linear because it simplifies to $3x + 3 > -5x + 4$.

3. $4x^2 + 1 < 3x + 2$ is nonlinear because the variable x has an exponent of 2.

5. $\frac{5}{7}(a+2) \geq \frac{3}{7}a + \frac{2}{3}$ is linear because it simplifies to $\frac{5}{7}a + \frac{10}{7} \geq \frac{3}{7}a + \frac{2}{3}$.

7. $0.6x + 2.7 \leq 5.2 - 1.9x$ is linear.

9. $x + 4(x - 8) > 2(3x + 1)$ is linear because it simplifies to $5x - 32 > 6x + 2$.

11. $\sqrt{2x-7} \leq x + 3$ is nonlinear because the radical expression contains a variable.

13. $\frac{5}{x} + 2x > 0$ is nonlinear because the variable x appears in the denominator.

15. $2p + 6 < 0$ is linear.

17. $x \geq 6$

0 1 2 3 4 5 6 7 8

The solution set is $[6, \infty)$.

19. $z < 12$

8 9 10 11 12 13 14 15 16

The solution set is $(-\infty, 12)$.

21. $b > -\frac{13}{5}$

$-\frac{13}{5}$

−8 −7 −6 −5 −4 −3 −2 −1 0

The solution set is $\left(-\frac{13}{5}, \infty\right)$.

23. $x \leq 4\frac{2}{3}$

$4\frac{2}{3}$

0 1 2 3 4 5 6 7 8

The solution set is $\left(-\infty, 4\frac{2}{3}\right]$.

25. $P \leq 12.59$

12.59

7 8 9 10 11 12 13 14 15

The solution set is $(-\infty, 12.59]$.

27. $q > -6.7$

−6.7

−8 −7 −6 −5 −4 −3 −2 −1 0

The solution set is $(-6.7, \infty)$.

29. $5 < x < 13$

5 6 7 8 9 10 11 12 13

The solution set is $(5, 13)$.

31. $8 > x > 2$

$2 < x < 8$

0 1 2 3 4 5 6 7 8

The solution set is $(2, 8)$.

33. $-4 < d \leq 0$

−4 −3 −2 −1 0 1 2 3 4

The solution set is $(-4, 0]$.

35. $-1 \leq m < 6$

−1 0 1 2 3 4 5 6 7

The solution set is $[-1, 6)$.

37. $2 \leq t \leq 7$

0 1 2 3 4 5 6 7 8

The solution set is $[2, 7]$.

39. $\frac{2}{5} < s \le 3\frac{1}{3}$

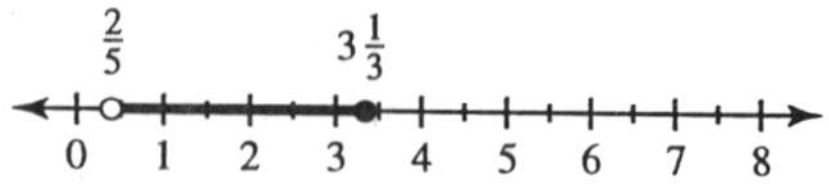

The solution set is $\left(\frac{2}{5}, 3\frac{1}{3}\right]$.

41. $-2.5 \le q < 3.5$

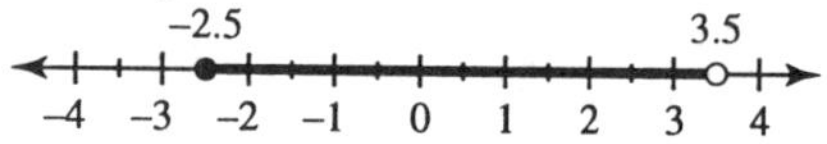

The solution set is $[-2.5, 3.5)$.

43. $\frac{4}{5} \le x \le 4.5$

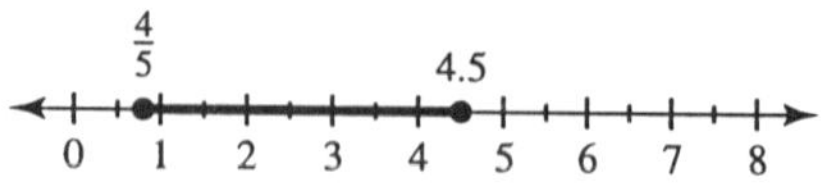

The solution set is $\left[\frac{4}{5}, 4.5\right]$.

45.

#	Inequality	Number Line	Interval Notation
a.	$x < 4.5$	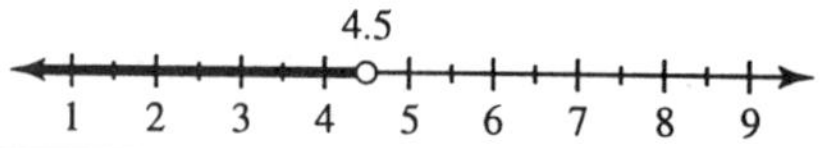	$(-\infty, 4.5)$
b.	$x \le 3$		$(-\infty, 3]$
c.	$x < -2$	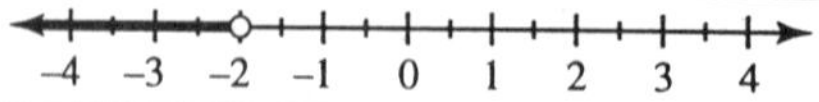	$(-\infty, -2)$
d.	$z \ge 5.7$	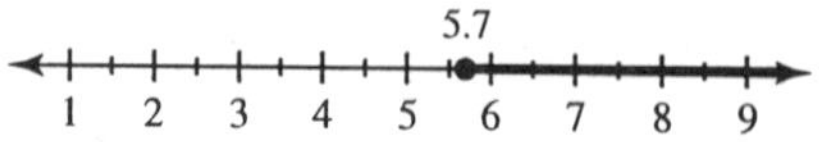	$[5.7, \infty)$
e.	$z > -7$	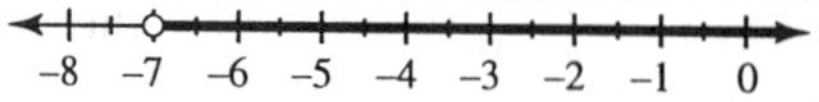	$(-7, \infty)$
f.	$z > 2$	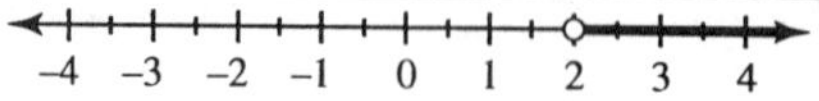	$(2, \infty)$
g.	$2 < y < 8$	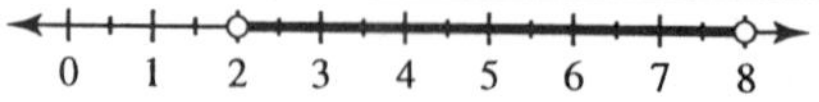	$(2, 8)$
h.	$-3 \le y < 2$	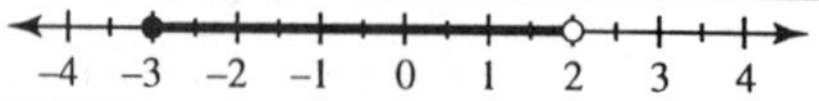	$[-3, 2)$
i.	$0 \le y \le 9$		$[0, 9]$

47. Let x = number of miles driven
$39.95 + 0.2x \le 150$

49. Let x = amount of weekly sales
$800 < 450 + 0.05x$

51. Let x = number of people
$150 \le 25 + 12.5x \le 200$

8.1 Experiencing Algebra the Calculator Way

1. $x < -2$

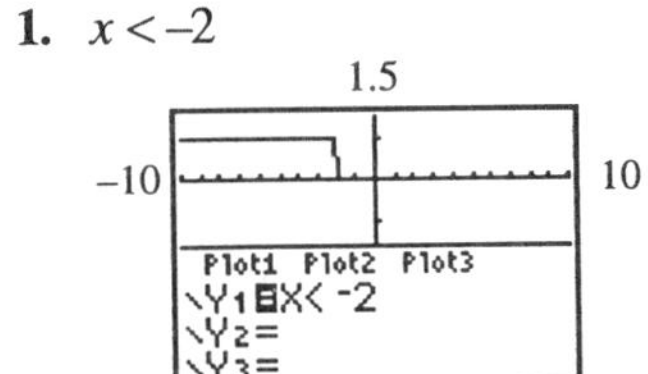

2. $x > -2$

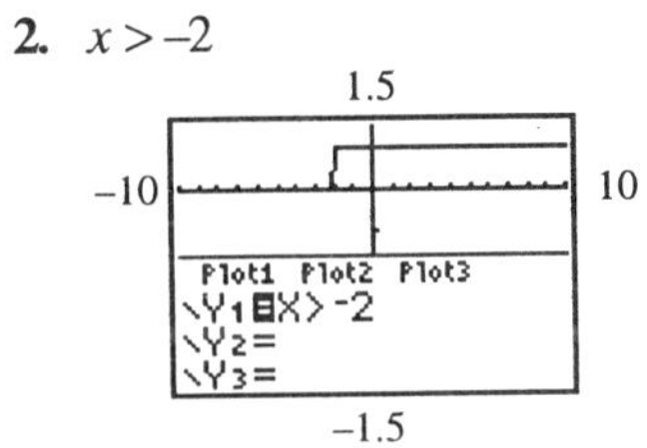

3. $x \le -2$

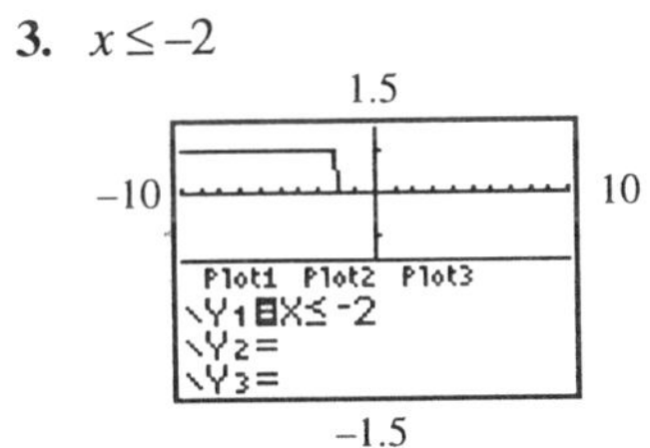

4. $x \ge -2$

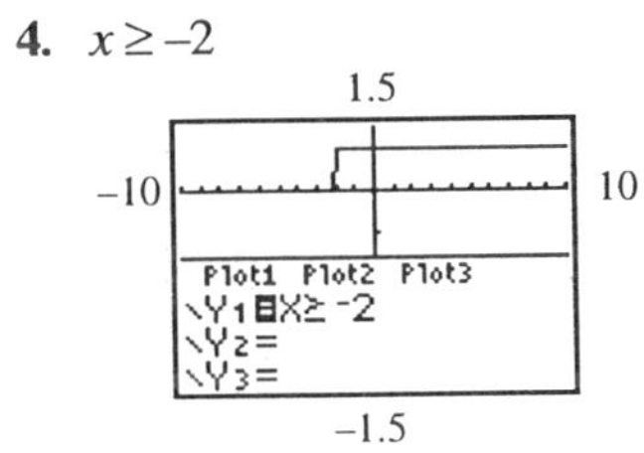

5. $-2.5 \le x \le 2.5$

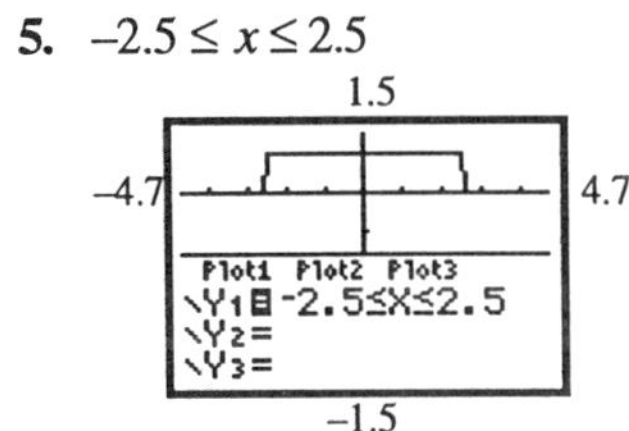

6. $-2.5 \le x < 2.5$

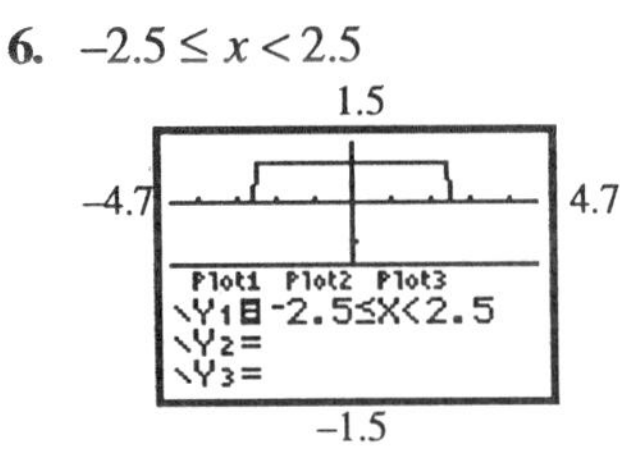

7. $-2.5 < x < 2.5$

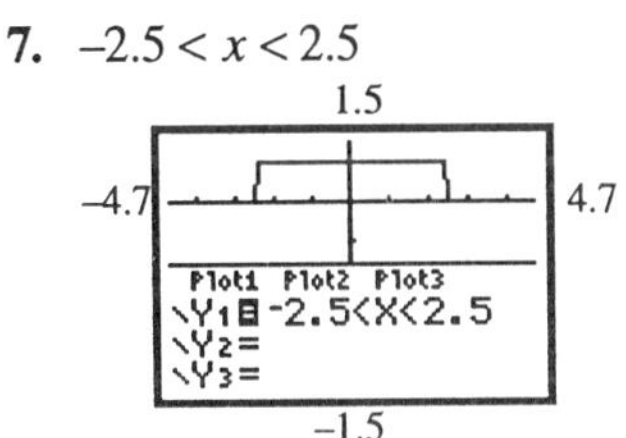

8.2 Experiencing Algebra the Exercise Way

1.

x	$6(x-12)$	$3x$	
23	$6(23-12)=66$	$3(23)=69$	$66 < 69$
24	$6(24-12)=72$	$3(24)=72$	$72 \nless 72$
25	$6(25-12)=78$	$3(25)=75$	$78 \nless 75$

The integers less than 24 are solutions of the inequality.

3.

x	$3(x+5)+3$	$3x+18$	
-1	$3(-1+5)+3=15$	$3(-1)+18=15$	$15 \ngtr 15$
0	$3(0+5)+3=18$	$3(0)+18=18$	$18 \ngtr 18$
1	$3(1+5)+3=21$	$3(1)+18=21$	$21 \ngtr 21$

The expressions are equal. The inequality is never true. There is no solution.

5.

x	$3x+3$	$x+2$	
-1	$3(-1)+3=0$	$-1+2=1$	$0 \ngtr 1$
0	$3(0)+3=3$	$0+2=2$	$3>2$
1	$3(1)+3=6$	$1+2=3$	$6>3$

The integers equal to and greater than 0 are solutions of the inequality.

7.

x	$2(4x+2)+2x-7$	$2(5x+4)$	
-1	$2[4(-1)+2]+2(-1)-7=-13$	$2[5(-1)+4]=-2$	$-13<-2$
0	$2[4(0)+2]+2(0)-7=-3$	$2[5(0)+4]=8$	$-3<8$
1	$2[4(1)+2]+2(1)-7=7$	$2[5(1)+4]=18$	$7<18$

The difference is always 11. The inequality is always true. The solutions are all integers.

9. $\frac{2}{3}x-\frac{2}{3}\geq-\frac{3}{4}x-\frac{7}{2}$

$Y1=\frac{2}{3}x-\frac{2}{3}$

$Y2=-\frac{3}{4}x-\frac{7}{2}$

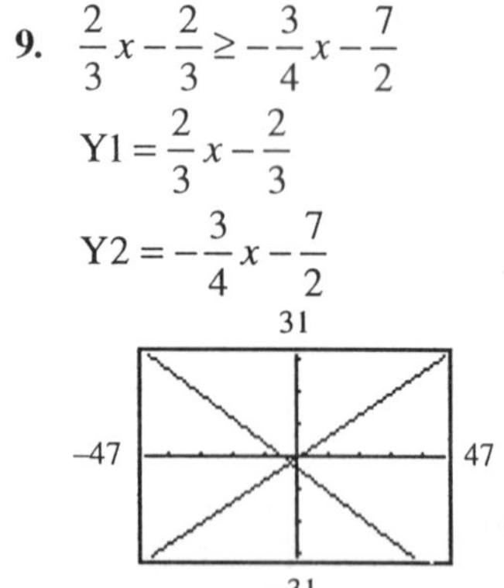

The intersection is $(-2, -2)$.
The solution set is all x that satisfies $x\geq-2$.

11. $2(x+3)-(x-1)>8-(x+9)$

$Y1=2(x+3)-(x-1)$

$Y2=8-(x+9)$

31

-47 47

-31

The intersection is $(-4, 3)$. The solution set is all x that satisfies $x>-4$.

13. $0.4x - 3.2 \le -0.6x - 0.2$
$Y1 = 0.4x - 3.2$
$Y2 = -0.6x - 0.2$

31
−47 47
−31

The intersection is (3, −2).
The solution set is all x that satisfies $x \le 3$.

15. $2(x + 1) > 3(x - 1) - x$
$Y1 = 2(x + 1)$
$Y2 = 3(x - 1)\,l - x$

31
−47 47
−31

The lines are parallel and the graph of Y1 is always above that of Y2.
The solution set is all real numbers.

17. $y - (3x + 1) > -x - (x + 1)$
$Y1 = x - (3x + 1)$
$Y2 = -x - (x + 1)$

31
−47 47
−31

The lines are coinciding. Since the inequality does not include the equal symbol, there is no solution.

19. $4x + 12 > 0$
$4x + 12 - 12 > 0 - 12$
$4x > -12$
$\frac{4x}{4} = \frac{-12}{4}$
$x > -3$
The solution set is $(-3, \infty)$.

21. $6x - 8 < -32$
$6x - 8 + 8 < -32 + 8$
$6x < -24$
$\frac{6x}{6} < \frac{-24}{6}$
$x < -4$
The solution set is $(-\infty, -4)$.

23. $-3x + 12 \ge 12$
$-3x + 12 - 12 \ge 12 - 12$
$-3x \ge 0$
$\frac{-3x}{-3} \le \frac{0}{-3}$
$x \le 0$
The solution set is $(-\infty, 0]$.

25. $-7x - 12 < -26$
$-7x - 12 + 12 < -26 + 12$
$-7x < -14$
$\frac{-7x}{-7} > \frac{-14}{-7}$
$x > 2$
The solution set is $(2, \infty)$.

27. $15.17 < 5.9x - 4.3$
$15.17 + 4.3 < 5.9x - 4.3 + 4.3$
$19.47 < 5.9x$
$\frac{19.47}{5.9} < \frac{5.9x}{5.9}$
$3.3 < x$
$x > 3.3$
The solution set is $(3.3, \infty)$.

29. $6.1 > -0.55a + 6.1$
$6.1 - 6.1 > -0.55a + 6.1 - 6.1$
$0 > -0.55a$
$\frac{0}{-0.55} < \frac{-0.55a}{-0.55}$
$0 < a$
$a > 0$
The solution set is $(0, \infty)$.

31. $2.07z + 4.12 \ge 16.54$
$2.07z + 4.12 - 4.12 \ge 16.54 - 4.12$
$2.07z \ge 12.42$
$\frac{2.07z}{2.07} \ge \frac{12.42}{2.07}$
$z \ge 6$
The solution set is $[6, \infty)$.

33. $-9.2b - 4.3 \le 70.6$
$-9.2b - 4.3 + 4.3 \le 70.6 + 4.3$
$-9.2b \le 74.9$
$\frac{-9.2b}{-9.2} \ge \frac{74.9}{-9.2}$
$b \ge -\frac{749}{92}$
The solution set is $\left[-\frac{749}{92}, \infty\right)$.

35. $-\frac{5}{9}b + \frac{11}{12} < \frac{23}{36}$
$36\left(-\frac{5}{9}b + \frac{11}{12}\right) < 36\left(\frac{23}{36}\right)$
$-20b + 33 < 23$
$-20b + 33 - 33 < 23 - 33$
$-20b < -10$
$\frac{-20b}{-20} > \frac{-10}{-20}$
$b > \frac{1}{2}$
The solution set is $\left(\frac{1}{2}, \infty\right)$.

37. $156z - 210 > 47z + 662$
$156z - 210 + 210 > 47z + 662 + 210$
$156z > 47z + 872$
$156z - 47z > 47z + 872 - 47z$
$109z > 872$
$\frac{109z}{109} > \frac{872}{109}$
$z > 8$
The solution set is $(8, \infty)$.

39. $-1.05x - 15.41 < 2.55x - 47.09$
$-1.05x - 15.41 + 15.41$
$< 2.55x - 47.09 + 15.41$
$-1.05x < 2.55x - 31.68$
$-1.05x - 2.55x < 2.55x \text{l}- 31.68 - 2.55x$
$-3.6x < -31.68$
$\frac{-3.6x}{-3.6} > \frac{-31.68}{-3.6}$
$x > 8.8$
The solution set is $(8.8, \infty)$.

41. $11x + \frac{1}{4} \le 2x + \frac{7}{36}$
$36\left(11x + \frac{1}{4}\right) \le 36\left(2x + \frac{7}{36}\right)$
$396x + 9 \le 72x + 7$
$396x + 9 - 9 \le 72x + 7 - 9$
$396x \le 72x - 2$
$396x - 72x \le 72x - 2 - 72x$
$324x \le -2$
$\frac{324x}{324} \le \frac{-2}{324}$
$x \le -\frac{1}{162}$
The solution set is $\left(-\infty, -\frac{1}{162}\right]$.

43. $\frac{2}{5}b - 12 < \frac{2}{3}b + 20$
$15\left(\frac{2}{5}b - 12\right) < 15\left(\frac{2}{3}b + 20\right)$
$6b - 180 < 10b + 300$
$6b - 180 + 180 < 10b \text{ l}+ 300 + 180$
$6b < 10b + 480$
$6b - 10b < 10b + 480 - 10b$
$-4b < 480$
$\frac{-4b}{-4} > \frac{480}{-4}$
$b > -120$
The solution set is $(-120, \infty)$.

45. $4x - (3x + 5) < x - 5$
$4x - 3x - 5 < x - 5$
$x - 5 < x - 5$
$x - 5 - x < x - 5 - x$
$-5 < -5$
Since this is a contradiction, there is no solution.

47. $4x - (3x + 5) \le x - 5$
$4x - 3x - 5 \le x - 5$
$x - 5 \le x - 5$
$x - 5 - x \le x - 5 - x$
$-5 \le -5$
Since this is an identity, the solutions et is all real numbers.

49. $0.05x + 10.5 < 0.15x - 0.25$
$0.05x\,l{+}\,10.5 - 10.5 < 0{,}.15x - 0.25 - 10.5$
$0.05x < 0.15x - 10.75$
$0.05x - 0.15x < 0.15x - 10.75 - 0.15x$
$-0.1x < -10.75$
$\frac{-0.1x}{-0.1} > \frac{-10.75}{-0.1}$
$x > 107.5$
The solution set is $(107.5, \infty)$.

51. Let x = sixth test score
$\frac{73+97+82+89+95+x}{6} \ge 87$
$\frac{436+x}{6} \ge 87$
$6\left(\frac{436+x}{6}\right) \ge 6(87)$
$436 + x \ge 522$
$436 + x - 436 \ge 522 - 436$
$x \ge 86$
His score must be 86 or greater.

53. Let x = number of hours worked
$9.75x + 5200 \le 7500$
$9.75x + 5200 - 5200 \le 7500 - 5200$
$9.75x \le 2300$
$\frac{9.75x}{9.75} \le \frac{2300}{9.75}$
$x \le 235.9$
He can work 235 hours or less.

55. Let x = number of people
$165x \le 2000$
$\frac{165x}{165} \le \frac{2000}{165}$
$x \le 12.1$
The number of people can be 12 or less.

57. Let x = number of calories for breakfast
$x + 150$ = number of calories for lunch
$2x + 50$ = number of calories for dinner
$x + x + 150 + 2x + 50 \le 1200$
$4x + 200 \le 1200$
$4x + 200 - 200 \le 1200 - 200$
$4x \le 1000$
$\frac{4x}{4} \le \frac{1000}{4}$
$x \le 250$
$x + 150 \le 400$
$2x + 50 \le 550$
She can have no more than 250 calories for breakfast, 400 calories for lunch, and 550 calories for dinner.

59. Let x = number of meat plates
$8.5x + 6.5(120 - x) \le 950$
$8.5x + 780 - 6.5x \le 950$
$2x + 780 - 780 \le 950 - 780$
$2x \le 170$
$\frac{2x}{2} \le \frac{170}{2}$
$x \le 85$
She can order 85 or less meat plates.

61. Let x = width
$\frac{3}{4}x + 30$ = length
$2x + \frac{3}{4}x + 30 - 4 \le 185$
$2.75x + 26 \le 185$
$2.75x + 26 - 26 \le 185 - 26$
$2.75x \le 159$
$\frac{2.75x}{2.75} \le \frac{159}{2.75}$
$x \le 57.8$
The width must be 57 or less feet.

63. Let x = amount of weekly sales
$300 + 0.05(x - 1000) > 500 + 0.05(x - 2000)$
$300 + 0.05x - 50 > 500 + 0.05x - 100$
$10.05x + 250 > 0.05x\,l{+}\,400$
$0.05x + 250 - 0.05x > 0.05x + 400 - 0.05x$
$250 > 400$
This is a contradiction. There is no solution. There are no values of sales that plan A will pay more than plan B.

65. $C(n) > 1\times 10^{12}$
$(4.73\times 10^{10})n + (7.66\times 10^{10}) > 1\times 10^{12}$
$(4.73\times 10^{10})n + (7.66\times 10^{10}) - (7.66\times 10^{10})$
$> 1\times 10^{12} - (7.66\times 10^{10})$
$(4.73\times 10^{10})n > 9.234\times 10^{11}$
$\dfrac{(4.73\times 10^{10})n}{4.73\times 10^{10}} > \dfrac{9.234\times 10^{11}}{4.73\times 10^{10}}$
$n > 1.9522\times 10^{1}$
$n > 19.522$
$1970 + 20 = 1990$
The years are 1990 to the present.

8.2 Experiencing Algebra the Calculator Way

Answers will vary.

8.3 Experiencing Algebra the Exercise Way

1. $2x + 1.7y > x - 4.6$ is linear.
In standard form,
$2x + 1.7y - x > x - 4.6 - x$
$x + 1.7y > -4.6$
$a = 1, b = 1.7, c = -4.6$

3. $y < x^2 + 2x - 3$ is nonlinear because the x term is squared.

5. $y > \sqrt{x} + 9$ is nonlinear because the radical expression contains a variable.

7. $\dfrac{x}{2} - \dfrac{y}{6} > \dfrac{1}{12}$ is linear. In standard form,
$12\left(\dfrac{x}{2} - \dfrac{y}{6}\right) > 12\left(\dfrac{1}{12}\right)$
$6x - 2y > 1$
$a = 6, b = -2, c = 1$

9. $4x + 16 \le y$ is linear. In standard form,
$4x + 16 - 16 \le y - 16$
$4x \le y - 16$
$4x - y \le y - 16 - y$
$4x - y \le -16$
$a = 4, b = -1, c = -16$

11. $y \le -\dfrac{2}{5}x + \dfrac{7}{15}$ is linear.
$15(y) \le 15\left(-\dfrac{2}{5}x + \dfrac{7}{15}\right)$
$15y \le -6x + 7$
$15y + 6x \le -6x + 7 + 6x$
$6x + 15y \le 7$
$a = 6, b = 15, c = 7$

13. $2x + y < 3$
$2x + y - 2x < 3 - 2x$
$y < 3 - 2x$
$y < -2x + 3$

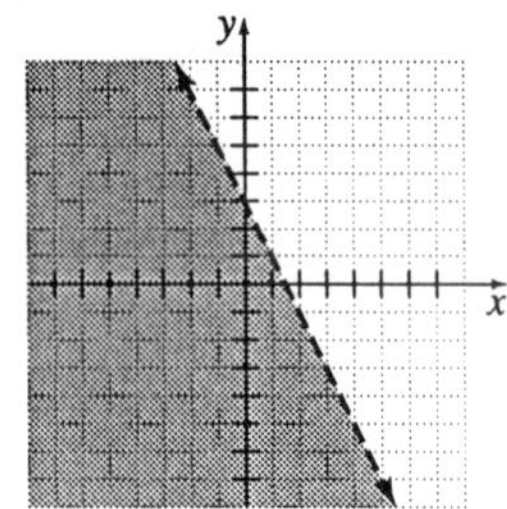

15. $x - 3y \ge 6$
$5x - 3y - 5x \ge 6 - 5x$
$-3y \ge 6 - 5x$
$\dfrac{-3y}{-3} \le \dfrac{-5x + 6}{-3}$
$y \le \dfrac{5}{3}x - 2$

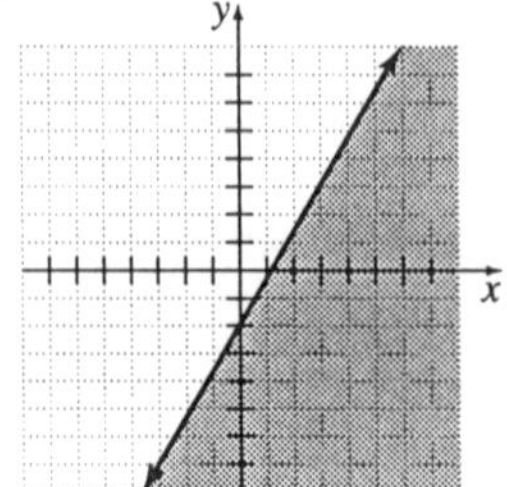

17. $y < -\frac{3}{4}x + 4$

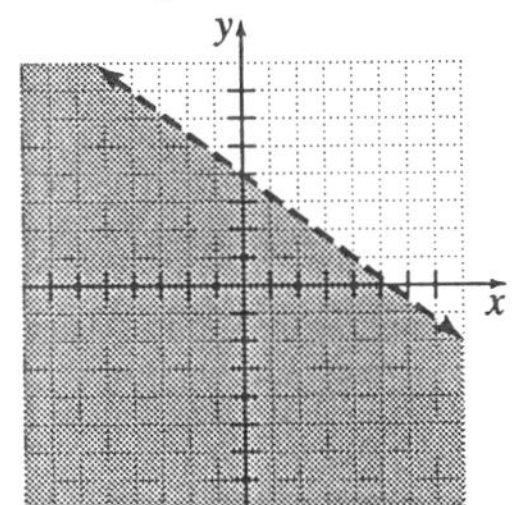

19. $y \geq 2.8x - 1.6$

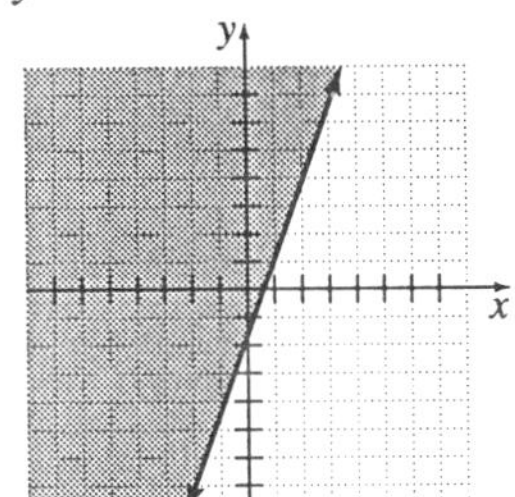

21. $2y > 3x - 5$

$y > \frac{3x-5}{2}$

$y > \frac{3}{2}x - \frac{5}{2}$

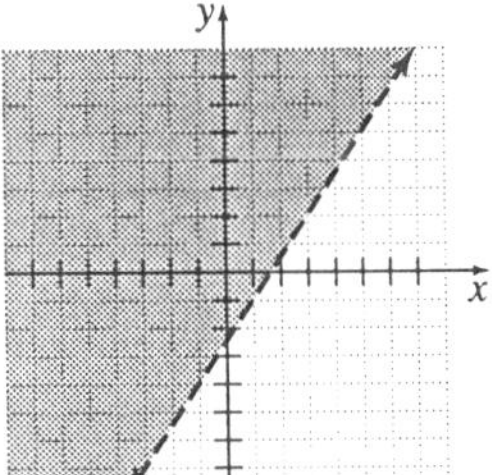

23. $3x + y - 4 \leq x + 2y - 3$

$3x + y - 4 + 4 \leq x + 2y - 3 + 4$

$3x + y \leq x + 2y + 1$

$3x + y - 3x \leq x + 2y + 1 - 3x$

$y \leq -2x + 2y + 1$

$y - 2y \leq -2x + 2y + 1 - 2y$

$-y \leq -2x + 1$

$\frac{-y}{-1} \geq \frac{-2x+1}{-1}$

$y \geq 2x - 1$

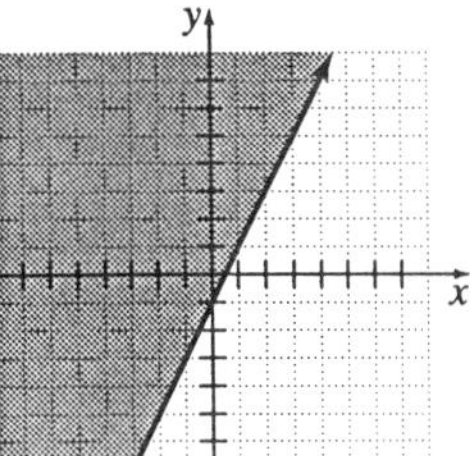

25. $5y > -20$

$\frac{5y}{5} > \frac{-20}{5}$

$y > -4$

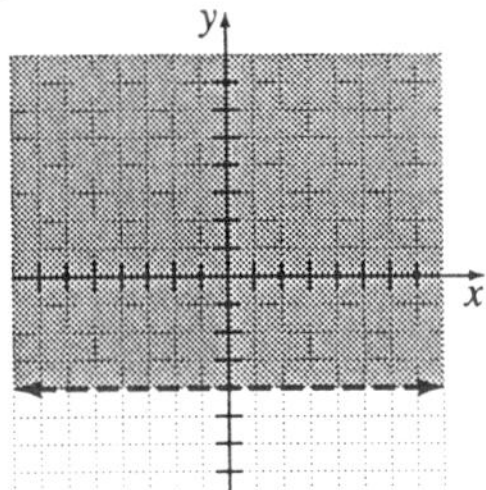

27. $3x + 6 \leq 9$

$3x + 6 - 6 \leq 9 - 6$

$3x \leq 3$

$\frac{3x}{3} \leq \frac{3}{3}$

$x \leq 1$

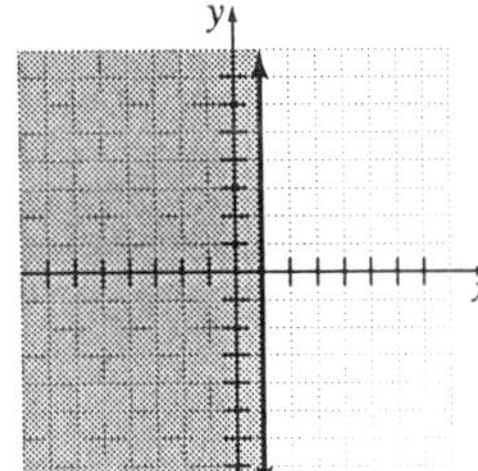

29. $3x+5y \ge 12$
$3x+5y-3x \ge 12-3x$
$5y \ge -3x+12$
$\frac{5y}{5} \ge \frac{-3x+12}{5}$
$y \ge -\frac{3}{5}x+\frac{12}{5}$

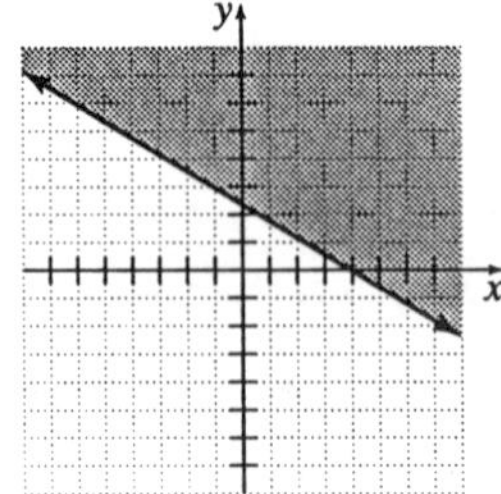

31. $-x-y<7$
$-x-y+x<7+x$
$-y<x+7$
$\frac{-y}{-1} > \frac{x+7}{-1}$
$y>-x-7$

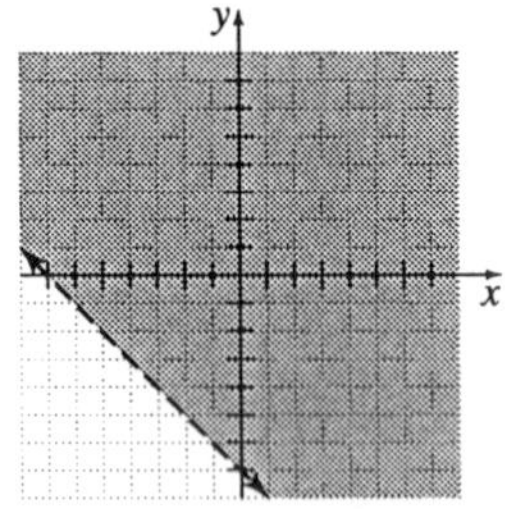

33. $-3x-9y>27$
$-3x-9y+3x>27+3x$
$-9y>3x+27$
$\frac{-9y}{-9} < \frac{3x+27}{-9}$
$y<-\frac{1}{3}x-3$

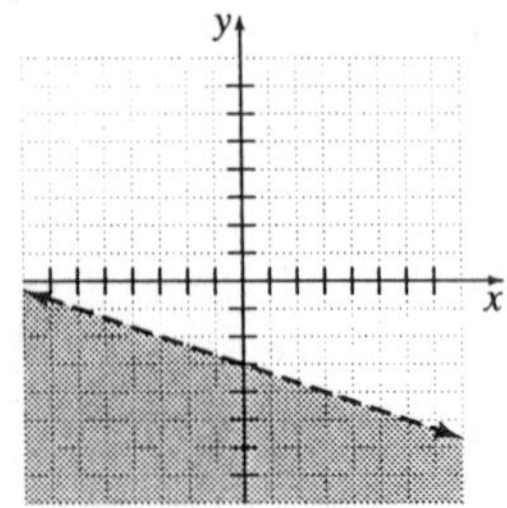

35. $\frac{x}{8}+\frac{y}{3} \le 1$
$\frac{x}{8}+\frac{y}{3}-\frac{x}{8} \le 1-\frac{x}{8}$
$\frac{y}{3} \le -\frac{x}{8}+1$
$3\left(\frac{y}{3}\right) \le 3\left(-\frac{x}{8}+1\right)$
$y \le -\frac{3}{8}x+3$

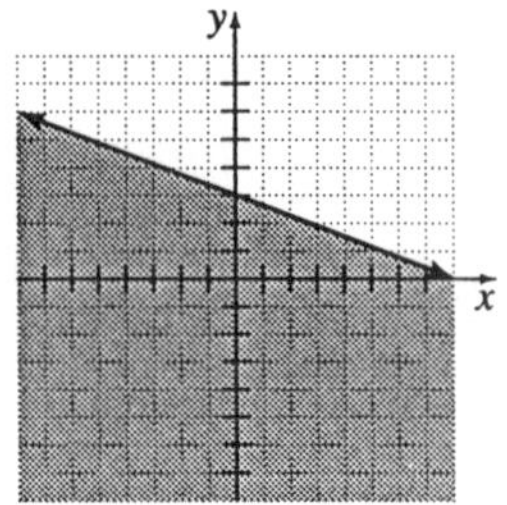

37. $-\frac{4}{7}x+\frac{2}{3}y \ge \frac{10}{21}$
$-\frac{4}{7}x+\frac{2}{3}y+\frac{4}{7}x \ge \frac{10}{21}+\frac{4}{7}x$
$\frac{2}{3}y \ge \frac{4}{7}x+\frac{10}{21}$
$\frac{3}{2}\left(\frac{2}{3}y\right) \ge \frac{3}{2}\left(\frac{4}{7}x+\frac{10}{21}\right)$
$y \ge \frac{6}{7}x+\frac{5}{7}$

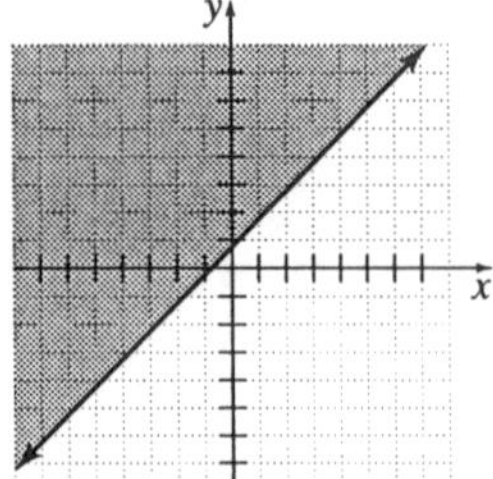

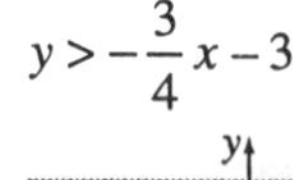

39. $1.8x - 3.2y > 0$
$1.8x - 3.2y - 1.8x > 0 - 1.8x$
$-3.2y > -1.8x$
$\frac{-3.2y}{-3.2} < \frac{-1.8x}{-3.2}$
$y < 0.5625x$

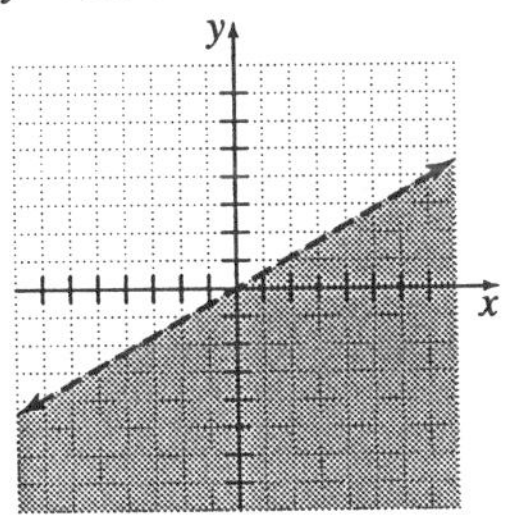

41. $4.6y < 3.5y + 5.94$
$4.6y - 3.5y < 3.5y + 5.94 - 3.5y$
$1.1y < 5.94$
$\frac{1.1y}{1.1} < \frac{5.94}{1.1}$
$y < 5.4$

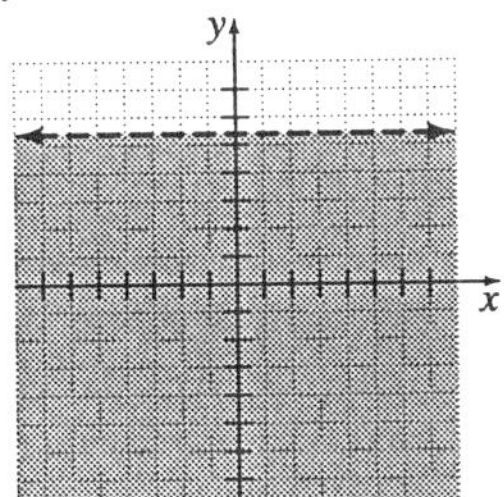

43. $x - y > 9$
$x - y - x > 9 - x$
$-y > -x + 9$
$\frac{-y}{-1} < \frac{-x+9}{-1}$
$y < x - 9$

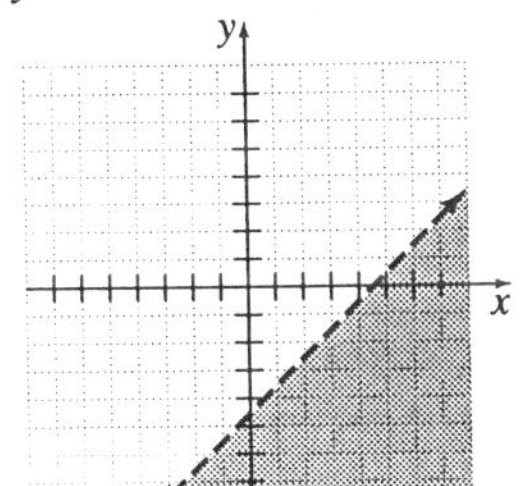

45. $-x - y > 9$
$-x - y + x > 9 + x$
$-y > x + 9$
$\frac{-y}{-1} < \frac{x+9}{-1}$
$y < -x - 9$

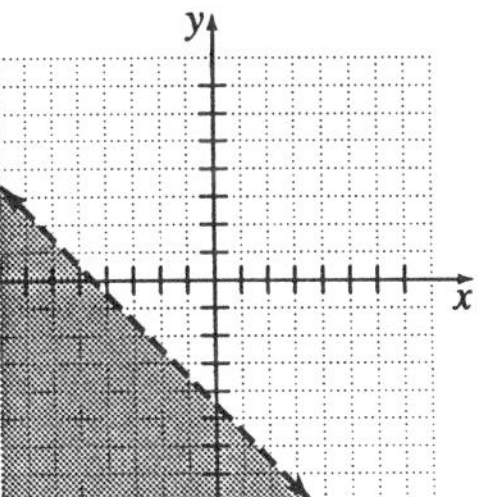

47. $y \leq x$

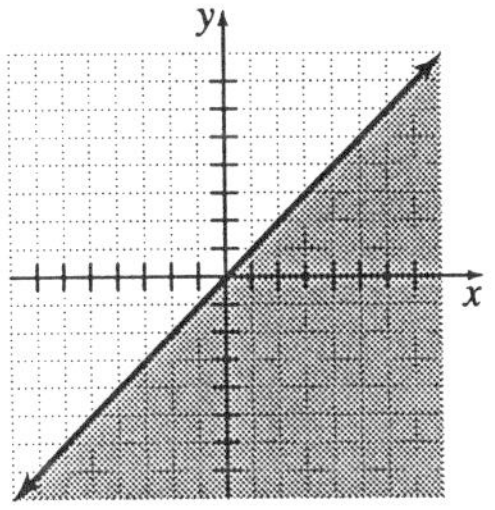

49. Let x = number of village pieces
y = number of angel ornaments
$25x + 12y \leq 225$
Solve for y.
$25x + 12y - 25x \leq 225 - 25x$
$12y \leq -25x + 225$
$\frac{12y}{12} \leq \frac{-25x + 225}{12}$
$y \leq -\frac{25}{12}x + \frac{75}{4}$

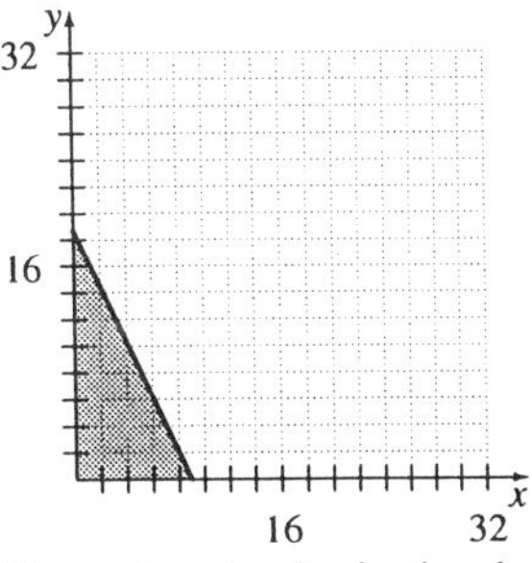

The ordered pairs in the shaded region represent the possible combinations.
Yes, the point (4, 7) is in the shaded region.

Therefore she can buy 4 village pieces and 7 angel ornaments.
No, the point (7, 9) is not in the shaded region. Therefore, she cannot buy 7 village pieces and 9 angel ornaments.

51. Let x = width
y = length
$2x + 2y \le 220$
Solve for y.
$2x + 2y - 2x \le 220 - 2x$
$2y \le -2x + 220$
$\frac{2y}{2} \le \frac{-2x+220}{2}$
$y \le -x + 110$

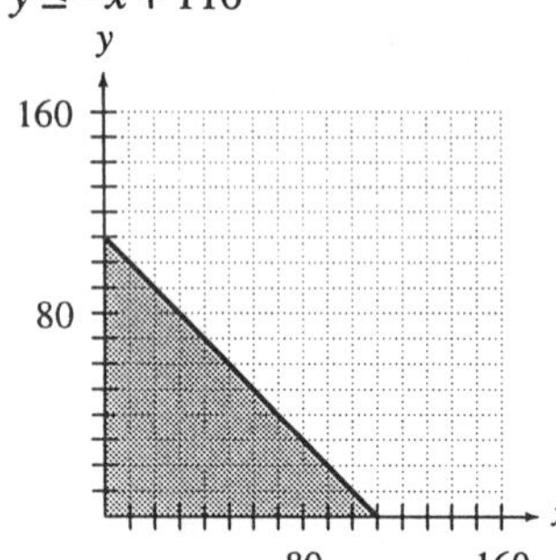

The ordered pairs in the shaded region represent the possible combinations.
No, the point (60, 70) is not in the shaded region. Therefore, she will not have enough for a 70 by 60 foot room.
Yes, the point (40, 60) is in the shaded region. Therefore, she will have enough for a 40 by 60 foot room.

53. Let x = number of days
y = number of items sold
$15x + 12y \ge 400$
Solve for y.
$15x + 12y - 15x \ge 400 - 15x$
$12y \ge -15x + 400$
$\frac{12y}{12} \ge \frac{-15x+400}{12}$
$y \ge -\frac{5}{4}x + \frac{100}{3}$

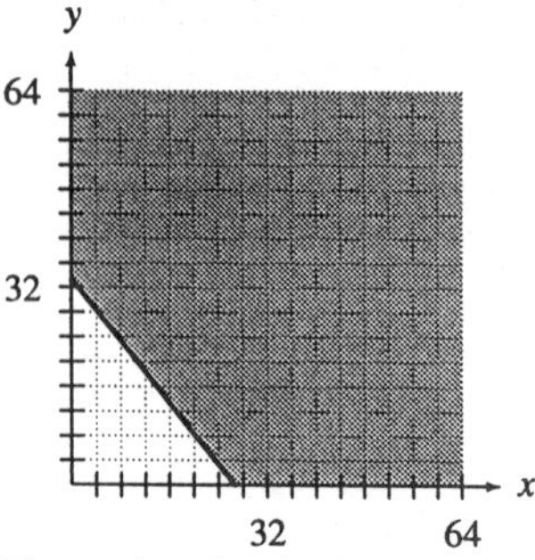

The ordered pairs in the shaded region represent the possible combinations.
No, the point (3, 20) is not in the shaded region. Therefore, he will not earn at least $400 working 3 days and selling 20 items.
Yes, the point (5, 30) is in the shaded region. Therefore, he will earn at least $400 working 5 days and selling 30 items.

55. Let x = number of adults
y = number of children
$4.5x + 2y \ge 250$
Solve for y.
$4.5x + 2y - 4.5x \ge 250 - 4.5x$
$2y \ge -4.5x + 250$
$\frac{2y}{2} \ge \frac{-4.5x+250}{2}$
$y \ge -2.25x + 125$

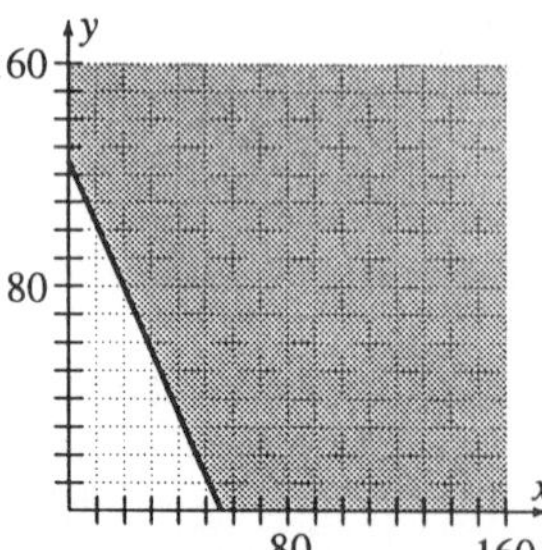

The ordered pairs in the shaded region represent the possible combinations.
No, the point (40, 25) is not in the shaded region. Therefore, they will not make their goal with 40 adults and 25 children.
Yes, the point (42, 45) is in the shaded region. Therefore, they will make their goal with 42 adults and 45 children.

8.3 Experiencing Algebra the Calculator Way

1. $6x-8<2x+32$
$6x-8+8<2x+32+8$
$6x<2x+40$
$6x-2x<2x+40-2x$
$4x<40$
$\frac{4x}{4}<\frac{40}{4}$
$x<10$

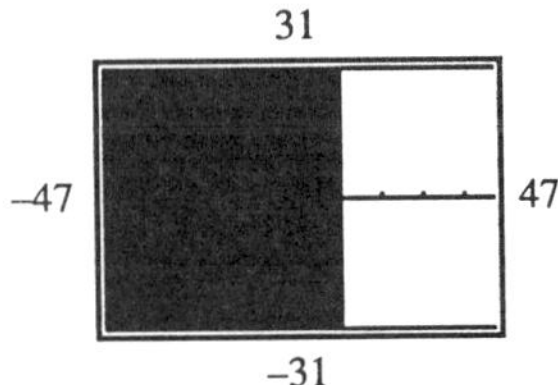

2. $3(x-4)<x-2$
$3x-12<x-2$
$3x-12+12<x-2+12$
$3x<x+10$
$3x-x<x+10-x$
$2x<10$
$\frac{2x}{2}<\frac{10}{2}$
$x<5$

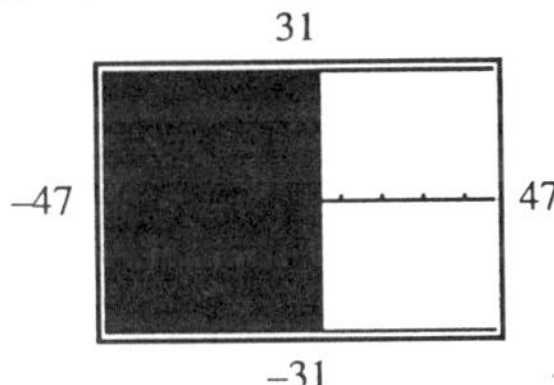

3. $5x>14$
$\frac{5x}{5}>\frac{14}{5}$
$x>\frac{14}{5}$

31
–47 47
–31

4. $2x-3.6\geq 1.8$
$2x-3.6+3.6\geq 1.8+3.6$
$2x\geq 5.4$
$\frac{2x}{2}\geq\frac{5.4}{2}$
$x\geq 2.7$

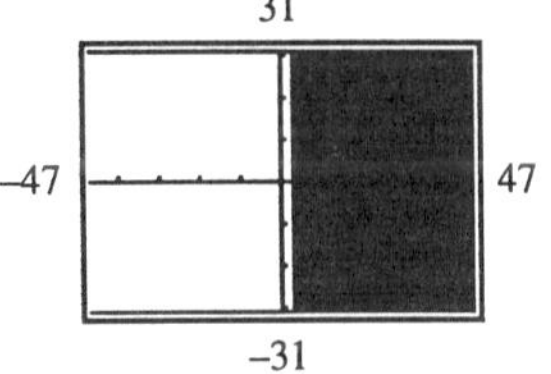

5. $x\leq\pi$

31
–47 47
–31

6. $x+\sqrt{2}\leq 3$
$x+\sqrt{2}-\sqrt{2}\leq 3-\sqrt{2}$
$x\leq 3-\sqrt{2}$

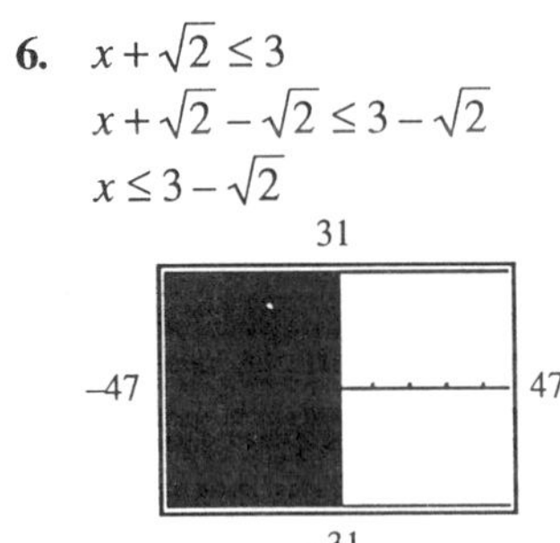

8.4 Experiencing Algebra the Exercise Way

1. Solve the first inequality for y.
$5x-3y>15$
$5x-3y-5x>15-5x$
$-3y>-5x+15$
$\frac{-3y}{-3}<\frac{-5x+15}{-3}$
$y<\frac{5}{3}x-5$
Solve the second inequality for y.
$4x+y>12$
$4x+y-4x>12-4x$
$y>-4x+12$

Find the intersection.

$$\begin{array}{l} 5x - 3y = 15 \\ \underline{12x - 3y = 36} \\ 17x \quad\quad = 51 \end{array}$$

$$\frac{17x}{17} = \frac{51}{17}$$

$x = 3$

$y = -4x + 12$

$y = -4(3) + 12$

$y = 0$

The intersection is (3, 0).

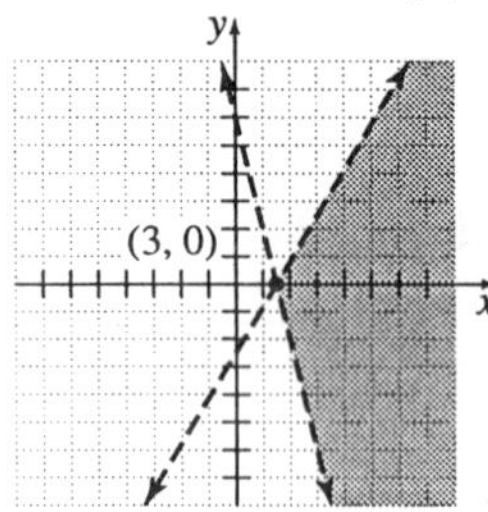

3. Solve the first inequality for y.

$2x + 4y \geq 8$

$2x + 4y - 2x \geq 8 - 2x$

$4y \geq -2x + 8$

$$\frac{4y}{4} \geq \frac{-2x + 8}{4}$$

$$y \geq -\frac{1}{2}x + 2$$

Solve the second inequality for y.

$3x + y \leq -8$

$3x + y - 3x \leq -8 - 3x$

$y \leq -3x - 8$

Find the intersection.

$$\begin{array}{l} 2x + 4y = \ 8 \\ \underline{-12x - 4y = 32} \\ -10x \quad\quad = 40 \end{array}$$

$$\frac{-10x}{-10} = \frac{40}{-10}$$

$x = -4$

$y = -3x - 8$

$y = -3(-4) - 8$

$y = 12 - 8$

$y = 4$

The intersection is (–4, 4).

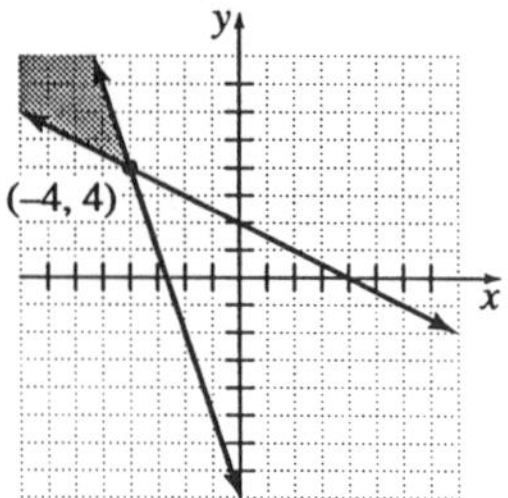

5. $y > 3x - 1$

$y > -2x + 4$

Find the intersection.

$3x - 1 = -2x + 4$

$3x - 1 + 1 = -2x + 4 + 1$

$3x = -2x + 5$

$3x + 2x = -2x + 5 + 2x$

$5x = 5$

$$\frac{5x}{5} = \frac{5}{5}$$

$x = 1$

$y = 3x - 1$

$y = 3(1) - 1$

$y = 2$

The intersection is (1, 2).

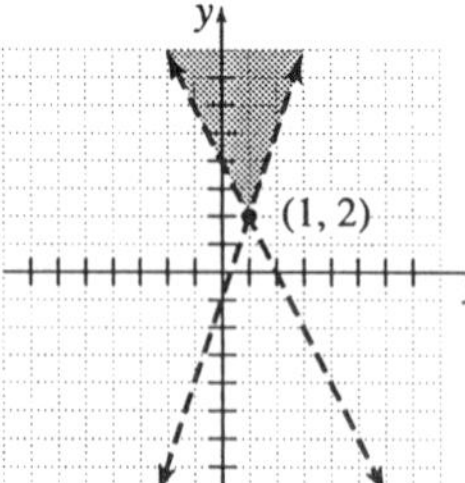

7. $y < \frac{3}{4}x + 1$

$y < -\frac{2}{3}x + 1$

Find the intersection.

$\frac{3}{4}x + 1 = -\frac{2}{3}x + 1$

$\frac{3}{4}x + 1 - 1 = -\frac{2}{3}x + 1 - 1$

$\frac{3}{4}x = -\frac{2}{3}x$

$\frac{3}{4}x + \frac{2}{3}x = -\frac{2}{3}x + \frac{2}{3}x$

$\frac{17}{12}x = 0$

$\frac{12}{17}\left(\frac{17}{12}x\right) = \frac{12}{17}(0)$

$x = 0$

$y = \frac{3}{4}x + 1$

$y = \frac{3}{4}(0) + 1$

$y = 1$

The intersection is (0, 1).

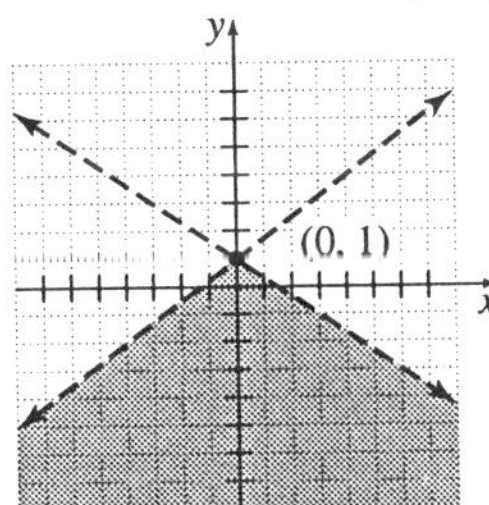

9. Solve the first inequality for y.

$3x + 2y < 12$

$3x + 2y - 3x < 12 - 3x$

$2y < -3x + 12$

$\frac{2y}{2} < \frac{-3x + 12}{2}$

$y < -\frac{3}{2}x + 6$

Solve the second inequality for y.

$y - 2 < 1$

$y - 2 + 2 < 1 + 2$

$y < 3$

Find the intersection.

$y = 3$

$3x + 2y = 12$

$3x + 2(3) = 12$

$3x + 6 - 6 = 12 - 6$

$3x = 6$

$\frac{3x}{3} = \frac{6}{3}$

$x = 2$

The intersection is (2, 3).

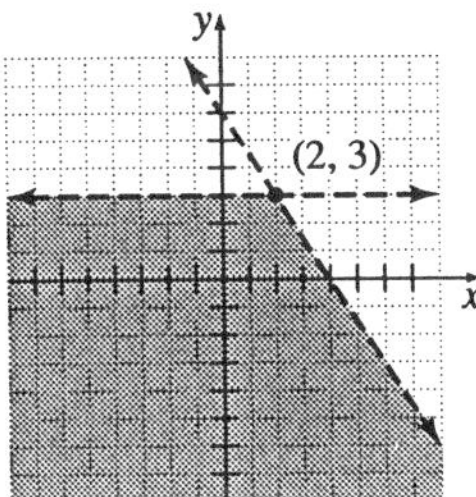

11. Solve the first inequality for y.

$3y - 2 > 2y + 2$

$3y - 2 + 2 > 2y + 2 + 2$

$3y > 2y + 4$

$3y - 2y > 2y + 4 - 2y$

$y > 4$

Solve the second inequality for x.

$x - 1 < 0$

$x - 1 + 1 < 0 + 1$

$x < 1$

Find the intersection.

$x = 1$

$y = 4$

The intersection is (1, 4).

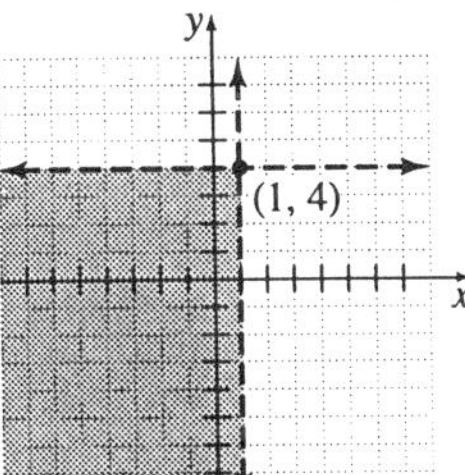

13. $y \le 3x - 5$

Solve the second inequality for *y*.

$x > 4 - 2y$

$x - 4 > 4 - 2y - 4$

$x - 4 > -2y$

$\frac{x-4}{-2} < \frac{-2y}{-2}$

$-\frac{1}{2}x + 2 < y$

or

$y > -\frac{1}{2}x + 2$

Find the intersection.

$3x - 5 = -\frac{1}{2}x + 2$

$3x - 5 + 5 = -\frac{1}{2}x + 2 + 5$

$3x = -\frac{1}{2}x + 7$

$3x + \frac{1}{2}x = -\frac{1}{2}x + 7 + \frac{1}{2}x$

$3.5x = 7$

$\frac{3.5x}{3.5} = \frac{7}{3.5}$

$x = 2$

$y = 3x - 5$

$y = 3(2) - 5$

$y = 1$

The intersection is (2, 1).

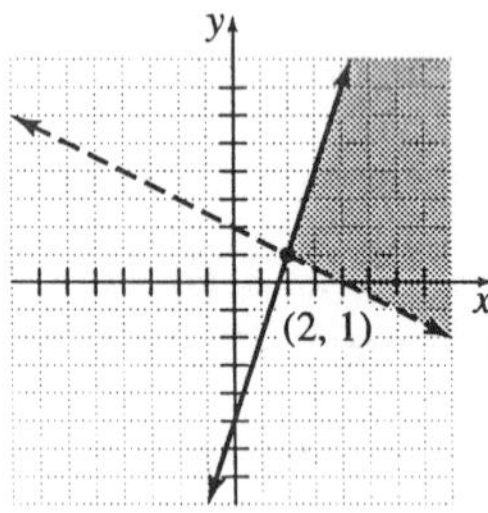

15. $y \le \frac{1}{2}x + 3$

Solve the second inequality for *y*.

$x + 2y < 6$

$x + 2y - x < 6 - x$

$2y < -x + 6$

$\frac{2y}{2} < \frac{-x+6}{2}$

$y < -\frac{1}{2}x + 3$

Find the intersection.

$\frac{1}{2}x + 3 = -\frac{1}{2}x + 3$

$\frac{1}{2}x + 3 - 3 = -\frac{1}{2}x + 3 - 3$

$\frac{1}{2}x = -\frac{1}{2}x$

$\frac{1}{2}x + \frac{1}{2}x = -\frac{1}{2}x + \frac{1}{2}x$

$x = 0$

$y = \frac{1}{2}x + 3$

$y = \frac{1}{2}(0) + 3$

$y = 3$

The intersection is (0, 3).

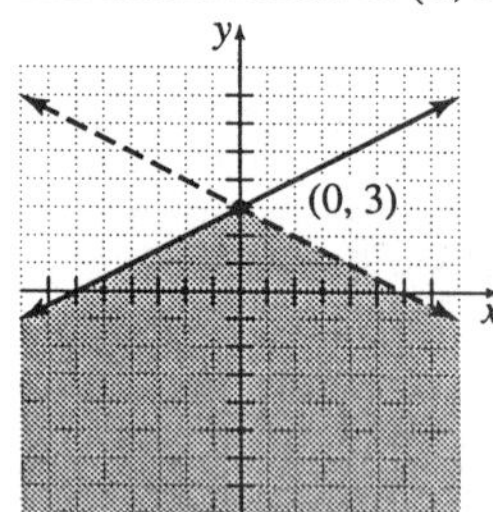

17. $y \le 10 - x$

Solve the second inequality for *y*.

$y > \frac{1}{2}x + 3$

$y > 0$

$x < 3$

Find the intersections and label the graph.

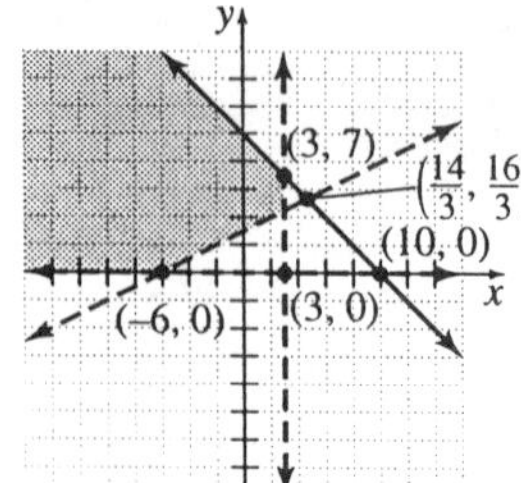

19. Solve the first inequality for y.

$x-2y>7$

$x-2y-x>7-x$

$-2y>-x+7$

$\frac{-2y}{-2}<\frac{-x+7}{-2}$

$y<\frac{1}{2}x-\frac{7}{2}$

Solve the second inequality for y.

$5x+10y<-3$

$5x+10y-5x<-3-5x$

$10y<-5x-3$

$\frac{10y}{10}<\frac{-5x-3}{10}$

$y<-\frac{1}{2}x-\frac{3}{10}$

$\text{Y1}=\frac{1}{2}x-\frac{7}{2}$

$\text{Y2}=-\frac{1}{2}-\frac{3}{10}$

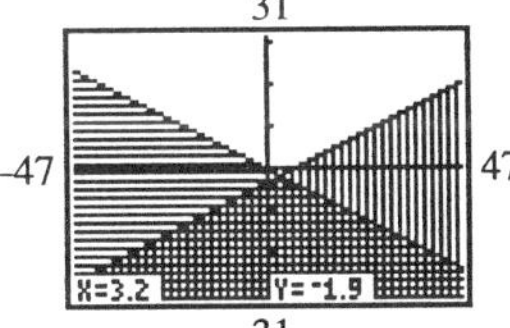

21. Solve the first inequality for y.

$3x-5y\geq 5$

$3x-5y-3x\geq 5-3x$

$-5y\geq -3x+5$

$\frac{-5y}{-5}\leq\frac{-3x+5}{-5}$

$y=\frac{3}{5}x-1$

Solve the second inequality for y.

$10y+15>2$

$10y+15-15>2-15$

$10y>-13$

$\frac{10y}{10}>\frac{-13}{10}$

$y>-1.3$

$\text{Y1}=\frac{3}{5}x-1$

$\text{Y2}=-1.3$

31
–47 47
–31

23. $y\geq x-3$

Solve the second inequality for y.

$10x-5y\geq 14$

$10x-5y-10x\geq 14-10x$

$-5y\geq -10x+14$

$\frac{-5y}{-5}\leq\frac{-10x+14}{-5}$

$y\leq 2x-\frac{14}{5}$

$\text{Y1}=x-3$

$\text{Y2}=2x-\frac{14}{5}$

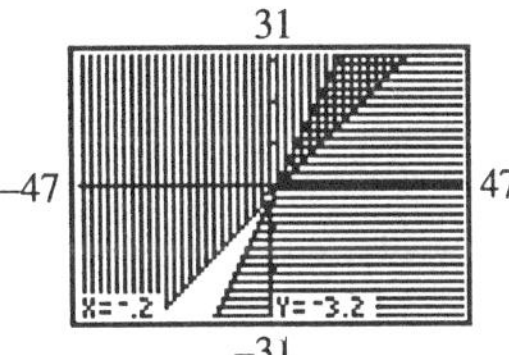

25. Solve the first inequality for y.

$2x+9y<102$

$2x+9y-2x<102-2x$

$9y<-2x+102$

$\frac{9y}{9}<\frac{-2x+102}{9}$

$y<-\frac{2}{9}x+\frac{34}{3}$

Solve the second inequality for y.

$5x-11y<-147$

$5x-11y-5x<-147-5x$

$-11y<-5x-147$

$\frac{-11y}{-11}>\frac{-5x-147}{-11}$

$y>\frac{5}{11}x+\frac{147}{11}$

$$\text{Y1} = -\frac{2}{9}x + \frac{34}{3}$$

$$\text{Y2} = \frac{5}{11}x + \frac{147}{11}$$

31

−47 47

X=-3 Y=12

−31

27. Solve the first inequality for y.

$$3.2x + 4.2y > 368$$

$$3.2x + 4.2y - 3.2x > 368 - 3.2x$$

$$4.2y > -3.2x + 368$$

$$\frac{4.2y}{4.2} > \frac{-3.2x + 368}{4.2}$$

$$y = -\frac{16}{21}x + \frac{1840}{21}$$

Solve the second inequality for y.

$$4.4x - 2.1y > 128$$

$$4.4x - 2.1y - 4.4x > 128 - 4.4x$$

$$-2.1y > -4.4x + 128$$

$$\frac{-2.1y}{-2.1} < \frac{-4.4x + 128}{-2.1}$$

$$y < \frac{44}{21}x - \frac{1280}{21}$$

$$\text{Y1} = -\frac{16}{21}x + \frac{1840}{21}$$

$$\text{Y2} = \frac{44}{21}x - \frac{1280}{21}$$

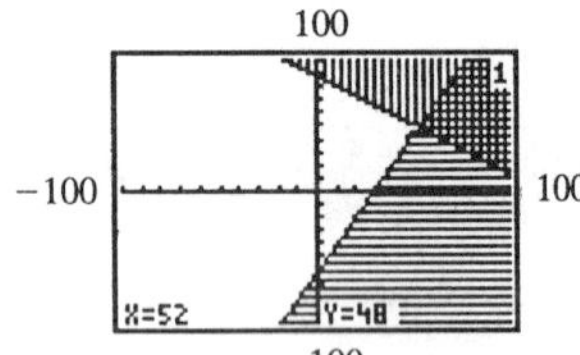

29. Solve the first inequality for y.

$$0.05x + 0.10y < 0.75$$

$$0.05x + 0.10y - 0.05x < 0.75 - 0.05x$$

$$0.10y < -0.05x + 0.75$$

$$\frac{0.10y}{0.10} < \frac{-0.05x + 0.75}{0.10}$$

$$y < -0.5x + 7.5$$

Solve the second inequality for y.

$$x + y > 11$$

$$x + y - x > 11 - x$$

$$y > -x + 11$$

$$\text{Y1} = -0.5x + 7.5$$

$$\text{Y2} = -x + 11$$

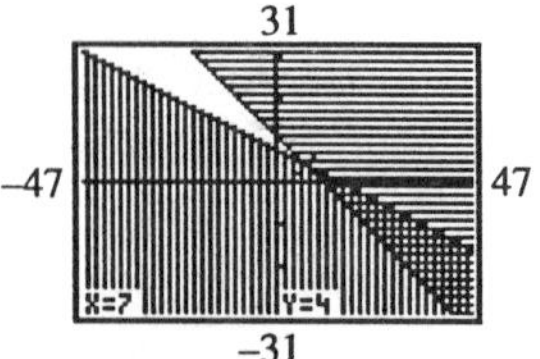

31. $2x - 3y > -6$

$-2x + 3y > -6$

or

$$y < \frac{2}{3}x + 2$$

$$y > \frac{2}{3}x - 2$$

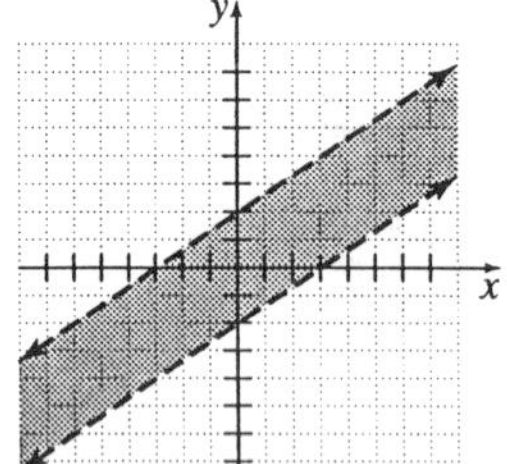

The solution is all ordered pairs contained in the shaded region.

33. $y > -3x + 2$

$y \leq -3x$

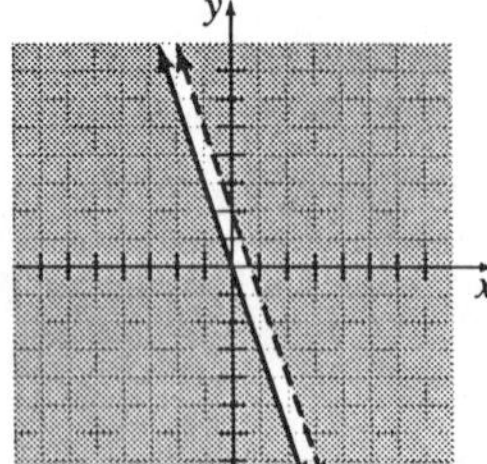

There are no ordered pairs that satisfy this system of linear inequalities.

35. $x + 4 > 0$
$3x + 2 < -4$
or
$x > -4$
$x < -2$

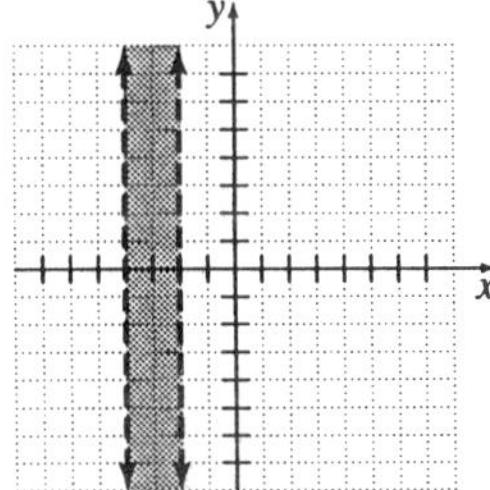

The solution is all ordered pairs contained in the shaded region.

37. $y < 4 - 2x$
$2x + y \le -1$
or
$y < -2x + 4$
$y \le -2x - 1$

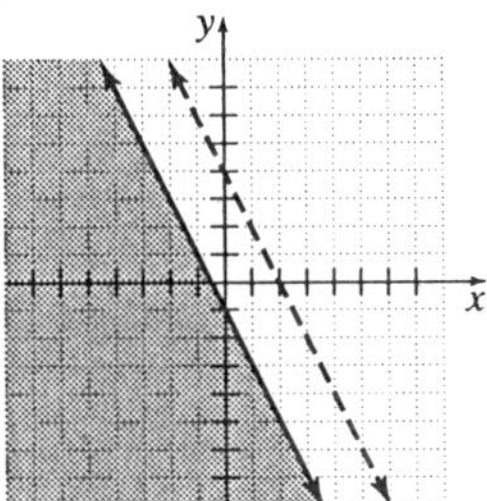

The solution is all ordered pairs below and including the boundary line $2x + y = -1$.

39. $3y \le x + 6$
$3y - x \ge 6$
or
$y \le \frac{1}{3}x + 2$
$y \ge \frac{1}{3}x + 2$

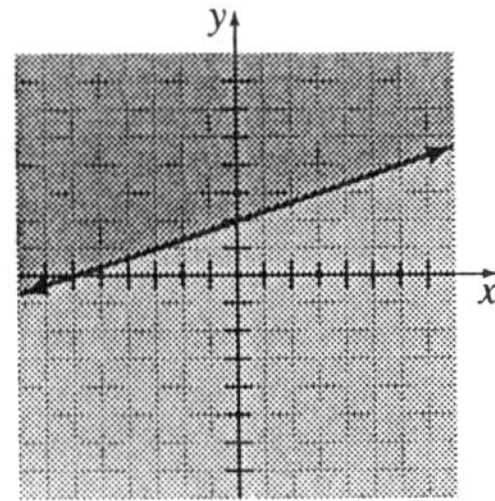

The solution is all ordered pairs on the line $3y = x + 6$ or $y = \frac{1}{3}x + 2$.

41. $y < 7$
$2y - 3 \le y + 4$
or
$y < 7$
$y \le 7$

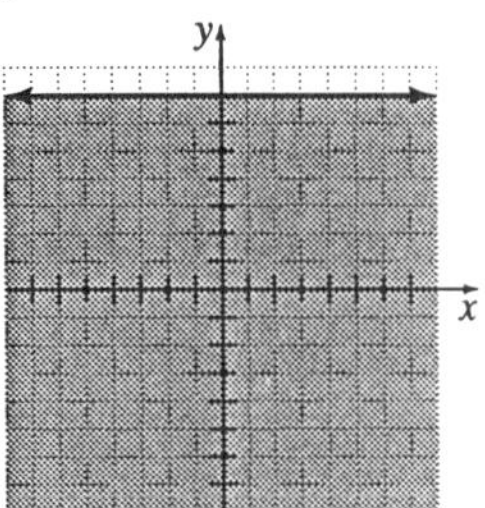

The solution is all ordered pairs below the line $y = 7$.

43. Let x = width
y = length
$2x + 2y \le 100$
$y \ge x + 10$
$x \ge 0$
$y \ge 0$
Solve the first inequality for y.
$y \le -x + 50$

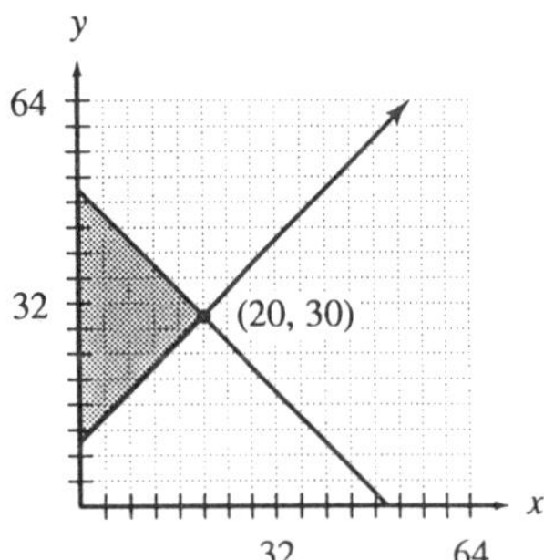

Possible answer:
The dimensions could be 10 feet by 30 feet.

45. Let x = amount invested at 6%
y = amount invested at 8%
$x + y \le 3000$
$0.06x + 0.08y \ge 200$
$x \ge 0$
$y \ge 0$
Solve the first two inequalities for y.
$y \le -x + 3000$
$y \ge -0.75x + 2500$

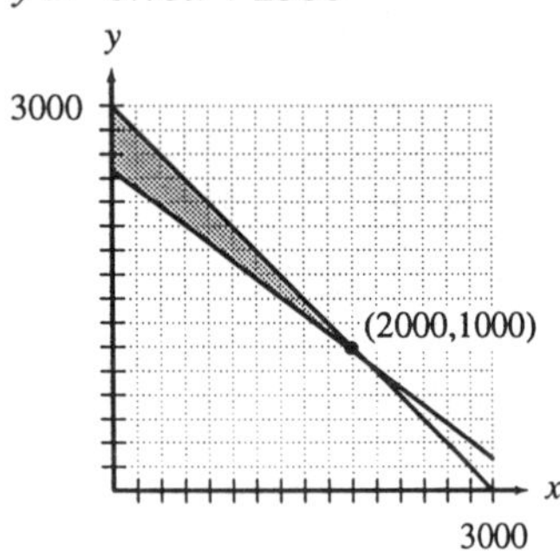

Possible answer:
The investments could be \$1000 at 6% and \$2000 at 8%.

47. Let x = number of hours at first job
y = number of hours at second job
$x + y \le 20$
$6.5x + 8.25y \ge 150$
$x \ge 0$
$y \ge 0$
Solve the first two inequalities for y.
$y \le -x + 20$
$$y \ge -\frac{6.5}{8.25}x + \frac{150}{8.25}$$

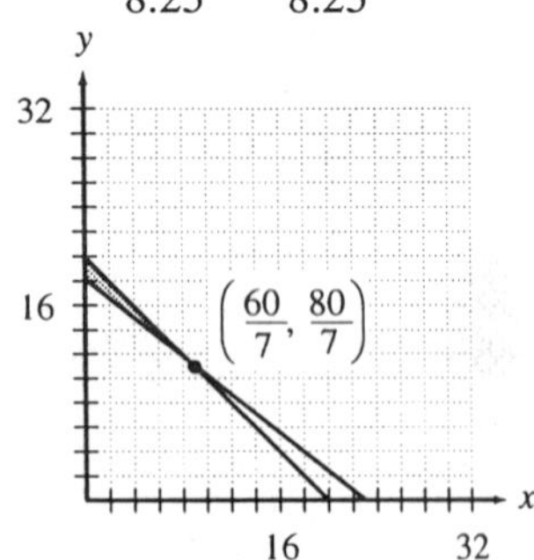

Possible answer:
He could work 6 hours on the first job and 14 hours on the second job.
No, the point (7, 10) is not in the shaded region. Therefore he could not work 7 hours on the first job and 10 hours on the second job.
Yes, the point (5, 15) is in the shaded region. Therefore, he could work 5 hours on the first job and 15 hours on the second job.

49. Let x = number of servings of lasagna
y = number of servings of veal
$1.75x + 2.25y \le 200$
$x \ge 50$
$y \ge 25$
Solve the first inequality for y.
$$y \le -\frac{7}{9}x + \frac{800}{9}$$

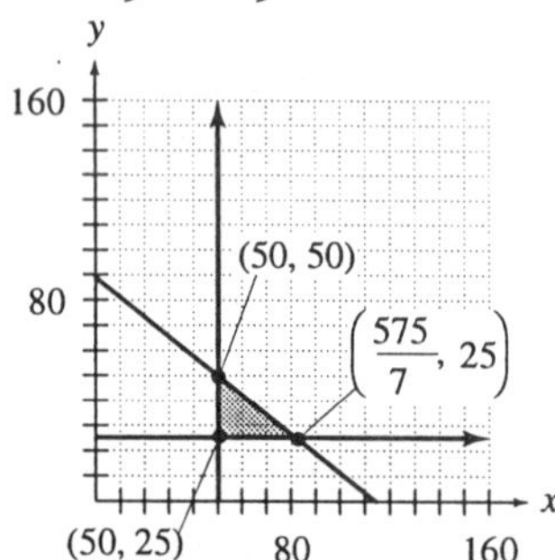

Possible answer:
A combination could be 75 servings of lasagna and 30 servings of veal.
Yes, the point (60, 35) is in the shaded region. Therefore, she could prepare 60 servings of lasagna and 35 servings of veal parmigiana.
No, the point (60, 50) is not in the shaded region. Therefore, she should not prepare 60 servings of lasagna and 50 servings of veal parmigiana.

8.4 Experiencing Algebra the Calculator Way

1. $3x + 4y > 12$
$2x + 5 < x + 7$
Solve the first inequality for y.
$3x + 4y - 3x > 12 - 3x$
$4y > -3x + 12$
$$\frac{4y}{4} > \frac{-3x + 12}{4}$$
$$y > -\frac{3}{4}x + 3$$

Solve the second inequality for y.
$2x+5<x+7$
$2x-x<7-5$
$x<2$
$\text{Y1}=-\dfrac{3}{4}x+3$

31

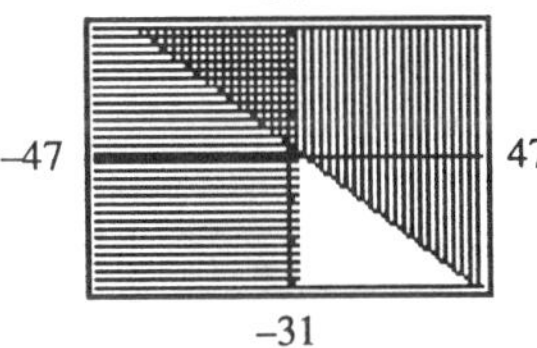

–47 47

–31

2. $x-4y\geq 8$
$3x-1\leq x-4$
Solve the first inequality for y.
$x-4y\geq 8$
$x-4y-x\geq 8-x$
$-4y\geq -x+8$
$\dfrac{-4y}{-4}\leq\dfrac{-x+8}{-4}$
$y\leq\dfrac{1}{4}x-2$
Solve the second inequality for x.
$3x-1\leq x-4$
$3x-x\leq -4+1$
$2x\leq -3$
$\dfrac{2x}{2}\leq\dfrac{-3}{2}$
$x\leq -\dfrac{3}{2}$
$\text{Y1}=\dfrac{1}{4}x-2$

31

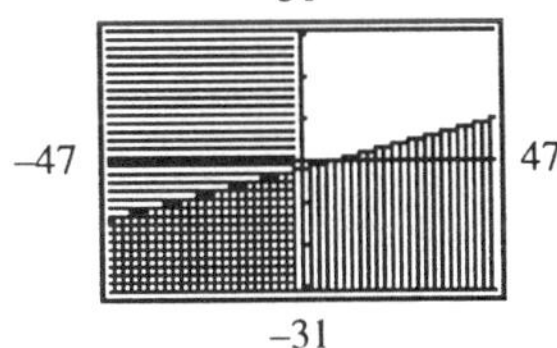

–47 47

–31

3. $x>3$
$x-3<2$
Solve the second inequality for x.
$x-3<2$
$x-3+3<2+3$
$x<5$

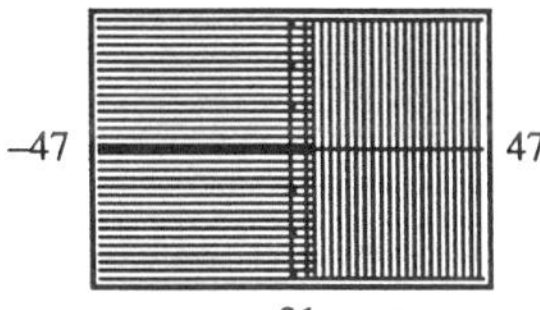

–47 47

–31

4. $2x+3>x$
$4y+3\geq -5$
Solve the first inequality for x and the second inequality for y.
$x>-3$
$y\geq -2$
$\text{Y1}=-2$

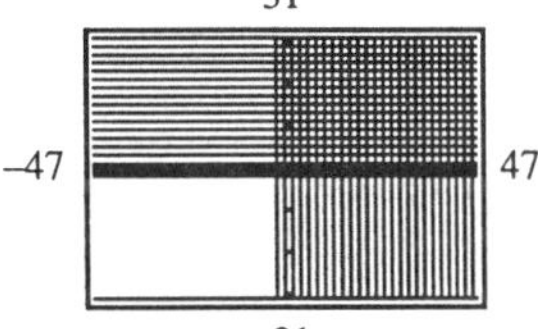

–47 47

–31

5. $4x-7<4$
$3x-9\leq 2-x$
Solve the first inequality for x and the second inequality for x.
$x<\dfrac{11}{4}$
$x\leq\dfrac{11}{4}$

31

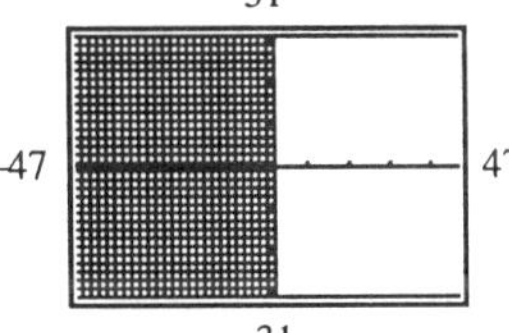

–47 47

–31

6. $3x+4\leq 2x+5$
$5x-9\geq 4(x-2)$
Solve the first and second inequalities for x.
$x\leq 1$
$x\geq 1$

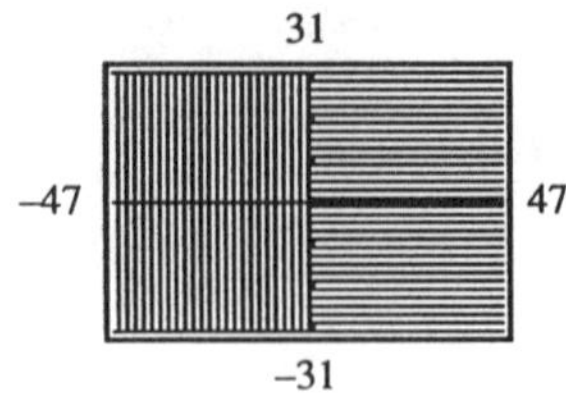

Chapter 8 Review

Reflections

1.–7. Answers will vary.

Exercises

1. $3x^2 - 2x + 1 < 0$ is nonlinear because the variable x is squared.

2. $5x - 4 > x + 7$ is linear.

3. $\frac{2}{3}x + \frac{4}{5} \le \frac{7}{15}$ is linear.

4. $\sqrt{x} - 3 \ge 6$ is nonlinear because the radical expression contains a variable.

5. $x - 3 \ge 6$ is linear.

6. $\frac{1}{x} - 3x \ge 6$ is non linear because the variable x appears in the denominator.

7. $1.5z - 12.6 < 14.7z$ is linear.

8. $3(a + 2) < 15a - (2a + 1)$ is linear because it simplifies to $3a + 6 < 13a - 1$.

9. $x < 3$

0 1 2 3 4 5 6 7 8

The solution set is $(-\infty, 3)$.

10. $x > -2$

−8 −7 −6 −5 −4 −3 −2 −1 0

The solution set is $(-2, \infty)$.

11. $x \le -5$

−8 −7 −6 −5 −4 −3 −2 −1 0

The solution set is $(-\infty, -5]$.

12. $x \ge -3.5$

−3.5

−8 −7 −6 −5 −4 −3 −2 −1 0

The solution set is $[-3.5, \infty)$.

13. $-2 < a < 4$

−4 −3 −2 −1 0 1 2 3 4

The solution set is $(-2, 4)$.

14. $-1 < b \le 0$

−4 −3 −2 −1 0 1 2 3 4

The solution set is $(-1, 0]$.

15. $3 \le c \le 5.5$

5.5

0 1 2 3 4 5 6 7 8

The solution set is $[3, 5.5]$.

16. $2\frac{1}{2} < d < 8$

$2\frac{1}{2}$

0 1 2 3 4 5 6 7 8

The solution set is $\left(2\frac{1}{2}, 8\right)$.

17. $-2.3 \le f \le -1\frac{1}{3}$

−2.3 $-1\frac{1}{3}$

−4 −3 −2 −1 0 1 2 3 4

The solution set is $\left[-2.3, -1\frac{1}{3}\right]$.

18. Let x = number of minutes
$(30 + 25)x \ge 300$

19. Let x = cost of base material
$x + 2x + \frac{1}{4}(2x) < 200,000$

20. Let x = number of pieces
$60 + 0.18x < 0.32x$

21.

x	$4x+7$	$2x-5$	
-7	$4(-7)+7=-21$	$2(-7)-5=-19$	$-21<-19$
-6	$4(-6)+7=-17$	$2(-6)-5=-17$	$-17 \not< -17$
-5	$4(-5)+7=-13$	$2(-5)-5=-15$	$-13 \not< -15$

The integers less than –6 are solutions of the inequality.

22.

x	$2.4x-9.6$	4.8	
5	$2.4(5)-9.6=2.4$	4.8	$2.4 \not> 4.8$
6	$2.4(6)-9.6=4.8$	4.8	$4.8 \not> 4.8$
7	$2.4(7)-9.6=7.2$	4.8	$7.2>4.8$

The integers greater than 6 are solutions of the inequality.

23.

x	$\frac{3}{5}x-\frac{7}{10}$	$\frac{1}{5}x+\frac{1}{2}$	
2	$\frac{3}{5}(2)-\frac{7}{10}=\frac{1}{2}$	$\frac{1}{5}(2)+\frac{1}{2}=\frac{9}{10}$	$\frac{1}{2}\le\frac{9}{10}$
3	$\frac{3}{5}(3)-\frac{7}{10}=\frac{11}{10}$	$\frac{1}{5}(3)+\frac{1}{2}=\frac{11}{10}$	$\frac{11}{10}\le\frac{11}{10}$
4	$\frac{3}{5}(4)-\frac{7}{10}=\frac{17}{10}$	$\frac{1}{5}(4)+\frac{1}{2}=\frac{13}{10}$	$\frac{17}{10}\not\le\frac{13}{10}$

The integers less than or equal to 3 are solutions of the inequality.

24.

x	$\frac{1}{2}x-2$	$-\frac{1}{3}x-\frac{11}{3}$	
-3	$\frac{1}{2}(-3)-2=\frac{7}{2}$	$-\frac{1}{3}(-3)-\frac{11}{3}=-\frac{8}{3}$	$-3 \not\ge -\frac{8}{3}$
-2	$\frac{1}{2}(-2)-2=-3$	$-\frac{1}{3}(-2)-\frac{11}{3}=-3$	$-3\ge-3$
-1	$\frac{1}{2}(-1)-2=-\frac{5}{2}$	$-\frac{1}{3}(-1)-\frac{11}{3}=-\frac{10}{3}$	$-\frac{5}{2}\ge-\frac{10}{3}$

The integers greater than or equal to –2 are solutions of the inequality.

25. $2x - 2 > -x + 4$
$Y1 = 2x - 2$
$Y2 = -x + 4$

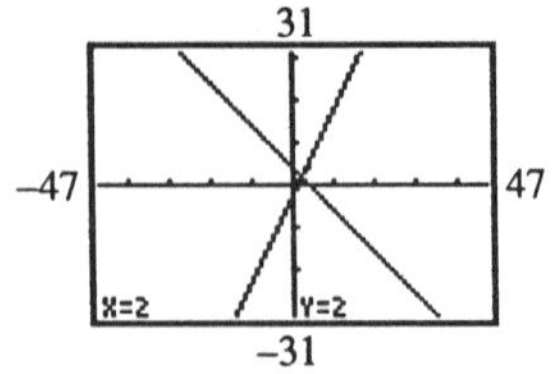

The intersection is (2, 2).
$x > 2$, $(2, \infty)$

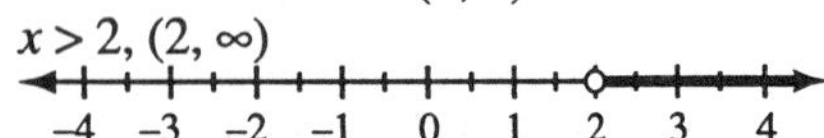

26. $1.2x + 0.72 \le -2.1x + 8.64$
$Y1 = 1.2x + 0.72$
$Y2 = -2.1x + 8.64$

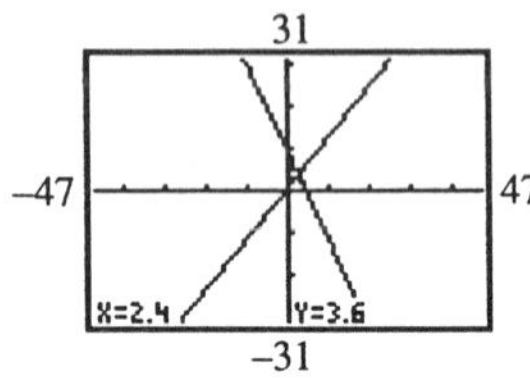

The intersection is (2.4, 3.6)
$x \le 2.4$
$(-\infty, 2.4]$

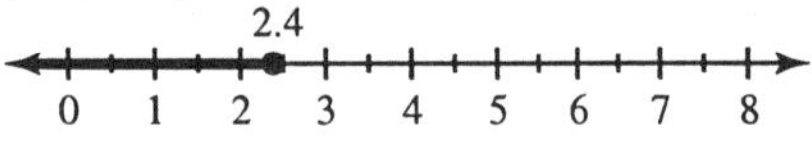

27. $(x + 6) - 3(x + 1) < (2x + 5) - 2(2x + 3)$
$Y1 = (x + 6) - 3(x + 1)$
$Y2 = (2x + 5) - 2(2x + 3)$

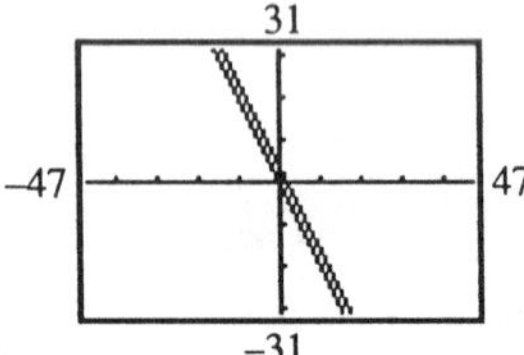

The lines are parallel and the graph of Y1 is always above the graph of Y2. Therefore, there is no solution.

28. $(x + 3) + (x + 1) \ge 3(x + 1) - (x - 1)$
$Y1 = (x + 3) + (x + 1)$
$Y2 = 3(x + 1) - (x - 1)$

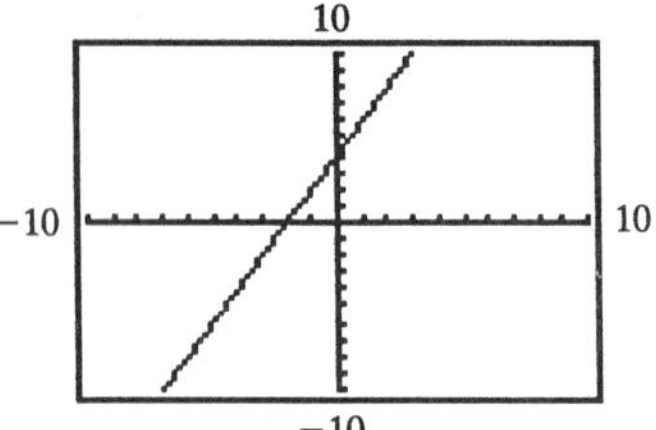

The lines are coinciding. Since the inequality includes the equal symbol the solution is all real numbers.
$(-\infty, \infty)$

29. $412 + x > 671$
$412 + x - 412 > 671 - 412$
$x > 259$
The solution set is $(259, \infty)$.

30. $y - \frac{7}{13} < \frac{11}{39}$
$y - \frac{7}{13} + \frac{7}{13} < \frac{11}{39} + \frac{7}{13}$
$y < \frac{11}{39} + \frac{21}{39}$
$y < \frac{32}{39}$
The solution set is $\left(-\infty, \frac{32}{39}\right)$.

31. $3x + 7 < 4x + 21$
$3x + 7 - 3x < 4x + 21 - 3x$
$7 < x + 21$
$7 - 21 < x + 21 - 21$
$-14 < x$
$x > -14$
The solution set is $(-14, \infty)$.

32. $14 + 2x < 2x$
$14 + 2x - 2x < 2x - 2x$
$14 < 0$
This is a contradiction.
There is no solution.

33. $8.7x + 4.33 \le -2.4x - 33.41$
$8.7x + 4.33 - 4.33 \le -2.4x - 3.41 - 4.33$
$8.7x \le -2.4x - 37.74$
$8.7x + 2.4x \le -2.4x - 37.74 + 2.4x$
$11.1x \le -37.74$
$\frac{11.1x}{11.1} \le \frac{-37.74}{11.1}$
$x \le -3.4$
The solution set is $(-\infty, -3.4]$.

34. $6.8z - 9.52 \ge 0$
$6.8z - 9.52 + 9.52 \ge 0 + 9.52$
$6.8z \ge 9.52$
$\frac{6.8z}{6.8} \ge \frac{9.52}{6.8}$
$z \ge 1.4$
The solution set is $[1.4, \infty)$.

35. $2(x + 5) - (x + 6) < 2(x + 2)$
$2x + 10 - x - 6 < 2x + 4$
$x + 4 < 2x + 4$
$x + 4 - 4 < 2x + 4 - 4$
$x < 2x$
$x - x < 2x - x$
$x < x$
$0 < x$
$x > 0$
The solution set is $(0, \infty)$.

36. $3(x - 4) + 2(x + 1) > 5x + 10$
$3x - 12 + 2x + 2 > 5x + 10$
$5x - 10 > 5x + 10$
$5x - 10 - 5x > 5x + 10 - 5x$
$-10 > 10$
This is a contradiction.
There is no solution.

37. Let x = number of miles driven
$49.95 + 0.18x \le 150$
$49.95 + 0.18x - 49.95 \le 150 - 49.95$
$0.18x \le 100.05$
$\frac{0.18x}{0.18} \le \frac{100.05}{0.18}$
$x \le 555.8\overline{3}$
The number of miles driven should be less than or equal to 555 miles.

38. Let x = amount of sixth month's sales
$\frac{2100 + 1300 + 1650 + 1250 + 1725 + x}{6} > 1500$
$\frac{8025 + x}{6} > 1500$
$6\left(\frac{8025 + x}{6}\right) > 6(1500)$
$8025 + x > 9000$
$8025 + x - 8025 > 9000 - 8025$
$x > 975$
His sales should be greater than \$975.

39. Let x = width
$2x + 2(x + 4) \le 40$
$2x + 2x + 8 \le 40$
$4x + 8 \le 40$
$4x + 8 - 8 \le 40 - 8$
$4x \le 32$
$\frac{4x}{4} \le \frac{32}{4}$
$x \le 8$
The width can be no more than 8 feet.

40. $x + 2y < 12$ is linear.
$x + 2y < 12$ is in standard form.
$a = 1, b = 2, c = 12$

41. $y < \frac{2}{3}x + \frac{5}{9}$ is linear.
$9(y) < 9\left(\frac{2}{3}x + \frac{5}{9}\right)$
$9y < 6x + 5$
$-5 < 6x - 9y$
$6x - 9y > -5$ is in standard form.
$a = 6, b = -9, c = -5$.

42. $x^2 + y^2 \ge 1$ is nonlinear because the x-variable and y-variable are squared.

43. $0.3x + 2.9 > 1.4y$ is linear.
$0.3x - 1.4y > -2.9$
$10(0.3x - 1.4y) > 10(-2.9)$
$3x - 14y > -29$ is in standard form.
$a = 3, b = -14, c = -29$

44. $y \ge \sqrt{x} - 1.44$ is nonlinear because the radical expression contains a variable.

45. $y < x^2 + 9$ is nonlinear because the x-variable is squared.

46. $12x + 6y < 48$
$12x + 16y - 12x < 48 - 12x$
$6y < -12x + 48$
$\frac{6}{6} < \frac{-12x + 48}{6}$
$y < -2x + 8$

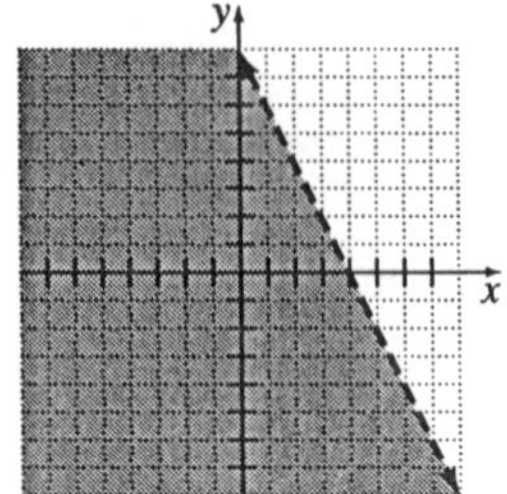

47. $y > \frac{3}{5}x - 6$

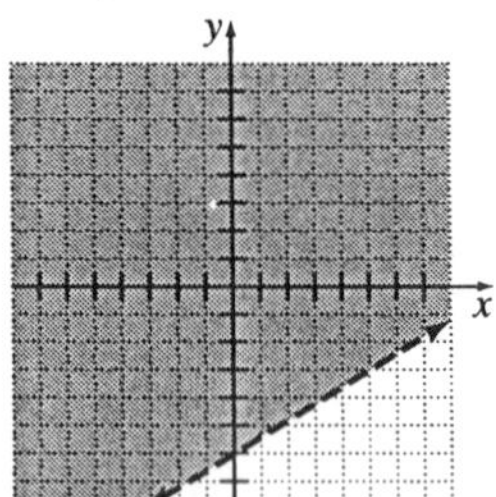

48. $4y \le x + 12$
$\frac{4y}{4} \le \frac{x + 12}{4}$
$y \le \frac{1}{4}x + 3$

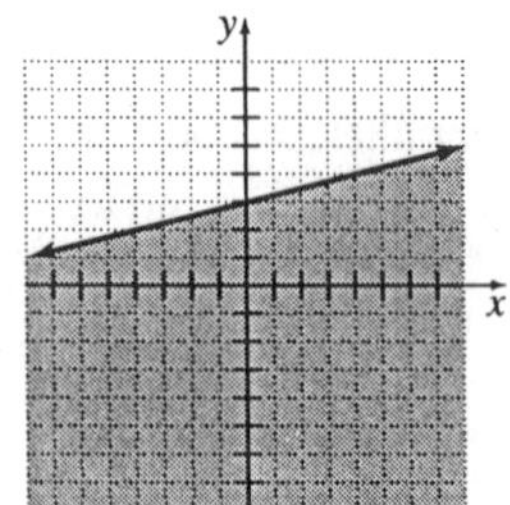

49. $y + 9 \ge 12$
$y + 9 - 9 \ge 12 - 9$
$y \ge 3$

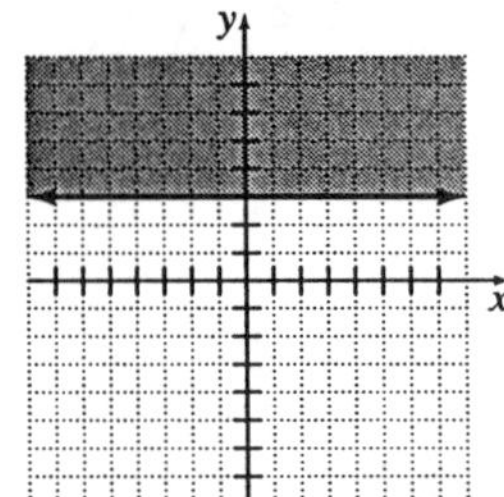

50. $5y - 2 < y - 10$
$5y - 2 + 2 < y - 10 + 2$
$5y < y - 8$
$5y - y < y - 8 - y$
$4y < -8$
$\frac{4y}{4} < \frac{-8}{4}$
$y < -2$

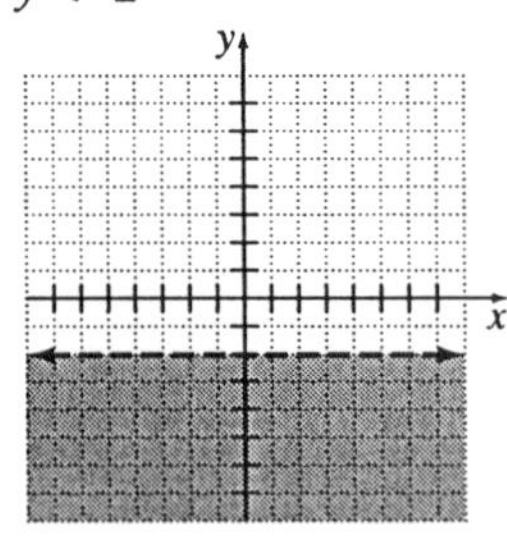

51. $2x + 16 > x + 18$
$2x + 16 - 16 > x + 18 - 16$
$2x > x + 2$
$2x - x > x + 2 - x$
$x > 2$

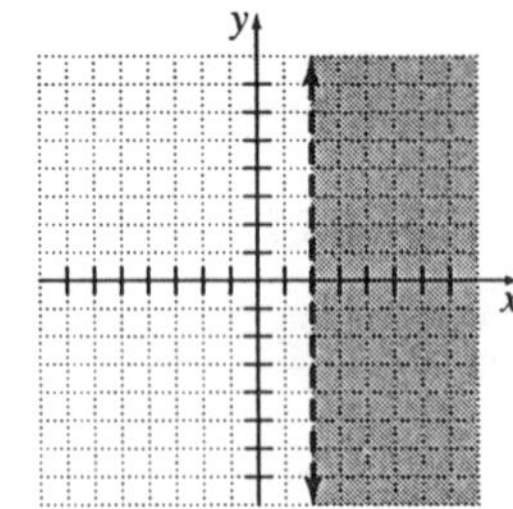

52. $8x - 12y > 24$
$8x - 12y - 8x > 24 - 8x$
$-12y > -8x + 24$
$\dfrac{-12y}{-12} < \dfrac{-8x + 24}{-12}$
$y < \dfrac{2}{3}x - 2$

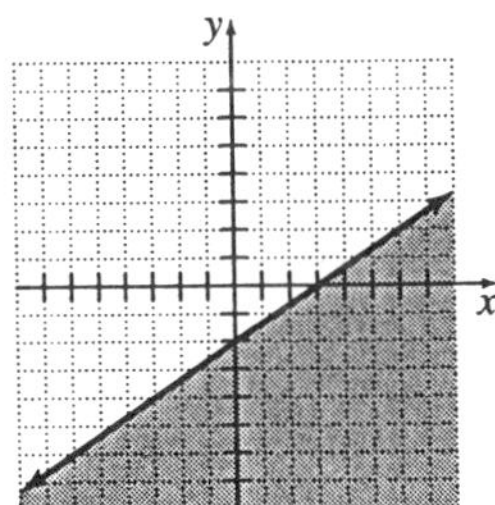

53. $4.4x + 1.1y \geq 12.1$
$4.4x + 1.1y - 4.4x \geq 12.1 - 4.4x$
$1.1y \geq -4.4x + 12.1$
$\dfrac{1.1y}{1.1} \geq \dfrac{-4.4x + 12.1}{1.1}$
$y \geq -4x + 11$

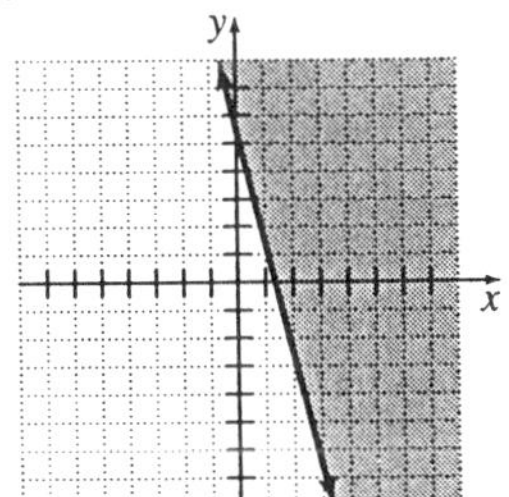

54. $7.4x - 14.8y \leq 29.6$
$7.4x - 14.8y - 7.4x \leq 29.6 - 7.4x$
$-14.8y \leq -7.4x + 29.6$
$\dfrac{-14.8y}{-14.8} \geq \dfrac{-7.4x + 29.6}{-14.8}$
$y \geq 0.5x - 2$

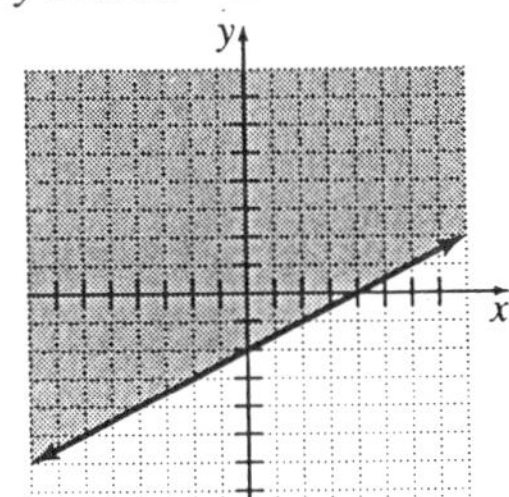

55. $\dfrac{x}{12} + \dfrac{y}{8} > -\dfrac{5}{4}$
$\dfrac{x}{12} + \dfrac{y}{8} - \dfrac{x}{12} > -\dfrac{5}{4} - \dfrac{x}{12}$
$\dfrac{y}{8} > -\dfrac{x}{12} - \dfrac{5}{4}$
$8\left(\dfrac{y}{8}\right) > 8\left(-\dfrac{x}{12} - \dfrac{5}{4}\right)$
$y > -\dfrac{2}{3}x - 10$

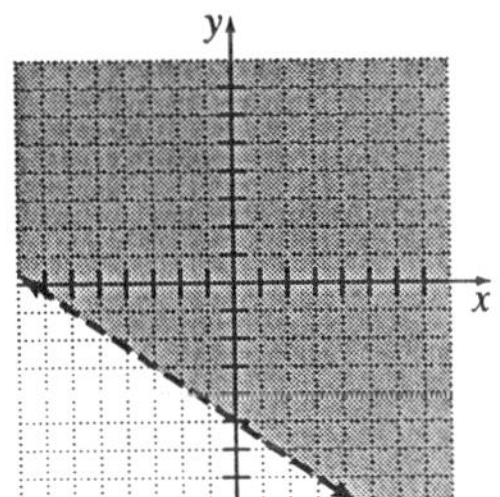

56. $-x - y < 2$
$-x - y + x < 2 + x$
$-y < x + 2$
$\dfrac{-y}{-1} > \dfrac{x + 2}{-1}$
$y > -x - 2$

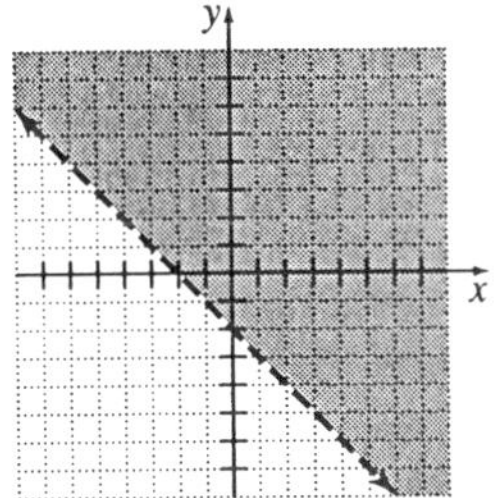

57. $y > -9x + 6$

58. Let x = number of rhododendrons
y = number of azaleas
$4x + 6y \le 85$
Solve the inequality for y.
$4x + 6y - 4x \le 85 - 4x$
$6y \le -4x + 85$
$\frac{6y}{6} \le \frac{-4x+85}{6}$
$y \le -\frac{2}{3}x + \frac{85}{6}$

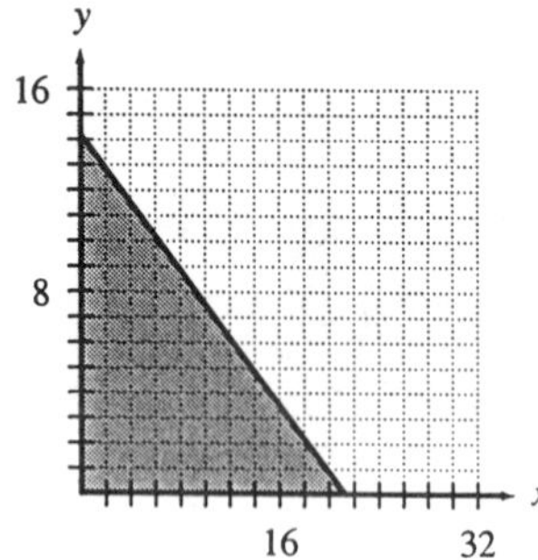

The possible combinations are in the shaded region.
Possible answer: She could buy 15 rhododendrons and 2 azaleas or 10 rhododendrons and 5 azaleas.

59. Let x = number correct
y = number incorrect
$5x - 3y \ge 80$
Solve the inequality for y.
$5x - 3y - 5x \ge 80 - 5x$
$-3y \ge -5x + 80$
$\frac{-3y}{-3} \le \frac{-5x+80}{-3}$
$y \le \frac{5}{3}x - \frac{80}{3}$

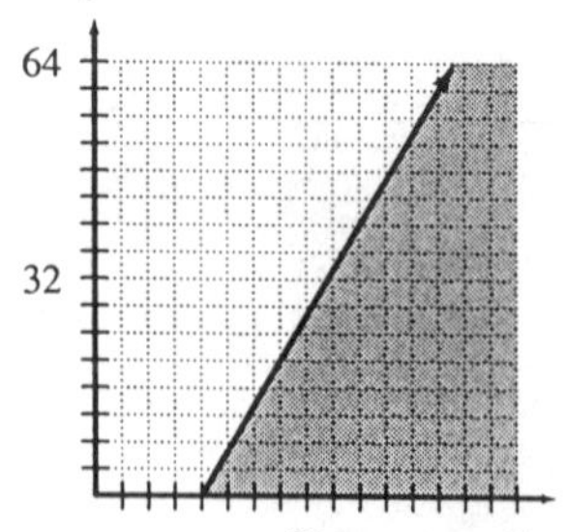

The possible combinations are in the shaded region.

Possible answer: She could get 40 correct and 3 incorrect or 30 correct and 0 wrong.

60. Solve the first inequality for y.
$2x + y > 10$
$2x + y - 2x > 10 - 2x$
$y > -2x + 10$
The second inequality is solved for y.
$y < 3x - 5$

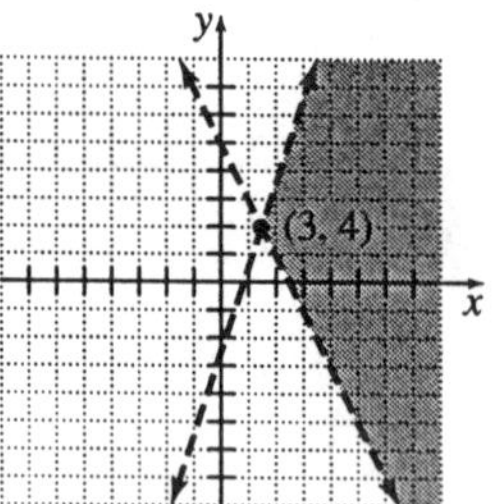

61. Solve the first inequality for x.
$3(x + 2) + 1 > -5$
$3x + 6 + 1 > -5$
$3x + 7 > -5$
$3x + 7 - 7 > -5 - 7$
$3x > -12$
$\frac{3x}{3} > \frac{-12}{3}$
$x > -4$
Solve the second inequality for y.
$2x - y < -8$
$2x - y - 2x < -8 - 2x$
$-y < -2x - 8$
$\frac{-y}{-1} > \frac{-2x-8}{-1}$
$y > 2x + 8$

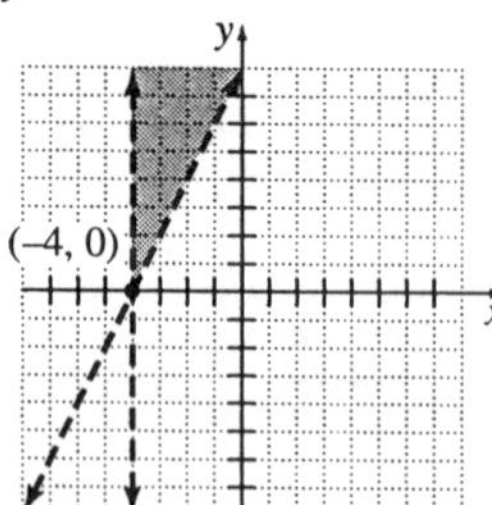

62. Solve the first inequality for x.
$x + 6 < 4$
$x + 6 - 6 < 4 - 6$
$x < -2$
Solve the second inequality for y.

$2(y+2) > y+3$
$2y+4 > y+3$
$2y+4-4 > y+3-4$
$2y > y-1-y$
$y > -1$

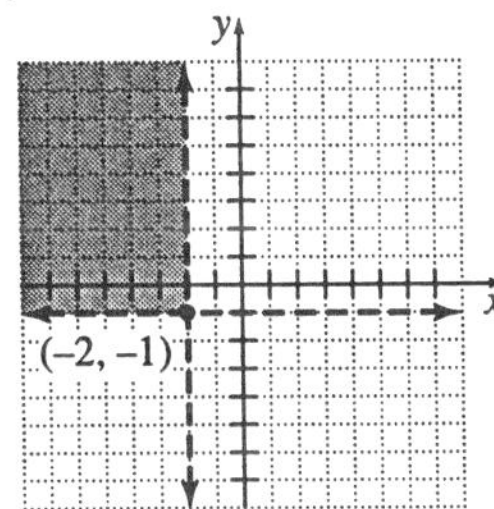

63. Solve the first inequality for y.
$x-2y \geq 12$
$x-2y-x \geq 12-x$
$-2y \geq -x+12$
$\frac{-2y}{-2} \leq \frac{-x+12}{-2}$
$y \leq \frac{1}{2}x-6$
Solve the second inequality for y.
$2x+3y < -6$
$2x+3y-2x < -6-2x$
$3y < -2x-6$
$\frac{3y}{3} < \frac{-2x-6}{3}$
$y < -\frac{2}{3}x-2$

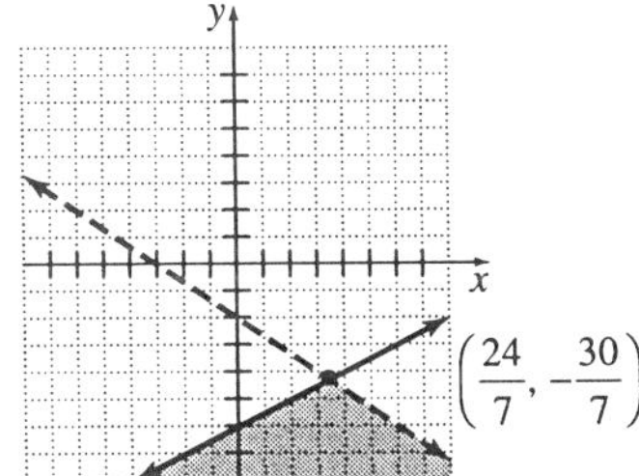

64. $y \geq 2(x+2)$
Solve the second inequality for y.
$2x-y \geq 5$
$2x-y-2x \geq 5-2x$
$-y \geq -2x+5$
$\frac{-y}{-1} \leq \frac{-2x+5}{-1}$
$y \leq 2x-5$

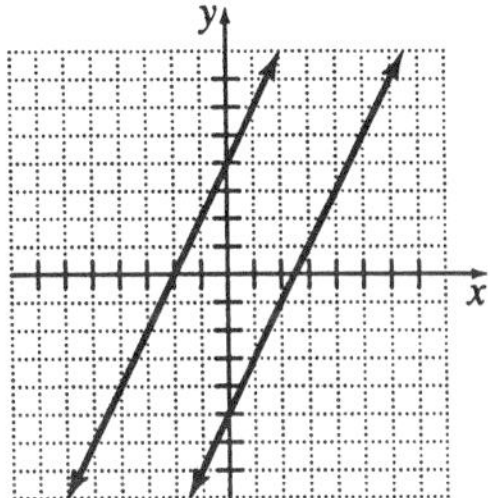

There is no solution.

65. $y > \frac{3}{2}x-6$
Solve the second inequality for y.
$3x-2y > 12$
$3x-2y-3x > 12-3x$
$-2y > -3x+12$
$\frac{-2y}{-2} < \frac{-3x+12}{-2}$
$y < \frac{3}{2}x-6$

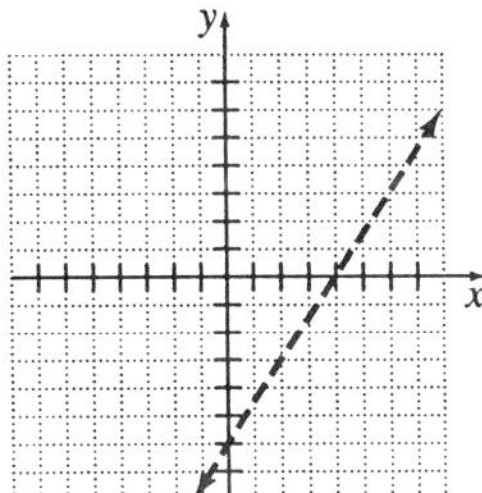

There is no solution.

66. $y < 3x-6$
$y < -3$

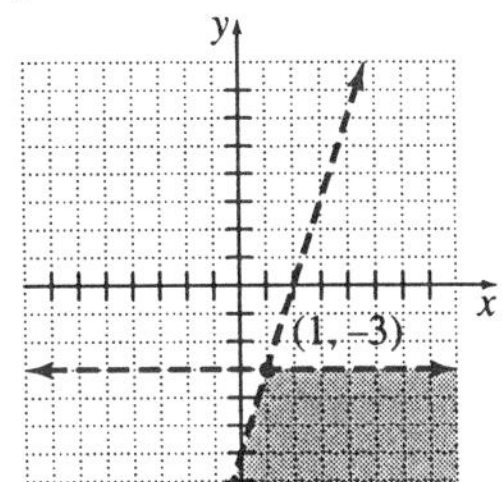

67. Solve the first inequality for y.
$2y-3>1$
$2y-3+3>1+3$
$2y>4$
$\frac{2y}{2}>\frac{4}{2}$
$y>2$
Solve the second inequality for y.
$5(y-4)>10$
$5y-20>10$
$5y-20+20>10+20$
$5y>30$
$\frac{5y}{5}>\frac{30}{5}$
$y>6$

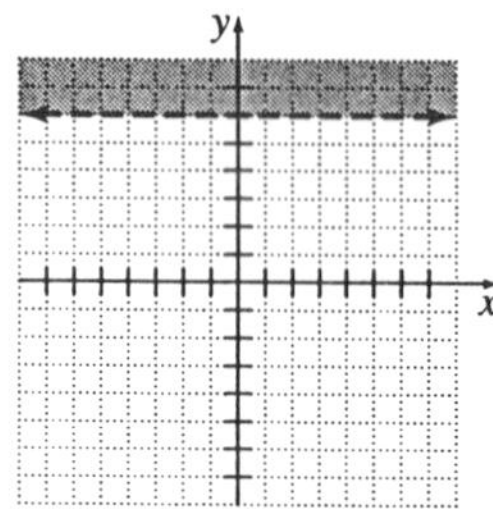

68. Solve the first inequality for x.
$2x-11<3$
$2x-11+11<3+11$
$2x<14$
$\frac{2x}{2}<\frac{14}{2}$
$x<7$
Solve the second inequality for x.
$14-2x>x-7$
$14-2x-14>x-7-14$
$-2x>x-21$
$-2x-x>x-21-x$
$-3x>-21$
$\frac{-3x}{-3}<\frac{-21}{-3}$
$x<7$

69. Solve the first inequality for y.
$y-1<4$
$y-1+1<4+1$
$y<5$
Solve the second inequality for y.
$3y-2<13$
$3y-2+2<13+2$
$3y<15$
$\frac{3y}{3}<\frac{15}{3}$
$y<5$

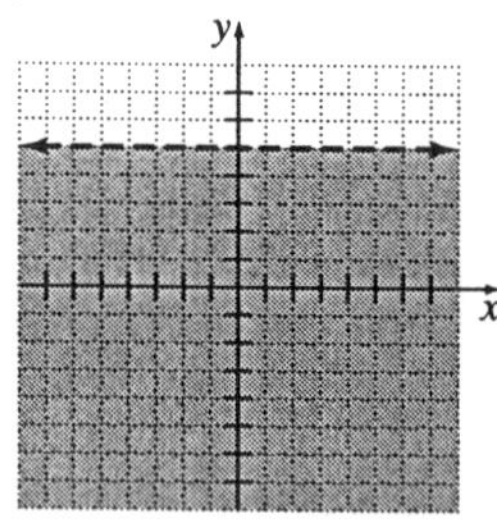

70. Solve the first inequality for y.
$2x+3y\le 6$
$2x+3y-2x\le 6-2x$
$3y\le -2x+6$
$\frac{3y}{3}\le\frac{-2x+6}{3}$
$y\le -\frac{2}{3}x+2$
Solve the second inequality for y.
$x+y\le 1$
$x+y-x\le 1-x$
$y\le -x+1$

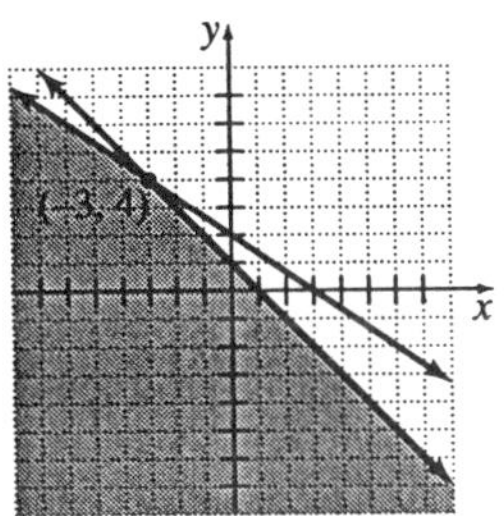

71. $y \le 2x - 15$
$y > -3x + 10$

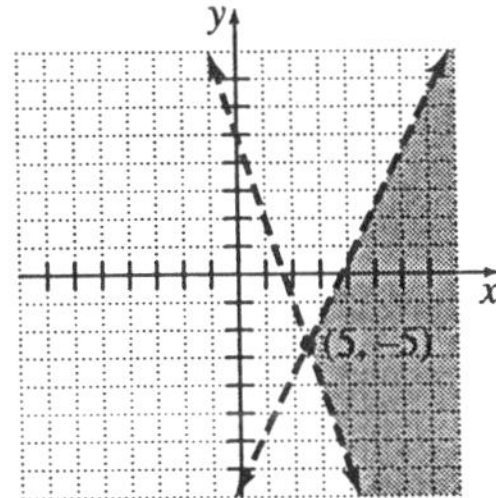

72. $y > 2x + 1$
Solve the second inequality for y.
$x > 2x + 7$
$x - 7 > 2y + 7 - 7$
$x - 7 > 2y$
$\frac{x-7}{2} > \frac{2y}{2}$
$\frac{1}{2}x - \frac{7}{2} > y$
$y < \frac{1}{2}x - \frac{7}{2}$

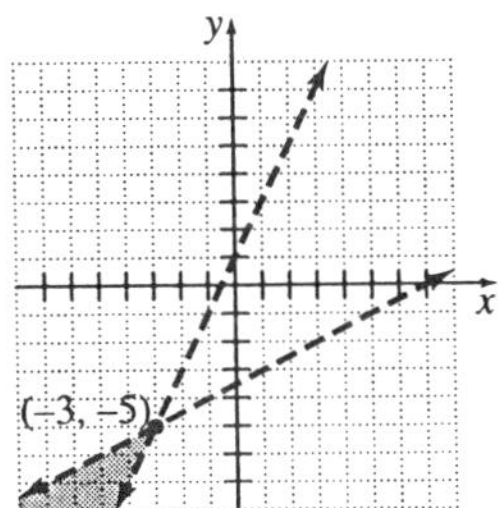

73. Solve the first inequality for y.
$2y + 6 \le 3y + 5$
$2y + 6 - 2y \le 3y + 5 - 2y$
$6 \le y + 5$
$6 - 5 \le y + 5 - 5$
$1 \le y$
$y \ge 1$

Solve the second inequality for x.
$2(x - 3) + 1 > 5$
$2x - 6 + 1 > 5$
$2x - 5 > 5$
$2x - 5 + 5 > 5 + 5$
$2x > 10$
$\frac{2x}{2} > \frac{10}{2}$
$x > 5$

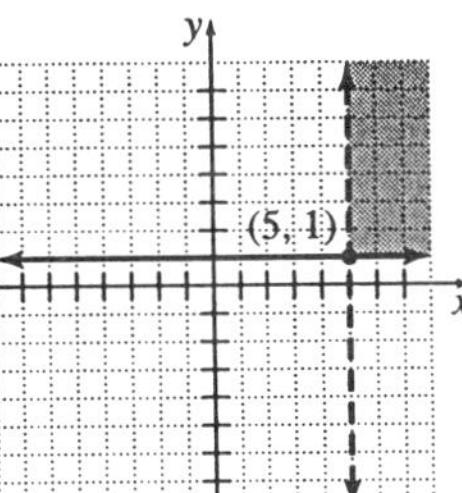

74. $y < 6 - x$
$y < 2x + 1$
$x \ge 0$
$y \ge 0$

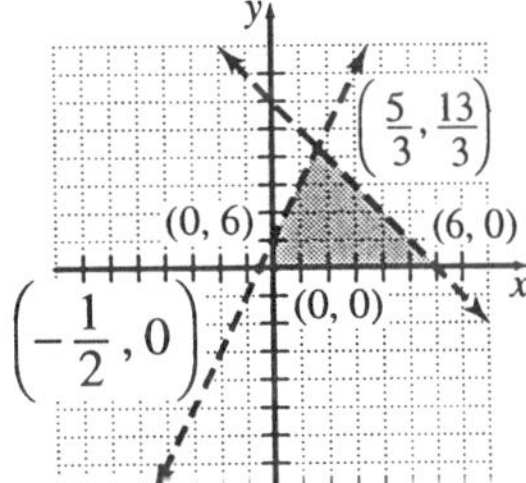

75. $y < 25 - \frac{1}{4}x$
$y < \frac{2}{3}x + 5$
$x \ge 0$
$y \ge 0$

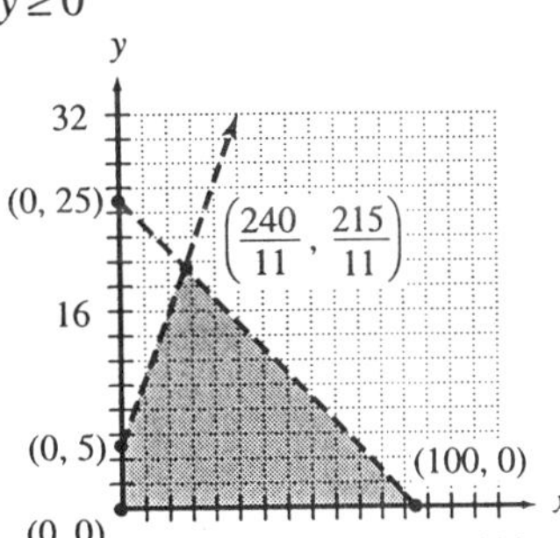

76. Let x = number of rhododendrons
y = number of azaleas
$4x + 6y \le 85$
$y + 4 \le x$
$x \ge 0$
$y \ge 0$
Solve the first two inequalities for y.
$y \le -\frac{2}{3}x + \frac{85}{6}$
$y \le x - 4$
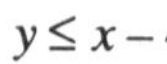

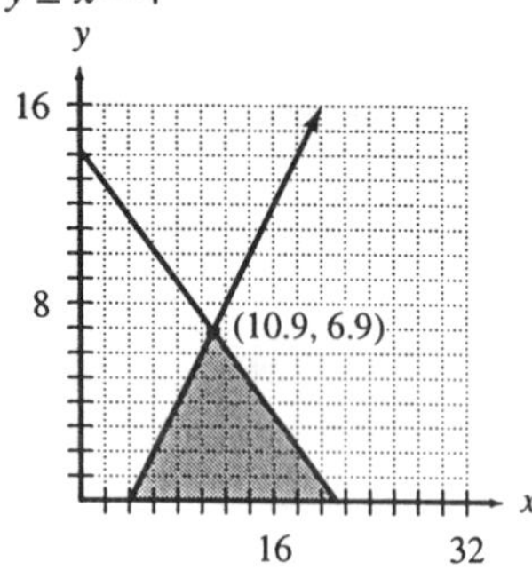

The possible combinations are in the shaded region.
Possible answer: She could purchase 15 rhododendrons and 2 azaleas.

77. Let x = amount invested at 5%
y = amount invested at 6%
$0.05x + 0.06y \ge 225$
$x + y \le 4000$
$x \ge 0$
$y \ge 0$
Solve the first two inequalities for y.
$y \ge -\frac{5}{6}x + 3750$
$y \le -x + 4000$

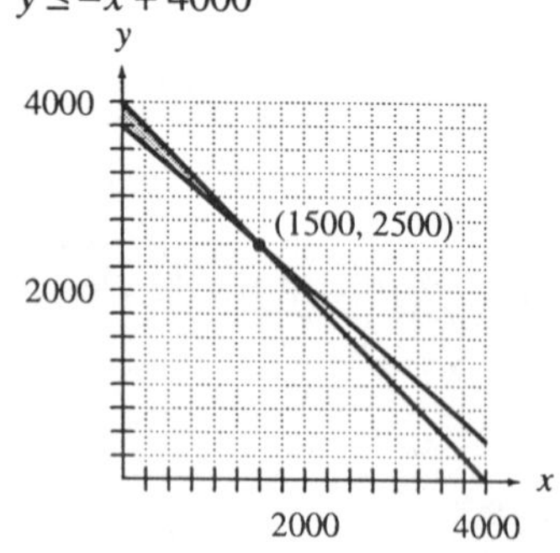

The possible combinations are in the shaded region.
Possible answer: You could invest $200 at 5% and $3800 at 6%.

Chapter 8 Mixed Review

1. $x < -2$

–4 –3 –2 –1 0 1 2 3 4

$(-\infty, -2)$

2. $x > 7$

0 1 2 3 4 5 6 7 8

$(7, \infty)$

3. $-1 < x < 3$

–4 –3 –2 –1 0 1 2 3 4

$(-1, 3)$

4. $x \ge 2.6$

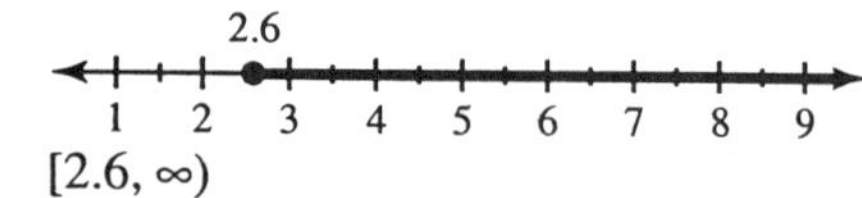

$[2.6, \infty)$

5. $x \le 3\frac{2}{3}$

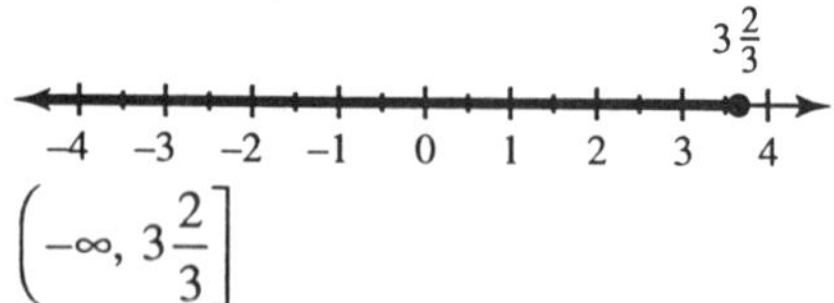

$\left(-\infty, 3\frac{2}{3}\right]$

6. $-2.4 < x \le 4\frac{1}{3}$

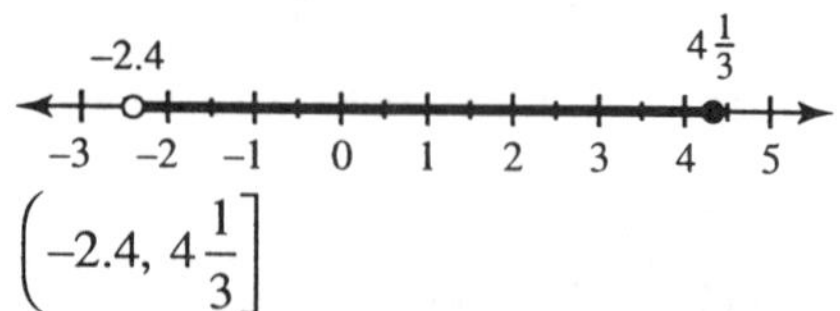

$\left(-2.4, 4\frac{1}{3}\right]$

7.

x	$5x+3$	$2x-9$	
-5	$5(-5)+3=-22$	$2(-5)-9=-19$	$-22<-19$
-4	$4(-4)+3=-17$	$2(-4)-9=-17$	$-17 \not< -17$
-3	$5(-3)+3=-12$	$2(-3)-9=-15$	$-12 \not< -15$

The integers less than –4 are solutions to the inequality.

8.

x	$5.6x-15.3$	$1.3x+19.1$	
7	$5.6(7)-15.3=23.9$	$1.3(7)+19.1=28.2$	$23.9 \not> 28.2$
8	$5.6(8)-15.3=29.5$	$1.3(8)+19.1=29.5$	$29.5 \not> 29.5$
9	$5.6(9)-15.3=35.1$	$1.3(9)+19.1=30.8$	$35.1>30.8$

The integers greater than 8 are solutions to the inequality.

9.

x	$\frac{1}{6}x+\frac{23}{3}$	$\frac{13}{6}-\frac{5}{3}x$	
-4	$\frac{1}{6}(-3)+\frac{23}{3}=7$	$\frac{13}{6}-\frac{5}{3}(-4)=8.8\overline{3}$	$7 \le 8.8\overline{3}$
-3	$\frac{1}{6}(-3)+\frac{23}{3}=7.1\overline{6}$	$\frac{13}{6}-\frac{5}{3}(-3)=7.1\overline{6}$	$7.1\overline{6} \le 7.1\overline{6}$
-2	$\frac{1}{6}(-2)+\frac{23}{3}=7.\overline{3}$	$\frac{13}{6}-\frac{5}{3}(-2)=5.5$	$7.\overline{3} \not\le 5.5$

The integers less than or equal to –3 are solutions to the inequality.

10.

x	$\frac{3}{7}x+\frac{9}{5}$	$\frac{4}{5}x-\frac{17}{5}$	
13	$\frac{3}{7}(13)+\frac{9}{5}=\frac{258}{35}$	$\frac{4}{5}(13)-\frac{17}{5}=7$	$\frac{258}{35} \ge 7$
14	$\frac{3}{7}(14)+\frac{9}{5}=\frac{39}{5}$	$\frac{4}{5}(14)-\frac{17}{5}=\frac{39}{5}$	$\frac{39}{5} \ge \frac{39}{5}$
15	$\frac{3}{7}(15)+\frac{9}{5}=\frac{288}{35}$	$\frac{4}{5}(15)-\frac{17}{5}=\frac{43}{5}$	$\frac{288}{35} \not\ge \frac{43}{5}$

The integers less than or equal to 14 are solutions to the inequality.

11. $5x-13>3(1-x)$
$Y1=5x-13$
$Y2=3(1-x)$

31
–47 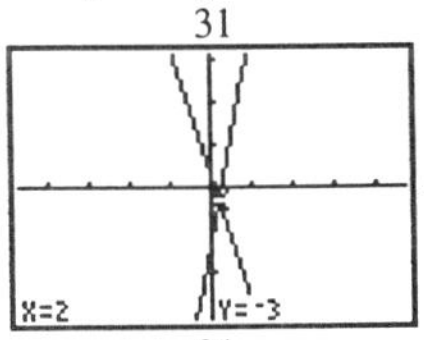 47
X=2 Y=-3
–31

The intersection is (2, –3).
$x>2$

$(2, \infty)$

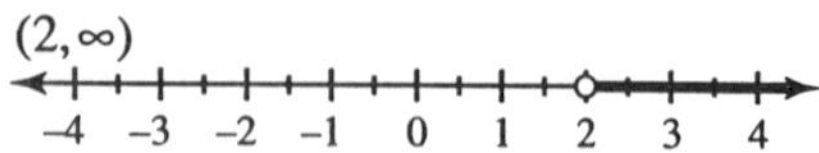

12. $2.1x + 31.71 \le 8.19 - 3.5x$
$Y1 = 2.1x + 31.71$
$Y2 = 8.19 - 3.5x$

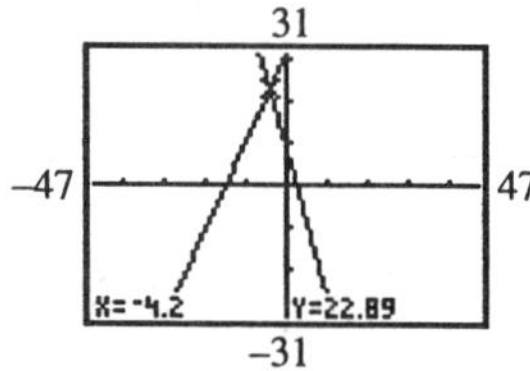

The intersection is (–4.2, 22.89).
$x \le -4.2$
$(-\infty, -4.2]$

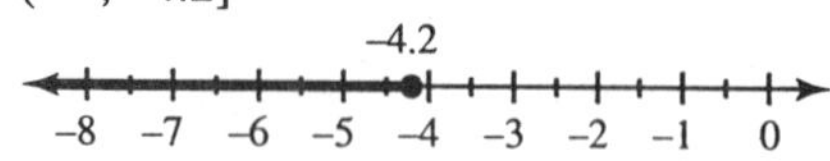

13. $3(x-1) + (x-1) < 2(x-1) + 2x - 1$
$Y1 = 3(x-1) + (x-1)$
$Y2 = 2(x-1) + 2x - 1$

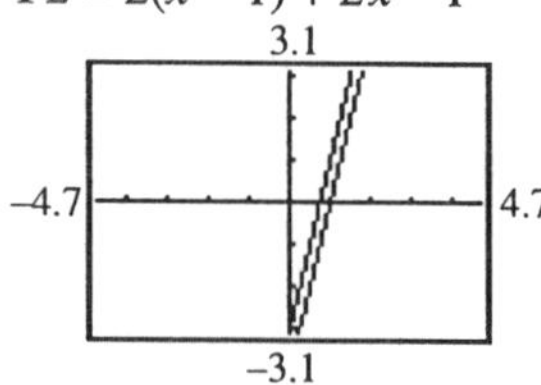

The lines are parallel and the graph of Y1 is always below the graph of Y2. Therefore, the solution is all real numbers.
$(-\infty, \infty)$

14. $4(x+2) - 3(x-5) \le 5x + 8 - 4(x+3)$
$Y1 = 4(x+2) - 3(x-5)$
$Y2 = 5x + 8 - 4(x+3)$

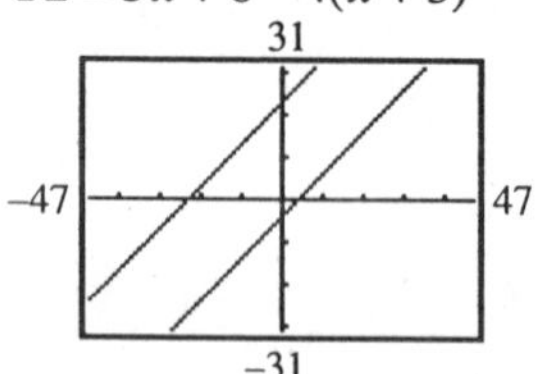

The lines are parallel and the graph of Y1 is always above the graph of Y2. Therefore, there is no solution.

15. $173 - x < 359$
$173 - x - 173 < 359 - 173$
$-x < 186$
$\frac{-x}{-1} > \frac{186}{-1}$
$x > -186$
The solution set is $(-186, \infty)$.

16. $z - \frac{4}{17} > \frac{15}{34}$
$z - \frac{4}{17} + \frac{4}{17} > \frac{15}{34} + \frac{4}{17}$
$z > \frac{15}{34} + \frac{8}{34}$
$z > \frac{23}{34}$
The solution set is $\left(\frac{23}{34}, \infty\right)$.

17. $5x + 4 > 3x + 18$
$5x + 4 - 4 > 3x + 18 - 4$
$5x > 3x + 14$
$5x - 3x > 3x + 18 - 3x$
$2x > 14$
$\frac{2x}{2} > \frac{14}{2}$
$x > 7$
The solution set is $(7, \infty)$.

18. $8x < 8x - 16$
$8x - 8x < 8x - 16 - 8x$
$0 < -16$
This is a contradiction. There is no solution.

19. $3.9x + 19.88 \ge -1.9x + 4.76$
$3.5x + 19.88 - 19.88 \ge -1.9x + 4.76 - 19.88$
$3.5x \ge -1.9x - 15.12$
$3.5x + 1.9x \ge -1.9x - 15.12 + 1.9x$
$5.4x \ge -15.12$
$\frac{5.4x}{5.4} \ge \frac{-15.12}{5.4}$
$x \ge -2.8$
The solution set is $[-2.8, \infty)$.

20. $2.6y + 9.62 \le 0$
$2.6y + 9.62 - 9.62 \le 0 - 9.62$
$2.6y \le -9.62$
$\frac{2.6y}{2.6} \le \frac{-9.62}{2.6}$
$y \le -3.7$
The solution set is $(-\infty, -3.7]$.

21. $3(x + 3) < 4(x + 1) - 2(x - 2)$
$3x + 9 < 4x + 4 - 2x + 4$
$3x + 9 < 2x + 8$
$3x + 9 - 9 < 2x + 8 - 9$
$3x < 2x - 1$
$3x - 2x < 2x - 1 - 2x$
$x < -1$
The solution set is $(-\infty, -1)$.

22. $3(x - 3) + 2(x + 2) < 5x + 7$
$5x - 9 + 2x + 4 < 5x + 7$
$5x - 5 < 5x + 7$
$5x - 5 - 5x < 5x + 7 - 5x$
$-5 < 7$
This is always true.
The solution set is $(-\infty, \infty)$.

23. Solve the first inequality for y.
$2y + 2 > y$
$2y + 2 - y > y - y$
$y + 2 > 0$
$y + 2 - 2 > 0 - 2$
$y > -2$
The second inequality is:
$y < -x + 1$

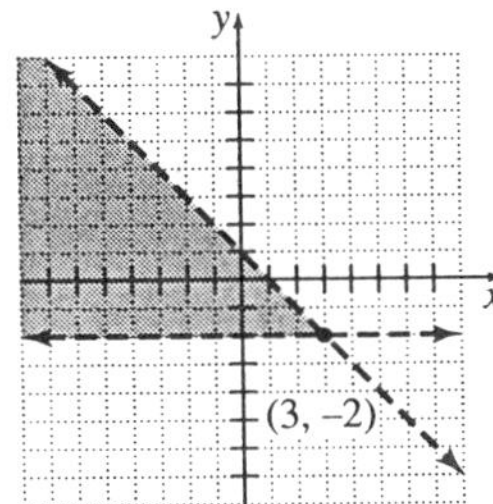

24. Solve the first inequality for x.
$2x + 3 > x$
$2x + 3 - 3 > x - 3$
$2x > x - 3$
$2x - x > x - 3 - x$
$x > -3$
Solve the second inequality for x.
$2x < 6$
$\frac{2x}{2} < \frac{6}{2}$
$x < 3$

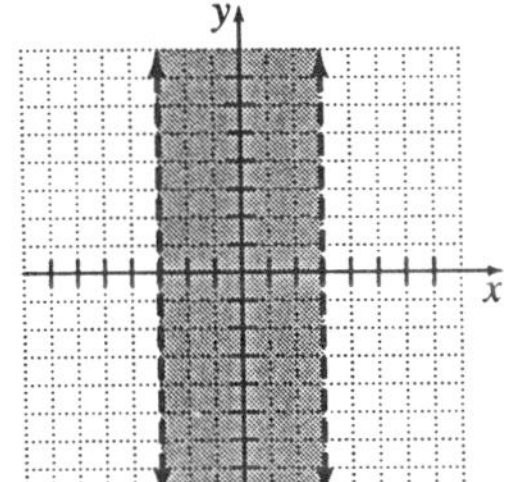

25. $y \ge 2x - 8$
Solve the second inequality for y.
$3x + y < 8$
$3x + y - 3x < 8 - 3x$
$y < -3x + 8$

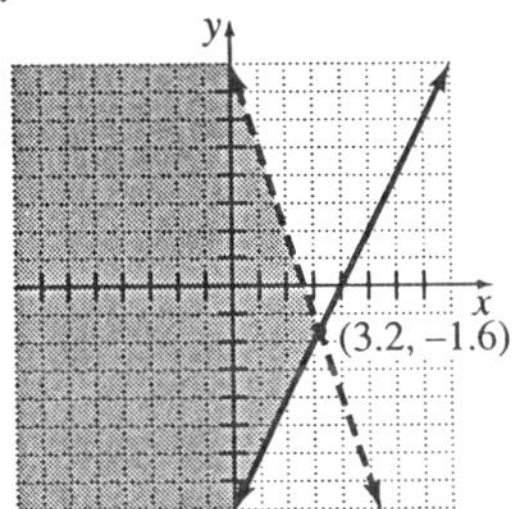

26. Solve the first inequality for x.
$2(x - 1) - 4 > 0$
$2x - 2 - 4 > 0$
$2x - 6 > 0$
$2x - 6 + 6 > 0 + 6$
$2x > 6$
$\frac{2x}{2} > \frac{6}{2}$
$x > 3$
Solve the second equation for y.
$2x + y \le 3$
$2x + y - 2x \le 3 - 2x$
$y \le -2x + 3$

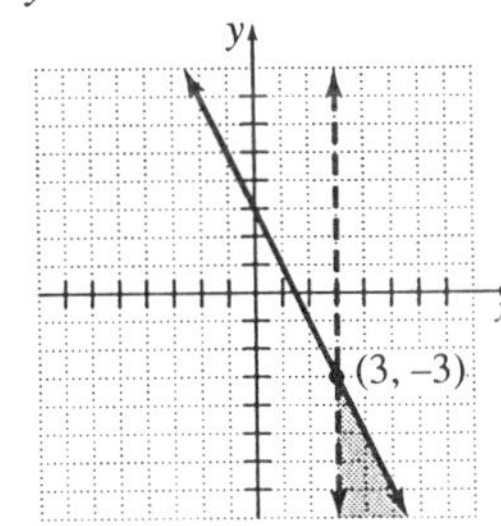

27. Solve the first inequality for y.
$x + y < 9 - x$
$x + y - x < 9 - x - x$
$y < -2x + 9$
Solve the second inequality for y.
$2x + y > 5$
$2x + y - 2x > 5 - 2x$
$y > -2x + 5$

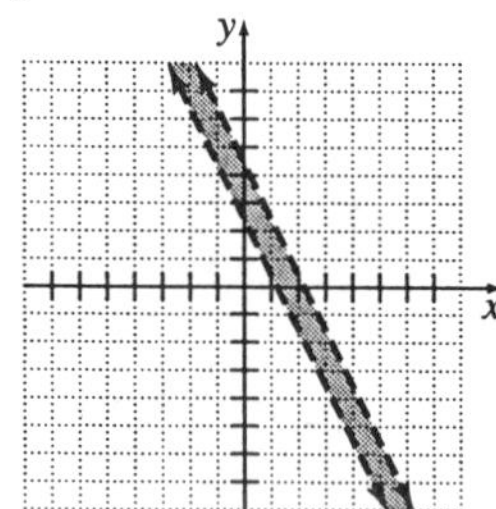

28. $y < -4x - 3$
Solve the second inequality for y.
$4x + 2y < y - 3$
$4x + 2y - 4x < y - 3 - 4x$
$2y < y - 4x - 3$
$2y - y < y - 4x - 3 - y$
$y < -4x - 3$

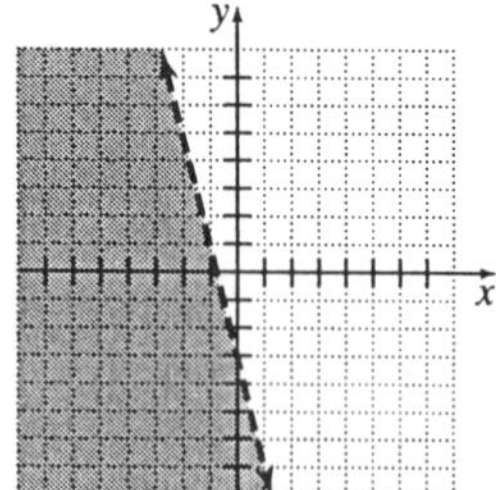

29. Solve the first inequality for x.
$3x + 4 \geq 2x$
$3x + 4 - 4 \geq 2x - 4$
$3x \geq 2x - 4$
$3x - 2x \geq 2x - 4 - 2x$
$x \geq -4$
Solve the second inequality for x.
$x + 7 \leq 3$
$x + 7 - 7 \leq 3 - 7$
$x \leq -4$

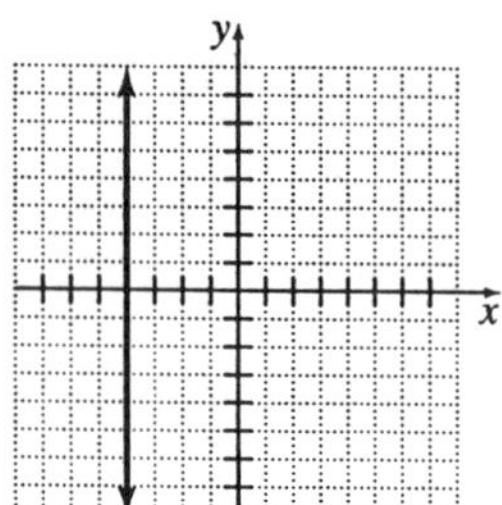

30. Solve the first equation for y.
$3y + 7 > 2y + 5$
$3y + 7 - 7 > 2y + 5 - 7$
$3y > 2y - 2$
$3y - 2y > 2y - 2 - 2y$
$y > -2$
Solve the second inequality for y.
$y + 9 < 7$
$y + 9 - 9 < 7 - 9$
$y < -2$

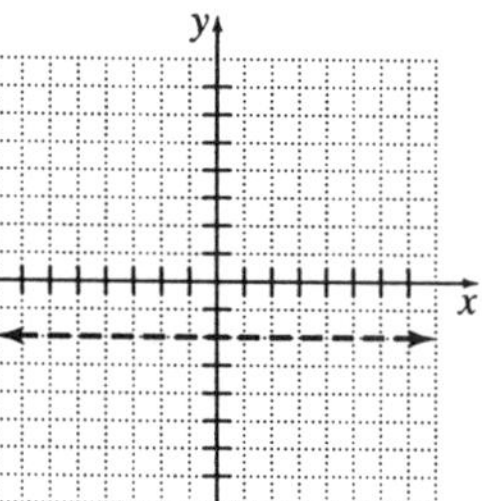

31. Solve the first inequality for x.
$3(x - 3) - 1 > 8$
$3x - 9 - 1 > 8$
$3x - 10 > 8$
$3x - 10 + 10 > 8 + 10$
$3x > 18$
$\frac{3x}{3} > \frac{18}{3}$
$x > 6$
Solve the second inequality for y.
$3y - 10 < 15 - 2y$
$3y - 10 + 10 < 15 - 2y + 10$
$3y < 25 - 2y$
$3y + 2y < 25 - 2y + 2y$
$5y < 25$
$\frac{5y}{5} < \frac{25}{5}$
$y < 5$

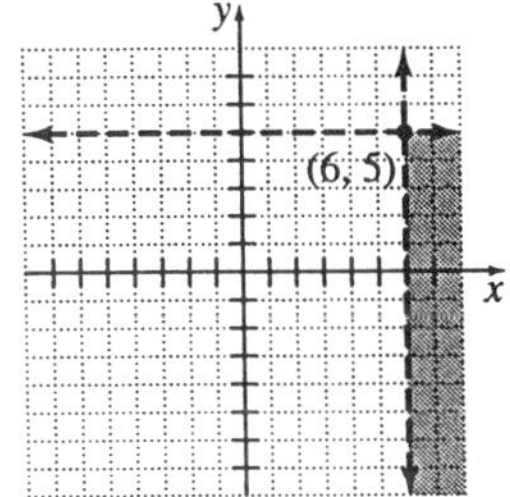

32. $y < 3x + 7$
Solve the second inequality for y.
$x - y > -1$
$x - y - x > -1 - x$
$-y > -x - 1$
$\frac{-y}{-1} < \frac{-x-1}{-1}$
$y < x + 1$

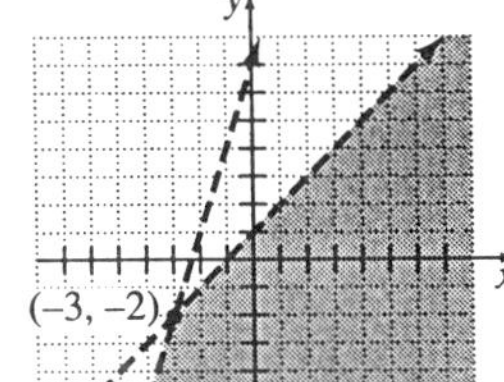

33. Solve the first inequality for y.
$x - 5y \le 15$
$x - 5y - x \le 15 - x$
$5y \le x + 15$
$\frac{-5y}{-5} \ge \frac{-x+15}{-5}$
$y \ge \frac{1}{5}x - 3$
Solve the second inequality for y.
$x + 5y \le -5$
$x + 5y - x \le -5 - x$
$5y \le -x - 5$
$\frac{5y}{5} \le \frac{-x-5}{5}$
$y \le -\frac{1}{5}x - 1$

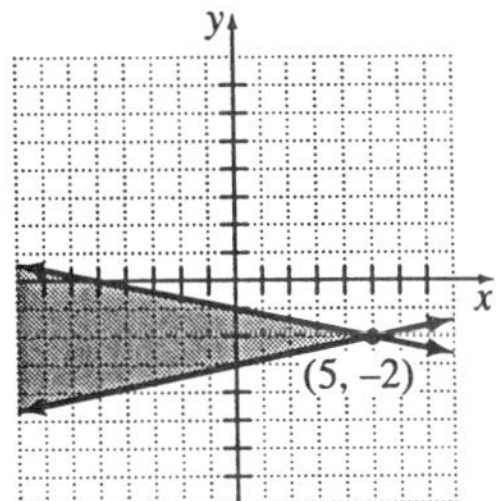

34. $y < -4x + 6$
$y > 3x - 8$

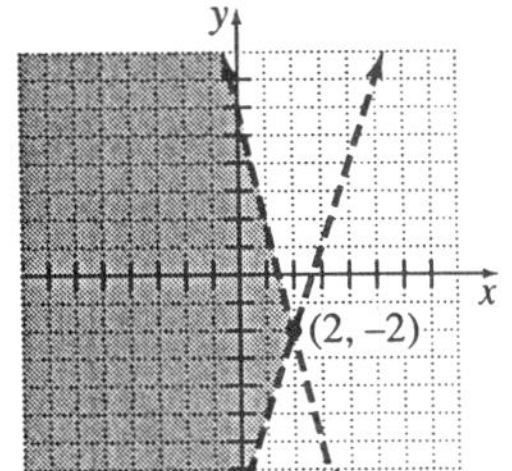

35. $y > x$
Solve the second inequality for y.
$x > 2 - y$
$x - 2 > 2 - y - 2$
$x - 2 > -y$
$\frac{x-2}{-1} < \frac{-y}{-1}$
$-x + 2 < y$
$y > -x + 2$

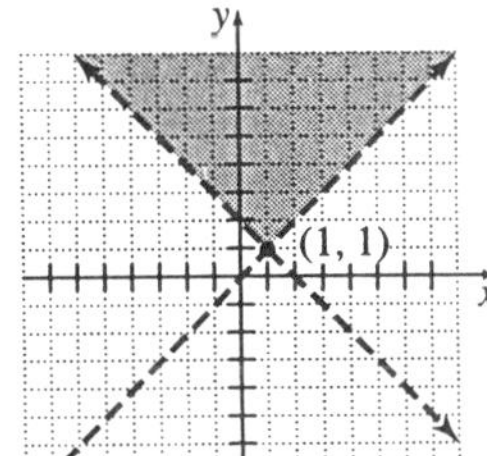

36. Solve the first inequality for x.
$x + 9 \le 3$
$x + 9 - 9 \le 3 - 9$
$x \le -6$
Solve the second inequality for y.
$3y > 2y - 1$
$3y - 2y > 2y - 1 - 2y$
$y > -1$

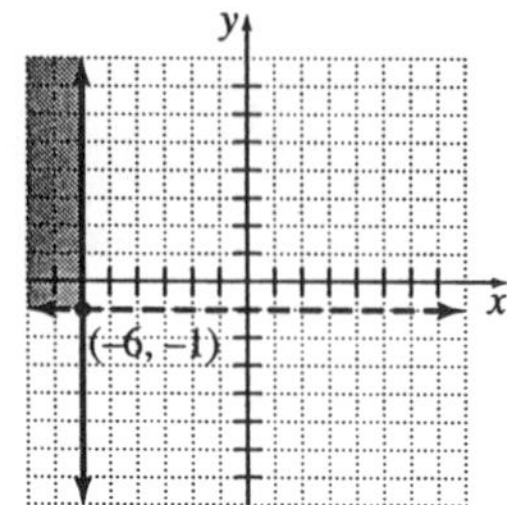

37. $y < -2x + 7$
$y < 2x$
$x \geq 0$
$y \geq 0$

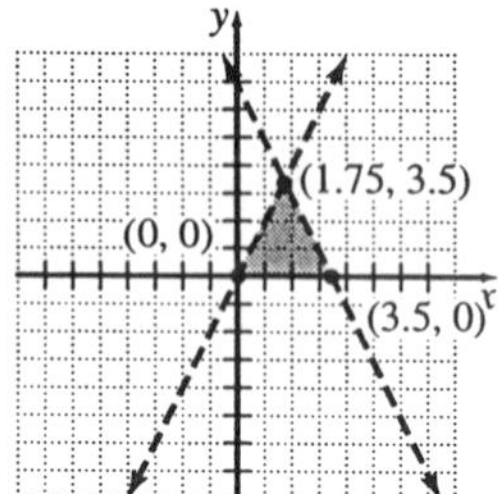

38. $y > \frac{3}{4}x - 4$

$y > -\frac{2}{3}x + 3$
$x \geq 0$
$y \geq 0$

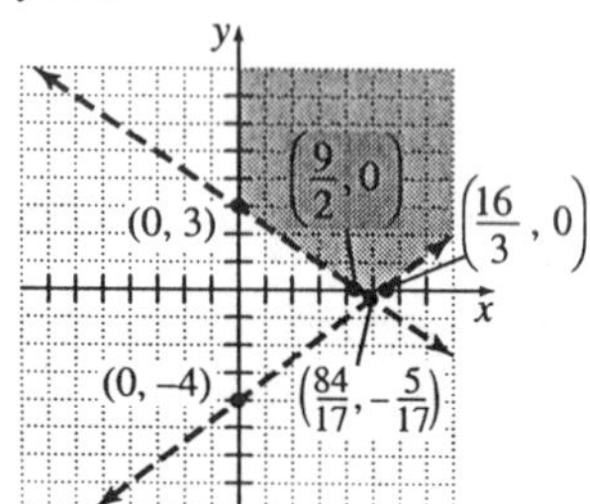

39. $9x - 5y < 45$
$9x - 5y - 9x < 45 - 9x$
$-5y < -9x + 45$
$\frac{-5y}{-5} > \frac{-9x + 45}{-5}$
$y > \frac{9}{5}x - 9$

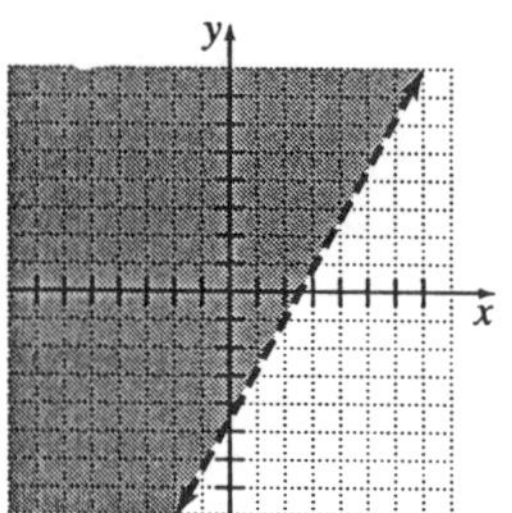

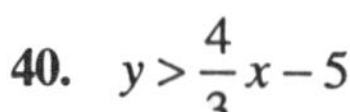

40. $y > \frac{4}{3}x - 5$

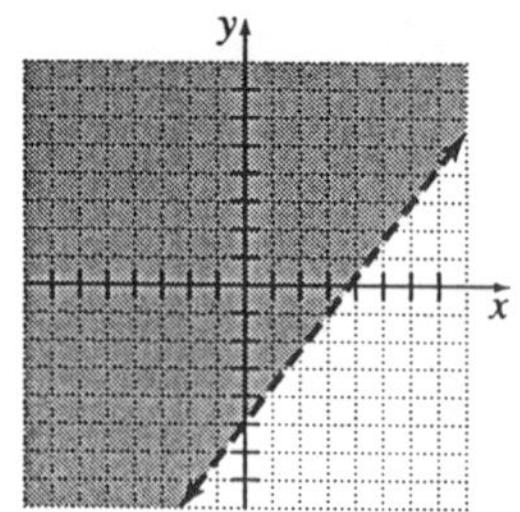

41. $3y \leq 2x + 9$
$\frac{3y}{3} \leq \frac{2x + 9}{3}$
$y \leq \frac{2}{3}x + 3$

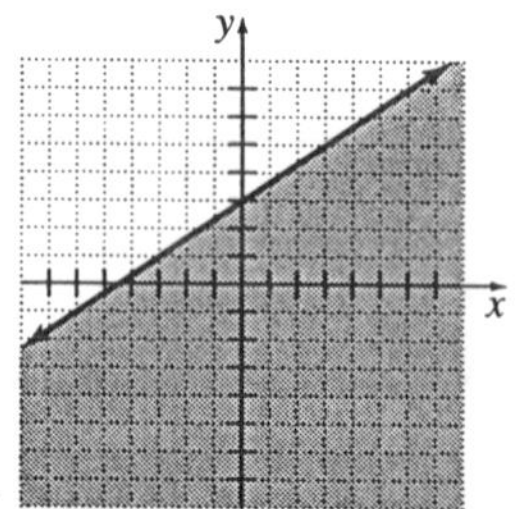

42. $y + 13 \geq 8$
$y + 13 - 13 \geq 8 - 13$
$y \geq -5$

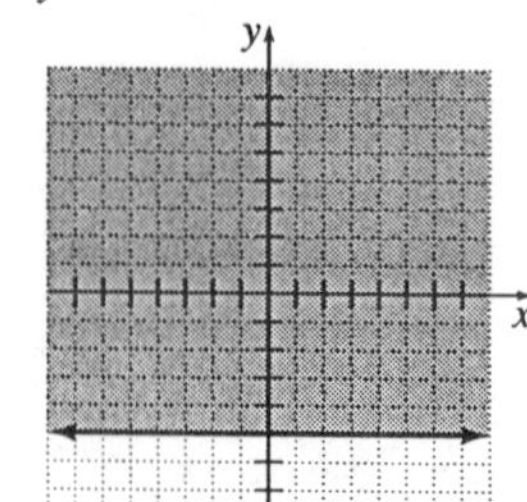

43. $-2y - 6 < y - 11$
$-2y - 6 + 6 < y - 11 + 6$
$-2y < y - 5$
$-2y - y < y - 5 - y$
$-3y < -5$
$\frac{-3y}{-3} > \frac{-5}{-3}$
$y > \frac{5}{3}$

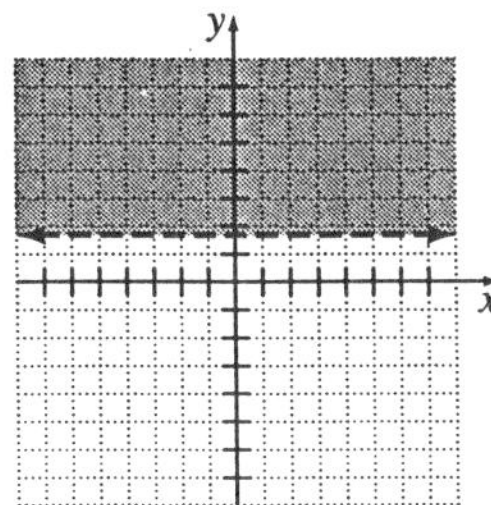

44. $x - 13 > 3x + 19$
$x - 13 + 13 > 3x + 19 + 13$
$x > 3x + 32$
$x - 3x > 3x + 32 - 3x$
$-2x > 32$
$\frac{-2x}{-2} < \frac{32}{-2}$
$x < -16$

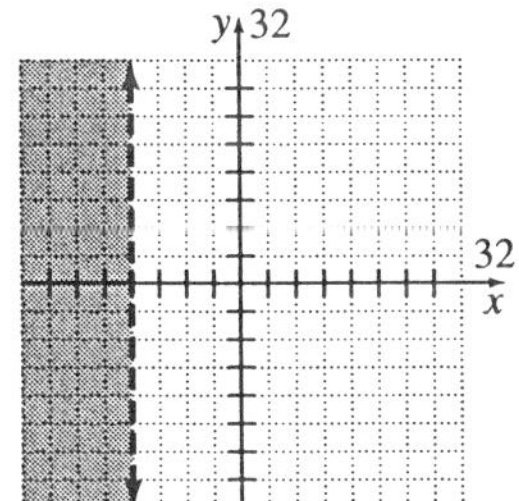

45. $x - 7y > 21$
$x - 7y - x > 21 - x$
$-7y > -x + 21$
$\frac{-7y}{-7} < \frac{-x + 21}{-7}$
$y < \frac{1}{7}x - 3$

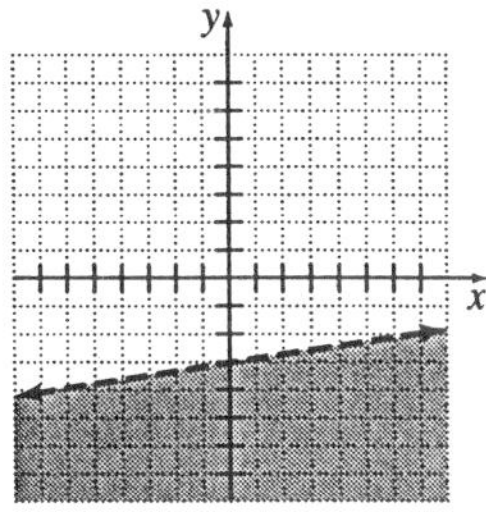

46. $2.7x + 5.4y \geq 16.2$
$2.7x + 5.4y - 2.7x \geq 16.2 - 2.7x$
$5.4y \geq -2.7x + 16.2$
$\frac{5.4y}{5.4} \geq \frac{-2.7x + 16.2}{5.4}$
$y \geq -0.5x + 3$

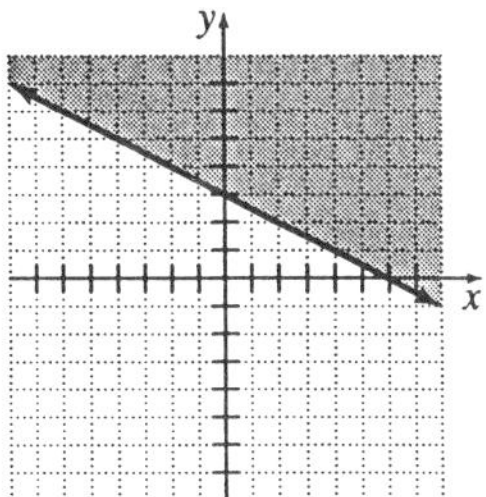

47. $4.8x - 1.8y \leq 14.4$
$4.8x - 1.8y - 4.8x \leq 14.4 - 4.8x$
$-1.8y \leq -4.8x + 14.4$
$\frac{-1.8y}{-1.8} \geq \frac{-4.8x + 14.4}{-1.8}$
$y \geq \frac{8}{3}x - 8$

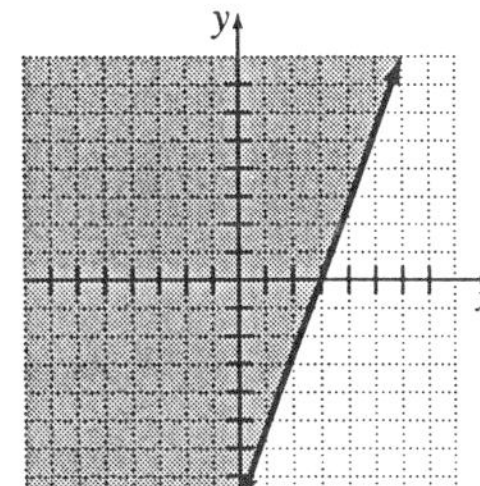

48. $\dfrac{x}{12}+\dfrac{y}{9}>-\dfrac{1}{3}$

$\dfrac{x}{12}+\dfrac{y}{9}-\dfrac{x}{12}>-\dfrac{1}{3}-\dfrac{x}{12}$

$\dfrac{y}{9}>-\dfrac{x}{12}-\dfrac{1}{3}$

$9\left(\dfrac{y}{9}\right)>9\left(-\dfrac{x}{12}-\dfrac{1}{3}\right)$

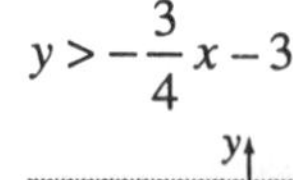

$y>-\dfrac{3}{4}x-3$

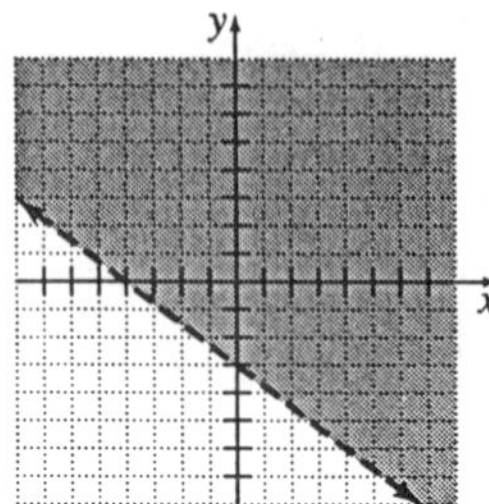

49. $x-y<7$

$x-y-x<7-x$

$-y<-x+7$

$\dfrac{-y}{-1}>\dfrac{-x+7}{-1}$

$y>x-7$

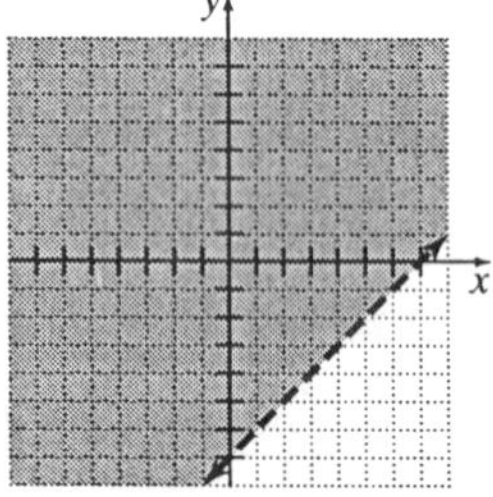

50. $y>-5x+7$

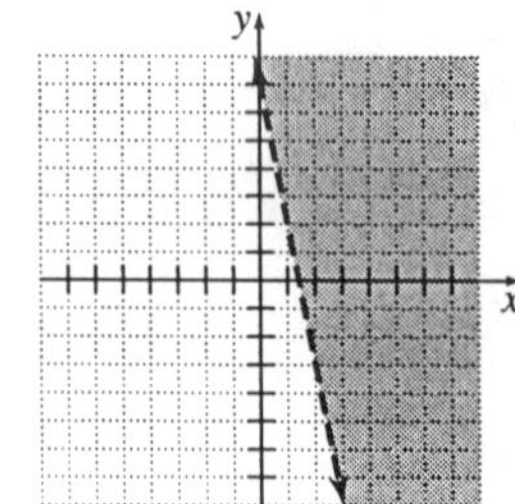

51. Let x = number of acres of oats
y = number of acres of wheat

$y\ge 2x$

$x+y\le 540$

$x\ge 0$

$y\ge 0$

Solve the second inequality for y.

$x+y\le 540$

$x+y-x\le 540-x$

$y\le -x+540$

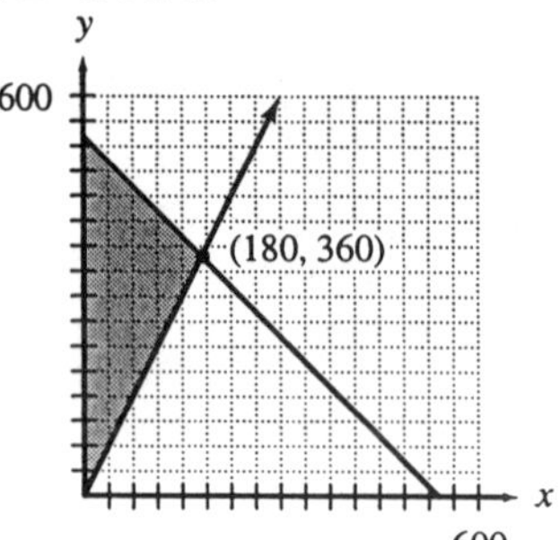

The possible combinations are represented by the shaded region.

Possible answer:

He could plant 100 acres of oats and 400 acres of wheat.

52. Let x = rent for efficiency apartment
y = rent for regular apartment

$3x+5y\ge 6000$

$y\ge x+75$

$x\ge 0$

$y\ge 0$

Solve the first inequality for y.

$3x+5y\ge 6000$

$3x+5y-3x\ge 6000-3x$

$5y\ge -3x+6000$

$\dfrac{5y}{5}\ge\dfrac{-3x+6000}{5}$

$y\ge -\dfrac{3}{5}x+1200$

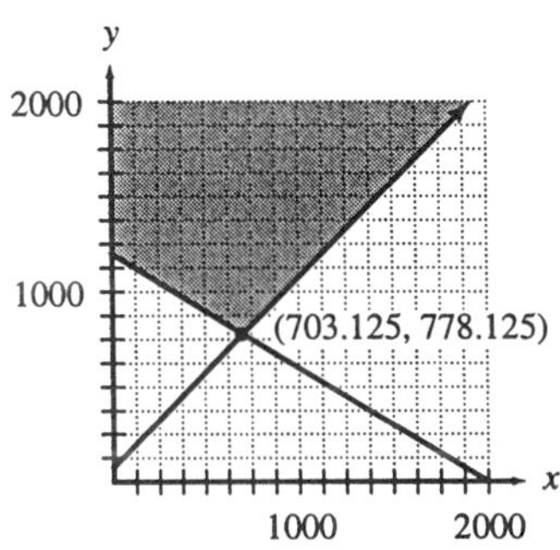

The possible combinations are represented by the shaded region.
Possible answer:
The rents could be \$1000 dollars for an efficiency and \$1500 for a regular apartment.

53. Let x = number of packs
$255 + 2.5x < 12.5x$
Solve the inequality for x.
$255 + 2.5x - 2.5x < 12.5x - 2.5x$
$255 < 10x$
$\frac{255}{10} < \frac{10x}{10}$
$25.5 < x$
$x > 25.5$
To make a profit at least 26 packs must be sold.
Possible answer:
There were 27 packs sold.

54. Let x = amount of bill for the sixth month
$\frac{45+36+52+48+31+x}{6} < 42$
Solve the inequality for x.
$\frac{212+x}{6} < 42$
$6\left(\frac{212+x}{6}\right) < 6(42)$
$212 + x < 252$
$212 + x - 212 < 252 - 212$
$x < 40$
Her sixth phone bill must be less than \$40.
Possible answer:
Her phone bill should be \$35.

55. Let x = length
$18x > 600$
Solve the inequality for x.
$\frac{18x}{18} > \frac{600}{18}$
$x > 33\frac{1}{3}$
The length must be at least $33\frac{1}{3}$ inches.
Possible answer:
The length is 40 inches.

56. Let x = number of wins
y = number of ties
$3x + y \geq 25$
Solve the inequality for y.
$3x + y - 3x \geq 25 - 3x$
$y \geq -3x + 25$

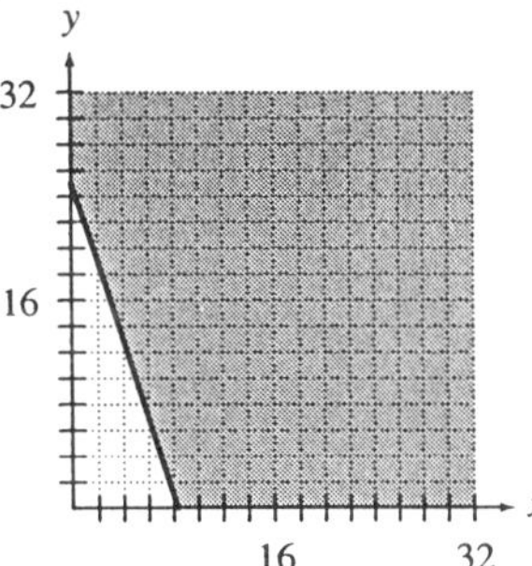

The possible combinations are represented in the shaded region
Possible answer:
They could have 10 wins and 5 ties.

57. Let x = number of liters of the 25% solution
y = number of liters of the 40% solution
$0.25x + 0.40y \leq 0.30(x + y)$
Solve the inequality for y.
$0.25x + 0.40y \leq 0.30x + 0.30y$
$0.25x + 0.40y - 0.25x$
$\leq 0.30x + 0.30y - 0.25x$
$0.40y \leq 0.05x + 0.30y$
$0.40y - 0.30y \leq 0.05x + 0.30y - 0.30y$
$0.1y \leq 0.05x$
$\frac{0.1y}{0.1} \leq \frac{0.05x}{0.1}$
$y \leq 0.5x$

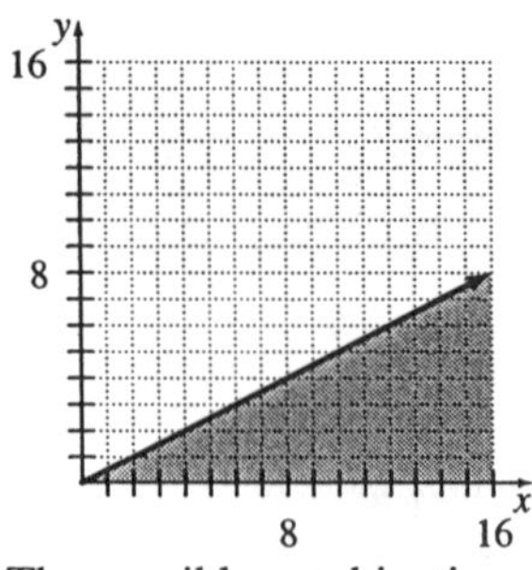

The possible combinations are represented by the shaded region.
Possible answer:
There should be 10 liters of 25% solution and 2 liters of 40% solution.

Chapter 8 Test

1. $2x - 3y > x + 8$ is linear because it simplifies to $y < \frac{1}{3}x - \frac{8}{3}$.

2. $4x^2 + 2x < x - 9$ is nonlinear because the x-variable is squared.

3. $5(x - 3) \geq 4 - (x + 1)$ is linear because it simplifies to $x \geq 3$.

4. $\frac{1}{2}x - 4 \geq y + \frac{3}{8}$ is linear.

5. $7(x + 2) - 3(x + 1) > 4(x + 8)$
$7x + 14 - 3x - 3 > 4x + 32$
$4x + 11 > 4x + 32$
$4x + 11 - 4x > 4x + 32 - 4x$
$11 > 32$
This is a contradiction. There is no solution.

6. $5x + 9 < 2x - 3$
$5x + 9 - 9 < 2x - 3 - 9$
$5x < 2x - 12$
$5x - 2x < 2x - 12 - 2x$
$3x < -12$
$\frac{3x}{3} < \frac{-12}{3}$
$x < -4$

–8 –7 –6 –5 –4 –3 –2 –1 0

The solution set is $(-\infty, -4)$.

7. $\frac{4}{5}(x - 10) < \frac{1}{5}(4x + 5) + 1$
$\frac{4}{5}x - 8 < \frac{4}{5}x + 2$
$\frac{4}{5}x - 8 - \frac{4}{5}x < \frac{4}{5}x + 2 - \frac{4}{5}x$
$-8 < 2$
This is always true. The solution is all real numbers.

The solution set is $(-\infty, \infty)$.

8. $5a - 7 \geq 8a + 1$
$5a - 7 + 7 \geq 8a + 1 + 7$
$5a \geq 8a + 8$
$5a - 8a \geq 8a + 8 - 8a$
$-3a \geq 8$
$\frac{-3a}{-3} \leq \frac{8}{-3}$
$a \leq -\frac{8}{3}$

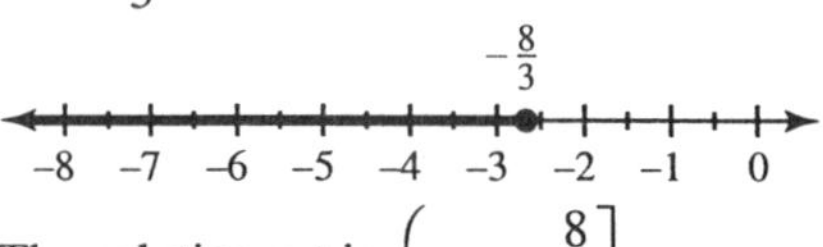

The solution set is $\left(-\infty, -\frac{8}{3}\right]$.

9. $y < -2x + 5$

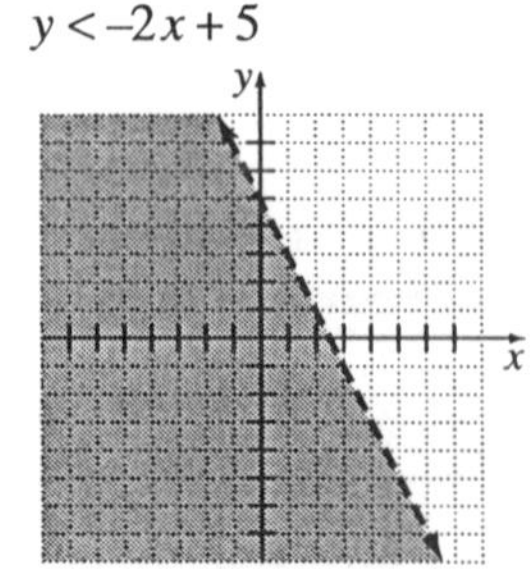

10. Solve the inequality for x.
$x + 3 > 2x - 1$
$x + 3 - x > 2x - 1 - x$
$3 > x - 1$
$3 + 1 > x - 1 + 1$
$4 > x$
$x < 4$

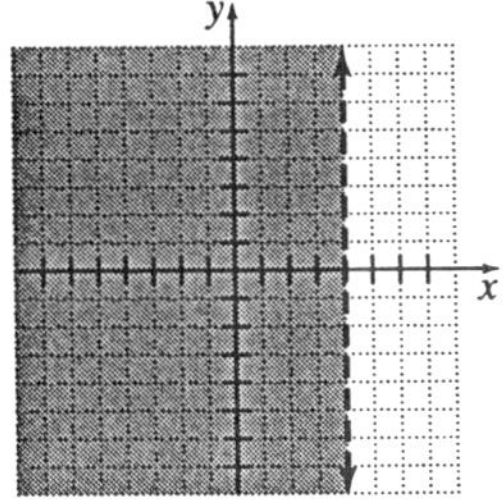

11. Solve the first inequality for y.
$x - y > 4$
$x - y - x > 4 - x$
$-y > -x + 4$
$\frac{-y}{-1} < \frac{-x+4}{-1}$
$y < x - 4$
Solve the second inequality for y.
$x + 2y > 4$
$x + 2y - x > 4 - x$
$2y > -x + 4$
$\frac{2y}{2} > \frac{-x+4}{2}$
$y > -\frac{1}{2}x + 2$

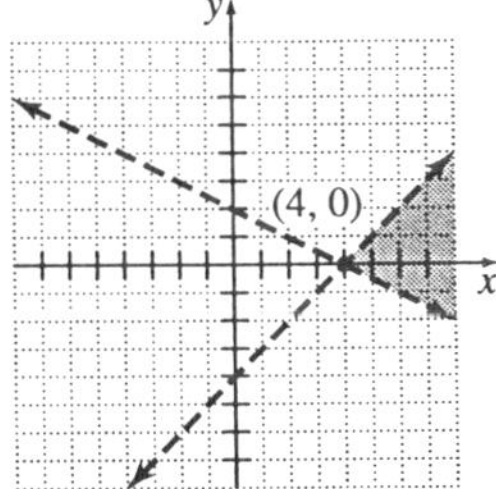

12. $y \geq 4x - 5$
$y > -3x + 4$

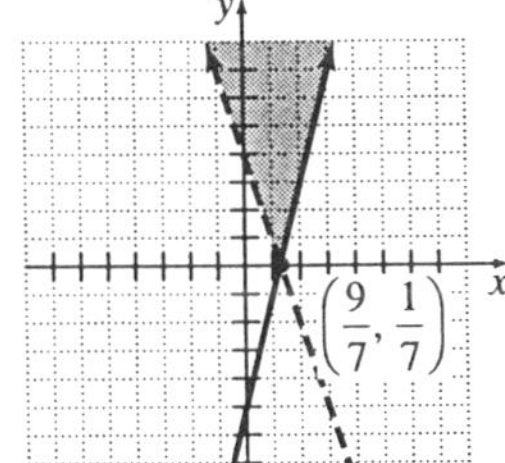

13. $y < 2x + 6$
$y < 2x - 1$

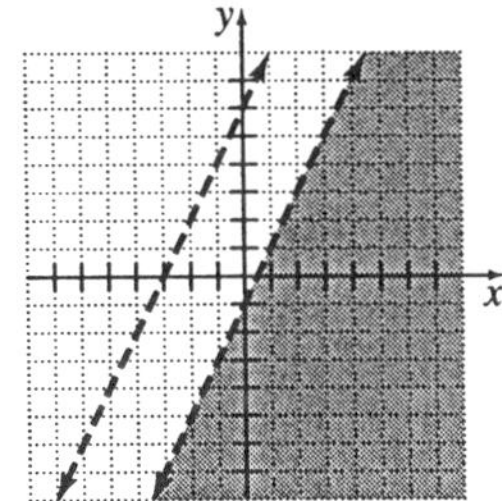

14. Solve the first inequality for y.
$3y - 1 > 2$
$3y - 1 + 1 > 2 + 1$
$3y > 3$
$\frac{3y}{3} > \frac{3}{3}$
$y > 1$
Solve the second inequality for x.
$x - 3 \leq -5$
$x - 3 + 3 \leq -5 + 3$
$x \leq -2$

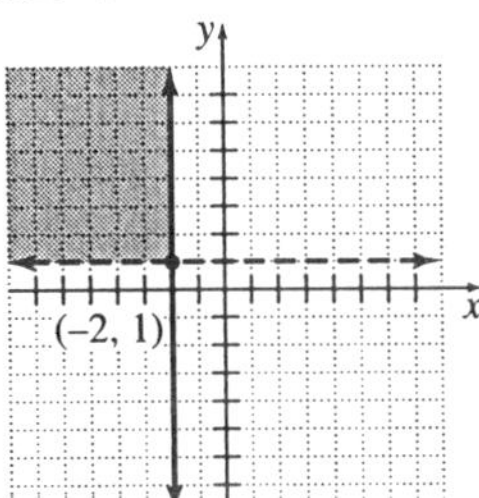

15. Let x = third score
$\frac{83 + 72 + x}{3} \geq 80$
Solve the inequality for x.
$\frac{155 + x}{3} \geq 80$
$3\left(\frac{155 + x}{3}\right) \geq 3(80)$
$155 + x \geq 240$
$155 + x - 155 \geq 240 - 155$
$x \geq 85$
The score must be at least 85.
Possible answer:
The score is 90 points.

16. Let x = number of good deeds
y = number of activity sheets
$5x + 2y \ge 20$
Solve the inequality for y.
$5x + 2y - 5x \ge 20 - 5x$
$2y \ge -5x + 20$
$\frac{2y}{2} \ge \frac{-5x+20}{2}$
$y \ge -\frac{5}{2}x + 10$

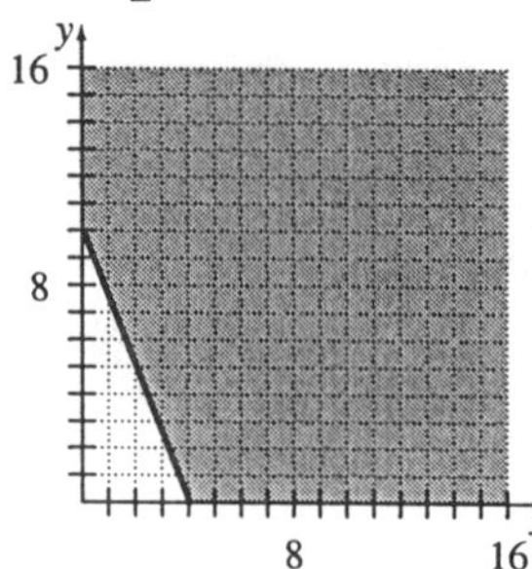

The solutions are represented in the shade region.
Possible answers:
The cub scout had 5 good deeds and 5 activity sheets.

17. Let x = amount invested at 4%
y = amount invested at 8%
$0.04x + 0.08y \ge 400$
$x + y \le 6000$
$x \ge 0$
$y \ge 0$
Solve the first two inequalities for y
$y \ge -0.5x + 5000$
$y \le -x + 6000$

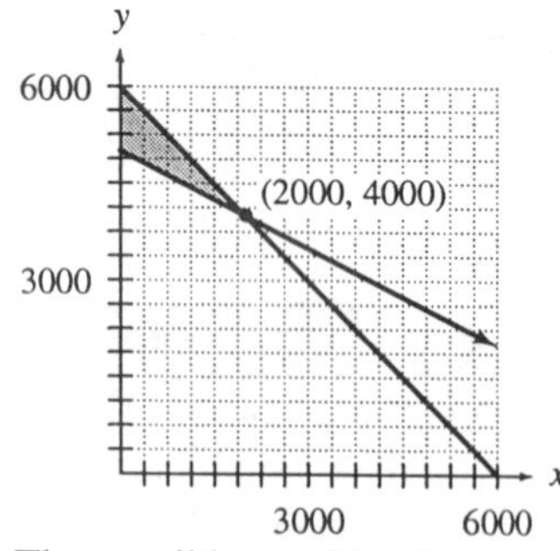

The possible combinations are represented by the shaded region.
Possible answer:
You could invest $500 at 4%
and $5500 at 8%

18. Answers will vary.

Chapters 1–8 Cumulative Review

1. $-(-9) = 9$

2. $-|-9| = -(9) = -9$

3. $\sqrt[3]{-\frac{27}{64}} = \frac{\sqrt[3]{-27}}{\sqrt[3]{64}} = \frac{-3}{4} = -\frac{3}{4}$

4. $\left(\frac{-3}{8}\right)^{-2} = \left(\frac{8}{-3}\right)^2 = \frac{(8)^2}{(-3)^2} = \frac{64}{9}$

5. $\left(-\frac{9}{14}\right)\left(\frac{7}{3}\right) = -\frac{3 \cdot 3 \cdot 7}{2 \cdot 7 \cdot 3} = -\frac{3}{2}$

6. $-\frac{3}{8} \div \left(-1\frac{2}{3}\right) = -\frac{3}{8} \div \left(-\frac{5}{3}\right)$
$= \left(-\frac{3}{8}\right)\left(-\frac{3}{5}\right) = \frac{9}{40}$

7. $12(-3) \div (-6)(2) \div (-2)$
$= (-36) \div (-6)(2) \div (-2)$
$= (6)(2) \div (-2)$
$= 12 \div (-2)$
$= -6$

8. $40 + 16 \div 8 - \sqrt{3^2 + 7 \cdot 5 + 5}$
$= 40 + 16 \div 8 - \sqrt{9 + 7 \cdot 5 + 5}$
$= 40 + 2 - \sqrt{9 + 35 + 5}$
$= 42 - \sqrt{49}$
$= 42 - 7$
$= 35$

9. $\dfrac{2(5^2-10)+4^2-1}{8^2-2(32)}$

$=\dfrac{2(25-10)+4^2-1}{8^2-2(32)}$

$=\dfrac{2(15)+16-1}{64-2(32)}$

$=\dfrac{30+16-1}{64-64}$

$=\dfrac{45}{0}=\text{undefined}$

10. $6x-2(4x-1)=6x-8x+2=-2x+2$

11. $4[2(x-3)+1]-[5(2x-4)-6]$
$=4(2x-5)-(10x-26)$
$=8x-20-10x+26$
$=-2x+6$

12.

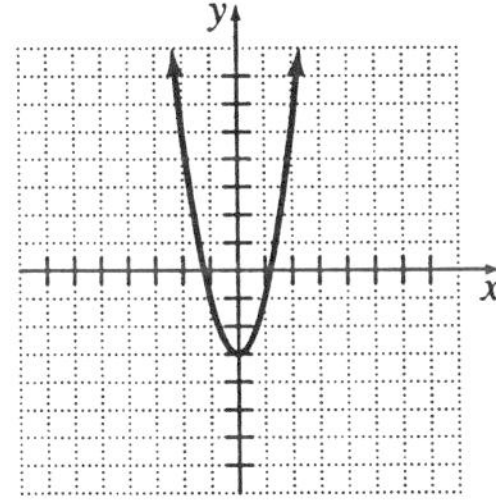

13. The domain of the relation is all real numbers. The range of the relation is all real numbers ≥ -3.

14. The relation is a function since all possible vertical lines cross the graph a maximum of one time.

15. The relative minimum value is -3 at $x = 0$. There is no relative maximum.

16. The relation is increasing for all $x > 0$, and decreasing for all $x < 0$.

17. For the x-intercept, solve $0=2x^2-3$.

$0=2x^2-3$
$3=2x^2$
$\dfrac{3}{2}=x^2$
$\pm\sqrt{\dfrac{3}{2}}=x,\ x\approx\pm1.225$

The x-intercepts are $(-1.225, 0)$ and $(1.225, 0)$.

For the y-intercept, solve $y=2(0)^2-3$.

$y=2(0)^2-3$
$y=2(0)-3$
$y=0-3$
$y=-3$

The y-intercept is $(0, -3)$.

18. $7x-3=5$
$7x=8$
$x=\dfrac{8}{7}$

19. $(x+3)-2(3x+4)=5$
$x+3-6x-8=5$
$-5x-5=5$
$-5x=10$
$x=\dfrac{10}{-5}$
$x=-2$

20. $1.2(x+3)-4(0.3x+0.15)=3$
$1.2x+3.6-1.2x-0.6=3$
$3=3$
All real numbers are solutions.

21. $|2x+6|=10$

$2x+6=10$	$2x+6=-10$
$2x=4$	$2x=-16$
$x=2$	$x=-8$

22. $5|x|+8=8$
$5|x|=0$
$|x|=0$
$x=0$

23. $5x + 4 > x - 8$
$5x > x - 12$
$4x > -12$
$x > -3$
$(-3, \infty)$

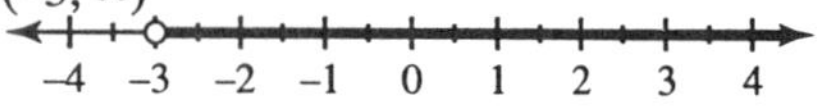

24. $-12x + 4 \geq -2x + 8$
$4 \geq 10x + 8$
$-4 \geq 10x$
$-\frac{4}{10} \geq x$
$-\frac{2}{5} \geq x$
$\left(-\infty, -\frac{2}{5}\right]$

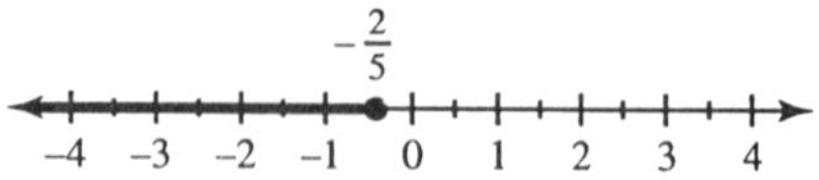

25. $4(x + 3) - 3(x - 2) \leq x - 5$
$4x + 12 - 3x + 6 \leq x - 5$
$x + 18 \leq x - 5$
$18 \leq -5$
The statement is false; there is no solution.

26.

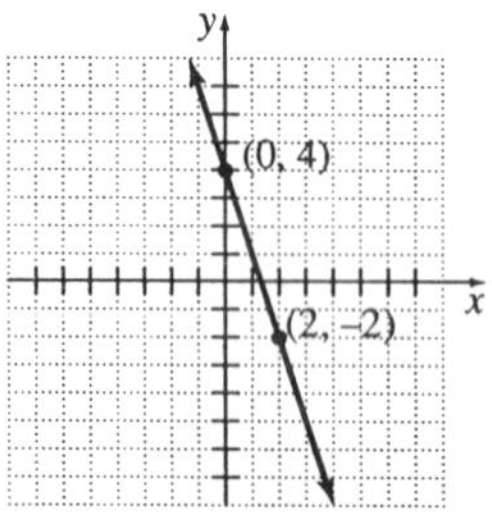

27.

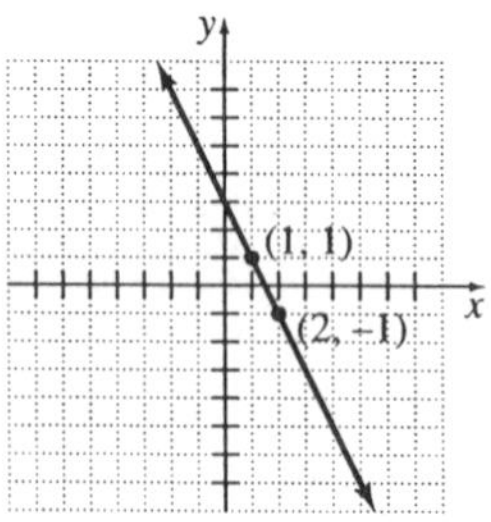

28.

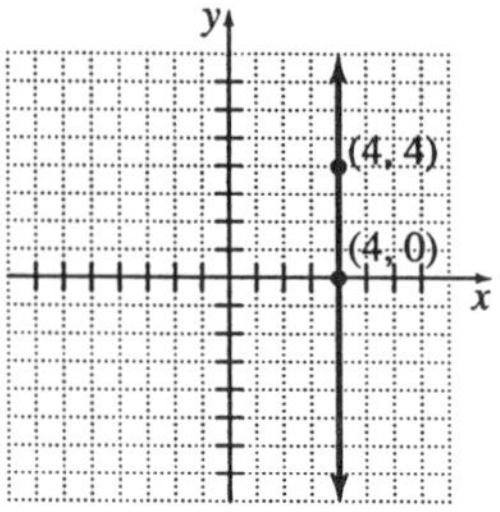

29.

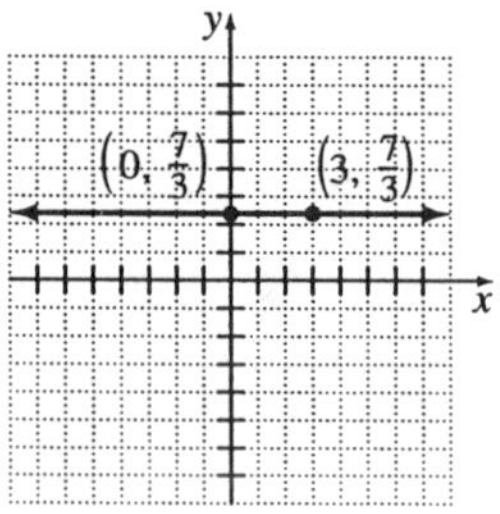

30. Graph $y = 4x - 3$ with a dashed line.
Use (0, 0) as a test point.
$0 < 4(0) - 3?$
$0 < -3$ False
Shade the half-plane not containing (0, 0).

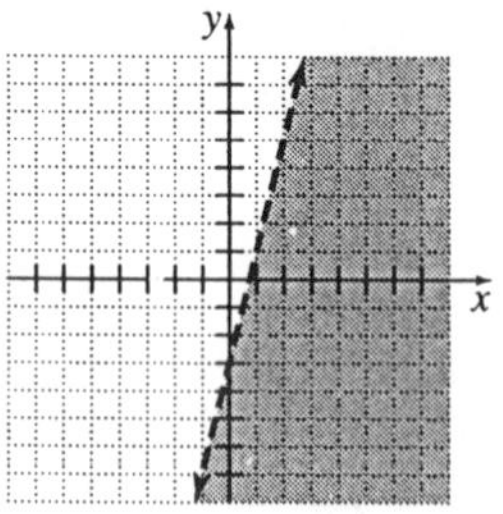

31. $2x + 3y = 21$
$3y = -2x + 21$
$y = -\frac{2}{3}x + 7$

$3x - 2y = 2$
$-2y = -3x + 2$
$y = \frac{3}{2}x - 1$

Since $\left(-\frac{2}{3}\right)\left(\frac{3}{2}\right) = -1$, the graph of the pair of equations is intersecting and perpendicular lines.

32. Subtract the equations.

$$\begin{array}{r} y=3x+4 \\ \underline{-y=-2x+5} \\ 0=x+9 \end{array}$$

$-9=x$

$y=3(-9)+4=-27=4=-23$

The solution is (–9, –23).

33. Substitute the expression $-2x+4$ for y in the first equation.

$4x+2(-2x+4)=8$

$4x-4x+8=8$

$8=8$

This is an identity. The solution is all ordered pairs which satisfy $y=-2x+4$.

34. Graph $y=-x$ with a solid line and $y=2x+4$ with a dashed line. Shade above the line $y=-x$ and below the line $y=2x+4$.

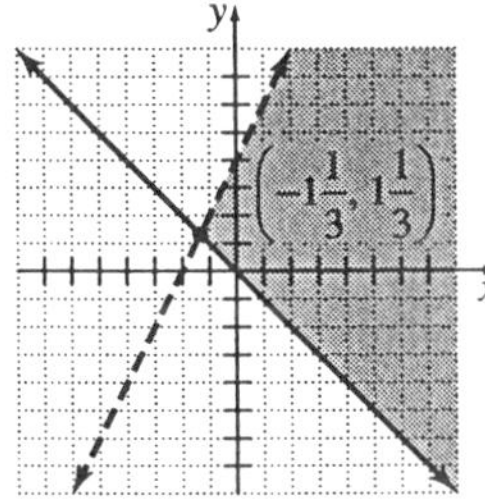

One solution is (1, 0).

35. $m=\dfrac{y_2-y_1}{x_2-x_1}=\dfrac{-2-3}{-1-(-2)}=\dfrac{-5}{1}=-5$

36. $m=\dfrac{y_2-y_1}{x_2-x_1}=\dfrac{3-(-5)}{6-6}=\dfrac{8}{0}$

The slope is undefined.

37. $y=mx+b$

$y=2x+3$

The slope is 2.

38. $y=mx+b$

$y=0x+1.4$

The slope is 0.

39. $m=\dfrac{y_2-y_1}{x_2-x_1}=\dfrac{-2-(-3)}{-1-2}=\dfrac{1}{-3}=-\dfrac{1}{3}$

$y-y_1=m(x-x_1)$

$y-(-3)=-\dfrac{1}{3}(x-2)$

$y+3=-\dfrac{1}{3}x+\dfrac{2}{3}$

$y=-\dfrac{1}{3}x-\dfrac{7}{3}$

40. $2x+3y=1$

$3y=-2x+1$

$y=-\dfrac{2}{3}x+\dfrac{1}{3}$

The slope of the graph is $-\dfrac{2}{3}$.

$y-y_1=m(x-x_1)$

$y-4=-\dfrac{2}{3}(x-5)$

$y-4=-\dfrac{2}{3}x+\dfrac{10}{3}$

$y=-\dfrac{2}{3}x+\dfrac{22}{3}$

41. $5{,}340{,}000=5.34\times10^6$

42. $1.2\times10^{-4}=0.00012$

43. $-4.783\text{ E}{-5}=-0.00004783$

44. $P=2L+2W$

$P-2W=2L$

$\dfrac{P-2W}{2}=L$

$L=\dfrac{P}{2}-W$

45. $f(x)=x^2+2x-1$

$f(-3)=(-3)^2+2(-3)-1$

$f(-3)=9+(-6)-1$

$f(-3)=9-6-1$

$f(-3)=2$

46. Let x be the amount that Lance invests.
$0.11x = 2775$
$x = \frac{2775}{0.11}$
$x \approx 25{,}227.273$
Lance should invest $25,227.28.

47. The sum of the measures of the angles is 90°. Let x be the measure of the smaller angle. Then the measure of the larger angle is $2x + 25$.
$x + (2x + 25) = 90$
$3x + 25 = 90$
$3x = 65$
$x = \frac{65}{3} = 21\frac{2}{3}$
$2x + 25 = 2\left(\frac{65}{3}\right) + 25 = \frac{205}{3} = 68\frac{1}{3}$
The smaller angle measures $21\frac{2}{3}°$ and the larger angle measures $68\frac{1}{3}°$.

48. Let x be the amount of Hazelnut Coffee that Mike uses. Since he is making 10 pounds total, the amount of Cinnamon Coffee that Mike uses is $10 - x$ pounds.
$7.5x + 6.75(10 - x) = 7(10)$
$100[7.5x + 6.75(10 - x)] = 100(70)$
$750x + 675(10 - x) = 7000$
$750x + 6750 - 675x = 7000$
$75x = 250$
$x = \frac{10}{3} = 3\frac{1}{3}$
$10 - \frac{10}{3} = \frac{20}{3} = 6\frac{2}{3}$
Mike should mix $3\frac{1}{3}$ pounds of Hazelnut Coffee and $6\frac{2}{3}$ pounds of Cinnamon Coffee.

49. Let x be April's score on her last test.
$\frac{82 + 88 + 80 + 95 + x}{5} \geq 85$
$345 + x \geq 425$
$x \geq 80$
April must score at least 80 on the last test to get a B in her Algebra class.

50. $c(x) = 35 + 1.50x$
To find the cost for 12 hours, we find $c(12)$.
$c(12) = 35 + 1.50(12)$
$c(12) = 35 + 18$
$c(12) = 53$
The cost of renting a chainsaw for 12 hours is $53.

Chapter 9

9.1 Experiencing Algebra the Exercise Way

1. Yes; $2a + 5$ is a polynomial.

3. Yes; $x^2 - 3x + 2$ is a polynomial.

5. No; $x^{1/2} - 6x - 7$ is not a polynomial because $x^{1/2}$ is not a monomial.

7. Yes; $5x^2 + 12xy + 2y^2$ is a polynomial.

9. Yes; $\sqrt{5}x - \sqrt{3}y$ is a polynomial.

11. No; $5\sqrt{x} - 3\sqrt{y}$ is not a polynomial because the first and second addends are not monomials.

13. Yes; $\frac{3}{5}x^3 - \frac{2}{3}x^2 + 4x - \frac{7}{10}$ is a polynomial.

15. No; $\frac{4}{x^2} + \frac{1}{x} - \frac{5}{7}$ is not a polynomial because the first and second addends are not monomials.

17. No; $a^{-2} + 17a^{-1} + 13$ is not a polynomial because the first and second addends are not monomials.

19. Yes; $0.07b^2 - 2.6b + 13.908$ is a polynomial.

21. $3a + 4b - 5c$ has three terms. Therefore, the expression is a trinomial.

23. $2z^2$ has one term. Therefore, it is a monomial.

25. $x - y$ has two terms. Therefore, the expression is a binomial.

27. $4p^4 - 2p^3 + 11p - 57$ has four terms. Therefore, the expression is a polynomial.

29. $6x^2 - 12 + 8x - 5x^2 + x - 17 = x^2 + 9x - 29$
There are three terms. Therefore, the expression is a trinomial.

31. $3b - 4 + 7b = 10b - 4$
There are two terms. Therefore, the expression is a binomial.

33. $x + 2x + 3x + 4x + 5x = 15x$
There is one term. Therefore, it is a monomial.

35. $\frac{1}{2}x + \frac{2}{3}y - \frac{3}{4}z$ has three terms. Therefore, the expression is a trinomial.

37. $2 - 15c$
2 or $2x^0$ has a degree of 0.
$-15c$ or $-15c^1$ has a degree of 1.
The degree of the polynomial is 1.

39. 123 or $123x^0$ has a degree of 0.
The degree of the polynomial is 0.

41. $5 + 5x - 4x - x = 5$
5 or $5x^0$ has a degree of 0.
The degree of the polynomial is 0.

43. $7x^5 + 2x^2y^2 - 12$
$7x^5$ has a degree of 5.
$2x^2y^2$ has a degree of 2 + 2 or 4.
-12 or $-12x^0$ has a degree of 0.
The degree of the polynomial is 5.

45. $\pi r^2 + 2\pi rh$
πr^2 has a degree of 2.
$2\pi rh$ or $2\pi r^1h^1$ has a degree of 1 + 1 or 2.
The degree of the polynomial is 2.

47. $4x^2yz^{12} - 8xy^2z^9 + 3x^3y^3z$
$4x^2yz^{12}$ has a degree of 2 + 1 + 12 or 15.
$-8xy^2z^9$ has a degree of 1 + 2 + 9 or 12.
$3x^3y^3z$ has a degree of 3 + 3 + 1 or 7.
The degree of the polynomial is 15.

49. $5-2a+3a^2+a^3=a^3+3a^2-2a+5$

51. $\frac{3}{5}x+\frac{4}{5}x^3-\frac{7}{15}x-\frac{8}{15}x^4$
$=-\frac{8}{15}x^4+\frac{4}{5}x^3+\frac{9}{15}x-\frac{7}{15}x$
$=-\frac{8}{15}x^4+\frac{4}{5}x^3+\frac{2}{15}x$

53. $0.1x^3-1.72+4.6x^2+3.06x^4$
$=3.06x^4+0.1x^3+4.6x^2-1.72$

55. $7b^2-6b^3+11-2b=11-2b+7b^2-6b^3$

57. $\frac{2}{7}x+\frac{1}{7}x^5-\frac{3}{14}x-\frac{5}{14}x^3$
$=\frac{4}{14}x-\frac{3}{14}x-\frac{5}{14}x^3+\frac{1}{7}x^5$
$=\frac{1}{14}x-\frac{5}{14}x^3+\frac{1}{7}x^5$

59. $0.5x^7-2.77+3.2x^3+9.76x^5$
$=-2.77+3.2x^3+9.76x^5+0.5x^7$

61. a. $3x^2-4x+1=3(4)^2-4(4)+1$
$=48-16+1$
$=33$
The result is 33.

b. $3x^2-4x+1=3(-2)^2-4(-2)+1$
$=12+8+1$
$=21$
The result is 21.

c. $3x^2-4x+1=3(0)^2-4(0)+1$
$=0-0+1$
$=1$
The result is 1.

63. a. $x^2+4xy+y^2=(2)^2+4(2)(3)+(3)^2$
$=4+24+9$
$=37$
The result is 37.

b. $x^2+4xy+y^2$
$=(-2)^2+4(-2)(-3)+(-3)^2$
$=4+24+9$
$=37$
The result is 37.

c. $x^2+4xy+y^2=(0)^2+4(0)(0)+(0)^2$
$=0+0+0$
$=0$
The result is 0.

65. a. $1.3m^3-2.5m^2+3.7m-4.9$
$=1.3(1)^3-2.5(1)^2+3.7(1)-4.9$
$=1.3-2.5+3.7-4.9$
$=-2.4$
The result is –2.4.

b. $1.3m^3-2.5m^2+3.7m-4.9$
$=1.3(2.5)^3-2.5(2.5)^2+3.7(2.5)-4.9$
$=20.3125-15.625+9.25-4.9$
$=9.0375$
The result is 9.0375.

c. $1.3m^3-2.5m^2+3.7m-4.9$
$=1.3(-1.5)^3-2.5(-1.5)^2$
$+3.7(-1.5)-4.9$
$=-4.3875-5.625-5.55-4.9$
$=-20.4625$
The result is –20.4625.

67. a. $\frac{2}{3}x^2-x-3=\frac{2}{3}\left(-\frac{3}{2}\right)^2-\left(-\frac{3}{2}\right)-3$
$=\frac{3}{2}+\frac{3}{2}-3$
$=0$
The result is 0.

b. $\frac{2}{3}x^2-x-3=\frac{2}{3}\left(\frac{1}{4}\right)^2-\frac{1}{4}-3$
$=\frac{1}{24}-\frac{1}{4}-3$
$=-\frac{77}{24}$
The result is $-\frac{77}{24}$.

c. $\frac{2}{3}x^2 - x - 3 = \frac{2}{3}(3)^2 - 3 - 3$
$= 6 - 3 - 3$
$= 0$
The result is 0.

69. Let x = measure of first side
$2x$ = measure of second side
$x + 1$ = measure of third side
$x + 2x + x + 1 = 4x + 1$
The perimeter measures $4x + 1$ units.
$4x + 1 = 4(8) + 1 = 33$
The perimeter measures 33 inches.

71. Let x = width
x = length
3 = height
$2x(x) + 2(x)(3) + 2(x)(3) = 2x^2 + 12x$
The surface area measures
$2x^2 + 12x$ square units.
$2(1.5)^2 + 12(1.5) = 22.5$
The surface area measures 22.5 m^2.

73. Area of $A = x(x)$
Area of $B = x(12)$
Area of $C = 3(7)$
$x(x) + x(12) + 3(7) = x^2 + 12x + 21$
The total area is $x^2 + 12x + 21$ m^2.

75. Let l = length of yard
w = width of yard
$lw - 15(20) = lw - 300$
The area not covered by the patio is
$lw - 300$ ft^2.
$120(75) - 300 = 8700$
The area not covered by the patio is
8700 ft^2.

77. Let x = length of one leg
y = length of other leg
$x^2 + y^2$
The square of the length of the hypotenuse is
$x^2 + y^2$.
$x^2 + y^2 = 4^2 + 7^2 = 65$
The square of the length of the hypotenuse is
65 ft^2. The hypotenuse is equal to $\sqrt{65}$ feet.

79. Let x = length of one side
$12x^2$
The replacement cost is $12x^2$ dollars.
$12x^2 = 12(12)^2 = 1728$
The replacement cost is \$1728.

81. Let x = length of one side

a. Revenue $= 200 + 12x^2$

b. Cost $= 75 + 2.75x^2$

c. $200 + 12x^2 = 200 + 12(15)^2 = 2900$
The revenue is \$2900.
$75 + 2.75x^2 = 75 + 2.75(15)^2 = 693.75$
The cost is \$693.75.
$2900 - 693.75 = 2206.25$
The net return is \$2206.25.

9.1 Experiencing Algebra the Calculator Way

1. $3x^3 - 8x^2 + 9x - 12$
when $x = \{-3, -1, 1, 3\}$
The calculator returns the list
$\{-192, -32, -8, 24\}$.

2. $4a^3 + 2a^2b + 2ab^2 + b^3$ for
$\{(-5, 2), (-3, 0), (-1, -2)\}$
The calculator returns the list
$\{-432, -108, -24\}$.

3. $\frac{1}{2}x^2 + \frac{3}{5}xy - \frac{1}{4}y^3$ for the (x, y) pairs
$\left(\frac{1}{2}, \frac{2}{5}\right), \left(\frac{1}{4}, \frac{3}{5}\right), \left(-\frac{1}{2}, 5\right), \left(0, \frac{2}{5}\right)$
The calculator returns the list
$\{0.229, 0.06725, -32.625, -0.016\}$, or, as
fractions, $\left\{\frac{229}{1000}, \frac{269}{4000}, -\frac{261}{8}, -\frac{2}{125}\right\}$.

9.2 Experiencing Algebra the Exercise Way

1. Possible answer:

x	$y = x^3 + 2x^2 - 5x - 6$	y
-2	$y = (-2)^3 + 2(-2)^2 - 5(-2) - 6 = 4$	4
-1	$y = (-1)^3 + 2(-1)^2 - 5(-1) - 6 = 0$	0
0	$y = (0)^3 + 2(0)^2 - 5(0) - 6 = -6$	-6
1	$y = (1)^3 + 2(1)^2 - 5(1) - 6 = -8$	-8
2	$y = (2)^3 + 2(2)^2 - 5(2) - 6 = 0$	0

3. Possible answer:

x	$y = x^2 + 4x + 1$	y
-2	$y = (-2)^2 + 4(-2) + 1 = -3$	-3
-1	$y = (-1)^2 + 4(-1) + 1 = -2$	-2
0	$y = (0)^2 + 4(0) + 1 = 1$	1
1	$y = (1)^2 + 4(1) + 1 = 6$	6
2	$y = (2)^2 + 4(2) + 1 = 13$	13

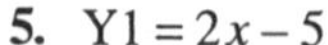

5. $Y1 = 2x - 5$

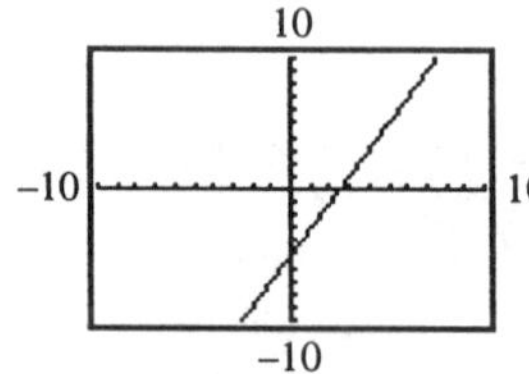

The range is all real numbers.

7. $Y1 = -x^2 + 6x - 4$

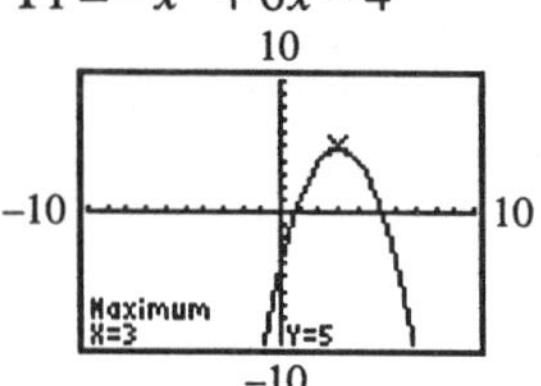

The range is all real numbers less than or equal to 5.

9. $Y1 = 2x^2 - 8x + 3$

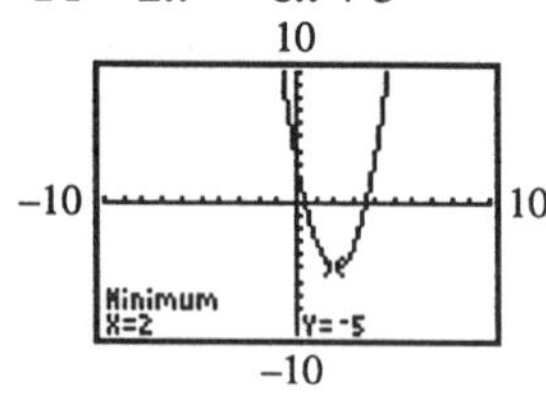

The range is all real numbers greater than or equal to –5.

11. $Y1 = x^3 + 3x^2 - 10x - 24$

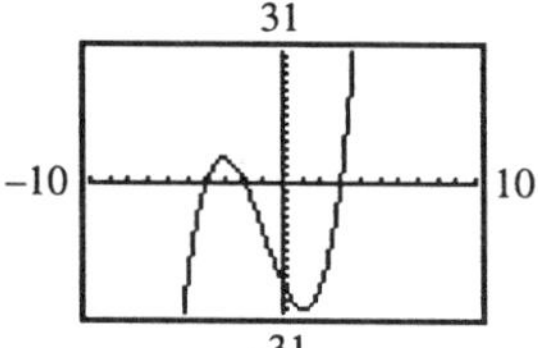

The y-coordinates have no absolute maximum or absolute minimum value. The range is all real numbers.

13. $Y1 = -x^4 - x^3 + 11x^2 + 9x - 18$

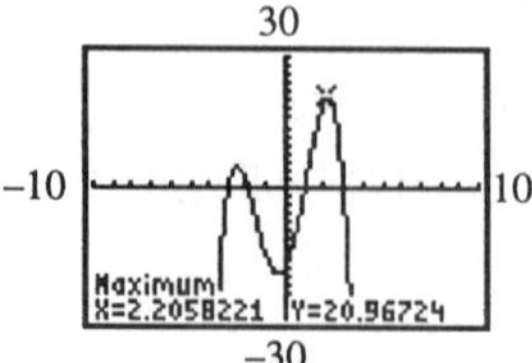

The y-coordinates have an absolute maximum at approximately 21. There is no absolute minimum value. The range is all real numbers less than 21.

15. $f(x) = -2x^3 + x^2 - 5x + 8$
$f(-2) = -2(-2)^3 + (-2)^2 - 5(-2) + 8$
$f(-2) = 16 + 4 + 10 + 8$
$f(-2) = 38$

17. $f(x) = -2x^3 + x^2 - 5x + 8$
$f(2) = -2(2)^3 + (2)^2 - 5(2) + 8$
$f(2) = -16 + 4 - 10 + 8$
$f(2) = -14$

19. $g(x) = 2.7x^3 - 1.5x^2 + 3.5x - 6.7$
$g(-1.7)$
$= 2.7(-1.7)^3 - 1.5(-1.7)^2 + 3.5(-1.7) - 6.7$
$g(-1.7) = -13.2651 - 4.335 - 5.95 - 6.7$
$g(-1.7) = -30.2501$

21. $g(x) = 2.7x^3 - 1.5x^2 + 3.5x - 6.7$
$g(2564.5) = 2.7(2564.5)^3 - 1.5(2564.5)^2 + 3.5(2564.5) - 6.7$
$g(2564.5) \approx 4.5528 \times 10^{10}$

23. $g(x) = 2.7x^3 - 1.5x^2 + 3.5x - 6.7$
$g(1.5)$
$= 2.7(1.5)^3 - 1.5(1.5)^2 + 3.5(1.5) - 6.7$
$g(1.5) = 9.1125 - 3.375 + 5.25 - 6.7$
$g(1.5) = 4.2875$

25. $g(x) = 2.7x^3 - 1.5x^2 + 3.5x - 6.7$
$g(1.1995) = 2.7(1.1995)^3 - 1.5(1.1995)^2 + 3.5(1.1995) - 6.7$
$g(1.1995) \approx -0.0001799$

27. $h(x) = \frac{1}{2}x^3 - \frac{3}{4}x^2 + \frac{3}{8}x - \frac{5}{8}$
$h\left(-\frac{3}{2}\right)$
$= \frac{1}{2}\left(-\frac{3}{2}\right)^3 - \frac{3}{4}\left(-\frac{3}{2}\right)^2 + \frac{3}{8}\left(-\frac{3}{2}\right) - \frac{5}{8}$
$h\left(-\frac{3}{2}\right) = -\frac{27}{16} - \frac{27}{16} - \frac{9}{16} - \frac{10}{16}$
$h\left(-\frac{3}{2}\right) = -\frac{73}{16}$

29. $h(x) = \frac{1}{2}x^3 - \frac{3}{4}x^2 + \frac{3}{8}x - \frac{5}{8}$
$h\left(\frac{5}{2}\right) = \frac{1}{2}\left(\frac{5}{2}\right)^3 - \frac{3}{4}\left(\frac{5}{2}\right)^2 + \frac{3}{8}\left(\frac{5}{2}\right) - \frac{5}{8}$
$h\left(\frac{5}{2}\right) = \frac{125}{16} - \frac{75}{16} + \frac{15}{16} - \frac{10}{16}$
$h\left(\frac{5}{2}\right) = \frac{55}{16}$

31. a. $R(x) = 150x - 5x^2$
$R(5) = 150(5) - 5(5)^2 = 625$
$R(10) = 150(10) - 5(10)^2 = 1000$
$R(15) = 150(15) - 5(15)^2 = 1125$
$R(20) = 150(20) - 5(20)^2 = 1000$
$R(25) = 150(25) - 5(25)^2 = 625$
$R(30) = 150(30) - 5(30)^2 = 0$
When 5, 10, 15, 20, 25, and 30 watches are ordered, the revenue is \$625, \$1000, \$1125, \$1000, \$625, and \$0 respectively.

b. $Y1 = 150x - 5x^2$

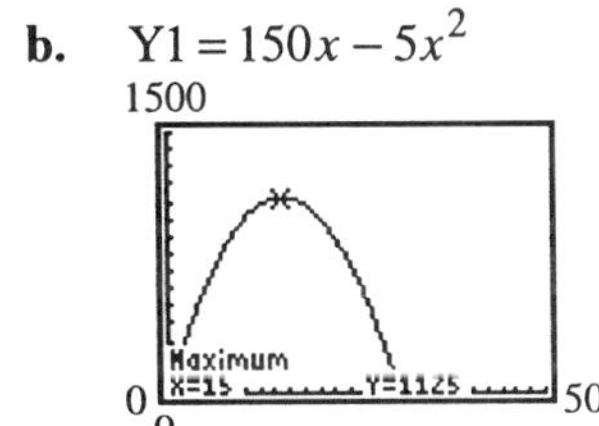

The range is all real numbers less than or equal to 1125.

c. This range shows us that the maximum revenue would be \$1125.

33. $S(x) = -0.011x^2 + 1.91x - 72.54$
$S(63) = -0.011(63)^2 + 1.91(63) - 72.54 = 4.131$
$S(65) = -0.011(65)^2 + 1.91(65) - 72.54 = 5.135$
$S(70) = -0.011(70)^2 + 1.91(70) - 72.54 = 7.26$
$S(75) = -0.011(75)^2 + 1.91(75) - 72.54 = 8.835$
$S(80) = -0.011(80)^2 + 1.91(80) - 72.54 = 9.86$
$S(85) = -0.011(85)^2 + 1.91(85) - 72.54 = 10.335$

Age (x)	63	65	70	75	80	85
Days ($S(x)$)	4.131	5.135	7.26	8.835	9.86	10.335

a. The function predicted a stay of 10 days for an 85-year-old woman. This is not a very good prediction of the actual data of an 8-day stay for an 85-year-old woman.

b. The function predicted a stay of 7 days for a 70-year-old woman. This is a fairly good prediction of the actual data of a 6-day stay, but not as good for a 9-day stay.

c. Answers will vary.

35. $s(t) = -16t^2 + 270t$
$Y1 = -16x^2 + 270x$

1500

Maximum
X=8.4375 Y=1139.0625

0 0 20

The domain is all real numbers between 0 and 8. This means the rocket is actually in the air for 8 seconds before it explodes. The range is all real numbers less than or equal to approximately 1139. That is, the maximum height of the rocket is 1139 ft.

9.2 Experiencing Algebra the Calculator Way

1. The domain is all real numbers greater than or equal to zero. The number of years, x, the salesperson spends in the same territory must be greater than or equal to zero.

2. The range is all real numbers between 0 and approximately 456,518.
$Y1 = -144,100 + 354,050x - 52,176x^2$

500,000

Maximum
X=3.392843 Y=456518.11

0 0 10

3. $S(x) = -144,100 + 354,050x - 52,176x^2$
$S(4.4)$
$= -144,100 + 354,050(4.4) - 52,176(4.4)^2$
$S(4.4) = 403,592.64$
Her estimated sales are $403,592.64. This is very close to the actual sales of $402,000.

4.

x	Actual Sales	Predicted Sales, $S(x)$	Difference ($)
0.8	$102,000	$105,747.36	–3747.36
1.1	$195,000	$182,222.04	12,777.96
1.4	$265,000	$249,305.04	15,694.96
1.5	$237,000	$269,579	–32,597
2.2	$345,000	$382,278.16	–37,278.16
3.0	$533,000	$448,466	84,534
3.6	$473,000	$454,279.04	18,720.96
4.4	$402,000	$403,592.64	–1592.64
4.5	$298,000	$392,561	–94,561
5.5	$263,000	$224,851	38,149

Answers will vary.

9.3 Experiencing Algebra the Exercise Way

1. $y = 1 - x - x^2 - x^3$ is nonquadratic because $1 - x - x^2 - x^3$ is a third-degree polynomial (x is cubed).

3. $f(x) = 8x + 11$ is nonquadratic because $a = 0$.

5. $g(x) = 0.5x^2 + 2.6x - 8.4$ is quadratic. It is in standard form.

7. $y = -2x^2 - 5x - 7$ is quadratic. It is written in standard form.

9. $a = \pi r^2$ is quadratic. It can be written in standard form $a = \pi r^2 + 0r + 0$.

11. $S = 6e^2$ is quadratic. It can be written in standard form $S = 6e^2 + 0e + 0$.

13. $y = \frac{7}{x^2} + 3x - 12$ is nonquadratic because $\frac{7}{x^2} + 3x - 12$ is not a polynomial (the squared variable term is in the denominator of a fraction).

15. $y = 2x^2 - 2x + 5$ is quadratic. It is written in standard form.

	function	a	b	c	graph wide/narrow	graph concave upward/ downward	graph vertex	axis of symmetry
17.	$y=-5x^2+10x+1$	–5	10	1	narrow	downward	(1, 6)	$x=1$
19.	$y=0.6x^2+6x-2$	0.6	6	–2	wide	upward	(–5, –17)	$x=-5$
21.	$y=2x^2+3x+5$	2	3	5	narrow	upward	(–0.75, 3.875)	$x=-0.75$
23.	$y=-\frac{1}{4}x^2+x-3$	$-\frac{1}{4}$	1	–3	wide	downward	(2, –2)	$x=2$
25.	$f(x)=x^2+8x+1$	1	8	1	neither	upward	(–4, –15)	$x=-4$

27. $f(x)=2x^2+5x-7$

Determine the vertex.

$$x=-\frac{b}{2a}=-\frac{5}{2(2)}=-\frac{5}{4}$$

$$f\left(-\frac{5}{4}\right)=2\left(-\frac{5}{4}\right)^2+5\left(-\frac{5}{4}\right)-7=-\frac{81}{8}$$

The vertex is $\left(-\frac{5}{4}, -\frac{81}{8}\right)$.

The axis of symmetry is the line $x=-\frac{5}{4}$.

x	$f(x)=2x^2+5x-7$	$f(x)$
–3	$f(-3)=2(-3)^2+5(-3)-7$ $f(-3)=-4$	–4
–2	$f(-2)=2(-2)^2+5(-2)-7$ $f(-2)=-9$	–9
0	$f(0)=2(0)^2+5(0)-7$ $f(0)=-7$	–7
1	$f(1)=2(1)^2+5(1)-7$ $f(1)=0$	0

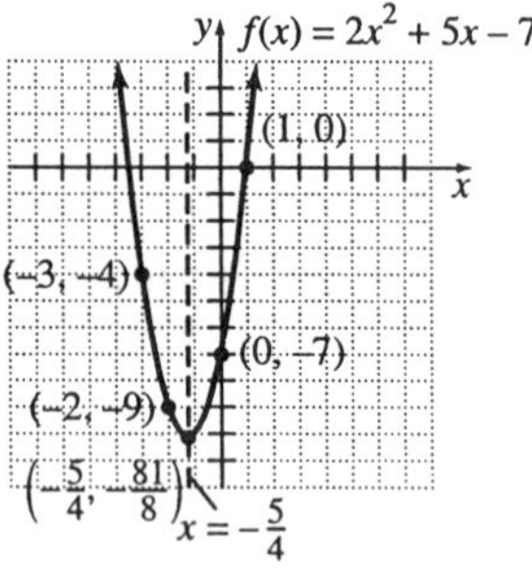

29. $y=\frac{1}{6}x^2+3x+12$

Determine the vertex.

$$x=-\frac{b}{2a}=-\frac{3}{2\left(\frac{1}{6}\right)}=-9$$

$$y=\frac{1}{6}(-9)^2+3(-9)+12=-\frac{3}{2}$$

The vertex is $\left(-9, -\frac{3}{2}\right)$.

The axis of symmetry is the line $x=-9$.

x	$y=\frac{1}{6}x^2+3x+12$	y
-14	$y=\frac{1}{6}(-14)^2+3(-14)+12$ $y=2\frac{2}{3}$	$2\frac{2}{3}$
-12	$y=\frac{1}{6}(-12)^2+3(-12)+12$ $y=0$	0
-6	$y=\frac{1}{6}(-6)^2+3(-6)+12$ $y=0$	0
-3	$y=\frac{1}{6}(-3)^2+3(-3)+12$ $y=4\frac{1}{2}$	$4\frac{1}{2}$

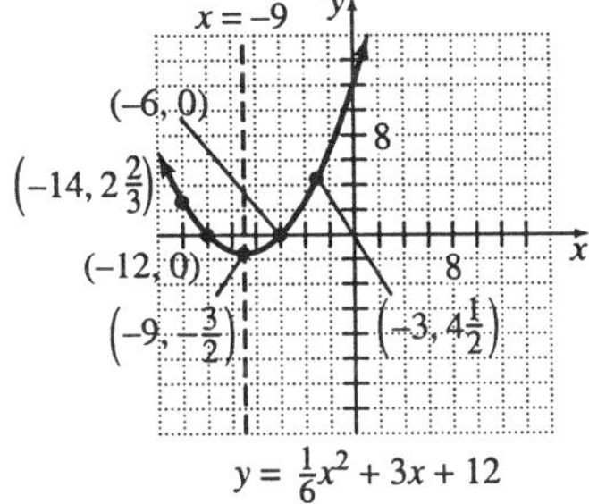

31. $h(x)=14+5x-x^2$

Determine the vertex.

$$x=-\frac{b}{2a}=-\frac{5}{2(-1)}=\frac{5}{2}$$

$$h\left(\frac{5}{2}\right)=14+5\left(\frac{5}{2}\right)-\left(\frac{5}{2}\right)^2=\frac{81}{4}$$

The vertex is $\left(\frac{5}{2}, \frac{81}{4}\right)$.

The axis of symmetry is the line $x=\frac{5}{2}$.

x	$h(x)=14+5x-x^2$	$h(x)$
0	$h(0)=14+5(0)-(0)^2$ $h(0)=14$	14
1	$h(1)=14+5(1)-(1)^2$ $h(1)=18$	18
4	$h(4)=14+5(4)-(4)^2$ $h(4)=18$	18
5	$h(5)=14+5(5)-(5)^2$ $h(5)=14$	14

33. $y=-2x^2+8x-3$

Determine the vertex.

$$x=-\frac{b}{2a}=-\frac{8}{2(-2)}=2$$

$$y=-2(2)^2+8(2)-3=5$$

The vertex is (2, 5).

The axis of symmetry is the line $x=2$.

x	$y=-2x^2+8x-3$	y
0	$y=-2(0)^2+8(0)-3$ $y=-3$	-3
1	$y=-2(1)^2+8(1)-3$ $y=3$	3
3	$y=-2(3)^2+8(3)-3$ $y=3$	3
4	$y=-2(4)^2+8(4)-3$ $y=-3$	-3

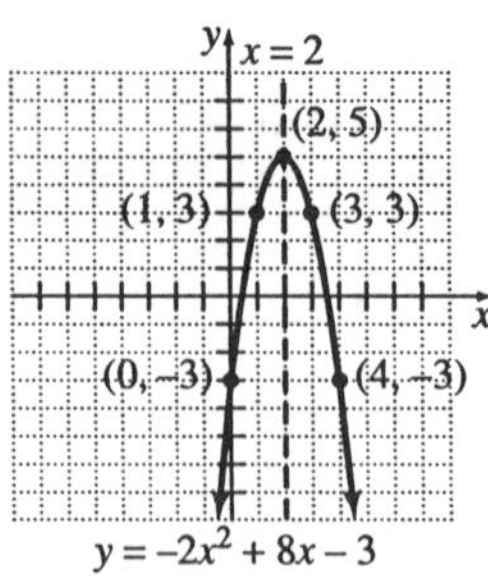

35. $g(x) = 0.8x^2 - 1.2x$

Determine the vertex.

$$x = -\frac{b}{2a} = -\frac{(-1.2)}{2(0.8)} = 0.75$$

$$g(0.75) = 0.8(0.75)^2 - 1.2(0.75) = -0.45$$

The vertex is (0.75, –0.45).
The axis of symmetry is the line $x = 0.75$.

x	$g(x) = 0.8x^2 - 1.2x$	$g(x)$
–1	$g(-1) = 0.8(-1)^2 - 1.2(-1)$ $g(-1) = 2$	2
0	$g(0) = 0.8(0)^2 - 1.2(0)$ $g(0) = 0$	0
2	$g(2) = 0.8(2)^2 - 1.2(2)$ $g(2) = 0.8$	0.8
4	$g(4) = 0.8(4)^2 - 1.2(4)$ $g(4) = 8$	8

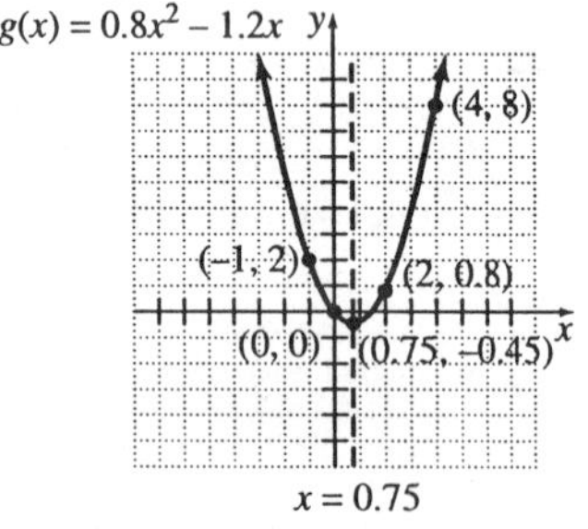

37. $y = 0.4x^2$

Determine the vertex.

$$x = -\frac{b}{2a} = -\frac{0}{2(0.4)} = 0$$

$$y = 0.4(0)^2 = 0$$

The vertex is (0, 0).
The axis of symmetry is the line $x = 0$.

x	$y = 0.4x^2$	y
–5	$y = 0.4(-5)^2$ $y = 10$	10
–2	$y = 0.4(-2)^2$ $y = 1.6$	1.6
2	$y = 0.4(2)^2$ $y = 1.6$	1.6
5	$y = 0.4(5)^2$ $y = 10$	10

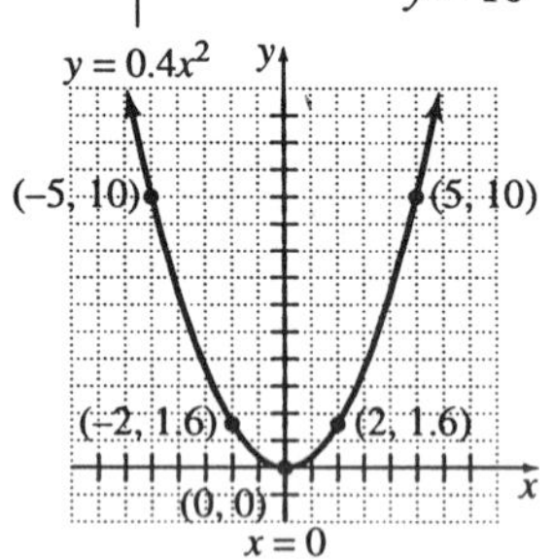

39. $y = 3x^2 - 3$

Determine the vertex.

$$x = -\frac{b}{2a} = -\frac{0}{2(3)} = 0$$

$$y = 3(0)^2 - 3 = -3$$

The vertex is (0, –3).
The axis of symmetry is the line $x = 0$.

x	$y = 3x^2 - 3$	y
−2	$y = 3(-2)^2 - 3$ $y = 9$	9
−1	$y = 3(-1)^2 - 3$ $y = 0$	0
1	$y = 3(1)^2 - 3$ $y = 0$	0
2	$y = 3(2)^2 - 3$ $y = 9$	9

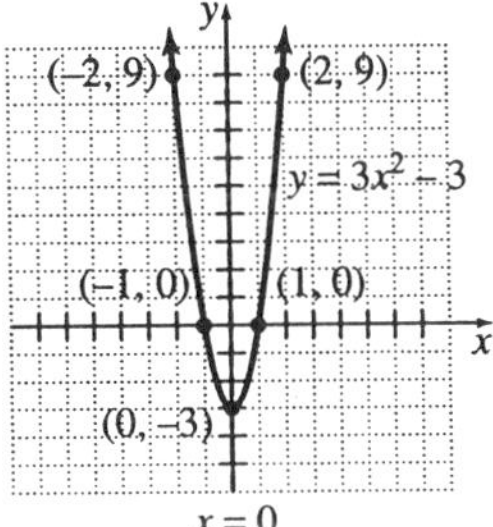

41. $f(x) = -0.5x^2 + 3x$

Determine the vertex.

$x = -\frac{b}{2a} = -\frac{3}{2(-0.5)} = 3$

$f(3) = -0.5(3)^2 + 3(3) = 4.5$

The vertex is (3, 4.5).

The axis of symmetry is the line $x = 3$.

x	$f(x) = -0.5x^2 + 3x$	$f(x)$
−2	$f(-2) = -0.5(-2)^2 + 3(-2)$ $f(-2) = -8$	−8
0	$f(0) = -0.5(0)^2 + 3(0)$ $f(0) = 0$	0
6	$f(6) = -0.5(6)^2 + 3(6)$ $f(6) = 0$	0
8	$f(8) = -0.5(8)^2 + 3(8)$ $f(8) = -8$	−8

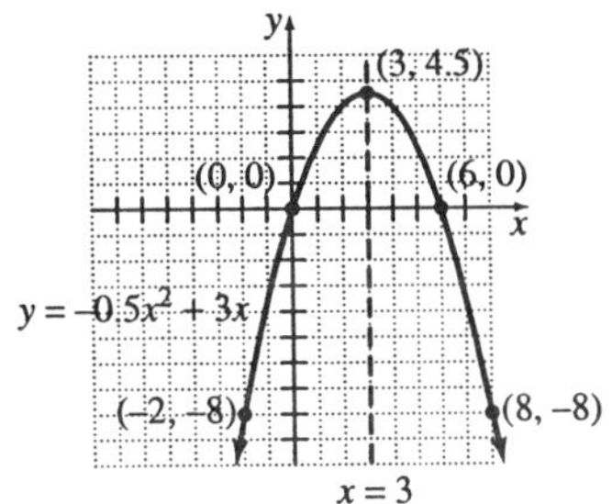

43. $s(t) = -16t^2 + 12t + 24$

$Y1 = -16x^2 + 12x + 24$

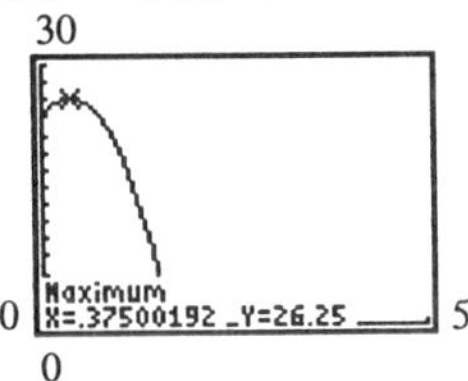

The vertex is approximately (0.375, 26.25). Therefore, the maximum height is 26.25 feet.

$s(t) = 0$ at $t \approx 1.7$

Therefore, the apple will hit the ground after approximately 1.7 seconds.

45. $A(x) = 140x - x^2$

$Y1 = 140x - x^2$

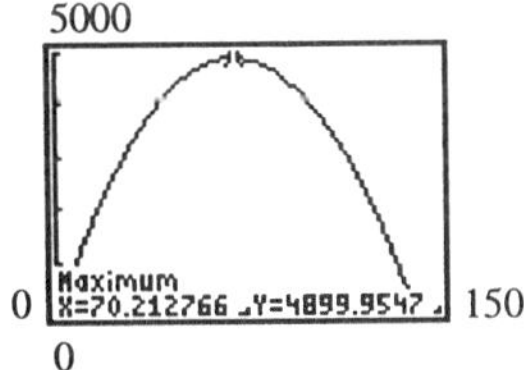

The vertex is (70, 4900).

The room has a maximum area of 4900 ft^2 when the width is 70 feet.

47. $s(x) = 5x^2 - 80x + 400$

$Y1 = 5x^2 - 80x + 400$

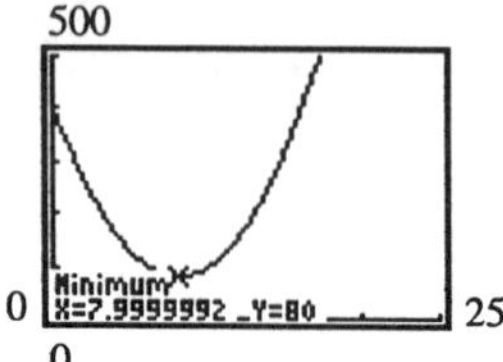

The vertex is (8, 80).
At a height of 8 inches the square of the hypotenuse will be minimized at 80 in^2.

49. $R(x) = 46x - x^2$

$Y1 = 46x - x^2$

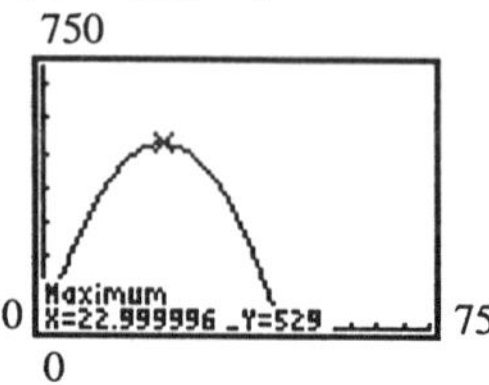

The vertex is (23, 529).
At x = $23 the revenue will be at a maximum of $529.

9.3 Experiencing Algebra the Calculator Way

1. $Y1 = x^2 - 2x$
$Y2 = x^2$
$Y3 = x^2 + 2x$

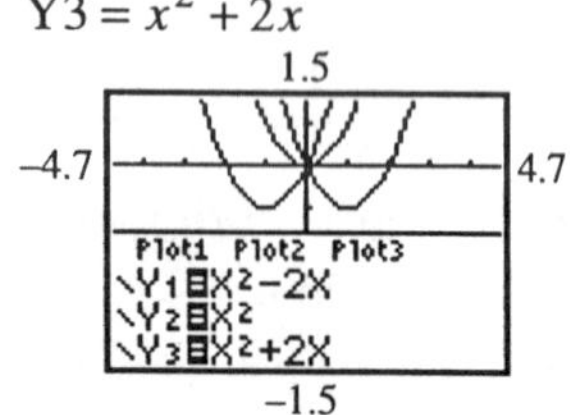

2. $Y1 = x^2 - 1$
$Y2 = x^2$
$Y3 = x^2 + 1$

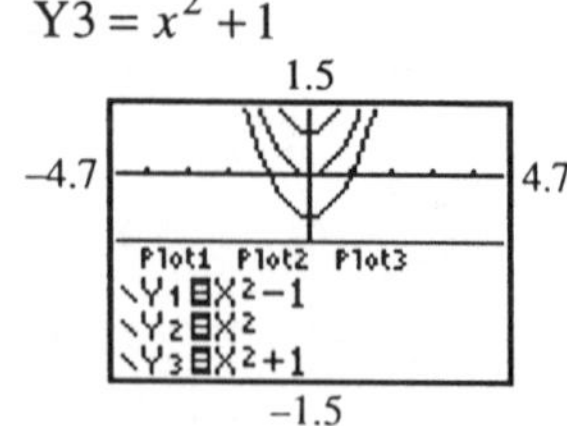

3. $Y1 = 0.3x^2$
$Y2 = x^2$
$Y3 = 3x^2$

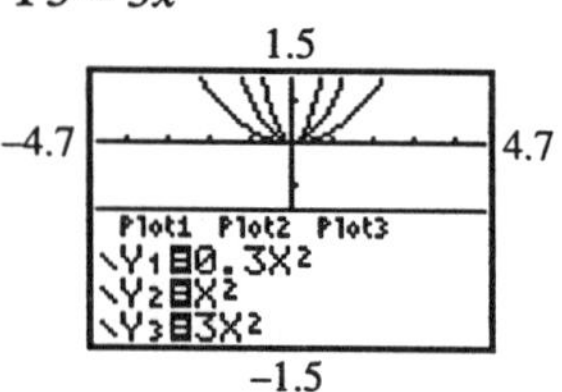

4. $Y1 = -3x^2$
$Y2 = 3x^2$

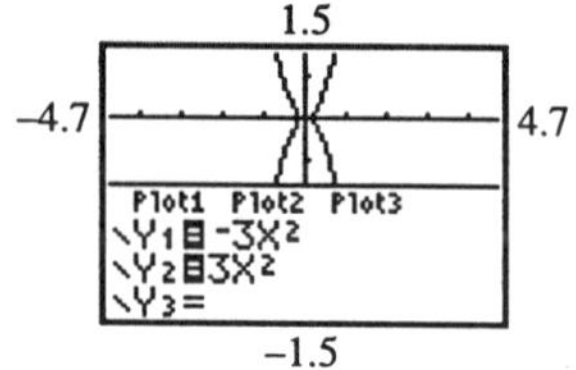

5. $Y1 = 3x^2$
$Y2 = 3x^2 + 1$
$Y3 = 3x^2 - 2$

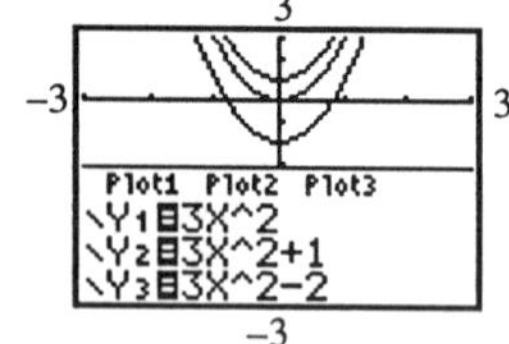

6. $Y1 = x^2 + 1$
$Y2 = x^2 + 2x + 1$
$Y3 = x^2 - 2x + 1$

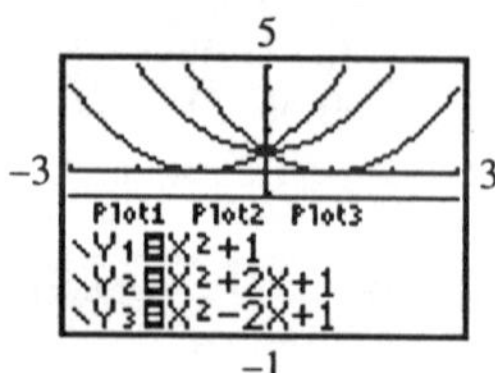

Answers may vary.

9.4 Experiencing Algebra the Exercise Way

1. Substitute (2, 0), (–4, 0), and (0, –8).

(1) $0 = a(2)^2 + b(2) + c$

(2) $0 = a(-4)^2 + b(-4) + c$

(3) $-8 = a(0)^2 + b(0) + c$

Simplify.

(1) $0 = 4a + 2b + c$

(2) $0 = 16a - 4b + c$

(3) $-8 = c$

Substitute –8 for c in (1) and (2).

(1) $8 = 4a + 2b$

(2) $8 = 16a - 4b$

Multiply equation (1) by 2 and solve for a and b.

(1) $16 = 8a + 4b$

(2) $\underline{8 = 16a - 4b}$

$24 = 24a$

$1 = a$

(1) $8 = 4(1) + 2b$

$4 = 2b$

$2 = b$

$a = 1,\ b = 2,\ c = -8$

$y = ax^2 + bx + c$

$y = x^2 + 2x - 8$

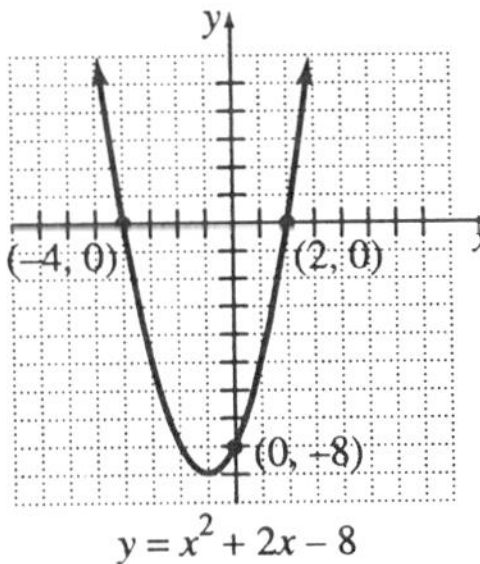

$y = x^2 + 2x - 8$

The graph passes through the given points.

3. Substitute (–4, 0), $\left(\frac{1}{2}, 0\right)$, and (0, –4).

(1) $0 = a(-4)^2 + b(-4) + c$

(2) $0 = a\left(\frac{1}{2}\right)^2 + b\left(\frac{1}{2}\right) + c$

(3) $-4 = a(0)^2 + b(0) + c$

Simplify.

(1) $0 = 16a - 4b + c$

(2) $0 = \frac{1}{4}a + \frac{1}{2}b + c$

(3) $-4 = c$

Substitute –4 for c in (1) and (2).

(1) $4 = 16a - 4b$

(2) $4 = \frac{1}{4}a + \frac{1}{2}b$

Multiply equation (2) by 8 and solve for a and b.

(1) $4 = 16a - 4b$

(2) $\underline{32 = 2a + 4b}$

$36 = 18a$

$2 = a$

(1) $4 = 16(2) - 4b$

$4 = 32 - 4b$

$-28 = -4b$

$7 = b$

$a = 2,\ b = 7,\ c = -4$

$y = ax^2 + bx + c$

$y = 2x^2 + 7x - 4$

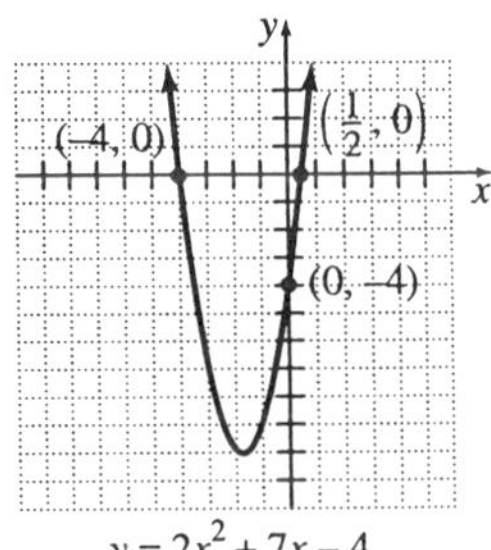

$y = 2x^2 + 7x - 4$

The graph passes through the given points.

5. Substitute (0, 3), (3, 0), and (–3, 0).

(1) $3 = a(0)^2 + b(0) + c$

(2) $0 = a(3)^2 + b(3) + c$

(3) $0 = a(-3)^2 + b(-3) + c$

Simplify.

(1) $3 = c$

(2) $0 = 9a + 3b + c$

(3) $0 = 9a - 3b + c$

Substitute 3 for c in (2) and (3).

(2) $-3 = 9a + 3b$

(3) $-3 = 9a - 3b$

Solve for a and b.

$$\begin{array}{ll} (2) & -3 = 9a + 3b \\ (3) & \underline{-3 = 9a - 3b} \\ & -6 = 18a \\ & -\frac{1}{3} = a \end{array}$$

(2) $-3 = 9\left(-\frac{1}{3}\right) + 3b$

$-3 = -3 + 3b$
$0 = 3b$
$0 = b$

$a = -\frac{1}{3},\ b = 0,\ c = 3$

$y = ax^2 + bx + c$

$y = -\frac{1}{3}x^2 + 3$

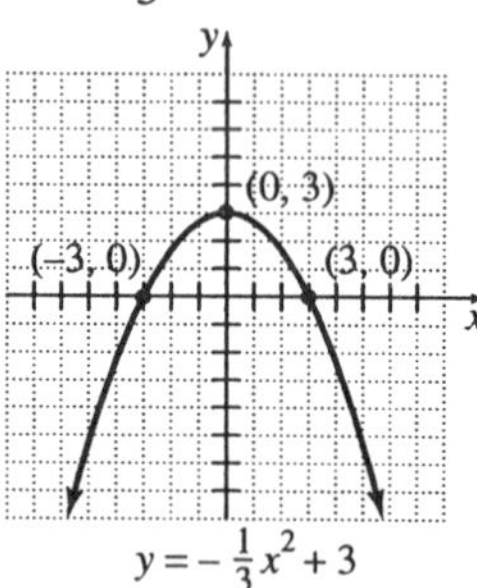

$y = -\frac{1}{3}x^2 + 3$

The graph passes through the given points.

7. Substitute (0, –4), (2, 0), and (–4, 0).

(1) $-4 = a(0)^2 + b(0) + c$

(2) $0 = a(2)^2 + b(2) + c$

(3) $0 = a(-4)^2 + b(-4) + c$

Simplify.

(1) $-4 = c$
(2) $0 = 4a + 2b + c$
(3) $0 = 16a - 4b + c$

Substitute –4 for c in (2) and (3).

(2) $4 = 4a + 2b$
(3) $4 = 16a - 4b$

Multiply equation (2) by 2 and solve for a and b.

$$\begin{array}{ll} (2) & 8 = 8a + 4b \\ (3) & \underline{4 = 16a - 4b} \\ & 12 = 24a \\ & \frac{1}{2} = a \end{array}$$

(2) $4 = 4\left(\frac{1}{2}\right) + 2b$

$4 = 2 + 2b$
$2 = 2b$
$1 = b$

$a = \frac{1}{2},\ b = 1,\ c = -4$

$y = ax^2 + bx + c$

$y = \frac{1}{2}x^2 + x - 4$

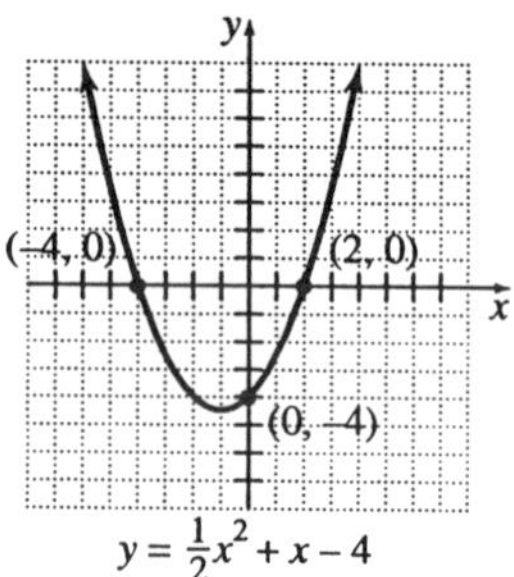

$y = \frac{1}{2}x^2 + x - 4$

The graph passes through the given points.

9. Substitute $\left(0, \frac{2}{3}\right)$, (–2, 0), and (2, 0).

(1) $\frac{2}{3} = a(0)^2 + b(0) + c$

(2) $0 = a(-2)^2 + b(-2) + c$

(3) $0 = a(2)^2 + b(2) + c$

Simplify.

(1) $\frac{2}{3} = c$

(2) $0 = 4a - 2b + c$
(3) $0 = 4a + 2b + c$

Substitute $\frac{2}{3}$ for c in (2) and (3).

(2) $-\frac{2}{3} = 4a - 2b$

(3) $-\frac{2}{3} = 4a + 2b$

Solve for a and b.

(2) $-\frac{2}{3} = 4a - 2b$

(3) $-\frac{2}{3} = 4a + 2b$

$-\frac{4}{3} = 8a$

$-\frac{1}{6} = a$

(2) $-\frac{2}{3} = 4\left(-\frac{1}{6}\right) - 2b$

$-\frac{2}{3} = -\frac{2}{3} - 2b$

$0 = -2b$

$0 = b$

$a = -\frac{1}{6},\ b = 0,\ c = \frac{2}{3}$

$y = ax^2 + bx + c$

$y = -\frac{1}{6}x^2 + \frac{2}{3}$

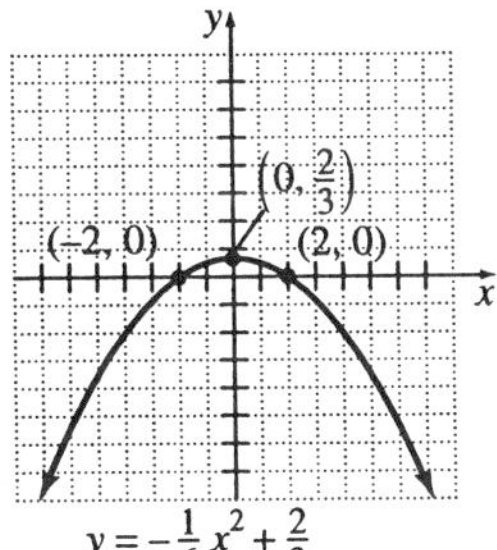

The graph passes through the given points.

11. Substitute (0, –2), $\left(\frac{1}{2}, 0\right)$, and $\left(\frac{9}{2}, 0\right)$.

(1) $-2 = a(0)^2 + b(0) + c$

(2) $0 = a\left(\frac{1}{2}\right)^2 + b\left(\frac{1}{2}\right) + c$

(3) $0 = a\left(\frac{9}{2}\right)^2 + b\left(\frac{9}{2}\right) + c$

Simplify.

(1) $-2 = c$

(2) $0 = \frac{1}{4}a + \frac{1}{2}b + c$

(3) $0 = \frac{81}{4}a + \frac{9}{2}b + c$

Substitute –2 for c in (2) and (3).

(2) $2 = \frac{1}{4}a + \frac{1}{2}b$

(3) $2 = \frac{81}{4}a + \frac{9}{2}b$

Multiply equation (2) by –9 and solve for a and b.

(2) $-18 = -\frac{9}{4}a - \frac{9}{2}b$

(3) $2 = \frac{81}{4}a + \frac{9}{2}b$

$-16 = \frac{72}{4}a$

$-\frac{8}{9} = a$

(2) $2 = \frac{1}{4}\left(-\frac{8}{9}\right) + \frac{1}{2}b$

$2 = -\frac{2}{9} + \frac{1}{2}b$

$\frac{20}{9} = \frac{1}{2}b$

$\frac{40}{9} = b$

$a = -\frac{8}{9},\ b = \frac{40}{9},\ c = -2$

$y = ax^2 + bx + c$

$y = -\frac{8}{9}x^2 + \frac{40}{9}x - 2$

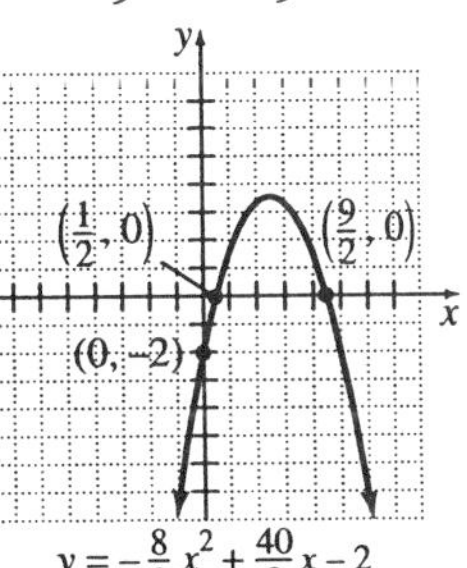

The graph passes through the given points.

13. Substitute (2, –15), (–2, –7), and (–4, 9).

(1) $-15 = a(2)^2 + b(2) + c$

(2) $-7 = a(-2)^2 + b(-2) + c$

(3) $9 = a(-4)^2 + b(-4) + c$

Simplify.
(1) $-15 = 4a + 2b + c$
(2) $-7 = 4a - 2b + c$
(3) $9 = 16a - 4b + c$
Solve for a, b, and c.

$$\begin{array}{ll} (1) & 15 = -4a - 2b - c \\ (2) & \underline{-7 = 4a - 2b + c} \\ & 8 = \quad -4b \\ & -2 = b \end{array}$$

(2) $-7 = 4a - 2(-2) + c$
(3) $9 = 16a - 4(-2) + c$

$$\begin{array}{ll} (2) & 11 = -4a - c \\ (3) & \underline{1 = 16a + c} \\ & 12 = 12a \\ & 1 = a \end{array}$$

(1) $-15 = 4(1) + 2(-2) + c$
$-15 = c$
$a = 1,\ b = -2,\ c = -15$
$y = ax^2 + bx + c$
$y = x^2 - 2x - 15$

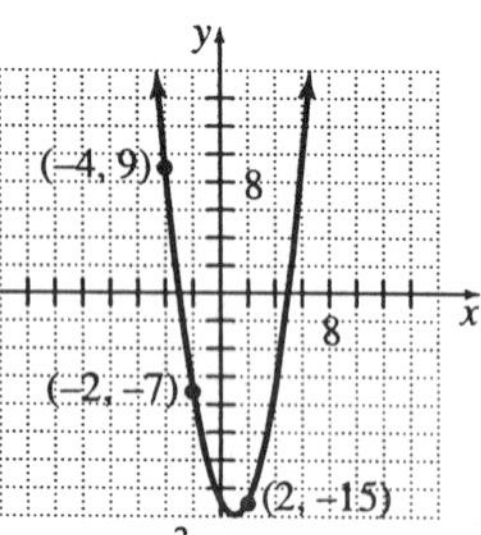

$y = x^2 - 2x - 15$

The graph passes through the given points.

15. Substitute (1, −10), (−2, 5), and (2, −7).
(1) $-10 = a(1)^2 + b(1) + c$
(2) $5 = a(-2)^2 + b(-2) + c$
(3) $-7 = a(2)^2 + b(2) + c$
Simplify.
(1) $-10 = a + b + c$
(2) $5 = 4a - 2b + c$
(3) $-7 = 4a + 2b + c$
Solve for a, b, and c.

$$\begin{array}{ll} (2) & -5 = -4a + 2b - c \\ (3) & \underline{-7 = 4a + 2b + c} \\ & -12 = \quad 4b \\ & -3 = b \end{array}$$

(1) $-10 = a + (-3) + c$
(2) $5 = 4a - 2(-3) + c$

$$\begin{array}{ll} (1) & 7 = -a - c \\ (2) & \underline{-1 = 4a + c} \\ & 6 = 3a \\ & 2 = a \end{array}$$

(1) $-10 = 2 - 3 + c$
$-9 = c$
$a = 2,\ b = -3,\ c = -9$
$y = ax^2 + bx + c$
$y = 2x^2 - 3x - 9$

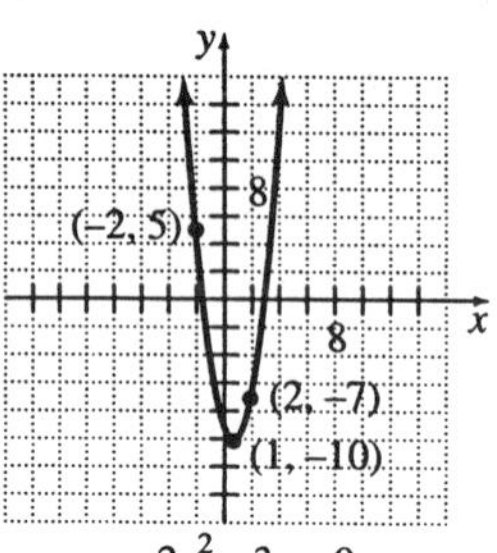

$y = 2x^2 - 3x - 9$

The graph passes through the given points.

17. Substitute (3, 2), (−3, 14), and (6, 5).
(1) $2 = a(3)^2 + b(3) + c$
(2) $14 = a(-3)^2 + b(-3) + c$
(3) $5 = a(6)^2 + b(6) + c$
Simplify.
(1) $2 = 9a + 3b + c$
(2) $14 = 9a - 3b + c$
(3) $5 = 36a + 6b + c$
Solve for a, b, and c.

$$\begin{array}{ll} (1) & -2 = -9a - 3b - c \\ (2) & \underline{14 = 9a - 3b + c} \\ & 12 = \quad -6b \\ & -2 = b \end{array}$$

(2) $14 = 9a - 3(-2) + c$
(3) $5 = 36a + 6(-2) + c$

$$\begin{array}{ll} (1) & -8 = -9a - c \\ (2) & \underline{17 = 36a + c} \\ & 9 = 27a \\ & \frac{1}{3} = a \end{array}$$

(1) $2 = 9\left(\frac{1}{3}\right) + 3(-2) + c$
$2 = 3 - 6 + c$
$2 = -3 + c$
$5 = c$

$a = \frac{1}{3}$, $b = -2$, $c = 5$

$y = ax^2 + bx + c$

$y = \frac{1}{3}x^2 - 2x + 5$

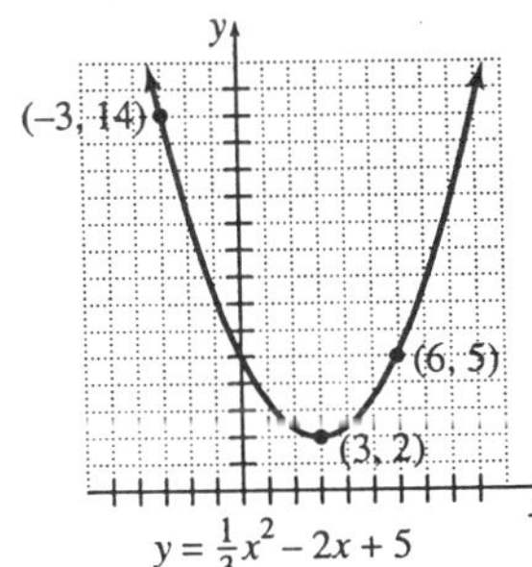

The graph passes through the given points.

19. Substitute (–1, –2), (1, 6), and (4, 3).

(1) $-2 = a(-1)^2 + b(-1) + c$
(2) $6 = a(1)^2 + b(1) + c$
(3) $3 = a(4)^2 + b(4) + c$

Simplify.

(1) $-2 = a - b + c$
(2) $6 = a + b + c$
(3) $3 = 16a + 4b + c$

Solve for a, b, and c.

(1) $2 = -a + b - c$
(2) $6 = a + b + c$
$8 = 2b$
$4 = b$

(2) $6 = a + 4 + c$
(3) $3 = 16a + 4(4) + c$

(2) $-2 = -a - c$
(3) $-13 = 16a + c$
$-15 = 15a$
$-1 = a$

(1) $-2 = -1 - 4 + c$
$-2 = -5 + c$
$3 = c$

$a = -1$, $b = 4$, $c = 3$

$y = ax^2 + bx + c$

$y = -x^2 + 4x + 3$

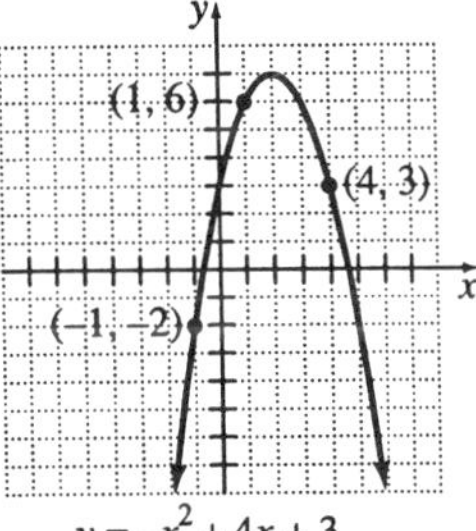

The graph passes through the given points.

21. Substitute (–1, –3), (–2, –5), and (1, –11).

(1) $-3 = a(-1)^2 + b(-1) + c$
(2) $-5 = a(-2)^2 + b(-2) + c$
(3) $-11 = a(1)^2 + b(1) + c$

Simplify.

(1) $-3 = a - b + c$
(2) $-5 = 4a - 2b + c$
(3) $-11 = a + b + c$

Solve for a, b, and c.

(1) $3 = -a + b - c$
(3) $-11 = a + b + c$
$-8 = 2b$
$-4 = b$

(1) $-3 = a - (-4) + c$
(2) $-5 = 4a - 2(-4) + c$

(1) $7 = -a - c$
(2) $-13 = 4a + c$
$-6 = 3a$
$-2 = a$

(1) $-3 = -2 - (-4) + c$
$-3 = 2 + c$
$-5 = c$

$a = -2$, $b = -4$, $c = -5$

$y = ax^2 + bx + c$

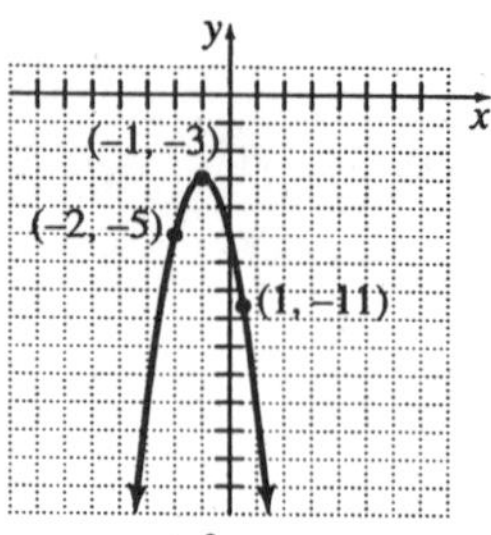

$y = -2x^2 - 4x - 5$

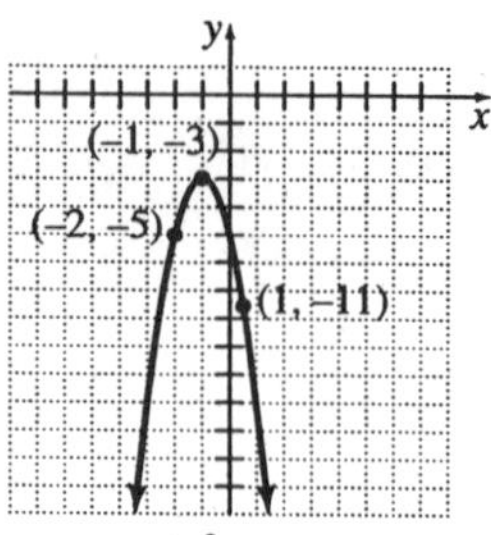

The graph passes through the given points.

23. Substitute (−2, 4), $\left(-3, \frac{5}{2}\right)$, and $\left(-1, \frac{13}{2}\right)$.

(1) $4 = a(-2)^2 + b(-2) + c$

(2) $\frac{5}{2} = a(-3)^2 + b(-3) + c$

(3) $\frac{13}{2} = a(-1)^2 + b(-1) + c$

Simplify.

(1) $4 = 4a - 2b + c$

(2) $\frac{5}{2} = 9a - 3b + c$

(3) $\frac{13}{2} = a - b + c$

Solve for a, b, and c.

(1) $-4 = -4a + 2b - c$

(2) $\frac{5}{2} = 9a - 3b + c$

(4) $-\frac{3}{2} = 5a - b$

(2) $-\frac{5}{2} = -9a + 3b - c$

(3) $\frac{13}{2} = a - b + c$

(5) $4 = -8a + 2b$

(4) $-3 = 10a - 2b$

(5) $4 = -8a + 2b$

$1 = 2a$

$\frac{1}{2} = a$

(1) $4 = 4\left(\frac{1}{2}\right) - 2b + c$

(3) $\frac{13}{2} = \frac{1}{2} - b + c$

(1) $-2 = 2b - c$

(3) $6 = -b + c$

$4 = b$

(1) $4 = 4\left(\frac{1}{2}\right) - 2(4) + c$

$4 = -6 + c$

$10 = c$

$a = \frac{1}{2},\ b = 4,\ c = 10$

$y = ax^2 + bx + c$

$y = \frac{1}{2}x^2 + 4x + 10$

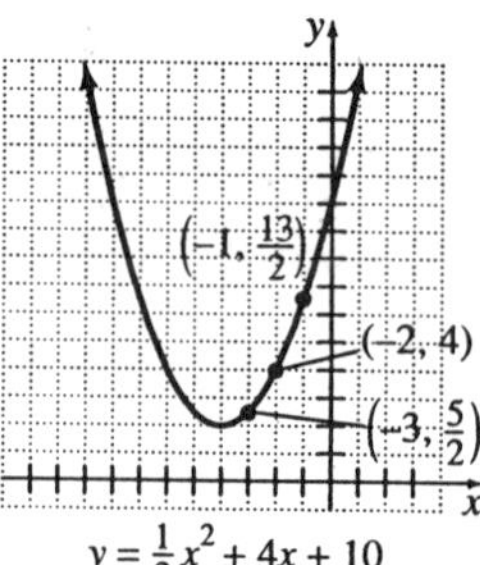

The graph passes through the given points.

25. Substitute (0, 150), (1, 146), and (2, 110).

(1) $150 = a(0)^2 + b(0) + c$

(2) $146 = a(1)^2 + b(1) + c$

(3) $110 = a(2)^2 + b(2) + c$

Simplify.

(1) $150 = c$

(2) $146 = a + b + c$

(3) $110 = 4a + 2b + c$

Substitute 150 for c in (2) and (3).

(2) $146 = a + b + 150$

(3) $110 = 4a + 2b + 150$

Simplify and solve for a and b.

(2) $-4 = a + b$

(3) $20 = -2a - b$

$16 = -a$

$-16 = a$

(2) $146 = -16 + b + 150$

$12 = b$

$a = -16,\ b = 12,\ c = 150$

$s(t) = at^2 + bt + c$

The position function is

$s(t) = -16t^2 + 12t + 150.$

$Y1 = -16x^2 + 12x + 150$
$Y2 = 0$

200

0 Intersection X=3.4597407 Y=0 5

0

The ball will hit the ground in approximately 3.46 seconds.

27. Substitute (0, 400), (2, 336), and (4, 144).

(1) $400 = a(0)^2 + b(0) + c$

(2) $336 = a(2)^2 + b(2) + c$

(3) $144 = a(4)^2 + b(4) + c$

Simplify.

(1) $400 = c$
(2) $336 = 4a + 2b + c$
(3) $144 = 16a + 4b + c$

Substitute 400 for c in (2) and (3).

(2) $336 = 4a + 2b + 400$
(3) $144 = 16a + 4b + 400$

Simplify and solve for a and b.

(2) $128 = -8a - 4b$
(3) $\underline{-256 = 16a + 4b}$
$-128 = 8a$
$-16 = a$

(2) $336 = 4(-16) + 2b + 400$
$0 = 2b$
$0 = b$

$a = -16,\ b = 0,\ c = 400$

$s(t) = at^2 + bt + c$

The position function is $s(t) = -16t^2 + 400$.

$Y1 = -16x^2 + 400$
$Y2 = 0$

500

0 Intersection X=5 Y=0 6

0

Trace the graph to see where $s(t) = 0$.
The dummy will hit the ground in 5 seconds.

29. Substitute (0, 2000), (10, 1730), and (20, 920).

(1) $2000 = a(0)^2 + b(0) + c$

(2) $1730 = a(10)^2 + b(10) + c$

(3) $920 = a(20)^2 + b(20) + c$

Simplify.

(1) $2000 = c$
(2) $1730 = 100a + 10b + c$
(3) $920 = 400a + 20b + c$

Substitute 2000 for c in (2) and (3).

(2) $1730 = 100a + 10b + 2000$
(3) $920 = 400a + 20b + 2000$

Simplify and solve for a and b.

(2) $540 = -200a - 20b$
(3) $\underline{-1080 = 400a + 20b}$
$-540 = 200a$
$-2.7 = a$

(2) $1730 = 100(-2.7) + 10b + 2000$
$0 = 10b$
$0 = b$

$a = -2.7,\ b = 0,\ c = 2000$

$s(t) = at^2 + bt + c$

The position function is
$s(t) = -2.7t^2 + 2000$.

$Y1 = -2.7x^2 + 2000$
$Y2 = 0$

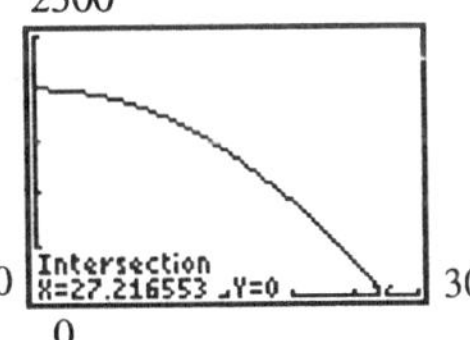

The package will touch the surface at approximately 27.22 seconds.

31. Substitute (0, 23.06), (5, 20.39), and (12, 23.20)

(1) $23.06 = a(0)^2 + b(0) + c$

(2) $20.39 = a(5)^2 + b(5) + c$

(3) $23.20 = a(12)^2 + b(12) + c$

Simplify.

(1) $23.06 = c$

(2) $20.39 = 25a^2 + 5b + c$

(3) $23.20 = 144a + 12b + c$

Substitute 23.06 for c in (2) and (3).
(2) $20.39 = 25a^2 + 5b + 23.06$
(3) $23.20 = 144a + 12b + 23.06$
Simplify and solve for a and b.
(2) $32.04 = -300a - 60b$
(3) $0.7 = 720a + 60b$
$32.74 = 420a$
$0.07795238095 \approx a$
(2) 20.39
$\approx 25(0.07795238095) + 5b + 23.06$
$-4.618809524 = 5b$
$-0.9237619048 = b$
$a \approx 0.07795,\ b \approx -0.92376,\ c = 23.06$
$f(x) = 0.07795x^2 - 0.92376x + 23.06$
$x = 2000 - 1980$
$x = 20$
$f(20) = 0.07795(20)^2 - 0.92376(20) + 23.06$
$f(20) = 35.766$
In the year 2000, the production is predicted to be 35.766 trillion cubic feet.
Answers will vary.

33. Let x = number of years after 1965
Substitute (0, 738), (10, 3352), and (20, 6972).
(1) $738 = a(0)^2 + b(0) + c$
(2) $3352 = a(10)^2 + b(10) + c$
(3) $6972 = a(20)^2 + b(20) + c$
Simplify.
(1) $738 = c$
(2) $3352 = 100a + 10b + c$
(3) $6972 = 400a + 20b + c$
Substitute 738 for c in (2) and (3).
(2) $3352 = 100a + 10b + 738$
(3) $6972 = 400a + 20b + 738$
Simplify and solve for a and b.
(2) $-5228 = -200a - 20b$
(3) $6234 = 400a + 20b$
$1006 = 200a$
$5.03 = a$
(2) $3352 = 100(5.03) + 10b + 738$
$2111 = 10b$
$211.1 = b$
$a = 5.03,\ b = 211.1,\ c = 738$
$f(x) = 5.03x^2 + 211.1x + 738$
$x = 1995 - 1965$
$x = 30$
$f(30) = 5.03(30)^2 + 211.1(30) + 738$
$f(30) = 11{,}598$
In 1995, the prediction is 11,598 franchises. This is fairly close to 10,953.
Answers will vary.

35. a. Substitute (5, 75), (10, 100), and (8, 96).
(1) $75 = a(5)^2 + b(5) + c$
(2) $100 = a(10)^2 + b(10) + c$
(3) $96 = a(8)^2 + b(8) + c$
Simplify.
(1) $75 = 25a + 5b + c$
(2) $100 = 100a + 10b + c$
(3) $96 = 64a + 8b + c$
Solve for a, b, and c.
(1) $-75 = -25a - 5b - c$
(2) $100 = 100a + 10b + c$
(4) $25 = 75a + 5b$

(2) $100 = 100a + 10b + c$
(3) $-96 = -64a - 8b - c$
(5) $4 = 36a + 2b$

(4) $-50 = -150a - 10b$
(5) $20 = 180a + 10b$
$-30 = 30a$
$-1 = a$
(4) $25 = 75(-1) + 5b$
$100 = 5b$
$20 = b$
(1) $75 = 25(-1) + 5(20) + c$
$0 = c$
$a = -1,\ b = 20,\ c = 0$
$f(x) = -x^2 + 20x$

b. $Y1 = -x^2 + 20x$

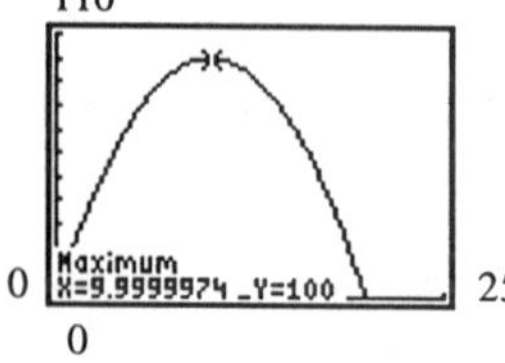

c. The vertex is at (10, 100). The maximum area that can be covered is 100 square yards when the width is 10 yards.

9.4 Experiencing Algebra the Calculator Way

1. (5, 15.8), (15, 130.8), (25, 353.8)
   $y = 0.54x^2 + 0.7x - 1.2$

2. (1, 767), (2, 1526), (4, 3344)
   $y = 50x^2 + 609x + 108$

3. (–2, 53), (–1, –66), (2, 297)
   $y = 60x^2 + 61x - 65$

Chapter 9 Review

Reflections

1.–8. Answers will vary.

Exercises

1. Yes; x is a monomial.

2. Yes; $5x - 3$ is a binomial.

3. No; $\sqrt{x} + 2$ is not a polynomial because $\sqrt{x}$ is not a monomial.

4. Yes; $3x^3 - 4x^2 + x - 1$ is a polynomial.

5. No; $\frac{3}{a} - 2a + 1$ is not a polynomial because $\frac{3}{a}$ is not a monomial.

6. Yes; $3a^4 - 2a^2 + 5$ is a trinomial.

7. $5x + 3x^3 - 2$
   $5x$ or $5x^1$ has a degree of 1.
   $3x^3$ has a degree of 3.
   -2 or $-2x^0$ has a degree of 0.
   The degree of the polynomial is 3.

8. $2x^2y + 3xy - 5$
   $2x^2y$ has a degree of 2 + 1 or 3.
   $3xy$ has a degree of 1 + 1 or 2.
   -5 or $-5x^0$ has a degree of 0.
   The degree of the polynomial is 3.

9. $x + 9$
   x has a degree of 1.
   9 or $9x^0$ has a degree of 0.
   The degree of the polynomial is 1.

10. $0.5a - 3.1a^2 + 9.6a + 3.1a^2 = 10.1a$
    $10.1a$ has a degree of 1.
    The degree of the polynomial is 1.

11. $12x^2 + 30x + 3$
    $12x^2$ has a degree of 2.
    $30x$ has a degree of 1.
    3 has a degree of 0.
    The degree of the polynomial is 2.

12. $5y^2 + 11y^4 + 12 - 6y + 9y^3$
    $= 11y^4 + 9y^3 + 5y^2 - 6y + 12$

13. $5 - p = -p + 5$

14. $\frac{1}{4}z^4 + \frac{1}{2}z^2 + z + \frac{1}{3}z^3 + 1$
    $= \frac{1}{4}z^4 + \frac{1}{3}z^3 + \frac{1}{2}z^2 + z + 1$

15. $0.6b - 2.3b^5 + 1.8 - 9.1b^3$
    $= -2.3b^5 - 9.1b^3 + 0.6b + 1.8$

16. $2x^3 + 11x^2 - 21x - 90$
    $= 2(3)^3 + 11(3)^2 - 21(3) - 90$
    $= 54 + 99 - 63 - 90$
    $= 0$
    The result is 0.

17. $2x^3 + 11x^2 - 21x - 90$
    $= 2(0)^3 + 11(0)^2 - 21(0) - 90$
    $= 0 + 0 - 0 - 90$
    $= -90$
    The result is –90.

18. $2x^3 + 11x^2 - 21x - 90$
    $= 2(1)^3 + 11(1)^2 - 21(1) - 90$
    $= 2 + 11 - 21 - 90$
    $= -98$
    The result is –98.

19. $2x^3 + 11x^2 - 21x - 90$
$= 2(-6)^3 + 11(-6)^2 - 21(-6) - 90$
$= -432 + 396 + 126 - 90$
$= 0$
The result is 0.

20. $2x^3 + 11x^2 - 21x - 90$
$= 2\left(-\frac{5}{2}\right)^3 + 11\left(-\frac{5}{2}\right)^2 - 21\left(-\frac{5}{2}\right) - 90$
$= -\frac{125}{4} + \frac{275}{4} + \frac{210}{4} - \frac{360}{4}$
$= 0$
The result is 0.

21. $a^3 + 2a^2b - 3ab^2 - b^3$
$= (0)^3 + 2(0)^2(1) - 3(0)(1)^2 - (1)^3$
$= 0 + 0 - 0 - 1$
$= -1$
The result is –1.

22. $a^3 + 2a^2b - 3ab^2 - b^3$
$= (-1)^3 + 2(-1)^2(0) - 3(-1)(0)^2 - 0^3$
$= -1 + 0 - 0 - 0$
$= -1$
The result is –1.

23. $a^3 + 2a^2b - 3ab^2 - b^3$
$= 0^3 + 2(0)^2(0) - 3(0)(0)^2 - 0^3$
$= 0 + 0 - 0 - 0$
$= 0$
The result is 0.

24. $a^3 + 2a^2b - 3ab^2 - b^3$
$= (1)^3 + 2(1)^2(1) - 3(1)(1)^2 - (1)^3$
$= 1 + 2 - 3 - 1$
$= -1$
The result is –1.

25. $a^3 + 2a^2b - 3ab^2 - b^3$
$= (-1)^3 + 2(-1)^2(1) - 3(-1)(1)^2 - (1)^3$
$= -1 + 2 + 3 - 1$
$= 3$
The result is 3.

26. $a^3 + 2a^2b - 3ab^2 - b^3$
$= (-1)^3 + 2(-1)^2(-1) - 3(-1)(-1)^2 - (-1)^3$
$= -1 - 2 + 3 + 1$
$= 1$
The result is 1.

27. Let w = width
w^2 = length
The perimeter is:
$2w + 2w^2$
$2(7) + 2(7)^2 = 112$
The perimeter is 112 yards.

28. Let x = length of first side
$3x$ = length of second side
$x^2 + 1$ = length of third side
The perimeter is:
$x + 3x + x^2 + 1 = x^2 + 4x + 1$
$(4)^2 + 4(4) + 1 = 33$
The perimeter is 33 inches.

29. Area of square $= a^2$
Area of rectangle $= 17a$
Area of triangle $= \frac{1}{2}(6)a = 3a$
The total area is
$a^2 + 17a + 3a = a^2 + 20a \text{ in}^2$.

30. Shaded area
= Rectangle area – Triangle area
$= 10x - \frac{1}{2}x(x)$
The shaded area is $-\frac{1}{2}x^2 + 10x \text{ in}^2$.

31. Area of lawn not covered by the garden
= Area of square – area of triangle
$= z^2 - \frac{1}{2}yx$
$z^2 - \frac{1}{2}xy = (80)^2 - \frac{1}{2}(10)(6) = 6370$
The area is 6370 ft^2.

32. Let w = width
l = length

a. $500 + 40lw$

b. $275 + 12lw$

c. Revenue = 500 + 40(25)(15) = \$15,500
Cost = 275 + 12(25)(15) = \$4775
Return = 15,500 – 4775 = \$10,725
The revenue is \$15,500, the cost is \$4775 and the net return is \$10,725.

33.

x	$y = x^3 - x^2 - 6x$	y
–3	$y = (-3)^3 - (-3)^2 - 6(-3) = -18$	–18
–2	$y = (-2)^3 - (-2)^2 - 6(-2) = 0$	0
–1	$y = (-1)^3 - (-1)^2 - 6(-1) = 4$	4
0	$y = 0^3 - 0^2 - 6(0) = 0$	0
1	$y = 1^3 - 1^2 - 6(1) = -6$	–6
2	$y = 2^3 - 2^2 - 6(2) = -8$	–8
3	$y = 3^3 - 3^2 - 6(3) = 0$	0

34. $Y1 = -3x + 5$

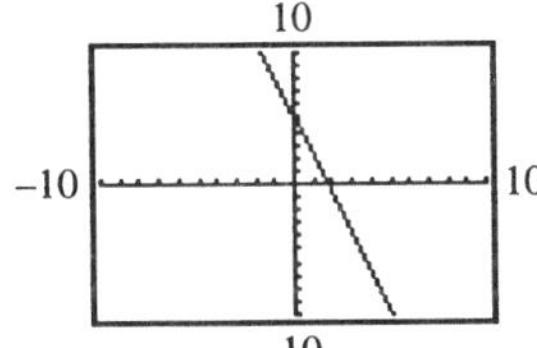

The range is all real numbers.

35. $Y1 = 2x^2 - 2x - 12$

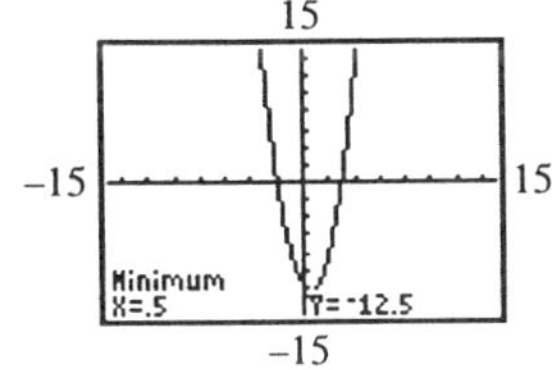

The range is all real numbers greater than or equal to –12.5.

36. $Y1 = x^3 + 2x^2 - 5x - 6$

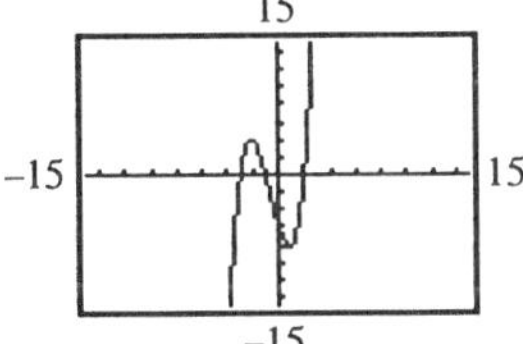

The y-coordinates have no absolute maximum or absolute minimum. Therefore, the range is all real numbers.

37. $Y1 = x^4 + 2x^3 - 5x^2 - 6x$

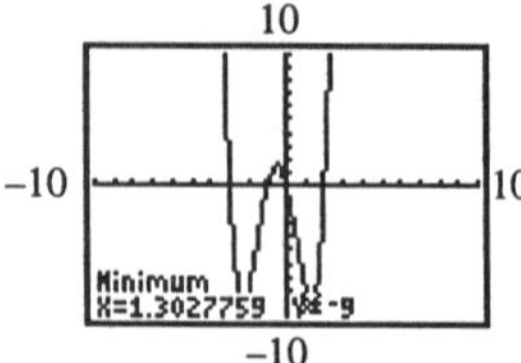

There is an absolute minimum at approximately $y = -9$. The range is all real numbers greater than or equal to –9.

38. $f(x) = 3x^3 - x^2 + 2x - 4$

$f(-2) = 3(-2)^3 - (-2)^2 + 2(-2) - 4$

$f(-2) = -24 - 4 - 4 - 4$

$f(-2) = -36$

39. $f(x) = 3x^3 - x^2 + 2x - 4$

$f(0) = 3(0)^3 - 0^2 + 2(0) - 4$

$f(0) = 0 - 0 + 0 - 4$

$f(0) = -4$

40. $f(x) = 3x^3 - x^2 + 2x - 4$

$f(2) = 3(2)^3 - 2^2 + 2(2) - 4$

$f(2) = 24 - 4 + 4 - 4$

$f(2) = 20$

41. $f(x) = 3x^3 - x^2 + 2x - 4$

$$f\left(-\frac{1}{2}\right) = 3\left(-\frac{1}{2}\right)^3 - \left(-\frac{1}{2}\right)^2 + 2\left(-\frac{1}{2}\right) - 4$$

$$f\left(-\frac{1}{2}\right) = -\frac{3}{8} - \frac{1}{4} - 1 - 4$$

$$f\left(-\frac{1}{2}\right) = -\frac{3}{8} - \frac{2}{8} - \frac{8}{8} - \frac{32}{8}$$

$$f\left(-\frac{1}{2}\right) = -\frac{45}{8}$$

42. $f(x) = 3x^3 - x^2 + 2x - 4$

$f(1.7) = 3(1.7)^3 - (1.7)^2 + 2(1.7) - 4$

$f(1.7) = 14.739 - 2.89 + 3.4 - 4$

$f(1.7) = 11.249$

43. a. $R(x) = 12x - 0.5x^2$

$R(5) = 12(5) - 0.5(5)^2 = 47.5$

$R(10) = 12(10) - 0.5(10)^2 = 70$

$R(15) = 12(15) - 0.5(15)^2 = 67.5$

$R(20) = 12(20) - 0.5(20)^2 = 40$

$R(25) = 12(25) - 0.5(25)^2 = -12.5$

The revenue for 5, 10, 15, 20, and 25 shirts is \$47.50, \$70, \$67.50, \$40, and –\$12.50, respectively.

Answers will vary.

b. $C(x) = 4x$

$C(5) = 4(5) = 20$

$C(10) = 4(10) = 40$

$C(15) = 4(15) = 60$

$C(20) = 4(20) = 80$

The cost for 5, 10, 15, and 20 shirts is \$20, \$40, \$60, and \$80, respectively.

c. Profit from

5 shirts: 47.50 – 20 = 27.50

10 shirts: 70 – 40 = 30

15 shirts: 67.50 – 60 = 7.50

20 shirts: 40 – 80 = –40

The profit from 5, 10, 15, and 20 shirts is \$27.50, \$30, \$7.50, and –\$40, respectively.

d. Answers will vary.

44. $y = 15,800 + 2.2x - 0.001x^2$

Gallons (x)	900	1000	1100	1200	1300	1400
Cost (y)	16,970	17,000	17,010	17,000	16,970	16,920

Answers will vary.

45. $y = x^2 + x + 1$ is quadratic.

46. $y = x^3 - x - 1$ is nonquadratic because $x^3 - x - 1$ is a third-degree polynomial (x is cubed).

47. $y = \frac{5}{x^2} + x + 1$ is nonquadratic because $\frac{5}{x^2} + x + 1$ is not a polynomial (the squared variable term is in the denominator of a fraction).

48. $y = x^2 + 4x + 4$.

	function	a	b	c	graph wide/narrow	graph concave upward/downward	graph vertex	axis of symmetry
49.	$y = -\frac{1}{4}x^2 + \frac{1}{2}x + 1$	$-\frac{1}{4}$	$\frac{1}{2}$	1	wide	downward	$\left(1, \frac{5}{4}\right)$	$x = 1$
50.	$f(x) = -2x^2 + 4x$	-2	4	0	narrow	downward	$(1, 2)$	$x = 1$
51.	$g(x) = \frac{1}{3}x^2 + x$	$\frac{1}{3}$	1	0	wide	upward	$\left(-\frac{3}{2}, -\frac{3}{4}\right)$	$x = -\frac{3}{2}$
52.	$y = 3x^2 - 3x + 1$	3	-3	1	narrow	upward	$\left(\frac{1}{2}, \frac{1}{4}\right)$	$x = \frac{1}{2}$

53. $f(x) = x^2 + 2x - 8$

Determine the vertex.

$x = -\frac{b}{2a} = -\frac{2}{2(1)} = -1$

$f(-1) = (-1)^2 + 2(-1) - 8 = -9$

The vertex is $(-1, -9)$.

The axis of symmetry is the line $x = -1$.

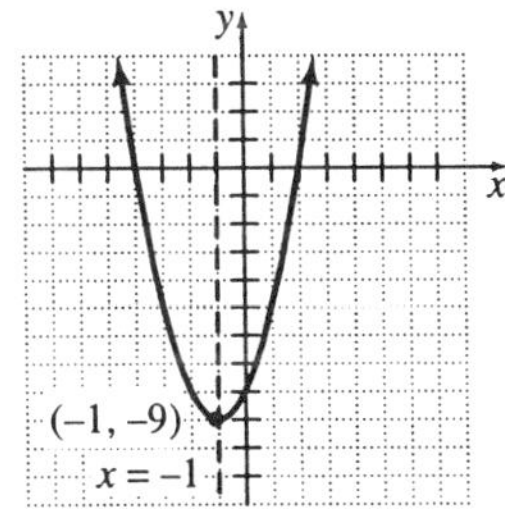

54. $y = -\frac{1}{2}x^2 + x - 2$

Determine the vertex.

$x = -\frac{b}{2a} = -\frac{1}{2\left(-\frac{1}{2}\right)} = 1$

$y = -\frac{1}{2}(1)^2 + 1 - 2 = -1\frac{1}{2}$

The vertex is $\left(1, -1\frac{1}{2}\right)$.

The axis of symmetry is the line $x = 1$.

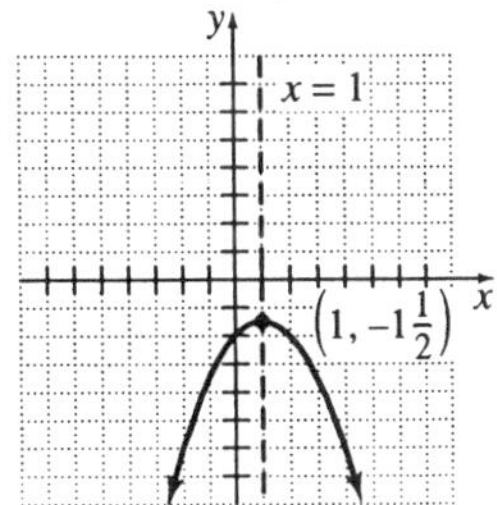

55. $h(x) = 2x^2 - 8$

Determine the vertex.

$x = -\frac{b}{2a} = -\frac{0}{2(2)} = 0$

$h(0) = 2(0)^2 - 8 = -8$

The vertex is (0, –8).

The axis of symmetry is the line $x = 0$.

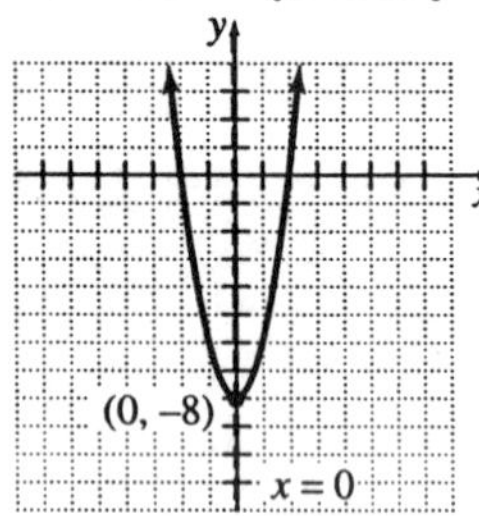

56. $A(w) = w^2 + 8w$

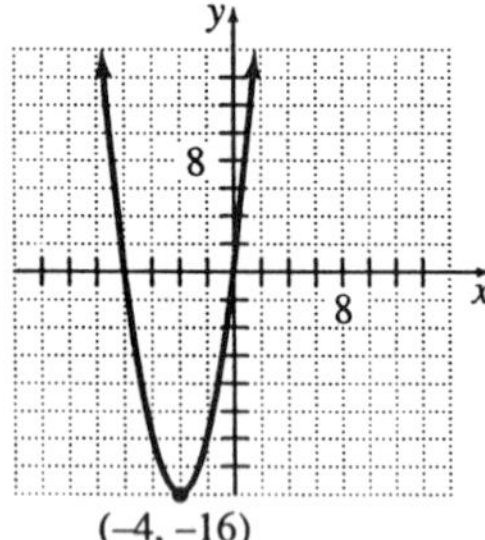

The vertex is (–4, –16).

No; the vertex has no physical meaning.

57. $R(x) = 30x - 0.50x^2$

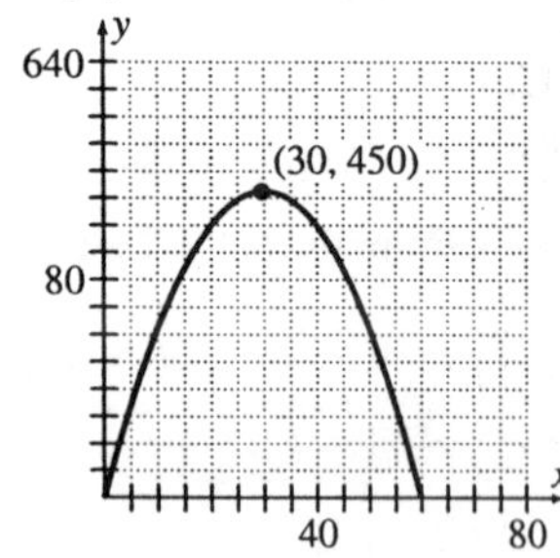

The vertex is (30, 450).

The revenue is at a maximum of $450 when 30 photos are ordered.

58. $s(t) = -16t^2 + 60t + 120$

Determine the vertex.

$x = -\frac{b}{2a} = -\frac{60}{2(-16)} = 1.875$

$s(1.875) = -16(1.875)^2 + 60(1.875) + 120$

$s(1.875) = 176.25$

The vertex is (1.875, 176.25).

The egg reaches a maximum height of 176.25 feet at 1.875 seconds.

$Y1 = -16x^2 + 60x + 120$

$Y2 = 0$

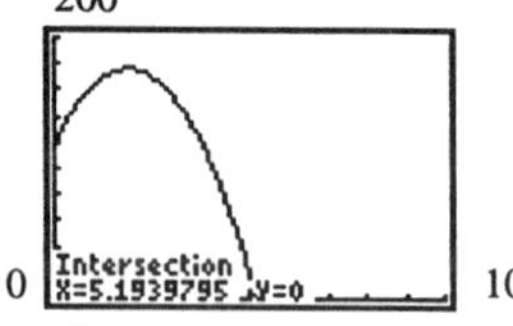

The egg will reach the ground in approximately 5.19 seconds.

59. Substitute (8, 0), (–3, 0), and (0, –24).

(1) $0 = a(8)^2 + b(8) + c$

(2) $0 = a(-3)^2 + b(-3) + c$

(3) $-24 = a(0)^2 + b(0) + c$

Simplify.

(1) $0 = 64a + 8b + c$

(2) $0 = 9a - 3b + c$

(3) $-24 = c$

Substitute –24 for c in (1) and (2).

(1) $24 = 64a + 8b$

(2) $24 = 9a - 3b$

Solve for a and b.

$$\begin{array}{ll} (1) & 72 = 192a + 24b \\ (2) & \underline{192 = 72a - 24b} \\ & 264 = 264a \\ & 1 = a \end{array}$$

(2) $0 = 9(1) - 3b - 24$

$15 = -3b$

$-5 = b$

$a = 1,\ b = -5,\ c = -24$

$y = ax^2 + bx + c$

$y = x^2 - 5x - 24$

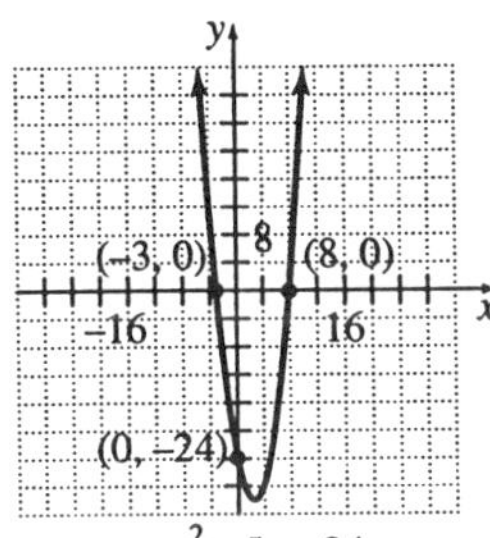

$y = x^2 - 5x - 24$

The graph passes through the given points.

60. Substitute (0, 5), (5, 0), and $\left(-\frac{1}{2}, 0\right)$.

(1) $5 = a(0)^2 + b(0) + c$

(2) $0 = a(5)^2 + b(5) + c$

(3) $0 = a\left(-\frac{1}{2}\right)^2 + b\left(-\frac{1}{2}\right) + c$

Simplify.

(1) $5 = c$

(2) $0 = 25a + 5b + c$

(3) $0 = \frac{1}{4}a - \frac{1}{2}b + c$

Substitute 5 for c in (2) and (3).

(2) $-5 = 25a + 5b$

(3) $-5 = \frac{1}{4}a - \frac{1}{2}b$

Solve for a and b.

$$\begin{array}{ll} (2) & -1 = 5a + b \\ (3) & -10 = \frac{1}{2}a - b \\ \hline & -11 = \frac{11}{2}a \\ & -2 = a \end{array}$$

(2) $0 = 25(-2) + 5b + 5$

$45 = 5b$

$9 = b$

$a = -2,\ b = 9,\ c = 5$

$y = ax^2 + bx + c$

$y = -2x^2 + 9x + 5$

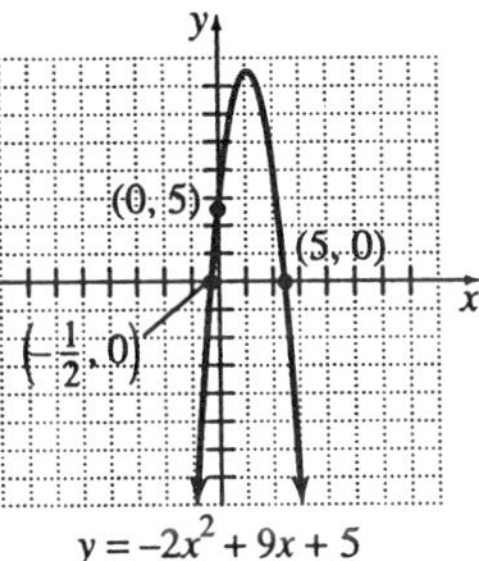

$y = -2x^2 + 9x + 5$

The graph passes through the given points.

61. Substitute (2, –4), (4, –6), and (7, 6).

(1) $-4 = a(2)^2 + b(2) + c$

(2) $-6 = a(4)^2 + b(4) + c$

(3) $6 = a(7)^2 + b(7) + c$

Simplify.

(1) $-4 = 4a + 2b + c$

(2) $-6 = 16a + 4b + c$

(3) $6 = 49a + 7b + c$

Solve for a, b, and c.

$$\begin{array}{ll} (1) & 4 = -4a - 2b - c \\ (2) & -6 = 16a + 4b + c \\ \hline (4) & -2 = 12a + 2b \end{array}$$

$$\begin{array}{ll} (1) & 4 = -4a - 2b - c \\ (3) & 6 = 49a + 7b + c \\ \hline (5) & 10 = 45a + 5b \end{array}$$

$$\begin{array}{ll} (4) & 10 = -60a - 10b \\ (5) & 20 = 90a + 10b \\ \hline & 30 = 30a \\ & 1 = a \end{array}$$

(4) $-2 = 12(1) + 2b$

$-14 = 2b$

$-7 = b$

(1) $-4 = 4(1) + 2(-7) + c$

$6 = c$

$a = 1,\ b = -7,\ c = 6$

$y = ax^2 + bx + c$

$y = x^2 - 7x + 6$

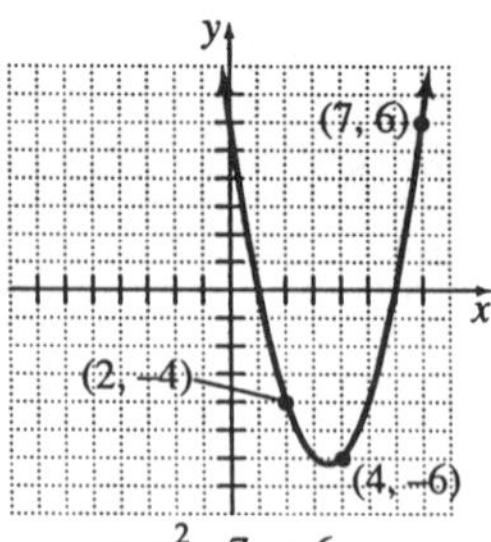

$y = x^2 - 7x + 6$

The graph passes through the given points.

62. $s(t) = at^2 + bt + c$

Substitute (0, 160), (1, 176), and (2, 160).

(1) $160 = a(0)^2 + b(0) + c$

(2) $176 = a(1)^2 + b(1) + c$

(3) $160 = a(2)^2 + b(2) + c$

Simplify.

(1) $160 = c$

(2) $176 = a + b + c$

(3) $160 = 4a + 2b + c$

Substitute 160 for c in (2) and (3).

(2) $176 = a + b + 160$

(3) $160 = 4a + 2b + 160$

Solve for a and b.

(2) $-32 = -2a - 2b$

(3) $\underline{\quad 0 = 4a + 2b}$

$-32 = 2a$

$-16 = a$

(2) $176 = -16 + b + 160$

$32 = b$

$a = -16,\ b = 32,\ c = 160$

$s(t) = at^2 + bt + c$

The position function is

$s(t) = -16t^2 + 32t + 160.$

$Y1 = -16x^2 + 32x + 160$

$Y2 = 0$

200

Intersection X=4.3166248 Y=0

0 5

0

The hammer will reach the ground at approximately 4.3 seconds.

63. Let x = number of years after 1900.

Substitute (10, 4), (50, 12.4), and (90, 31.2)

(1) $4 = a(10)^2 + b(10) + c$

(2) $12.4 = a(50)^2 + b(50) + c$

(3) $31.2 = a(90)^2 + b(90) + c$

Simplify.

(1) $4 = 100a + 10b + c$

(2) $12.4 = 2500a + 50b + c$

(3) $31.2 = 8100a + 90b + c$

Solve for a, b, and c.

(1) $-4 = -100a - 10b - c$

(2) $\underline{12.4 = 2500a + 50b + c}$

(4) $8.4 = 2400a + 40b$

(1) $-4 = -100a - 10b - c$

(3) $\underline{31.2 = 8100a + 90b + c}$

(5) $27.2 = 8000a + 80b$

(4) $-16.8 = -4800a - 80b$

$\underline{27.2 = 8000a + 80b}$

$10.4 = 3200a$

$0.00325 = a$

(4) $8.4 = 2400(0.00325) + 40b$

$0.6 = 40b$

$0.015 = b$

(1) $4 = 100(0.00325) + 10(0.015) + c$

$3.525 = c$

$a = 0.00325,\ b = 0.015,\ c = 3.525$

The quadratic function is

$f(x) = 0.00325x^2 + 0.015x + 3.525.$

In 1980, $x = 1980 - 1900 = 80.$

$f(80) = 0.00325(80)^2 + 0.015(80) + 3.525$

$f(80) = 25.525$

In 1970, $x = 1970 - 1900 = 70.$

$f(70) = 0.00325(70)^2 + 0.015(70) + 3.525$

$f(70) = 20.5$

In 1960, $x = 1960 - 1900 = 60.$

$f(60) = 0.00325(60)^2 + 0.015(60) + 3.525$

$f(60) = 16.125$

In 1980, 1970, and 1960, the predictions are 25.525 million, 20.5 million, and 16.125 million, respectively.

Answers will vary.

In 2000, $x = 2000 - 1900 = 100.$

$f(100)$
$= 0.00325(100)^2 + 0.015(100) + 3.525$
$f(100) = 37.525$
In the year 2000, the prediction is 37.525 million.

Chapter 9 Mixed Review

1. $2x^3 - 3x^2 - 29x - 30$
$= 2(5)^3 - 3(5)^2 - 29(5) - 30$
$= 250 - 75 - 145 - 30$
$= 0$
The result is 0.

2. $2x^3 - 3x^2 - 29x - 30$
$= 2(0)^3 - 3(0)^2 - 29(0) - 30$
$= 0 - 0 - 0 - 30$
$= -30$
The result is –30.

3. $2x^3 - 3x^2 - 29x - 30$
$= 2(1)^3 - 3(1)^2 - 29(1) - 30$
$= 2 - 3 - 29 - 30$
$= -60$
The result is –60.

4. $2x^3 - 3x^2 - 29x - 30$
$= 2(-2)^3 - 3(-2)^2 - 29(-2) - 30$
$= -16 - 12 + 58 - 30$
$= 0$
The result is 0.

5. $2x^3 - 3x^2 - 29x - 30$
$= 2\left(-\frac{3}{2}\right)^3 - 3\left(-\frac{3}{2}\right)^2 - 29\left(-\frac{3}{2}\right) - 30$
$= -\frac{27}{4} - \frac{27}{4} + \frac{174}{4} - \frac{120}{4}$
$= 0$
The result is 0.

6. $2a^3 + 4a^2b - 2ab^2 + b^3$
$= 2(-1)^3 + 4(-1)^2(0) - 2(-1)(0)^2 + 0^3$
$= -2 + 0 - 0 + 0$
$= -2$
The result is –2.

7. $2a^3 + 4a^2b - 2ab^2 + b^3$
$= 2(1)^3 + 4(1)^2(1) - 2(1)(1)^2 + 1^3$
$= 2 + 4 - 2 + 1$
$= 5$
The result is 5.

8. $2a^3 + 4a^2b - 2ab^2 + b^3$
$= 2(-1)^3 + 4(-1)^2(1) - 2(-1)(1)^2 + 1^3$
$= -2 + 4 + 2 + 1$
$= 5$
The result is 5.

9. $2a^3 + 4a^2b - 2ab^2 + b^3$
$= 2(-1)^3 + 4(-1)^2(-1) - 2(-1)(-1)^2 + (-1)^3$
$= -2 - 4 + 2 - 1$
$= -5$
The result is –5.

10. **a.** binomial; degree of each term is 0, 2; degree of polynomial is 2; $3x^2 + 5$

b. polynomial; degree of each term is 2, 3, 0, 1; degree of polynomial is 3; $-5a^3 + 15a^2 + a + 4$

c. polynomial; degree of each term is 4, 1, 0, 5, 2; degree of polynomial is 5; $x^5 + 5x^4 - 3x^2 + x - 2$

11. **a.** $3b^2 + 13b - 4$; trinomial; degree of each term is 2, 1, 0; degree of polynomial is 2.

b. $3x^2y - 4xy + 3xy^2 + 5 - 4x^2y^2$; polynomial; degree of each term is 3, 2, 3, 0, 4; degree of polynomial is 4.

c. $6xyz$; monomial; degree of term is 3; degree of polynomial is 3.

12. $f(x) = 2x^3 - 3x^2 - 23x + 12$
$f(-3) = 2(-3)^3 - 3(-3)^2 - 23(-3) + 12$
$f(-3) = -54 - 27 + 69 + 12$
$f(-3) = 0$
The result is 0.

13. $f(x) = 2x^3 - 3x^2 - 23x + 12$
$f(0) = 2(0)^3 - 3(0)^2 - 23(0) + 12$
$f(0) = 0 - 0 - 0 + 12$
$f(0) = 12$
The result is 12.

14. $f(x) = 2x^3 - 3x^2 - 23x + 12$
$f(4) = 2(4)^3 - 3(4)^2 - 23(4) + 12$
$f(4) = 128 - 48 - 92 + 12$
$f(4) = 0$
The result is 0.

15. $f(x) = 2x^3 - 3x^2 - 23x + 12$
$f\left(\frac{1}{2}\right) = 2\left(\frac{1}{2}\right)^3 - 3\left(\frac{1}{2}\right)^2 - 23\left(\frac{1}{2}\right) + 12$
$f\left(\frac{1}{2}\right) = \frac{1}{4} - \frac{3}{4} - \frac{46}{4} + \frac{48}{4}$
$f\left(\frac{1}{2}\right) = 0$
The result is 0.

16. $f(x) = 2x^3 - 3x^2 - 23x + 12$
$f(2.2) = 2(2.2)^3 - 3(2.2)^2 - 23(2.2) + 12$
$f(2.2) = 21.296 - 14.52 - 50.6 + 12$
$f(2.2) = -31.824$
The result is –31.824.

17.

x	$y = 2x^3 + 2x^2 - 12x$	y
–3	$y = 2(-3)^3 + 2(-3)^2 - 12(-3) = 0$	0
–2	$y = 2(-2)^3 + 2(-2)^2 - 12(-2) = 16$	16
–1	$y = 2(-1)^3 + 2(-1)^2 - 12(-1) = 12$	12
0	$y = 2(0)^3 + 2(0)^2 - 12(0) = 0$	0
1	$y = 2(1)^3 + 2(1)^2 - 12(1) = -8$	–8
2	$y = 2(2)^3 + 2(2)^2 - 12(2) = 0$	0
3	$y = 2(3)^3 + 2(3)^2 - 12(3) = 36$	36

18. $Y1 = 4x - 2$

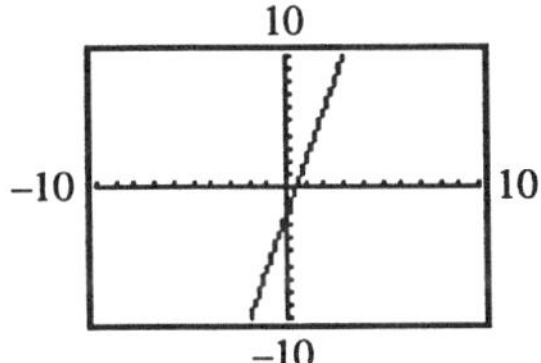

The range is all real numbers.

19. $Y1 = 3x^2 + 3x - 6$

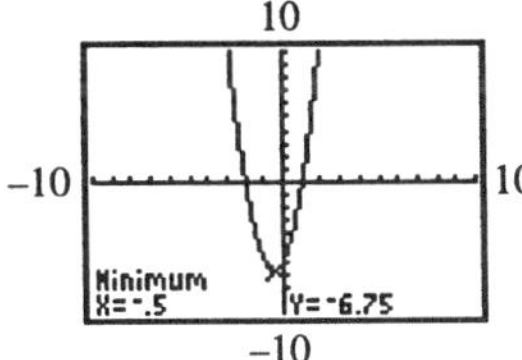

The range is all real numbers greater than or equal to –6.75.

20. $Y1 = x^3 - 3x^2 - 13x + 15$

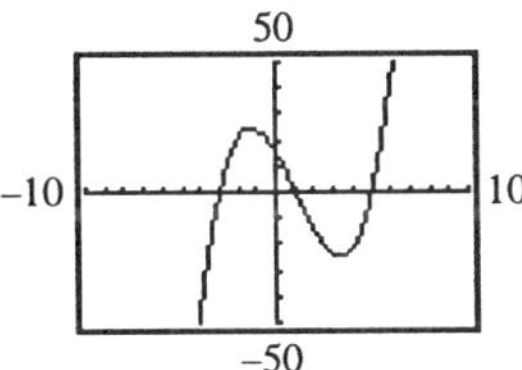

The y-coordinates have no absolute maximum and no absolute minimum. The range is all real numbers.

21. $Y1 = x^4 - 3x^3 - 13x^2 + 15x$

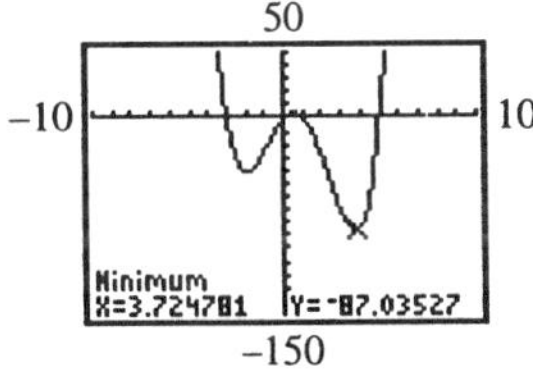

The y-coordinates have an absolute minimum at approximately –87.04. The range is all real numbers greater than or equal to –87.04.

22. $f(x) = 2x^2 + 7x - 4$

Determine the vertex.

$$x = -\frac{b}{2a} = -\frac{7}{2(2)} = -\frac{7}{4}$$

$$f\left(-\frac{7}{4}\right) = 2\left(-\frac{7}{4}\right)^2 + 7\left(-\frac{7}{4}\right) - 4 = -\frac{81}{8}$$

The vertex is $\left(-\frac{7}{4}, -\frac{81}{8}\right)$.

The axis of symmetry is the line $x = -\frac{7}{4}$.

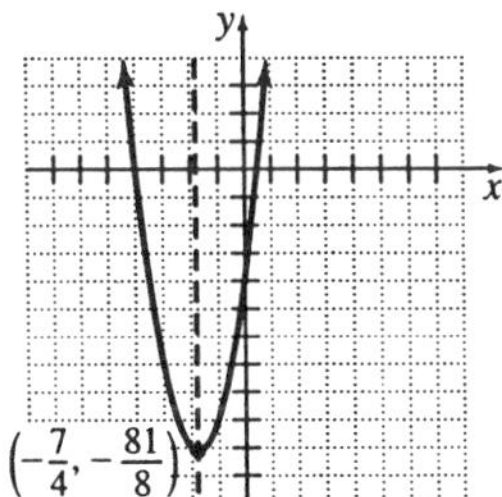

23. $y = \frac{1}{4}x^2 - x + 3$

Determine the vertex.

$$x = -\frac{b}{2a} = -\frac{(-1)}{2\left(\frac{1}{4}\right)} = 2$$

$$y = \frac{1}{4}(2)^2 - 2 + 3 = 2$$

The vertex is (2, 2).

The axis of symmetry is the line $x = 2$.

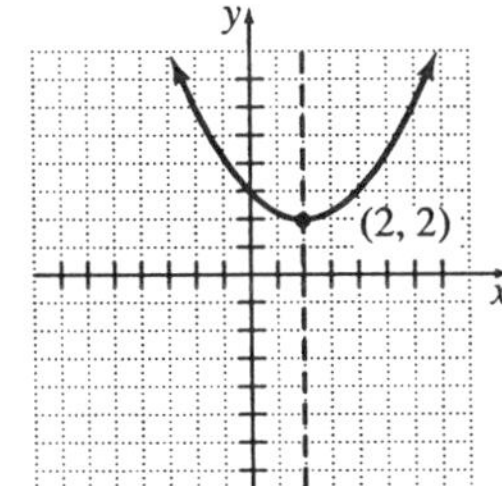

24. $h(x) = -x^2 + 9$
Determine the vertex.
$$x = -\frac{b}{2a} = -\frac{0}{2(-1)} = 0$$
$h(0) = -0^2 + 9 = 9$
The vertex is (0, 9).
The axis of symmetry is the line $x = 0$.

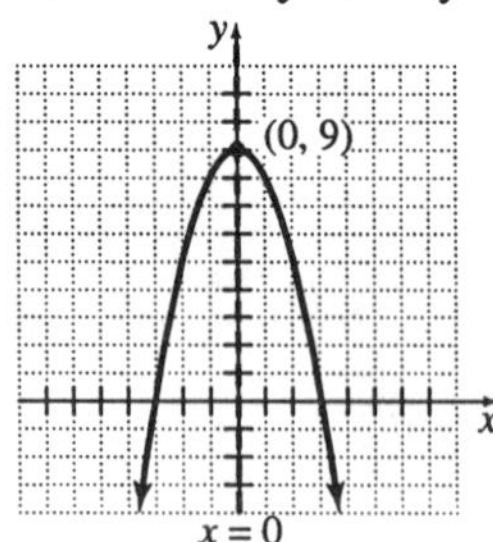

	function	a	b	c	graph wide/narrow	graph concave upward/downward	graph vertex	axis of symmetry
25.	$y = \frac{1}{3}x^2 + \frac{2}{3}x + 1$	$\frac{1}{3}$	$\frac{2}{3}$	1	wide	upward	$\left(-1, \frac{2}{3}\right)$	$x = -1$
26.	$f(x) = -3x^2 + 6x$	–3	6	0	narrow	downward	(1, 3)	$x = 1$
27.	$y = -\frac{1}{4}x^2 + x + 3$	$-\frac{1}{4}$	1	3	wide	downward	(2, 4)	$x = 2$
28.	$g(x) = 2x^2 + 4x - 6$	2	4	–6	narrow	upward	(–1, –8)	$x = -1$

29. Substitute (–2, 12), (–4, 8), and (–7, 17).
(1) $12 = a(-2)^2 + b(-2) + c$
(2) $8 = a(-4)^2 + b(-4) + c$
(3) $17 = a(-7)^2 + b(-7) + c$
Simplify.
(1) $12 = 4a - 2b + c$
(2) $8 = 16a - 4b + c$
(3) $17 = 49a - 7b + c$
Solve for a, b, and c.
(1) $-12 = -4a + 2b - c$
(2) $\underline{8 = 16a - 4b + c}$
(4) $-4 = 12a - 2b$

(1) $-12 = -4a + 2b - c$
(3) $\underline{17 = 49a - 7b + c}$
(5) $5 = 45a - 5b$

(4) $20 = -60a + 10b$
(5) $10 = 90a - 10b$
$30 = 30a$
$1 = a$
(4) $-4 = 12(1) - 2b$
$-16 = -2b$
$8 = b$
(1) $12 = 4(1) - 2(8) + c$
$24 = c$
$a = 1, b = 8, c = 24$
$y = ax^2 + bx + c$
$y = x^2 + 8x + 24$

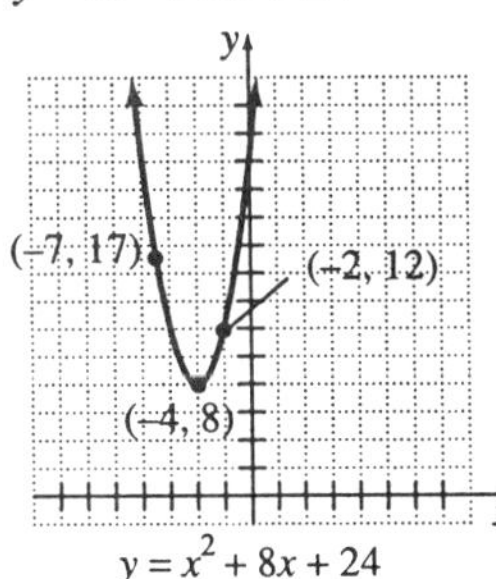

The graph passes through the given points.

30. Substitute (9, 0), (−4, 0), and (0, −36).
(1) $0 = a(9)^2 + b(9) + c$
(2) $0 = a(-4)^2 + b(-4) + c$
(3) $-36 = a(0)^2 + b(0) + c$
Simplify.
(1) $0 - 81a + 9b + c$
(2) $0 = 16a - 4b + c$
(3) $-36 = c$
Substitute −36 for c in (1) and (2).
(1) $36 = 81a + 9b$
(2) $36 = 16a - 4b$
Solve for a and b.
(1) $4 = 9a + b$
(2) $9 = 4a - b$
$13 = 13a$
$1 = a$
(1) $0 = 81(1) + 9b - 36$
$-45 = 9b$
$-5 = b$
$a = 1, b = -5, c = -36$
$y = ax^2 + bx + c$
$y = x^2 - 5x - 36$

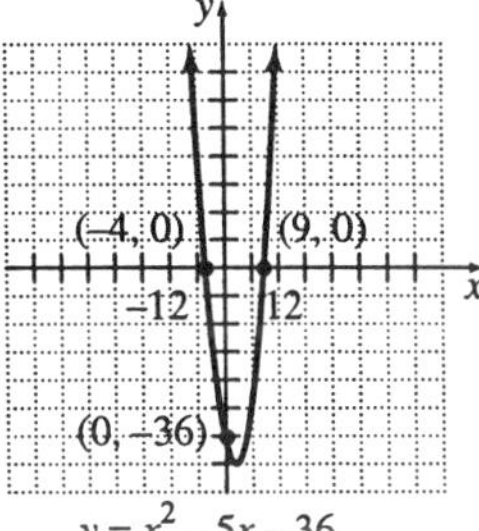

The graph passes through the given points.

31. Substitute (0, −12), (3, 0), and $\left(-\frac{4}{3}, 0\right)$.
(1) $-12 = a(0)^2 + b(0) + c$
(2) $0 = a(3)^2 + b(3) + c$
(3) $0 = a\left(-\frac{4}{3}\right)^2 + b\left(-\frac{4}{3}\right) + c$
Simplify.
(1) $-12 = c$
(2) $0 = 9a + 3b + c$
(3) $0 = \frac{16}{9}a - \frac{4}{3}b + c$
Substitute −12 for c in (2) and (3).
(2) $12 = 9a + 3b$
(3) $12 = \frac{16}{9}a - \frac{4}{3}b$
Solve for a and b.
(2) $48 = 36a + 12b$
(3) $108 = 16a - 12b$
$156 = 52a$
$3 = a$
(2) $0 = 9(3) + 3b - 12$
$-15 = 3b$
$-5 = b$
$a = 3, b = -5, c = -12$
$y = ax^2 + bx + c$
$y = 3x^2 - 5x - 12$

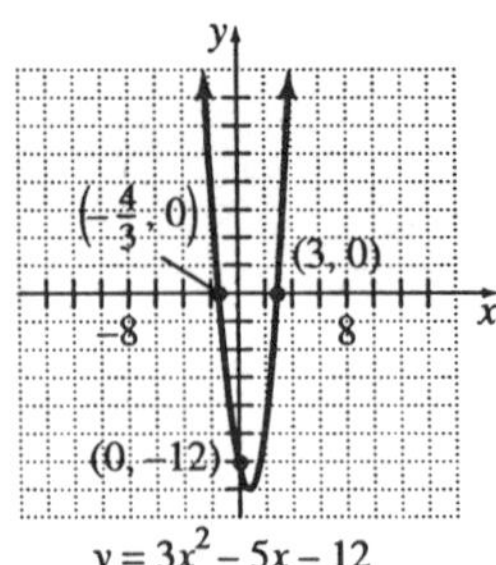

$y = 3x^2 - 5x - 12$

The graph passes through the given points.

32. Let $w = \text{width}$
$w^3 + 5 = \text{length}$
The perimeter is:
$2w + 2(w^3 + 5) = 2w^3 + 2w + 10$
The polynomial for the perimeter is
$2w^3 + 2w + 10.$
$2(3)^3 + 2(3) + 10 = 70$
The perimeter is 70 feet.

33. Let $x = \text{length of first side}$
$2x - 3 = \text{length of second side}$
$x^2 - 7 = \text{length of third side}$
The perimeter is:
$x + 2x - 3 + x^2 - 7 = x^2 + 3x - 10$
The polynomial for the perimeter is
$x^2 + 3x - 10.$
$(10)^2 + 3(10) - 10 = 120$
The perimeter is 120 cm.

34. Area of first $= \frac{1}{2}(5)x$
Area of second $= x(x+2)$
Area of third $= \frac{1}{2}x(5+8)$
Total area is:
$\frac{5}{2}x + x(x+2) + \frac{13}{2}x = x^2 + 11x$
The polynomial for the total area is
$x^2 + 11x$ square units.

35. Shaded area
= Area of rectangle + area of semicircle
$= 3x(x) + \frac{1}{2}\pi\left(\frac{x}{2}\right)^2$
$= 3x^2 + \frac{\pi}{8}x^2$
$= \frac{24+\pi}{8}x^2$
The polynomial for the shaded area is
$\frac{24+\pi}{8}x^2$ square units.

36. Area of yard not covered by pool
= Area of rectangle – area of circle
$= xy - \pi z^2$
The polynomial is $xy - \pi z^2$.
$xy - \pi z^2 = 80(50) - \pi(8)^2$
$= 4000 - 64\pi$
≈ 3798.9
The area measures approximately 3798.9 ft^2.

37. a. Let $l = \text{length}$
$w = \text{width}$
The cost is $1.5lw$ dollars.

b. $800 - 1.5lw$ dollars

c. $800 - 1.5(15)(10) = 575$
The net return will be \$575.

38. a.

x	$P(x) = 10x + 2x^2$	$P(x)$
0	$P(0) = 10(0) + 2(0)^2 = 0$	0
1	$P(1) = 10(1) + 2(1)^2 = 12$	12
2	$P(2) = 10(2) + 2(2)^2 = 28$	28
3	$P(3) = 10(3) + 2(3)^2 = 48$	48
4	$P(4) = 10(4) + 2(4)^2 = 72$	72
5	$P(5) = 10(5) + 2(5)^2 = 100$	100
6	$P(6) = 10(6) + 2(6)^2 = 132$	132
7	$P(7) = 10(7) + 2(7)^2 = 168$	168

b. Answers will vary.

39. a. $y = -0.25 + 0.5x + 0.0082z - 0.0000081z^2$

$y = -0.25 + 0.5(3.71) + 0.0082(750) - 0.0000081(750)^2$

$y = 3.19875$

The predicted GPA is 3.19875.

b.

GMAT score, z	500	550	600	650	700	750
GPA in MBA, y	3.575	3.55975	3.504	3.40775	3.271	3.09375

c. As the entrance exam score goes up, the student's predicted performance goes down.

40. $s(t) = -0.8t^2 + 1500$

$Y1 = -0.8x^2 + 1500$

$Y2 = 0$

2000

Intersection
X=43.30127 Y=0

0 50

0

The object will reach the ground in approximately 43.3 seconds.

41. Let x = number of years after 1900.

Substitute (20, 1.6), (40, 2.0), (70, 3.5).

(1) $1.6 = a(20)^2 + b(20) + c$

(2) $2 = a(40)^2 + b(40) + c$

(3) $3.5 = a(70)^2 + b(70) + c$

Simplify.

(1) $1.6 = 400a + 20b + c$

(2) $2 = 1600a + 40b + c$

(3) $3.5 = 4900a + 70b + c$

Solve for a, b, and c.

(1) $-1.6 = -400a - 20b - c$

(2) $2 = 1600a + 40b + c$

(4) $0.4 = 1200a + 20b$

$$\begin{aligned}(1)\quad -1.6 &= -400a - 20b - c\\ (3)\quad 3.5 &= 4900a + 70b + c\\ \hline (5)\quad 1.9 &= 4500a + 50b\end{aligned}$$

$$\begin{aligned}(4)\quad -20 &= -60{,}000a - 1000b\\ (5)\quad 38 &= 90{,}000a + 1000b\\ \hline 18 &= 30{,}000a\\ 0.0006 &= a\end{aligned}$$

(4) $0.4 = 1200(0.006) + 20b$
$-0.32 = 20b$
$-0.016 = b$
(1) $1.6 = 400(0.0006) + 20(-0.016) + c$
$1.68 = c$
$a = 0.0006$, $b = -0.016$, $c = 1.68$
$f(x) = ax^2 + bx + c$
The quadratic function is
$f(x) = 0.0006x^2 - 0.016x + 1.68.$
In 1930, $x = 1930 - 1900 = 30$
$f(30) = 0.0006(30)^2 - 0.016(30) + 1.68$
$f(30) = 1.74$
In 1990, $x = 1990 - 1900 = 90$
$f(90) = 0.0006(90)^2 - 0.016(90) + 1.68$
$f(90) = 5.1$
In 1930 and 1990, the predicted divorce rate per 1000 population is 1.74 and 5.1, respectively.
Answers will vary.
In 2000, $x = 2000 - 1900 = 100$.
$f(100) = 0.0006(100)^2 - 0.016(100) + 1.68$
$f(100) = 6.08$
In the year 2000, the prediction is 6.08 per 1000 population.

Chapter 9 Test

1. $123x^2y^3z$ is a monomial because it has one term.

2. $3a^3 + 5a^2b + 7ab^2 + 9b^3$ is a polynomial because it has 4 terms.

3. $2 - c$ is a binomial because it has 2 terms.

4. The term with the largest degree is $-0.5x^5$. Therefore, the degree of the polynomial is 5.

5. The term with the largest degree is $5x^2y^3$. Therefore the degree of the polynomial is $2 + 3$ or 5.

6. $15 + 3x^4 - 7x + x^5 + 9x^2 + 21x$
$= x^5 + 3x^4 + 9x^2 + 14x + 15$

7. $a - \frac{2}{3}a^3 - \frac{5}{6}a^2 - \frac{4}{9} = -\frac{4}{9} + a - \frac{5}{6}a^2 - \frac{2}{3}a^3$

8. $a^2 + 3ab^3 - 7b^2 - b - 6$
$= (0)^2 + 3(0)(3)^3 - 7(3)^2 - 3 - 6$
$= 0 + 0 - 63 - 3 - 6$
$= -72$
The result is –72.

9. $a^2 + 3ab^3 - 7b^2 - b - 6$
$= (-2)^2 + 3(-2)(0)^3 - 7(0)^2 - 0 - 6$
$= 4 - 0 - 0 - 0 - 6$
$= -2$
The result is –2.

10. $a^2 + 3ab^3 - 7b^2 - b - 6$
$= (2)^2 + 3(2)(-3)^3 - 7(-3)^2 - (-3) - 6$
$= 4 - 162 - 63 + 3 - 6$
$= -224$
The result is –224.

11. Let l = length
w = width
The total cost is \$16.50 multiplied by the area and added to \$75.
The total cost will be $16.5lw + 75$ dollars.
$16.5(5)\left(7\frac{1}{3}\right) + 75 = \$680$
The total cost will be \$680.

12. $g(x) = 3x^2 + 7x - 6$
$g(-3) = 3(-3)^2 + 7(-3) - 6$
$g(-3) = 27 - 21 - 6$
$g(-3) = 0$
The result is 0.

13. $g(x) = 3x^2 + 7x - 6$
$g(0) = 3(0)^2 + 7(0) - 6$
$g(0) = 0 + 0 - 6$
$g(0) = -6$
The result is –6.

14. $g(x) = 3x^2 + 7x - 6$
$g(1) = 3(1)^2 + 7(1) - 6$
$g(1) = 3 + 7 - 6$
$g(1) = 4$
The result is 4.

15. $y = \frac{1}{2}x^2 - 2x - 6$

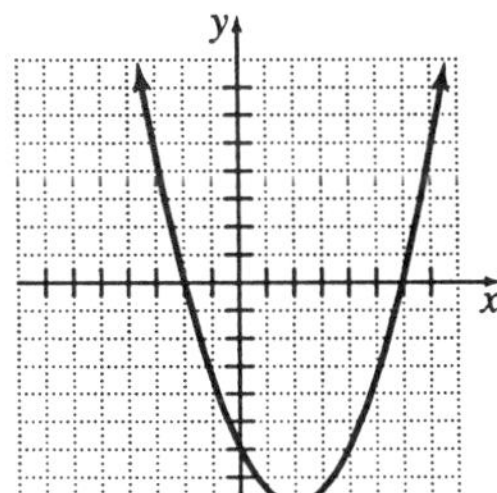

16. $x = -\frac{b}{2a}$
$x = -\frac{(-2)}{2\left(\frac{1}{2}\right)}$
$x = 2$
$y = \frac{1}{2}(2)^2 - 2(2) - 6$
$y = 2 - 4 - 6$
$y = -8$
The vertex is (2, –8).

17. The y-coordinates have an absolute minimum at –8. The range is all real numbers greater than or equal to –8.

18. Yes; the graph of the relation passes the vertical line test.

	function	a	b	c	graph wide/narrow	graph concave upward/downward	graph vertex	axis of symmetry
19.	$y=\frac{1}{2}x^2+2x+3$	$\frac{1}{2}$	2	3	wide	upward	(–2, 1)	$x=-2$
20.	$y=3x^2-3x+\frac{1}{4}$	3	–3	$\frac{1}{4}$	narrow	upward	$\left(\frac{1}{2}, -\frac{1}{2}\right)$	$x=\frac{1}{2}$

21. Substitute (0, 8), (2, 0), and (4, 0).

(1) $8=a(0)^2+b(0)+c$

(2) $0=a(2)^2+b(2)+c$

(3) $0=a(4)^2+b(4)+c$

Simplify.

(1) $8=c$

(2) $0=4a+2b+c$

(3) $0=16a+4b+c$

Substitute 8 for c in (2) and (3).

(2) $-8=4a+2b$

(3) $-8=16a+4b$

Solve for a and b.

(2) $16=-8a-4b$

(3) $\underline{-8=16a+4b}$

$8=8a$

$1=a$

(2) $0=4(1)+2b+8$

$-12=2b$

$-6=b$

$a=1,\ b=-6,\ c=8$

$y=ax^2+bx+c$

$y=x^2-6x+8$

22. Substitute (–1, 6), (1, 4), and (2, 9).

(1) $6=a(-1)^2+b(-1)+c$

(2) $4=a(1)^2+b(1)+c$

(3) $9=a(2)^2+b(2)+c$

Simplify.

(1) $6=a-b+c$

(2) $4=a+b+c$

(3) $9=4a+2b+c$

Solve for a, b, and c.

(1) $-6=-a+b-c$

(2) $\underline{4=a+b+c}$

$-2=2b$

$-1=b$

(1) $6=a+-(-1)+c$

(3) $9=4a+2(-1)+c$

(1) $-5=-a-c$

(3) $\underline{11=4a+c}$

$6=3a$

$2=a$

(1) $6=2-(-1)+c$

$3=c$

$a=2,\ b=-1,\ c=3$

$y=ax^2+bx+c$

$y=2x^2-x+3$

23. Answers will vary.

Chapter 10

10.1 Experiencing Algebra the Exercise Way

1. $4^3a^2 = 4\cdot 4\cdot 4\cdot a\cdot a$

3. $(-3x)^4 = (-3x)(-3x)(-3x)(-3x)$

5. $a^3b^0c^5 = a^3\cdot 1\cdot c^5 = a\cdot a\cdot a\cdot c\cdot c\cdot c\cdot c\cdot c$

7. $\left(\frac{3x}{4}\right)^3 = \left(\frac{3x}{4}\right)\left(\frac{3x}{4}\right)\left(\frac{3x}{4}\right)$

9. $5(x+y)^2 = 5(x+y)(x+y)$

11. $p^{-3} = \frac{1}{p^3}$

13. $\frac{1}{q^{-5}} = q^5$

15. $\frac{p^{-3}}{q^{-5}} = \frac{q^5}{p^3}$

17. $p^{-3}q^5 = \frac{q^5}{p^3}$

19. $\frac{4^3m^{-2}}{3^4n^{-3}} = \frac{4^3n^3}{3^4m^2} = \frac{64n^3}{81m^2}$

21. $\frac{4^{-3}m^{-2}}{3^{-4}n^{-3}} = \frac{3^4n^3}{4^3m^2} = \frac{81n^3}{64m^2}$

23. $\frac{-4(m-n)^{-1}}{5(m+n)^{-2}} = \frac{-4(m+n)^2}{5(m-n)}$

25. $\frac{4^{-1}(m-n)}{5^{-1}(m+n)^2} = \frac{5(m-n)}{4(m+n)^2}$

27. $a^{-3} + 2a^{-2} - 3a^{-1} + 4a^0 = \frac{1}{a^3} + \frac{2}{a^2} - \frac{3}{a} + 4$

29. $x^5\cdot x^8 = x^{5+8} = x^{13}$

31. $y^{-5}\cdot y^{13} = y^{-5+13} = y^8$

33. $z^{-9}\cdot z^4 = z^{-9+4} = z^{-5} = \frac{1}{z^5}$

35. $p^{-2}\cdot p^{-7} = p^{-2+(-7)} = p^{-9} = \frac{1}{p^9}$

37. $(x+y)^4(x+y)^{-4} = (x+y)^{4+(-4)}$
$= (x+y)^0 = 1$

39. $(x+3)^2(x+3) = (x+3)^{2+1} = (x+3)^3$

41. $\frac{p^{11}}{p^6} = p^{11-6} = p^5$

43. $\frac{y^{-4}}{y^{-5}} = y^{-4-(-5)} = y^1 = y$

45. $\frac{b^6}{b^8} = b^{6-8} = b^{-2} = \frac{1}{b^2}$

47. $\frac{(2x-3)^8}{(2x-3)^3} = (2x-3)^{8-3} = (2x-3)^5$

49. $\frac{(p+q)^2}{(p+q)^{-1}} = (p+q)^{2-(-1)} = (p+q)^3$

51. $\frac{(4-x)^{-3}}{(4-x)^2}$
$= (4-x)^{-3-2}$
$= (4-x)^{-5}$
$= \frac{1}{(4-x)^5}$

53. $\dfrac{(z-5)^{-4}}{(z-5)^{-1}}$
$=(z-5)^{-4-(-1)}$
$=(z-5)^{-3}$
$=\dfrac{1}{(z-5)^3}$

55. $(a^5)^6 = a^{5\cdot 6} = a^{30}$

57. $(c^{-4})^{-2} = c^{(-4)(-2)} = c^8$

59. $[(x+y)^3]^2 = (x+y)^{3\cdot 2} = (x+y)^6$

61. $[(a-b)^4]^{-1}$
$=(a-b)^{4(-1)}$
$=(a-b)^{-4}$
$=\dfrac{1}{(a-b)^4}$

63. $(x^2)^0 = x^{2\cdot 0} = x^0 = 1$

65. $(5m)^4 = 5^4 m^4 = 625m^4$

67. $\left(\dfrac{b}{d}\right)^4 = \dfrac{b^4}{d^4}$

69. $(pq)^{21} = p^{21}q^{21}$

71. $\left(\dfrac{3b}{c}\right)^4 = \dfrac{3^4 b^4}{c^4} = \dfrac{81b^4}{c^4}$

73. $(-2c)^6 = (-2)^6 c^6 = 64c^6$

75. $\left(\dfrac{4x}{y}\right)^{-3} = \left(\dfrac{y}{4x}\right)^3 = \dfrac{y^3}{4^3 x^3} = \dfrac{y^3}{64x^3}$

77. $\left(\dfrac{-3p}{q}\right)^{-4} = \left(\dfrac{q}{-3p}\right)^4 = \dfrac{q^4}{(-3)^4 p^4} = \dfrac{q^4}{81p^4}$

79. $(p^5q^7)^3 = p^{5\cdot 3}q^{7\cdot 3} = p^{15}q^{21}$

81. $[(2a)^2]^5$
$=(2a)^{2\cdot 5}$
$=(2a)^{10}$
$=2^{10}a^{10}$
$=1024a^{10}$

83. $\left[\left(\dfrac{x}{2y}\right)^2\right]^3$
$=\left(\dfrac{x}{2y}\right)^{2\cdot 3}$
$=\left(\dfrac{x}{2y}\right)^6$
$=\dfrac{x^6}{2^6 y^6}$
$=\dfrac{x^6}{64y^6}$

85. Let x = length of original side
$5x$ = length of enlarged side
The original area is x^2 square units.
The enlarged area is $(5x)^2 = 5^2x^2 = 25x^2$ square units.
The enlarged area is 25 times bigger.
$x^2 = 6^2 = 36$
$25x^2 = 25(6)^2 = 25(36) = 900$
If the original side is 6 feet, the original area and enlarged area are 36 ft^2 and 900 ft^2, respectively.

87. Let x = length of original side
$4x$ = length of enlarged side
The volume of the original bin is x^3 cubic units.
The volume of the enlarged bin is $(4x)^3 = 4^3x^3 = 64x^3$ cubic units.
The volume of the enlarged bin is 64 times greater.
$x^3 = (1.5)^3 = 3.375$
$64x^3 = 64(3.375) = 216$
If the original side measures 1.5 feet, then the volumes of the original bin and enlarged bin are 3.375 ft^3 and 216 ft^3, respectively.

10.1 Experiencing Algebra the Calculator Way

1. $x^3 \cdot x^2 = x^6$
   $Y1 = x^3 \cdot x^2$
   $Y2 = x^6$

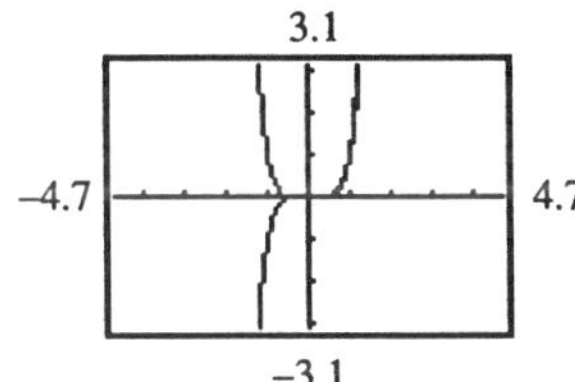

They are not equivalent.

2. $x^3 \cdot x^2 = x^5$
   $Y1 = x^3 \cdot x^2$
   $Y2 = x^5$

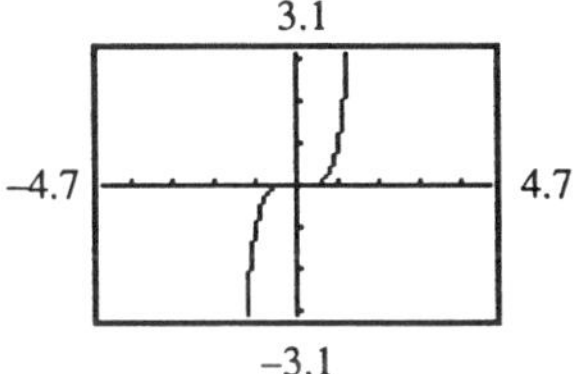

They are equivalent.

3. $\frac{x^3}{x^{-2}} = x^5$
   $Y1 = \frac{x^3}{x^{-2}}$
   $Y2 = x^5$

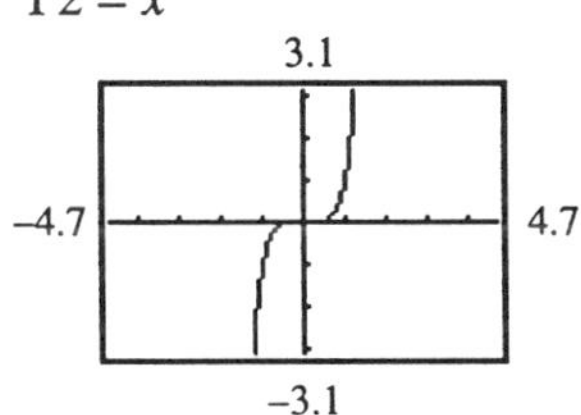

They are equivalent.

4. $\frac{x^3}{x^{-2}} = x$
   $Y1 = \frac{x^3}{x^{-2}}$
   $Y2 = x$

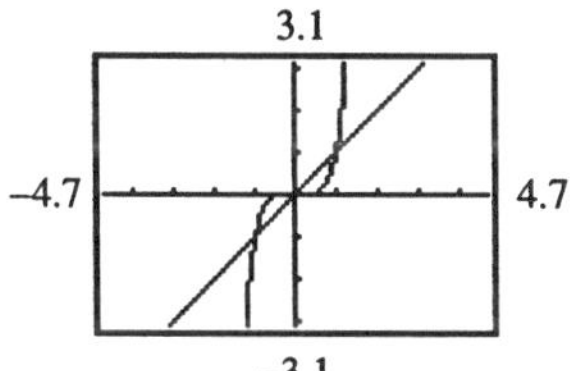

They are not equivalent.

5. $\frac{x^3}{x^2} = x^5$
   $Y1 = \frac{x^3}{x^2}$
   $Y2 = x^5$

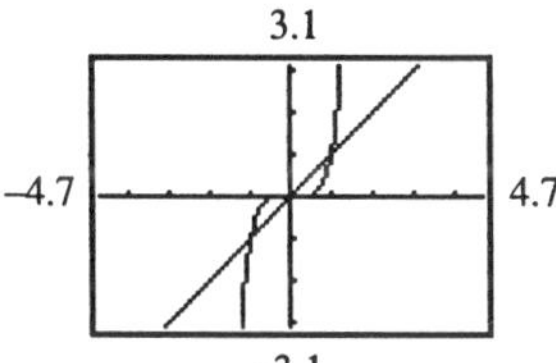

They are not equivalent.

6. $\frac{x^3}{x^2} = x$
   $Y1 = \frac{x^3}{x^2}$
   $Y2 = x$

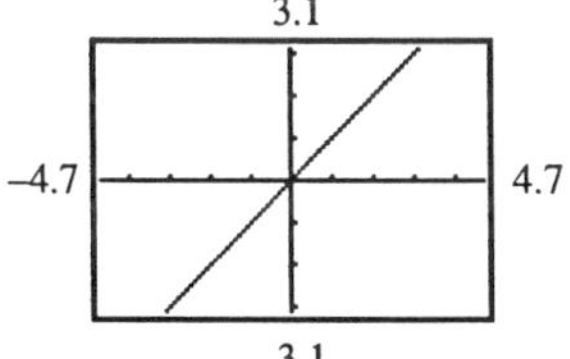

They are equivalent.

10.2 Experiencing Algebra the Exercise Way

1.
$$\begin{array}{r} 9x^2 - 17x + 31 \\ 2x^4 \quad + 3x^2 \quad + 12 \\ 5x^3 \quad - 17x + 11 \\ \underline{2x^4 + 4x^3 - 7x^2 - \ 8x - 26} \\ 4x^4 + 9x^3 + 5x^2 - 42x + 28 \end{array}$$

3. $(5x^4 + 6x + 3x^3 - 2x^2 - 12) + (4x^4 + 21 - 8x^2 - 9x)$
$= 5x^4 + 6x + 3x^3 - 2x^2 - 12 + 4x^4 + 21 - 8x^2 - 9x$
$= 9x^4 + 3x^3 - 10x^2 - 3x + 9$

5. $(5x^2y - 3xy^2 + 6y^3) + (15x^3 - 8x^2y + 3xy^2)$
$= 5x^2y - 3xy^2 + 6y^3 + 15x^3 - 8x^2y + 3xy^2$
$= -3x^2y + 6y^3 + 15x^3$

7. $(6 - 7a^3 + 3a^2 - 5a) + (6a + 8a^3 + 2) + (5a^2 - 8a - 9)$
$= 6 - 7a^3 + 3a^2 - 5a + 6a + 8a^3 + 2 + 5a^2 - 8a - 9$
$= a^3 + 8a^2 - 7a - 1$

9. $\left(\frac{2}{3}y^4 + \frac{1}{6}y^3 + 3y^2 - \frac{1}{3}y + \frac{5}{9}\right) + \left(\frac{7}{3}y^3 - \frac{8}{9}y^2 + \frac{5}{6}y - 3\right)$
$= \frac{2}{3}y^4 + \frac{1}{6}y^3 + 3y^2 - \frac{1}{3}y + \frac{5}{9} + \frac{7}{3}y^3 - \frac{8}{9}y^2 + \frac{5}{6}y - 3$
$= \frac{2}{3}y^4 + \left(\frac{1}{6} + \frac{7}{3}\right)y^3 + \left(3 - \frac{8}{9}\right)y^2 + \left(-\frac{1}{3} + \frac{5}{6}\right)y + \left(\frac{5}{9} - 3\right)$
$= \frac{2}{3}y^4 + \frac{5}{3}y^3 + \frac{19}{9}y^2 + \frac{1}{2}y - \frac{22}{9}$

11. $(12.07x^3 + 8.6x^2 - 3.19x + 14) + (6.7x^3 - 9.83x^2 + 7x - 4.265)$
$= 12.07x^3 + 8.6x^2 - 3.19x + 14 + 6.7x^3 - 9.83x^2 + 7x - 4.265$
$= 18.77x^3 - 1.23x^2 + 3.81x + 9.735$

13.
$$\begin{array}{r} 4756a^3 - \ 3219a^2 - 1816a + 2083 \\ \underline{361a^3 + 54217a^2 + \quad 0a + \ 12} \\ 5117a^3 + 50998a^2 - 1816a + 2095 \end{array}$$

15. $(3a + 4b) + (5c + 6d)$
$= 3a + 4b + 5c + 6d$

17. $(5z^3 + 27z^2 - 35z + 42) - (3z^3 + 16z - 72)$
$= 5z^3 + 27z^2 - 35z + 42 - 3z^3 - 16z + 72$
$= 2z^3 + 27z^2 - 51z + 114$

19. $(a^5 - 9) - (a^5 - a^4 + a^3 - a^2 + a - 9)$
$= a^5 - 9 - a^5 + a^4 - a^3 + a^2 - a + 9$
$= a^4 - a^3 + a^2 - a$

21. $(16x^2-32+9x)-(12x+7+9x^2)$
$=16x^2-32+9x-12x-7-9x^2$
$=7x^2-3x-39$

23. $(13a^3-6a^2+11)-(12a-3+18a^3)$
$=13a^3-6a^2+11-12a+3-18a^3$
$=-5a^3-6a^2-12a+14$

25.
$$\begin{array}{r}42x^3+17x^2y+3xy^2+23y^3\\ \underline{0x^3-47x^2y+0xy^2-12y^3}\\ 42x^3-30x^2y+3xy^2+11y^3\end{array}$$

27. $(4a+7c)-(2b+6d)$
$=4a+7c-2b-6d$
$=4a-2b+7c-6d$

29. $\left(\frac{5}{7}x^2+\frac{8}{21}x-\frac{11}{14}\right)-\left(\frac{1}{2}x^2+\frac{5}{6}x+\frac{19}{42}\right)$
$=\frac{5}{7}x^2+\frac{8}{21}x-\frac{11}{14}-\frac{1}{2}x^2-\frac{5}{6}x-\frac{19}{42}$
$=\left(\frac{5}{7}-\frac{1}{2}\right)x^2+\left(\frac{8}{21}-\frac{5}{6}\right)x+\left(-\frac{11}{14}-\frac{19}{42}\right)$
$=\frac{3}{14}x^2-\frac{19}{42}x-\frac{26}{21}$

31.
$$\begin{array}{r}21.2x^3+0.9x^2y-13.22xy^2+81.07y^3\\ \underline{-12.2x^3-0.1x^2y-\ 0.78xy^2-13.07y^3}\\ 9x^3+0.8x^2y-\quad 14xy^2+\quad 68y^3\end{array}$$

33.
$$\begin{array}{r}5062z^2-\ 106z+8295\\ \underline{-\ 379z^2-4297z-1108}\\ 4683z^2-4403z+7187\end{array}$$

35. $-6p^3q\cdot 7p^{-3}q^{-4}$
$=-42p^{3+(-3)}q^{1+(-4)}$
$=-42p^0q^{-3}$
$=-\frac{42}{q^3}$

37. $(1.4x^2)\cdot(4.3x^3y^{-2})$
$=6.02x^{2+3}y^{-2}$
$=\frac{6.02x^5}{y^2}$

39. $\left(\frac{3}{7}x^2y^{-1}\right)\cdot\left(\frac{14}{15}x^{-1}y^4\right)$
$=\frac{2}{5}x^{2-1}y^{-1+4}$
$=\frac{2}{5}xy^3$

41. $4a^2(3a^2-5ab+b^2)$
$=4a^2(3a^2)-4a^2(5ab)+4a^2(b^2)$
$=12a^{2+2}\quad 20a^{2+1}b+4a^2b^2$
$=12a^4-20a^3b+4a^2b^2$

43. $-3x^{-1}(-2x^3+3x^2-x+1)$
$=-3x^{-1}(-2x^3)-3x^{-1}(3x^2)-3x^{-1}(-x)-3x^{-1}(1)$
$=6x^{-1+3}-9x^{-1+2}+3x^{-1+1}-3x^{-1}$
$=6x^2-9x^1+3x^0-3\left(\frac{1}{x}\right)$
$=6x^2-9x+3-\frac{3}{x}$

45. $(3a^2+2ab+b^2)\cdot(-ab)$
$=3a^2(-ab)+2ab(-ab)+b^2(-ab)$
$=-3a^{2+1}b-2a^{1+1}b^{1+1}-ab^{2+1}$
$=-3a^3b-2a^2b^2-ab^3$

47. $\frac{3}{5}a^2(5a-15a^2+25a^3)$
$=\frac{3}{5}a^2(5a)-\frac{3}{5}a^2(15a^2)+\frac{3}{5}a^2(25a^3)$
$=3a^{2+1}-9a^{2+2}+15a^{2+3}$
$=3a^3-9a^4+15a^5$

49. $\frac{24x^5}{3x^3}$
$=\frac{24}{3}\cdot\frac{x^5}{x^3}$
$=\frac{24}{3}\cdot x^{5-3}$
$=8x^2$

51. $\frac{13x^4y^2}{2x^4y^7}$
$=\frac{13}{2}\cdot\frac{x^4}{x^4}\cdot\frac{y^2}{y^7}$
$=\frac{13}{2}\cdot x^{4-4}\cdot y^{2-7}$
$=\frac{13}{2}x^0y^{-5}$
$=\frac{13}{2}(1)\left(\frac{1}{y^5}\right)$
$=\frac{13}{2y^5}$

53. $\frac{122a^{-3}b^3}{11a^2b^{-2}}$
$=\frac{122}{11}\cdot\frac{a^{-3}}{a^2}\cdot\frac{b^3}{b^{-2}}$
$=\frac{122}{11}\cdot a^{-3-2}\cdot b^{3-(-2)}$
$=\frac{122}{11}a^{-5}b^5$
$=\frac{122}{11}\cdot\frac{1}{a^5}\cdot b^5$
$=\frac{122b^5}{11a^5}$

55. $\frac{(2x^2)^3}{6x}$
$=\frac{2^3x^{2\cdot3}}{6x}$
$=\frac{8x^6}{6x}$
$=\frac{8}{6}\cdot\frac{x^6}{x}$
$=\frac{4x^5}{3}$

57. $\left(\frac{2x^2}{6x}\right)^3$
$=\left(\frac{2}{6}\cdot\frac{x^2}{x}\right)^3$
$=\left(\frac{x}{3}\right)^3$
$=\frac{x^3}{3^3}$
$=\frac{x^3}{27}$

59. $\dfrac{(6x)^6}{(6x)^4}$
$= (6x)^{6-4}$
$= (6x)^2$
$= 6^2 x^2$
$= 36x^2$

61. $\dfrac{(3a^{-2}b)^3}{6ab^2}$
$= \dfrac{3^3 a^{-2\cdot 3} b^3}{6ab^2}$
$= \dfrac{27a^{-6}b^3}{6ab^2}$
$= \dfrac{27}{6}\cdot\dfrac{a^{-6}}{a}\cdot\dfrac{b^3}{b^2}$
$= \dfrac{9b}{2a^7}$

63. $\left(\dfrac{3a^2b}{6ab^2}\right)^3$
$= \left(\dfrac{3}{6}\cdot\dfrac{a^2}{a}\cdot\dfrac{b}{b^2}\right)^3$
$= \left(\dfrac{a}{2b}\right)^3$
$= \dfrac{a^3}{2^3 b^3}$
$= \dfrac{a^3}{8b^3}$

65. $\left(\dfrac{3a^2b}{6ab^2}\right)^{-3}$
$= \left(\dfrac{6ab^2}{3a^2b}\right)^3$
$= \left(\dfrac{6}{3}\cdot\dfrac{a}{a^2}\cdot\dfrac{b^2}{b}\right)^3$
$= \left(\dfrac{2b}{a}\right)^3$
$= \dfrac{2^3 b^3}{a^3}$
$= \dfrac{8b^3}{a^3}$

67. $\dfrac{(x^2y^{-3}z)^4}{(2xyz^2)^2}$
$= \dfrac{x^{2\cdot 4}y^{-3\cdot 4}z^4}{2^2x^2y^2z^{2\cdot 2}}$
$= \dfrac{x^8y^{-12}z^4}{4x^2y^2z^4}$
$= \dfrac{1}{4}\cdot\dfrac{x^8}{x^2}\cdot\dfrac{y^{-12}}{y^2}\cdot\dfrac{z^4}{z^4}$
$= \dfrac{x^6}{4y^{14}}$

69. $\dfrac{7{,}309{,}800a^9}{0.000031a^3}$
$= \dfrac{7{,}309{,}800}{0.000031}\cdot\dfrac{a^9}{a^3}$
$= (2.358 \times 10^{11})a^6$

71. $\dfrac{2.35\cdot 10^{-4}x^5}{4.7\cdot 10^8 x^2}$
$= \dfrac{2.35\cdot 10^{-4}}{4.7\cdot 10^8}\cdot\dfrac{x^5}{x^2}$
$= (5\times 10^{-13})x^3$

73. $\dfrac{6x^3+12x^2-18x}{3x}$

$=\dfrac{6x^3}{3x}+\dfrac{12x^2}{3x}-\dfrac{18x}{3x}$

$=2x^{3-1}+4x^{2-1}-6x^{1-1}$

$=2x^2+4x-6$

75. $\dfrac{4x^2-2x+12}{8x}$

$=\dfrac{4x^2}{8x}-\dfrac{2x}{8x}+\dfrac{12}{8x}$

$=\dfrac{1}{2}x^{2-1}-\dfrac{1}{4}x^{1-1}+\dfrac{3}{2x}$

$=\dfrac{1}{2}x-\dfrac{1}{4}+\dfrac{3}{2x}$

77. $\dfrac{3.72x^4-6.96x^2+1.08}{1.2x^2}$

$=\dfrac{3.72x^4}{1.2x^2}-\dfrac{6.96x^2}{1.2x^2}+\dfrac{1.08}{1.2x^2}$

$=3.1x^{4-2}-5.8x^{2-2}+\dfrac{0.9}{x^2}$

$=3.1x^2-5.8+\dfrac{0.9}{x^2}$

79. $\dfrac{9x^4+6x^3y-18x^2y^2-24xy^3+72y^4}{-3xy}$

$=\dfrac{9x^4}{-3xy}+\dfrac{6x^3y}{-3xy}-\dfrac{18x^2y^2}{-3xy}-\dfrac{24xy^3}{-3xy}+\dfrac{72y^4}{-3xy}$

$=-3x^{4-1}y^{-1}-2x^{3-1}y^{1-1}+6x^{2-1}y^{2-1}+8x^{1-1}y^{3-1}-24x^{-1}y^{4-1}$

$=-\dfrac{3x^3}{y}-2x^2+6xy+8y^2-\dfrac{24y^3}{x}$

81. Let x = number of pots

a. $C(x)=200+(2.50+2.00)x-0.05x$
$C(x)=200+4.45x$

b. $R(x)=13.5x$

c. $P(x)=R(x)-C(x)$
$P(x)=13.5x-(200+4.45x)$
$P(x)=9.05x-200$

d. $A(x) = \frac{9.05x - 200}{x}$

$A(x) = 9.05 - \frac{200}{x}$

e. $A(20) = 9.05 - \frac{200}{20} = -0.95$

$A(30) = 9.05 - \frac{200}{30} \approx 2.38$

The average profit per pot if 20 and 30 are produced is –\$0.95 and \$2.38, respectively.

83. Let width = 15 feet
length = $(15 + x)$ feet
The polynomial for the perimeter is:
$2(15) + 2(15 + x)$
$= 30 + 30 + 2x$
$= 60 + 2x$
$60 + 2x = 60 + 2(8) = 76$
The perimeter is 76 feet.

85. Surface area of cube: $6x^2$

Surface area of sphere: $4\pi\left(\frac{x}{2}\right)^2 = \pi x^2$

Difference: $6x^2 - \pi x^2$
$= 6x^2 - \pi x^2 = 6(8)^2 - \pi(8)^2$
≈ 182.9
The difference is approximately 182.9 in^2.

87. Area of rectangle: $8(2x) = 16x$
Area of square: x^2 square inches
Area of shade: $16x - x^2$ square inches

89. Let r = radius
$4r + 5$ = length
Area of two semi-circles: πr^2
Area of rectangle: $2r(4r + 5)$
Total area:
$\pi r^2 + 2r(4r + 5)$
$= \pi r^2 + 8r^2 + 10r$
$= (\pi + 8)r^2 + 10r$ square feet

r (feet)	$(\pi+8)r^2+10r$	Area $(\text{feet})^2$
8	$(\pi+8)8^2+10(8)\approx 793.06$	793.06
9	$(\pi+8)9^2+10(9)\approx 992.47$	992.47
10	$(\pi+8)10^2+10(10)\approx 1214.16$	1214.16
11	$(\pi+8)11^2+10(11)\approx 1458.13$	1458.13
12	$(\pi+8)12^2+10(12)\approx 1724.39$	1724.39

91. a. x = height
$4x$ = base
Area is:
$\frac{1}{2}(4x)x = 2x^2$ square feet

b. $2x$ = height
$4x + 10$ = base
Area is:
$\frac{1}{2}(4x+10)(2x) = 4x^2+10x$ square feet

c. Ratio $= \dfrac{\text{Planned area}}{\text{Current area}}$
$= \dfrac{4x^2+10x}{2x^2}$
$= \dfrac{4x^2}{2x^2}+\dfrac{10x}{2x^2}$
$= 2+\dfrac{5}{x}$

d. Current Area $= 2x^2$
$= 2(10)^2$
$= 200$
The current area is $200\ \text{ft}^2$.
Planned area $= 4x^2+10x$
$= 4(10)^2+10(10)$
$= 400 + 100$
$= 500$
The planned area is $500\ \text{ft}^2$.
Ratio $= 2+\dfrac{5}{x}$
$= 2+\dfrac{5}{10}$
$= 2+\dfrac{1}{2}$
$= \dfrac{5}{2}$
The ratio is $\dfrac{5}{2}$.

93. Let r = radius of heel

Area of two heels = $2\pi r^2$

Pressure from each heel = $\dfrac{110}{2\pi r^2}$

$= \dfrac{55}{\pi r^2}$ pounds per square inch

r (inch)	$\dfrac{55}{\pi r^2}$	Pounds per square inch
0.25	$\dfrac{55}{\pi(0.25)^2} \approx 280.1$	280.1
0.50	$\dfrac{55}{\pi(0.5)^2} \approx 70.0$	70.0
0.75	$\dfrac{55}{\pi(0.75)^2} \approx 31.1$	31.1
1.00	$\dfrac{55}{\pi(1)^2} \approx 17.5$	17.5

For a radius of 1 inch, the woman's pressure on the floor approximates the elephant's.

95. Let w = width

$2w$ = length

Area = $2w^2$

Cost per square foot = $\dfrac{80{,}000}{2w^2} = \dfrac{40{,}000}{w^2}$

w (feet)	$\dfrac{40{,}000}{w^2}$	Cost per square foot
20	$\dfrac{40{,}000}{(20)^2} = 100$	100
25	$\dfrac{40{,}000}{(25)^2} = 64$	64
30	$\dfrac{40{,}000}{(30)^2} \approx 44.44$	44.44
35	$\dfrac{40{,}000}{(35)^2} \approx 32.65$	32.65
40	$\dfrac{40{,}000}{(40)^2} = 25$	25
45	$\dfrac{40{,}000}{(45)^2} \approx 19.75$	19.75
50	$\dfrac{40{,}000}{(50)^2} = 16$	16

A home with a size of 35 ft by 70 ft would cost approximately $33 per square foot.

10.2 Experiencing Algebra the Calculator Way

1. $Y1 = (3x^4 - x^3 + 2x^2 - 5) - (2x^4 + x^3 + x^2 - 4)$

$Y2 = x^4 + x^2 - 9$

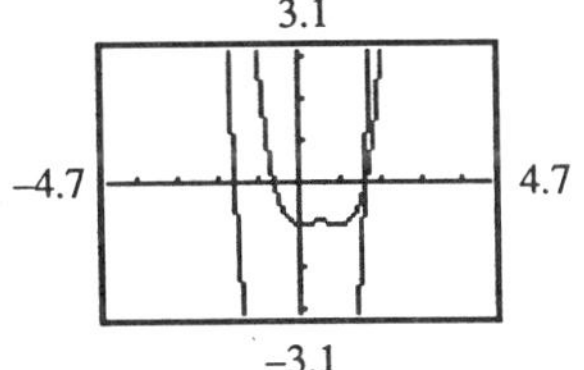

No; they are not equivalent.

2. $Y1 = (3x^4 - x^3 + 2x^2 - 5) - (2x^4 + x^3 + x^2 - 4)$

$Y2 = x^4 - 2x^3 + x^2 - 1$

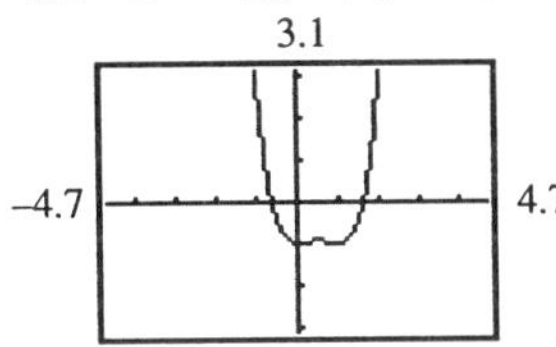

Yes; they are equivalent.

3. $Y1 = 2x^2(x^2 - 3x - 2)$

$Y2 = 2x^4 - 6x^3 - 4x^2$

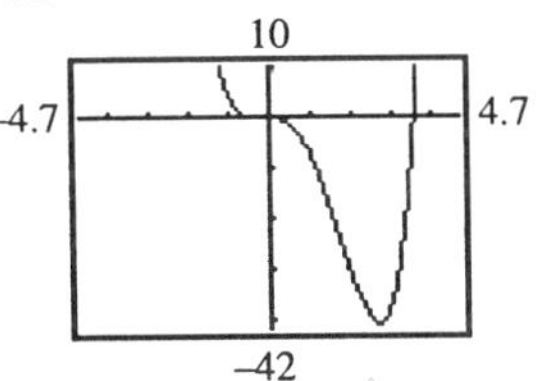

Yes, they are equivalent.

4. $Y1 = \left(\frac{1}{2}x\right)\left(2x + \frac{1}{3}\right)$

$Y2 = x^2 - \frac{1}{6}x$

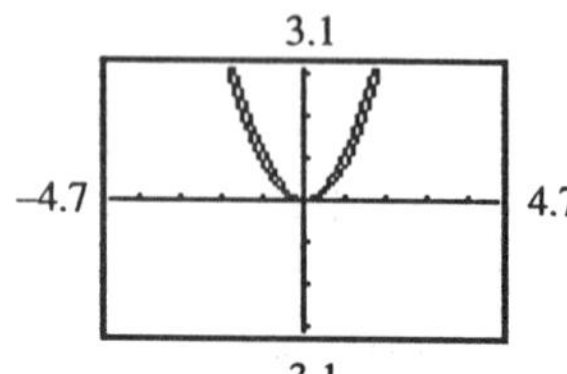

No; they are not equivalent.

5. $Y1 = -0.4x(7 - 5x + x^2)$

$Y2 = 2.8x + 2x^2 - 0.4x^3$

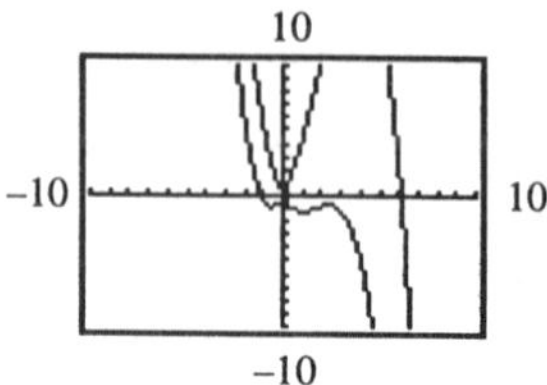

No; they are not equivalent.

6. $Y1 = \frac{2x^3 - 6x^2 + 2x}{2x}$

$Y2 = x^2 - 3x + 1$

10

−10 10

−10

Yes; they are equivalent.

7. $Y1 = \frac{-6x^2 + 9x}{-3x}$

$Y2 = 2x + 3$

10

−10 10

−10

No; they are not equivalent.

8. $Y1 = \frac{-8x^4 + 4x^3 - 8x^2 + 4x}{4x}$

$Y2 = -2x^3 + x^2 - 2x + 4$

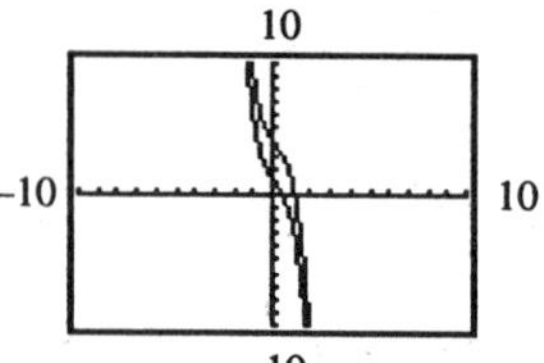

No; they are not equivalent.

9. $Y1 = \frac{3x^4 - 3x^3 + 3x}{3x}$

$Y2 = x^3 - x^2 + 1$

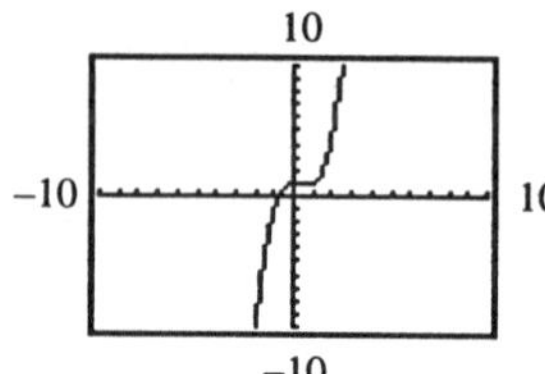

Yes; they are equivalent.

10.3 Experiencing Algebra the Exercise Way

1. $(x + 3)(x + 8)$
$= x(x + 8) + 3(x + 8)$
$= x^2 + 8x + 3x + 24$
$= x^2 + 11x + 24$

3. $(2x + 1)(3x - 4)$
$= 2x(3x - 4) + 1(3x - 4)$
$= 6x^2 - 8x + 3x - 4$
$= 6x^2 - 5x - 4$

5. $(4 - x)(x + 2)$
$= 4(x + 2) - x(x + 2)$
$= 4x + 8 - x^2 - 2x$
$= -x^2 + 2x + 8$

7. $(5 + x)(3 + 2x)$
$= 5(3 + 2x) + x(3 + 2x)$
$= 15 + 10x + 3x + 2x^2$
$= 2x^2 + 13x + 15$

9. $(2x+5)(3y-2)$
$=2x(3y-2)+5(3y-2)$
$=6xy-4x+15y-10$

11. $(3x+4y)(x-2y)$
$=3x(x-2y)+4y(x-2y)$
$=3x^2-6xy+4xy-8y^2$
$=3x^2-2xy-8y^2$

13. $(a-2.4)(5a+3.8)$
$=a(5a+3.8)-2.4(5a+3.8)$
$=5a^2+3.8a-12a-9.12$
$=5a^2-8.2a-9.12$

15. $(2x+1.1)(3y+3.2)$
$=2x(3y+3.2)+1.1(3y+3.2)$
$=6xy+6.4x+3.3y+3.52$

17. $\left(a+\frac{2}{3}\right)\left(a+\frac{1}{3}\right)$
$=a\left(a+\frac{1}{3}\right)+\frac{2}{3}\left(a+\frac{1}{3}\right)$
$=a^2+\frac{1}{3}a+\frac{2}{3}a+\frac{2}{9}$
$=a^2+a+\frac{2}{9}$

19. $\left(2x+\frac{1}{4}\right)\left(8x-\frac{1}{6}\right)$
$=2x\left(8x-\frac{1}{6}\right)+\frac{1}{4}\left(8x-\frac{1}{6}\right)$
$=16x^2-\frac{1}{3}x+2x-\frac{1}{24}$
$=16x^2+\frac{5}{3}x-\frac{1}{24}$

21. $(2x^2-3)(x^2+4)$
$=2x^2(x^2+4)-3(x^2+4)$
$=2x^4+8x^2-3x^2-12$
$=2x^4+5x^2-12$

23. $(4x^2+3)(2x+1)$
$=4x^2(2x+1)+3(2x+1)$
$=8x^3+4x^2+6x+3$

25. $(2x+3y^2)(3x^2-5y)$
$=2x(3x^2-5y)+3y^2(3x^2-5y)$
$=6x^3-10xy+9x^2y^2-15y^3$

27. $(3a^2+5b^3)(a^2+b^3)$
$=3a^2(a^2+b^3)+5b^3(a^2+b^3)$
$=3a^4+3a^2b^3+5a^2b^3+5b^6$
$=3a^4+8a^2b^3+5b^6$

29. $(x+1)(x^2-x+1)$
$=x(x^2-x+1)+1(x^2-x+1)$
$=x^3-x^2+x+x^2-x+1$
$=x^3+1$

31. $(3x-2)(2x^2-5x-3)$
$=3x(2x^2-5x-3)-2(2x^2-5x-3)$
$=6x^3-15x^2-9x-4x^2+10x+6$
$=6x^3-19x^2+x+6$

33. $(x^2+x+1)(x^2+2x+3)$
$=x^2(x^2+2x+3)+x(x^2+2x+3)+1(x^2+2x+3)$
$=x^4+2x^3+3x^2+x^3+2x^2+3x+x^2+2x+3$
$=x^4+3x^3+6x^2+5x+3$

35. $(a+b+c)^2$
$=(a+b+c)(a+b+c)$
$=a(a+b+c)+b(a+b+c)+c(a+b+c)$
$=a^2+ab+ac+ab+b^2+bc+ac+bc+c^2$
$=a^2+b^2+c^2+2ab+2bc+2ac$

37. $(z+3)^3$
$=(z+3)(z+3)(z+3)$
$=[z(z+3)+3(z+3)](z+3)$
$=(z^2+3z+3z+9)(z+3)$
$=(z^2+6z+9)(z+3)$
$=(z^2+6z+9)(z)+(z^2+6z+9)(3)$
$=z^3+6z^2+9z+3z^2+18z+27$
$=z^3+9z^2+27z+27$

39. $(3a-2b)^3$
$=(3a-2b)(3a-2b)(3a-2b)$
$=[3a(3a-2b)-2b(3a-2b)](3a-2b)$
$=(9a^2-6ab-6ab+4b^2)(3a-2b)$
$=(9a^2-12ab+4b^2)(3a-2b)$
$=(9a^2-12ab+4b^2)(3a)+(9a^2-12ab+4b^2)(-2b)$
$=27a^3-36a^2b+12ab^2-18a^2b+24ab^2-8b^3$
$=27a^3-54a^2b+36ab^2-8b^3$

41. $(x-5)(x+5)$
$=x^2-5^2$
$=x^2-25$

43. $(3m+7)(3m-7)$
$=(3m)^2-7^2$
$=9m^2-49$

45. $(2a+3b)(2a-3b)$
$=(2a)^2-(3b)^2$
$=4a^2-9b^2$

47. $(4x-1.5)(4x+1.5)$
$=(4x)^2-(1.5)^2$
$=16x^2-2.25$

49. $\left(\frac{2}{5}x-1\right)\left(\frac{2}{5}x+1\right)$
$=\left(\frac{2}{5}x\right)^2-1^2$
$=\frac{4}{25}x^2-1$

51. $\left(\frac{1}{3}x+\frac{4}{5}\right)\left(\frac{1}{3}x-\frac{4}{5}\right)$
$=\left(\frac{1}{3}x\right)^2-\left(\frac{4}{5}\right)^2$
$=\frac{1}{9}x^2-\frac{16}{25}$

53. $(x^2+7)(x^2-7)$
$=(x^2)^2-7^2$
$=x^4-49$

55. $(2x^3+5y)(2x^3-5y)$
$=(2x^3)^2-(5y)^2$
$=4x^6-25y^2$

57. $(m+7)^2$
$=m^2+2(m)(7)+7^2$
$=m^2+14m+49$

59. $(x-y)^2$
$=x^2-2xy+y^2$

61. $(2p+9q)^2$
$=(2p)^2+2(2p)(9q)+(9q)^2$
$=4p^2+36pq+81q^2$

63. $(6c-5)^2$
$=(6c)^2-2(6c)(5)+(5)^2$
$=36c^2-60c+25$

65. $(3x^3+2)^2$
$=(3x^3)^2+2(3x^3)(2)+2^2$
$=9x^6+12x^3+4$

67. $(2x^2-3y^3)^2$
$=(2x^2)^2-2(2x^2)(3y^3)+(3y^3)^2$
$=4x^4-12x^2y^3+9y^6$

69. a. length: $(18-2x)$ in.
width: $(12-2x)$ in.
height: (x) in.

b. Volume: $x(18-2x)(12-2x)$
$=(18x-2x^2)(12-2x)$
$=18x(12-2x)-2x^2(12-2x)$
$=216x-36x^2-24x^2+4x^3$
$=4x^3-60x^2+216x$
The volume is $4x^3-60x^2+216x$ in^3.

c. Surface Area:
$2(x)(12-2x)+2(x)(18-2x)+(12-2x)(18-2x)$
$=24x-4x^2+36x-4x^2+216-24x-36x+4x^2$
$=-4x^2+216$
The surface area is $-4x^2+216$ in^2.

d. Volume:
$4(2)^3-60(2)^2+216(2)=224$
Surface area:
$-4(2)^2+216=200$
The volume is 224 in^3 and the surface area is 200 in^2.

71. a. Area of outer circle:
$(\pi x^2)\ \text{ft}^2$

b. Area of inner circle:
$[\pi(x-5)^2]\ \text{ft}^2$

c. Difference of areas:
$\pi x^2 - \pi(x-5)^2$
$= \pi x^2 - \pi(x^2 - 10x + 25)$
$= \pi x^2 - \pi x^2 + 10\pi x - 25\pi$
$= 10\pi x - 25\pi$
The area of the deck is
$(10\pi x - 25\pi)\ \text{ft}^2$.

d. Area:
$10\pi(22) - 25\pi$
$= 220\pi - 25\pi$
$= 195\pi$
The area is $195\pi\ \text{ft}^2$.

73. a. The old volume is $x^3\ \text{ft}^3$.

b. New volume:
$x(x+5)(x+8)$
$= (x^2+5x)(x+8)$
$= x^2(x+8) + 5x(x+8)$
$= x^3 + 8x^2 + 5x^2 + 40x$
$= x^3 + 13x^2 + 40x$
The new volume is
$(x^3 + 13x^2 + 40x)\ \text{ft}^3$.

c. Ratio of new volume to old volume:
$$\frac{x^3+13x^2+40x}{x^3}$$
$$= \frac{x(x^2+13x+40)}{x(x^2)}$$
$$= \frac{x^2+13x+40}{x^2}$$
The ratio is $\dfrac{x^2+13x+40}{x^2}$

d. Ratio:
$$\frac{(12)^2 + 13(12) + 40}{(12)^2}$$
$$= \frac{340}{144}$$
$$= \frac{85}{36}$$
The ratio is $\dfrac{85}{36}$.

10.3 Experiencing Algebra the Calculator Way

1. $\text{Y1} = (2x-1)(2x+1)$
$\text{Y2} = 2x^2 - 1$

They are not equivalent.
$(2x-1)(2x+1)$
$= (2x)^2 - 1^2$
$= 4x^2 - 1$
The correct expression is $4x^2 - 1$.

2. $\text{Y1} = (x+1)(x-1)$
$\text{Y2} = x^2 + x + 2$

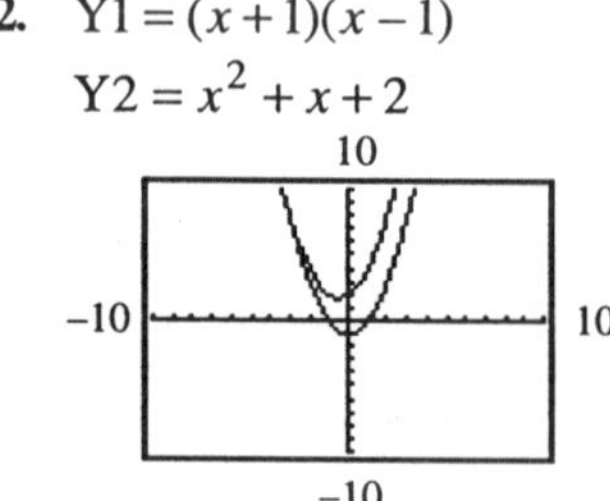

They are not equivalent.
$(x+1)(x-1)$
$= x^2 - 1^2$
$= x^2 - 1$
The correct expression is $x^2 - 1$.

3. $Y1 = (x-2)(x-1)$
$Y2 = x^2 - 3x + 2$

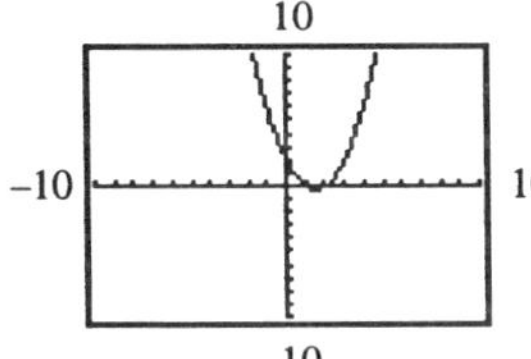

They are equivalent.

4. $Y1 = (0.5x+1)(4x-0.8)$
$Y2 = 2x^2 + 3.6x - 0.8$

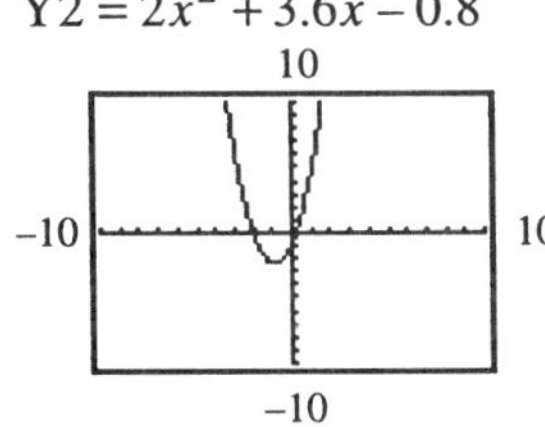

They are equivalent.

5. $Y1 = \left(\frac{1}{2}x - 3\right)\left(2x + \frac{1}{3}\right)$
$Y2 = x^2 - \frac{35}{6}x + 1$

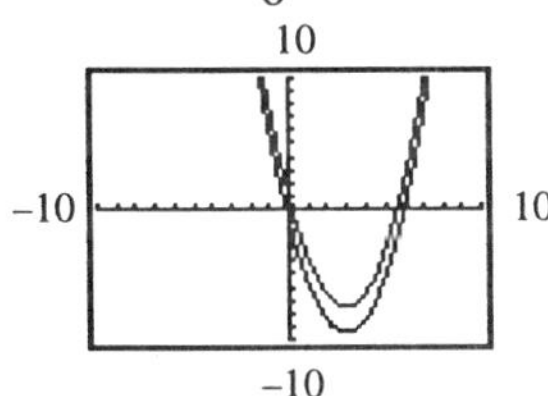

They are not equivalent.

$\left(\frac{1}{2}x - 3\right)\left(2x + \frac{1}{3}\right)$
$= \frac{1}{2}x\left(2x + \frac{1}{3}\right) - 3\left(2x + \frac{1}{3}\right)$
$= x^2 + \frac{1}{6}x - 6x - 1$
$= x^2 - \frac{35}{6}x - 1$

The correct expression is $x^2 - \frac{35}{6}x - 1$.

6. $Y1 = (x^2+4)(x^2-1)$
$Y2 = x^4 + 3x^2 - 4$

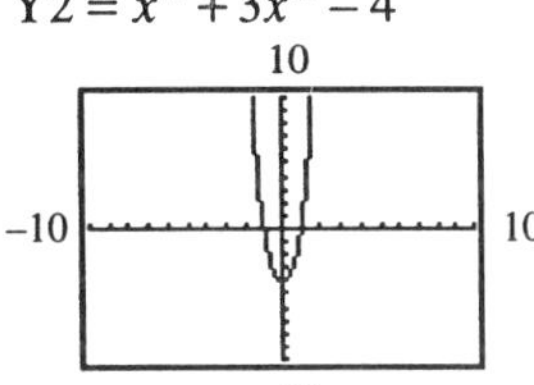

They are equivalent.

7. $Y1 = (x^2+1)(2x-1)$
$Y2 = 2x^3 - x^2 + 2x - 1$

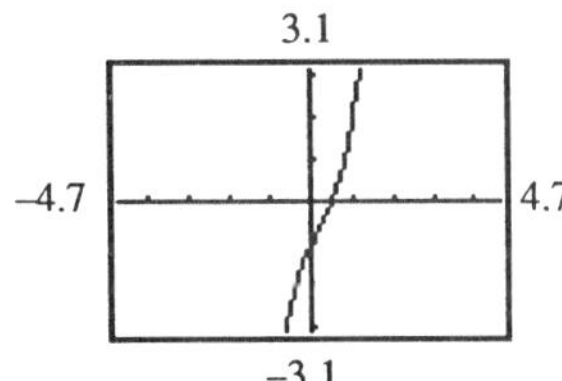

They are equivalent.

8. $Y1 = (x^2-1)(0.3x-1.4)$
$Y2 = 0.3x^3 - 1.4x^2 - 0.3x - 1.4$

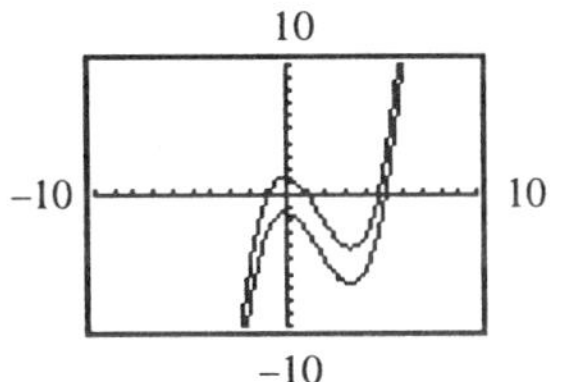

They are not equivalent.

$(x^2-1)(0.3x-1.4)$
$= x^2(0.3x-1.4) - 1(0.3x-1.4)$
$= 0.3x^3 - 1.4x^2 - 0.3x + 1.4$

The correct expression is $0.3x^3 - 1.4x^2 - 0.3x + 1.4$.

10.4 Experiencing Algebra the Exercise Way

1. $60a^2b^4c^3$ and $50a^3bc^2$
$60 = 2^2 \cdot 3 \cdot 5$
$50 = 2 \cdot 5^2$
The GCF for the coefficients is $2 \cdot 5 = 10$.
The GCF for the variable factors is a^2bc^2.
The GCF for $60a^2b^4c^3$ and $50a^3bc^2$ is $10a^2bc^2$.

3. $252x^3y^4$ and $180x^2z$
$252 = 2^2 \cdot 3^2 \cdot 7$
$180 = 2^2 \cdot 3^2 \cdot 5$
The GCF for the coefficients is $2^2 \cdot 3^2 = 36$.
The GCF for the variable factors is x^2.
The GCF for $252x^3y^4$ and $180x^2z$ is $36x^2$.

5. $45pq^4$ and $135r^4s$
$45 = 1 \cdot 3^2 \cdot 5$
$135 = 3^3 \cdot 5$
The GCF for the coefficients is $3^2 \cdot 5 = 45$. There are no common variable factors. The GCF for $45pq^4$ and $135r^4s$ is 45.

7. $63xyz$ and $98xyz$
$63 = 3^2 \cdot 7$
$98 = 2 \cdot 7^2$
The GCF for the coefficients is 7. The GCF for the variable factors is xyz. The GCF for $63xyz$ and $98xyz$ is $7xyz$.

9. $60a^2b^3c^3$, $90ab^2c$, and $150a^2b^4c^2$
$60 = 2^2 \cdot 3 \cdot 5$
$90 = 2 \cdot 3^2 \cdot 5$
$150 = 2 \cdot 3 \cdot 5^2$
The GCF for the coefficients is $2 \cdot 3 \cdot 5 = 30$.
The GCF for the variable factors is ab^2c.
The GCF for $60a^2b^3c^3$, $90ab^2c$, and $150a^2b^4c^2$ is $30ab^2c$.

11. $4x + 12y$
$= 4 \cdot x + 4 \cdot 3y$
$= 4(x + 3y)$

13. $8x^3 - 4x^2 + 12x - 24$
$= 4 \cdot 2x^3 - 4 \cdot x^2 + 4 \cdot 3x - 4 \cdot 6$
$= 4(2x^3 - x^2 + 3x - 6)$

15. $3a^4 - 5a^3 + 7a^2$
$= a^2 \cdot 3a^2 - a^2 \cdot 5a + a^2 \cdot 7$
$= a^2(3a^2 - 5a + 7)$

17. $-3x^5 - 9x^4 - 12x^3$
$= -3x^3 \cdot x^2 + (-3x^3) \cdot 3x + (-3x^3) \cdot 4$
$= -3x^3(x^2 + 3x + 4)$

19. $7x^4y^2 - 3x^2y^2 + 9x^2y^4$
$= x^2y^2 \cdot 7x^2 - x^2y^2 \cdot 3 + x^2y^2 \cdot 9y^2$
$= x^2y^2(7x^2 - 3 + 9y^2)$

21. $8a^5b^3c + 4a^4b^2 + 16a^3c$
$= 4a^3 \cdot 2a^2b^3c + 4a^3 \cdot ab^2 + 4a^3 \cdot 4c$
$= 4a^3(2a^2b^3c + ab^2 + 4c)$

23. $66u^3v^4 - 88u^4v^3$
$= 22u^3v^3 \cdot 3v + 22u^3v^3 \cdot 4u$
$= 22u^3v^3(3v + 4u)$

25. $3x^3 + 5y^4$
There is no greatest common factor, therefore the binomial does not factor.

27. $5x(x + 3) - 4(x + 3)$
$= (x + 3)(5x - 4)$

29. $x(2x + y) + 2y(2x + y)$
$= (2x + y)(x + 2y)$

31. $6x^2 + 10x + 21x + 35$
$= (6x^2 + 10x) + (21x + 35)$
$= 2x(3x + 5) + 7(3x + 5)$
$= (3x + 5)(2x + 7)$

33. $x^2 + 8x + x + 8$
$= (x^2 + 8x) + (x + 8)$
$= x(x + 8) + 1(x + 8)$
$= (x + 8)(x + 1)$

35. $2a^2 + 3a - 2a - 3$
$= (2a^2 + 3a) + (-2a - 3)$
$= a(2a + 3) - 1(2a + 3)$
$= (2a + 3)(a - 1)$

37. $2x^2 + xy + 4xy + 2y^2$
$= (2x^2 + xy) + (4xy + 2y^2)$
$= x(2x + y) + 2y(2x + y)$
$= (2x + y)(x + 2y)$

39. $x^2 + xy - xy - y^2$
$= (x^2 + xy) + (-xy - y^2)$
$= x(x + y) - y(x + y)$
$= (x + y)(x - y)$

41. $10xy - 55y + 24x - 132$
$= (10xy - 55y) + (24x - 132)$
$= 5y(2x - 11) + 12(2x - 11)$
$= (2x - 11)(5y + 12)$

43. $12ac + 3bc + 4ad + bd$
$= (12ac + 3bc) + (4ad + bd)$
$= 3c(4a + b) + d(4a + b)$
$= (4a + b)(3c + d)$

45. $2x^2y^2 + 3xy - 8xy - 12$
$= (2x^2y^2 + 3xy) + (-8xy - 12)$
$= xy(2xy + 3) - 4(2xy + 3)$
$= (2xy + 3)(xy - 4)$

47. $-x^2 - 3x - xy - 3y$
$= -1(x^2 + 3x + xy + 3y)$
$= -1[(x^2 + 3x) + (xy + 3y)]$
$= -1[x(x + 3) + y(x + 3)]$
$= -1(x + 3)(x + y)$

49. $x^4 + x^2y^2 + 2x^2y^2 + 2y^4$
$= (x^4 + x^2y^2) + (2x^2y^2 + 2y^4)$
$= x^2(x^2 + y^2) + 2y^2(x^2 + y^2)$
$= (x^2 + y^2)(x^2 + 2y^2)$

51. $ac + bc + ad + bd$
$= (ac + bc) + (ad + bd)$
$= c(a + b) + d(a + b)$
$= (a + b)(c + d)$

53. $8x^2 + 4x + 24x + 12$
$= 4(2x^2 + x + 6x + 3)$
$= 4[(2x^2 + x) + (6x + 3)]$
$= 4[x(2x + 1) + 3(2x + 1)]$
$= 4(2x + 1)(x + 3)$

55. $u^4 + u^3v - 2u^3v - 2u^2v^2$
$= u^2(u^2 + uv - 2uv - 2v^2)$
$= u^2[(u^2 + uv) + (-2uv - 2v^2)]$
$= u^2[u(u + v) - 2v(u + v)]$
$= u^2(u + v)(u - 2v)$

57. $6a^4 + 6a^3b^2 + 6a^3b + 6a^2b^3$
$= 6a^2(a^2 + ab^2 + ab + b^3)$
$= 6a^2[(a^2 + ab^2) + (ab + b^3)]$
$= 6a^2[a(a + b^2) + b(a + b^2)]$
$= 6a^2(a + b^2)(a + b)$

59. $-4x^3 - 12x^2 - 2x^2 - 6x$
$= -2x(2x^2 + 6x + x + 3)$
$= -2x[(2x^2 + 6x) + (x + 3)]$
$= -2x[2x(x + 3) + 1(x + 3)]$
$= -2x(x + 3)(2x + 1)$

61. $4x^2 + 6xz + 6xz + 9z^2$
$= (4x^2 + 6xz) + (6xz + 9z^2)$
$= 2x(2x + 3z) + 3z(2x + 3z)$
$= (2x + 3z)(2x + 3z)$
$= (2x + 3z)^2$

63. $36a^2 - 30ab - 30ab + 25b^2$
$= (36a^2 - 30ab) + (-30ab + 25b^2)$
$= 6a(6a - 5b) - 5b(6a - 5b)$
$= (6a - 5b)(6a - 5b)$
$= (6a - 5b)^2$

65. $5m^3 + 5m^2n + 5m^2n + 5mn^2$
$= 5m(m^2 + mn + mn + n^2)$
$= 5m[(m^2 + mn) + (mn + n^2)]$
$= 5m[m(m+n) + n(m+n)]$
$= 5m(m+n)(m+n)$
$= 5m(m+n)^2$

67. $2x^3 + 3x + 8x^2 + 12$
$= (2x^3 + 3x) + (8x^2 + 12)$
$= x(2x^2 + 3) + 4(2x^2 + 3)$
$= (2x^2 + 3)(x + 4)$

69. $5a^2x + 2b^2x + 15a^2y + 6b^2y$
$= (5a^2x + 2b^2x) + (15a^2y + 6b^2y)$
$= x(5a^2 + 2b^2) + 3y(5a^2 + 2b^2)$
$= (5a^2 + 2b^2)(x + 3y)$

71. **a.** Let x = amount received on the first goal
$x + (x+1) + (x+2) + (x+3)$
$+ (x+4) + (x+5) + (x+6) + (x+7)$
$+ (x+8)$
$= 9x + 36$
She will receive $(9x + 36)$ dollars.

b. $9x + 36 = 9(x + 4)$

c. The binomial $(x + 4)$ is the average amount in dollars that she will receive for each of her 9 goals.

d. $9(x + 4) = 9(10 + 4) = 126$
She will receive \$126.
Check:
$9 \text{ days} \cdot \frac{\$14}{\text{day}} = \$126$

73. **a.** $n^2 - n = n(n-1)$

b. $n^2 - n = (21)^2 - 21 = 420$
$n(n-1) = 21(21-1) = 420$
Each equals 420 times.

c. The factored expression was easier to evaluate. Answers will vary.

75. $x^2 + 7x - 3x - 21$
$= (x^2 + 7x) + (-3x - 21)$
$= x(x+7) - 3(x+7)$
$= (x+7)(x-3)$
The rectangle's length is $(x + 7)$ units and width is $(x - 3)$ units.

10.4 Experiencing Algebra the Calculator Way

1. $30 = 2 \cdot 3 \cdot 5$

2. $108 = 2^2 \cdot 3^3$

3. $525 = 3 \cdot 5^2 \cdot 7$

4. $1287 = 3^2 \cdot 11 \cdot 13$

5. $1547 = 7 \cdot 13 \cdot 17$

6. $4500 = 2^2 \cdot 3^2 \cdot 5^3$

10.5 Experiencing Algebra the Exercise Way

1. $x^2 + 14x + 45$
$a = 1,\ b = 14,\ c = 45$

factor	factor	sum of factors	
1	45	46	
3	15	18	
5	9	14	$\leftarrow b$

$x^2 + 14x + 45 = (x+5)(x+9)$

3. $y^2 - 15y + 56$
$a = 1,\ b = -15,\ c = 56$

factor	factor	sum of factors	
−1	−56	−57	
−2	−28	−30	
−4	−14	−18	
−7	−8	−15	$\leftarrow b$

$y^2 - 15y + 56 = (y-7)(y-8)$

5. $p^2 - 9p - 36$
$a = 1,\ b = -9,\ c = -36$

factor	factor	sum of factors	
–1	36	35	
1	–36	–35	
–2	18	16	
2	–18	–16	
–3	12	9	
3	–12	–9	$\leftarrow b$
–4	9	5	
4	–9	–5	
–6	6	0	

$p^2 - 9p - 36 = (p+3)(p-12)$

7. $z^2 + 6z + 12$
$a = 1,\ b = 6,\ c = 12$

factor	factor	sum of factors
1	12	13
2	6	8
3	4	7

The possible factors of 12 do not add to 6. Therefore $z^2 + 6z + 12$ does not factor.

9. $x^4 + 25x^2 + 144$
$a = 1,\ b = 25,\ c = 144$

factor	factor	sum of factors	
1	144	145	
2	72	74	
3	48	51	
4	36	40	
6	24	30	
8	18	26	
9	16	25	$\leftarrow b$
12	12	24	

$x^4 + 25x^2 + 144 = (x^2+9)(x^2+16)$

11. $x^4 - 2x^2 - 3$
$a = 1,\ b = -2,\ c = -3$

factor	factor	sum of factors	
–1	3	2	
1	–3	–2	$\leftarrow b$

$x^4 - 2x^2 - 3 = (x^2+1)(x^2-3)$

13. $3a^2 + 48a + 165 = 3(a^2 + 16a + 55)$
$a = 1,\ b = 16,\ c = 55$

factor	factor	sum of factors	
1	55	56	
5	11	16	$\leftarrow b$

$3a^2 + 48a + 165 = 3(a+5)(a+11)$

15. $4c^2 + 44c - 104$
$= 4(c^2 + 11c - 26)$
$a = 1,\ b = 11,\ c = -26$

factor	factor	sum of factors	
–1	26	25	
1	–26	–25	
–2	13	11	$\leftarrow b$
2	–13	–11	

$4c^2 + 44c - 104 = 4(c-2)(c+13)$

17. $x^2 - 11xy + 24y^2$
$a = 1,\ b = -11,\ c = 24$

factor	factor	sum of factors	
–1	–24	–25	
–2	–12	–14	
–3	–8	–11	$\leftarrow b$
–4	–6	–10	

$x^2 - 11xy + 24 = (x-3y)(x-8y)$

19. $x^2+11xy-12y^2$
$a=1,\ b=11,\ c=-12$

factor	factor	sum of factors	
–1	12	11	$\leftarrow b$
1	–12	–11	
–2	6	4	
2	–6	–4	
–3	4	1	
3	–4	–1	

$x^2+11xy-12y^2=(x-y)(x+12y)$

21. $-3a^2-15ab-18b^2=-3(a^2+5ab+6b^2)$
$a=1,\ b=5,\ c=6$

factor	factor	sum of factors	
1	6	7	
2	3	5	$\leftarrow b$

$-3a^2-15ab-18b^2=-3(a+2b)(a+3b)$

23. $-2x^2+14xy+36y^2$
$=-2(x^2-7xy-18y^2)$
$a=1,\ b=-7,\ c=-18$

factor	factor	sum of factors	
–1	18	17	
1	–18	–17	
–2	9	7	
2	–9	–7	$\leftarrow b$
–3	6	3	
3	–6	–3	

$-2x^2+14xy+36y^2=-2(x+2y)(x-9y)$

25. $a=3,\ b=10,\ c=3,\ ac=9$
$3x^2+10x+3$
$=3x^2+x+9x+3$
$=x(3x+1)+3(3x+1)$
$=(3x+1)(x+3)$

27. $a=2,\ b=-15,\ c=7,\ ac=14$
$2x^2-15x+7$
$=2x^2-14x-x+7$
$=2x(x-7)-1(x-7)$
$=(x-7)(2x-1)$

29. $a=3,\ b=-1,\ c=-2,\ ac=-6$
$3x^2-x-2$
$=3x^2-3x+2x-2$
$=3x(x-1)+2(x-1)$
$=(x-1)(3x+2)$

31. $a=5,\ b=9,\ c=-2,\ ac=-10$
$5m^2+9m-2$
$=5m^2+10m-m-2$
$=5m(m+2)-1(m+2)$
$=(m+2)(5m-1)$

33. $a=2,\ b=7,\ c=-3,\ ac=-6$
$2m^2+7m-3$ does not factor because there are no factors of –6 that add to 7.

35. $a=4,\ b=25,\ c=6,\ ac=24$
$4a^2+25a+6$
$=4a^2+24a+a+6$
$=4a(a+6)+1(a+6)$
$=(a+6)(4a+1)$

37. $a=9,\ b=-13,\ c=4,\ ac=36$
$9d^2-13d+4$
$=9d^2-9d-4d+4$
$=9d(d-1)-4(d-1)$
$=(d-1)(9d-4)$

39. $a=6,\ b=-23,\ c=-4,\ ac=-24$
$6x^2-23x-4$
$=6x^2-24x+x-4$
$=6x(x-4)+1(x-4)$
$=(x-4)(6x+1)$

41. $a=8,\ b=7,\ c=-18,\ ac=-144$
$8y^2+7y-18$
$=8y^2+16y-9y-18$
$=8y(y+2)-9(y+2)$
$=(y+2)(8y-9)$

43. $a = 6,\ b = 17,\ c = 12,\ ac = 72$
$$6b^2 + 17b + 12$$
$$= 6b^2 + 9b + 8b + 12$$
$$= 3b(2b + 3) + 4(2b + 3)$$
$$= (2b + 3)(3b + 4)$$

45. $a = 20,\ b = -31,\ c = 12,\ ac = 240$
$$20x^2 - 31x + 12$$
$$= 20x^2 - 16x - 15x + 12$$
$$= 4x(5x - 4) - 3(5x - 4)$$
$$= (5x - 4)(4x - 3)$$

47. $a = 18,\ b = -9,\ c = -20,\ ac = -360$
$$18x^2 - 9x - 20$$
$$= 18x^2 - 24x + 15x - 20$$
$$= 6x(3x - 4) + 5(3x - 4)$$
$$= (3x - 4)(6x + 5)$$

49. $a = 6,\ b = -19,\ c = -7,\ ac = -42$
$$18p^2 - 57p - 21$$
$$= 3(6p^2 - 19p - 7)$$
$$= 3(6p^2 - 21p + 2p - 7)$$
$$= 3[3p(2p - 7) + 1(2p - 7)]$$
$$= 3(2p - 7)(3p + 1)$$

51. $a = 2,\ b = 11,\ c = 9,\ ac = 18$
$$2x^4 + 11x^2 + 9$$
$$= 2x^4 + 9x^2 + 2x^2 + 9$$
$$= x^2(2x^2 + 9) + 1(2x^2 + 9)$$
$$= (2x^2 + 9)(x^2 + 1)$$

53. $a = 4,\ b = 13,\ c = -12,\ ac = -48$
$$4m^4 + 13m^2 - 12$$
$$= 4m^4 + 16m^2 - 3m^2 - 12$$
$$= 4m^2(m^2 + 4) - 3(m^2 + 4)$$
$$= (m^2 + 4)(4m^2 - 3)$$

55. $a = 6,\ b = -5,\ c = -56,\ ac = -336$
$$-6x^2 + 5x + 56$$
$$= -1(6x^2 - 5x - 56)$$
$$= -1(6x^2 - 21x + 16x - 56)$$
$$= -1[3x(2x - 7) + 8(2x - 7)]$$
$$= -1(2x - 7)(3x + 8)$$

57. $a = 6,\ b = 5,\ c = -6,\ ac = -36$
$$6x^2 + 5xy - 6y^2$$
$$= 6x^2 + 9xy - 4xy - 6y^2$$
$$= 3x(2x + 3y) - 2y(2x + 3y)$$
$$= (2x + 3y)(3x - 2y)$$

59. $a = 4,\ b = -39,\ c = 56,\ ac = 224$
$$4u^2 - 39uv + 56v^2$$
$$= 4u^2 - 32uv - 7uv + 56v^2$$
$$= 4u(u - 8v) - 7v(u - 8v)$$
$$= (u - 8v)(4u - 7v)$$

61. $a = 9,\ b = 13,\ c = 4,\ ac = 36$
$$9x^4 + 13x^2y^2 + 4y^4$$
$$= 9x^4 + 9x^2y^2 + 4x^2y^2 + 4y^4$$
$$= 9x^2(x^2 + y^2) + 4y^2(x^2 + y^2)$$
$$= (x^2 + y^2)(9x^2 + 4y^2)$$

63. $a = 10,\ b = 1,\ c = -21,\ ac = -210$
$$10x^2y^2 + xy - 21$$
$$= 10x^2y^2 + 15xy - 14xy - 21$$
$$= 5xy(2xy + 3) - 7(2xy + 3)$$
$$= (2xy + 3)(5xy - 7)$$

65. Let w = original width
$w + 8$ = original length

a. $w^2 + 14w + 24$
$$= w^2 + 12w + 2w + 24$$
$$= w(w + 12) + 2(w + 12)$$
$$= (w + 12)(w + 2)$$
The increased width is $(w + 2)$ in. and the increased length is $(w + 12)$ in.

b. $(w + 2) - w = 2$
The width was increased by 2 in.

c. $(w + 12) - (w + 8) = 4$
The length was increased by 4 in.

67. Let x = length of original triangle's small side

a. $x^2 + \frac{21}{2}x + 20$

$= \frac{1}{2}(2x^2 + 21x + 40)$

$= \frac{1}{2}(2x^2 + 16x + 5x + 40)$

$= \frac{1}{2}[2x(x+8) + 5(x+8)]$

$= \frac{1}{2}(x+8)(2x+5)$

The base is $(x + 8)$ in. and the height is $(2x + 5)$ in.

The lengths of the legs are $(x + 8)$ in. and $(2x + 5)$ in.

b. They were each increased by 8 in.

c. $(2x + 5) - 8 = 2x - 3$

The expression is $(2x - 3)$ in.

10.5 Experiencing Algebra the Calculator Way

1. $a = 96,\ b = -16,\ c = -2,$

```
96→A: -16→B: -2→C
                -2
```

$Y1 = \frac{AC}{x}$

$Y2 = x + Y1 = B$

X	Y1	Y2
7	-27.43	0
8	-24	1
9	-21.33	0
10	-19.2	0
11	-17.45	0
12	-16	0
13	-14.77	0

X=7

$96x^2 - 16x - 2$

$= 96x^2 - 24x + 8x - 2$

$= 24x(4x - 1) + 2(4x - 1)$

$= (4x - 1)(24x + 2)$

$= 2(4x - 1)(12x + 1)$

2. $a = 24,\ b = 7,\ c = -55$

```
24→A:7→B: -55→C
               -55
```

$Y1 = \frac{AC}{x}$

$Y2 = x + Y1 = B$

X	Y1	Y2
39	-33.85	0
40	-33	1
41	-32.2	0
42	-31.43	0
43	-30.7	0
44	-30	0
45	-29.33	0

X=39

$24x^2 + 7x - 55$

$= 24x^2 + 40x - 33x - 55$

$= 8x(3x + 5) - 11(3x + 5)$

$= (3x + 5)(8x - 11)$

3. $a = 32,\ b = 102,\ c = 81$

```
32→A:102→B:81→C
                81
```

$Y1 = \frac{AC}{x}$

$Y2 = x + Y1 = B$

X	Y1	Y2
47	55.149	0
48	54	1
49	52.898	0
50	51.84	0
51	50.824	0
52	49.846	0
53	48.906	0

X=47

$32x^2 + 102x + 81$

$= 32x^2 + 48x + 54x + 81$

$= 16x(2x + 3) + 27(2x + 3)$

$= (2x + 3)(16x + 27)$

4. $a = 72,\ b = -99,\ c = 34$

```
72→A: -99→B:34→C
                34
```

$Y1 = \frac{AC}{x}$

$Y2 = x + Y1 = B$

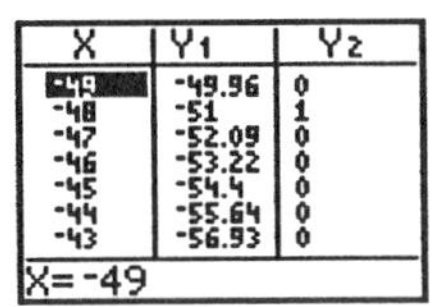

$72x^2 - 99x + 34$
$= 72x^2 - 48x - 51x + 34$
$= 24x(3x - 2) - 17(3x - 2)$
$= (3x - 2)(24x - 17)$

5. $a = 4,\ b = 109,\ c = 225$

```
4→A:109→B:225→C
                225
```

$\text{Y1} = \dfrac{AC}{x}$
$\text{Y2} = x + \text{Y1} = B$

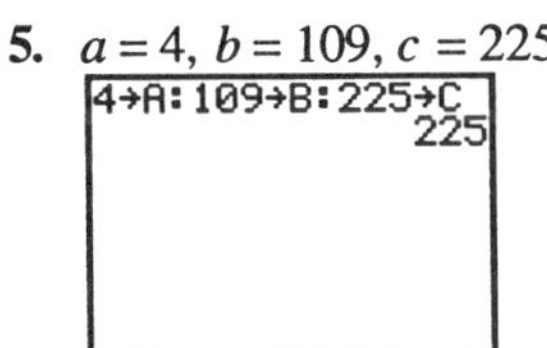

$4p^4 + 109p^2 + 225$
$= 4p^4 + 100p^2 + 9p^2 + 225$
$= 4p^2(p^2 + 25) + 9(p^2 + 25)$
$= (p^2 + 25)(4p^2 + 9)$

10.6 Experiencing Algebra the Exercise Way

1. $x^2 - 100$
$= x^2 - (10)^2$
$= (x + 10)(x - 10)$

3. $121 - c^2$
$= (11)^2 - c^2$
$= (11 + c)(11 - c)$

5. $49a^2 - 4$
$= (7a)^2 - 2^2$
$= (7a + 2)(7a - 2)$

7. $25 - 4y^2$
$= 5^2 - (2y)^2$
$= (5 + 2y)(5 - 2y)$

9. $16u^2 - 9v^2$
$= (4u)^2 - (3v)^2$
$= (4u + 3v)(4u - 3v)$

11. $7z^2 - 28$
$= 7(z^2 - 4)$
$= 7(z^2 - 2^2)$
$= 7(z + 2)(z - 2)$

13. $25 + 4p^2$ will not factor. This is not a difference of squares, but a sum of squares.

15. $x^4 - 625$
$= (x^2)^2 - (25)^2$
$= (x^2 + 25)(x^2 - 25)$
$= (x^2 + 25)(x^2 - 5^2)$
$= (x^2 + 25)(x + 5)(x - 5)$

17. $256 - z^4$
$= (16)^2 - (z^2)^2$
$= (16 + z^2)(16 - z^2)$
$= (16 + z^2)(4^2 - z^2)$
$= (16 + z^2)(4 + z)(4 - z)$

19. $x^8 - 1$
$= (x^4)^2 - 1^2$
$= (x^4 + 1)(x^4 - 1)$
$= (x^4 + 1)[(x^2)^2 - 1^2]$
$= (x^4 + 1)(x^2 + 1)(x^2 - 1)$
$= (x^4 + 1)(x^2 + 1)(x^2 - 1^2)$
$= (x^4 + 1)(x^2 + 1)(x + 1)(x - 1)$

21. $x^2 + 4x + 4$
$= x^2 + 2(x)(2) + 2^2$
$= (x + 2)^2$

23. $16z^2 + 40z + 25$
$= (4z)^2 + 2(4z)(5) + 5^2$
$= (4z + 5)^2$

25. $x^2+13x+169$ does not factor. The first and last terms are perfect squares, x^2 and 13^2. However, the middle term is not $2(x)(13)$.

27. $x^2-10x+25$
$=x^2-2(x)(5)+5^2$
$=(x-5)^2$

29. $36z^2-60z+25$
$=(6z)^2-2(6z)(5)+5^2$
$=(6z-5)^2$

31. $c^2-16d+16d^2$ does not factor. The first and last terms are perfect squares, c^2 and $(4d)^2$. However, the middle term is not $2(c)(4d)$.

33. a^4-32a^2+256
$=(a^2)^2-2(a^2)(16)+(16)^2$
$=(a^2-16)^2$
$=(a^2-16)(a^2-16)$
$=(a^2-4^2)(a^2-4^2)$
$=(a+4)(a-4)(a+4)(a-4)$
$=(a+4)^2(a-4)^2$

35. $16x^4-72x^2+81$
$=(4x^2)^2-2(4x^2)(9)+9^2$
$=(4x^2-9)^2$
$=(4x^2-9)(4x^2-9)$
$=[(2x)^2-3^2][(2x)^2-3^2]$
$=(2x+3)(2x-3)(2x+3)(2x-3)$
$=(2x+3)^2(2x-3)^2$

37. $3x^2+24x+48$
$=3(x^2+8x+16)$
$=3[x^2+2(x)(4)+4^2]$
$=3(x+4)^2$

39. $2a^2-12a+18$
$=2(a^2-6a+9)$
$=2[a^2-2(1)(3)+3^2]$
$=2(a-3)^2$

41. $p^3+2p^2q+pq^2$
$=p(p^2+2pq+q^2)$
$=p[p^2+2(p)(q)+q^2]$
$=p(p+q)^2$

43. $2p^5-4p^3q^2+2pq^4$
$=2p(p^4-2p^2q^2+q^4)$
$=2p[(p^2)^2-2(p^2)(q^2)+(q^2)^2]$
$=2p(p^2-q^2)^2$
$=2p(p^2-q^2)(p^2-q^2)$
$=2p(p+q)(p-q)(p+q)(p-q)$
$=2p(p+q)^2(p-q)^2$

45. $m^5+2m^3n^2+mn^4$
$=m(m^4+2m^2n^2+n^4)$
$=m[(m^2)^2+2(m^2)(n^2)+(n^2)^2]$
$=m(m^2+n^2)^2$

47. x^3-27
$=x^3-3^3$
$=(x-3)(x^2+3x+3^2)$
$=(x-3)(x^2+3x+9)$

49. a^3+64
$=a^3+4^3$
$=(a+4)(a^2-4a+4^2)$
$=(a+4)(a^2-4a+16)$

51. $27x^3+64y^3$
$=(3x)^3+(4y)^3$
$=(3x+4y)[(3x)^2-(3x)(4y)+(4y)^2]$
$=(3x+4y)(9x^2-12xy+16y^2)$

53. $8p^3 - 125q^3$
$= (2p)^3 - (5q)^3$
$= (2p - 5q)[(2p)^2 + (2p)(5q) + (5q)^2]$
$= (2p - 5q)(4p^2 + 10pq + 25q^2)$

55. $p^4 + 64pq^3$
$= p(p^3 + 64q^3)$
$= p[p^3 + (4q)^3]$
$= p(p + 4q)[p^2 - p(4q) + (4q)^2]$
$= p(p + 4q)(p^2 - 4pq + 16q^2)$

57. $81u^4 - 3uv^3$
$= 3u(27u^3 - v^3)$
$= 3u[(3u)^3 - v^3]$
$= 3u(3u - v)[(3u)^2 + (3u)(v) + v^2]$
$= 3u(3u - v)(9u^2 + 3uv + v^2)$

59. $3y^3 + 3y^2 + 3y$
$= 3y(y^2 + y + 1)$

61. $10abc^2 + 15abc - 20ab$
$= 5ab(2c^2 + 3c - 4)$

63. $-40a^2 - 24ab + 48ac$
$= -8a(5a + 3b - 6c)$

65. $108x^5 - 75x^3$
$= 3x^3(36x^2 - 25)$
$= 3x^3[(6x)^2 - 5^2]$
$= 3x^3(6x + 5)(6x - 5)$

67. $-200x^3y + 32xy^3$
$= -8xy(25x^2 - 4y^2)$
$= -8xy[(5x)^2 - (2y)^2]$
$= -8xy(5x + 2y)(5x - 2y)$

69. $x^3 - 16x^2 + 64x$
$= x(x^2 - 16x + 64)$
$= x[x^2 - 2(x)(8) + 8^2]$
$= x(x - 8)^2$

71. $12u^2v + 36uv^2 + 27v^3$
$= 3v(4u^2 + 12uv + 9v^2)$
$= 3v[(2u)^2 + 2(2u)(3v) + (3v)^2]$
$= 3v(2u + 3v)^2$

73. $3x^4 - 48x^2 + 7x^2 - 112$
$= (3x^4 - 48x^2) + (7x^2 - 112)$
$= 3x^2(x^2 - 16) + 7(x^2 - 16)$
$= (x^2 - 16)(3x^2 + 7)$
$= (x^2 - 4^2)(3x^2 + 7)$
$= (x + 4)(x - 4)(3x^2 + 7)$

75. $4p^4 - 37p^2 + 9$
$ac = 36$

factors	sum
–36, –1	–37

$= 4p^4 - 36p^2 - p^2 + 9$
$= 4p^2(p^2 - 9) - 1(p^2 - 9)$
$= (p^2 - 9)(4p^2 - 1)$
$= (p^2 - 3^2)[(2p)^2 - 1^2]$
$= (p + 3)(p - 3)(2p + 1)(2p - 1)$

77. $2 + 8y - 42y^2$
$= -42y^2 + 8y + 2$
$= -2(21y^2 - 4y - 1)$
$ac = -21$

factors	sum
–7, 3	–4

$= -2(21y^2 - 7y + 3y - 1)$
$= -2[7y(3y - 1) + 1(3y - 1)]$
$= -2(3y - 1)(7y + 1)$

79. $4u^2v^2 + 36uv + 56$
$= 4(u^2v^2 + 9uv + 14)$
$ac = 14$

factors	sum
2, 7	9

$= 4(u^2v^2 + 2uv + 7uv + 14)$
$= 4[uv(uv + 2) + 7(uv + 2)]$
$= 4(uv + 2)(uv + 7)$

81. $32x^3 - 64x^2 - 28x^2 + 56x$
$= 4x(8x^2 - 16x - 7x + 14)$
$= 4x[8x(x - 2) - 7(x - 2)]$
$= 4x(x - 2)(8x - 7)$

83. $12x^4 + 26x^3 - 30x^2$
$= 2x^2(6x^2 + 13x - 15)$
$ac = -90$

factors	sum
18, –5	13

$= 2x(6x^2 + 18x - 5x - 15)$
$= 2x[6x(x + 3) - 5(x + 3)]$
$= 2x(x + 3)(6x - 5)$

85. $x^6 - 5x^3y^3 + 3x^3y^3 - 15y^6$
$= x^3(x^3 - 5y^3) + 3y^3(x^3 - 5y^3)$
$= (x^3 - 5y^3)(x^3 + 3y^3)$

87. $x^4 - 5x^3 + 8x^2 - 40x$
$= x(x^3 - 5x^2 + 8x - 40)$
$= x[x^2(x - 5) + 8(x - 5)]$
$= x(x - 5)(x^2 + 8)$

89. $1 - k^8$
$= 1^2 - (k^4)^2$
$= (1 + k^4)(1 - k^4)$
$= (1 + k^4)[1 - (k^2)^2]$
$= (1 + k^4)(1 + k^2)(1 - k^2)$
$= (1 + k^4)(1 + k^2)(1 + k)(1 - k)$

91. a. Large area – small area
$= x^2 - (15)^2$
$= x^2 - 225$
The garden area is $(x^2 - 225)$ ft^2.

b. $x^2 - 225$
$= x^2 - (15)^2$
$= (x + 15)(x - 15)$
The rectangular plot has dimensions of $(x + 15)$ ft by $(x - 15)$ ft.

c. $x^2 - 225 = (100)^2 - 225 = 9775$
$85(100) = 8500$
Yes, the plot that is 100 ft on each side has a larger garden area than the 85 by 100 foot plot.

10.6 Experiencing Algebra the Calculator Way

Students should check exercises using the computer program.
Answers will vary.

Chapter 10 Review

Reflections

1.–5. Answers will vary.

Exercises

1. $-5c^2 = -5 \cdot c \cdot c$

2. $(-5c)^2 = (-5c)(-5c)$

3. $4(x + y)^2 z^0 = 4(x + y)^2(1)$
$= 4(x + y)(x + y)$

4. $\left(\frac{2}{3x}\right)^4 = \left(\frac{2}{3x}\right)\left(\frac{2}{3x}\right)\left(\frac{2}{3x}\right)\left(\frac{2}{3x}\right)$

5. $\frac{x(y - z)^{-1}}{z^{-3}} = \frac{xz^3}{y - z}$

6. $3y^{-3}+2y^{-2}+y^{-1}+10y^{0}$
$=\dfrac{3}{y^3}+\dfrac{2}{y^2}+\dfrac{1}{y}+10$

7. $a^9\cdot a^{-4}=a^{9+(-4)}=a^5$

8. $(p+q)^{-3}(p+q)=(p+q)^{-3+1}$
$=(p+q)^{-2}=\dfrac{1}{(p+q)^2}$

9. $\dfrac{t^{12}}{t^9}=t^{12-9}=t^3$

10. $\dfrac{m^{-2}}{m^{-4}}=m^{-2-(-4)}=m^2$

11. $\dfrac{(x+3y)^{-5}}{(x+3y)^{-4}}=(x+3y)^{-5-(-4)}$
$=(x+3y)^{-1}=\dfrac{1}{x+3y}$

12. $\dfrac{72x^4y^5z}{24x^2y^4z^3}=\dfrac{72}{24}\cdot\dfrac{x^4}{x^2}\cdot\dfrac{y^5}{y^4}\cdot\dfrac{z}{z^3}$
$=3x^{4-2}y^{5-4}z^{1-3}=\dfrac{3x^2y}{z^2}$

13. $(z^4)^3=z^{4\cdot 3}=z^{12}$

14. $(w^{-3})^2=w^{(-3)(2)}=w^{-6}=\dfrac{1}{w^6}$

15. $(p^{-2})^{-4}=p^{(-2)(-4)}=p^8$

16. $(a^3)^0=a^{3\cdot 0}=a^0=1$

17. $(-2a)^4=(-2)^4a^4=16a^4$

18. $(-2a)^5=(-2)^5a^5=-32a^5$

19. $\left(\dfrac{-2x}{3z}\right)^3=\dfrac{(-2)^3x^3}{3^3z^3}=-\dfrac{8x^3}{27z^3}$

20. $\left(\dfrac{4a}{5b}\right)^{-2}=\left(\dfrac{5b}{4a}\right)^2=\dfrac{5^2b^2}{4^2a^2}=\dfrac{25b^2}{16a^2}$

21. $(a^2b^3c)^3=a^{2\cdot 3}b^{3\cdot 3}c^3=a^6b^9c^3$

22. $\left[(3x)^3\right]^2=(3^3x^3)^2=(27x^3)^2=(27)^2x^{3\cdot 2}$
$=729x^6$

23. $\left[\left(\dfrac{m}{4n}\right)^2\right]^2=\left(\dfrac{m^2}{4^2n^2}\right)^2=\left(\dfrac{m^2}{16n^2}\right)^2$
$=\dfrac{m^{2\cdot 2}}{(16)^2n^{2\cdot 2}}=\dfrac{m^4}{256n^4}$

24. Let x = length of original side
$\dfrac{1}{4}x$ = length of reduced side
Area of original square: x^2 square units
Area of reduced square:
$\left(\dfrac{1}{4}x\right)^2=\dfrac{x^2}{16}$ square units
The smaller area is $\dfrac{1}{16}$ of the original area.

25. Let r = original radius
$2r$ = increased radius
Area of original garden: πr^2 square units
Area of enlarged garden:
$\pi(2r)^2=4\pi r^2$ square units
The enlarged garden has an area that is 4 times that of the original garden.
$4\pi r^2=4\pi(6)^2=144\pi$
The garden has an area of 144π ft^2.

26.
$$\begin{array}{r} 5x^4+3x^3 \qquad\quad +6x-3 \\ 4x^3+5x^2 \qquad +7 \\ x^4 \qquad -3x^2+\ x-9 \\ x^3+2x^2-7x+1 \\ \hline 6x^4+8x^3+4x^2+0x-4 \end{array}$$
or
$$6x^4+8x^3+4x^2-4$$

27. $(4y^2+3y-7)+(2y+8)+(5y^2-4y+1)$
$=4y^2+3y-7+2y+8+5y^2-4y+1$
$=9y^2+y+2$

28. $(3.57z^3-2.08z^2+8.77z-1.99)+(4.73-2.98z+5.64z^2)$
$=3.57z^3-2.08z^2+8.77z-1.99+4.73-2.98z+5.64z^2$
$=3.57z^3+3.56z^2+5.79z+2.74$

29. $(4x^3+8x^2-6x+2)-(2x^3+12x^2-x+9)$
$=4x^3+8x^2-6x+2-2x^3-12x^2+x-9$
$=2x^3-4x^2-5x-7$

30. $(5a^4+a^3+a^2+a+1)-(a^4+2a^3+3a^2+4a+5)$
$=5a^4+a^3+a^2+a+1-a^4-2a^3-3a^2-4a-5$
$=4a^4-a^3-2a^2-3a-4$

31. $(65z^4+27z^2+36)-(16z^3+8z+12)$
$=65z^4+27z^2+36-16z^3-8z-12$
$=65z^4-16z^3+27z^2-8z+24$

32. $\left(\frac{5}{8}b^4+\frac{7}{8}b^3-\frac{3}{4}b^2+\frac{1}{2}b-\frac{1}{4}\right)-\left(\frac{1}{2}b^4+\frac{3}{8}b^2+\frac{1}{8}b\right)$
$=\frac{5}{8}b^4+\frac{7}{8}b^3-\frac{3}{4}b^2+\frac{1}{2}b-\frac{1}{4}-\frac{1}{2}b^4-\frac{3}{8}b^2-\frac{1}{8}b$
$=\left(\frac{5}{8}-\frac{1}{2}\right)b^4+\frac{7}{8}b^3+\left(-\frac{3}{4}-\frac{3}{8}\right)b^2+\left(\frac{1}{2}-\frac{1}{8}\right)b-\frac{1}{4}$
$=\frac{1}{8}b^4+\frac{7}{8}b^3-\frac{9}{8}b^2+\frac{3}{8}b-\frac{1}{4}$

33. $-3a^3b(5a^{-2}b^3)$
$=-3(5)a^{3-2}b^{1+3}$
$=-15ab^4$

34. $6x^3(3x^2+2x-7)$
$=6x^3(3x^2)+6x^3(2x)+6x^3(-7)$
$=18x^{3+2}+12x^{3+1}-42x^3$
$=18x^5+12x^4-42x^3$

35. $7a^{-2}(3a^6+a^4-2a^2)$
$=7a^{-2}(3a^6)+7a^{-2}(a^4)+7a^{-2}(-2a^2)$
$=21a^{-2+6}+7a^{-2+4}-14a^{-2+2}$
$=21a^4+7a^2-14$

36. $(-6.9x^3z^{-4})(3.4xz^3)$
$=(-6.9)(3.4)x^{3+1}z^{-4+3}$
$=-23.46x^4z^{-1}$
$=-\dfrac{23.46x^4}{z}$

37. $\dfrac{144x^6y^3z^4}{2x^5y^2z^6}$
$=\dfrac{144}{2}x^{6-5}y^{3-2}z^{4-6}$
$=72x^1y^1z^{-2}$
$=\dfrac{72xy}{z^2}$

38. $\dfrac{(2a^2)^3}{24a^5}$
$=\dfrac{2^3a^{2\cdot3}}{24a^5}$
$=\dfrac{8a^6}{24a^5}$
$=\dfrac{8}{24}a^{6-5}$
$=\dfrac{a}{3}$ or $\dfrac{1}{3}a$

39. $\left(\dfrac{27a^5b^3}{9a^2b^7}\right)^2$
$=\left(\dfrac{27}{9}a^{5-2}b^{3-7}\right)^4$
$=\left(\dfrac{3a^3}{b^4}\right)^2$
$=\dfrac{3^2a^{3\ 2}}{b^{4\cdot2}}$
$=\dfrac{9a^6}{b^8}$

40. $\dfrac{824,500,000y^{-5}}{0.0097y^7}$
$=85,000,000,000y^{-5-7}$
$=\dfrac{85,000,000,000}{y^{12}}$ or $\dfrac{8.5\times10^{10}}{y^{12}}$

41. $\dfrac{15b^4 - 10b^3 - 25b^2 + 5b}{-5b}$
$= \dfrac{15b^4}{-5b} - \dfrac{10b^3}{-5b} - \dfrac{25b^2}{-5b} + \dfrac{5b}{-5b}$
$= -3b^{4-1} + 2b^{3-1} + 5b^{2-1} - 1b^{1-1}$
$= -3b^3 + 2b^2 + 5b - 1$

42. $\dfrac{6ab^2 + 18a^2b}{3a^2b^2}$
$= \dfrac{6ab^2}{3a^2b^2} + \dfrac{18a^2b}{3a^2b^2}$
$= 2a^{1-2}b^{2-2} + 6a^{2-2}b^{1-2}$
$= 2a^{-1}b^0 + 6a^0b^{-1}$
$= \dfrac{2}{a} + \dfrac{6}{b}$

43. **a.** $C(x) = 10 + 3.5x$

b. $R(x) = (10 - 0.25x)x$

c. $P(x) = R(x) - C(x)$
$P(x) = (10 - 0.25x)x - (10 + 3.5x)$
$P(x) = 10x - 0.25x^2 - 10 - 3.5x$
$P(x) = -0.25x^2 + 6.5x - 10$

d. $A(x) = \dfrac{-0.25x^2 + 6.5x - 10}{x}$
$A(x) = -\dfrac{0.25x^2}{x} + \dfrac{6.5x}{x} - \dfrac{10}{x}$
$A(x) = -0.25x + 6.5 - \dfrac{10}{x}$

e. $x = 13$

44. Let length = 20 meters
width = $(20 - x)$ meters
Perimeter: $2(20) + 2(20 - x)$
$= 40 + 40 - 2x$
$= 80 - 2x$
The perimeter is $(80 - 2x)$ meters.

45. Area = Patio area – Box area
Area = $x(x+8) - 3^2$
Area = $x^2 + 8x - 9$
$9^2 + 8(9) - 9 = 144$
The patio needs 144 ft^2 of carpeting.

46. $(y+9)^2$
$= y^2 + 2y(9) + 9^2$
$= y^2 + 18y + 81$

47. $(x^3 - 5)^2$
$= (x^3)^2 - 2(x^3)(5) + 5^2$
$= x^6 - 10x^3 + 25$

48. $(5x - 2)(x + 11)$
$= 5x(x + 11) - 2(x + 11)$
$= 5x^2 + 55x - 2x - 22$
$= 5x^2 + 53x - 22$

49. $(7 - z)(7 + z)$
$= 7^2 - z^2$
$= 49 - z^2$

50. $(y - 1.8)(y + 3.4)$
$= y(y + 3.4) - 1.8(y + 3.4)$
$= y^2 + 3.4y - 1.8y - 1.8(3.4)$
$= y^2 + 1.6y - 6.12$

51. $(z^2 - 10)(z^2 + 10)$
$= (z^2)^2 - (10)^2$
$= z^4 - 100$

52. $\left(\dfrac{4}{5}x - \dfrac{1}{2}\right)\left(\dfrac{4}{5}x + \dfrac{1}{2}\right)$
$= \left(\dfrac{4}{5}x\right)^2 - \left(\dfrac{1}{2}\right)^2$
$= \dfrac{16}{25}x^2 - \dfrac{1}{4}$

53. $(b-4)^3$
$= (b-4)(b-4)(b-4)$
$= [b(b-4)-4(b-4)](b-4)$
$= (b^2-4b-4b+16)(b-4)$
$= (b^2-8b+16)(b-4)$
$= (b^2-8b+16)b+(b^2-8b+16)(-4)$
$= b^3-8b^2+16b-4b^2+32b-64$
$= b^3-12b^2+48b-64$

54. $(2x+1)(3x^2+5x-4)$
$= 2x(3x^2+5x-4)+1(3x^2+5x-4)$
$= 6x^3+10x^2-8x+3x^2+5x-4$
$= 6x^3+13x^2-3x-4$

55. $(a+b+c)^2$
$= (a+b+c)(a+b+c)$
$= a(a+b+c)+b(a+b+c)+c(a+b+c)$
$= a^2+ab+ac+ab+b^2+bc+ac+bc+c^2$
$= a^2+b^2+c^2+2ab+2bc+2ac$

56. $(a-3)(a^2+3a+9)$
$= a(a^2+3a+9)-3(a^2+3a+9)$
$= a^3+3a^2+9a-3a^2-9a-27$
$= a^3-27$

57. a. length: $(x+3)$ in.
width: $(x-3)$ in.
height: x in.

b. Volume:
$x(x+3)(x-3)$
$= x(x^2-3^2)$
$= x(x^2-9)$
$= x^3-9x$
The volume is (x^3-9x) in^3.

c. Surface area:
$2(x+3)(x-3)+2x(x+3)+2x(x-3)$
$= 2(x^2-3^2)+2x(x)+2x(3)+2x(x)-2x(3)$
$= 2x^2-18+2x^2+6x+2x^2-6x$
$= 6x^2-18$
The surface area is $(6x^2-18)$ in^2.

58. **a.** current length = $3x$
current width = x
Area = $x(3x) = 3x^2$
The current area is $3x^2$ ft^2.

b. new length = $3x + 9$
new width = $2x$
Area:
$2x(3x+9)$
$= 2x(3x) + 2x(9)$
$= 6x^2 + 18x$
The new area is $(6x^2 + 18x)$ ft^2.

c. Ratio of planned area to current area:
$$\frac{6x^2+18x}{3x^2} = \frac{3x(2x+6)}{3x(x)} = \frac{2x+6}{x}$$
The ratio is $\frac{2x+6}{x}$.

59. $20a^6 - 28a^4 + 44a^2$
$= 4a^2(5a^4 - 7a^2 + 11)$

60. $22u^3v^2 + 22u^2v^3$
$= 22u^2v^2(u + v)$

61. $3x^3 + 3x + x^2 + 1$
$= (3x^3 + 3x) + (x^2 + 1)$
$= 3x(x^2 + 1) + 1(x^2 + 1)$
$= (x^2 + 1)(3x + 1)$

62. $7a^4 + 7a^2b^2 + 7a^2b^2 + 7b^4$
$= 7(a^4 + a^2b^2 + a^2b^2 + b^4)$
$= 7[(a^4 + a^2b^2) + (a^2b^2 + b^4)]$
$= 7[a^2(a^2 + b^2) + b^2(a^2 + b^2)]$
$= 7(a^2 + b^2)(a^2 + b^2)$
$= 7(a^2 + b^2)^2$

63. $15ac + 18ad + 20bc + 24bd$
$= (15ac + 18ad) + (20bc + 24bd)$
$= 3a(5c + 6d) + 4b(5c + 6d)$
$= (5c + 6d)(3a + 4b)$

64. **a.** $\frac{1}{2}n^2 + \frac{1}{2}n$
$= \frac{1}{2}n(n+1)$

b. $\frac{1}{2}n^2 + \frac{1}{2}n = \frac{1}{2}(12)^2 + \frac{1}{2}(12) = 78$
$\frac{1}{2}n(n+1) = \frac{1}{2}(12)(12+1) = 78$
The sum is 78.

c. The factored expression is easier to evaluate. Answers will vary.

65. $2x^2 - 6x + 5x - 15$
$= (2x^2 - 6x) + (5x - 15)$
$= 2x(x - 3) + 5(x - 3)$
$= (x - 3)(2x + 5)$
The width is $(x - 3)$ units and the length is $(2x + 5)$ units.

66. Total pay:
$x+(x+10)+(x+20)+(x+30)+(x+40)+(x+50)+(x+60)$
$=7x+210$
$=7(x+30)$
The employee earns an average of
$(x+30)$ dollars for each of 7 months.
$7(960+30)=6930$
The salary is $6930.

67. $z^2+2z-99$
$a=1$, $b=2$, $c=-99$

factor	factor	sum of factors	
–1	99	98	
1	–99	–98	
–3	33	30	
3	–33	–30	
–9	11	2	$\leftarrow b$
9	–11	–2	

$z^2+2z-99=(z-9)(z+11)$

68. $p^2+5pq-66q^2$
$a=1$, $b=5$, $c=-66$

factor	factor	sum of factors	
–1	66	65	
1	–66	–65	
–2	33	31	
2	–33	–31	
–3	22	19	
3	–22	–19	
–6	11	5	$\leftarrow b$
6	–11	–5	

$p^2+5pq-66q^2=(p-6q)(p+11q)$

69. $6a^2+96a+234$
$=6(a^2+16a+39)$
$a=1$, $b=16$, $c=39$

factor	factor	sum of factors	
1	39	40	
3	13	16	$\leftarrow b$

$6a^2+96a+234=6(a+3)(a+13)$

70. x^4+8x^2+15
$a=1$, $b=8$, $c=15$

factor	factor	sum of factors	
1	15	16	
3	5	8	$\leftarrow b$

$x^4+8x^2+15=(x^2+3)(x^2+5)$

71. $4q^3 - 28q^2 - 240q$
$= 4q(q^2 - 7q - 60)$
$a = 1,\ b = -7,\ c = -60$

factor	factor	sum of factors	
–1	60	59	
1	–60	–59	
–2	30	28	
2	–30	–28	
–3	20	17	
3	–20	–17	
–4	15	11	
4	–15	–11	
–5	12	7	
5	–12	–7	$\leftarrow b$
–6	10	4	
6	–10	–4	

$4q^3 - 28q^2 - 240q = 4q(q+5)(q-12)$

72. $x^2y^2 - 4xy - 117$
$a = 1,\ b = -4,\ c = -117$

factor	factor	sum of factors	
–1	117	116	
1	–117	–116	
–3	39	36	
3	–39	–36	
–9	13	4	
9	–13	–4	$\leftarrow b$

$x^2y^2 - 4xy - 117 = (xy+9)(xy-13)$

73. $-7x^2 + 98x - 168$
$= -7(x^2 - 14x + 24)$
$a = 1,\ b = -14,\ c = 24$

factor	factor	sum of factors	
–1	–24	–25	
–2	–12	–14	$\leftarrow b$
–3	–8	–11	
–4	–6	–10	

$-7x^2 + 98x - 168 = -7(x-2)(x-12)$

74. $a = 2,\ b = -11,\ c = 5,\ ac = 10$
$2x^2 - 11x + 5$
$= 2x^2 - 10x - x + 5$
$= 2x(x-5) - 1(x-5)$
$= (x-5)(2x-1)$

75. $a = 6,\ b = 17,\ c = 5,\ ac = 30$
$6x^2 + 17x + 5$
$= 6x^2 + 15x + 2x + 5$
$= 3x(2x+5) + 1(2x+5)$
$= (2x+5)(3x+1)$

76. $a = 4,\ b = 13,\ c = 3,\ ac = 12$
$28a^2b^2 + 91ab + 21$
$= 7(4a^2b^2 + 13ab + 3)$
$= 7(4a^2b^2 + 12ab + ab + 3)$
$= 7[4ab\,(ab+3) + 1(ab+3)]$
$= 7(ab+3)(4ab+1)$

77. $a = 15,\ b = 34,\ c = -16,\ ac = -240$
$-45x^3 - 102x^2 + 48x$
$= -3x(15x^2 + 34x - 16)$
$= -3x(15x^2 + 40x - 6x - 16)$
$= -3x[5x(3x+8) - 2(3x+8)]$
$= -3x(3x+8)(5x-2)$

78. a. $2x^2 + 10x + \frac{21}{2}$
$= \frac{1}{2}(4x^2 + 20x + 21)$
$= \frac{1}{2}(4x^2 + 6x + 14x + 21)$
$= \frac{1}{2}[2x(2x+3) + 7(2x+3)]$
$= \frac{1}{2}(2x+3)(2x+7)$
The base is $(2x + 7)$ units and the height is $(2x + 3)$ units.

b. $2x + 7 - 2x = 7$
The base was increased by 7 units.

c. $2x + 3 - x = x + 3$
The height was increased by $(x + 3)$ units.

79. a. $x^2 + 17x + 30$
$= x^2 + 2x + 15x + 30$
$= x(x+2) + 15(x+2)$
$= (x+2)(x+15)$
The new length is $(x + 15)$ units and the new width is $(x + 2)$ units.

b. $x + 15 - x = 15$
It was increased by 15 units.

c. $x + 2 - (x - 6) = 8$
It was increased by 8 units.

80. $x^2 - 169$
$= x^2 - (13)^2$
$= (x + 13)(x - 13)$

81. $625 - a^2$
$= (25)^2 - a^2$
$= (25 + a)(25 - a)$

82. $12x^2 - 75$
$= 3(4x^2 - 25)$
$= 3[(2x)^2 - 5^2]$
$= 3(2x + 5)(2x - 5)$

83. $p^2 - q^2$
$= (p + q)(p - q)$

84. $p^2 + q^2$ will not factor.
This is not the difference of squares, but the sum of squares.

85. $9x^2 - 25y^2$
$= (3x)^2 - (5y)^2$
$= (3x + 5y)(3x - 5y)$

86. $16x^4 - 81$
$= (4x^2)^2 - 9^2$
$= (4x^2 + 9)(4x^2 - 9)$
$= (4x^2 + 9)[(2x)^2 - 3^2]$
$= (4x^2 + 9)(2x + 3)(2x - 3)$

87. $p^2 + 12p + 36$
$= p^2 + 2p(6) + 6^2$
$= (p + 6)^2$

88. $q^2 - 16q + 64$
$= q^2 - 2q(8) + 8^2$
$= (q - 8)^2$

89. $c^3 + 27$
$= c^3 + 3^3$
$= (c + 3)(c^2 - 3c + 3^2)$
$= (c + 3)(c^2 - 3c + 9)$

90. $8z^3 - 125$
$= (2z)^3 - 5^3$
$= (2z - 5)[(2z)^2 + (2z)(5) + 5^2]$
$= (2z - 5)(4z^2 + 10z + 25)$

91. $5h^3 + 40k^3$
$= 5(h^3 + 8k^3)$
$= 5[h^3 + (2k)^3]$
$= 5(h + 2k)[h^2 + h(2k) + (2k)^2]$
$= 5(h + 2k)(h^2 + 2hk + 4k^2)$

92. $-12x^3+60x^2y-75xy^2$
$=-3x(4x^2-20xy+25y^2)$
$=-3x[(2x)^2-2(2x)(5y)+(5y)^2]$
$=-3x(2x-5y)^2$

93. $7x^4+7x^3+7x^2$
$=7x^2(x^2+x+1)$

94. $12x^3-243x$
$=3x(4x^2-81)$
$=3x[(2x)^2-9^2]$
$=3x(2x+9)(2x-9)$

95. $32x^3+32x^2+8x$
$=8x(4x^2+4x+1)$
$=8x[(2x)^2+2(2x)(1)+1^2]$
$=8x(2x+1)^2$

96. $24x^3-14x^2-90x$
$=2x(12x^2-7x-45)$
$=2x(12x^2-27x+20x-45)$
$=2x[3x(4x-9)+5(4x-9)]$
$=2x(4x-9)(3x+5)$

97. $256x^4-288x^2+81$
$=(16x^2)^2-2(16x^2)(9)+9^2$
$=(16x^2-9)^2$
$=(16x^2-9)(16x^2-9)$
$=[(4x)^2-3^2][(4x)^2-3^2]$
$=(4x+3)(4x-3)(4x+3)(4x-3)$
$=(4x+3)^2(4x-3)^2$

98. $36x^4-25x^2+4$
$=36x^4-16x^2-9x^2+4$
$=4x^2(9x^2-4)-1(9x^2-4)$
$=(9x^2-4)(4x^2-1)$
$=[(3x)^2-(2)^2][(2x)^2-(1)^2]$
$=(3x+2)(3x-2)(2x+1)(2x-1)$

99. $2x^4+14x^3-8x^2-56x$
$=2x(x^3+7x^2-4x-28)$
$=2x[x^2(x+7)-4(x+7)]$
$=2x(x+7)(x^2-4)$
$=2x(x+7)(x-2)(x+2)$

100. a. Area of land – Area of base
$=x(4x)-5^2$
$=4x^2-25$
The land area not covered is
$(4x^2-25)$ ft^2.

b. $4x^2-25$
$=(2x)^2-5^2$
$=(2x+5)(2x-5)$
The dimensions would be $(2x+5)$ ft by $(2x-5)$ ft.

c. $(2\cdot 80+5)$ ft by $(2\cdot 80-5)$ ft
165 ft by 155 ft
The dimensions would be 165 ft by 155 ft.

Chapter 10 Mixed Review

1. $z^2+9z-90$
$=z^2+15z-6z-90$
$=z(z+15)-6(z+15)$
$=(z+15)(z-6)$

2. $a^2-18a+72$
$=a^2-6a-12a+72$
$=a(a-6)-12(a-6)$
$=(a-6)(a-12)$

3. $x^2+14xy+45y^2$
$=x^2+9xy+5xy+45y^2$
$=x(x+9y)+5y(x+9y)$
$=(x+9y)(x+5y)$

4. $5a^2+70a+245$
$=5(a^2+14a+49)$
$=5[a^2+2a(7)+7^2]$
$=5(a+7)^2$

5. $2a^2 + 8ab + 12b^2$
$= 2(a^2 + 4ab + 6b^2)$

6. $x^4 + 10x^2 + 21$
$= x^4 + 3x^2 + 7x^2 + 21$
$= x^2(x^2 + 3) + 7(x^2 + 3)$
$= (x^2 + 3)(x^2 + 7)$

7. $3q^3 - 33q^2 - 126q$
$= 3q(q^2 - 11q - 42)$
$= 3q(q^2 - 3q + 14q - 42)$
$= 3q[q(q-3) + 14(q-3)]$
$= 3q(q-3)(q+14)$

8. $-6x^2 + 42x + 360$
$= -6(x^2 - 7x - 60)$
$= -6(x^2 + 5x - 12x - 60)$
$= -6[x(x+5) - 12(x+5)]$
$= -6(x+5)(x-12)$

9. $10 + 7x + x^2$
$= x^2 + 7x + 10$
$= x^2 + 2x + 5x + 10$
$= x(x+2) + 5(x+2)$
$= (x+2)(x+5)$

10. $x^2 - 289$
$= x^2 - (17)^2$
$= (x+17)(x-17)$

11. $x^3 - 1$
$= x^3 - 1^3$
$= (x-1)[x^2 + x(1) + 1^2]$
$= (x-1)(x^2 + x + 1)$

12. $4x^2 - 64$
$= 4(x^2 - 16)$
$= 4(x^2 - 4^2)$
$= 4(x+4)(x-4)$

13. $u^2 + v^2$ does not factor. It is the sum of two squares, not the difference of two squares.

14. $36x^2 - 49y^2$
$= (6x)^2 - (7y)^2$
$= (6x + 7y)(6x - 7y)$

15. $81x^4 - 1$
$= (9x^2)^2 - 1^2$
$= (9x^2 + 1)(9x^2 - 1)$
$= (9x^2 + 1)[(3x)^2 - 1^2]$
$= (9x^2 + 1)(3x + 1)(3x - 1)$

16. $p^2 + 22p + 121$
$= p^2 + 2p(11) + (11)^2$
$= (p + 11)^2$

17. $q^2 - 30q + 225$
$= q^2 - 2q(15) + (15)^2$
$= (q - 15)^2$

18. $27a^3 + 64b^3$
$= (3a)^3 + (4b)^3$
$= (3a + 4b)[(3a)^2 + 3a(4b) + (4b)^2]$
$= (3a + 4b)(9a^2 + 12ab + 16b^2)$

19. $27a^2b^2 - 72ab + 48$
$= 3(9a^2b^2 - 24ab + 16)$
$= 3[(3ab)^2 - 2(3ab)(4) + 4^2]$
$= 3(3ab - 4)^2$

20. $-50x^3 + 120x^2y - 72xy^2$
$= -2x(25x^2 - 60xy + 36y^2)$
$= -2x[(5x)^2 - 2(5x)(6y) + (6y)^2]$
$= -2x(5x - 6y)^2$

21. $8x^4 - 2x^3 + 6x^2 - 12x$
$= 2x(4x^3 - x^2 + 3x - 6)$

22. $35u^3v^2 + 25u^2v^3$
$= 5u^2v^2(7u + 5v)$

23. $2x^3+10x+x^2+5$
$=2x(x^2+5)+1(x^2+5)$
$=(x^2+5)(2x+1)$

24. $m^2-2mn-8mn+16n^2$
$=m(m-2n)-8n(m-2n)$
$=(m-2n)(m-8n)$

25. $4a^4+8a^2b^2+8a^2b^2+16b^4$
$=4(a^4+2a^2b^2+2a^2b^2+4b^4)$
$=4[a^2(a^2+2b^2)+2b^2(a^2+2b^2)]$
$=4(a^2+2b^2)(a^2+2b^2)$
$=4(a^2+2b^2)^2$

26. $2x^2-13x+11$
$=2x^2-2x-11x+11$
$=2x(x-1)-11(x-1)$
$=(x-1)(2x-11)$

27. $24ac+20ad+18bc+15bd$
$=4a(6c+5d)+3b(6c+5d)$
$=(6c+5d)(4a+3b)$

28. $7x^2-19x-6$
$=7x^2-21x+2x-6$
$=7x(x-3)+2(x-3)$
$=(x-3)(7x+2)$

29. $10x^2-11x-6$
$=10x^2-15x+4x-6$
$=5x(2x-3)+2(2x-3)$
$=(2x-3)(5x+2)$

30. $36a^2+66a+24$
$=6(6a^2+11a+4)$
$=6(6a^2+3a+8a+4)$
$=6[3a(2a+1)+4(2a+1)]$
$=6(2a+1)(3a+4)$

31. $-30x^2-28x+32$
$=-2(15x^2+14x-16)$
$=-2(15x^2+10x-24x-16)$
$=-2[5x(3x+2)-8(3x+2)]$
$=-2(3x+2)(5x-8)$

32. $12x^4+13x^2+3$
$=12x^4+4x^2+9x^2+3$
$=4x^2(3x^2+1)+3(3x^2+1)$
$=(3x^2+1)(4x^2+3)$

33. $54x^3+36x^2+6x$
$=6x(9x^2+6x+1)$
$=6x\left[(3x)^2+2(3x)(1)+1^2\right]$
$=6x(3x+1)^2$

34. $81x^4-72x^2+16$
$=(9x^2)^2-2(9x^2)(4)+4^2$
$=(9x^2-4)^2$
$=\left[(3x)^2-2^2\right]^2$
$=[(3x+2)(3x-2)]^2$
$=(3x+2)^2(3x-2)^2$

35. $4x^4-61x^2+225$
$=(4x^4-36x^2)-(25x^2+225)$
$=4x^2(x^2-9)-25(x^2-9)$
$=(4x^2-25)(x^2-9)$
$=[(2x)^2-(5)^2][x^2-(3)^2]$
$=(2x+5)(2x-5)(x+3)(x-3)$

36. $12x^3+18x^2-30x^2-45x$
$=3x(4x^2+6x-10x-15)$
$=3x[2x(2x+3)-5(2x+3)]$
$=3x(2x+3)(2x-5)$

37. $3x^4+15x^3-27x^2-135x$
$=3x(x^3+5x^2-9x-45)$
$=3x[x^2(x+5)-9(x+5)]$
$=3x(x+5)(x^2-9)$
$=3x(x+5)(x+3)(x-3)$

38. $m^{-7}n^5=\dfrac{n^5}{m^7}$

39. $\dfrac{5^{-3}t^3}{4^{-3}s^4}=\dfrac{4^3t^3}{5^3s^4}=\dfrac{64t^3}{125s^4}$

40. $\dfrac{(d+2e)^3}{(d-2e)^{-2}} = (d+2e)^2(d-2e)^2$

41. $2y^{-4} + 4y^{-3} + 6y^{-2} + 8y^{-1} + 10y^0$
$= \dfrac{2}{y^4} + \dfrac{4}{y^3} + \dfrac{6}{y^2} + \dfrac{8}{y} + 10$

42. $(2y^2 + 4y - 3) + (9y + 7) + (8y^2 - 12y + 15)$
$= 2y^2 + 4y - 3 + 9y + 7 + 8y^2 - 12y + 15$
$= 10y^2 + y + 19$

43. $\left(\dfrac{3}{8}a^3 + \dfrac{3}{4}a^2 - \dfrac{5}{8}a + \dfrac{1}{4}\right) + \left(\dfrac{3}{4}a^3 + \dfrac{7}{16}a - \dfrac{5}{8}a^2 + \dfrac{11}{16}\right)$
$= \dfrac{3}{8}a^3 + \dfrac{3}{4}a^2 - \dfrac{5}{8}a + \dfrac{1}{4} + \dfrac{3}{4}a^3 + \dfrac{7}{16}a - \dfrac{5}{8}a^2 + \dfrac{11}{16}$
$= \left(\dfrac{3}{8} + \dfrac{3}{4}\right)a^3 + \left(\dfrac{3}{4} - \dfrac{5}{8}\right)a^2 + \left(-\dfrac{5}{8} + \dfrac{7}{16}\right)a + \left(\dfrac{1}{4} + \dfrac{11}{16}\right)$
$= \dfrac{9}{8}a^3 + \dfrac{1}{8}a^2 - \dfrac{3}{16}a + \dfrac{15}{16}$

44. $(4.9z^3 - 6.82z^2 + 12z - 11.07) + (4.6 - 1.83z + 4.9z^2)$
$= 4.9z^3 - 6.82z^2 + 12z - 11.07 + 4.6 - 1.83z + 4.9z^2$
$= 4.9z^3 - 1.92z^2 + 10.17z - 6.47$

45. $(2x^3 + 6x^2 - 9x + 13) - (4x^3 + 17x^2 - x + 6)$
$= 2x^3 + 6x^2 - 9x + 13 - 4x^3 - 17x^2 + x - 6$
$= -2x^3 - 11x^2 - 8x + 7$

46. $(6a^4 + 3a^2 + 4a + 5) - (5a^4 + 4a^3 + 3a^2 + 2a + 1)$
$= 6a^4 + 3a^2 + 4a + 5 - 5a^4 - 4a^3 - 3a^2 - 2a - 1$
$= a^4 - 4a^3 + 2a + 4$

47. $(117z^4 + 43z^2 + 88) - (18z^3 + 50z + 32)$
$= 117z^4 + 43z^2 + 88 - 18z^3 - 50z - 32$
$= 117z^4 - 18z^3 + 43z^2 - 50z + 56$

48. $11x^5y^6(13x^{-2}y^{-6})$
$= (11)(13)x^{5-2}y^{6-6}$
$= 143x^3y^0$
$= 143x^3$

49. $-6x(2x^4 - 4x^3 + 6x^2 + 8x - 10)$
$= -12x^{1+4} + 24x^{1+3} - 36x^{1+2} - 48x^{1+1} + 60x$
$= -12x^5 + 24x^4 - 36x^3 - 48x^2 + 60x$

50. $9a^{-1}(4a^5 + 2a^3 - 3a)$
$= 36a^{-1+5} + 18a^{-1+3} - 27a^{-1+1}$
$= 36a^4 + 18a^2 - 27a^0$
$= 36a^4 + 18a^2 - 27$

51. $(m - 11)(m + 11)$
$= m(m + 11) - 11(m + 11)$
$= m^2 + 11m - 11m - 121$
$= m^2 - 121$

52. $(z - 8)^2$
$= z^2 - 2(z)(8) + 8^2$
$= z^2 - 16z + 64$

53. $(4a - 7)(a + 13)$
$= 4a(a + 13) - 7(a + 13)$
$= 4a^2 + 52a - 7a - 91$
$= 4a^2 + 45a - 91$

54. $(13 - x)(13 + x)$
$= (13)^2 - x^2$
$= 169 - x^2$

55. $(b^3 + 4)^2$
$= (b^3)^2 + 2(b^3)(4) + 4^2$
$= b^6 + 8b^3 + 16$

56. $(t + 2)^3$
$= (t + 2)(t + 2)(t + 2)$
$= [t(t + 2) + 2(t + 2)](t + 2)$
$= (t^2 + 2t + 2t + 4)(t + 2)$
$= (t^2 + 4t + 4)(t + 2)$
$= (t^2 + 4t + 4)t + (t^2 + 4t + 4)(2)$
$= t^3 + 4t^2 + 4t + 2t^2 + 8t + 8$
$= t^3 + 6t^2 + 12t + 8$

57. $(7x - 2)(x^2 + 4x - 3)$
$= 7x(x^2 + 4x - 3) - 2(x^2 + 4x - 3)$
$= 7x^3 + 28x^2 - 21x - 2x^2 - 8x + 6$
$= 7x^3 + 26x^2 - 29x + 6$

58. $(p + q + r)^2$
$= (p + q + r)(p + q + r)$
$= p(p + q + r) + q(p + q + r) + r(p + q + r)$
$= p^2 + pq + pr + pq + q^2 + qr + pr + qr + r^2$
$= p^2 + q^2 + r^2 + 2pq + 2qr + 2pr$

59. $(b - 4)(b^2 + 4b + 16)$
$= b(b^2 + 4b + 16) - 4(b^2 + 4b + 16)$
$= b^3 + 4b^2 + 16b - 4b^2 - 16b - 64$
$= b^3 - 64$

60. $\dfrac{c^9}{c^6} = c^{9-6} = c^3$

61. $\dfrac{(a+4b)^{-6}}{(a+4b)^{-2}} = (a+4b)^{-6-(-2)} = (a+4b)^{-4}$
$= \dfrac{1}{(a+4b)^4}$

62. $\dfrac{81x^5y^7z}{27x^2y^6z^2} = \dfrac{81}{27}x^{5-2}y^{7-6}z^{1-2} = \dfrac{3x^3y}{z}$

63. $\dfrac{(3a^3)^3}{54a^7} = \dfrac{3^3a^{3\cdot3}}{54a^7} = \dfrac{27a^9}{54a^7} = \dfrac{27}{54}a^{9-7} = \dfrac{a^2}{2}$

64. $\left(\dfrac{21a^6b^2}{7a^3b^6}\right)^2 = \left(\dfrac{3a^3}{b^4}\right)^2 = \dfrac{3^2a^{3\cdot2}}{b^{4\cdot2}} = \dfrac{9a^6}{b^8}$

65. $\dfrac{14b^4 - 21b^3 - 35b^2 + 28b}{-7b}$
$= \dfrac{14b^4}{-7b} - \dfrac{21b^3}{-7b} - \dfrac{35b^2}{-7b} + \dfrac{28b}{-7b}$
$= -2b^{4-1} + 3b^{3-1} + 5b^{2-1} - 4b^{1-1}$
$= -2b^3 + 3b^2 + 5b - 4$

66. $\dfrac{16cd^2+8c^2d}{4c^2d^2}$
$=\dfrac{16cd^2}{4c^2d^2}+\dfrac{8c^2d}{4c^2d^2}$
$=4c^{1-2}d^{2-2}+2c^{2-2}d^{1-2}$
$=\dfrac{4}{c}+\dfrac{2}{d}$

67. $s^2\cdot s^8=s^{2+8}=s^{10}$

68. $y^5\cdot y^{-8}=y^{5-8}=y^{-3}=\dfrac{1}{y^3}$

69. $(m+n)^5(m+n)=(m+n)^{5+1}=(m+n)^6$

70. $(b^5)^7=b^{5\cdot 7}=b^{35}$

71. $(d^6)^{-3}=d^{6\cdot(-3)}=d^{-18}=\dfrac{1}{d^{18}}$

72. $(x^0)^8=x^{0\cdot 8}=x^0=1$

73. $(-3d)^6=(-3)^6d^6=729d^6$

74. $(-3d)^3=(-3)^3d^3=-27d^3$

75. $\left(\dfrac{m}{n}\right)^{-6}=\left(\dfrac{n}{m}\right)^6=\dfrac{n^6}{m^6}$

76. $(mn)^{-6}=\left(\dfrac{1}{mn}\right)^6=\dfrac{1}{m^6n^6}$

77. $(x^4y^5z)^4=x^{4\cdot 4}y^{5\cdot 4}z^4=x^{16}y^{20}z^4$

78. $[(4x)^2]^3=(4^2x^2)^3=(16x^2)^3=(16)^3x^{2\cdot 3}$
$=4096x^6$

79. $\left[\left(\dfrac{c}{2d}\right)^3\right]^2=\left(\dfrac{c^3}{2^3d^3}\right)^2$
$=\left(\dfrac{c^3}{8d^3}\right)^2=\dfrac{c^{3\cdot 2}}{8^2d^{3\cdot 2}}$
$=\dfrac{c^6}{64d^6}$

80. $8x^2-2x-3$
$=8x^2-6x+4x-3$
$=2x(4x-3)+1(4x-3)$
$=(4x-3)(2x+1)$
The width is $(2x+1)$ units and the length is $(4x-3)$ units.

81. a. $6x^2-10x-4$
$=6x^2+2x-12x-4$
$=2x(3x+1)-4(3x+1)$
$=(3x+1)(2x-4)$
The new length is $(3x+1)$ inches and the new width is $(2x-4)$ inches.

b. $3x+1-x=2x+1$
The length was increased by $(2x+1)$ inches.

c. $2x-4-(x-2)=x-2$
The width was increased by $(x-2)$ inches.

82. a. $x^2+12x+32$
$=\dfrac{1}{2}(2x^2+24x+64)$
$=\dfrac{1}{2}(2x^2+8x+16x+64)$
$=\dfrac{1}{2}[2x(x+4)+16(x+4)]$
$=\dfrac{1}{2}(x+4)(2x+16)$
The base is $(2x+16)$ units and the height is $(x+4)$ units.

b. $2x+16-2x=16$
The base was increased by 16 units.

c. $x+4-x=4$
The height was increased by 4 units.

83. a. length: $(2x+3)$ in.
width: x in.
height: x in.

b. Volume:
$x(x)(2x+3)$
$=x^2(2x+3)$
$=2x^3+3x^2$
The volume is $(2x^3+3x^2)$ in^3.

c. Surface area:
$2x(2x+3)+2x(2x+3)+2x^2$
$=4x^2+6x+4x^2+6x+2x^2$
$=10x^2+12x$
The surface area is $(10x^2+12x)$ in^2.

d. Volume:
$2x^3+3x^2$
$=2(8)^3+3(8)^2$
$=1024+192$
$=1216$
The volume is 1216 in^3.

e. Surface area:
$10x^2+12x$
$=10(8)^2+12(8)$
$=640+96$
$=736$
The surface area is 736 in^2.

84. a. height: x cm
base: $(2x+5)$ cm
Area:
$\frac{1}{2}(2x+5)(x)$
$=\frac{1}{2}x(2x)+\frac{1}{2}x(5)$
$=x^2+\frac{5}{2}x$
The area is $\left(x^2+\frac{5}{2}x\right)$ cm^2.

b. $x^2+\frac{5}{2}x$
$=(12)^2+\frac{5}{2}(12)$
$=144+30$
$=174$
The area is 174 cm^2.

85. a. old width: x ft
old length: x ft
Area $=x^2$
The current area is x^2 ft^2.

b. new width: $2x$
new length: $2x+5$
Area $=2x(2x+5)$
Area $=4x^2+10x$
The new area will be $(4x^2+10x)$ ft^2.

c. Ratio of enlarged area to original area:
$\frac{4x^2+10x}{x^2}$
$=\frac{x(4x+10)}{x(x)}$
$=\frac{4x+10}{x}$
The ratio is $\frac{4x+10}{x}$.

Chapter 10 Test

1. $81a^3+54a^2+9a$
$=9a(9a^2+6a+1)$
$=9a[(3a)^2+2(3a)(1)+1^2]$
$=9a(3a+1)^2$

2. p^3+125
$=p^3+5^3$
$=(p+5)(p^2-5p+5^2)$
$=(p+5)(p^2-5p+25)$

3. $-8a^4b^2 - 36a^3b^3 - 16a^2b^4$
$= -4a^2b^2(2a^2 + 9ab + 4b^2)$
$= -4a^2b^2(2a^2 + 8ab + 1ab + 4b^2)$
$= -4a^2b^2[2a(a + 4b) + b(a + 4b)]$
$= -4a^2b^2(a + 4b)(2a + b)$

4. $a(a^2 + b^2) - 5b(a^2 + b^2)$
$= (a^2 + b^2)(a - 5b)$

5. $15x^2 - 21xy + 10xy - 14y^2$
$= 3x(5x - 7y) + 2y(5x - 7y)$
$= (5x - 7y)(3x + 2y)$

6. $64a^2 - 49b^2$
$= (8a)^2 - (7b)^2$
$= (8a + 7b)(8a - 7b)$

7. $25x^2 - 70x + 49$
$= (5x)^2 - 2(5x)(7) + 7^2$
$= (5x - 7)^2$

8. $3x^3 - 27x^2 + 24x$
$= 3x(x^2 - 9x + 8)$
$= 3x(x^2 - x - 8x + 8)$
$= 3x[x(x - 1) - 8(x - 1)]$
$= 3x(x - 1)(x - 8)$

9. $x^2 - 4xy - 21y^2$
$= x^2 + 3xy - 7xy - 21y^2$
$= x(x + 3y) - 7y(x + 3y)$
$= (x + 3y)(x - 7y)$

10. $14x^2 + 25x + 9$
$= 14x^2 + 7x + 18x + 9$
$= 7x(2x + 1) + 9(2x + 1)$
$= (2x + 1)(7x + 9)$

11. $4x^4 + 27x^2 - 7$
$= 4x^4 + 28x^2 - x^2 - 7$
$= 4x^2(x^2 + 7) - 1(x^2 + 7)$
$= (x^2 + 7)(4x^2 - 1)$
$= (x^2 + 7)[(2x)^2 - 1^2]$
$= (x^2 + 7)(2x + 1)(2x - 1)$

12. $x^2 + 8x + 14$ will not factor.

13. $(2x - 1)(2x - 1)^8 = (2x - 1)^{1+8} = (2x - 1)^9$

14. $3x^2y^0z^{-3} = \dfrac{3x^2(1)}{z^3} = \dfrac{3x^2}{z^3}$

15. $(a + b)^3(a - b)^{-5} = \dfrac{(a + b)^3}{(a - b)^5}$

16. $(-2x^2y^{-1})^3$
$= (-2)^3 x^{2\cdot 3} y^{(-1)(3)}$
$= -8x^6y^{-3}$
$= \dfrac{-8x^6}{y^3}$

17. $\dfrac{(3c - 7)^3}{(3c - 7)^5}$
$= (3c - 7)^{3-5}$
$= (3c - 7)^{-2}$
$= \dfrac{1}{(3c - 7)^2}$

18. $\left[\dfrac{(2p)^3}{3q}\right]^{-2} = \left(\dfrac{2^3p^3}{3q}\right)^{-2} = \left(\dfrac{8p^3}{3q}\right)^{-2}$
$= \left(\dfrac{3q}{8p^3}\right)^2 = \dfrac{3^2q^2}{8^2p^{3\cdot 2}} = \dfrac{9q^2}{64p^6}$

19. $(3y^2+16-7y+5y^3)+(7y+6y^2+4y^4-11)$
$=3y^2+16-7y+5y^3+7y+6y^2+4y^4-11$
$=4y^4+5y^3+9y^2+5$

20. $(7x^5+23x^3+17x^2-39)-(2x^4-4x^2-2x-9)$
$=7x^5+23x^3+17x^2-39-2x^4+4x^2+2x+9$
$=7x^5-2x^4+23x^3+21x^2+2x-30$

21. $(-2p^3q^{-5}r^2)(5.7p^6q^7r)$
$=(-2)(5.7)p^{3+6}q^{-5+7}r^{2+1}$
$=-11.4p^9q^2r^3$

22. $-4t(2t^3-3t^2-8t+6)$
$=-8t^{1+3}+12t^{1+2}+32t^{1+1}-24t$
$=-8t^4+12t^3+32t^2-24t$

23. $(9-5d)(9+5d)$
$=(9)^2-(5d)^2$
$=81-25d^2$

24. $(3x+4)(5x-7)$
$=3x(5x-7)+4(5x-7)$
$=15x^2-21x+20x-28$
$=15x^2-x-28$

25. $(4z-3)(2z^2-z+5)$
$=4z(2z^2-z+5)-3(2z^2-z+5)$
$=8z^3-4z^2+20z-6z^2+3z-15$
$=8z^3-10z^2+23z-15$

26. $(x+3)^2$
$=x^2+2(x)(3)+3^2$
$=x^2+6x+9$

27. $\dfrac{24a^2b^{-3}c}{3ab^2c^{-2}}$
$=\dfrac{24}{3}a^{2-1}b^{-3-2}c^{1-(-2)}$
$=8ab^{-5}c^3$
$=\dfrac{8ac^3}{b^5}$

28. $\dfrac{(a^{-2}b^4)^3}{a^2b^5c}$
$=\dfrac{a^{(-2)\cdot 3}b^{4\cdot 3}}{a^2b^5c}$
$=\dfrac{a^{-6}b^{12}}{a^2b^5c}$
$=a^{-6-2}b^{12-5}c^{-1}$
$=a^{-8}b^7c^{-1}$
$=\dfrac{b^7}{a^8c}$

29. $\dfrac{15x^6+25x^5-5x^3}{5x^2}$
$=\dfrac{15x^6}{5x^2}+\dfrac{25x^5}{5x^2}-\dfrac{5x^3}{5x^2}$
$=3x^{6-2}+5x^{5-2}-1x^{3-2}$
$=3x^4+5x^3-x$

30. a. width: $(x+4)$ in.
height: x in.
length: $(2x)$ in.
volume:
$x(2x)(x+4)$
$= 2x^2(x+4)$
$= 2x^3+8x^2$
The volume is $(2x^3+8x^2)$ in^3.

b. Surface area:
$2x(x+4)+2x(2x)+2(2x)(x+4)$
$= 2x^2+8x+4x^2+4x^2+16x$
$= 10x^2+24x$
The surface area is $(10x^2+24x)$ in^2.

c. Volume:
$2x^3+8x^2$
$= 2(5)^3+8(5)^2$
$= 250+200$
$= 450$
The volume is 450 in^3.
Surface area:
$10x^2+24x$
$= 10(5)^2+24(5)$
$= 250+120$
$= 370$
The surface area is 370 in^2.

31. a. Factor by grouping

b. He did not factor completely.
$(4x^2-9)$ can be factored further.

c. $12x^3+28x^2-27x-63$
$= (12x^3+28x^2)+(-27x-63)$
$= 4x^2(3x+7)-9(3x+7)$
$= (3x+7)(4x^2-9)$
$= (3x+7)[(2x)^2-3^2]$
$= (3x+7)(2x+3)(2x-3)$

Chapter 11

11.1 Experiencing Algebra the Exercise Way

1. $3x^3 - 2x^2 + x = 5$ or $3x^3 - 2x^2 + x - 5 = 0$ is a cubic polynomial equation.

3. $3\sqrt{y} + y - 4 = 0$ is not a polynomial because y is the radicand of $\sqrt{y}$.

5. $\frac{1}{4}x^4 + 3x^2 - \frac{3}{4} = 0$ is a polynomial equation.

7. $4(x - 2)(x + 7) = 16$ or $4x^2 + 20x - 72 = 0$ is a quadratic polynomial equation.

9. $3x^{-2} - 5x = 4x^2$ is not a polynomial because x has an exponent of -2.

11. $1.7x^2 + 3.2x = 5.7$ or $1.7x^2 + 3.2x - 5.7 = 0$ is a quadratic polynomial equation.

13. $x^2 + 8 = 6x$

$Y1 = x^2 + 8$

$Y2 = 6x$

X	Y1	Y2
0	8	0
1	9	6
2	12	12
3	17	18
4	24	24
5	33	30
6	44	36

X=0

The solutions are 2 and 4.

15. $x^3 = x$

$Y1 = x^3$

$Y2 = x$

X	Y1	Y2
-4	-64	-4
-3	-27	-3
-2	-8	-2
-1	-1	-1
0	0	0
1	1	1
2	8	2

X= -4

The solutions are -1, 1, and 0.

17. $x^2 - 7 = x^2 + 3$

$Y1 = x^2 - 7$

$Y2 = x^2 + 3$

X	Y1	Y2
0	-7	3
1	-6	4
2	-3	7
3	2	12
4	9	19
5	18	28
6	29	39

X=0

The expression on the left is always 10 less than the expression on the right. There is no solution.

19. $x^2 + 5x + 1 = 1 + x(5 + x)$

$Y1 = x^2 + 5x + 1$

$Y2 = 1 + x(5 + x)$

X	Y1	Y2
0	1	1
1	7	7
2	15	15
3	25	25
4	37	37
5	51	51
6	67	67

X=0

The expressions are equal for all x-values in the table. The solution is the set of all real numbers.

21. $x^2 - 2x + 6 = 12 - 4x + 2x^2$

$Y1 = x^2 - 2x + 6$

$Y2 = 12 - 4x + 2x^2$

X	Y1	Y2
0	6	12
1	5	10
2	6	12
3	9	18
4	14	28
5	21	42
6	30	60

X=0

The expression on the left is always less than the expression on the right. In standard form, the equation is $x^2 - 2x + 6 = 0$. There is no real-number solution.

23. $\frac{1}{2}x^2 - x = 6 - 3x$

$Y1 = \frac{1}{2}x^2 - x$

$Y2 = 6 - 3x$

X	Y1	Y2
-7	31.5	27
-6	24	24
-5	17.5	21
-4	12	18
-3	7.5	15
-2	4	12
-1	1.5	9

X=-7

X	Y1	Y2
0	0	6
1	-.5	3
2	0	0
3	1.5	-3
4	4	-6
5	7.5	-9
6	12	-12

X=0

The solutions are –6 and 2.

25. $2 - 0.2x^2 = 0.6x$

$Y1 = 2 - 0.2x^2$

$Y2 = 0.6x$

X	Y1	Y2
-6	-5.2	-3.6
-5	-3	-3
-4	-1.2	-2.4
-3	.2	-1.8
-2	1.2	-1.2
-1	1.8	-.6
0	2	0

X=-6

X	Y1	Y2
1	1.8	.6
2	1.2	1.2
3	.2	1.8
4	-1.2	2.4
5	-3	3
6	-5.2	3.6
7	-7.8	4.2

X=1

The solutions are –5 and 2.

27. $x^2 - 3 = 6$

$Y1 = x^2 - 3$

$Y2 = 6$

31

–47 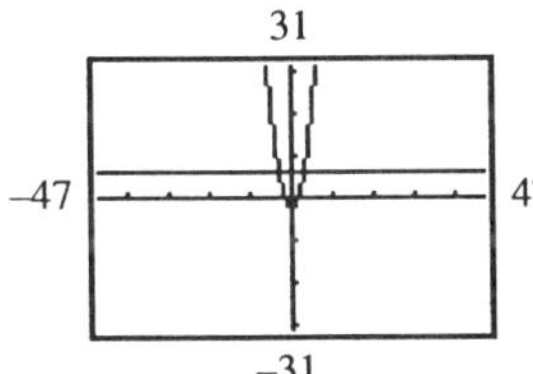 47

–31

The graphs intersect at (–3, 6) and (3, 6). The solutions are –3 and 3.

29. $x^2 - 3 = 2x$

$Y1 = x^2 - 3$

$Y2 = 2x$

31

–47 47

–31

The graphs intersect at (–1, –2) and (3, 6). The solutions are –1 and 3.

31. $x^3 = 4x$

$Y1 = x^3$

$Y2 = 4x$

10

–10 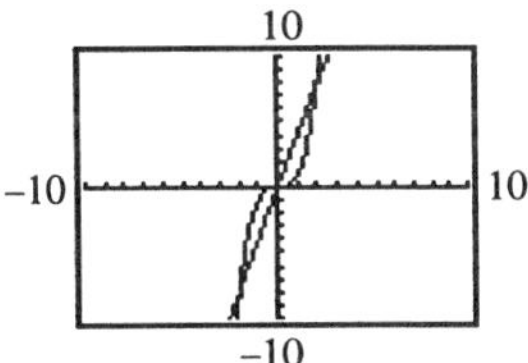 10

–10

The graphs intersect at (–2, –8), (0, 0), and (2, 8). The solutions are –2, 0, and 2.

33. $x^2 - 3x - 10 = 0$

$Y1 = x^2 - 3x - 10$

$Y2 = 0$

31

–47 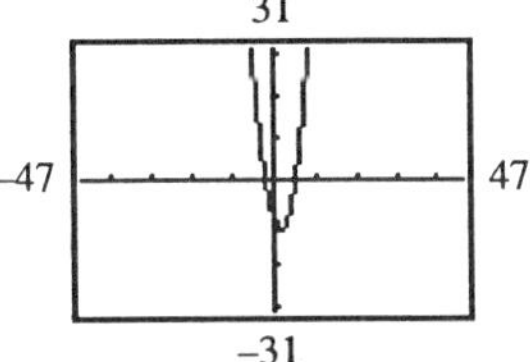 47

–31

The graphs intersect at (–2, 0) and (5, 0). The solutions are –2 and 5.

35. $x^2 - 2x + 1 = x^2 - 2x - 3$

$Y1 = x^2 - 2x + 1$

$Y2 = x^2 - 2x - 3$

20

–10 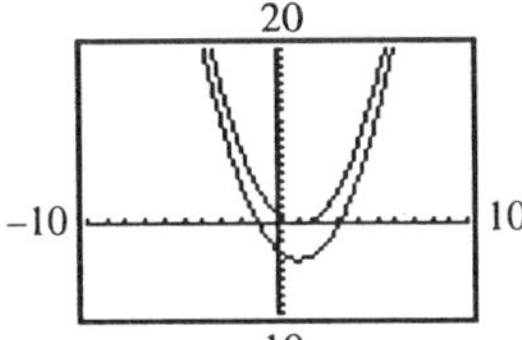 10

–10

The graphs do not intersect. There is no solution.

37. $x^2 + 1 = 3x^2 + 3$
$Y1 = x^2 + 1$
$Y2 = 3x^2 + 3$

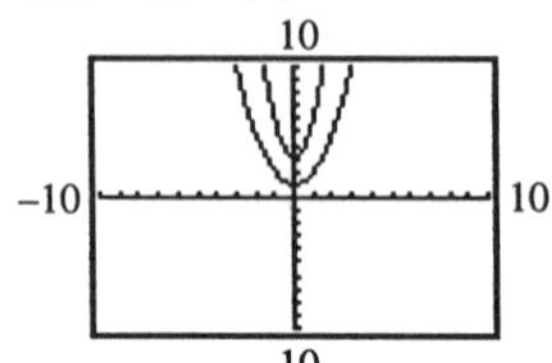

The graphs do not intersect. There is no real-number solution.

39. $x(x + 3) = x^2 + 3x$
$Y1 = x(x + 3)$
$Y1 = x^2 + 3x$

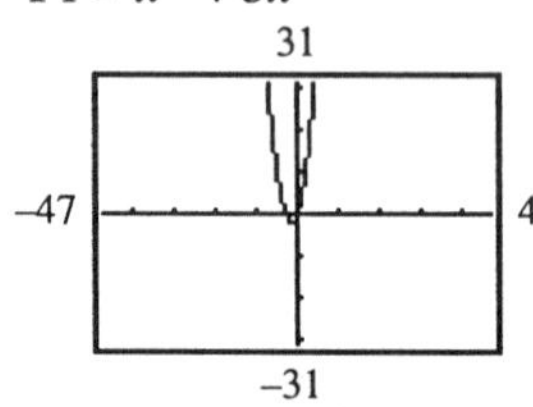

The graphs are the same. The solution is the set of all real numbers.

41. $4x^2 - x^3 = x^2 - 4x$
$Y1 = 4x^2 - x^3$
$Y2 = x^2 - 4x$

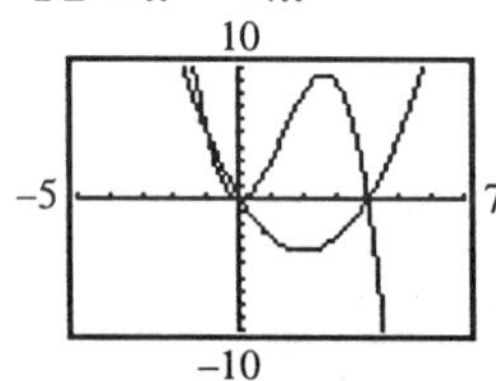

The graphs intersect at (–1, 5), (0, 0) and (4, 0). The solutions are –1, 0, and 4.

43. $x^2 - 10x + 30 = \frac{1}{2}x^2 - 5x + 15$
$Y1 = x^2 - 10x + 30$
$Y2 = \frac{1}{2}x^2 - 5x + 15$

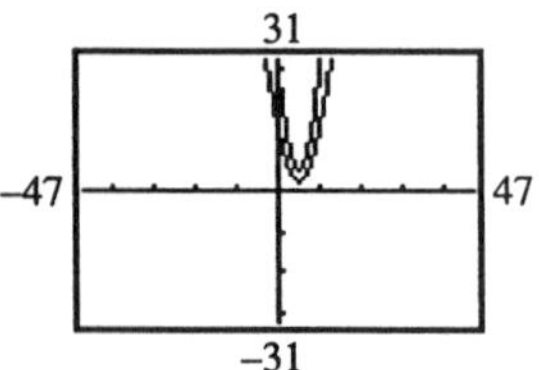

The graphs do not intersect.
There is no real-number solution.

45. $x^3 - 2x^2 + 1 = x^3 - 2x^2 + 9$
$Y1 = x^3 - 2x^2 + 1$
$Y2 = x^3 - 2x^2 + 9$

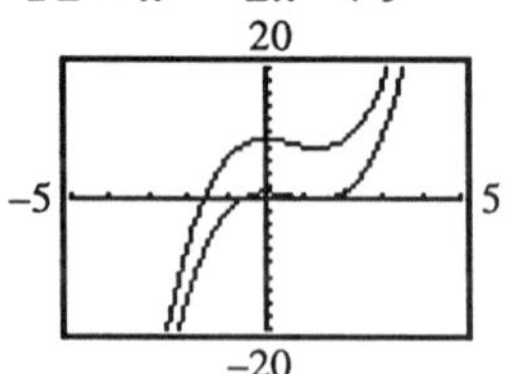

The graphs do not intersect. There is no solution.

47. $x(x^2 - 3) - 5(x + 1) = x^3 - 8x - 5$
$Y1 = x(x^2 - 3) - 5(x + 1)$
$Y2 = x^3 - 8x - 5$

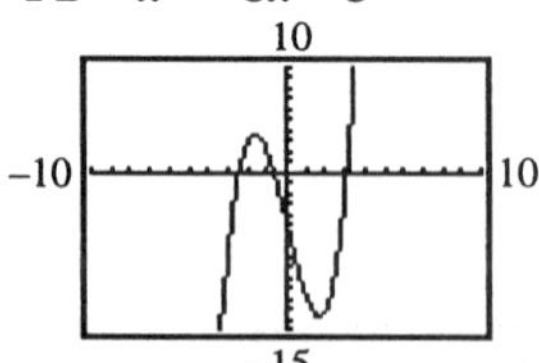

The graphs are the same. The solution is all real numbers.

49. $4x^2 = 9$
$Y1 = 4x^2$
$Y2 = 9$

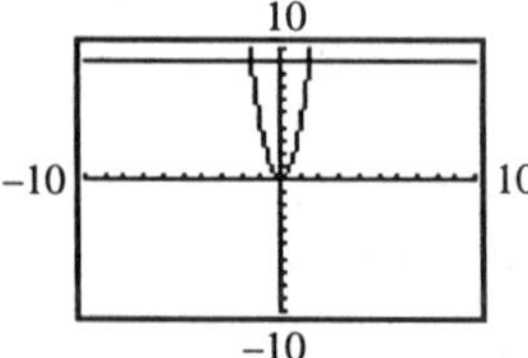

The graphs intersect at $\left(-\frac{3}{2}, 9\right)$ and $\left(\frac{3}{2}, 9\right)$. The solutions are $-\frac{3}{2}$ and $\frac{3}{2}$.

51. $10x^3 - 7x^2 - 4x = 3x - 4$
$Y1 = 10x^3 - 7x^2 - 4x$
$Y2 = 3x - 4$

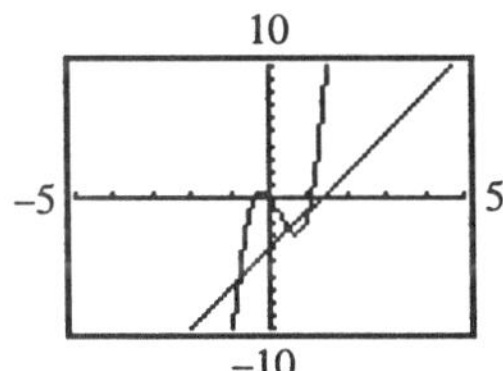

The graphs intersect at (–0.8, –6.4), (0.5, –2.5), and (1, –1). The solutions are –0.8, 0.5, and 1.

53. $x^2 - 0.9x - 10.36 = 0$
$Y1 = x^2 - 0.9x - 10.36$
$Y2 = 0$

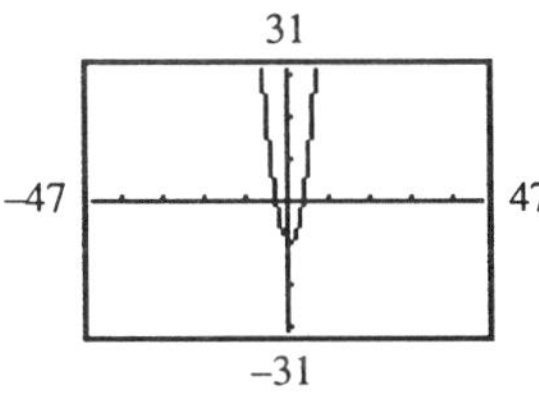

The graphs intersect at (–2.8, 0) and (3.7, 0). The solutions are –2.8 and 3.7.

55. $x^3 + 3.7x^2 = 1.74x + 7.56$
$Y1 = x^3 + 3.7x^2$
$Y2 = 1.74x + 7.56$

15
−10 10
−10

The graphs intersect at (–3.6, 1.296), (–1.5, 4.95), and (1.4, 9.996). The solutions are –3.6, –1.5, and 1.4.

57. $s = -16t^2 + v_0t + s_0$
$0 = -16t^2 + 0t + 40$
$0 = -16t^2 + 40$
$Y1 = 0$
$Y2 = -16x^2 + 40$

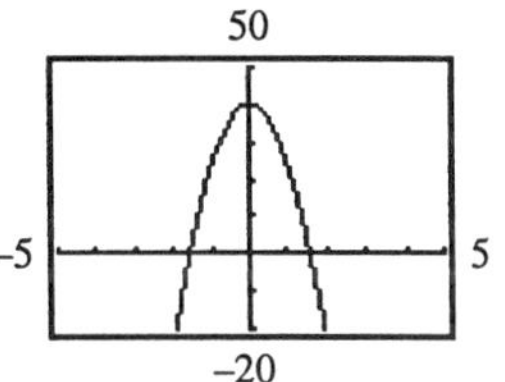

The graphs intersect at about (–1.58, 0) and (1.58, 0). There cannot be a negative amount of time. Therefore, it will hit the ground in approximately 1.58 seconds.

59. $s = -16t^2 + v_0t + s_0$
$0 = -16t^2 - 5t + 40$
$Y1 = 0$
$Y2 = -16x^2 - 5x + 40$

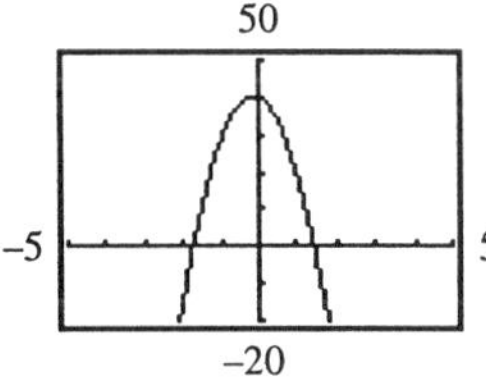

The graphs intersect at approximately (–1.75, 0) and (1.43, 0). There cannot be a negative amount of time. Therefore, the dagger will hit the ground in approximately 1.43 seconds.

61. $s = -16t^2 + v_0t + s_0$
$0 = -16t^2 + 5t + 40$
$Y1 = 0$
$Y2 = -16x^2 + 5x + 40$

50
−5 5
−20

The graphs intersect at approximately (–1.43, 0) and (1.75, 0). Disregard a negative time. It will hit the ground in approximately 1.75 seconds.

63. **a.** $30 + x$

b. $40 - x$

c. 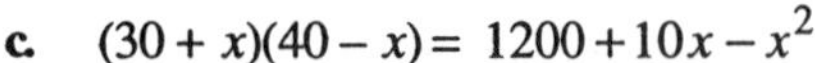 $(30+x)(40-x)=1200+10x-x^2$

d. $1100=1200+10x-x^2$

e. $Y1=1100$
$Y2=1200+10x-x^2$

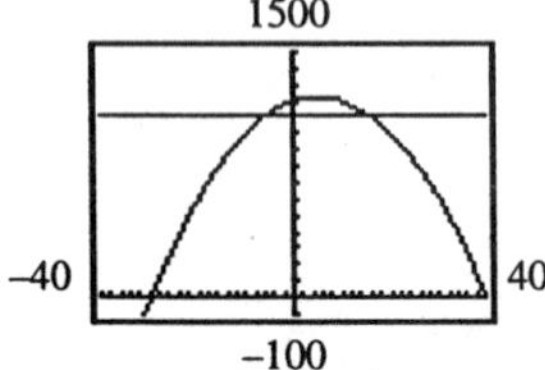

The graphs intersect at approximately (–6.18, 1100) and (16.18, 1100). Disregard a negative number of seats. There should be 16 unsold seats and 24 sold seats to meet this goal.

65. $y=11.55+20.67x-1.35x^2$
$100=11.55+20.67x-1.35x^2$
$Y1=100$
$Y2=11.55+20.67x-1.35x^2$

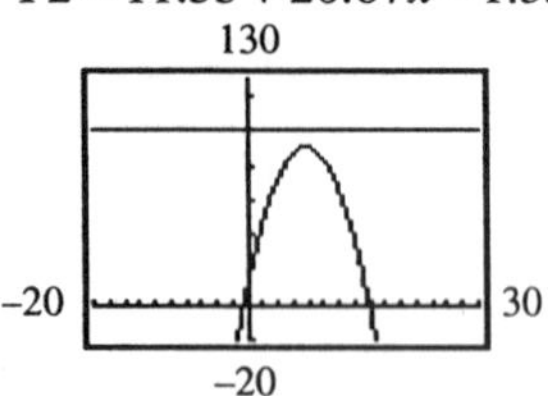

The graphs do not intersect. There are no number of sales reps that can be assigned and meet the goal. There is no real-number solution.

67. Volume = length × width × height
Volume = 2000
height = x
length = $2x$
width = $x+1$
$2000=2x(x+1)(x)$
$2000=2x^3+2x^2$
$Y1=2000$
$Y2=2x^3+2x^2$

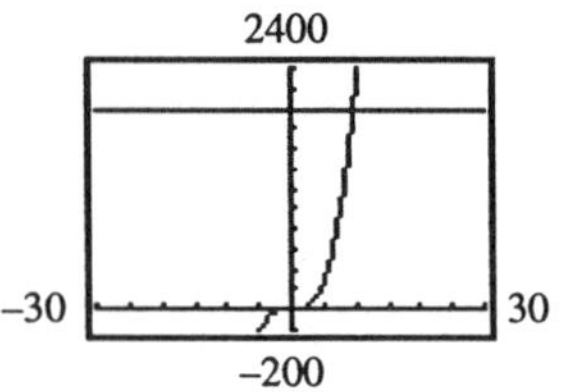

The graphs intersect at about (9.68, 2000). The dimensions are about 20 in. by 11 in. by 10 in.

11.1 Experiencing Algebra the Calculator Way

Students should verify solutions with the calculator.

11.2 Experiencing Algebra the Exercise Way

1. $(x+6)(x+11)=0$
$x+6=0$ or $x+11=0$
$x=-6$ $\quad$ $x=-11$
The solutions are –11 and –6.

3. $\left(\frac{3}{5}x-\frac{9}{20}\right)\left(x+\frac{2}{3}\right)=0$
$\frac{3}{5}x-\frac{9}{20}=0$ or $x+\frac{2}{3}=0$
$\frac{3}{5}x=\frac{9}{20}$ $\quad$ $x=-\frac{2}{3}$
$x=\frac{3}{4}$

The solutions are $-\frac{2}{3}$ and $\frac{3}{4}$.

5. $3x(x+9)(2x-5)=0$
$3x=0$ or $x+9=0$ or $2x-5=0$
$x=0$ $\quad$ $x=-9$ $\quad$ $2x=5$
$x=\frac{5}{2}$

The solutions are –9, 0 and $\frac{5}{2}$.

7. $(7x-49)(49x-7)=0$
$7x-49=0$ or $49x-7=0$
$7x=49$ $\quad$ $49x=7$
$x=7$ $\quad$ $x=\frac{1}{7}$

The solutions are $\frac{1}{7}$ and 7.

9. $(4x+3)(2x-9)(x+6)=0$
$4x+3=0$ or $2x-9=0$ or $x+6=0$
$4x=-3$ $\quad 2x=9$ $\quad x=-6$
$x=-\frac{3}{4}$ $\quad x=\frac{9}{2}$
The solutions are -6, $-\frac{3}{4}$, and $\frac{9}{2}$.

11. $(0.2x+6.8)(1.3x-1.69)=0$
$0.2x+6.8=0$ or $1.3x-1.69=0$
$0.2x=-6.8$ $\quad 1.3x=1.69$
$x=-34$ $\quad x=1.3$
The solutions are -34 and 1.3.

13. $0=x^2+10x+24$
$0=(x+6)(x+4)$
$x+6=0$ or $x+4=0$
$x=-6$ $\quad x=-4$
The solutions are -6 and -4.

15. $x^2+33=14x$
$x^2-14x+33=0$
$(x-3)(x-11)=0$
$x-3=0$ or $x-11=0$
$x=3$ $\quad x=11$
The solutions are 3 and 11.

17. $4x^2+5x+24x+30=0$
$(4x^2+5x)+(24x+30)=0$
$x(4x+5)+6(4x+5)=0$
$(4x+5)(x+6)=0$
$4x+5=0$ or $x+6=0$
$4x=-5$ $\quad x=-6$
$x=-\frac{5}{4}$
The solutions are -6 and $-\frac{5}{4}$.

19. $5x^2+3x=8$
$5x^2+3x-8=0$
$(5x+8)(x-1)=0$
$5x+8=0$ or $x-1=0$
$5x=-8$ $\quad x=1$
$x=-\frac{8}{5}$
The solutions are $-\frac{8}{5}$ and 1.

21. $15x^2=35x$
$15x^2-35x=0$
$5x(3x-7)=0$
$5x=0$ or $3x-7=0$
$x=0$ $\quad 3x=7$
$x=\frac{7}{3}$
The solutions are 0 and $\frac{7}{3}$.

23. $18x^2-3x=5-30x$
$18x^2+27x-5=0$
$(6x-1)(3x+5)=0$
$6x-1=0$ or $3x+5=0$
$6x=1$ $\quad 3x=-5$
$x=\frac{1}{6}$ $\quad x=-\frac{5}{3}$
The solutions are $-\frac{5}{3}$ and $\frac{1}{6}$.

25. $16x^2+72x+81=0$
$(4x+9)^2=0$
$4x+9=0$
$4x=-9$
$x=-\frac{9}{4}$
The solution is $-\frac{9}{4}$.

27. $4x^2+25x+18=5x-7$
$4x^2+20x+25=0$
$(2x+5)^2=0$
$2x+5=0$
$2x=-5$
$x=-\frac{5}{2}$
The solution is $-\frac{5}{2}$.

29. $b^2+7=71$
$b^2-64=0$
$(b+8)(b-8)=0$
$b+8=0$ or $b-8=0$
$b=-8$ $\quad b=8$
The solutions are -8 and 8.

31. $9z^2 = 25$
$9z^2 - 25 = 0$
$(3z + 5)(3z - 5) = 0$
$3z + 5 = 0$ or $3z - 5 = 0$
$3z = -5$ $\quad 3z = 5$
$z = -\frac{5}{3}$ $\quad z = \frac{5}{3}$
The solutions are $-\frac{5}{3}$ and $\frac{5}{3}$

33. $x^2 + 10x + 20 = 0$
The equation does not factor.

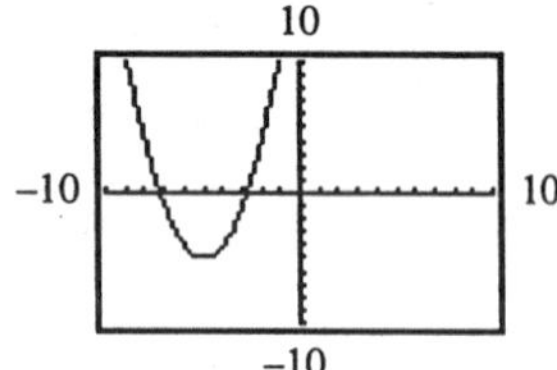

The solutions (the x-coordinates of the intersections) are approximately −7.24 and −2.76.

35. $(x+1)^2 = 49$
$(x+1)^2 - 49 = 0$
$[(x + 1) + 7][(x + 1) - 7] = 0$
$(x + 8)(x - 6) = 0$
$x + 8 = 0$ or $x - 6 = 0$
$x = -8$ $\quad x = 6$
The solutions are −8 and 6.

37. $(x - 3)(x - 2) = 42$
$x^2 - 5x + 6 = 42$
$x^2 - 5x - 36 = 0$
$(x - 9)(x + 4) = 0$
$x - 9 = 0$ or $x + 4 = 0$
$x = 9$ $\quad x = -4$
The solutions are −4 and 9.

39. $(x - 7)(2x + 3) = 2(x^2 - 5x - 10) - (x + 1)$
$2x^2 - 11x - 21 = 2x^2 - 10x - 20 - x - 1$
$2x^2 - 11x - 21 = 2x^2 - 11x - 21$
$0 = 0$
This is an identity. The solution is the set of all real numbers.

41. $(x+3)^2 = x^2 + 6x + 11$
$x^2 + 6x + 9 = x^2 + 6x + 11$
$9 = 11$
This is a contradiction.
There is no solution.

43. $x^2 + (x+3)^2 = 225$
$x^2 + x^2 + 6x + 9 = 225$
$2x^2 + 6x - 216 = 0$
$2(x^2 + 3x - 108) = 0$
$2(x + 12)(x - 9) = 0$
$x + 12 = 0$ or $x - 9 = 0$
$x = -12$ $\quad x = 9$
The solutions are −12 and 9.

45. $x^3 + 7x^2 - 9x - 63 = 0$
$x^2(x+7) - 9(x+7) = 0$
$(x+7)(x^2 - 9) = 0$
$(x + 7)(x + 3)(x - 3) = 0$
$x + 7 = 0$ or $x + 3 = 0$ or $x - 3 = 0$
$x = -7$ $\quad x = -3$ $\quad x = 3$
The solutions are −7, −3, and 3.

47. $18x^3 + 45x^2 - 50x - 125 = 0$
$9x^2(2x+5) - 25(2x+5) = 0$
$(2x+5)(9x^2 - 25) = 0$
$(2x + 5)(3x + 5)(3x - 5) = 0$
$2x + 5 = 0$ or $3x + 5 = 0$ or $3x - 5 = 0$
$2x = -5$ $\quad 3x = -5$ $\quad 3x = 5$
$x = -\frac{5}{2}$ $\quad x = -\frac{5}{3}$ $\quad x = \frac{5}{3}$
The solutions are $-\frac{5}{2}$, $-\frac{5}{3}$, and $\frac{5}{3}$.

49. $3x^3 - 3x^2 + 12x - 12 = 0$
$3x^2(x-1) + 12(x-1) = 0$
$(x-1)(3x^2 + 12) = 0$
$(x-1)3(x^2 + 4) = 0$
$x - 1 = 0$ or $x^2 + 4 = 0$
$x = 1$ $\quad x^2 = -4$
The solution is 1.

51. a. length = x ft
height = 3 ft
width = $\left(\frac{1}{2}x+1\right)$ ft
volume = length × width × height
volume = $x(3)\left(\frac{1}{2}x+1\right)=3x\left(\frac{1}{2}x+1\right)$
$=\frac{3}{2}x^2+3x$
The volume is $\left(\frac{3}{2}x^2+3x\right)$ ft^3.

b. $72=\frac{3}{2}x^2+3x$
$0=3x^2+6x-144$
$0=(3x-18)(x+8)$
$3x-18=0$ or $x+8=0$
$3x=18$ $x=-8$
$x=6$
Disregard a negative length. The dimensions are 6 ft by 3 ft by 4 ft.

53. a. height = x in.
width = $(3x)$ in.
length = $(4x+2)$ in.
surface area
$=2x(3x)+2x(4x+2)+3x(4x+2)$
$=6x^2+8x^2+4x+12x^2+6x$
$=26x^2+10x$
The surface area is $(26x^2+10x)$ in^2.

b. $700=26x^2+10x$
$0=26x^2+10x-700$
$0=2(13x^2+5x-350)$
$0=2(13x+70)(x-5)$
$13x+70=0$ or $x-5=0$
$13x=-70$ $x=5$
$x=-\frac{70}{13}$
Disregard a negative height. The dimensions are 15 in. by 5 in. by 22 in.

55. a. base = x ft
height = $(x+4)$ ft
area = $\frac{1}{2}x(x+4)=\frac{1}{2}x^2+2x$
The area is $\left(\frac{1}{2}x^2+2x\right)$ ft^2.

b. $30=\frac{1}{2}x^2+2x$
$0=x^2+4x-60$
$0=(x+10)(x-6)$
$x+10=0$ or $x-6=0$
$x=-10$ $x=6$
Disregard a negative base.
The dimensions are 6 ft base and 10 ft height.

57. height = x ft
base = $(x+6)$ ft
hypotenuse = $(x+12)$ ft
$(x+12)^2=x^2+(x+6)^2$
$x^2+24x+144=x^2+x^2+12x+36$
$0=x^2-12x-108$
$0=(x-18)(x+6)$
$x-18=0$ or $x+6=0$
$x=18$ $x=-6$
Disregard a negative height.
The measurements are: 18 ft height, 24 ft base, and 30 ft hypotenuse.

59. $V^2=30FS$
$V^2=30(0.40)(243)$
$V^2=2916$
$V=\sqrt{2916}$
$V=54$
The vehicle was traveling at 54 mph.

11.2 Experiencing Algebra the Calculator Way

1. $x^2-133x+4350=0$
$\text{Y1}=x^2-133x+4350$
$\text{Y2}=0$

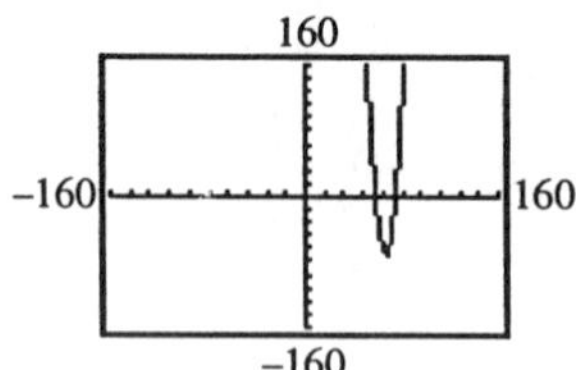

The solutions (the x-coordinates of the intersection) are 58 and 75.

2. $12x^2 - 12x - 3672 = 0$
   $Y1 = 12x^2 - 12x - 3672$
   $Y2 = 0$

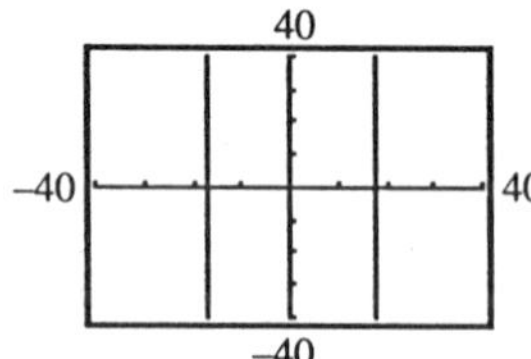

The solutions (the x-coordinates of the intersection) are –17 and 18.

3. $x^2 + 250x + 15,625 = 0$
   $Y1 = x^2 + 250x + 15,625$
   $Y2 = 0$

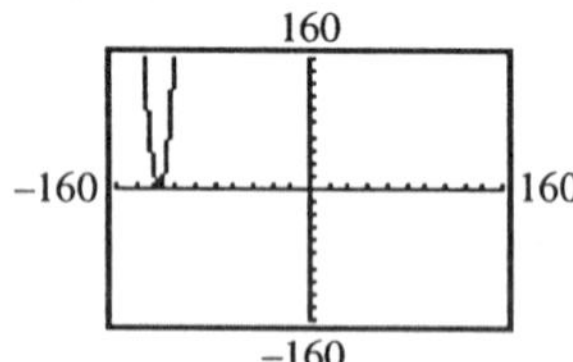

The solution (the x-coordinate of the intersection) is –125.

4. $x^2 - 186x + 8649 = 0$
   $Y1 = x^2 - 186x + 8649$
   $Y2 = 0$

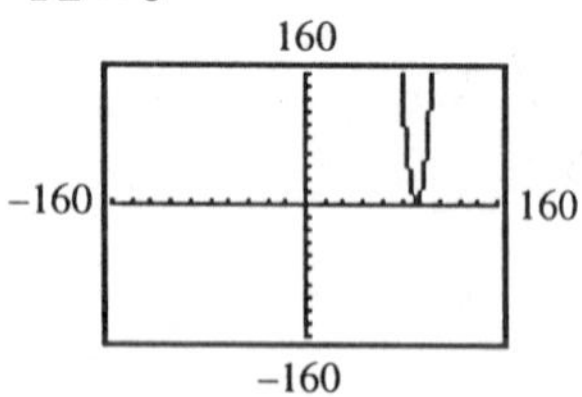

The solution (the x-coordinate of the intersection) is 93.

5. $13x^2 + 195x = 377x + 5655$
   $Y1 = 13x^2 + 195x$
   $Y2 = 377x + 5655$

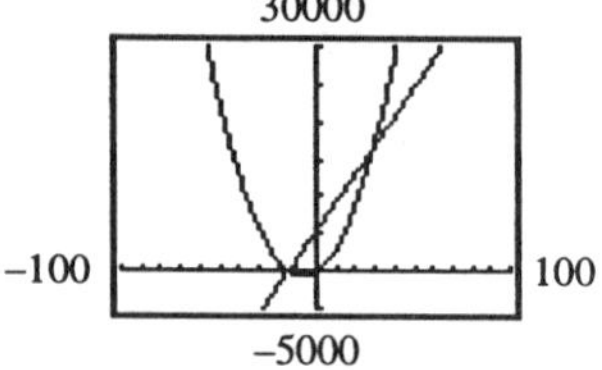

The solutions (the x-coordinates of the intersections) are –15 and 29.

6. Answers will vary.

11.3 Experiencing Algebra the Exercise Way

1. $\sqrt{81} = \sqrt{9^2} = 9$

3. $-\sqrt{100} = -\sqrt{(10)^2} = -10$

5. $\sqrt{3}\sqrt{27} = \sqrt{3 \cdot 27} = \sqrt{81} = 9$

7. $\sqrt{6}\sqrt{21} = \sqrt{6 \cdot 21} = \sqrt{126} = \sqrt{9 \cdot 14} = 3\sqrt{14}$

9. $\sqrt{147} = \sqrt{49 \cdot 3} = 7\sqrt{3}$

11. $-\sqrt{125} = -\sqrt{25 \cdot 5} = -5\sqrt{5}$

13. $\sqrt{\dfrac{36}{49}} = \dfrac{\sqrt{36}}{\sqrt{49}} = \dfrac{6}{7}$

15. $\sqrt{\dfrac{11}{144}} = \dfrac{\sqrt{11}}{\sqrt{144}} = \dfrac{\sqrt{11}}{12}$

17. $\sqrt{\dfrac{64}{13}} = \dfrac{\sqrt{64}}{\sqrt{13}} = \dfrac{8}{\sqrt{13}} \cdot \dfrac{\sqrt{13}}{\sqrt{13}} = \dfrac{8\sqrt{13}}{13}$

19. $-\dfrac{\sqrt{28}}{\sqrt{9}} = -\dfrac{\sqrt{4 \cdot 7}}{3} = -\dfrac{2\sqrt{7}}{3}$

21. $\dfrac{\sqrt{45}}{\sqrt{24}} = \dfrac{\sqrt{9 \cdot 5}}{\sqrt{4 \cdot 6}} = \dfrac{3\sqrt{5}}{2\sqrt{6}} \cdot \dfrac{\sqrt{6}}{\sqrt{6}} = \dfrac{3\sqrt{30}}{2 \cdot 6} = \dfrac{\sqrt{30}}{4}$

23. $-\sqrt{\frac{18}{48}}$

$= -\sqrt{\frac{9}{24}}$

$= -\frac{\sqrt{9}}{\sqrt{24}}$

$= -\frac{3}{\sqrt{4 \cdot 6}}$

$= -\frac{3}{2\sqrt{6}} \cdot \frac{\sqrt{6}}{\sqrt{6}}$

$= -\frac{3\sqrt{6}}{12} = -\frac{\sqrt{6}}{4}$

25. $x^2 = 144$

$x = \sqrt{144}$ or $x = -\sqrt{144}$

$x = 12$ or $x = -12$

The solutions are ± 12.

27. $a^2 = 13$

$a = \sqrt{13}$ or $a = -\sqrt{13}$

The solutions are $\pm\sqrt{13}$.

29. $q^2 = 98$

$q = \sqrt{98}$ or $q = -\sqrt{98}$

$q = \sqrt{49 \cdot 2}$ $q = -\sqrt{49 \cdot 2}$

$q = 7\sqrt{2}$ $q = -7\sqrt{2}$

The solutions are $\pm 7\sqrt{2}$

31. $2x^2 - 32 = 0$

$2x^2 = 32$

$x^2 = 16$

$x = \sqrt{16}$ or $x = -\sqrt{16}$

$x = 4$ $x = -4$

The solutions are ± 4.

33. $4x^2 - 25 = 0$

$4x^2 = 25$

$x^2 = \frac{25}{4}$

$x = \sqrt{\frac{25}{4}}$ or $x = -\sqrt{\frac{25}{4}}$

$x = \frac{5}{2}$ $x = -\frac{5}{2}$

The solutions are $\pm\frac{5}{2}$.

35. $9x^2 = 2$

$x^2 = \frac{2}{9}$

$x = \sqrt{\frac{2}{9}}$ or $x = -\sqrt{\frac{2}{9}}$

$x = \frac{\sqrt{2}}{3}$ $x = -\frac{\sqrt{2}}{3}$

The solutions are $\pm\frac{\sqrt{2}}{3}$.

37. $x^2 + 4 = 6$

$x^2 = 2$

$x = \sqrt{2}$ or $x = -\sqrt{2}$

The solutions are $\pm\sqrt{2}$.

39. $m^2 + 7 = 5$

$m^2 = -2$

There is no real-number solution because there is no real number whose square is negative.

41. $(x-5)^2 = 0$

$x - 5 = \sqrt{0}$

$x - 5 = 0$

$x = 5$

The solution is 5.

43. $(z-7)^2 = 4$

$z - 7 = \sqrt{4}$ or $z - 7 = -\sqrt{4}$

$z - 7 = 2$ $z - 7 = -2$

$z = 9$ $z = 5$

The solutions are 5 and 9.

45. $(4a-3)^2=4$

$4a-3=\sqrt{4}$ or $4a-3=-\sqrt{4}$

$4a-3=2$ $\quad$ $4a-3=-2$

$4a=5$ $\quad$ $4a=1$

$a=\frac{5}{4}$ $\quad$ $a=\frac{1}{4}$

The solutions are $\frac{1}{4}$ and $\frac{5}{4}$.

47. $x^2=1.69$

$x=\sqrt{1.69}$ or $x=-\sqrt{1.69}$

$x=1.3$ $\quad$ $x=-1.3$

The solutions are ±1.3.

49. $(x+3)^2-1=3$

$(x+3)^2=4$

$x+3=\sqrt{4}$ or $x+3=-\sqrt{4}$

$x+3=2$ $\quad$ $x+3=-2$

$x=-1$ $\quad$ $x=-5$

The solutions are –5 and –1.

51. $2(m-4)^2-6=12$

$2(m-4)^2=18$

$(m-4)^2=9$

$m-4=\sqrt{9}$ or $m-4=-\sqrt{9}$

$m-4=3$ $\quad$ $m-4=-3$

$m=7$ $\quad$ $m=1$

The solutions are 1 and 7.

53. $x^2+10x+25=9$

$(x+5)^2=9$

$x+5=\sqrt{9}$ or $x+5=-\sqrt{9}$

$x+5=3$ $\quad$ $x+5=-3$

$x=-2$ $\quad$ $x=-8$

The solutions are –8 and –2.

55. $9x^2-6x+1=144$

$(3x-1)^2=144$

$3x-1=\sqrt{144}$ or $3x-1=-\sqrt{144}$

$3x-1=12$ $\quad$ $3x-1=-12$

$3x=13$ $\quad$ $3x=-11$

$x=\frac{13}{3}$ $\quad$ $x=-\frac{11}{3}$

The solutions are $-\frac{11}{3}$ and $\frac{13}{3}$.

57. $(x-7)^2-5=1$

$(x-7)^2=6$

$x-7=\sqrt{6}$ or $x-7=-\sqrt{6}$

$x=7+\sqrt{6}$ $\quad$ $x=7-\sqrt{6}$

The solutions are $7\pm\sqrt{6}$.

59. $(2x+1)^2-3=7$

$(2x+1)^2=10$

$2x+1=\sqrt{10}$ or $2x+1=-\sqrt{10}$

$2x=-1+\sqrt{10}$ $\quad$ $2x=-1-\sqrt{10}$

$x=\frac{-1+\sqrt{10}}{2}$ $\quad$ $x=\frac{-1-\sqrt{10}}{2}$

The solutions are $\frac{-1\pm\sqrt{10}}{2}$.

61. $(x+3)^2-6=6$

$(x+3)^2=12$

$x+3=\sqrt{12}$ or $x+3=-\sqrt{12}$

$x+3=2\sqrt{3}$ $\quad$ $x+3=-2\sqrt{3}$

$x=-3+2\sqrt{3}$ $\quad$ $x=-3-2\sqrt{3}$

The solutions are $-3\pm2\sqrt{3}$.

63. $s=-16t^2+v_0t+s_0$

$0=-16t^2+0t+400$

$16t^2=400$

$t^2=\frac{400}{16}$

$t^2=25$

$t=\pm\sqrt{25}$

$t=5$

Use the positive square root. It will take 5 seconds for the balloon to hit the ground.

65. $s=-16t^2+v_0t+s_0$

$5000=-16t^2+0t+12{,}000$

$16t^2=7000$

$t^2=\frac{7000}{16}$

$t=\pm\sqrt{\frac{7000}{16}}$

$t=\frac{10\sqrt{70}}{4}$

$t = \frac{5\sqrt{70}}{2} \approx 20.9$
Use the positive square root.
It will take $\frac{5\sqrt{70}}{2}$ or approximately 20.9 seconds.

67. $a^2 + b^2 = c^2$
$a^2 + (60)^2 = (61)^2$
$a^2 = 121$
$a = \pm\sqrt{121}$
$a = 11$
Use the positive square root.
It is 11 inches high.

69. $a^2 + b^2 = c^2$
$a^2 + (3000)^2 = (5000)^2$
$a^2 = 16,000,000$
$a = \pm\sqrt{16,000,000}$
$a = 4000$
Use the positive square root.
The distance is 4000 ft.

71. $a^2 + b^2 = c^2$
$x^2 + x^2 = (50)^2$
$2x^2 = 2500$
$x^2 = 1250$
$x = \pm\sqrt{1250}$
$x = \pm\sqrt{625 \cdot 2}$
$x = 25\sqrt{2}$
Use the positive square root.
The distance is $25\sqrt{2}$ or approximately 35.36 feet.

73. $C(x) = 0.044x^2 + 4700$
$9000 = 0.044x^2 + 4700$
$4300 = 0.044x^2$
$\frac{4300}{0.044} = x^2$
$\pm\sqrt{\frac{4300}{0.044}} = x$
$312.61 \approx x$
Use the positive square root.
She manages about 313 accounts.

11.3 Experiencing Algebra the Calculator Way

Original Pizza Diameter	Twice as Large Pizza Diameter
5 inches	7 inches
6 inches	8 inches
7 inches	10 inches
8 inches	11 inches
9 inches	13 inches
10 inches	14 inches
11 inches	16 inches
12 inches	17 inches

11.4 Experiencing Algebra the Exercise Way

1. x^2+6x

$b=\frac{6}{2}=3$

$b^2=3^2=9$

We need to add 9.

3. x^2-3x

$b=-\frac{3}{2}$

$b^2=\left(-\frac{3}{2}\right)^2=\frac{9}{4}$

We need to add $\frac{9}{4}$.

5. $x^2+\frac{3}{4}x$

$b=\frac{3}{4}\cdot\frac{1}{2}=\frac{3}{8}$

$b^2=\left(\frac{3}{8}\right)^2=\frac{9}{64}$

We need to add $\frac{9}{64}$.

7. x^2+x

$b=\frac{1}{2}$

$b^2=\left(\frac{1}{2}\right)^2=\frac{1}{4}$

We need to add $\frac{1}{4}$.

9. x^2-6x

$b=-\frac{6}{2}=-3$

$b^2=(-3)^2=9$

We need to add 9.

11. x^2+9x

$b=\frac{9}{2}$

$b^2=\left(\frac{9}{2}\right)^2=\frac{81}{4}$

We need to add $\frac{81}{4}$.

13. $x^2+\frac{8}{9}x$

$b=\frac{8}{9}\cdot\frac{1}{2}=\frac{4}{9}$

$b^2=\left(\frac{4}{9}\right)^2=\frac{16}{81}$

We need to add $\frac{16}{81}$.

15. x^2-14x

$b=-\frac{14}{2}=-7$

$b^2=(-7)^2=49$

We need to add 49.

17. $x^2+6x-20=35$

$x^2+6x=55$

$x^2+6x+9=55+9$

$(x+3)^2=64$

$x+3=\sqrt{64}$ or $x+3=-\sqrt{64}$

$x+3=8$ $\quad$ $x+3=-8$

$x=5$ $\quad$ $x=-11$

The solutions are −11 and 5.

19. $x^2-3x=28$

$x^2-3x+\frac{9}{4}=28+\frac{9}{4}$

$\left(x-\frac{3}{2}\right)^2=\frac{121}{4}$

$x-\frac{3}{2}=\sqrt{\frac{121}{4}}$ or $x-\frac{3}{2}=-\sqrt{\frac{121}{4}}$

$x-\frac{3}{2}=\frac{11}{2}$ $\quad$ $x-\frac{3}{2}=-\frac{11}{2}$

$x=\frac{14}{2}$ $\quad$ $x=-\frac{8}{2}$

$x=7$ $\quad$ $x=-4$

The solutions are −4 and 7.

21. $x^2 + \frac{4}{7}x + \frac{3}{49} = 0$

$x^2 + \frac{4}{7}x = -\frac{3}{49}$

$x^2 + \frac{4}{7}x + \frac{4}{49} = -\frac{3}{49} + \frac{4}{49}$

$\left(x + \frac{2}{7}\right)^2 = \frac{1}{49}$

$x + \frac{2}{7} = \sqrt{\frac{1}{49}}$ or $x + \frac{2}{7} = -\sqrt{\frac{1}{49}}$

$x + \frac{2}{7} = \frac{1}{7}$ $\qquad$ $x + \frac{2}{7} = -\frac{1}{7}$

$x = -\frac{1}{7}$ $\qquad$ $x = -\frac{3}{7}$

The solutions are $-\frac{3}{7}$ and $-\frac{1}{7}$.

23. $x^2 + x - 30 = 60$

$x^2 + x = 90$

$x^2 + x + \frac{1}{4} = 90 + \frac{1}{4}$

$\left(x + \frac{1}{2}\right)^2 = \frac{361}{4}$

$x + \frac{1}{2} = \sqrt{\frac{361}{4}}$ or $x + \frac{1}{2} = -\sqrt{\frac{361}{4}}$

$x + \frac{1}{2} = \frac{19}{2}$ $\qquad$ $x + \frac{1}{2} = -\frac{19}{2}$

$x = \frac{18}{2}$ $\qquad$ $x = -\frac{20}{2}$

$x = 9$ $\qquad$ $x = -10$

The solutions are –10 and 9.

25. $x^2 - 6x = 2$

$x^2 - 6x + 9 = 2 + 9$

$(x - 3)^2 = 11$

$x - 3 = \sqrt{11}$ or $x - 3 = -\sqrt{11}$

$x = 3 + \sqrt{11}$ $\qquad$ $x = 3 - \sqrt{11}$

The solutions are $3 \pm \sqrt{11}$.

27. $x^2 + 9x = 1$

$x^2 + 9x + \frac{81}{4} = 1 + \frac{81}{4}$

$\left(x + \frac{9}{2}\right)^2 = \frac{85}{4}$

$x + \frac{9}{2} = \sqrt{\frac{85}{4}}$ or $x + \frac{9}{2} = -\sqrt{\frac{85}{4}}$

$x + \frac{9}{2} = \frac{\sqrt{85}}{2}$ $\qquad$ $x + \frac{9}{2} = -\frac{\sqrt{85}}{2}$

$x = \frac{-9 + \sqrt{85}}{2}$ $\qquad$ $x = \frac{-9 - \sqrt{85}}{2}$

The solutions are $\frac{-9 \pm \sqrt{85}}{2}$.

29. $x^2 + \frac{8}{9}x = 2$

$x^2 + \frac{8}{9}x + \frac{16}{81} = 2 + \frac{16}{81}$

$\left(x + \frac{4}{9}\right)^2 = \frac{178}{81}$

$x + \frac{4}{9} = \sqrt{\frac{178}{81}}$ or $x + \frac{4}{9} = -\sqrt{\frac{178}{81}}$

$x + \frac{4}{9} = \frac{\sqrt{178}}{9}$ $\qquad$ $x + \frac{4}{9} = -\frac{\sqrt{178}}{9}$

$x = \frac{-4 + \sqrt{178}}{9}$ $\qquad$ $x = \frac{-4 - \sqrt{178}}{9}$

The solutions are $\frac{-4 \pm \sqrt{178}}{9}$

31. $x^2 - x - 5 = 0$

$x^2 - x = 5$

$x^2 - x + \frac{1}{4} = 5 + \frac{1}{4}$

$\left(x - \frac{1}{2}\right)^2 = \frac{21}{4}$

$x-\frac{1}{2}=\sqrt{\frac{21}{4}}$ or $x-\frac{1}{2}=-\sqrt{\frac{21}{4}}$

$x-\frac{1}{2}=\frac{\sqrt{21}}{2}$ $\quad$ $x-\frac{1}{2}=-\frac{\sqrt{21}}{2}$

$x=\frac{1+\sqrt{21}}{2}$ $\quad$ $x=\frac{1-\sqrt{21}}{2}$

The solutions are $\frac{1\pm\sqrt{21}}{2}$.

33. $x^2+x+10=0$

$x^2+x=-10$

$x^2+x+\frac{1}{4}=-10+\frac{1}{4}$

$\left(x+\frac{1}{2}\right)^2=-\frac{39}{4}$

There is no real-number solution because there is no real number whose square is negative.

35. $2x^2+6x-1=0$

$2x^2+6x=1$

$x^2+3x=\frac{1}{2}$

$x^2+3x+\frac{9}{4}=\frac{1}{2}+\frac{9}{4}$

$\left(x+\frac{3}{2}\right)^2=\frac{11}{4}$

$x+\frac{3}{2}=\sqrt{\frac{11}{4}}$ or $x+\frac{3}{2}=-\sqrt{\frac{11}{4}}$

$x+\frac{3}{2}=\frac{\sqrt{11}}{2}$ $\quad$ $x+\frac{3}{2}=-\frac{\sqrt{11}}{2}$

$x=\frac{-3+\sqrt{11}}{2}$ $\quad$ $x=\frac{-3-\sqrt{11}}{2}$

The solutions are $\frac{-3\pm\sqrt{11}}{2}$.

37. $3x^2+x-7=0$

$3x^2+x=7$

$x^2+\frac{1}{3}x=\frac{7}{3}$

$x^2+\frac{1}{3}x+\frac{1}{36}=\frac{7}{3}+\frac{1}{36}$

$\left(x+\frac{1}{6}\right)^2=\frac{85}{36}$

$x+\frac{1}{6}=\sqrt{\frac{85}{36}}$ or $x+\frac{1}{6}=-\sqrt{\frac{85}{36}}$

$x+\frac{1}{6}=\frac{\sqrt{85}}{6}$ $\quad$ $x+\frac{1}{6}=-\frac{\sqrt{85}}{6}$

$x=\frac{-1+\sqrt{85}}{6}$ $\quad$ $x=\frac{-1-\sqrt{85}}{6}$

The solutions are $\frac{-1\pm\sqrt{85}}{6}$.

39. $x^2-14x+55=6$

$x^2-14x=-49$

$x^2-14x+49=-49+49$

$(x-7)^2=0$

$x-7=0$

$x=7$

The solution is 7.

41. $4x^2-20x+30=5$

$4x^2-20x=-25$

$x^2-5x=-\frac{25}{4}$

$x^2-5x+\frac{25}{4}=-\frac{25}{4}+\frac{25}{4}$

$\left(x-\frac{5}{2}\right)^2=0$

$x-\frac{5}{2}=0$

$x=\frac{5}{2}$

The solution is $\frac{5}{2}$.

43. $\frac{1}{2}x^2+5x-2=0$

$\frac{1}{2}x^2+5x=2$

$x^2+10x=4$

$x^2+10x+25=4+25$

$(x+5)^2=29$

$x+5=\sqrt{29}$ or $x+5=-\sqrt{29}$

$x=-5+\sqrt{29}$ $\quad$ $x=-5-\sqrt{29}$

The solutions are $-5\pm\sqrt{29}$.

45. $\frac{1}{3}x^2+\frac{1}{9}x-\frac{1}{6}=0$

$\frac{1}{3}x^2+\frac{1}{9}x=\frac{1}{6}$

$x^2+\frac{1}{3}x=\frac{1}{2}$

$x^2+\frac{1}{3}x+\frac{1}{36}=\frac{1}{2}+\frac{1}{36}$

$\left(x+\frac{1}{6}\right)^2=\frac{19}{36}$

$x+\frac{1}{6}=\sqrt{\frac{19}{36}}$ or $x+\frac{1}{6}=-\sqrt{\frac{19}{36}}$

$x+\frac{1}{6}=\frac{\sqrt{19}}{6}$ $\quad$ $x+\frac{1}{6}=-\frac{\sqrt{19}}{6}$

$x=\frac{-1+\sqrt{19}}{6}$ $\quad$ $x=\frac{-1-\sqrt{19}}{6}$

The solutions are $\frac{-1\pm\sqrt{19}}{6}$.

47. $\frac{2}{3}x^2-2x-\frac{5}{6}=0$

$\frac{2}{3}x^2-2x=\frac{5}{6}$

$x^2-3x=\frac{5}{4}$

$x^2-3x+\frac{9}{4}=\frac{5}{4}+\frac{9}{4}$

$\left(x-\frac{3}{2}\right)^2=\frac{14}{4}$

$x-\frac{3}{2}=\sqrt{\frac{14}{4}}$ or $x-\frac{3}{2}=-\sqrt{\frac{14}{4}}$

$x-\frac{3}{2}=\frac{\sqrt{14}}{2}$ $\quad$ $x-\frac{3}{2}=-\frac{\sqrt{14}}{2}$

$x=\frac{3+\sqrt{14}}{2}$ $\quad$ $x=\frac{3-\sqrt{14}}{2}$

The solutions are $\frac{3\pm\sqrt{14}}{2}$.

49. Let x = number of cars purchased

$20-x$ = cost per car

$x(20-x)=75$

$20x-x^2=75$

$x^2-20x=-75$

$x^2-20x+100=-75+100$

$(x-10)^2=25$

$x-10=\sqrt{25}$ or $x-10=-\sqrt{25}$

$x-10=5$ $\quad$ $x-10=-5$

$x=15$ $\quad$ $x=5$

The number of cars is 15 because the dealer will not sell cars for less than \$12.

51. $D(x)=\frac{1}{3}x^2-64x+3100$

$2000=\frac{1}{3}x^2-64x+3100$

$-1100=\frac{1}{3}x^2-64x$

$-3300=x^2-192x$

$-3300+9216=x^2-192x+9216$

$5916=(x-96)^2$

$x-96=\sqrt{5916}$ or $x-96=-\sqrt{5916}$

$x=96+\sqrt{5916}$ $\quad$ $x=96-\sqrt{5916}$

$x\approx 173$ $\quad$ $x\approx 19$

Disregard the price over \$50. The price should be set at \$19.

53. $C(x) = 0.11x^2 + 3.08x + 11$
$55 = 0.11x^2 + 3.08x + 11$
$44 = 0.11x^2 + 3.08x$
$400 = x^2 + 28x$
$400 + 196 = x^2 + 28x + 196$
$596 = (x + 14)^2$
$x + 14 = \sqrt{596}$ or $x + 14 = -\sqrt{596}$
$x = -14 + \sqrt{596}$ $x = -14 - \sqrt{596}$
$x \approx 10.4$ $x = -38.4$
Disregard a negative amount. The contract should be about \$10,400,000.

55. Let x = width
$x + 4$ = length
$x(x + 4) = 117$
$x^2 + 4x = 117$
$x^2 + 4x + 4 = 117 + 4$
$(x + 2)^2 = 121$
$x + 2 = \sqrt{121}$ or $x + 2 = -\sqrt{121}$
$x + 2 = 11$ $x + 2 = -11$
$x = 9$ $x = -13$
Disregard a negative width. The width is 9 in. and the length is 13 in.

57. Let x = length
$x - 5$ = width
$x(x - 5) = 85$
$x^2 - 5x = 85$
$x^2 - 5x + \frac{25}{4} = 85 + \frac{25}{4}$
$\left(x - \frac{5}{2}\right)^2 = \frac{365}{4}$
$x - \frac{5}{2} = \sqrt{\frac{365}{4}}$ or $x - \frac{5}{2} = -\sqrt{\frac{365}{4}}$
$x = \frac{5 + \sqrt{365}}{2}$ $x = \frac{5 - \sqrt{365}}{2}$
$x \approx 12.1$ $x \approx -7.1$
Disregard a negative length. The length is 12.1 ft and the width is 7.1 ft.

59. **a.** $s = -16t^2 + v_0t + s_0$
$0 = -16t^2 + 32t + 16$

b. $16t^2 - 32t - 16 = 0$
$t^2 - 2t - 1 = 0$
$t^2 - 2t = 1$
$t^2 - 2t + 1 = 1 + 1$
$(t - 1)^2 = 2$
$t - 1 = \sqrt{2}$ or $t - 1 = -\sqrt{2}$
$t = 1 + \sqrt{2}$ $t = 1 - \sqrt{2}$
Disregard a negative time. It will take $1 + \sqrt{2}$ seconds.

c. $1 + \sqrt{2} \approx 2.4$
It will take approximately 2.4 seconds.

11.4 Experiencing Algebra the Calculator Way

1. $x^2 + 8x + 15 = 0$
$x^2 + 8x = -15$
$x^2 + 8x + 16 = -15 + 16$
$(x + 4)^2 = 1$
$x + 4 = 1$ or $x + 4 = -1$
$x = -3$ or $x = -5$
$Y1 = x^2 + 8x + 15$
$Y2 = 0$

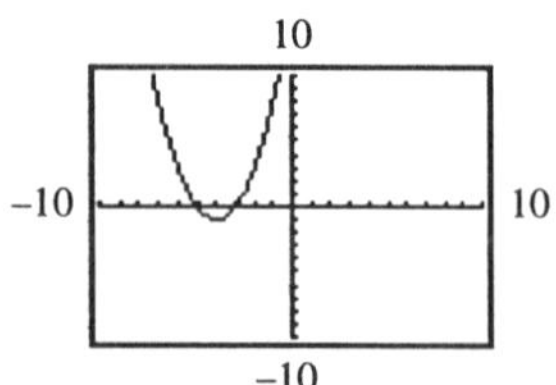

The solutions are –5 and –3.

2. $x^2 + 4x + 7 = 0$
$x^2 + 4x = -7$
$x^2 + 4x + 4 = -7 + 4$
$(x + 2)^2 = -3$
There is no real number solution.
$Y1 = x^2 + 4x + 7$
$Y2 = 0$

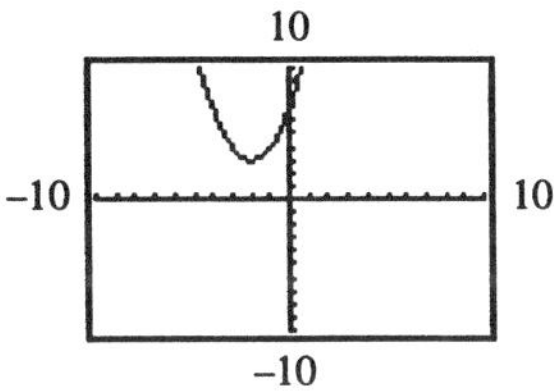

There is no real-number solution.

3. $x^2 + x - 1 = 0$

$x^2 + x = 1$

$x^2 + x + \frac{1}{4} = 1 + \frac{1}{4}$

$\left(x + \frac{1}{2}\right)^2 = \frac{5}{4}$

$x + \frac{1}{2} = \sqrt{\frac{5}{4}}$ or $x + \frac{1}{2} = -\sqrt{\frac{5}{4}}$

$x = \frac{-1 + \sqrt{5}}{2}$ $\quad$ $x = \frac{-1 - \sqrt{5}}{2}$

$\text{Y1} = x^2 + x - 1$

$\text{Y2} = 0$

The solutions are approximately −1.618 and 0.618.

4. $2x^2 + 11x + 12 = 0$

$2x^2 + 11x = -12$

$x^2 + \frac{11}{2}x = -6$

$x^2 + \frac{11}{2}x + \frac{121}{16} = -6 + \frac{121}{16}$

$\left(x + \frac{11}{4}\right)^2 = \frac{25}{16}$

$x + \frac{11}{4} = \sqrt{\frac{25}{16}}$ or $x + \frac{11}{4} = -\sqrt{\frac{25}{16}}$

$x = -\frac{6}{4}$ $\quad$ $x = -\frac{16}{4}$

$x = -\frac{3}{2}$ $\quad$ $x = -4$

$\text{Y1} = 2x^2 + 11x + 12$

$\text{Y2} = 0$

The solutions are −4 and −1.5.

5. $2x^2 + 6x + 7 = 0$

$2x^2 + 6x = -7$

$x^2 + 3x = -\frac{7}{2}$

$x^2 + 3x + \frac{9}{4} = -\frac{7}{2} + \frac{9}{4}$

$\left(x + \frac{3}{2}\right)^2 = -\frac{5}{4}$

$\text{Y1} = 2x^2 + 6x + 7$

$\text{Y2} = 0$

There is no real-number solution.

6. $8x^2 + 8x - 5 = 0$

$8x^2 + 8x = 5$

$x^2 + x = \frac{5}{8}$

$x^2 + x + \frac{1}{4} = \frac{5}{8} + \frac{1}{4}$

$\left(x + \frac{1}{2}\right)^2 = \frac{7}{8}$

$x + \frac{1}{2} = \sqrt{\frac{7}{8}}$ or $x + \frac{1}{2} = -\sqrt{\frac{7}{8}}$

$x = -\frac{1}{2} + \frac{\sqrt{7}}{2\sqrt{2}} \cdot \frac{\sqrt{2}}{\sqrt{2}}$ $\quad$ $x = -\frac{1}{2} - \frac{\sqrt{7}}{2\sqrt{2}} \cdot \frac{\sqrt{2}}{\sqrt{2}}$

$x = \frac{-2 + \sqrt{14}}{4}$ $\quad$ $x = \frac{-2 - \sqrt{14}}{4}$

$Y1 = 8x^2 + 8x - 5$
$Y2 = 0$

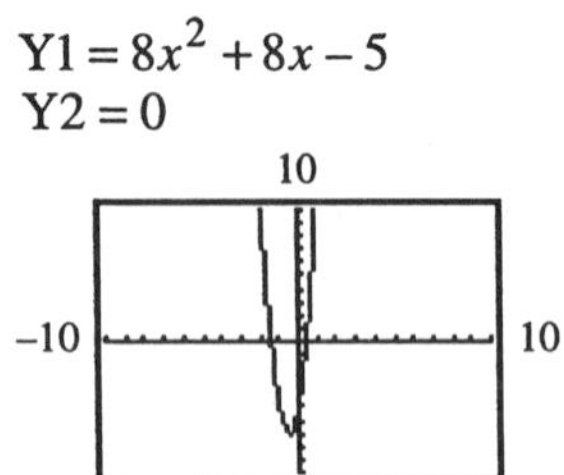

The solutions are approximately –1.435 and 0.435.

11.5 Experiencing Algebra the Exercise Way

1. $x^2 - 12x + 27 = 0$

$x = \frac{-b \pm \sqrt{b^2 - 4ac}}{2a}$

$x = \frac{-(-12) \pm \sqrt{(-12)^2 - 4(1)(27)}}{2(1)}$

$x = \frac{12 \pm \sqrt{144 - 108}}{2}$

$x = \frac{12 \pm \sqrt{36}}{2}$

$x = \frac{12 \pm 6}{2}$

$x = \frac{12 - 6}{2}$ $\quad$ $x = \frac{12 + 6}{2}$

$x = 3$ $\quad$ $x = 9$

The solutions are 3 and 9.

3. $2x^2 + 3x - 15 = 2x + 6$

$2x^2 + x - 21 = 0$

$x = \frac{-b \pm \sqrt{b^2 - 4ac}}{2a}$

$x = \frac{-1 \pm \sqrt{1^2 - 4(2)(-21)}}{2(2)}$

$x = \frac{-1 \pm \sqrt{1 + 168}}{4}$

$x = \frac{-1 \pm \sqrt{169}}{4}$

$x = \frac{-1 \pm 13}{4}$

$x = \frac{-1 - 13}{4}$ $\quad$ $x = \frac{-1 + 13}{4}$

$x = -\frac{7}{2}$ $\quad$ $x = 3$

The solutions are $-\frac{7}{2}$ and 3.

5. $2z^2 + 11z + 5 = 0$

$z = \frac{-b \pm \sqrt{b^2 - 4ac}}{2a}$

$z = \frac{-11 \pm \sqrt{(11)^2 - 4(2)(5)}}{2(2)}$

$z = \frac{-11 \pm \sqrt{121 - 40}}{4}$

$z = \frac{-11 \pm \sqrt{81}}{4}$

$z = \frac{-11 \pm 9}{4}$

$z = \frac{-11 - 9}{4}$ $\quad$ $z = \frac{-11 + 9}{4}$

$z = -5$ $\quad$ $z = -\frac{1}{2}$

The solutions are –5 and $-\frac{1}{2}$.

7. $8(2p^2 - p + 1) = 7$

$16p^2 - 8p + 1 = 0$

$p = \frac{-b \pm \sqrt{b^2 - 4ac}}{2a}$

$p = \frac{-(-8) \pm \sqrt{(-8)^2 - 4(16)(1)}}{2(16)}$

$p = \frac{8 \pm \sqrt{64 - 64}}{32}$

$p = \frac{8 \pm \sqrt{0}}{32}$

$p = \frac{8}{32}$

$p = \frac{1}{4}$

The solution is $\frac{1}{4}$.

9. $x^2+3x+4=0$

$$x=\frac{-b\pm\sqrt{b^2-4ac}}{2a}$$
$$x=\frac{-3\pm\sqrt{3^2-4(1)(4)}}{2(1)}$$
$$x=\frac{-3\pm\sqrt{9-16}}{2}$$
$$x=\frac{-3\pm\sqrt{-7}}{2}$$

There is no real-number solution.

11. $-5a^2+4a-7=0$

$$a=\frac{-b\pm\sqrt{b^2-4ac}}{2a}$$
$$a=\frac{-4\pm\sqrt{4^2-4(-5)(-7)}}{2(-5)}$$
$$a=\frac{-4\pm\sqrt{16-140}}{-10}$$
$$a=\frac{-4\pm\sqrt{-124}}{-10}$$

There is no real-number solution.

13. $v^2-5v+2=0$

$$v=\frac{-b\pm\sqrt{b^2-4ac}}{2a}$$
$$v=\frac{-(-5)\pm\sqrt{(-5)^2-4(1)(2)}}{2(1)}$$
$$v=\frac{5\pm\sqrt{25-8}}{2}$$
$$v=\frac{5\pm\sqrt{17}}{2}$$

The solutions are $\frac{5\pm\sqrt{17}}{2}$.

15. $x(x-4)+1=0$

$x^2-4x+1=0$

$$x=\frac{-b\pm\sqrt{b^2-4ac}}{2a}$$
$$x=\frac{-(-4)\pm\sqrt{(-4)^2-4(1)(1)}}{2(1)}$$
$$x=\frac{4\pm\sqrt{16-4}}{2}$$
$$x=\frac{4\pm\sqrt{12}}{2}$$
$$x=\frac{4\pm2\sqrt{3}}{2}$$
$$x=2\pm\sqrt{3}$$

The solutions are $2\pm\sqrt{3}$.

17. $4x(x+1)+6=6-x$

$4x^2+5x+0=0$

$$x=\frac{-b\pm\sqrt{b^2-4ac}}{2a}$$
$$x=\frac{-5\pm\sqrt{5^2-4(4)(0)}}{2(4)}$$
$$x=\frac{-5\pm\sqrt{25}}{8}$$
$$x=\frac{-5\pm5}{8}$$

$x=\frac{-5-5}{8}$ $x=\frac{-5+5}{8}$

$x=-\frac{5}{4}$ $x=0$

The solutions are $-\frac{5}{4}$ and 0.

19. $3d^2+10=17$

$3d^2+0d-7=0$

$$d=\frac{-b\pm\sqrt{b^2-4ac}}{2a}$$
$$d=\frac{-0\pm\sqrt{0^2-4(3)(-7)}}{2(3)}$$
$$d=\frac{0\pm\sqrt{84}}{6}$$
$$d=\frac{\pm2\sqrt{21}}{6}$$
$$d=\pm\frac{\sqrt{21}}{3}$$

The solutions are $\pm\frac{\sqrt{21}}{3}$.

21. $16m = m^2 + 55$

$m^2 - 16m + 55 = 0$

$m = \dfrac{-b \pm \sqrt{b^2 - 4ac}}{2a}$

$m = \dfrac{-(-16) \pm \sqrt{(-16)^2 - 4(1)(55)}}{2(1)}$

$m = \dfrac{16 \pm \sqrt{256 - 220}}{2}$

$m = \dfrac{16 \pm \sqrt{36}}{2}$

$m = \dfrac{16 \pm 6}{2}$

$m = \dfrac{16 - 6}{2}$ or $m = \dfrac{16 + 6}{2}$

$m = 5$ $\quad$ $m = 11$

The solutions are 5 and 11.

23. $(x - 4)(x + 4) = 2(x - 4)$

$x^2 - 2x - 8 = 0$

$x = \dfrac{-b \pm \sqrt{b^2 - 4ac}}{2a}$

$x = \dfrac{-(-2) \pm \sqrt{(-2)^2 - 4(1)(-8)}}{2(1)}$

$x = \dfrac{2 \pm \sqrt{4 + 32}}{2}$

$x = \dfrac{2 \pm \sqrt{36}}{2}$

$x = \dfrac{2 \pm 6}{2}$

$x = \dfrac{2 - 6}{2}$ or $x = \dfrac{2 + 6}{2}$

$x = -2$ $\quad$ $x = 4$

The solutions are –2 and 4.

25. $x^2 - 6.3x + 7.2 = 0$

$x = \dfrac{-b \pm \sqrt{b^2 - 4ac}}{2a}$

$x = \dfrac{-(-6.3) \pm \sqrt{(-6.3)^2 - 4(1)(7.2)}}{2(1)}$

$x = \dfrac{6.3 \pm \sqrt{39.69 - 28.8}}{2}$

$x = \dfrac{6.3 \pm \sqrt{10.89}}{2}$

$x = \dfrac{6.3 \pm 3.3}{2}$

$x = \dfrac{6.3 - 3.3}{2}$ or $x = \dfrac{6.3 + 3.3}{2}$

$x = 1.5$ $\quad$ $x = 4.8$

The solutions are 1.5 and 4.8.

27. $1.8 - 5.6x - x^2 = 0$

$x^2 + 5.6x - 1.8 = 0$

$x = \dfrac{-b \pm \sqrt{b^2 - 4ac}}{2a}$

$x = \dfrac{-5.6 \pm \sqrt{(5.6)^2 - 4(1)(-1.8)}}{2(1)}$

$x = \dfrac{-5.6 \pm \sqrt{31.36 + 7.2}}{2}$

$x = \dfrac{-5.6 \pm \sqrt{38.56}}{2}$

$x = \dfrac{-5.6 \pm 2\sqrt{9.64}}{2}$

$x = -2.8 \pm \sqrt{9.64}$

The solutions are $-2.8 \pm \sqrt{9.64}$

29. $a^2 + 24.01 = 9.8a$

$a^2 - 9.8a + 24.01 = 0$

$a = \dfrac{-(-9.8) \pm \sqrt{(-9.8)^2 - 4(1)(24.01)}}{2(1)}$

$a = \dfrac{9.8 \pm \sqrt{96.04 - 96.04}}{2}$

$a = \dfrac{9.8 \pm 0}{2}$

$a = \dfrac{9.8}{2}$

$a = 4.9$

The solution is 4.9.

31. $1.7z^2 + 1.3z + 5.6 = 0$

$z = \dfrac{-b \pm \sqrt{b^2 - 4ac}}{2a}$

$z = \dfrac{-1.3 \pm \sqrt{(1.3)^2 - 4(1.7)(5.6)}}{2(1.7)}$

$z = \dfrac{-1.3 \pm \sqrt{1.69 - 38.08}}{3.4}$

$z = \dfrac{-1.3 \pm \sqrt{-36.39}}{3.4}$

There is no real-number solution.

33. $x^2 - 11x + 24 = 0$

$b^2 - 4ac = (-11)^2 - 4(1)(24) = 25$

There are two rational solutions because 25 is a perfect square.

35. $a^2 + 12a + 36 = 0$

$b^2 - 4ac = (12)^2 - 4(1)(36) = 0$

There is one rational solution because the discriminant is 0.

37. $z^2 = 4z - 5$

$z^2 - 4z + 5 = 0$

$b^2 - 4ac = (-4)^2 - 4(1)(5) = -4$

There is no real-number solution because the discriminant is negative.

39. $6x^2 - 11x - 7 = 0$

$b^2 - 4ac = (-11)^2 - 4(6)(-7) = 289$

There are two rational solutions because 289 is a perfect square.

41. $7p^2 - 15 = 0$

$b^2 - 4ac = 0^2 - 4(7)(-15) = 420$

There are two irrational solutions because 420 is positive but not a perfect square.

43. $1 - 5x - 4x^2 = 0$

$4x^2 + 5x - 1 = 0$

$b^2 - 4ac = (5)^2 - 4(4)(-1) = 41$

There are two irrational solutions because 41 is positive but not a perfect square.

45. $6.25z - 2.1 = 2.5z^2$

$2.5z^2 - 6.25z + 2.1 = 0$

$b^2 - 4ac = (-6.25)^2 - 4(2.5)(2.1) = 18.0625$

There are two rational solutions because 18.0625 is a perfect square.

47. $0.3x - 2.8 = 1.7x^2$

$1.7x^2 - 0.3x + 2.8 = 0$

$b^2 - 4ac = (-0.3)^2 - 4(1.7)(2.8) = -18.95$

There are no real-number solutions because the discriminant is negative.

49. $A = P(1 + r)^t$

$6000 = 5000(1 + r)^2$

$\dfrac{6}{5} = 1 + 2r + r^2$

$r^2 + 2r - \dfrac{1}{5} = 0$

$r = \dfrac{-2 \pm \sqrt{(2)^2 - 4(1)\left(-\frac{1}{5}\right)}}{2(1)}$

$r = \dfrac{-2 \pm \sqrt{4 + \frac{4}{5}}}{2}$

$r = \dfrac{-2 \pm \sqrt{4.8}}{2}$

$r = \dfrac{-2 - \sqrt{4.8}}{2}$ or $r = \dfrac{-2 + \sqrt{4.8}}{2}$

$r \approx -2.095$ $\quad r \approx 0.095$

Disregard a negative value. The interest rate must be about 9.5%.

51. $R(x) = 6000 + 70x - x^2$

$7125 = 6000 + 70x - x^2$

$x^2 - 70x + 1125 = 0$

$x = \dfrac{70 \pm \sqrt{(-70)^2 - 4(1)(1125)}}{2(1)}$

$x = \dfrac{70 \pm \sqrt{400}}{2}$

$x = \dfrac{70 \pm 20}{2}$

$x = \dfrac{70 - 20}{2}$ or $x = \dfrac{70 + 20}{2}$

$x = 25$ $\quad x = 45$

The values of x are 25 and 45.

Answers will vary.

53. $y = 0.086x^2 - 9.34x + 277.78$
$100 = 0.086x^2 - 9.34x + 277.78$
$0 = 0.086x^2 - 9.34x + 177.78$
$x = \dfrac{9.34 \pm \sqrt{(-9.34)^2 - 4(0.086)(177.78)}}{2(0.086)}$
$x = \dfrac{9.34 \pm \sqrt{26.07928}}{0.172}$
$x = \dfrac{9.34 - \sqrt{26.07928}}{0.172} \approx 24.61$ or
$x = \dfrac{9.34 + \sqrt{26.07928}}{0.172} \approx 83.99$
$1950 + 25 = 1975$
The year was 1975.

11.5 Experiencing Algebra the Calculator Way

Equation	Value of Discriminant	Type of Roots	Number of Unlike Roots	Roots
1. $x^2 + 6 = 5x$	1	rational	2	2, 3
2. $9x^2 + 6x = -1$	0	rational	1	$-\frac{1}{3}$
3. $2x^2 + 1 = 7x$	41	irrational	2	0.149, 3.351
4. $x^2 + 6x = -10$	–4	not real	2	not real
5. $x^2 = 6 - x$	25	rational	2	–3, 2
6. $5x^2 - 6x = 0$	36	rational	2	0, 1.2
7. $x^2 + 0.36 = 1.2x$	0	rational	1	0.6
8. $1.7x^2 + x + 1.9 = 0$	–11.92	not real	2	not real
9. $1.5x^2 + 1.2x = 3.6$	23.04	rational	2	–2, 1.2
10. $\frac{1}{4}x^2 + x = \frac{1}{8}$	1.125	irrational	2	–4.121, 0.121
11. $x^2 - \frac{1}{6}x = \frac{1}{6}$	$0.69\overline{4}$	rational	2	$-0.\overline{3}, 0.5$
12. $\frac{1}{5}x^2 + \frac{2}{3}x = -\frac{7}{8}$	$-0.2\overline{5}$	not real	2	not real

11.6 Experiencing Algebra the Exercise Way

1. Let x = length of one leg
   $x + 3$ = length of other leg
   $a^2 + b^2 = c^2$
   $x^2 + (x+3)^2 = (15)^2$
   $x^2 + x^2 + 6x + 9 = 225$
   $2x^2 + 6x - 216 = 0$
   $2(x^2 + 3x - 108) = 0$
   $2(x+12)(x-9) = 0$
   $x + 12 = 0$ or $x - 9 = 0$
   $x = -12$ $x = 9$
   Disregard a negative length.
   The legs are 9 inches and 12 inches.

3. Let x = length of one leg
   $2x + 2$ = length of other leg
   $a^2 + b^2 = c^2$
   $x^2 + (2x+2)^2 = (13)^2$
   $x^2 + 4x^2 + 8x + 4 = 169$
   $5x^2 + 8x - 165 = 0$
   $(5x + 33)(x - 5) = 0$
   $5x + 33 = 0$ or $x - 5 = 0$
   $x = -\frac{33}{5}$ $x = 5$
   Disregard a negative length.
   The legs are 5 feet and 12 feet.

5. Let x = length of one leg
   $2x - 3$ = length of hypotenuse
   $a^2 + b^2 = c^2$
   $x^2 + (45)^2 = (2x-3)^2$
   $x^2 + 2025 = 4x^2 - 12x + 9$
   $3x^2 - 12x - 2016 = 0$
   $3(x^2 - 4x - 672) = 0$
   $3(x + 24)(x - 28) = 0$
   $x + 24 = 0$ or $x - 28 = 0$
   $x = -24$ $x = 28$
   Disregard a negative length.
   The leg is 28 cm and the hypotenuse is 53 cm.

7. $a^2 + b^2 = c^2$
   $(99)^2 + x^2 = (195)^2$
   $9801 + x^2 = 38{,}025$
   $x^2 = 28{,}224$
   $x = \sqrt{28{,}224}$ or $x = -\sqrt{28{,}224}$
   $x = 168$ $x = -168$
   Disregard a negative distance.
   The distance is 168 yards.

9. Let x = length of side
   $a^2 + b^2 = c^2$
   $x^2 + x^2 = (20)^2$
   $2x^2 = 400$
   $x^2 = 200$
   $x = \pm\sqrt{200}$
   $x \approx \pm 14.1$
   Disregard a negative length.
   They should each measure approximately 14.1 feet.

11. Let x = width
    $3x - 5$ = length
    $a^2 + b^2 = c^2$
    $x^2 + (3x-5)^2 = (70)^2$
    $x^2 + 9x^2 - 30x + 25 = 4900$
    $10x^2 - 30x - 4875 = 0$
    $5\left(2x^2 - 6x - 975\right) = 0$
    $x = \frac{-(-6) \pm \sqrt{(-6)^2 - 4(2)(-975)}}{2(2)}$
    $x = \frac{6 \pm \sqrt{7836}}{4}$
    $x = \frac{6 - \sqrt{7836}}{4}$ $x = \frac{6 + \sqrt{7836}}{4}$
    $x \approx -20.6$ $x \approx 23.6$
    Disregard a negative length.
    The width is about 23.6 feet and the length is about 65.8 feet.

13. $a^2 + b^2 = c^2$
    $(65)^2 + (65)^2 = c^2$
    $8450 = c^2$
    $\pm\sqrt{8450} = c$
    $\pm 91.9 \approx c$
    Disregard a negative distance.
    The distance is about 91.9 feet.

15. Let x = distance beyond first base
$a^2+b^2=c^2$
$(65+x)^2+(65)^2=(97)^2$
$4225+130x+x^2+4225=9409$
$x^2+130x-959=0$
$(x+137)(x-7)=0$
$x+137=0$ or $x-7=0$
$x=-137$ $\quad x=7$
Disregard a negative distance.
It was 7 feet beyond first base.

17. $a^2+b^2=c^2$
$(2640)^2+x^2=(2640.5)^2$
$x^2=2640.25$
$x=\pm\sqrt{2640.5}$
$x\approx\pm 51.4$
Disregard a negative distance.
It would rise approximately 51.4 inches.

19. $c=a\sqrt{2}$
$c=3.75\sqrt{2}$
$c\approx 5.30$
The hypotenuse measures approximately 5.30 m.

21. $c=a\sqrt{2}$
$7.6=a\sqrt{2}$
$\frac{7.6}{\sqrt{2}}=a$
$a=\frac{7.6\sqrt{2}}{2}$
$a=3.8\sqrt{2}$
$a\approx 5.37$
The lengths are approximately 5.37 ft.

23. $c=a\sqrt{2}$
$c=20\sqrt{2}$
$c\approx 28.28$
The wire is approximately 28.28 ft.

25. $c=a\sqrt{2}$
$42.5=a\sqrt{2}$
$\frac{42.5}{\sqrt{2}}=a$
$a=\frac{42.5\sqrt{2}}{2}$
$a=21.25\sqrt{2}$
$a\approx 30.05$
The building is approximately 30.05 ft tall.

27. $c=a\sqrt{2}$
$40=a\sqrt{2}$
$\frac{40}{\sqrt{2}}=a$
$a=\frac{40\sqrt{2}}{2}$
$a=20\sqrt{2}$
$a\approx 28.28$
Yes, 28.28 ft is less than 30 ft.

29. $b=a\sqrt{3}$
$18.6=a\sqrt{3}$
$\frac{18.6}{\sqrt{3}}=a$
$a=\frac{18.6\sqrt{3}}{3}$
$a=6.2\sqrt{3}\approx 10.7$
$2a=12.4\sqrt{3}\approx 21.5$
The lengths are approximately 10.7 in. and 21.5 in.

31. $c=2a$
$22=2a$
$a=11$
$a\sqrt{3}=11\sqrt{3}\approx 19.1$
The lengths are 11 in. and approximately 19.1 in.

33. $a=21$
$2a=42$
$a\sqrt{3}=21\sqrt{3}\approx 36.4$
The lengths are 42 cm and approximately 36.4 cm.

35. $a=4000$
$a\sqrt{3}=4000\sqrt{3}\approx 6928.2$
$2a=8000$
The horizontal distance is about 6928.2 ft and the slanted distance is 8000 ft.

37. $a = 8$
$2a = 16$
$a\sqrt{3} = 8\sqrt{3} \approx 13.9$
The height of the pole is approximately 13.9 ft and the length of the wire is 16 ft.

39. $E = mc^2$ for c
$$\frac{E}{m} = c^2$$
$$c = \pm\sqrt{\frac{E}{m}}$$
$$c = \pm\frac{\sqrt{Em}}{m}$$

41. $V = \pi r^2 h$ for r
$$\frac{V}{\pi h} = r^2$$
$$r = \pm\sqrt{\frac{V}{\pi h}}$$
$$r = \pm\frac{\sqrt{V\pi h}}{\pi h}$$

43. $E = \frac{1}{2}mv^2$ for v
$$\frac{2E}{m} = v^2$$
$$v = \pm\sqrt{\frac{2E}{m}}$$
$$v = \pm\frac{\sqrt{2Em}}{m}$$

45. $C = \frac{n(n-1)}{2}$ for n
$$2C = n^2 - n$$
$$n^2 - n - 2C = 0$$
$$n = \frac{-(-1) \pm \sqrt{(-1)^2 - 4(1)(-2C)}}{2(1)}$$
$$n = \frac{1 \pm \sqrt{1+8C}}{2}$$

47. $x^2 - y^2 = c^2$ for x
$$x^2 = y^2 + c^2$$
$$x = \pm\sqrt{y^2 + c^2}$$

49. $A = P(1+r)^2$ for r
$$\frac{A}{P} = (1+r)^2$$
$$1 + r = \pm\sqrt{\frac{A}{P}}$$
$$1 + r = \pm\frac{\sqrt{AP}}{P}$$
$$r = -1 \pm \frac{\sqrt{AP}}{P}$$

11.6 Experiencing Algebra the Calculator Way

1. $c = 34.45$, $a = 23.87$
$b \approx 24.84$ cm

2. $a = 184.8$, $b = 5927.56$
$c = 5930.44$ in.

3. $c = 5769$, $a = 1800$
$b = 5481$ ft

4. $a = 9639$, $b = 9720$
$c = 13{,}689$ dm

5. $c = \frac{5}{64}$, $a = \frac{1}{16}$
$b = \frac{3}{64}$ ft

11.7 Experiencing Algebra the Exercise Way

1. $2x^2 - 5 < 4x + 1$ is a quadratic inequality because it can be written in standard form $2x^2 - 4x - 6 < 0$.

3. $1.3z - 2.8 > 4.3z^2$ is a quadratic inequality because it can be written in standard form $4.3z^2 - 1.3z + 2.8 < 0$.

5. $4x^{-2} + 5x^{-1} \geq 2x + 1$ is not a quadratic inequality because it has variables with negative exponents.

7. $\frac{1}{3}x^2 - \frac{5}{6}x > \frac{3}{4}$ is a quadratic inequality because it can be written in standard form $\frac{1}{3}x^2 - \frac{5}{6}x - \frac{3}{4} > 0$.

9. $2x^3 + 4 \le x^2 + x$ is not a quadratic inequality because it has a third-degree term.

11. $3\sqrt{x} + 5x < x^2 - 5$ is not a quadratic inequality because it has a square root of a variable.

13. $x^2 + 3x - 7 \ge x + 8$

$\mathrm{Y1} = x^2 + 3x - 7$

$\mathrm{Y2} = x + 8$

X	Y1	Y2
-7	21	1
-6	11	2
-5	3	3
-4	-3	4
-3	-7	5
-2	-9	6
-1	-9	7

X=-7

X	Y1	Y2
0	-7	8
1	-3	9
2	3	10
3	11	11
4	21	12
5	33	13
6	47	14

X=0

The solution set is all integers less than or equal to –5 and greater than or equal to 3.

15. $3x^2 + 9x + 7 < 11 - 2x$

$\mathrm{Y1} = 3x^2 + 9x + 7$

$\mathrm{Y2} = 11 - 2x$

X	Y1	Y2
-7	91	25
-6	61	23
-5	37	21
-4	19	19
-3	7	17
-2	1	15
-1	1	13

X=-7

X	Y1	Y2
0	7	11
1	19	9
2	37	7
3	61	5
4	91	3
5	127	1
6	169	-1

X=0

The solution set is all integers greater than –4 and less than 1.

17. $4x^2 - 13x - 7 > 18x + 1$

$\mathrm{Y1} = 4x^2 - 13x - 7$

$\mathrm{Y2} = 18x + 1$

X	Y1	Y2
-2	35	-35
-1	10	-17
0	-7	1
1	-16	19
2	-17	37
3	-10	55
4	5	73

X=-2

X	Y1	Y2
5	28	91
6	59	109
7	98	127
8	145	145
9	200	163
10	263	181
11	334	199

X=5

The solution set is all integers less than 0 or greater than 8.

19. $x^2 + 2x + 1 > 4$

$\mathrm{Y1} = x^2 + 2x + 1$

$\mathrm{Y2} = 4$

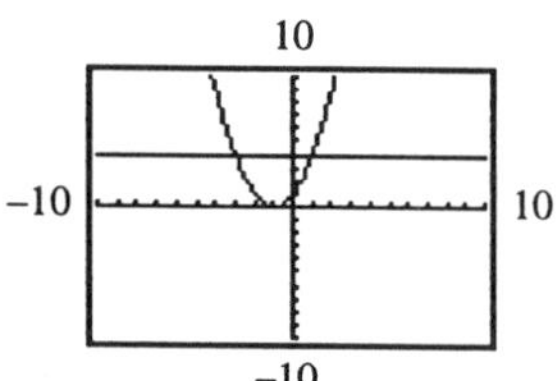

The intersections are (–3, 4) and (1, 4). The solution is $(-\infty, -3) \cup (1, \infty)$.

21. $2x^2 + x - 3 < 7$

$\mathrm{Y1} = 2x^2 + x - 3$

$\mathrm{Y2} = 7$

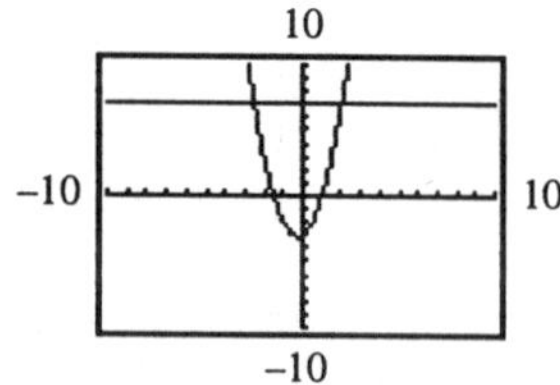

The intersections are (–2.5, 7) and (2, 7). The solution is (–2.5, 2).

23. $24+5x-x^2<-x+8$
$Y1=24+5x-x^2$
$Y2=-x+8$

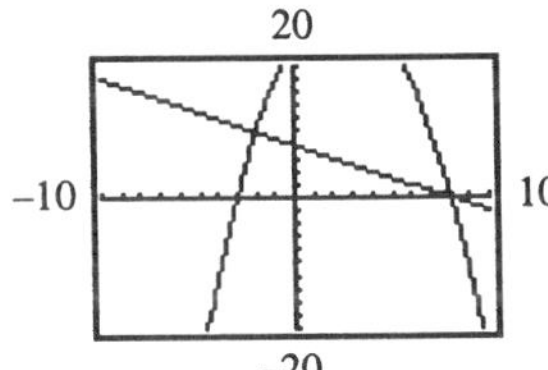

The interstctions are (–2, 10) and (8, 0).
The solution is $(-\infty,-2)\cup(8,\infty)$

25. $2x^2-5\le x^2+4$
$Y1=2x^2-5$
$Y2=x^2+4$

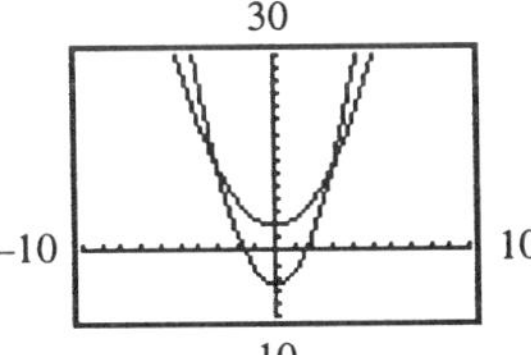

The intersections are (–3, 13) and (3, 13).
The solution is [–3, 3].

27. $\frac{1}{4}x^2-3\ge x^2+2$
$Y1=\frac{1}{4}x^2-3$
$Y2=x^2+2$

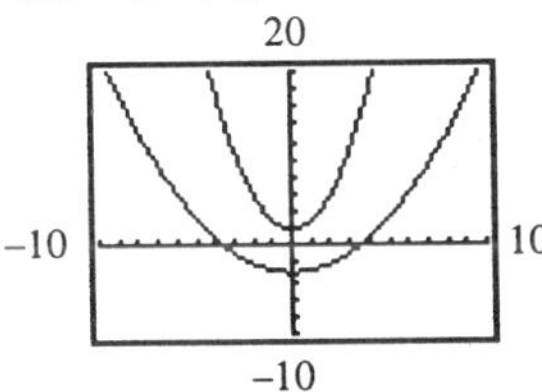

The graphs do not intersect. Y1 is always less than Y2.
There is no solution.

29. $x^2+6x+10\ge -x^2-6x-8$
$Y1=x^2+6x+10$
$Y2=-x^2-6x-8$

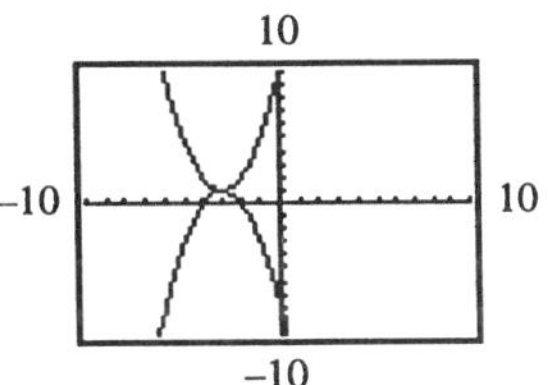

The intersection is (–3, 1). Y1 is always greater than or equl to Y2. The solution is $(-\infty, \infty)$.

31. $x^2-4x+8\ge\frac{1}{4}x^2-x+5$
$Y1=x^2-4x+8$
$Y2=\frac{1}{4}x^2-x+5$

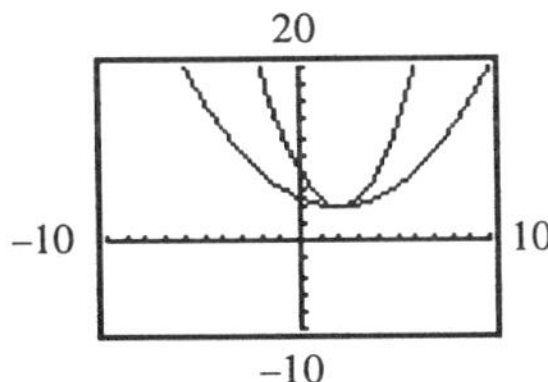

The intersection is (2, 4).
Y1 is always greater than or equal to Y2.
The solution is $(-\infty,\infty)$.

33. $x^2-2x-15\ge 9$
Determine the critical points.
$x^2-2x-15=9$
$x^2-2x-24=0$
$(x-6)(x+4)=0$
$x=6$ or $x=-4$
Possible intervals:
$(-\infty, -4]$, $[-4, 6]$, $[6, \infty)$
Use a test point in each interval to determine the solution is $(-\infty, -4]\cup[6, \infty)$.

35. $8+2x-x^2>-7$
Determine the critical points.
$8+2x-x^2=-7$
$x^2-2x-15=0$
$(x-5)(x+3)=0$
$x=5$ or $x=-3$
Possible intervals: $(-\infty, -3)$, $(-3, 5)$, $(5, \infty)$
Use a test point in each interval to determine the solution is $(-3, 5)$.

37. $4x^2 - 12x + 9 \le 4$
Determine the critical points.
$4x^2 - 12x + 9 = 4$
$4x^2 - 12x + 5 = 0$
$(2x - 1)(2x - 5) = 0$
$x = \frac{1}{2}$ or $x = \frac{5}{2}$
Possible intervals:
$\left(-\infty, \frac{1}{2}\right], \left[\frac{1}{2}, \frac{5}{2}\right], \left[\frac{5}{2}, \infty\right)$
Use a test point in each interval to determine the solution is $\left[\frac{1}{2}, \frac{5}{2}\right]$.

39. $6 + 5x - x^2 < -8$
Determine the critical points.
$6 + 5x - x^2 = -8$
$x^2 - 5x - 14 = 0$
$(x - 7)(x + 2) = 0$
$x = 7$ or $x = -2$
Possible intervals:
$(-\infty, -2), (-2, 7), (7, \infty)$
Use a test point in each interval to determine the solution is $(-\infty, -2) \cup (7, \infty)$.

41. $x^2 + 3 \ge -x + 15$
Determine the critical points.
$x^2 + 3 = -x + 15$
$x^2 + x - 12 = 0$
$(x + 4)(x - 3) = 0$
$x = -4$ or $x = 3$
Possible intervals:
$(-\infty, -4], [-4, 3], [3, \infty)$.
Use a test point in each interval to determine the solution is $(-\infty, -4] \cup [3, \infty)$.

43. $10 - 2x^2 < 4x + 4$
Determine the critical points.
$10 - 2x^2 = 4x + 4$
$2x^2 + 4x - 6 = 0$
$2(x + 3)(x - 1) = 0$
$x = -3$ or $x = 1$
Possible intervals:
$(-\infty, -3), (-3, 1), (1, \infty)$
Use a test point in each interval to determine the solution is $(-\infty, -3) \cup (1, \infty)$.

45. $x^2 - x - 6 \le 12 + 2x - 2x^2$
Determine the critical points.
$x^2 - x - 6 = 12 + 2x - 2x^2$
$3x^2 - 3x - 18 = 0$
$3(x + 2)(x - 3) = 0$
$x = -2$ or $x = 3$
Possible intervals:
$(-\infty, -2], [-2, 3], [3, \infty)$
Use a test point in each interval to determine the solution is [–2, 3].

47. $x^2 - 3x + \frac{21}{4} \ge 2x^2 - 6x + \frac{15}{2}$
Determine the critical points.
$x^2 - 3x + \frac{21}{4} = 2x^2 - 6x + \frac{15}{2}$
$x^2 - 3x + \frac{9}{4} = 0$
$\left(x - \frac{3}{2}\right)^2 = 0$
$x = \frac{3}{2}$
Possible intervals:
$\left(-\infty, \frac{3}{2}\right], \left[\frac{3}{2}, \infty\right)$
Use a test point in each interval to determine neither interval is a solution.
So the solution is only $\frac{3}{2}$.

49. $x^2 - x - \frac{7}{4} > -2x^2 + 2x - \frac{5}{2}$
Determine the critical points.
$x^2 - x - \frac{7}{4} = -2x^2 + 2x - \frac{5}{2}$
$3x^2 - 3x + \frac{3}{4} = 0$
$3\left(x^2 - x + \frac{1}{4}\right) = 0$
$3\left(x - \frac{1}{2}\right)^2 = 0$
$x = \frac{1}{2}$

Possible intervals:

$\left(-\infty, \frac{1}{2}\right), \left(\frac{1}{2}, \infty\right)$

Use a test point in each interval to determine both are solutions. The solution is

$\left(-\infty, \frac{1}{2}\right) \cup \left(\frac{1}{2}, \infty\right).$

51. $x^2 + 4x + 8 \geq 4$

Determine the critical points.

$x^2 + 4x + 8 = 4$

$x^2 + 4x + 4 = 0$

$(x+2)^2 = 0$

$x = -2$

Possible intervals:

$(-\infty, -2], [-2, \infty)$

Use a test point in each interval to determine that both intervals are solutions. Therefore, the solution is $(-\infty, \infty)$.

53. $2 + 4x - x^2 > 10 - 4x + x^2$

Determine the critical points.

$2 + 4x - x^2 = 10 - 4x + x^2$

$2x^2 - 8x + 8 = 0$

$2(x-2)^2 = 0$

$x = 2$

Possible intervals:

$(-\infty, 2), (2, \infty)$

Use a test point in each interval to determine neither interval is a solution. Therefore, there is no solution.

55. $s(t) = -16t^2 + v_0t + s_0$

$s(t) = -16t^2 + 30t + 25$

Solve:

$25 < -16t^2 + 30t + 25$

Determine the critical points.

$25 = -16t^2 + 30t + 25$

$0 = -16t^2 + 30t$

$0 = -2t(8t - 15)$

$t = 0$ or $t = \frac{15}{8}$

Possible intervals:

$(-\infty, 0), \left(0, \frac{15}{8}\right), \left(\frac{15}{8}, \infty\right)$

Use a test point in each interval to determine the solution is $\left(0, \frac{15}{8}\right)$.

The egg is higher than the stand between 0 and $\frac{15}{8}$ seconds.

57. $s(t) = -16t^2 + v_0t + s_0$

$s(t) = -16t^2 + 75t + 1200$

Solve:

$800 \geq -16t^2 + 75t + 1200$

Determine the intersection points.

$Y1 = 800$

$Y2 = -16x^2 + 75x + 1200$

$Y3 = 0$

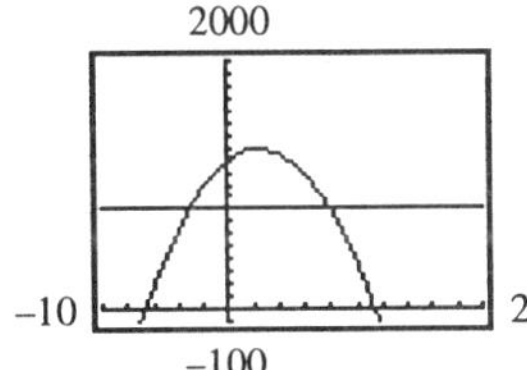

The intersections are about (–3.178, 800), (7.866, 800) and (11.316, 0).

Since time cannot be negative, the possible intervals are:

[0, 7.866] or [7.866, 11.316]

The graph of Y1 is above the graph of Y2 on the interval [7.866, 11.316].

The rock is no more than 800 feet from about 7.866 seconds to 11.316 seconds.

59. $R(x) = (5 + x)(78 - 3x)$

$R(x) = -3x^2 + 63x + 390$

$600 \leq -3x^2 + 63x + 390$

Determine the intersection points.

$Y1 = 600$

$Y2 = -3x^2 + 63x + 390$

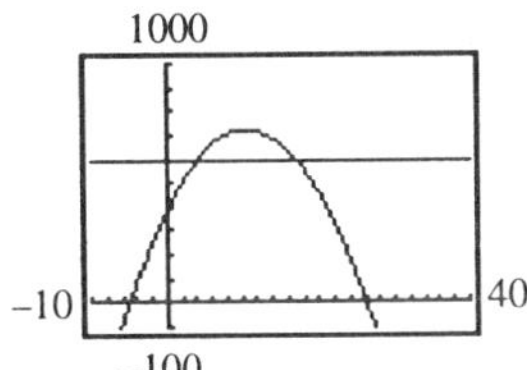

The intersections are about (4.2, 600) and (16.8, 600).
The possible intervals are [0, 4.2], [4.2, 16.8], $[16.8, \infty)$.
The graph of Y1 is below the graph of Y2 on the interval [4.2, 16.8].
Since x is the number of stereos over 5 ordered, the number ordered must be greater than 9 and less than 22 stereos.

61. $C(x) = 3x^2 + x + 35$
$100 \geq 3x^2 + x + 35$
Determine the intersection points.
$\text{Y1} = 100$
$\text{Y2} = 3x^2 + x + 35$

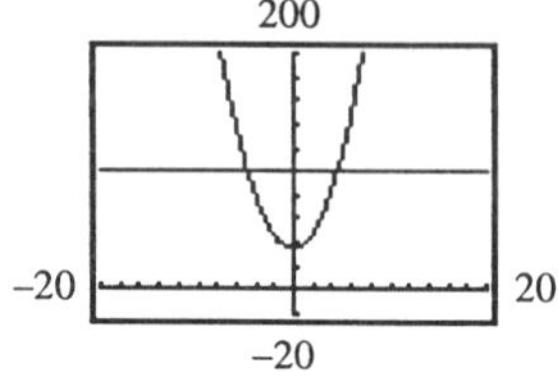

The intersections are about (–4.8, 100) and (4.5, 100).
Since the number of days must be positive, the possible intervals are:
$(0, 4.5], [4.5, \infty)$.
The graph of Y1 is above the graph of Y2 on the interval (0, 4.5].
The booth can be operated from 1 day to 4 days.

63. $y = 0.058x^2 - 0.79x + 6.125$
$10 < 0.058x^2 - 0.79x + 6.125$
Determine the intersection points.
$\text{Y1} = 10$
$\text{Y2} = 0.058x^2 - 0.79x + 6.125$

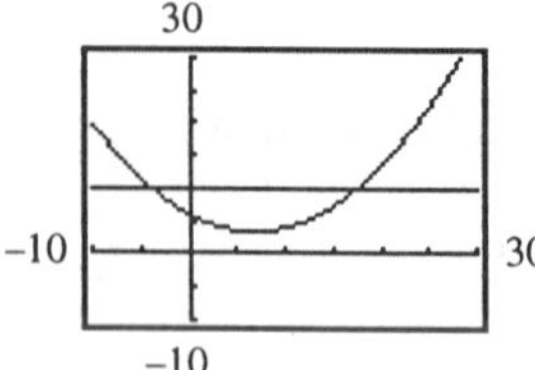

The intersections are about (–3.8, 10) and (17.4, 10).
Since time can only be positive, the possible intervals are (0, 17.4) and $(17.4, \infty)$.
The graph of Y1 is below the graph of Y2 on the interval $(17.4, \infty)$.
1970 + 18 = 1988
The years are from 1988 and on.

11.7 Experiencing Algebra the Calculator Way

Answers will vary. Students should check the solutions of the previous exercises.

Chapter 11 Review

Reflections

1.–6. Answers will vary.

Exercises

1. $2x^2 + 5x - 16 = 7x + 8$
$\text{Y1} = 2x^2 + 5x - 16$
$\text{Y2} = 7x + 8$

X	Y1	Y2
-5	9	-27
-4	-4	-20
-3	-13	-13
-2	-18	-6
-1	-19	1
0	-16	8
1	-9	15

X=-5

X	Y1	Y2
2	2	22
3	17	29
4	36	36
5	59	43
6	86	50
7	117	57
8	152	64

X=2

The solutions are –3 and 4.

2. $\frac{1}{3}x^2 + 5x + 6 = x - 3$
$\text{Y1} = \frac{1}{3}x^2 + 5x + 6$
$\text{Y2} = x - 3$

X	Y1	Y2
-12	-6	-15
-11	-8.667	-14
-10	-10.67	-13
-9	-12	-12
-8	-12.67	-11
-7	-12.67	-10
-6	-12	-9

X=-12

X	Y1	Y2
-5	-10.67	-8
-4	-8.667	-7
-3	-6	-6
-2	-2.667	-5
-1	1.3333	-4
0	6	-3
1	11.333	-2

X=-5

The solutions are –9 and –3.

3. $0.7x^2 - 2.5x + 4.6 = 1.7x + 1.1$

$Y1 = 0.7x^2 - 2.5x + 4.6$

$Y2 = 1.7x + 1.1$

X	Y1	Y2
0	4.6	1.1
1	2.8	2.8
2	2.4	4.5
3	3.4	6.2
4	5.8	7.9
5	9.6	9.6
6	14.8	11.3

X=0

The solutions are 1 and 5.

4. $x^2 - 4x + 4 = 9$

$Y1 = x^2 - 4x + 4$

$Y2 = 9$

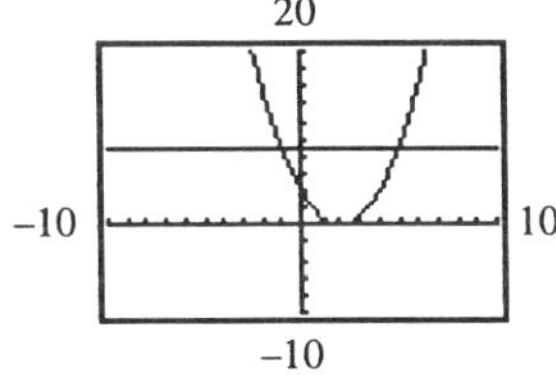

The graphs intersect at (–1, 9) and (5, 9).
The solutions are –1 and 5.

5. $x^2 - 6 = \frac{1}{4}x^2 + 6$

$Y1 = x^2 - 6$

$Y2 = \frac{1}{4}x^2 + 6$

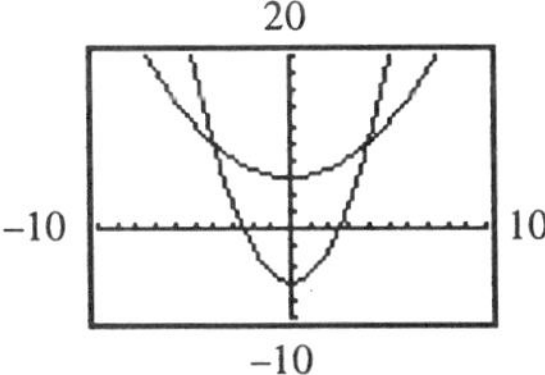

The graphs intersect at (–4, 10) and (4, 10).
The solutions are –4 and 4.

6. $x^2 + 3x - 5 = 7 - 2x - x^2$

$Y1 = x^2 + 3x - 5$

$Y2 = 7 - 2x - x^2$

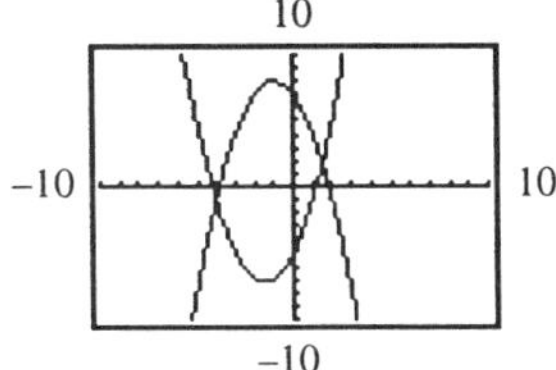

The graphs intersect at (–4, –1) and (1.5, 1.75).
The solutions are –4 and 1.5.

7. $-0.3x^2 + 4x + 5 = 2.2x - 11.5$

$Y1 = -0.3x^2 + 4x + 5$

$Y2 = 2.2x - 11.5$

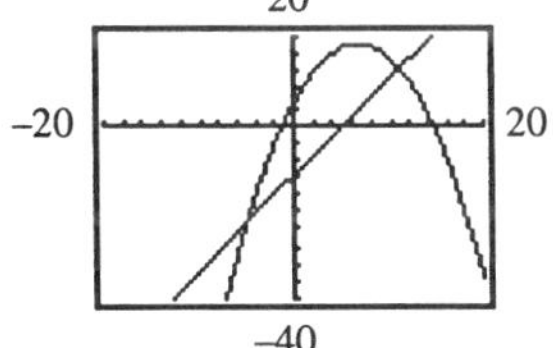

The graphs intersect at (–5, –22.5) and (11, 12.7).
The solutions are –5 and 11.

8. $x^2 + 8x + 8 = 0$

$Y1 = x^2 + 8x + 8$

$Y2 = 0$

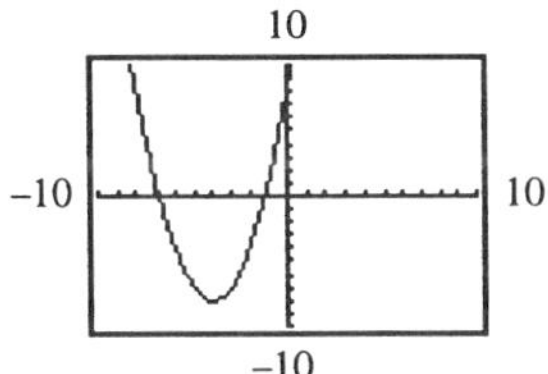

The graphs intersect at about (–6.83, 0) and (–1.17, 0).
The solutions are approximately 6.83 and –1.17.

9. $\frac{1}{5}x^2 - 5x + 4 = 8 - \frac{1}{3}x^2$

$Y1 = \frac{1}{5}x^2 - 5x + 4$

$Y2 = 8 - \frac{1}{3}x^2$

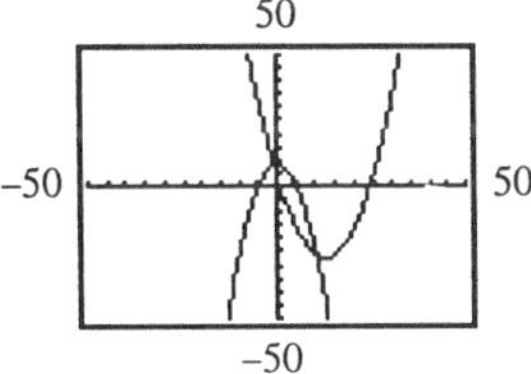

The graphs intersect at about (–0.74, 7.82) and (10.12, –26.11).
The solutions are approximately –0.74 and 10.12.

10. $s = -16t^2 + v_0t + s_0$
$0 = -16t^2 + 0t + 96$
$0 = -16t^2 + 96$
$Y1 = 0$
$Y2 = -16x^2 + 96$

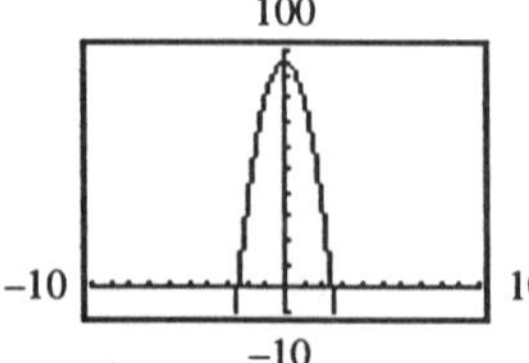

The graphs intersect at about (–2.45, 0) and (2.45, 0).
Disgregard a negative time.
It will reach the ground in approximately 2.45 seconds.

11. $C(x) = 8x^2 + 20x + 450$
$1850 = 8x^2 + 20x + 450$
$Y1 = 1850$
$Y2 = 8x^2 + 20x + 450$

2000
–50 50
–100

The graphs intersect at about (–14.54, 1850) and (12.04, 1850).
Disregard a negative number of items.
12 items can be produced.

12. $x^2 - 5 = x + 1$
$x^2 - x - 6 = 0$
$(x - 3)(x + 2) = 0$
$x - 3 = 0$ or $x + 2 = 0$
$x = 3$ $\quad x = -2$
The solutions are –2 and 3.

13. $6x^2 - x - 77 = 0$
$(2x + 7)(3x - 11) = 0$
$2x + 7 = 0$ or $3x - 11 = 0$
$x = -\frac{7}{2}$ $\quad x = \frac{11}{3}$
The solutions are $-\frac{7}{2}$ and $\frac{11}{3}$.

14. $2x + 10 = x^2 - 5x + 2$
$x^2 - 7x - 8 = 0$
$(x - 8)(x + 1) = 0$
$x - 8 = 0$ or $x + 1 = 0$
$x = 8$ $\quad x = -1$
The solutions are –1 and 8.

15. $x^2 - 7x - 60 = 0$
$(x + 5)(x - 12) = 0$
$x + 5 = 0$ or $x - 12 = 0$
$x = -5$ $\quad x = 12$
The solutions are –5 and 12.

16. $3x^2 + 5x - 3 = 1 - 6x$
$3x^2 + 11x - 4 = 0$
$(3x - 1)(x + 4) = 0$
$3x - 1 = 0$ or $x + 4 = 0$
$x = \frac{1}{3}$ $\quad x = -4$
The solutions are –4 and $\frac{1}{3}$.

17. $6x^2 - 8x = 9x - 12$
$6x^2 - 17x + 12 = 0$
$(3x - 4)(2x - 3) = 0$
$3x - 4 = 0$ or $2x - 3 = 0$
$x = \frac{4}{3}$ $\quad x = \frac{3}{2}$
The solutions are $\frac{4}{3}$ and $\frac{3}{2}$.

18. $\frac{1}{4}x^2 + \frac{3}{2}x - \frac{19}{8} = \frac{3}{8}x + 2$
$\frac{1}{4}x^2 + \frac{9}{8}x - \frac{35}{8} = 0$
$8\left(\frac{1}{4}x^2 + \frac{9}{8}x - \frac{35}{8}\right) = 8 \cdot 0$
$2x^2 + 9x - 35 = 0$
$(2x - 5)(x + 7) = 0$
$2x - 5 = 0$ or $x + 7 = 0$
$x = \frac{5}{2}$ $\quad x = -7$
The solutions are $\frac{5}{2}$ and –7.

19. $9x^2 - 49 = 0$
$(3x - 7)(3x + 7) = 0$
$3x - 7 = 0$ or $3x + 7 = 0$
$x = \frac{7}{3}$ $\quad x = -\frac{7}{3}$
The solutions are $\pm\frac{7}{3}$.

20. $s = -16t^2 + v_0 t + s_0$
$0 = -16t^2 + 48t + 16$
$0 = -16(t^2 - 3t - 1)$
$t^2 - 3t - 1 = 0$
The equation does not factor.
$Y1 = x^2 - 3x - 1$
$Y2 = 0$

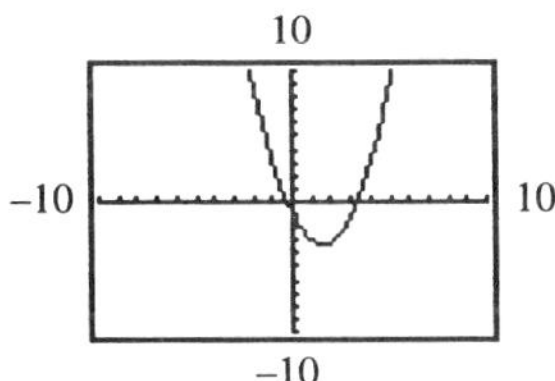

The graphs intersect at about (–0.303, 0) and (3.303, 0).
Disregard a negative time.
The ball will hit the ground in approximately 3.3 seconds.

21. $\sqrt{25} = \sqrt{5^2} = 5$

22. $-\sqrt{49} = -\sqrt{7^2} = -7$

23. $\sqrt{5}\sqrt{45} = \sqrt{5 \cdot 45} = \sqrt{225} = \sqrt{(15)^2} = 15$

24. $\sqrt{15}\sqrt{21} = \sqrt{315} = \sqrt{9 \cdot 35} = 3\sqrt{35}$

25. $\sqrt{8} = \sqrt{4 \cdot 2} = 2\sqrt{2}$

26. $\sqrt{\frac{64}{81}} = \frac{\sqrt{64}}{\sqrt{81}} = \frac{8}{9}$

27. $\sqrt{\frac{13}{25}} = \frac{\sqrt{13}}{\sqrt{25}} = \frac{\sqrt{13}}{5}$

28. $\sqrt{\frac{25}{3}} = \frac{\sqrt{25}}{\sqrt{3}} = \frac{5}{\sqrt{3}} \cdot \frac{\sqrt{3}}{\sqrt{3}} = \frac{5\sqrt{3}}{3}$

29. $\sqrt{\frac{2}{5}} = \frac{\sqrt{2}}{\sqrt{5}} \cdot \frac{\sqrt{5}}{\sqrt{5}} = \frac{\sqrt{10}}{5}$

30. $\sqrt{\frac{15}{21}} = \sqrt{\frac{5}{7}} = \frac{\sqrt{5}}{\sqrt{7}} \cdot \frac{\sqrt{7}}{\sqrt{7}} = \frac{\sqrt{35}}{7}$

31. $\sqrt{\frac{6}{50}} = \sqrt{\frac{3}{25}} = \frac{\sqrt{3}}{\sqrt{25}} = \frac{\sqrt{3}}{5}$

32. $\frac{\sqrt{75}}{\sqrt{3}} = \sqrt{\frac{75}{3}} = \sqrt{25} = 5$

33. $4x^2 - 100 = 0$
$4x^2 = 100$
$x^2 = 25$
$x = \sqrt{25}$ or $x = -\sqrt{25}$
$x = 5$ $\quad x = -5$
The solutions are ± 5.

34. $a^2 + 5 = 9$
$a^2 = 4$
$a = \sqrt{4}$ or $a = -\sqrt{4}$
$a = 2$ $\quad a = -2$
The solutions are ± 2.

35. $x^2 + 3 = 15$
$x^2 = 12$
$x = \sqrt{12}$ or $x = -\sqrt{12}$
$x = \sqrt{4 \cdot 3}$ $\quad x = -\sqrt{4 \cdot 3}$
$x = 2\sqrt{3}$ $\quad x = -2\sqrt{3}$
The solutions are $\pm 2\sqrt{3}$.

36. $15 + b^2 = 7$
$b^2 = -8$
There is no real-number solution because there is no real number whose square is negative.

37. $(x-4)^2 = 9$

$x-4=\sqrt{9}$ or $x-4=-\sqrt{9}$

$x-4=3$ $\quad x-4=-3$

$x=7$ $\quad x=1$

The solutions are 1 and 7.

38. $x^2+18x+81=16$

$(x+9)^2=16$

$x+9=\sqrt{16}$ or $x+9=-\sqrt{16}$

$x+9=4$ $\quad x+9=-4$

$x=-5$ $\quad x=-13$

The solutions are –13 and –5.

39. $(z-2)^2+5=7$

$(z-2)^2=2$

$z-2=\sqrt{2}$ $\quad z-2=-\sqrt{2}$

$z=2+\sqrt{2}$ $\quad z=2-\sqrt{2}$

The solutions are $2\pm\sqrt{2}$.

40. $s=-16t^2+v_0t+s_0$

$6000=-16t^2+0t+10,000$

$-4000=-16t^2$

$250=t^2$

$t=\sqrt{250}$ or $t=-\sqrt{250}$

$t=5\sqrt{10}$ $\quad t=-5\sqrt{10}$

Disregard a negative time.

It will take $5\sqrt{10}$ seconds or approximately 15.8 seconds.

41. $a^2+b^2=c^2$

$a^2+(15)^2=(35)^2$

$a^2+225=1225$

$a^2=1000$

$a=\sqrt{1000}$ or $a=-\sqrt{1000}$

$a=10\sqrt{10}$ $\quad a=-10\sqrt{10}$

Disregard a negative length.

The horizontal distance is $10\sqrt{10}$ feet or approximately 31.6 feet.

42. $s(x)=28x^2+2000$

$3200=28x^2+2000$

$1200=28x^2$

$\frac{1200}{28}=x^2$

$x=\sqrt{\frac{1200}{28}}$ or $x=-\sqrt{\frac{1200}{28}}$

$x=\sqrt{\frac{300}{7}}$ $\quad x=-\sqrt{\frac{300}{7}}$

$x=\frac{10\sqrt{3}}{\sqrt{7}}\cdot\frac{\sqrt{7}}{\sqrt{7}}$ $\quad x=\frac{-10\sqrt{3}}{\sqrt{7}}\cdot\frac{\sqrt{7}}{\sqrt{7}}$

$x=\frac{10\sqrt{21}}{7}$ $\quad x=\frac{-10\sqrt{21}}{7}$

$x\approx 6.55$ $\quad x=-6.55$

Disregard a negative time.

1991 + 7 = 1998

Sales will reach $3200 million in 1998.

43. $p^2-4p-96=0$

$p^2-4p=96$

$p^2-4p+4=96+4$

$(p-2)^2=100$

$p-2=\sqrt{100}$ or $p-2=-\sqrt{100}$

$p-2=10$ $\quad p-2=-10$

$p=12$ $\quad p=-8$

The solutions are –8 and 12.

44. $z^2+12z=-3$

$z^2+12z+36=-3+36$

$(z+6)^2=33$

$z+6=\sqrt{33}$ or $z+6=-\sqrt{33}$

$z=-6+\sqrt{33}$ $\quad z=-6-\sqrt{33}$

The solutions are $-6\pm\sqrt{33}$.

45. $2x^2-5x-12=0$

$2x^2-5x=12$

$x^2-\frac{5}{2}x=6$

$x^2-\frac{5}{2}x+\frac{25}{16}=6+\frac{25}{16}$

$\left(x-\frac{5}{4}\right)^2=\frac{121}{16}$

$x-\frac{5}{4}=\sqrt{\frac{121}{16}}$ or $x-\frac{5}{4}=-\sqrt{\frac{121}{16}}$

$x-\frac{5}{4}=\frac{11}{4}$ $\quad$ $x-\frac{5}{4}=-\frac{11}{4}$

$x=\frac{5}{4}+\frac{11}{4}$ $\quad$ $x=\frac{5}{4}-\frac{11}{4}$

$x=4$ $\quad$ $x=-\frac{3}{2}$

The solutions are $-\frac{3}{2}$ and 4.

46. $x^2+\frac{1}{2}x=1$

$x^2+\frac{1}{2}x+\frac{1}{16}=1+\frac{1}{16}$

$\left(x+\frac{1}{4}\right)^2=\frac{17}{16}$

$x+\frac{1}{4}=\sqrt{\frac{17}{16}}$ or $x+\frac{1}{4}=-\sqrt{\frac{17}{16}}$

$x=-\frac{1}{4}+\frac{\sqrt{17}}{4}$ $\quad$ $x=-\frac{1}{4}-\frac{\sqrt{17}}{4}$

$x=\frac{-1+\sqrt{17}}{4}$ $\quad$ $x=\frac{-1-\sqrt{17}}{4}$

The solutions are $\frac{-1\pm\sqrt{17}}{4}$.

47. $C(x)=60+x(20+2x)$
$150=60+x(20+2x)$
$150=60+20x+2x^2$
$2x^2+20x-90=0$
$x^2+10x-45=0$
$x^2+10x=45$
$x^2+10x+25=45+25$
$(x+5)^2=70$

$x+5=\sqrt{70}$ or $x+5=-\sqrt{70}$

$x=-5+\sqrt{70}$ $\quad$ $x=-5-\sqrt{70}$

$x\approx 3.4$ $\quad$ $x\approx -13.4$

Disregard a negative amount. Since x is the number of figurines over 3, a total of 6 figurines can be purchased for \$150.

48. Let $x=$ width
$2x+8=$ length
$x(2x+8)=90$
$2x^2+8x-90=0$
$x^2+4x-45=0$
$x^2+4x+4=45+4$
$(x+2)^2=49$

$x+2=\sqrt{49}$ or $x+2=-\sqrt{49}$

$x+2=7$ $\quad$ $x+2=-7$

$x=5$ $\quad$ $x=-9$

Disregard a negative width. The dimensions are 5 inches by 18 inches.

49. $x^2-2x-63=0$

$x=\frac{-b\pm\sqrt{b^2-4ac}}{2a}$

$x=\frac{-(-2)\pm\sqrt{(-2)^2-4(1)(-63)}}{2(1)}$

$x=\frac{2\pm\sqrt{4+252}}{2}$

$x=\frac{2\pm\sqrt{256}}{2}$

$x=\frac{2\pm 16}{2}$

$x=\frac{2-16}{2}$ or $x=\frac{2+16}{2}$

$x=-7$ $\quad$ $x=9$

The solutions are –7 and 9.

50. $x^2=3x+3$

$x^2-3x-3=0$

$x=\frac{-b\pm\sqrt{b^2-4ac}}{2a}$

$x=\frac{-(-3)\pm\sqrt{(-3)^2-4(1)(-3)}}{2(1)}$

$x=\frac{3\pm\sqrt{9+12}}{2}$

$x=\frac{3\pm\sqrt{21}}{2}$

The solutions are $\frac{3\pm\sqrt{21}}{2}$.

51. $25x^2+1=10x$

$25x^2-10x+1=0$

$x=\dfrac{-b\pm\sqrt{b^2-4ac}}{2a}$

$x=\dfrac{-(-10)\pm\sqrt{(-10)^2-4(25)(1)}}{2(25)}$

$x=\dfrac{10\pm\sqrt{100-100}}{50}$

$x=\dfrac{10\pm 0}{50}$

$x=\dfrac{1}{5}$

The solution is $\dfrac{1}{5}$.

52. $z^2+\dfrac{7}{20}z-\dfrac{3}{10}=0$

$20\left(z^2+\dfrac{7}{20}z-\dfrac{3}{10}\right)=0\cdot 20$

$20z^2+7z-6=0$

$z=\dfrac{-b\pm\sqrt{b^2-4ac}}{2a}$

$z=\dfrac{-7\pm\sqrt{7^2-4(20)(-6)}}{2(20)}$

$z=\dfrac{-7\pm\sqrt{529}}{40}$

$z=\dfrac{-7\pm 23}{40}$

$z=\dfrac{-7+23}{40}$ or $z=\dfrac{-7-23}{40}$

$z=\dfrac{16}{40}$ $\qquad z=-\dfrac{30}{40}$

$z=\dfrac{2}{5}$ $\qquad z=-\dfrac{3}{4}$

The solutions are $-\dfrac{3}{4}$ and $\dfrac{2}{5}$.

53. $x^2+2.1x-10.8=0$

$x=\dfrac{-b\pm\sqrt{b^2-4ac}}{2a}$

$x=\dfrac{-2.1\pm\sqrt{(2.1)^2-4(1)(-10.8)}}{2(1)}$

$x=\dfrac{-2.1\pm\sqrt{4.41+43.2}}{2}$

$x=\dfrac{-2.1\pm\sqrt{47.61}}{2}$

$x=\dfrac{-2.1\pm 6.9}{2}$

$x=\dfrac{-2.1-6.9}{2}$ or $x=\dfrac{-2.1+6.9}{2}$

$x=-4.5$ $\qquad x=2.4$

The solutions are –4.5 and 2.4.

54. $y^2-5y+12=0$

$y=\dfrac{-b\pm\sqrt{b^2-4ac}}{2a}$

$y=\dfrac{-(-5)\pm\sqrt{(-5)^2-4(1)(12)}}{2(1)}$

$y=\dfrac{5\pm\sqrt{25-48}}{2}$

$y=\dfrac{5\pm\sqrt{-23}}{2}$

There is no real-number solution.

55. $x^2-10x+6=0$

$x=\dfrac{-b\pm\sqrt{b^2-4ac}}{2a}$

$x=\dfrac{-(-10)\pm\sqrt{(-10)^2-4(1)(6)}}{2(1)}$

$x=\dfrac{10\pm\sqrt{76}}{2}$

$x=\dfrac{10\pm 2\sqrt{19}}{2}$

$x=5\pm\sqrt{19}$

56. $3a^2 - 4a - 12 = 0$

$a = \dfrac{-b \pm \sqrt{b^2 - 4ac}}{2a}$

$a = \dfrac{-(-4) \pm \sqrt{(-4)^2 - 4(3)(-12)}}{2(3)}$

$a = \dfrac{4 \pm \sqrt{160}}{6} = \dfrac{4 \pm 4\sqrt{10}}{6} = \dfrac{2 \pm 2\sqrt{10}}{3}$

57. $x^2 + 2x + 10 = 0$

$b^2 - 4ac = 2^2 - 4(1)(10) = -36$

There is no real-number solution because the discriminant is negative.

58. $x^2 + 20x + 55 = 2x - 26$

$x^2 + 18x + 81 = 0$

$b^2 - 4ac = (18)^2 - 4(1)(81) = 0$

There is one rational solution because the discriminant is 0.

59. $x^2 = 10x + 75$

$x^2 - 10x - 75 = 0$

$b^2 - 4ac = (-10)^2 - 4(1)(-75) = 400$

There are two rational solutions because 400 is a perfect square.

60. $x = x^2 - 11$

$x^2 - x - 11 = 0$

$b^2 - 4ac = (-1)^2 - 4(1)(-11) = 45$

There are two irrational solutions because 45 is postive but not a perfect square.

61. $A = P(1+r)^t$

$1166.40 = 1000(1+r)^2$

$1.1664 = (1+r)^2$

$1 + r = \sqrt{1.1664}$ or $1 + r = -\sqrt{1.1664}$

$r = -1 + \sqrt{1.1664}$ $r = -1 - \sqrt{1.1664}$

$r = -1 + 1.08$ $r = -1 - 1.08$

$r = 0.08$ $r = -2.08$

Disregard a negative interest. The interest rate would be 8%.

62. $A = P(1+r)^t$

$1200 = 1000(1+r)^2$

$1.2 = (1+r)^2$

$1 + r = \sqrt{1.2}$ or $1 + r = -\sqrt{1.2}$

$r = -1 + \sqrt{1.2}$ $r = -1 - \sqrt{1.2}$

$r \approx 0.095$ $r \approx -2.095$

Disregard a negative interest. The interest is about 9.5%.

63. $R(x) = 600 + 26x - 2x^2$

$450 = 600 + 26x - 2x^2$

$2x^2 - 26x - 150 = 0$

$2\left(x^2 - 13x - 75\right) = 0$

$x = \dfrac{-(-13) \pm \sqrt{(-13)^2 - 4(1)(-75)}}{2(1)}$

$x = \dfrac{13 \pm \sqrt{469}}{2}$

$x = \dfrac{13 - \sqrt{469}}{2}$ or $x = \dfrac{13 + \sqrt{469}}{2}$

$x \approx -4.33$ $x \approx 17.33$

Disregard a negative amount. For a value of 17.33, the revenue will be approximately $450.

64. $a^2 + b^2 - c^2$

$(11)^2 + (60)^2 = c^2$

$3721 = c^2$

$c = \sqrt{3721}$ or $c = -\sqrt{3721}$

$c = 61$ $c = -61$

Disregard a negative length. The hypotenuse is 61 in.

65. $a^2 + b^2 = c^2$

$a^2 + (240)^2 = (250)^2$

$a^2 = 4900$

$a = \sqrt{4900}$ or $a = -\sqrt{4900}$

$a = 70$ $a = -70$

Disregard a negative length. The pond is 70 feet wide.

66. Let x = length
$x-9$ = width.
$$a^2+b^2=c^2$$
$$(x)^2+(x-9)^2=(45)^2$$
$$x^2+x^2-18x+81=2025$$
$$2x^2-18x-1944=0$$
$$2\left(x^2-9x-972\right)=0$$
$$2(x-36)(x+27)=0$$
$$x=36 \quad \text{or} \quad x=-27$$
Disregard the negative length. The dimensions are 36 ft by 27 ft. Now check
$$(27.5)^2+(35.4)^2=(45)^2$$
$$756.25+1253.16=2025$$
$$2009.41=2025$$
This is a contradiction. The corners are not square.

67. $c=a\sqrt{2}$
$$16.85=a\sqrt{2}$$
$$\frac{16.85}{\sqrt{2}}=a$$
$$a=\frac{16.85\sqrt{2}}{2}$$
$$a=8.425\sqrt{2}$$
$$a\approx 11.91$$
The length of the legs is $8.425\sqrt{2}$ cm or approximately 11.91 cm.

68. $c=a\sqrt{2}$
$$c=7\sqrt{2}$$
$$c\approx 9.9$$
The rope is $7\sqrt{2}$ feet or approximately 9.9 feet.

69. $2a=20$
$$a=10$$
$$a\sqrt{3}=10\sqrt{3}$$
The lengths of the sides are 10 in. and $10\sqrt{3}$ in.

70. $A=\frac{1}{4}\pi d^2$
$$4A=\pi d^2$$
$$d^2=\frac{4A}{\pi}$$
$$d=\sqrt{\frac{4A}{\pi}}=\frac{2\sqrt{A}}{\sqrt{\pi}}\cdot\frac{\sqrt{\pi}}{\sqrt{\pi}}=\frac{2\sqrt{\pi A}}{\pi}$$

71. $a=bx^2+c$ for x
$$a-c=bx^2$$
$$\frac{a-c}{b}=x^2$$
$$x=\sqrt{\frac{a-c}{b}} \quad \text{or} \quad x=-\sqrt{\frac{a-c}{b}}$$
$$x=\frac{\sqrt{b(a-c)}}{b} \qquad x=-\frac{\sqrt{b(a-c)}}{b}$$
$$x=\pm\frac{\sqrt{b(a-c)}}{b}$$

72. $s=-16t^2+v_0t+s_0$ for t
$$0=-16t^2+v_0t+(s_0-s)$$
$$t=\frac{-b\pm\sqrt{b^2-4ac}}{2a}$$
$$t=\frac{-v_0\pm\sqrt{(v_0)^2-4(-16)(s_0-s)}}{2(-16)}$$
$$t=\frac{-v_0\pm\sqrt{(v_0)^2+64(s_0-s)}}{-32}$$

73. $x^2+6x-9\le 4x+15$
$$\text{Y1}=x^2+6x-9$$
$$\text{Y2}=4x+15$$

X	Y1	Y2
-8	7	-17
-7	-2	-13
-6	-9	-9
-5	-14	-5
-4	-17	-1
-3	-18	3
-2	-17	7

X=-8

X	Y1	Y2
-1	-14	11
0	-9	15
1	-2	19
2	7	23
3	18	27
4	31	31
5	46	35

X=-1

The solution set is all integers greater than or equal to –6 and less than or equal to 4.

74. $2x^2+3x-8>6+4x-x^2$

$Y1=2x^2+3x-8$

$Y2=6+4x-x^2$

X	Y1	Y2
-3	1	-15
-2	-6	-6
-1	-9	1
0	-8	6
1	-3	9
2	6	10
3	19	9

X=-3

The solution set is all integers less than –2 or greater than 2.

75. $x^2+2x-3\ge 5$

$Y1=x^2+2x-3$

$Y2=5$

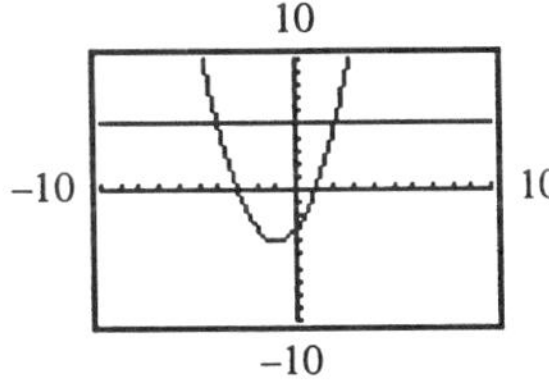

The intersections are (–4, 5) and (2, 5).
The solution is $(-\infty,-4]\cup[2,\infty)$.

76. $5-x^2\ge -4$

$Y1=5-x^2$

$Y2=-4$

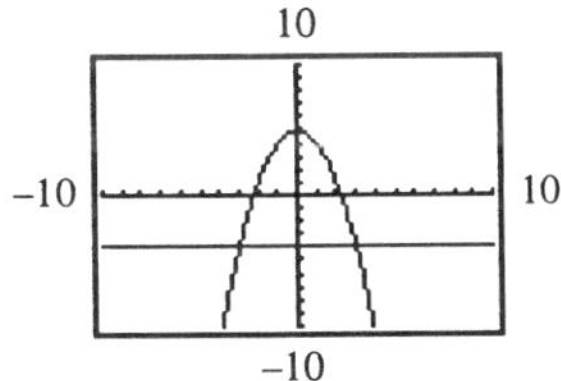

The intersections are (–3, –4) and (3, –4).
The solution is [–3, 3].

77. $x^2-2x-1<x+3$

$Y1=x^2-2x-1$

$Y2=x+3$

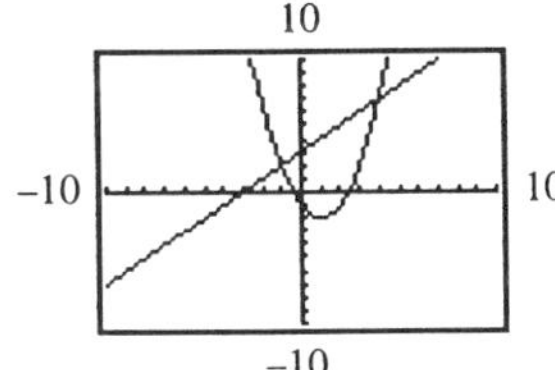

The intersections are (–1, 2) and (4, 7).
The solution is (–1, 4).

78. $x^2+3<1-x^2$

$Y1=x^2+3$

$Y2=1-x^2$

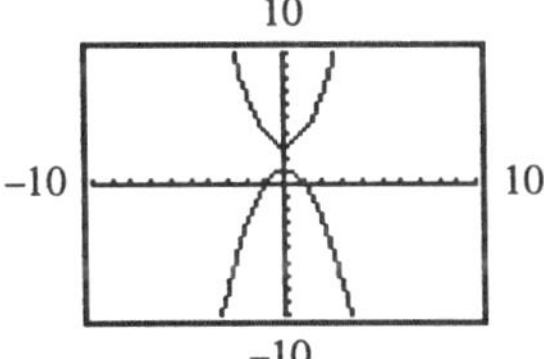

There are no intersections. Y1 is always greater than Y2.
There is no solution.

79. $x^2-6x+11>-\frac{1}{2}x^2+3x-\frac{5}{2}$

$Y1=x^2-6x+11$

$Y2=-\frac{1}{2}x^2+3x-\frac{5}{2}$

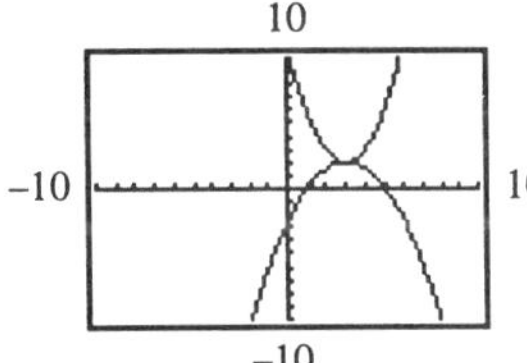

The intersection is (3, 2). Y1 is always greater than Y2 except for $x=3$.
The solution is $(-\infty,3)\cup(3,\infty)$.

80. $x^2-8x+18\le\frac{1}{4}x^2-2x+6$

$Y1=x^2-8x+18$

$Y2=\frac{1}{4}x^2-2x+6$

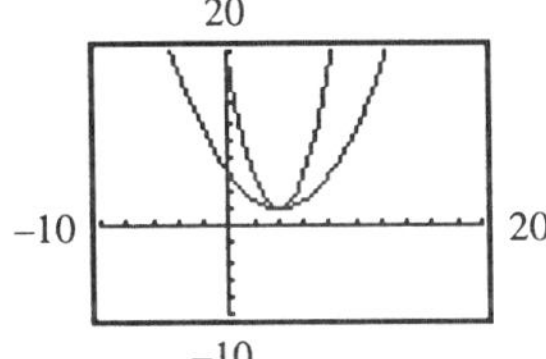

The intersection is (4, 2).
Y1 is always greater than Y2 except at $x=4$.
Therefore, the solution is 4.

81. $10-6x+x^2 < -4+6x-x^2$

$Y1 = 10-6x+x^2$

$Y2 = -4+6x-x^2$

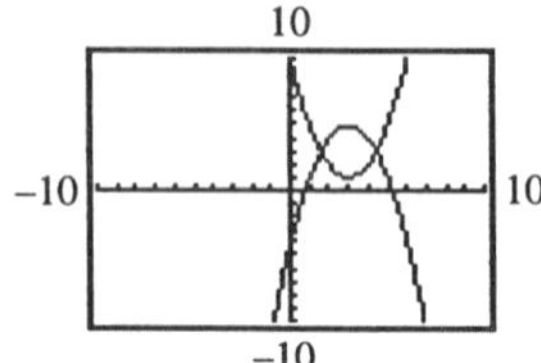

The intersections are about (1.59, 3) and (4.41, 3).

The solution is $(1.59, 4.41)$

82. $x^2-4x+4 < \frac{1}{4}x^2 - x + 3$

Determine the critical points.

$x^2-4x+4 = \frac{1}{4}x^2 - x + 3$

$4x^2-16x+16 = x^2-4x+12$

$3x^2-12x+4=0$

$x = \dfrac{-b \pm \sqrt{b^2-4ac}}{2a}$

$x = \dfrac{-(-12) \pm \sqrt{(-12)^2-4(3)(4)}}{2(3)}$

$x = \dfrac{12 \pm \sqrt{96}}{6}$

$x = \dfrac{12 \pm 4\sqrt{6}}{6}$

$x = 2 \pm \frac{2}{3}\sqrt{6}$

Possible intervals:

$\left(-\infty, 2-\frac{2}{3}\sqrt{6}\right), \left(2-\frac{2}{3}\sqrt{6}, 2+\frac{2}{3}\sqrt{6}\right),$

$\left(2+\frac{2}{3}\sqrt{6}, \infty\right)$

Use a test point in each interval to determine the solution is $\left(2-\frac{2}{3}\sqrt{6}, 2+\frac{2}{3}\sqrt{6}\right)$.

83. $x^2+x-6 \le 12-2x-2x^2$

Determine the critical points.

$x^2+x-6 = 12-2x-2x^2$

$3x^2+3x-18=0$

$3(x^2+x-6)=0$

$3(x+3)(x-2)=0$

$x=-3$ or $x=2$

Possible intervals:

$(-\infty,-3], [-3,2], [2,\infty)$

Use a test point in each interval to determine the solution is [–3, 2].

84. $3-z^2 > -2z$

Determine the critical points.

$3-z^2 = -2z$

$z^2-2z-3=0$

$(z+1)(z-3)=0$

$z=-1$ or $z=3$

Possible intervals:

$(-\infty,-1), (-1,3), (3,\infty)$

Use a test point in each interval to determine the solution is (–1, 3).

85. $x^2+6x+11 > -x^2-6x-7$

Determine the critical points.

$2x^2+12x+18=0$

$2(x^2+6x+9)=0$

$2(x+3)^2=0$

$x=-3$

Possible intervals:

$(-\infty,-3), (-3,\infty)$

Use a test point in each interval to determine the solution is $(-\infty,-3) \cup (-3,\infty)$.

86. $1-2x^2 \le 1-x^2$

Determine the critical points.

$1-2x^2 = 1-x^2$

$x^2=0$

$x=0$

Possible intervals:

$(-\infty,0], [0,\infty)$

Use a test point in each interval to determermie the solution is $(-\infty,\infty)$.

87. $x^2+4x-2<-\frac{1}{2}x^2-2x-8$

Determine the critical points.

$x^2+4x-2=-\frac{1}{2}x^2-2x-8$

$2x^2+8x-4=-x^2-4x-16$

$3x^2+12x+12=0$

$3(x^2+4x+4)=0$

$3(x+2)^2=0$

$x=-2$

Possible intervals:

$(-\infty,-2),\ (-2,\infty)$

Use a test point in each interval to determine there is no solution.

88. $p^2-6p-1<0$

Determine the critical points.

$p=\frac{-b\pm\sqrt{b^2-4ac}}{2a}$

$p=\frac{-(-6)\pm\sqrt{(-6)^2-4(1)(-1)}}{2(1)}$

$p=\frac{6\pm\sqrt{40}}{2}$

$p=\frac{6\pm2\sqrt{10}}{2}$

$p=3\pm\sqrt{10}$

Possible intervals:

$\left(-\infty,\ 3-\sqrt{10}\right),\ \left(3-\sqrt{10},\ 3+\sqrt{10}\right),$

$\left(3+\sqrt{10},\ \infty\right)$

Use a test point in each interval to determine the solution is $\left(3-\sqrt{10},3+\sqrt{10}\right)$.

89. $R(x)=(5+x)(65-2x)$

$R(x)=-2x^2+55x+325$

$500\le-2x^2+55x+325$

Determine the critical points.

$Y1=500$

$Y2=-2x^2+55x+325$

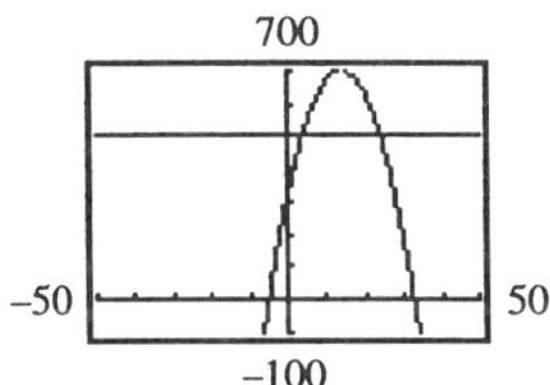

The intersections are about (3.672, 500) and (23.828, 500).

The solution is [3.672, 23.828].

Since x is the number of items over 5, the order must be greater than 8 items but less than 29 items.

90. $C(x)=3x^2+25x+50$

$200\ge3x^2+25x+50$

Determine the critical points.

$Y1=200$

$Y2=3x^2+25x+50$

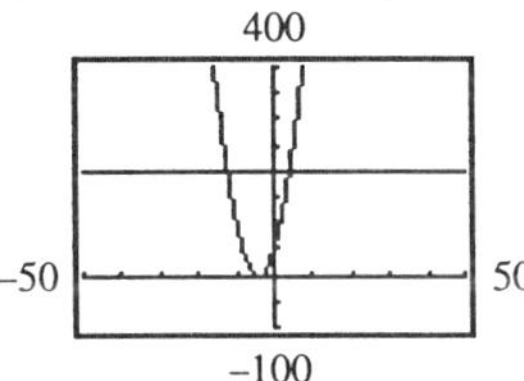

The intersections are about (–12.37, 200) and (4.04, 200). Disregard negative values.

The solution is (0, 4.04].

The booth can be rented up to 4 days.

91. $s(t)=-16t^2+v_0t+s_0$

$0\le-16t^2+25t+30$

Determine the critical points.

$Y1=0$

$Y2=-16x^2+25x+30$

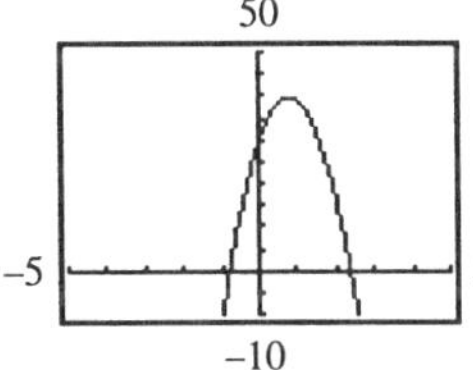

The intersections are about (–0.795, 0) and (2.358, 0).

The solution is [–0.795, 2.358].

Disregard negative times.

The dart is above the ground from 0 seconds to approximately 2.358 seconds.

Chapter 11 Mixed Review

1. $-\sqrt{144} = -\sqrt{(12)^2} = -12$

2. $\sqrt{125} = \sqrt{25 \cdot 5} = 5\sqrt{5}$

3. $\sqrt{7}\sqrt{28} = \sqrt{7 \cdot 28} = \sqrt{196} = 14$

4. $\sqrt{\frac{121}{225}} = \frac{\sqrt{121}}{\sqrt{225}} = \frac{11}{15}$

5. $\sqrt{\frac{24}{135}} = \frac{\sqrt{24}}{\sqrt{135}} = \frac{\sqrt{4 \cdot 6}}{\sqrt{9 \cdot 15}} = \frac{2\sqrt{6}}{3\sqrt{15}} \cdot \frac{\sqrt{15}}{\sqrt{15}}$
$= \frac{2\sqrt{90}}{45} = \frac{6\sqrt{10}}{45} = \frac{2\sqrt{10}}{15}$

6. $\frac{\sqrt{63}}{\sqrt{28}} = \sqrt{\frac{63}{28}} = \sqrt{\frac{9}{4}} = \frac{\sqrt{9}}{\sqrt{4}} = \frac{3}{2}$

7. $x^2 - 4 = 2x - 1$
$Y1 = x^2 - 4$
$Y2 = 2x - 1$

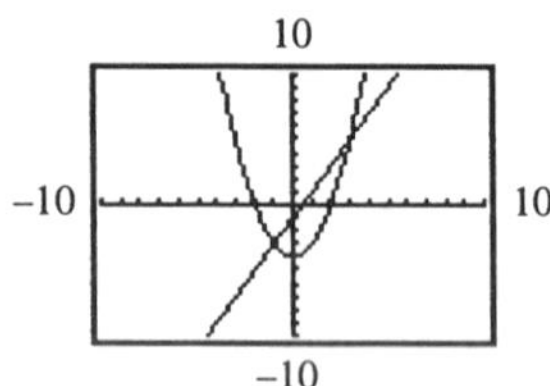

The intersections are (–1, –3) and (3, 5).
The solutions are –1 and 3.

8. $3 - x^2 = -3 - \frac{1}{3}x^2$
$Y1 = 3 - x^2$
$Y2 = -3 - \frac{1}{3}x^2$

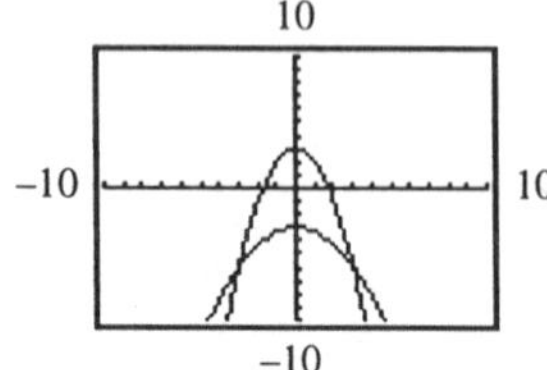

The intersections are (–3, –6) and (3, –6).
The solutions are –3 and 3.

9. $x^2 + 6x + 9 = 4$
$Y1 = x^2 + 6x + 9$
$Y2 = 4$

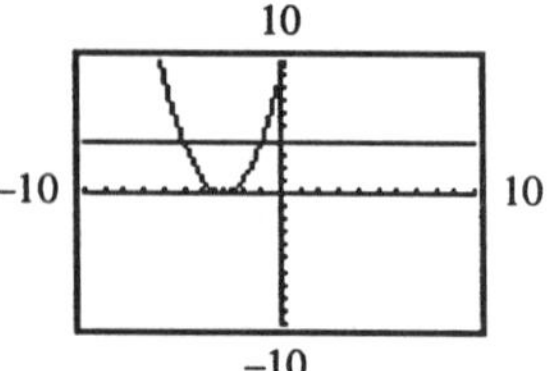

The intersections are (–5, 4) and (–1, 4).
The solutions are –5 and –1.

10. $15x^2 + 13x - 72 = 0$
$Y1 = 15x^2 + 13x - 72$
$Y2 = 0$

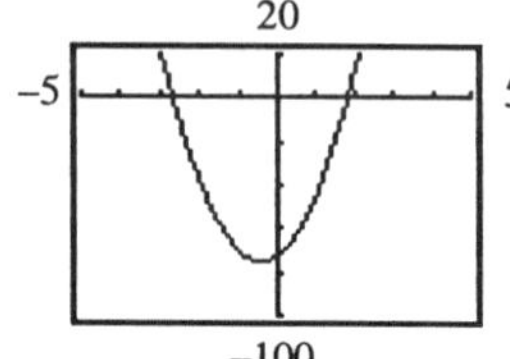

The intersections are
$\left(-2\frac{2}{3}, 0\right)$ and $\left(1\frac{4}{5}, 0\right)$

The solutions are $-2\frac{2}{3}$ and $1\frac{4}{5}$.

11. $2x + 10 = x^2 + 4x + 7$
$Y1 = 2x + 10$
$Y2 = x^2 + 4x + 7$

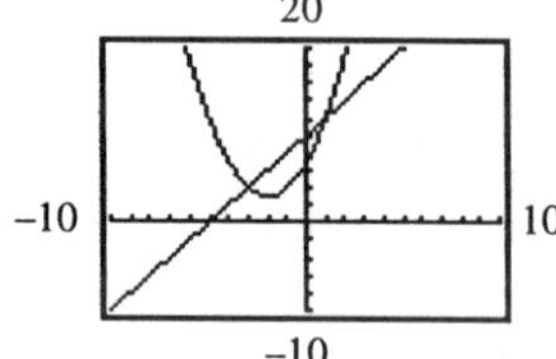

The intersections are (–3, 4) and (1, 12).
The solutions are –3 and 1.

12. $x^2 + 9x + 7 = 0$
$Y1 = x^2 + 9x + 7$
$Y2 = 0$

20

–10 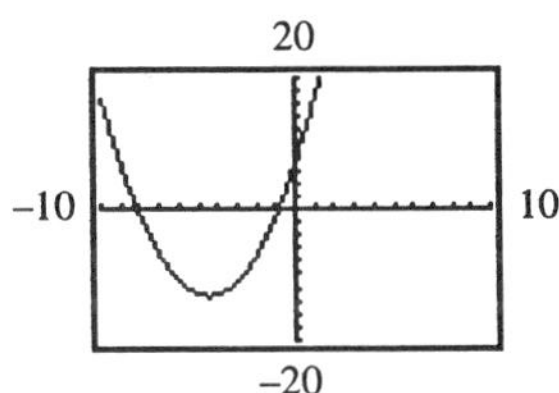 10

–20

The intersections are about (–8.14, 0) and (–0.86, 0).
The solutions are approximately –8.14 and –0.86.

13. $\frac{1}{6}x^2 - 2x + 7 = 9 - \frac{1}{4}x^2$

$Y1 = \frac{1}{6}x^2 - 2x + 7$

$Y2 = 9 - \frac{1}{4}x^2$

20

–20 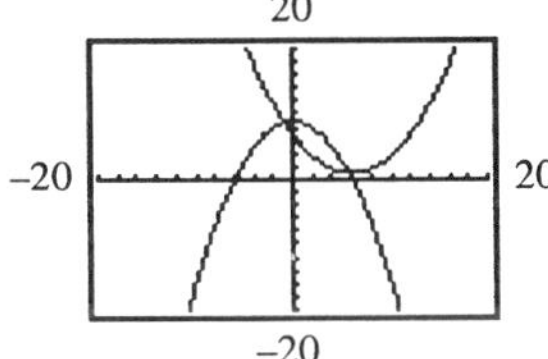 20

–20

The intersections are about (–0.85, 8.82) and (5.65, 1.02).
The solutions are approximately –0.85 and 5.65.

14. $x^2 - 6x - 40 = 0$
$Y1 = x^2 - 6x - 40$
$Y2 = 0$

20

–20 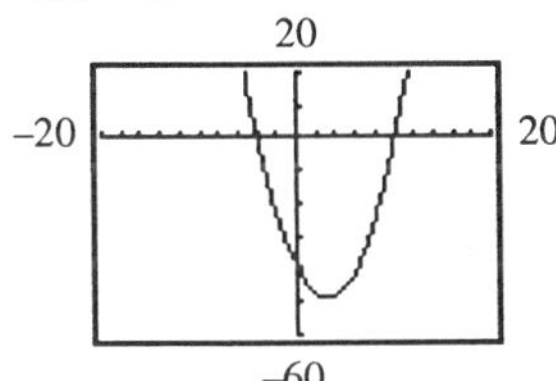 20

–60

The intersections are (–4, 0) and (10, 0).
The solutions are –4 and 10.

15. $2x^2 - x - 28 = x^2 + 3x + 17$
$Y1 = 2x^2 - x - 28$
$Y2 = x^2 + 3x + 17$

X	Y1	Y2
[illegible]	50	35
-5	27	27
-4	8	21
-3	-7	17
-2	-18	15
-1	-25	15
0	-28	17
X= -6		

X	Y1	Y2
[illegible]	-27	21
2	-22	27
3	-13	35
4	0	45
5	17	57
6	38	71
7	63	87
X=1		

X	Y1	Y2
[illegible]	92	105
9	125	125
10	162	147
11	203	171
12	248	197
13	297	225
14	350	255
X=8		

The solutions are –5 and 9.

16. $3x^2 - 5x - 10 = 7x + 5$
$Y1 = 3x^2 - 5x - 10$
$Y2 = 7x + 5$

X	Y1	Y2
[illegible]	-2	-2
0	-10	5
1	-12	12
2	-8	19
3	2	26
4	18	33
5	40	40
X= -1		

The solutions are –1 and 5.

17. $12x^2 + 9x = 8x + 6$
$12x^2 + x - 6 = 0$
$(3x - 2)(4x + 3) = 0$
$3x - 2 = 0$ or $4x + 3 = 0$
$x = \frac{2}{3}$ $\quad$ $x = -\frac{3}{4}$

The solutions are $-\frac{3}{4}$ and $\frac{2}{3}$.

18. $x^2 + 3x - 88 = 0$
$(x + 11)(x - 8) = 0$
$x + 11 = 0$ or $x - 8 = 0$
$x = -11$ $\quad$ $x = 8$
The solutions are –11 and 8.

19. $5x^2+14x-6=8-19x$
$5x^2+33x-14=0$
$(5x-2)(x+7)=0$
$5x-2=0$ or $x+7=0$
$x=\frac{2}{5}$ $x=-7$
The solutions are –7 and $\frac{2}{5}$.

20. $49x^2-16=0$
$49x^2=16$
$x^2=\frac{16}{49}$
$x=\sqrt{\frac{16}{49}}$ or $x=-\sqrt{\frac{16}{49}}$
$x=\frac{4}{7}$ $x=-\frac{4}{7}$
The solutions are $\pm\frac{4}{7}$.

21. $5x^2-180=0$
$5x^2=180$
$x^2=36$
$x=\sqrt{36}$ or $x=-\sqrt{36}$
$x=6$ $x=-6$
The solutions are ± 6.

22. $11+b^2=2$
$b^2=-9$
There is no real-number solution.

23. $(x-9)^2=16$
$x-9=\sqrt{16}$ or $x-9=-\sqrt{16}$
$x-9=4$ $x-9=-4$
$x=9+4$ $x=9-4$
$x=13$ $x=5$
The solutions are 5 and 13.

24. $(p-7)^2+6=9$
$(p-7)^2=3$
$p-7=\sqrt{3}$ or $p-7=-\sqrt{3}$
$p=7+\sqrt{3}$ $p=7-\sqrt{3}$
The solutions are $7\pm\sqrt{3}$.

25. $z^2+12z=-33$
$z^2+12z+33=0$
$z=\frac{-b\pm\sqrt{b^2-4ac}}{2a}$
$z=\frac{-12\pm\sqrt{(12)^2-4(1)(33)}}{2(1)}$
$z=\frac{-12\pm\sqrt{144-132}}{2}$
$z=\frac{-12\pm\sqrt{12}}{2}$
$z=\frac{-12\pm 2\sqrt{3}}{2}$
$z=-6\pm\sqrt{3}$
The solutions are $-6\pm\sqrt{3}$.

26. $m^2+7m+12=9$
$m^2+7m+3=0$
$m=\frac{-b\pm\sqrt{b^2-4ac}}{2a}$
$m=\frac{-7\pm\sqrt{7^2-4(1)(3)}}{2(1)}$
$m=\frac{-7\pm\sqrt{37}}{2}$
The solutions are $\frac{-7\pm\sqrt{37}}{2}$.

27. $b^2+2.4b-4.32=0$
$100(b^2+2.4b-4.32)=0\cdot 100$
$100b^2+240b-432=0$
$b=\frac{-240\pm\sqrt{(240)^2-4(100)(-432)}}{2(100)}$
$b=\frac{-240\pm\sqrt{230,400}}{200}$
$b=\frac{-240\pm 480}{200}$
$b=\frac{-240+480}{200}$ or $b=\frac{-240-480}{200}$
$b=1.2$ $b=-3.6$
The solutions are –3.6 and 1.2.

28. $q^2 - 7q - 78 = 0$
$(q + 6)(q - 13) = 0$
$q + 6 = 0$ or $q - 13 = 0$
$q = -6$ $\quad q = 13$
The solutions are –6 and 13.

29. $x^2 + 20x + 84 = 0$
$(x + 14)(x + 6) = 0$
$x + 14 = 0$ or $x + 6 = 0$
$x = -14$ $\quad x = -6$
The solutions are –14 and –6.

30. $2y^2 + 3y - 5 = 0$
$(2y + 5)(y - 1) = 0$
$2y + 5 = 0$ or $y - 1 = 0$
$y = -\frac{5}{2}$ $\quad y = 1$
The solutions are $-\frac{5}{2}$ and 1.

31. $4x^2 + 16x - 2 = 0$
$2(2x^2 + 8x - 1) = 0$
$$x = \frac{-b \pm \sqrt{b^2 - 4ac}}{2a}$$
$$x = \frac{-8 \pm \sqrt{8^2 - 4(2)(-1)}}{2(2)}$$
$$x = \frac{-8 \pm \sqrt{72}}{4}$$
$$x = \frac{-8 \pm 6\sqrt{2}}{4}$$
$$x = \frac{-4 \pm 3\sqrt{2}}{2}$$
The solutions are $\frac{-4 \pm 3\sqrt{2}}{2}$.

32. $4 + 8x + x^2 = 0$
$x^2 + 8x + 4 = 0$
$$x = \frac{-b \pm \sqrt{b^2 - 4ac}}{2a}$$
$$x = \frac{-8 \pm \sqrt{(8)^2 - 4(1)(4)}}{2(1)}$$
$$x = \frac{-8 \pm \sqrt{48}}{2}$$
$$x = \frac{-8 \pm 4\sqrt{3}}{2}$$
$$x = -4 \pm 2\sqrt{3}$$
The solutions are $4 \pm 2\sqrt{5}$.

33. $x^2 = 7x + 2$
$x^2 - 7x - 2 = 0$
$$x = \frac{-b \pm \sqrt{b^2 - 4ac}}{2a}$$
$$x = \frac{-(-7) \pm \sqrt{(-7)^2 - 4(1)(-2)}}{2(1)}$$
$$x = \frac{7 \pm \sqrt{57}}{2}$$
The solutions are $\frac{7 \pm \sqrt{57}}{2}$.

34. $36q^2 + 25 = 60q$
$36q^2 - 60q + 25 = 0$
$$q = \frac{-b \pm \sqrt{b^2 - 4ac}}{2a}$$
$$q = \frac{-(-60) \pm \sqrt{(-60)^2 - 4(36)(25)}}{2(36)}$$
$$q = \frac{60 \pm \sqrt{0}}{72}$$
$$q = \frac{60}{72}$$
$$q = \frac{5}{6}$$
The solution is $\frac{5}{6}$.

35. $b^2 - 7b + 16 = 0$

$$b = \frac{-b \pm \sqrt{b^2 - 4ac}}{2a}$$

$$b = \frac{-(-7) \pm \sqrt{(-7)^2 - 4(1)(16)}}{2(1)}$$

$$b = \frac{7 \pm \sqrt{-15}}{2}$$

There is no real-number solution.

36. $z^2 + 6z - 55 = 0$

$$z = \frac{-b \pm \sqrt{b^2 - 4ac}}{2a}$$

$$z = \frac{-6 \pm \sqrt{6^2 - 4(1)(-55)}}{2(1)}$$

$$z = \frac{-6 \pm \sqrt{256}}{2}$$

$$z = \frac{-6 \pm 16}{2}$$

$z = \dfrac{-6 - 16}{2}$ or $z = \dfrac{-6 + 16}{2}$

$z = -11$ $\qquad$ $z = 5$

The solutions are -11 and 5.

37. $4m^2 + 7m = 36$

$$4m^2 + 7m - 36 = 0$$

$$m = \frac{-b \pm \sqrt{b^2 - 4ac}}{2a}$$

$$m = \frac{-7 \pm \sqrt{7^2 - 4(4)(-36)}}{2(4)}$$

$$m = \frac{-7 \pm \sqrt{625}}{8}$$

$$m = \frac{-7 \pm 25}{8}$$

$m = \dfrac{-7 - 25}{8}$ or $m = \dfrac{-7 + 25}{8}$

$m = -4$ $\qquad$ $m = \dfrac{9}{4}$

The solutions are -4 and $\dfrac{9}{4}$.

38. $a = \dfrac{x(x+1)}{2}$

$$2a = x^2 + x$$

$$x^2 + x - 2a = 0$$

$$x = \frac{-b \pm \sqrt{b^2 - 4ac}}{2a}$$

$$x = \frac{-1 \pm \sqrt{1^2 - 4(1)(-2a)}}{2(1)}$$

$$x = \frac{-1 \pm \sqrt{1 + 8a}}{2}$$

39. $x^2 + y^2 + z^2 = r^2$

$$y^2 = r^2 - x^2 - z^2$$

$$y = \sqrt{r^2 - x^2 - z^2}$$

40. $A = 4\pi r^2$

$$\frac{A}{4\pi} = r^2$$

$$r = \pm\sqrt{\frac{A}{4\pi}}$$

$$r = \pm\frac{\sqrt{A\pi}}{2\pi}$$

41. $x^2 + 2x - 7 \le 4x + 8$

$\mathrm{Y1} = x^2 + 2x - 7$

$\mathrm{Y2} = 4x + 8$

X	Y1	Y2
-5	8	-12
-4	1	-8
-3	-4	-4
-2	-7	0
-1	-8	4
0	-7	8
1	-4	12

X=-5

X	Y1	Y2
2	1	16
3	8	20
4	17	24
5	28	28
6	41	32
7	56	36
8	73	40

X=2

The solutions are all integers greater than or equal to -3 and less than or equal to 5.

42. $x^2+9x-11 \le 4x+3$
$Y1 = x^2+9x-11$
$Y2 = 4x+3$

X	Y1	Y2
-9	-11	-33
-8	-19	-29
-7	-25	-25
-6	-29	-21
-5	-31	-17
-4	-31	-13
-3	-29	-9

X=-9

X	Y1	Y2
-2	-25	-5
-1	-19	-1
0	-11	3
1	-1	7
2	11	11
3	25	15
4	41	19

X=-2

The solutions are all integers greater than or equal to –7 and less than or equal to 2.

43. $z^2+5z-6<12-10z-2z^2$
Determine the critical points.
$3z^2+15z-18=0$
$3(z^2+5z-6)=0$
$3(z+6)(z-1)=0$
$z=-6$ or $z=1$
Possible intervals:
$(-\infty,-6)$, $(-6,1)$, $(1,\infty)$
Use a test point in each interval to determine the solution is $(-6, 1)$.

44. $m^2-15>2m$
Determine the critical points.
$m^2-2m-15=0$
$(m-5)(m+3)=0$
$m=5$ or $m=-3$
Possible intervals:
$(-\infty,-3)$, $(-3, 5)$, $(5, \infty)$
Use a test point in each interval to determine the solution is $(-\infty,-3)\cup(5,\infty)$.

45. $x^2+2x-1>-x^2-2x+1$
Determine the critical points.
$2x^2+4x-2=0$
$2(x^2+2x-1)=0$

$$x=\frac{-b\pm\sqrt{b^2-4ac}}{2a}$$
$$x=\frac{-2\pm\sqrt{2^2-4(1)(-1)}}{2(1)}$$
$$x=\frac{-2\pm\sqrt{8}}{2}$$
$$x=\frac{-2\pm2\sqrt{2}}{2}$$
$$x=-1\pm\sqrt{2}$$

Possible intervals:
$\left(-\infty,-1-\sqrt{2}\right)$, $\left(-1-\sqrt{2},-1+\sqrt{2}\right)$, $\left(-1+\sqrt{2},\infty\right)$
Use a test point in each interval to determine the solution is $\left(-\infty,-1-\sqrt{2}\right)\cup\left(-1+\sqrt{2},\infty\right)$.

46. $p^2-8p-2>0$
Determine the critical points.
$$p=\frac{-b\pm\sqrt{b^2-4ac}}{2a}$$
$$p=\frac{8\pm\sqrt{(-8)^2-4(1)(-2)}}{2(1)}$$
$$p=\frac{8\pm\sqrt{72}}{2}$$
$$p=\frac{8\pm6\sqrt{2}}{2}$$
$$p=4\pm3\sqrt{2}$$

Possible intervals:
$\left(-\infty,4-3\sqrt{2}\right)$, $\left(4-3\sqrt{2},4+3\sqrt{2}\right)$, $\left(4+3\sqrt{2},\infty\right)$
Use a test point in each interval to determine the solution is $\left(-\infty,4-3\sqrt{2}\right)\cup\left(4+3\sqrt{2},\infty\right)$.

47. $2x^2-5x-11\le19+8x-x^2$
Determine the critical points.
$3x^2-13x-30=0$
$(3x+5)(x-6)=0$

$3x+5=0$ or $x-6=0$

$x=-\frac{5}{3}$ $x=6$

Possible intervals:

$\left(-\infty, -\frac{5}{3}\right], \left[-\frac{5}{3}, 6\right], [6, \infty)$

Use a test point in each interval to determine the solution is $\left[-\frac{5}{3}, 6\right]$.

48. $3q^2+2q-17 \ge 5q+19$

Determine the critical points.

$3q^2-3q-36=0$

$3(q^2-q-12)=0$

$3(q-4)(q+3)=0$

$q=4$ $q=-3$

Possible intervals:

$(-\infty,-3], [-3,4], [4,\infty)$

Use a test point in each interval to determine the solution is $(-\infty,-3]\cup[4,\infty)$.

49. $x^2-4x-1 \ge x+5$

$Y1=x^2-4x-1$

$Y2=x+5$

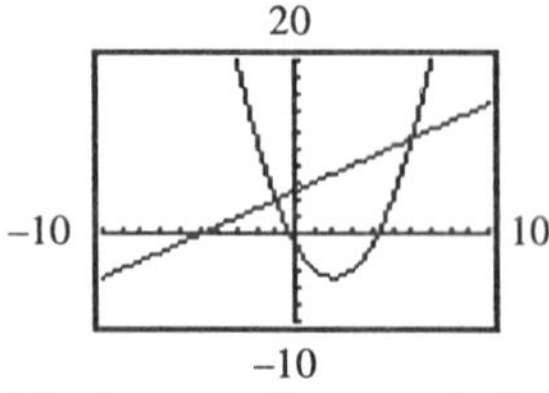

The intersections are (–1, 4) and (6, 11).

The solution is $(-\infty,-1]\cup[6,\infty)$.

50. $2-x^2 > 5+x^2$

$Y1=2-x^2$

$Y2=5+x^2$

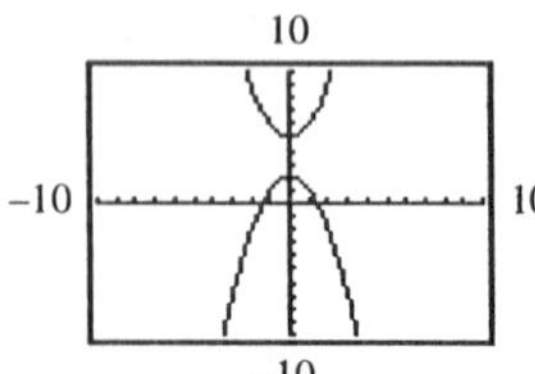

There are no intersections. Y1 is always less than Y2.

There is no solution.

51. $x^2-2x-3<12$

$Y1=x^2-2x-3$

$Y2=12$

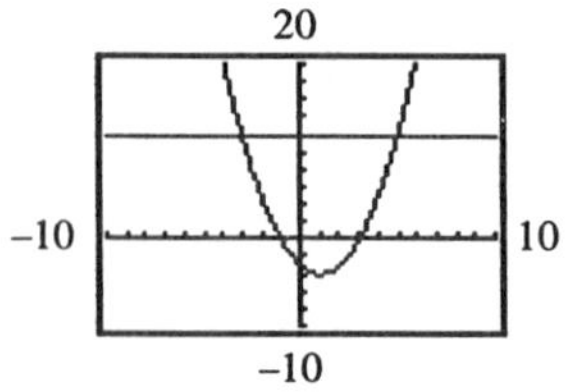

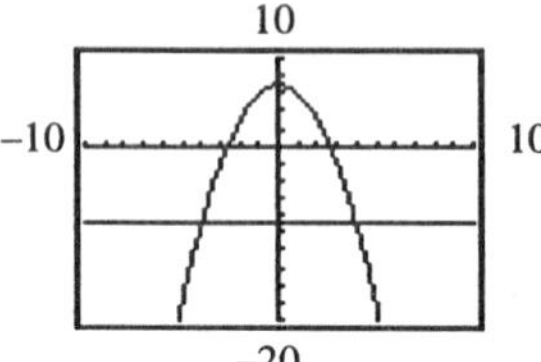

The intersections are (–3, 12) and (5, 12)

The solution is (–3, 5).

52. $7-x^2>-9$

$Y1=7-x^2$

$Y2=-9$

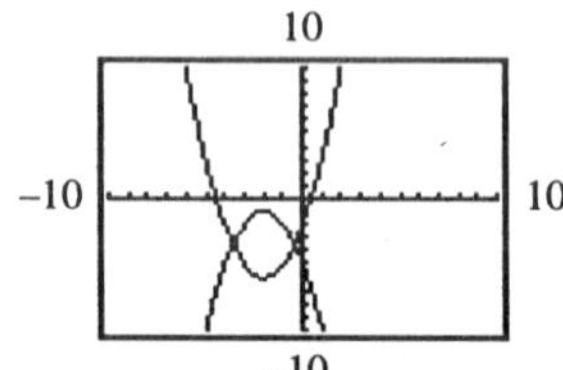

The intersections are (–4, –9) and (4, –9).

The solution is (–4, 4).

53. $x^2+4x-2>-5-4x-x^2$

$Y1=x^2+4x-2$

$Y2=-5-4x-x^2$

10

–10 10

–10

The intersections are about (–3.58, –3.5) and (–0.42, –3.5).

The solution is $(-\infty,-3.58)\cup(-0.42,\infty)$.

54. $-x^2-10x-27<x^2+10x+23$

$Y1=-x^2-10x-27$

$Y2=x^2+10x+23$

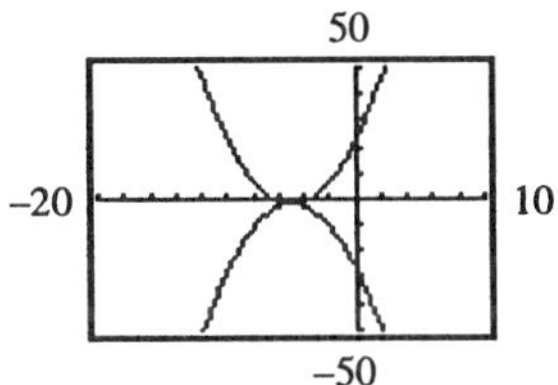

The intersection is (–5, –2).
Y1 is less than Y2 except at $x = -5$.
The solution is $(-\infty, -5) \cup (-5, \infty)$.

55. $-x^2 + 6x - 7 > -\frac{1}{3}x^2 + 2x - 4$

$Y1 = -x^2 + 6x - 7$

$Y2 = -\frac{1}{3}x^2 + 2x - 4$

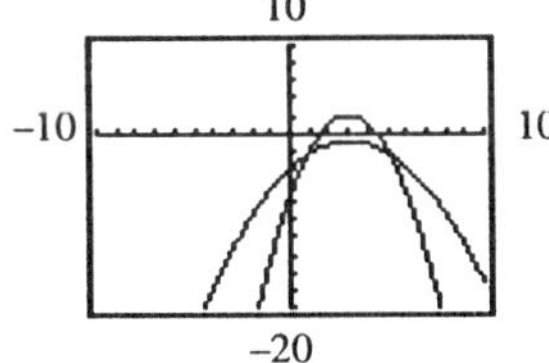

The intersections are about (0.88, –2.5) and (5.12, –2.5).
The solution is (0.88, 5.12).

56. $-x^2 + 6x - 6 \geq -\frac{1}{3}x^2 + 2x$

$Y1 = -x^2 + 6x - 6$

$Y2 = -\frac{1}{3}x^2 + 2x$

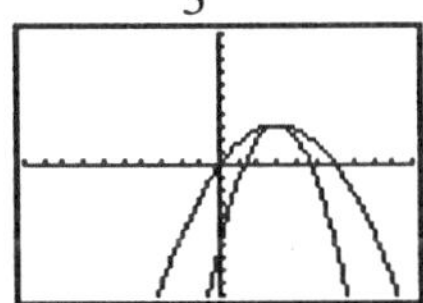

The intersection is (3, 3). Y1 is always less than Y2 except at $x = 3$.
Therefore, the solution is only 3.

57. $s(t) = -16t^2 + v_0t + s_0$

$0 = -16t^2 + 0t + 160$

$0 = -16t^2 + 160$

$t^2 = 10$

$t = \pm\sqrt{10}$

$t \approx \pm\ 3.16$

Disregard a negative time.
The notebook will reach the ground in $\sqrt{10}$ seconds or approximately 3.16 seconds.

58. $C(x) = 5x^2 + 75x + 875$

$3500 = 5x^2 + 75x + 875$

$5x^2 + 75x - 2625 = 0$

$5(x^2 + 15x - 525) = 0$

$x = \frac{-15 \pm \sqrt{(15)^2 - 4(1)(-525)}}{2(1)}$

$x = \frac{-15 \pm \sqrt{2325}}{2}$

$x = \frac{-15 \pm 5\sqrt{93}}{2}$

$x = \frac{-15 + 5\sqrt{93}}{2} \approx 16.6$

Disregard a negative value.
The number is about 16 items.

59. $s(t) = -16t^2 + v_0t + s_0$

$0 = -16t^2 + 64t + 32$

$0 = -16(t^2 - 4t - 2)$

$t = \frac{-(-4) \pm \sqrt{(-4)^2 - 4(1)(-2)}}{2(1)}$

$t = \frac{4 \pm \sqrt{24}}{2}$

$t = \frac{4 \pm 2\sqrt{6}}{2}$

$t = 2 \pm \sqrt{6}$

$t = 2 + \sqrt{6} \approx 4.45$

Disregard a negative value of time.
It will reach the ground in $2 + \sqrt{6}$ seconds or approximately 4.45 seconds.

60. $s(t) = -16t^2 + v_0t + s_0$

$4000 = -16t^2 + 0t + 11{,}000$

$16t^2 = 7000$

$t^2 = 437.5$

$t = \pm\sqrt{437.5}$

$t = \sqrt{437.5} \approx 20.92$

Disregard a negative value of time.
It will take $\sqrt{437.5}$ seconds or approximately 20.92 seconds.

61. $a^2 + b^2 = c^2$
$(34)^2 + b^2 = (42.5)^2$
$b^2 = 650.25$
$b = \pm\sqrt{650.25}$
$b = \sqrt{650.25}$
$b = 25.5$
The other leg is 25.5 meters.

62. Let $x = \text{width}$
$3x - 6 = \text{length}$
$x(3x - 6) = 144$
$3x^2 - 6x - 144 = 0$
$3(x^2 - 2x - 48) = 0$
$3(x - 8)(x + 6) = 0$
$x = 8$ or $x = -6$
Disregard a negative width. The dimensions are 8 cm by 18 cm.

63. $A = P(1+r)^t$
$1323 = 1200(1+r)^2$
$1.1025 = (1+r)^2$
$1 + r = \pm\sqrt{1.1025}$
$r = -1 \pm 1.05$
$r = -1 + 1.05 = 0.05$
Disregard a negative value.
The interest rate is 5%.

64. $A = P(1+r)^t$
$1500 = 1200(1+r)^2$
$1.25 = (1+r)^2$
$1 + r = \pm\sqrt{1.25}$
$r = -1 \pm \sqrt{1.25}$
$r \approx -1 + 1.118$
$r \approx 0.118$
Disregard a negative value.
The interest rate is approximately 11.8%.

65. $R(x) = 850 + 35x - 5x^2$
$650 = 850 + 35x - 5x^2$
$5x^2 - 35x - 200 = 0$
$5(x^2 - 7x - 40) = 0$
$$x = \frac{-(-7) \pm \sqrt{(-7)^2 - 4(1)(-40)}}{2(1)}$$
$$x = \frac{7 \pm \sqrt{209}}{2}$$
$$x = \frac{7 + \sqrt{209}}{2} \approx 10.7$$
Disregard a negative value.
A value of about 11 will yield a revenue of about $650.

66. $a^2 + b^2 = c^2$
$(20)^2 + (48)^2 = c^2$
$c^2 = 2704$
$c = \pm\sqrt{2704}$
$c = \sqrt{2704}$
$c = 52$
Disregard a negative length.
The hypotenuse is 52 inches.

67. $a^2 + b^2 = c^2$
$a^2 + (350)^2 = (370)^2$
$a^2 = 14,400$
$a = \pm\sqrt{14,400}$
$a = \sqrt{14,400}$
$a = 120$
Disregard a negative distance.
The pond is 120 ft wide.

68. Let x = length of one leg
$3x - 1$ = length of other leg
$a^2 + b^2 = c^2$
$x^2 + (3x - 1)^2 = (37)^2$
$x^2 + 9x^2 - 6x + 1 = 1369$
$10x^2 - 6x - 1368 = 0$
$2(5x^2 - 3x - 684) = 0$
$2(5x + 57)(x - 12) = 0$
$5x + 57 = 0$ or $x - 12 = 0$
$x = -\frac{57}{5}$ $\quad x = 12$
Disregard a negative length. The legs measure 12 in. and 35 in.

69. $c = a\sqrt{2}$

$14.21 = a\sqrt{2}$

$\dfrac{14.21}{\sqrt{2}} = a$

$a = \dfrac{14.21\sqrt{2}}{2}$

$a = 7.105\sqrt{2}$

They each measure $7.105\sqrt{2}$ cm or approximately 10 cm.

70. $c = a\sqrt{2}$

$c = 11\sqrt{2}$

The rope is $11\sqrt{2}$ feet long or approximately 15.6 feet.

71. $2a = 32$

$a = 16$

$a\sqrt{3} = 16\sqrt{3}$

The sides measure 16 in. and $16\sqrt{3}$ in.

72. $C(x) = 5x^2 + 20x + 100$

$500 \ge 5x^2 + 20x + 100$

$\text{Y1} = 500$

$\text{Y2} = 5x^2 + 20x + 100$

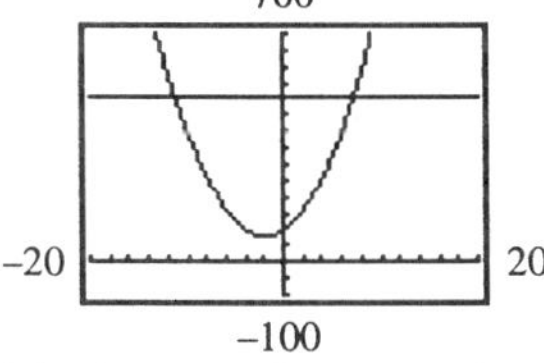

The intersections are about (–11.2, 500) and (7.2, 500). The number of days can only be positive. Therefore, the solution is (0, 7.2]. The booth can be rented up to 7 days.

Chapter 11 Test

1. $x^2 - 4x - 9 = 2x + 7$

$\text{Y1} = x^2 - 4x - 9$

$\text{Y2} = 2x + 7$

X	Y1	Y2
-3	12	1
-2	3	3
-1	-4	5
0	-9	7
1	-12	9
2	-13	11
3	-12	13

X=-3

X	Y1	Y2
4	-9	15
5	-4	17
6	3	19
7	12	21
8	23	23
9	36	25
10	51	27

X=4

The solutions are –2 and 8.

2. $2x^3 + 9x^2 - 23x - 66 = 0$

$\text{Y1} = 2x^3 + 9x^2 - 23x - 66$

$\text{Y2} = 0$

50
–20 20
–85

The graphs intersect at (3, 0), (–2, 0), and (–5.5, 0). The solutions are –5.5, –2, and 3.

3. $4(x-5)^2 + 7 = 71$

$4(x-5)^2 = 64$

$(x-5)^2 = 16$

$x - 5 = \sqrt{16}$ or $x - 5 = -\sqrt{16}$

$x - 5 = 4$ $\quad x - 5 = -4$

$x = 9$ $\quad x = 1$

The solutions are 1 and 9.

4. $x^2 - 6x + 13 = 0$

$x = \dfrac{-b \pm \sqrt{b^2 - 4ac}}{2a}$

$x = \dfrac{-(-6) \pm \sqrt{(-6)^2 - 4(1)(13)}}{2(1)}$

$x = \dfrac{6 \pm \sqrt{-16}}{2}$

There is no real-number solution.

5. $2x^2 - 7x + 3 = 2x - 1$
   $Y1 = 2x^2 - 7x + 3$
   $Y2 = 2x - 1$

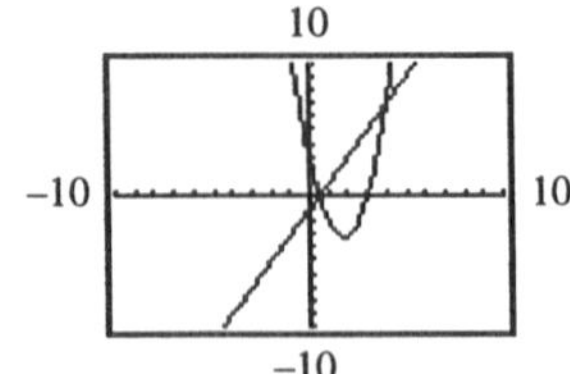

   The graphs intersect at (0.5, 0) and (4, 7). The solutions are 0.5 and 4.

6. $x^2 - 5x + 4 = 7 - 3x - x^2$
   $Y1 = x^2 - 5x + 4$
   $Y2 = 7 - 3x - x^2$

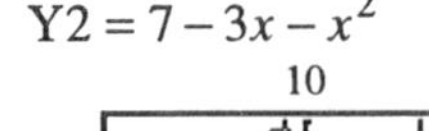

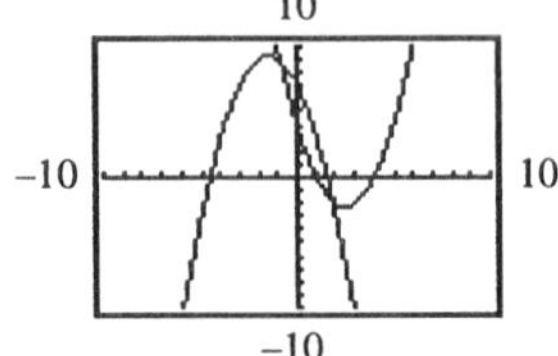

   The graphs intersect at about (–0.82, 8.79) and (1.82, –1.79). The solutions are –0.82 and 1.82.

7. $x^2 - x - 6 \le 6$
   $Y1 = x^2 - x - 6$
   $Y2 = 6$

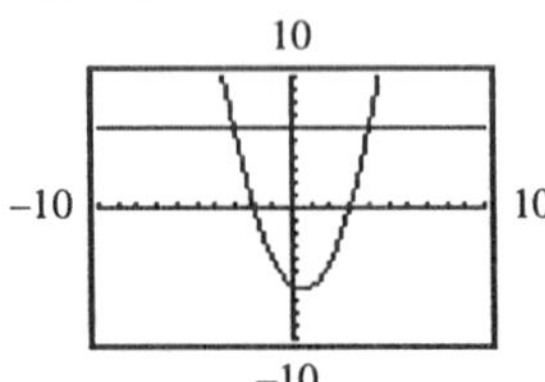

   The graphs intersect at (–3, 6) and (4, 6). The solution is [–3, 4].

8. $(x + 3)(x - 2) < 3(x + 3)$
   $Y1 = (x + 3)(x - 2)$
   $Y2 = 3(x + 3)$

X	Y1	Y2
-5	14	-6
-4	6	-3
-3	0	0
-2	-4	3
-1	-6	6
0	-6	9
1	-4	12

X=-5

X	Y1	Y2
2	0	15
3	6	18
4	14	21
5	24	24
6	36	27
7	50	30
8	66	33

X=2

   The solutions are all integers greater than –3 and less than 5.

9. $2x^2 + 7x - 9 \ge 4x + 11$
   Determine the critical points.
   $2x^2 + 3x - 20 = 0$
   $(2x - 5)(x + 4) = 0$
   $x = \frac{5}{2}$ $\quad x = -4$
   Possible intervals:
   $(-\infty, -4], \left[-4, \frac{5}{2}\right], \left[\frac{5}{2}, \infty\right)$
   Use a test point in each interval to determine the solution is $(-\infty, -4] \cup \left[\frac{5}{2}, \infty\right)$.

10. $x^2 - 9x + 9 = 7x - 5$
    $x^2 - 16x + 14 = 0$
    $b^2 - 4ac = (-16)^2 - 4(1)(14) = 200$
    There are two irrational solutions because 200 is positive but not a perfect square.
    $x = \frac{-b \pm \sqrt{b^2 - 4ac}}{2a}$
    $x = \frac{16 \pm \sqrt{200}}{2}$
    $x = \frac{16 \pm 10\sqrt{2}}{2}$
    $x = 8 \pm 5\sqrt{2}$
    The roots are $8 \pm 5\sqrt{2}$.

11. $x^2 - 3x + 12 = 0$
    $b^2 - 4ac = (-3)^2 - 4(1)(12) = -39$
    There are no real solutions because the discriminant is negative.

12. $s = at^2 + c$

$s - c = at^2$

$\frac{s-c}{a} = t^2$

$t = \pm\sqrt{\frac{s-c}{a}}$

$t = \pm\frac{\sqrt{a(s-c)}}{a}$

13. $x^2 - 2 = x + 4$

$Y1 = x^2 - 2$

$Y2 = x + 4$

a.

X	Y1	Y2
-3	7	1
-2	2	2
-1	-1	3
0	-2	4
1	-1	5
2	2	6
3	7	7

X= -3

The solutions are –2 and 3.

b.

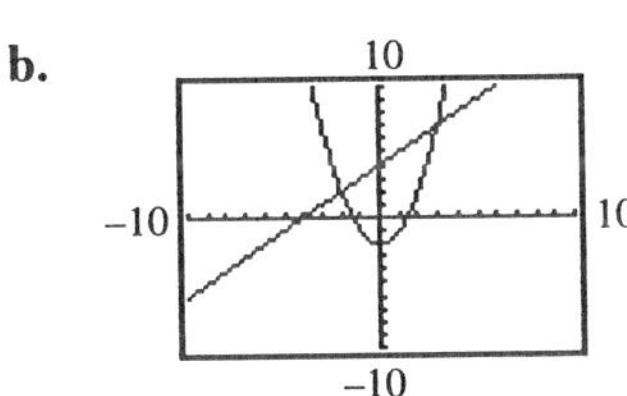

The graphs intersect at (–2, 2) and (3, 7). The solutions are –2 and 3.

c. $x^2 - 2 = x + 4$

$x^2 - x - 6 = 0$

$(x - 3)(x + 2) = 0$

$x - 3 = 0$ or $x + 2 = 0$

$x = 3$ $\quad x = -2$

The solutions are –2 and 3.

14. Let x = length of one leg

$x + 4$ = length of other leg

$a^2 + b^2 = c^2$

$x^2 + (x+4)^2 = (20)^2$

$x^2 + x^2 + 8x + 16 = 400$

$2x^2 + 8x - 384 = 0$

$2(x^2 + 4x - 192) = 0$

$2(x + 16)(x - 12) = 0$

$x = -16$ or $x = 12$

Disregard a negative length. The legs are 12 in. and 16 in.

15. $s = -16t^2 + v_0t + s_0$

$2600 = -16t^2 + 0 \cdot t + 11{,}500$

$-16t^2 = -8900$

$t^2 = 556.25$

$t = \pm\sqrt{556.25}$

$t \approx 23.6$

Disregard a negative value of time. Keanu was free-falling for about 24 seconds.

16. $A = P(1+r)^t$

$1573 = 1400(1+r)^2$

$\frac{1573}{1400} = (1+r)^2$

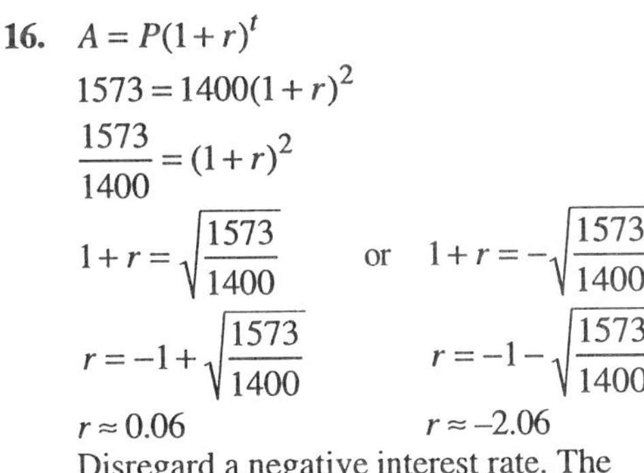

$1 + r = \sqrt{\frac{1573}{1400}}$ or $1 + r = -\sqrt{\frac{1573}{1400}}$

$r = -1 + \sqrt{\frac{1573}{1400}}$ $\quad r = -1 - \sqrt{\frac{1573}{1400}}$

$r \approx 0.06$ $\quad r \approx -2.06$

Disregard a negative interest rate. The interest rate must be approximately 6%.

17. $C(x) = 40 + 25x - 2x^2$

$100 \geq 40 + 25x - 2x^2$

$Y1 = 100$

$Y2 = 40 + 25x - 2x^2$

120

–20 20

–10

The graphs intersect at about (3.2, 100) and (9.3, 100).

The solution is (0, 3.2] ∪ [9.3, ∞). Since the rental is limited to 7 days, x = 1, 2, or 3. You can rent for up to 3 additional days.

18. Answers will vary.

Chapter 12

12.1 Experiencing Algebra the Exercise Way

1. $\dfrac{x^2+10x+25}{x+5}$ is a rational expression and an algebraic fraction because $(x^2+10x+25)$ and $(x+5)$ are both polynomials.

3. $z-4-\dfrac{25}{z-4}$ is a rational expression because it can be written as $\left(\dfrac{z}{1}-\dfrac{4}{1}-\dfrac{25}{z-4}\right)$.

5. $\dfrac{x+3}{\sqrt{x}-2}$ is not a rational expression because $\left(\sqrt{x}-2\right)$ is not a polynomial.

7. $6x^2+17x-3$ is a rational expression.

9. $\dfrac{11}{3xy}$ is a rational expression and an algebraic fraction because 11 and $3xy$ are both polynomials.

11. $3x^{-2}+4x^{-1}+3-x$ is a rational expression because it can be written as $\left(\dfrac{3}{x^2}+\dfrac{4}{x}+\dfrac{3}{1}-\dfrac{x}{1}\right)$.

13. $\dfrac{0.5x+7}{1.2x-4}$ is a rational expression and an algebraic fraction because $(0.5x+7)$ and $(1.2x-4)$ are both polynomials.

15. $\dfrac{2-3\sqrt{x}}{2x}$ is not a rational expression because $\left(2-3\sqrt{x}\right)$ is not a polynomial.

17. $y=\dfrac{3-x}{x}$
$x=0$
The restricted value is 0.
Domain: $x\neq 0$
$(-\infty, 0)\cup(0, \infty)$

19. $h(x)=\dfrac{x-5}{x^2-11x+30}$
$x^2-11x+30=0$
$(x-5)(x-6)=0$
$x=5 \quad x=6$
The restricted values are 5 and 6.
Domain: $x\neq 5, x\neq 6$
$(-\infty, 5)\cup(5, 6)\cup(6, \infty)$

21. $C(x)=\dfrac{x+7}{x^2-4}$
$x^2-4=0$
$(x+2)(x-2)=0$
$x=-2 \quad x=2$
The restricted values are –2 and 2.
Domain: $x\neq -2, x\neq 2$
$(-\infty, -2)\cup(-2, 2)\cup(2, \infty)$

23. $y=\dfrac{2x^2+7}{x^2-5x}$
$x^2-5x=0$
$x(x-5)=0$
$x=0 \quad x=5$
The restricted values are 0 and 5.
Domain: $x\neq 0, x\neq 5$
$(-\infty, 0)\cup(0, 5)\cup(5, \infty)$

25. $f(x)=\dfrac{x+2}{x^3-2x^2+4x-8}$
$x^3-2x^2+4x-8=0$
$x^2(x-2)+4(x-2)=0$
$(x-2)(x^2+4)=0$
$x=2$
The restricted value is 2.
Domain: $x\neq 2$
$(-\infty, 2)\cup(2, \infty)$

27. $f(x)=\dfrac{7}{2x^2+17x+35}$

$2x^2+17x+35=0$
$(2x+7)(x+5)=0$
$x=-\dfrac{7}{2}\quad x=-5$

The restricted values are –5 and $-\dfrac{7}{2}$.

Domain: $x\neq -5,\ x\neq -\dfrac{7}{2}$

$(-\infty,-5)\cup\left(-5,-\dfrac{7}{2}\right)\cup\left(-\dfrac{7}{2},\infty\right)$

29. $y=\dfrac{x^2-3x}{8x^2+6x-9}$

$8x^2+6x-9=0$
$(4x-3)(2x+3)=0$
$x=\dfrac{3}{4}\quad x=-\dfrac{3}{2}$

The restricted values are $-\dfrac{3}{2}$ and $\dfrac{3}{4}$.

Domain: $x\neq -\dfrac{3}{2},\ x\neq \dfrac{3}{4}$

$\left(-\infty,-\dfrac{3}{2}\right)\cup\left(-\dfrac{3}{2},\dfrac{3}{4}\right)\cup\left(\dfrac{3}{4},\infty\right)$

31. $y=\dfrac{4x}{4x^2+36x+81}$

$4x^2+36x+81=0$
$(2x+9)^2=0$
$x=-\dfrac{9}{2}$

The restricted value is $-\dfrac{9}{2}$.

Domain: $x\neq -\dfrac{9}{2}$

$\left(-\infty,-\dfrac{9}{2}\right)\cup\left(-\dfrac{9}{2},\infty\right)$

33. $h(x)=\dfrac{6-x}{x^2-22x+121}$

$x^2-22x+121=0$
$(x-11)^2=0$
$x=11$

The restricted value is 11.
Domain: $x\neq 11$
$(-\infty,11)\cup(11,\infty)$

35. $y=\dfrac{3x^2}{9x^2-25}$

$9x^2-25=0$
$x^2=\dfrac{25}{9}$
$x=\pm\sqrt{\dfrac{25}{9}}$
$x=\pm\dfrac{5}{3}$

The restricted values are $-\dfrac{5}{3}$ and $\dfrac{5}{3}$.

Domain: $x\neq -\dfrac{5}{3},\ x\neq \dfrac{5}{3}$

$\left(-\infty,-\dfrac{5}{3}\right)\cup\left(-\dfrac{5}{3},\dfrac{5}{3}\right)\cup\left(\dfrac{5}{3},\infty\right)$

37. $C(x)=\dfrac{5x^3+1}{2x^3+4x^2-18x-36}$

$2x^3+4x^2-18x-36=0$
$2x^2(x+2)-18(x+2)=0$
$(x+2)(2x^2-18)=0$
$2(x+2)(x+3)(x-3)=0$
$x=-2\quad x=-3\quad x=3$
The restricted values are –3, –2, and 3.
Domain: $x\neq -3,\ x\neq -2,\ x\neq 3$
$(-\infty,-3)\cup(-3,-2)\cup(-2,3)\cup(3,\infty)$

39. $y=\dfrac{3x+8}{2x^2+5x+7}$

$2x^2+5x+7=0$
$x=\dfrac{-5\pm\sqrt{5^2-4(2)(7)}}{2(2)}$
$x=\dfrac{-5\pm\sqrt{-31}}{4}$
There are no restricted values.
Domain: all real numbers
$(-\infty,\infty)$

41. $y = \dfrac{x+4}{2x+3}$

$2x+3=0$

$x = -\dfrac{3}{2}$

The restricted value is $-\dfrac{3}{2}$.

Domain: $x \neq -\dfrac{3}{2}$

$\left(-\infty, -\dfrac{3}{2}\right) \cup \left(-\dfrac{3}{2}, \infty\right)$

43.

x	$y = \frac{6}{x} + \frac{9}{x^2}$	y
–3	$y = \frac{6}{-3} + \frac{9}{(-3)^2} = -1$	–1
–1	$y = \frac{6}{-1} + \frac{9}{(-1)^2} = 3$	3
0	$y = \frac{6}{0} + \frac{9}{0^2}$ is undefined	undefined
1	$y = \frac{6}{1} + \frac{9}{1^2} = 15$	15
3	$y = \frac{6}{3} + \frac{9}{(3)^2} = 3$	3

The solutions are (–3, –1), (–1, 3), (0, undefined), (1, 15), and (3, 3).

45.

x	$h(x) = \frac{x^2-49}{x-7}$	$h(x)$
–3	$h(-3) = \frac{(-3)^2-49}{-3-7} = 4$	4
–1	$h(-1) = \frac{(-1)^2-49}{-1-7} = 6$	6
0	$h(0) = \frac{0^2-49}{0-7} = 7$	7
1	$h(1) = \frac{1^2-49}{1-7} = 8$	8
3	$h(3) = \frac{3^2-49}{3-7} = 10$	10

The solutions are (–3, 4), (–1, 6), (0, 7), (1, 8), and (3, 10).

47.

x	$y = \frac{2x}{x^2+3}$	y
–3	$y = \frac{2(-3)}{(-3)^2+3} = -\frac{1}{2}$	$-\frac{1}{2}$
–1	$y = \frac{2(-1)}{(-1)^2+3} = -\frac{1}{2}$	$-\frac{1}{2}$
0	$y = \frac{2(0)}{0^2+3} = 0$	0
1	$y = \frac{2(1)}{1^2+3} = \frac{1}{2}$	$\frac{1}{2}$
3	$y = \frac{2(3)}{3^2+3} = \frac{1}{2}$	$\frac{1}{2}$

The solutions are $\left(-3, -\dfrac{1}{2}\right)$, $\left(-1, -\dfrac{1}{2}\right)$, (0, 0), $\left(1, \dfrac{1}{2}\right)$, and $\left(3, \dfrac{1}{2}\right)$.

49.

x	$y = \frac{50}{x^2} + \frac{15}{x}$	y
–3	$y = \frac{50}{(-3)^2} + \frac{15}{-3} = \frac{5}{9}$	$\frac{5}{9}$
–1	$y = \frac{50}{(-1)^2} + \frac{15}{-1} = 35$	35
0	$y = \frac{5^0}{0^2} + \frac{15}{0}$ is undefined	undefined
1	$y = \frac{50}{1^2} + \frac{15}{1} = 65$	65
3	$y = \frac{50}{3^2} + \frac{15}{3} = \frac{95}{9}$	$\frac{95}{9}$

The solutions are $\left(-3, \dfrac{5}{9}\right)$, (–1, 35), (0, undefined), (1, 65), and $\left(3, \dfrac{95}{9}\right)$.

51.

x	$f(x)=\frac{x^2+1}{x-3}$	$f(x)$
−3	$f(-3)=\frac{(-3)^2+1}{-3-3}=-\frac{5}{3}$	$-\frac{5}{3}$
−1	$f(-1)=\frac{(-1)^2+1}{-1-3}=-\frac{1}{2}$	$-\frac{1}{2}$
0	$f(0)=\frac{0^2+1}{0-3}=-\frac{1}{3}$	$-\frac{1}{3}$
1	$f(1)=\frac{1^2+1}{1-3}=-1$	−1
3	$f(3)=\frac{3^2+1}{3-3}$ is undefined	undefined

The solutions are $\left(-3,-\frac{5}{3}\right)$, $\left(-1,-\frac{1}{2}\right)$, $\left(0,-\frac{1}{3}\right)$, (1, −1), and (3, undefined).

53.

x	$R(x)=\dfrac{2x^2+7x+6}{x+2}$	$R(x)$
−3	$R(-3)=\frac{2(-3)^2+7(-3)+6}{-3+2}=-3$	−3
−1	$R(-1)=\frac{2(-1)^2+7(-1)+6}{-1+2}=1$	1
0	$R(0)=\frac{2(0)^2+7(0)+6}{0+2}=3$	3
1	$R(1)=\frac{2(1)^2+7(1)+6}{1+2}=5$	5
3	$R(3)=\frac{2(3)^2+7(3)+6}{3+2}=9$	9

The solutions are (−3, −3), (−1, 1), (0, 3), (1, 5), and (3, 9).

55. $y=\dfrac{8}{x}$

The restricted value is 0. Set up a table of values using x-values less than and greater than 0. Connect the points with a smooth curve.

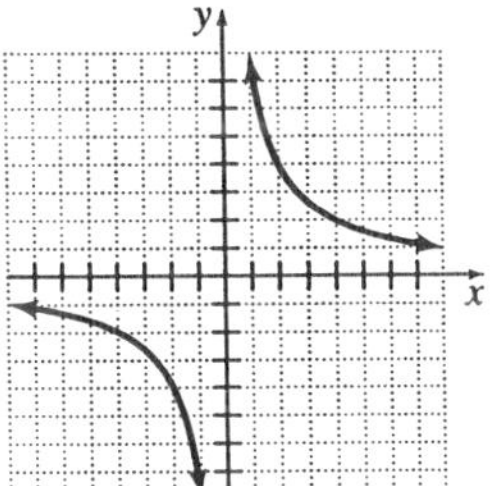

57. $y=\dfrac{x-5}{x+3}$

The restricted value is −3. Set up a table of values using x-values less than and greater than −3. Connect the points with a smooth curve.

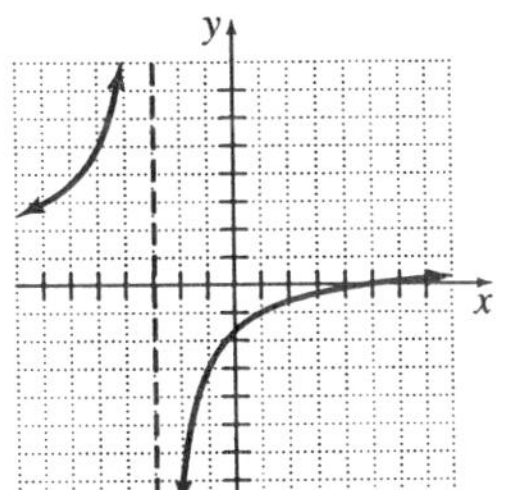

59. $f(x)=\dfrac{2x+3}{x^2-2x-3}$

$x^2-2x-3=0$

$(x-3)(x+1)=0$

$x=3 \quad x=-1$

The restricted values are −1 and 3. Set up a table of x-values less than −1, between −1 and 3, and greater than 3. Connect the points with a smooth curve.

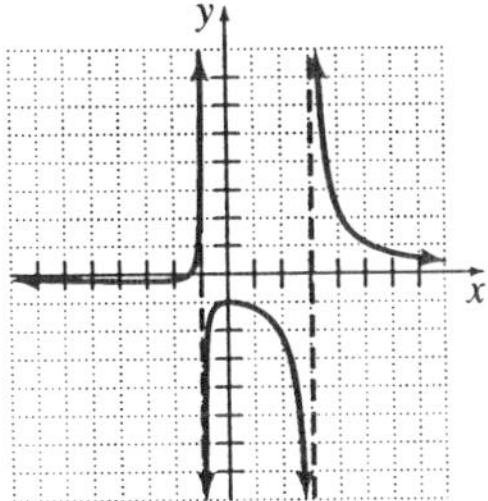

61. $g(x) = \dfrac{2x^2 - 5x + 2}{x - 2}$
The restricted value is 2. Set up a table of x-values less than and greater than 2. Connect the points with a smooth curve.

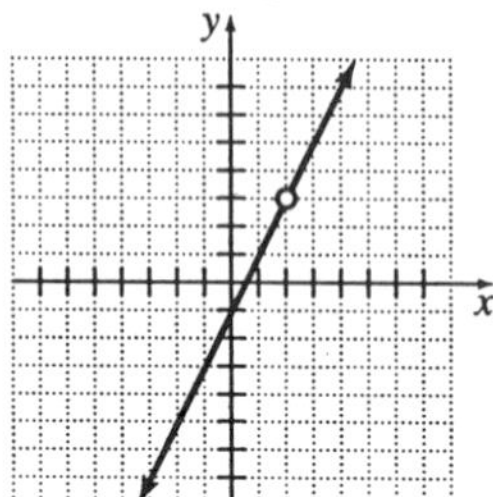

63. **a.** $t = \dfrac{d}{r}$

running time: $\dfrac{26}{x}$

bicycle time: $\dfrac{110}{x+20}$

The total time: $T(x) = \dfrac{26}{x} + \dfrac{110}{x+20}$

b. $\text{Y1} = \dfrac{26}{x} + \dfrac{110}{x+20}$

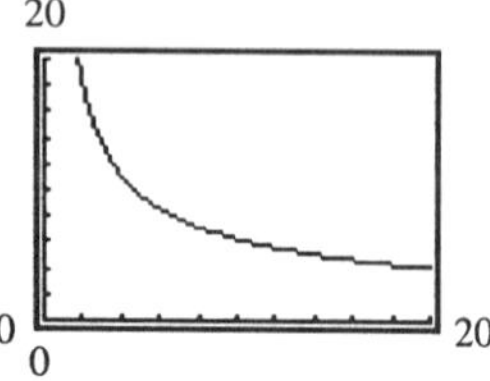

c. At $x = 9$, $y \approx 6.7$
The total time is approximately 6.7 hours.

d. At $y = 8$, $x \approx 6.6$
The average speeds for running and bicycling are approximately 6.6 mph and 26.6 mph, respectively.

65. $F(s) = \dfrac{10}{(s+3)^2}$

$\text{Y1} = \dfrac{10}{(x+3)^2}$

X	Y1
-4	10
-3	ERROR
-2	10
-1	2.5
0	1.1111
1	.625
2	.4

X=-4

X	Y1
-2	10
-1	2.5
0	1.1111
1	.625
2	.4
3	.27778
4	.20408

X=-2

12.1 Experiencing Algebra the Calculator Way

1. $y = \dfrac{6}{x-3} + \dfrac{6}{x+3}$
(4, 6.8571), (8, 1.7455), and (10, 1.3187)

2. $y = \dfrac{x^2 - 9}{x - 3}$
(–6, –3), (–1, 2), (0, 3),
(2, 5), (3, undefined), (5, 8)

3. $y = \dfrac{x^2 + 9}{x - 3}$
(–6, –5), (–1, –2.5), (0, –3),
(2, –13), (3, undefined), (5, 17)

4. $y = \dfrac{12x^2}{3x}$
(–5, –20), (–2, –8), (0, undefined),
(1, 4), (3, 12), (6, 24)

5. $y = \dfrac{3x}{12x^2}$
(–5, –0.05), (–2, –0.125), (0, undefined),
(1, 0.25), (3, 0.08333), (6, 0.04167)

6. $y = \dfrac{x^2 + x - 20}{x^2 + 6x + 5}$
(–2.5, 4.3333), (–1.3, 17.667), (0, –4),
(1.8, –0.7857), (2.7, –0.3514), (5.9, 0.27536)

7. $y = \dfrac{x - 4}{x + 1}$
(–2.5, 4.3333), (–1.3, 17.667)
(0, –4), (1.8, –0.7857),
(2.7, –0.3514), (5.9, 0.27536)

12.2 Experiencing Algebra the Exercise Way

1. $\dfrac{36x^3y^2z}{54x^2y^5}$
$= \dfrac{(18x^2y^2)(2xz)}{(18x^2y^2)(3y^3)}$
$= \dfrac{2xz}{3y^3}$

3. $\dfrac{-56a^3b^2}{84a^2b^2c}$
$= \dfrac{(28a^2b^2)(-2a)}{(28a^2b^2)(3c)}$
$= -\dfrac{2a}{3c}$

5. $\dfrac{9x-27}{18x-6}$
$= \dfrac{9(x-3)}{6(3x-1)}$
$= \dfrac{3\cdot 3(x-3)}{3\cdot 2(3x-1)}$
$= \dfrac{3(x-3)}{2(3x-1)}$

7. $\dfrac{8-z}{z-8}$
$= \dfrac{-1(z-8)}{z-8}$
$= -1$

9. $\dfrac{5x^2y^3+10xy^4}{15xy^2-20x^2y^2}$
$= \dfrac{5xy^3(x+2y)}{5xy^2(3-4x)}$
$= \dfrac{y(x+2y)}{3-4x}$

11. $\dfrac{-6x+9y-3z}{3x-21y+3z}$
$= \dfrac{3(-2x+3y-z)}{3(x-7y+z)}$
$= \dfrac{-2x+3y-z}{x-7y+z}$

13. $\dfrac{x^2-8x+12}{x^2-10x+24}$
$= \dfrac{(x-6)(x-2)}{(x-6)(x-4)}$
$= \dfrac{x-2}{x-4}$

15. $\dfrac{x^2-x-6}{x^2+9x+14}$
$= \dfrac{(x+2)(x-3)}{(x+2)(x+7)}$
$= \dfrac{x-3}{x+7}$

17. $\dfrac{3x^2+7x+2}{3x^2-11x-4}$
$= \dfrac{(3x+1)(x+2)}{(3x+1)(x-4)}$
$= \dfrac{x+2}{x-4}$

19. $\dfrac{6x^2-x-2}{4x^2-4x-3}$
$= \dfrac{(2x+1)(3x-2)}{(2x+1)(2x-3)}$
$= \dfrac{3x-2}{2x-3}$

21. $\frac{8x^2+12x+4}{24x^2-40x-16}$
$=\frac{4(2x^2+3x+1)}{8(3x^2-5x-2)}$
$=\frac{4(x+1)(2x+1)}{8(x-2)(3x+1)}$
$=\frac{(x+1)(2x+1)}{2(x-2)(3x+1)}$

23. $\frac{2x^2+xy-y^2}{3x^2+4xy+y^2}$
$=\frac{(2x-y)(x+y)}{(3x+y)(x+y)}$
$=\frac{2x-y}{3x+y}$

25. $\frac{64-x^2}{x^3+2x^2-64x-128}$
$=\frac{64-x^2}{x^2(x+2)-64(x+2)}$
$=\frac{64-x^2}{(x+2)(x^2-64)}$
$=\frac{-1(x^2-64)}{(x+2)(x^2-64)}$
$=-\frac{1}{x+2}$

27. $\frac{2x^2-7x-15}{x^3-5x^2+11x-55}$
$=\frac{(2x+3)(x-5)}{x^2(x-5)+11(x-5)}$
$=\frac{(x-5)(2x+3)}{(x-5)(x^2+11)}$
$=\frac{2x+3}{x^2+11}$

29. $\frac{12a^2b^2}{35cd^2}\cdot\frac{49c^2d^2}{27ab^3}$
$=\frac{12a^2b^2\cdot 49c^2d^2}{35cd^2\cdot 27ab^3}$
$=\frac{588a^2b^2c^2d^2}{945ab^3cd^2}$
$=\frac{28ac}{45b}$

31. $\frac{-2x^3}{5y^2}\cdot\frac{15xy^2}{8x^2}$
$=\frac{-2x^3\cdot 15xy^2}{5y^2\cdot 8x^2}$
$=\frac{-30x^4y^2}{40x^2y^2}$
$=-\frac{3x^2}{4}$

33. $(2.6x^2)\cdot\frac{3.1x}{y^2}$
$=\frac{(2.6x^2)\cdot 3.1x}{y^2}$
$=\frac{8.06x^3}{y^2}$

35. $\frac{1}{3}\cdot\frac{2a^2}{3b}$
$=\frac{1.2a^2}{3\cdot 3b}$
$=\frac{2a^2}{9b}$

37. $\frac{x+5}{x}\cdot\frac{x^3}{x+4}$
$=\frac{(x+5)x^3}{x(x+4)}$
$=\frac{x^2(x+5)}{x+4}$

39. $\dfrac{3x+2}{x+4}\cdot\dfrac{x-4}{3x+2}$
$=\dfrac{(3x+2)(x-4)}{(3x+2)(x+4)}$
$=\dfrac{x-4}{x+4}$

41. $\dfrac{x+7}{2x+1}\cdot\dfrac{5x+2}{x}$
$=\dfrac{(x+7)(5x+2)}{x(2x+1)}$

43. $\dfrac{x^2-x-6}{2x^2+3x+1}\cdot\dfrac{2x^2-x-1}{x^2+5x+6}$
$=\dfrac{(x^2-x-6)(2x^2-x-1)}{(2x^2+3x+1)(x^2+5x+6)}$
$=\dfrac{(x-3)(x+2)(2x+1)(x-1)}{(2x+1)(x+1)(x+3)(x+2)}$
$=\dfrac{(x-3)(x-1)}{(x+1)(x+3)}$

45. $\dfrac{x^2-xy-2y^2}{x^2-9y^2}\cdot\dfrac{x^2-3xy}{x^2-y^2}$
$=\dfrac{(x^2-xy-2y^2)(x^2-3xy)}{(x^2-9y^2)(x^2-y^2)}$
$=\dfrac{(x-2y)(x+y)(x)(x-3y)}{(x-3y)(x+3y)(x+y)(x-y)}$
$=\dfrac{x(x-2y)}{(x+3y)(x-y)}$

47. $\dfrac{54a^3b^2}{c}\div\dfrac{36ab^2}{c}$
$=\dfrac{54a^3b^2}{c}\cdot\dfrac{c}{36ab^2}$
$=\dfrac{54a^3b^2c}{36ab^2c}$
$=\dfrac{3a^2}{2}$

49. $\dfrac{-21x^2y^3}{z}\div\dfrac{7xy^2}{z^2}$
$=\dfrac{-21x^2y^3}{z}\cdot\dfrac{z^2}{7xy^2}$
$=\dfrac{-21x^2y^3z^2}{7xy^2z}$
$=-3xyz$

51. $(32ab^2)\div\dfrac{8ab^3}{c}$
$=(32ab^2)\cdot\dfrac{c}{8ab^3}$
$=\dfrac{32ab^2c}{8ab^3}$
$=\dfrac{4c}{b}$

53. $\dfrac{21x^2y^2}{z^3}\div(-3xy^3)$
$=\dfrac{21x^2y^2}{z^3}\cdot\dfrac{1}{-3xy^3}$
$=\dfrac{21x^2y^2}{-3xy^3z^3}$
$=-\dfrac{7x}{yz^3}$

55. $\dfrac{a+4}{9a^2-1}\div\dfrac{5a+20}{3a+1}$
$=\dfrac{a+4}{9a^2-1}\cdot\dfrac{3a+1}{5a+20}$
$=\dfrac{(a+4)(3a+1)}{(3a+1)(3a-1)(5)(a+4)}$
$=\dfrac{1}{5(3a-1)}$

57. $\frac{x^2-16}{5y-2} \div \frac{x+4}{25y^2-4}$
$= \frac{x^2-16}{5y-2} \cdot \frac{25y^2-4}{x+4}$
$= \frac{(x+4)(x-4)(5y-2)(5y+2)}{(5y-2)(x+4)}$
$= (x-4)(5y+2)$

59. $\frac{2x^2-5x-3}{x^2-5x-14} \div \frac{3x^2+10x+3}{x^2-4x-21}$
$= \frac{2x^2-5x-3}{x^2-5x-14} \cdot \frac{x^2-4x-21}{3x^2+10x+3}$
$= \frac{(2x+1)(x-3)(x+3)(x-7)}{(x+2)(x-7)(x+3)(3x+1)}$
$= \frac{(2x+1)(x-3)}{(x+2)(3x+1)}$

61. $\frac{8x^2+2x-3}{4x^2-1} \div (4x+3)$
$= \frac{8x^2+2x-3}{4x^2-1} \cdot \frac{1}{4x+3}$
$= \frac{(4x+3)(2x-1)}{(2x+1)(2x-1)(4x+3)}$
$= \frac{1}{2x+1}$

63. $(x+9) \div \frac{2x^2+19x+9}{3x+4}$
$= (x+9) \cdot \frac{3x+4}{2x^2+19x+9}$
$= \frac{(x+9)(3x+4)}{(2x+1)(x+9)}$
$= \frac{3x+4}{2x+1}$

65. $\frac{1}{3x-5} \div \frac{1}{x+7}$
$= \frac{1}{3x-5} \div \frac{x+7}{1}$
$= \frac{x+7}{3x-5}$

67. $\frac{x^2+3xy}{7xy+2y^2} \div \frac{x^2+4xy+3y^2}{7x^2-5xy-2y^2}$
$= \frac{x^2+3xy}{7xy+2y^2} \cdot \frac{7x^2-5xy-2y^2}{x^2+4xy+3y^2}$
$= \frac{x(x+3y)(7x+2y)(x-y)}{y(7x+2y)(x+3y)(x+y)}$
$= \frac{x(x-y)}{y(x+y)}$

69. $\frac{4x+8y}{3x-12y} \div \frac{4x+4y}{6x+6y}$
$= \frac{4x+8y}{3x-12y} \cdot \frac{6x+6y}{4x+4y}$
$= \frac{4(x+2y)(6)(x+y)}{3(x-4y)(4)(x+y)}$
$= \frac{2(x+2y)}{(x-4y)}$

71. $t = \frac{d}{r}$
$t = \frac{170}{x}$
The expression for the time traveled is $\frac{170}{x}$ hours.
$d = rt$
$d = (x-10)\left(\frac{170}{x}\right)$
The expression for the new distance is $(x-10)\left(\frac{170}{x}\right)$ miles.
$d = (x-10)\left(\frac{170}{x}\right)$
$d = (50-10)\left(\frac{170}{50}\right)$
$d = 136$
The distance is 136 miles.

73. $V = LWH$

$15x^3 + 7x^2 - 2x$

$= x(15x^2 + 7x - 2)$

$= x(3x+2)(5x-1)$

The length is $(5x - 1)$ inches.

height: 4 in.

width: $(3\cdot 4 + 2) = 14$ in.

length: $(5\cdot 4 - 1) = 19$ in.

75. **a.** $C(x) = 300 + 5x$

$R(x) = 25x$

b. $\dfrac{25x}{300+5x}$

$= \dfrac{5(5x)}{5(60+x)}$

$= \dfrac{5x}{60+x}$

c. $\dfrac{5x}{60+x}$

$= \dfrac{5(40)}{60+40}$

$= \dfrac{200}{100}$

$= 2$

The ratio is 2.

12.2 Experiencing Algebra the Calculator Way

Expression	Restricted Values	Algebraically Simplified Expression	Is the simplified function linear?
1. $\frac{x^2-9}{x-3}$	$x=3$	$x+3$	yes
2. $\frac{x^2+9}{x-3}$	$x=3$	$\frac{x^2+9}{x-3}$	no
3. $\frac{12x^2}{3x}$	$x=0$	$4x$	yes
4. $\frac{3x}{12x^2}$	$x=0$	$\frac{1}{4x}$	no
5. $\frac{5x^2-10x}{5x}$	$x=0$	$x-2$	yes
6. $\frac{2x^2-3x}{2x^2-5x}$	$x=0, \frac{5}{2}$	$\frac{2x-3}{2x-5}$	no
7. $\frac{4-x}{x-4}$	$x=4$	-1	yes
8. $\frac{x^2+x-20}{x^2+6x+5}$	$x=-5, -1$	$\frac{x-4}{x+1}$	no
9. $\frac{x^3-6x^2-16x}{2x}$	$x=0$	$\frac{x^2-6x-16}{2}$	no
10. $\frac{2x^2-3x-20}{x-4}$	$x=4$	$2x+5$	yes

12.3 Experiencing Algebra the Exercise Way

1. $\frac{5}{x}+\frac{8}{x}$

$=\frac{5+8}{x}$

$=\frac{13}{x}$

3. $\frac{5}{x}+\frac{8}{-x}$

$=\frac{5-8}{x}$

$=-\frac{3}{x}$

5. $\frac{2x}{5xy}+\frac{3x}{5xy}$
$=\frac{2x+3x}{5xy}$
$=\frac{5x}{5xy}$
$=\frac{1}{y}$

7. $\frac{x-1}{x+6}+\frac{x+1}{x+6}$
$=\frac{x-1+x+1}{x+6}$
$=\frac{2x}{x+6}$

9. $\frac{2b}{b^2-25}+\frac{10}{b^2-25}$
$=\frac{2b+10}{b^2-25}$
$=\frac{2(b+5)}{(b-5)(b+5)}$
$=\frac{2}{b-5}$

11. $\frac{x-5}{x+3}+\frac{x+5}{x+3}+\frac{x+9}{x+3}$
$=\frac{x-5+x+5+x+9}{x+3}$
$=\frac{3x+9}{x+3}$
$=\frac{3(x+3)}{x+3}$
$=3$

13. $\frac{3x-2}{x-5}+\frac{x+8}{5-x}$
$=\frac{3x-2+(-x-8)}{x-5}$
$=\frac{2x-10}{x-5}$
$=\frac{2(x-5)}{x-5}$
$=2$

15. $\frac{a^2}{a-3}+\frac{9}{3-a}$
$=\frac{a^2-9}{a-3}$
$=\frac{(a-3)(a+3)}{a-3}$
$=a+3$

17. $\frac{5x}{x^2-9}+\frac{15}{9-x^2}$
$=\frac{5x-15}{x^2-9}$
$=\frac{5(x-3)}{(x+3)(x-3)}$
$=\frac{5}{x+3}$

19. $\frac{3}{2x}+\frac{7}{6x^2}$
$=\frac{3(3x)+7}{6x^2}$
$=\frac{9x+7}{6x^2}$

21. $\frac{3x+1}{x-6}+\frac{x-4}{2x-12}$
$=\frac{2(3x+1)+x-4}{2(x-6)}$
$=\frac{6x+2+x-4}{2(x-6)}$
$=\frac{7x-2}{2(x-6)}$

23. $\frac{2x}{2x-3}+\frac{6x+9}{4x^2-9}$
$=\frac{2x(2x+3)+6x+9}{(2x-3)(2x+3)}$
$=\frac{4x^2+6x+6x+9}{(2x-3)(2x+3)}$
$=\frac{4x^2+12x+9}{(2x-3)(2x+3)}$

$$= \frac{(2x+3)(2x+3)}{(2x-3)(2x+3)}$$
$$= \frac{2x+3}{2x-3}$$

25. $\frac{7}{3x-5} + \frac{4}{3x+5}$
$$= \frac{7(3x+5)+4(3x-5)}{(3x-5)(3x+5)}$$
$$= \frac{21x+35+12x-20}{(3x-5)(3x+5)}$$
$$= \frac{33x+15}{(3x-5)(3x+5)}$$
$$= \frac{3(11x+5)}{(3x-5)(3x+5)}$$

27. $\frac{7}{x-5} + \frac{x+1}{x^2-10x+25}$
$$= \frac{7(x-5)+x+1}{(x-5)^2}$$
$$= \frac{7x-35+x+1}{(x-5)^2}$$
$$= \frac{8x-34}{(x-5)^2}$$
$$= \frac{2(4x-17)}{(x-5)^2}$$

29. $\frac{b+8}{b} + \frac{b}{b+8}$
$$= \frac{(b+8)^2+b^2}{b(b+8)}$$
$$= \frac{b^2+16b+64+b^2}{b(b+8)}$$
$$= \frac{2b^2+16b+64}{b(b+8)}$$
$$= \frac{2(b^2+8b+32)}{b(b+8)}$$

31. $\frac{x-2}{4x+12} + \frac{2x+1}{x^2-9}$
$$= \frac{(x-2)(x-3)+4(2x+1)}{4(x+3)(x-3)}$$
$$= \frac{x^2-5x+6+8x+4}{4(x+3)(x-3)}$$
$$= \frac{x^2+3x+10}{4(x+3)(x-3)}$$

33. $\frac{x+7}{x^2+2x-15} + \frac{8-x}{x^2+9x+20}$
$$= \frac{x+7}{(x+5)(x-3)} + \frac{8-x}{(x+5)(x+4)}$$
$$= \frac{(x+7)(x+4)+(8-x)(x-3)}{(x+5)(x-3)(x+4)}$$
$$= \frac{x^2+11x+28+11x-x^2-24}{(x+5)(x-3)(x+4)}$$
$$= \frac{22x+4}{(x+5)(x-3)(x+4)}$$
$$= \frac{2(11x+2)}{(x+5)(x-3)(x+4)}$$

35. $\frac{3a-b}{a^2b} + \frac{a+2b}{ab^2}$
$$= \frac{b(3a-b)+a(a+2b)}{a^2b^2}$$
$$= \frac{3ab-b^2+a^2+2ab}{a^2b^2}$$
$$= \frac{a^2+5ab-b^2}{a^2b^2}$$

37. $\frac{3x}{x-2y} + \frac{4y}{2y-x}$
$$= \frac{3x+(-1)(4y)}{x-2y}$$
$$= \frac{3x-4y}{x-2y}$$

39. $\dfrac{5x+2y}{x^2-9y^2}+\dfrac{6}{x+3y}$
$=\dfrac{5x+2y+6(x-3y)}{(x-3y)(x+3y)}$
$=\dfrac{5x+2y+6x-18y}{(x-3y)(x+3y)}$
$=\dfrac{11x-16y}{(x-3y)(x+3y)}$

41. $\dfrac{8}{x^2+6xz+9z^2}+\dfrac{2}{x^2-9z^2}$
$=\dfrac{8}{(x+3z)(x+3z)}+\dfrac{2}{(x+3z)(x-3z)}$
$=\dfrac{8(x-3z)+2(x+3z)}{(x+3z)^2(x-3z)}$
$=\dfrac{8x-24z+2x+6z}{(x+3z)^2(x-3z)}$
$=\dfrac{10x-18z}{(x+3z)^2(x-3z)}$
$=\dfrac{2(5x-9z)}{(x+3z)^2(x-3z)}$

43. $\dfrac{4}{d}-\dfrac{9}{d}$
$=\dfrac{4-9}{d}$
$=-\dfrac{5}{d}$

45. $\dfrac{9xy}{7x^2y}-\dfrac{2xy}{7x^2y}$
$=\dfrac{9xy-2xy}{7x^2y}$
$=\dfrac{7xy}{7x^2y}$
$=\dfrac{7xy\cdot(1)}{7xy\cdot(x)}$
$=\dfrac{1}{x}$

47. $\dfrac{2x-3}{x+9}-\dfrac{x-3}{x+9}$
$=\dfrac{2x-3-(x-3)}{x+9}$
$=\dfrac{2x-3-x+3}{x+9}$
$=\dfrac{x}{x+9}$

49. $\dfrac{4x-13}{x-5}-\dfrac{2x-3}{x-5}$
$=\dfrac{4x-13-(2x-3)}{x-5}$
$=\dfrac{4x-13-2x+3}{x-5}$
$=\dfrac{2x-10}{x-5}$
$=\dfrac{2(x-5)}{x-5}$
$=2$

51. $\dfrac{x+3}{x+11}-\dfrac{2x-8}{x+11}$
$=\dfrac{x+3-(2x-8)}{x+11}$
$=\dfrac{x+3-2x+8}{x+11}$
$=\dfrac{-x+11}{x+11}$

53. $\dfrac{z^2}{z+4}-\dfrac{16}{z+4}$
$=\dfrac{z^2-16}{z+4}$
$=\dfrac{(z+4)(z-4)}{z+4}$
$=z-4$

55. $\frac{3c}{c^2-9}-\frac{9}{c^2-9}$
$=\frac{3c-9}{c^2-9}$
$=\frac{3(c-3)}{(c+3)(c-3)}$
$=\frac{3}{c+3}$

57. $\frac{5x+1}{2x+3}-\frac{x+2}{2x+3}-\frac{2x-5}{2x+3}$
$=\frac{5x+1-x-2-2x+5}{2x+3}$
$=\frac{2x+4}{2x+3}$
$=\frac{2(x+2)}{2x+3}$

59. $\frac{7x+5}{x-2}-\frac{x-1}{2-x}$
$=\frac{7x+5+x-1}{x-2}$
$=\frac{8x+4}{x-2}$
$=\frac{4(2x+1)}{x-2}$

61. $\frac{3x}{x^2-4}-\frac{6}{4-x^2}$
$=\frac{3x+6}{x^2-4}$
$=\frac{3(x+2)}{(x-2)(x+2)}$
$=\frac{3}{x-2}$

63. $\frac{6x-3}{3x+15}-\frac{x-1}{x+5}$
$=\frac{6x-3-3(x-1)}{3(x+5)}$
$=\frac{6x-3-3x+3}{3(x+5)}$
$=\frac{3x}{3(x+5)}$
$=\frac{x}{x+5}$

65. $\frac{7}{2x-5}-\frac{9}{2x+5}$
$=\frac{7(2x+5)-9(2x-5)}{(2x-5)(2x+5)}$
$=\frac{14x+35-18x+45}{(2x-5)(2x+5)}$
$=\frac{-4x+80}{(2x-5)(2x+5)}$
$=\frac{-4(x-20)}{(2x-5)(2x+5)}$

67. $\frac{6}{x^2+12x+36}-\frac{2x+1}{x+6}$
$=\frac{6-(x+6)(2x+1)}{(x+6)^2}$
$=\frac{6-(2x^2+13x+6)}{(x+6)^2}$
$=\frac{6-2x^2-13x-6}{(x+6)^2}$
$=\frac{-2x^2-13x}{(x+6)^2}$
$=\frac{-x(2x+13)}{(x+6)^2}$

69. $\frac{x}{8-x}-\frac{8-x}{x}$
$=\frac{x(x)-(8-x)(8-x)}{x(8-x)}$
$=\frac{x^2-(64-16x+x^2)}{x(8-x)}$
$=\frac{x^2-64+16x-x^2}{x(8-x)}$
$=\frac{16x-64}{x(8-x)}$
$=\frac{16(x-4)}{x(8-x)}$

71. $\dfrac{x+2}{3x+12}-\dfrac{x-4}{x^2-16}$
$=\dfrac{(x+2)(x-4)-(x-4)(3)}{3(x+4)(x-4)}$
$=\dfrac{x^2-2x-8-3x+12}{3(x+4)(x-4)}$
$=\dfrac{x^2-5x+4}{3(x+4)(x-4)}$
$=\dfrac{(x-4)(x-1)}{3(x+4)(x-4)}$
$=\dfrac{x-1}{3(x+4)}$

73. $\dfrac{11}{b^2-9}-\dfrac{7}{2b^2-b-15}$
$=\dfrac{11}{(b+3)(b-3)}-\dfrac{7}{(2b+5)(b-3)}$
$=\dfrac{11(2b+5)-7(b+3)}{(b+2)(b-3)(2b+5)}$
$=\dfrac{22b+55-7b-21}{(b+3)(b-3)(2b+5)}$
$=\dfrac{15b+34}{(b+3)(b-3)(2b+5)}$

75. $\dfrac{7}{5ab^2}-\dfrac{15}{10a^2b}-\dfrac{3}{2b}$
$=\dfrac{7(2a)-15(b)-3(5a^2b)}{10a^2b^2}$
$=\dfrac{14a-15b-15a^2b}{10a^2b^2}$

77. $\dfrac{4p-q}{p+3q}-\dfrac{p+2q}{p+3q}$
$=\dfrac{4p-q-(p+2q)}{p+3q}$
$=\dfrac{4p-q-p-2q}{p+3q}$
$=\dfrac{3p-3q}{p+3q}$
$=\dfrac{3(p-q)}{p+3q}$

79. $\dfrac{7}{x^2+6xy+9y^2}-\dfrac{2}{x^2-9y^2}$
$=\dfrac{7(x-3y)-2(x+3y)}{(x+3y)(x+3y)(x-3y)}$
$=\dfrac{7x-21y-2x-6y}{(x+3y)^2(x-3y)}$
$=\dfrac{5x-27y}{(x+3y)^2(x-3y)}$

81. $\dfrac{x}{x+5}-\dfrac{x}{x-5}+\dfrac{10x}{x^2-25}$
$=\dfrac{x(x-5)-4(x+5)+10x}{(x+5)(x-5)}$
$=\dfrac{x^2-5x-4x-20+10x}{(x+5)(x-5)}$
$=\dfrac{x^2+x-20}{(x+5)(x-5)}$
$=\dfrac{(x+5)(x-4)}{(x+5)(x-5)}$
$=\dfrac{x-4}{x-5}$

83. $1+\dfrac{x-9}{x+10}$
$=\dfrac{x+10+x-9}{x+10}$
$=\dfrac{2x+1}{x+10}$

85. $5-\dfrac{4}{2x+7}$
$=\dfrac{5(2x+7)-4}{2x+7}$
$=\dfrac{10x+35-4}{2x+7}$
$=\dfrac{10x+31}{2x+7}$

87. $\dfrac{2x-3}{x+7}-1$
$=\dfrac{2x-3-(x+7)}{x+7}$
$=\dfrac{2x-3-x-7}{x+7}$
$=\dfrac{x-10}{x+7}$

89. $\dfrac{3}{x+3}+x+5$
$=\dfrac{3+x(x+3)+5(x+3)}{x+3}$
$=\dfrac{3+x^2+3x+5x+15}{x+3}$
$=\dfrac{x^2+8x+18}{x+3}$

91. $2x+1-\dfrac{3}{x-4}$
$=\dfrac{2x(x-4)+1(x-4)-3}{x-4}$
$=\dfrac{2x^2-8x+x-4-3}{x-4}$
$=\dfrac{2x^2-7x-7}{x-4}$

93. a. width of smaller rectangle $=\dfrac{20}{L}$

width of larger rectangle $=\dfrac{45}{L}$

$\dfrac{20}{L}+\dfrac{45}{L}=\dfrac{20+45}{L}=\dfrac{65}{L}$

The total width is $\dfrac{65}{L}$ feet.

b. width of smaller rectangle $=\dfrac{20}{L}$

width of larger rectangle $=\dfrac{45}{2L-5}$

$\dfrac{20}{L}+\dfrac{45}{2L-5}$
$=\dfrac{20(2L-5)+45(L)}{L(2L-5)}$
$=\dfrac{40L-100+45L}{L(2L-5)}$
$=\dfrac{85L-100}{L(2L-5)}$
$=\dfrac{5(17L-20)}{L(2L-5)}$

The total width is $\dfrac{5(17L-20)}{L(2L-5)}$ feet.

95. $2\left(\dfrac{24}{x-1}\right)+2\left(\dfrac{72}{x+1}\right)$
$=\dfrac{48(x+1)+144(x-1)}{(x-1)(x+1)}$
$=\dfrac{48x+48+144x-144}{(x-1)(x+1)}$
$=\dfrac{192x-96}{(x-1)(x+1)}$
$=\dfrac{96(2x-1)}{(x-1)(x+1)}$

The perimeter is $\dfrac{96(2x-1)}{(x-1)(x+1)}$ feet.

97. a. $R(x)=55x$
$C(x)=100+25x$
$\dfrac{R(x)}{C(x)}:\ \dfrac{55x}{100+25x}=\dfrac{5(11x)}{5(20+5x)}$
$=\dfrac{11x}{20+5x}$

The ratio is $\dfrac{11x}{20+5x}$.

b. $R(x)=40x$
$C(x)=800+20x$
$\dfrac{R(x)}{C(x)}:\ \dfrac{40x}{800+20x}=\dfrac{20(2x)}{20(40+x)}$
$=\dfrac{2x}{40+x}$

The ratio is $\dfrac{2x}{40+x}$.

c. Difference:

$$\frac{11x}{20+5x}-\frac{2x}{40+x}$$
$$=\frac{11x(40+x)-2x(5)(4+x)}{5(4+x)(40+x)}$$
$$=\frac{440x+11x^2-40x-10x^2}{5(4+x)(40+x)}$$
$$=\frac{x^2+400x}{5(4+x)(40+x)}$$
$$=\frac{x(x+400)}{5(4+x)(40+x)}$$

d. $$\frac{100(100+400)}{5(4+100)(40+100)}$$
$$=\frac{100(500)}{5(104)(140)}$$
$$=\frac{125}{182}$$

99. $$\frac{2}{s}+\frac{-1}{s+1}$$
$$=\frac{2(s+1)-1(s)}{s(s+1)}$$
$$=\frac{2s+2-s}{s(s+1)}$$
$$=\frac{s+2}{s(s+1)}$$

The sum is $\frac{s+2}{s(s+1)}$.

12.3 Experiencing Algebra the Calculator Way

1. $$\frac{x+3}{x}+\frac{x-2}{x-1}$$
$$=\frac{(x-1)(x+3)+x(x-2)}{x(x-1)}$$
$$=\frac{x^2+2x-3+x^2-2x}{x(x-1)}$$
$$=\frac{2x^2-3}{x(x-1)}$$

The graphs of $Y1=\frac{x+3}{x}+\frac{x-2}{x-1}$ and $Y2=\frac{2x^2-3}{x(x-1)}$ are the same.

2. $$\frac{x+3}{x}-\frac{x-2}{x-1}$$
$$=\frac{(x-1)(x+3)-x(x-2)}{x(x-1)}$$
$$=\frac{x^2+2x-3-x^2+2x}{x(x-1)}$$
$$=\frac{4x-3}{x(x-1)}$$

The graphs of $Y1=\frac{x+3}{x}-\frac{x-2}{x-1}$ and $Y2=\frac{4x-3}{x(x-1)}$ are the same.

3. $$\frac{x+2}{x-2}+\frac{x-2}{x^2-4}$$
$$=\frac{(x+2)(x+2)+(x-2)}{(x-2)(x+2)}$$
$$=\frac{x^2+4x+4+x-2}{(x-2)(x+2)}$$
$$=\frac{x^2+5x+2}{(x-2)(x+2)}$$

The graphs of $Y1=\frac{x+2}{x-2}+\frac{x-2}{x^2-4}$ and $Y2=\frac{x^2+5x+2}{(x-2)(x+2)}$ are the same.

4. $$\frac{x+2}{x-2}-\frac{x-2}{x^2-4}$$
$$=\frac{(x+2)(x+2)-(x-2)}{(x+2)(x-2)}$$
$$=\frac{x^2+4x+4-x+2}{(x+2)(x-2)}$$
$$=\frac{x^2+3x+6}{(x+2)(x-2)}$$

The graphs of $Y1=\frac{x+2}{x-2}-\frac{x-2}{x^2-4}$ and $Y2=\frac{x^2+3x+6}{(x+2)(x-2)}$ are the same.

5. $\frac{x+3}{x^2+x-6}+\frac{x-3}{x^2-5x+6}$
$=\frac{(x+3)(x-3)+(x-3)(x+3)}{(x+3)(x-2)(x-3)}$
$=\frac{x^2-9+x^2-9}{(x+3)(x-2)(x-3)}$
$=\frac{2(x^2-9)}{(x+3)(x-2)(x-3)}$
$=\frac{2(x+3)(x-3)}{(x+3)(x-2)(x-3)}$
$=\frac{2}{x-2}$
The graphs of
$Y1=\frac{x+3}{x^2+x-6}+\frac{x-3}{x^2-5x+6}$ and
$Y2=\frac{2}{x-2}$ are the same.

6. $\frac{x+3}{x^2+x-6}-\frac{x-3}{x^2-5x+6}$
$=\frac{(x+3)(x-3)-(x-3)(x+3)}{(x+3)(x-2)(x-3)}$
$=\frac{x^2-9-x^2+9}{(x+3)(x-2)(x-3)}$
$=\frac{0}{(x+3)(x-2)(x-3)}$
$=0$
The graphs of
$Y1=\frac{x+3}{x^2+x-6}-\frac{x-3}{x^2-5x+6}$ and $Y2=0$
are the same.

12.4 Experiencing Algebra the Exercise Way

1. $x^{-3}+2x^{-2}-x^{-1}=5$ is a rational equation because it can be written as
$\frac{1}{x^3}+\frac{2}{x^2}-\frac{1}{x}-5=0.$

3. $\frac{1}{x+7}+\frac{1}{x}=5$ is a rational equation because it can be written as $\frac{1}{x+7}+\frac{1}{x}-5=0.$

5. $v^{1/2}+3v=10$ is not rational because a variable to a fractional exponent cannot be written as a rational expression.

7. $\frac{\sqrt{2x+7}}{x}=\frac{1}{3-x}$ is not rational because $\sqrt{2x+7}$ cannot be written as a rational expression.

9. $\frac{2.7}{z^2}-\frac{5.3}{z}+0.9=0$ is a rational equation.

11. $\frac{1}{3x^2}+\frac{5}{6x}=\frac{8}{9}$ is a rational equation because it can be written as
$\frac{1}{3x^2}+\frac{5}{6x}-\frac{8}{9}=0.$

13. $\frac{5}{\sqrt{3m}}+\frac{7}{\sqrt{m}}=\frac{1}{\sqrt{5}}$ is not rational because there are variables in the radicand.

15. $\frac{\sqrt{3}}{x^2}+\frac{\sqrt{2}}{x}=7$ is a rational equation because it can be written as $\frac{\sqrt{3}}{x^2}+\frac{\sqrt{2}}{x}-7=0.$

17. $\frac{q}{7}=7q^{-1}$
The restricted value is 0. Set up a table of values on the calculator letting $Y1=\frac{x}{7}$ and $Y2=7x^{-1}$. The solutions are –7 and 7 because when –7 and 7 are substituted for the variable in both expressions, the resulting values are equal.

19. $\dfrac{3(x-3)}{2x+1}=\dfrac{x-3}{x+1}$

The restricted values are –1 and $-\dfrac{1}{2}$. Set up a table of values on the calculator letting $Y1=\dfrac{3(x-3)}{2x+1}$ and $Y2=\dfrac{x-3}{x+1}$. The solutions are –2 and 3.

21. $\dfrac{5}{6}=\dfrac{7}{x}-\dfrac{1}{3}$

The restricted value is 0. Set up a table of values on the calculator letting $Y1=\dfrac{5}{6}$ and $Y2=\dfrac{7}{x}-\dfrac{1}{3}$. The solution is 6.

23. $\dfrac{5}{2s}=\dfrac{4}{s}-\dfrac{1}{2}$

The restricted value is 0. Set up a table of values on the calculator letting $Y1=\dfrac{5}{2x}$ and $Y2=\dfrac{4}{x}-\dfrac{1}{2}$. The solution is 3.

25. $\dfrac{4}{x+8}-\dfrac{3}{x+2}=-\dfrac{5}{56}$

The restricted values are –8 and –2. Set up a table of values on the calculator letting $Y1=\dfrac{4}{x+8}-\dfrac{3}{x+2}$ and $Y2=-\dfrac{5}{56}$. The integer solution is 6.

27. $\dfrac{2G+5}{G-1}=9$

The restricted value is 1. Set up a table of values letting $Y1=\dfrac{2x+5}{x-1}$ and $Y2=9$. The solution is 2.

29. $\dfrac{b-4}{b+2}=\dfrac{1}{4}$

The restricted value is –2. Set up a table of values letting $Y1=\dfrac{x-4}{x+2}$ and $Y2=\dfrac{1}{4}$. The solution is 6.

31. $\dfrac{z-3}{z-4}=\dfrac{7}{z-4}$

The restricted value is 4. Set up a table of values letting $Y1=\dfrac{x-3}{x-4}$ and $Y2=\dfrac{7}{x-4}$. The solution is 10.

33. $\dfrac{1}{x+6}+\dfrac{1}{x-6}=\dfrac{6}{x^2-36}$

The restricted values are –6 and 6. Set up a table of values letting $Y1=\dfrac{1}{x+6}+\dfrac{1}{x-6}$ and $Y2=\dfrac{6}{x^2-36}$. The solution is 3.

35. $\dfrac{-5}{a-2}=\dfrac{1+a}{2-a}$

The restricted value is 2. Set up a table of values letting $Y1=-\dfrac{5}{x-2}$ and $Y2=\dfrac{1+x}{2-x}$. The solution is 4.

37. $\dfrac{1}{x-1}+\dfrac{4}{x^2-1}=1$

The restricted values are –1 and 1. Set up a table of values letting $Y1=\dfrac{1}{x-1}+\dfrac{4}{x^2-1}$ and Y2 = 1. The solutions are –2 and 3.

39. $\dfrac{a-3}{a-1}+\dfrac{3}{a+1}=\dfrac{(a+3)(a-2)}{a^2-1}$

The restricted values are –1 and 1. Set up a table of values letting $Y1=\dfrac{x-3}{x-1}+\dfrac{3}{x+1}$ and $Y2=\dfrac{(x+3)(x-2)}{x^2-1}$. The solutions are all integers not equal to –1 or 1.

41. $\dfrac{3}{x-5}+\dfrac{3}{x^2-25}=\dfrac{3x+1}{x^2-25}$

The restricted values are –5 and 5. Set up a table of values letting $Y1=\dfrac{3}{x-5}+\dfrac{3}{x^2-25}$ and $Y2=\dfrac{3x+1}{x^2-25}$. There is no solution.

43. $\frac{1}{3}=\frac{x+1}{x+2}-\frac{1}{2}$

$Y1=\frac{1}{3}$

$Y2=\frac{x+1}{x+2}-\frac{1}{2}$

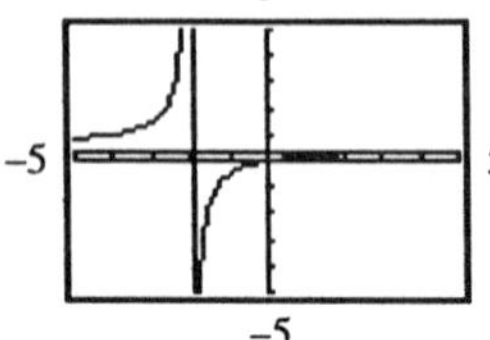

The solution is the x-coordinate of the intersection, 4.

45. $\frac{5}{c}+c=\frac{21}{2}$

$Y1=\frac{5}{x}+x$

$Y2=\frac{21}{2}$

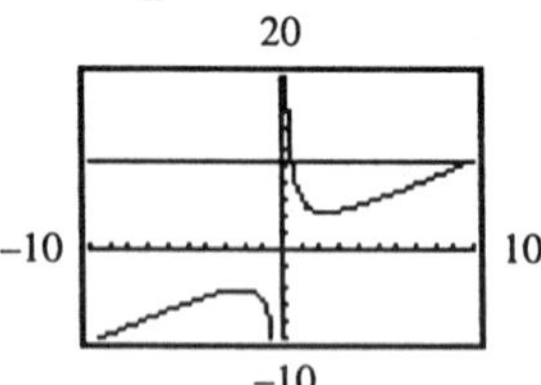

The solutions are the x-coordinates of the intersections, $\frac{1}{2}$ or 0.5 and 10.

47. $\frac{9}{x}-\frac{2}{x}=21$

$Y1=\frac{9}{x}-\frac{2}{x}$

$Y2=21$

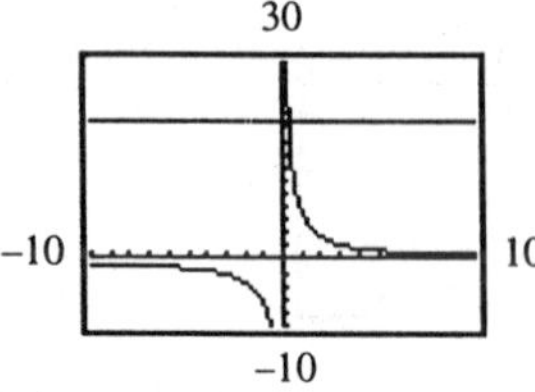

The solution is the x-coordinate of the intersection, $\frac{1}{3}$ or $0.\overline{3}$.

49. $\frac{3}{k}=\frac{17}{4}-\frac{2}{5k}$

$Y1=\frac{3}{x}$

$Y2=\frac{17}{4}-\frac{2}{5x}$

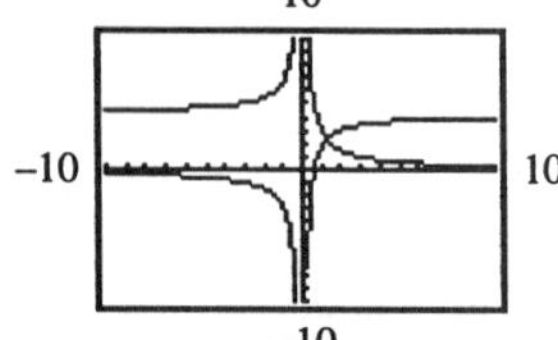

The solution is the x-coordinate of the intersection, 0.8.

51. $\frac{x}{3}=\frac{x}{7}+\frac{2}{3}$

$Y1=\frac{x}{3}$

$Y2=\frac{x}{7}+\frac{2}{3}$

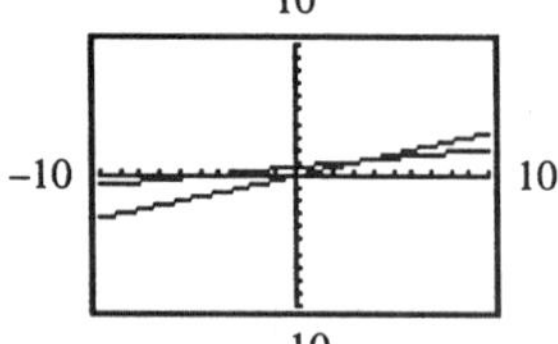

The solution is the x-coordinate of the intersection, 3.5.

53. $\frac{x+5}{x-3}=\frac{1+x}{2-x}$

$Y1=\frac{x+5}{x-3}$

$Y2=\frac{1+x}{2-x}$

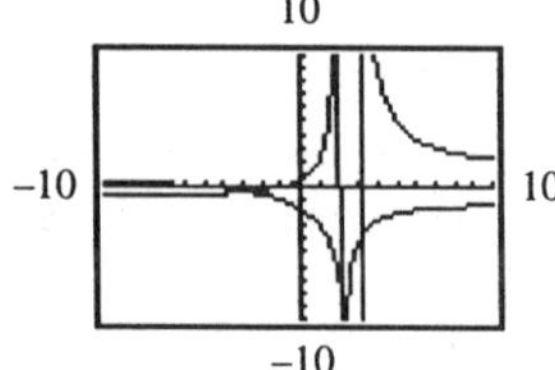

The solutions are the x-coordinates of the intersections, approximately −2.8117 and 2.3117.

55. $\frac{3-z}{7-z} = \frac{15}{z-5}$

$Y1 = \frac{3-x}{7-x}$

$Y2 = \frac{15}{x-5}$

10

−10 20

−10

The solutions are the x-coordinates of the intersections, 8 and 15.

57. $\frac{1}{x+3} + \frac{1}{x} = 2$

$Y1 = \frac{1}{x+3} + \frac{1}{x}$

$Y2 = 2$

10

−10 10

−10

The solutions are the x-coordinates of the intersections, approximately −2.5811 and 0.5811.

59. $w^{-3} + 2w^{-2} - w^{-1} = 1$

$Y1 = x^{-3} + 2x^{-2} - x^{-1}$

$Y2 = 1$

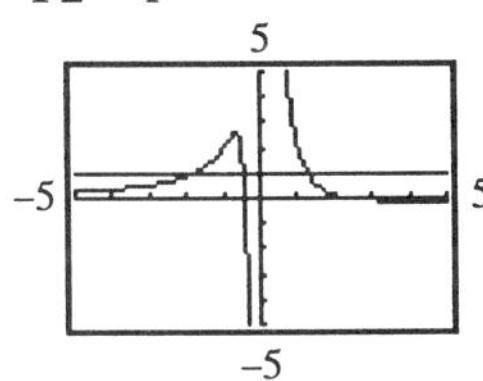

The solutions are the x-coordinates of the intersections, approximately −1.8019, −0.4450, and 1.2470.

61. $2.7x = 5.3 - \frac{0.9}{x}$

$Y1 = 2.7x$

$Y2 = 5.3 - \frac{0.9}{x}$

10

−10 10

−10

The solutions are the x-coordinates of the intersections, approximately 0.1878 and 1.7752.

63. $\frac{1}{3u^2} + \frac{5}{6u} = \frac{8}{9}$

$Y1 = \frac{1}{3x^2} + \frac{5}{6x}$

$Y2 = \frac{8}{9}$

5

−5 5

−5

The solutions are the x-coordinates of the intersections, approximately −0.3024 and 1.2399.

65. $\frac{\sqrt{3}}{x^2} + \frac{\sqrt{2}}{x} = 7$

$Y1 = \frac{\sqrt{3}}{x^2} + \frac{\sqrt{2}}{x}$

$Y2 = 7$

10

−10 10

−10

The solutions are the x-coordinates of the intersections, approximately −0.4066 and 0.6086.

67. $\dfrac{3x+7}{x-1}=\dfrac{3x-3}{x-1}$

$Y1=\dfrac{3x+7}{x-1}$

$Y2=\dfrac{3x-3}{x-1}$

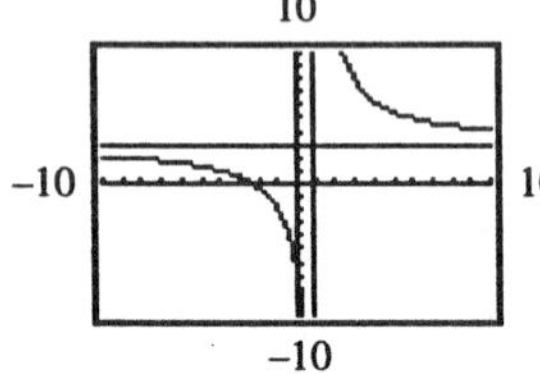

The graphs do not appear to meet. There is no solution.

69. $\dfrac{b+4}{b-2}=\dfrac{b^2+3b-4}{b^2-3b+2}$

$Y1=\dfrac{x+4}{x-2}$

$Y2=\dfrac{x^2+3x-4}{x^2-3x+2}$

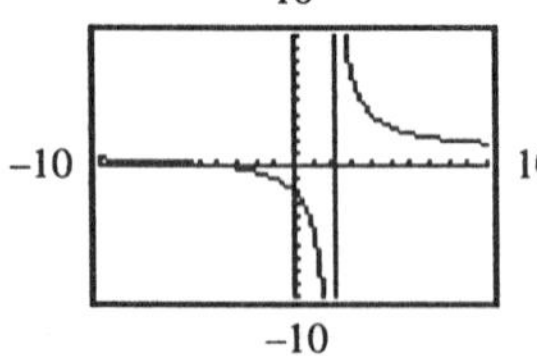

The graphs are the same. The solution is all real numbers except $b=2$ and $b=1$.

71. Let x = monthly gross pay before the reduction

$\dfrac{85.56}{x}=\dfrac{53.22}{x-1200}$

$Y1=\dfrac{85.56}{x}$

$Y2=\dfrac{53.22}{x-1200}$

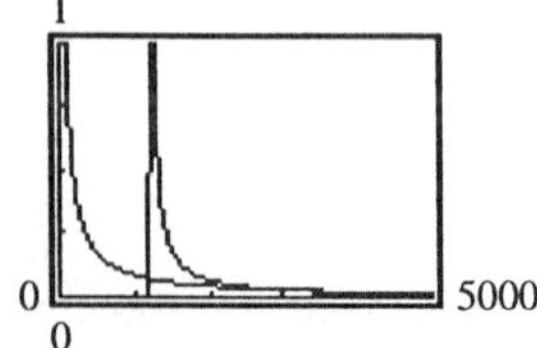

The solution is the x-coordinate of the intersection, approximately 3174.77. His monthly salary is approximately \$3175.

73. Let x = length of first time period

$\dfrac{132}{x}=\dfrac{225}{x+1.6}$

$Y1=\dfrac{132}{x}$

$Y2=\dfrac{225}{x+1.6}$

100
0 5
0

The solution is the x-coordinate of the intersection, 2.3. His first time is 2.3 hours and his second time is 3.9 hours.

75. Let x = interest on second account

$\dfrac{286}{(x+0.02)2}+\dfrac{117}{x(2)}=3500$

$Y1=\dfrac{286}{2(x+0.02)}+\dfrac{117}{2x}$

$Y2=3500$

4000
0 1
0

The solution is $x=0.045$.
The interest on the first account is 6.5% and on the second account is 4.5%.

77. $\dfrac{1}{R}=\dfrac{1}{r_1}+\dfrac{1}{r_2}$

$\dfrac{1}{11.5}=\dfrac{1}{20.5}+\dfrac{1}{r_2}$

$Y1=\dfrac{1}{11.5}$

$Y2=\dfrac{1}{20.5}+\dfrac{1}{x}$

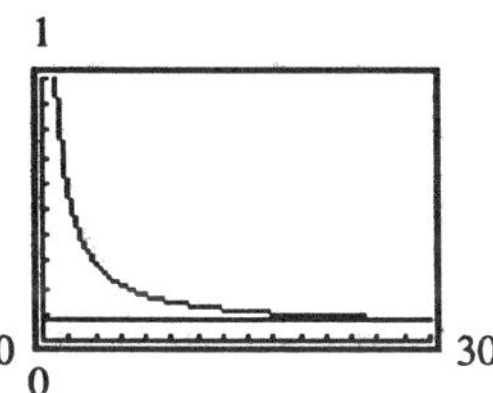

The solution is approximately 26.2. The resistance of the second vessel is 26.2 dynes.

12.4 Experiencing Algebra the Calculator Way

Students should check the answers to exercises in text using the methods outlined here. Answers will vary.

12.5 Experiencing Algebra the Exercise Way

1. $\frac{1}{a}=\frac{5}{9}+\frac{2}{3}$

$9a\left(\frac{1}{a}\right)=9a\left(\frac{5}{9}\right)+9a\left(\frac{2}{3}\right)$

$9=5a+6a$

$9=11a$

$\frac{9}{11}=a$

The solution is $\frac{9}{11}$.

3. $\frac{2}{z}=\frac{4}{7}-\frac{2}{3}$

$21z\left(\frac{2}{z}\right)=21z\left(\frac{4}{7}\right)-21z\left(\frac{2}{3}\right)$

$42=12z-14z$

$42=-2z$

$-21=z$

The solution is –21.

5. $\frac{11}{x}-\frac{3}{4}=\frac{5}{8}$

$8x\left(\frac{11}{x}\right)-8x\left(\frac{3}{4}\right)=8x\left(\frac{5}{8}\right)$

$88-6x=5x$

$88=11x$

$8=x$

The solution is 8.

7. $1-\frac{8}{x}=\frac{10}{x}$

$x(1)-x\left(\frac{8}{x}\right)=x\left(\frac{10}{x}\right)$

$x-8=10$

$x=18$

The solution is 18.

9. $1+\frac{7}{x}=\frac{19}{x}$

$x(1)+x\left(\frac{7}{x}\right)=x\left(\frac{19}{x}\right)$

$x+7=19$

$x=12$

The solution is 12.

11. $\frac{1}{3}=\frac{5}{x}+\frac{7}{3x}$

$3x\left(\frac{1}{3}\right)=3x\left(\frac{5}{x}\right)+3x\left(\frac{7}{3x}\right)$

$x=15+7$

$x=22$

The solution is 22.

13. $\frac{1}{2}=\frac{5}{2x}-\frac{6}{x}$

$2x\left(\frac{1}{2}\right)=2x\left(\frac{5}{2x}\right)-2x\left(\frac{6}{x}\right)$

$x=5-12$

$x=-7$

The solution is –7.

15. $\frac{9}{x-3}=1$

$(x-3)\left(\frac{9}{x-3}\right)=1(x-3)$

$9=x-3$

$12=x$

The solution is 12.

17. $2=\dfrac{6}{x-2}$

$(x-2)(2)=(x-2)\left(\dfrac{6}{x-2}\right)$

$2x-4=6$

$2x=10$

$x=5$

The solution is 5.

19. $\dfrac{x-4}{2x+3}=\dfrac{5}{21}$

$5(2x+3)=21(x-4)$

$10x+15=21x-84$

$99=11x$

$9=x$

The solution is 9.

21. $\dfrac{1}{x-6}=\dfrac{17}{2x+3}$

$1(2x+3)=17(x-6)$

$2x+3=17x-102$

$105=15x$

$7=x$

The solution is 7.

23. $\dfrac{8}{2x-3}+\dfrac{4}{x}=0$

$\dfrac{8}{2x-3}=\dfrac{-4}{x}$

$-4(2x-3)=8x$

$-8x+12=8x$

$12=16x$

$\dfrac{3}{4}=x$

The solution is $\dfrac{3}{4}$.

25. $\dfrac{8}{x-2}=\dfrac{6}{x}$

$8x=6(x-2)$

$8x=6x-12$

$2x=-12$

$x=-6$

The solution is –6.

27. $\dfrac{2x+3}{x+4}=\dfrac{5}{x+4}$

$(x+4)\left(\dfrac{2x+3}{x+4}\right)=(x+4)\left(\dfrac{5}{x+4}\right)$

$2x+3=5$

$2x=2$

$x=1$

The solution is 1.

29. $\dfrac{16}{x}=x$

$x^2=16$

$x=\pm\sqrt{16}$

$x=\pm 4$

The solutions are ±4.

31. $\dfrac{z}{7}-\dfrac{7}{z}=0$

$\dfrac{z}{7}=\dfrac{7}{z}$

$z^2=49$

$z=\pm\sqrt{49}$

$z=\pm 7$

The solutions are ±7.

33. $\dfrac{10}{k}+3=k$

$k\left(\dfrac{10}{k}\right)+k(3)=k(k)$

$10+3k=k^2$

$k^2-3k-10=0$

$(k-5)(k+2)=0$

$k=5 \quad \text{or} \quad k=-2$

The solutions are –2 and 5.

35. $b=8-\dfrac{16}{b}$

$b(b)=b(8)-b\left(\dfrac{16}{b}\right)$

$b^2=8b-16$

$b^2-8b+16=0$

$(b-4)(b-4)=0$

$b=4$

The solution is 4.

37. $a+\frac{10}{a}=\frac{1}{a}-6$

$a(a)+a\left(\frac{10}{a}\right)=a\left(\frac{1}{a}\right)-a(6)$

$a^2+10=1-6a$

$a^2+6a+9=0$

$(a+3)^2=0$

$a=-3$

The solution is –3.

39. $\frac{x}{3}+\frac{5}{x}=\frac{1}{x}-\frac{8}{3}$

$3x\left(\frac{x}{3}\right)+3x\left(\frac{5}{x}\right)=3x\left(\frac{1}{x}\right)-3x\left(\frac{8}{3}\right)$

$x^2+15=3-8x$

$x^2+8x+12=0$

$(x+6)(x+2)=0$

$x=-6$ or $x=-2$

The solutions are –6 and –2.

41. $\frac{x}{4}+\frac{4}{x}=3-\frac{4}{x}$

$4x\left(\frac{x}{4}\right)+4x\left(\frac{4}{x}\right)=4x(3)-4x\left(\frac{4}{x}\right)$

$x^2+16=12x-16$

$x^2-12x+32=0$

$(x-4)(x-8)=0$

$x=4$ or $x=8$

The solutions are 4 and 8.

43. $\frac{y}{5}-\frac{21}{5y}=1+\frac{3}{y}$

$5y\left(\frac{y}{5}\right)-5y\left(\frac{21}{5y}\right)=5y(1)+5y\left(\frac{3}{y}\right)$

$y^2-21=5y+15$

$y^2-5y-36=0$

$(y-9)(y+4)=0$

$y=9$ or $y=-4$

The solutions are –4 and 9.

45. $\frac{x+7}{x+9}=\frac{x}{x+1}$

$(x+7)(x+1)=x(x+9)$

$x^2+8x+7=x^2+9x$

$7=x$

The solution is 7.

47. $\frac{x-6}{x+5}=\frac{x-4}{3x+2}$

$(x-6)(3x+2)=(x+5)(x-4)$

$3x^2-16x-12=x^2+x-20$

$2x^2-17x+8=0$

$(2x-1)(x-8)=0$

$x=\frac{1}{2}$ or $x=8$

The solutions are $\frac{1}{2}$ and 8.

49. $x^{-1}-5=7$

$x^{-1}=12$

$\frac{1}{x}=\frac{12}{1}$

$12x=1$

$x=\frac{1}{12}$

The solution is $\frac{1}{12}$.

51. $x^{-1}+\frac{2}{3}=\frac{3}{5}$

$\frac{1}{x}=\frac{3}{5}-\frac{2}{3}$

$\frac{1}{x}=\frac{1}{-15}$

$x=-15$

The solution is –15.

53. $x^{-2}-6=3$

$\frac{1}{x^2}=\frac{9}{1}$

$9x^2=1$

$x^2=\frac{1}{9}$

$x=\pm\sqrt{\frac{1}{9}}$

$x=\pm\frac{1}{3}$

The solutions are $\pm\frac{1}{3}$.

55. $1-4x^{-1}=60x^{-2}$

$1-\frac{4}{x}=\frac{60}{x^2}$

$x^2(1)-x^2\left(\frac{4}{x}\right)=x^2\left(\frac{60}{x^2}\right)$

$x^2-4x=60$

$x^2-4x-60=0$

$(x+6)(x-10)=0$

$x=-6$ or $x=10$

The solutions are –6 and 10.

57. $\frac{9}{x-7}+3=\frac{2x-5}{x-7}$

$(x-7)\left(\frac{9}{x-7}\right)+(x-7)(3)$

$=(x-7)\left(\frac{2x-5}{x-7}\right)$

$9+3x-21=2x-5$

$x=7$

(extraneous)

There is no solution.

59. $\frac{x+2}{3}=\frac{x^2-x+7}{3x-6}$

$3(x-2)\left(\frac{x+2}{3}\right)=3(x-2)\left(\frac{x^2-x+7}{3x-6}\right)$

$x^2-4=x^2-x+7$

$x=11$

The solution is 11.

61. $\frac{x-3}{x+4}=\frac{14}{x^2+6x+8}$

$(x+4)(x+2)\left(\frac{x-3}{x+4}\right)$

$=(x+4)(x+2)\left(\frac{14}{x^2+6x+8}\right)$

$x^2-x-6=14$

$x^2-x-20=0$

$(x-5)(x+4)=0$

$x=5$ or $x=-4$ (extraneous)

The solution is 5.

63. $\frac{25}{p^2+p-12}=\frac{p+4}{p-3}$

$(p+4)(p-3)\frac{25}{(p+4)(p-3)}$

$=\frac{p+4}{p-3}(p+4)(p-3)$

$25=(p+4)^2$

$p^2+8p-9=0$

$(p+9)(p-1)=0$

$p=-9$ or $p=1$

The solutions are –9 and 1.

65. $\frac{m-2}{m+1}=\frac{7m+22}{m^2+6m+5}$

$(m+1)(m+5)\left(\frac{m-2}{m+1}\right)$

$=\frac{7m+22}{(m+1)(m+5)}(m+1)(m+5)$

$(m+5)(m-2)=7m+22$

$m^2+3m-10=7m+22$

$m^2-4m-32=0$

$(m-8)(m+4)=0$

$m=8$ or $m=-4$

The solutions are –4 and 8.

67. $\frac{1-4x}{1-x}+2=\frac{6x-3}{x-1}$

$(x-1)\left(\frac{1-4x}{1-x}+2\right)=(x-1)\left(\frac{6x-3}{x-1}\right)$

$-1+4x+2x-2=6x-3$

$6x-3=6x-3$

$-3=-3$

This is an identity. The solution is the set of all real numbers not equal to 1.

69. $2+\frac{z+2}{2z+1}=\frac{3z+6}{2z+1}+1$

$(2z+1)\left(2+\frac{z+2}{2z+1}\right)=(2z+1)\left(\frac{3z+6}{2z+1}+1\right)$

$2(2z+1)+z+2=3z+6+2z+1$

$5z+4=5z+7$

$4=7$

This is a contradiction. There is no solution.

71. $\frac{2}{x-4}+\frac{5x}{x^2-16}=\frac{1}{x+4}$

$(x-4)(x+4)\left(\frac{2}{x-4}+\frac{5x}{x^2-16}\right)$

$=(x-4)(x+4)\left(\frac{1}{x+4}\right)$

$2(x+4)+5x=x-4$

$7x+8=x-4$

$6x=-12$

$x=-2$

The solution is –2.

73. $\frac{1}{x+5}+\frac{21}{x^2-25}=2$

$(x+5)(x-5)\left(\frac{1}{x+5}+\frac{21}{x^2-25}\right)$

$=2(x+5)(x-5)$

$x-5+21=2x^2-50$

$2x^2-x-66=0$

$(2x+11)(x-6)=0$

$x=-\frac{11}{2}$ or $x=6$

The solutions are $-\frac{11}{2}$ and 6.

75. Let r = resistance of one resistor

$2r-5$ = resistance of other resistor

$\frac{1}{R}=\frac{1}{r_1}+\frac{1}{r_2}$

$\frac{1}{6}=\frac{1}{r}+\frac{1}{2r-5}$

$6r(2r-5)\left(\frac{1}{6}\right)=6r(2r-5)\left(\frac{1}{r}+\frac{1}{2r-5}\right)$

$r(2r-5)=6(2r-5)+6r$

$2r^2-5r=12r-30+6r$

$2r^2-23r+30=0$

$(2r-3)(r-10)=0$

$r=\frac{3}{2}$ or $r=10$

$2r-5=-2$ $\quad$ $2r-5=15$

The resistance of the resistors is 10 ohms and 15 ohms, respectively.

77. Let x = additional money raised and spent

$\frac{x+120}{x+300}=0.45$

$x+120=0.45(x+300)$

$x+120=0.45x+135$

$0.55x=15$

$\frac{0.55x}{0.55}=\frac{15}{0.55}$

$x=27.27$

They should raise and spend $27.27.

79. Let x = number of feet added to the height

$A=\frac{1}{2}bh \Rightarrow b=\frac{2A}{h}=\frac{2(60)}{8}=15$

The current base is 15.

$15=\frac{2(60+45)}{8+x}$

$15(8+x)=2(60+45)$

$120+15x=210$

$15x=90$

$x=6$

The height should be increased by 6 feet.

12.5 Experiencing Algebra the Calculator Way

Students should try the approach outlined here to solve some of the exercises in this section. Answers will vary.

12.6 Experiencing Algebra the Exercise Way

1. Let x = time to complete the report working together.

$\frac{1}{3}$ is Miriam's rate.

$\frac{1}{4.2}$ is Saul's rate.

$\frac{1}{x}$ is the combined rate.

$\frac{1}{3}+\frac{1}{4.2}=\frac{1}{x}$

$\frac{4.2+3}{12.6}=\frac{1}{x}$

$x(4.2+3)=12.6(1)$

$7.2x=12.6$

$x=1.75$

It will take them 1.75 hours working together.

3. Let x = time it takes Simone to clean the house.

$\frac{1}{2x}$ is Jacque's rate.

$\frac{1}{x}$ is Simone's rate.

$\frac{1}{3\frac{1}{3}} = \frac{1}{\frac{10}{3}} = \frac{3}{10}$ is the combined rate.

$\frac{1}{2x} + \frac{1}{x} = \frac{3}{10}$

$\frac{1+2}{2x} = \frac{3}{10}$

$3(10) = 3(2x)$

$30 = 6x$

$x = 5$

$2x = 10$

It takes Jacques 10 hours and Simone 5 hours.

5. Let x = time to do the job together.

$\frac{1}{9}$ is Felix's rate.

$\frac{1}{6}$ is Oscar's rate.

$\frac{1}{x}$ is the combined rate.

$\frac{1}{9} + \frac{1}{6} = \frac{1}{x}$

$\frac{15}{54} = \frac{1}{x}$

$15x = 54$

$x = 3.6$

It will take 3.6 hours working together.

7. Let x = time for both pipelines to fill the tank.

$\frac{1}{4}$ is the red pipe's rate.

$\frac{1}{7}$ is the blue pipe's rate.

$\frac{1}{x}$ is the combined rate.

$\frac{1}{7} + \frac{1}{4} = \frac{1}{x}$

$\frac{11}{28} = \frac{1}{x}$

$11x = 28$

$x \approx 2.55$

It will take approxiamtely 2.55 hours for both pipes together.

9. Let x = time it takes line B.

$\frac{1}{\frac{2x}{3}} = \frac{3}{2x}$ is line A's rate.

$\frac{1}{x}$ is line B's rate.

$\frac{1}{4\frac{4}{5}} = \frac{1}{\frac{24}{5}} = \frac{5}{24}$ is the combined rate

$\frac{3}{2x} + \frac{1}{x} = \frac{5}{24}$

$\frac{3+2}{2x} = \frac{5}{24}$

$10x = 120$

$x = 12$

$\frac{2}{3}x = 8$

It takes Line A 8 hours and line B 12 hours.

11. Let x = number of shares.

$\frac{15.75}{1000} = \frac{x}{1800}$

$1000x = 28{,}350$

$x = 28.35$

He can buy 28.35 shares for \$1800.

13. Let x = number of additional pionts.

$\frac{45 + x}{60 + x} = \frac{80}{100}$

$100(45 + x) = 80(60 + x)$

$4500x + 100x = 4800 + 80x$

$20x = 300$

$x = 15$

There must be an additional 15 points.

15. $\frac{PQ}{XY}=\frac{QR}{YZ}$
$\frac{24}{x}=\frac{12}{2}$
$12x=48$
$x=4$
$\overline{XY}$ measures 4 inches.
$\frac{PR}{XZ}=\frac{QR}{YZ}$
$\frac{x}{3}=\frac{12}{2}$
$2x=36$
$x=18$
$\overline{PR}$ measures 18 inches.

17. $\frac{PQ}{XY}=\frac{PR}{XZ}$
$\frac{x}{4.03}=\frac{5.875}{2.35}$
$2.35x=(4.03)(5.875)$
$x=10.075$
$\overline{PQ}$ measures 10.075 cm.
$\frac{QR}{YZ}=\frac{PR}{XZ}$
$\frac{4.4}{x}=\frac{5.875}{2.35}$
$5.875x=(4.4)(2.35)$
$x=1.76$
$\overline{YZ}$ measures 1.76 cm.

19. $\frac{QR}{YZ}=\frac{PQ}{XY}$
$\frac{x}{1\frac{3}{4}}=\frac{10\frac{5}{16}}{4\frac{1}{8}}$
$4\frac{1}{8}(x)=\left(1\frac{3}{4}\right)\left(10\frac{5}{16}\right)$
$x=4\frac{3}{8}$
$\overline{QR}$ measures $4\frac{3}{8}$ feet.

$\frac{PR}{XZ}=\frac{PQ}{XY}$
$\frac{5\frac{5}{8}}{x}=\frac{10\frac{5}{16}}{4\frac{1}{8}}$
$10\frac{5}{16}(x)=\left(5\frac{5}{8}\right)\left(4\frac{1}{8}\right)$
$x=2\frac{1}{4}$
$\overline{XZ}$ measures $2\frac{1}{4}$ feet.

21. Let x = height of cliff
$\frac{x}{35}=\frac{5.5}{4}$
$4x=192.5$
$x=48.125$
The estimated height is 48.125 feet.

12.6 Experiencing Algebra the Calculator Way

1. Calculator methods will vary.
$A=lw\Rightarrow w=\frac{A}{l}=\frac{25}{x}$
$P=2l+2w$
$P(x)=2x+2\left(\frac{25}{x}\right)$
$25=2x+\frac{50}{x}$
$25x=2x^2+50$
$2x^2-25x+50=0$
$(2x-5)(x-10)=0$
$x=\frac{5}{2}$ $x=10$
$\frac{25}{x}=\frac{25}{\frac{5}{2}}=10$ $\frac{25}{x}=\frac{25}{10}=\frac{5}{2}$
The dimensions are $\frac{5}{2}$ feet by 10 feet.

2. Calculator methods will vary.
   Let x = time for Tom to paint the fence.
   $\frac{1}{2x}$ is Becky's rate.
   $\frac{1}{x}$ is Tom's rate.
   $\frac{1}{4}$ is the combined rate.
   $\frac{1}{2x}+\frac{1}{x}=\frac{1}{4}$
   $\frac{3}{2x}=\frac{1}{4}$
   $2x=12$
   $x=6$
   It will take Becky 12 hours and Tom 6 hours.

3. Calculator methods will vary.
   Let x = time to complete the trip.
   $\frac{24}{144}=\frac{x}{200-144}$
   $144x=1344$
   $x=9\frac{1}{3}$
   It will take an additional $9\frac{1}{3}$ hours.

4. Calculator methods will vary.
   Let x = number of dozen of cookies
   $f(x)=\frac{9+0.55x}{x}$
   $0.75=\frac{9+0.55x}{x}$
   $0.75x=9+0.55x$
   $0.2x=9$
   $x=45$
   She should bake 45 dozen.

Chapter 12 Review

Reflections

1.–6. Answers will vary.

Exercises

1. $3+\frac{x-5}{x^2+6x+9}$ is a rational expression because it can be written as $\left(\frac{3}{1}+\frac{x-5}{x^2+6x+9}\right)$.

2. $x^2-\frac{\sqrt{x}+3x}{x+1}$ is a non-rational expression because $(\sqrt{x}+3x)$ is not a polynomial.

3. $5x^2-3x+2-4x^{-1}$ is a rational expression because it can be written as $\left(5x^2-3x+2-\frac{4}{x}\right)$.

4. $2x+5$ is a rational expression because it can be written as $\left(\frac{2x}{1}+\frac{5}{1}\right)$.

5. $y=\frac{-3}{x^2}$
   $x^2=0$
   $x=0$
   The restricted value is 0.
   Domain: $x\neq 0$; $(-\infty, 0)\cup(0, \infty)$

6. $f(x)=\frac{2x+7}{x^2+5x-24}$
   $x^2+5x-24=0$
   $(x+8)(x-3)=0$
   $x=-8$ $\quad$ $x=3$
   The restricted values are –8 and 3.
   Domain: $x\neq -8$, $x\neq 3$;
   $(-\infty, -8)\cup(-8, 3)\cup(3, \infty)$

7. $y=\frac{5}{4x^2+9}$
   $4x^2+9=0$
   $4x^2=-9$
   $x^2=-\frac{9}{4}$
   There are no restricted values.
   Domain: All real numbers; $(-\infty, \infty)$

8. $y = \dfrac{5}{4x^2 - 9}$

$4x^2 - 9 = 0$

$x^2 = \dfrac{9}{4}$

$x = \pm\sqrt{\dfrac{9}{4}}$

$x = \pm\dfrac{3}{2}$

The restricted values are $\pm\dfrac{3}{2}$.

Domain: $x \neq -\dfrac{3}{2}$, $x \neq \dfrac{3}{2}$;

$\left(-\infty, -\dfrac{3}{2}\right) \cup \left(-\dfrac{3}{2}, \dfrac{3}{2}\right) \cup \left(\dfrac{3}{2}, \infty\right)$

9.

x	$y = \frac{180}{x^2} + \frac{30}{x}$	y
−2	$y = \frac{180}{(-2)^2} + \frac{30}{-2} = 30$	30
−1	$y = \frac{180}{(-1)^2} + \frac{30}{-1} = 150$	150
0	$y = \frac{180}{0^2} + \frac{30}{0}$ is undefined	undefined
1	$y = \frac{180}{1^2} + \frac{30}{1} = 210$	210
2	$y = \frac{180}{2^2} + \frac{30}{2} = 60$	60

The solutions are (−2, 30), (−1, 150), (0, undefined), (1, 210), and (2, 60).

10.

x	$g(x)=\frac{x+3}{x^2-x-2}$	$g(x)$
−2	$g(-2)=\frac{-2+3}{(-2)^2-(-2)-2}=\frac{1}{4}$	$\frac{1}{4}$
−1	$g(-1)=\frac{-1+3}{(-1)^2-(-1)-2}=\text{undefined}$	undefined
0	$g(0)=\frac{0+3}{0^2-0-2}=-\frac{3}{2}$	$-\frac{3}{2}$
1	$g(1)=\frac{1+3}{1^2-1-2}=-2$	−2
2	$g(2)=\frac{2+3}{2^2-2-2}=\text{undefined}$	undefined

The solutions are $\left(-2,\frac{1}{4}\right)$, (−1, undefined), $\left(0,-\frac{3}{2}\right)$, (1, −2), and (2, undefined).

11. $p(x)=1-\frac{4}{x}-\frac{21}{x^2}$

The restricted value is 0.

Set up a table of values using x-values less than and greater than 0. Connect the points with a smooth curve.

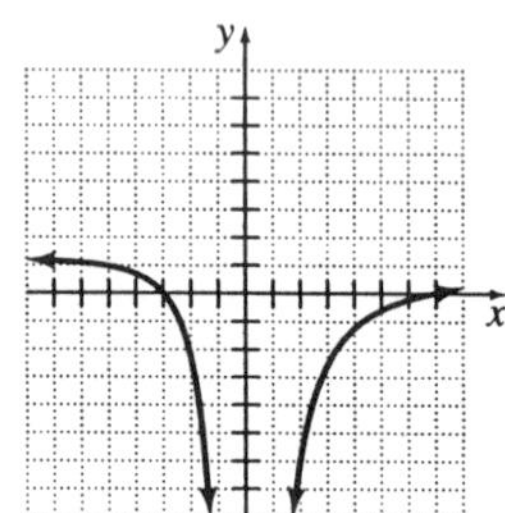

12. $y=\frac{x^2-3x-10}{x^2-x-12}$

$x^2-x-12=0$

$(x-4)(x+3)=0$

$x=4$ or $x=-3$

The restricted values are −3 and 4. Set up a table of values. Connect with a smooth curve.

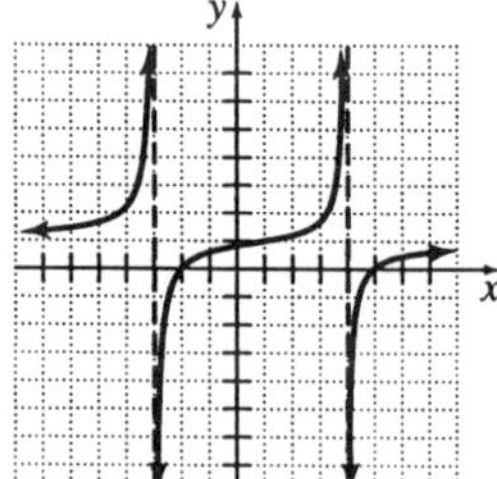

13. a. $d=rt$

$t=\frac{d}{r}$

$t(x)=\frac{125}{10+x}+\frac{80}{x}$

b.

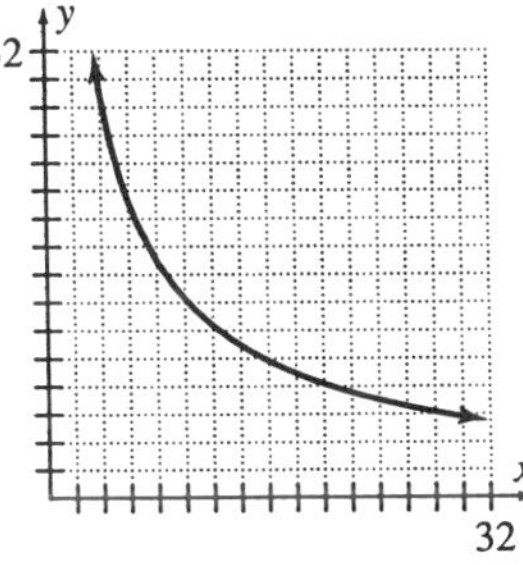

c. At $x = 30$, $y \approx 5.8$
The trip took 5.8 hours.

d. At $y = 4.5$, $x = 40$.
Gretchen's speed was 40 mph and Bob's speed was 50 mph.

14. $\dfrac{-56x^3yz}{126x^2y^3z}$
$= \dfrac{(14x^2yz)(-4x)}{(14x^2yz)(9y^2)}$
$= -\dfrac{4x}{9y^2}$

15. $\dfrac{5p-15}{15p+25}$
$= \dfrac{5(p-3)}{5(3p+5)}$
$= \dfrac{p-3}{3p+5}$

16. $\dfrac{15x-3x^2}{6x^3-30x^2}$
$= \dfrac{-3x(x-5)}{3x(2x)(x-5)}$
$= -\dfrac{1}{2x}$

17. $\dfrac{2x^2+x-15}{4x^2+13x+3}$
$= \dfrac{(2x-5)(x+3)}{(4x+1)(x+3)}$
$= \dfrac{2x-5}{4x+1}$

18. $\dfrac{b^2-49}{b^3+7b^2+2b+14}$
$= \dfrac{(b-7)(b+7)}{(b^2+2)(b+7)}$
$= \dfrac{b-7}{b^2+2}$

19. $\dfrac{9x^3}{35y} \cdot \dfrac{-5y^2}{6x^5}$
$= \dfrac{-45x^3y^2}{210x^5y}$
$= \dfrac{(15x^3y)(-3y)}{(15x^3y)(14x^2)}$
$= -\dfrac{3y}{14x^2}$

20. $\dfrac{a+5}{a-4} \cdot (a+1)$
$= \dfrac{a^2+6a+5}{a-4}$

21. $\dfrac{4m-4}{m^2-25} \cdot \dfrac{m-5}{2m^2+5m-7}$
$= \dfrac{4(m-1)(m-5)}{(m-5)(m+5)(2m+7)(m-1)}$
$= \dfrac{4}{(m+5)(2m+7)}$

22. $\dfrac{a-5}{a+4} \cdot \dfrac{a+4}{5-a}$
$= \dfrac{(a-5)(a+4)}{-1(a+4)(a-5)}$
$= -1$

23. $\dfrac{7x^2y}{5y-3x} \cdot \dfrac{3x-5y}{42xy^3}$
$= \dfrac{7x^2y(3x-5y)}{-42xy^3(3x-5y)}$
$= -\dfrac{x}{6y^2}$

24. $\dfrac{2m+6}{9m+45}\cdot\dfrac{-3m-15}{10m+30}$
$=\dfrac{2(m+3)(-3)(m+5)}{9(m+5)(10)(m+3)}$
$=-\dfrac{1}{15}$

25. $\dfrac{14z^5}{25x^2y^3}\div\dfrac{-7z^3}{15xy^4}$
$=\dfrac{14z^5}{25x^2y^3}\cdot\dfrac{15xy^4}{-7z^3}$
$=\dfrac{210xy^4z^5}{-175x^2y^3z^3}$
$=-\dfrac{6yz^2}{5x}$

26. $\dfrac{27a^2b}{8cd}\div\dfrac{9c^2d}{16ab^2}$
$=\dfrac{27a^2b}{8cd}\cdot\dfrac{16ab^2}{9c^2d}$
$=\dfrac{432a^3b^3}{72c^3d^2}$
$=\dfrac{6a^3b^3}{c^3d^2}$

27. $\dfrac{x^2+4x+3}{5x^2+5}\div\dfrac{x^2-3x+2}{x^4-1}$
$=\dfrac{(x+3)(x+1)}{5(x^2+1)}\cdot\dfrac{(x-1)(x+1)(x^2+1)}{(x-2)(x-1)}$
$=\dfrac{(x+3)(x+1)^2}{5(x-2)}$

28. $\dfrac{3x^2y}{z^3}\div(2xyz)$
$=\dfrac{3x^2y}{z^3}\cdot\dfrac{1}{2xyz}$
$=\dfrac{3x^2y}{2xyz^4}$
$=\dfrac{3x}{2z^4}$

29. $(21a^2bc)\div\dfrac{3ab}{2c}$
$=21a^2bc\cdot\dfrac{2c}{3ab}$
$=\dfrac{42a^2bc^2}{3ab}$
$=14ac^2$

30. $\dfrac{x^2-4}{x^2+4}\div(x-2)$
$=\dfrac{(x-2)(x+2)}{x^2+4}\cdot\dfrac{1}{(x-2)}$
$=\dfrac{x+2}{x^2+4}$

31. $(2x^2-x-3)\div\dfrac{2x^2-7x+6}{2x^2-x-6}$
$=(2x^2-x-3)\cdot\dfrac{2x^2-x-6}{2x^2-7x+6}$
$=\dfrac{(2x-3)(x+1)(2x+3)(x-2)}{(2x-3)(x-2)}$
$=(x+1)(2x+3)$

32. $\dfrac{a+3}{a+6}\div\dfrac{a+2}{a+4}$
$=\dfrac{a+3}{a+6}\cdot\dfrac{a+4}{a+2}$
$=\dfrac{(a+3)(a+4)}{(a+6)(a+2)}$

33. $\frac{x-11}{x-3} \div \frac{11-x}{3-x}$
$= \frac{x-11}{x-3} \cdot \frac{3-x}{11-x}$
$= \frac{-1(11-x)(3-x)}{-1(3-x)(11-x)}$
$= 1$

34. $d = rt$
$r = \frac{d}{t}$
$r = \frac{220}{t}$
The expression of his average speed is $\frac{220}{t}$ mph.
$d = rt$
$= \left(\frac{220}{t} \cdot \frac{1}{2}\right)(t+2)$
$= \frac{110}{t}(t+2)$
$= \frac{110(t+2)}{t}$
His new distance is $\frac{110(t+2)}{t}$ miles.

35. $V = lwh$
$4x^3 + 12x^2 = (2x)^2 h$
$h = \frac{4x^3 + 12x^2}{4x^2}$
$h = \frac{4x^2(x+3)}{4x^2}$
$h = x + 3$
The height is $(x+3)$ inches.

36. a. $C(x) = 500 + 40x$
$R(x) = 80x$

b. $\frac{500+40x}{80x}$
$= \frac{20(25+2x)}{20(4x)}$
$= \frac{25+2x}{4x}$
The ratio is $\frac{25+2x}{4x}$.

37. $\frac{3y}{5x} + \frac{7y^2}{5x}$
$= \frac{3y+7y^2}{5x}$
$= \frac{y(3+7y)}{5x}$

38. $\frac{5b^3}{7a} - \frac{2b}{7a}$
$= \frac{5b^3 - 2b}{7a}$
$= \frac{b\left(5b^2 - 2\right)}{7a}$

39. $\frac{x-8}{2x-5} + \frac{3x-2}{2x-5}$
$= \frac{x-8+3x-2}{2x-5}$
$= \frac{4x-10}{2x-5}$
$= \frac{2(2x-5)}{2x-5}$
$= 2$

40. $\frac{10x^2+3x+5}{4x^2-9} - \frac{2-7x}{4x^2-9}$
$= \frac{10x^2+3x+5-2+7x}{4x^2-9}$
$= \frac{10x^2+10x+3}{4x^2-9}$
$= \frac{10x^2+10x+3}{(2x-3)(2x+3)}$

41. $\dfrac{4}{5x^3y}+\dfrac{7}{15x^2y^2}$
$=\dfrac{4(3y)+7(x)}{15x^3y^2}$
$=\dfrac{12y+7x}{15x^3y^2}$

42. $\dfrac{5x}{x-8}+\dfrac{2x}{x+4}$
$=\dfrac{5x(x+4)+2x(x-8)}{(x-8)(x+4)}$
$=\dfrac{5x^2+20x+2x^2-16x}{(x-8)(x+4)}$
$=\dfrac{7x^2+4x}{(x-8)(x+4)}$
$=\dfrac{x(7x+4)}{(x-8)(x+4)}$

43. $\dfrac{8}{2x-1}-\dfrac{4}{x+5}$
$=\dfrac{8(x+5)-4(2x-1)}{(2x-1)(x+5)}$
$=\dfrac{8x+40-8x+4}{(2x-1)(x+5)}$
$=\dfrac{44}{(2x-1)(x+5)}$

44. $\dfrac{7}{x-y}+\dfrac{4}{y-x}$
$=\dfrac{7-4}{x-y}$
$=\dfrac{3}{x-y}$

45. $\dfrac{4x}{9x^2-16}+\dfrac{2}{15x-20}$
$=\dfrac{4x}{(3x-4)(3x+4)}+\dfrac{2}{5(3x-4)}$
$=\dfrac{4x(5)+2(3x+4)}{5(3x-4)(3x+4)}$
$=\dfrac{20x+6x+8}{5(3x-4)(3x+4)}$
$=\dfrac{26x+8}{5(3x-4)(3x+4)}$
$=\dfrac{2(13x+4)}{5(3x-4)(3x+4)}$

46. $\dfrac{5}{2x^2-5x-3}+\dfrac{7}{3x^2-11x+6}$
$=\dfrac{5}{(2x+1)(x-3)}+\dfrac{7}{(3x-2)(x-3)}$
$=\dfrac{5(3x-2)+7(2x+1)}{(2x+1)(x-3)(3x-2)}$
$=\dfrac{15x-10+14x+7}{(2x+1)(x-3)(3x-2)}$
$=\dfrac{29x-3}{(2x+1)(x-3)(3x-2)}$

47. $\dfrac{7}{2x-18}-\dfrac{3x+4}{x^2-81}$
$=\dfrac{7}{2(x-9)}-\dfrac{3x+4}{(x+9)(x-9)}$
$=\dfrac{7(x+9)-(3x+4)(2)}{2(x-9)(x+9)}$
$=\dfrac{7x+63-6x-8}{2(x-9)(x+9)}$
$=\dfrac{x+55}{2(x-9)(x+9)}$

48. $\dfrac{8}{2x^2+x-6}-\dfrac{2}{3x^2+4x-4}$

$=\dfrac{8}{(2x-3)(x+2)}-\dfrac{2}{(3x-2)(x+2)}$

$=\dfrac{8(3x-2)-2(2x-3)}{(2x-3)(x+2)(3x-2)}$

$=\dfrac{24x-16-4x+6}{(2x-3)(x+2)(3x-2)}$

$=\dfrac{20x-10}{(2x-3)(x+2)(3x-2)}$

$=\dfrac{10(2x-1)}{(2x-3)(x+2)(3x-2)}$

49. $\dfrac{7}{x+5}+\dfrac{6}{x-5}+\dfrac{8}{5-x}$

$=\dfrac{7(x-5)+6(x+5)-8(x+5)}{(x+5)(x-5)}$

$=\dfrac{7x-35+6x+30-8x-40}{(x+5)(x-5)}$

$=\dfrac{5x-45}{(x+5)(x-5)}$

$=\dfrac{5(x-9)}{(x+5)(x-5)}$

50. $\dfrac{7x}{x+2}-\dfrac{11}{3x}-\dfrac{5}{6}$

$=\dfrac{7x(6x)-11(2)(x+2)-5(x)(x+2)}{6x(x+2)}$

$=\dfrac{42x^2-22x-44-5x^2-10x}{6x(x+2)}$

$=\dfrac{37x^2-32x-44}{6x(x+2)}$

51. $\dfrac{x-1}{x+1}+\dfrac{x-3}{x}+\dfrac{x+1}{2x}$

$=\dfrac{2x(x-1)+(x-3)(2)(x+1)+(x+1)(x+1)}{2x(x+1)}$

$=\dfrac{2x^2-2x+2x^2-4x-6+x^2+2x+1}{2x(x+1)}$

$=\dfrac{5x^2-4x-5}{2x(x+1)}$

The perimeter is $\dfrac{5x^2-4x-5}{2x(x+1)}$ units.

52. $\dfrac{14x^2+14x+6}{2x(x+1)}-\dfrac{3x}{x+1}$

$=\dfrac{14x^2+14x+6-3x(2x)}{2x(x+1)}$

$=\dfrac{14x^2+14x+6-6x^2}{2x(x+1)}$

$=\dfrac{8x^2+14x+6}{2x(x+1)}$

$=\dfrac{2(4x^2+7x+3)}{2x(x+1)}$

$=\dfrac{2(4x+3)(x+1)}{2x(x+1)}$

$=\dfrac{4x+3}{x}$

The length is $\dfrac{4x+3}{x}$ units.

53. $\dfrac{2x+3}{6}=\dfrac{5}{\sqrt{x}-1}$ is not rational because $\left(\sqrt{x}-1\right)$ is not a polynomial.

54. $\dfrac{x^2+6x-7}{x^2-6x+5}=\dfrac{2x+1}{x^2}$ is a rational equation because it can be written as $\dfrac{x^2+6x-7}{x^2-6x+5}-\dfrac{2x+1}{x^2}=0.$

55. $2x^2+5x-15=0$ is a rational equation.

56. $3x-7=4x^{1/3}$ is not rational because a variable to a fractional exponent cannot be written as a rational expression.

57. $5x^{-2}+3x^{-1}+4=18$ is a rational equation because it can be written as $\frac{5}{x^2}+\frac{3}{x}-14=0$

58. $\frac{2}{x}-\frac{5}{x-3}=17$ is a rational equation because it can be written as
$\frac{2}{x}-\frac{5}{x-3}-17=0$

59. $x+3=\frac{3x-26}{2-x}$
The restricted value is 2. Set up a table of values on the calculator letting
$Y1=x+3$ and $Y2=\frac{3x-26}{2-x}$
The solutions are –8 and 4.

60. $\frac{3}{x+3}=\frac{6}{2x+6}$
The restricted value is –3. Set up a table of values on the calculator letting
$Y1=\frac{3}{x+3}$ and $Y2=\frac{6}{2x+6}$.
The solutions are all integers not equal to –3.

61. $\frac{x}{8}=\frac{2}{x}$
The restricted value is 0. Set up a table of values on the calculator letting
$Y1=\frac{x}{8}$ and $Y2=\frac{2}{x}$.
The solutions are –4 and 4.

62. $\frac{x}{x+5}=\frac{x+8}{x+5}$
The restricted value is –5. Set up a table of values on the calculator letting
$Y1=\frac{x}{x+5}$ and $Y2=\frac{x+8}{x+5}$. There is no solution.

63. $\frac{x}{4}=\frac{16}{x}$
$Y1=\frac{x}{4}$
$Y2=\frac{16}{x}$

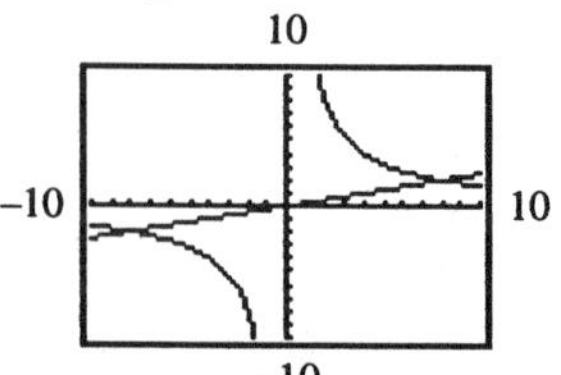

The solutions are the x-coordinates of the intersections, –8 and 8.

64. $\frac{7-x}{2-x}=\frac{x+9}{x-2}$
$Y1=\frac{7-x}{2-x}$
$Y2=\frac{x+9}{x-2}$

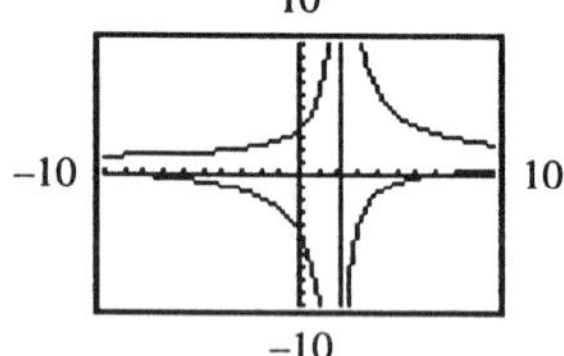

The graphs do not appear to intersect. There is no solution.

65. $x+2=\frac{6x+19}{x+2}$
$Y1=x+2$
$Y2=\frac{6x+19}{x+2}$

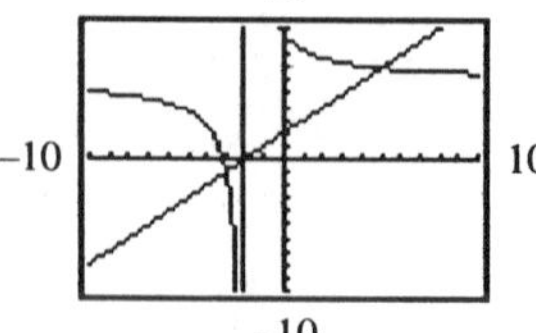

The solutions are the x-coordinates of the intersections, –3 and 5.

66. $1+\frac{5}{x}=\frac{3x+10}{3x}+\frac{5}{3x}$

$Y1=1+\frac{5}{x}$

$Y2=\frac{3x+10}{3x}+\frac{5}{3x}$

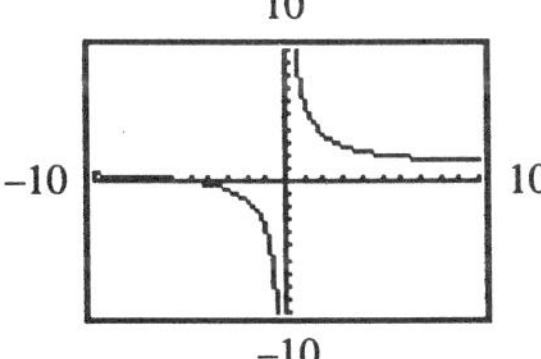

The graphs are the same. The solutions are all real numbers except 0.

67. $4x=\frac{12-x}{x+3}$

$Y1=4x$

$Y2=\frac{12-x}{x+3}$

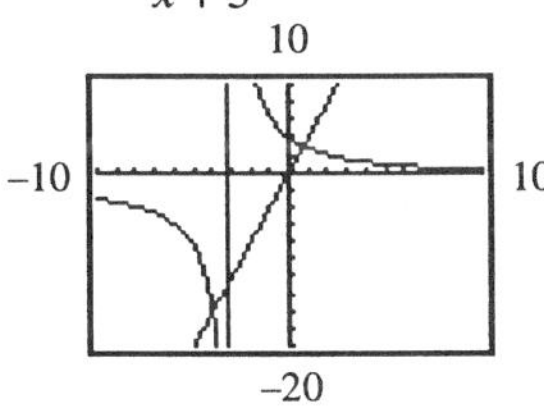

The solutions are the x-coordinates of the intersections, −4 and 0.75.

68. $\frac{70}{x}=\frac{91}{x+0.6}$

$Y1=\frac{70}{x}$

$Y2=\frac{91}{x+0.6}$

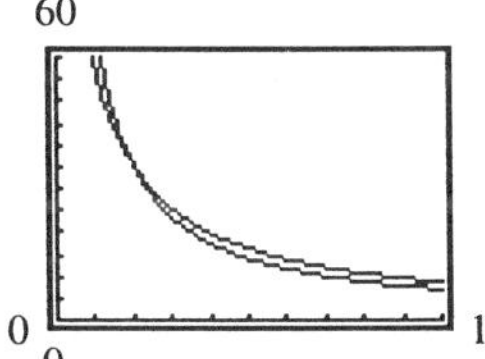

The solution is the x-coordinate of the intersection, 2. The times were 2 hours and 2.6 hours.

69. Let x = resistance of one resistor

$2x+5$ = resistance of second resistor

$\frac{1}{R}=\frac{1}{r_1}+\frac{1}{r_2}$

$\frac{1}{30}=\frac{1}{x}+\frac{1}{2x+5}$

$Y1=\frac{1}{30}$

$Y2=\frac{1}{x}+\frac{1}{2x+5}$

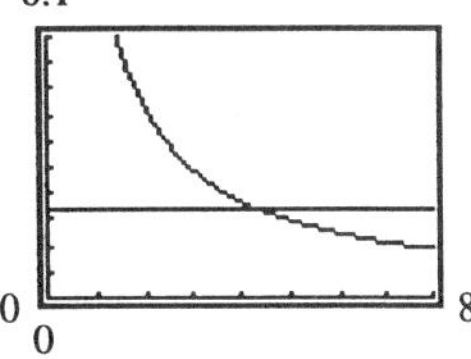

The solution is the x-coordinate of the intersection.

$x\approx 44$

$2x+5\approx 93$

The resistances of the two resistors are 44 ohms and 93 ohms.

70. $\frac{4}{3x}=1-\frac{17}{3x}$

$\frac{4}{3x}=\frac{3x-17}{3x}$

$4=3x-17$

$21=3x$

$7=x$

The solution is 7.

71. $\frac{5}{2x+4}=\frac{2x+1}{2}$

$2(5)=(2x+4)(2x+1)$

$10=4x^2+10x+4$

$4x^2+10x-6=0$

$2(2x^2+5x-3)=0$

$2(2x-1)(x+3)=0$

$x=\frac{1}{2}$　　or　　$x=-3$

The solutions are −3 and $\frac{1}{2}$.

72. $\dfrac{3x+7}{2x-4}=\dfrac{11}{3}$
$3(3x+7)=11(2x-4)$
$9x+21=22x-44$
$65=13x$
$5=x$
The solution is 5.

73. $\dfrac{z}{50}-\dfrac{2}{z}=0$
$\dfrac{z}{50}=\dfrac{2}{z}$
$z^2=100$
$z=\pm\sqrt{100}$
$z=\pm 10.$
The solutions are ± 10.

74. $b=1+\dfrac{30}{b}$
$b-1=\dfrac{30}{b}$
$b(b-1)=30$
$b^2-b-30=0$
$(b-6)(b+5)=0$
$b=6$ or $b=-5$
The solutions are –5 and 6.

75. $a+\dfrac{13}{a}=-\dfrac{3}{a}-8$
$a\left(a+\dfrac{13}{a}\right)=a\left(-\dfrac{3}{a}-8\right)$
$a^2+13=-3-8a$
$a^2+8a+16=0$
$(a+4)^2=0$
$a=-4$
The solution is –4.

76. $x^{-2}-18=7$
$\dfrac{1}{x^2}=25$
$25x^2=1$
$x^2=\dfrac{1}{25}$
$x=\pm\sqrt{\dfrac{1}{25}}$
$x=\pm\dfrac{1}{5}$
The solutions are $\pm\dfrac{1}{5}$.

77. $1=12p^{-2}+p^{-1}$
$1=\dfrac{12}{p^2}+\dfrac{1}{p}$
$p^2(1)=p^2\left(\dfrac{12}{p^2}+\dfrac{1}{p}\right)$
$p^2=12+p$
$p^2-p-12=0$
$(p+3)(p-4)=0$
$p=-3$ or $p=4$
The solutions are –3 and 4.

78. $\dfrac{x-2}{x-6}+3=\dfrac{3x-14}{x-6}$
$(x-6)\left(\dfrac{x-2}{x-6}+3\right)=(x-6)\left(\dfrac{3x-14}{x-6}\right)$
$x-2+3x-18=3x-14$
$4x-20=3x-14$
$x=6$ (extraneous)
There is no solution.

79. $\dfrac{3}{m+4}+\dfrac{7}{m^2-16}=2$
$(m+4)(m-4)\left(\dfrac{3}{m+4}+\dfrac{7}{m^2-16}\right)$
$=(m+4)(m-4)(2)$
$3(m-4)+7=2(m^2-16)$
$3m-12+7=2m^2-32$
$2m^2-3m-27=0$

$(2m-9)(m+3)=0$

$m=\frac{9}{2}$ or $m=-3$

The solutions are –3 and $\frac{9}{2}$.

80. $\frac{x+5}{x-2}=\frac{x+8}{2(x+3)-(x+8)}$

$\frac{x+5}{x-2}=\frac{x+8}{2x+6-x-8}$

$\frac{x+5}{x-2}=\frac{x+8}{x-2}$

$x+5=x+8$

$5=8$

This is a contradiction. There is no solution.

81. $\frac{a}{a+5}=\frac{a-4}{a-3}$

$a(a-3)=(a+5)(a-4)$

$a^2-3a=a^2+a-20$

$20=4a$

$5=a$

The solution is 5.

82. $\frac{2x}{x+3}=1-\frac{7}{x+3}$

$(x+3)\left(\frac{2x}{x+3}\right)=(x+3)\left(1-\frac{7}{x+3}\right)$

$2x=x+3-7$

$x=-4$

The solution is –4.

83. $\frac{2}{(3x+5)-(2x+1)}=\frac{8}{3(x+5)+(x+1)}$

$\frac{2}{3x+5-2x-1}=\frac{8}{3x+15+x+1}$

$\frac{2}{x+4}=\frac{8}{4x+16}$

$4(x+4)\left(\frac{2}{x+4}\right)=4(x+4)\left(\frac{8}{4x+16}\right)$

$8=8$

This is an identity. The solutions are all real numbers except –4.

84. $\frac{7}{x+3}+\frac{2x-1}{x^2+x-6}=\frac{5x-2}{x^2-4}$

$\frac{7}{x+3}+\frac{2x-1}{(x+3)(x-2)}=\frac{5x-2}{(x+2)(x-2)}$

$(x+3)(x+2)(x-2)\left[\frac{7}{x+3}+\frac{2x-1}{(x+3)(x-2)}\right]$

$=(x+3)(x+2)(x-2)\left[\frac{5x-2}{(x+2)(x-2)}\right]$

$7(x^2-4)+(x+2)(2x-1)=(x+3)(5x-2)$

$7x^2-28+2x^2+3x-2=5x^2+13x-6$

$9x^2+3x-30=5x^2+13x-6$

$4x^2-10x-24=0$

$2(2x^2-5x-12)=0$

$2(2x+3)(x-4)=0$

$x=-\frac{3}{2}$ or $x=4$

The solutions are $-\frac{3}{2}$ and 4.

85. Let x = resistance of one resistor

$x-10$ = resistance of other resistor

$\frac{1}{R}=\frac{1}{r_1}+\frac{1}{r_2}$

$\frac{1}{12}=\frac{1}{x}+\frac{1}{x-10}$

$12x(x-10)\left(\frac{1}{12}\right)$

$=12x(x-10)\left(\frac{1}{x}+\frac{1}{x-10}\right)$

$x(x-10)=12(x-10)+12x$

$x^2-10x=12x-120+12x$

$x^2-34x+120=0$

$(x-30)(x-4)=0$

$x=30$ or $x=4$

$x-10=20$ $x-10=-6$

The resistances of the two resistors are 20 ohms and 30 ohms.

86. Let x = additional amount Maya should add

$\frac{x+85}{x+210} = \frac{50}{100}$

$100(x + 85) = 50(x + 210)$

$100x + 8500 = 50x + 10500$

$50x = 2000$

$x = 40$

She should add another $40.

87. Let x = time to do the room together.

$\frac{1}{6}$ is Norman's rate.

$\frac{1}{8}$ is his assistant's rate.

$\frac{1}{x}$ is the combined rate.

$\frac{1}{6} + \frac{1}{8} = \frac{1}{x}$

$\frac{8+6}{48} = \frac{1}{x}$

$14x = 48$

$x = \frac{48}{14}$

$x = \frac{24}{7}$

$x \approx 3.4$

It will take them $\frac{24}{7}$ hours or approximately 3.4 hours.

88. Let x = time it takes Lucy to pack a crate.

$\frac{1}{x}$ is Viv's rate.

$\frac{1}{x+10}$ is Lucy's rate.

$\frac{1}{30}$ is the combined rate.

$\frac{1}{x} + \frac{1}{x+10} = \frac{1}{30}$

$30x(x+10)\left(\frac{1}{x} + \frac{1}{x+10}\right)$

$= 30x(x+10)\left(\frac{1}{30}\right)$

$30(x + 10) + 30x = x(x + 10)$

$30x + 300 + 30x = x^2 + 10x$

$x^2 - 50x - 300 = 0$

$x = \frac{50 \pm \sqrt{(-50)^2 - 4(1)(-300)}}{2(1)}$

$x = \frac{50 \pm \sqrt{3700}}{2}$

$x \approx 55.4$

It will take Viv 55.4 minutes and Lucy 65.4 minutes.

89. Let x = time to fill the barrel together.

$\frac{1}{10}$ is one line's rate.

$\frac{1}{15}$ is another line's rate.

$\frac{1}{x}$ is the combined rate.

$\frac{1}{10} + \frac{1}{15} = \frac{1}{x}$

$\frac{3+2}{30} = \frac{1}{x}$

$5x = 30$

$x = 6$

It will take 6 minutes working together.

90. Let x = time for high speed line to produce one order.

$\frac{1}{x}$ is the high speed line's rate.

$\frac{1}{2x}$ is the other line's rate.

$\frac{1}{3\frac{1}{2}} = \frac{1}{\frac{7}{2}} = \frac{2}{7}$ is the combined rate.

$$\frac{1}{x}+\frac{1}{2x}=\frac{2}{7}$$
$$14x\left(\frac{1}{x}+\frac{1}{2x}\right)=14x\left(\frac{2}{7}\right)$$
$$14+7=4x$$
$$\frac{21}{4}=x$$
$$5\frac{1}{4}=x$$

It will take the high speed line $5\frac{1}{4}$ hours and the other line $10\frac{1}{2}$ hours working alone.

91. Let x = Solomon's earnings.
$$\frac{18.50}{30}=\frac{x}{100}$$
$$100(18.50)=30x$$
$$61.67\approx x$$
He will earn \$61.67.

92. Let x = additional cups
$$\frac{1.5+x}{12+x}=\frac{25}{100}$$
$$100(1.5+x)=25(12+x)$$
$$150+100x=300+25x$$
$$75x=150$$
$$x=2$$
Another 2 cups should be added.

93. a.
$$\frac{GI}{JL}=\frac{GH}{JK}$$
$$\frac{x}{21}=\frac{4}{7}$$
$$7x=84$$
$$x=12$$
$\overline{GI}$ measures 12 inches.
$$\frac{HI}{KL}=\frac{GH}{JK}$$
$$\frac{8}{x}=\frac{4}{7}$$
$$4x=56$$
$$x=14$$
$\overline{KL}$ measures 14 inches.

b.
$$\frac{HI}{KL}=\frac{GH}{JK}$$
$$\frac{x}{5\frac{13}{16}}=\frac{3\frac{1}{2}}{2\frac{5}{8}}$$
$$2\frac{5}{8}(x)=\left(5\frac{13}{16}\right)\left(3\frac{1}{2}\right)$$
$$x=7\frac{3}{4}$$
$\overline{HI}$ measures $7\frac{3}{4}$ feet.
$$\frac{GI}{JL}=\frac{GH}{JK}$$
$$\frac{13}{x}=\frac{3\frac{1}{2}}{2\frac{5}{8}}$$
$$3\frac{1}{2}(x)=13\left(2\frac{5}{8}\right)$$
$$x=9\frac{3}{4}$$
$\overline{JL}$ measures $9\frac{3}{4}$ feet.

c.
$$\frac{GH}{JK}=\frac{HI}{KL}$$
$$\frac{x}{1.88}=\frac{9.22}{2.305}$$
$$2.305(x)=(9.22)(1.88)$$
$$x=7.52$$
$\overline{GH}$ measures 7.52 cm.
$$\frac{GI}{UL}=\frac{HI}{KL}$$
$$\frac{14.08}{x}=\frac{9.22}{2.305}$$
$$9.22(x)=(14.08)(2.305)$$
$$x=3.52$$
$\overline{JL}$ measures 3.52 cm.

94. Let x = height of bluff
$$\frac{x}{120}=\frac{6}{15}$$
$$15x=720$$
$$x=48$$
The bluff is 48 feet high.

95. Let x = length of tree's shadow

$\frac{x}{9} = \frac{64}{6}$

$6x = 576$

$x = 96$

The tree's shadow was 96 feet long.

Chapter 12 Mixed Review

1.

x	$y = \frac{45}{x^2} + \frac{15}{x}$	y
–2	$y = \frac{45}{(-2)^2} + \frac{15}{-2} = 3.75$	3.75
–1	$y = \frac{45}{(-1)^2} + \frac{15}{-1} = 30$	30
0	$y = \frac{45}{0^2} + \frac{15}{0}$ is undefined	undefined
1	$y = \frac{45}{1^2} + \frac{15}{1} = 60$	60
2	$y = \frac{45}{2^2} + \frac{15}{2} = 18.75$	18.75

The solutions are (–2, 3.75), (–1, 30), (0, undefined). (1, 60), and (2, 18.75).

2.

x	$g(x) = \frac{x+5}{x^2+5x+4}$	$g(x)$
–2	$g(-2) = \frac{-2+5}{(-2)^2+5(-2)+4} = -\frac{3}{2}$	$-\frac{3}{2}$
–1	$g(-1) = \frac{-1+5}{(-1)^2+5(-1)+4}$ is undefined	undefined
0	$g(0) = \frac{0+5}{0^2+5(0)+4} = \frac{5}{4}$	$\frac{5}{4}$
1	$g(1) = \frac{1+5}{1^2+5(1)+4} = \frac{3}{5}$	$\frac{3}{5}$
2	$g(2) = \frac{2+5}{2^2+5(2)+4} = \frac{7}{18}$	$\frac{7}{18}$

The solutions are $\left(-2, -\frac{3}{2}\right)$, (–1, undefined), $\left(0, \frac{5}{4}\right)$, $\left(1, \frac{3}{5}\right)$, and $\left(2, \frac{7}{18}\right)$.

3. $p(x) = 2 - \frac{3}{x} - \frac{2}{x^2}$

The restricted value is 0. Set up a table of x-values less than and greater than 0. Connect with a smooth curve.

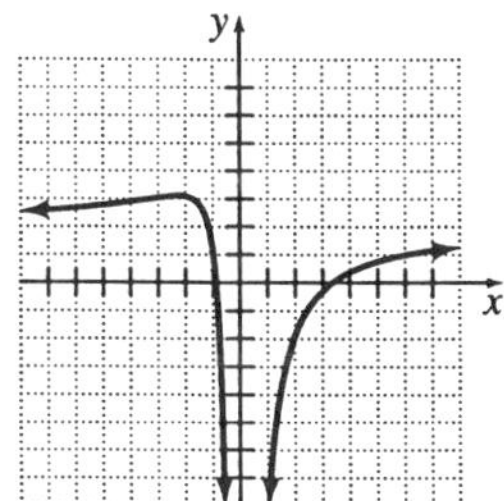

4. $y = \frac{x^2 - 5x - 6}{x^2 + 4x - 5}$

$x^2 + 4x - 5 = 0$

$(x + 5)(x - 1) = 0$

$x = -5$ or $x = 1$

The restricted values are –5 and 1. Set up a table of x-values less than –5, betwen –5 and 1, and greater than 1. Connect with a smooth curve.

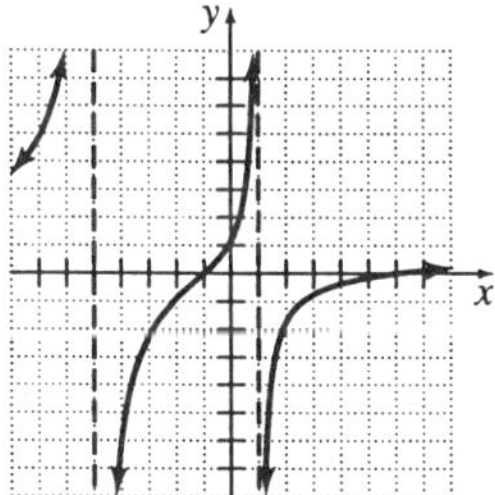

5. $x + 5 = \frac{x - 11}{2 - x}$

Set up a table of values on the calculator letting $\text{Y1} = x + 5$ and $\text{Y2} = \frac{x - 11}{2 - x}$. The solutions are –7 and 3.

6. $\frac{x + 4}{2x + 1} = \frac{x - 4}{(x + 1) + x}$

Set up a table of values on the calculator letting $\text{Y1} = \frac{x + 4}{2x + 1}$ and $\text{Y2} = \frac{x - 4}{(x + 1) + x}$.

There is no solution.

7. $5 = \frac{3(x + 5)}{x + 3} + \frac{4x}{2x + 6}$

Set up a table of values letting $\text{Y1} = 5$ and $\text{Y2} = \frac{3(x + 5)}{x + 3} + \frac{4x}{2x + 6}$

The solutions are all real numbers except –3.

8. $\frac{x}{16} = \frac{4}{x}$

Set up a table of values on the calculator letting $\text{Y1} = \frac{x}{16}$ and $\text{Y2} = \frac{4}{x}$.

The solutions are ±8.

9. $\frac{1}{2} - \frac{3}{2x} = \frac{1}{2} + \frac{1}{2x} - \frac{2}{x}$

$\text{Y1} = \frac{1}{2} - \frac{3}{2x}$

$\text{Y2} = \frac{1}{2} + \frac{1}{2x} - \frac{2}{x}$

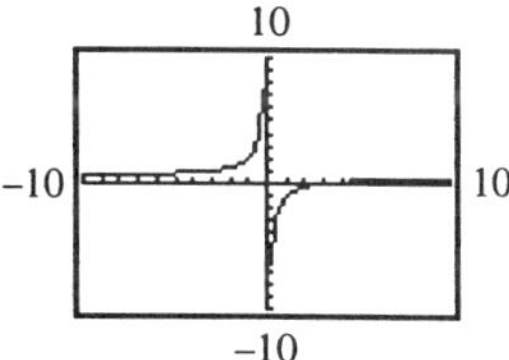

The graphs are the same. The solution is all real numbers except 0.

10. $5x = \frac{9x + 4}{x + 2}$

$\text{Y1} = 5x$

$\text{Y2} = \frac{9x + 4}{x + 2}$

5

–5 5

–7

The solutions are the x-coordinates of the intersections, –1 and $\frac{4}{5}$.

11. $\frac{x}{3}=\frac{27}{x}$

$Y1=\frac{x}{3}$

$Y2=\frac{27}{x}$

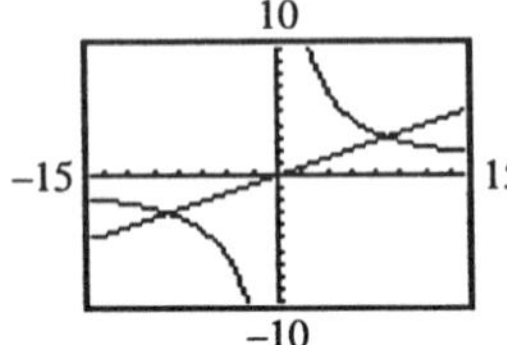

The solutions are the x-coordinates of the intersections, –9 and 9.

12. $\frac{4-x}{3-x}=\frac{x+11}{x-3}$

$Y1=\frac{4-x}{3-x}$

$Y2=\frac{x+11}{x-3}$

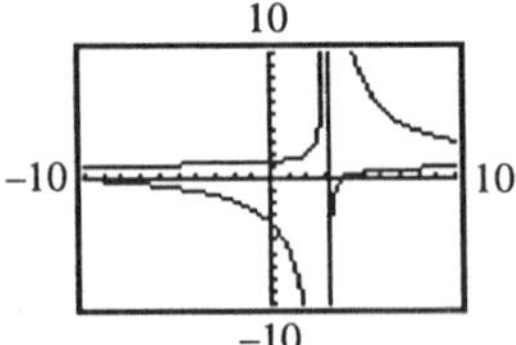

The graphs do not appear to intersect. There is no solution.

13. $\frac{x+8}{2}=\frac{2x+11}{x+1}$

$Y1=\frac{x+8}{2}$

$Y2=\frac{2x+11}{x+1}$

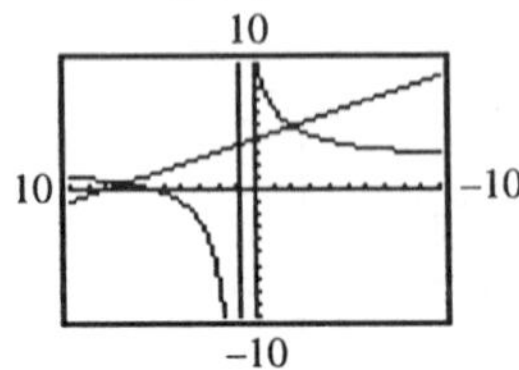

The solutions are the x-coordinates of the intersections, –7 and 2.

14. $y=\frac{5}{9x^2+25}$

$9x^2+25=0$

$x^2=-\frac{25}{9}$

There are no restricted values.
Domain: all real numbers; $(-\infty, \infty)$

15. $y=\frac{5}{9x^2-25}$

$9x^2-25=0$

$x^2=\frac{25}{9}$

$x=\pm\sqrt{\frac{25}{9}}$

$x=\pm\frac{5}{3}$

The restricted values are $\pm\frac{5}{3}$.

Domain: $x\neq-\frac{5}{3}, x\neq\frac{5}{3}$;

$\left(-\infty, -\frac{5}{3}\right)\cup\left(-\frac{5}{3}, \frac{5}{3}\right)\cup\left(\frac{5}{3}, \infty\right)$

16. $g(x)=\frac{x+5}{3x}$

$3x=0$

$x=0$

The restricted value is 0.
Domain: $x\neq 0$; $(-\infty, 0)\cup(0, \infty)$

17. $h(x)=\frac{5x-7}{3x^2+2x-8}$

$3x^2+2x-8=0$

$(3x-4)(x+2)=0$

$x=\frac{4}{3}$ or $x=-2$

The restricted values are –2 and $\frac{4}{3}$.

Domain: $x\neq-2, x\neq\frac{4}{3}$;

$(0, -2)\cup\left(-2, \frac{4}{3}\right)\cup\left(\frac{4}{3}, \infty\right)$

18. $\dfrac{3x^2+19x+20}{2x^2+13x+15}$
$=\dfrac{(3x+4)(x+5)}{(2x+3)(x+5)}$
$=\dfrac{3x+4}{2x+3}$

19. $\dfrac{k^2-25}{k^3-5k^2+6k-30}$
$=\dfrac{(k+5)(k-5)}{(k^2+6)(k-5)}$
$=\dfrac{k+5}{k^2+6}$

20. $\dfrac{-72x^4y^2z}{96xy^6z}$
$=\dfrac{(24xy^2z)(-3x^3)}{(24xy^2z)(4y^4)}$
$=-\dfrac{3x^3}{4y^4}$

21. $\dfrac{7q-21}{14q+28}$
$=\dfrac{7(q-3)}{14(q+2)}$
$=\dfrac{q-3}{2(q+2)}$

22. $\dfrac{12x^2-6x^3}{9x^3-36x}$
$=\dfrac{6x^2(2-x)}{9x(x^2-4)}$
$=\dfrac{-2x(3x)(x-2)}{3(3x)(x-2)(x+2)}$
$=\dfrac{-2x}{3(x+2)}$

23. $\dfrac{7}{3x+2}-\dfrac{2}{x-3}$
$=\dfrac{7(x-3)-2(3x+2)}{(3x+2)(x-3)}$
$=\dfrac{7x-21-6x-4}{(3x+2)(x-3)}$
$=\dfrac{x-25}{(3x+2)(x-3)}$

24. $\dfrac{14}{a-b}+\dfrac{9}{b-a}$
$=\dfrac{14-9}{a-b}$
$=\dfrac{5}{a-b}$

25. $\dfrac{7b}{3a}+\dfrac{5b^2}{3a}$
$=\dfrac{7b+5b^2}{3a}$
$=\dfrac{b(7+5b)}{3a}$

26. $\dfrac{3p^2}{4m}-\dfrac{2p}{4m}$
$=\dfrac{3p^2-2p}{4m}$
$=\dfrac{p(3p-2)}{4m}$

27. $\dfrac{5}{4x^2y^3}+\dfrac{3}{6x^2y}$
$=\dfrac{5(3)+3(2y^2)}{12x^2y^3}$
$=\dfrac{3(5+2y^2)}{3(4x^2y^3)}$
$=\dfrac{5+2y^2}{4x^2y^3}$

28. $\dfrac{x}{x+3}+\dfrac{4x}{x-5}$
$=\dfrac{x(x-5)+4x(x+3)}{(x+3)(x-5)}$
$=\dfrac{x^2-5x+4x^2+12x}{(x+3)(x-5)}$
$=\dfrac{5x^2+7x}{(x+3)(x-5)}$
$=\dfrac{x(5x+7)}{(x+3)(x-5)}$

29. $\dfrac{2x}{4x^2-25}+\dfrac{1}{6x-15}$
$=\dfrac{2x}{(2x-5)(2x+5)}+\dfrac{1}{3(2x-5)}$
$=\dfrac{2x(3)+(2x+5)}{3(2x-5)(2x+5)}$
$=\dfrac{6x+2x+5}{3(2x-5)(2x+5)}$
$=\dfrac{8x+5}{3(2x-5)(2x+5)}$

30. $\dfrac{7}{2x^2+5x-3}+\dfrac{5}{3x^2+11x+6}$
$=\dfrac{7}{(2x-1)(x+3)}+\dfrac{5}{(3x+2)(x+3)}$
$=\dfrac{7(3x+2)+5(2x-1)}{(2x-1)(x+3)(3x+2)}$
$=\dfrac{21x+14+10x-5}{(2x-1)(x+3)(3x+2)}$
$=\dfrac{31x+9}{(2x-1)(x+3)(3x+2)}$

31. $\dfrac{3z-5}{2z-7}+\dfrac{z-9}{2z-7}$
$=\dfrac{3z-5+z-9}{2z-7}$
$=\dfrac{4z-14}{2z-7}$
$=\dfrac{2(2z-7)}{2z-7}$
$=2$

32. $\dfrac{10x^2-8x}{25x^2-16}-\dfrac{5x-4}{25x^2-16}$
$=\dfrac{10x^2-8x-5x+4}{25x^2-16}$
$=\dfrac{10x^2-13x+4}{25x^2-16}$
$=\dfrac{(5x-4)(2x-1)}{(5x-4)(5x+4)}$
$=\dfrac{2x-1}{5x+4}$

33. $\dfrac{3}{x+6}+\dfrac{2}{x-6}+\dfrac{4}{6-x}$
$=\dfrac{3(x-6)+2(x+6)-4(x+6)}{(x+6)(x-6)}$
$=\dfrac{3x-18+2x+12-4x-24}{(x+6)(x-6)}$
$=\dfrac{x-30}{(x+6)(x-6)}$

34. $\dfrac{x}{x+1}-\dfrac{3}{2x}-\dfrac{3}{4}$
$=\dfrac{4x(x)-3(2)(x+1)-3(x)(x+1)}{4x(x+1)}$
$=\dfrac{4x^2-6x-6-3x^2-3x}{4x(x+1)}$
$=\dfrac{x^2-9x-6}{4x(x+1)}$

35. $\dfrac{5}{2x+10}-\dfrac{2x+1}{x^2-25}$
$=\dfrac{5(x-5)-(2x+1)(2)}{2(x+5)(x-5)}$
$=\dfrac{5x-25-4x-2}{2(x+5)(x-5)}$
$=\dfrac{x-27}{2(x+5)(x-5)}$

36. $\dfrac{3}{4x^2-17x+15}-\dfrac{2}{5x^2-19x+12}$
$=\dfrac{3}{(4x-5)(x-3)}-\dfrac{2}{(5x-4)(x-3)}$
$=\dfrac{3(5x-4)-2(4x-5)}{(4x-5)(x-3)(5x-4)}$
$=\dfrac{15x-12-8x+10}{(4x-5)(x-3)(5x-4)}$
$=\dfrac{7x-2}{(4x-5)(x-3)(5x-4)}$

37. $\dfrac{m^2-36}{2m^2+9m+10}\cdot\dfrac{3m+6}{m-6}$
$=\dfrac{(m-6)(m+6)(3)(m+2)}{(2m+5)(m+2)(m-6)}$
$=\dfrac{3(m+6)}{2m+5}$

38. $\dfrac{z-3}{z-7}\cdot\dfrac{7-z}{z-3}$
$=\dfrac{(z-3)(-1)(z-7)}{(z-7)(z-3)}$
$=-1$

39. $\dfrac{8a^4}{21b}\cdot\dfrac{-35b^3}{12a^7}$
$=\dfrac{-280a^4b^3}{252a^7b}$
$=-\dfrac{10b^2}{9a^3}$

40. $(2c+3)\cdot\dfrac{c+7}{c+3}$
$=\dfrac{(2c+3)(c+7)}{c+3}$

41. $\dfrac{3b-4a}{24ab^3}\cdot\dfrac{8a^2b}{4a-3b}$
$=\dfrac{-1(4a-3b)(8a^2b)}{24ab^3(4a-3b)}$
$=-\dfrac{a}{3b^2}$

42. $\dfrac{5p+20}{12p+96}\cdot\dfrac{-9p-72}{15p+60}$
$=\dfrac{5(p+4)(-9)(p+8)}{12(p+8)(15)(p+4)}$
$=-\dfrac{1}{4}$

43. $\dfrac{22g^8}{15h^3k^2}\div\dfrac{-11g^4}{10hk^3}$
$=\dfrac{22g^8}{15h^3k^2}\cdot\dfrac{10hk^3}{-11g^4}$
$=\dfrac{220g^8hk^3}{-165g^4h^3k^2}$
$=-\dfrac{4g^4k}{3h^2}$

44. $\dfrac{18p^2q^2}{25m^2n}\div\dfrac{3mn^3}{5pq^5}$
$=\dfrac{18p^2q^2}{25m^2n}\cdot\dfrac{5pq^5}{3mn^3}$
$=\dfrac{90p^3q^7}{75m^3n^4}$
$=\dfrac{6p^3q^7}{5m^3n^4}$

45. $(15m^3np^2)\div\dfrac{2mn}{5p}$
$=15m^3np^2\cdot\dfrac{5p}{2mn}$
$=\dfrac{75m^3np^3}{2mn}$
$=\dfrac{75m^2p^3}{2}$

46. $\dfrac{4x^2-9}{4x^2+9}\div(2x+3)$
$=\dfrac{(2x-3)(2x+3)}{4x^2+9}\cdot\dfrac{1}{2x+3}$
$=\dfrac{2x-3}{4x^2+9}$

47. $\frac{x^2-x-6}{2x^2+8} \div \frac{x^2-5x+6}{x^4-16}$
$= \frac{(x-3)(x+2)}{2(x^2+4)} \cdot \frac{(x^2+4)(x-2)(x+2)}{(x-3)(x-2)}$
$= \frac{(x+2)^2}{2}$

48. $\frac{5a^2b^3}{c^5} \div (3a^2bc^2)$
$= \frac{5a^2b^3}{c^5} \cdot \frac{1}{3a^2bc^2}$
$= \frac{5a^2b^3}{3a^2bc^7}$
$= \frac{5b^2}{3c^7}$

49. $\frac{z-3}{z-6} \div \frac{z+4}{z+8}$
$= \frac{z-3}{z-6} \cdot \frac{z+8}{z+4}$
$= \frac{(z-3)(z+8)}{(z-6)(z+4)}$

50. $\frac{k-7}{k-12} \div \frac{7-k}{12-k}$
$= \frac{k-7}{k-12} \cdot \frac{12-k}{7-k}$
$= \frac{-1(7-k)(12-k)}{-1(12-k)(7-k)}$
$= 1$

51. $(2x^2+9x-5) \div \frac{2x^2+3x-2}{2x^2+5x+2}$
$= (2x-1)(x+5) \cdot \frac{(2x+1)(x+2)}{(2x-1)(x+2)}$
$= (x+5)(2x+1)$

52. $\frac{4}{7} - \frac{1}{y} = \frac{9}{28}$
$\frac{4y-7}{7y} = \frac{9}{28}$
$28(4y-7) = 9(7y)$
$112y - 196 = 63y$
$49y = 196$
$y = 4$
The solution is 4.

53. $\frac{7}{3x} = 2 - \frac{17}{3x}$
$3x\left(\frac{7}{3x}\right) = 3x\left(2 - \frac{17}{3x}\right)$
$7 = 6x - 17$
$24 = 6x$
$4 = x$
The solution is 4.

54. $\frac{9}{2x+3} = -3$
$-3(2x+3) = 9$
$-6x - 9 = 9$
$-6x = 18$
$x = -3$
The solution is –3.

55. $\frac{23}{18} = \frac{4}{9} + \frac{5}{m}$
$18m\left(\frac{23}{18}\right) = 18m\left(\frac{4}{9} + \frac{5}{m}\right)$
$23m = 8m + 90$
$15m = 90$
$m = 6$
The solution is 6.

56. $\frac{4}{x-1} = \frac{5x-16}{7}$
$4(7) = (x-1)(5x-16)$
$28 = 5x^2 - 21x + 16$
$5x^2 - 21x - 12 = 0$
$x = \frac{21 \pm \sqrt{(-21)^2 - 4(5)(-12)}}{2(5)}$
$x = \frac{21 \pm \sqrt{681}}{10}$
$x \approx -0.51$ and $x \approx 4.71$
The solutions are approxiamtely –0.51 and 4.71.

57. $\frac{7x-2}{3x+5}=16$
$16(3x+5)=7x-2$
$48x+80=7x-2$
$41x=-82$
$x=-2$
The solution is –2.

58. $2-\frac{4}{x+3}=\frac{x+2}{x+3}$
$(x+3)\left(2-\frac{4}{x+3}\right)=(x+3)\left(\frac{x+2}{x+3}\right)$
$2(x+3)-4=x+2$
$2x+6-4=x+2$
$2x+2=x+2$
$x=0$
The solution is 0.

59. $\frac{5}{m+3}+\frac{8}{m^2-9}=\frac{3}{5}$
$5(m+3)(m-3)\left[\frac{5}{m+3}+\frac{8}{(m+3)(m-3)}\right]$
$=5(m+3)(m-3)\left(\frac{3}{5}\right)$
$5(5)(m-3)+5(8)=3(m^2-9)$
$25m-75+40=3m^2-27$
$3m^2-25m+8=0$
$(3m-1)(m-8)=0$
$m=\frac{1}{3}$ $\quad m=8$
The solutions are $\frac{1}{3}$ and 8.

60. $\frac{16}{2x+6}=\frac{5}{x}$
$16x=5(2x+6)$
$16x=10x+30$
$6x=30$
$x=5$
The solution is 5.

61. $35=p^{-2}+2p^{-1}$
$35=\frac{1}{p^2}+\frac{2}{p}$
$p^2(35)=p^2\left(\frac{1}{p^2}+\frac{2}{p}\right)$
$35p^2=1+2p$
$35p^2-2p-1=0$
$(7p+1)(5p-1)=0$
$p=-\frac{1}{7}$ $\quad p=\frac{1}{5}$
The solutions are $-\frac{1}{7}$ and $\frac{1}{5}$.

62. $\frac{x-6}{x-4}=\frac{x+7}{2(x+3)-(x+10)}$
$\frac{x-6}{x-4}=\frac{x+7}{2x+6-x-10}$
$\frac{x-6}{x-4}=\frac{x+7}{x-4}$
$x-6=x+7$
$-6=7$
This is a contradiction.
There is no solution.

63. $\frac{a}{a-7}=\frac{a+1}{a-2}$
$a(a-2)=(a-7)(a+1)$
$a^2-2a=a^2-6a-7$
$4a=-7$
$a=-\frac{7}{4}$
The solution is $-\frac{7}{4}$.

64. $1-\frac{6}{x}=\frac{6}{x}-\frac{1}{2}+\frac{21}{2x}$
$2x\left(1-\frac{6}{x}\right)=2x\left(\frac{6}{x}-\frac{1}{2}+\frac{21}{2x}\right)$
$2x-12=12-x+21$
$3x=45$
$x=15$
The solution is 15.

65. $\frac{4x+7}{3}-\frac{x+6}{5}=\frac{4x-1}{5}$
$15\left(\frac{4x+7}{3}-\frac{x+6}{5}\right)=15\left(\frac{4x-1}{5}\right)$
$5(4x+7)-3(x+6)=3(4x-1)$
$20x+35-3x-18=12x-3$
$17x+17=12x-3$
$5x=-20$
$x=-4$
The solution is – 4.

66. $1+7x^{-1}=3$
$1+\frac{7}{x}=3$
$\frac{7}{x}=2$
$2x=7$
$x=\frac{7}{2}$
The solution is $\frac{7}{2}$.

67. $\frac{2x}{x-4}=1+\frac{9}{x-4}$
$(x-4)\left(\frac{2x}{x-4}\right)=(x-4)\left(1+\frac{9}{x-4}\right)$
$2x=x-4+9$
$x=5$
The solution is 5.

68. $\frac{10}{x+5}-\frac{24}{x^2+x-20}=\frac{1}{x^2-16}$
$\frac{10}{x+5}-\frac{24}{(x+5)(x-4)}=\frac{1}{(x-4)(x+4)}$
$(x+5)(x-4)(x+4)\left[\frac{10}{x+5}-\frac{24}{(x+5)(x-4)}\right]$
$=(x+5)(x-4)(x+4)\left[\frac{1}{(x-4)(x+4)}\right]$
$10(x^2-16)-24(x+4)=x+5$
$10x^2-160-24x-96=x+5$
$10x^2-25x-261=0$
$x=\frac{-(-25)\pm\sqrt{(-25)^2-4(10)(-261)}}{2(10)}$
$x=\frac{25\pm\sqrt{11065}}{20}$
$x\approx-4.01$ and $x\approx6.51$
The solutions are approximately –4.01 and 6.51.

69. $\frac{2}{(x+3)+(x-9)}=\frac{3}{2(x+4)+(x-17)}$
$\frac{2}{2x-6}=\frac{3}{2x+8+x-17}$
$\frac{2}{2(x-3)}=\frac{3}{3(x-3)}$
$\frac{1}{x-3}=\frac{1}{x-3}$
$1=1$
This is an identity.
The solutions are all real numbers except 3.

70. Let x = time to paper the room together.
$\frac{1}{10}$ is Dave's rate.
$\frac{1}{7}$ is Joan's rate.
$\frac{1}{x}$ is the combined rate.
$\frac{1}{10}+\frac{1}{7}=\frac{1}{x}$
$\frac{17}{70}=\frac{1}{x}$
$17x=70$
$x=\frac{70}{17}$
$x\approx4.1$
It will take them $\frac{70}{17}$ hours or approximately 4.1 hours.

71. Let x = additional pints of glue

$\frac{2+x}{7+x}=\frac{30}{100}$

$100(2+x)=30(7+x)$

$200+100x=210+30x$

$70x=10$

$x=\frac{10}{70}$

$x=\frac{1}{7}$

Another $\frac{1}{7}$ pint of glue is needed.

72. Let x = amount of tips.

$\frac{42}{9}=\frac{x}{15}$

$(42)(15)=9x$

$70=x$

She would earn $70.

73. His average rate of speed is $\frac{120}{t}$ mph.

The new distance is

$\left(\frac{2}{3}\right)\left(\frac{120}{t}\right)(t+1)=\frac{80(t+1)}{t}$ miles.

74. Let x = height of tree

$\frac{x}{4}=\frac{18}{3}$

$3x=72$

$x=24$

The tree is 24 feet high.

75. Let x = width

$5x$ = length

$v=lwh$

$10x^3+5x^2=(5x)(x)h$

$h=\frac{10x^3+5x^2}{5x^2}$

$=\frac{5x^2(2x+1)}{5x^2}$

$=2x+1$

The height is $(2x+1)$ inches.

76. Let x = time it takes the first line to fill an order.

$\frac{1}{x}$ is the first line's rate

$\frac{1}{x+1.5}$ is the second line's rate

$\frac{1}{3\frac{1}{3}}=\frac{1}{\frac{10}{3}}=\frac{3}{10}$ is the combined rate.

$\frac{1}{x}+\frac{1}{x+1.5}=\frac{3}{10}$

$10x(x+1.5)\left(\frac{1}{x}+\frac{1}{x+1.5}\right)$

$=10x(x+1.5)\left(\frac{3}{10}\right)$

$10(x+1.5)+10x=3x(x+1.5)$

$10x+15+10x=3x^2+4.5x$

$3x^2-15.5x-15=0$

$x=\frac{-(-15.5)\pm\sqrt{(-15.5)^2-4(3)(-15)}}{2(3)}$

$x=\frac{15.5\pm 20.5}{6}$

$x=6$

$x+1.5=7.5$

It would take the first line 6 hours and the second line 7.5 hours working alone.

77. $\frac{6}{x-3}+3+\frac{x-1}{x}$

$=\frac{6x+3x(x-3)+(x-1)(x-3)}{x(x-3)}$

$=\frac{6x+3x^2-9x+x^2-4x+3}{x(x-3)}$

$=\frac{4x^2-7x+3}{x(x-3)}$

$=\frac{(4x-3)(x-1)}{x(x-3)}$

The perimeter is $\frac{(4x-3)(x-1)}{x(x-3)}$ units.

78. Let x = additional amount

$\frac{70+x}{250+x}=\frac{1}{3}$

$3(70+x)=250+x$

$210+3x=250+x$

$2x=40$

$x=20$

She needs an additional \$20.

79. Let x = number of miles.

$\frac{65.50}{x}=\frac{0.24}{1}$

$0.24x=65.5$

$x\approx 272.9$

She drove 273 miles.

80. Let x = resistance of one resistor

$12+x$ = resistance of other resistor

$\frac{1}{R}=\frac{1}{r_1}+\frac{1}{r_2}$

$\frac{1}{25}=\frac{1}{x}+\frac{1}{12+x}$

$\frac{1}{25}=\frac{12+x+x}{x(12+x)}$

$x(12+x)=25(12+2x)$

$12x+x^2=300+50x$

$x^2-38x-300=0$

$x=\frac{-(-38)\pm\sqrt{(-38)^2-4(1)(-300)}}{2(1)}$

$x=\frac{38\pm\sqrt{2644}}{2}$

$x\approx 44.7$

$12+x\approx 56.7$

Each resistor is approximately 45 ohms and 57 ohms.

Chapter 12 Test

1. $y=\frac{x-7}{2x^2+6x}$

$2x^2+6x=0$

$2x(x+3)=0$

$x=0$ or $x=-3$

The restricted values are –3 and 0.

Domain: $x\neq -3,\ x\neq 0$;

$(-\infty,-3)\cup(-3,0)\cup(0,\infty)$

2. $g(x)=\frac{x^2-4x-21}{x^2-5x-36}$

$x^2-5x-36=0$

$(x-9)(x+4)=0$

$x=9$ or $x=-4$

The restricted values are –4 and 9.

Domain: $x\neq -4,\ x\neq 9$;

$(-\infty,-4)\cup(-4,9)\cup(9,\infty)$

3. $f(x)=\frac{x+4}{x-2}$

The restricted value is 2.

Construct a table of x-values less than and greater than 2. Connect with a smooth curve.

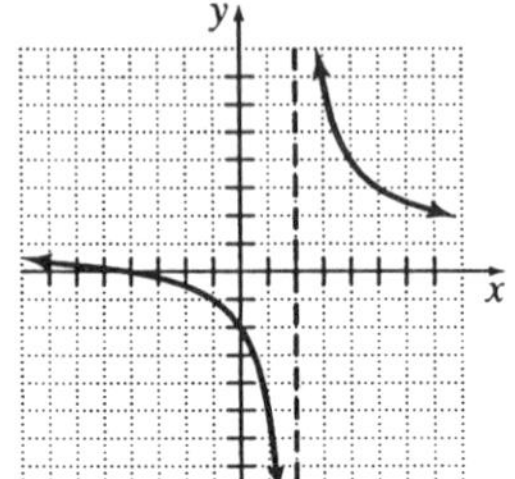

4. $\frac{24x+56}{9x^2-49}$

$=\frac{8(3x+7)}{(3x-7)(3x+7)}$

$=\frac{8}{3x-7}$

5. $\frac{2z^3-18z}{z^3+3z^2+9z+27}$

$=\frac{2z(z^2-9)}{z^2(z+3)+9(z+3)}$

$=\frac{2z(z+3)(z-3)}{(z^2+9)(z+3)}$

$=\frac{2z(z-3)}{z^2+9}$

6. $\dfrac{5}{x^2+4x}+\dfrac{2}{x^2+x-12}$
$=\dfrac{5}{x(x+4)}+\dfrac{2}{(x+4)(x-3)}$
$=\dfrac{5(x-3)+2(x)}{x(x+4)(x-3)}$
$=\dfrac{5x-15+2x}{x(x+4)(x-3)}$
$=\dfrac{7x-15}{x(x+4)(x-3)}$

7. $\dfrac{5x^2-40x}{20x^2+30x}\div\dfrac{x^2-64}{2x^2+19x+24}$
$=\dfrac{5x^2-40x}{20x^2+30x}\cdot\dfrac{2x^2+19x+24}{x^2-64}$
$=\dfrac{(5x)(x-8)(2x+3)(x+8)}{(10x)(2x+3)(x-8)(x+8)}$
$=\dfrac{1}{2}$

8. $\dfrac{2x+1}{x^2+5x}-\dfrac{x-4}{x^2+10x+25}$
$=\dfrac{2x+1}{x(x+5)}-\dfrac{x-4}{(x+5)(x+5)}$
$=\dfrac{(2x+1)(x+5)-x(x-4)}{x(x+5)^2}$
$=\dfrac{2x^2+11x+5-x^2+4x}{x(x+5)^2}$
$=\dfrac{x^2+15x+5}{x(x+5)^2}$

9. $\dfrac{24x^2y}{x^2-xy-2y^2}\cdot\dfrac{3x^2-12xy+12y^2}{16x^3y}$
$=\dfrac{24x^2y(3)(x-2y)(x-2y)}{(x+y)(x-2y)16x^3y}$
$=\dfrac{9(x-2y)}{2x(x+y)}$

10. $x-5=\dfrac{5-x}{x}$

a. Numerically:
The restricted value is 0. Set up a table of values on the calculator letting
$Y1 = x-5$ and $Y2=\dfrac{5-x}{x}$.
The solutions are -1 and 5.

b. Graphically:

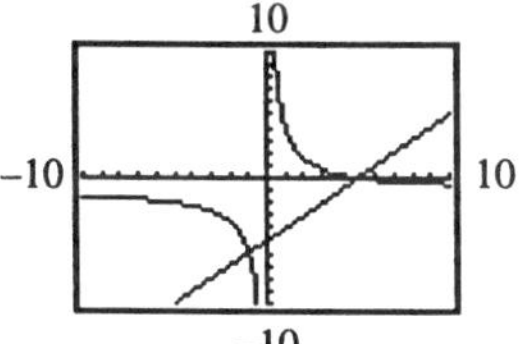

The solutions are the x-coordinates of the intersections, -1 and 5.

c. Algebraically:
$x-5=\dfrac{5-x}{x}$
$x(x-5)=5-x$
$x^2-5x=5-x$
$x^2-4x-5=0$
$(x+1)(x-5)=0$
$x=-1$ and $x=5$
The solutions are -1 and 5.

11. $\dfrac{16}{x-3}=\dfrac{x-3}{x}$
$16x=(x-3)^2$
$16x=x^2-6x+9$
$x^2-22x+9=0$
$x=\dfrac{-(-22)\pm\sqrt{(-22)^2-4(1)(9)}}{2(1)}$
$x=\dfrac{22\pm\sqrt{448}}{2}$
$x=\dfrac{22\pm8\sqrt{7}}{2}$
$x=11\pm4\sqrt{7}$
The solutions are $11\pm4\sqrt{7}$.

12. $\frac{b}{5}+\frac{1}{2}=\frac{1}{5}-\frac{11}{10b}$

$10b\left(\frac{b}{5}+\frac{1}{2}\right)=10b\left(\frac{1}{5}-\frac{11}{10b}\right)$

$2b^2+5b=2b-11$

$2b^2+3b+11=0$

$b=\frac{-3\pm\sqrt{3^2-4(2)(11)}}{2(2)}$

$b=\frac{-3\pm\sqrt{-79}}{4}$

There are no real-number solutions.

13. $\frac{3}{m}-\frac{m}{75}=0$

$\frac{3}{m}=\frac{m}{75}$

$3(75)=m^2$

$\pm\sqrt{225}=m$

$\pm 15=m$

The solutions are ±15.

14. $6+14x^{-1}+4x^{-2}=0$

$6+\frac{14}{x}+\frac{4}{x^2}=0$

$x^2\left(6+\frac{14}{x}+\frac{4}{x^2}\right)=x^2(0)$

$6x^2+14x+4=0$

$2(3x^2+7x+2)=0$

$2(3x+1)(x+2)=0$

$x=-\frac{1}{3}$ or $x=-2$

The solutions are –2 and $-\frac{1}{3}$.

15. $\frac{x^2+6x+9}{4x+20}=\frac{x+6}{x+5}$

$\frac{(x+3)^2}{4(x+5)}=\frac{x+6}{x+5}$

$4(x+5)\left[\frac{(x+3)^2}{4(x+5)}\right]=4(x+5)\left(\frac{x+6}{x+5}\right)$

$(x+3)^2=4(x+6)$

$x^2+6x+9=4x+24$

$x^2+2x-15=0$

$(x+5)(x-3)=0$

$x=-5$ (extraneous) or $x=3$

The solution is 3.

16. $\frac{2}{x+2}=\frac{3}{x+3}$

$2(x+3)=3(x+2)$

$2x+6=3x+6$

$0=x$

The solution is 0.

17. $2+\frac{10.68}{x}+\frac{10.72}{x^2}=\frac{1}{x}+\frac{5.34}{x^2}+\frac{5.36}{x^3}$

$Y1=2+\frac{10.68}{x}+\frac{10.72}{x^2}$

$Y2=\frac{1}{x}+\frac{5.34}{x^2}+\frac{5.36}{x^3}$

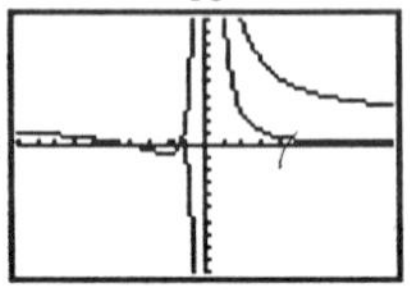

The solutions are the x-coordinates of the intersections, –4, –1.34, and 0.5.

18. Let x = height of tree

$\frac{x}{7}=\frac{32.5}{10.5}$

$10.5x=7(32.5)$

$x=21\frac{2}{3}$

The height is $21\frac{2}{3}$ feet.

19. Let x = time it takes Jill to rake the leaves.

$\frac{1}{6}$ is Jack's rate.

$\frac{1}{x}$ is Jill's rate.

$\frac{1}{2\frac{2}{5}} = \frac{1}{\frac{12}{5}} = \frac{5}{12}$ is the combined rate.

$$\frac{1}{6} + \frac{1}{x} = \frac{5}{12}$$

$$12x\left(\frac{1}{6} + \frac{1}{x}\right) = 12x\left(\frac{5}{12}\right)$$

$$2x + 12 = 5x$$

$$12 = 3x$$

$$4 = x$$

Jill can rake the leaves in 4 hours.

20. Answers will vary.

Chapter 13

13.1 Experiencing Algebra the Exercise Way

1. $\sqrt{81} = 9$ because $9^2 = 81$.

3. $\sqrt{1.44} = 1.2$ because $(1.2)^2 = 1.44$.

5. $\sqrt{\frac{4}{9}} = \frac{2}{3}$ because $\left(\frac{2}{3}\right)^2 = \frac{4}{9}$.

7. $\sqrt[3]{\frac{-1}{27}} = \frac{-1}{3}$ because $\left(\frac{-1}{3}\right)^3 = \frac{-1}{27}$.

9. $-\sqrt[4]{625} = -(5) = -5$

11. $\sqrt[4]{-625}$ is not a real number because the radicand is negative and the root is even.

13. $\sqrt[3]{216} = 6$ because $6^3 = 216$.

15. $\sqrt[3]{-216} = -6$ because $(-6)^3 = -216$.

17. $-\sqrt[3]{216} = -(6) = -6$.

19. $\sqrt{110} \approx 10.488$

21. $-\sqrt{110} \approx -10.488$

23. $\sqrt{-110}$ is not a real number because the radicand is negative and the root is even.

25. $\sqrt[3]{110} \approx 4.791$

27. $-\sqrt[3]{110} \approx -4.791$

29. $\sqrt[5]{20} \approx 1.821$

31. $\sqrt[3]{12.5} \approx 2.321$

33. $36^{1/2} = \sqrt{36} = 6$

35. $216^{1/3} = \sqrt[3]{216} = 6$

37. $-81^{1/4} = -\sqrt[4]{81} = -3$

39. $(-81)^{1/4} = \sqrt[4]{-81}$ is not a real number.

41. $-8^{1/3} = -\sqrt[3]{8} = -2$

43. $(-8)^{1/3} = \sqrt[3]{-8} = -2$

45. $36^{-1/2} = \frac{1}{36^{1/2}} = \frac{1}{\sqrt{36}} = \frac{1}{6}$

47. $216^{-1/3} = \frac{1}{216^{1/3}} = \frac{1}{\sqrt[3]{216}} = \frac{1}{6}$

49. $81^{-1/4} = \frac{1}{81^{1/4}} = \frac{1}{\sqrt[4]{81}} = \frac{1}{3}$

51. $-81^{-1/4} = -\frac{1}{81^{1/4}} = -\frac{1}{\sqrt[4]{81}} = -\frac{1}{3}$

53. $8^{-1/3} = \frac{1}{8^{1/3}} = \frac{1}{\sqrt[3]{8}} = \frac{1}{2}$

55. $(-8)^{-1/3} = \frac{1}{(-8)^{1/3}} = \frac{1}{\sqrt[3]{-8}} = \frac{1}{-2} = -\frac{1}{2}$

57. $522^{1/2} = \sqrt{522} \approx 22.847$

59. $522^{1/4} = \sqrt[4]{522} \approx 4.780$

61. $522^{-1/3} = \frac{1}{522^{1/3}} = \frac{1}{\sqrt[3]{522}} \approx 0.124$

63. $522^{-1/5} = \frac{1}{522^{1/5}} = \frac{1}{\sqrt[5]{522}} \approx 0.286$

65. $27^{4/3} = \left(\sqrt[3]{27}\right)^4 = 3^4 = 81$

67. $(-27)^{4/3} = \left(\sqrt[3]{-27}\right)^4 = (-3)^4 = 81$

69. $27^{-4/3} = \frac{1}{27^{4/3}} = \frac{1}{(\sqrt[3]{27})^4} = \frac{1}{3^4} = \frac{1}{81}$

71. $-27^{-4/3} = -\frac{1}{27^{4/3}} = -\frac{1}{(\sqrt[3]{27})^4} = -\frac{1}{3^4}$

$= -\frac{1}{81}$

73. $4^{5/2} = (\sqrt{4})^5 = 2^5 = 32$

75. $8^{7/3} = (\sqrt[3]{8})^7 = 2^7 = 128$

77. $4^{-7/2} = \frac{1}{4^{7/2}} = \frac{1}{(\sqrt{4})^7} = \frac{1}{2^7} = \frac{1}{128}$

79. $-16^{3/2} = -(\sqrt{16})^3 = -4^3 = -64$

81. $(-9)^{3/2} = (\sqrt{-9})^3$ is not a real number.

83. $-81^{-3/2} = -\frac{1}{81^{3/2}} = -\frac{1}{(\sqrt{81})^3} = -\frac{1}{9^3}$

$= -\frac{1}{729}$

85. $28^{5/4} = (\sqrt[4]{28})^5 \approx 64.409$

87. $(-21)^{4/3} = (\sqrt[3]{-21})^4 \approx 57.937$

89. $5^{-2/3} = \frac{1}{5^{2/3}} = \frac{1}{(\sqrt[3]{5})^2} \approx 0.342$

91. $-42^{-2/3} = -\frac{1}{42^{2/3}} = -\frac{1}{(\sqrt[3]{42})^2} \approx -0.083$

93. $(-88)^{3/8} = (\sqrt[8]{-88})^3$ is not a real number.

95. $s = \sqrt{A}$

a. $s = \sqrt{289}$
$s = 17$
The length of the side is 17 feet.

b. $s = \sqrt{90.25}$
$s = 9.5$
The length of the side is 9.5 inches.

c. $s = \sqrt{115}$
$s \approx 10.724$
The length of the side is approximately 10.724 mm.

97. $r = \sqrt{\frac{A}{\pi}}$

a. $r = \sqrt{\frac{196\pi}{\pi}}$
$r = 14$
The length of the radius is 14 inches.

b. $r = \sqrt{\frac{49}{\pi}}$
$r = \frac{7}{\sqrt{\pi}}$

The length of the radius is $\frac{7}{\sqrt{\pi}} \approx 3.949$ yards.

c. $r = \sqrt{\frac{108}{\pi}}$
$r \approx 5.863$
The length of the radius is
$\sqrt{\frac{108}{\pi}} \approx 5.863$ meters

99. $V = s^3$
$\sqrt[3]{V} = s$
$s = \sqrt[3]{V}$

a. $s = \sqrt[3]{1728}$
$s = 12$
The length of the side is 12 inches.

b. $s = \sqrt[3]{120}$
$s \approx 4.93$
The length of the side is 4.93 meters.

101. $V = \frac{4}{3}\pi r^3$
$r = \sqrt[3]{\frac{3V}{4\pi}}$

a. $r = \sqrt[3]{\frac{3(288\pi)}{4\pi}}$
$r = 6$
$d = 2r$
$d = 2(6)$
$d = 12$
The radius is 6 dm and the diameter is 12 dm.

b. $r = \sqrt[3]{\frac{3(250)}{4\pi}}$
$r \approx 3.908$
$d = 2r$
$d \approx 7.816$
The radius is approximately 3.908 yards and the diameter is approximately 7.816 yards.

103. Area of sphere $= 4\pi r^2$
Area of $\frac{1}{2}$ of a sphere $= 2\pi r^2$
Solve the equation for r:
$A = 2\pi r^2$
$r = \sqrt{\frac{A}{2\pi}}$
$r = \sqrt{\frac{8200}{2\pi}}$
$r \approx 36.126$
$d \approx 72.252$
The radius and diameter are approximately 36.126 ft and 72.252 ft, respectively.

13.1 Experiencing Algebra the Calculator Way

x	Year	Number of Employees
1	1989	6257
2	1990	7883
3	1991	9024
4	1992	9932
5	1993	10,699
6	1994	11,370
7	1995	11,969
8	1996	12,514
9	1997	13,015
10	1998	13,480
11	1999	13,915
12	2000	14,325

Answers will vary.

13.2 Experiencing Algebra the Exercise Way

1. $y = \sqrt{5}x + 7$ is a nonradical function because the radicand does not contain x.

3. $f(x) = 2\sqrt{3x+1} - 5$ is a radical function because the radicand contains x.

5. $y = 5\sqrt[3]{x} - 7$ is a radical function because the radicand contains x.

7. $y = (3x-5)^{4/5}$ is a radical function because when it is written in the form $y = \left(\sqrt[5]{3x-5}\right)^4$ it can be seen that the radicand contains x.

9. $g(x) = 7^{1/2}x + 5$ is a nonradical function because when it is written in the form $g(x) = \sqrt{7}x + 5$ it can be seen that the radicand does not contain x.

11. $y = \sqrt{x-3}$
$x - 3 < 0$
$x < 3$
The restricted values are all real numbers less than 3.
domain: $[3, \infty)$

13. $y = \sqrt[3]{x-3}$
There are no restricted values because the index of the radical is odd.
domain: $(-\infty, \infty)$

15. $f(x) = (7-2x)^{2/3}$
There are no restricted values because the denominator of the rational exponent is odd.
domain: $(-\infty, \infty)$

17. $h(x) = (7-2x)^{3/4}$
$7 - 2x < 0$
$-2x < -7$
$x > \frac{7}{2}$
The restricted values are all real numbers greater than $\frac{7}{2}$.
domain: $\left(-\infty, \frac{7}{2}\right]$

19. $y = 2\sqrt{x} + 3$
$x < 0$
The restricted values are all real numbers less than 0.
domain: $[0, \infty)$

21. $y = \sqrt{2x+1} - \sqrt{x-3}$
$2x + 1 < 0$ or $x - 3 < 0$
$2x < -1$ $\quad x < 3$
$x < -\frac{1}{2}$
The restricted values are all real numbers less than 3.
domain: $[3, \infty)$

23. $y = \left(\sqrt[3]{5x+9}\right)^4$
There are no restricted values because the index of the radical is odd.
domain: $(-\infty, \infty)$

25. $y = \left(\sqrt[4]{3x+2}\right)^5$
$3x + 2 < 0$
$3x < -2$
$x < -\frac{2}{3}$
The restricted values are all real numbers less than $-\frac{2}{3}$.
domain: $\left[-\frac{2}{3}, \infty\right)$

27.–39. Answers will vary. Possible answers follow.

27.

x	$y = \sqrt{x+7}$	y
–3	$y = \sqrt{-3+7} = 2$	2
2	$y = \sqrt{2+7} = 3$	3
9	$y = \sqrt{9+7} = 4$	4
10	$y = \sqrt{10+7} \approx 4.123$	4.123

Therefore, (–3, 2), (2, 3), (9, 4), and (10, 4.123) are four solutions.

29.

x	$f(x) = \sqrt[3]{5x+21}$	$f(x)$
–2	$f(-2) = \sqrt[3]{5(-2)+21} \approx 2.224$	2.224
–1	$f(-1) = \sqrt[3]{5(-1)+21} \approx 2.520$	2.5198
0	$f(0) = \sqrt[3]{5(0)+21} \approx 2.759$	2.7589
1	$f(1) = \sqrt[3]{5(1)+21} \approx 2.962$	2.9625

Therefore, (–2, 2.224), (–1, 2.5198), (0, 2.7589), and (1, 2.9625) are four solutions.

31.

x	$y=1+6\sqrt{x-9}$	y
9	$y=1+6\sqrt{9-9}=1$	1
10	$y=1+6\sqrt{10-9}=7$	7
13	$y=1+6\sqrt{13-9}=13$	13
18	$y=1+6\sqrt{18-9}=19$	19

Therefore, (9, 1), (10, 7), (13, 13), and (18, 19) are four solutions.

33.

x	$h(x)=2\sqrt{2-3x}$	$h(x)$
–4	$h(-4)=2\sqrt{2-3(-4)}\approx 7.483$	7.4833
–3	$h(-3)=2\sqrt{2-3(-3)}\approx 6.633$	6.6332
–2	$h(-2)=2\sqrt{2-3(-2)}\approx 5.657$	5.6569
0	$h(0)=2\sqrt{2-3(0)}\approx 2.828$	2.8284

Therefore, (–4, 7.4833), (–3, 6.6332), (–2, 5.6569), and (0, 2.8284) are four solutions.

35.

x	$g(x)=\sqrt{2x-1}+2$	$g(x)$
$\frac{1}{2}$	$g\left(\frac{1}{2}\right)=\sqrt{2\left(\frac{1}{2}\right)-1}+2=2$	2
1	$g(1)=\sqrt{2(1)-1}+2=3$	3
3	$g(3)=\sqrt{2(3)-1}+2\approx 4.236$	4.2361
5	$g(5)=\sqrt{2(5)-1}+2=5$	5

Therefore $\left(\frac{1}{2}, 2\right)$, (1, 3), (3, 4.2361), and (5, 5) are four solutions.

37.

x	$y=5(x+2)^{2/3}$	y
–3	$y=5(-3+2)^{2/3}=5$	5
–2	$y=5(-2+2)^{2/3}=0$	0
–1	$y=5(-1+2)^{2/3}=5$	5
6	$y=5(6+2)^{2/3}=20$	20

Therefore, (–3, 5), (–2, 0), (–1, 5) and (6, 20) are four solutions.

39.

x	$y=-2(x-1)^{3/2}$	y
1	$y=-2(1-1)^{3/2}=0$	0
2	$y=-2(2-1)^{3/2}=-2$	–2
5	$y=-2(5-1)^{3/2}=-16$	–16
10	$y=-2(10-1)^{3/2}=-54$	–54

Therefore, (1, 0), (2, –2), (5, –16), and (10, –54) are four solutions.

41. $y=\sqrt{x}$

domain: $[0, \infty)$

Plot and connect ordered pairs that satisfy the equation.

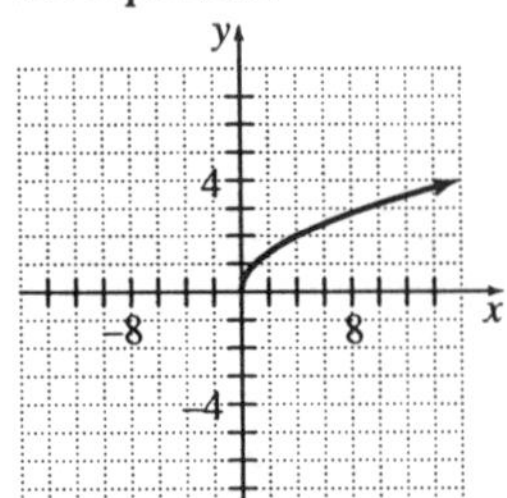

43. $y = \sqrt[3]{x} - 10$
domain: $(-\infty, \infty)$
Plot and connect ordered pairs that satisfy the equation.

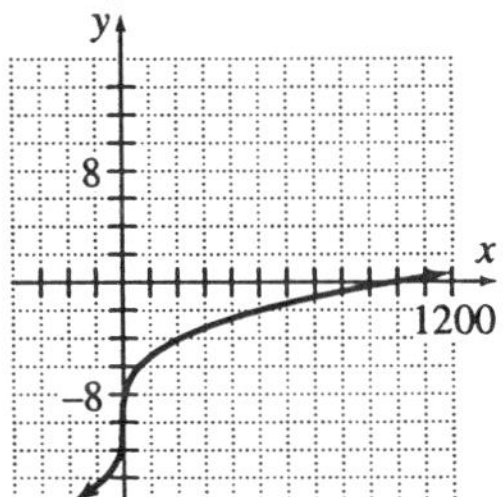

45. $y = 5\sqrt{x} + 10$
domain: $[0, \infty)$
Plot and connect ordered pairs that satisfy the equation.

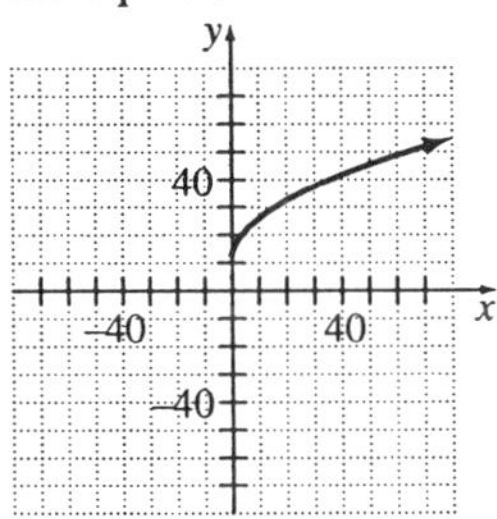

47. $f(x) = \sqrt{x - 5}$
domain: $[5, \infty)$
Plot and connect ordered pairs that satisfy the equation.

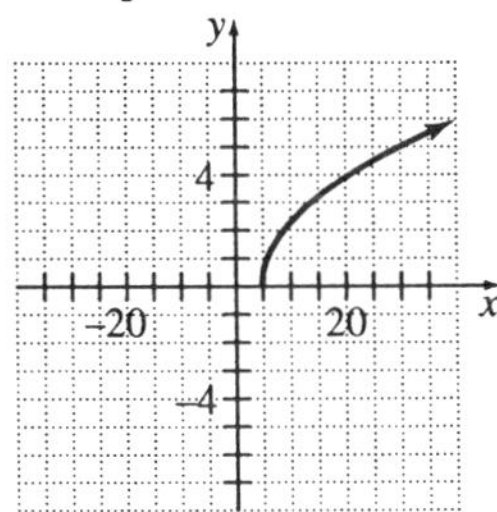

49. $g(x) = \sqrt[3]{8x}$
domain: $(-\infty, \infty)$
Plot and connect ordered pairs that satisfy the equation.

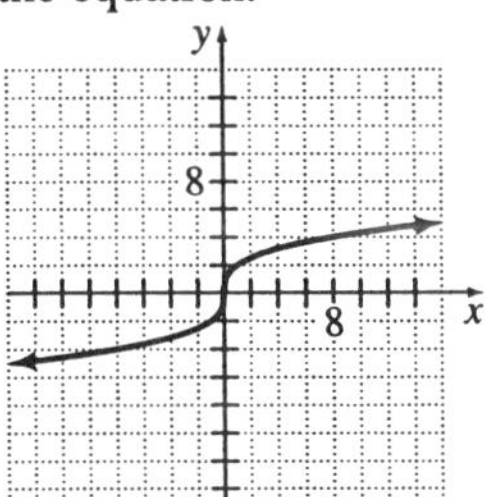

51. $y = 5 - x^{3/2}$
domain: $[0, \infty)$
Plot and connect ordered pairs that satisfy the equations.

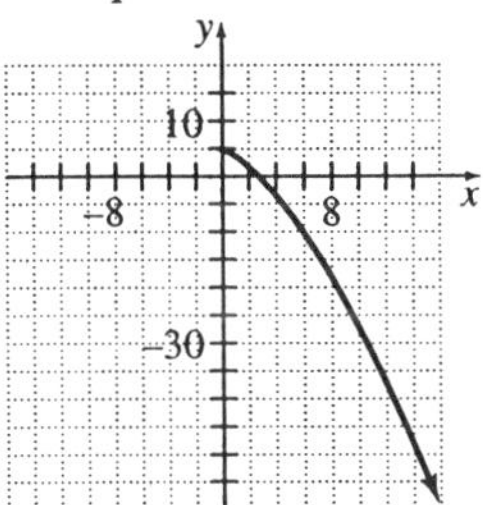

53. $y = x^{3/4} + 2$
domain: $[0, \infty)$
Plot and connect ordered pairs that satisfy the equation.

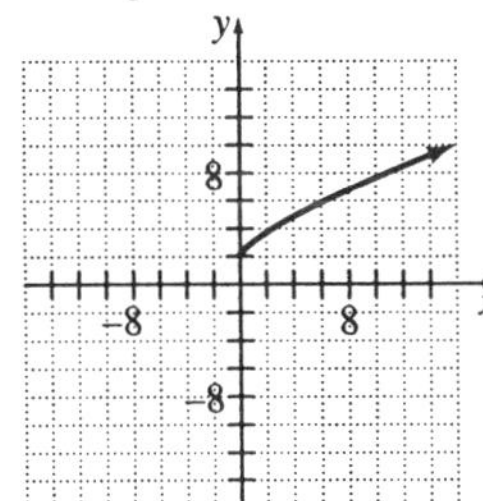

55. $y = 3 - x^{2/3}$
domain: $(-\infty, \infty)$
Plot and connect ordered pairs that satisfy the equation.

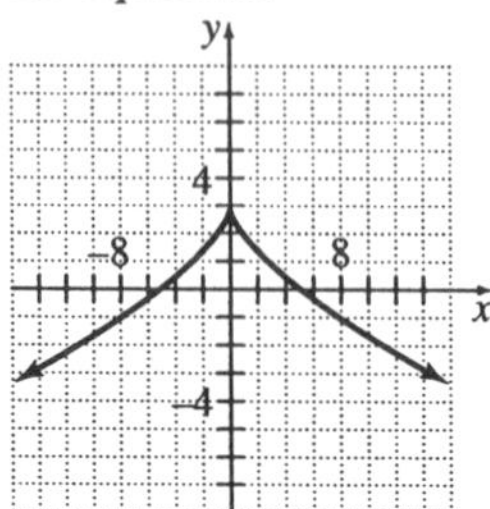

57. $y = x^{4/5} + 1$
domain: $(-\infty, \infty)$
Plot and connect ordered pairs that satisfy the equation.

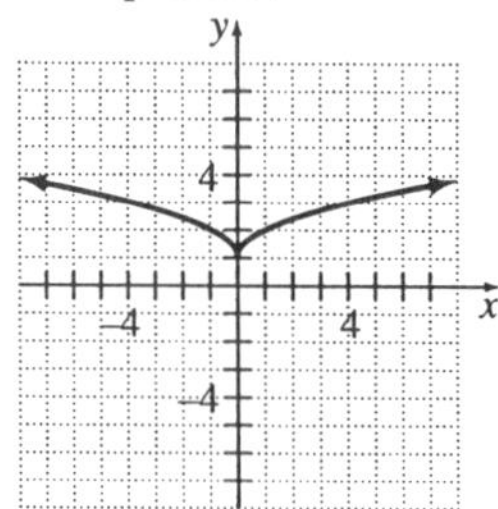

59. $s = A^{1/2}$

A (ft^2)	s (ft)
10	3.1623
20	4.4721
30	5.4772
40	6.3246
50	7.0711
60	7.746
70	8.3666
80	8.9443
90	9.4868
100	10

The lengths of the sides (in inches) for the given areas are 3.1623, 4.4721, 5.4772, 6.3246, 7.0711, 7.746, 8.3666, 8.9443, 9.4868, and 10, respectively.

61. $r = \sqrt{n}$

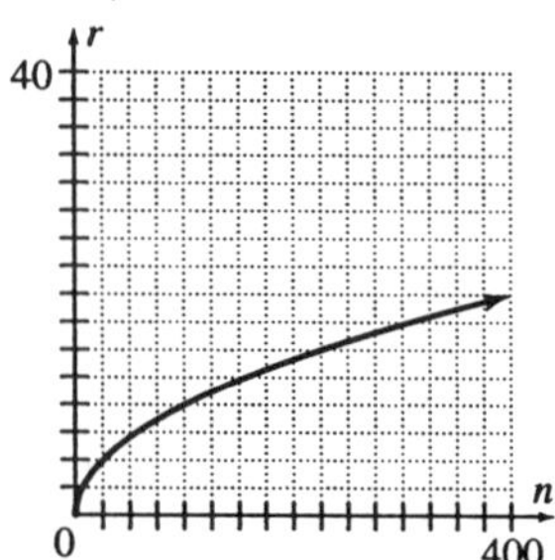

At $n = 250$, $r \approx 15.8$.
There should be 16 rows in the table.

63. $V = \sqrt{12S}$

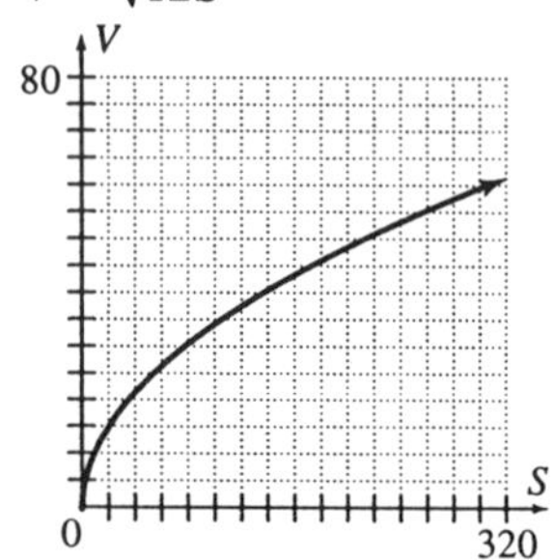

At $S = 250$, $V \approx 54.8$
The velocity would be about 55 mph.

13.2 Experiencing Algebra the Calculator Way

Students should follow the instructions and determine how their calculator graphs the functions. Answers will vary.

13.3 Experiencing Algebra the Exercise Way

1. $9^{-1/2} = \dfrac{1}{9^{1/2}} = \dfrac{1}{3}$

3. $\dfrac{1}{8^{-1/3}} = 8^{1/3} = 2$

5. $27^{1/3} \cdot 27^{4/3}$
$= 27^{\frac{1}{3}+\frac{4}{3}}$
$= 27^{5/3}$
$= 243$

7. $\dfrac{32^{4/5}}{32^{1/5}}$
$= 32^{\frac{4}{5}-\frac{1}{5}}$
$= 32^{3/5}$
$= 8$

9. $(125^{1/2})^{2/3}$
$= 125^{\frac{1}{2}\cdot\frac{2}{3}}$
$= 125^{1/3}$
$= 5$

11. $3^{1/2} \cdot 12^{1/2}$
$= (3 \cdot 12)^{1/2}$
$= 36^{1/2}$
$= 6$

13. $\left(\dfrac{8}{27}\right)^{1/3}$
$= \dfrac{8^{1/3}}{27^{1/3}}$
$= \dfrac{2}{3}$

15. $\left(\dfrac{9}{16}\right)^{-1/2}$
$= \left(\dfrac{16}{9}\right)^{1/2}$
$= \dfrac{16^{1/2}}{9^{1/2}}$
$= \dfrac{4}{3}$

17. $[(2^{1/2})(8^{1/2})]^{-3/2}$
$= [(2 \cdot 8)^{1/2}]^{-3/2}$
$= (16^{1/2})^{-3/2}$
$= 4^{-3/2}$
$= \dfrac{1}{4^{3/2}}$
$= \dfrac{1}{8}$

19. $x^{-2/3} = \dfrac{1}{x^{2/3}}$

21. $\dfrac{12}{y^{-3/4}} = 12y^{3/4}$

23. $z^{2/3} \cdot z^{3/4}$
$= z^{\frac{2}{3}+\frac{3}{4}}$
$= z^{\frac{8}{12}+\frac{9}{12}}$
$= z^{17/12}$

25. $\dfrac{p^{5/6}}{p^{2/3}}$
$= p^{\frac{5}{6}-\frac{2}{3}}$
$= p^{\frac{5}{6}-\frac{4}{6}}$
$= p^{1/6}$

27. $(b^{3/4})^{8/9}$
$= b^{\frac{3}{4}\cdot\frac{8}{9}}$
$= b^{2/3}$

29. $(x^2y)^{3/4}$
$= x^{2\cdot 3/4}y^{3/4}$
$= x^{3/2}y^{3/4}$

31. $\left(\dfrac{x^2}{y^3}\right)^{5/6}$
$= \dfrac{x^{2\cdot 5/6}}{y^{3\cdot 5/6}}$
$= \dfrac{x^{5/3}}{y^{5/2}}$

33. $\left(\dfrac{a}{b}\right)^{-1/3}$
$= \left(\dfrac{b}{a}\right)^{1/3}$
$= \dfrac{b^{1/3}}{a^{1/3}}$

35. $c^{3/5} \cdot c^{-4/5}$
$= c^{\frac{3}{5}-\frac{4}{5}}$
$= c^{-1/5}$
$= \dfrac{1}{c^{1/5}}$

37. $(8a^5b^6)^{2/3}$
$= 8^{2/3}a^{5\cdot 2/3}b^{6\cdot 2/3}$
$= 4a^{10/3}b^4$

39. $(5a^{3/4}b^{2/5})(2a^{1/3}b^{2/5})$
$= 5\cdot 2a^{\frac{3}{4}+\frac{1}{3}}b^{\frac{2}{5}+\frac{2}{5}}$
$= 10a^{13/12}b^{4/5}$

41. $\left(\dfrac{m^{3/7}}{2m^{-2/7}}\right)^2$
$= \left(\dfrac{m^{\frac{3}{7}+\frac{2}{7}}}{2}\right)^2$
$= \left(\dfrac{m^{5/7}}{2}\right)^2$
$= \dfrac{m^{\frac{5}{7}\cdot 2}}{2^2}$
$= \dfrac{m^{10/7}}{4}$

43. $(3a^{1/3}b^2c^{1/6})^{-2}(2a^{4/3}b^{1/4}c^{5/6})^3$
$= \dfrac{(2a^{4/3}b^{1/4}c^{5/6})^3}{(3a^{1/3}b^2c^{1/6})^2}$
$= \dfrac{2^3a^{\frac{4}{3}\cdot 3}b^{\frac{1}{4}\cdot 3}c^{\frac{5}{6}\cdot 3}}{3^2a^{\frac{1}{3}\cdot 2}b^{2\cdot 2}c^{\frac{1}{6}\cdot 2}}$
$= \dfrac{8a^4b^{3/4}c^{5/2}}{9a^{2/3}b^4c^{1/3}}$
$= \dfrac{8}{9}a^{4-\frac{2}{3}}b^{\frac{3}{4}-4}c^{\frac{5}{2}-\frac{1}{3}}$
$= \dfrac{8}{9}a^{10/3}b^{-13/4}c^{13/6}$
$= \dfrac{8a^{10/3}c^{13/6}}{9b^{13/4}}$

45. $x^{1/4}(x^{2/3} - x^{4/5})$
$= x^{\frac{1}{4}+\frac{2}{3}} - x^{\frac{1}{4}+\frac{4}{5}}$
$= x^{11/12} - x^{21/20}$

47. $2x^{2/5}(x^{1/5} + 3y^{2/5})$
$= 2x^{\frac{2}{5}+\frac{1}{5}} + 2\cdot 3x^{2/5}y^{2/5}$
$= 2x^{3/5} + 6x^{2/5}y^{2/5}$

49. $a^{1/4}b^{1/3}(3 - a^{1/2}b^{5/6})$
$= 3a^{1/4}b^{1/3} - a^{\frac{1}{4}+\frac{1}{2}}b^{\frac{1}{3}+\frac{5}{6}}$
$= 3a^{1/4}b^{1/3} - a^{3/4}b^{7/6}$

51. $(x^{1/2} - y^{1/2})(x^{1/3} + y^{1/3})$
$= x^{\frac{1}{2}+\frac{1}{3}} + x^{1/2}y^{1/3} - y^{1/2}x^{1/3} - y^{\frac{1}{2}+\frac{1}{3}}$
$= x^{5/6} + x^{1/2}y^{1/3} - x^{1/3}y^{1/2} - y^{5/6}$

53. $(x^{1/4} + y^{1/4})(x^{1/4} - y^{1/4})$
$= x^{\frac{1}{4}+\frac{1}{4}} - x^{1/4}y^{1/4} + y^{1/4}x^{1/4} - y^{\frac{1}{4}+\frac{1}{4}}$
$= x^{1/2} - y^{1/2}$

55. $(x^2 + y^2)(x^{1/2} - y^{1/2})$
$= x^{2+\frac{1}{2}} - x^2y^{1/2} + x^{1/2}y^2 - y^{2+\frac{1}{2}}$
$= x^{5/2} - x^2y^{1/2} + x^{1/2}y^2 - y^{5/2}$

57. $(x^{1/3} + 2)(x^{1/3} - 2)$
$= x^{\frac{1}{3}+\frac{1}{3}} - 2x^{1/3} + 2x^{1/3} - 2\cdot 2$
$= x^{2/3} - 4$

59. $(x^{1/4} + 2)^2$
$= x^{\frac{1}{4}\cdot 2} + 2\cdot 2x^{1/4} + 2^2$
$= x^{1/2} + 4x^{1/4} + 4$

61. $(x - y^{1/2})^2$
$= x^2 - 2\cdot xy^{1/2} - y^{\frac{1}{2}\cdot 2}$
$= x^2 - 2xy^{1/2} - y$

63. $(a^{1/2}+b^{1/2})^2$
$=a^{\frac{1}{2}\cdot 2}+2a^{1/2}b^{1/2}+b^{\frac{1}{2}\cdot 2}$
$=a+2a^{1/2}b^{1/2}+b$

65. Let s_1 = length of the original side.
Let s_2 = length of the larger side.
$s_1=V^{1/3}$
$s_2=(8V)^{1/3}$
$\frac{s_2}{s_1}=\frac{(8V)^{1/3}}{V^{1/3}}=2$
The side length should be doubled.

67. Let t_1 = time to fall from the original height.
Let t_2 = time to fall from three times the original height.
$t_1=\left(\frac{d}{16.1}\right)^{1/2}$
$t_2=\left(\frac{3d}{16.1}\right)^{1/2}$
$t_2-t_1=\left(\frac{3d}{16.1}\right)^{1/2}-\left(\frac{d}{16.1}\right)^{1/2}$
$=\frac{3^{1/2}d^{1/2}}{(16.1)^{1/2}}-\frac{d^{1/2}}{(16.1)^{1/2}}$
$=d^{1/2}\left(\frac{3^{1/2}-1}{(16.1)^{1/2}}\right)$
$\approx 0.1824d^{1/2}$
The time is increased by about $0.1824d^{1/2}$ seconds.

13.3 Experiencing Algebra the Calculator Way

1. $(x^3)^{2/3}$
$=x^{3\cdot 2/3}$
$=x^2$
$Y1=(x^3)^{2/3}$
$Y2=x^2$

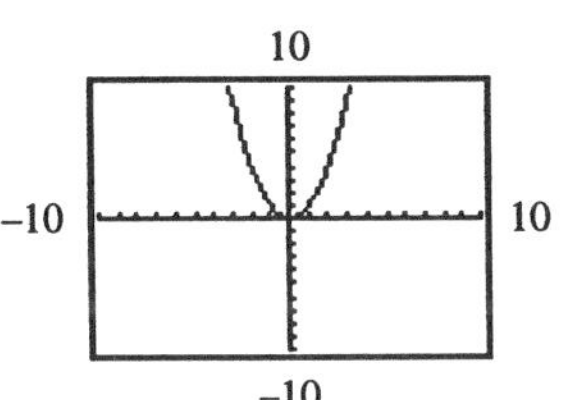

2. $(x^3)^{-2/3}$
$=\left(\frac{1}{x^3}\right)^{2/3}$
$=\frac{1^{2/3}}{x^{3\cdot 2/3}}$
$=\frac{1}{x^2}$
$Y1=(x^3)^{-2/3}$
$Y2=\frac{1}{x^2}$

10
-5
5
-10

3. $\frac{x^4}{x^{9/4}}$
$=x^{4-\frac{9}{4}}$
$=x^{7/4}$
$Y1=\frac{x^4}{x^{9/4}}$
$Y2=x^{7/4}$

10
-10
10
-10

4. $\frac{x^2}{x^{7/2}}$

$= x^{2-\frac{7}{2}}$

$= x^{-3/2}$

$= \frac{1}{x^{3/2}}$

$\text{Y1} = \frac{x^2}{x^{7/2}}$

$\text{Y2} = \frac{1}{x^{3/2}}$

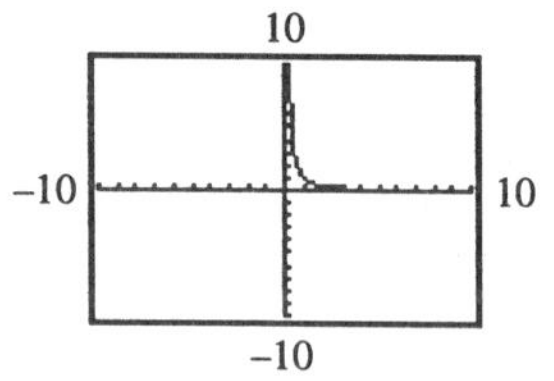

10
−10 10
−10

5. $x^{3/4}(x^{3/2} - x^{2/3})$

$= x^{\frac{3}{4}+\frac{3}{2}} - x^{\frac{3}{4}+\frac{2}{3}}$

$= x^{9/4} - x^{17/12}$

$\text{Y1} = x^{3/4}(x^{3/2} - x^{2/3})$

$\text{Y2} = x^{9/4} - x^{17/12}$

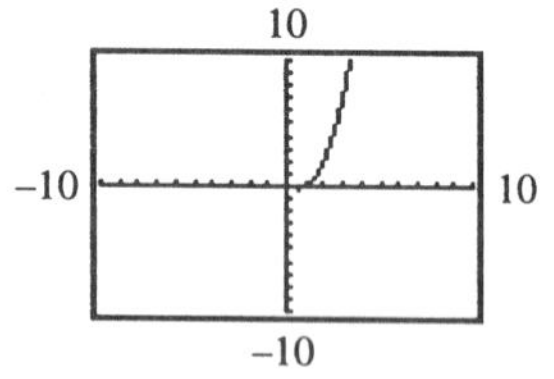

10
−10 10
−10

6. $(x^{1/3}+1)(x^{1/3}-1)$

$= x^{\frac{1}{3}+\frac{1}{3}} - x^{1/3} + x^{1/3} - 1$

$= x^{2/3} - 1$

$\text{Y1} = (x^{1/3}+1)(x^{1/3}-1)$

$\text{Y2} = x^{2/3} - 1$

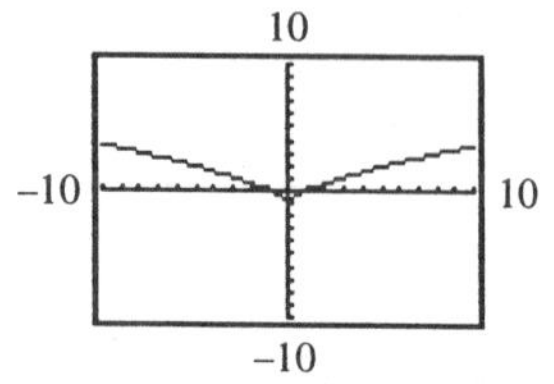

10
−10 10
−10

13.4 Experiencing Algebra the Exercise Way

1. $\sqrt{28}$

$= \sqrt{4 \cdot 7}$

$= \sqrt{2^2 \cdot 7}$

$= 2\sqrt{7}$

3. $\sqrt[3]{-686}$

$= \sqrt[3]{-343 \cdot 2}$

$= \sqrt[3]{(-7)^3 \cdot 2}$

$= -7\sqrt[3]{2}$

5. $\sqrt[4]{112}$

$= \sqrt[4]{16 \cdot 7}$

$= \sqrt[4]{2^4 \cdot 7}$

$= 2\sqrt[4]{7}$

7. $\sqrt[5]{-6250}$

$= \sqrt[5]{(-3125) \cdot 2}$

$= \sqrt[5]{(-5)^5 \cdot 2}$

$= -5\sqrt[5]{2}$

9. $\sqrt{20x^4y^3z^2}$

$= \sqrt{4 \cdot 5x^4y^2 \cdot yz^2}$

$= \sqrt{4}\sqrt{5}\sqrt{x^4}\sqrt{y^2}\sqrt{y}\sqrt{z^2}$

$= 2\sqrt{5}x^2y\sqrt{yz}$

$= 2x^2yz\sqrt{5y}$

11. $\sqrt[3]{72m^5n^9}$

$= \sqrt[3]{8 \cdot 9m^3 \cdot m^2n^9}$

$= \sqrt[3]{8}\sqrt[3]{9}\sqrt[3]{m^3}\sqrt[3]{m^2}\sqrt[3]{n^9}$

$= 2\sqrt[3]{9}m\sqrt[3]{m^2}n^3$

$= 2mn^3\sqrt[3]{9m^2}$

13. $\sqrt[4]{162x^4y^5}$
$= \sqrt[4]{81 \cdot 2}\sqrt[4]{x^4}\sqrt[4]{y^4 \cdot y}$
$= 3\sqrt[4]{2}xy\sqrt[4]{y}$
$= 3xy\sqrt[4]{2y}$

15. $\sqrt[5]{486a^6b^{10}c^2}$
$\sqrt[5]{243 \cdot 2}\sqrt[5]{a^5 \cdot a}\sqrt[5]{b^{10}}\sqrt[5]{c^2}$
$= 3\sqrt[5]{2}a\sqrt[5]{a}b^2\sqrt[5]{c^2}$
$= 3ab^2\sqrt[5]{2ac^2}$

17. $\sqrt{5x^2 + 30x + 45}$
$= \sqrt{5(x^2 + 6x + 9)}$
$= \sqrt{5(x+3)^2}$
$= (x+3)\sqrt{5}$

19. $\sqrt{7x} \cdot \sqrt{14y}$
$= \sqrt{7x \cdot 14y}$
$= \sqrt{98xy}$
$= \sqrt{49 \cdot 2xy}$
$= 7\sqrt{2xy}$

21. $2\sqrt{7xy} \cdot x\sqrt{14y}$
$= 2x\sqrt{7xy \cdot 14y}$
$= 2x\sqrt{98xy^2}$
$= 2x\sqrt{49 \cdot 2 \cdot x \cdot y^2}$
$= 2x(7y)\sqrt{2x}$
$= 14xy\sqrt{2x}$

23. $\sqrt{7x + 14y} \cdot \sqrt{3x + 6y}$
$= \sqrt{(7x + 14y)(3x + 6y)}$
$= \sqrt{7(x+y)(3)(x+y)}$
$= \sqrt{21(x+y)^2}$
$= (x+y)\sqrt{21}$

25. $\sqrt{x+2} \cdot \sqrt{x^2 + 3x + 2}$
$= \sqrt{(x+2)(x^2 + 3x + 2)}$
$= \sqrt{(x+2)(x+2)(x+1)}$
$= (x+2)\sqrt{x+1}$

27. $\sqrt{x^2 + x - 2} \cdot \sqrt{x^2 + 3x - 4}$
$= \sqrt{(x^2 + x - 2)(x^2 + 3x - 4)}$
$= \sqrt{(x+2)(x-1)(x+4)(x-1)}$
$= (x-1)\sqrt{(x+2)(x+4)}$

29. $\sqrt[3]{-4x^4y^2} \cdot \sqrt[3]{6xy}$
$= \sqrt[3]{(-4x^4y^2)(6xy)}$
$= \sqrt[3]{-24x^5y^3}$
$= \sqrt[3]{-8 \cdot 3x^3 \cdot x^2y^3}$
$= -2xy\sqrt[3]{3x^2}$

31. $-\sqrt{\dfrac{36}{49}}$
$= -\dfrac{\sqrt{36}}{\sqrt{49}}$
$= -\dfrac{6}{7}$

33. $\sqrt{\dfrac{1210}{1440}}$
$= \dfrac{\sqrt{1210}}{\sqrt{1440}}$
$= \dfrac{\sqrt{121 \cdot 10}}{\sqrt{144 \cdot 10}}$
$= \dfrac{11\sqrt{10}}{12\sqrt{10}}$
$= \dfrac{11}{12}$

35. $-\sqrt{\dfrac{8}{27}}$

$=-\dfrac{\sqrt{8}}{\sqrt{27}}$

$=-\dfrac{\sqrt{4\cdot 2}}{\sqrt{9\cdot 3}}$

$=-\dfrac{2\sqrt{2}}{3\sqrt{3}}$

$=-\dfrac{2\sqrt{2}}{3\sqrt{3}}\cdot\dfrac{\sqrt{3}}{\sqrt{3}}$

$=-\dfrac{2\sqrt{2\cdot 3}}{3\cdot 3}$

$=-\dfrac{2\sqrt{6}}{9}$

37. $\sqrt[3]{\dfrac{4}{125}}$

$=\dfrac{\sqrt[3]{4}}{\sqrt[3]{125}}$

$=\dfrac{\sqrt[3]{4}}{5}$

39. $\sqrt[3]{-\dfrac{9}{25}}$

$=\dfrac{\sqrt[3]{-9}}{\sqrt[3]{25}}$

$=\dfrac{\sqrt[3]{-9}}{\sqrt[3]{25}}\cdot\dfrac{\sqrt[3]{5}}{\sqrt[3]{5}}$

$=\dfrac{\sqrt[3]{-9\cdot 5}}{\sqrt[3]{125}}$

$=\dfrac{-\sqrt[3]{45}}{5}$

41. $\sqrt[5]{\dfrac{1}{16}}$

$=\dfrac{\sqrt[5]{1}}{\sqrt[5]{16}}$

$=\dfrac{1}{\sqrt[5]{16}}\cdot\dfrac{\sqrt[5]{2}}{\sqrt[5]{2}}$

$=\dfrac{\sqrt[5]{2}}{\sqrt[5]{32}}$

$=\dfrac{\sqrt[5]{2}}{2}$

43. $\sqrt{\dfrac{4x^2}{9y^2}}$

$=\dfrac{\sqrt{4x^2}}{\sqrt{9y^2}}$

$=\dfrac{2x}{3y}$

45. $\sqrt{\dfrac{3x}{25y^2}}$

$=\dfrac{\sqrt{3x}}{\sqrt{25y^2}}$

$=\dfrac{\sqrt{3x}}{5y}$

47. $\sqrt{\dfrac{4xy}{5z}}$

$=\dfrac{\sqrt{4xy}}{\sqrt{5z}}$

$=\dfrac{2\sqrt{xy}}{\sqrt{5z}}\cdot\dfrac{\sqrt{5z}}{\sqrt{5z}}$

$=\dfrac{2\sqrt{5xyz}}{5z}$

49. $\sqrt{\dfrac{2x^2}{6}}$

$=\sqrt{\dfrac{x^2}{3}}$

$=\dfrac{\sqrt{x^2}}{\sqrt{3}}$

$=\dfrac{x}{\sqrt{3}}\cdot\dfrac{\sqrt{3}}{\sqrt{3}}$

$=\dfrac{x\sqrt{3}}{3}$

51. $\sqrt[3]{\dfrac{27x^3}{y^6}}$

$= \dfrac{\sqrt[3]{27x^3}}{\sqrt[3]{y^6}}$

$= \dfrac{3x}{y^2}$

53. $\sqrt[3]{\dfrac{3}{25x^2}}$

$= \dfrac{\sqrt[3]{3}}{\sqrt[3]{25x^2}}$

$= \dfrac{\sqrt[3]{3}}{\sqrt[3]{25x^2}} \cdot \dfrac{\sqrt[3]{5x}}{\sqrt[3]{5x}}$

$= \dfrac{\sqrt[3]{3 \cdot 5x}}{\sqrt[3]{125x^3}}$

$= \dfrac{\sqrt[3]{15x}}{5x}$

55. $\sqrt[3]{\dfrac{3x^2y}{5xy^2z}}$

$= \sqrt[3]{\dfrac{3x}{5yz}}$

$= \dfrac{\sqrt[3]{3x}}{\sqrt[3]{5yz}}$

$= \dfrac{\sqrt[3]{3x}}{\sqrt[3]{5yz}} \cdot \dfrac{\sqrt[3]{25y^2z^2}}{\sqrt[3]{25y^2z^2}}$

$= \dfrac{\sqrt[3]{75xy^2z^2}}{5yz}$

57. $\dfrac{\sqrt{5}}{\sqrt{x}}$

$= \dfrac{\sqrt{5}}{\sqrt{x}} \cdot \dfrac{\sqrt{x}}{\sqrt{x}}$

$= \dfrac{\sqrt{5x}}{x}$

59. $\dfrac{\sqrt{4x}}{\sqrt{8xy}}$

$= \sqrt{\dfrac{4x}{8xy}}$

$= \sqrt{\dfrac{1}{2y}}$

$= \dfrac{\sqrt{1}}{\sqrt{2y}} \cdot \dfrac{\sqrt{2y}}{\sqrt{2y}}$

$= \dfrac{\sqrt{2y}}{2y}$

61. $\dfrac{\sqrt{z^3}}{\sqrt{8}}$

$= \dfrac{z\sqrt{z}}{2\sqrt{2}}$

$= \dfrac{z\sqrt{z}}{2\sqrt{2}} \cdot \dfrac{\sqrt{2}}{\sqrt{2}}$

$= \dfrac{z\sqrt{2z}}{4}$

63. $\dfrac{\sqrt{ab^4}}{\sqrt{a^2b}}$

$= \sqrt{\dfrac{ab^4}{a^2b}}$

$= \sqrt{\dfrac{b^3}{a}}$

$= \dfrac{\sqrt{b^3}}{\sqrt{a}}$

$= \dfrac{b\sqrt{b}}{\sqrt{a}} \cdot \dfrac{\sqrt{a}}{\sqrt{a}}$

$= \dfrac{b\sqrt{ab}}{a}$

65. $\dfrac{3x\sqrt{5x^2y}}{6\sqrt{15x^3y}}$

$= \dfrac{x}{2}\sqrt{\dfrac{5x^2y}{15x^3y}}$

$= \dfrac{x}{2}\sqrt{\dfrac{1}{3x}}$

$= \dfrac{x}{2\sqrt{3x}} \cdot \dfrac{\sqrt{3x}}{\sqrt{3x}}$

$= \dfrac{x\sqrt{3x}}{6x}$

$= \dfrac{\sqrt{3x}}{6}$

67. $\dfrac{\sqrt[3]{3}}{\sqrt[3]{x^2}}$

$= \dfrac{\sqrt[3]{3}}{\sqrt[3]{x^2}} \cdot \dfrac{\sqrt[3]{x}}{\sqrt[3]{x}}$

$= \dfrac{\sqrt[3]{3x}}{x}$

69. $\dfrac{\sqrt[3]{3x}}{\sqrt[3]{12xy}}$

$= \sqrt[3]{\dfrac{3x}{12xy}}$

$= \sqrt[3]{\dfrac{1}{4y}}$

$= \dfrac{\sqrt[3]{1}}{\sqrt[3]{4y}} \cdot \dfrac{\sqrt[3]{2y^2}}{\sqrt[3]{2y^2}}$

$= \dfrac{\sqrt[3]{2y^2}}{2y}$

71. $\dfrac{\sqrt[3]{-8xy^2z^2}}{\sqrt[3]{3x^2y^2z}}$

$= \sqrt[3]{\dfrac{-8xy^2z^2}{3x^2y^2z}}$

$= \sqrt[3]{\dfrac{-8z}{3x}}$

$= \dfrac{\sqrt[3]{-8z}}{\sqrt[3]{3x}}$

$= \dfrac{-2\sqrt[3]{z}}{\sqrt[3]{3x}} \cdot \dfrac{\sqrt[3]{9x^2}}{\sqrt[3]{9x^2}}$

$= \dfrac{-2\sqrt[3]{9x^2z}}{3x}$

73. $T = 2\pi\sqrt{\dfrac{L}{32}}$

$T = 2\pi\sqrt{\dfrac{7}{32}}$

$T \approx 2.9$

The period is approximately 2.9 seconds.

75. $T = 2\pi\sqrt{\dfrac{L}{32}}$

$T = 2\pi\sqrt{\dfrac{3}{32}}$

$T \approx 1.92$

It takes approximately 1.92 seconds.

77. $I = \sqrt{\dfrac{P}{R}}$

$I = \sqrt{\dfrac{10,800}{300}}$

$I = \sqrt{36}$

$I = 6$

The current is 6 amperes.

79. $I = \sqrt{\dfrac{P}{R}}$

$I = \sqrt{\dfrac{425}{10}}$

$I = \sqrt{42.5}$

$I \approx 6.52$

The current is approximately 6.52 amperes.

13.4 Experiencing Algebra the Calculator Way

1. $\sqrt[3]{2x^2} \cdot \sqrt[3]{108x}$

$= \sqrt[3]{2x^2 \cdot 108x}$

$= \sqrt[3]{216x^3}$

$= 6x$

$\text{Y1} = \sqrt[3]{2x^2} \cdot \sqrt[3]{108x}$

$\text{Y2} = 6x$

X	Y1	Y2
-3	-18	-18
-2	-12	-12
-1	-6	-6
0	0	0
1	6	6
2	12	12
3	18	18
X=-3		

2. $3\sqrt{12a^2} \cdot \sqrt{6a^4}$

$= 3\sqrt{12a^2 \cdot 6a^4}$

$= 3\sqrt{72a^6}$

$= 3 \cdot 6a^3\sqrt{2}$

$= 18a^3\sqrt{2}$

$\text{Y1} = 3\sqrt{12x^2} \cdot \sqrt{6x^4}$

$\text{Y2} = 18x^3\sqrt{2}$

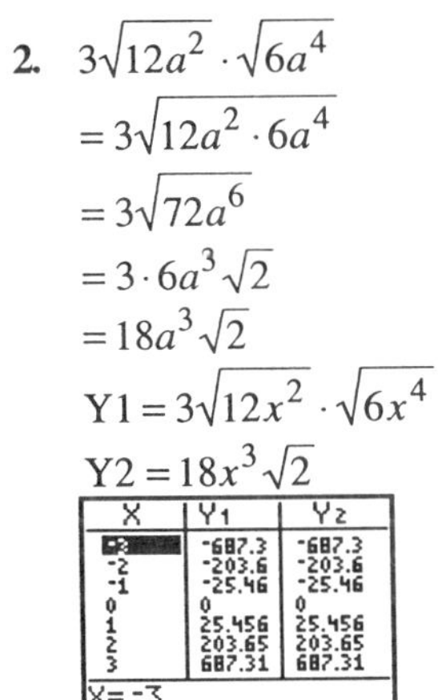

X	Y1	Y2
-3	-687.3	-687.3
-2	-203.6	-203.6
-1	-25.46	-25.46
0	0	0
1	25.456	25.456
2	203.65	203.65
3	687.31	687.31
X=-3		

3. $\sqrt{x^2 + 10x + 25}$

$= \sqrt{(x+5)^2}$

$= x + 5$

$\text{Y1} = \sqrt{x^2 + 10x + 25}$

$\text{Y2} = x + 5$

X	Y1	Y2
-3	2	2
-2	3	3
-1	4	4
0	5	5
1	6	6
2	7	7
3	8	8
X=-3		

4. $\sqrt{b^3 + 6b^2 + 9b}$

$= \sqrt{b(b^2 + 6b + 9)}$

$= \sqrt{b(b+3)^2}$

$= (b+3)\sqrt{b}$

$\text{Y1} = \sqrt{x^3 + 6x^2 + 9x}$

$\text{Y2} = (x+3)\sqrt{x}$

X	Y1	Y2
-3	0	ERROR
-2	ERROR	ERROR
-1	ERROR	ERROR
0	0	0
1	4	4
2	7.0711	7.0711
3	10.392	10.392
X=-3		

5. $\sqrt{2x^2 - 4x + 2}$

$= \sqrt{2(x^2 - 2x + 1)}$

$= \sqrt{2(x-1)^2}$

$= (x-1)\sqrt{2}$

$\text{Y1} = \sqrt{2x^2 - 4x + 2}$

$\text{Y2} = (x-1)\sqrt{2}$

X	Y1	Y2
-2	4.2426	-4.243
-1	2.8284	-2.828
0	1.4142	-1.414
1	0	0
2	1.4142	1.4142
3	2.8284	2.8284
4	4.2426	4.2426
X=4		

6. $\sqrt{\dfrac{36x^2}{27x^3}}$

$= \sqrt{\dfrac{4}{3x}}$

$= \dfrac{\sqrt{4}}{\sqrt{3x}}$

$= \dfrac{2}{\sqrt{3x}} \cdot \dfrac{\sqrt{3x}}{\sqrt{3x}}$

$= \dfrac{2\sqrt{3x}}{3x}$

$$Y1=\sqrt{\frac{36x^2}{27x^3}}$$

$$Y2=\frac{2\sqrt{3x}}{3x}$$

X	Y1	Y2
-2	ERROR	ERROR
-1	ERROR	ERROR
0	ERROR	ERROR
1	1.1547	1.1547
2	.8165	.8165
3	.66667	.66667
4	.57735	.57735

X=-2

13.5 Experiencing Algebra the Exercise Way

1. $2\sqrt{28}-\sqrt{63}$
$=4\sqrt{7}-3\sqrt{7}$
$=(4-3)\sqrt{7}$
$=\sqrt{7}$

3. $\sqrt{10}-\sqrt{\frac{1}{10}}$
$=\sqrt{10}-\frac{\sqrt{1}}{\sqrt{10}}\cdot\frac{\sqrt{10}}{\sqrt{10}}$
$=\sqrt{10}-\frac{\sqrt{10}}{10}$
$=\left(1-\frac{1}{10}\right)\sqrt{10}$
$=\frac{9}{10}\sqrt{10}$

5. $\sqrt[3]{24}-4\sqrt[3]{3}$
$=2\sqrt[3]{3}-4\sqrt[3]{3}$
$=(2-4)\sqrt[3]{3}$
$=-2\sqrt[3]{3}$

7. $5\sqrt{75}-2\sqrt{27}+\sqrt{48}$
$=25\sqrt{3}-6\sqrt{3}+4\sqrt{3}$
$=(25-6+4)\sqrt{3}$
$=23\sqrt{3}$

9. $5\sqrt{x}+9\sqrt{x}$
$=(5+9)\sqrt{x}$
$=14\sqrt{x}$

11. $\sqrt{25x}+\sqrt{36x}$
$=5\sqrt{x}+6\sqrt{x}$
$=(5+6)\sqrt{x}$
$=11\sqrt{x}$

13. $\sqrt{8x^3}-\sqrt{50x^3}$
$=2x\sqrt{2x}-5x\sqrt{2x}$
$=(2x-5x)\sqrt{2x}$
$=-3x\sqrt{2x}$

15. $\sqrt{9a}+\sqrt{16a^3}$
$=3\sqrt{a}+4a\sqrt{a}$
$=(3+4a)\sqrt{a}$

17. $7a\sqrt{b}+9a\sqrt{b}-2a\sqrt{b}$
$=(7a+9a-2a)\sqrt{b}$
$=14a\sqrt{b}$

19. $5\sqrt{pq}-4\sqrt[3]{pq}+2\sqrt{pq}+11\sqrt[3]{pq}$
$=(5+2)\sqrt{pq}+(-4+11)\sqrt[3]{pq}$
$=7\sqrt{pq}+7\sqrt[3]{pq}$

21. $7y\sqrt{x^3y}+3x\sqrt{xy^3}-4xy\sqrt{xy}$
$=7xy\sqrt{xy}+3xy\sqrt{xy}-4xy\sqrt{xy}$
$=(7xy+3xy-4xy)\sqrt{xy}$
$=6xy\sqrt{xy}$

23. $2x\sqrt[3]{x^2y^4z}-3y\sqrt[3]{x^5yz}$
$=2xy\sqrt[3]{x^2yz}-3xy\sqrt[3]{x^2yz}$
$=(2xy-3xy)\sqrt[3]{x^2yz}$
$=-xy\sqrt[3]{x^2yz}$

25. $\sqrt{7}\left(\sqrt{5}-\sqrt{7}\right)$
$=\sqrt{35}-\sqrt{49}$
$=\sqrt{35}-7$

27. $\sqrt{3}\left(\sqrt{x}-\sqrt{5}\right)$
$=\sqrt{3x}-\sqrt{15}$

29. $3\sqrt{a}\left(2\sqrt{a}-5\right)$
$=6\sqrt{a^2}-15\sqrt{a}$
$=6a-15\sqrt{a}$

31. $2\sqrt[3]{x}\left(4\sqrt[3]{x^2}-6\sqrt[3]{x}\right)$
$=8\sqrt[3]{x^3}-12\sqrt[3]{x^2}$
$=8x-12\sqrt[3]{x^2}$

33. $\left(\sqrt{3}-5\sqrt{6}\right)\left(2\sqrt{3}+\sqrt{8}\right)$
$=2\sqrt{3^2}+\sqrt{24}-10\sqrt{18}-5\sqrt{48}$
$=6+2\sqrt{6}-30\sqrt{2}-20\sqrt{3}$

35. $\left(\sqrt{3}-\sqrt{x}\right)\left(\sqrt{2}+\sqrt{x}\right)$
$=\sqrt{6}+\sqrt{3x}-\sqrt{2x}-x$

37. $\left(5-\sqrt{6}\right)\left(5+\sqrt{6}\right)$
$=5^2-\left(\sqrt{6}\right)^2$
$=25 \quad 6$
$=19$

39. $\left(12+\sqrt{p}\right)\left(12-\sqrt{p}\right)$
$=(12)^2-\left(\sqrt{p}\right)^2$
$=144-p$

41. $\left(\sqrt{2x}+\sqrt{3y}\right)\left(\sqrt{2x}-\sqrt{3y}\right)$
$=\left(\sqrt{2x}\right)^2-\left(\sqrt{3y}\right)^2$
$=2x-3y$

43. $\left(\sqrt{a}+4\right)^2$
$=\left(\sqrt{a}\right)^2+2\left(\sqrt{a}\right)(4)+4^2$
$=a+8\sqrt{a}+16$

45. $\left(3\sqrt{b}-2\right)^2$
$=\left(3\sqrt{b}\right)^2-2\left(3\sqrt{b}\right)(2)+2^2$
$=9b-12\sqrt{b}+4$

47. $\left(\sqrt{x}-\sqrt{y}\right)^2$
$=\left(\sqrt{x}\right)^2-2\left(\sqrt{x}\right)\left(\sqrt{y}\right)+\left(\sqrt{y}\right)^2$
$=x-2\sqrt{xy}+y$

49. $\dfrac{\sqrt{21x}-\sqrt{14}}{\sqrt{7}}$
$=\dfrac{\sqrt{21x}}{\sqrt{7}}-\dfrac{\sqrt{14}}{\sqrt{7}}$
$=\sqrt{\dfrac{21x}{7}}-\sqrt{\dfrac{14}{7}}$
$=\sqrt{3x}-\sqrt{2}$

51. $\dfrac{\sqrt{a}-12}{\sqrt{a}}$
$=\dfrac{\sqrt{a}}{\sqrt{a}}-\dfrac{12}{\sqrt{a}}$
$=\sqrt{\dfrac{a}{a}}-\dfrac{12}{\sqrt{a}}\cdot\dfrac{\sqrt{a}}{\sqrt{a}}$
$=1-\dfrac{12\sqrt{a}}{a}$

53. $\dfrac{\sqrt{x}+\sqrt{y}+\sqrt{z}}{\sqrt{x}}$
$=\dfrac{\sqrt{x}}{\sqrt{x}}+\dfrac{\sqrt{y}}{\sqrt{x}}+\dfrac{\sqrt{z}}{\sqrt{x}}$
$=\sqrt{\dfrac{x}{x}}+\dfrac{\sqrt{y}}{\sqrt{x}}\cdot\dfrac{\sqrt{x}}{\sqrt{x}}+\dfrac{\sqrt{z}}{\sqrt{x}}\cdot\dfrac{\sqrt{x}}{\sqrt{x}}$
$=1+\dfrac{\sqrt{xy}}{x}+\dfrac{\sqrt{xz}}{x}$

55. $\dfrac{18}{\sqrt{6}+\sqrt{3}}$

$= \dfrac{18}{\sqrt{6}+\sqrt{3}} \cdot \dfrac{\sqrt{6}-\sqrt{3}}{\sqrt{6}-\sqrt{3}}$

$= \dfrac{18\sqrt{6}-18\sqrt{3}}{\left(\sqrt{6}\right)^2-\left(\sqrt{3}\right)^2}$

$= \dfrac{18\left(\sqrt{6}-\sqrt{3}\right)}{6-3}$

$= \dfrac{18\left(\sqrt{6}-\sqrt{3}\right)}{3}$

$= 6\left(\sqrt{6}-\sqrt{3}\right)$

57. $\dfrac{\sqrt{3}+\sqrt{2}}{\sqrt{3}-\sqrt{2}}$

$= \dfrac{\sqrt{3}+\sqrt{2}}{\sqrt{3}-\sqrt{2}} \cdot \dfrac{\sqrt{3}+\sqrt{2}}{\sqrt{3}+\sqrt{2}}$

$= \dfrac{\left(\sqrt{3}\right)^2+2\sqrt{3}\sqrt{2}+\left(\sqrt{2}\right)^2}{\left(\sqrt{3}\right)^2-\left(\sqrt{2}\right)^2}$

$= \dfrac{3+2\sqrt{6}+2}{3-2}$

$= 5+2\sqrt{6}$

59. $\dfrac{3x}{\sqrt{x}-2}$

$= \dfrac{3x}{\sqrt{x}-2} \cdot \dfrac{\sqrt{x}+2}{\sqrt{x}+2}$

$= \dfrac{3x\left(\sqrt{x}+2\right)}{\left(\sqrt{x}\right)^2-2^2}$

$= \dfrac{3x\sqrt{x}+6x}{x-4}$

61. $\dfrac{3b-4}{\sqrt{3b}-2}$

$= \dfrac{3b-4}{\sqrt{3b}-2} \cdot \dfrac{\sqrt{3b}+2}{\sqrt{3b}+2}$

$= \dfrac{(3b-4)\left(\sqrt{3b}+2\right)}{\left(\sqrt{3b}\right)^2-2^2}$

$= \dfrac{(3b-4)\left(\sqrt{3b}+2\right)}{3b-4}$

$= \sqrt{3b}+2$

63. $\dfrac{\sqrt{x}+3}{\sqrt{x}-3}$

$= \dfrac{\sqrt{x}+3}{\sqrt{x}-3} \cdot \dfrac{\sqrt{x}+3}{\sqrt{x}+3}$

$= \dfrac{\left(\sqrt{x}\right)^2+2(3)\sqrt{x}+3^2}{\left(\sqrt{x}\right)^2-3^2}$

$= \dfrac{x+6\sqrt{x}+9}{x-9}$

65. $\dfrac{3\sqrt{x}-4}{4-3\sqrt{x}}$

$= \dfrac{-1\left(4-3\sqrt{x}\right)}{4-3\sqrt{x}}$

$= -1$

67. Determine the length of the sides of each painting.
Let s_1 = length of the side of the first painting.
$s_1 = \sqrt{A_1}$
$s_1 = \sqrt{490}$
$s_1 = 7\sqrt{10}$
Let s_2 = length of the side of the second painting.
$s_2 = \sqrt{A_2}$
$s_2 = \sqrt{810}$
$s_2 = 9\sqrt{10}$
Determine how much more framing material

will be needed for the bigger painting.

$$\begin{aligned}4s_2 - 4s_1 &= 4\left(9\sqrt{10}\right) - 4\left(7\sqrt{10}\right)\\ &= 36\sqrt{10} - 28\sqrt{10}\\ &= 8\sqrt{10}\\ &\approx 25.3\end{aligned}$$

It will require $8\sqrt{10} \approx 25.3$ inches more of material.

69. Determine the length of the sides of each of the cubes.
Let s_1 = length of the side of the first box.

$$s_1 = \sqrt[3]{V_1}$$
$$s_1 = \sqrt[3]{1750}$$
$$s_1 = 5\sqrt[3]{14}$$

Let s_2 = length of the side of the second box.

$$s_2 = \sqrt[3]{V_2}$$
$$s_2 = \sqrt[3]{896}$$
$$s_2 = 4\sqrt[3]{14}$$

Determine the difference in the perimeters of the bases.

$$\begin{aligned}4s_1 - 4s_2 &= 4\left(5\sqrt[3]{14}\right) - 4\left(4\sqrt[3]{14}\right)\\ &= 20\sqrt[3]{14} - 16\sqrt[3]{14}\\ &= 4\sqrt[3]{14}\\ &\approx 9.6\end{aligned}$$

The difference is $4\sqrt[3]{14} \approx 9.6$ inches.

13.5 Experiencing Algebra the Calculator Way

Students should check results by one of the methods presented.

1. $\left(3\sqrt[3]{x} + 5\sqrt{x}\right) + \left(\sqrt[3]{x} - 3\sqrt{x}\right) + \left(4\sqrt{x} + 7\right)$

$$= \left(3\sqrt[3]{x} + \sqrt[3]{x}\right) + \left(5\sqrt{x} - 3\sqrt{x} + 4\sqrt{x}\right) + 7$$
$$= 4\sqrt[3]{x} + 6\sqrt{x} + 7$$

X	Y_1	Y_2
0	7	7
1	17	17
2	20.525	20.525
3	23.161	23.161
4	25.35	25.35
5	27.256	27.256
6	28.965	28.965

X=0

2. $\left(2\sqrt{x} + 7\right) - \left(\sqrt{x} + 9\right)$

$$\begin{aligned}&= 2\sqrt{x} + 7 - \sqrt{x} - 9\\ &= 2\sqrt{x} - \sqrt{x} + 7 - 9\\ &= \sqrt{x} - 2\end{aligned}$$

10

−10 10

−10

3. $2\sqrt[3]{x}\left(3\sqrt[3]{2x} + 5\right)$

$$\begin{aligned}&= 2\sqrt[3]{x}\left(3\sqrt[3]{2x}\right) + 2\sqrt[3]{x}(5)\\ &= 6\sqrt[3]{2x^2} + 10\sqrt[3]{x}\end{aligned}$$

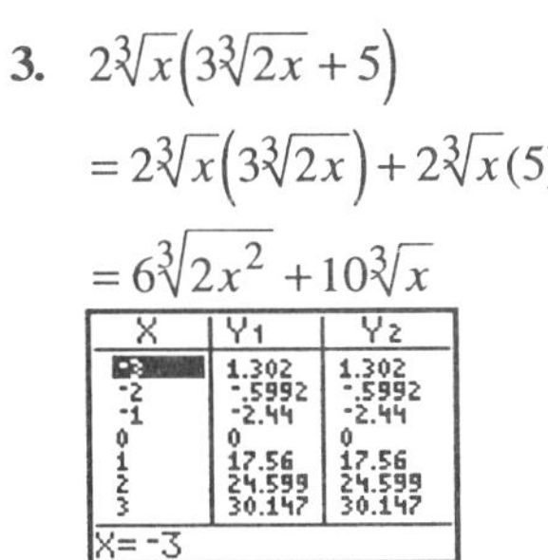

X	Y_1	Y_2
-3	1.302	1.302
-2	-.5992	-.5992
-1	-2.44	-2.44
0	0	0
1	17.56	17.56
2	24.599	24.599
3	30.147	30.147

X= -3

4. $\left(2\sqrt{x} + 3\sqrt{2x}\right)\left(\sqrt{x} + \sqrt{2x}\right)$

$$\begin{aligned}&= 2\sqrt{x}\left(\sqrt{x}\right) + 2\sqrt{x}\left(\sqrt{2x}\right) + 3\sqrt{2x}\left(\sqrt{x}\right)\\ &\quad + 3\sqrt{2x}\left(\sqrt{2x}\right)\\ &= 2x + 2x\sqrt{2} + 3x\sqrt{2} + 6x\\ &= 8x + 5x\sqrt{2}\end{aligned}$$

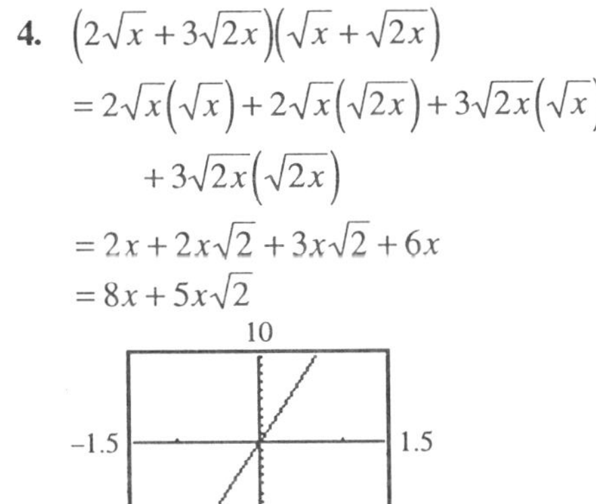

5. $\dfrac{x-16}{\sqrt{x}+4}$

$=\dfrac{x-16}{\sqrt{x}+4}\cdot\dfrac{\sqrt{x}-4}{\sqrt{x}-4}$

$=\dfrac{(x-16)\left(\sqrt{x}-4\right)}{\left(\sqrt{x}\right)^2-4^2}$

$=\dfrac{(x-16)\left(\sqrt{x}-4\right)}{x-16}$

$=\sqrt{x}-4$

10

–10 10

–10

6. $\dfrac{21}{\sqrt{x}-7}$

$=\dfrac{21}{\sqrt{x}-7}\cdot\dfrac{\sqrt{x}+7}{\sqrt{x}+7}$

$=\dfrac{21\left(\sqrt{x}+7\right)}{\left(\sqrt{x}\right)^2-7^2}$

$=\dfrac{21\sqrt{x}+147}{x-49}$

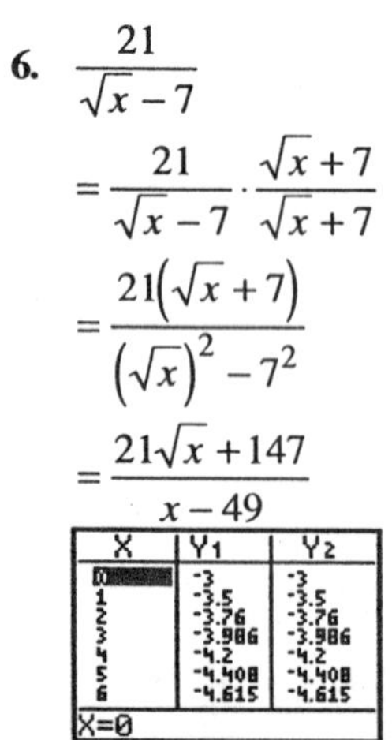

X	Y1	Y2
0	-3	-3
1	-3.5	-3.5
2	-3.76	-3.76
3	-3.986	-3.986
4	-4.2	-4.2
5	-4.408	-4.408
6	-4.615	-4.615

X=0

7. $\left(3\sqrt{x}+5\right)\left(\sqrt{x}-7\right)+3\sqrt{x}-6$

$=3\sqrt{x}\left(\sqrt{x}\right)-3\sqrt{x}(7)+5\sqrt{x}+5(-7)+3\sqrt{x}-6$

$=3x-21\sqrt{x}+5\sqrt{x}-35+3\sqrt{x}-6$

$=3x-13\sqrt{x}-41$

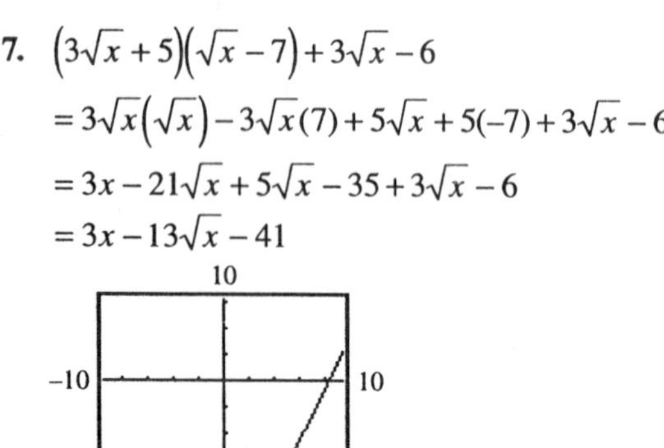

13.6 Experiencing Algebra the Exercise Way

1. $\sqrt{2x+3}=x+1$ is a radical equation because $\sqrt{2x+3}-x-1=0$.

3. $3x-\sqrt{7}=2x+1$ is a nonradical equation because the radicand does not contain a variable.

5. $x^{2/3}+\sqrt{3}=0$ is a radical equation because $\left(\sqrt[3]{x}\right)^2+\sqrt{3}=0$.

7. $x+5^{3/4}=0$ is a nonradical equation because when it is written as $x+\left(\sqrt[4]{5}\right)^3$ it can be seen that the radical does not contain a variable.

9. $x^2-3x+2=0$ is a nonradical equation because there is no radicand.

11. $x^{1/3}+5\sqrt[3]{x}=12$ is a radical equation because $x^{1/3}+5\sqrt[3]{x}-12=0$.

13.–17. Students can set up a table to solve the radical equations numerically on a calculator.

13. $\sqrt{5x+4}=x-4$

$Y1=\sqrt{5x+4}$

$Y2=x-4$

The restricted values are all real numbers less than $-\dfrac{4}{5}$. The solution is 12.

15. $(2x+1)^{2/3}=\sqrt[3]{4x^2+4x+1}$

$Y1=(2x+1)^{2/3}$

$Y2=\sqrt[3]{4x^2+4x+1}$

There are no restricted values. The expressions are always equal. The solutions are all real numbers.

17. $(5-2x)^{1/2} = \sqrt[4]{4x^2-20x+25}$
$Y1 = (5-2x)^{1/2}$
$Y2 = \sqrt[4]{4x^2-20x+25}$
The restricted values are all real numbers greater than $\frac{5}{2}$. The expressions are always equal. The solutions are all real numbers less than or equal to $\frac{5}{2}$.

19. $\sqrt{x} = 3$
$Y1 = \sqrt{x}$
$Y2 = 3$

10
-10 10
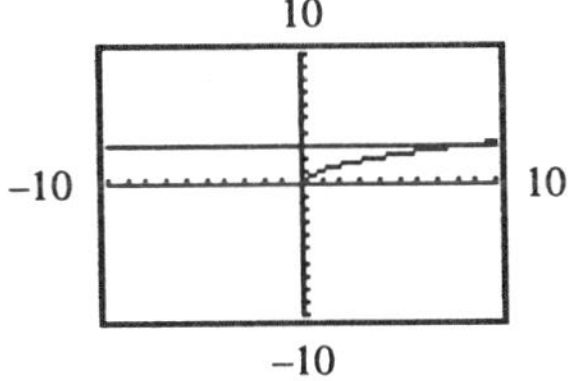
-10

The solution is the x-coordinate of the intersection, 9.

21. $\sqrt{x-2} = 3$
$Y1 = \sqrt{x-2}$
$Y2 = 3$

10
-5 25
-5

The solution is the x-coordinate of the intersection, 11.

23. $\sqrt{3x+1} = \sqrt{26-2x}$
$Y1 = \sqrt{3x+1}$
$Y2 = \sqrt{26-2x}$

10
-10 10
-10

The solution is the x-coordinate of the intersection, 5.

25. $\sqrt{3x+7} = x-1$
$Y1 = \sqrt{3x+7}$
$Y2 = x-1$

10
-10 10
-10

The solution is the x-coordinate of the intersection, 6.

27. $x^3 = \sqrt[3]{x}$
$Y1 = x^3$
$Y2 = \sqrt[3]{x}$

5
-5 5
-5

The solutions are the x-coordinates of the intersections, –1, 0, and 1.

29. $(x+2)^2 = \sqrt{x+2}$
$Y1 = (x+2)^2$
$Y2 = \sqrt{x+2}$

5
-5 5
-5

The solutions are the x-coordinates of the intersections, –2 and –1.

31. $\sqrt{2x+1} = 2x+1$
$Y1 = \sqrt{2x+1}$
$Y2 = 2x+1$

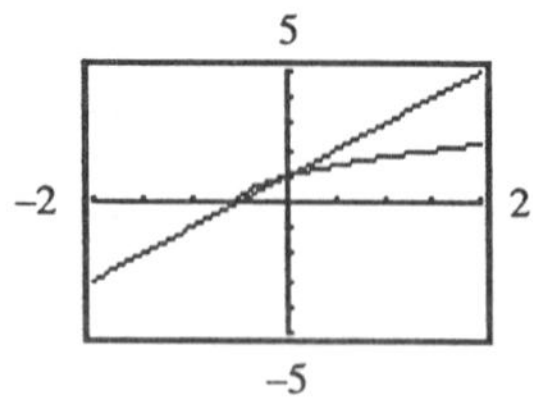

The solutions are the x-coordinates of the intersections, $-\frac{1}{2}$ and 0.

33. $(3x+5)^{1/2} = \sqrt{3x-5}$

$Y1 = (3x+5)^{1/2}$

$Y2 = \sqrt{3x-5}$

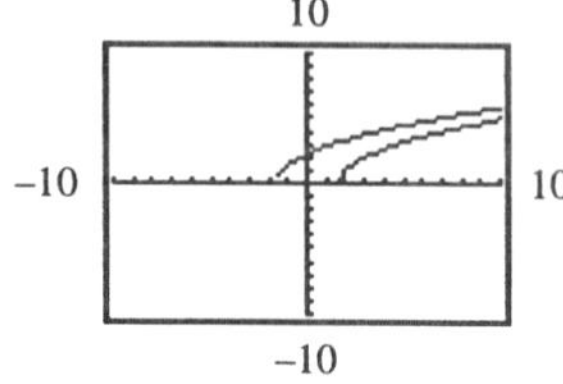

The graphs do not appear to intersect. There is no solution.

35. $\sqrt[3]{x^2+6x+9} = (x+3)^{2/3}$

$Y1 = \sqrt[3]{x^2+6x+9}$

$Y2 = (x+3)^{2/3}$

10
−10 10
−10

The graphs are the same. The solutions are all real numbers.

37. $\sqrt{x-3} = \sqrt[4]{x^2-6x+9}$

$Y1 = \sqrt{x-3}$

$Y2 = \sqrt[4]{x^2-6x+9}$

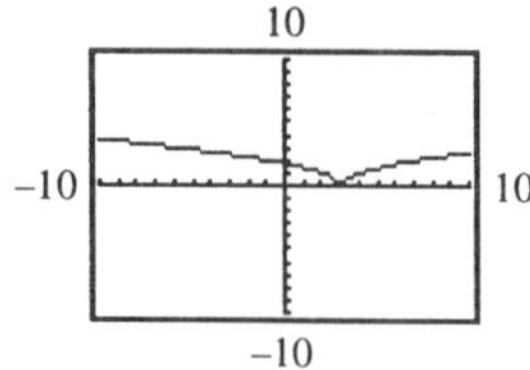

The solutions are all real numbers greater than or equal to 3.

39. $\sqrt[3]{(x+1)^2} = \sqrt[3]{x^2-2x+1}$

$Y1 = \sqrt[3]{(x+1)^2}$

$Y2 = \sqrt[3]{x^2-2x+1}$

10
−10 10
−10

The solution is the x-coordinate of the intersection, 0.

41. $(x+2)^{3/2} = x+20$

$Y1 = (x+2)^{3/2}$

$Y2 = x+20$

50
−10 10
−10

The solution is the x-coordinate of the intersection, 7.

43. $\sqrt[3]{x^2-6x+9} = x+9$

$Y1 = \sqrt[3]{x^2-6x+9}$

$Y2 = x+9$

10
−10 10
−10

The solution is the x-coordinate of the intersection, −5.

45. $x+5 = \sqrt[3]{4x+5}$

$Y1 = x+5$

$Y2 = \sqrt[3]{4x+5}$

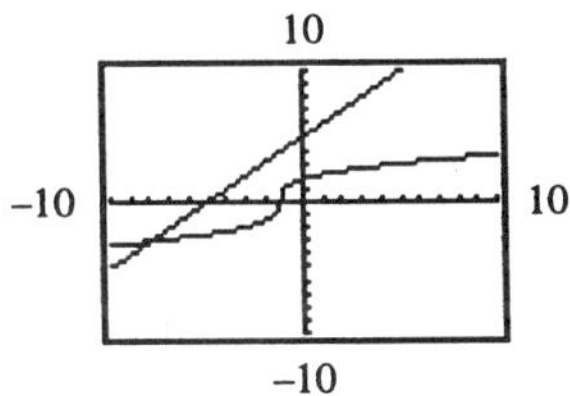

The solution is the x-coordinate of the intersection, –8.

47. $\sqrt{x} = 4.23$
$Y1 = \sqrt{x}$
$Y2 = 4.23$

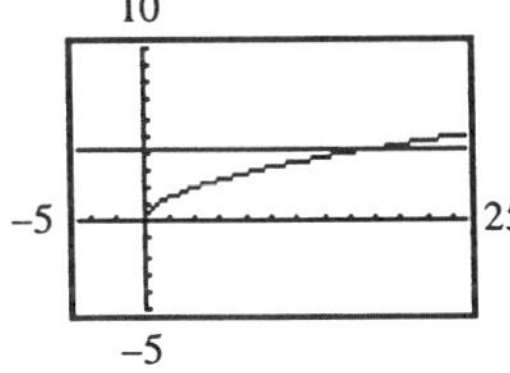

The solution is the x-coordinate of the intersection, 17.89.

49. $3x + 1 = \sqrt[3]{2x - 5}$
$Y1 = 3x + 1$
$Y2 = \sqrt[3]{2x - 5}$

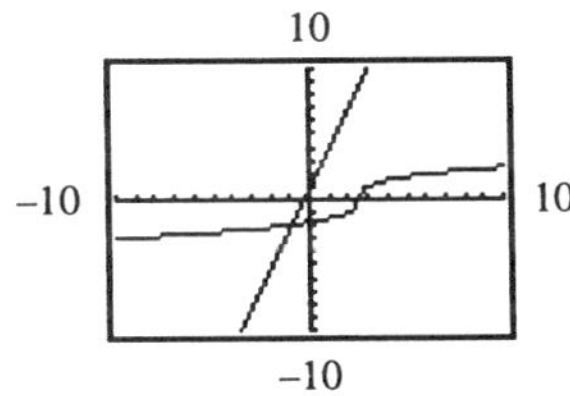

The approximate solution is the x-coordinate of the intersection, –0.97.

51. $\sqrt[3]{2x + 3} = \sqrt{4 - x}$
$Y1 = \sqrt[3]{2x + 3}$
$Y2 = \sqrt{4 - x}$

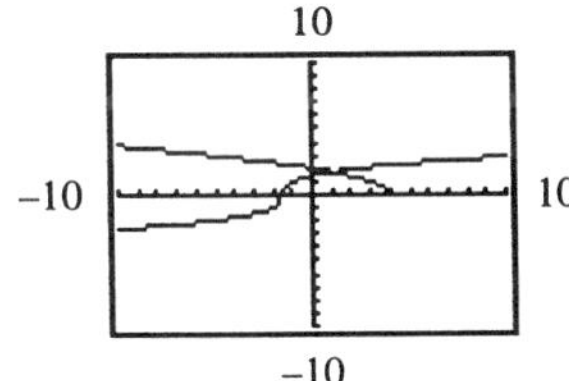

The approximate solution is the x-coordinate of the intersection, 1.04.

53. $(5x + 7)^{2/3} = (x - 6)^{3/4}$
$Y1 = (5x + 7)^{2/3}$
$Y2 = (x - 6)^{3/4}$

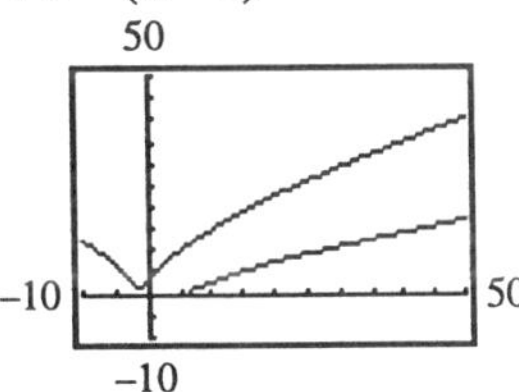

The graphs do not appear to intersect. There is no solution.

55. $3x + 2 = \sqrt{9x^2 + 12x + 4}$
$Y1 = 3x + 2$
$Y2 = \sqrt{9x^2 + 12x + 4}$

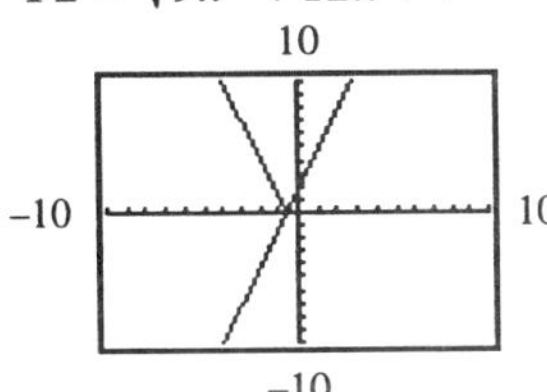

The graphs are the same for $x \geq -\frac{2}{3}$. The solutions are all real numbers greater than or equal to $-\frac{2}{3}$.

57. $2L + 2W = P$
$2\sqrt{x + 4} + 2\sqrt{3x} = 20$
$Y1 = 2\sqrt{x + 4} + 2\sqrt{3x}$
$Y2 = 20$

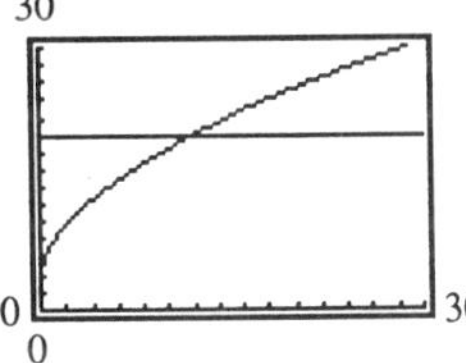

The x-coordinate of the intersection is 12.
width $= \sqrt{x + 4} = \sqrt{12 + 4} = 4$
length $= \sqrt{3x} = \sqrt{3(12)} = 6$
The width is 4 inches and the length is 6 inches.

59. $t=\sqrt{\dfrac{2d}{32.2}}$

$5.845=\sqrt{\dfrac{2d}{32.2}}$

$\text{Y1}=5.845$

$\text{Y2}=\sqrt{\dfrac{2x}{32.2}}$

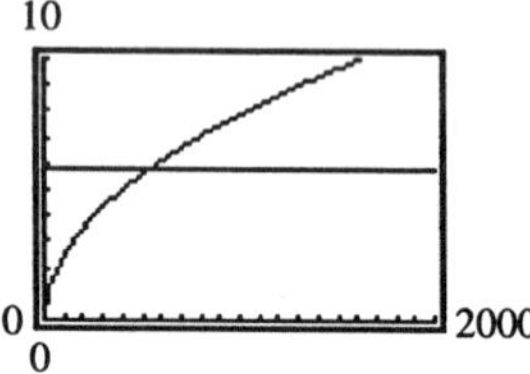

The x-coordinate of the intersection is about 550.0. The height is approximately 550 feet.

13.6 Experiencing Algebra the Calculator Way

1. $\sqrt{x-2}+1=\sqrt{x-1}$

$\text{Y1}=\sqrt{x-2}+1$

$\text{Y2}=\sqrt{x-1}$

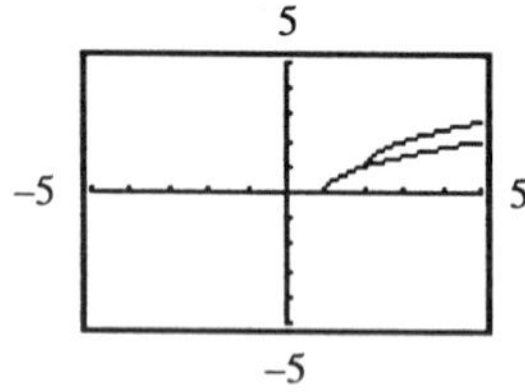

The solution is 2.

2. $\sqrt{x-1}=\sqrt{x}-1$

$\text{Y1}=\sqrt{x-1}$

$\text{Y2}=\sqrt{x}-1$

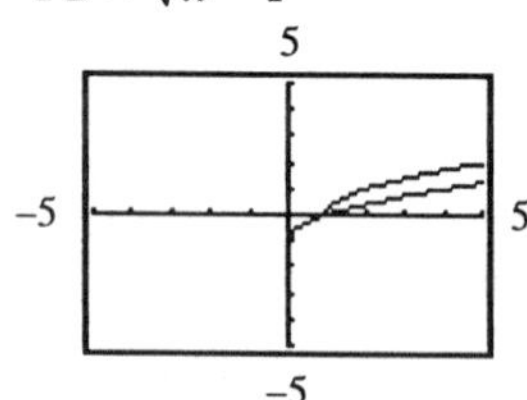

The solution is 1.

3. $\sqrt[3]{x-6}=x$

$\text{Y1}=\sqrt[3]{x-6}$

$\text{Y2}=x$

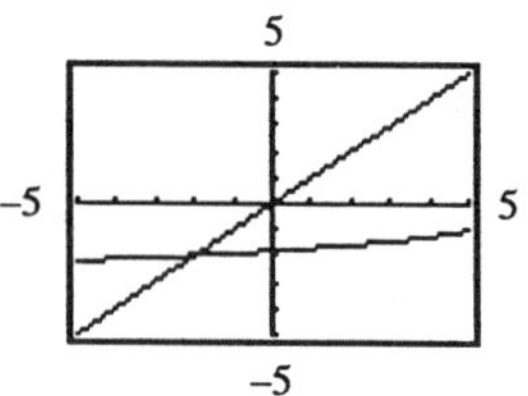

The solution is –2.

4. $\sqrt[4]{x+4}=1$

$\text{Y1}=\sqrt[4]{x+4}$

$\text{Y2}=1$

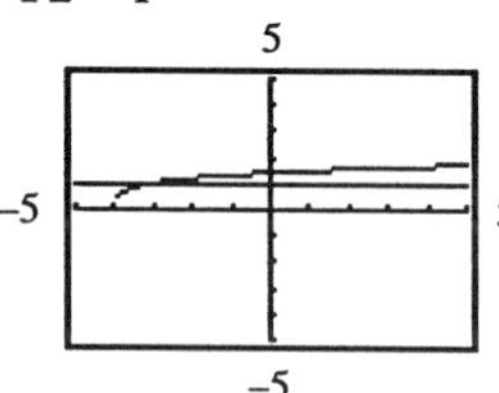

The solution is –3.

5. $\sqrt{x-3}+1=\sqrt{x-1}$

$\text{Y1}=\sqrt{x-3}+1$

$\text{Y2}=\sqrt{x-1}$

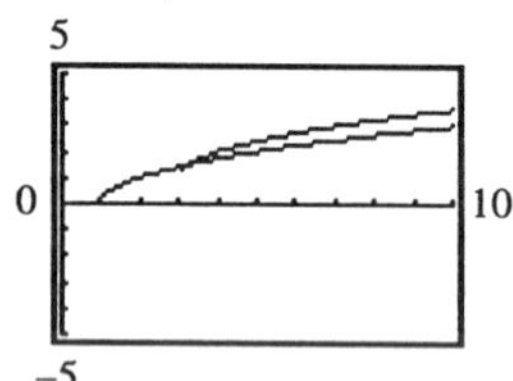

The solution is 3.25.

6. $\sqrt[3]{x^2-5}-3=-1$

$\text{Y1}=\sqrt[3]{x^2-5}-3$

$\text{Y2}=-1$

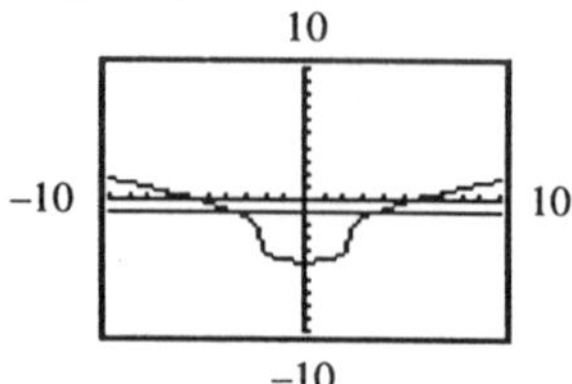

The approximately solutions are –3.606 and 3.606.

7. $\sqrt{x-1}+x=\sqrt{x}+1$

$Y1=\sqrt{x-1}+x$

$Y2=\sqrt{x}+1$

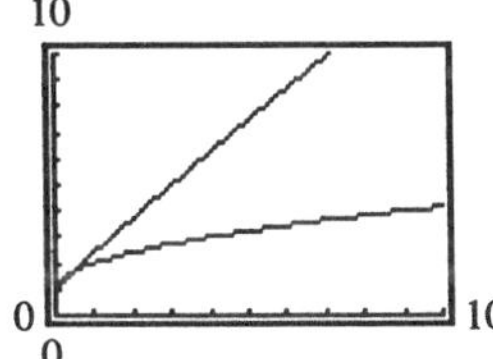

The approximate solution is 1.514.

8. $\sqrt[4]{x^2-x+2}=2$

$Y1=\sqrt[4]{x^2-x+2}$

$Y2=2$

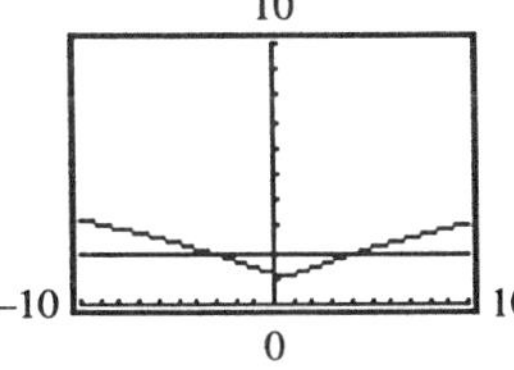

The approximate solutions are −3.275 and 4.275.

9. $\sqrt{2(2x+1)}-1=\sqrt{2(x+1)}$

$Y1=\sqrt{2(2x+1)}-1$

$Y2=\sqrt{2(x+1)}$

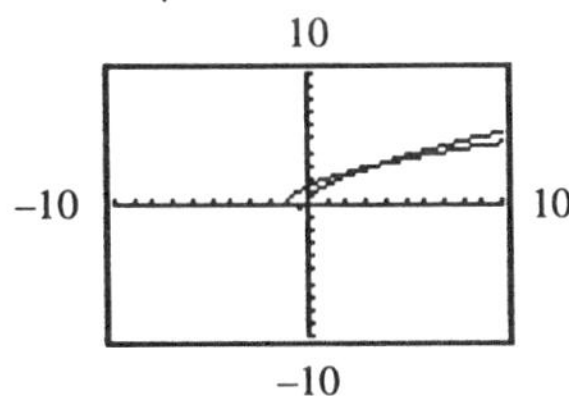

The solution is 3.5.

10. $1+3\sqrt{2x-3}=3\sqrt{15}+1$

$Y1=1+3\sqrt{2x-3}$

$Y2=3\sqrt{15}+1$

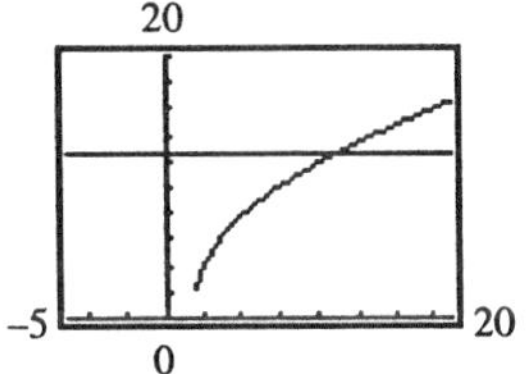

The solution is 9.

11. $\sqrt{4-18x}=2x^2-5x-18$

$Y1=\sqrt{4-18x}$

$Y2=2x^2-5x-18$

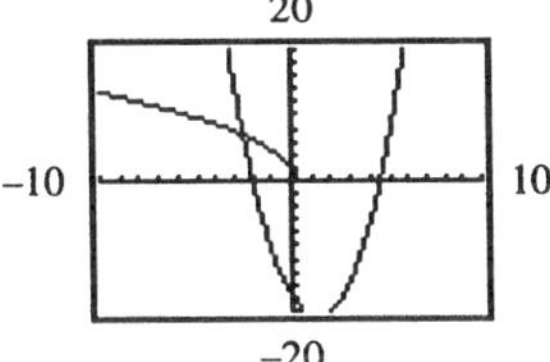

The solution is −2.5.

12. $\sqrt{\frac{5x+9}{3x-8}}=\frac{x-1}{x-4}$

$Y1=\sqrt{\frac{5x+9}{3x-8}}$

$Y2=\frac{x-1}{x-4}$

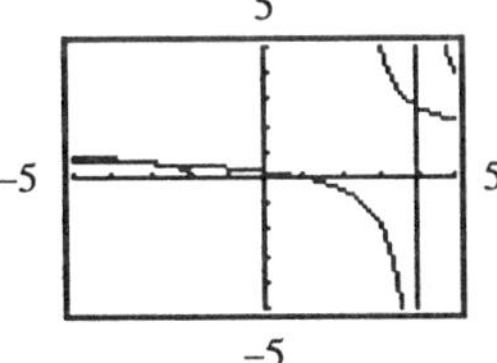

The solutions are approximately −2.84 and 8.

13. $\sqrt[3]{3x+4}=\sqrt{36-x}$

$Y1=\sqrt[3]{3x+4}$

$Y2=\sqrt{36-x}$

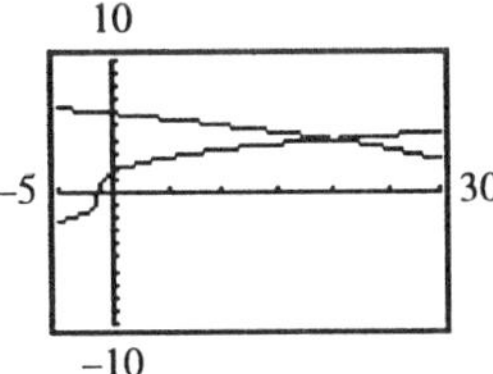

The solution is 20.

14. $\sqrt[4]{3x+4} = \sqrt[3]{12-x}$
$Y1 = \sqrt[4]{3x+4}$
$Y2 = \sqrt[3]{12-x}$

5
–10 10
–5

The solution is 4.

15. $\sqrt[4]{1-5x} = x^2 + x - 4$
$Y1 = \sqrt[4]{1-5x}$
$Y2 = x^2 + x - 4$

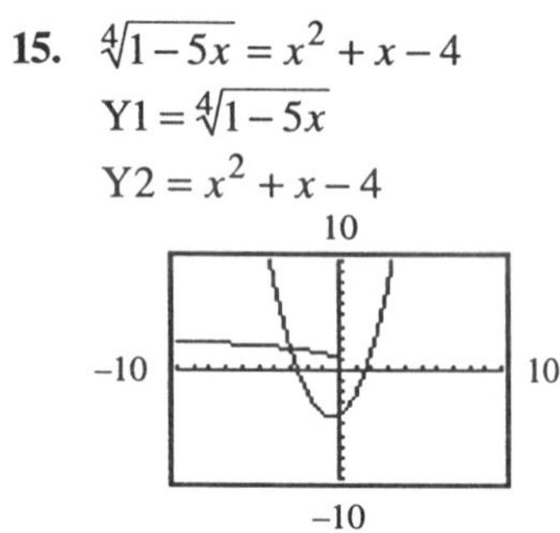

The solution is –3.

13.7 Experiencing Algebra the Exercise Way

1. $\sqrt{x} - 3 = 5$
$\sqrt{x} = 8$
$\left(\sqrt{x}\right)^2 = 8^2$
$x = 64$
The solution is 64.

3. $\sqrt{x} + 1.7 = 4.5$
$\sqrt{x} = 2.8$
$\left(\sqrt{x}\right)^2 = (2.8)^2$
$x = 7.84$
The solution is 7.84.

5. $\sqrt{x} + 8 = 5$
$\sqrt{x} = -3$
There is no solution.

7. $\sqrt{3x} - 4 = 2$
$\sqrt{3x} = 6$
$\left(\sqrt{3x}\right)^2 = 6^2$
$3x = 36$
$x = 12$
The solution is 12.

9. $9 - 3\sqrt{2z} = 0$
$9 = 3\sqrt{2z}$
$3 = \sqrt{2z}$
$3^2 = \left(\sqrt{2z}\right)^2$
$9 = 2z$
$\frac{9}{2} = z$
The solution is $\frac{9}{2}$.

11. $3\sqrt{6x} + 1 = 10$
$3\sqrt{6x} = 9$
$\sqrt{6x} = 3$
$\left(\sqrt{6x}\right)^2 = 3^2$
$6x = 9$
$x = \frac{3}{2}$
The solution is $\frac{3}{2}$.

13. $\sqrt{x+5} = 3$
$\left(\sqrt{x+5}\right)^2 = 3^2$
$x + 5 = 9$
$x = 4$
The solution is 4.

15. $\sqrt{2x+5} + 4 = 9$
$\sqrt{2x+5} = 5$
$\left(\sqrt{2x+5}\right)^2 = 5^2$
$2x + 5 = 25$
$2x = 20$
$x = 10$
The solution is 10.

17. $\sqrt{3x+4}-2=6$
$\sqrt{3x+4}=8$
$\left(\sqrt{3x+4}\right)^2=8^2$
$3x+4=64$
$3x=60$
$x=20$
The solution is 20.

19. $\sqrt{5-4x}=x$
$\left(\sqrt{5-4x}\right)^2=x^2$
$5-4x=x^2$
$x^2+4x-5=0$
$(x+5)(x-1)=0$
$x=-5$ (extraneous) $\quad x=1$
The solution is 1.

21. $\sqrt{6x-5}=\sqrt{4x+5}$
$\left(\sqrt{6x-5}\right)^2=\left(\sqrt{4x+5}\right)^2$
$6x-5=4x+5$
$2x=10$
$x=5$
The solution is 5.

23. $3\sqrt{x+2}=\sqrt{x+10}$
$\left(3\sqrt{x+2}\right)^2=\left(\sqrt{x+10}\right)^2$
$9(x+2)=x+10$
$8x=-8$
$x=-1$
The solution is –1.

25. $x=3+2\sqrt{x-4}$
$x-3=2\sqrt{x-4}$
$(x-3)^2=\left(2\sqrt{x-4}\right)^2$
$x^2-6x+9=4(x-4)$
$x^2-10x+25=0$
$(x-5)^2=0$
$x=5$
The solution is 5.

27. $x=\sqrt{21-x-x^2}+3$
$x-3=\sqrt{21-x-x^2}$
$(x-3)^2=\left(\sqrt{21-x-x^2}\right)^2$
$x^2-6x+9=21-x-x^2$
$2x^2-5x-12=0$
$(2x+3)(x-4)=0$
$x=-\frac{3}{2}$ (extraneous) $\quad x=4$
The solution is 4.

29. $\sqrt{x^2-14x+49}=7-x$
$\left(\sqrt{x^2-14x+49}\right)^2=(7-x)^2$
$x^2-14x+49=49-14x+x^2$
$49=49$
The solutions are all real numbers less than or equal to 7.

31. $x=\sqrt{2x+7}$
$x^2=\left(\sqrt{2x+7}\right)^2$
$x^2=2x+7$
$x^2-2x-7=0$
$x=\dfrac{-(-2)\pm\sqrt{(-2)^2-4(1)(-7)}}{2(1)}$
$x\approx-1.83$ (extraneous) $\quad x\approx3.83$
The approximate solution is 3.83.

33. $5+\sqrt{25-2x}=x$
$\sqrt{25-2x}=x-5$
$\left(\sqrt{25-2x}\right)^2=(x-5)^2$
$25-2x=x^2-10x+25$
$x^2-8x=0$
$x(x-8)=0$
$x=0$ (extraneous) $\quad x=8$
The solution is 8.

35. $\sqrt{x}-3=\sqrt{x-27}$
$\left(\sqrt{x}-3\right)^2=\left(\sqrt{x-27}\right)^2$
$x-6\sqrt{x}+9=x-27$
$\sqrt{x}=6$
$\left(\sqrt{x}\right)^2=6^2$
$x=36$
The solution is 36.

37. $\sqrt{x-7}=7-\sqrt{x}$
$\left(\sqrt{x-7}\right)^2=\left(7-\sqrt{x}\right)^2$
$x-7=49-14\sqrt{x}+x$
$4=\sqrt{x}$
$4^2=\left(\sqrt{x}\right)^2$
$16=x$
The solution is 16.

39. $\sqrt{2x+5}-7=\sqrt{5+2x}+3$
$-7=3$
This is a contradiction. There is no solution.

41. $\sqrt{2x+3}=1-\sqrt{x+5}$
$\left(\sqrt{2x+3}\right)^2=\left(1-\sqrt{x+5}\right)^2$
$2x+3=1-2\sqrt{x+5}+x+5$
$2\sqrt{x+5}=-x+3$
$\left(2\sqrt{x+5}\right)^2=(-x+3)^2$
$4(x+5)=x^2-6x+9$
$x^2-10x-11=0$
$(x-11)(x+1)=0$
$x=11$, $x=-1$ (both extraneous)
There is no solution.

43. $\sqrt[3]{3x}=-6$
$\left(\sqrt[3]{3x}\right)^3=(-6)^3$
$3x=-216$
$x=-72$
The solution is –72.

45. $\sqrt[4]{2x}=10$
$\left(\sqrt[4]{2x}\right)^4=(10)^4$
$2x=10{,}000$
$x=5000$
The solution is 5000.

47. $\sqrt[5]{2x}=-2$
$\left(\sqrt[5]{2x}\right)^5=(-2)^5$
$2x=-32$
$x=-16$
The solution is –16.

49. $\sqrt[3]{x}+3=1$
$\sqrt[3]{x}=-2$
$\left(\sqrt[3]{x}\right)^3=(-2)^3$
$x=-8$
The solution is –8.

51. $\sqrt[4]{x}+1=5$
$\sqrt[4]{x}=4$
$\left(\sqrt[4]{x}\right)^4=(4)^4$
$x=256$
The solution is 256.

53. $\sqrt[3]{2x+1}=3$
$\left(\sqrt[3]{2x+1}\right)^3=3^3$
$2x+1=27$
$2x=26$
$x=13$
the solution is 13.

55. $\sqrt[4]{x-5}=2$
$\left(\sqrt[4]{x-5}\right)^4=2^4$
$x-5=16$
$x=21$
The solution is 21.

57. $\sqrt[5]{2x-5}=3$
$\left(\sqrt[5]{2x-5}\right)^5=3^5$
$2x-5=243$
$2x=248$
$x=124$
The solution is 124.

59. $\sqrt[3]{x^2+7x}+7=9$
$\sqrt[3]{x^2+7x}=2$
$\left(\sqrt[3]{x^2+7x}\right)^3=2^3$
$x^2+7x=8$
$x^2+7x-8=0$
$(x+8)(x-1)=0$
$x=-8 \qquad x=1$
The solutions are –8 and 1.

61. $\sqrt[4]{3x-5}=\sqrt[4]{2x+4}$
$\left(\sqrt[4]{3x-5}\right)^4=\left(\sqrt[4]{2x+4}\right)^4$
$3x-5=2x+4$
$x=9$
The solution is 9.

63. $\sqrt[4]{5x-2}=2\sqrt[4]{3}$
$\left(\sqrt[4]{5x-2}\right)^4=\left(2\sqrt[4]{3}\right)^4$
$5x-2=16(3)$
$5x=50$
$x=10$
The solution is 10.

65. $x=\sqrt[4]{18x^2-81}$
$x^4=\left(\sqrt[4]{18x^2-81}\right)^4$
$x^4=18x^2-81$
$x^4-18x^2+81=0$
$(x^2-9)^2=0$
$x^2=9$
$x=-3$ (extraneous) $\qquad x=3$
The solution is 3.

67. $\sqrt[4]{2x}+\sqrt[4]{3x}=0$
$\sqrt[4]{2x}=-\sqrt[4]{3x}$
$\left(\sqrt[4]{2x}\right)^4=\left(-\sqrt[4]{3x}\right)^4$
$2x=3x$
$0=x$
The solution is 0.

69. $\sqrt[4]{3x-5}+\sqrt[4]{x-3}=0$
$\sqrt[4]{3x-5}=-\sqrt[4]{x-3}$
$\left(\sqrt[4]{3x-5}\right)^4=\left(-\sqrt[4]{x-3}\right)^4$
$3x-5=x-3$
$2x=2$
$x=1$ (restricted value)
There is no solution.

71. $\sqrt[3]{3x-5}+\sqrt[3]{x-3}=0$
$\sqrt[3]{3x-5}=-\sqrt[3]{x-3}$
$\left(\sqrt[3]{3x-5}\right)^3=\left(-\sqrt[3]{x-3}\right)^3$
$3x-5=-(x-3)$
$4x=8$
$x=2$
The solution is 2.

73. $\sqrt[4]{7x-1}-\sqrt[4]{x+11}=0$
$\sqrt[4]{7x-1}=\sqrt[4]{x+11}$
$\left(\sqrt[4]{7x-1}\right)^4=\left(\sqrt[4]{x+11}\right)^4$
$7x-1=x+11$
$6x=12$
$x=2$
The solution is 2.

75. $x^{4/3}=16$
$(x^{4/3})^3=(16)^3$
$x^4=4096$
$x=\pm\sqrt[4]{4096}$
$x=\pm 8$
The solutions are ±8.

77. $(x+6)^{2/5}=4$
$[(x+6)^{2/5}]^5=4^5$
$(x+6)^2=1024$
$x+6=\pm\sqrt{1024}$
$x=-6\pm 32$
$x=-38$ $x=26$
The solutions are –38 and 26.

79. $(x-4)^{3/4}=27$
$[(x-4)^{3/4}]^4=(27)^4$
$(x-4)^3=531,441$
$x-4=\sqrt[3]{531,441}$
$x=4+81$
$x=85$
The solution is 85.

81. $x^{-2/3}=4$
$(x^{-2/3})^3=4^3$
$\frac{1}{x^2}=64$
$x=\pm\sqrt{\frac{1}{64}}$
$x=\pm\frac{1}{8}$
The solutions are $\pm\frac{1}{8}$.

83. $(x-5)^{-3/4}=27$
$[(x-5)^{-3/4}]^4=(27)^4$
$\frac{1}{(x-5)^3}=531,441$
$(x-5)^3=\frac{1}{531,441}$
$x-5=\sqrt[3]{\frac{1}{531,441}}$
$x=5+\frac{1}{81}$
$x=\frac{406}{81}$
The solution is $\frac{406}{81}$.

85. $(5x-3)^{-2/3}=\frac{1}{25}$
$(5x-3)^{2/3}=25$
$[(5x-3)^{2/3}]^3=(25)^3$
$(5x-3)^2=15,625$
$5x-3=\pm\sqrt{15,625}$
$5x=3\pm 125$
$x=\frac{3\pm125}{5}$
$x=-\frac{122}{5}$ $x=\frac{128}{5}$
The solutions are $-\frac{122}{5}$ and $\frac{128}{5}$.

87. $x^{2/3}+5=3$
$x^{2/3}=-2$
There is no real-number solution.

89. $x^{3/4}-7=1$
$x^{3/4}=8$
$[x^{3/4}]^4=8^4$
$x^3=4096$
$x=\sqrt[3]{4096}$
$x=16$
The solution is 16.

91. $d=\sqrt{(x_2-x_1)^2+(y_2-y_1)^2}$
$d=\sqrt{(8-8)^2+(2+3)^2}$
$d=\sqrt{5^2}$
$d=5$
The distance is 5 units.

93. $d=\sqrt{(x_2-x_1)^2+(y_2-y_1)^2}$
$d=\sqrt{(5+3)^2+(6-6)^2}$
$d=\sqrt{8^2}$
$d=8$
The distance is 8 units.

95. $d = \sqrt{(x_2 - x_1)^2 + (y_2 - y_1)^2}$

$d = \sqrt{(2+3)^2 + (5+4)^2}$

$d = \sqrt{106}$

The distance is $\sqrt{106}$ units.

97. $d = \sqrt{(x_2 - x_1)^2 + (y_2 - y_1)^2}$

$13 = \sqrt{(3-8)^2 + (4-y)^2}$

$(13)^2 = \left(\sqrt{(-5)^2 + 16 - 8y + y^2}\right)^2$

$169 = y^2 - 8y + 41$

$0 = y^2 - 8y - 128$

$0 = (y+8)(y-16)$

$y = -8 \quad y = 16$

The y-coordinates are –8 and 16.
The two possible points are (8, –8), (8, 16).

99. $d = \sqrt{(x_2 - x_1)^2 + (y_2 - y_1)^2}$

$10 = \sqrt{(-2-x)^2 + (5+1)^2}$

$(10)^2 = \left(\sqrt{4 + 4x + x^2 + 36}\right)^2$

$100 = 40 + 4x + x^2$

$0 = x^2 + 4x - 60$

$0 = (x+10)(x-6)$

$x = -10,\ x = 6$

The y-coordinates are –10 and 6.
The two possible points are (–10, –1), (6, –1).

101. $d = \sqrt{(x_2 - x_1)^2 + (y_2 - y_1)^2}$

$5 = \sqrt{(-2-6)^2 + (1-y)^2}$

$5 = \sqrt{64 + (1-y)^2}$

$25 = 64 + (1-y)^2$

$-39 = (1-y)^2$

There is no solution.

103. $d = kE^{1/3}$

$13{,}200 = 0.02E^{1/3}$

$660{,}000 = E^{1/3}$

$(660{,}000)^3 = (E^{1/3})^3$

$2.875 \times 10^{17} \approx E$

The energy was approximately 2.875×10^{17} ft-lb.

105. $T = \left(\frac{LH^2}{25}\right)^{1/4}$

$4 = \left(\frac{L8^2}{25}\right)^{1/4}$

$4^4 = \left[\left(\frac{64L}{25}\right)^{1/4}\right]^4$

$256 = \frac{64L}{25}$

$6400 = 64L$

$100 = L$

The crushing load was 100 tons.

13.7 Experiencing Algebra the Calculator Way

1. $\left(\sqrt[4]{2x+1}\right)\left(\sqrt[3]{x-13}\right) = 9$

$Y1 = \left(\sqrt[4]{2x+1}\right)\left(\sqrt[3]{x-13}\right)$

$Y2 = 9$

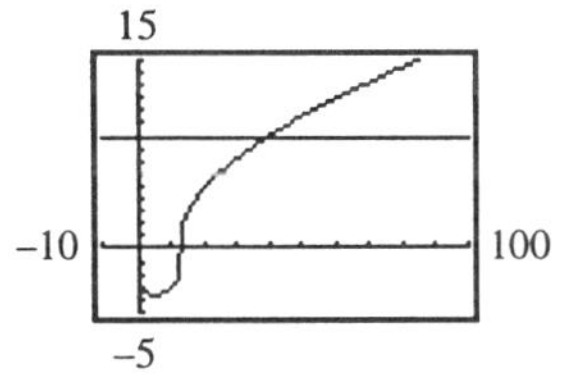

The solution is 40.

2. $\sqrt[5]{3x+2} - \sqrt[4]{8x+1} = -1$

$Y1 = \sqrt[5]{3x+2} - \sqrt[4]{8x+1}$

$Y2 = -1$

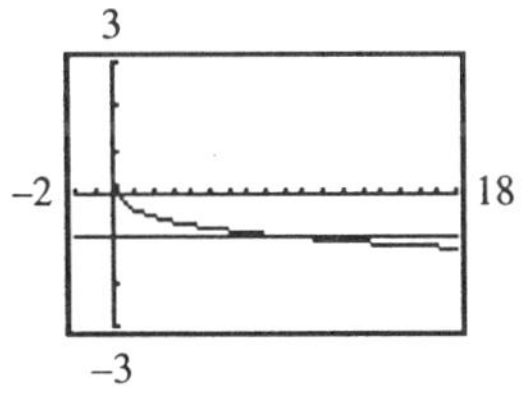

The solution is 10.

3. $\dfrac{\sqrt{3x+4}}{\sqrt[4]{x-4}} = \sqrt[3]{4x-16}$

$Y1 = \dfrac{\sqrt{3x+4}}{\sqrt[4]{x-4}}$

$Y2 = \sqrt[3]{4x-16}$

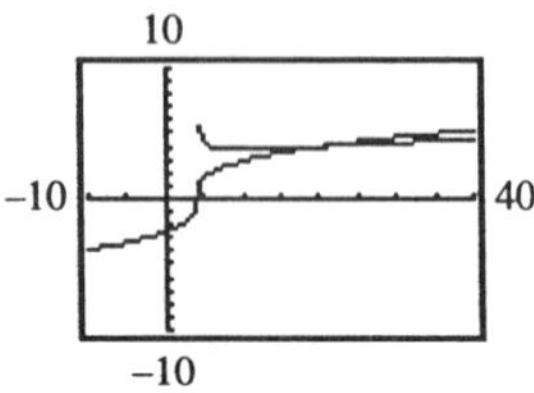

The solution is 20.

4. $\sqrt[3]{2x+1} = \sqrt[4]{3x^2}$

$Y1 = \sqrt[3]{2x+1}$

$Y2 = \sqrt[4]{3x^2}$

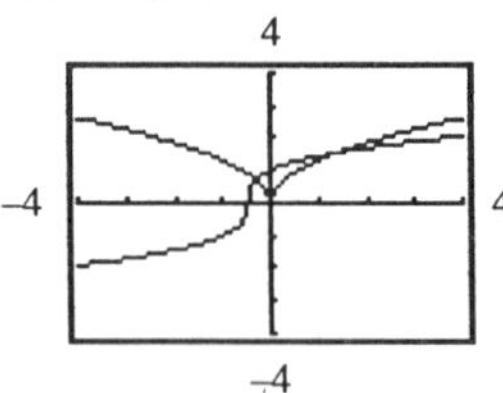

The approximate solutions are −0.307 and 1.412.

5. $x^{3/2} + 2x^{1/2} - 7 = 0$

$Y1 = x^{3/2} + 2x^{1/2} - 7$

$Y2 = 0$

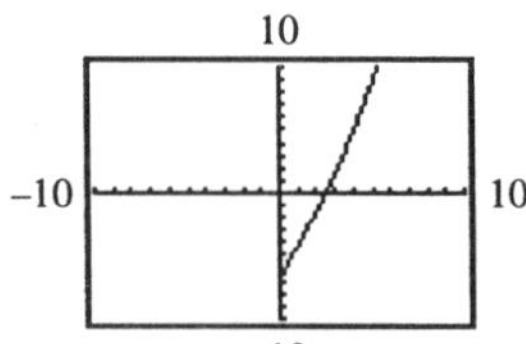

The approximate solution is 2.462.

6. $3x^{2/3} - 5x^{1/3} - 9 = 0$

$Y1 = 3x^{2/3} - 5x^{1/3} - 9$

$Y2 = 0$

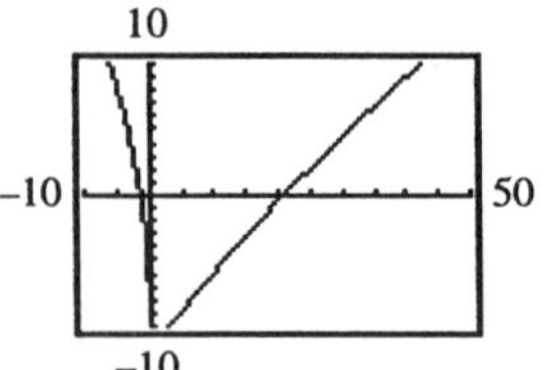

The approximate solutions are 20.920 and −1.291.

Chapter 13 Review

Reflections

1.–7. Answers will vary.

Exercises

1. $\sqrt{225}$

$= \sqrt{(15)^2}$

$= 15$

2. $\sqrt{2.89}$

$= \sqrt{(1.7)^2}$

$= 1.7$

3. $\sqrt{\dfrac{49}{64}}$

$= \sqrt{\dfrac{7^2}{8^2}}$

$= \dfrac{7}{8}$

4. $\sqrt[3]{-\dfrac{27}{125}}$

$= \sqrt[3]{\left(-\dfrac{3}{5}\right)^3}$

$= -\dfrac{3}{5}$

5. $-\sqrt[4]{1296}$

$= -\sqrt[4]{6^4}$

$= -6$

6. $\sqrt[4]{-1296}$ is not a real number.

7. $\sqrt{150}$
≈ 12.247

8. $\sqrt[3]{-35}$
≈ -3.271

9. $\sqrt[5]{125}$
≈ 2.627

10. $(-64)^{1/3}$
$= \sqrt[3]{-64}$
$= -4$

11. $-16^{1/4}$
$= -\sqrt[4]{16}$
$= -2$

12. $(-16)^{1/4}$
$= \sqrt[4]{-16}$ is not a real number.

13. $-64^{-1/3}$
$= -\dfrac{1}{64^{1/3}}$
$= -\dfrac{1}{\sqrt[3]{64}}$
$= -\dfrac{1}{4}$

14. $-64^{4/3}$
$= -\left(\sqrt[3]{64}\right)^4$
$= -4^4$
$= -256$

15. $(-64)^{3/2}$
$= \left(\sqrt{-64}\right)^3$ is not a real number.

16. $16^{-3/4}$
$= \dfrac{1}{16^{3/4}}$
$= \dfrac{1}{\left(\sqrt[4]{16}\right)^3}$
$= \dfrac{1}{2^3}$
$= \dfrac{1}{8}$

17. $(-16)^{-3/4}$
$= \dfrac{1}{(-16)^{3/4}}$
$= \dfrac{1}{\left(\sqrt[4]{-16}\right)^3}$ is not a real number.

18. $s = \sqrt{A}$
$s = \sqrt{121}$
$s = 11$
The side is 11 inches.

19. $s = \sqrt[3]{V}$
$s = \sqrt[3]{\dfrac{125}{216}}$
$s = \dfrac{5}{6}$
The side is $\dfrac{5}{6}$ yard.

20. $V = \dfrac{4}{3}\pi r^3$
$r = \sqrt[3]{\dfrac{3V}{4\pi}}$
$r = \sqrt[3]{\dfrac{3(65.45)}{4\pi}}$
$r \approx 2.5$
$d = 2r \approx 5$
The approximate radius is 2.5 cm and the diameter is 5 cm.

21. $y = \sqrt{2x - 7}$

$2x - 7 < 0$
$2x < 7$
$x < \frac{7}{2}$

The restricted values are all real numbers less than $\frac{7}{2}$.

domain: $\left[\frac{7}{2}, \infty\right)$

22. $y = \sqrt[3]{2x - 7}$

There are no restricted values because the index of the radical is odd.
domain: $(-\infty, \infty)$

23. $y = (4x + 9)^{3/4}$

$y = \left(\sqrt[4]{4x + 9}\right)^3$
$4x + 9 < 0$
$4x < -9$
$x < -\frac{9}{4}$

The restricted values are all real numbers less than $-\frac{9}{4}$.

domain: $\left[-\frac{9}{4}, \infty\right)$

24. $y = 2\sqrt{x}$

domain: $[0, \infty)$
Plot and connect ordered pairs that satisfy the equation.

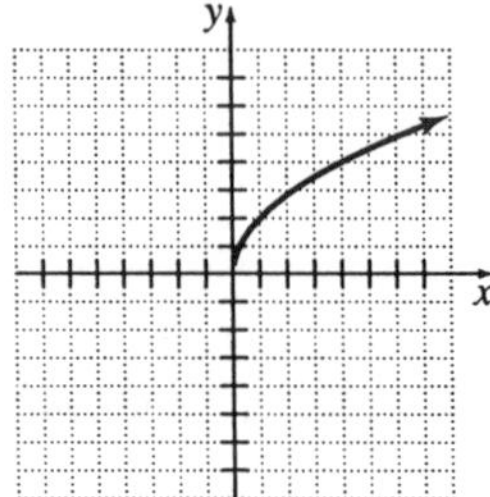

25. $y = x^{3/2} + 1$

domain: $[0, \infty)$
Plot and connect ordered pairs that satisfy the equation.

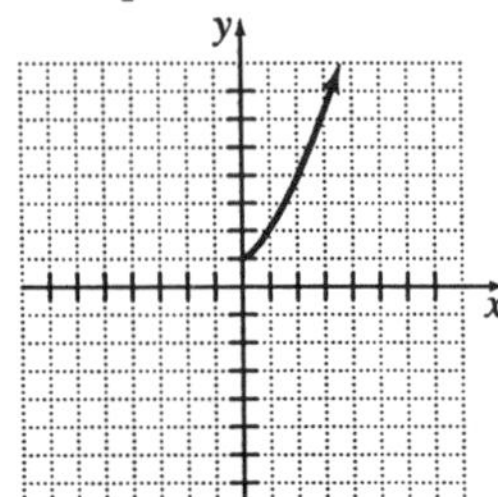

26. $s = \sqrt{\frac{n(n+1)}{12}}$

$s = \sqrt{\frac{23(23+1)}{12}}$
$s = \sqrt{46}$
$s \approx 6.78$

The variability is about 6.78.

27. $V = \sqrt{9S}$

$Y1 = \sqrt{9x}$

50

0 200

0

At $x = 150$, $y \approx 36.7$.
The approximate speed is 36.7 mph.

28. $\frac{1}{4^{-1/2}}$

$= 4^{1/2}$
$= \sqrt{4}$
$= 2$

29. $8^{2/3} \cdot 8^{5/3}$

$= 8^{\frac{2}{3}+\frac{5}{3}}$
$= 8^{7/3}$
$= \left(\sqrt[3]{8}\right)^7$
$= 2^7$
$= 128$

30. $\dfrac{9^{7/3}}{9^{2/3}}$

$=9^{\frac{7}{3}-\frac{2}{3}}$

$=9^{5/3}$

$=\left(\sqrt[3]{9}\right)^5$

≈ 38.9

31. $\left(\dfrac{4}{9}\right)^{-3/2}$

$=\left(\dfrac{9}{4}\right)^{3/2}$

$=\left(\sqrt{\dfrac{9}{4}}\right)^3$

$=\left(\dfrac{3}{2}\right)^3$

$=\dfrac{27}{8}$

32. $[(3^{1/2})(27^{1/2})]^{-3/2}$

$=[(3\cdot 27)^{1/2}]^{-3/2}$

$=[(81)^{1/2}]^{-3/2}$

$=(9)^{-3/2}$

$=\dfrac{1}{9^{3/2}}$

$=\dfrac{1}{\left(\sqrt{9}\right)^3}$

$=\dfrac{1}{3^3}$

$=\dfrac{1}{27}$

33. $(64^{2/3})^{3/4}$

$=\left[\left(\sqrt[3]{64}\right)^2\right]^{3/4}$

$=[(4)^2]^{3/4}$

$=(16)^{3/4}$

$=\left(\sqrt[4]{16}\right)^3$

$=2^3$

$=8$

34. $\dfrac{x^{2/3}}{x^{5/6}}$

$=x^{\frac{2}{3}-\frac{5}{6}}$

$=x^{\frac{4}{6}-\frac{5}{6}}$

$=x^{-1/6}$

$=\dfrac{1}{x^{1/6}}$

35. $y^{2/5}\cdot y^{-3/10}$

$=y^{\frac{2}{5}-\frac{3}{10}}$

$=y^{\frac{4}{10}-\frac{3}{10}}$

$=y^{1/10}$

36. $(z^{3/5})^{5/9}$

$=z^{\frac{3}{5}\cdot\frac{5}{9}}$

$=z^{1/3}$

37. $\left(\dfrac{a^3}{b^6}\right)^{-5/12}$

$=\left(\dfrac{b^6}{a^3}\right)^{5/12}$

$=\dfrac{b^{6\cdot 5/12}}{a^{3\cdot 5/12}}$

$=\dfrac{b^{5/2}}{a^{5/4}}$

38. $(8x^6y^9)^{4/3}$
$= 8^{4/3}x^{6\cdot 4/3}y^{9\cdot 4/3}$
$= \left(\sqrt[3]{8}\right)^4 x^8y^{12}$
$= 16x^8y^{12}$

39. $(2a^{3/4}b^{1/3})(3a^{1/3}b^{2/3})$
$= 2\cdot 3a^{\frac{3}{4}+\frac{1}{3}}b^{\frac{1}{3}+\frac{2}{3}}$
$= 6a^{13/12}b$

40. Let s_1 = length of the original side.
Let s_2 = length of the new side.
$s_1 = \sqrt{A_1}$
$s_2 = \sqrt{5A_1}$
$\frac{s_2}{s_1} = \frac{\sqrt{5A_1}}{\sqrt{A_1}} = \sqrt{5}$
The new side should be $\sqrt{5}$ times larger than the old side.

41. $\sqrt{6300}$
$= \sqrt{900\cdot 7}$
$= \sqrt{(30)^2\cdot 7}$
$= 30\sqrt{7}$

42. $\sqrt[3]{-320}$
$= \sqrt[3]{(-64)\cdot 5}$
$= \sqrt[3]{(-4)^3\cdot 5}$
$= -4\sqrt[3]{5}$

43. $\sqrt[4]{162}$
$= \sqrt[4]{81\cdot 2}$
$= \sqrt[4]{(3)^4\cdot 2}$
$= 3\sqrt[4]{2}$

44. $\sqrt{45x^4y^7z^2}$
$= \sqrt{9\cdot 5x^4y^6\cdot y\cdot z^2}$
$= 3x^2y^3z\sqrt{5y}$

45. $\sqrt[3]{-64x^2y^7}$
$= \sqrt[3]{-64x^2y^6\cdot y}$
$= -4y^2\sqrt[3]{x^2y}$

46. $\sqrt{12x^2-36x+27}$
$= \sqrt{3(4x^2-12x+9)}$
$= \sqrt{3(2x-3)^2}$
$= (2x-3)\sqrt{3}$

47. $\sqrt{2x}\cdot\sqrt{8x^3}$
$= \sqrt{(2x)(8x^3)}$
$= \sqrt{16x^4}$
$= 4x^2$

48. $\sqrt[3]{-2a^2b}\cdot\sqrt[3]{20a^2b^5}$
$= \sqrt[3]{(-2)(20)a^{2+2}b^{1+5}}$
$= \sqrt[3]{-40a^4b^6}$
$= -2ab^2\sqrt[3]{5a}$

49. $\sqrt{x^2-2xy}\cdot\sqrt{3x-6y}$
$= \sqrt{(x^2-2xy)(3x-6y)}$
$= \sqrt{3x^3-6x^2y-6x^2y+12xy^2}$
$= \sqrt{3x^3-12x^2y+12xy^2}$
$= \sqrt{3x(x^2-4xy+4y^2)}$
$= \sqrt{3x(x-2y)^2}$
$= (x-2y)\sqrt{3x}$

50. $\sqrt{\frac{25}{64}}$
$= \frac{\sqrt{25}}{\sqrt{64}}$
$= \frac{5}{8}$

51. $\sqrt{\frac{13}{289}}$
$= \frac{\sqrt{13}}{\sqrt{289}}$
$= \frac{\sqrt{13}}{17}$

52. $\sqrt{\frac{27}{343}}$
$= \frac{\sqrt{27}}{\sqrt{343}}$
$= \frac{3\sqrt{3}}{7\sqrt{7}}$
$= \frac{3\sqrt{3}}{7\sqrt{7}} \cdot \frac{\sqrt{7}}{\sqrt{7}}$
$= \frac{3\sqrt{21}}{7 \cdot 7}$
$= \frac{3\sqrt{21}}{49}$

53. $\sqrt[3]{\frac{6}{25}}$
$= \frac{\sqrt[3]{6}}{\sqrt[3]{25}}$
$= \frac{\sqrt[3]{6}}{\sqrt[3]{25}} \cdot \frac{\sqrt[3]{5}}{\sqrt[3]{5}}$
$= \frac{\sqrt[3]{30}}{5}$

54. $\frac{\sqrt{50}}{\sqrt{60}}$
$= \frac{5\sqrt{2}}{2\sqrt{15}}$
$= \frac{5\sqrt{2}}{2\sqrt{15}} \cdot \frac{\sqrt{15}}{\sqrt{15}}$
$= \frac{5\sqrt{30}}{2 \cdot 15}$
$= \frac{\sqrt{30}}{6}$

55. $\sqrt{\frac{25a^2}{64b^4}}$
$= \frac{\sqrt{25a^2}}{\sqrt{64b^4}}$
$= \frac{5a}{8b^2}$

56. $\sqrt{\frac{16m}{5}}$
$= \frac{\sqrt{16m}}{\sqrt{5}}$
$= \frac{4\sqrt{m}}{\sqrt{5}} \cdot \frac{\sqrt{5}}{\sqrt{5}}$
$= \frac{4\sqrt{5m}}{5}$

57. $\sqrt[3]{\frac{3z^3}{4x^2y}}$
$= \frac{\sqrt[3]{3z^3}}{\sqrt[3]{4x^2y}}$
$= \frac{z\sqrt[3]{3}}{\sqrt[3]{4x^2y}} \cdot \frac{\sqrt[3]{2xy^2}}{\sqrt[3]{2xy^2}}$
$= \frac{z\sqrt[3]{6xy^2}}{2xy}$

58. $\frac{4a\sqrt{5ab^2}}{12\sqrt{10ab^3}}$
$= \left(\frac{4a}{12}\right)\sqrt{\frac{5ab^2}{10ab^3}}$
$= \frac{a}{3}\sqrt{\frac{1}{2b}}$
$= \frac{a}{3\sqrt{2b}} \cdot \frac{\sqrt{2b}}{\sqrt{2b}}$
$= \frac{a\sqrt{2b}}{6b}$

59. $T = 2\pi\sqrt{\frac{L}{32}}$
$T = 2\pi\sqrt{\frac{3}{32}}$
$T \approx 1.9$
The period is approximately 1.9 seconds.

60. $3\sqrt{5} - 2\sqrt{5} + 7\sqrt{5} - \sqrt{5}$
$= \sqrt{5}(3 - 2 + 7 - 1)$
$= \sqrt{5}(7)$
$= 7\sqrt{5}$

61. $7\sqrt{11} + 2\sqrt{44}$
$= 7\sqrt{11} + 4\sqrt{11}$
$= \sqrt{11}(7 + 4)$
$= 11\sqrt{11}$

62. $\sqrt{15} + \sqrt{\frac{1}{15}}$
$= \sqrt{15} + \frac{1}{\sqrt{15}} \cdot \frac{\sqrt{15}}{\sqrt{15}}$
$= \sqrt{15} + \frac{\sqrt{15}}{15}$
$= \sqrt{15}\left(1 + \frac{1}{15}\right)$
$= \frac{16}{15}\sqrt{15}$

63. $5\sqrt[3]{16} - \sqrt[3]{54}$
$= 5 \cdot 2\sqrt[3]{2} - 3\sqrt[3]{2}$
$= \sqrt[3]{2}(10 - 3)$
$= 7\sqrt[3]{2}$

64. $\sqrt{49x} - \sqrt{25x}$
$= 7\sqrt{x} - 5\sqrt{x}$
$= \sqrt{x}(7 - 5)$
$= 2\sqrt{x}$

65. $4b\sqrt{a^3b} - 7a\sqrt{ab^3} + 8ab\sqrt{ab}$
$= 4ab\sqrt{ab} - 7ab\sqrt{ab} + 8ab\sqrt{ab}$
$= \sqrt{ab}(4ab - 7ab + 8ab)$
$= 5ab\sqrt{ab}$

66. $\sqrt{7}\left(\sqrt{14} - \sqrt{7}\right)$
$= \sqrt{7} \cdot \sqrt{14} - \left(\sqrt{7}\right)^2$
$= \sqrt{98} - 7$
$= 7\sqrt{2} - 7$

67. $\sqrt{2a}\left(6 + \sqrt{2a}\right)$
$= \sqrt{2a}(6) + \left(\sqrt{2a}\right)^2$
$= 6\sqrt{2a} + 2a$

68. $\left(2 - \sqrt{5}\right)\left(4 + \sqrt{5}\right)$
$= 2 \cdot 4 + 2\sqrt{5} - 4\sqrt{5} - \left(\sqrt{5}\right)^2$
$= 8 - 2\sqrt{5} - 5$
$= 3 - 2\sqrt{5}$

69. $\left(\sqrt{5x} + \sqrt{7y}\right)\left(\sqrt{5x} - \sqrt{7y}\right)$
$= \left(\sqrt{5x}\right)^2 - \left(\sqrt{7y}\right)^2$
$= 5x - 7y$

70. $\left(\sqrt{x} + 8\right)^2$
$= \left(\sqrt{x}\right)^2 + 2 \cdot 8\sqrt{x} + 8^2$
$= x + 16\sqrt{x} + 64$

71. $3\sqrt[3]{x}\left(2\sqrt[3]{x^2} + \sqrt[3]{x}\right)$
$= 6\sqrt[3]{x^3} + 3\sqrt[3]{x^2}$
$= 6x + 3\sqrt[3]{x^2}$

72. $\dfrac{\sqrt{15x}-\sqrt{30}}{\sqrt{5}}$

$=\dfrac{\sqrt{15x}}{\sqrt{5}}-\dfrac{\sqrt{30}}{\sqrt{5}}$

$=\sqrt{\dfrac{15x}{5}}-\sqrt{\dfrac{30}{5}}$

$=\sqrt{3x}-\sqrt{6}$

73. $\dfrac{\sqrt{z}-5}{\sqrt{z}}$

$=\dfrac{\sqrt{z}}{\sqrt{z}}-\dfrac{5}{\sqrt{z}}$

$=\sqrt{\dfrac{z}{z}}-\dfrac{5}{\sqrt{z}}\cdot\dfrac{\sqrt{z}}{\sqrt{z}}$

$=1-\dfrac{5\sqrt{z}}{z}$

74. $\dfrac{24}{\sqrt{5}+2}$

$=\dfrac{24}{\sqrt{5}+2}\cdot\dfrac{\sqrt{5}-2}{\sqrt{5}-2}$

$=\dfrac{24\left(\sqrt{5}-2\right)}{\left(\sqrt{5}\right)^2-2^2}$

$=\dfrac{24\sqrt{5}-48}{5-4}$

$=24\sqrt{5}-48$

75. $\dfrac{\sqrt{x}+2}{\sqrt{x}-2}$

$=\dfrac{\sqrt{x}+2}{\sqrt{x}-2}\cdot\dfrac{\sqrt{x}+2}{\sqrt{x}+2}$

$=\dfrac{\left(\sqrt{x}\right)^2+2\cdot 2\sqrt{x}+2^2}{\left(\sqrt{x}\right)^2-2^2}$

$=\dfrac{x+4\sqrt{x}+4}{x-4}$

76. $\dfrac{2\sqrt{x}-5}{5-2\sqrt{x}}$

$=\dfrac{-1\left(5-2\sqrt{x}\right)}{5-2\sqrt{x}}$

$=-1$

77. $\dfrac{2x-9}{\sqrt{2x}-3}$

$=\dfrac{2x-9}{\sqrt{2x}-3}\cdot\dfrac{\sqrt{2x}+3}{\sqrt{2x}+3}$

$=\dfrac{(2x-9)\left(\sqrt{2x}+3\right)}{\left(\sqrt{2x}\right)^2-3^2}$

$=\sqrt{2x}+3$

78. Let s_1 = length of the side of the first mirror

$s_1=\sqrt{A_1}$

$s_1=\sqrt{720}$

$s_1=12\sqrt{5}$

Let s_2 = length of the side of the second mirror.

$s_2=\sqrt{A_2}$

$s_2=\sqrt{1620}$

$s_2=18\sqrt{5}$

$4s_2-4s_1=4\left(18\sqrt{5}\right)-4\left(12\sqrt{5}\right)$

$=72\sqrt{5}-48\sqrt{5}$

$=\sqrt{5}(72-48)$

$=24\sqrt{5}$

It will require $24\sqrt{5}$ inches more material.

79. $\sqrt{3x-5}=x-3$

$Y1=\sqrt{3x-5}$

$Y2=x-3$

Using the table feature on the calculator, the solution is 7.

80. $\sqrt{6x}-1=(6x)^{1/2}$

$Y1=\sqrt{6x}-1$

$Y2=(6x)^{1/2}$

Using the table feature on the calculator, the first expression is always less than the second expression. There is no solution.

81. $\sqrt{x^2+10x+25}=x+5$

$Y1=\sqrt{x^2+10x+25}$

$Y2=x+5$

Using the table feature on the calculator, the expressions are always equal. The solutions are all real numbers greater than or equal to −5.

82. $\sqrt{2x-7}=2x-7$

$Y1=\sqrt{2x-7}$

$Y2=2x-7$

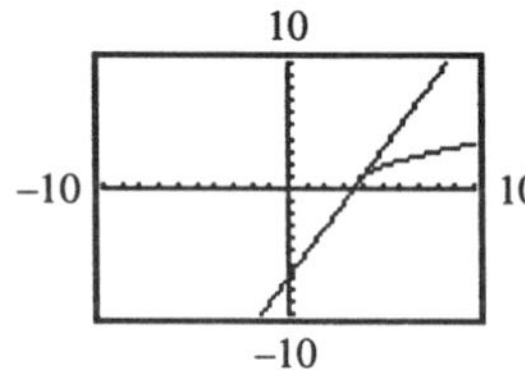

The solutions are the x-coordinates of the intersections, $\frac{7}{2}$ and 4.

83. $\sqrt{x^2+2x+1}=x+1$

$Y1=\sqrt{x^2+2x+1}$

$Y2=x+1$

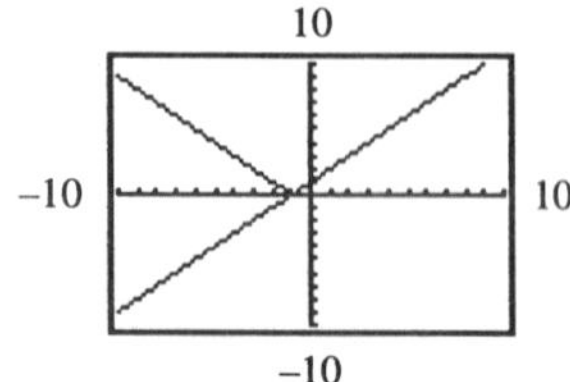

The graphs are the same for $x\ge -1$. The solutions are all real numbers greater than or equal to −1.

84. $(x+4)^{3/2}=\sqrt[3]{2x+3}$

$Y1=(x+4)^{3/2}$

$Y2=\sqrt[3]{2x+3}$

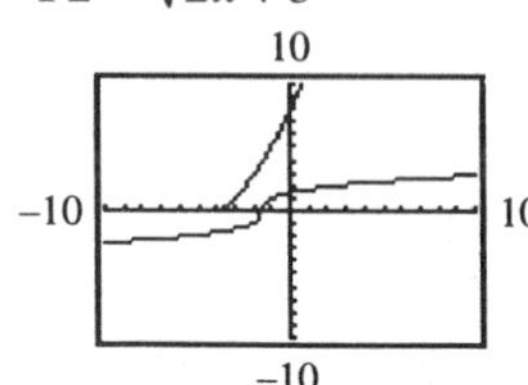

The graphs do not appear to intersect. There is no solution.

85. $(x+5)^{3/5}=(4x-1)^{2/3}$

$Y1=(x+5)^{3/5}$

$Y2=(4x-1)^{2/3}$

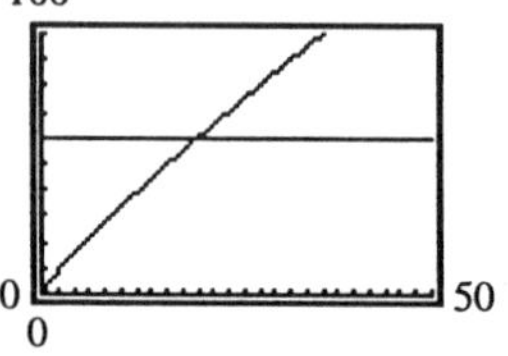

The approximate solutions are the x-coordinates of the intersections, −0.683 and 1.62.

86. $2L+2W=P$

$2\sqrt{7x+4}+2(x-2)=60$

$Y1=2\sqrt{7x+4}+2(x-2)$

$Y2=60$

100

0 50

0

The solution is 20.

$\sqrt{7(20)+4}=12$ $\quad$ $20-2=18$

The width is 12 in. and the length is 18 in.

87. $t=\sqrt{\frac{2d}{32.2}}$

$6=\sqrt{\frac{2d}{32.2}}$

$Y1=6$

$Y2=\sqrt{\frac{2x}{32.2}}$

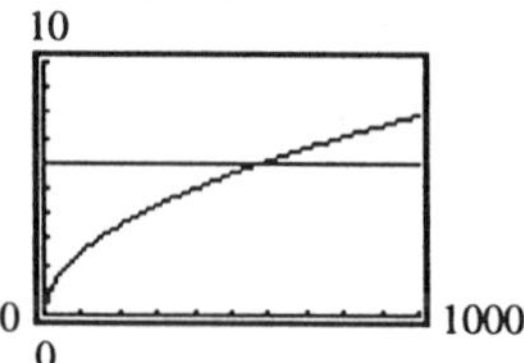

The solution is 579.6. It fell from a height of 579.6 feet.

88. $2\sqrt{3x}+3=15$
$2\sqrt{3x}=12$
$\sqrt{3x}=6$
$\left(\sqrt{3x}\right)^2=6^2$
$3x=36$
$x=12$
The solution is 12.

89. $\sqrt{4x-3}=x-2$
$\left(\sqrt{4x-3}\right)^2=(x-2)^2$
$4x-3=x^2-4x+4$
$0=x^2-8x+7$
$0=(x-7)(x-1)$
$x=7$ $\quad$ $x=1$ (extraneous)
The solution is 7.

90. $2\sqrt{x+5}=\sqrt{5x+9}$
$\left(2\sqrt{x+5}\right)^2=\left(\sqrt{5x+9}\right)^2$
$4(x+5)=5x+9$
$4x+20=5x+9$
$11=x$
The solution is 11.

91. $\sqrt{x+6}+3=\sqrt{5x-1}$
$\left(\sqrt{x+6}+3\right)^2-\left(\sqrt{5x-1}\right)^2$
$x+6+6\sqrt{x+6}+9=5x-1$
$6\sqrt{x+6}=4x-16$
$\left(6\sqrt{x+6}\right)^2=(4x-16)^2$
$36(x+6)=16x^2-128x+256$
$0=16x^2-164x+40$
$0=4(4x^2-41x+10)$
$0=4(4x-1)(x-10)$
$x=\frac{1}{4}$ (extraneous) $\quad$ $x=10$
The solution is 10.

92. $\sqrt{x+8}=\sqrt{x-2}$
$\left(\sqrt{x+8}\right)^2=\left(\sqrt{x-2}\right)^2$
$x+8=x-2$
$8=-2$
This is a contradiction.
There is no solution.

93. $\sqrt[3]{2x}=-4$
$\left(\sqrt[3]{2x}\right)^3=(-4)^3$
$2x=-64$
$x=-32$
The solution is -32.

94. $\sqrt[4]{x+3}=3$
$\left(\sqrt[4]{x+3}\right)^4=3^4$
$x+3=81$
$x=78$
The solution is 78.

95. $(x+7)^{2/5}=4$
$[(x+7)^{2/5}]^5=4^5$
$(x+7)^2=1024$
$\sqrt{(x+7)^2}=\pm\sqrt{1024}$
$x+7=\pm32$
$x=-7\pm32$
$x=-39$ $\quad$ $x=25$
The solutions are -39 and 25.

96. $x^{-3/2}=8$
$x^{3/2}=\frac{1}{8}$
$(x^{3/2})^2=\left(\frac{1}{8}\right)^2$
$x^3=\frac{1}{64}$
$(x^3)^{1/3}=\left(\frac{1}{64}\right)^{1/3}$
$x=\frac{1}{4}$
The solution is $\frac{1}{4}$.

97. $2x^{2/3} - 5 = 1$
$2x^{2/3} = 6$
$x^{2/3} = 3$
$(x^{2/3})^3 = 3^3$
$x^2 = 27$
$\sqrt{x^2} = \pm\sqrt{27}$
$x = \pm 3\sqrt{3}$
The solutions are $\pm 3\sqrt{3}$.

98. $d = \sqrt{(x_2 - x_1)^2 + (y_2 - y_1)^2}$
$d = \sqrt{(4-4)^2 + (7+3)^2}$
$d = \sqrt{10^2}$
$d = 10$
The distance is 10 units.

99. $d = \sqrt{(x_2 - x_1)^2 + (y_2 - y_1)^2}$
$d = \sqrt{(7+2)^2 + (-4+4)^2}$
$d = \sqrt{9^2}$
$d = 9$
The distance is 9 units.

100. $d = \sqrt{(x_2 - x_1)^2 + (y_2 - y_1)^2}$
$d = \sqrt{(9-6)^2 + (6-2)^2}$
$d = \sqrt{3^2 + 4^2}$
$d = \sqrt{9+16}$
$d = \sqrt{25}$
$d = 5$
The distance is 5 units.

101. $d = \sqrt{(x_2 - x_1)^2 + (y_2 - y_1)^2}$
$d = \sqrt{(4+3)^2 + (8+1)^2}$
$d = \sqrt{7^2 + 9^2}$
$d = \sqrt{63+81}$
$d = \sqrt{144}$
$d = 12$
The distance is 12 units.

102. $d = \sqrt{(x_2 - x_1)^2 + (y_2 - y_1)^2}$
$5 = \sqrt{(6-3)^2 + (y+2)^2}$
$5 = \sqrt{3^2 + y^2 + 4y + 4}$
$(5)^2 = \left(\sqrt{y^2 + 4y + 13}\right)^2$
$25 = y^2 + 4y + 13$
$0 = y^2 + 4y - 12$
$0 = (y+6)(y-2)$
$y = -6 \quad y = 2$
The y-coordinates are –6 and 2.
The two possible points are (6, –6) and (6, 2).

103. $V = \sqrt{10.5S}$
$40 = \sqrt{10.5S}$
$(40)^2 = \left(\sqrt{10.5S}\right)^2$
$1600 = 10.5S$
$\frac{1600}{10.5} = \frac{10.5S}{10.5}$
$152.4 \approx S$
The skid mark would be approximately 152.4 feet long.

Chapter 13 Mixed Review

1. $\sqrt{196}$
$= \sqrt{(14)^2}$
$= 14$

2. $\sqrt{4.41}$
$= \sqrt{(2.1)^2}$
$= 2.1$

3. $\sqrt[3]{-\frac{27}{64}}$
$= \sqrt[3]{\left(-\frac{3}{4}\right)^3}$
$= -\frac{3}{4}$

4. $\sqrt[4]{256}$
$= \sqrt[4]{4^4}$
$= 4$

5. $\sqrt[4]{-256}$ is not a real number.

6. $\sqrt[3]{-29}$
≈ -3.072

7. $121^{1/2}$
$= \sqrt{121}$
$= 11$

8. $(-125)^{1/3}$
$= \sqrt[3]{-125}$
$= -5$

9. $-81^{1/4}$
$= -\sqrt[4]{81}$
$= -3$

10. $(-81)^{1/4}$
$= \sqrt[4]{-81}$ is not a real number.

11. $729^{5/6}$
$= \left(\sqrt[6]{729}\right)^5$
$= 3^5$
$= 243$

12. $(-729)^{5/6}$
$= \left(\sqrt[6]{-729}\right)^5$ is not a real number.

13. $(-8)^{-2/3}$
$= \dfrac{1}{-8^{2/3}}$
$= \dfrac{1}{\left(\sqrt[3]{-8}\right)^2}$
$= \dfrac{1}{(-2)^2}$
$= \dfrac{1}{4}$

14. $(-81)^{-3/4}$
$= \dfrac{1}{(-81)^{3/4}}$
$= \dfrac{1}{\left(\sqrt[4]{-81}\right)^3}$ is not a real number.

15. $\dfrac{1}{9^{-1/2}}$
$= 9^{1/2}$
$= \sqrt{9}$
$= 3$

16. $27^{2/3} \cdot 27^{5/3}$
$= 27^{\frac{2}{3}+\frac{5}{3}}$
$= 27^{7/3}$
$= \left(\sqrt[3]{27}\right)^7$
$= 3^7$
$= 2187$

17. $\dfrac{6^{7/3}}{6^{2/3}}$
$= 6^{\frac{7}{3}-\frac{2}{3}}$
$= 6^{5/3}$
$= \left(\sqrt[3]{6}\right)^5$
≈ 19.812

18. $\left(36^{2/3}\right)^{3/4}$
$= 36^{\frac{2}{3}\cdot\frac{3}{4}}$
$= 36^{1/2}$
$= \sqrt{36}$
$= 6$

19. $\left(\dfrac{9}{16}\right)^{-3/2}$
$= \left(\dfrac{16}{9}\right)^{3/2}$
$= \left(\sqrt{\dfrac{16}{9}}\right)^3$

$$= \left(\frac{4}{3}\right)^3$$
$$= \frac{64}{27}$$

20. $\left[\left(3^{1/2}\right)\left(12^{1/2}\right)\right]^{-3/2}$
$$= \left[(3 \cdot 12)^{1/2}\right]^{-3/2}$$
$$= \left[(36)^{1/2}\right]^{-3/2}$$
$$= 6^{-3/2}$$
$$= \frac{1}{6^{3/2}}$$
$$= \frac{1}{\left(\sqrt{6}\right)^3}$$
$$\approx 0.068$$

21. $-\sqrt{\frac{72}{98}}$
$$= -\frac{\sqrt{72}}{\sqrt{98}}$$
$$= -\frac{6\sqrt{2}}{7\sqrt{2}}$$
$$= -\frac{6}{7}$$

22. $\sqrt{\frac{15}{144}}$
$$= \frac{\sqrt{15}}{\sqrt{144}}$$
$$= \frac{\sqrt{15}}{12}$$

23. $\sqrt{\frac{50}{338}}$
$$= \sqrt{\frac{25}{169}}$$
$$= \frac{\sqrt{25}}{\sqrt{169}}$$
$$= \frac{5}{13}$$

24. $\sqrt[3]{-\frac{64}{125}}$
$$= \sqrt[3]{\left(-\frac{4}{5}\right)^3}$$
$$= -\frac{4}{5}$$

25. $\sqrt{150}$
$$= \sqrt{25 \cdot 6}$$
$$= 5\sqrt{6}$$

26. $\sqrt[3]{-448}$
$$= \sqrt[3]{(-64)(7)}$$
$$= -4\sqrt[3]{7}$$

27. $\sqrt[4]{48}$
$$= \sqrt[4]{16 \cdot 3}$$
$$= 2\sqrt[4]{3}$$

28. $\sqrt[5]{-486}$
$$= \sqrt[5]{(-243)(2)}$$
$$= -3\sqrt[5]{2}$$

29. $\sqrt{15} \cdot \sqrt{35}$
$$= \sqrt{(15)(35)}$$
$$= \sqrt{525}$$
$$= \sqrt{25 \cdot 21}$$
$$= 5\sqrt{21}$$

30. $\sqrt{\frac{3}{7}} - \sqrt{\frac{7}{3}}$
$$= \frac{\sqrt{3}}{\sqrt{7}} - \frac{\sqrt{7}}{\sqrt{3}}$$
$$= \frac{\sqrt{3}}{\sqrt{7}} \cdot \frac{\sqrt{7}}{\sqrt{7}} - \frac{\sqrt{7}}{\sqrt{3}} \cdot \frac{\sqrt{3}}{\sqrt{3}}$$
$$= \frac{\sqrt{21}}{7} - \frac{\sqrt{21}}{3}$$
$$= \sqrt{21}\left(\frac{1}{7} - \frac{1}{3}\right)$$

$=\sqrt{21}\left(\frac{3}{21}-\frac{7}{21}\right)$
$=-\frac{4}{21}\sqrt{21}$

31. $6\sqrt{13}+3\sqrt{52}$
$=6\sqrt{13}+6\sqrt{13}$
$=\sqrt{13}(6+6)$
$=12\sqrt{13}$

32. $4\sqrt{7}-3\sqrt{7}+11\sqrt{7}-\sqrt{7}$
$=\sqrt{7}(4-3+11-1)$
$=11\sqrt{7}$

33. $\sqrt{3}\left(\sqrt{6}-\sqrt{3}\right)$
$=\sqrt{18}-\left(\sqrt{3}\right)^2$
$=3\sqrt{2}-3$

34. $\left(7-\sqrt{3}\right)\left(9+\sqrt{3}\right)$
$=63+7\sqrt{3}-9\sqrt{3}-(\sqrt{3})^2$
$=63-2\sqrt{3}-3$
$=60-2\sqrt{3}$

35. $\left(\sqrt{10}-\sqrt{5}\right)\left(\sqrt{10}+\sqrt{5}\right)$
$=\left(\sqrt{10}\right)^2-\left(\sqrt{5}\right)^2$
$=10-5$
$=5$

36. $6\sqrt[3]{375}+2\sqrt[3]{24}$
$=6\sqrt[3]{125\cdot 3}+2\sqrt[3]{8\cdot 3}$
$=30\sqrt[3]{3}+4\sqrt[3]{3}$
$=\sqrt[3]{3}(30+4)$
$=34\sqrt[3]{3}$

37. $\frac{x^{2/3}}{x^{3/4}}$
$=x^{\frac{2}{3}-\frac{3}{4}}$
$=x^{\frac{8}{12}-\frac{9}{12}}$
$=x^{-1/12}$
$=\frac{1}{x^{1/12}}$

38. $z^{3/5}\cdot z^{-3/2}$
$=z^{\frac{3}{5}-\frac{3}{2}}$
$=z^{\frac{6}{10}-\frac{15}{10}}$
$=z^{-9/10}$
$=\frac{1}{z^{9/10}}$

39. $\left(\frac{p^4}{q^8}\right)^{-5/16}$
$=\left(\frac{q^8}{p^4}\right)^{5/16}$
$=\frac{q^{8\cdot 5/16}}{p^{4\cdot 5/16}}$
$=\frac{q^{5/2}}{p^{5/4}}$

40. $\left(27x^9y^{12}\right)^{4/3}$
$=27^{4/3}x^{9\cdot 4/3}y^{12\cdot 4/3}$
$=81x^{12}y^{16}$

41. $\sqrt{72x^6y^3z^4}$
$=\sqrt{36\cdot 2x^6y^2\cdot yz^4}$
$=6x^3yz^2\sqrt{2y}$

42. $\sqrt[3]{-64x^2y^7}$
$=\sqrt[3]{-64x^2y^6\cdot y}$
$=-4y^2\sqrt[3]{x^2y}$

43. $2\sqrt{3xy}\cdot y\sqrt{6x}$
$=2y\sqrt{(3xy)(6x)}$
$=2y\sqrt{18x^2y}$
$=6xy\sqrt{2y}$

44. $\sqrt[3]{-3a^5b^5}\cdot\sqrt[3]{18ab^2}$
$=\sqrt[3]{\left(-3a^5b^5\right)\left(18ab^2\right)}$
$=\sqrt[3]{-54a^6b^7}$
$=-3a^2b^2\sqrt[3]{2b}$

45. $\sqrt{10x+35y}\cdot\sqrt{2x+7y}$
$=\sqrt{(10x+35y)(2x+7y)}$
$=\sqrt{20x^2+140xy+245y^2}$
$\sqrt{5\left(4x^2+28xy+49y^2\right)}$
$=\sqrt{5(2x+7y)^2}$
$=(2x+7y)\sqrt{5}$

46. $\sqrt{\dfrac{36x^2}{49y^4}}$
$=\dfrac{\sqrt{36x^2}}{\sqrt{49y^4}}$
$=\dfrac{6x}{7y^2}$

47. $\sqrt{\dfrac{25z}{3}}$
$=\dfrac{\sqrt{25z}}{\sqrt{3}}$
$=\dfrac{5\sqrt{z}}{\sqrt{3}}\cdot\dfrac{\sqrt{3}}{\sqrt{3}}$
$=\dfrac{5\sqrt{3z}}{3}$

48. $\sqrt[3]{\dfrac{-27a}{b^3}}$
$=\dfrac{\sqrt[3]{-27a}}{\sqrt[3]{b^3}}$
$=\dfrac{-3\sqrt[3]{a}}{b}$

49. $\dfrac{\sqrt{6a}}{\sqrt{12ab}}$
$=\sqrt{\dfrac{6a}{12ab}}$
$=\sqrt{\dfrac{1}{2b}}$
$=\dfrac{1}{\sqrt{2b}}\cdot\dfrac{\sqrt{2b}}{\sqrt{2b}}$
$=\dfrac{\sqrt{2b}}{2b}$

50. $\dfrac{\sqrt{x^3y^2}}{\sqrt{x^4y^5}}$
$=\sqrt{\dfrac{x^3y^2}{x^4y^5}}$
$=\sqrt{\dfrac{1}{xy^3}}$
$=\dfrac{\sqrt{1}}{\sqrt{xy^3}}$
$=\dfrac{1}{y\sqrt{xy}}\cdot\dfrac{\sqrt{xy}}{\sqrt{xy}}$
$=\dfrac{\sqrt{xy}}{xy^2}$

51. $\sqrt{121x}-\sqrt{81x}$
$=11\sqrt{x}-9\sqrt{x}$
$=\sqrt{x}(11-9)$
$=2\sqrt{x}$

52. $16d\sqrt{c^3d} - 11c\sqrt{cd^3} + 9cd\sqrt{cd}$
$= 16cd\sqrt{cd} - 11cd\sqrt{cd} + 9cd\sqrt{cd}$
$= \sqrt{cd}(16cd - 11cd + 9cd)$
$= 14cd\sqrt{cd}$

53. $\left(\sqrt{a}+9\right)^2$
$= \left(\sqrt{a}\right)^2 + 2\cdot 9\sqrt{a} + 9^2$
$= a + 18\sqrt{a} + 81$

54. $\sqrt{6a}\left(3+\sqrt{6a}\right)$
$= 3\sqrt{6a} + \left(\sqrt{6a}\right)^2$
$= 3\sqrt{6a} + 6a$

55. $\left(\sqrt{3x}+\sqrt{2y}\right)\left(\sqrt{3x}-\sqrt{2y}\right)$
$= \left(\sqrt{3x}\right)^2 - \left(\sqrt{2y}\right)^2$
$= 3x - 2y$

56. $2\sqrt[3]{x^2}\left(3\sqrt[3]{x} + 5\sqrt[3]{x^2}\right)$
$= 2\sqrt[3]{x^2}\left(3\sqrt[3]{x}\right) + 2\sqrt[3]{x^2}\left(5\sqrt[3]{x^2}\right)$
$= 6\sqrt[3]{x^3} + 10\sqrt[3]{x^4}$
$= 6x + 10x\sqrt[3]{x}$

57. $\dfrac{\sqrt{18x}-\sqrt{42}}{\sqrt{6}}$
$= \dfrac{\sqrt{18x}}{\sqrt{6}} - \dfrac{\sqrt{42}}{\sqrt{6}}$
$= \sqrt{\dfrac{18x}{6}} - \sqrt{\dfrac{42}{6}}$
$= \sqrt{3x} - \sqrt{7}$

58. $\dfrac{\sqrt{m}-9}{\sqrt{m}}$
$= \dfrac{\sqrt{m}}{\sqrt{m}} - \dfrac{9}{\sqrt{m}}$
$= 1 - \dfrac{9}{\sqrt{m}}\cdot\dfrac{\sqrt{m}}{\sqrt{m}}$
$= 1 - \dfrac{9\sqrt{m}}{m}$

59. $\dfrac{18}{\sqrt{7}+2}$
$= \dfrac{18}{\sqrt{7}+2}\cdot\dfrac{\sqrt{7}-2}{\sqrt{7}-2}$
$= \dfrac{18\left(\sqrt{7}-2\right)}{\left(\sqrt{7}\right)^2 - 2^2}$
$= \dfrac{18\left(\sqrt{7}-2\right)}{7-4}$
$= \dfrac{18\left(\sqrt{7}-2\right)}{3}$
$= 6\left(\sqrt{7}-2\right)$

60. $\dfrac{3+\sqrt{x}}{3-\sqrt{x}}$
$= \dfrac{3+\sqrt{x}}{3-\sqrt{x}}\cdot\dfrac{3+\sqrt{x}}{3+\sqrt{x}}$
$= \dfrac{3^2 + 2\cdot 3\sqrt{x} + \left(\sqrt{x}\right)^2}{3^2 - \left(\sqrt{x}\right)^2}$
$= \dfrac{9 + 6\sqrt{x} + x}{9 - x}$

61. $\dfrac{7-3\sqrt{a}}{3\sqrt{a}-7}$
$= \dfrac{-1\left(3\sqrt{a}-7\right)}{3\sqrt{a}-7}$
$= -1$

62. $\frac{25-7x}{5-\sqrt{7x}}$

$=\frac{25-7x}{5-\sqrt{7x}}\cdot\frac{5+\sqrt{7x}}{5+\sqrt{7x}}$

$=\frac{(25-7x)(5+\sqrt{7x})}{(5)^2-(\sqrt{7x})^2}$

$=\frac{(25-7x)(5+\sqrt{7x})}{25-7x}$

$=5+\sqrt{7x}$

63. $f(x)=5-\sqrt{x}$

The restricted values are all real numbers less than 0.

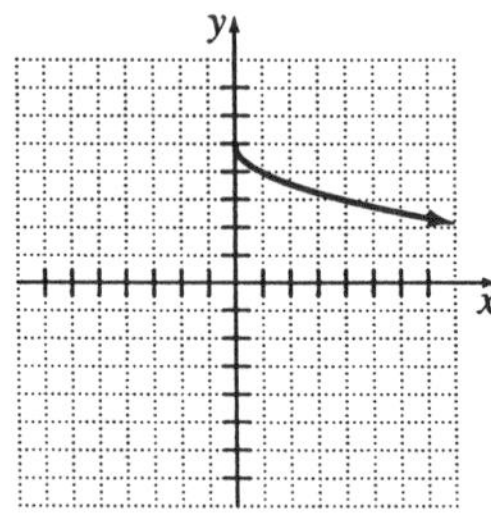

64. $y=x^{3/2}-3$

The restricted values are all real numbers less than 0.

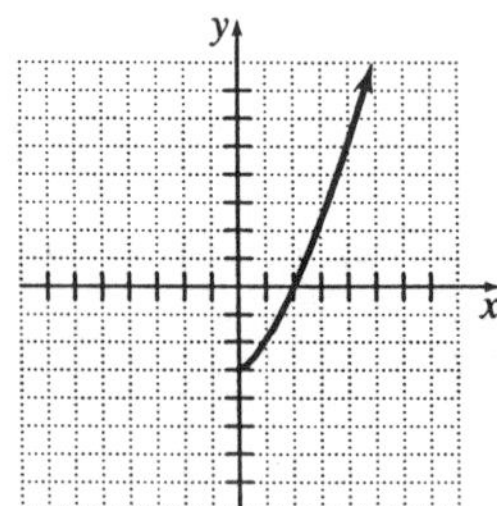

65. $\sqrt{x^2+8x+16}=x+4$

$\left(\sqrt{x^2+8x+16}\right)^2=(x+4)^2$

$x^2+8x+16=x^2+8x+16$

$16=16$

The solution is all real numbers greater than or equal to –4.

66. $\sqrt{x-2}=x-2$

$\left(\sqrt{x-2}\right)^2=(x-2)^2$

$x-2=x^2-4x+4$

$0=x^2-5x+6$

$0=(x-3)(x-2)$

$x=3 \qquad x=2$

The solutions are 2 and 3.

67. $\sqrt{x^2+6x+9}=x+3$

$\left(\sqrt{x^2+6x+9}\right)^2=(x+3)^2$

$x^2+6x+9=x^2+6x+9$

$9=9$

The solutions are all real numbers greater than or equal to –3.

68. $(x-1)^{3/2}=\sqrt[3]{2x+5}$

$\text{Y1}=(x-1)^{3/2}$

$\text{Y2}=\sqrt[3]{2x+5}$

10

–10 10

–10

The approximate solution is the x-coordinate of the intersection, 2.681.

69. $(x+4)^{3/5}=(5x-1)^{2/3}$

$\text{Y1}=(x+4)^{3/5}$

$\text{Y2}=(5x-1)^{2/3}$

10

–10 10

–10

The approximate solutions are the x-coordinates of the intersections, –0.429 and 1.061.

70. $\sqrt{x}+2=11$
$\sqrt{x}=9$
$\left(\sqrt{x}\right)^2=9^2$
$x=81$
The solution is 81.

71. $3\sqrt{5x}+7=52$
$3\sqrt{5x}=45$
$\sqrt{5x}=15$
$\left(\sqrt{5x}\right)^2=(15)^2$
$5x=225$
$x=45$
The solution is 45.

72. $\sqrt{3x+1}=x-3$
$\left(\sqrt{3x+1}\right)^2=(x-3)^2$
$3x+1=x^2-6x+9$
$0=x^2-9x+8$
$0=(x-8)(x-1)$
$x=8$ $\quad$ $x=1$ (extraneous)
The solution is 8.

73. $2\sqrt{x+9}=\sqrt{9x+1}$
$\left(2\sqrt{x+9}\right)^2=\left(\sqrt{9x+1}\right)^2$
$4(x+9)=9x+1$
$4x+36=9x+1$
$35=5x$
$7=x$
The solution is 7.

74. $\sqrt{2x+1}+2=\sqrt{6x+1}$
$\left(\sqrt{2x+1}+2\right)^2=\left(\sqrt{6x+1}\right)^2$
$2x+1+4\sqrt{2x+1}+4=6x+1$
$\sqrt{2x+1}=x-1$
$\left(\sqrt{2x+1}\right)^2=(x-1)^2$
$2x+1=x^2-2x+1$
$0=x^2-4x$
$0=x(x-4)$
$x=0$ (extraneous) $\quad$ $x=4$
The solution is 4.

75. $\sqrt{10-x}=\sqrt{3-x}$
$\left(\sqrt{10-x}\right)^2=\left(\sqrt{3-x}\right)^2$
$10-x=3-x$
$10=3$
This is a contradiction. There is no solution.

76. $\sqrt[3]{3x}=-6$
$\left(\sqrt[3]{3x}\right)^3=(-6)^3$
$3x=-216$
$x=-72$
The solution is -72.

77. $\sqrt[4]{x+2}=3$
$\left(\sqrt[4]{x+2}\right)^4=3^4$
$x+2=81$
$x=79$
The solution is 79.

78. $(x+5)^{2/5}=9$
$[(x+5)^{2/5}]^5=9^5$
$(x+5)^2=59,049$
$\sqrt{(x+5)^2}=\pm\sqrt{59,049}$
$x+5=\pm 243$
$x=-5\pm 243$
$x=-248 \quad x=238$
The solutions are -248 and 238.

79. $x^{-3/2}=27$
$x^{3/2}=\frac{1}{27}$
$(x^{3/2})^2=\left(\frac{1}{27}\right)^2$
$x^3=\frac{1}{729}$
$(x^3)^{1/3}=\left(\frac{1}{729}\right)^{1/3}$
$x=\frac{1}{9}$
The solution is $\frac{1}{9}$.

80. $3x^{2/3} - 7 = 2$
$3x^{2/3} = 9$
$x^{2/3} = 3$
$(x^{2/3})^3 = 3^3$
$x^2 = 27$
$\sqrt{x^2} = \pm\sqrt{27}$
$x = \pm 3\sqrt{3}$
The solutions are $\pm 3\sqrt{3}$.

81. $s = \sqrt{A}$
$s = \sqrt{529}$
$s = 23$
The side is 23 inches.

82. $s = \sqrt{\frac{n(n+1)}{12}}$
$s = \sqrt{\frac{27(27+1)}{12}}$
$s = \sqrt{63}$
$s = 3\sqrt{7}$
$s \approx 7.937$
The variability is approximately 7.937.

83. $t = \sqrt{\frac{2d}{32.2}}$
$4 = \sqrt{\frac{2d}{32.2}}$
$16 = \frac{2d}{32.2}$
$515.2 = 2d$
$257.6 = d$
It fell from a height of 257.6 feet.

84. $T = 2\pi\sqrt{\frac{L}{32}}$
$T = 2\pi\sqrt{\frac{3.5}{32}}$
$T \approx 2.08$
The period is approximately 2.08 seconds.

85. $d = \sqrt{(x_2 - x_1)^2 + (y_2 - y_1)^2}$
$d = \sqrt{(6-6)^2 + (7+3)^2}$
$d = \sqrt{10^2}$
$d = 10$
The distance is 10 units.

86. $d = \sqrt{(x_2 - x_1)^2 + (y_2 - y_1)^2}$
$d = \sqrt{(2+4)^2 + (8-8)^2}$
$d = \sqrt{6^2}$
$d = 6$
The distance is 6 units.

87. $d = \sqrt{(x_2 - x_1)^2 + (y_2 - y_1)^2}$
$d = \sqrt{(9-5)^2 + (2+1)^2}$
$d = \sqrt{16+9}$
$d = \sqrt{25}$
$d = 5$
The distance is 5 units.

88. $d = \sqrt{(x_2 - x_1)^2 + (y_2 - y_1)^2}$
$d = \sqrt{(-2-5)^2 + (10-7)^2}$
$d = \sqrt{49+9}$
$d = \sqrt{58}$
The distance is $\sqrt{58}$ units.

89. $d = \sqrt{(x_2 - x_1)^2 + (y_2 - y_1)^2}$
$4 = \sqrt{(2-x)^2 + (8-[-2])^2}$
$4 = \sqrt{(2-x)^2 + 100}$
$16 = (2-x)^2 + 100$
$-84 = (2-x)^2$
There is no solution.

90. $d = \sqrt{(x_2 - x_1)^2 + (y_2 - y_1)^2}$
$5 = \sqrt{(3-7)^2 + (2-y)^2}$
$25 = 16 + (2-y)^2$
$9 = (2-y)^2$
$\pm 3 = 2 - y$

$y = 2 \pm 3$
$y = 5 \quad y = -1$
The y-coordinates are 5 and -1. The possible points are (7, 5) and (7, -1).

Chapter 13 Test

1. $\sqrt{196}$
$= \sqrt{(14)^2}$
$= 14$

2. $\sqrt{2.25}$
$= \sqrt{(1.5)^2}$
$= 1.5$

3. $\sqrt[4]{21}$
≈ 2.141

4. $\sqrt{-25}$ is not a real number.

5. $\sqrt[3]{-\frac{8}{125}}$
$= \sqrt[3]{\left(-\frac{2}{5}\right)^3}$
$= -\frac{2}{5}$

6. $16^{5/2} \cdot 16^{-1/4}$
$= 16^{\frac{5}{2}-\frac{1}{4}}$
$= 16^{\frac{10}{4}-\frac{1}{4}}$
$= 16^{9/4}$
$= \left(\sqrt[4]{16}\right)^9$
$= 2^9$
$= 512$

7. $(81^{4/3})^{3/8}$
$= 81^{\frac{4}{3}\cdot\frac{3}{8}}$
$= 81^{1/2}$
$= \sqrt{81}$
$= 9$

8. $64^{2/3}$
$= \left(\sqrt[3]{64}\right)^2$
$= 4^2$
$= 16$

9. $\left(\frac{8}{27}\right)^{-2/3}$
$= \left(\frac{27}{8}\right)^{2/3}$
$= \left(\sqrt[3]{\frac{27}{8}}\right)^2$
$= \left(\frac{3}{2}\right)^2$
$= \frac{9}{4}$

10. $\frac{x^{4/5}}{x^{3/10}}$
$= x^{\frac{4}{5}-\frac{3}{10}}$
$= x^{\frac{8}{10}-\frac{3}{10}}$
$= x^{1/2}$
$= \sqrt{x}$

11. $\left(\sqrt{2x} - \sqrt{5y}\right)\left(\sqrt{2x} + \sqrt{5y}\right)$
$= \left(\sqrt{2x}\right)^2 - \left(\sqrt{5y}\right)^2$
$= 2x - 5y$

12. $\sqrt{6x^3y} \cdot \sqrt{15xy^2}$
$= \sqrt{(6x^3y)(15xy^2)}$
$= \sqrt{90x^4y^3}$
$= \sqrt{9 \cdot 10x^4y^2 \cdot y}$
$= 3x^2y\sqrt{10y}$

13. $\sqrt{\frac{3z^3}{2xy^2}}$

$= \frac{\sqrt{3z^3}}{\sqrt{2xy^2}}$

$= \frac{z\sqrt{3z}}{y\sqrt{2x}} \cdot \frac{\sqrt{2x}}{\sqrt{2x}}$

$= \frac{z\sqrt{6xz}}{2xy}$

14. $\frac{\sqrt{50xy^3}}{\sqrt{98x^3y^2}}$

$= \sqrt{\frac{50xy^3}{98x^3y^2}}$

$= \sqrt{\frac{25y}{49x^2}}$

$= \frac{\sqrt{25y}}{\sqrt{49x^2}}$

$= \frac{5\sqrt{y}}{7x}$

15. $\sqrt{81x} + 2\sqrt{25x} - 3\sqrt{16x}$

$= 9\sqrt{x} + 10\sqrt{x} - 12\sqrt{x}$

$= \sqrt{x}(9 + 10 - 12)$

$= 7\sqrt{x}$

16. $5\sqrt[3]{x}\left(2\sqrt[3]{x^2} + 3\sqrt[3]{x}\right)$

$= 5\sqrt[3]{x} \cdot 2\sqrt[3]{x^2} + 5\sqrt[3]{x} \cdot 3\sqrt[3]{x}$

$= 10\sqrt[3]{x^3} + 15\sqrt[3]{x^2}$

$= 10x + 15\sqrt[3]{x^2}$

17. $\frac{\sqrt{x}+4}{\sqrt{x}+1}$

$= \frac{\sqrt{x}+4}{\sqrt{x}+1} \cdot \frac{\sqrt{x}-1}{\sqrt{x}-1}$

$= \frac{\left(\sqrt{x}\right)^2 - \sqrt{x} + 4\sqrt{x} - 4}{\left(\sqrt{x}\right)^2 - 1^2}$

$= \frac{x + 3\sqrt{x} - 4}{x - 1}$

18. **a.**

x	$(x+4)^{1/3}+3$	2
–8	1.4126	2
–7	1.5578	2
–6	1.7401	2
–5	2	2
–4	3	2
–3	4	2
–2	4.2599	2

The solution is $x = -5$.

b. $Y1 = (x+4)^{1/3} + 3$
$Y2 = 2$

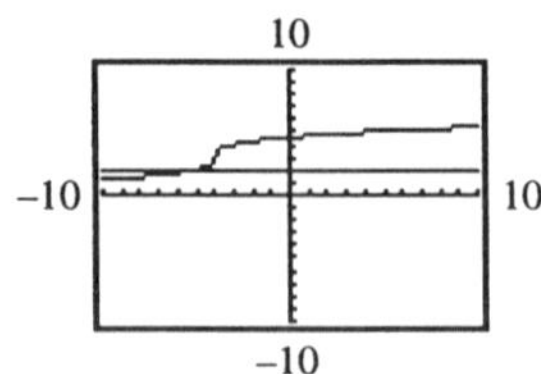

The solution is $x = -5$.

c. $(x+4)^{1/3} + 3 = 2$

$(x+4)^{1/3} = -1$

$(x+4) = -1$

$x = -5$

19. $3\sqrt{x} - 5 = 13$

$3\sqrt{x} = 18$

$\sqrt{x} = 6$

$\left(\sqrt{x}\right)^2 = 6^2$

$x = 36$

The solution is 36.

20. $\sqrt{3x+1} = x-1$
$\left(\sqrt{3x+1}\right)^2 = (x-1)^2$
$3x+1 = x^2 - 2x + 1$
$0 = x^2 - 5x$
$0 = x(x-5)$
$x = 0$ (extraneous) $\quad x = 5$
The solution is 5.

21. $\sqrt{x^2+10x+25} = x+5$
$\left(\sqrt{x^2+10x+25}\right)^2 = (x+5)^2$
$x^2+10x+25 = x^2+10x+25$
$25 = 25$
The solutions are all real numbers greater than or equal to –5.

22. $\sqrt{5x+1} - 2 = \sqrt{x+1}$
$\left(\sqrt{5x+1}-2\right)^2 = \left(\sqrt{x+1}\right)^2$
$5x+1-4\sqrt{5x+1}+4 = x+1$
$\sqrt{5x+1} = x+1$
$\left(\sqrt{5x+1}\right)^2 = (x+1)^2$
$5x+1 = x^2+2x+1$
$0 = x^2 - 3x$
$0 = x(x-3)$
$x = 0$ (extraneous) $\quad x = 3$
The solution is 3.

23. $f(x) = \sqrt{x+7} - 3$
$x + 7 < 0$
$x < -7$
The restricted values are all real numbers less than –7.

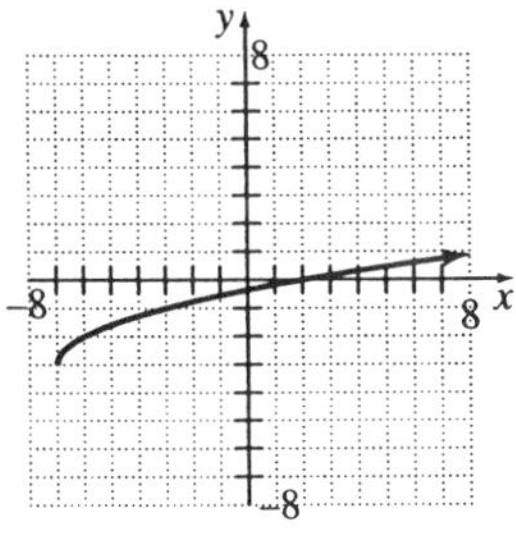

24. $d = \sqrt{(x_2-x_1)^2 + (y_2-y_1)^2}$
$d = \sqrt{(3-3)^2 + (5+2)^2}$
$d = \sqrt{0^2 + 7^2}$
$d = \sqrt{7^2}$
$d = 7$
The distance is 7 units.

25. $d = \sqrt{(x_2-x_1)^2 + (y_2-y_1)^2}$
$d = \sqrt{(5-2)^2 + (4+3)^2}$
$d = \sqrt{9+49}$
$d = \sqrt{58}$
The distance is $\sqrt{58}$ units.

26. Let s_1 = length of the side of the smaller tablecloth.
$s_1 = \sqrt{A_1}$
$s_1 = \sqrt{2000}$
$s_1 = 20\sqrt{5}$
Let s_2 = length of the side of the larger tablecloth.
$s_2 = \sqrt{A_2}$
$s_2 = \sqrt{3125}$
$s_2 = 25\sqrt{5}$
$4s_2 - 4s_1 = 4\left(25\sqrt{5}\right) - 4\left(20\sqrt{5}\right)$
$= 100\sqrt{5} - 80\sqrt{5}$
$= \sqrt{5}(100-80)$
$= 20\sqrt{5}$
It will require $20\sqrt{5}$ inches more border.

27. Answers will vary.

Chapters 1–13 Cumulative Review

1. $2^0 x^{-1} y^2 = \dfrac{y^2}{x}$

2. $(3x^{1/2}y^{3/4})^2 = 3^2(x^{1/2})^2(y^{3/4})^2 = 9xy^{3/2}$

3. $\left(\frac{(2s)^2}{3t}\right)^{-3} = \left(\frac{2^2 s^2}{3t}\right)^{-3} = \left(\frac{4s^2}{3t}\right)^{-3}$
$= \left(\frac{3t}{4s^2}\right)^3 = \frac{(3t)^3}{(4s^2)^3} = \frac{3^3 t^3}{4^3 (s^2)^3} = \frac{27t^3}{64s^6}$

4. $(2x^3 + 6x^2y - 2xy^2 + y^3) + (3x^3 + 4xy^2 - y^3)$
$= 2x^3 + 3x^3 + 6x^2y - 2xy^2 + 4xy^2 + y^3 - y^3$
$= (2+3)x^3 + 6x^2y + (-2+4)xy^2 + (1-1)y^3$
$= 5x^3 + 6x^2y + 2xy^2 + 0y^3$
$= 5x^3 + 6x^2y + 2xy^2$

5. $(1.2a^2 - 3.6ab + b^2) - (4a^2 + 2.71ab - 3.4b^2)$
$= (1.2a^2 - 3.6ab + b^2) + (-4a^2 - 2.71ab + 3.4b^2)$
$= 1.2a^2 - 4a^2 - 3.6ab - 2.71ab + b^2 + 3.4b^2$
$= (1.2 - 4)a^2 + (-3.6 - 2.71)ab + (1 + 3.4)b^2$
$= -2.8a^2 - 6.31ab + 4.4b^2$

6. $(6.8m^2n)(-2mn^2p)$
$= (6.8)(-2)m^{2+1}n^{1+2}p$
$= -13.6m^3n^3p$

7. $(3a - b)(2a + 4b)$
$= (3a)(2a) + (3a)(4b) + (-b)(2a) + (-b)(4b)$
$= 6a^2 + 12ab - 2ab - 4b^2$
$= 6a^2 + 10ab - 4b^2$

8. $(2x + 3)(2x - 3) = (2x)^2 - 3^2 = 4x^2 - 9$

9. $(2x + 3)^2 = (2x)^2 + 2(2x)(3) + 3^2$
$= 4x^2 + 12x + 9$

10. $\frac{-15x^2y^3z}{3xyz} = -5x^{2-1}y^{3-1}z^{1-1} = -5xy^2z^0$
$= -5xy^2$

11. $\frac{2m^2n + 4mn - 8n^2}{2m^2n}$
$= \frac{2m^2n}{2m^2n} + \frac{4mn}{2m^2n} - \frac{8n^2}{2m^2n}$
$= 1 + \frac{2}{m} - \frac{4n}{m^2}$

12. $\dfrac{x^2+2x-3}{x^2-9}=\dfrac{(x+3)(x-1)}{(x+3)(x-3)}=\dfrac{x-1}{x-3}$

13. $\dfrac{x}{x+2}+\dfrac{3}{x-2}$

$=\dfrac{x(x-2)}{(x+2)(x-2)}+\dfrac{3(x+2)}{(x+2)(x-2)}$

$=\dfrac{x^2-2x+3x+6}{(x+2)(x-2)}=\dfrac{x^2+x+6}{(x+2)(x-2)}$

14. $\dfrac{1}{x+2}-\dfrac{3}{x-3}-\dfrac{x+3}{x^2+x-12}$

$=\dfrac{1}{x+2}-\dfrac{3}{x-3}-\dfrac{x+3}{(x+4)(x-3)}$

$=\dfrac{1(x-3)(x+4)}{(x+2)(x-3)(x+4)}-\dfrac{3(x+2)(x+4)}{(x+2)(x-3)(x+4)}-\dfrac{(x+3)(x+2)}{(x+2)(x-3)(x+4)}$

$=\dfrac{(x^2+x-12)-3(x^2+6x+8)-(x^2+5x+6)}{(x+2)(x-3)(x+4)}$

$=\dfrac{x^2+x-12-3x^2-18x-24-x^2-5x-6}{(x+2)(x-3)(x+4)}$

$=\dfrac{-3x^2-22x-42}{(x+2)(x-3)(x+4)}=-\dfrac{3x^2+22x+42}{(x+2)(x-3)(x+4)}$

15. $\dfrac{2a-5}{a+2}\cdot\dfrac{2a^2+3a-2}{4a^2-25}$

$=\dfrac{2a-5}{a+2}\cdot\dfrac{(2a-1)(a+2)}{(2a+5)(2a-5)}$

$=\dfrac{(2a-5)(2a-1)(a+2)}{(a+2)(2a+5)(2a-5)}=\dfrac{2a-1}{2a+5}$

16. $\dfrac{2x}{2x^2+9x+4}\div\dfrac{4x^2y}{x^2+9x+20}$

$=\dfrac{2x}{(x+4)(2x+1)}\div\dfrac{4x^2y}{(x+5)(x+4)}$

$=\dfrac{2x}{(x+4)(2x+1)}\cdot\dfrac{(x+5)(x+4)}{4x^2y}$

$=\dfrac{2x(x+5)(x+4)}{4x^2y(x+4)(2x+1)}=\dfrac{x+5}{2xy(2x+1)}$

17. $\sqrt{8y}+\sqrt{2y}-2\sqrt{18y}$

$=\sqrt{4\cdot 2y}+\sqrt{2y}-2\sqrt{9\cdot 2y}$

$=2\sqrt{2y}+\sqrt{2y}-6\sqrt{2y}=-3\sqrt{2y}$

18. $\sqrt[3]{25a^4b}\cdot\sqrt[3]{5ab}=\sqrt[3]{25a^4b\cdot 5ab}$

$=\sqrt[3]{125a^5b^2}=\sqrt[3]{5^3a^3\cdot a^2\cdot b^2}$

$=5a\sqrt[3]{a^2b^2}$

19. $\left(\sqrt{3}-\sqrt{5}\right)\left(\sqrt{3}+\sqrt{5}\right)=\left(\sqrt{3}\right)^2-\left(\sqrt{5}\right)^2$

$=3-5=-2$

20. $\dfrac{\sqrt{21abc^3}}{\sqrt{7a^3b}}=\sqrt{\dfrac{21abc^3}{7a^3b}}=\sqrt{\dfrac{3c^3}{a^2}}$

$=\sqrt{\dfrac{3c^2\cdot c}{a^2}}=\dfrac{c}{a}\sqrt{3c}$

21. $\dfrac{\sqrt{x}+2}{\sqrt{x}-1}=\dfrac{\sqrt{x}+2}{\sqrt{x}-1}\cdot\dfrac{\sqrt{x}+1}{\sqrt{x}+1}$

$=\dfrac{\left(\sqrt{x}\right)^2+\sqrt{x}+2\sqrt{x}+2}{\left(\sqrt{x}\right)^2-1^2}=\dfrac{x+3\sqrt{x}+2}{x-1}$

22. $x^{1/2}(x^{2/3}-x^{3/4})=x^{1/2+2/3}-x^{1/2+3/4}$
$=x^{7/6}-x^{5/4}$

23. $16a^2-25b^2=(4a)^2-(5b)^2$
$=(4a+5b)(4a-5b)$

24. $x^2-2x-8=(x-4)(x+2)$

25. $3x^2-9x-30=3(x^2-3x+10)$
$=3(x-5)(x+2)$

26. $f(x)=-x^2+3x-1$
$f(-2)=-(-2)^2+3(-2)-1$
$f(-2)=-4+(-6)-1$
$f(-2)=-11$

27.

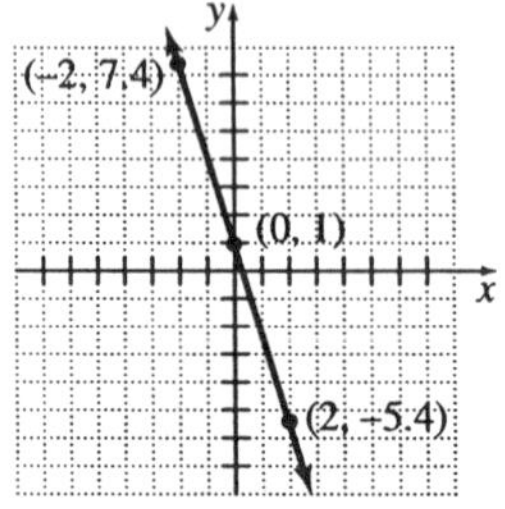

Domain: all real numbers
Range: all real numbers

28. $y=x^2+4x+6$
$a=1,\ b=4,\ c=6$
$-\dfrac{b}{2a}=-\dfrac{4}{2(1)}=-2$

x	$y=x^2+4x+6$	y
-2	$y=(-2)^2+4(-2)+6$ $y=2$	2
-1	$y=(-1)^2+4(-1)+6$ $y=3$	3
0	$y=0^2+4(0)+6$ $y=6$	6

The vertex is (–2, 2); the axis of symmetry is $x=-2$.

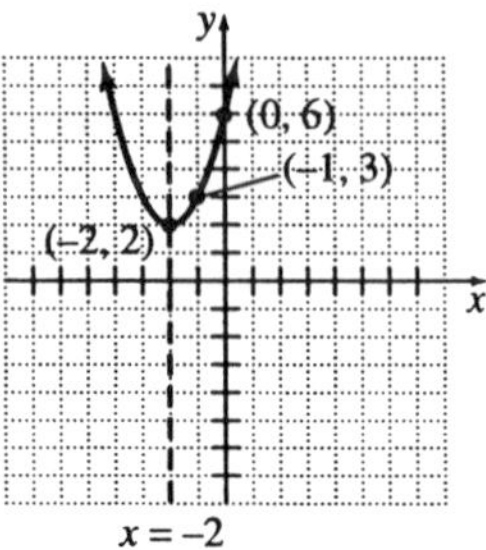

Domain: all real numbers
Range: all real numbers ≥ 2

29. $y = g(x) = -2x^2 + x + 1$

$a = -2,\ b = 1,\ c = 1$

$\frac{-b}{2a} = \frac{-1}{2(-2)} = \frac{1}{4}$

x	$g(x) = -2x^2 + x + 1$	y
0	$g(0) = -2(0)^2 + 0 + 1$ $g(0) = 1$	1
$\frac{1}{4}$	$g\left(\frac{1}{4}\right) = -2\left(\frac{1}{4}\right)^2 + \frac{1}{4} + 1$ $g\left(\frac{1}{4}\right) = \frac{9}{8}$	$\frac{9}{8}$
1	$g(1) = -2(1)^2 + 1 + 1$ $g(1) = 0$	0
2	$g(2) = -2(2)^2 + 2 + 1$ $g(2) = -5$	-5
$-\frac{1}{2}$	$g\left(-\frac{1}{2}\right) = -2\left(-\frac{1}{2}\right)^2 + \left(-\frac{1}{2}\right) + 1$ $g\left(-\frac{1}{2}\right) = 0$	0

The vertex is $\left(\frac{1}{4}, \frac{9}{8}\right)$; the axis of symmetry is $x = \frac{1}{4}$.

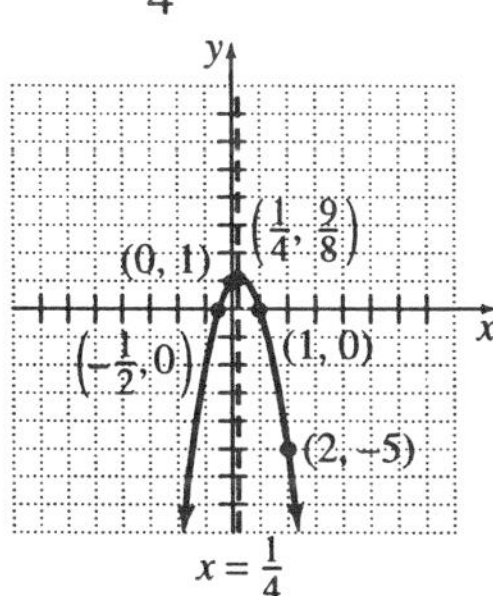

Domain: all real numbers

Range: all real numbers $\leq \frac{9}{8}$

30. $y = \frac{x^2 - 4}{x + 2} = \frac{(x+2)(x-2)}{x+2}$

When $x \neq -2$, $y = x - 2$

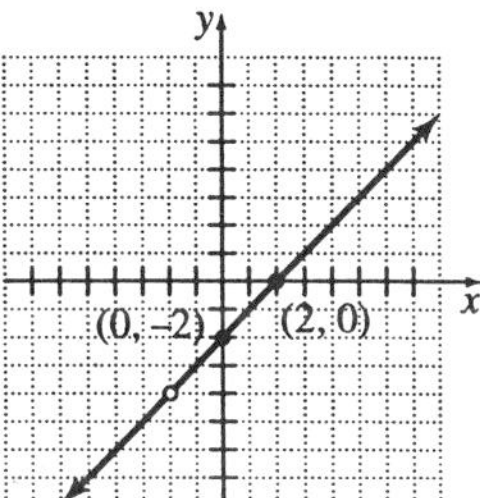

Domain: all real numbers $\neq -2$

Range: all real numbers $\neq -4$

31.

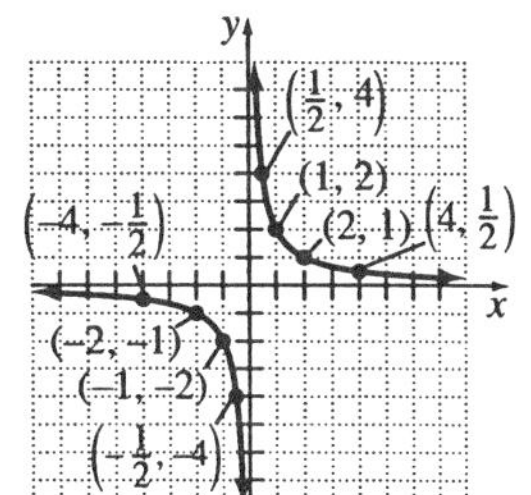

Domain: all real numbers $\neq 0$

Range: all real numbers $\neq 0$

32.

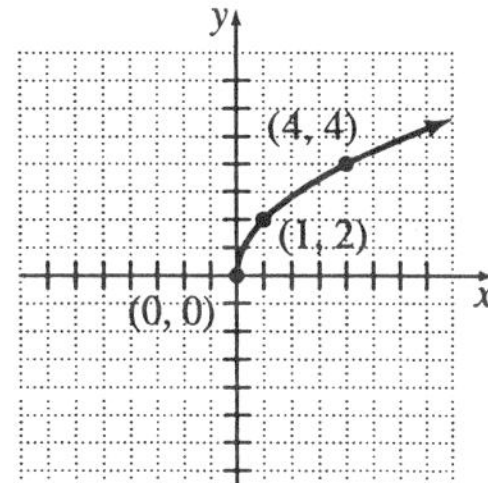

Domain: all real numbers ≥ 0

Range: all real numbers ≥ 0

33.

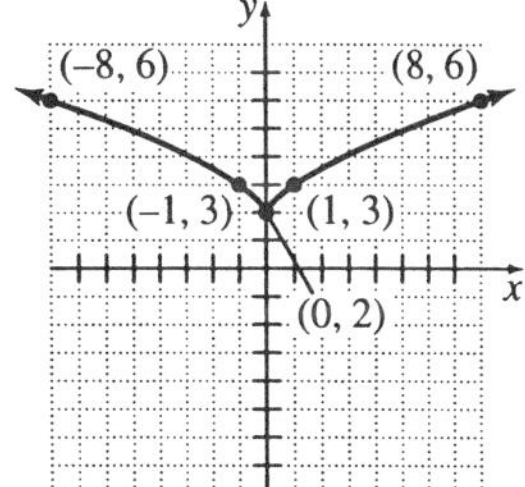

Domain: all real numbers

Range: all real numbers ≥ 2

34. $2(x+3)-4(x-1)=x-2$
$2x+6-4x+4=x-2$
$-2x+10=x-2$
$12=3x$
$4=x$

35. $2.4x-1.2(2x-3)=3.6$
$2.4x-2.4x+3.6=3.6$
$3.6=3.6$
This is an identity. The solution is all real numbers.

36. $x^2+2x-15=0$
$(x+5)(x-3)=0$
$x+5=0$ or $x-3=0$
$x=-5$ or $x=3$

37. $2t^2+3t=4$
$2t^2+3t-4=0$
$t=\dfrac{-3\pm\sqrt{3^2-4(2)(-4)}}{2(2)}=\dfrac{-3\pm\sqrt{41}}{4}$

38. $\dfrac{x+1}{x-1}=\dfrac{x+3}{x+2}$
$(x-1)(x+2)\dfrac{x+1}{x-1}=(x-1)(x+2)\dfrac{x+3}{x+2}$
$(x+2)(x+1)=(x-1)(x+3)$
$x^2+3x+2=x^2+2x-3$
$x=-5$

39. $\dfrac{2}{x^2}+\dfrac{3}{2x^2}=\dfrac{1}{8}$
$(8x^2)\dfrac{2}{x^2}+(8x^2)\dfrac{3}{2x^2}=(8x^2)\dfrac{1}{8}$
$16+12=x^2$
$28=x^2$
$\pm\sqrt{28}=x$
$\pm2\sqrt{7}=x$

40. $\sqrt{2x-3}+5=9$
$\sqrt{2x-3}=4$
$2x-3=16$
$2x=19$
$x=\dfrac{19}{2}$

41. $2x^2-9x-18<0$
$(x-6)(2x+3)<0$
$\left(-\dfrac{3}{2}, 6\right)$

42. From the second equation, $x=8-y$. Substituting $8-y$ for x in the first equation gives
$3(8-y)-4y=3$
$24-3y-4y=3$
$-7y=-21$
$y=3$
$x=8-3=5$
The solution is (5, 3).

43. $m=\dfrac{y_2-y_1}{x_2-x_1}=\dfrac{2-(-5)}{-4-3}=\dfrac{7}{-7}=-1$
$y-y_1=m(x-x_1)$
$y-(-5)=-1(x-3)$
$y+5=-x+3$
$y=-x-2$

44. $2x+3y=5$
$3y=-2x+5$
$y=-\dfrac{2}{3}x+\dfrac{5}{3}$
The line we want has slope $\dfrac{3}{2}$.
$y-y_1=m(x-x_1)$
$y-1=\dfrac{3}{2}[x-(-2)]$
$y-1=\dfrac{3}{2}x+3$
$y=\dfrac{3}{2}x+4$

45. A general quadratic equation has the form $y=ax^2+bx+c$. From the given points we have:
$-4=a(0)^2+b(0)+c$
$0=a(-4)^2+b(-4)+c$
$0=a(1)^2+b(1)+c$
or
$-4=c$
$0=16a-4b+c$
$0=a+b+c$

Substitute the value of –4 for c into the second and third equations.
$0 = 16a - 4b - 4$
$0 = a + b - 4$
Multiply the second equation above by 4 and add
$0 = 16a - 4b - 4$
$\underline{0 = 4a + 4b - 16}$
$0 = 20a - 20$
$20 = 20a$
$1 = a$
Since $a + b + c = 0$ and $a = 1$, $c = -4$, we have $1 + b - 4 = 0$ or $b = 3$. The quadratic equation is $y = x^2 + 3x - 4$.

46. Let l be the length of the rectangle and w be the width. Since the diagonal measures 20 feet, we have $\sqrt{l^2 + w^2} = 20$ or $l^2 + w^2 = 400$. Substitute $w + 4$ for l in this equation.
$(w + 4)^2 + w^2 = 400$
$w^2 + 8w + 16 + w^2 = 400$
$2w^2 + 8w - 384 = 0$
$w^2 + 4w - 192 = 0$
$(w + 16)(w - 12) = 0$
$w + 16 = 0$ or $w - 12 = 0$
$w = -16$ or $w = 12$
Since length cannot be negative, the width is 12 feet and the length is $12 + 4 = 16$ feet. The rectangle is 12 feet by 16 feet.

47. If the point $(0, y)$ is 5 units from the point $(-3, 4)$, then, by the distance formula,
$\sqrt{(-3-0)^2 + (4-y)^2} = 5$
$(-3-0)^2 + (4-y)^2 = 25$
$9 + 16 - 8y + y^2 = 25$
$y^2 - 8y = 0$
$y(y - 8) = 0$
$y = 0$ or $y - 8 = 0$
$y = 8$
The possible y-coordinates are 0 or 8.

48. Since Kevin needs 5 hours to rake the yard, he rakes $\frac{1}{5}$ of the yard per hour. Similarly, Elizabeth rakes $\frac{1}{4}$ of the yard in 1 hour. Let x be the number of hours that they work together raking leaves. The proportion of the yard that Kevin rakes is $\frac{1}{5}x$ and the proportion that Elizabeth rakes is $\frac{1}{4}x$.
$\frac{1}{5}x + \frac{1}{4}x = 1$
$20\left(\frac{1}{5}x\right) + 20\left(\frac{1}{4}x\right) = 20(1)$
$4x + 5x = 20$
$9x = 20$
$x = \frac{20}{9} = 2\frac{2}{9}$
It will take Kevin and Elizabeth $2\frac{2}{9}$ hours or about 2 hours, 13 minutes to rake the leaves, thus they will be done before 2:30. Yes, they can finish raking the leaves.

49. Let x be the number of items produced and sold. The cost of producing x items is $1000 + 0.55x$, while the revenue from selling x items is $1.25x$.
$1.25x \geq 1000 + 0.55x$
$0.70x \geq 1000$
$x \geq \frac{1000}{0.7} = \frac{10,000}{7} \approx 1428.6$
The company must sell at least 1429 items in order to break even.

50. The position equation is $s = -16t^2 + v_0 t + s_0$. We have to find t when s is 2500 if v_0 is 0 and s_0 is 10,000.

$2500 = -16t^2 + 0t + 10{,}000$

$2500 = -16t^2 + 10{,}000$

$16t^2 = 7500$

$t^2 = \frac{1875}{4}$

$t = \pm\sqrt{\frac{1875}{4}} = \pm\frac{25\sqrt{3}}{2}$

Since the amount of time will be positive, $t = \frac{25\sqrt{3}}{2} \approx 21.65$. The skydiver was free-falling for about 21.65 seconds.

Chapter 14

14.1 Experiencing Algebra the Exercise Way

1. $\sqrt{-100}$
$= \sqrt{100}i$
$= 10i$

3. $\sqrt{-\frac{16}{49}}$
$= \sqrt{\frac{16}{49}}i$
$= \frac{4}{7}i$

5. $\sqrt{-32}$
$= \sqrt{32}i$
$= 4\sqrt{2}i$

7. $2\sqrt{-50}$
$= 2\sqrt{50}i$
$= 2 \cdot 5\sqrt{2}i$
$= 10\sqrt{2}i$

9. $\sqrt{-9} + \sqrt{-16}$
$= \sqrt{9}i + \sqrt{16}i$
$= 3i + 4i$
$= 7i$

11. $\sqrt{-25} - \sqrt{-100}$
$= \sqrt{25}i - \sqrt{100}i$
$= 5i - 10i$
$= -5i$

13. $7\sqrt{-36} + 9\sqrt{-4}$
$= 7\sqrt{36}i + 9\sqrt{4}i$
$= 7 \cdot 6i + 9 \cdot 2i$
$= 42i + 18i$
$= 60i$

15. $2\sqrt{-121} - 3\sqrt{-9}$
$= 2\sqrt{121}i - 3\sqrt{9}i$
$= 2 \cdot 11i - 3 \cdot 3i$
$= 22i - 9i$
$= 13i$

17. $\frac{\sqrt{-144}}{\sqrt{-225}}$
$= \frac{\sqrt{144}i}{\sqrt{225}i}$
$= \frac{12i}{15i}$
$= \frac{4}{5}$

19. $\sqrt{-5}\sqrt{-8}\sqrt{-10}$
$= \sqrt{5}i\sqrt{8}i\sqrt{10}i$
$= \sqrt{400}i^3$
$= 20(-i)$
$= -20i$

21. $\left(\sqrt{-6} + \sqrt{-4}\right)\left(\sqrt{-3} - \sqrt{-8}\right)$
$= \left(\sqrt{6}i + \sqrt{4}i\right)\left(\sqrt{3}i - \sqrt{8}i\right)$
$= \sqrt{18}i^2 - \sqrt{48}i^2 + \sqrt{12}i^2 - \sqrt{32}i^2$
$= 3\sqrt{2}(-1) - 4\sqrt{3}(-1) + 2\sqrt{3}(-1) - 4\sqrt{2}(-1)$
$= -3\sqrt{2} + 4\sqrt{3} - 2\sqrt{3} + 4\sqrt{2}$
$= \sqrt{2} + 2\sqrt{3}$

23. $\left(\sqrt{-5} - \sqrt{-7}\right)\left(\sqrt{-5} + \sqrt{-7}\right)$
$= \left(\sqrt{5}i - \sqrt{7}i\right)\left(\sqrt{5}i + \sqrt{7}i\right)$
$= \left(\sqrt{5}i\right)^2 - \left(\sqrt{7}i\right)^2$
$= 5i^2 - 7i^2$
$= 5(-1) - 7(-1)$
$= -5 + 7$
$= 2$

25. $\frac{5}{7}$ is a real number. The real part is $\frac{5}{7}$ and the imaginary part is 0.

27. $2i$ is a pure imaginary number. The real part is 0 and the imaginary part is 2.

29. $\sqrt{3}-5i$ is an imaginary number. The real part is $\sqrt{3}$ and the imaginary part is –5.

31. –6 is a real number. The real part is –6 and the imaginary part is 0.

33. $2\sqrt{5}i$ is a pure imaginary number. The real part is 0 and the imaginary part is $2\sqrt{5}$.

35. $-4\sqrt{7}+\sqrt{11}$ is a real number. The real part is $-4\sqrt{7}+\sqrt{11}$ and the imaginary part is 0.

37. $(3+5i)+(-8-i)$
$=3+5i-8-i$
$=-5+4i$

39. $(6+2i)-(3+3i)$
$=6+2i-3-3i$
$=3-i$

41. $8i-(5-6i)$
$=8i-5+6i$
$=-5+14i$

43. $(4-5i)-(3-5i)$
$=4-5i-3+5i$
$=1$

45. $(7+i)-6$
$=7+i-6$
$=1+i$

47. $10+(3+5i)$
$=10+3+5i$
$=13+5i$

49. $\left(\frac{1}{2}+5i\right)+\left(3-\frac{2}{3}i\right)$
$=\frac{1}{2}+5i+3-\frac{2}{3}i$
$=\frac{1}{2}+3+\left(5-\frac{2}{3}\right)i$
$=\frac{7}{2}+\frac{13}{3}i$

51. $(4.5+6.7i)-(2.88-4.68i)$
$=4.5+6.7i-2.88+4.68i$
$=4.5-2.88+(6.7+4.68)i$
$=1.62+11.38i$

53. $\left(2\sqrt{3}-\sqrt{2}i\right)+\left(5\sqrt{3}+2\sqrt{2}i\right)$
$=2\sqrt{3}-\sqrt{2}i+5\sqrt{3}+2\sqrt{2}i$
$=2\sqrt{3}+5\sqrt{3}+\left(-\sqrt{2}+2\sqrt{2}\right)i$
$=7\sqrt{3}+\sqrt{2}i$

55. $-7(4+9i)$
$=-28-63i$

57. $5i(8-2i)$
$=40i-10i^2$
$=40i-10(-1)$
$=10+40i$

59. $(4-3i)(6+5i)$
$=24+20i-18i-15i^2$
$=24+20i-18i-15(-1)$
$=24+20i-18i+15$
$=39+2i$

61. $(5+7i)(5-7i)$
$=(5)^2-(7i)^2$
$=25-49i^2$
$=25+49$
$=74$

63. $(-6-i)(-6+i)$
$=(-6)^2-(i)^2$
$=36-(-1)$
$=36+1$
$=37$

65. $5.6(2.1 - 8.7i)$
$= 11.76 - 48.72i$

67. $(0.4 - 3.1i)(5.7 + 0.8i)$
$= 2.28 + 0.32i - 17.67i - 2.48i^2$
$= 2.28 + 0.32i - 17.67i - 2.48(-1)$
$= 2.28 + 0.32i - 17.67i + 2.48$
$= 4.76 - 17.35i$

69. $\left(\frac{3}{5} + \frac{1}{2}i\right)\left(\frac{2}{5} - \frac{2}{3}i\right)$
$= \frac{6}{25} - \frac{6}{15}i + \frac{2}{10}i - \frac{2}{6}i^2$
$= \frac{6}{25} - \frac{6}{15}i + \frac{1}{5}i - \frac{1}{3}(-1)$
$= \frac{18}{75} - \frac{6}{15}i + \frac{3}{15}i + \frac{25}{75}$
$= \frac{43}{75} - \frac{1}{5}i$

71. $\sqrt{2}i\left(\sqrt{2} + \sqrt{3}i\right)$
$= \sqrt{4}i + \sqrt{6}i^2$
$= 2i + \sqrt{6}(-1)$
$= -\sqrt{6} + 2i$

73. $\left(\sqrt{3} - \sqrt{5}i\right)\left(\sqrt{3} + \sqrt{5}i\right)$
$= \left(\sqrt{3}\right)^2 - \left(\sqrt{5}i\right)^2$
$= 3 - 5i^2$
$= 3 - 5(-1)$
$= 3 + 5$
$= 8$

75. $\frac{7 + 21i}{7}$
$= \frac{7}{7} + \frac{21i}{7}$
$= 1 + 3i$

77. $\frac{-4 + 7i}{3 - i}$
$= \frac{(-4 + 7i)(3 + i)}{(3 - i)(3 + i)}$
$= \frac{-12 - 4i + 21i + 7i^2}{9 + 1}$
$= \frac{-12 - 4i + 21i - 7}{9 + 1}$
$= \frac{-19 + 17i}{10}$
$= -\frac{19}{10} + \frac{17}{10}i$

79. $\frac{6 - 5i}{2i}$
$= \frac{(6 - 5i)(-2i)}{(2i)(-2i)}$
$= \frac{-12i + 10i^2}{4}$
$= \frac{-12i - 10}{4}$
$= -\frac{10}{4} - \frac{12i}{4}$
$= -\frac{5}{2} - 3i$

81. $\frac{16.2 - 13.5i}{2.7i}$
$= \frac{(16.2 - 13.5i)(-2.7i)}{(2.7i)(-2.7i)}$
$= \frac{-43.74i + 36.45i^2}{7.29}$
$= \frac{-43.74i - 36.45}{7.29}$
$= \frac{-36.45}{7.29} - \frac{43.74i}{7.29}$
$= -5 - 6i$

83. $\dfrac{21.2-10.4i}{4.5+i}$

$=\dfrac{(21.2-10.4i)(4.5-i)}{(4.5+i)(4.5-i)}$

$=\dfrac{95.4-21.2i-46.8i+10.4i^2}{20.25+1}$

$=\dfrac{95.4-21.2i-46.8i-10.4}{21.25}$

$=\dfrac{85-68i}{21.25}$

$=4-3.2i$

85. $\dfrac{\sqrt{6}+\sqrt{14}i}{\sqrt{2}i}$

$=\dfrac{\sqrt{6}\cdot i}{\sqrt{2}i\cdot i}+\dfrac{\sqrt{14}\cdot i}{\sqrt{2}\cdot i}$

$=-\dfrac{\sqrt{6}i}{\sqrt{2}}+\dfrac{\sqrt{14}}{\sqrt{2}}$

$=-\sqrt{\dfrac{6}{2}}i+\sqrt{\dfrac{14}{2}}$

$=-\sqrt{3}i+\sqrt{7}$

$=\sqrt{7}-\sqrt{3}i$

87. $\dfrac{2+5\sqrt{6}i}{2\sqrt{3}+\sqrt{2}i}$

$=\dfrac{\left(2+5\sqrt{6}i\right)\left(2\sqrt{3}-\sqrt{2}i\right)}{\left(2\sqrt{3}+\sqrt{2}i\right)\left(2\sqrt{3}-\sqrt{2}i\right)}$

$=\dfrac{4\sqrt{3}-2\sqrt{2}i+10\sqrt{18}i-5\sqrt{12}i^2}{\left(2\sqrt{3}\right)^2+2}$

$=\dfrac{4\sqrt{3}-2\sqrt{2}i+30\sqrt{2}i+10\sqrt{3}}{12+2}$

$=\dfrac{14\sqrt{3}+28\sqrt{2}i}{14}$

$=\dfrac{14\sqrt{3}}{14}+\dfrac{28\sqrt{2}i}{14}$

$=\sqrt{3}+2\sqrt{2}i$

89. $V=IZ$

$V=(3+5i)(5+4i)$

$V=15+12i+25i+20i^2$

$V=15+12i+25i-20$

$V=-5+37i$

The total voltage is $-5+37i$ volts.

$|V|=\sqrt{(-5)^2+(37)^2}\approx 37.336$

The magnitude of the total voltage is approximately 37.336 volts.

91. $I=\dfrac{V}{Z}$

$I=\dfrac{18+i}{2+3i}$

$I=\dfrac{(18+i)(2-3i)}{(2+3i)(2-3i)}$

$I=\dfrac{36-54i+2i-3i^2}{4+9}$

$I=\dfrac{36-54i+2i+3}{13}$

$I=\dfrac{39-52i}{13}$

$I=\dfrac{39}{13}-\dfrac{52}{13}i$

$I=3-4i$

The current is $3-4i$ amperes.

The magnitude of the current is

$|I|=\sqrt{(3)^2+(-4)^2}=5.$

The magnitude of the current is 5 amperes.

14.1 Experiencing Algebra the Calculator Way

1.–3.

```
√(-25)
                 5i
(-27)^(1/3)
                 -3
(-64)^(1/4)
               2+2i
```

4.–6.

```
(-32)^(1/5)
                 -2
(-324)^(1/4)
               3+3i
(-40000)^(1/4)
             10+10i
```

7.–9.

```
1.5-3.5i▸Frac
        3/2-7/2i
0.4+3.2i▸Frac
       2/5+16/5i
-1.8i▸Frac
           -9/5i
```

10.–12.

```
abs(6+8i)
              10
abs(6-8i)
              10
abs(2-5i)
     5.385164807
```

13.–15.

```
abs(22)
              22
abs(-7i)
               7
abs(√(3)i)
     1.732050808
```

14.2 Experiencing Algebra the Exercise Way

1. $x^2+9=0$
$x^2=-9$
$x=\pm\sqrt{9}i$
$x=\pm 3i$

3. $a^2+7=0$
$a^2=-7$
$a=\pm\sqrt{7}i$

5. $z^2+5=1$
$z^2=-4$
$z=\pm\sqrt{4}i$
$z=\pm 2i$

7. $3p^2+75=0$
$3p^2=-75$
$p^2=-25$
$p=\pm\sqrt{25}i$
$p=\pm 5i$

9. $4d^2+12=0$
$4d^2=-12$
$d^2=-3$
$d=\pm\sqrt{3}i$

11. $(t+1)^2+9=0$
$(t+1)^2=-9$
$t+1=\pm\sqrt{9}i$
$t+1=\pm 3i$
$t=-1\pm 3i$

13. $4(x-5)^2+22=2$
$4(x-5)^2=-20$
$(x-5)^2=-5$
$x-5=\pm\sqrt{5}i$
$x=5\pm\sqrt{5}i$

15. $4(2x+5)^2=-16$
$(2x+5)^2=-4$
$2x+5=\pm\sqrt{4}i$
$2x+5=\pm 2i$
$2x=-5\pm 2i$
$x=\dfrac{-5\pm 2i}{2}$
$x=-\dfrac{5}{2}\pm i$

17. $y^2+1.44=0$
$y^2=-1.44$
$y=\pm\sqrt{1.44}i$
$y=\pm 1.2i$

19. $(x-1.2)^2=-0.36$
$x-1.2=\pm\sqrt{0.36}i$
$x-1.2=\pm 0.6i$
$x=1.2\pm 0.6i$

21. $2(z+2.5)^2+10.58=0$
$2(z+2.5)^2=-10.58$
$(z+2.5)^2=-5.29$
$z+2.5=\pm\sqrt{5.29}i$
$z+2.5=\pm 2.3i$
$z=-2.5\pm 2.3i$

23. $x^2+\frac{9}{16}=0$

$x^2=-\frac{9}{16}$

$x=\pm\sqrt{\frac{9}{16}}i$

$x=\pm\frac{3}{4}i$

25. $\left(b-\frac{1}{2}\right)^2+\frac{1}{4}=0$

$\left(b-\frac{1}{2}\right)^2=-\frac{1}{4}$

$b-\frac{1}{2}=\pm\sqrt{\frac{1}{4}}i$

$b-\frac{1}{2}=\pm\frac{1}{2}i$

$b=\frac{1}{2}\pm\frac{1}{2}i$

27. $x^2+2x+4=0$

$x=\frac{-b\pm\sqrt{b^2-4ac}}{2a}$

$x=\frac{-2\pm\sqrt{2^2-4(1)(4)}}{2(1)}$

$x=\frac{-2\pm\sqrt{-12}}{2}$

$x=\frac{-2\pm2\sqrt{3}i}{2}$

$x=\frac{-2}{2}\pm\frac{2\sqrt{3}i}{2}$

$x=-1\pm\sqrt{3}i$

29. $b^2-10b+27=0$

$b=\frac{-(-10)\pm\sqrt{(-10)^2-4(1)(27)}}{2(1)}$

$b=\frac{10\pm\sqrt{-8}}{2}$

$b=\frac{10\pm2\sqrt{2}i}{2}$

$b=\frac{10}{2}\pm\frac{2\sqrt{2}}{2}i$

$b=5\pm\sqrt{2}i$

31. $4y^2+4y+5=0$

$y=\frac{-b\pm\sqrt{b^2-4ac}}{2a}$

$y=\frac{-4\pm\sqrt{4^2-4(4)(5)}}{2(4)}$

$y=\frac{-4\pm\sqrt{-64}}{8}$

$y=\frac{-4\pm8i}{8}$

$y=\frac{-4}{8}\pm\frac{8i}{8}$

$y=-\frac{1}{2}\pm i$

33. $9p^2-12p+8=0$

$p=\frac{-b\pm\sqrt{b^2-4ac}}{2a}$

$p=\frac{-(-12)\pm\sqrt{(-12)^2-4(9)(8)}}{2(9)}$

$p=\frac{12\pm\sqrt{-144}}{18}$

$p=\frac{12\pm12i}{18}$

$p=\frac{12}{18}\pm\frac{12i}{18}$

$p=\frac{2}{3}\pm\frac{2}{3}i$

35. $x^2 - 2.4x + 3.44 = 0$

$$x = \frac{-b \pm \sqrt{b^2 - 4ac}}{2a}$$
$$x = \frac{-(-2.4) \pm \sqrt{(-2.4)^2 - 4(1)(3.44)}}{2(1)}$$
$$x = \frac{2.4 \pm \sqrt{-8}}{2}$$
$$x = \frac{2.4 \pm 2\sqrt{2}i}{2}$$
$$x = \frac{2.4}{2} \pm \frac{2\sqrt{2}i}{2}$$
$$x = 1.2 \pm \sqrt{2}i$$

37. $2y^2 - 2y + 1.22 = 0$

$$y = \frac{-b \pm \sqrt{b^2 - 4ac}}{2a}$$
$$y = \frac{-(-2) \pm \sqrt{(-2)^2 - 4(2)(1.22)}}{2(2)}$$
$$y = \frac{2 \pm \sqrt{-5.76}}{4}$$
$$y = \frac{2 \pm 2.4i}{4}$$
$$y = \frac{2}{4} \pm \frac{2.4i}{4}$$
$$y = \frac{1}{2} \pm \frac{3}{5}i$$

39. $x^2 - \frac{2}{3}x + \frac{13}{36} = 0$

$$36\left(x^2 - \frac{2}{3}x + \frac{13}{36}\right) = 0 \cdot 36$$
$$36x^2 - 24x + 13 = 0$$
$$x = \frac{24 \pm \sqrt{(-24)^2 - 4(36)(13)}}{2(36)}$$
$$x = \frac{24 \pm \sqrt{-1296}}{72}$$
$$x = \frac{24}{72} \pm \frac{36}{72}i = \frac{1}{3} \pm \frac{1}{2}i$$

41. $4z^2 + \frac{4}{3}z + \frac{2}{9} = 0$

$$9\left(4z^2 + \frac{4}{3}z + \frac{2}{9}\right) = 0 \cdot 9$$
$$36z^2 + 12z + 2 = 0$$
$$z = \frac{-12 \pm \sqrt{12^2 - 4(36)(2)}}{72}$$
$$z = \frac{-12 \pm \sqrt{-144}}{72}$$
$$z = -\frac{12}{72} \pm \frac{12}{72}i$$
$$z = -\frac{1}{6} \pm \frac{1}{6}i$$

43. $\frac{x}{x-2} = \frac{5}{x+3}$

$$x(x+3) = 5(x-2)$$
$$x^2 + 3x = 5x - 10$$
$$x^2 - 2x + 10 = 0$$
$$x = \frac{-b \pm \sqrt{b^2 - 4ac}}{2a}$$
$$x = \frac{-(-2) \pm \sqrt{(-2)^2 - 4(1)(10)}}{2(1)}$$
$$x = \frac{2 \pm \sqrt{-36}}{2}$$
$$x = \frac{2 \pm 6i}{2}$$
$$x = \frac{2}{2} \pm \frac{6i}{2}$$
$$x = 1 \pm 3i$$

45. $\dfrac{y+7}{y-3} = \dfrac{6}{y+4}$

$(y+7)(y+4) = 6(y-3)$

$y^2 + 11y + 28 = 6y - 18$

$y^2 + 5y + 46 = 0$

$y = \dfrac{-b \pm \sqrt{b^2 - 4ac}}{2a}$

$y = \dfrac{-5 \pm \sqrt{5^2 - 4(1)(46)}}{2(1)}$

$y = \dfrac{-5 \pm \sqrt{-159}}{2}$

$y = -\dfrac{5}{2} \pm \dfrac{\sqrt{159}}{2} i$

47. $\dfrac{z-5}{z(z+3)} = \dfrac{4}{5}$

$5(z-5) = 4z(z+3)$

$5z - 25 = 4z^2 + 12z$

$0 = 4z^2 + 7z + 25$

$z = \dfrac{-b \pm \sqrt{b^2 - 4ac}}{2a}$

$z = \dfrac{-7 \pm \sqrt{7^2 - 4(4)(25)}}{2(4)}$

$z = \dfrac{-7 \pm \sqrt{-351}}{8}$

$z = \dfrac{-7 \pm 3\sqrt{39} i}{8}$

$z = -\dfrac{7}{8} \pm \dfrac{3\sqrt{39}}{8} i$

49. $G(s) = \dfrac{3(s+2)}{s(s+1)}$

$-3 = \dfrac{3(s+2)}{s(s+1)}$

$-3s^2 - 3s = 3s + 6$

$3s^2 + 6s + 6 = 0$

$\dfrac{3(s^2 + 2s + 2)}{3} = \dfrac{0}{3}$

$s^2 + 2s + 2 = 0$

$s = \dfrac{-2 \pm \sqrt{2^2 - 4(1)(2)}}{2}$

$s = \dfrac{-2 \pm \sqrt{-4}}{2}$

$s = \dfrac{-2 \pm 2i}{2}$

$s = -1 \pm i$

51. $K = 1.00 - 2.67r + r^2$

$-1.75 = 1 - 2.67r + r^2$

$0 = 2.75 - 2.67r + r^2$

$0 = r^2 - 2.67r + 2.75$

$r = \dfrac{2.67 \pm \sqrt{(-2.67)^2 - 4(1)(2.75)}}{2}$

$r = \dfrac{2.67 \pm \sqrt{-3.8711}}{2}$

$r = \dfrac{2.67}{2} \pm \dfrac{\sqrt{3.8711}}{2} i$

$r = 1.335 \pm \dfrac{\sqrt{3.8711}}{2} i$

14.2 Experiencing Algebra the Calculator Way

1. $x^2 - 10x + 34 = 0$

$x = \dfrac{-b \pm \sqrt{b^2 - 4ac}}{2a}$

$x = \dfrac{-(-10) \pm \sqrt{(-10)^2 - 4(1)(34)}}{2(1)}$

$x = \dfrac{10 \pm \sqrt{-36}}{2}$

$x = \dfrac{10 \pm 6i}{2}$

$x = 5 \pm 3i$

```
5+3i→X:X²-10X+34
                0
5-3i→X:X²-10X+34
                0
```

2. $x^2 + 14x + 65 = 0$

$$x = \frac{-b \pm \sqrt{b^2 - 4ac}}{2a}$$
$$x = \frac{-14 \pm \sqrt{(14)^2 - 4(1)(65)}}{2(1)}$$
$$x = \frac{-14 \pm \sqrt{-64}}{2}$$
$$x = \frac{-14 \pm 8i}{2}$$
$$x = -7 \pm 4i$$

```
-7+4i→X:X²+14X+6
5
               0
-7-4i→X:X²+14X+6
5
               0
```

3. $4x^2 + 20x + 26 = 0$

$$x = \frac{-20 \pm \sqrt{(20)^2 - 4(4)(26)}}{2(4)}$$
$$x = \frac{-20 \pm \sqrt{-16}}{8}$$
$$x = \frac{-20 \pm 4i}{8}$$
$$x = -\frac{5}{2} \pm \frac{1}{2}i$$

```
-5/2+1/2i→X:4X^2
+20X+26
               0
-5/2-1/2i→X:4X^2
+20X+26
               0
```

4. $16x^2 - 8x + 10 = 0$

$$x = \frac{-(-8) \pm \sqrt{(-8)^2 - 4(16)(10)}}{2(16)}$$
$$x = \frac{8 \pm \sqrt{-576}}{32}$$
$$x = \frac{8 \pm 24i}{32}$$
$$x = \frac{1}{4} \pm \frac{3}{4}i$$

```
1/4+3/4i→X:16X²-
8X+10
               0
1/4-3/4i→X:16X²-
8X+10
               0
```

Chapter 14 Review

Reflections

1.–7. Answers will vary.

Exercises

1. $\sqrt{-64}$
$= \sqrt{64}i$
$= 8i$

2. $\sqrt{-\frac{25}{49}}$
$= \sqrt{\frac{25}{49}}i$
$= \frac{5}{7}i$

3. $\sqrt{-6.25}$
$= \sqrt{6.25}i$
$= 2.5i$

4. $\sqrt{-50}$
$= \sqrt{50}i$
$= 5\sqrt{2}i$

5. $\sqrt{-\frac{81}{2}}$
$= \sqrt{\frac{81}{2}}i$
$= \frac{\sqrt{81}}{\sqrt{2}}i$
$= \frac{9}{\sqrt{2}} \cdot \frac{\sqrt{2}}{\sqrt{2}}i$
$= \frac{9\sqrt{2}}{2}i$

6. $-5\sqrt{-32}$
$= -5\sqrt{32}i$
$= -5 \cdot 4\sqrt{2}i$
$= -20\sqrt{2}i$

7. $\sqrt{-25} + \sqrt{-49}$
$= \sqrt{25}i + \sqrt{49}i$
$= 5i + 7i$
$= 12i$

8. $\sqrt{-144} - \sqrt{-64}$
$= \sqrt{144}i - \sqrt{64}i$
$= 12i - 8i$
$= 4i$

9. $2\sqrt{-36} + 6\sqrt{-81}$
$= 2\sqrt{36}i + 6\sqrt{81}i$
$= 2 \cdot 6i + 6 \cdot 9i$
$= 12i + 54i$
$= 66i$

10. $11\sqrt{-4} - 3\sqrt{-36}$
$= 11\sqrt{4}i - 3\sqrt{36}i$
$= 11 \cdot 2i - 3 \cdot 6i$
$= 22i - 18i$
$= 4i$

11. $\dfrac{\sqrt{-169}}{\sqrt{-225}}$
$= \dfrac{\sqrt{169}i}{\sqrt{225}i}$
$= \dfrac{13i}{15i}$
$= \dfrac{13}{15}$

12. $\sqrt{-2}\sqrt{-6}\sqrt{-75}$
$= \sqrt{2}i\sqrt{6}i\sqrt{75}i$
$= \sqrt{900}i^3$
$= 30(-i)$
$= -30i$

13. $\sqrt{-2}\left(\sqrt{-6} + \sqrt{75}\right)$
$= \sqrt{2}i\left(\sqrt{6}i + \sqrt{75}\right)$
$= \sqrt{12}i^2 + \sqrt{150}i$
$= 2\sqrt{3}(-1) + 5\sqrt{6}i$
$= -2\sqrt{3} + 5\sqrt{6}i$

14. $\left(\sqrt{-2} + \sqrt{-18}\right)\left(\sqrt{-4} - \sqrt{-9}\right)$
$= \left(\sqrt{2}i + \sqrt{18}i\right)\left(\sqrt{4}i - \sqrt{9}i\right)$
$= \sqrt{8}i^2 - \sqrt{18}i^2 + \sqrt{72}i^2 - \sqrt{162}i^2$
$= 2\sqrt{2}(-1) - 3\sqrt{2}(-1) + 6\sqrt{2}(-1) - 9\sqrt{2}(-1)$
$= -2\sqrt{2} + 3\sqrt{2} - 6\sqrt{2} + 9\sqrt{2}$
$= 4\sqrt{2}$

15. $\left(\sqrt{-7} + \sqrt{-11}\right)\left(\sqrt{-7} - \sqrt{-11}\right)$
$= \left(\sqrt{7}i + \sqrt{11}i\right)\left(\sqrt{7}i - \sqrt{11}i\right)$
$= \left(\sqrt{7}i\right)^2 - \left(\sqrt{11}i\right)^2$
$= 7i^2 - 11i^2$
$= -7 + 11$
$= 4$

16. $-\sqrt{3}i$ is a pure imaginary number. The real part is 0 and the imaginary part is $-\sqrt{3}$.

17. $\dfrac{\sqrt{11}}{3}$ is a real number. The real part is $\dfrac{\sqrt{11}}{3}$ and the imaginary part is 0.

18. $\sqrt{13} - 2i$ is an imaginary number. The real part is $\sqrt{13}$ and the imaginary part is –2.

19. $4\sqrt{2} + 1$ is a real number. The real part is $4\sqrt{2} + 1$ and the imaginary part is 0.

20. $3 - 17i$ is an imaginary number. The real part is 3 and the imaginary part is –17.

21. –43 is a real number. The real part is –43 and the imaginary part is 0.

22. $(17+4i)+(12-6i)$
$=17+4i+12-6i$
$=29-2i$

23. $-15i(3-6i)$
$=-45i+90i^2$
$=-45i+90(-1)$
$=-90-45i$

24. $(12-3i)-(1+i)$
$=12-3i-1-i$
$=11-4i$

25. $(2+5i)(-3+4i)$
$=-6+8i-15i+20i^2$
$=-6+8i-15i-20$
$=-26-7i$

26. $(3+11i)(3-11i)$
$=3^2-(11i)^2$
$=9-121i^2$
$=9-121(-1)$
$=9+121$
$=130$

27. $6-(2+12i)$
$=6-2-12i$
$=4-12i$

28. $\dfrac{12+15i}{3}$
$=\dfrac{12}{3}+\dfrac{15i}{3}$
$=4+5i$

29. $\dfrac{14-21i}{-7i}$
$=\dfrac{(14-21i)(7i)}{(-7i)(7i)}$
$=\dfrac{98i-147i^2}{49}$
$=\dfrac{98i-147(-1)}{49}$
$=\dfrac{147+98i}{49}$
$=3+2i$

30. $\dfrac{41+i}{5-2i}$
$=\dfrac{(41+i)(5+2i)}{(5-2i)(5+2i)}$
$=\dfrac{205+82i+5i+2i^2}{25+4}$
$=\dfrac{205+82i+5i-2}{29}$
$=\dfrac{203+87i}{29}$
$=7+3i$

31. $\left(\dfrac{3}{5}-5i\right)+\left(\dfrac{2}{5}+2i\right)$
$=\dfrac{3}{5}-5i+\dfrac{2}{5}+2i$
$=1-3i$

32. $(13.7+0.29i)-(14.92+1.5i)$
$=13.7+0.29i-14.92-1.5i$
$=-1.22-1.21i$

33. $\left(5\sqrt{7}+8\sqrt{13}i\right)+\left(3\sqrt{7}-2\sqrt{13}i\right)$
$=5\sqrt{7}+8\sqrt{13}i+3\sqrt{7}-2\sqrt{13}i$
$=8\sqrt{7}+6\sqrt{13}i$

34. $\sqrt{6}i\left(\sqrt{3}-\sqrt{6}i\right)$
$=\sqrt{18}i-\sqrt{36}i^2$
$=3\sqrt{2}i-6(-1)$
$=6+3\sqrt{2}i$

35. $\left(\dfrac{2}{3}-\dfrac{3}{5}i\right)\left(\dfrac{1}{2}+\dfrac{1}{3}i\right)$
$=\dfrac{1}{3}+\dfrac{2}{9}i-\dfrac{3}{10}i-\dfrac{1}{5}i^2$
$=\dfrac{5}{15}+\dfrac{20}{90}i-\dfrac{27}{90}i+\dfrac{3}{15}$
$=\dfrac{8}{15}-\dfrac{7}{90}i$

36. $5.8i(1.9-2i)$
$=11.02i-11.6i^2$
$=11.02i-11.6(-1)$
$=11.6+11.02i$

37. $\dfrac{\sqrt{10}+5\sqrt{3}i}{\sqrt{5}}$
$=\dfrac{\sqrt{10}}{\sqrt{5}}+\dfrac{5\sqrt{3}}{\sqrt{5}}i$
$=\sqrt{\dfrac{10}{5}}+\dfrac{5\sqrt{3}}{\sqrt{5}}\cdot\dfrac{\sqrt{5}}{\sqrt{5}}i$
$=\sqrt{2}+\dfrac{5\sqrt{15}}{5}i$
$=\sqrt{2}+\sqrt{15}i$

38. $\dfrac{18-11\sqrt{6}i}{3\sqrt{2}+\sqrt{3}i}$
$=\dfrac{\left(18-11\sqrt{6}i\right)\left(3\sqrt{2}-\sqrt{3}i\right)}{\left(3\sqrt{2}+\sqrt{3}i\right)\left(3\sqrt{2}-\sqrt{3}i\right)}$
$=\dfrac{54\sqrt{2}-18\sqrt{3}i-33\sqrt{12}i+11\sqrt{18}i^2}{\left(3\sqrt{2}\right)^2-\left(\sqrt{3}i\right)^2}$
$=\dfrac{54\sqrt{2}-18\sqrt{3}i-66\sqrt{3}i-33\sqrt{2}}{18+3}$
$=\dfrac{21\sqrt{2}-84\sqrt{3}i}{21}$
$=\sqrt{2}-4\sqrt{3}i$

39. $V=IZ$
$V=(5+4i)(9-2i)$
$V=45-10i+36i-8i^2$
$V=45-10i+36i+8$
$V=53+26i$
The total voltage is $53+26i$ volts.
$|V|=\sqrt{53^2+26^2}\approx 59.0$
The magnitude of the total voltage is 59.0 volts.

40. $Z=\dfrac{V}{I}$
$Z=\dfrac{33+19i}{5+2i}$
$Z=\dfrac{(33+19i)(5-2i)}{(5+2i)(5-2i)}$
$=\dfrac{165-66i+95i-38i^2}{25+4}$
$=\dfrac{165-66i+95i+38}{29}$
$=\dfrac{203+29i}{29}$
$=7+i$
The impedance is $7+i$ ohms.
$|Z|=\sqrt{7^2+1^2}=\sqrt{50}=5\sqrt{2}$
The magnitude of the impedance is $5\sqrt{2}$ ohms.

41. $x^2+100=0$
$x^2=-100$
$x=\pm\sqrt{100}i$
$x=\pm 10i$

42. $y^2+15=0$
$y^2=-15$
$y=\pm\sqrt{15}i$

43. $z^2+10=1$
$z^2=-9$
$z=\pm\sqrt{9}i$
$z=\pm 3i$

44. $5t^2+125=0$
$5t^2=-125$
$t^2=-25$
$t=\pm\sqrt{25}i$
$t=\pm 5i$

45. $3a^2+33=0$
$3a^2=-33$
$a^2=-11$
$a=\pm\sqrt{11}i$

46. $(r-5)^2+36=0$
$(r-5)^2=-36$
$r-5=\pm\sqrt{36}i$
$r-5=\pm 6i$
$r=5\pm 6i$

47. $7(x+2)^2+23=2$
$7(x+2)^2=-21$
$(x+2)^2=-3$
$x+2=\pm\sqrt{3}i$
$x=-2\pm\sqrt{3}i$

48. $3(4x+1)^2=-27$
$(4x+1)^2=-9$
$4x+1=\pm\sqrt{9}i$
$4x+1=\pm 3i$
$4x=-1\pm 3i$
$x=\dfrac{-1\pm 3i}{4}$
$x=-\dfrac{1}{4}\pm\dfrac{3}{4}i$

49. $u^2+4.84=0$
$u^2=-4.84$
$u=\pm\sqrt{4.84}i$
$u=\pm 2.2i$

50. $(v-3.4)^2+3.24=0$
$(v-3.4)^2=-3.24$
$v-3.4=\pm\sqrt{3.24}i$
$v-3.4=\pm 1.8i$
$v=3.4\pm 1.8i$

51. $c^2+\dfrac{36}{121}=0$
$c^2=-\dfrac{36}{121}$
$c=\pm\sqrt{\dfrac{36}{121}}i$
$c=\pm\dfrac{6}{11}i$

52. $\left(m+\dfrac{2}{5}\right)^2+\dfrac{16}{25}=0$
$\left(m+\dfrac{2}{5}\right)^2=-\dfrac{16}{25}$
$m+\dfrac{2}{5}=\pm\sqrt{\dfrac{16}{25}}i$
$m+\dfrac{2}{5}=\pm\dfrac{4}{5}i$
$m=-\dfrac{2}{5}\pm\dfrac{4}{5}i$

53. $x^2-10x+29=0$
$x=\dfrac{-b\pm\sqrt{b^2-4ac}}{2a}$
$x=\dfrac{-(-10)\pm\sqrt{(-10)^2-4(1)(29)}}{2(1)}$
$x=\dfrac{10\pm\sqrt{-16}}{2}$
$x=\dfrac{10\pm 4i}{2}$
$x=5\pm 2i$

54. $4y^2+4y+10=0$
$y=\dfrac{-b\pm\sqrt{b^2-4ac}}{2a}$
$y=\dfrac{-4\pm\sqrt{4^2-4(4)(10)}}{2(4)}$
$y=\dfrac{-4\pm\sqrt{-144}}{8}$
$y=\dfrac{-4\pm 12i}{8}$
$y=-\dfrac{1}{2}\pm\dfrac{3}{2}i$

55. $z^2+16z+17=0$

$$z=\frac{-b\pm\sqrt{b^2-4ac}}{2a}$$
$$z=\frac{-16\pm\sqrt{(16)^2-4(1)(17)}}{2(1)}$$
$$z=\frac{-16\pm\sqrt{188}}{2}$$
$$z=\frac{-16\pm 2\sqrt{47}}{2}$$
$$z=-8\pm\sqrt{47}$$

56. $9x^2-12x+40=0$

$$x=\frac{-b\pm\sqrt{b^2-4ac}}{2a}$$
$$x=\frac{-(-12)\pm\sqrt{(-12)^2-4(9)(40)}}{2(9)}$$
$$x=\frac{12\pm\sqrt{-1296}}{18}$$
$$x=\frac{12\pm 36i}{18}$$
$$x=\frac{2}{3}\pm 2i$$

57. $x^2-2.4x+2.65=0$

$$x=\frac{-b\pm\sqrt{b^2-4ac}}{2a}$$
$$x=\frac{-(-2.4)\pm\sqrt{(-2.4)^2-4(1)(2.65)}}{2(1)}$$
$$x=\frac{2.4\pm\sqrt{-4.84}}{2}$$
$$x=\frac{2.4\pm 2.2i}{2}$$
$$x=1.2\pm 1.1i$$

58. $4x^2+4x+5.84=0$

$$x=\frac{-b\pm\sqrt{b^2-4ac}}{2a}$$
$$x=\frac{-4\pm\sqrt{4^2-4(4)(5.84)}}{2(4)}$$
$$x=\frac{-4\pm\sqrt{-77.44}}{8}$$
$$x=\frac{-4\pm 8.8i}{8}$$
$$x=-0.5\pm 1.1i$$

59. $x^2-\frac{3}{2}x+\frac{5}{8}=0$

$$8\left(x^2-\frac{3}{2}x+\frac{5}{8}\right)=0\cdot 8$$
$$8x^2-12x+5=0$$
$$x=\frac{12\pm\sqrt{(-12)^2-4(8)(5)}}{16}$$
$$x=\frac{12\pm\sqrt{-16}}{16}$$
$$x=\frac{12}{16}\pm\frac{4}{16}i$$
$$x=\frac{3}{4}\pm\frac{1}{4}i$$

60. $4x^2-12x+\frac{85}{9}=0$

$$9\left(4x^2-12x+\frac{85}{9}\right)=0\cdot 9$$
$$36x^2-108x+85=0$$
$$x=\frac{108\pm\sqrt{(-108)^2-4(36)(85)}}{72}$$
$$x=\frac{108\pm\sqrt{-576}}{72}$$
$$x=\frac{108}{72}\pm\frac{24}{72}i$$
$$x=\frac{3}{2}\pm\frac{1}{3}i$$

61. $\frac{x}{x-5} = \frac{10}{x+1}$
$x(x+1) = 10(x-5)$
$x^2 + x = 10x - 50$
$x^2 - 9x + 50 = 0$
$x = \frac{-(-9) \pm \sqrt{(-9)^2 - 4(1)(50)}}{2(1)}$
$x = \frac{9 \pm \sqrt{-119}}{2}$
$x = \frac{9}{2} \pm \frac{\sqrt{119}}{2} i$

62. $\frac{y+4}{y-8} = \frac{9}{y+6}$
$(y+4)(y+6) = 9(y-8)$
$y^2 + 10y + 24 = 9y - 72$
$y^2 + y + 96 = 0$
$y = \frac{-1 \pm \sqrt{1^2 - 4(1)(96)}}{2(1)}$
$y = \frac{-1 \pm \sqrt{-383}}{2}$
$y = -\frac{1}{2} \pm \frac{\sqrt{383}}{2} i$

63. $\frac{b-8}{b(b+7)} = \frac{2}{3}$
$3(b-8) = 2b(b+7)$
$3b - 24 = 2b^2 + 14b$
$0 = 2b^2 + 11b + 24$
$b = \frac{-11 \pm \sqrt{(11)^2 - 4(2)(24)}}{2(2)}$
$b = \frac{-11 \pm \sqrt{-71}}{4}$
$b = -\frac{11}{4} \pm \frac{\sqrt{71}}{4} i$

64. $G(s) = \frac{3(s+10)}{s(s+5)}$
$-1 = \frac{3(s+10)}{s(s+5)}$
$-s(s+5) = 3(s+10)$
$-s^2 - 5s = 3s + 30$
$0 = s^2 + 8s + 30$
$s = \frac{-8 \pm \sqrt{(8)^2 - 4(1)(30)}}{2(1)}$
$s = \frac{-8 \pm \sqrt{64 - 120}}{2}$
$s = \frac{-8 \pm \sqrt{-56}}{2}$
$s = \frac{-8 \pm 2\sqrt{14}i}{2}$
$s = -4 \pm \sqrt{14}i$

Chapter 14 Mixed Review

1. $-\frac{\sqrt{23}}{5}$ is a real number. The real part is $-\frac{\sqrt{23}}{5}$ and the imaginary part is 0.

2. $\sqrt{29} + 5i$ is an imaginary number. The real part is $\sqrt{29}$ and the imaginary part is 5.

3. $16 + 22i$ is an imaginary number. The real part is 16 and the imaginary part is 22.

4. $-3\sqrt{5} - 7$ is a real number. The real part is $-3\sqrt{5} - 7$ and the imaginary part is 0.

5. $\sqrt{31}i$ is a pure imaginary number. The real part is 0 and the imaginary part is $\sqrt{31}$.

6. $-\pi$ is a real number. The real part is $-\pi$ and the imaginary part is 0.

7. $\sqrt{-169}$
$= \sqrt{169}i$
$= 13i$

8. $\sqrt{-\frac{64}{81}}$
$= \sqrt{\frac{64}{81}}i$
$= \frac{8}{9}i$

9. $\sqrt{-\frac{25}{6}}$
$= \sqrt{\frac{25}{6}}i$
$= \frac{\sqrt{25}}{\sqrt{6}}i$
$= \frac{5}{\sqrt{6}} \cdot \frac{\sqrt{6}}{\sqrt{6}}i$
$= \frac{5\sqrt{6}}{6}i$

10. $10\sqrt{-80}$
$= 10\sqrt{80}i$
$= 10 \cdot 4\sqrt{5}i$
$= 40\sqrt{5}i$

11. $\sqrt{-6.25}$
$= \sqrt{6.25}i$
$= 2.5i$

12. $-\sqrt{-108}$
$= -\sqrt{108}i$
$= -6\sqrt{3}i$

13. $(22.3 - 1.33i) + (8.55 - 2.9i)$
$= 22.3 - 1.33i + 8.55 - 2.9i$
$= 30.85 - 4.23i$

14. $\sqrt{-64} + \sqrt{-4}$
$= \sqrt{64}i + \sqrt{4}i$
$= 8i + 2i$
$= 10i$

15. $\sqrt{-196} - \sqrt{-100}$
$= \sqrt{196}i - \sqrt{100}i$
$= 14i - 10i$
$= 4i$

16. $\left(\frac{3}{4} + \frac{1}{2}i\right)\left(\frac{4}{9} - 2i\right)$
$= \frac{1}{3} - \frac{3}{2}i + \frac{2}{9}i - i^2$
$= \frac{1}{3} - \frac{27}{18}i + \frac{4}{18}i + 1$
$= \frac{4}{3} - \frac{23}{18}i$

17. $-1.5i(2.9 - 4i)$
$= -4.35i + 6i^2$
$= -4.35i + 6(-1)$
$= -6 - 4.35i$

18. $7\sqrt{-25} + 3\sqrt{-64}$
$= 7\sqrt{25}i + 3\sqrt{64}i$
$= 7 \cdot 5i + 3 \cdot 8i$
$= 35i + 24i$
$= 59i$

19. $\sqrt{-2}\sqrt{-6}\sqrt{-48}$
$= \sqrt{2}i\sqrt{6}i\sqrt{48}i$
$= \sqrt{576}i^3$
$= 24(-i)$
$= -24i$

20. $\left(\sqrt{-3} - \sqrt{-10}\right)\left(\sqrt{-5} + \sqrt{-6}\right)$
$= \left(\sqrt{3}i - \sqrt{10}i\right)\left(\sqrt{5}i + \sqrt{6}i\right)$
$= \sqrt{15}i^2 - \sqrt{18}i^2 - \sqrt{50}i^2 - \sqrt{60}i^2$
$= \sqrt{15}(-1) - 3\sqrt{2}(-1) - 5\sqrt{2}(-1) - 2\sqrt{15}(-1)$
$= -\sqrt{15} + 3\sqrt{2} + 5\sqrt{2} + 2\sqrt{15}$
$= \sqrt{15} + 8\sqrt{2}$

21. $\left(\sqrt{-13}+\sqrt{-17}\right)\left(\sqrt{-13}-\sqrt{-17}\right)$
$=\left(\sqrt{13}i+\sqrt{17}i\right)\left(\sqrt{13}i-\sqrt{17}i\right)$
$=\left(\sqrt{13}i\right)^2-\left(\sqrt{17}i\right)^2$
$=13i^2-17i^2$
$=13(-1)-17(-1)$
$=4$

22. $(21-i)-(7+i)$
$=21-i-7-i$
$=14-2i$

23. $(9-3i)(-9+3i)$
$=-81+27i+27i-9i^2$
$=-81+27i+27i-9(-1)$
$=-72+54i$

24. $(-2+12i)(-2-12i)$
$=(-2)^2-(12i)^2$
$=4-144i^2$
$=4+144$
$=148$

25. $3i-(5+7i)$
$=3i-5-7i$
$=-5-4i$

26. $\dfrac{24-16i}{-8}$
$=\dfrac{24}{-8}-\dfrac{16i}{-8}$
$=-3+2i$

27. $\dfrac{18-27i}{9i}$
$=\dfrac{18}{9i}-\dfrac{27i}{9i}$
$=\dfrac{2}{i}\cdot\dfrac{i}{i}-3$
$=\dfrac{2i}{-1}-3$
$=-3-2i$

28. $\dfrac{32-9i}{2-3i}$
$=\dfrac{(32-9i)(2+3i)}{(2-3i)(2+3i)}$
$=\dfrac{64+96i-18i-27i^2}{4+9}$
$=\dfrac{64+96i-18i+27}{13}$
$=\dfrac{91+78i}{13}$
$=7+6i$

29. $\left(\dfrac{5}{6}-7i\right)+\left(\dfrac{5}{6}+i\right)$
$=\dfrac{5}{6}-7i+\dfrac{5}{6}+i$
$=\dfrac{5}{3}-6i$

30. $\sqrt{10}i\left(\sqrt{5}-\sqrt{2}i\right)$
$=\sqrt{50}i-\sqrt{20}i^2$
$=5\sqrt{2}i-2\sqrt{5}(-1)$
$=2\sqrt{5}+5\sqrt{2}i$

31. $\dfrac{\sqrt{30}+6\sqrt{2}i}{\sqrt{6}}$
$=\dfrac{\sqrt{30}}{\sqrt{6}}+\dfrac{6\sqrt{2}}{\sqrt{6}}i$
$=\sqrt{\dfrac{30}{6}}+\dfrac{6\sqrt{2}}{\sqrt{6}}\cdot\dfrac{\sqrt{6}}{\sqrt{6}}i$
$=\sqrt{5}+\dfrac{6\sqrt{12}}{6}i$
$=\sqrt{5}+2\sqrt{3}i$

32. $\dfrac{12i+\sqrt{10}}{2\sqrt{5}+\sqrt{2}i}$

$= \dfrac{(12i+\sqrt{10})(2\sqrt{5}-\sqrt{2}i)}{(2\sqrt{5}+\sqrt{2}i)(2\sqrt{5}-\sqrt{2}i)}$

$= \dfrac{24\sqrt{5}i-12\sqrt{2}i^2+2\sqrt{50}-\sqrt{20}i}{(2\sqrt{5})^2-(\sqrt{2}i)^2}$

$= \dfrac{24\sqrt{5}i+12\sqrt{2}+10\sqrt{2}-2\sqrt{5}i}{20+2}$

$= \dfrac{22\sqrt{2}+22\sqrt{5}i}{22}$

$= \sqrt{2}+\sqrt{5}i$

33. $\dfrac{\sqrt{-289}}{\sqrt{-841}}$

$= \dfrac{\sqrt{289}i}{\sqrt{841}i}$

$= \dfrac{17i}{29i}$

$= \dfrac{17}{29}$

34. $\dfrac{d-11}{d(d+9)} = \dfrac{5}{8}$

$8(d-11) = 5d(d+9)$

$8d-88 = 5d^2+45d$

$0 = 5d^2+37d+88$

$d = \dfrac{-b \pm \sqrt{b^2-4ac}}{2a}$

$d = \dfrac{-(37) \pm \sqrt{(37)^2-4(5)(88)}}{2(5)}$

$d = \dfrac{-37 \pm \sqrt{-391}}{10}$

$d = -\dfrac{37}{10} \pm \dfrac{\sqrt{391}}{10}i$

35. $x^2+10.24 = 0$

$x^2 = -10.24$

$x = \pm\sqrt{10.24}i$

$x = \pm 3.2i$

36. $(m+7)^2+25 = 0$

$(m+7)^2 = -25$

$m+7 = \pm\sqrt{25}i$

$m+7 = \pm 5i$

$m = -7 \pm 5i$

37. $4(y-3)^2+29 = 5$

$4(y-3)^2 = -24$

$(y-3)^2 = -6$

$y-3 = \pm\sqrt{6}i$

$y = 3 \pm \sqrt{6}i$

38. $x^2-16x+67 = 0$

$x = \dfrac{-b \pm \sqrt{b^2-4ac}}{2a}$

$x = \dfrac{-(-16) \pm \sqrt{(-16)^2-4(1)(67)}}{2(1)}$

$x = \dfrac{16 \pm \sqrt{-12}}{2}$

$x = \dfrac{16 \pm 2\sqrt{3}i}{2}$

$x = 8 \pm \sqrt{3}i$

39. $9y^2+30y+32 = 0$

$y = \dfrac{-b \pm \sqrt{b^2-4ac}}{2a}$

$y = \dfrac{-30 \pm \sqrt{(30)^2-4(9)(32)}}{2(9)}$

$y = \dfrac{-30 \pm \sqrt{-252}}{18}$

$y = \dfrac{-30 \pm 6\sqrt{7}i}{18}$

$y = -\dfrac{30}{18} \pm \dfrac{6\sqrt{7}}{18}i$

$y = -\dfrac{5}{3} \pm \dfrac{\sqrt{7}}{3}i$

40. $a^2 + 100 = 19$
$a^2 = -81$
$a = \pm\sqrt{81}i$
$a = \pm 9i$

41. $6b^2 + 78 = 0$
$6b^2 = -78$
$b^2 = -13$
$b = \pm\sqrt{13}i$

42. $4x^2 + 12x + 10.96 = 0$
$x = \frac{-b \pm \sqrt{b^2 - 4ac}}{2a}$
$x = \frac{-12 \pm \sqrt{(12)^2 - 4(4)(10.96)}}{2(4)}$
$x = \frac{-12 \pm \sqrt{-31.36}}{8}$
$x = \frac{-12 \pm 5.6i}{8}$
$x = -1.5 \pm 0.7i$

43. $x^2 + \frac{5}{4}x + \frac{7}{16} = 0$
$16\left(x^2 + \frac{5}{4}x + \frac{7}{16}\right) = 0 \cdot 16$
$16x^2 + 20x + 7 = 0$
$x = \frac{-20 \pm \sqrt{20^2 - 4(16)(7)}}{32}$
$x = \frac{-20 \pm \sqrt{-48}}{32}$
$x = \frac{-20}{32} \pm \frac{4\sqrt{3}}{32}i$
$x = -\frac{5}{8} \pm \frac{\sqrt{3}}{8}i$

44. $\frac{y+9}{y+1} = \frac{5}{y-7}$
$(y + 9)(y - 7) = 5(y + 1)$
$y^2 + 2y - 63 = 5y + 5$
$y^2 - 3y - 68 = 0$
$y = \frac{-b \pm \sqrt{b^2 - 4ac}}{2a}$
$y = \frac{-(-3) \pm \sqrt{(-3)^2 - 4(1)(-68)}}{2(1)}$
$y = \frac{3 \pm \sqrt{281}}{2}$
$y = \frac{3}{2} \pm \frac{\sqrt{281}}{2}$

45. $\left(y + \frac{3}{7}\right)^2 + \frac{25}{49} = 0$
$\left(y + \frac{3}{7}\right)^2 = -\frac{25}{49}$
$y + \frac{3}{7} = \pm\sqrt{\frac{25}{49}}i$
$y + \frac{3}{7} = \pm\frac{5}{7}i$
$y = -\frac{3}{7} \pm \frac{5}{7}i$

46. $Z = \frac{V}{I}$ or $I = \frac{V}{Z}$
$I = \frac{30 + 52i}{9 + 5i}$
$I = \frac{(30 + 52i)(9 - 5i)}{(9 + 5i)(9 - 5i)}$
$I = \frac{270 - 150i + 468i - 260i^2}{81 + 25}$
$I = \frac{530 + 318i}{106}$
$I = 5 + 3i$
The current is $5 + 3i$ amperes.
$|I| = \sqrt{5^2 + 3^2}$
$= \sqrt{34}$
≈ 5.8
The magnitude of the current is approximately 5.8 amperes.

47. $G(s)=\dfrac{3(s+10)}{s(s+5)}$

$-2=\dfrac{3(s+10)}{s(s+5)}$

$-2s(s+5)=3(s+10)$

$-2s^2-10s=3s+30$

$0=2s^2+13s+30$

$s=\dfrac{-13\pm\sqrt{13^2-4(2)(30)}}{4}$

$s=\dfrac{-13}{4}\pm\dfrac{\sqrt{71}}{4}i$

Chapter 14 Test

1. $\sqrt{-64}+\sqrt{-9}$
$=\sqrt{64}i+\sqrt{9}i$
$=8i+3i$
$=11i$

2. $3\sqrt{-100}-2\sqrt{-49}$
$=3\sqrt{100}i-2\sqrt{49}i$
$=3\cdot 10i-2\cdot 7i$
$=30i-14i$
$=16i$

3. $\sqrt{-2}\sqrt{-7}\sqrt{-56}$
$=\sqrt{2}i\sqrt{7}i\sqrt{56}i$
$=\sqrt{784}i^3$
$=28(-i)$
$=-28i$

4. $\sqrt{-2}\left(\sqrt{-7}-\sqrt{56}\right)$
$=\sqrt{2}i\left(\sqrt{7}i-\sqrt{56}\right)$
$=\sqrt{14}i^2-\sqrt{112}i$
$=\sqrt{14}(-1)-4\sqrt{7}i$
$=-\sqrt{14}-4\sqrt{7}i$

5. $\dfrac{\sqrt{-225}}{\sqrt{-289}}$
$=\dfrac{\sqrt{225}i}{\sqrt{289}i}$
$=\dfrac{15i}{17i}$
$=\dfrac{15}{17}$

6. $\left(\sqrt{-6}-\sqrt{-13}\right)\left(\sqrt{-6}+\sqrt{-13}\right)$
$=\left(\sqrt{6}i-\sqrt{13}i\right)\left(\sqrt{6}i+\sqrt{13}i\right)$
$=\left(\sqrt{6}i\right)^2-\left(\sqrt{13}i\right)^2$
$=6i^2-13i^2$
$=6(-1)-13(-1)$
$=-6+13$
$=7$

7. $(8+2i)(-5+i)$
$=-40+8i-10i+2i^2$
$=-40+8i-10i+2(-1)$
$=-40+8i-10i-2$
$=-42-2i$

8. $\dfrac{29+29i}{5-2i}$
$=\dfrac{(29+29i)(5+2i)}{(5-2i)(5+2i)}$
$=\dfrac{145+58i+145i+58i^2}{25+4}$
$=\dfrac{145+58i+145i-58}{29}$
$=\dfrac{87+203i}{29}$
$=3+7i$

9. $(22.47-13.6i)-(12.2-7.32i)$
$=22.47-13.6i-12.2+7.32i$
$=10.27-6.28i$

10. $\sqrt{15}i\left(\sqrt{3}+\sqrt{5}i\right)$
$=\sqrt{45}i+\sqrt{75}i^2$
$=3\sqrt{5}i+5\sqrt{3}(-1)$
$=-5\sqrt{3}+3\sqrt{5}i$

11. $\frac{44+16i}{2i}$
$=\frac{44}{2i}+\frac{16i}{2i}$
$=\frac{22}{i}\cdot\frac{i}{i}+8$
$=\frac{22i}{-1}+8$
$=8-22i$

12. $\left(\frac{1}{2}+\frac{3}{7}i\right)\left(\frac{14}{15}-\frac{7}{9}i\right)$
$=\frac{7}{15}-\frac{7}{18}i+\frac{2}{5}i-\frac{1}{3}i^2$
$=\frac{7}{15}-\frac{35}{90}i+\frac{36}{90}i+\frac{5}{15}$
$=\frac{4}{5}+\frac{1}{90}i$

13. $x^2+196=0$
$x^2=-196$
$x=\pm\sqrt{196}i$
$x=\pm 14i$

14. $(t+2)^2+15=0$
$(t+2)^2=-15$
$t+2=\pm\sqrt{15}i$
$t=-2\pm\sqrt{15}i$

15. $d^2+\frac{4}{49}=0$
$d^2=-\frac{4}{49}$
$d=\pm\sqrt{\frac{4}{49}}i$
$d=\pm\frac{2}{7}i$

16. $2z^2+5=2.12$
$2z^2=-2.88$
$z^2=-1.44$
$z=\pm\sqrt{1.44}i$
$z=\pm 1.2i$

17. $x^2-6x+17=0$
$x=\frac{-b\pm\sqrt{b^2-4ac}}{2a}$
$x=\frac{-(-6)\pm\sqrt{(-6)^2-4(1)(17)}}{2(1)}$
$x=\frac{6\pm\sqrt{-32}}{2}$
$x=\frac{6\pm 4\sqrt{2}i}{2}$
$x=3\pm 2\sqrt{2}i$

18. $9x^2-6x+7=0$
$x=\frac{-b\pm\sqrt{b^2-4ac}}{2a}$
$x=\frac{-(-6)\pm\sqrt{(-6)^2-4(9)(7)}}{2(9)}$
$x=\frac{6\pm\sqrt{-216}}{18}$
$x=\frac{6\pm 6\sqrt{6}i}{18}$
$x=\frac{1}{3}\pm\frac{\sqrt{6}}{3}i$

19. $\frac{y-2}{y(y+1)}=\frac{4}{7}$
$7(y-2)=4y(y+1)$
$7y-14=4y^2+4y$
$0=4y^2-3y+14$

$$y = \frac{-b \pm \sqrt{b^2 - 4ac}}{2a}$$
$$y = \frac{-(-3) \pm \sqrt{(-3)^2 - 4(4)(14)}}{2(4)}$$
$$y = \frac{3 \pm \sqrt{-215}}{8}$$
$$y = \frac{3 \pm \sqrt{215}i}{8}$$
$$y = \frac{3}{8} \pm \frac{\sqrt{215}}{8}i$$

20. $V = IZ$
$V = (8 + 3i)(11 - 3i)$
$V = 88 - 24i + 33i - 9i^2$
$V = 88 - 24i + 33i + 9$
$V = 97 + 9i$
The total voltage is $97 + 9i$ volts.
$|V| = \sqrt{97^2 + 9^2}$
$= \sqrt{9490}$
≈ 97.4
The magnitude of the total voltage is approximately 97.4 volts.

21. Answers will vary.

Chapter 15

15.1 Experiencing Algebra the Exercise Way

1. $h = \{(-3, 5), (-2, 4), (-1, 3), (0, 2), (1, 1), (2, 0)\}$
   $h^{-1} = \{(5, -3), (4, -2), (3, -1), (2, 0), (1, 1), (0, 2)\}$

3. $A = \{(3, 6), (2, 7), (1, 8), (0, 8), (-1, 7), (-2, 6)\}$
   $A^{-1} = \{(6, 3), (7, 2), (8, 1), (8, 0), (7, -1), (6, -2)\}$

5. $M = \left\{\left(\frac{1}{5}, 1.3\right), \left(\frac{3}{5}, 2.3\right), (1, 3.3), \left(\frac{7}{5}, 4.3\right), \left(\frac{9}{5}, 5.3\right), \left(\frac{11}{5}, 6.3\right)\right\}$
   $M^{-1} = \left\{\left(1.3, \frac{1}{5}\right), \left(2.3, \frac{3}{5}\right), (3.3, 1), \left(4.3, \frac{7}{5}\right), \left(5.3, \frac{9}{5}\right), \left(6.3, \frac{11}{5}\right)\right\}$

7. $y = 2x - 8$
   $x = 2y - 8$
   $x + 8 = 2y$
   $\frac{x+8}{2} = y$
   $y = \frac{x+8}{2}$
   $y = \frac{1}{2}x + 4$

9. $y = -3x + 2$
   $x = -3y + 2$
   $3y = 2 - x$
   $y = \frac{2-x}{3}$
   $y = -\frac{1}{3}x + \frac{2}{3}$

11. $y = \frac{3}{4}x + 9$
    $x = \frac{3}{4}y + 9$
    $x - 9 = \frac{3}{4}y$
    $\frac{4}{3}(x - 9) = \frac{4}{3}\left(\frac{3}{4}y\right)$
    $y = \frac{4}{3}(x - 9)$
    $y = \frac{4}{3}x - 12$

13. $y = 0.125x - 2.5$
    $x = 0.125y - 2.5$
    $x + 2.5 = 0.125y$
    $\frac{x+2.5}{0.125} = y$
    $y = \frac{x+2.5}{0.125}$
    $y = 8x + 20$

15. $y = x^2 - 2$
    $x = y^2 - 2$
    $x + 2 = y^2$
    $\pm\sqrt{x+2} = y$
    $y = \pm\sqrt{x+2}$

17. $k = \{(5, 2), (10, 4), (15, 8), (20, 16), (25, 32)\}$
    The function k is a one-to-one function because no two ordered pairs have the same first coordinate or the same second coordinate. The inverse is a function.

19. $Q = \{(-2, 0), (-4, 2), (-6, 4), (-8, 2), (-10, 0)\}$
    The function Q is not a one-to-one function because the second coordinates 0 and 2 are repeated in two ordered pairs. The inverse is not a function.

21. Yes, the graph represents a one-to-one function because all possible vertical as well as horizontal lines cross the graph only once.

23. No, the graph is a function because it passes the vertical line test, but a horizontal line can be drawn that intersects the graph more than once so it is not a one-to-one function.

25. Yes, the graph represents a one-to-one function because all possible vertical as well as horizontal lines cross the graph only once.

27. $g(x) = 3x - 6$

$$y = 3x - 6$$
$$x = 3y - 6$$
$$x + 6 = 3y$$
$$\frac{x+6}{3} = y$$
$$y = \frac{1}{3}x + 2$$

Therefore,

$$g^{-1}(x) = \frac{1}{3}x + 2$$

$Y1 = 3x - 6$

$Y2 = \frac{1}{3}x + 2$

$Y3 = x$

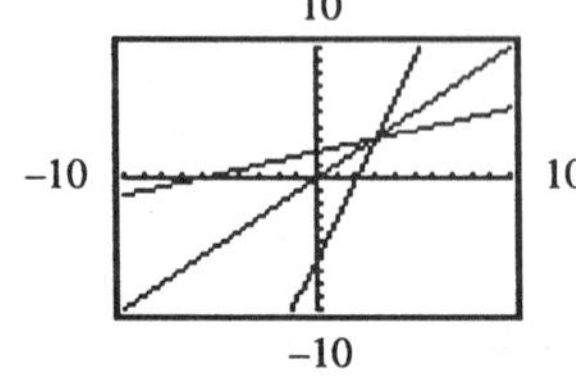

29. $y = \frac{2}{3}x - 4$

$$y = \frac{2}{3}x - 4$$
$$x = \frac{2}{3}y - 4$$
$$x + 4 = \frac{2}{3}y$$
$$\frac{3}{2}(x+4) = \frac{3}{2}\left(\frac{2}{3}y\right)$$
$$y = \frac{3}{2}x + 6$$

Therefore,

$$y^{-1} = \frac{3}{2}x + 6$$

$Y1 = \frac{2}{3}x - 4$

$Y2 = \frac{3}{2}x + 6$

$Y3 = x$

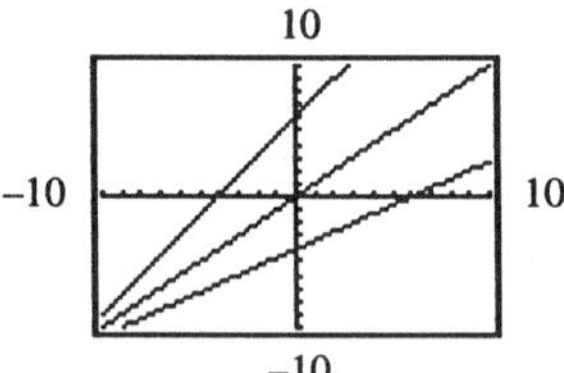

31. $h(x) = x^2 - 1$

$$y = x^2 - 1$$
$$x = y^2 - 1$$
$$x + 1 = y^2$$
$$\pm\sqrt{x+1} = y$$

Therefore,

$$h^{-1}(x) = \pm\sqrt{x+1}$$

$Y1 = x^2 - 1$

$Y2 = \sqrt{x+1}$

$Y3 = -\sqrt{x+1}$

$Y4 = x$

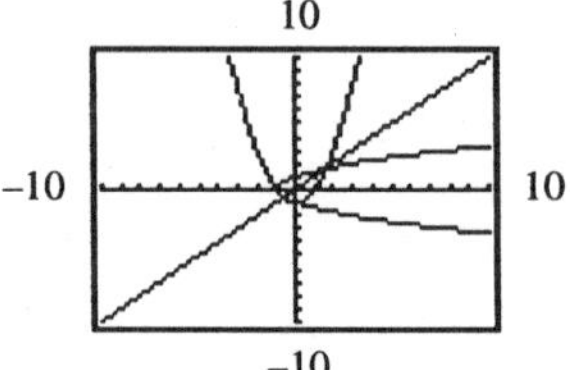

33. $y = \frac{1}{3}x^3 - 4$

$y = \frac{1}{3}x^3 - 4$

$x = \frac{1}{3}y^3 - 4$

$x + 4 = \frac{1}{3}y^3$

$3(x+4) = y^3$

$\sqrt[3]{3(x+4)} = y$

Therefore,

$y^{-1} = \sqrt[3]{3x+12}$

$Y1 = \frac{1}{3}x^3 - 4$

$Y2 = \sqrt[3]{3x+12}$

$Y3 = x$

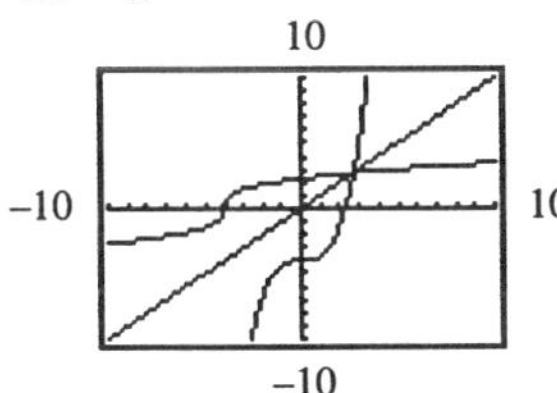

35. Let x = value of stock transactions.
Her monthly income can be represented by
$f(x) = 1500 + 0.02x$

$y = 1500 + 0.02x$

$x = 1500 + 0.02y$

$x - 1500 = 0.02y$

$\frac{x - 1500}{0.02} = y$

$y = 50x - 75{,}000$

Therefore, $f^{-1}(x) = 50x - 75{,}000$.

The inverse function represents the amount of Julia's transactions in terms of her monthly income.

37. Let x = amount of time
The distance traveled can be represented by
$f(x) = 55x$

$y = 55x$

$x = 55y$

$\frac{x}{55} = y$

Therefore, $f^{-1}(x) = \frac{x}{55}$.

The inverse function represents the time in terms of the distance traveled.

39. Let x = number of months to repay loan
$I = Prt = 2000(0.005)x = 10x$
Total to repay = $P + Prt = 2000 + 10x$
The total to repay can be represented by
$f(x) = 2000 + 10x$

$y = 2000 + 10x$

$x = 2000 + 10y$

$x - 2000 = 10y$

$\frac{x - 2000}{10} = y$

$y = \frac{1}{10}x - 200$

Therefore, $f^{-1}(x) = \frac{1}{10}x - 200$.

The inverse function represents the number of months in terms of the total amount to repay.

15.1 Experiencing Algebra the Calculator Way

1. $y = 0.3x^2 - 4$

$x = 0.3y^2 - 4$

$x + 4 = 0.3y^2$

$\frac{x+4}{0.3} = y^2$

$\pm\sqrt{\frac{x+4}{0.3}} = y$

$y^{-1} = \pm\sqrt{\frac{x+4}{0.3}}$

$Y1 = 0.3x^2 - 4$

$Y2 = \sqrt{\frac{x+4}{0.3}}$

$Y3 = -\sqrt{\frac{x+4}{0.3}}$

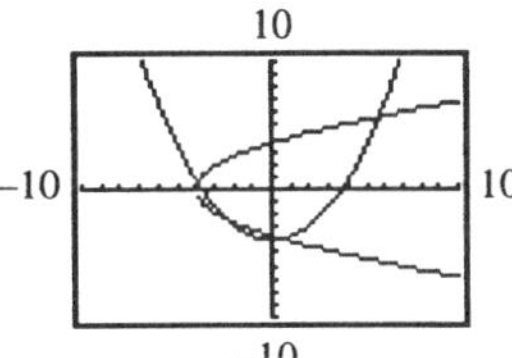

2. $f(x) = 0.1x^3 + 2$

$y = 0.1x^3 + 2$

$x = 0.1y^3 + 2$

$x - 2 = 0.1y^3$

$\frac{x-2}{0.1} = y^3$

$\sqrt[3]{\frac{x-2}{0.1}} = y$

$f^{-1}(x) = \sqrt[3]{\frac{x-2}{0.1}}$

$\text{Y1} = 0.1x^3 + 2$

$\text{Y2} = \sqrt[3]{\frac{x-2}{0.1}}$

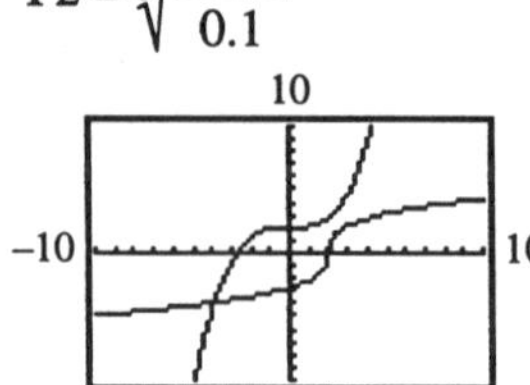

3. $g(x) = 5\sqrt{x} - 1.5$

$y = 5\sqrt{x} - 1.5$

$x = 5\sqrt{y} - 1.5$

$x + 1.5 = 5\sqrt{y}$

$\frac{x+1.5}{5} = \sqrt{y}$

$\left(\frac{x+1.5}{5}\right)^2 = y$

$g^{-1}(x) = \left(\frac{x+1.5}{5}\right)^2$

$\text{Y1} = 5\sqrt{x} - 1.5$

$\text{Y2} = \left(\frac{x+1.5}{5}\right)^2$

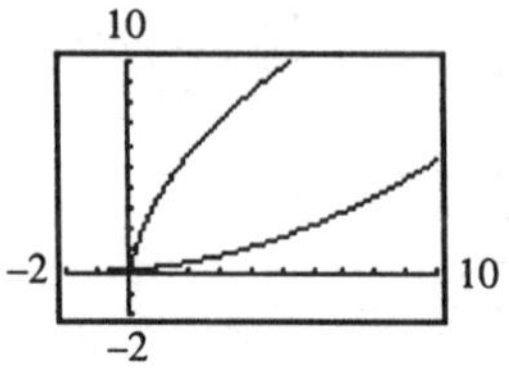

4. $y = \frac{1}{x} + 2$

$x = \frac{1}{y} + 2$

$x - 2 = \frac{1}{y}$

$y = \frac{1}{x-2}$

$y^{-1} = \frac{1}{x-2}$

$\text{Y1} = \frac{1}{x} + 2$

$\text{Y2} = \frac{1}{x-2}$

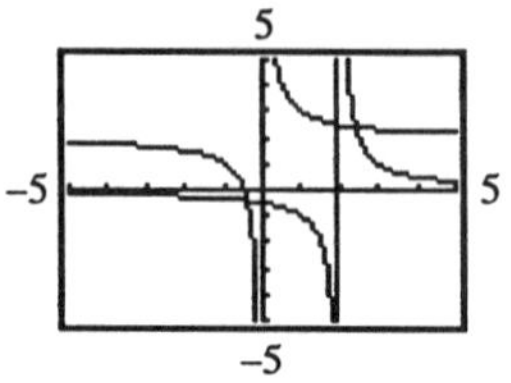

15.2 Experiencing Algebra the Exercise Way

1. $y = 7^x$ is an exponential function because it has a rational number base and a variable exponent.

3. $f(x) = 0.3^x$ is an exponential function because it has a rational number base and a variable exponent.

5. $g(x) = x^{0.3}$ is not exponential because the base is a variable.

7. $y = 1.57^x$ is an exponential function because it has a rational number base and a variable exponent.

9. $y = x^{-3}$ is not exponential because the base is a variable.

11. $R(x) = \left(\frac{2}{3}\right)^x$ is an exponential function because it has a rational number base and a variable exponent.

13. $g(x) = 16^x$
$g(3) = 16^3 = 4096$

15. $g(x) = 16^x$
$g(-2) = 16^{-2} = \dfrac{1}{16^2} = \dfrac{1}{256}$

17. $g(x) = 16^x$
$g\left(\dfrac{1}{2}\right) = 16^{1/2} = \sqrt{16} = 4$

19. $g(x) = 16^x$
$g\left(\sqrt{2}\right) = 16^{\sqrt{2}} \approx 50.453$

21. $h(x) = 0.64^x$
$h(0.3) = 0.64^{0.3} \approx 0.875$

23. $h(x) = 0.64^x$
$h\left(-\dfrac{1}{3}\right) = 0.64^{-1/3} = \dfrac{1}{0.64^{1/3}} \approx 1.160$

25. $h(x) = 0.64^x$
$h\left(-\sqrt{2}\right) = 0.64^{-\sqrt{2}} = \dfrac{1}{0.64^{\sqrt{2}}} \approx 1.880$

27. $f(x) = 4^x$
$Y1 = 4^x$

X	Y1
-3	.01563
-2	.0625
-1	.25
0	1
1	4
2	16
3	64

Y1■4^X

3.1

-4.7 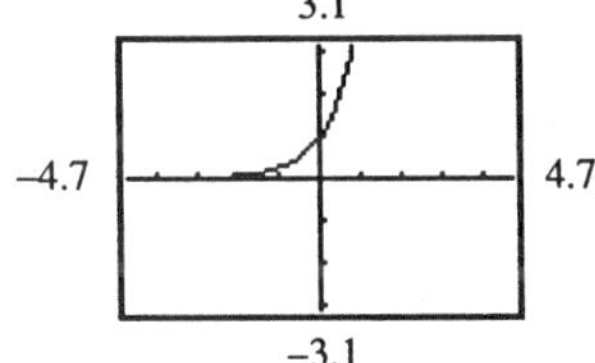 4.7

-3.1

29. $g(x) = 4^{-x}$
$Y1 = 4^{-x}$

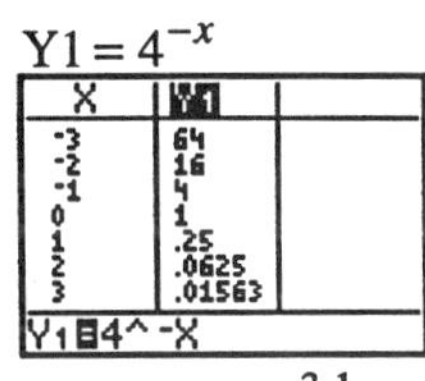

X	Y1
-3	64
-2	16
-1	4
0	1
1	.25
2	.0625
3	.01563

Y1■4^-X

3.1

-4.7 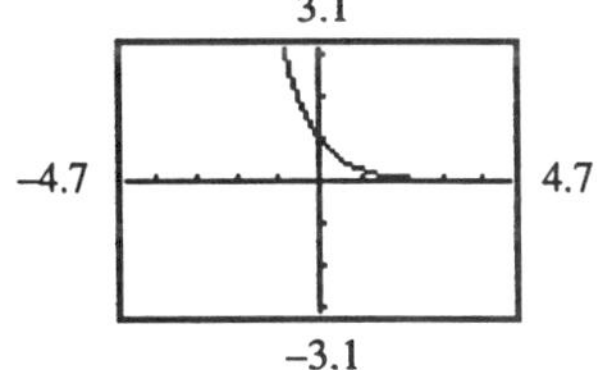 4.7

-3.1

31. $h(x) = 4^{2x}$
$Y1 = 4^{2x}$

X	Y1
-3	2.4E-4
-2	.00391
-1	.0625
0	1
1	16
2	256
3	4096

Y1■4^(2X)

3.1

-4.7 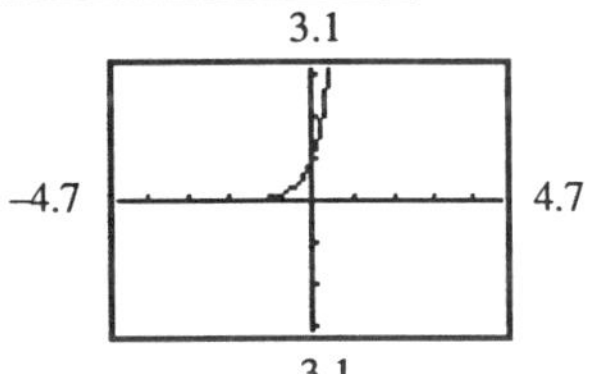 4.7

-3.1

33. $j(x) = 4^{\frac{1}{2}x}$
$Y1 = 4^{\frac{1}{2}x}$

X	Y1
-3	.125
-2	.25
-1	.5
0	1
1	2
2	4
3	8

Y1■4^(X/2)

3.1

-4.7 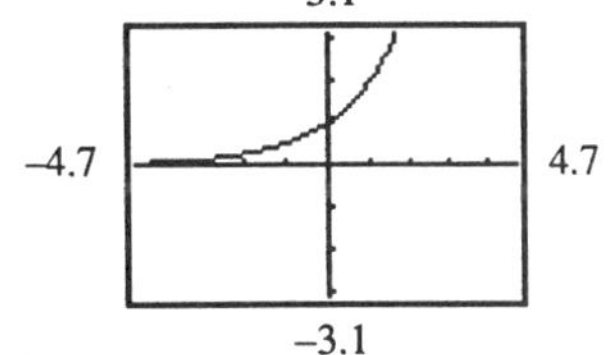 4.7

-3.1

35. $k(x) = 4^{x-1}$

$Y1 = 4^{x-1}$

X	Y1
-3	.00391
-2	.01563
-1	.0625
0	.25
1	1
2	4
3	16

Y1=4^(X−1)

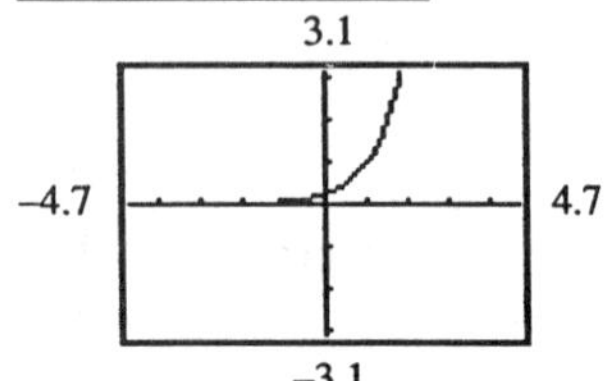

37. $m(x) = 4^x - 1$

$Y1 = 4^x - 1$

X	Y1
-3	-.9844
-2	-.9375
-1	-.75
0	0
1	3
2	15
3	63

Y1=4^X−1

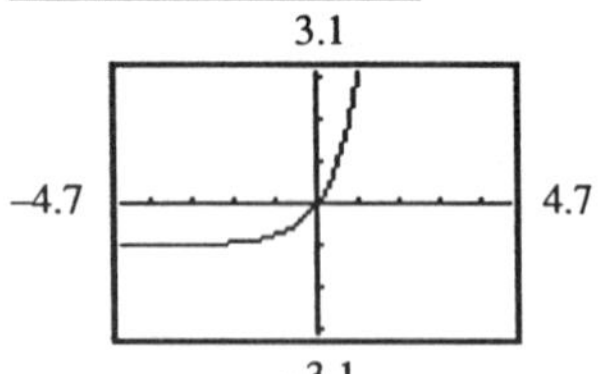

39. $y = e^{\frac{1}{2}x}$

$Y1 = e^{\frac{1}{2}x}$

X	Y1
-3	.22313
-2	.36788
-1	.60653
0	1
1	1.6487
2	2.7183
3	4.4817

Y1=e^(X/2)

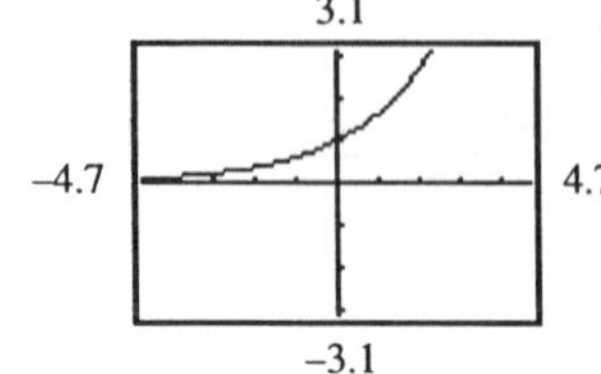

41. $y = \frac{1}{2}e^x$

$Y1 = \frac{1}{2}e^x$

X	Y1
-3	.02489
-2	.06767
-1	.18394
0	.5
1	1.3591
2	3.6945
3	10.043

Y1=1/2e^(X)

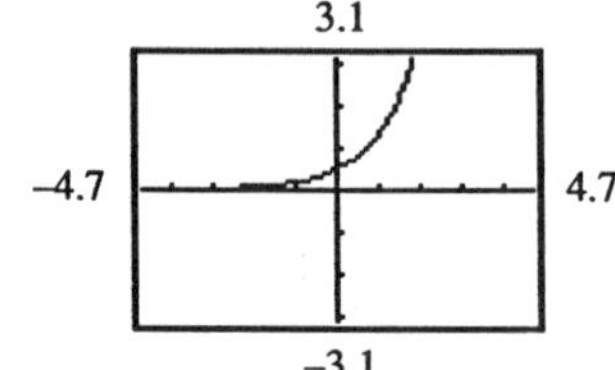

43. $y = e^{-\frac{1}{2}x}$

$Y1 = e^{-\frac{1}{2}x}$

X	Y1
-3	4.4817
-2	2.7183
-1	1.6487
0	1
1	.60653
2	.36788
3	.22313

Y1=e^((−1/2)X)

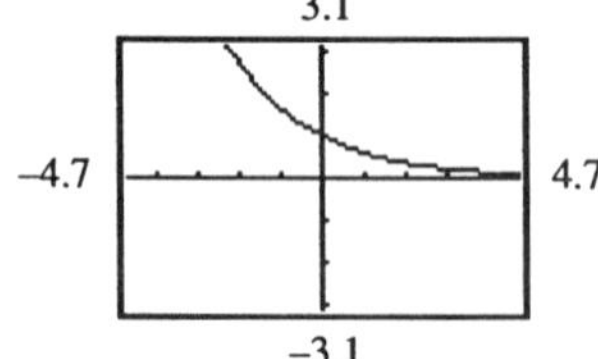

45. $y = e^x + \frac{1}{2}$

$Y1 = e^x + \frac{1}{2}$

X	Y1
-3	.54979
-2	.63534
-1	.86788
0	1.5
1	3.2183
2	7.8891
3	20.586

Y1=e^(X)+1/2

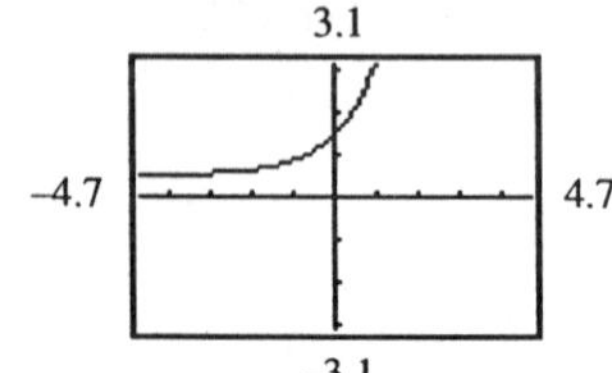

47. $y = \frac{1}{2}e^{\frac{1}{2}x}$

$\text{Y1} = \frac{1}{2}e^{\frac{1}{2}x}$

X	Y1
-3	.11157
-2	.18394
-1	.30327
0	.5
1	.82436
2	1.3591
3	2.2408

Y1■1/2e^((1/2)X)

3.1

−4.7 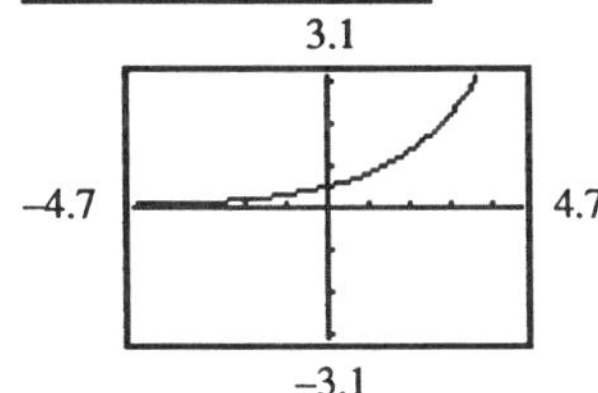 4.7

−3.1

49. $A = P(1+r)^t$

$A = 6000(1+0.055)^t$

$\text{Y1} = 6000(1.055)^x$

10000

0 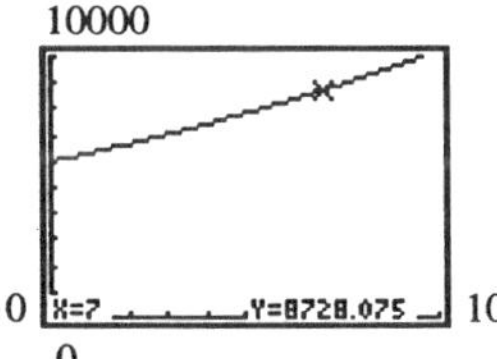 10

0

$A \approx 8728.08$

The investment was worth $8728.08.

51. $A = Pe^{rt}$

$A = 8000e^{0.048t}$

$\text{Y1} = 8000e^{0.048x}$

20,000

0 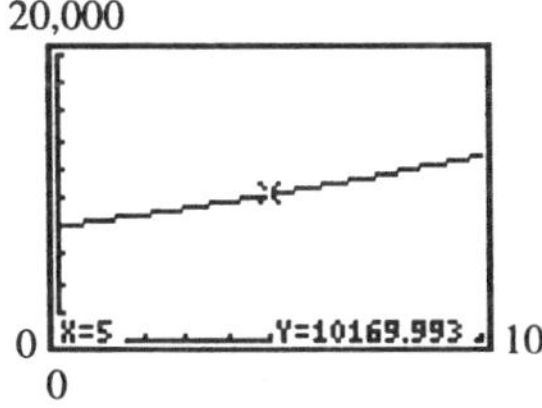 10

0

$A \approx 10{,}169.99$

$10{,}169.99 - 8000 = 2169.99$

She will have earned $2169.99.

53. $y = (281.49)1.047^x$

$\text{Y1} = (281.49)1.047^x$

1000

0 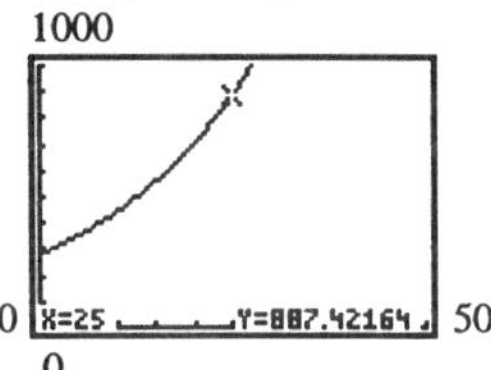 50

0

$x = 2005 - 1980 = 25$

At $x = 25$, $y \approx 887.42$

The number of passenger miles will be 887.42 billion.

15.2 Experiencing Algebra the Calculator Way

1. a. $\text{Y1} = e^x$

3.1

−4.7 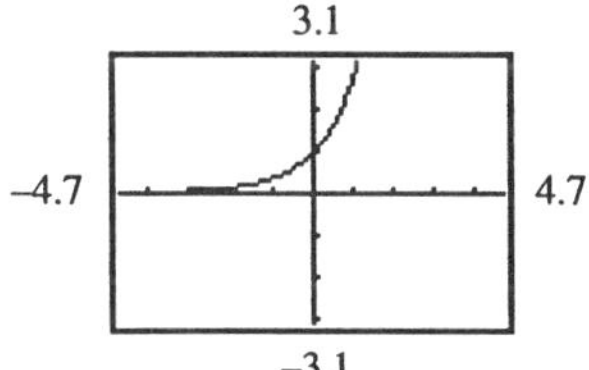 4.7

−3.1

b. $\text{Y1} = e^{-x}$

3.1

−4.7 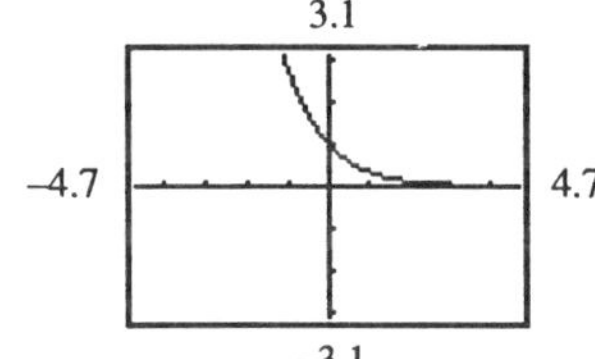 4.7

−3.1

2. a. $\text{Y1} = e^x + e^{-x}$

6.2

−4.7 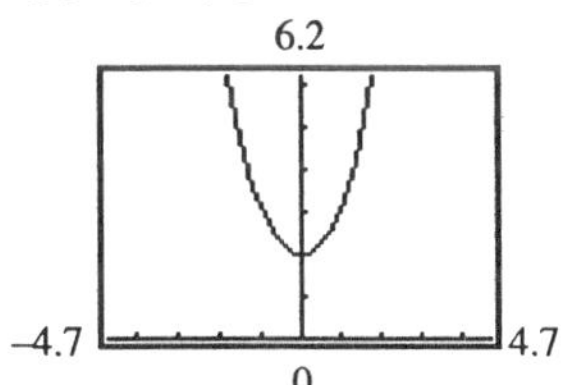 4.7

0

b. $Y1 = e^x - e^{-x}$

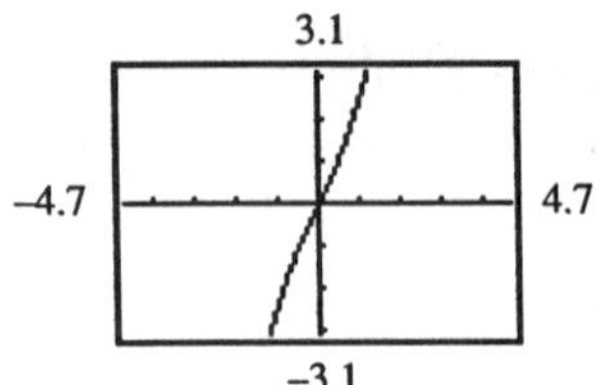

3. a. $Y1 = (e^x)(e^{-x})$

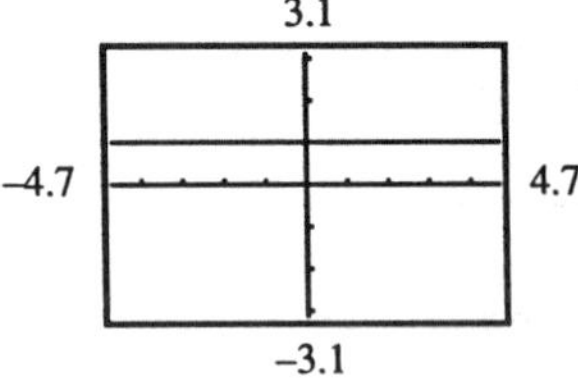

b. $Y1 = e^x \div e^{-x}$

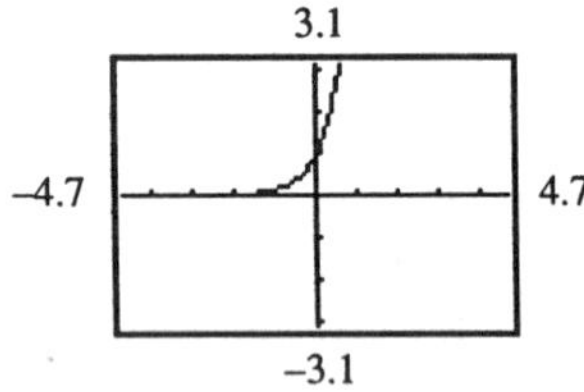

Answers will vary.

4. a. $Y1 = e^{\frac{1}{2}x}$

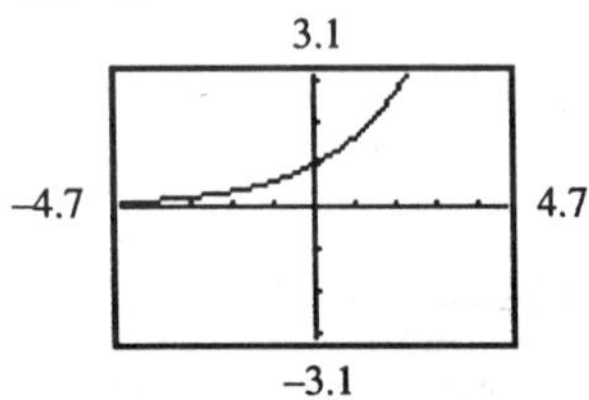

b. $Y1 = e^{-\frac{1}{2}x}$

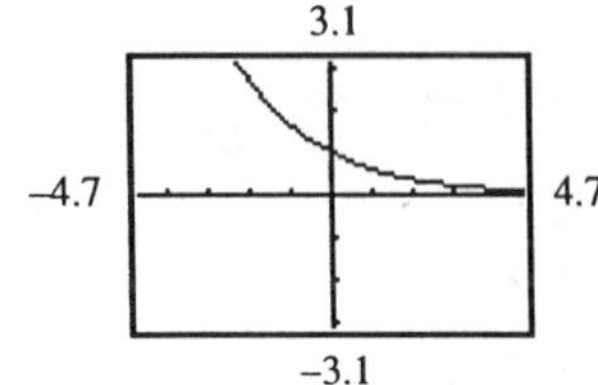

5. a. $Y1 = e^{\frac{1}{2}x} + e^{-\frac{1}{2}x}$

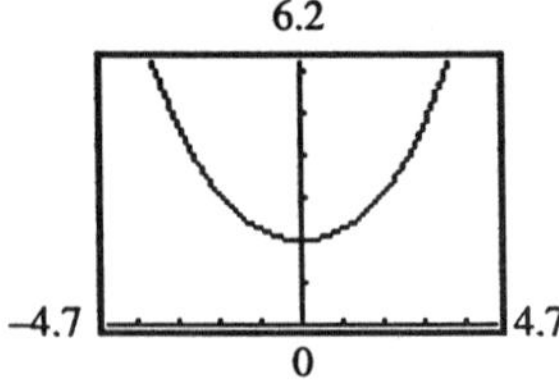

b. $Y1 = e^{\frac{1}{2}x} - e^{-\frac{1}{2}x}$

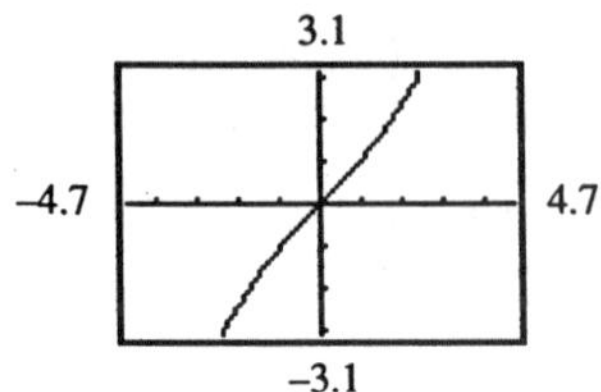

6. a. $Y1 = \left(e^{\frac{1}{2}x}\right)\left(e^{-\frac{1}{2}x}\right)$

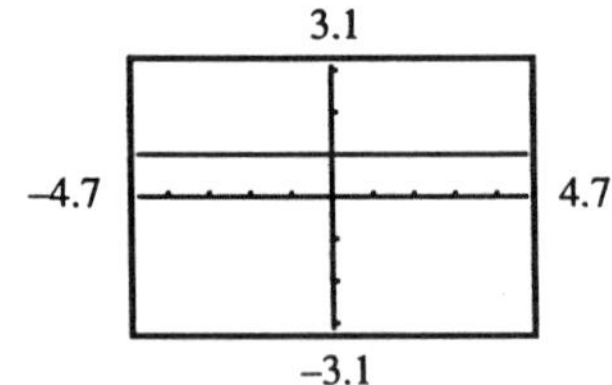

b. $Y1 = \left(e^{\frac{1}{2}x}\right) \div \left(e^{-\frac{1}{2}x}\right)$

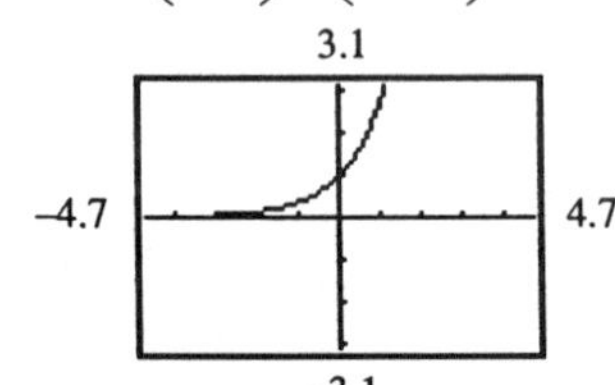

Answers will vary.

15.3 Experiencing Algebra the Exercise Way

1. $f(x) = 11^x$
$f^{-1}(x) = \log_{11} x$

3. $g(x) = 6^x$
$g^{-1}(x) = \log_6 x$

5. $H(x) = k^x$
$H^{-1}(x) = \log_k x$

7. $2^6 = 64$
$\log_2 64 = 6$

9. $3^5 = 243$
$\log_3 243 = 5$

11. $4^4 = 256$
$\log_4 256 = 4$

13. $2^{-3} = \left(\frac{1}{2}\right)^3 = \frac{1}{8}$
$\log_2 \frac{1}{8} = -3$

15. $4^{-1} = \frac{1}{4} = 0.25$
$\log_4 0.25 = -1$

17. $3^{-4} = \left(\frac{1}{3}\right)^4 = \frac{1}{81}$
$\log_3 \frac{1}{81} = -4$

19. $\log 10 = 1$

21. $\log 0.0001 = -4$

23. $\ln e^3 = 3$

25. $\ln e^{-5} = -5$

27. $\ln \frac{1}{e^5} = \ln e^{-5} = -5$

29. $\log 15 \approx 1.176$

31. $\log \frac{1}{12} \approx -1.079$

33. $\log 1.35 \approx 0.130$

```
log(15)
      1.176091259
log(1/12)
     -1.079181246
log(1.35)
      .1303337685
```

35. $\ln 14 \approx 2.639$

37. $\ln 2.85 \approx 1.047$

39. $\ln \frac{1}{5} \approx -1.609$

```
ln(14)
       2.63905733
ln(2.85)
      1.047318994
ln(1/5)
     -1.609437912
```

41. $\log_4 12 = \frac{\log 12}{\log 4} \approx 1.792$

43. $\log_2 10 = \frac{\log 10}{\log 2} \approx 3.322$

45. $\log_5 2.88 = \frac{\log 2.88}{\log 5} \approx 0.657$

47. $\log_5 \frac{2}{3} = \frac{\log \frac{2}{3}}{\log 5} \approx -0.252$

49. $\log_8 15 = \frac{\ln 15}{\ln 8} \approx 1.302$

51. $\log_3 5.9 = \frac{\ln 5.9}{\ln 3} \approx 1.616$

53. $\log_5 \frac{1}{7} = \frac{\ln \frac{1}{7}}{\ln 5} \approx -1.209$

55. $f(x) = \log_5 x$
$Y1 = \frac{\log x}{\log 5}$

X	Y1
.001	-4.292
.01	-2.861
.1	-1.431
1	0
1.5	.25193
2	.43068
3	.68261

Y1=log(X)/log(5)

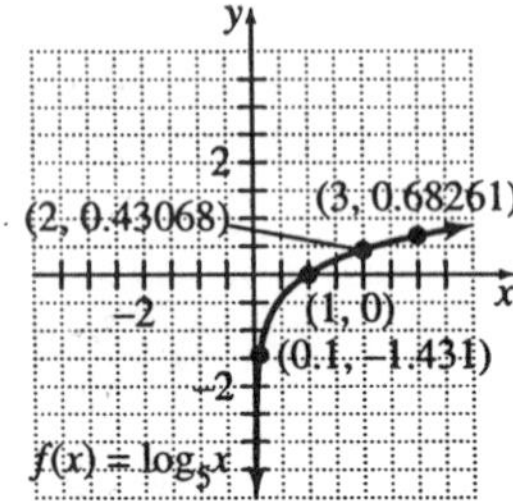

57. $g(x) = \log(x+2)$
$Y1 = \log(x+2)$

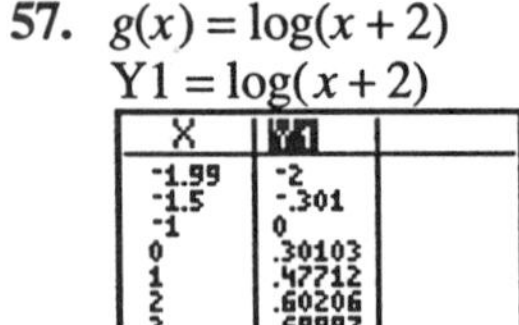

X	Y1
-1.99	-2
-1.5	-.301
-1	0
0	.30103
1	.47712
2	.60206
3	.69897

Y1=log(X+2)

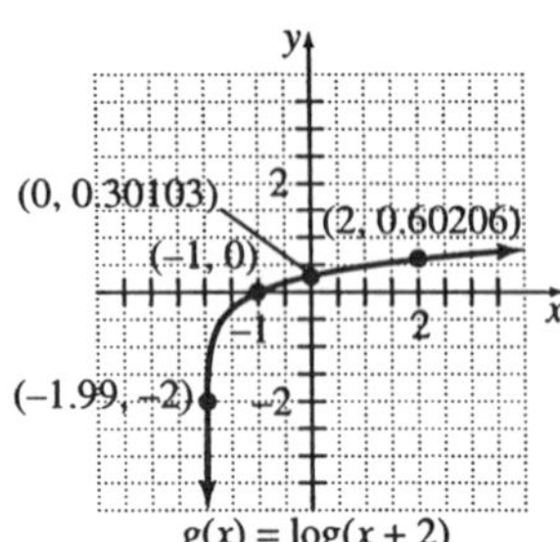

59. $h(x) = \ln(x+2)$
$Y1 = \ln(x+2)$

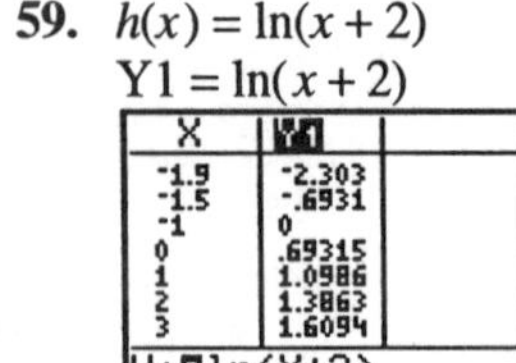

X	Y1
-1.9	-2.303
-1.5	-.6931
-1	0
0	.69315
1	1.0986
2	1.3863
3	1.6094

Y1=ln(X+2)

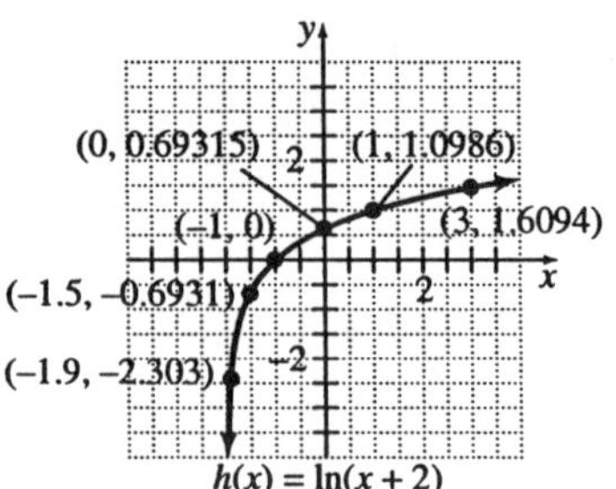

61. $\text{pH} = -\log[\text{H}^+]$
$6.2 = -\log[\text{H}^+]$
$\log[\text{H}^+] = -6.2$
$10^{-6.2} = [\text{H}^+]$
$0.0000006 \approx [\text{H}^+]$
The hydrogen ion concentration is 0.0000006 moles per liter.

63. $\text{pH} = -\log\left[\text{H}^+\right]$
$\text{pH} = -\log\left(3.2 \times 10^{-9}\right)$
$\text{pH} \approx 8.49485$
The pH is approximately 8.5.

65. $R = \log \dfrac{I}{I_0}$

$3.5 = \log \dfrac{I}{I_0}$ $\quad$ $8.3 = \log \dfrac{I}{I_0}$

$10^{3.5} = \dfrac{I}{I_0}$ $\quad$ $10^{8.3} = \dfrac{I}{I_0}$

$I = 10^{3.5} I_0$ $\quad$ $I = 10^{8.3} I_0$

$$\frac{10^{8.3} I_0}{10^{3.5} I_0} = \frac{10^{8.3}}{10^{3.5}} \approx 63{,}095.7$$

The larger earthquake's intensity was 63,095.7 times as great as the smaller earthquake's intensity.

15.3 Experiencing Algebra the Calculator Way

1. 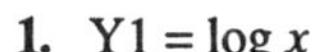 $Y1 = \log x$

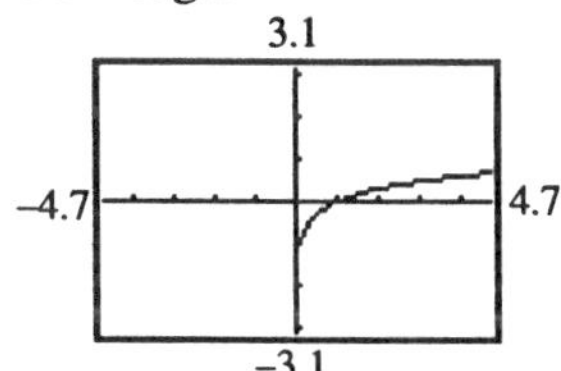

2. $Y1 = \log(x + 1)$

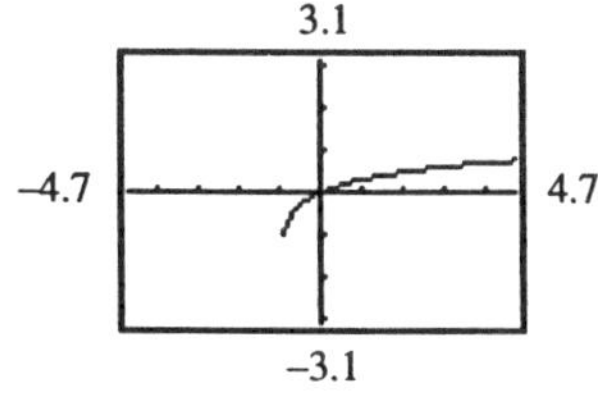

3. $Y1 = \log x + \log(x + 1)$

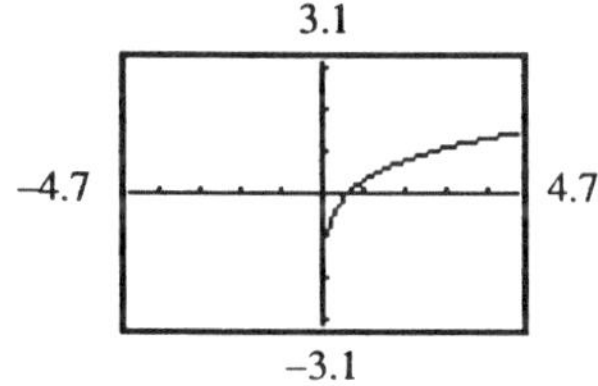

4. $Y1 = \log[x(x + 1)]$

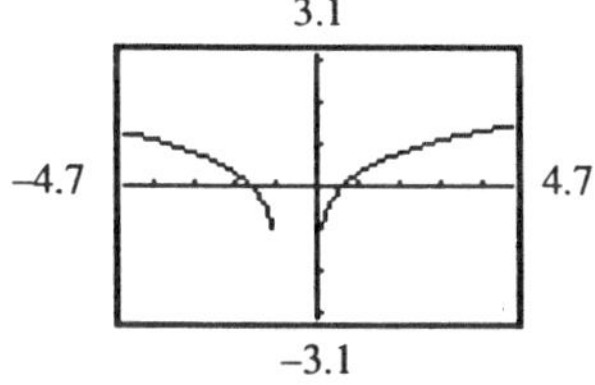

5. $Y1 = \log x - \log(x + 1)$

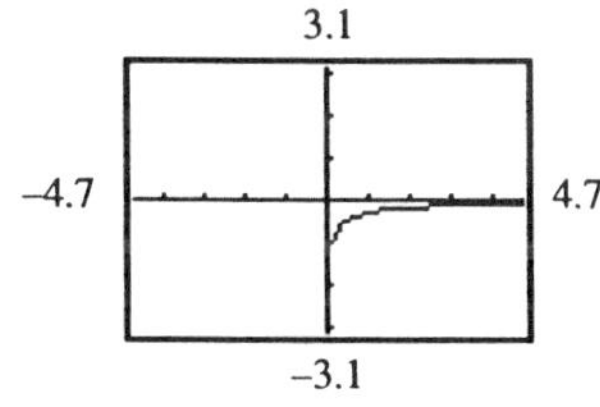

6. $Y1 = \log\left(\frac{x}{x+1}\right)$

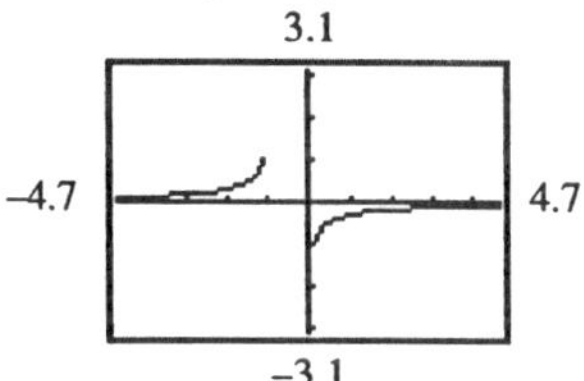

Answers will vary.

15.4 Experiencing Algebra the Exercise Way

1. $\log 12a = \log 12 + \log a$

3. $\ln x^3 = 3\ln x$

5. $\log_5 \frac{x}{5}$

$= \log_5 x - \log_5 5$

$= \log_5 x - 1$

7. $\log \frac{2x^2}{y}$

$= \log 2x^2 - \log y$

$= \log 2 + \log x^2 - \log y$

$= \log 2 + 2\log x - \log y$

9. $\log_3 x^3 y^2$

$= \log_3 x^3 + \log_3 y^2$

$= 3\log_3 x + 2\log_3 y$

11. $\ln\sqrt[3]{xy^2}$

$= \ln\left(xy^2\right)^{1/3}$

$= \frac{1}{3}\ln\left(xy^2\right)$

$= \frac{1}{3}\left(\ln x + \ln y^2\right)$

$= \frac{1}{3}(\ln x + 2\ln y)$

$= \frac{1}{3}\ln x + \frac{2}{3}\ln y$

13. $\log \frac{\sqrt{2x}}{\sqrt[3]{y}}$
$= \log \frac{(2x)^{1/2}}{y^{1/3}}$
$= \log(2x)^{1/2} - \log y^{1/3}$
$= \frac{1}{2}\log(2x) - \frac{1}{3}\log y$
$= \frac{1}{2}(\log 2 + \log x) - \frac{1}{3}\log y$
$= \frac{1}{2}\log 2 + \frac{1}{2}\log x - \frac{1}{3}\log y$

15. $\log_3 3a$
$= \log_3 3 + \log_3 a$
$= 1 + \log_3 a$

17. $\log_5 10xy$
$= \log_5 10 + \log_5 x + \log_5 y$

19. $\log_a\left(ab^2\right)$
$= \log_a a + \log_a b^2$
$= 1 + 2\log_a b$

21. $\log x + \log(x+5)$
$= \log x(x+5)$

23. $2\ln x + 3\ln y$
$= \ln x^2 + \ln y^3$
$= \ln x^2 y^3$

25. $2\log_3(x+3) - \log_3(x-1)$
$= \log_3(x+3)^2 - \log_3(x-1)$
$= \log_3 \frac{(x+3)^2}{x-1}$

27. $\frac{1}{2}\ln x - \frac{1}{5}\ln(x+1)$
$= \ln x^{1/2} - \ln(x+1)^{1/5}$
$= \ln\sqrt{x} - \ln\sqrt[5]{x+1}$
$= \ln\frac{\sqrt{x}}{\sqrt[5]{x+1}}$

29. $\log xy - \log xz$
$= \log\frac{xy}{xz}$
$= \log\frac{y}{z}$

31. $G = \log\left(\frac{P_0}{P_i}\right)^{10}$
$G = \log\left(\frac{20}{1.5}\right)^{10}$
$G \approx 11.249$
The power gain is approximately 11.249 decibels.

33. $y = c\ln\frac{I_0}{I}$
$y = 12\ln 3.5$
$y \approx 15.033$
The depth is approximately 15.033 units.

35. $t = \frac{\ln k}{r}$
$t = \frac{\ln 3}{0.05}$
$t \approx 21.972$
It will take approximately 22 years.

37. $t = \frac{\ln k}{r}$
$t = \frac{\ln 4}{0.09}$
$t \approx 15.403$
It will take approximately 15.4 years.

15.4 Experiencing Algebra the Calculator Way

1. $Y1 = \log 3x$

3.1
−4.7 4.7
−3.1

2. $Y1 = \log x + \log 2x$

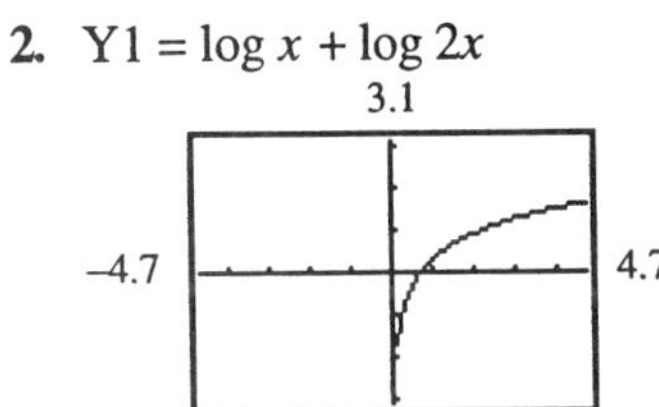

3. $Y1 = \log(x^2 - x)$

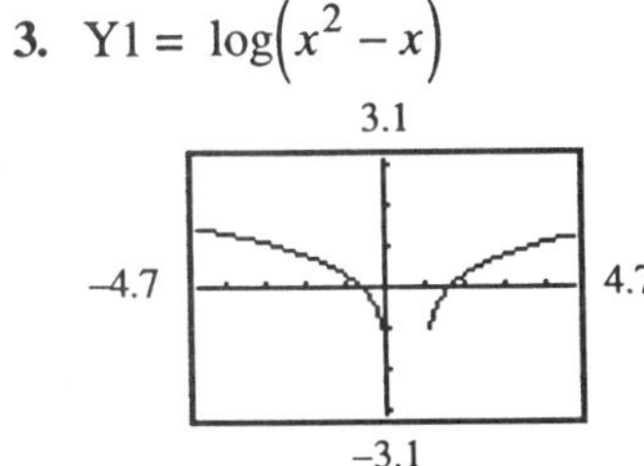

4. $Y1 = \log x^2 - \log x$

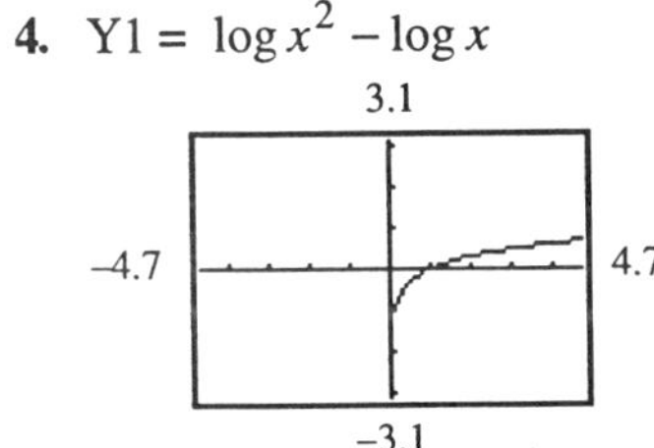

5. No; they are not equivalent.

6. No; they are not equivalent.

7. The logarithm of a sum is not equivalent to the sum of the logarithms.

8. The logarithm of a difference is not equivalent to the difference of the logarithms.

15.5 Experiencing Algebra the Exercise Way

1. $-3 + \log x^2 = 1$ is a logarithmic equation because the logarithmic expression contains a variable.

3. $e^3 + x = 2x - 1$ is neither an exponential nor a logarithmic equation. The exponential expression does not contain a variable.

5. $e^{2x} - 1 = 5$ is an exponential equation because the exponential expression contains a variable.

7. $x = \log 4 + \log 7$ is neither an exponential nor a logarithmic equation. The logarithmic expressions do not contain a variable.

9. $5^{2x} = 625$
$5^{2x} = 5^4$
$2x = 4$
$x = 2$
The solution is 2.

11. $e^t = 2$
$\ln e^t = \ln 2$
$t = \ln 2$
$t \approx 0.693$
The solution is ≈ 0.693.

13. $5^x = 1$
$5^0 = 1$
$x = 0$
The solution is 0.

15. $3e^{-x} = 2$
$\ln 3e^{-x} = \ln 2$
$\ln 3 + \ln e^{-x} = \ln 2$
$\ln e^{-x} = \ln 2 - \ln 3$
$-x \ln e = \ln 2 - \ln 3$
$-x = \ln 2 - \ln 3$
$x = \ln 3 - \ln 2$
$x \approx 0.405$
The solution is ≈ 0.405.

17. $(2^{3x})(2^5) = 2^8$
$2^{3x+5} = 2^8$
$3x + 5 = 8$
$3x = 3$
$x = 1$
The solution is 1.

19. $2^x = 2^{3x-6}$

$x = 3x - 6$

$6 = 2x$

$3 = x$

The solution is 3.

21. $5^{2x(x-2)} = 5^{3(x-1)}$

$2x(x-2) = 3(x-1)$

$2x^2 - 4x = 3x - 3$

$2x^2 - 7x + 3 = 0$

$(2x-1)(x-3) = 0$

$x = \frac{1}{2} \quad x = 3$

The solutions are $\frac{1}{2}$ and 3.

23. $e^{x(x-2)} = e^{x+10}$

$x(x-2) = x + 10$

$x^2 - 2x = x + 10$

$x^2 - 3x - 10 = 0$

$(x-5)(x+2) = 0$

$x = 5 \qquad x = -2$

The solutions are –2 and 5.

25. $3^{x+x(x-3)} = 3^{(x-1)^2}$

$x + x(x-3) = (x-1)^2$

$x + x^2 - 3x = x^2 - 2x + 1$

$0 = 1$

This is a contradiction. There is no solution.

27. $8^{5(x-3)} = 8^{4(x-2)}8^{(x-7)}$

$8^{5(x-3)} = 8^{4(x-2)+(x-7)}$

$5(x-3) = 4(x-2) + (x-7)$

$5x - 15 = 4x - 8 + x - 7$

$-15 = -15$

This is an identity. The solutions are all real numbers.

29. $3^{3x} = 9^{x+1}$

$Y1 = 3^{3x}$

$Y2 = 9^{x+1}$

X	Y1	Y2
0	1	9
1	27	81
2	729	729
3	19683	6561
4	531441	59049
5	1.43E7	531441
6	3.87E8	4.78E6

Y1=3^(3X)

The solution is 2.

31. $5^{2b+5} = 5^4$

$2b + 5 = 4$

$2b = -1$

$b = -\frac{1}{2}$

The solution is $-\frac{1}{2}$.

33. $3^{2x^2} \cdot 3^{5x} = 27$

$Y1 = 3^{2x^2} \cdot 3^{5x}$

$Y2 = 27$

31

–4.7   4.7

Intersection X=.5 Y=27

–31

The solutions are –3 and $\frac{1}{2}$.

35. $e^{0.5x} = 4$

$\ln e^{0.5x} = \ln 4$

$0.5x = \ln 4$

$x = \frac{\ln 4}{0.5}$

$x \approx 2.773$

The solution is ≈ 2.773.

37. $(3.5)^x = 12$

$\log(3.5)^x = \log 12$

$x \log 3.5 = \log 12$

$x = \frac{\log 12}{\log 3.5}$

$x \approx 1.984$

The solution is ≈ 1.984.

39. $\log_5 c = 5$

$5^{\log_5 c} = 5^5$

$c = 5^5$

$c = 3125$

The solution is 3125.

41. $\log x = 0$

$10^0 = x$

$1 = x$

The solution is 1.

43. $\ln z = 0$

$e^0 = z$

$1 = z$

The solution is 1.

45. $\log_3 k = 2$

$3^2 = k$

$9 = k$

The solution is 9.

47. $\log_5 x = -3$

$5^{-3} = x$

$\frac{1}{5^3} = x$

$\frac{1}{125} = x$

The solution is $\frac{1}{125}$.

49. $\log_3(a+2) = 2$

$3^2 = a+2$

$9 = a+2$

$7 = a$

The solution is 7.

51. $2\ln x = 1$

$\ln x = \frac{1}{2}$

$e^{\frac{1}{2}} = x$

$\sqrt{e} = x$

$1.649 \approx x$

The solution is ≈ 1.649.

53. $\log x + 4 = 2$

$\log x = -2$

$10^{-2} = x$

$\frac{1}{10^2} = x$

$\frac{1}{100} = x$

The solution is $\frac{1}{100}$.

55. $5\ln z + 4 = 4$

$5\ln z = 0$

$\ln z = 0$

$e^0 = z$

$1 = z$

The solution is 1.

57. $\ln x + \ln(x+1) = \ln 12$

$Y1 = \ln x + \ln(x+1)$

$Y2 = \ln 12$

3.1

–4.7 4.7

Intersection X=3 Y=2.4849066

–3.1

The solution is 3.

59. $2\ln(p-5) = \ln 9$

$\ln(p-5)^2 = \ln 9$

$(p-5)^2 = 9$

$p^2 - 10p + 25 = 9$

$p^2 - 10p + 16 = 0$

$(p-8)(p-2) = 0$

$p = 8 \quad p = 2$ (restricted)

The solution is 8.

61. $\log 2x + \log 3x = \log 6 + 2\log x$

$Y1 = \log 2x + \log 3x$

$Y2 = \log 6 + 2\log x$

X	Y1	Y2
0	ERROR	ERROR
1	.77815	.77815
2	1.3802	1.3802
3	1.7324	1.7324
4	1.9823	1.9823
5	2.1761	2.1761
6	2.3345	2.3345

Y1=log(2X)+log(...

This is an identity. The solution is all real numbers greater than 0.

63. $\ln x + \ln(x+2) = 2\ln(x+2)$
$\ln x(x+2) = \ln(x+2)^2$
$x(x+2) = (x+2)^2$
$x^2 + 2x = x^2 + 4x + 4$
$-2x = 4$
$x = -2$ (restricted)
There is no solution.

65. $\log(x+4) - \log x = \log 2$
$\log\frac{x+4}{x} = \log 2$
$\frac{x+4}{x} = 2$
$x + 4 = 2x$
$4 = x$
The solution is 4.

67. $\log x - \log(x+6) = -1$
$\log\frac{x}{x+6} = -1$
$10^{-1} = \frac{x}{x+6}$
$\frac{1}{10} = \frac{x}{x+6}$
$x + 6 = 10x$
$6 = 9x$
$\frac{2}{3} = x$
The solution is $\frac{2}{3}$.

69. $\log_3(x+16) - \log_3 x = 2$
$\log_3\frac{x+16}{x} = 2$
$3^2 = \frac{x+16}{x}$
$9 = \frac{x+16}{x}$
$9x = x + 16$
$8x = 16$
$x = 2$
The solution is 2.

71. Let t = number of years since 1970.
$A(t) = A_0e^{kt}$
$131.005 = 121.6e^{k(7)}$
$\frac{131.005}{121.6} = e^{7k}$
$\ln\frac{131.005}{121.6} = 7k$
$\frac{1}{7}\ln\frac{131.005}{121.6} = k$
$0.011 \approx k$
$A(t) = 121.6e^{0.011t}$
$t = 2025 - 1990 = 35$
$A(35) = 121.6e^{0.011(35)}$
$A(35) \approx 178.7$
The prediction is 178.7 million.

73. $A(t) = A_0e^{0.0307t}$
$2 = e^{0.0307t}$
$\ln 2 = 0.0307t$
$\frac{\ln 2}{0.0307} = t$
$22.58 = t$
$1990 + 22.58 = 2012.58$
It will double its value in the year 2012.

75. $A(t) = A_0e^{kt}$
$\frac{1}{2} = e^{k(4.5)}$
$\ln\frac{1}{2} = 4.5k$
$\frac{\ln\frac{1}{2}}{4.5} = k$
$-0.154 \approx k$
The decay factor is approximately –0.154.
$0.98 = e^{-0.154t}$
$\ln 0.98 = -0.154t$
$\frac{\ln 0.98}{-0.154} = t$
$0.131 = t$
It will take approximately 0.131 billion years, or 131 million years.

15.5 Experiencing Algebra the Calculator Way

Students should read graphing techniques.

Chapter 15 Review

Reflections

1.–7. Answers will vary.

Exercises

1. $h = \{(2, 4.5), (4, 3.5), (6, 2.5), (8, 1.5), (10, 0.5)\}$
$h^{-1} = \{(4.5, 2), (3.5, 4), (2.5, 6), (1.5, 8), (0.5, 10)\}$

2.
$$y = 3x - 9$$
$$x = 3y - 9$$
$$x + 9 = 3y$$
$$\frac{x+9}{3} = y$$
$$y = \frac{1}{3}x + 3$$
Therefore, $y^{-1} = \frac{1}{3}x + 3.$

3.
$$y = x^2 + 1$$
$$x = y^2 + 1$$
$$x - 1 = y^2$$
$$\pm\sqrt{x-1} = y$$
$$y = \pm\sqrt{x-1}$$
Therefore, $y^{-1} = \pm\sqrt{x-1}.$

4.
$$y = x^3 - 1$$
$$x = y^3 - 1$$
$$x + 1 = y^3$$
$$\sqrt[3]{x+1} = y$$
$$y = \sqrt[3]{x+1}$$
Therefore, $y^{-1} = \sqrt[3]{x+1}.$

5. $P = \{(9, 1), (8, 2), (7, 3), (6, 1), (5, 2), (4, 3)\}$
The function P is not a one-to-one function because the second coordinates 1, 2, and 3 are each repeated in two ordered pairs. The inverse is not a function.

6. $Q = \{(1, 9), (2, 8), (3, 7), (4, 6), (5, 5), (6, 4), (7, 3)\}$
The function Q is a one-to-one function because no two ordered pairs have the same first coordinate or the same second coordinate. The inverse is a function.

7. No, the graph does not represent a function because a vertical line intersects the graph more than once.

8. Yes, the graph represents a one-to-one function because all possible vertical as well as horizontal lines cross the graph only once.

9. No, the graph is a function because it passes the vertical line test, but a horizontal line can be drawn that intersects the graph more than once so it is not a one-to-one function.

10. $f(x) = \frac{3}{4}x - 3$
$$y = \frac{3}{4}x - 3$$
$$x = \frac{3}{4}y - 3$$
$$x + 3 = \frac{3}{4}y$$
$$\frac{4}{3}(x+3) = \frac{4}{3}\left(\frac{3}{4}y\right)$$
$$y = \frac{4}{3}x + 4$$
Therefore,
$$f^{-1}(x) = \frac{4}{3}x + 4$$

$$Y1 = \frac{3}{4}x - 3$$
$$Y2 = \frac{4}{3}x + 4$$
$$Y3 = x$$

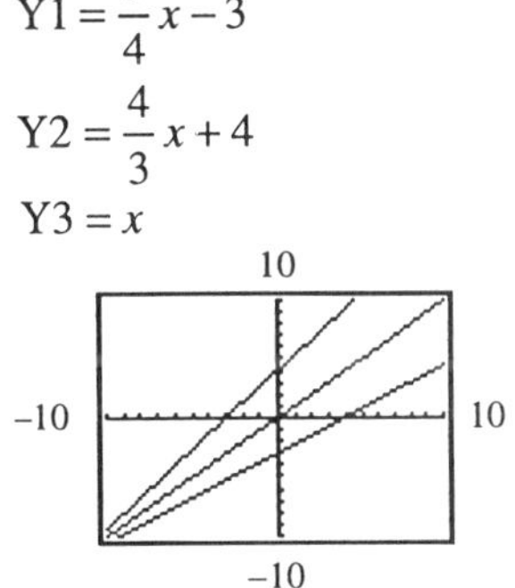

11. $y = x^3 + 8$

$$y = x^3 + 8$$
$$x = y^3 + 8$$
$$x - 8 = y^3$$
$$\sqrt[3]{x-8} = y$$

Therefore,

$$y^{-1} = \sqrt[3]{x-8}$$

$\text{Y1} = x^3 + 8$
$\text{Y2} = \sqrt[3]{x-8}$
$\text{Y3} = x$

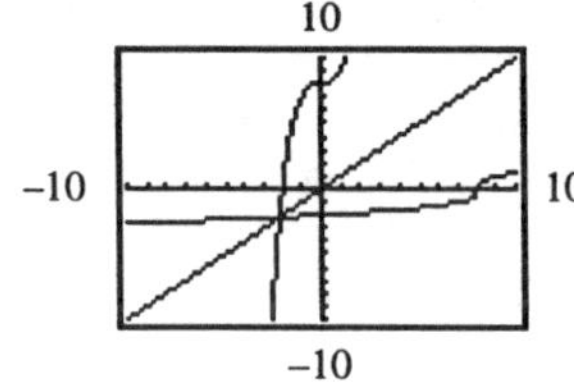

12. Let x = value of all sales
The function that represents her weekly income is $f(x) = 570 + 0.03x$.

$$y = 570 + 0.03x$$
$$x = 570 + 0.03y$$
$$x - 570 = 0.03y$$
$$\frac{x-570}{0.03} = y$$
$$y = 33.\overline{3}x - 19{,}000$$

The inverse function is
$f^{-1}(x) = 33.\overline{3}x - 19{,}000$ which represents the value of all sales in terms of Motomo's weekly salary.

13. Let x = number of hours
The function that represents the number of problems in x hours is $f(x) = 15x$.

$$y = 15x$$
$$x = 15y$$
$$\frac{x}{15} = y$$

The inverse function is $f^{-1}(x) = \frac{x}{15}$ which represents the number of hours in terms of the number of problems solved.

14. $f(x) = 2^x$ is an exponential function because it has a rational number base and a variable exponent.

15. $y = x^2$ is not exponential because the base is a variable.

16. $g(x) = 1.21^x$
$g(2) = 1.21^2 = 1.4641$

17. $g(x) = 1.21^x$
$g(-1) = 1.21^{-1} = \frac{1}{1.21} \approx 0.826$

18. $g(x) = 1.21^x$
$g(0) = 1.21^0 = 1$

19. $g(x) = 1.21^x$
$g\left(\frac{1}{2}\right) = 1.21^{1/2} = \sqrt{1.21} = 1.1$

20. $g(x) = 1.21^x$
$g(0.3) = 1.21^{0.3} \approx 1.059$

21. $g(x) = 1.21^x$
$g\left(\sqrt{2}\right) = 1.21^{\sqrt{2}} \approx 1.309$

22. $f(x) = 5^x - 1$
$\text{Y1} = 5^x - 1$

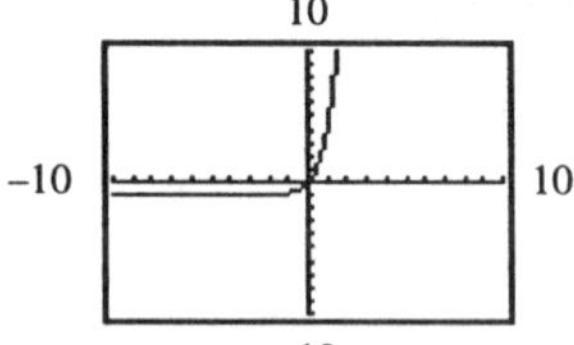

23. 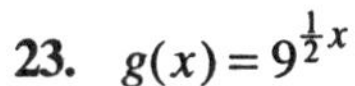
$g(x) = 9^{\frac{1}{2}x}$

$\text{Y1} = 9^{\frac{1}{2}x}$

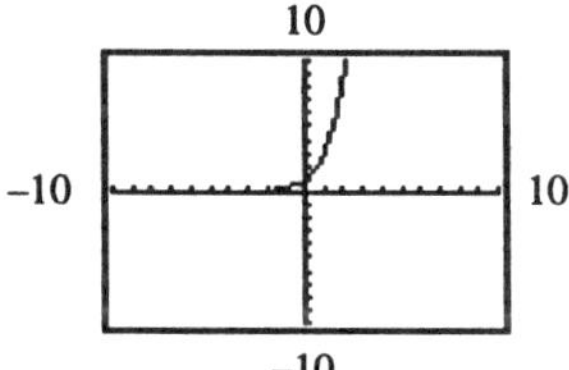

24. $y = 2e^{0.5x}$

$\text{Y1} = 2e^{0.5x}$

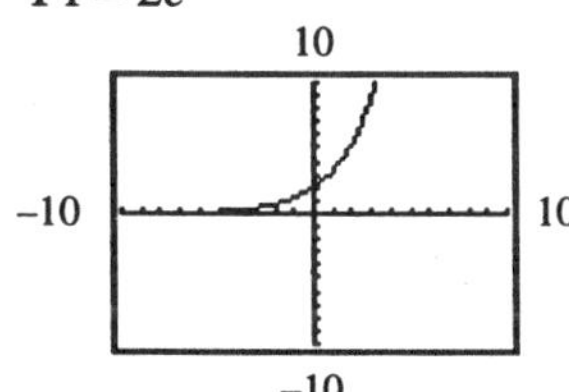

25. $A = P(1+r)^t$

$A = 1000(1+0.04)^t$

$A = 1000(1.04)^t$

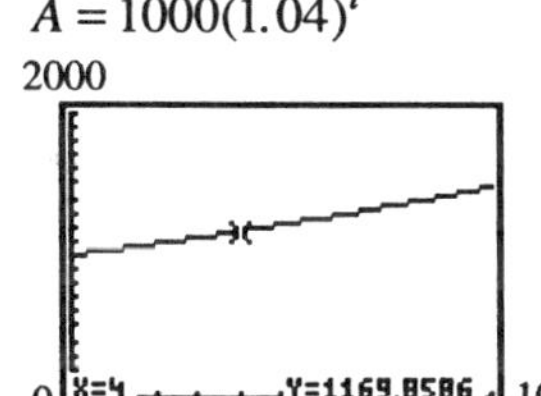

At $x = 4$, $y \approx 1169.86$
After 4 years the investment was worth \$1169.86.

26. $y = (78.16)1.115^t$

$t = 2005 - 1970 = 35$

$y = (78.16)1.115^{35}$

$y \approx 3528.6$

The expenditures are expected to be \$3528.6 billion.

27. $h(x) = 8^x$

$h^{-1}(x) = \log_8 x$

28. $A(x) = Ae^{kx}$

$y = Ae^{kx}$

$x = Ae^{ky}$

$\frac{x}{A} = e^{ky}$

$\ln\frac{x}{A} = \ln e^{ky}$

$\ln\frac{x}{A} = ky$

$\frac{1}{k}\ln\frac{x}{A} = y$

Therefore, $A^{-1}(x) = \frac{1}{k}\ln\frac{x}{A}$.

29. $\log 100 = 2$ because $10^2 = 100$

30. $\log_2 16 = 4$ because $2^4 = 16$

31. $\log_3 \frac{1}{27} = -3$ because $3^{-3} = \frac{1}{27}$

32. $\log 1.5 \approx 0.176$

33. $\ln e^4 = 4$

34. $\ln 10 \approx 2.303$

35. $\ln e^{-2} = -2$

36. $\log\frac{3}{5} \approx -0.222$

37. $\log_5 15 = \frac{\log 15}{\log 5} \approx 1.683$

38. $y = \log_3 x$

$\text{Y1} = \frac{\log x}{\log 3}$

X	Y1
.001	-6.288
.01	-4.192
.1	-2.096
1	0
2	.63093
3	1
4	1.2619

Y1=log(X)/log(3)

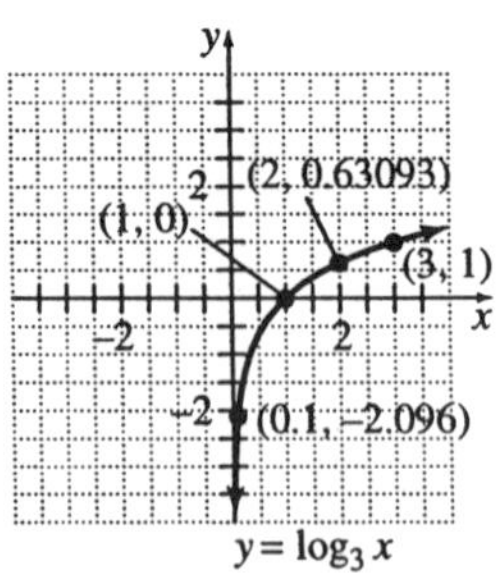

39. $f(x) = \ln(x-2)$
$Y1 = \ln(x-2)$

X	Y1
2.0001	-9.21
2.01	-4.605
2.1	-2.303
3	0
4	.69315
5	1.0986
6	1.3863

Y1 = ln(X−2)

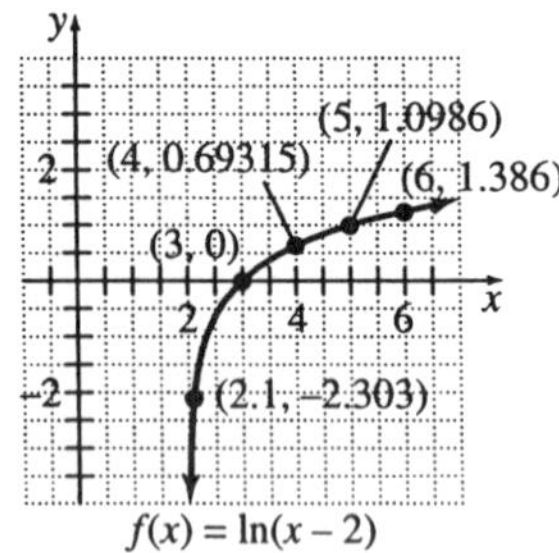

40. $\text{pH} = -\log\left[H^+\right]$
$\text{pH} = -\log\left[6.2 \times 10^{-3}\right]$
$\text{pH} \approx 2.208$
The pH is approximately 2.208.

41. $\text{pH} = -\log[H^+]$
$7.4 = -\log[H^+]$
$10^{-7.4} = [H^+]$
$3.98 \times 10^{-8} \approx [H^+]$
The hydrogen ion concentration is 3.98×10^{-8} moles per liter.

42. $\log 2x^4$
$= 4 \log 2x$
$= 4(\log 2 + \log x)$
$= 4 \log 2 + 4 \log x$

43. $\log_7 \frac{7x}{y}$
$= \log_7 7x - \log_7 y$
$= \log_7 7 + \log_7 x - \log_7 y$
$= 1 + \log_7 x - \log_7 y$

44. $\ln 25x^2y^3z$
$= \ln 25 + \ln x^2 + \ln y^3 + \ln z$
$= \ln 25 + 2 \ln x + 3 \ln y + \ln z$

45. $\ln \frac{\sqrt[3]{2x^2y}}{\sqrt{yz}}$
$= \ln \frac{\left(2x^2y\right)^{1/3}}{(yz)^{1/2}}$
$= \ln\left(2x^2y\right)^{1/3} - \ln(yz)^{1/2}$
$= \frac{1}{3}\ln\left(2x^2y\right) - \frac{1}{2}\ln(yz)$
$= \frac{1}{3}\left(\ln 2 + \ln x^2 + \ln y\right) - \frac{1}{2}(\ln y + \ln z)$
$= \frac{1}{3}\ln 2 + \frac{2}{3}\ln x + \frac{1}{3}\ln y - \frac{1}{2}\ln y - \frac{1}{2}\ln z$

46. $\log(x-3) + \log(x+3)$
$= \log(x-3)(x+3)$
$= \log(x^2 - 9)$

47. $5 \ln x - 3 \ln y$
$= \ln x^5 - \ln y^3$
$= \ln \frac{x^5}{y^3}$

48. $\frac{1}{3}\log x - \frac{1}{2}\log y$
$= \log x^{1/3} - \log y^{1/2}$
$= \log \frac{x^{1/3}}{y^{1/2}}$
$= \log \frac{\sqrt[3]{x}}{\sqrt{y}}$

49. $\log ab - \log bc + \log cd$
$= \log \frac{ab}{bc} + \log cd$
$= \log \frac{abcd}{bc}$
$= \log ad$

50. $t = \frac{\ln k}{r}$
$t = \frac{\ln 3}{0.06}$
$t \approx 18.3$
It will take approximately 18.3 years.

51. $e^x + 5 = 7$ is an exponential function.

52. $\ln 7 - x = \ln 3$ is neither because the logarithmic expressions do not contain a variable.

53. $\log x^2 + 2 = \log x$ is a logarithmic function.

54. $x = e^3 - 1$ is neither because the exponential expression does not contain a variable.

55. $2^x + 7 = 2^{3x}$ is an exponential function.

56. $y = \log_2 5 - e^2$ is neither because there is no variable present in either the logarithmic expression or the exponential expression.

57. $3^{x+1} = 243$
$3^{x+1} = 3^5$
$x + 1 = 5$
$x = 4$
The solution is 4.

58. $e^{-5t} - 2 = 3$
$e^{-5t} = 5$
$\ln e^{-5t} = \ln 5$
$-5t = \ln 5$
$t = \frac{\ln 5}{-5}$
$t \approx -0.322$
The solution is ≈ -0.322.

59. $12^a = 1$
$12^0 = 1$
$a = 0$
The solution is 0.

60. $(3^{2x})(3^5) = 3^8$
$3^{2x+5} = 3^8$
$2x + 5 = 8$
$2x = 3$
$x = \frac{3}{2}$
The solution is $\frac{3}{2}$.

61. $e^{2x(x-3)} = e^{2(x-2)(x-1)}$
$2x(x-3) = 2(x-2)(x-1)$
$2x^2 - 6x = 2x^2 - 6x + 4$
$0 = 4$
This is a contradiction. There is no solution.

62. $4^{x^2+2x} = 64$
$4^{x^2+2x} = 4^3$
$x^2 + 2x = 3$
$x^2 + 2x - 3 = 0$
$(x + 3)(x - 1) = 0$
$x = -3 \quad x = 1$
The solutions are –3 and 1.

63. $5^{2x-3} = 5^{x+9} \cdot 5^{x-12}$
$5^{2x-3} = 5^{x+9+x-12}$
$2x - 3 = x + 9 + x - 12$
$2x - 3 = 2x - 3$
$-3 = -3$
This is an identity. The solutions are all real numbers.

64. $10^{2.3x} = 4$
$\log 10^{2.3x} = \log 4$
$2.3x = \log 4$
$x = \frac{\log 4}{2.3}$
$x \approx 0.262$
The solution is ≈ 0.262.

65. $\log_7 a = 49$
$7^{\log_7 a} = 7^{49}$
$a = 7^{49}$
$a = 2.569 \times 10^{41}$
The solution is 2.569×10^{41}.

66. $5\log_3 x = 45$
$\log_3 x = 9$
$3^{\log_3 x} = 3^9$
$x = 3^9$
$x = 19{,}683$
The solution is 19,683.

67. $\ln z = 1$
$e^{\ln z} = e^1$
$z = e^1$
$z \approx 2.718$
The solution is ≈ 2.718.

68. $\log b = 0$
$10^{\log b} = 10^0$
$b = 10^0$
$b = 1$
The solution is 1.

69. $\log k = -3$
$10^{\log k} = 10^{-3}$
$k = \frac{1}{10^3}$
$k = \frac{1}{1000}$
The solution is $\frac{1}{1000}$.

70. $5 + \log_2 x = 3$
$\log_2 x = -2$
$x = 2^{-2}$
$x = \frac{1}{2^2}$
$x = \frac{1}{4}$
The solution is $\frac{1}{4}$.

71. $3\ln x = 1$
$\ln x = \frac{1}{3}$
$x = e^{1/3}$
$x \approx 1.396$
The solution is ≈ 1.396.

72. $\ln\left(2x^2 + 7x\right) = \ln 15$
$2x^2 + 7x = 15$
$2x^2 + 7x - 15 = 0$
$(2x - 3)(x + 5) = 0$
$x = \frac{3}{2} \quad x = -5$
The solutions are -5 and $\frac{3}{2}$.

73. $2 \log (x - 2) = \log 9$
$\log(x-2)^2 = \log 9$
$(x-2)^2 = 9$
$x^2 - 4x + 4 = 9$
$x^2 - 4x - 5 = 0$
$(x - 5)(x + 1) = 0$
$x = 5 \quad x = -1$ (restricted)
The solution is 5.

74. $\log x + \log (x + 4) = 2 \log (x + 2)$
$\log x(x+4) = \log(x+2)^2$
$x(x+4) = (x+2)^2$
$x^2 + 4x = x^2 + 4x + 4$
$0 = 4$
This is a contradiction. There is no solution.

75. $7\ln x - \ln x^3 = 4\ln x$

$\ln \frac{x^7}{x^3} = \ln x^4$

$x^4 = x^4$

This is an identity. The solutions are all real numbers greater than 0.

76. Let t = number of years since 1990.

$A(t) = A_0 e^{kt}$

$949.4 = 697.5e^{k(4)}$

$\frac{949.4}{697.5} = e^{4k}$

$\ln \frac{949.4}{697.5} = 4k$

$\frac{1}{4}\ln \frac{949.4}{697.5} = k$

$k \approx 0.077$

$A(t) = 697.5e^{0.077t}$

$t = 2000 - 1990 = 10$

$A(10) = 697.5e^{0.077(10)}$

$A(10) \approx 1506.44$

$t = 2005 - 1990 = 15$

$A(15) = 697.5e^{0.077(15)}$

$A(15) \approx 2213.88$

$t = 2010 - 1990 = 20$

$A(20) = 697.5e^{0.077(20)}$

$A(20) \approx 3253.55$

The estimated expenditures in the years 2000, 2005, and 2010 are
\$1506.44 billion,
\$2213.88 billion, and
\$3253.55 billion, respectively.

77. $A(t) = A_0 e^{kt}$

$\frac{2}{3} = e^{-0.000121t}$

$\ln \frac{2}{3} = -0.000121t$

$\frac{\ln \frac{2}{3}}{-0.000121} = t$

$t \approx 3351$

The time since the skeleton's demise is 3351 years.

Chapter 15 Mixed Review

1. $\log 0.001 = -3$ because $10^{-3} = 0.001$

2. $\log_7 343 = 3$ because $7^3 = 343$

3. $\log_2 \frac{1}{64} = -6$ because $2^{-6} = \frac{1}{64}$

4. $\log 5.1 \approx 0.708$

5. $\ln e^{-3} = -3$

6. $\ln 100 \approx 4.605$

7. $\ln e = 1$

8. $\ln \frac{4}{7} \approx -0.560$

9. $\log_3 36 = \frac{\log 36}{\log 3} \approx 3.262$

10. $\log 3x^2 y$

$= \log 3 + \log x^2 + \log y$

$= \log 3 + 2\log x + \log y$

11. $\log_3 \frac{9a}{b}$

$= \log_3 9a - \log_3 b$

$= \log_3 9 + \log_3 a - \log_3 b$

$= 2 + \log_3 a - \log_3 b$

12. $\ln 100p^3q^2r$

$= \ln 100 + \ln p^3 + \ln q^2 + \ln r$

$= \ln 100 + 3\ln p + 2\ln q + \ln r$

13. $\log \frac{\sqrt[5]{6x^3}}{\sqrt{xy}}$

$= \log \frac{(6x^3)^{1/5}}{(xy)^{1/2}}$

$= \log(6x^3)^{1/5} - \log(xy)^{1/2}$

$= \frac{1}{5}\log(6x^3) - \frac{1}{2}\log xy$

$= \frac{1}{5}(\log 6 + \log x^3) - \frac{1}{2}(\log x + \log y)$

$= \frac{1}{5}\log 6 + \frac{3}{5}\log x - \frac{1}{2}\log x - \frac{1}{2}\log y$

14. $\log(x^2 - 9) - \log(x+3)$

$= \log \frac{x^2 - 9}{x+3}$

$= \log \frac{(x+3)(x-3)}{x+3}$

$= \log(x-3)$

15. $2\log c - 5\log d$

$= \log c^2 - \log d^5$

$= \log \frac{c^2}{d^5}$

16. $\frac{1}{2}\ln x - 3\ln y$

$= \ln x^{1/2} - \ln y^3$

$= \ln \frac{\sqrt{x}}{y^3}$

17. $\log 2xy - \log xz + \log 3yz$

$= \log \frac{2xy}{xz} + \log 3yz$

$= \log \frac{2xy(3yz)}{xz}$

$= \log 6y^2$

18. $y = 5 - x$

$x = 5 - y$

$y = 5 - x$

Therefore $y^{-1} = 5 - x$.

19. $y = 1 - x^2$

$x = 1 - y^2$

$y^2 = 1 - x$

$y = \pm\sqrt{1-x}$

Therefore $y^{-1} = \pm\sqrt{1-x}$.

20. $y = 4 - x^3$

$x = 4 - y^3$

$y^3 = 4 - x$

$y = \sqrt[3]{4-x}$

Therefore $y^{-1} = \sqrt[3]{4-x}$.

21. $y = 2 - \frac{4}{5}x$

$x = 2 - \frac{4}{5}y$

$\frac{4}{5}y = 2 - x$

$\frac{5}{4}\left(\frac{4}{5}y\right) = \frac{5}{4}(2-x)$

$y = \frac{5}{2} - \frac{5}{4}x$

Therefore, $y^{-1} = -\frac{5}{4}x + \frac{5}{2}$.

22. $m(x) = 5^x$

$m^{-1}(x) = \log_5 x$

23. $G(x) = ab^x$

$y = ab^x$

$x = ab^y$

$\frac{x}{a} = b^y$

$\log_b\left(\frac{x}{a}\right) = \log_b b^y$

$\log_b\left(\frac{x}{a}\right) = y$

Therefore, $G^{-1}(x) = \log_b\left(\frac{x}{a}\right)$.

24. $g(t)=1+\left(\frac{4}{9}\right)^t$

$g(3)=1+\left(\frac{4}{9}\right)^3$

$g(3)\approx 1.088$

25. $g(t)=1+\left(\frac{4}{9}\right)^t$

$g(-1)=1+\left(\frac{4}{9}\right)^{-1}$

$g(-1)=1+\frac{9}{4}$

$g(-1)=\frac{13}{4}$

26. $g(t)=1+\left(\frac{4}{9}\right)^t$

$g(0)=1+\left(\frac{4}{9}\right)^0$

$g(0)=1+1$

$g(0)=2$

27. $g(t)=1+\left(\frac{4}{9}\right)^t$

$g\left(\frac{1}{2}\right)=1+\left(\frac{4}{9}\right)^{\frac{1}{2}}$

$g\left(\frac{1}{2}\right)=1+\frac{2}{3}$

$g\left(\frac{1}{2}\right)=1\frac{2}{3}$

28. $g(t)=1+\left(\frac{4}{9}\right)^t$

$g(1.5)=1+\left(\frac{4}{9}\right)^{1.5}$

$g(1.5)\approx 1.296$

29. $g(t)=1+\left(\frac{4}{9}\right)^t$

$g\left(-\sqrt{3}\right)=1+\left(\frac{4}{9}\right)^{-\sqrt{3}}$

$g\left(-\sqrt{3}\right)\approx 5.074$

30. $m(x)=3^x-2$

$\text{Y1}=3^x-2$

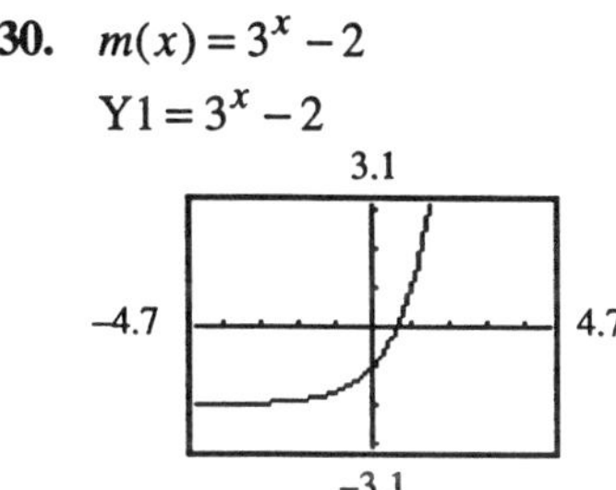

31. $k(x)=8^{\frac{1}{3}x}$

$\text{Y1}=8^{\frac{1}{3}x}$

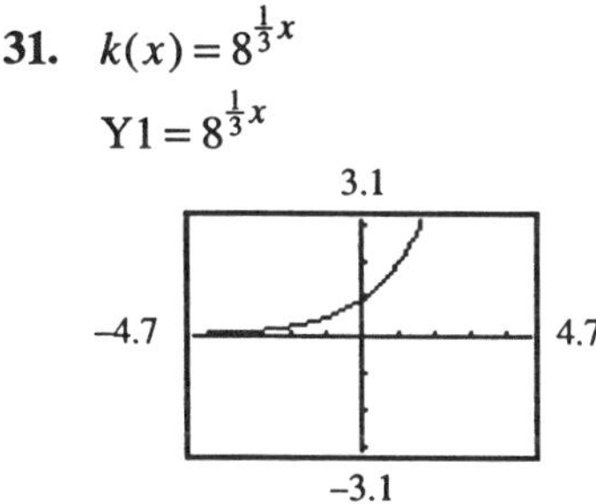

32. $y=-2e^{0.4x}$

$\text{Y1}=-2e^{0.4x}$

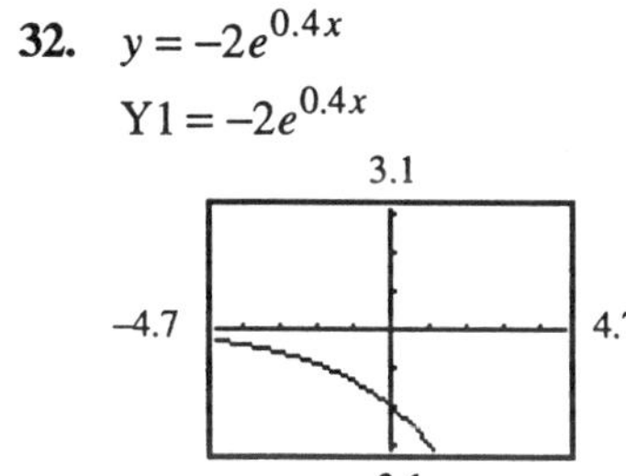

33. $y=\log_2 x-2$

$\text{Y1}=\frac{\log x}{\log 2}-2$

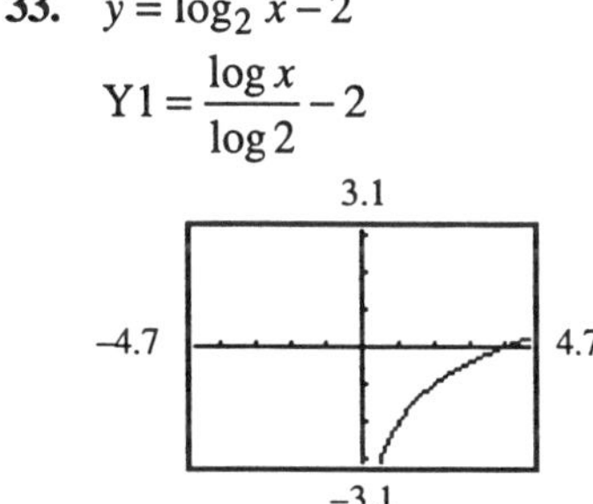

34. $g(x) = \log(x + 1)$
$Y1 = \log(x+1)$

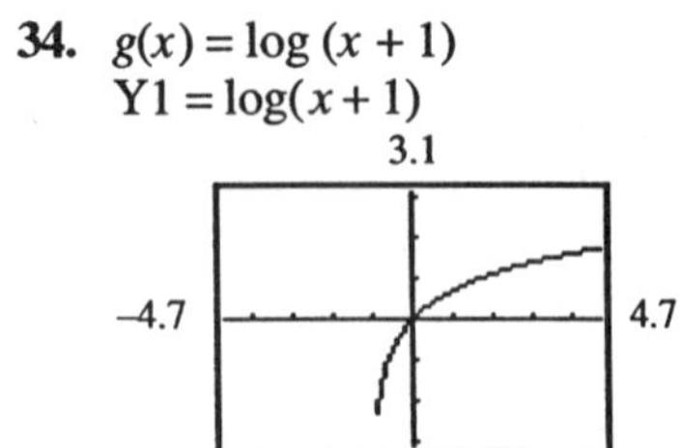

35. $\log_6 b = 216$
$b = 6^{216}$

The solution is 6^{216}.

36. $2\log_3 z = 6$
$\log_3 z = 3$
$z = 3^3$
$z = 27$
The solution is 27.

37. $\log z = 1$
$z = 10^1$
$z = 10$
The solution is 10.

38. $\ln b = 0$
$b = e^0$
$b = 1$
The solution is 1.

39. $\log 2m = -4$
$2m = 10^{-4}$
$m = \dfrac{1}{2 \cdot 10^4}$
$m = \dfrac{1}{20{,}000}$

The solution is $\dfrac{1}{20{,}000}$.

40. $-1 + \log_3 x = 4$
$\log_3 x = 5$
$x = 3^5$
$x = 243$
The solution is 243.

41. $2\log x = 1$
$\log x = \dfrac{1}{2}$
$x = 10^{1/2}$
$x = \sqrt{10}$
$x \approx 3.162$
The solution is ≈ 3.162.

42. $\log\left(2x^2 + 3x\right) = \log 9$
$2x^2 + 3x = 9$
$2x^2 + 3x - 9 = 0$
$(2x - 3)(x + 3) = 0$
$x = \dfrac{3}{2} \quad x = -3$

The solutions are –3 and $\dfrac{3}{2}$.

43. $2\ln(x+3) = \ln 4$
$\ln(x+3)^2 = \ln 4$
$(x+3)^2 = 4$
$x^2 + 6x + 5 = 0$
$(x + 1)(x + 5) = 0$
$x = -1 \quad x = -5$ (restricted).
The solution is –1.

44. $\log 4x + \log(x + 1) = 2\log(2x + 1)$
$\log 4x(x+1) = \log(2x+1)^2$
$4x(x+1) = (2x+1)^2$
$4x^2 + 4x = 4x^2 + 4x + 1$
$0 = 1$
This is a contradiction. There is no solution.

45. $2\log a + \log a^4 = 6\log a$
$\log a^2 + \log a^4 = \log a^6$
$\log a^2 \cdot a^4 = \log a^6$
$a^6 = a^6$
This is an identity. The solutions are all real numbers greater than 0.

46. $5^{x-3}=125$
$5^{x-3}=5^3$
$x-3=3$
$x=6$
The solution is 6.

47. $e^{-t}+4=5$
$e^{-t}=1$
$\ln e^{-t}=\ln 1$
$t=0$
The solution is 0.

48. $7^a=e$
$\ln 7^a=\ln e$
$a\ln 7=1$
$a=\dfrac{1}{\ln 7}$
$a\approx 0.514$
The solution is ≈ 0.514.

49. $(6^7)(6^{3x})=6^2$
$6^{7+3x}=6^2$
$7+3x=2$
$3x=-5$
$x=-\dfrac{5}{3}$
The solution is $-\dfrac{5}{3}$.

50. $e^{x(x+6)}=e^{(x+5)(x+1)}$
$x(x+6)=(x+5)(x+1)$
$x^2+6x=x^2+6x+5$
$0=5$
This is a contradiction. There is no solution.

51. $2^{x^2}\cdot 2^{-3x}=16$
$2^{x^2-3x}=2^4$
$x^2-3x=4$
$x^2-3x-4=0$
$(x-4)(x+1)=0$
$x=4 \quad x=-1$
The solutions are -1 and 4.

52. $7^{2x-6}=(7^{3x-11})(7^{5-x})$
$7^{2x-6}=7^{3x-11+5-x}$
$2x-6=3x-11+5-x$
$2x-6=2x-6$
$-6=-6$
This is an identity. The solutions are all real numbers.

53. $10^{1.8x}=9$
$\log 10^{1.8x}=\log 9$
$1.8x=\log 9$
$x=\dfrac{\log 9}{1.8}$
$x\approx 0.530$
The solution is ≈ 0.530.

54. Let x = amount of sales
The situation can be represented by
$f(x)=1200+0.025x$.
$y=1200+0.025x$
$x=1200+0.025y$
$x-1200=0.025y$
$\dfrac{x-1200}{0.025}=y$
$y=40x-48{,}000$
Therefore, $f^{-1}(x)=40x-48{,}000$.
The inverse function represents her total sales in terms of her monthly income.

55. Let x = number of hours
The situation can be represented by
$f(x)=95x$.
$y=95x$
$x=95y$
$\dfrac{x}{95}=y$
Therefore, $f^{-1}(x)=\dfrac{x}{95}$.
The inverse function represents the number of hours in terms of the distance traveled.

56. $A=P(1+r)^t$
$A=8500(1.065)^t$
$Y1=8500(1.065)^x$

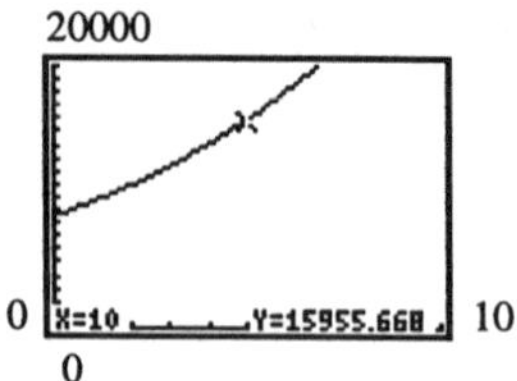

At $x = 10$, $y \approx 15955.67$
The investment was worth $15,955.67.

57. $y = 2.490e^{0.0673t}$
$t = 2010 - 1970 = 40$
$y = 2.490e^{0.0673(40)}$
$y \approx 36.755$
The model predicts that there will be 36.755 thousand (or 36,755) cable systems in the year 2010.
Answers will vary.

58. $\text{pH} = -\log\left[\text{H}^+\right]$
$\text{pH} = -\log\left(1.6 \times 10^{-4}\right)$
$\text{pH} \approx 3.796$
The pH is approximately 3.796.

59. $\text{pH} = -\log[\text{H}^+]$
$1.6 = -\log[\text{H}^+]$
$[\text{H}^+] = 10^{-1.6}$
$[\text{H}^+] \approx 2.5 \times 10^{-2}$
The hydrogen ion concentration is approximately 2.5×10^{-2} moles per liter.

60. $t = \dfrac{\ln k}{r}$
$t = \dfrac{\ln 2}{0.056}$
$t \approx 12.4$
It will take approximately 12.4 years.

61. Let t = number of years since 1980.
$A(t) = A_0e^{kt}$
$207.8 = 96.9e^{k(10)}$
$\dfrac{207.8}{96.9} = e^{10k}$
$\ln\dfrac{207.8}{96.9} = \ln e^{10k}$
$\ln\dfrac{207.8}{96.9} = 10k$
$k = \dfrac{1}{10}\ln\dfrac{207.8}{96.9}$
$k \approx 0.076$
$A(t) = 96.9e^{0.076t}$
$t = 2000 - 1980 = 20$
$A(20) = 96.9e^{0.076(20)}$
$A(20) \approx 443.0$
$t = 2005 - 1980 = 25$
$A(25) = 96.9e^{0.076(25)}$
$A(25) \approx 647.9$
$t = 2010 - 1980 = 30$
$A(30) = 96.9e^{0.076(30)}$
$A(30) \approx 947.4$
The estimated revenues for 2000, 2005, and 2010 are (in billions) $443.0, $647.9, and $947.4, respectively.

62. $A(t) = A_0e^{kt}$
$A(t) = A_0e^{-0.000121t}$
$0.4 = e^{-0.000121t}$
$\ln 0.4 = -0.000121t$
$t = \dfrac{\ln 0.4}{-0.000121}$
$t \approx 7573$
It has been about 7573 years since the organism died.

Chapter 15 Test

1. $f(x) = x^3$

$y = x^3$
$x = y^3$
$\sqrt[3]{x} = y$
Therefore
$f^{-1}(x) = \sqrt[3]{x}$

$Y1 = x^3$
$Y2 = \sqrt[3]{x}$
$Y3 = x$

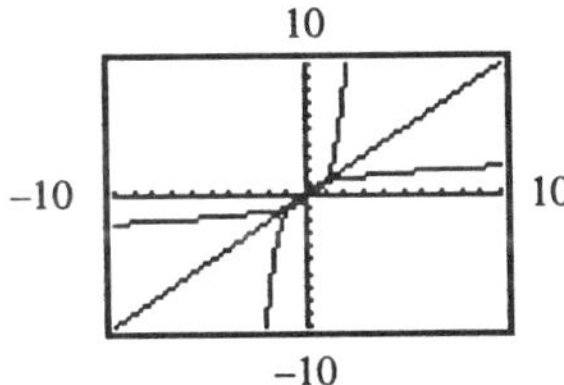

2. $f(x) = 2^x + 1$
   $Y1 = 2^x + 1$

3.1
−4.7 4.7
−3.1

3. $g(x) = \ln(x + 2)$
   $Y1 = \ln(x + 2)$

3.1
−4.7 4.7
−3.1

4. $F(x) = 2x - 8$

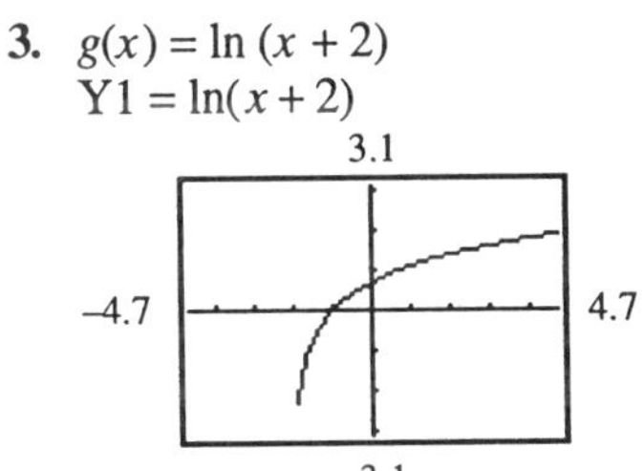

$$y = 2x - 8$$
$$x = 2y - 8$$
$$x + 8 = 2y$$
$$\frac{x+8}{2} = y$$
$$y = \frac{1}{2}x + 4$$

Therefore, $F^{-1}(x) = \frac{1}{2}x + 4$.

5. $g(x) = 16^{\frac{1}{2}x}$

$$g(-1) = 16^{\frac{1}{2}(-1)}$$
$$g(-1) = \frac{1}{\sqrt{16}}$$
$$g(-1) = \frac{1}{4}$$

6. $g(x) = 16^{\frac{1}{2}x}$

$$g(0) = 16^{\frac{1}{2}\cdot 0}$$
$$g(0) = 1$$

7. $g(x) = 16^{\frac{1}{2}x}$

$$g(4) = 16^{\frac{1}{2}\cdot 4}$$
$$g(4) = 256$$

8. $g(x) = 16^{\frac{1}{2}x}$

$$g(0.3) = 16^{\frac{1}{2}(0.3)}$$
$$g(0.3) \approx 1.516$$

9. $g(x) = 16^{\frac{1}{2}x}$

$$g\left(\sqrt{3}\right) = 16^{\frac{1}{2}\cdot\sqrt{3}}$$
$$g\left(\sqrt{3}\right) \approx 11.036$$

10. $\log 3x^2y^3z$

$$= \log 3 + \log x^2 + \log y^3 + \log z$$
$$= \log 3 + 2\log x + 3\log y + \log z$$

11. $\log_3 \frac{9a^2}{b}$

$$= \log_3 9a^2 - \log_3 b$$
$$= \log_3 9 + \log_3 a^2 - \log_3 b$$
$$= 2 + 2\log_3 a - \log_3 b$$

12. $2\log x + \log(x - 4)$

$$= \log x^2 + \log(x - 4)$$
$$= \log x^2(x - 4)$$

13. $\frac{1}{2}\ln x - 2\ln y$
$= \ln x^{1/2} - \ln y^2$
$= \ln \frac{\sqrt{x}}{y^2}$

14. a. $Y1 = 9^x$
$Y2 = 3^{x+1}$

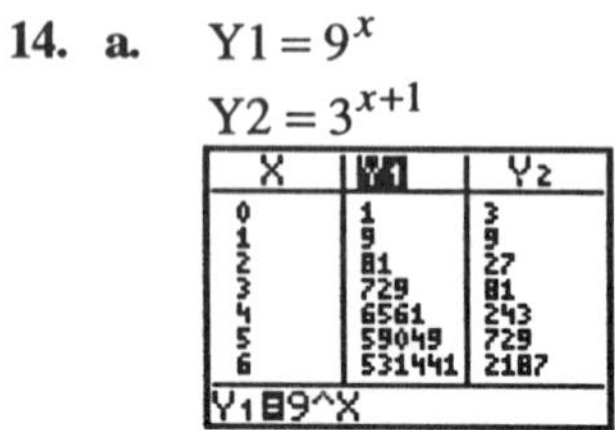

X	Y1	Y2
0	1	3
1	9	9
2	81	27
3	729	81
4	6561	243
5	59049	729
6	531441	2187

Y1=9^X

The solution is $x = 1$.

b. $Y1 = 9^x$
$Y2 = 3^{x+1}$

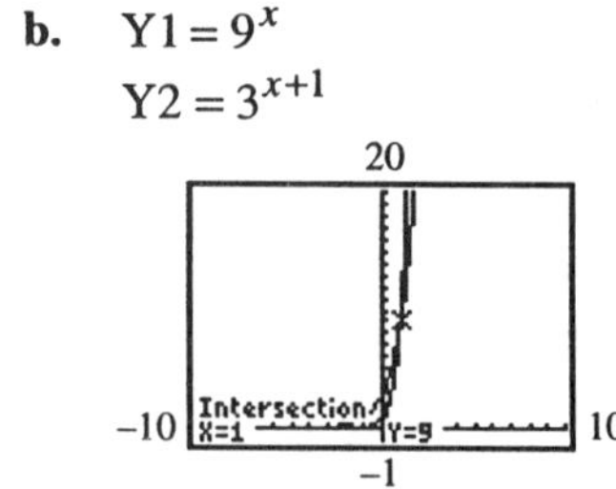

The solution is $x = 1$.

c. $9^x = 3^{x+1}$
$(3^2)^x = 3^{x+1}$
$3^{2x} = 3^{x+1}$
$2x = x + 1$
$x = 1$
The solution is $x = 1$.

15. $(2^{3x})(2^{-2}) = 8$
$2^{3x-2} = 2^3$
$3x - 2 = 3$
$3x = 5$
$x = \frac{5}{3}$
The solution is $\frac{5}{3}$.

16. $2\ln a = 3.1$
$\ln a = 1.55$
$e^{\ln a} = e^{1.55}$
$a = e^{1.55}$
$a \approx 4.711$
The solution is ≈ 4.771.

17. $2\log x + \log x^4 = 6\log x$
$\log x^2 + \log x^4 = \log x^6$
$\log x^2 \cdot x^4 = \log x^6$
$x^6 = x^6$
This is an identity. The solutions are all real numbers greater than 0.

18. $9^x = 5$
$\log 9^x = \log 5$
$x\log 9 = \log 5$
$x = \frac{\log 5}{\log 9}$
$x \approx 0.732$
The solution is ≈ 0.732.

19. Let x = amount of weekly sales
$f(x)$ = total weekly income

a. $f(x) = 400 + 0.03x$

b. $y = 400 + 0.03x$
$x = 400 + 0.03y$
$x - 400 = 0.03y$
$\frac{x - 400}{0.03} = y$
$y = 33.\overline{3}x - 13,333.\overline{3}$
Therefore, $f^{-1}(x) = 33.\overline{3}x - 13,333.\overline{3}$.

c. The inverse function represents the value of weekly sales in terms of his weekly income.

20. $A = P(1+r)^t$
$f(x) = 2000(1.05)^x$
$f(20) = 2000(1.05)^{20}$
$f(20) \approx 5306.60$
The investment was worth approximately $5306.60.

21. $t = \dfrac{\ln k}{r}$
$t = \dfrac{\ln 2}{0.075}$
$t \approx 9.24$
It will take approximately 9.24 years.

22. Answers will vary.

Chapters 1–15 Cumulative Review

1. $5^{-2}x^0y^2 = \dfrac{y^2}{5^2} = \dfrac{y^2}{25}$

2. $(-2x^{-3}y^2)^2 = (-2)^2(x^{-3})^2(y^2)^2$
$= 4x^{-6}y^4$
$= \dfrac{4y^4}{x^6}$

3. $\left(\dfrac{4x^{3/4}}{x^{-1/4}}\right)^2 = (4x^{3/4-(-1/4)})^2$
$= (4x)^2$
$= 4^2x^2 = 16x^2$

4. $(3.2a^2 - 2.6ab + 1.7b^2) - (4a^2 + 2.6ab - b^2)$
$= (3.2a^2 - 2.6ab + 1.7b^2) + (-4a^2 - 2.6ab + b^2)$
$= (3.2 - 4)a^2 + (-2.6 - 2.6)ab + (1.7 + 1)b^2$
$= -0.8a^2 - 5.2ab + 2.7b^2$

5. $(4.5x^2y)(-2y^2z) = (4.5)(-2)x^2y^{1+2}z$
$= -9x^2y^3z$

6. $(3x + 2y)(3x - 2y) = (3x)^2 - (2y)^2$
$= 9x^2 - 4y^2$

7. $(2x + 6)(2x - 3)$
$= (2x)(2x) + (2x)(-3) + (6)(2x) + (6)(-3)$
$= 4x^2 - 6x + 12x - 18 = 4x^2 + 6x - 18$

8. $(x-4)^2 = x^2 - 2(x)(4) + 4^2 = x^2 - 8x + 16$

9. $\dfrac{25x^2y^4z^2}{-5xyz} = -5x^{2-1}y^{4-1}z^{2-1} = -5xy^3z$

10. $\dfrac{2r^2s+4rs-8s^2}{2rs} = \dfrac{2r^2s}{2rs} + \dfrac{4rs}{2rs} - \dfrac{8s^2}{2rs}$
$= r + 2 - \dfrac{4s}{r}$

11. $\dfrac{x^2-x-6}{x^2-9} = \dfrac{(x-3)(x+2)}{(x-3)(x+3)} = \dfrac{x+2}{x+3}$

12. $\dfrac{x}{x+2} + \dfrac{3}{x-2} + \dfrac{x-1}{x^2-4}$
$= \dfrac{x}{x+2} + \dfrac{3}{x-2} + \dfrac{x-1}{(x+2)(x-2)}$
$= \dfrac{x(x-2)}{(x+2)(x-2)} + \dfrac{3(x+2)}{(x+2)(x-2)} + \dfrac{x-1}{(x+2)(x-2)}$
$= \dfrac{x^2-2x+3x+6+x-1}{(x+2)(x-2)}$
$= \dfrac{x^2+2x+5}{(x+2)(x-2)}$

13. $\dfrac{1}{x+4} - \dfrac{x}{x-2}$
$= \dfrac{1(x-2)}{(x+4)(x-2)} - \dfrac{x(x+4)}{(x+4)(x-2)}$
$= \dfrac{x-2-(x^2+4x)}{(x+4)(x-2)} = \dfrac{x-2-x^2-4x}{(x+4)(x-2)}$
$= \dfrac{-x^2-3x-2}{(x+4)(x-2)} = -\dfrac{x^2+3x+2}{(x+4)(x-2)}$
$= -\dfrac{(x+2)(x+1)}{(x+4)(x-2)}$

14. $\dfrac{2m-3}{m+1} \cdot \dfrac{2m^2+5m+3}{4m^2-9}$
$= \dfrac{2m-3}{m+1} \cdot \dfrac{(m+1)(2m+3)}{(2m+3)(2m-3)}$
$= \dfrac{(2m-3)(m+1)(2m+3)}{(m+1)(2m+3)(2m-3)} = 1$

15. $\dfrac{16x}{y^3+y^2-12y} \div \dfrac{4x^2y}{y^2-y-20}$
$= \dfrac{16x}{y(y+4)(y-3)} \div \dfrac{4x^2y}{(y-5)(y+4)}$
$= \dfrac{16x}{y(y+4)(y-3)} \cdot \dfrac{(y-5)(y+4)}{4x^2y}$
$= \dfrac{16x(y-5)(y+4)}{y(y+4)(y-3) \cdot 4x^2y} = \dfrac{4(y-5)}{xy^2(y-3)}$

16. $\sqrt{18x} - \sqrt{50x} + 4\sqrt{8x}$
$= \sqrt{9 \cdot 2x} - \sqrt{25 \cdot 2x} + 4\sqrt{4 \cdot 2x}$
$= 3\sqrt{2x} - 5\sqrt{2x} + 8\sqrt{2x} = 6\sqrt{2x}$

17. $\sqrt[3]{16a^2b^2} \cdot \sqrt[3]{8ab^2}$
$= \sqrt[3]{16a^2b^2 \cdot 8ab^2} = \sqrt[3]{128a^3b^4}$
$= \sqrt[3]{64a^3b^3 \cdot 2b} = 4ab\sqrt[3]{2b}$

18. $\left(\sqrt{2}-3\right)\left(\sqrt{2}+\sqrt{3}\right) = \left(\sqrt{2}\right)^2 - \left(\sqrt{3}\right)^2$
$= 2 - 3 = -1$

19. $\dfrac{\sqrt{32xy^2z}}{\sqrt{8x^3y}} = \sqrt{\dfrac{32xy^2z}{8x^3y}} = \sqrt{\dfrac{4yz}{x^2}} = \dfrac{2}{x}\sqrt{yz}$

20. $\dfrac{\sqrt{x}+4}{\sqrt{x}-3} = \dfrac{\left(\sqrt{x}+4\right)\left(\sqrt{x}+3\right)}{\left(\sqrt{x}-3\right)\left(\sqrt{x}+3\right)} = \dfrac{x+7\sqrt{x}+12}{x-9}$

21. $x^{2/3}(x^{3/4} - x^{1/3}) = x^{2/3+3/4} - x^{2/3+1/3}$
$= x^{17/12} - x$

22. $25m^2 - 36n^2 = (5m)^2 - (6n)^2$
$= (5m - 6n)(5m + 6n)$

23. $2x^2 - 4x - 16 = 2(x^2 - 2x - 8)$
$= 2(x - 4)(x + 2)$

24. $x^2 + 6x - 6$ does not factor.

25.

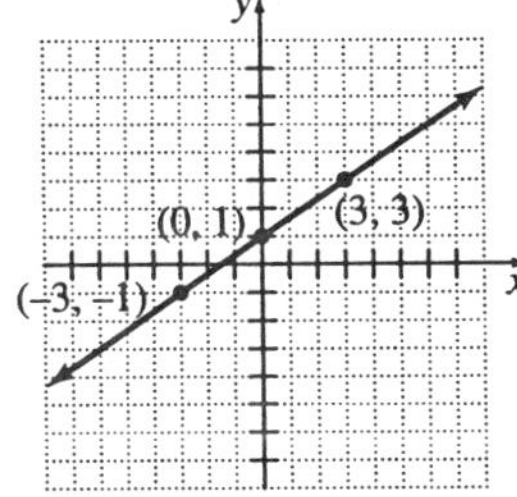

Domain: all real numbers
Range: all real numbers

26. $y = x^2 + 5x + 4$

$a = 1,\ b = 5,\ c = 4$

$\frac{-b}{2a} = \frac{-5}{2(1)} = -\frac{5}{2}$

x	$y = x^2 + 5x + 4$	y
−4	$y = (-4)^2 + 5(-4) + 4$ $y = 0$	0
$-\frac{5}{2}$	$y = \left(-\frac{5}{2}\right)^2 + 5\left(-\frac{5}{2}\right) + 4$ $y = -\frac{9}{4}$	$-\frac{9}{4}$
−1	$y = (-1)^2 + 5(-1) + 4$ $y = 0$	0
0	$y = 0^2 + 5(0) + 4$ $y = 4$	4

The vertex is $\left(-\frac{5}{2}, -\frac{9}{4}\right)$. The axis of symmetry is $x = -\frac{5}{2}$.

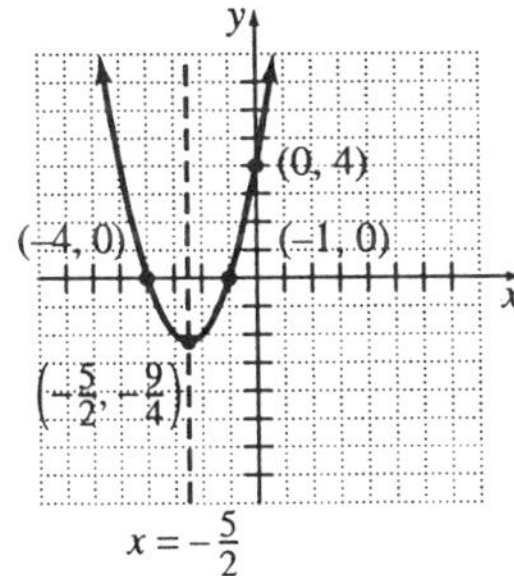

Domain: all real numbers

Range: all real numbers $\geq -\frac{9}{4}$

27.

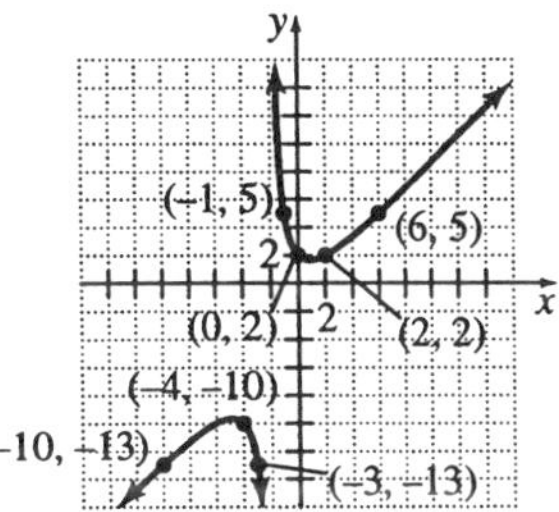

Domain: all real numbers $\neq -2$
Range: all real numbers ≤ -9.66 or ≥ 1.66

28.

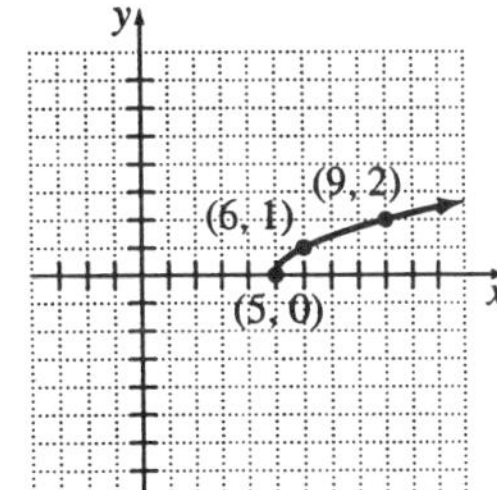

Domain: all real numbers $x \geq 5$
Range: all real numbers ≥ 0

29.

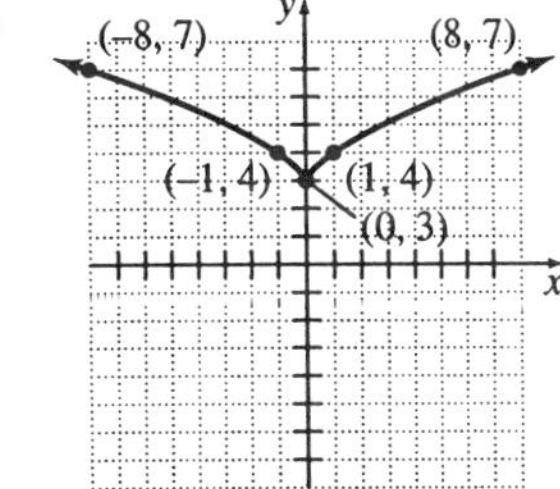

Domain: all real numbers
Range: all real numbers ≥ 3

30.

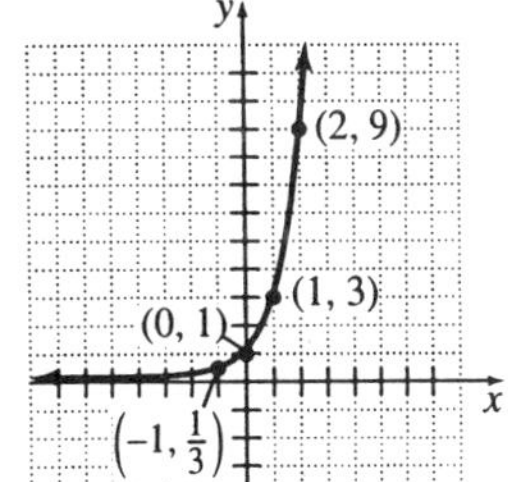

Domain: all real numbers
Range: all real numbers > 0

31.

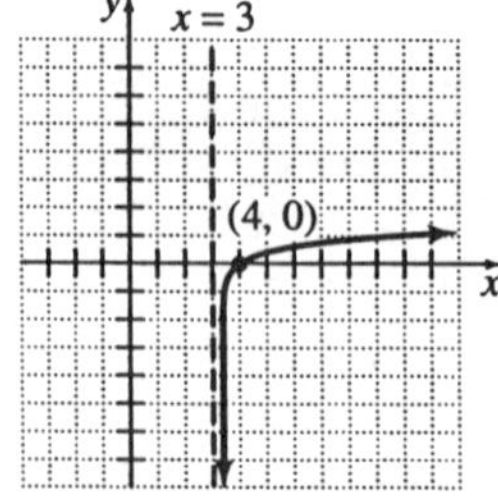

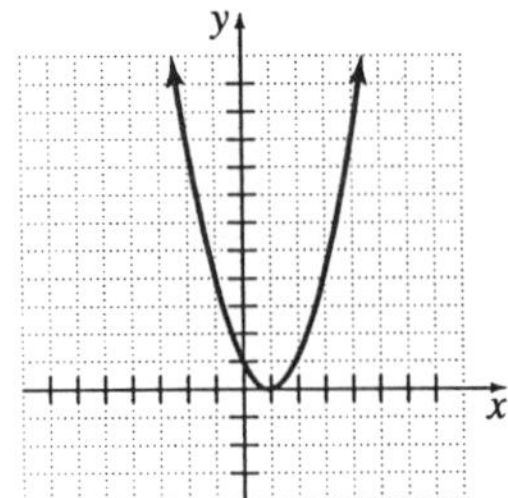

Domain: all real numbers > 3
Range: all real numbers

32. $2(x-5)-(x+4)=2x-8$
$2x-10-x-4=2x-8$
$x-14=2x-8$
$-6=x$

33. $2(x+3.1)=(x-4.2)+(x-6)$
$2x+6.2=x-4.2+x-6$
$2x+6.2=2x-10.2$
$6.2=-10.2$
This is a contradiction. There is no solution.

34. $x^2-2x-12=12$
$x^2-2x-24=0$
$(x-6)(x+4)=0$
$x-6=0$ or $x+4=0$
$x=6$ or $x=-4$

35. $x^2-5x+9=0$
$$x=\frac{-(-5)\pm\sqrt{(-5)^2-4(1)(9)}}{2(1)}$$
$$x=\frac{5\pm\sqrt{25-36}}{2}$$
$$x=\frac{5\pm\sqrt{-11}}{2}$$
$$x=\frac{5}{2}\pm\frac{\sqrt{11}}{2}i$$

36. $\frac{x+1}{x+2}=\frac{x-1}{x+3}$
$$(x+2)(x+3)\frac{x+1}{x+2}=(x+2)(x+3)\frac{x-1}{x+3}$$
$(x+3)(x+1)=(x+2)(x-1)$
$x^2+4x+3=x^2+x-2$
$3x=-5$
$x=-\frac{5}{3}$

37. $4\sqrt{x+1}-14=-2$
$4\sqrt{x+1}=12$
$\sqrt{x+1}=3$
$x+1=9$
$x=8$

38. $e^{2x-4}=e^{x(x-2)}$
$e^{2x-4}=e^{x^2-2x}$
$2x-4=x^2-2x$
$0=x^2-4x+4$
$0=(x-2)^2$
$0=x-2$
$x=2$

39. $2\log_3 x=\log_3 5$
$\log_3 x^2=\log_3 5$
$x^2=5$
$x=\pm\sqrt{5}$
$x=\sqrt{5}$

40. $2x-3<3(x-4)$
$2x-3<3x-12$
$9<x$
$(9,\infty)$

41. $x^2+2x\ge 2$
$x^2+2x-2\ge 0$
Solve $x^2+2x-2=0$.
$$x=\frac{-2\pm\sqrt{2^2-4(1)(-2)}}{2(1)}$$
$$x=\frac{-2\pm\sqrt{4+8}}{2}$$
$$x=\frac{-2\pm2\sqrt{3}}{2}$$
$x=-1\pm\sqrt{3}$
$\left(-\infty,-1-\sqrt{3}\right]\cup\left[-1+\sqrt{3},\infty\right)$

42. $3x - 2y = 4$
$x + y = 8$
From the second equation, $y = 8 - x$.
Substitute $8 - x$ for y in the first equation.
$3x - 2(8 - x) = 4$
$3x - 16 + 2x = 4$
$5x = 20$
$x = 4$

$y = 8 - x$
$y = 8 - 4$
$y = 4$
The solution is (4, 4).

43. $m = \frac{y_2 - y_1}{x_2 - x_1} = \frac{2-(-1)}{-2-3} = \frac{2+1}{-2-3}$
$= \frac{3}{-5} = -\frac{3}{5}$
$y - y_1 = m(x - x_1)$
$y - (-1) = -\frac{3}{5}(x - 3)$
$y + 1 = -\frac{3}{5}x + \frac{9}{5}$
$y = -\frac{3}{5}x + \frac{4}{5}$

44. $x + 4y = 2.3$
$4y = -x + 2.3$
$y = -\frac{1}{4}x + \frac{2.3}{4}$
The slope of the desired line is $\frac{-1}{\left(-\frac{1}{4}\right)}$ or 4.
$y - y_1 = m(x - x_1)$
$y - 1 = 4[x - (-2)]$
$y - 1 = 4(x + 2)$
$y - 1 = 4x + 8$
$y = 4x + 9$

45. The general form of a quadratic equation is
$y = ax^2 + bx + c$.
$-6 = a(0)^2 + b(0) + c$
$0 = a(-4)^2 + b(-4) + c$
$0 = a(1)^2 + b(1) + c$
These equations simplify to
$-6 = c$
$0 = 16a - 4b + c$
$0 = a + b + c$
Substitute -6 for c in the second and third equations.
$0 = 16a - 4b - 6$
$0 = a + b - 6$
Multiply the second equation by 4 and add.

$$\begin{array}{r} 0 = 16a - 4b - 6 \\ \underline{0 = 4a + 4b - 24} \\ 0 = 20a + 0b - 30 \end{array}$$

$30 = 20a$
$\frac{3}{2} = a$
Now we find b.
$0 = a + b + c$
$0 = \frac{3}{2} + b - 6$
$\frac{9}{2} = b$
The quadratic equation is
$y = \frac{3}{2}x^2 + \frac{9}{2}x - 6$.

46. $f(x) = \frac{2}{3}x + 6$
$y = \frac{2}{3}x + 6$
$x = \frac{2}{3}y + 6$
$x - 6 = \frac{2}{3}y$
$\frac{3}{2}(x - 6) = y$
$\frac{3}{2}x - 9 = y$
$f^{-1}(x) = \frac{3}{2}x - 9$

47. Let x be the number of books printed.
The cost of printing x books is $5x + 525$.
The revenue from selling x books is $15x$.
$15x \geq 5x + 525$
$10x \geq 525$
$x \geq 52.5$
The company must sell 53 books in order to break even.

48. Let w be the width of Nathan's garden. Then the length is $w + 3$. The length of the diagonal is $\sqrt{w^2+(w+3)^2}$.

$15=\sqrt{w^2+(w+3)^2}$
$15^2=w^2+w^2+6w+9$
$225=2w^2+6w+9$
$0=2w^2+6w-216$
$0=w^2+3w-108$
$0=(w+12)(w-9)$
$w+12=0$ or $w-9=0$
$w=-12$ or $w=9$

Since the width cannot be negative, the width of the garden is 9 feet and the length is $9 + 3 = 12$ feet. The dimensions are 9 feet by 12 feet.

49. The left drain drains $\frac{1}{4}$ of the tank per hour, while the right drain drains $\frac{1}{6.5}$ or $\frac{2}{13}$ of the tank per hour. Let x be the number of hours it takes for both drains to drain the tank.

$\frac{1}{4}x+\frac{2}{13}x=1$
$52\left(\frac{1}{4}x\right)+52\left(\frac{2}{13}x\right)=52(1)$
$13x+8x=52$
$21x=52$
$x=\frac{52}{21}$
$x\approx 2.48$

It will take the drains about 2.5 hours to drain the tank.

50. $t=\frac{\ln k}{r}$
$t=\frac{\ln 3}{0.045}$
$t\approx 24.41$

It will take about 24.4 years for the money to triple.

Calculator Appendix

Absolute value

Absolute value expressions are entered in the calculator using the ABS function.

To evaluate an absolute value expression, enter MATH ▶ 1 followed by the expression and) .

EXAMPLE Evaluate $|-16|$.

```
abs( -16)
                16
abs( -16+2)
                14
```

MATH ▶ 1 (−) 1 6) ENTER

The result is 16.

Evaluate $|-16 + 2|$.

MATH ▶ 1 (−) 1 6 + 2) ENTER

The result is 14.

CALC function

Intersect under the CALC function determines the coordinates of the intersection point of two graphs on the graphics screen.

To find the intersection of two graphs, enter 2nd CALC 5 ; use the arrow keys to move the cursor as close as possible to the intersection on the first graph and press ENTER ; move the cursor as close as possible to the intersection on the second graph and press ENTER ; then use the arrow keys to move the cursor close to the intersection for a guess and press ENTER. The coordinates of the intersection are displayed at the bottom of the screen.

EXAMPLE Determine the intersection point of the graphs of the equations $y = 3x + 5$ and $y = -x$.

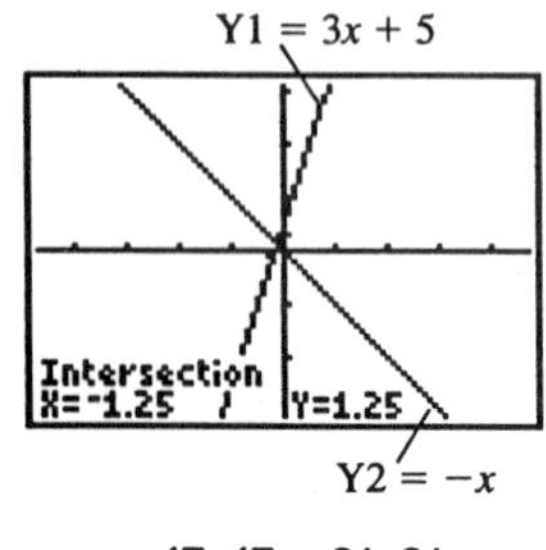

−47, 47, −31, 31

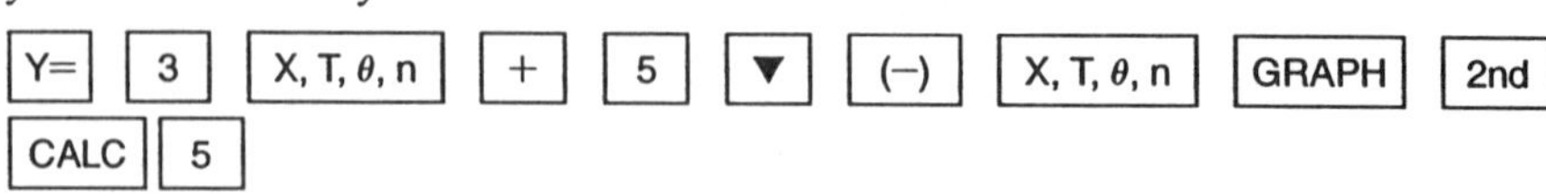

Use the arrow keys to move the cursor as close as possible to the intersection on the first graph and press ENTER; move the cursor as close as possible to the intersection on the second graph and press ENTER; then make sure the cursor is near the intersection for a guess and press ENTER.

The result is $x = -1.25$ and $y = 1.25$.

Maximum under the CALC function determines the coordinates of the maximum point of a graph between two given points on the graph that have been chosen by the user.

To find the maximum point of a graph, enter 2nd CALC 4; use the arrow keys to move the cursor to the left of the maximum point and press ENTER; move the cursor to the right of the maximum point and press ENTER; then move the cursor as close as possible to the maximum point (this should be between the previously two chosen points) and press ENTER. The coordinates of the maximum point will be displayed at the bottom of the screen.

EXAMPLE Determine the coordinates of the maximum point of the graph of the equation $y = -x^2 - 5$.

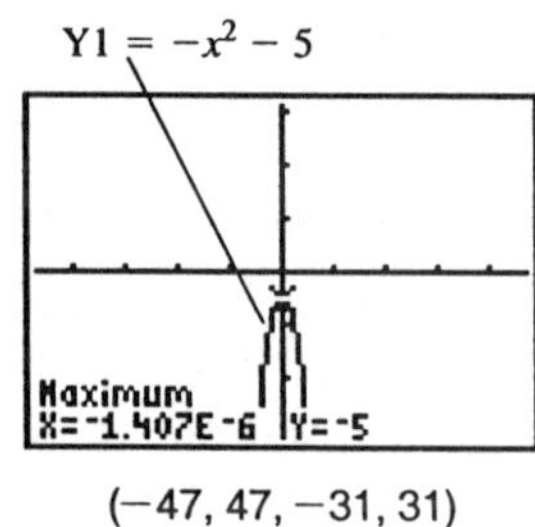

(−47, 47, −31, 31)

Y= (−) X, T, θ, n x^2 − 5 GRAPH 2nd CALC 4

Use the arrow keys to move the cursor to the left of the maximum point and press ENTER; the move the cursor to the right of the maximum point and press ENTER; then move the cursor as close as possible to the maximum point (this should be between the previously two chosen points) and press ENTER.

The result is $x = -1.407\text{E-}6$ (this is very close to 0) and $y = -5$.

Minimum under the CALC function determines the coordinates of the minimum point of a graph between two given points on the graph that have been chosen by the user.

To find the minimum point of a graph, enter 2nd CALC 3 ; use the arrow keys to move the cursor to the left of the minimum point and press ENTER ; move the cursor to the right of the minimum point and press ENTER ; then move the cursor as close as possible to the minimum point (this should be between the previously two chosen points) and press ENTER . The coordinates of the minimum point will be displayed at the bottom of the screen.

EXAMPLE Determine the coordinates of the minimum point of the graph of the equation $y = x^2 - 5$.

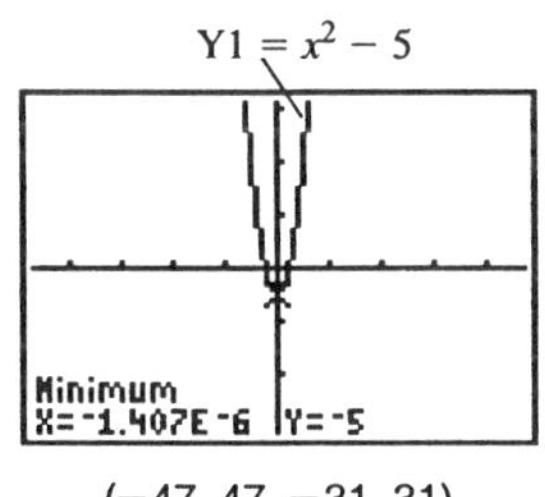

(−47, 47, −31, 31)

Y= X, T, θ, n x² − 5 GRAPH 2nd CALC 3

Use the arrow keys to move the cursor to the left of the minimum point and-press ENTER ; move the cursor to the right of the minimum point and press ENTER ; then move the cursor as close as possible to the minimum point (this should be between the previously two chosen points) and press ENTER .

The result is $x = -1.407\text{E-}6$ (this is very close to 0) and $y = -5$.

Value under the CALC function determines the y-coordinate of a point on a graph on the graphics screen given an x-coordinate chosen by the user.

To find the y-coordinate for a given x-coordinate on a graph, enter 2nd CALC 1 ; then enter the given x-coordinate and press ENTER . The x-coordinate and the y-coordinate are displayed at the bottom of the screen.

EXAMPLE Determine the y-coordinate for the x-coordinate 3 on the graph of the equation $y = x^2 - 5$.

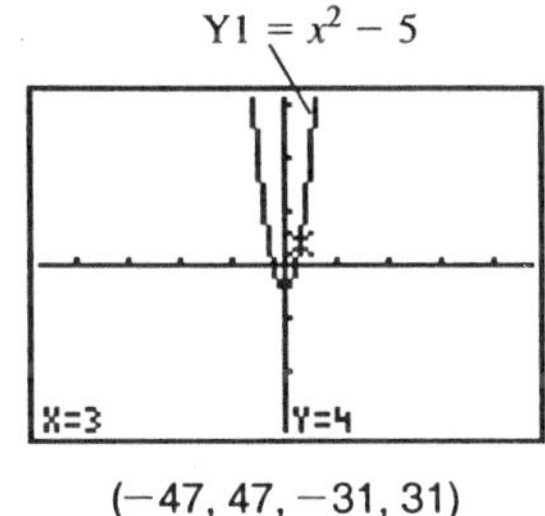

(−47, 47, −31, 31)

Y= X, T, θ, n x² − 5 GRAPH 2nd CALC 1 3 ENTER

The x-coordinate, 3, and the y-coordinate, 4, are displayed at the bottom of the screen.

Zero under the CALC function determines the coordinates of the x-intercept of a graph on the graphics screen.

To find the x-intercept of a graph, enter 2nd CALC 2 ; use the arrow keys to move the cursor to the left of the intercept and press ENTER ; move the cursor to the right of the intercept and press ENTER ; then move the cursor as close as possible to the intercept and press ENTER . The coordinates of the x-intercept are displayed at the bottom of the screen.

EXAMPLE Determine the coordinates of the x-intercept of the graph of the equation $y = 3x + 5$.

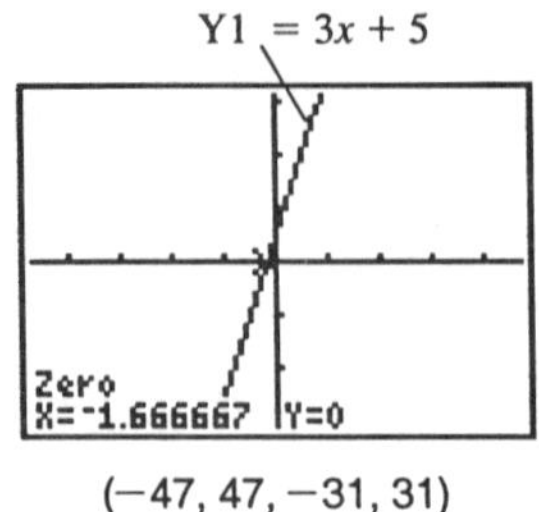

(−47, 47, −31, 31)

Y= 3 X, T, θ, n + 5 GRAPH 2nd CALC 2

Use the ◄ to move to the left of the x-intercept and press ENTER ; use the ► to move to the right of the x-intercept and press ENTER ; then use the ◄ to move close to the x-intercept (between the last two entries) and press ENTER .

The result is $x = -1.666667$ and $y = 0$.

Colon

To enter more than one command within one entry, enter ALPHA : .

EXAMPLE Evaluate $2x + 5$ for $x = 3$.

3→X:2X+5
11

3 STO▶ X, T, θ, n ALPHA : 2 X, T, θ, n + 5 ENTER

The result is 11.

Evaluate

To evaluate an expression involving variables, first store the value of each variable. Enter the value, followed by STO▶ , followed by the name of the variable (ALPHA and the letter of the variable name), followed by a line entry (ALPHA :). Repeat this process until all variables are entered with their values. After the last : , enter the expression to be evaluated.

EXAMPLE Evaluate $4a$ for $a = 3$.

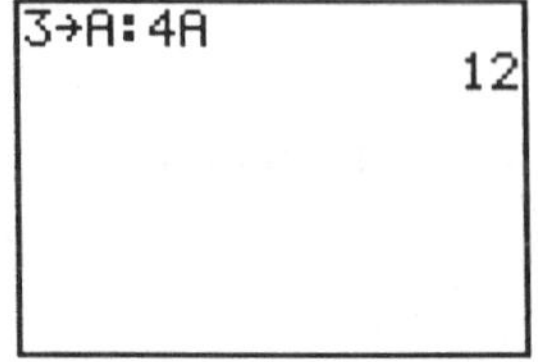

3 STO▶ ALPHA A ALPHA : 4 ALPHA A ENTER

The result is 12.

To change an entry without reentering the entire entry, use 2nd ENTRY to recall the last entry. Change the entry by using the arrow keys to locate the position of the desired change and enter the change. DEL and 2nd

[INS] will allow you to delete and insert characters for entries that are shorter and longer then the original entry.

EXAMPLE Recall the previous entry and evaluate the expression $4a$ for $a = -2$.

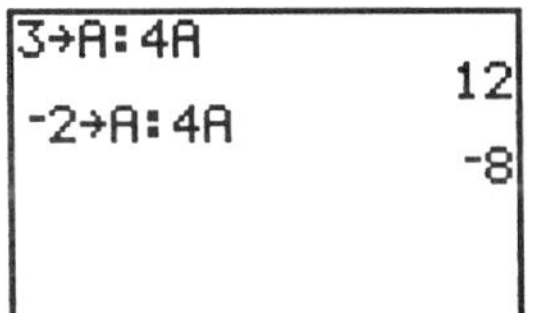

Enter [2nd] [ENTRY]; then use the left arrow key to move over the 3. Enter [(–)] and then insert the 2: [2nd] [INS] [2] [ENTER].

The result is −8.

Exponent

To evaluate a squared expression, enter the expression followed by [x^2]. Be sure to use parentheses correctly.

EXAMPLE Evaluate 9^2.

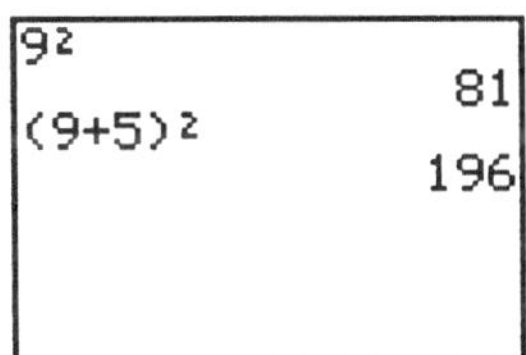

[9] [x^2] [ENTER]

The result is 81.

Evaluate $(9 + 5)^2$.

[(] [9] [+] [5] [)] [x^2] [ENTER]

The result is 196.

To evaluate a cubed expression, enter the expression followed by [MATH] [3]. Be sure to use parentheses correctly.

EXAMPLE Evaluate 9^3.

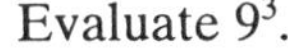

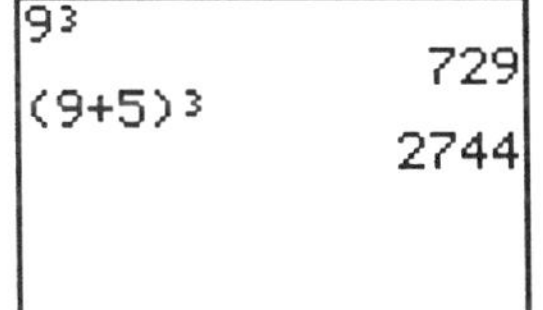

[9] [MATH] [3] [ENTER]

The result is 729.

Evaluate $(9 + 5)^3$.

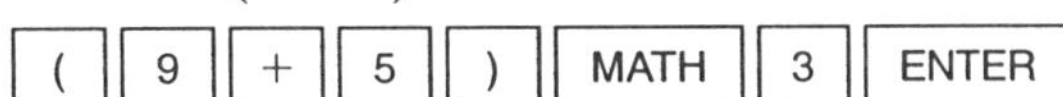

[(] [9] [+] [5] [)] [MATH] [3] [ENTER]

The result is 2744.

To evaluate an exponential expression other than a squared or cubed expression, enter the expression followed by [^] followed by the value of the exponent. Be sure to use parentheses correctly.

EXAMPLE Evaluate 9^5.

```
9^5
                59049
9^(1/2)
                    3
(9+5)^5
               537824
```

[9] [^] [5] [ENTER]

The result is 59,049.

Evaluate $9^{1/2}$.

[9] [^] [(] [1] [÷] [2] [)] [ENTER]

The result is 3.

Evaluate $(9 + 5)^5$.

[(] [9] [+] [5] [)] [^] [5] [ENTER]

The result is 537,824.

Fraction

Proper fractions are entered as division expressions. To obtain an answer in fractional notation, enter the expression followed by [MATH] [1] [ENTER].

To ensure the correct order of operation, place fractions in parentheses.

EXAMPLE Enter $\frac{3}{4}$.

```
(3/4)▸Frac
                  3/4
```

[(] [3] [÷] [4] [)] [MATH] [1] [ENTER]

The result is $\frac{3}{4}$.

Negative proper fractions are entered in one of two ways. The negative sign may be entered within the parentheses in front of the numerator, or an opposite sign may be entered outside the parentheses of a positive fraction.

EXAMPLE Enter $-\frac{3}{4}$.

```
(-3/4)▸Frac
                 -3/4
-(3/4)▸Frac
                 -3/4
```

[(] [(−)] [3] [÷] [4] [)] [MATH] [1] [ENTER]

or [(−)] [(] [3] [÷] [4] [)] [MATH] [1] [ENTER]

The result is $-\frac{3}{4}$.

Mixed numbers are entered as a sum of an integer and a proper fraction. The entire sum should be in parentheses to ensure proper order of operation. There is no need to place the proper-fraction addend in a second set of parentheses.

EXAMPLE Enter $1\frac{2}{3}$.

```
(1+2/3)▸Frac
            5/3
```

[( ] [1] [+] [2] [÷] [3] [ )] [MATH] [1] [ENTER]

The result is $\frac{5}{3}$ which is equivalent to $1\frac{2}{3}$.

Negative mixed numbers are entered as the opposite of a sum of an integer and a proper fraction. The entire sum should be in parentheses to ensure proper order of operation. The opposite sign precedes the parentheses. There is no need to place the proper-fraction addend in a second set of parentheses.

EXAMPLE Enter $-1\frac{2}{3}$.

```
-(1+2/3)▸Frac
           -5/3
```

[(−)] [( ] [1] [+] [2] [÷] [3] [ )] [MATH] [1] [ENTER]

The result is $-\frac{5}{3}$ which is equivalent to $-1\frac{2}{3}$.

To evaluate an expression and obtain the results in fractional notation, enter the expression followed by [MATH] [1] [ENTER].

EXAMPLE Evaluate $\frac{3}{4} \div \frac{2}{3}$.

```
(3/4)/(2/3)▸Frac
             9/8
```

[( ] [3] [÷] [4] [ )] [÷] [( ] [2] [÷] [3] [ )] [MATH] [1] [ENTER]

The result is $\frac{9}{8}$.

To change a real number to a fraction in lowest terms, enter the number followed by [MATH] [1] [ENTER].

EXAMPLE Write 0.25 as a fraction in lowest terms.

```
.25▸Frac
             1/4
```

[.] [2] [5] [MATH] [1] [ENTER]

The result is $\frac{1}{4}$.

Irrational numbers will not convert to a fraction. Therefore, the result is a decimal approximation of the number.

EXAMPLE Write π as a fraction in lowest terms.

```
π▸Frac
     3.141592654
```

[2nd] [π] [MATH] [1] [ENTER]

The result is 3.141592654.

Graphing

To graph points on the calculator, first set the screen to the desired setting (see *Screens*). Enter 2nd QUIT to return to the home or text screen. Enter 2nd DRAW ▶ 1 , followed by the x-coordinate of the point and , , followed by the y-coordinate of the point and) ENTER . To clear the screen of the graphed points, enter 2nd DRAW 1 .

EXAMPLE Graph the point (3, 2) on a decimal screen.

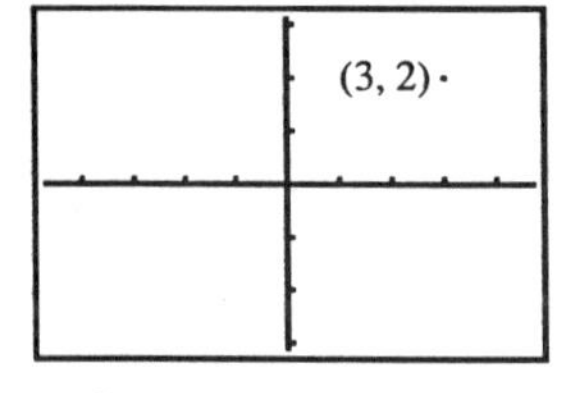

(−4.7, 4.7, −3.1, 3.1)

ZOOM 4 2nd QUIT 2nd DRAW ▶ 1 3 , 2) ENTER

To graph an equation on the calculator, first set the screen to the desired setting (see *Screens*). Enter the equation that is solved for y by entering Y= , followed by the expression in terms of x. If more than one equation is to be entered, use ENTER or ▼ to move to the next y. Finally, enter GRAPH .

EXAMPLE Graph $y = 2x + 5$ on an integer screen.

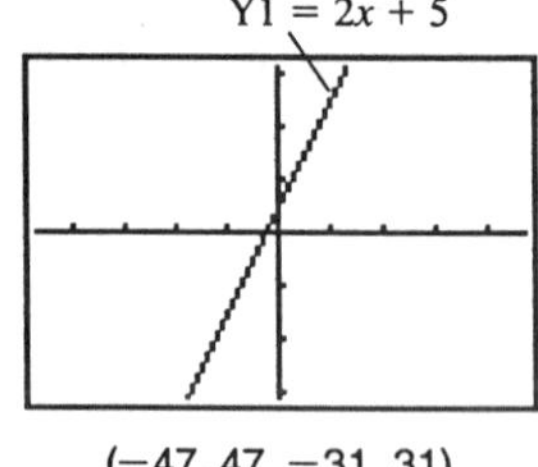

(−47, 47, −31, 31)

ZOOM 6 ZOOM 8 ENTER Y= 2 X, T, θ, n + 5 GRAPH

To determine the coordinates of points on the screen, use the arrow keys. The coordinates will be displayed on the bottom of the screen.

To determine the coordinates of the points graphed on the screen, enter TRACE followed by the arrow keys. The coordinates will be displayed on the bottom of the screen, and the equation will be displayed in the upper left-hand corner of the screen.

EXAMPLE Determine whether (4, 13) is on the graph of the equation $y = 2x + 5$.

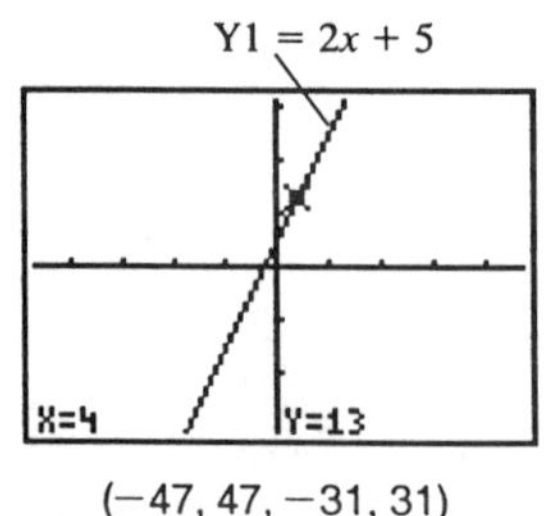

(−47, 47, −31, 31)

ZOOM 6 ZOOM 8 ENTER Y= 2 X, T, θ, n + 5 GRAPH TRACE ▶ ▶ ▶ ▶

To graph an inequality on the calculator, first set the screen to the desired setting (see *Screens*). Solve the inequality for y, exchange the inequality symbol for an equal sign, and enter the equation. Press Y= followed by the expression in terms of x. Move the cursor to the left of Y1 with the left arrow key. For a "greater than" inequality, press ENTER twice: ENTER ENTER . For a "less than" inequality, press ENTER three times: ENTER ENTER ENTER . Then enter GRAPH .

EXAMPLE 

Graph $y > 2x + 5$ on an integer screen.

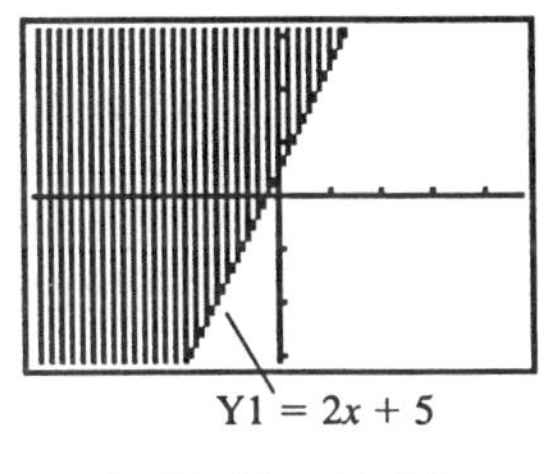

(−47, 47, −31, 31)

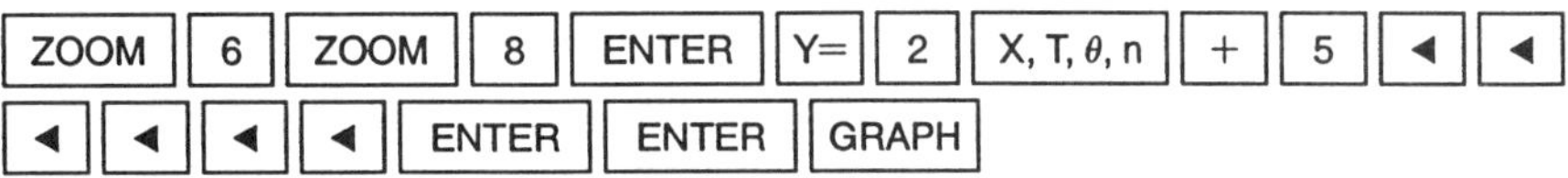

Grouping

To enter parentheses, brackets, and braces, use [( ] and [ )].

EXAMPLE Evaluate $2 + \{3[2 + (5 - 1)]\}$.

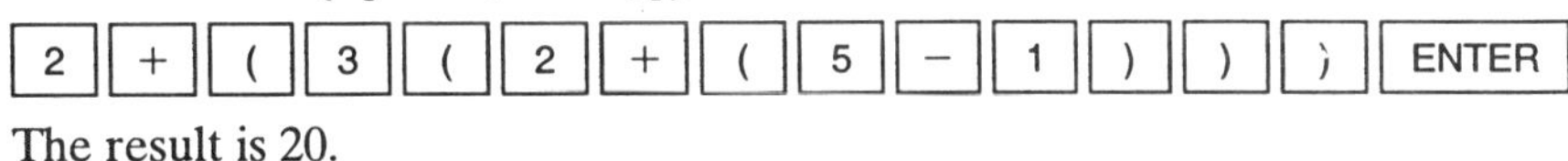

The result is 20.

To evaluate division problems with an expression for the dividend and/or divisor, enter the expressions in parentheses.

EXAMPLE Evaluate $\frac{9 + 5}{3 + 4}$.

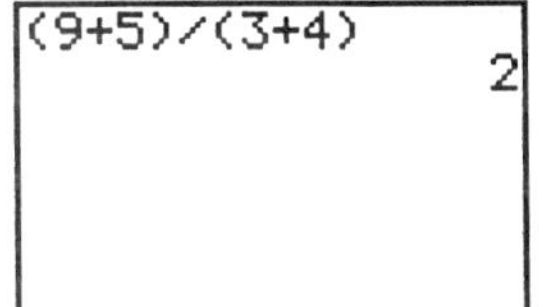

[( ] [9] [+] [5] [ )] [÷] [( ] [3] [+] [4] [ )] [ENTER]

The result is 2.

Negative sign

[(−)] means *negative* or *opposite of.*

Note: Do **not** use the subtraction operation symbol, [−], for a negative sign, [(−)].

EXAMPLE Evaluate $-5 - 3$.

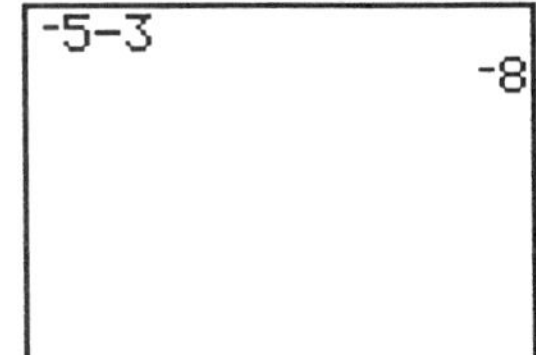

[(−)] [5] [−] [3] [ENTER]

The result is −8.

Opposite

Opposite expressions are entered in the calculator using the [(–)] key.

To evaluate an opposite expression, enter [(–)] followed by the expression enclosed in parentheses.

EXAMPLE Evaluate $-(16)$.

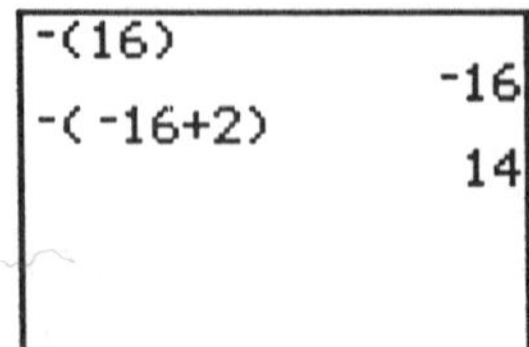

[(–)] [( ] [1] [6] [ )] [ENTER]

The result is -16.

Evaluate $-(-16 + 2)$.

[(–)] [( ] [(–)] [1] [6] [+] [2] [ )] [ENTER]

The result is 14.

Radical

To evaluate a square-root expression, enter [2nd] [√] followed by the expression and [ )].

EXAMPLE Evaluate $\sqrt{9}$.

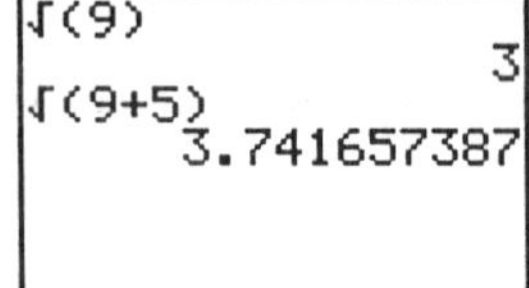

[2nd] [√] [9] [ )] [ENTER]

The result is 3.

Evaluate $\sqrt{9 + 5}$.

[2nd] [√] [9] [+] [5] [ )] [ENTER]

The result is 3.741657387.

To evaluate a cube-root expression, enter [MATH] [4] followed by the term and [ )].

EXAMPLE Evaluate $\sqrt[3]{27}$.

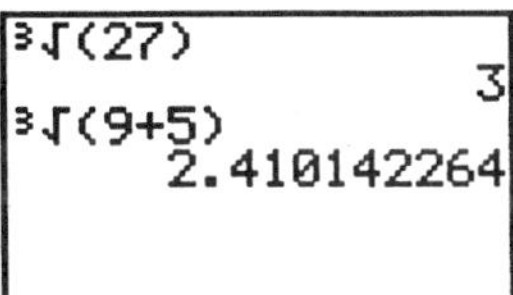

MATH 4 2 7) ENTER

The result is 3.

Evaluate $\sqrt[3]{9 + 5}$.

MATH 4 9 + 5) ENTER

The result is 2.410142264.

To evaluate a radical expression other than a square or cube root, enter the value of the index, followed by MATH 5, followed by the term. Be sure to enclose the expression in parentheses, as needed.

EXAMPLE Evaluate $\sqrt[4]{81}$.

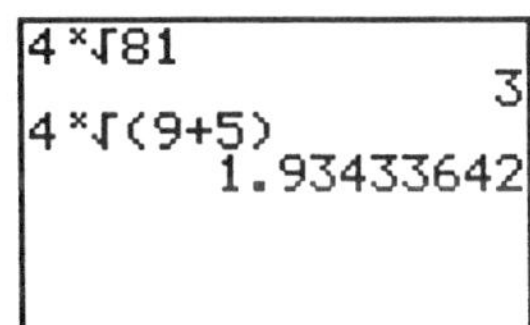

4 MATH 5 8 1 ENTER

The result is 3.

Evaluate $\sqrt[4]{9 + 5}$.

4 MATH 5 (9 + 5) ENTER

The result is 1.93433642.

Screens

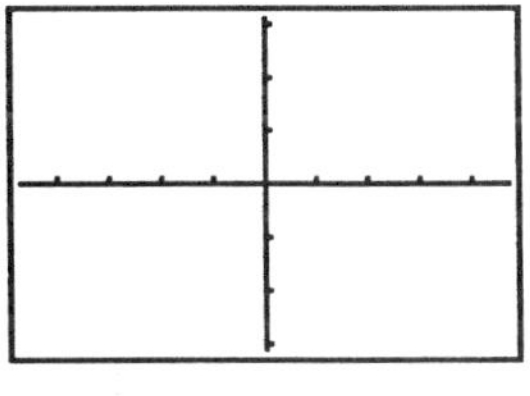

(−47, 47, −31, 31)

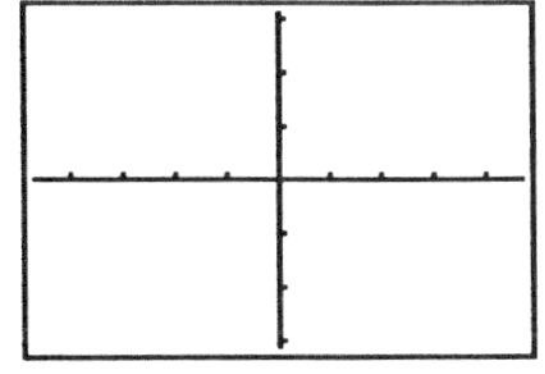

(−4.7, 4.7, −3.1, 3.1)

The **home** or **text screen** is a blank screen. If you want to return to this screen from another screen, enter 2nd QUIT as many times as needed.

The **integer graphics screen** used in this text is a centered integer screen set up by entering ZOOM 6 ZOOM 8 ENTER. *Note:* Do **not** forget the ENTER.

The window setting is (−47, 47, −31, 31).

The **decimal graphics screen** is the window setting (−4.7, 4.7, −3.1, 3.1) and is set up by entering ZOOM 4.

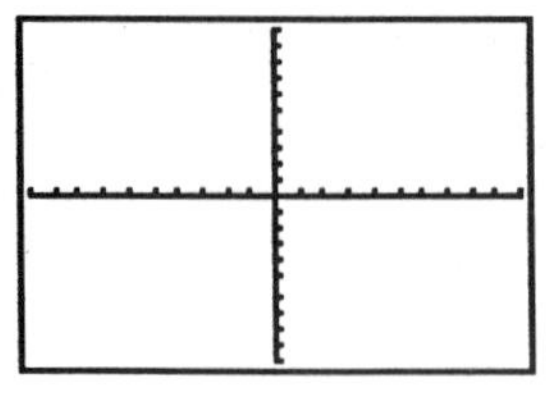
(−10, 10, −10, 10)

The **standard graphics screen** is the window setting (−10, 10, −10, 10) and is set up by entering ZOOM 6.

A **graphics screen of your choosing** is set up by entering WINDOW ▼, followed by reentering values and using the down arrow until the desired setting is entered. To view the screen, enter GRAPH.

EXAMPLE Enter the window (0, 10, 1, 0, 25, 10, 1).

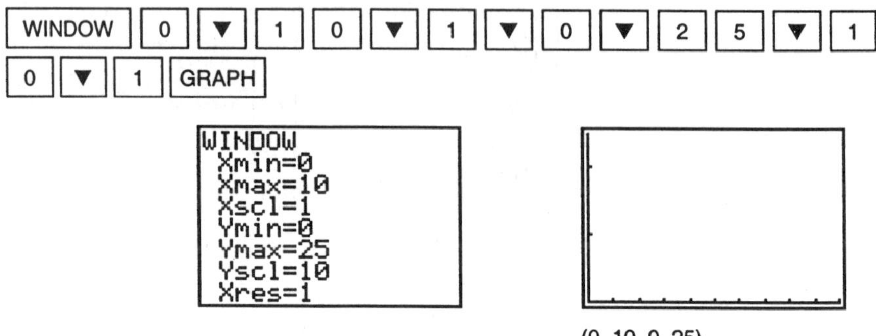

(0, 10, 0, 25)

Subtraction

− means subtraction.

Note: Do **not** use the subtraction operation symbol, −, for a negative sign, (−).

EXAMPLE Evaluate −5 − 3.

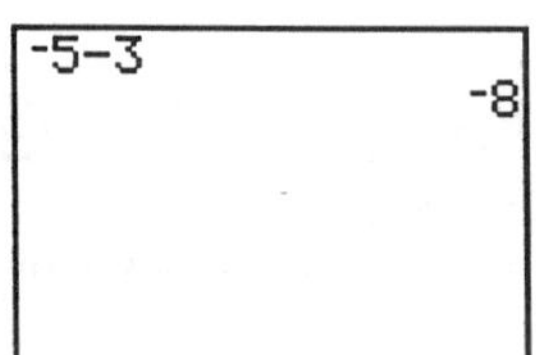

(−) 5 − 3 ENTER

The result is −8.

Table of values

The calculator will display a table of values for expressions entered in the Y= menu. First, enter Y= followed by the expression. To set up the table, enter 2nd TBLSET, followed by the least desired number wanted in the

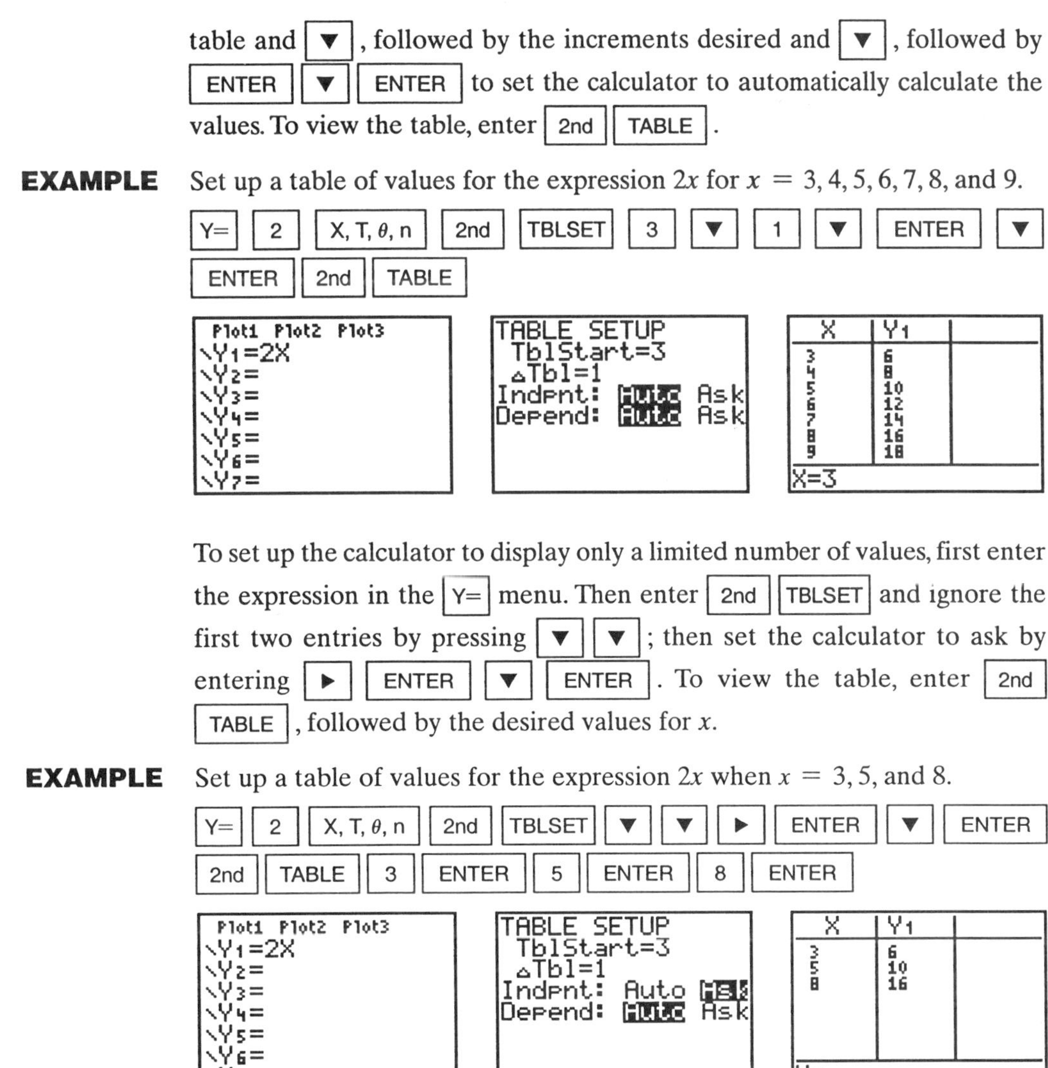

table and [▼], followed by the increments desired and [▼], followed by [ENTER] [▼] [ENTER] to set the calculator to automatically calculate the values. To view the table, enter [2nd] [TABLE].

EXAMPLE Set up a table of values for the expression $2x$ for $x = 3, 4, 5, 6, 7, 8,$ and 9.

[Y=] [2] [X, T, θ, n] [2nd] [TBLSET] [3] [▼] [1] [▼] [ENTER] [▼] [ENTER] [2nd] [TABLE]

To set up the calculator to display only a limited number of values, first enter the expression in the [Y=] menu. Then enter [2nd] [TBLSET] and ignore the first two entries by pressing [▼] [▼]; then set the calculator to ask by entering [▶] [ENTER] [▼] [ENTER]. To view the table, enter [2nd] [TABLE], followed by the desired values for x.

EXAMPLE Set up a table of values for the expression $2x$ when $x = 3, 5,$ and 8.

[Y=] [2] [X, T, θ, n] [2nd] [TBLSET] [▼] [▼] [▶] [ENTER] [▼] [ENTER] [2nd] [TABLE] [3] [ENTER] [5] [ENTER] [8] [ENTER]

TEST

The calculator tests whether an inequality or equation is true or false and returns a 0 for a false statement and a 1 for a true statement. Enter the statement in the calculator and then press [ENTER]. The inequality symbols and the equal symbol are under the TEST function. You will find these by entering [2nd] [TEST] and then entering the number in front of the symbol.

EXAMPLE Determine whether $3 < 4$ is true or false.

```
3<4
                    1
```

[3] [2nd] [TEST] [5] [4] [ENTER]

The result is 1 for true.

Since the calculator rounds decimal numbers, a 0 for false may result when the statement entered is true and a 1 for true when the statement is false. To recheck, enter each expression separately and check the results yourself.

EXAMPLE Determine whether $2.00000000001 = 2$ is true or false.

```
2.00000000001=2
                    1
```

[2] [.] [0] [0] [0] [0] [0] [0] [0] [0] [0] [0] [1] [2nd] [TEST] [1] [2] [ENTER]

The result is 1 for true.

However, we can see without a calculator that 2.00000000001 and 2 are not equal values.